Prealgebra

Julie Miller
Daytona State College

Molly O'Neill
Daytona State College

Nancy Hyde
Professor Emeritus
Broward College

McGraw Hill Higher Education

Boston Burr Ridge, IL Dubuque, IA New York San Francisco St. Louis
Bangkok Bogotá Caracas Kuala Lumpur Lisbon London Madrid Mexico City
Milan Montreal New Delhi Santiago Seoul Singapore Sydney Taipei Toronto

The McGraw-Hill Companies

Higher Education

PREALGEBRA

Published by McGraw-Hill, a business unit of The McGraw-Hill Companies, Inc., 1221 Avenue of the Americas, New York, NY 10020.

This book is printed on acid-free paper.

2 3 4 5 6 7 8 9 0 DOW/DOW 1 0 9 8 7 6 5 4 3 2 1 0

ISBN 978–0–07–338431–3
MHID 0–07–338431–3

ISBN 978–0–07–723893–3 (Annotated Instructor's Edition)
MHID 0–07–723893–1

Editorial Director: *Stewart K. Mattson*
Executive Editor: *David Millage*
Director of Development: *Kristine Tibbetts*
Senior Developmental Editor: *Emilie J. Berglund*
Marketing Manager: *Victoria Anderson*
Lead Project Manager: *Peggy J. Selle*
Senior Production Supervisor: *Sherry L. Kane*
Lead Media Project Manager: *Stacy A. Patch*
Senior Designer: *Laurie B. Janssen*
Cover Illustration: *Imagineering Media Services, Inc.*
Lead Photo Research Coordinator: *Carrie K. Burger*
Supplement Producer: *Mary Jane Lampe*
Compositor: *Aptara®, Inc.*
Typeface: *10/12 Times Ten Roman*
Printer: *R. R. Donnelley Willard, OH*

Library of Congress Cataloging-in-Publication Data

Miller, Julie, 1962-
Prealgebra / Julie Miller, Molly O'Neill, Nancy Hyde.—1st ed.
p. cm.
Includes index.
ISBN 978–0–07–338431–3—ISBN 0–07–338431–3 (hard copy : alk. paper)
I. Mathematics—Textbooks. I. O'Neill, Molly, 1953 II. Hyde, Nancy. III. Title.

QA39.3.M56 2011
510—dc22

2009022051

www.mhhe.com

Letter from the Authors

Dear Colleagues,

We originally embarked on this textbook project because we were seeing a lack of student success in our developmental math sequence. In short, we were not getting the results we wanted from our students with the materials and textbooks that we were using at the time. The primary goal of our project was to create teaching and learning materials that would get better results.

At Daytona State College, our students were instrumental in helping us develop the clarity of writing; the step-by-step examples; and the pedagogical elements, such as Avoiding Mistakes, Concept Connections, and Problem Recognition Exercises, found in our textbooks. They also helped us create the content for the McGraw-Hill video exercises that accompany this text. Using our text with a course redesign at Daytona State College, our student success rates in developmental courses have improved by 20% since 2006 (for further information, see *The Daytona Beach News Journal,* December 18, 2006). We think you will agree that these are the kinds of results we are all striving for in developmental mathematics courses.

This project has been a true collaboration with our Board of Advisors and colleagues in developmental mathematics around the country. We are sincerely humbled by those of you who adopted the first edition and the over 400 colleagues around the country who partnered with us providing valuable feedback and suggestions through reviews, symposia, focus groups, and being on our Board of Advisors. You partnered with us to create materials that will help students get better results. For that we are immeasurably grateful.

As an author team, we have an ongoing commitment to provide the best possible text materials for instructors and students. With your continued help and suggestions we will continue the quest to help all of our students get better results.

Sincerely,

Julie Miller
julie.miller.math@gmail.com

Molly O'Neill
molly.s.oneill@gmail.com

Nancy Hyde
nhyde@montanasky.com

About the Authors

Julie Miller

Julie Miller has been on the faculty in the School of Mathematics at Daytona State College for 20 years, where she has taught developmental and upper-level courses. Prior to her work at DSC, she worked as a software engineer for General Electric in the area of flight and radar simulation. Julie earned a bachelor of science in applied mathematics from Union College in Schenectady, New York, and a master of science in mathematics from the University of Florida. In addition to this textbook, she has authored several course supplements for college algebra, trigonometry, and precalculus, as well as several short works of fiction and nonfiction for young readers.

"My father is a medical researcher, and I got hooked on math and science when I was young and would visit his laboratory. I can remember using graph paper to plot data points for his experiments and doing simple calculations. He would then tell me what the peaks and features in the graph meant in the context of his experiment. I think that applications and hands-on experience made math come alive for me and I'd like to see math come alive for my students."

—Julie Miller

Molly O'Neill

Molly O'Neill is also from Daytona State College, where she has taught for 22 years in the School of Mathematics. She has taught a variety of courses from developmental mathematics to calculus. Before she came to Florida, Molly taught as an adjunct instructor at the University of Michigan–Dearborn, Eastern Michigan University, Wayne State University, and Oakland Community College. Molly earned a bachelor of science in mathematics and a master of arts and teaching from Western Michigan University in Kalamazoo, Michigan. Besides this textbook, she has authored several course supplements for college algebra, trigonometry, and precalculus and has reviewed texts for developmental mathematics.

"I differ from many of my colleagues in that math was not always easy for me. But in seventh grade I had a teacher who taught me that if I follow the rules of mathematics, even I could solve math problems. Once I understood this, I enjoyed math to the point of choosing it for my career. I now have the greatest job because I get to do math every day and I have the opportunity to influence my students just as I was influenced. Authoring these texts has given me another avenue to reach even more students."

—Molly O'Neill

Nancy Hyde

Nancy Hyde served as a full-time faculty member of the Mathematics Department at Broward College for 24 years. During this time she taught the full spectrum of courses from developmental math through differential equations. She received a bachelor of science degree in math education from Florida State University and a master's degree in math education from Florida Atlantic University. She has conducted workshops and seminars for both students and teachers on the use of technology in the classroom. In addition to this textbook, she has authored a graphing calculator supplement for *College Algebra*.

"I grew up in Brevard County, Florida, with my father working at Cape Canaveral. I was always excited by mathematics and physics in relation to the space program. As I studied higher levels of mathematics I became more intrigued by its abstract nature and infinite possibilities. It is enjoyable and rewarding to convey this perspective to students while helping them to understand mathematics."

—Nancy Hyde

Dedication

To Susan and Jim Conway

—Julie Miller

To my father, Richard

—Molly O'Neill

To the "Group," Chris, Jim, Kathy, and Dennis

—Nancy Hyde

Get Better Results with Miller/O'Neill/Hyde

About the Cover

A mosaic is made up of pieces placed together to create a unified whole. Similarly, a prealgebra course provides an array of topics that together create a solid mathematical foundation for the developmental mathematics student.

The Miller/O'Neill/Hyde developmental mathematics series helps students see the whole picture through better pedagogy and supplemental materials. In this *Prealgebra* textbook, Julie Miller, Molly O'Neill, and Nancy Hyde focused their efforts on guiding students successfully through core topics, building mathematical proficiency, and getting better results!

"We originally embarked on this textbook project because we were seeing a lack of student success in courses beyond our developmental sequence. We wanted to build a better bridge between developmental algebra and higher level math courses. Our goal has been to develop pedagogical features to help students achieve better results in mathematics."

—Julie Miller, Molly O'Neill, Nancy Hyde

How Will Miller/O'Neill/Hyde Help Your Students *Get Better Results?*

Better Clarity, Quality, and Accuracy

Julie Miller, Molly O'Neill, and Nancy Hyde know what students need to be successful in mathematics. Better results come from clarity in their exposition, quality of step-by-step worked examples, and accuracy of their exercises sets; but it takes more than just great authors to build a textbook series to help students achieve success in mathematics. Our authors worked with a strong mathematical team of instructors from around the country to ensure that the clarity, quality, and accuracy you expect from the Miller/O'Neill/Hyde series was included in this edition.

"I would describe it as an excellent text written by teachers! It has easy to understand explanations for students. The examples are good and the mathematics is solid. The students should find it "easy to read."
—Teresa Hasenauer, *Indian River State College*

Better Exercise Sets!

Comprehensive sets of exercises are available for every student level. Julie Miller, Molly O'Neill, and Nancy Hyde worked with a national board of advisors from across the country to offer the appropriate depth and breadth of exercises for your students. **Problem Recognition Exercises** were created to improve student performance while testing.

Our practice exercise sets help students progress from skill development to conceptual understanding. Student tested and instructor approved, the Miller/O'Neill/Hyde exercise sets will help your student *get better results.*

- **Problem Recognition Exercises**
- **Skill Practice Exercises**
- **Study Skills Exercises**
- **Mixed Exercises**
- **Expanding Your Skills Exercises**

"Extremely interesting and effective practice exercises. I love the study skills component that reminds students of essential habits that will help students achieve success. Review is a great reinforcement technique for recently mastered skills. The exercises are innovative and there are plenty of relevant applications from several fields and disciplines."
—Corinna Goehring, *Jackson State Community College*

Better Step-By-Step Pedagogy!

Prealgebra provides enhanced step-by-step learning tools to help students *get better results.*

- **Worked Examples** provide an "easy-to-understand" approach, clearly guiding each student through a step-by-step approach to master each practice exercise for better comprehension.
- **TIPs** offer students extra cautious direction to help improve understanding through hints and further insight.
- **Avoiding Mistakes** boxes alert students to common errors and provide practical ways to avoid them. Both of these learning aids will help students get better results by showing how to work through a problem using a clearly defined step-by-step methodology that has been class tested and student approved.

"MOH seems to look into the heads of students and see what mistakes they make and then help the students to avoid them—not only do the Avoiding Mistakes help with this, but the TIPs along the way also help."
—Linda Schott, *Ozark Technical Community College*

Formula for Student Success

Step-by-Step Worked Examples

- Do you get the feeling that there is a disconnection between your students' class work and homework?
- Do your students have trouble finding worked examples that match the practice exercises?
- Do you prefer that your students see examples in the textbook that match the ones you use in class?

Miller/O'Neill/Hyde's *Worked Examples* offer a clear, concise methodology that replicates the mathematical processes used in the authors' classroom lectures!

"I really like the visuals you use on the worked problems; there is just so much more help for students. I wouldn't be very worried about a student who missed a class and had to catch up by reading and following the text."
—Terry Kidd, *Salt Lake Community College*

Skill Practice

10. D.J. signs up for a new credit card that earns travel miles with a certain airline. She initially earns 15,000 travel miles by signing up for the new card. Then for each dollar spent she earns 2.5 travel miles. If at the end of one year she has 38,500 travel miles, how many dollars did she charge on the credit card?

Example 8 Using a Linear E

Joanne has a cellular phone plan in
pays $39.95 per month for 450 min of air time. Additional minutes beyond 450 are charged at a rate of $0.40 per minute. If Joanne's bill comes to $87.95, how many minutes did she use beyond 450 min?

Solution:

	Step 1: Read the problem.
Let x represent the number of minutes beyond 450. Then $0.40x$ represents the cost for x additional minutes.	Step 2: Label the variable.
(Monthly fee) + (Cost of additional minutes) = (Total cost)	Step 3: Write an equation in words.
$39.95 + 0.40x = 87.95$	Step 4: Write a mathematical equation.
$39.95 + 0.40x = 87.95$	Step 5: Solve the equation.
$39.95 - 39.95 + 0.40x = 87.95 - 39.95$	Subtract 39.95.
$0.40x = 48.00$	

"The Worked Examples are very easy for the students to follow with great step-by-step detailed explanations. Miller/O'Neill/Hyde excels with their Worked Examples."
—Kelli Hammer, *Broward College*

Joanne talked for 120 min b

Answer

10. D.J. charged $9400.

"The authors do a great job of explaining their thinking as they work from step to step in the examples. This helps to demystify the process of mathematics. The "avoiding mistakes" also help to reduce common student misconceptions."
—Nicole Lloyd, *Lansing Community College*

To ensure that the classroom experience also matches the examples in the text and the practice exercises, we have included references to even-numbered exercises to be used as Classroom Examples. These exercises are highlighted in the Practice Exercises at the end of each section.

Get Better Results

Better Learning Tools

Chapter Openers

Tired of students not being prepared? The Miller/O'Neill/Hyde *Chapter Openers* help students get better results through engaging *Puzzles and Games* that introduce the chapter concepts and ask "Are You Prepared?"

Chapter 7

In this chapter, we present the concept of percent. Percents are used to measure the number of parts per hundred of some whole amount. As a consumer, it is important to have a working knowledge of percents.

Are You Prepared?

To prepare for your work with percents, take a minute to review multiplying and dividing by a power of 10. Also practice solving equations and proportions. Work the problems on the left. Record the answers in the spaces on the right, according to the number of digits to the left and right of the decimal point. Then write the letter of each choice to complete the sentence below. If you need help, review Sections 5.3, 5.4, and 3.4.

1. 0.582×100 **2.** 0.002×100 **S.** _ • _ _

4. 318×0.01 **F.** _ _ • _ _

6. Solve. $\frac{34}{160} = \frac{x}{100}$ **P.** _ _ _ _ •

$= \frac{36}{100}$ **O.** • _

8. Solve. $0.3x = 16.2$ **I.** _ _ •

E. • _ _ _

H. _ _ • _

U. • _ _

...ician's bakery was called _ _ _ _ _ (1 2 3 4 5) _ _ (2 6) _ _ (7 8).

> "I really like the puzzle idea! I like for the chapter opener to be an activity instead of just reading. That way, students don't even realize they are preparing themselves for the concepts ahead and it is not nearly as boring."
>
> —Jacqui Fields, *Wake Technical Community College*

TIP and Avoiding Mistakes Boxes

TIP and Avoiding Mistakes boxes have been created based on the authors' classroom experiences—they have also been integrated into the **Worked Examples.** These pedagogical tools will help students get better results by learning how to work through a problem using a clearly defined step-by-step methodology.

Example 1 Adding and Subtracting Like Fractions

Add. Write the answer as a fraction or whole number.

a. $\frac{1}{4} + \frac{5}{4}$ **b.** $\frac{2}{15} + \frac{1}{15} - \frac{13}{15}$

Solution:

a. $\frac{1}{4} + \frac{5}{4} = \frac{1+5}{4}$ Add the numerators.

$= \frac{6}{4}$ Write the sum over the common denominator.

$= \frac{\cancel{6}^{3}}{\cancel{4}_{2}}$ Simplify to lowest terms.

Avoiding Mistakes

Notice that when adding fractions, we do not add the denominators. We add *only* the numerators.

Avoiding Mistakes Boxes:

Avoiding Mistakes boxes are integrated throughout the textbook to alert students to common errors and how to avoid them.

> "The MOH text does a better job of pointing out the common mistakes students make."
>
> —Kaye Black, *Bluegrass Community & Technical College*

TIP Boxes

Teaching tips are usually revealed only in the classroom. Not anymore! TIP boxes offer students helpful hints and extra direction to help improve understanding and further insight.

TIP: Example 1(a) can also be solved by using a percent proportion.

$\frac{5.5}{100} = \frac{x}{20{,}000}$ What is 5.5% of 20,000?

$(5.5)(20{,}000) = 100x$

$110{,}000 = 100x$

$\frac{110{,}000}{100} = \frac{100x}{100}$ Divide both sides by 100.

$1100 = x$ The sales tax is \$1100.

> "I think that one of the best features of this chapter (and probably will continue throughout the text) is the TIP section."
>
> —Ena Salter, *Manatee Community College*

Better Exercise Sets! Better Practice! Better Results!

- Do your students have trouble with problem solving?
- Do you want to help students overcome math anxiety?
- Do you want to help your students improve performance on math assessments?

Problem Recognition Exercises

Problem Recognition Exercises present a collection of problems that look similar to a student upon first glance, but are actually quite different in the manner of their individual solutions. Students sharpen critical thinking skills and better develop their "solution recall" to help them distinguish the method needed to solve an exercise—an essential skill in developmental mathematics.

Problem Recognition Exercises, tested in a developmental mathematics classroom, were created to improve student performance while testing.

> "I like how this author doesn't title all the sections within this PRE. I believe that would be important during testing-anxiety situations. How many times do our students say they did not know what to do and (are) not sure what they were being asked?"
>
> —Christine Baade, *San Juan College*

Problem Recognition Exercises

Operations on Fractions versus Solving Proportions

For Exercises 1–6, identify the problem as a proportion or as a product of fractions. Then solve the proportion or multiply the fractions.

1. a. $\frac{x}{4} = \frac{15}{8}$ **b.** $\frac{1}{4} \cdot \frac{15}{8}$ **2. a.** $\frac{2}{5} \cdot \frac{3}{10}$ **b.** $\frac{2}{5} = \frac{y}{10}$

3. a. $\frac{2}{7} \times \frac{3}{14}$ **b.** $\frac{2}{7} = \frac{n}{14}$ **4. a.** $\frac{m}{5} = \frac{6}{15}$ **b.** $\frac{3}{5} \times \frac{6}{15}$

5. a. $\frac{48}{p} = \frac{16}{3}$ **b.** $\frac{48}{8} \cdot \frac{16}{3}$ **6. a.** $\frac{10}{7} \cdot \frac{28}{5}$ **b.** $\frac{10}{7} = \frac{28}{t}$

For Exercises 7–10, solve the proportion or perform the indicated operation on fractions.

7. a. $\frac{3}{7} = \frac{6}{z}$ **b.** $\frac{3}{7} \div \frac{6}{35}$ **c.** $\frac{3}{7} + \frac{6}{35}$ **d.** $\frac{3}{7} \cdot \frac{6}{35}$

8. a. $\frac{4}{5} \div \frac{20}{3}$ **b.** $\frac{4}{v} = \frac{20}{3}$ **c.** $\frac{4}{5} \times \frac{20}{3}$ **d.** $\frac{4}{5} - \frac{20}{3}$

9. a. $\frac{14}{5} \cdot \frac{10}{7}$ **b.** $\frac{14}{5} = \frac{x}{7}$ **c.** $\frac{14}{5} - \frac{10}{7}$ **d.** 14 ÷ [illegible]

10. a. 11 [illegible] 6/y **b.** $\frac{11}{3} + \frac{66}{11}$ **c.** $\frac{11}{3} \div \frac{66}{11}$

> "These are brilliant! I often do such things in class to get across a point, and I haven't seen them in a text before."
>
> —Russell Penner, *Mohawk Valley Community College*

> "This book does a much better job of pairing similar problems for students to be able to practice recognizing different exercises—and as an instructor, I can use these exercises as part of a review and lecture about the need to understand a problem versus just memorizing a process."
>
> —Vicki Lucido, *St. Louis Community College–Florissant Valley*

Student Centered Applications!

The Miller/O'Neill/Hyde Board of Advisors partnered with our authors to bring the *best applications* from every region in the country! These applications include real data and topics that are more relevant and interesting to today's student.

20. The number of new jobs for selected industries are given in the table. (*Source:* Bureau of Labor Statistics)

a. Which category has the greatest number of new jobs? How many new jobs is this?

b. Draw a bar graph with vertical bars to illustrate these data.

Industry	Number of New Jobs
Health care	219,400
Temporary help	212,000
Construction	173,000
Food service	167,600
Retail	78,600

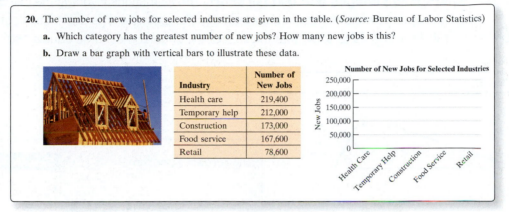

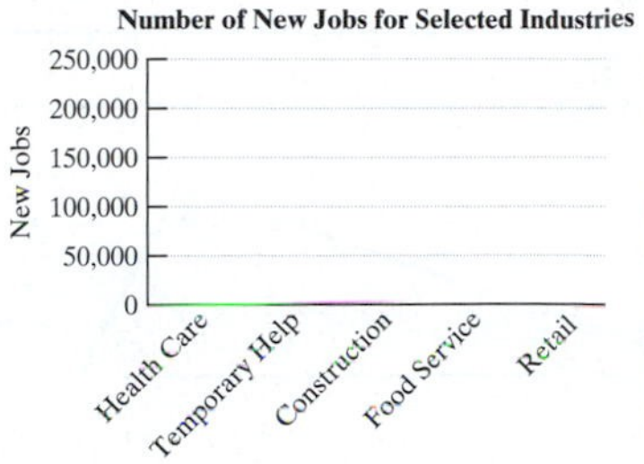

Group Activities!

Each chapter concludes with a Group Activity to promote classroom discussion and collaboration— helping students not only to solve problems but to explain their solutions for better mathematical mastery. Group Activities are great for instructors and adjuncts— bringing a more interactive approach to teaching mathematics! All required materials, activity time, and suggested group sizes are provided in the end-of-chapter material. Activities include Investigating Probability, Tracking Stocks, Card Games with Fractions, and more!

Group Activity

Remodeling the Classroom

Materials: A tape measure for measuring the size of the classroom.
Advertisements online or from the newspaper for carpet anc

Estimated time: 30 minutes

Group Size: 3–4

In this activity, your group will determine the cost for updating your classroom with new paint and new carpet.

1. Measure and record the dimensions of the room and also the height of the walls. You may want to sketch the floor and walls and then label their dimensions on the figure.

2. Calculate and record the area of the floor.

3. Ca culate and record the total area of the walls. Subtract any area taken up by doors, windows, and ialkboards.

at would be suitable for your classroom. You may have to look

"I love this group activity. It gives students the answer to the commonly asked question "When will I ever use this?" while it reinforces the material they have been studying."
—Karen Walsh, *Broward College*

"I really like the hands-on nature of taking actual measurements in the MOH activity. It is also something that translates to a project that students may actually have to do someday as homeowners—very practical."
—Kristi Laird, *Jackson Community College*

"MOH's group activity involves true participation and interaction; fun with fractions!"
—Monika Bender, *Central Texas College*

Dynamic Math Animations

The Miller/O'Neill/Hyde author team has developed a series of Flash animations to illustrate difficult concepts where static images and text fall short. The animations leverage the use of on-screen movement and morphing shapes to enhance conceptual learning. For example, one animation "cuts" a triangle into three pieces and rotates the pieces to show that the sum of the angular measures equals 180° (below).

Sum of the Angles in a Triangle

180°

This animation shows that the sum of the measures of the angles in a triangle is 180°.

Pythagorean Theorem | **Section 8.6**

Objectives

1. Triangles
2. Square Roots
3. Pythagorean Theorem

...gon. Furthermore, the sum of the measures of the ... Teachers often demonstrate this fact by tearing a ...nown in Figure 8-24. Then they align the **vertices** ... a straight angle.

22° 60° 98° 22°

Figure 8-24

Animation

PROPERTY Angles of a Triangle
The sum of the measures of the angles of a triangle equals 180°.

Example 1 Finding the Measure of Angles Within a Triangle

Find the measure of angles a and b.

a.

a

38°

b.

b

43° a 130°

Solution:

a. Recall that the □ symbol represents a 90° angle.

$38° + 90° + m(\angle a) = 180°$ The sum of the angles within a triangle is 180°.

$128° + m(\angle a) = 180°$ Add the measures of the two known angles.

$128° - 128° + m(\angle a) = 180° - 128°$ Solve for $m(\angle a)$.

Skill Practice

Find the measures of angles a and b.

1.

a

42°

2.

39° b

a

100°

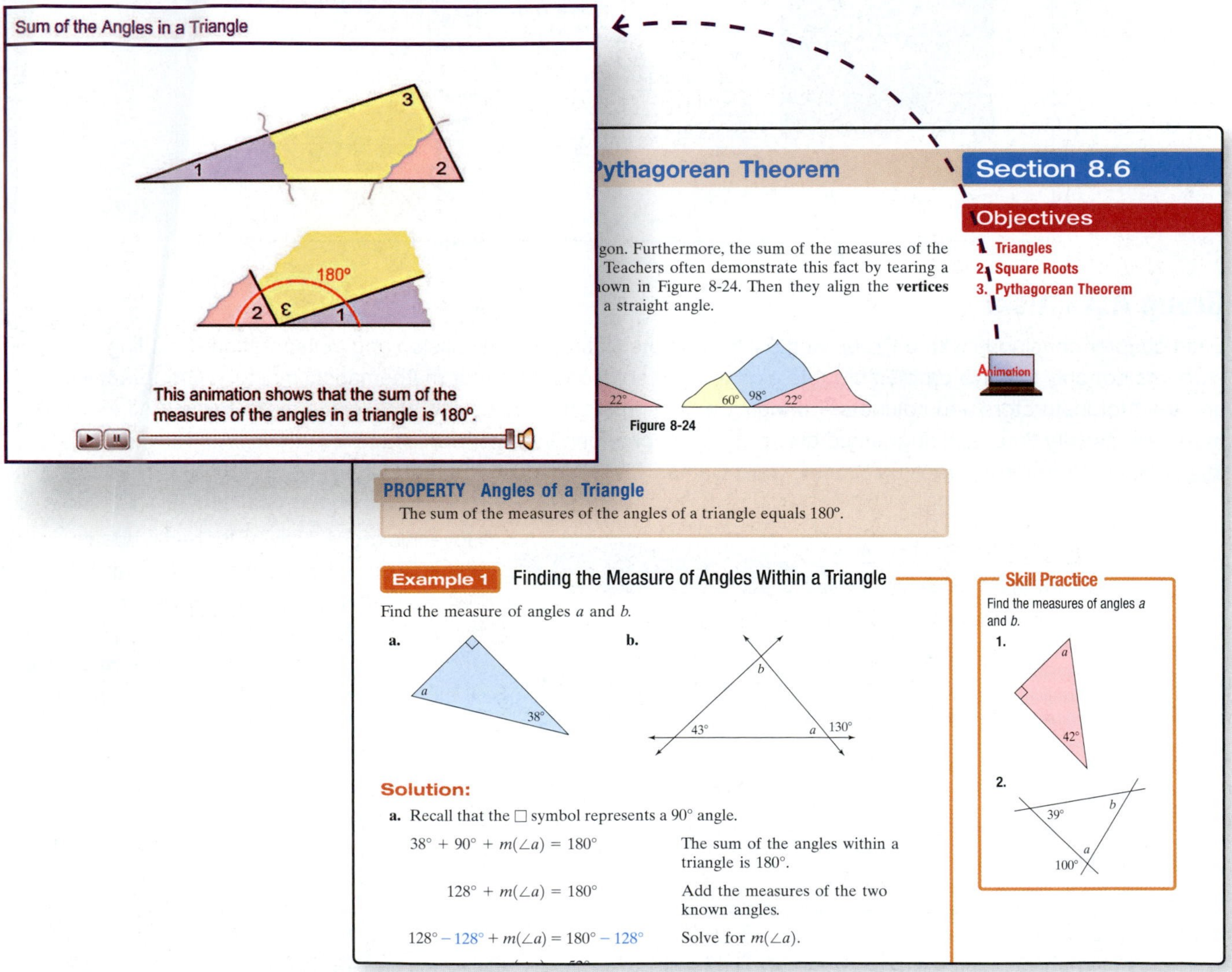

Through their classroom experience, the authors recognize that such media assets are great teaching tools for the classroom and excellent for online learning. The Miller/O'Neill/Hyde animations are interactive and quite diverse in their use. Some provide a virtual laboratory for which an application is simulated and where students can collect data points for analysis and modeling. Others provide interactive question-and-answer sessions to test conceptual learning. For word problem applications, the animations ask students to estimate answers and practice "number sense."

N

O

P

Q

R

S

Geometry Formulas

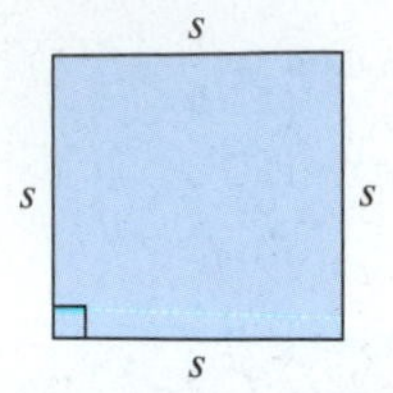

Perimeter of a Square

$P = s + s + s + s$

$P = 4s$

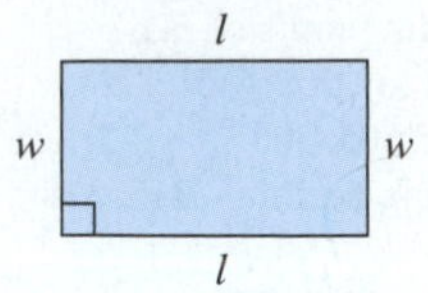

Perimeter of a Rectangle

$P = l + l + w + w$

$P = 2l + 2w$

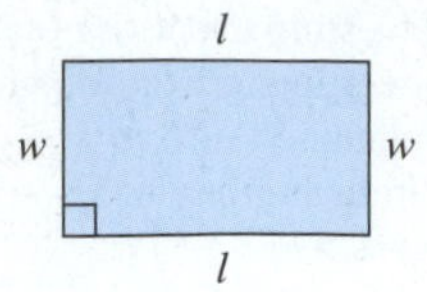

Area of a Rectangle

$A =$ length $\times$ width

$A = lw$

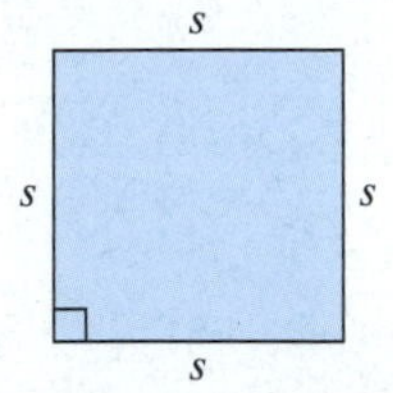

Area of a Square

$A =$ length $\times$ width

$A = s \cdot s$

$A = s^2$

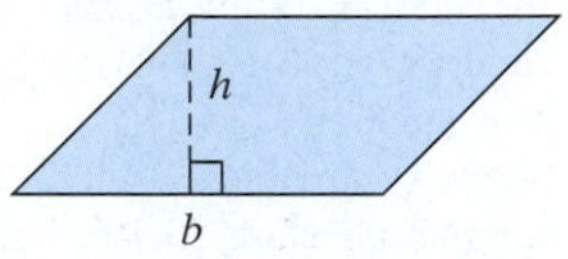

Area of a Parallelogram

$A =$ base $\times$ height

$A = bh$

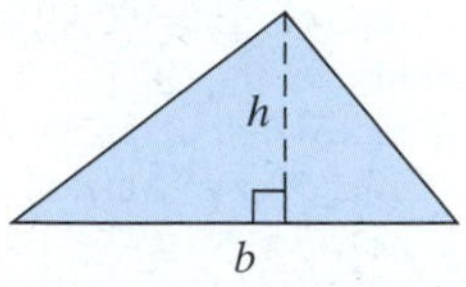

Area of a Triangle

$A = \frac{1}{2} \times$ base $\times$ height

$A = \frac{1}{2}bh$

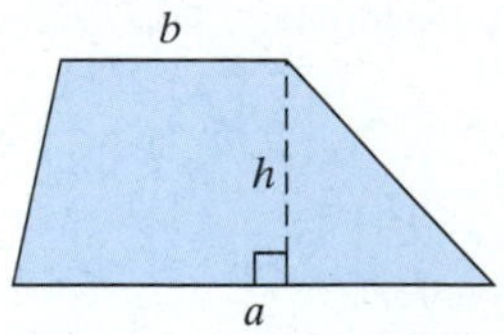

Area of a Trapezoid

$A = \frac{1}{2} \times$ (sum of the parallel sides) $\times$ height

$A = \frac{1}{2} \cdot (a + b) \cdot h$

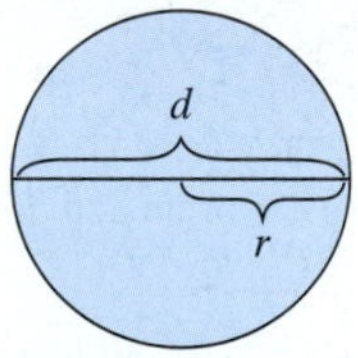

Circumference of a Circle

The circumference, C, of a circle is given by: $C = \pi d$ or $C = 2\pi r$

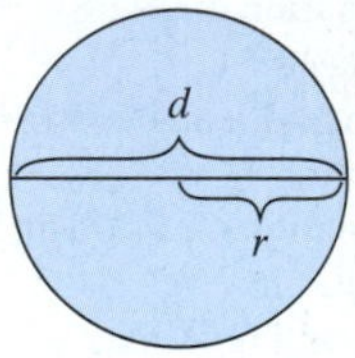

Area of a Circle

The area, A, of a circle is given by: $A = \pi r^2$

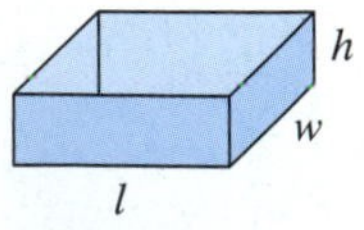

Rectangular Solid

$V = lwh$

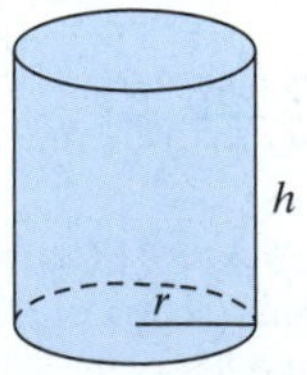

Right Circular Cylinder

$V = \pi r^2 h$

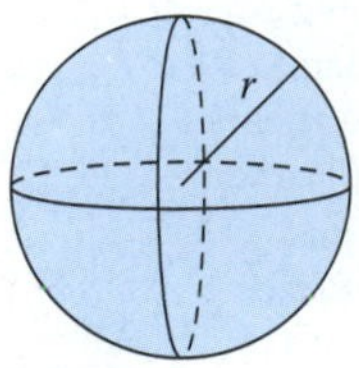

Sphere

$V = \frac{4}{3}\pi r^3$

360° Development Process

McGraw-Hill's 360° Development Process is an ongoing, never-ending, market-oriented approach to building accurate and innovative print and digital products. It is dedicated to continual large-scale and incremental improvement that is driven by multiple customer-feedback loops and checkpoints. This is initiated during the early planning stages of our new products, and intensifies during the development and production stages—then begins again upon publication, in anticipation of the next edition.

A key principle in the development of any mathematics text is its ability to adapt to teaching specifications in a universal way. The only way to do so is by contacting those universal voices—and learning from their suggestions. We are confident that our book has the most current content the industry has to offer, thus pushing our desire for accuracy to the highest standard possible. In order to accomplish this, we have moved through an arduous road to production. Extensive and open-minded advice is critical in the production of a superior text.

Here is a brief overview of the initiatives included in the *Prealgebra*, 360° Development Process:

Board of Advisors

A hand-picked group of trusted teachers' active in the Prealgebra course served as chief advisors and consultants to the author and editorial team with regards to manuscript development. The Board of Advisors reviewed parts of the manuscript; served as a sounding board for pedagogical, media, and design concerns; consulted on organizational changes; and attended a focus group to confirm the manuscript's readiness for publication.

Would you like to inquire about becoming a BOA member?
If so, email the editor, David Millage at david_millage@mcgraw-hill.com.

Prealgebra

Vanetta Grier-Felix, *Seminole Community College*
Teresa Hasenauer, *Indian River State College*
Shelbra Jones, *Wake Technical Community College*
Nicole Lloyd, *Lansing Community College*
Kausha Miller, *Bluegrass Community and Technical College*
Linda Schott, *Ozarks Technical Community College*
Renee Sundrud, *Harrisburg Area Community College*

Beginning Algebra

Anabel Darini, *Suffolk County Community College*
Sabine Eggleston, *Edison State College*
Brandie Faulkner, *Tallahassee Community College*
Kelli Hammer, *Broward College–South*
Joseph Howe, *St. Charles Community College*
Laura Lossi, *Broward College–Central*
DiDi Quesada, *Miami Dade College*

Intermediate Algebra

Connie Buller, *Metropolitan Community College*
Nancy Carpenter, *Johnson County Community College*
Pauline Chow, *Harrisburg Area Community College*
Donna Gerken, *Miami Dade College*
Gayle Krzemien, *Pikes Peak Community College*
Judy McBride, *Indiana University–Purdue University at Indianapolis*
Patty Parkison, *Ball State University*

Beginning and Intermediate Algebra

Annette Burden, *Youngstown State University*
Lenore Desilets, *DeAnza College*
Gloria Guerra, *St. Philip's College*
Julie Turnbow, *Collin County Community College*
Suzanne Williams, *Central Piedmont Community College*
Janet Wyatt, *Metropolitan Community College–Longview*

With ALEKS®

New ALEKS Instructor Module

Enhanced Functionality and Streamlined Interface Help to Save Instructor Time

ALEKS® The new ALEKS Instructor Module features enhanced functionality and a streamlined interface based on research with ALEKS instructors and homework management instructors. Paired with powerful assignment-driven features, textbook integration, and extensive content flexibility, the new ALEKS Instructor Module simplifies administrative tasks and makes ALEKS more powerful than ever.

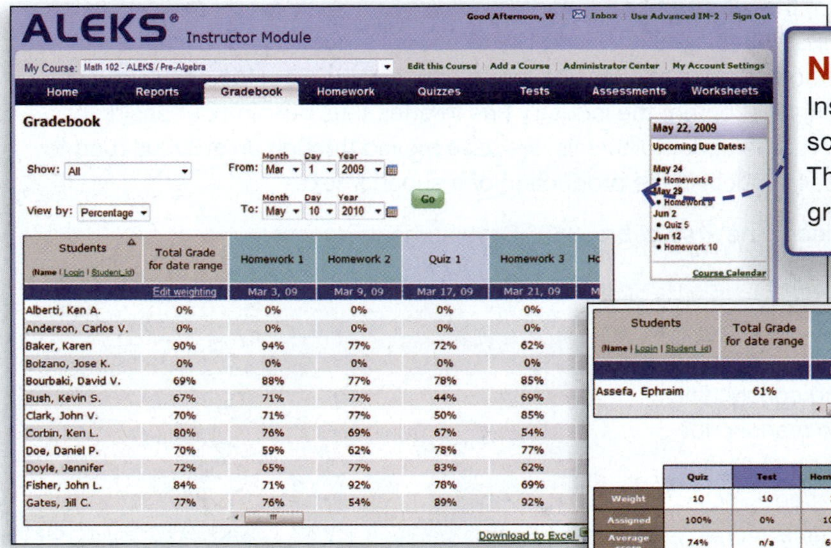

Gradebook view for all students

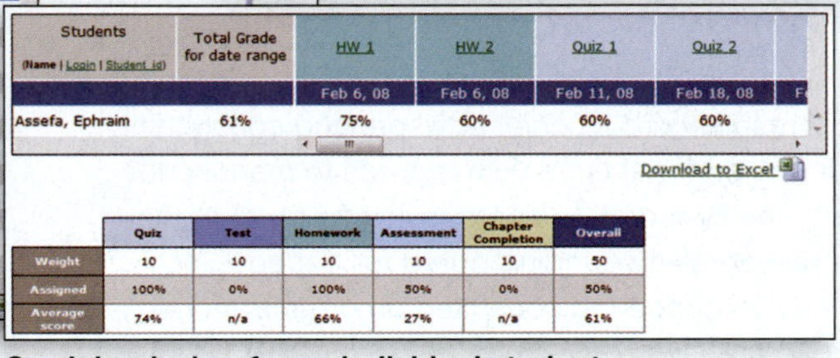

Gradebook view for an individual student

New Gradebook!

Instructors can seamlessly track student scores on automatically graded assignments. They can also easily adjust the weighting and grading scale of each assignment.

Track Student Progress Through Detailed Reporting

Instructors can track student progress through automated reports and robust reporting features.

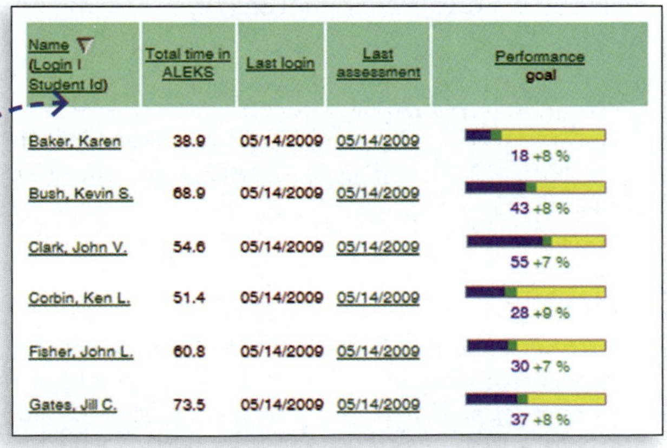

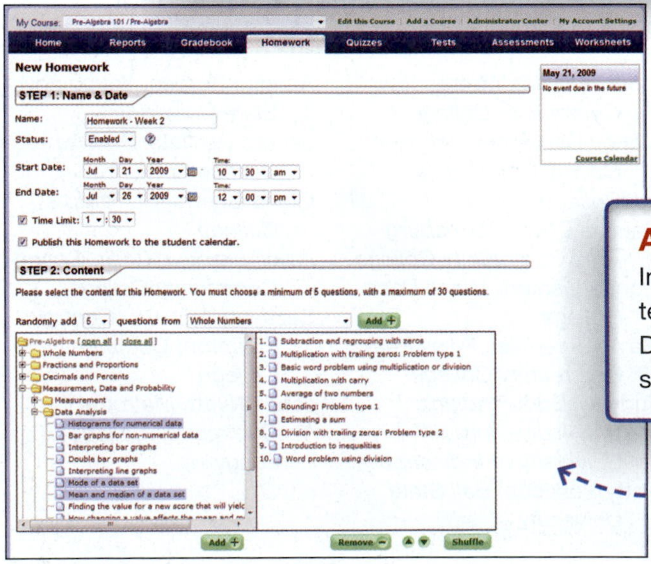

Select topics for each assignment

Automatically Graded Assignments

Instructors can easily assign homework, quizzes, tests, and assessments to all or select students. Deadline extensions can also be created for select students.

Learn more about ALEKS by visiting **www.aleks.com/highered/math** or contact your McGraw-Hill representative.

Get Better Results

Experience Student Success!

ALEKS® ALEKS is a unique online math tool that uses adaptive questioning and artificial intelligence to correctly place, prepare, and remediate students . . . all in one product! Institutional case studies have shown that **ALEKS has improved pass rates by over 20% versus traditional online homework, and by over 30% compared to using a text alone.**

By offering each student an individualized learning path, ALEKS directs students to work on the math topics that they are ready to learn. Also, to help students keep pace in their course, instructors can correlate ALEKS to their textbook or syllabus in seconds.

To learn more about how ALEKS can be used to boost student performance, please visit **www.aleks.com/highered/math** or contact your McGraw-Hill representative.

ALEKS Pie

Each student is given her or his own individualized learning path.

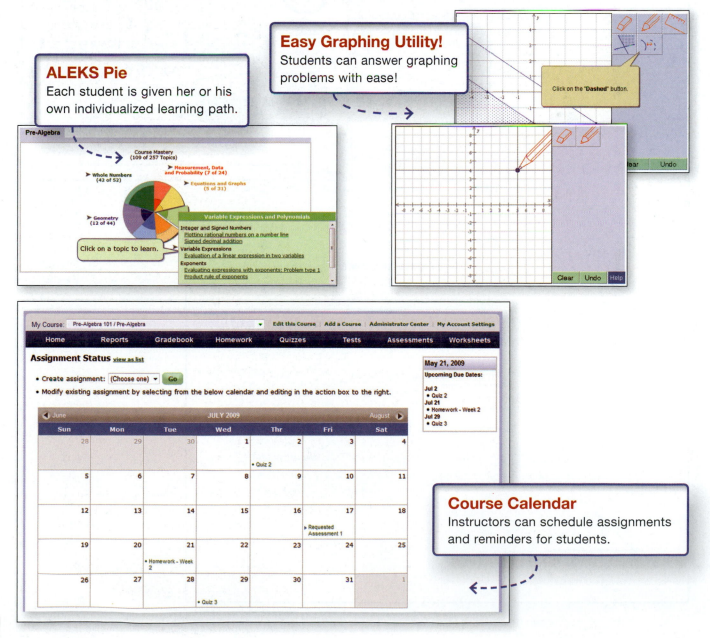

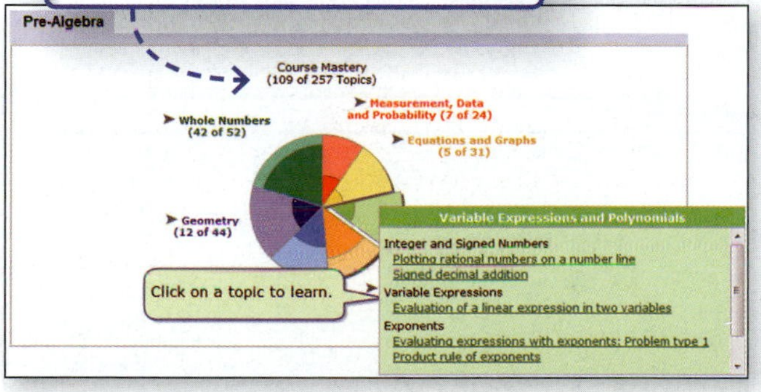

Easy Graphing Utility!

Students can answer graphing problems with ease!

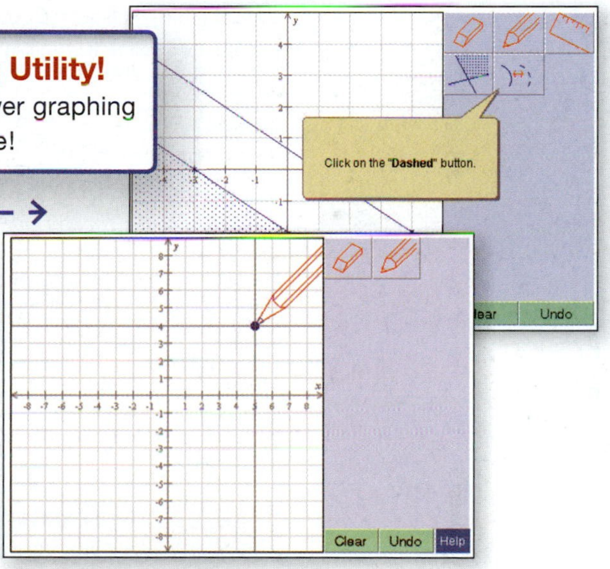

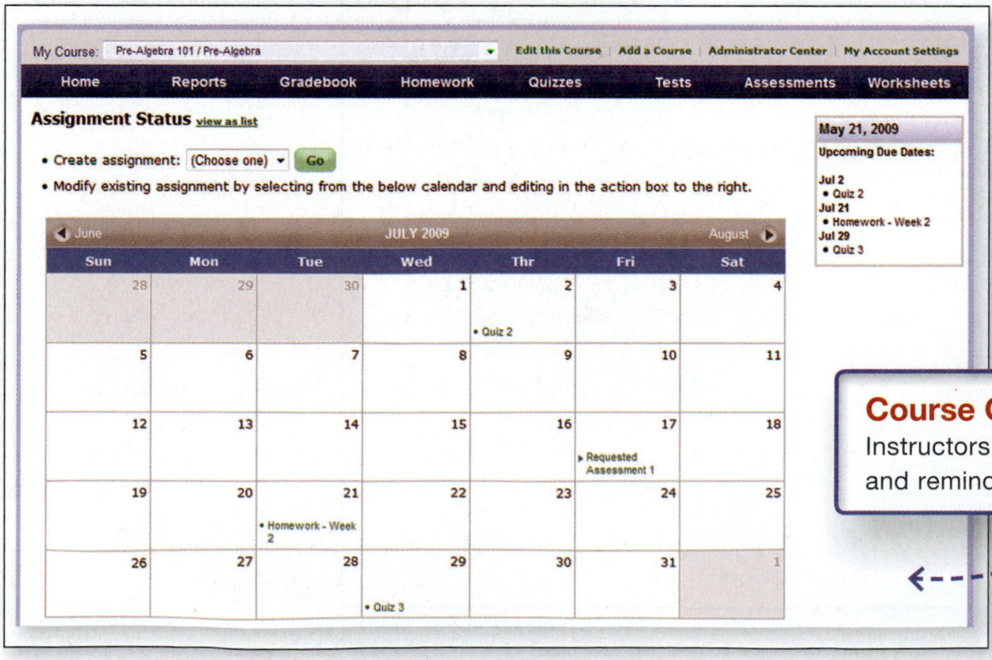

Course Calendar

Instructors can schedule assignments and reminders for students.

Get Better Results

The animations were created by the authors based on over 75 years of combined teaching experience! To facilitate the use of the animations, the authors have placed icons in the text to indicate where animations are available. Students and instructors can access these assets online in ALEKS.

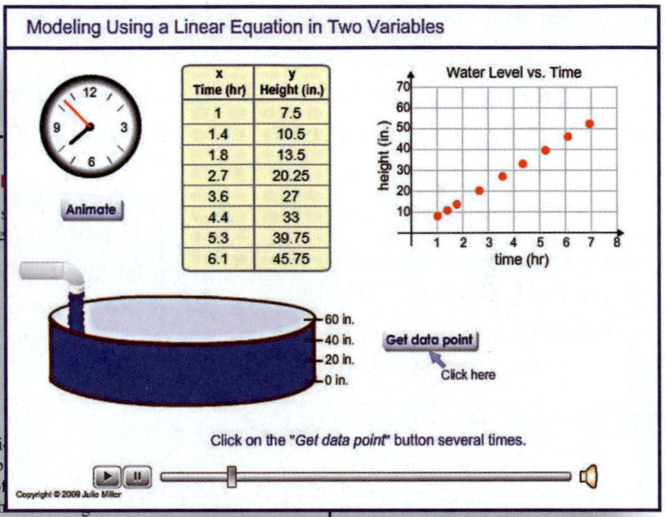

Animation

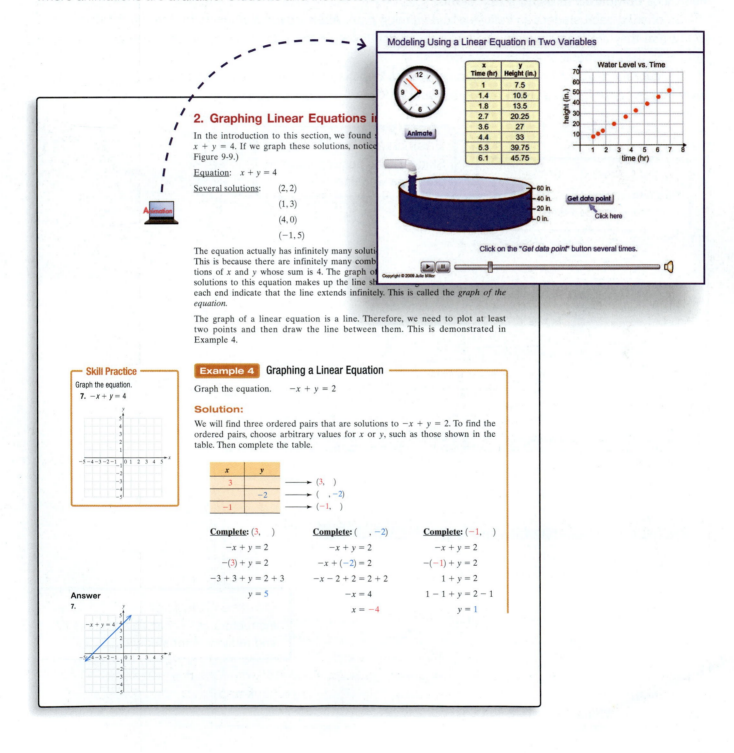

2. Graphing Linear Equations i

In the introduction to this section, we found $x + y = 4$. If we graph these solutions, notice Figure 9-9.)

Equation: $x + y = 4$

Several solutions: $(2, 2)$

$(1, 3)$

$(4, 0)$

$(-1, 5)$

The equation actually has infinitely many soluti This is because there are infinitely many comb tions of x and y whose sum is 4. The graph o solutions to this equation makes up the line sh each end indicate that the line extends infinitely. This is called the *graph of the equation.*

The graph of a linear equation is a line. Therefore, we need to plot at least two points and then draw the line between them. This is demonstrated in Example 4.

Skill Practice

Graph the equation.

7. $-x + y = 4$

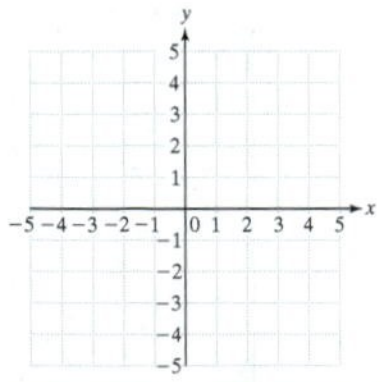

Example 4 Graphing a Linear Equation

Graph the equation. $-x + y = 2$

Solution:

We will find three ordered pairs that are solutions to $-x + y = 2$. To find the ordered pairs, choose arbitrary values for x or y, such as those shown in the table. Then complete the table.

x	y	
3		→ (3,)
	−2	→ (, −2)
−1		→ (−1,)

Complete: (3,)

$-x + y = 2$

$-(3) + y = 2$

$-3 + 3 + y = 2 + 3$

$y = 5$

Complete: (, −2)

$-x + y = 2$

$-x + (-2) = 2$

$-x - 2 + 2 = 2 + 2$

$-x = 4$

$x = -4$

Complete: (−1,)

$-x + y = 2$

$-(-1) + y = 2$

$1 + y = 2$

$1 - 1 + y = 2 - 1$

$y = 1$

Answer

7.

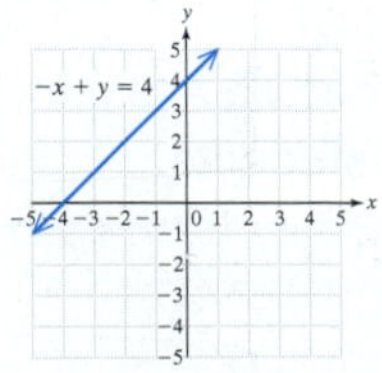

Better Development!

Question: How do you build a better developmental mathematics textbook series?

Answer: Employ a developmental mathematics instructor from the classroom to become a McGraw-Hill editor!

Emilie Berglund joined the developmental mathematics team at McGraw-Hill, bringing her extensive classroom experience to the Miller/O'Neill/Hyde textbook series. A former developmental mathematics instructor at Utah Valley State College, Ms. Berglund has won numerous teaching awards and has served as the beginning algebra course coordinator for the department. Ms. Berglund's experience teaching developmental mathematics students from the Miller/O'Neill/Hyde translates into more well-developed pedagogy throughout the textbook series and can be seen in everything from the updated Worked Examples to the Exercise Sets.

Listening to You . . .

This textbook has been reviewed by over 300 teachers across the country. Our textbook is a commitment to your students, providing a clear explanation, a concise writing style, step-by-step learning tools, and the best exercises and applications in developmental mathematics. **How do we know? You told us so!**

Teachers *Just Like You* are saying great things about the Miller/O'Neill/Hyde developmental mathematics series:

"I would say that the authors have definitely been in a classroom and know how to say things in a simple manner (but mathematically sound and correct). They often write things exactly as I say them in class."
—Teresa Hasenauer, *Indian River State College*

"A text with exceptional organization and presentation."
—Shelbra Jones, *Wake Technical Community College*

"I really like the 'avoiding mistakes' and 'tips' areas. I refer to these in class all the time."
—Joe Howe, *St. Charles Community College*

Acknowledgments and Reviewers

The development of this textbook series would never have been possible without the creative ideas and feedback offered by many reviewers. We are especially thankful to the following instructors for their careful review of the manuscript.

Symposia

Every year McGraw-Hill conducts general mathematics symposia that are attended by instructors from across the country. These events provide opportunities for editors from McGraw-Hill to gather information about the needs and challenges of instructors teaching these courses. This information helped to create the book plan for *Prealgebra.* A forum is also offered for the attendees to exchange ideas and experiences with colleagues they otherwise might not have met.

Advisors Symposium—Barton Creek, Texas

Connie Buller, *Metropolitan Community College*
Pauline Chow, *Harrisburg Area Community College*
Anabel Darini, *Suffolk County Community College*
Maria DeLucia, *Middlesex County College*
Sabine Eggleston, *Edison State College*
Brandie Faulkner, *Tallahassee Community College*
Vanetta Grier-Felix, *Seminole Community College*
Gloria Guerra, *St. Philip's College*
Joseph Howe, *St. Charles Community College*
Laura Lossi, *Broward College-Central*
Gayle Krzemien, *Pikes Peak Community College*
Nicole Lloyd, *Lansing Community College*
Judy McBride, *Indiana University–Purdue University at Indianapolis*
Kausha Miller, *Bluegrass Community and Technical College*
Patty Parkison, *Ball State University*
Linda Schott, *Ozarks Technical and Community College*
Renee Sundrud, *Harrisburg Area Community College*
Janet Wyatt, *Metropolitan Community College–Longview*

Napa Valley Symposium

Antonio Alfonso, *Miami Dade College*
Lynn Beckett-Lemus, *El Camino College*
Kristin Chatas, *Washtenaw Community College*
Maria DeLucia, *Middlesex County College*
Nancy Forrest, *Grand Rapids Community College*
Michael Gibson, *John Tyler Community College*
Linda Horner, *Columbia State College*
Matthew Hudock, *St. Philip's College*
Judith Langer, *Westchester Community College*
Kathryn Lavelle, *Westchester Community College*
Scott McDaniel, *Middle Tennessee State University*
Adelaida Quesada, *Miami Dade College*
Susan Shulman, *Middlesex County College*
Stephen Toner, *Victor Valley College*
Chariklia Vassiliadis, *Middlesex County College*
Melanie Walker, *Bergen Community College*

Myrtle Beach Symposium

Patty Bonesteel, *Wayne State University*
Zhixiong Chen, *New Jersey City University*
Latonya Ellis, *Bishop State Community College*
Bonnie Filer, *Tubaugh University of Akron*
Catherine Gong, *Citrus College*
Marcia Lambert, *Pitt Community College*
Katrina Nichols, *Delta College*
Karen Stein, *The University of Akron*
Walter Wang, *Baruch College*

La Jolla Symposium

Darryl Allen, *Solano Community College*
Yvonne Aucoin, *Tidewater Community College-Norfolk*
Sylvia Carr, *Missouri State University*
Elizabeth Chu, *Suffolk County Community College*
Susanna Crawford, *Solano Community College*
Carolyn Facer, *Fullerton College*
Terran Felter, *California State University–Bakersfield*
Elaine Fitt, *Bucks County Community College*
John Jerome, *Suffolk County Community College*
Sandra Jovicic, *The University of Akron*
Carolyn Robinson, *Mt. San Antonio College*
Carolyn Shand-Hawkins, *Missouri State University*

Class Tests

Multiple class tests provided the editorial team with an understanding of how content and the design of a textbook impact a student's homework and study habits in the general mathematics course area.

Special "*thank you*" to our Manuscript Class-Testers

Manuscript Review Panels

Over 200 teachers and academics from across the country reviewed the various drafts of the manuscript to give feedback on content, design, pedagogy, and organization. This feedback was summarized by the book team and used to guide the direction of the text.

Reviewers of Miller/O'Neill/Hyde Prealgebra

Ken Aeschliman, *Oakland Community College*
Joyce Ahlgren, *California State University–San Bernardino*
Khadija Ahmed, *Monroe County Community College*
Victoria Anemelu, *San Bernardino Valley College*
Carla Arriola, *Broward College–North*
Christine Baade, *San Juan College*
Carlos Barron, *Mountain View College*
Monika Bender, *Central Texas College*
Kaye Black, *Bluegrass Community and Technical College*
E. Deronn Bowen, *Broward College–Central*
Susan Caldiero, *Cosumnes River College*
Julie Chung, *American River College*
John Close, *Salt Lake Community College*
William Coe, *Montgomery College*
Lois Colpo, *Harrisburg Area Community College*
Greg Cripe, *Spokane Falls Community College*
Ann Davis, *Pasadena Area Community College*
Marcial Echenique, *Broward College–North*
Monette Elizalde, *Palo Alto College*
Jacqui Fields, *Wake Technical Community College*
Rhoderick Fleming, *Wake Technical Community College*
Lori Fuller, *Tunxis Community College*
Corinna Goehring, *Jackson State Community College*
Susan Grody, *Broward College–North*
Kathryn Gundersen, *Three Rivers Community College*
Safa Hamed, *Oakton Community College*
Kelli Hammer, *Broward College–South*
Teresa Hasenauer, *Indian River State College*
Mary Beth Headlee, *Manatee Community College*
Rebecca Heiskell, *Mountain View College*
Paul Hernandez, *Palo Alto College*
Glenn Jablonski, *Triton College*
Erin Jacob, *Corning Community College*
Ted Jenkins, *Chaffey College*
Yolanda Johnson, *Tarrant County College–South*
Shelbra Jones, *Wake Technical Community College*
Ryan Kasha, *Valencia Community College–West*
Susan Kautz, *Cy Fair College*
Joanne Kendall, *Cy Fair College*
Terry Kidd, *Salt Lake Community College*
Barbara Kistler, *Lehigh Carbon Community College*

Reviewers of the Miller/O'Neill/Hyde Developmental Mathematics Series *(continued)*

Bernadette Kocyba, *J. Sargent Reynolds Community College*
Jeff Koleno, *Lorain County Community College*
Kathy Kopelousos, *Lewis and Clark Community College*
Gayle Krzemien, *Pikes Peak Community College*
Carla Kulinsky, *Salt Lake Community College*
Myrna La Rosa, *Triton College*
Kristi Laird, *Jackson State Community College*
Lider Lamar, *Seminole Community College–Lake Mary*
Alice Lawson-Johnson, *Palo Alto College*
Jeanine Lewis, *Aims Community College–Main*
Lisa Lindloff, *McLennan Community College*
Nicole Lloyd, *Lansing Community College*
Barbara Lott, *Seminole Community College–Lake Mary*
Ann Loving, *J. Sargeant Reynolds Community College*
Vicki Lucido, *St. Louis Community College–Florissant Valley*
Judy Maclaren, *Trinidad State Junior College*
Joseph Maire, *San Jacinto College–Pasadena*
J Robert Malena, *Community College of Allegheny County-South*
Barbara Manley, *Jackson State Community College*
Linda Marable, *Nashville State Technical Community College*
Diane Martling, *William Rainey Harper College*
Victoria Mcclendon, *Northwest Arkansas Community College*
Ruth McGowan, *St. Louis Community College–Florissant Valley*
Trudy Meyer, *El Camino College*
Kausha Miller, *Bluegrass Community and Technical College*
Angel Miranda, *Valencia Community College–Osceola*
Danielle Morgan, *San Jacinto College–South*
Kathleen Offenholley, *Brookdale Community College*
Maria Parker, *Oxnard College*
Russell Penner, *Mohawk Valley Community College*
Sara Pries, *Sierra College*
Kumars Ranjbaran, *Mountain View College*
Ali Ravandi, *College of the Mainland*
Linda Reist, *Macomb Community College*
Nancy Ressler, *Oakton Community College*
Natalie Rivera, *Estrella Mountain Community College*
Pat Rowe, *Columbus State Community College*
Kristina Sampson, *Cy Fair College*
Rainer Schochat, *Triton College*
Linda Schott, *Ozarks Technical Community College*
Sally Sestini, *Cerritos College*
Kathleen Shepherd, *Monroe County Community College*
Rose Shirey, *College of the Mainland*
Domingo Soria-Martin, *Solano Community College*
Joel Spring, *Broward College–South*
Melissa Spurlock, *Anne Arundel Community College*
Shirley Stewart, *Pikes Peak Community College*
Barbara Strauch, *Devry University–Tinley Park*
Jennifer Strehler, *Oakton Community College*
Renee Sundrud, *Harrisburg Area Community College*
Shae Thompson, *Montana State University–Bozeman*
John Thoo, *Yuba College*
Joseph Tripp, *Ferris State University*
Laura Van Husen, *Midland College*
Karen Walsh, *Broward College–North*
Jennifer Wilson, *Tyler Junior College*
Rick Woodmansee, *Sacramento City College*
Michael Yarbrough, *Cosumnes River College*
Kevin Yokoyama, *College of the Redwoods*

Special thanks go to Jon Weerts for preparing the *Instructor's Solutions Manual* and the *Student's Solution Manual* and to Carrie Green, Rebecca Hubiak, and Hal Whipple for their work ensuring accuracy. Many thanks to Cindy Reed for her work in the video series, and to Kelly Jackson for advising us on the Instructor Notes.

Finally, we are forever grateful to the many people behind the scenes at McGraw-Hill without whom we would still be on page 1. To our developmental editor (and math instructor extraordinaire), Emilie Berglund, thanks for your day-to-day support and understanding of the world of developmental mathematics. To David Millage, our executive editor and overall team captain, thanks for keeping the train on the track. Where did you find enough hours in the day? To Torie Anderson and Sabina Navsariwala, we greatly appreciate your countless hours of support and creative ideas promoting all of our efforts. To our director of development and champion, Kris Tibbetts, thanks for being there in our time of need. To Pat Steele, where would we be without your watchful eye over our manuscript? To our publisher, Stewart Mattson, we're grateful for your experience and energizing new ideas. Thanks for believing in us. To Jeff Huettman and Amber Bettcher, we give our greatest appreciation for the exciting technology so critical to student success, and to Peggy Selle, thanks for keeping watch over the whole team as the project came together.

Most importantly, we give special thanks to all the students and instructors who use *Prealgebra* in their classes.

Get Better Results

A COMMITMENT TO ACCURACY

You have a right to expect an accurate textbook, and McGraw-Hill invests considerable time and effort to make sure that we deliver one. Listed below are the many steps we take to make sure this happens.

Our Accuracy Verification Process

First Round

Step 1: Numerous **college math instructors** review the manuscript and report on any errors that they may find. Then the authors make these corrections in their final manuscript.

Second Round

Step 2: Once the manuscript has been typeset, the **authors** check their manuscript against the first page proofs to ensure that all illustrations, graphs, examples, exercises, solutions, and answers have been correctly laid out on the pages, and that all notation is correctly used.

Step 3: An outside, **professional mathematician** works through every example and exercise in the page proofs to verify the accuracy of the answers.

Step 4: A **proofreader** adds a triple layer of accuracy assurance in the first pages by hunting for errors, then a second, corrected round of page proofs is produced.

Third Round

Step 5: The **author team** reviews the second round of page proofs for two reasons: (1) to make certain that any previous corrections were properly made, and (2) to look for any errors they might have missed on the first round.

Step 6: A **second proofreader** is added to the project to examine the new round of page proofs to double check the author team's work and to lend a fresh, critical eye to the book before the third round of paging.

Fourth Round

Step 7: A **third proofreader** inspects the third round of page proofs to verify that all previous corrections have been properly made and that there are no new or remaining errors.

Step 8: Meanwhile, in partnership with **independent mathematicians,** the text accuracy is verified from a variety of fresh perspectives:

- The **test bank authors** check for consistency and accuracy as they prepare the computerized test item file.
- The **solutions manual author** works every exercise and verifies his/her answers, reporting any errors to the publisher.
- A **consulting group of mathematicians,** who write material for the text's MathZone site, notifies the publisher of any errors they encounter in the page proofs.
- A video production company employing **expert math instructors** for the text's videos will alert the publisher of any errors it might find in the page proofs.

Final Round

Step 9: The **project manager,** who has overseen the book from the beginning, performs a **fourth proofread** of the textbook during the printing process, providing a final accuracy review.

What results is a mathematics textbook that is as accurate and error-free as is humanly possible, and our authors and publishing staff are confident that our many layers of quality assurance have produced textbooks that are the leaders in the industry for their integrity and correctness.

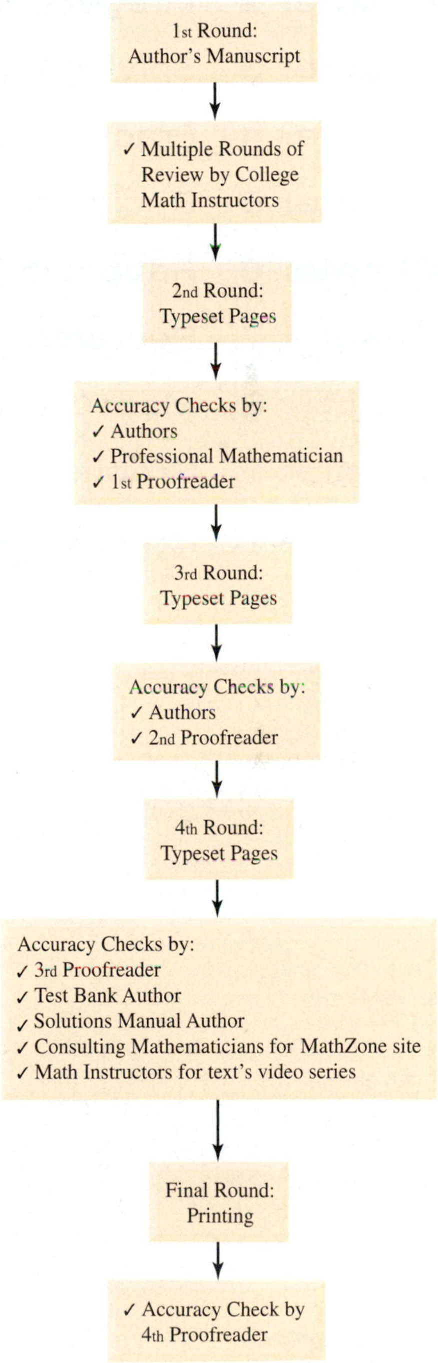

Brief Contents

Contents

Whole Numbers

1

CHAPTER OUTLINE

Chapter 1

Chapter 1 begins with adding, subtracting, multiplying, and dividing whole numbers. We also include rounding, estimating, and applying the order of operations.

Are You Prepared?

Before you begin this chapter, check your skill level by trying the exercises in the puzzle. If you have trouble with any of these problems, come back to the puzzle as you work through the chapter, and fill in the answers that gave you trouble.

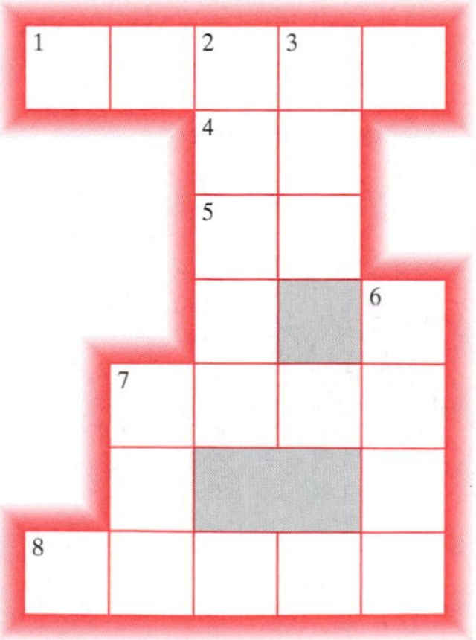

Across

1. $24{,}159 - 3168$

4. 8^2

5. 2^5

7. $1600 \div 4 \times 8$

8. $3^3 \cdot 1000$

Down

2. $39{,}504 + 56{,}798$

3. $11{,}304 \div 12$

6. $(13 - \sqrt{9})^3$

7. $100 - 5 \cdot 4 + 277$

Section 1.1 Study Tips

Objectives

1. Before the Course
2. During the Course
3. Preparation for Exams
4. Where to Go for Help

In taking a course in algebra, you are making a commitment to yourself, your instructor, and your classmates. Following some or all of the study tips below can help you be successful in this endeavor. The features of this text that will assist you are printed in blue.

1. Before the Course

- Purchase the necessary materials for the course before the course begins or on the first day.
- Obtain a three-ring binder to keep and organize your notes, homework, tests, and any other materials acquired in the class. We call this type of notebook a *portfolio*.
- Arrange your schedule so that you have enough time to attend class and to do homework. A common rule is to set aside at least 2 hours for homework for every hour spent in class. That is, if you are taking a 4-credit-hour course, plan on at least 8 hours a week for homework. If you experience difficulty in mathematics, plan for more time. A 4-credit-hour course will then take *at least* 12 hours each week—about the same as a part-time job.
- Communicate with your employer and family members the importance of your success in this course so that they can support you.
- Be sure to find out the type of calculator (if any) that your instructor requires.

2. During the Course

- Read the section in the text *before* the lecture to familiarize yourself with the material and terminology.
- Attend every class, and be on time.
- Take notes in class. Write down all of the examples that the instructor presents. Read the notes after class, and add any comments to make your notes clearer to you. Use a tape recorder to record the lecture if the instructor permits the recording of lectures.
- Ask questions in class.
- Read the section in the text *after* the lecture, and pay special attention to the Tip boxes and Avoiding Mistakes boxes.
- After you read an example, try the accompanying Skill Practice problem in the margin. The skill practice problem mirrors the example and tests your understanding of what you have read.
- Do homework every night. Even if your class does not meet every day, you should still do some work every night to keep the material fresh in your mind.
- Check your homework with the answers that are supplied in the back of this text. Correct the exercises that do not match, and circle or star those that you cannot correct yourself. This way you can easily find them and ask your instructor the next day.
- Write the definition and give an example of each Key Term found at the beginning of the Practice Exercises.
- The Problem Recognition Exercises are located in most chapters. These provide additional practice distinguishing among a variety of problem types. Sometimes the most difficult part of learning mathematics is retaining all that you learn. These exercises are excellent tools for retention of material.

- Form a study group with fellow students in your class, and exchange phone numbers. You will be surprised by how much you can learn by talking about mathematics with other students.
- If you use a calculator in your class, read the Calculator Connections boxes to learn how and when to use your calculator.
- Ask your instructor where you might obtain extra help if necessary.

3. Preparation for Exams

- Look over your homework. Pay special attention to the exercises you have circled or starred to be sure that you have learned that concept.
- Read through the Summary at the end of the chapter. Be sure that you understand each concept and example. If not, go to the section in the text and reread that section.
- Give yourself enough time to take the Chapter Test uninterrupted. Then check the answers. For each problem you answered incorrectly, go to the Review Exercises and do all of the problems that are similar.
- To prepare for the final exam, complete the Cumulative Review Exercises at the end of each chapter, starting with Chapter 2. If you complete the cumulative reviews after finishing each chapter, then you will be preparing for the final exam throughout the course. The Cumulative Review Exercises are another excellent tool for helping you retain material.

4. Where to Go for Help

- At the first sign of trouble, see your instructor. Most instructors have specific office hours set aside to help students. Don't wait until after you have failed an exam to seek assistance.
- Get a tutor. Most colleges and universities have free tutoring available.
- When your instructor and tutor are unavailable, use the Student Solutions Manual for step-by-step solutions to the odd-numbered problems in the exercise sets.
- Work with another student from your class.
- Work on the computer. Many mathematics tutorial programs and websites are available on the Internet, including the website that accompanies this text: www.mhhe.com/moh

Group Activity

Becoming a Successful Student

Materials: Computer with Internet access and textbook

Estimated time: 15 minutes

Group Size: 4

Good time management, good study skills, and good organization will help you be successful in this course. Answer the following questions and compare your answers with your group members.

1. To motivate yourself to complete a course, it is helpful to have clear reasons for taking the course. List your goals for taking this course and discuss them with your group.

2. For the following week, write down the times each day that you plan to study math.

Monday	Tuesday	Wednesday	Thursday	Friday	Saturday	Sunday

3. Write down the date of your next math test. ____________

4. Taking 12 credit-hours is the equivalent of a full-time job. Often students try to work too many hours while taking classes at school.

   a. Write down the number of hours you work per week and the number of credit hours you are taking this term.

   number of hours worked per week ____________

   number of credit-hours this term ____________

   b. The table gives a recommended limit to the number of hours you should work for the number of credit hours you are taking at school. (Keep in mind that other responsibilities in your life such as your family might also make it necessary to limit your hours at work even more.) How do your numbers from part (a) compare to those in the table? Are you working too many hours?

Number of Credit-Hours	Maximum Number of Hours of Work per Week
3	40
6	30
9	20
12	10
15	0

5. Look through the book in Chapter 2 and find the page number corresponding to each feature in the book. Discuss with your group members how you might use each feature.

   Problem Recognition Exercises: page ____________

   Chapter Summary: page ____________

   Chapter Review Exercises: page ____________

   Chapter Test: page ____________

   Cumulative Review Exercises: page ____________

6. Look at the Skill Practice exercises in the margin (for example, find Skill Practice exercises 2–4 in Section 1.2). Where are the answers to these exercises located? Discuss with your group members how you might use the Skill Practice exercises.

7. Discuss with your group members places where you can go for extra help in math. Then write down three of the suggestions.

____

____

____

8. Do you keep an organized notebook for this class? Can you think of any suggestions that you can share with your group members to help them keep their materials organized?

9. Do you think that you have math anxiety? Read the following list for some possible solutions. Check the activities that you can realistically try to help you overcome this problem.

_____ Read a book on math anxiety.

_____ Search the Web for helpful tips on handling math anxiety.

_____ See a counselor to discuss your anxiety.

_____ See your instructor to inform him or her about your situation.

_____ Evaluate your time management to see if you are trying to do too much. Then adjust your schedule accordingly.

10. Some students favor different methods of learning over others. For example, you might prefer:

- Learning through listening and hearing.
- Learning through seeing images, watching demonstrations, and visualizing diagrams and charts.
- Learning by experience through a hands-on approach by doing things.
- Learning through reading and writing.

Most experts believe that the most effective learning comes when a student engages in *all* of these activities. However, each individual is different and may benefit from one activity more than another. You can visit a number of different websites to determine your "learning style." Try doing a search on the Internet with the key words "*learning styles assessment.*" Once you have found a suitable website, answer the questionnaire and the site will give you feedback on what method of learning works best for you.

Introduction to Whole Numbers

Section 1.2

Objectives

1. Place Value
2. Standard Notation and Expanded Notation
3. Writing Numbers in Words
4. The Number Line and Order

1. Place Value

Numbers provide the foundation that is used in mathematics. We begin this chapter by discussing how numbers are represented and named. All numbers in our numbering system are composed from the **digits** 0, 1, 2, 3, 4, 5, 6, 7, 8, and 9. In mathematics, the numbers 0, 1, 2, 3, 4, 5, 6, 7, 8, 9, 10, 11, 12, . . . are called the *whole numbers.* (The three dots are called *ellipses* and indicate that the list goes on indefinitely.)

For large numbers, commas are used to separate digits into groups of three called **periods**. For example, the number of live births in the United States in a recent year was 4,058,614 (*Source: The World Almanac*). Numbers written in this way are said to be in **standard form**. The position of each digit within a number determines the place value of the digit. To interpret the number of births in the United States, refer to the place value chart (Figure 1-1).

Concept Connections

1. Determine the place value of each 3 in the number 303.

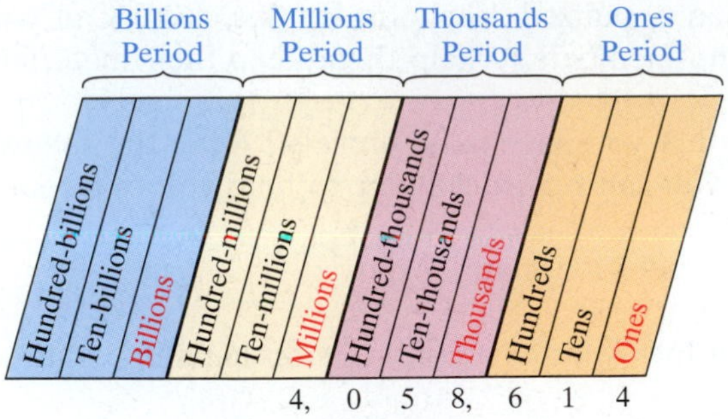

Figure 1-1

The digit 5 in the number 4,058,614 represents 5 ten-thousands because it is in the ten-thousands place. The digit 4 on the left represents 4 millions, whereas the digit 4 on the right represents 4 ones.

Skill Practice

Determine the place value of the digit 4 in each number.

2. 547,098,632
3. 1,659,984,036
4. 6420

Example 1 Determining Place Value

Determine the place value of the digit 2 in each number.

a. 417,216,900 **b.** 724 **c.** 502,000,700

Solution:

a. 417,216,900 hundred-thousands

b. 724 tens

c. 502,000,700 millions

Skill Practice

5. Alaska is the largest state geographically. Its land area is 571,962 square miles (mi^2). Give the place value for each digit.

Example 2 Determining Place Value

The altitude of Mount Everest, the highest mountain on earth, is 29,035 feet (ft). Give the place value for each digit in this number.

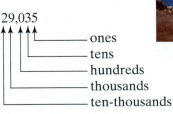

Solution:

29,035

- 5 — ones
- 3 — tens
- 0 — hundreds
- 9 — thousands
- 2 — ten-thousands

Answers

1. First 3 (on the left) represents 3 hundreds, while the second 3 (on the right) represents 3 ones.
2. Ten-millions
3. Thousands
4. Hundreds
5. 5: hundred-thousands
   7: ten-thousands
   1: thousands 9: hundreds
   6: tens 2: ones

2. Standard Notation and Expanded Notation

A number can also be written in an expanded form by writing each digit with its place value unit. For example, the number 287 can be written as

$$287 = 2 \text{ hundreds} + 8 \text{ tens} + 7 \text{ ones}$$

This is called **expanded form**.

Example 3 Converting Standard Form to Expanded Form

Convert to expanded form.

a. 4672 **b.** 257,016

Solution:

a. 4672 4 thousands + 6 hundreds + 7 tens + 2 ones

b. 257,016 2 hundred-thousands + 5 ten-thousands + 7 thousands + 1 ten + 6 ones

Skill Practice

Convert to expanded form.

6. 837

7. 4,093,062

Example 4 Converting Expanded Form to Standard Form

Convert to standard form.

a. 2 hundreds + 5 tens + 9 ones

b. 1 thousand + 2 tens + 5 ones

Solution:

a. 2 hundreds + 5 tens + 9 ones = 259

b. Each place position from the thousands place to the ones place must contain a digit. In this problem, there is no reference to the hundreds place digit. Therefore, we assume 0 hundreds. Thus,

1 thousand + 0 hundreds + 2 tens + 5 ones = 1025

Skill Practice

Convert to standard form.

8. 8 thousands + 5 hundreds + 5 tens + 1 one

9. 5 hundred-thousands + 4 thousands + 8 tens + 3 ones

3. Writing Numbers in Words

The word names of some two-digit numbers appear with a hyphen, while others do not. For example:

Number	Number Name
12	twelve
68	sixty-eight
40	forty
42	forty-two

To write a three-digit or larger number, begin at the leftmost group of digits. The number named in that group is followed by the period name, followed by a comma. Then the next period is named, and so on.

Answers

6. 8 hundreds + 3 tens + 7 ones

7. 4 millions + 9 ten-thousands + 3 thousands + 6 tens + 2 ones

8. 8551 **9.** 504,083

Skill Practice

10. Write the number 1,450,327,214 in words.

Example 5 Writing a Number in Words

Write the number 621,417,325 in words.

Solution:

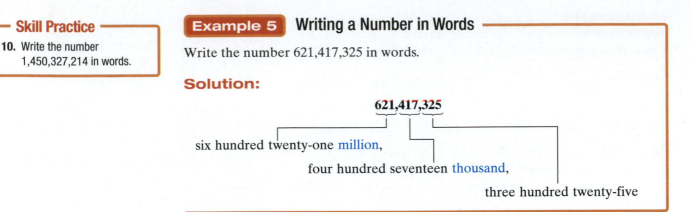

621,417,325

six hundred twenty-one million,

four hundred seventeen thousand,

three hundred twenty-five

Notice from Example 5 that when naming numbers, the name of the ones period is not attached to the last group of digits. Also note that for whole numbers, the word *and* should not appear in word names. For example, the number 405 should be written as four hundred five.

Skill Practice

11. Write the number in standard form: fourteen thousand, six hundred nine.

Example 6 Writing a Number in Standard Form

Write the number in standard form.

Six million, forty-six thousand, nine hundred three

Solution:

six million nine hundred three

6,046,903

forty-six thousand

We have seen several examples of writing a number in standard form, in expanded form, and in words. Standard form is the most concise representation. Also note that when we write a four-digit number in standard form, the comma is often omitted. For example, the number 4,389 is often written as 4389.

4. The Number Line and Order

Whole numbers can be visualized as equally spaced points on a line called a *number line* (Figure 1-2).

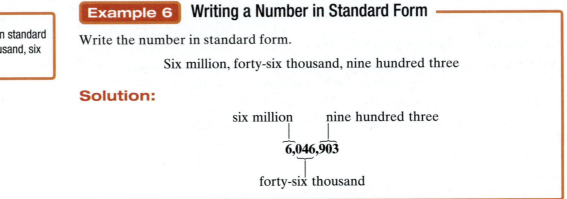

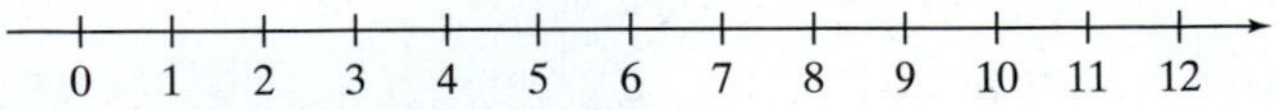

Figure 1-2

The whole numbers begin at 0 and are ordered from left to right by increasing value.

A number is graphed on a number line by placing a dot at the corresponding point. For any two numbers graphed on a number line, the number to the

Answers

10. One billion, four hundred fifty million, three hundred twenty-seven thousand, two hundred fourteen

11. 14,609

left is less than the number to the right. Similarly, a number to the right is greater than the number to the left. In mathematics, the symbol $<$ is used to denote "is less than," and the symbol $>$ means "is greater than." Therefore,

$3 < 5$	means	3 is less than 5
$5 > 3$	means	5 is greater than 3

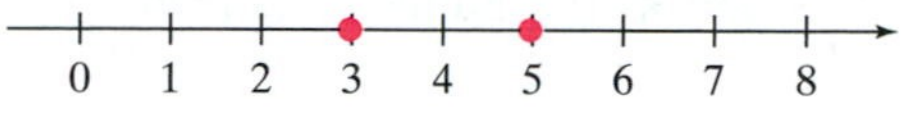

Example 7 Determining Order of Two Numbers

Fill in the blank with the symbol $<$ or $>$.

a. 7 ☐ 0 **b.** 30 ☐ 82

Solution:

a. $7 \boxed{>} 0$

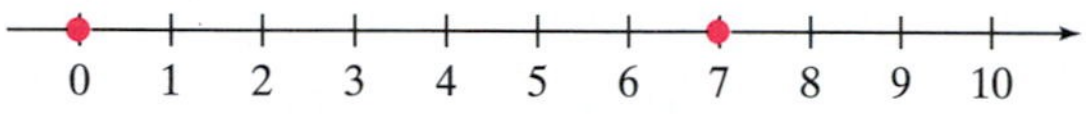

b. $30 \boxed{<} 82$

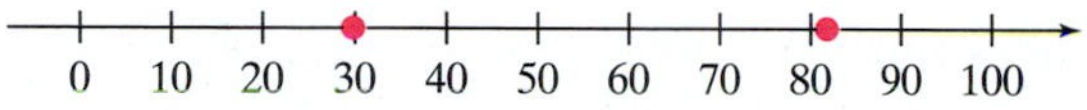

To visualize the numbers 82 and 30 on the number line, it may be necessary to use a different scale. Rather than setting equally spaced marks in units of 1, we can use units of 10. The number 82 must be somewhere between 80 and 90 on the number line.

Skill Practice

Fill in the blank with the symbol $<$ or $>$.

12. 9 ☐ 5

13. 8 ☐ 18

Answers

12. $>$ **13.** $<$

Section 1.2 Practice Exercises

Boost *your* GRADE at ALEKS.com!

ALEKS version 3.0

- Practice Problems
- Self-Tests
- NetTutor
- e-Professors
- Videos

Study Skills Exercises

In this text, we provide skills for you to enhance your learning experience. Many of the practice exercises begin with an activity that focuses on one of seven areas: learning about your course, using your text, taking notes, doing homework, taking an exam (test and math anxiety), managing your time, and studying for the final exam.

Each activity requires only a few minutes and will help you pass this class and become a better math student. Many of these skills can be carried over to other disciplines and help you become a model college student.

1. To begin, write down the following information.

a. Instructor's name

b. Instructor's office number

c. Instructor's telephone number

d. Instructor's email address

e. Instructor's office hours

f. Days of the week that the class meets

g. The room number in which the class meets

h. Is there a lab requirement for this course? If so, where is the lab located and how often must you go?

2. Define the key terms.

a. Digit **b. Standard form** **c. Periods** **d. Expanded form**

Objective 1: Place Value

3. Name the place value for each digit in the number 8,213,457.

4. Name the place value for each digit in the number 103,596.

For Exercises 5–24, determine the place value for each underlined digit. **(See Example 1.)**

5. 321 **6.** 689 **7.** 214 **8.** 738

9. 8710 **10.** 2293 **11.** 1430 **12.** 3101

13. 452,723 **14.** 655,878 **15.** 1,023,676,207 **16.** 3,111,901,211

17. 22,422 **18.** 58,106 **19.** 51,033,201 **20.** 93,971,224

21. The number of U.S. travelers abroad in a recent year was 10,677,881. **(See Example 2.)**

22. The area of Lake Superior is 31,820 square miles (mi^2).

23. For a recent year, the total number of U.S. $1 bills in circulation was 7,653,468,440.

24. For a certain flight, the cruising altitude of a commercial jet is 31,000 ft.

Objective 2: Standard Notation and Expanded Notation

For Exercises 25–32, convert the numbers to expanded form. **(See Example 3.)**

25. 58 **26.** 71 **27.** 539 **28.** 382

29. 5203 **30.** 7089 **31.** 10,241 **32.** 20,873

For Exercises 33–40, convert the numbers to standard form. **(See Example 4.)**

33. 5 hundreds + 2 tens + 4 ones

34. 3 hundreds + 1 ten + 8 ones

35. 1 hundred + 5 tens

36. 6 hundreds + 2 tens

37. 1 thousand + 9 hundreds + 6 ones

38. 4 thousands + 2 hundreds + 1 one

39. 8 ten-thousands + 5 thousands + 7 ones

40. 2 ten-thousands + 6 thousands + 2 ones

41. Name the first four periods of a number (from right to left).

42. Name the first four place values of a number (from right to left).

Objective 3: Writing Numbers in Words

For Exercises 43–50, write the number in words. **(See Example 5.)**

43. 241

44. 327

45. 603

46. 108

47. 31,530

48. 52,160

49. 100,234

50. 400,199

51. The Shuowen jiezi dictionary, an ancient Chinese dictionary that dates back to the year 100, contained 9535 characters. Write the number 9535 in words.

52. Interstate I-75 is 1377 miles (mi) long. Write the number 1377 in words.

53. The altitude of Mt. McKinley in Alaska is 20,320 ft. Write the number 20,320 in words.

54. There are 1800 seats in the Regal Champlain Theater in Plattsburgh, New York. Write the number 1800 in words.

55. Researchers calculate that about 590,712 stone blocks were used to construct the Great Pyramid. Write the number 590,712 in words.

56. In the United States, there are approximately 60,000,000 cats living in households. Write the number 60,000,000 in words.

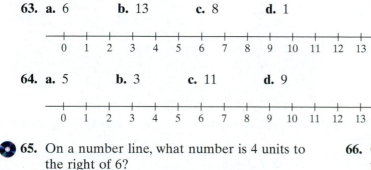

For Exercises 57–62, convert the number to standard form. **(See Example 6.)**

57. Six thousand, five

58. Four thousand, four

59. Six hundred seventy-two thousand

60. Two hundred forty-eight thousand

61. One million, four hundred eighty-four thousand, two hundred fifty

62. Two million, six hundred forty-seven thousand, five hundred twenty

Objective 4: The Number Line and Order

For Exercises 63–64, graph the numbers on the number line.

63. **a.** 6 **b.** 13 **c.** 8 **d.** 1

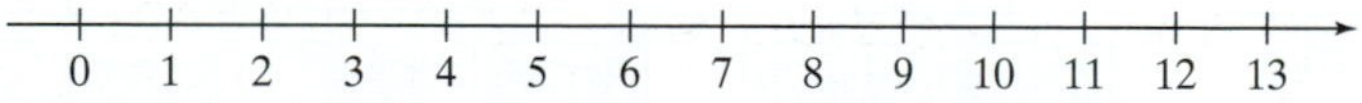

64. **a.** 5 **b.** 3 **c.** 11 **d.** 9

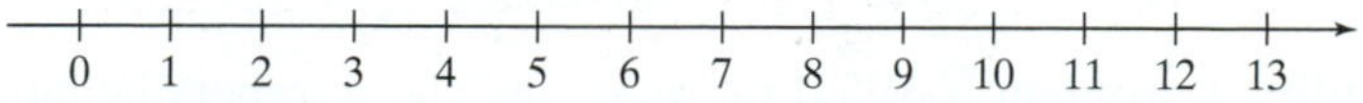

65. On a number line, what number is 4 units to the right of 6?

66. On a number line, what number is 8 units to the left of 11?

67. On a number line, what number is 3 units to the left of 7?

68. On a number line, what number is 5 units to the right of 0?

For Exercises 69–72, translate the inequality to words.

69. $8 > 2$ **70.** $6 < 11$ **71.** $3 < 7$ **72.** $14 > 12$

For Exercises 73–84, fill in the blank with the inequality symbol $<$ or $>$. **(See Example 7.)**

73. 6 ☐ 11 **74.** 14 ☐ 13 **75.** 21 ☐ 18 **76.** 5 ☐ 7

77. 3 ☐ 7 **78.** 14 ☐ 24 **79.** 95 ☐ 89 **80.** 28 ☐ 30

81. 0 ☐ 3 **82.** 8 ☐ 0 **83.** 90 ☐ 91 **84.** 48 ☐ 47

Expanding Your Skills

85. Answer true or false. The number 12 is a digit.

86. Answer true or false. The number 26 is a digit.

87. What is the greatest two-digit number?

88. What is the greatest three-digit number?

89. What is the greatest whole number?

90. What is the least whole number?

91. How many zeros are there in the number ten million?

92. How many zeros are there in the number one hundred billion?

93. What is the greatest three-digit number that can be formed from the digits 6, 9, and 4? Use each digit only once.

94. What is the greatest three-digit number that can be formed from the digits 0, 4, and 8? Use each digit only once.

Section 1.3 Addition and Subtraction of Whole Numbers and Perimeter

Objectives

1. Addition of Whole Numbers
2. Properties of Addition
3. Subtraction of Whole Numbers
4. Translations and Applications Involving Addition and Subtraction
5. Perimeter

1. Addition of Whole Numbers

We use addition of whole numbers to represent an increase in quantity. For example, suppose Jonas typed 5 pages of a report before lunch. Later in the afternoon he typed 3 more pages. The total number of pages that he typed is found by adding 5 and 3.

$$5 \text{ pages} + 3 \text{ pages} = 8 \text{ pages}$$

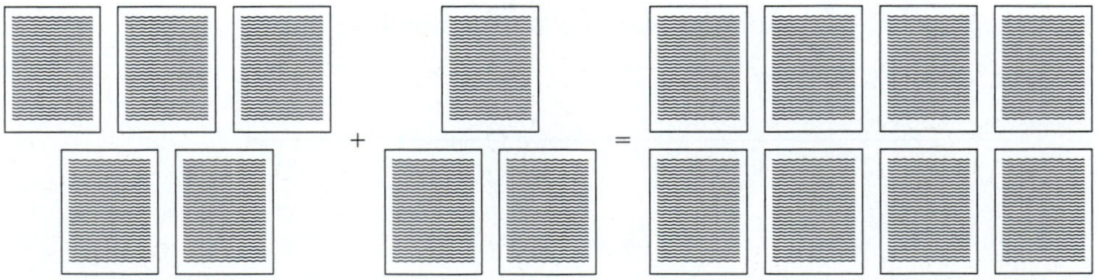

The result of an addition problem is called the **sum**, and the numbers being added are called **addends**. Thus,

$$5 + 3 = 8$$

addends (5, 3) sum (8)

Concept Connections

1. Identify the addends and the sum.
$3 + 7 + 12 = 22$

Answer

1. Addends: 3, 7, and 12; sum: 22

The number line is a useful tool to visualize the operation of addition. To add 5 and 3 on a number line, begin at 5 and move 3 units to the right. The final location indicates the sum.

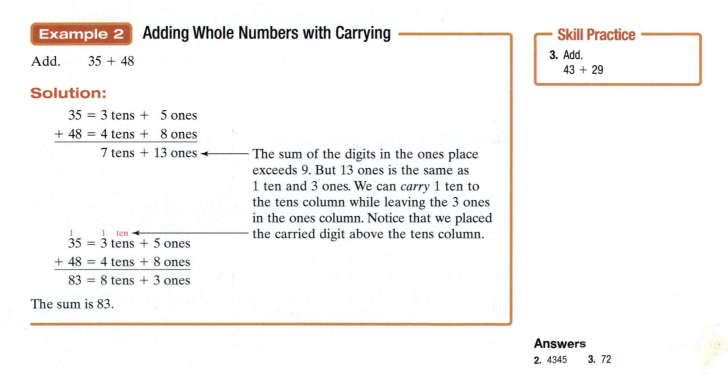

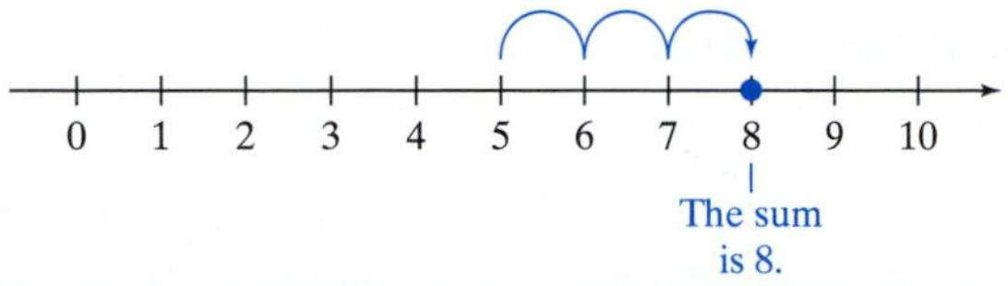

You can use a number line to find the sum of any pair of digits. The sums for all possible pairs of one-digit numbers should be memorized (see Exercise 7). Memorizing these basic addition facts will make it easier for you to add larger numbers.

To add whole numbers with several digits, line up the numbers vertically by place value. Then add the digits in the corresponding place positions.

Example 1 Adding Whole Numbers

Add. 261 + 28

Solution:

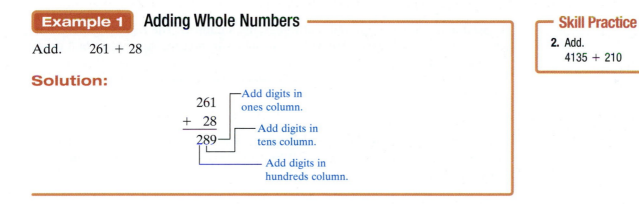

$$\begin{array}{r} 261 \\ +\ \ 28 \\ \hline 289 \end{array}$$

Add digits in ones column.

Add digits in tens column.

Add digits in hundreds column.

> **Skill Practice**
>
> **2.** Add.
> 4135 + 210

Sometimes when adding numbers, the sum of the digits in a given place position is greater than 9. If this occurs, we must do what is called *carrying* or *regrouping*. Example 2 illustrates this process.

Example 2 Adding Whole Numbers with Carrying

Add. 35 + 48

Solution:

$$\begin{array}{rl} 35 = & 3 \text{ tens} + 5 \text{ ones} \\ +\ 48 = & 4 \text{ tens} + 8 \text{ ones} \\ \hline & 7 \text{ tens} + 13 \text{ ones} \end{array}$$

The sum of the digits in the ones place exceeds 9. But 13 ones is the same as 1 ten and 3 ones. We can *carry* 1 ten to the tens column while leaving the 3 ones in the ones column. Notice that we placed the carried digit above the tens column.

$$\begin{array}{rl} \overset{1}{3}5 = & \overset{1 \text{ ten}}{3} \text{ tens} + 5 \text{ ones} \\ +\ 48 = & 4 \text{ tens} + 8 \text{ ones} \\ \hline 83 = & 8 \text{ tens} + 3 \text{ ones} \end{array}$$

The sum is 83.

> **Skill Practice**
>
> **3.** Add.
> 43 + 29

Answers

2. 4345 **3.** 72

Addition of numbers may include more than two addends.

Skill Practice

4. Add.

$$\begin{array}{r} 57{,}296 \\ 4{,}089 \\ +\ 9{,}762 \\ \hline \end{array}$$

Example 3 **Adding Whole Numbers**

Add. 21,076 + 84,158 + 2419

Solution:

$$\begin{array}{r} \scriptstyle 1\,2 \\ 21{,}076 \\ 84{,}158 \\ +\ \ 2{,}419 \\ \hline 107{,}653 \end{array}$$

In this example, the sum of the digits in the ones column is 23. Therefore, we write the 3 and carry the 2.

In the tens column, the sum is 15. Write the 5 in the tens place and carry the 1.

2. Properties of Addition

A **variable** is a letter or symbol that represents a number. The following are examples of variables: a, b, and c. We will use variables to present three important properties of addition.

Most likely you have noticed that 0 added to any number is that number. For example:

$6 + 0 = 6$ $527 + 0 = 527$ $0 + 88 = 88$ $0 + 15 = 15$

In each example, the number in red can be replaced with any number that we choose, and the statement would still be true. This fact is stated as the addition property of 0.

PROPERTY Addition Property of 0

For any number a,

$$a + 0 = a \quad \text{and} \quad 0 + a = a$$

The sum of any number and 0 is that number.

The order in which we add two numbers does not affect the result. For example: $11 + 20 = 20 + 11$. This is true for any two numbers and is stated in the next property.

PROPERTY Commutative Property of Addition

For any numbers a and b,

$$a + b = b + a$$

Changing the order of two addends does not affect the sum.

In mathematics we use parentheses ( ) as grouping symbols. To add more than two numbers, we can group them and then add. For example:

$(2 + 3) + 8$ Parentheses indicate that $2 + 3$ is added first. Then 8 is added to the result.

$= 5 + 8$

$= 13$

Answer

4. 71,147

$2 + (3 + 8)$ Parentheses indicate that 3 + 8 is added first. Then the result is added to 2.

$= 2 + 11$

$= 13$

PROPERTY Associative Property of Addition

For any numbers a, b, and c,

$$(a + b) + c = a + (b + c)$$

The manner in which addends are grouped does not affect the sum.

Example 4 Applying the Properties of Addition

a. Rewrite 9 + 6, using the commutative property of addition.

b. Rewrite (15 + 9) + 5, using the associative property of addition.

Solution:

a. $9 + 6 = 6 + 9$ Change the order of the addends.

b. $(15 + 9) + 5 = 15 + (9 + 5)$ Change the grouping of the addends.

Skill Practice

5. Rewrite 3 + 5, using the commutative property of addition.

6. Rewrite (1 + 7) + 12, using the associative property of addition.

3. Subtraction of Whole Numbers

Jeremy bought a case of 12 sodas, and on a hot afternoon he drank 3 of the sodas. We can use the operation of subtraction to find the number of sodas remaining.

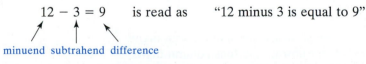

12 sodas − 3 sodas = 9 sodas

The symbol "−" between two numbers is a subtraction sign, and the result of a subtraction is called the **difference**. The number being subtracted (in this case, 3) is called the **subtrahend**. The number 12 from which 3 is subtracted is called the **minuend**.

$12 - 3 = 9$ is read as "12 minus 3 is equal to 9"

minuend subtrahend difference

Subtraction is the reverse operation of addition. To find the number of sodas that remain after Jeremy takes 3 sodas away from 12 sodas, we ask the following question:

"3 added to what number equals 12?"

That is,

$12 - 3 = ?$ is equivalent to $? + 3 = 12$

Answers

5. $3 + 5 = 5 + 3$

6. $(1 + 7) + 12 = 1 + (7 + 12)$

Subtraction can also be visualized on the number line. To evaluate $7 - 4$, start from the point on the number line corresponding to the minuend (7 in this case). Then move to the *left* 4 units. The resulting position on the number line is the difference.

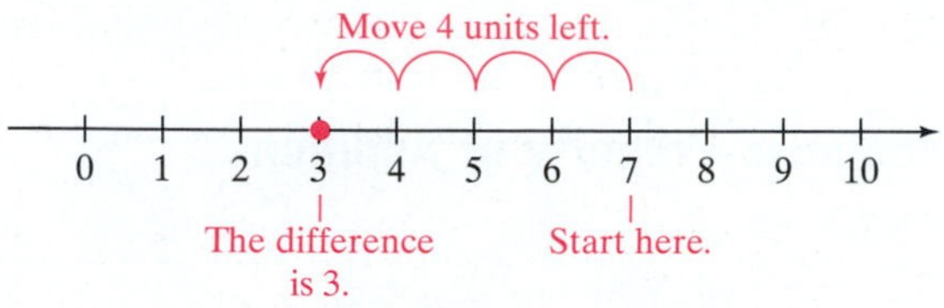

To check the result, we can use addition.

$$7 - 4 = 3 \quad \text{because} \quad 3 + 4 = 7$$

Skill Practice

Subtract. Check by using addition.

7. $11 - 5$ **8.** $8 - 0$

9. $7 - 2$ **10.** $5 - 5$

Example 5 Subtracting Whole Numbers

Subtract and check the answer by using addition.

a. $8 - 2$ **b.** $10 - 6$ **c.** $5 - 0$ **d.** $3 - 3$

Solution:

a. $8 - 2 = 6$ because $6 + 2 = 8$

b. $10 - 6 = 4$ because $4 + 6 = 10$

c. $5 - 0 = 5$ because $5 + 0 = 5$

d. $3 - 3 = 0$ because $0 + 3 = 3$

When subtracting large numbers, it is usually more convenient to write the numbers vertically. We write the minuend on top and the subtrahend below it. Starting from the ones column, we subtract digits having corresponding place values.

Skill Practice

Subtract. Check by using addition.

11. $472 - 261$ **12.** $3947 - 137$

Example 6 Subtracting Whole Numbers

Subtract and check the answer by using addition.

a. $976 - 124$ **b.** $2498 - 197$

Solution:

a.

$$\begin{array}{r} 976 \\ -\ 124 \\ \hline 852 \end{array} \qquad \underline{\text{Check}}: \begin{array}{r} 852 \\ +\ 124 \\ \hline 976\ \checkmark \end{array}$$

- Subtract the ones column digits.
- Subtract the tens column digits.
- Subtract the hundreds column digits.

b.

$$\begin{array}{r} 2498 \\ -\ 197 \\ \hline 2301 \end{array} \qquad \underline{\text{Check}}: \begin{array}{r} 2301 \\ +\ 197 \\ \hline 2498\ \checkmark \end{array}$$

Answers

7. 6 **8.** 8 **9.** 5 **10.** 0

11. 211 **12.** 3810

When a digit in the subtrahend (bottom number) is larger than the corresponding digit in the minuend (top number), we must "regroup" or borrow a value from the column to the left.

$$\begin{array}{r} 92 = 9 \text{ tens} + 2 \text{ ones} \\ -\ 74 = 7 \text{ tens} + 4 \text{ ones} \\ \hline \end{array}$$

In the ones column, we cannot take 4 away from 2. We will regroup by borrowing 1 ten from the minuend. Furthermore, 1 ten = 10 ones.

$$\begin{array}{r} \overset{8\ +10}{\not{9}\ 2} = \overset{8}{\not{9}} \text{ tens} + \overset{+10 \text{ ones}}{2 \text{ ones}} \\ -\ 7\ 4 = 7 \text{ tens} + 4 \text{ ones} \\ \hline \end{array}$$

We now have 12 ones in the minuend.

$$\begin{array}{r} \overset{8\ 12}{\not{9}\not{2}} = \overset{8}{\not{9}} \text{ tens} + 12 \text{ ones} \\ -\ 7\,4 = 7 \text{ tens} + \ 4 \text{ ones} \\ \hline 1\,8 = 1 \text{ ten} + \ 8 \text{ ones} \end{array}$$

> **TIP:** The process of *borrowing* in subtraction is the reverse operation of *carrying* in addition.

Example 7 Subtracting Whole Numbers with Borrowing

Subtract and check the result with addition.

$$\begin{array}{r} 134{,}616 \\ -\ 53{,}438 \\ \hline \end{array}$$

Solution:

$$\begin{array}{r} 134{,}6\overset{0}{\not{1}}\overset{16}{\not{6}} \\ -\ 53{,}438 \\ \hline 8 \end{array}$$

In the ones place, 8 is greater than 6. We borrow 1 ten from the tens place.

$$\begin{array}{r} 134{,}\overset{5}{\not{6}}\overset{\overset{10}{\not{0}}}{\not{1}}\overset{16}{\not{6}} \\ -\ 53{,}438 \\ \hline 78 \end{array}$$

In the tens place, 3 is greater than 0. We borrow 1 hundred from the hundreds place.

$$\begin{array}{r} \overset{0}{\not{1}}\overset{13}{\not{3}}4{,}\overset{5}{\not{6}}\overset{\overset{10}{\not{0}}}{\not{1}}\overset{16}{\not{6}} \\ -\ 53{,}438 \\ \hline 81{,}178 \end{array}$$

In the ten-thousands place, 5 is greater than 3. We borrow 1 hundred-thousand from the hundred-thousands place.

Check:

$$\begin{array}{r} \overset{1\,1}{81{,}178} \\ +\ 53{,}438 \\ \hline 134{,}616\ \checkmark \end{array}$$

Skill Practice

Subtract. Check by addition.

13. $\begin{array}{r} 23{,}126 \\ -\ 6{,}048 \\ \hline \end{array}$

Answer

13. 17,078

Skill Practice

Subtract. Check by addition.

14. 700 − 531

Example 8 Subtracting Whole Numbers with Borrowing

Subtract and check the result with addition. 500 − 247

Solution:

$$\begin{array}{r} 500 \\ -\,247 \\ \hline \end{array}$$

In the ones place, 7 is greater than 0. We try to borrow 1 ten from the tens place. However, the tens place digit is 0. Therefore we must first borrow from the hundreds place.

$$\begin{array}{r} {\scriptstyle 4\ 10}\quad \\ \not{5}\,\not{0}\,0 \\ -\,2\ 4\ 7 \\ \hline \end{array}$$

← 1 hundred = 10 tens

$$\begin{array}{r} {\scriptstyle 9}\quad\; \\ {\scriptstyle 4\ \not{10}\ 10} \\ \not{5}\,\not{0}\,\not{0} \\ -\,2\ 4\ 7 \\ \hline 2\ 5\ 3 \end{array}$$

← Now we can borrow 1 ten to add to the ones place.

Subtract.

Check:

$$\begin{array}{r} {\scriptstyle 1\,1}\; \\ 253 \\ +\,247 \\ \hline 500 \end{array}$$ ✓

4. Translations and Applications Involving Addition and Subtraction

In the English language, there are many different words and phrases that imply addition. A partial list is given in Table 1-1.

Table 1-1

Word/Phrase	Example	In Symbols
Sum	The sum of 6 and x	$6 + x$
Added to	3 added to 8	$8 + 3$
Increased by	y increased by 2	$y + 2$
More than	10 more than 6	$6 + 10$
Plus	8 plus 3	$8 + 3$
Total of	The total of a and b	$a + b$

Skill Practice

Translate and simplify.

15. 50 more than 80

16. 12 increased by 14

Example 9 Translating an English Phrase to a Mathematical Statement

Translate each phrase to an equivalent mathematical statement and simplify.

a. 12 added to 109

b. The sum of 1386 and 376

Answers

14. 169 **15.** 80 + 50; 130

16. 12 + 14; 26

Solution:

a. 109 + 12

$$\begin{array}{r} \overset{1}{1}09 \\ +\ 12 \\ \hline 121 \end{array}$$

b. 1386 + 376

$$\begin{array}{r} 1\overset{1}{3}\overset{1}{8}6 \\ +\ 376 \\ \hline 1762 \end{array}$$

Table 1-2 gives several key phrases that imply subtraction.

Table 1-2

Word/Phrase	Example	In Symbols
Minus	15 minus x	$15 - x$
Difference	The difference of 10 and 2	$10 - 2$
Decreased by	a decreased by 1	$a - 1$
Less than	5 less than 12	$12 - 5$
Subtract . . . from	Subtract 3 from 8	$8 - 3$
Subtracted from	6 subtracted from 10	$10 - 6$

In Table 1-2, make a note of the last three entries. The phrases *less than*, *subtract . . . from* and *subtracted from* imply a specific order in which the subtraction is performed. In all three cases, begin with the second number listed and subtract the first number listed.

Example 10 **Translating an English Phrase to a Mathematical Statement**

Translate the English phrase to a mathematical statement and simplify.

a. The difference of 150 and 38 **b.** 30 subtracted from 82

Solution:

a. From Table 1-2, the *difference* of 150 and 38 implies $150 - 38$.

$$\begin{array}{r} 1\overset{4}{\cancel{5}}\overset{10}{0} \\ -\ \ 38 \\ \hline 112 \end{array}$$

b. The phrase "30 subtracted from 82" implies that 30 is taken away from 82. We have $82 - 30$.

$$\begin{array}{r} 82 \\ -\ 30 \\ \hline 52 \end{array}$$

Skill Practice

Translate the English phrase to a mathematical statement and simplify.

17. Twelve decreased by eight

18. Subtract three from nine.

We noted earlier that addition is commutative. That is, the order in which two numbers are added does not affect the sum. This is *not* true for subtraction. For example, $82 - 30$ is not equal to $30 - 82$. The symbol $\neq$ means "is not equal to." Thus, $82 - 30 \neq 30 - 82$.

Answers

17. $12 - 8$; 4

18. $9 - 3$; 6

In Examples 11 and 12, we use addition and subtraction of whole numbers to solve application problems.

Skill Practice

19. Refer to the table in Example 11.
a. Find the total number of bronze medals won.
b. Find the number of medals won by the United States.

Example 11 Solving an Application Problem Involving a Table

The table gives the number of gold, silver, and bronze medals won in the 2006 Winter Olympics for selected countries.

a. Find the total number of medals won by Canada.

b. Determine the total number of silver medals won by these three countries.

	Gold	Silver	Bronze
Germany	11	12	6
United States	9	9	7
Canada	7	10	7

Solution:

a. The number of medals won by Canada appears in the last row of the table. The word "total" implies addition.

$7 + 10 + 7 = 24$ Canada won 24 medals.

b. The number of silver medals is given in the middle column. The total is

$12 + 9 + 10 = 31$ There were 31 silver medals won by these countries.

Skill Practice

Refer to the graph for Example 12.
20. a. Has the number of robberies increased or decreased between 1995 and 2005?
b. Determine the amount of increase or decrease.

Example 12 Solving an Application Problem

The number of reported robberies in the United States has fluctuated for selected years, as shown in the graph.

a. Find the increase in the number of robberies from the year 2000 to 2005.

b. Find the decrease in the number of robberies from the year 1995 to 2000.

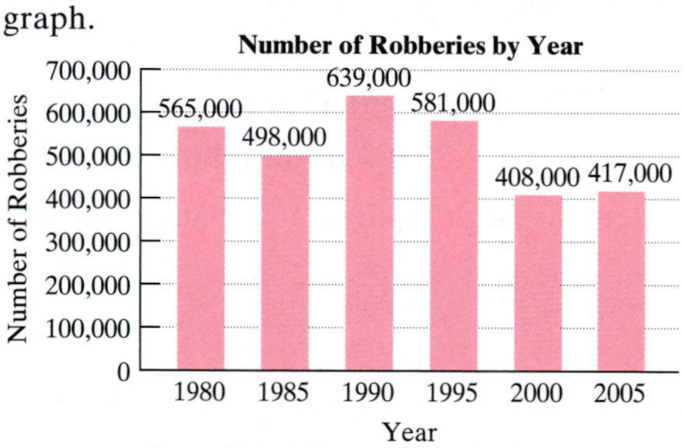

Source: Federal Bureau of Investigation

Solution:

For the purpose of finding an amount of increase or decrease, we will subtract the smaller number from the larger number.

a. Because the number of robberies went *up* from 2000 to 2005, there was an *increase*. To find the amount of increase, subtract the smaller number from the larger number.

$$\begin{array}{r} 4\overset{0}{\cancel{1}}\overset{17}{\cancel{7}},000 \\ -\ 408,000 \\ \hline 9,000 \end{array}$$

From 2000 to 2005, there was an increase of 9000 robberies in the United States.

Answers
19. a. 20 **b.** 25
20. a. decreased **b.** 164,000 robberies

b. Because the number of robberies went *down* from 1995 to 2000, there was a *decrease*. To find the amount of decrease, subtract the smaller number from the larger number.

$$\begin{array}{r} \scriptstyle 7\ 11 \\ 5\not{8}\not{1},000 \\ -\ 408,000 \\ \hline 173,000 \end{array}$$

From 1995 to 2000 there was a decrease of 173,000 robberies in the United States.

5. Perimeter

One special application of addition is to find the perimeter of a polygon. A **polygon** is a flat closed figure formed by line segments connected at their ends. Familiar figures such as triangles, rectangles, and squares are examples of polygons. See Figure 1-3.

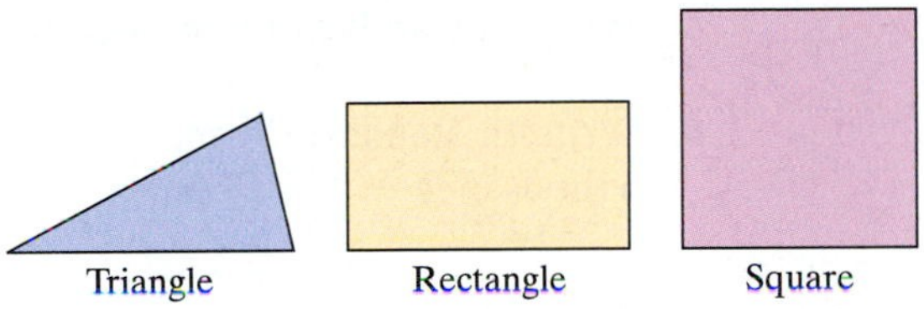

Figure 1-3

Animation

The **perimeter** of any polygon is the distance around the outside of the figure. To find the perimeter, add the lengths of the sides.

Example 13 **Finding Perimeter**

A paving company wants to edge the perimeter of a parking lot with concrete curbing. Find the perimeter of the parking lot.

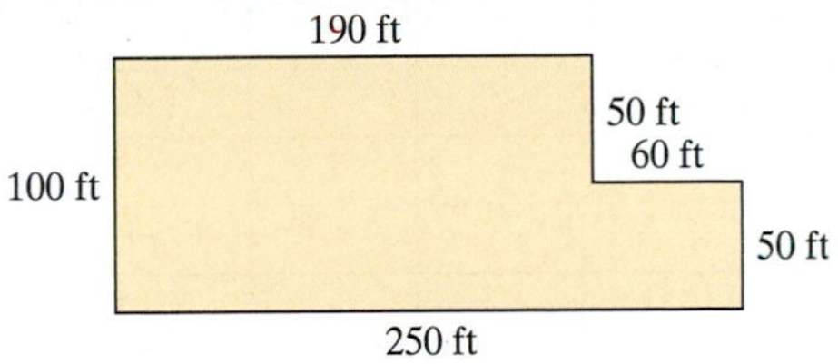

Solution:

The perimeter is the sum of the lengths of the sides.

$$\begin{array}{r} \scriptstyle 3 \\ 190\text{ ft} \\ 50\text{ ft} \\ 60\text{ ft} \\ 50\text{ ft} \\ 250\text{ ft} \\ +\ 100\text{ ft} \\ \hline 700\text{ ft} \end{array}$$

The distance around the parking lot (the perimeter) is 700 ft.

Skill Practice

21. Find the perimeter of the garden.

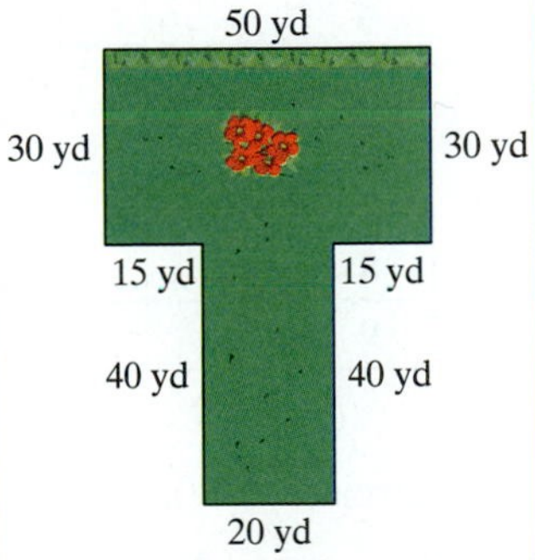

Answer

21. 240 yd

Section 1.3 Practice Exercises

Study Skills Exercises

1. It is very important to attend class every day. Math is cumulative in nature, and you must master the material learned in the previous class to understand the lesson for the next day. Because this is so important, many instructors tie attendance to the final grade. Write down the attendance policy for your class.

2. Define the key terms.

a. Sum **b. Addends** **c. Polygon** **d. Perimeter**

e. Difference **f. Subtrahend** **g. Minuend** **h. Variable**

Review Exercises

For Exercises 3–6, write the number in the form indicated.

3. Write 351 in expanded form.

4. Write in standard form: two thousand, four

5. Write in standard form: four thousand, twelve

6. Write in standard form: 6 thousands + 2 hundreds + 6 ones

Objective 1: Addition of Whole Numbers

7. Fill in the table.

+	0	1	2	3	4	5	6	7	8	9
0										
1										
2										
3										
4										
5										
6										
7										
8										
9										

For Exercises 8–10, identify the addends and the sum.

8. 11 + 10 = 21

9. 1 + 13 + 4 = 18

10. 5 + 8 + 2 = 15

For Exercises 11–18, add. **(See Example 1.)**

11.
$$\begin{array}{r} 42 \\ +\ 33 \\ \hline \end{array}$$

12.
$$\begin{array}{r} 21 \\ +\ 53 \\ \hline \end{array}$$

13.
$$\begin{array}{r} 12 \\ 15 \\ +\ 32 \\ \hline \end{array}$$

14.
$$\begin{array}{r} 10 \\ 8 \\ +\ 30 \\ \hline \end{array}$$

15. 890 + 107

16. 444 + 354

17. 4 + 13 + 102

18. 11 + 221 + 5

For Exercises 19–32, add the whole numbers with carrying. **(See Examples 2–3.)**

19. $\begin{array}{r} 76 \\ +\ 45 \\ \hline \end{array}$ **20.** $\begin{array}{r} 25 \\ +\ 59 \\ \hline \end{array}$ **21.** $\begin{array}{r} 87 \\ +\ 24 \\ \hline \end{array}$ **22.** $\begin{array}{r} 38 \\ +\ 77 \\ \hline \end{array}$

23. $\begin{array}{r} 658 \\ +\ 231 \\ \hline \end{array}$ **24.** $\begin{array}{r} 642 \\ +\ 295 \\ \hline \end{array}$ **25.** $\begin{array}{r} 152 \\ +\ 549 \\ \hline \end{array}$ **26.** $\begin{array}{r} 462 \\ +\ 388 \\ \hline \end{array}$

27. $79 + 112 + 12$ **28.** $62 + 907 + 34$ **29.** $4980 + 10{,}223$ **30.** $23{,}112 + 892$

31. $10{,}223 + 25{,}782 + 4980$ **32.** $92{,}377 + 5622 + 34{,}659$

Objective 2: Properties of Addition

For Exercises 33–36, rewrite the addition problem, using the commutative property of addition. **(See Example 4.)**

33. $101 + 44 = \square + \square$ **34.** $8 + 13 = \square + \square$ **35.** $x + y = \square + \square$ **36.** $t + q = \square + \square$

For Exercises 37–40, rewrite the addition problem using the associative property of addition, by inserting a pair of parentheses. **(See Example 4.)**

37. $(23 + 9) + 10 = 23 + 9 + 10$ **38.** $7 + (12 + 8) = 7 + 12 + 8$

39. $r + (s + t) = r + s + t$ **40.** $(c + d) + e = c + d + e$

41. Explain the difference between the commutative and associative properties of addition.

42. Explain the addition property of 0. Then simplify the expressions.

a. $423 + 0$ **b.** $0 + 25$ **c.** $\begin{array}{r} 67 \\ +\ 0 \\ \hline \end{array}$ **d.** $0 + x$

Objective 3: Subtraction of Whole Numbers

For Exercises 43–44, identify the minuend, subtrahend, and the difference.

43. $12 - 8 = 4$ **44.** $\begin{array}{r} 9 \\ -\ 6 \\ \hline 3 \end{array}$

For Exercises 45–48, write the subtraction problem as a related addition problem. For example, $19 - 6 = 13$ can be written as $13 + 6 = 19$.

45. $27 - 9 = 18$ **46.** $20 - 8 = 12$ **47.** $102 - 75 = 27$ **48.** $211 - 45 = 166$

For Exercises 49–52, subtract, then check the answer by using addition. **(See Example 5.)**

49. $8 - 3$ Check: $\square + 3 = 8$ **50.** $7 - 2$ Check: $\square + 2 = 7$

51. $4 - 1$ Check: $\square + 1 = 4$ **52.** $9 - 1$ Check: $\square + 1 = 9$

For Exercises 53–56, subtract and check the answer by using addition. **(See Example 6.)**

53. $\begin{array}{r} 1347 \\ -\ 221 \\ \hline \end{array}$ **54.** $\begin{array}{r} 4865 \\ -\ 713 \\ \hline \end{array}$ **55.** $14{,}356 - 13{,}253$ **56.** $34{,}550 - 31{,}450$

For Exercises 57–72, subtract the whole numbers involving borrowing. **(See Examples 7–8.)**

57. $76 - 59$

58. $64 - 48$

59. $710 - 189$

60. $850 - 303$

61. $6002 - 1238$

62. $3000 - 2356$

63. $10{,}425 - 9{,}022$

64. $23{,}901 - 8{,}064$

65. $62{,}088 - 59{,}871$

66. $32{,}112 - 28{,}334$

67. 3700 − 2987

68. 8000 − 3788

69. 32,439 − 1498

70. 21,335 − 4123

71. 8,007,234 − 2,345,115

72. 3,045,567 − 1,871,495

73. Use the expression 7 – 4 to explain why subtraction is not commutative.

74. Is subtraction associative? Use the numbers 10, 6, 2 to explain.

Objective 4: Translations and Applications Involving Addition and Subtraction

For Exercises 75–92, translate the English phrase to a mathematical statement and simplify. **(See Examples 9–10.)**

75. The sum of 13 and 7

76. The sum of 100 and 42

77. 45 added to 7

78. 81 added to 23

79. 5 more than 18

80. 2 more than 76

81. 1523 increased by 90

82. 1320 increased by 448

83. The total of 5, 39, and 81

84. 78 decreased by 6

85. Subtract 100 from 422.

86. Subtract 42 from 89.

87. 72 less than 1090

88. 60 less than 3111

89. The difference of 50 and 13

90. The difference of 405 and 103

91. Subtract 35 from 103.

92. Subtract 14 from 91.

93. Three top television shows entertained the following number of viewers in one week: 26,548,000 for *American Idol,* 26,930,000 for *Grey's Anatomy,* and 20,805,000 for *House.* Find the sum of the viewers for these shows.

94. To schedule enough drivers for an upcoming week, a local pizza shop manager recorded the number of deliveries each day from the previous week: 38, 54, 44, 61, 97, 103, 124. What was the total number of deliveries for the week?

95. The table gives the number of desks and chairs delivered each quarter to an office supply store. Find the total number of desks delivered for the year. **(See Example 11.)**

	Chairs	Desks
1st Quarter	220	115
2nd Quarter	185	104
3rd Quarter	201	93
4th Quarter	198	111

96. A portion of Jonathan's checking account register is shown. What is the total amount of the three checks written?

Check No.	Description	Debit	Credit	Balance
1871	Electric	$60		$180
1872	Groceries	82		98
	Payroll		$1256	1354
1874	Restaurant	58		1296
	Deposit		150	1446

97. The altitude of White Mountain Peak in California is 14,246 ft. Denali in Alaska is 20,320 ft. How much higher is Denali than White Mountain Peak?

98. There are 55 DVDs to shelve one evening at a video rental store. If Jason puts away 39 before leaving for the day, how many are left for Patty to put away?

For Exercises 99–102, use the information from the graph. **(See Example 12.)**

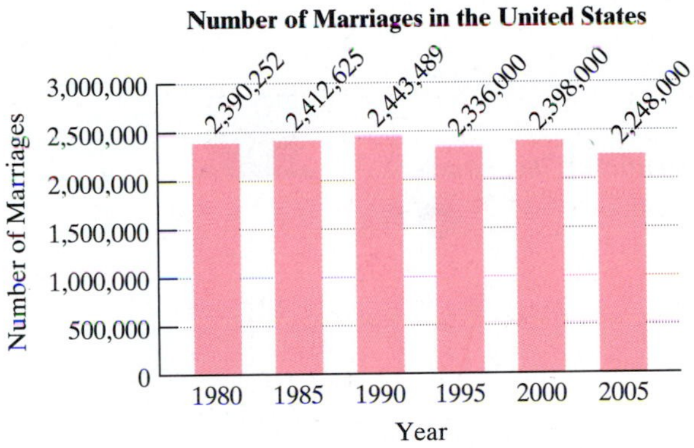

Source: National Center for Health Statistics

Figure for Exercises 99–102

99. What is the difference in the number of marriages between 1980 and 2000?

100. Find the decrease in the number of marriages in the United States between the years 2000 and 2005.

101. What is the difference in the number of marriages between the year having the greatest and the year having the least?

102. Between which two 5-year periods did the greatest increase in the number of marriages occur? What is the increase?

103. The staff for U.S. public schools is categorized in the graph. Determine the number of staff other than teachers.

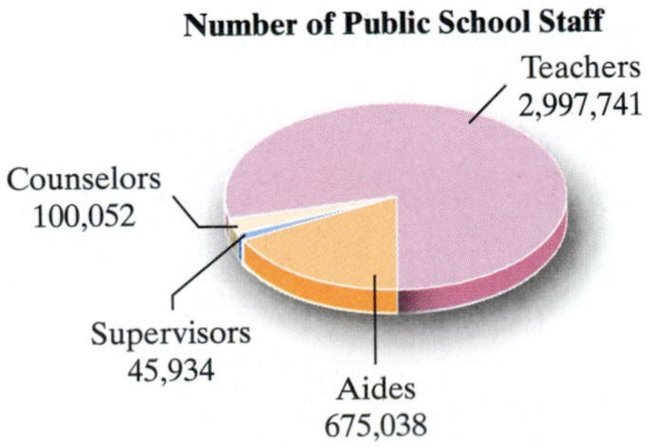

Source: National Center for Education Statistics

104. The pie graph shows the costs incurred in managing Sub-World sandwich shop for one month. From this information, determine the total cost for one month.

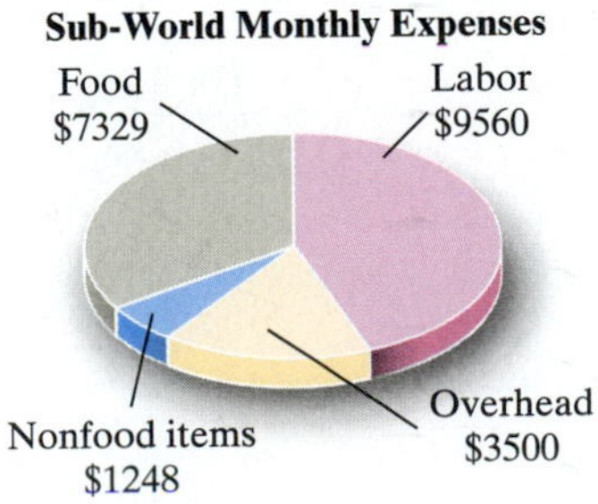

105. Pinkham Notch Visitor Center in the White Mountains of New Hampshire has an elevation of 2032 ft. The summit of nearby Mt. Washington has an elevation of 6288 ft. What is the difference in elevation?

106. A recent report in *USA Today* indicated that the play *The Lion King* had been performed 4043 times on Broadway. At that time, the play *Hairspray* had been performed 2064 times on Broadway. How many more times had *The Lion King* been performed than *Hairspray*?

107. Jeannette has two children who each attended college in Boston. Her son Ricardo attended Bunker Hill Community College where the yearly tuition and fees came to $2600. Her daughter Ricki attended M.I.T. where the yearly tuition and fees totaled $26,960. If Jeannette paid the full amount for both children to go to school, what was her total expense for tuition and fees for 1 year?

108. Clyde and Mason each leave a rest area on the Florida Turnpike. Clyde travels north and Mason travels south. After 2 hr, Clyde has gone 138 mi and Mason, who ran into heavy traffic, traveled only 96 mi. How far apart are they?

Objective 5: Perimeter

Animation

For Exercises 109–112, find the perimeter. **(See Example 13.)**

109.

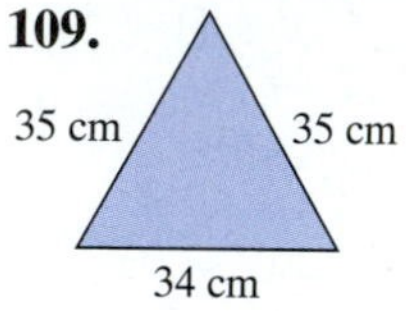

110.

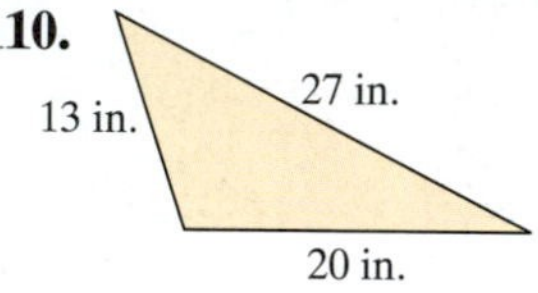

111.

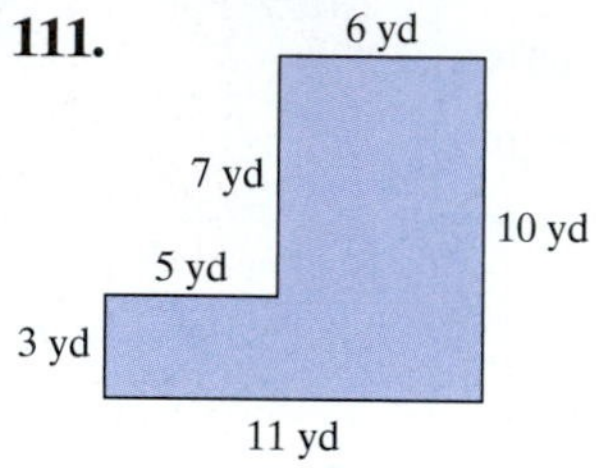

112.

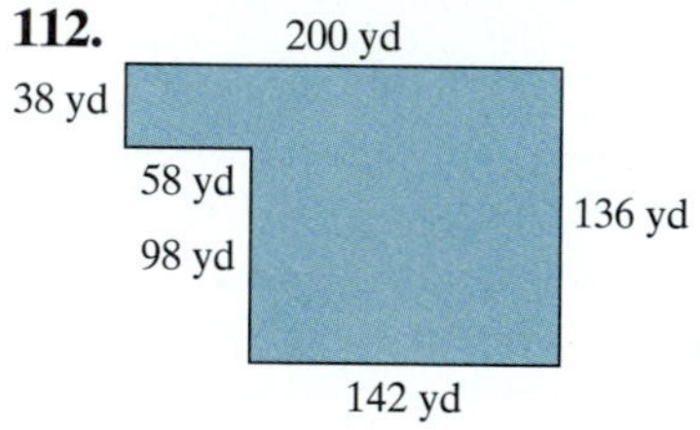

113. Find the perimeter of an NBA basketball court.

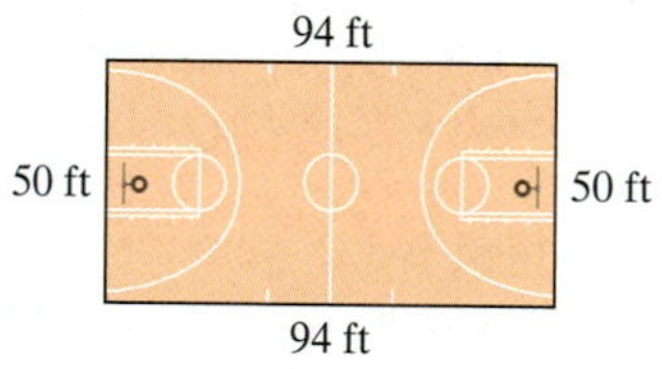

114. A major league baseball diamond is in the shape of a square. Find the distance a batter must run if he hits a home run.

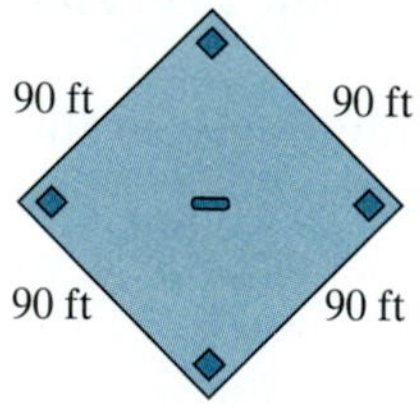

For Exercises 115–116, find the missing length.

115. The perimeter of the triangle is 39 m.

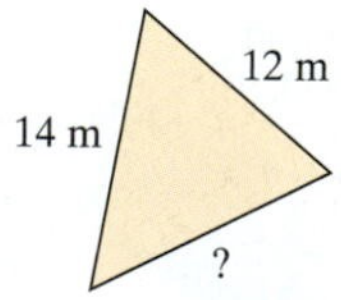

116. The perimeter of the figure is 547 cm.

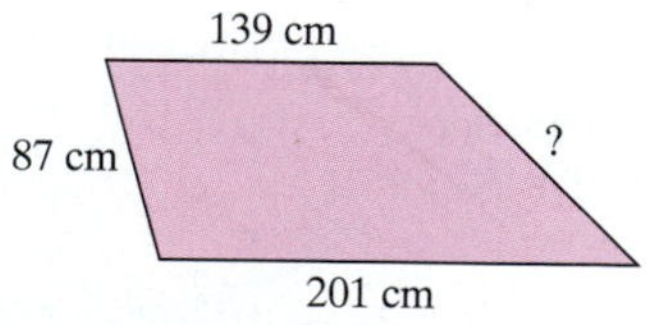

Calculator Connections

Topic: Adding and Subtracting Whole Numbers

The following keystrokes demonstrate the procedure to add numbers on a calculator. The ENTER key (or on some calculators, the = key or EXE key) tells the calculator to complete the calculation. Notice that commas used in large numbers are not entered into the calculator.

Expression	Keystrokes	Result
92,406 + 83,168	92406 + 83168 ENTER	175574

Your calculator may use the = key or EXE key.

To subtract numbers on a calculator, use the subtraction key, −. Do not confuse the subtraction key with the (−) key. The (−) will be presented later to enter negative numbers.

Expression	Keystrokes	Result
345,899 − 43,018	345899 − 43018 ENTER	302881

Calculator Exercises

For Exercises 117–122, perform the indicated operation by using a calculator.

117.
45,418
81,990
9,063
+ 56,309

118.
9,300,050
7,803,513
3,480,009
+ 907,822

119.
3,421,019
822,761
1,003,721
+ 9,678

120.
4,905,620
− 458,318

121.
953,400,415
− 56,341,902

122.
82,025,160
−79,118,705

For Exercises 123–126, refer to the table showing the land area for five states.

State	Land Area (mi^2)
Rhode Island	1045
Tennessee	41,217
West Virginia	24,078
Wisconsin	54,310
Colorado	103,718

123. Find the difference in the land area between Colorado and Wisconsin.

124. Find the difference in the land area between Tennessee and West Virginia.

125. What is the combined land area for Rhode Island, Tennessee, and Wisconsin?

126. What is the combined land area for all five states?

Section 1.4 Rounding and Estimating

Objectives

1. Rounding
2. Estimation
3. Using Estimation in Applications

1. Rounding

Rounding a whole number is a common practice when we do not require an exact value. For example, Madagascar lost 3956 mi^2 of rainforest between 1990 and 2008. We might round this number to the nearest thousand and say that approximately 4000 mi^2 was lost. In mathematics, we use the symbol $\approx$ to read "is approximately equal to." Therefore, 3956 $mi^2 \approx$ 4000 mi^2.

A number line is a helpful tool to understand rounding. For example, the number 48 is closer to 50 than it is to 40. Therefore, 48 rounded to the nearest ten is 50.

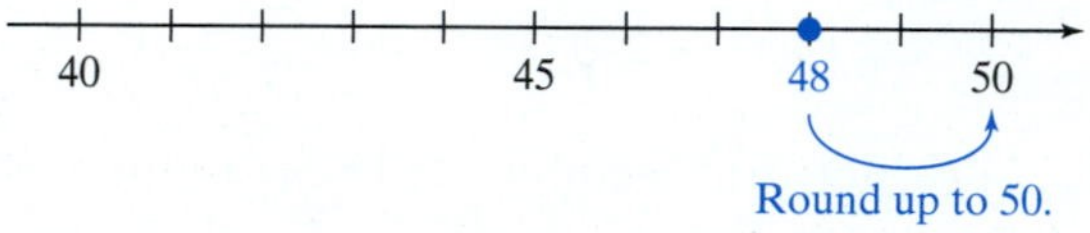

Concept Connections

1. Is the number 82 closer to 80 or to 90? Round 82 to the nearest ten.
2. Is the number 65 closer to 60 or to 70? Round the number to the nearest ten.

The number 43, on the other hand, is closer to 40 than to 50. Therefore, 43 rounded to the nearest ten is 40.

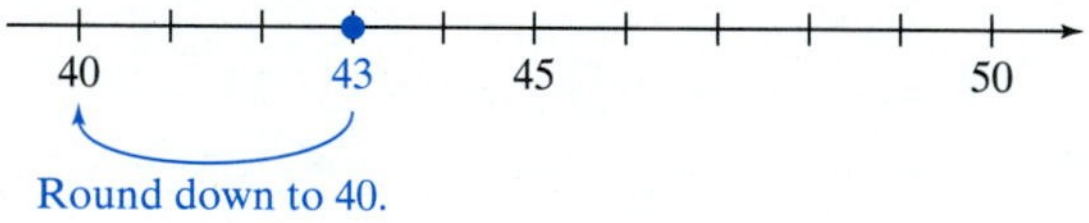

The number 45 is halfway between 40 and 50. In such a case, our convention will be to round *up* to the next-larger ten.

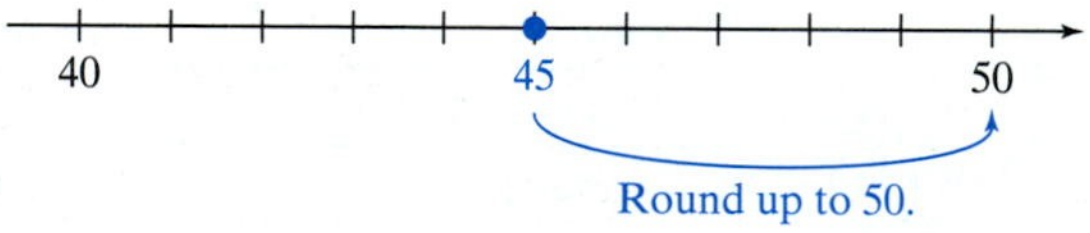

The decision to round up or down to a given place value is determined by the digit to the *right* of the given place value. The following steps outline the procedure.

PROCEDURE Rounding Whole Numbers

Step 1 Identify the digit one position to the right of the given place value.

Step 2 If the digit in Step 1 is a 5 or greater, then add 1 to the digit in the given place value. If the digit in Step 1 is less than 5, leave the given place value unchanged.

Step 3 Replace each digit to the right of the given place value by 0.

Answers

1. Closer to 80; 80
2. The number 65 is the same distance from 60 and 70; round up to 70.

Example 1 Rounding a Whole Number

Round 3741 to the nearest hundred.

Solution:

$3\,7\,\boxed{4}\,1 \approx 3700$

hundreds place (arrow to 7)

This is the digit to the right of the hundreds place. Because 4 is less than 5, leave the hundreds digit unchanged. Replace the digits to its right by zeros.

> **Skill Practice**
>
> 3. Round 12,461 to the nearest thousand.

Example 1 could also have been solved by drawing a number line. Use the part of a number line between 3700 and 3800 because 3741 lies between these numbers.

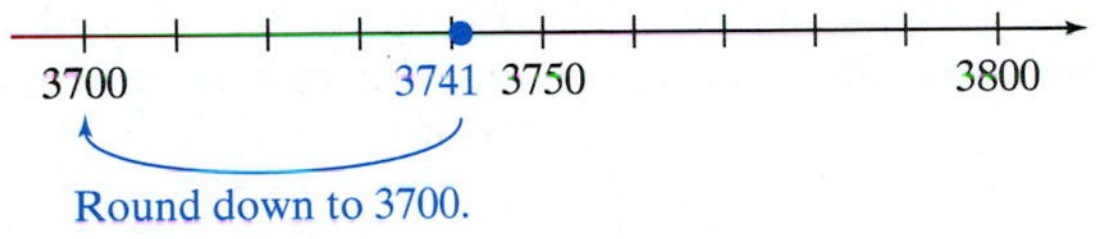

Example 2 Rounding a Whole Number

Round 1,790,641 to the nearest hundred-thousand.

Solution:

$1,7\,\boxed{9}\,0,6\,4\,1 \approx 1{,}800{,}000$

hundred-thousands place (arrow to 7)

This is the digit to the right of the given place value. Because 9 is greater than 5, add 1 to the hundred-thousands place, add: $7 + 1 = 8$. Replace the digits to the right of the hundred-thousands place by zeros.

> **Skill Practice**
>
> 4. Round 147,316 to the nearest ten-thousand.

Example 3 Rounding a Whole Number

Round 1503 to the nearest thousand.

Solution:

$1\,\boxed{5}\,0\,3 \approx 2000$

thousands place (arrow to 1)

This is the digit to the right of the thousands place. Because this digit is 5, we round up. We increase the thousands place digit by 1. That is, $1 + 1 = 2$. Replace the digits to its right by zeros.

> **Skill Practice**
>
> 5. Round 7,521,460 to the nearest million.

Answers

3. 12,000
4. 150,000
5. 8,000,000

Skill Practice

6. Round 39,823 to the nearest thousand.

Example 4 Rounding a Whole Number

Round the number 24,961 to the hundreds place.

Solution:

$$2\,4,\overset{+1}{9}\,\boxed{6}\,1 \approx 25{,}000$$

This is the digit to the right of the hundreds place. Because 6 is greater than 5, add 1 to the hundreds place digit. Replace the digits to the right of the hundreds place with 0.

2. Estimation

We use the process of rounding to estimate the result of numerical calculations. For example, to estimate the following sum, we can round each addend to the nearest ten.

31	rounds to ⟶	30
12	rounds to ⟶	10
+ 49	rounds to ⟶	+ 50
		90

The estimated sum is 90 (the actual sum is 92).

Skill Practice

7. Estimate the sum by rounding each number to the nearest hundred.

3162 + 4931 + 2206

Example 5 Estimating a Sum

Estimate the sum by rounding to the nearest thousand.

$$6109 + 976 + 4842 + 11{,}619$$

Solution:

6,109	rounds to ⟶	$\overset{1}{}$6,000
976	rounds to ⟶	1,000
4,842	rounds to ⟶	5,000
+ 11,619	rounds to ⟶	+ 12,000
		24,000

The estimated sum is 24,000 (the actual sum is 23,546).

Skill Practice

8. Estimate the difference by rounding each number to the nearest million.

35,264,000 − 21,906,210

Example 6 Estimating a Difference

Estimate the difference 4817 − 2106 by rounding each number to the nearest hundred.

Solution:

4817	rounds to ⟶	4800
− 2106	rounds to ⟶	− 2100
		2700

The estimated difference is 2700 (the actual difference is 2711).

Answers

6. 40,000 7. 10,300
8. 13,000,000

3. Using Estimation in Applications

Example 7 Estimating a Sum in an Application

A driver for a delivery service must drive from Chicago, Illinois, to Dallas, Texas, and make several stops on the way. The driver follows the route given on the map. Estimate the total mileage by rounding each distance to the nearest hundred miles.

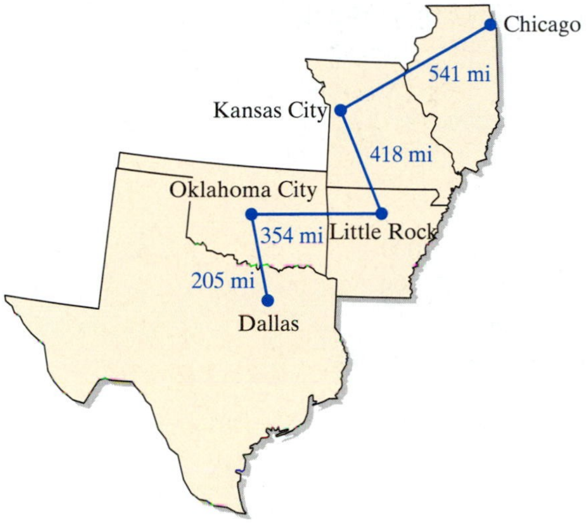

Solution:

541	rounds to ⟶	500
418	rounds to ⟶	400
354	rounds to ⟶	400
+ 205	rounds to ⟶	+ 200
		1500

The driver traveled approximately 1500 mi.

Skill Practice

9. The two countries with the largest areas of rainforest are Brazil and the Democratic Republic of Congo. The rainforest areas are 4,776,980 square kilometers (km^2) and 1,336,100 km^2, respectively. Round the values to the nearest hundred-thousand. Estimate the total area of rainforest for these two countries.

Example 8 Estimating a Difference in an Application

In a recent year, the U.S. Census Bureau reported that the number of males over the age of 18 was 100,994,367. The same year, the number of females over 18 was 108,133,727. Round each value to the nearest million. Estimate how many more females over 18 there were than males over 18.

Solution:

The number of males was approximately 101,000,000. The number of females was approximately 108,000,000.

$$\begin{array}{r} 108{,}000{,}000 \\ -\ 101{,}000{,}000 \\ \hline 7{,}000{,}000 \end{array}$$

There were approximately 7 million more women over age 18 in the United States than men.

Skill Practice

10. In a recent year, there were 135,073,000 persons over the age of 16 employed in the United States. During the same year, there were 6,742,000 persons over the age of 16 who were unemployed. Approximate each value to the nearest million. Use these values to approximate how many more people were employed than unemployed.

Answers

9. 6,100,000 km^2
10. 128,000,000

Section 1.4 Practice Exercises

Study Skills Exercises

1. After doing a section of homework, check the odd-numbered answers in the back of the text. Choose a method to identify the exercises that gave you trouble (i.e., circle the number or put a star by the number). List some reasons why it is important to label these problems.

2. Define the key term **rounding**.

Review Exercises

For Exercises 3–6, add or subtract as indicated.

3. $\begin{array}{r} 59 \\ -\ 33 \\ \hline \end{array}$

4. $\begin{array}{r} 130 \\ -\ 98 \\ \hline \end{array}$

5. $\begin{array}{r} 4009 \\ +\ 998 \\ \hline \end{array}$

6. $\begin{array}{r} 12{,}033 \\ +\ 23{,}441 \\ \hline \end{array}$

7. Determine the place value of the digit 6 in the number 1,860,432.

8. Determine the place value of the digit 4 in the number 1,860,432.

Objective 1: Rounding

9. Explain how to round a whole number to the hundreds place.

10. Explain how to round a whole number to the tens place.

For Exercises 11–28, round each number to the given place value. **(See Examples 1–4.)**

11. 342; tens

12. 834; tens

13. 725; tens

14. 445; tens

15. 9384; hundreds

16. 8363; hundreds

17. 8539; hundreds

18. 9817; hundreds

19. 34,992; thousands

20. 76,831; thousands

21. 2578; thousands

22. 3511; thousands

23. 9982; hundreds

24. 7974; hundreds

25. 109,337; thousands

26. 437,208; thousands

27. 489,090; ten-thousands

28. 388,725; ten-thousands

29. In the first weekend of its release, the movie *Spider-Man 3* grossed $148,431,020. Round this number to the millions place.

30. The average per capita personal income in the United States in a recent year was $33,050. Round this number to the nearest thousand.

31. The average center-to-center distance from the Earth to the Moon is 238,863 mi. Round this to the thousands place.

32. A shopping center in Edmonton, Alberta, Canada, covers an area of 492,000 square meters (m^2). Round this number to the hundred-thousands place.

Objective 2: Estimation

For Exercises 33–36, estimate the sum or difference by first rounding each number to the nearest ten. **(See Examples 5–6.)**

33.
```
   57
   82
 + 21
```

34.
```
   33
   78
 + 41
```

35.
```
  639
- 422
```

36.
```
  851
- 399
```

For Exercises 37–40, estimate the sum or difference by first rounding each number to the nearest hundred. **(See Examples 5–6.)**

37.
```
  892
+ 129
```

38.
```
  347
+ 563
```

39.
```
  3412
- 1252
```

40.
```
  9771
- 4544
```

Objective 3: Using Estimation in Applications

For Exercises 41–42, refer to the table.

Brand	Manufacturer	Sales ($)
M&Ms	Mars	97,404,576
Hershey's Milk Chocolate	Hershey Chocolate	81,296,784
Reese's Peanut Butter Cups	Hershey Chocolate	54,391,268
Snickers	Mars	53,695,428
KitKat	Hershey Chocolate	38,168,580

41. Round the individual sales to the nearest million to estimate the total sales brought in by the Mars company. **(See Example 7.)**

42. Round the sales to the nearest million to estimate the total sales brought in by the Hershey Chocolate Company.

43. Neil Diamond earned $71,339,710 in U.S. tours in one year while Paul McCartney earned $59,684,076. Round each value to the nearest million dollars to estimate how much more Neil Diamond earned. **(See Example 8.)**

44. The number of women in the 45–49 age group who gave birth in 1981 is 1190. By 2001 this number increased to 4844. Round each value to the nearest thousand to estimate how many more women in the 45–49 age group gave birth in 2001.

For Exercises 45–48, use the given table.

45. Round each revenue to the nearest hundred-thousand to estimate the total revenue for the years 2000 through 2002.

46. Round the revenue to the nearest hundred-thousand to estimate the total revenue for the years 2003 through 2005.

47. a. Determine the year with the greatest revenue. Round this revenue to the nearest hundred-thousand.

b. Determine the year with the least revenue. Round this revenue to the nearest hundred-thousand.

48. Estimate the difference between the year with the greatest revenue and the year with the least revenue.

Beach Parking Revenue for Daytona Beach, Florida

Year	Revenue
2000	$3,316,897
2001	3,272,028
2002	3,360,289
2003	3,470,295
2004	3,173,050
2005	1,970,380

Source: Daytona Beach News Journal

Table for Exercises 45–48

For Exercises 49–52, use the graph provided.

49. Determine the state with the greatest number of students enrolled in grades 6–12. Round this number to the nearest thousand.

50. Determine the state with the least number of students enrolled in grades 6–12. Round this number to the nearest thousand.

51. Use the information in Exercises 49 and 50 to estimate the difference between the number of students in the state with the greatest enrollment and that of the least enrollment.

52. Estimate the total number of students enrolled in grades 6–12 in the selected states by first rounding the number of students to the thousands place.

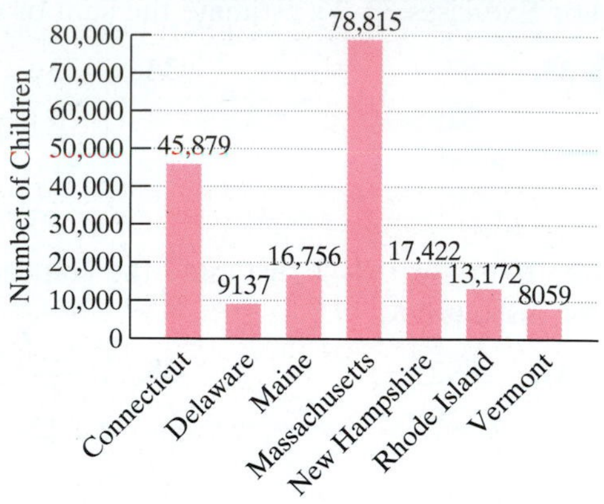

Source: National Center for Education Statistics

Figure for Exercises 49–52

Expanding Your Skills

For Exercises 53–56, round the numbers to estimate the perimeter of each figure. (Answers may vary.)

53.

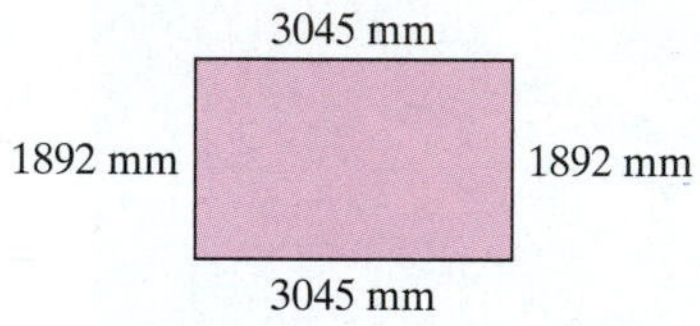

54.

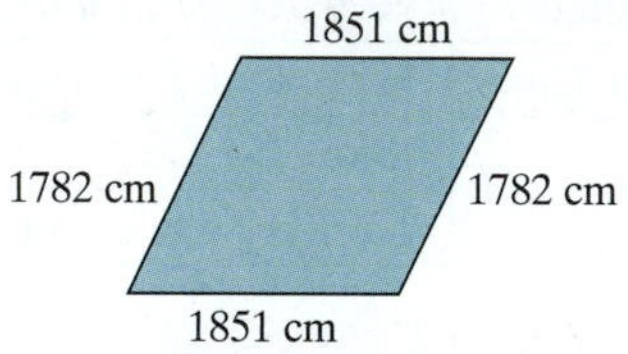

55.

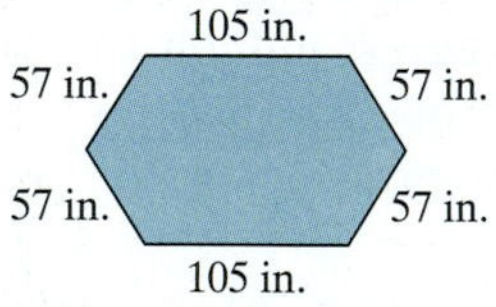

56.

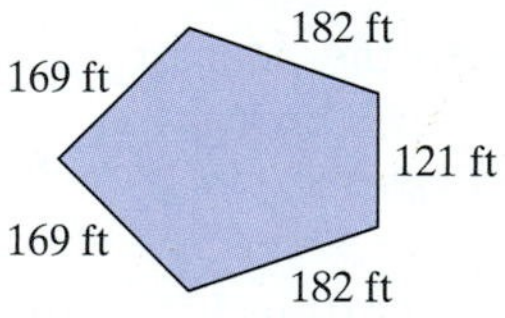

Section 1.5 Multiplication of Whole Numbers and Area

Objectives

1. Introduction to Multiplication
2. Properties of Multiplication
3. Multiplying Many-Digit Whole Numbers
4. Estimating Products by Rounding
5. Applications Involving Multiplication
6. Area of a Rectangle

1. Introduction to Multiplication

Suppose that Carmen buys three cartons of eggs to prepare a large family brunch. If there are 12 eggs per carton, then the total number of eggs can be found by adding three 12s.

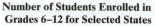

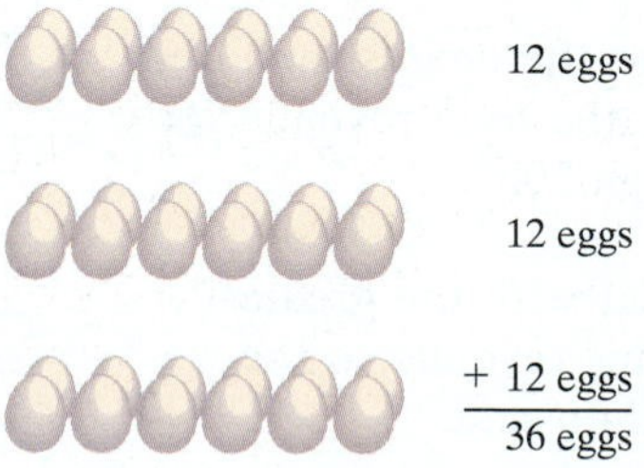

When each addend in a sum is the same, we have what is called *repeated* addition. Repeated addition is also called **multiplication**. We use the multiplication sign $\times$ to express repeated addition more concisely.

$$12 + 12 + 12 \quad \text{is equal to} \quad 3 \times 12$$

The expression 3×12 is read "3 times 12" to signify that the number 12 is added 3 times. The numbers 3 and 12 are called **factors**, and the result, 36, is called the **product**.

The symbol $\cdot$ may also be used to denote multiplication such as in the expression $3 \cdot 12 = 36$. Two factors written adjacent to each other with no other operator between them also implies multiplication. The quantity $2y$, for example, is understood to be 2 times y. If we use this notation to multiply two numbers, parentheses are used to group one or both factors. For example:

$$3(12) = 36 \qquad (3)12 = 36 \qquad \text{and} \qquad (3)(12) = 36$$

all represent the product of 3 and 12.

Concept Connections

1. How can multiplication be used to compute the sum $4+4+4+4+4+4+4$?

TIP: In the expression 3(12), the parentheses are necessary because two adjacent factors written together with no grouping symbol would look like the number 312.

The products of one-digit numbers such as $4 \times 5 = 20$ and $2 \times 7 = 14$ are basic facts. All products of one-digit numbers should be memorized (see Exercise 6).

Example 1 Identifying Factors and Products

Identify the factors and the product.

a. $6 \times 3 = 18$ **b.** $5 \cdot 2 \cdot 7 = 70$

Solution:

a. Factors: 6, 3; product: 18 **b.** Factors: 5, 2, 7; product: 70

Skill Practice

Identify the factors and the product.

2. $3 \times 11 = 33$
3. $2 \cdot 5 \cdot 8 = 80$

2. Properties of Multiplication

Recall from Section 1.3 that the order in which two numbers are added does not affect the sum. The same is true for multiplication. This is stated formally as the *commutative property of multiplication.*

PROPERTY Commutative Property of Multiplication

For any numbers, a and b,

$$a \cdot b = b \cdot a$$

Changing the order of two factors does not affect the product.

Answers

1. 7×4
2. Factors: 3 and 11; product: 33
3. Factors 2, 5, and 8; product: 80

The following rectangular arrays help us visualize the commutative property of multiplication. For example $2 \cdot 5 = 5 \cdot 2$.

$2 \cdot 5 = 10$ 2 rows of 5

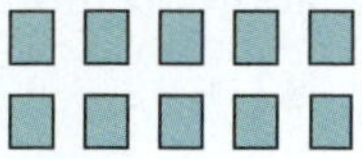

$5 \cdot 2 = 10$ 5 rows of 2

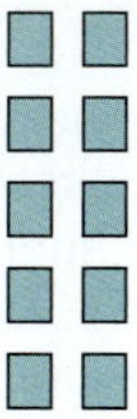

Multiplication is also an associative operation. For example,

$$(3 \cdot 5) \cdot 2 = 3 \cdot (5 \cdot 2).$$

PROPERTY Associative Property of Multiplication

For any numbers, a, b, and c,

$$(a \cdot b) \cdot c = a \cdot (b \cdot c)$$

The manner in which factors are grouped under multiplication does not affect the product.

TIP: The variable "x" is commonly used in algebra and can be confused with the multiplication symbol "×." For this reason it is customary to limit the use of "×" for multiplication.

Skill Practice

4. Rewrite the expression $6 \cdot 5$, using the commutative property of multiplication. Then find the product.
5. Rewrite the expression $3 \cdot (1 \cdot 7)$, using the associative property of multiplication. Then find the product.

Example 2 Applying Properties of Multiplication

a. Rewrite the expression $3 \cdot 9$, using the commutative property of multiplication. Then find the product.

b. Rewrite the expression $(4 \cdot 2) \cdot 3$, using the associative property of multiplication. Then find the product.

Solution:

a. $3 \cdot 9 = 9 \cdot 3$. The product is 27.

b. $(4 \cdot 2) \cdot 3 = 4 \cdot (2 \cdot 3)$.

To find the product, we have

$$\begin{aligned} &4 \cdot (2 \cdot 3) \\ &= 4 \cdot (6) \\ &= 24 \end{aligned}$$

The product is 24.

Two other important properties of multiplication involve factors of 0 and 1.

Answers

4. $5 \cdot 6$; product is 30
5. $(3 \cdot 1) \cdot 7$; product is 21

PROPERTY Multiplication Property of 0

For any number a,

$$a \cdot 0 = 0 \quad \text{and} \quad 0 \cdot a = 0$$

The product of any number and 0 is 0.

The product $5 \cdot 0 = 0$ can easily be understood by writing the product as repeated addition.

$$\underbrace{0 + 0 + 0 + 0 + 0}_{\text{Add 0 five times.}} = 0$$

PROPERTY Multiplication Property of 1

For any number a,

$$a \cdot 1 = a \quad \text{and} \quad 1 \cdot a = a$$

The product of any number and 1 is that number.

The next property of multiplication involves both addition and multiplication. First consider the expression $2(4 + 3)$. By performing the operation within parentheses first, we have

$$2(4 + 3) = 2(7) = 14$$

We get the same result by multiplying 2 times each addend within the parentheses:

$$2(4 + 3) = (2 \cdot 4) + (2 \cdot 3) = 8 + 6 = 14$$

This result illustrates the distributive property of multiplication over addition (sometimes we simply say *distributive property* for short).

PROPERTY Distributive Property of Multiplication over Addition

For any numbers a, b, and c,

$$a(b + c) = a \cdot b + a \cdot c$$

The product of a number and a sum can be found by multiplying the number by each addend.

Example 3 Applying the Distributive Property of Multiplication Over Addition

Apply the distributive property and simplify.

a. $3(4 + 8)$ **b.** $7(3 + 0)$

Solution:

a. $3(4 + 8) = (3 \cdot 4) + (3 \cdot 8) = 12 + 24 = 36$

b. $7(3 + 0) = (7 \cdot 3) + (7 \cdot 0) = 21 + 0 = 21$

Skill Practice

Apply the distributive property and simplify.

6. $2(6 + 4)$

7. $5(0 + 8)$

Answers

6. $(2 \cdot 6) + (2 \cdot 4)$; 20

7. $(5 \cdot 0) + (5 \cdot 8)$; 40

3. Multiplying Many-Digit Whole Numbers

When multiplying numbers with several digits, it is sometimes necessary to carry. To see why, consider the product $3 \cdot 29$. By writing the factors in expanded form, we can apply the distributive property. In this way, we see that 3 is multiplied by both 20 and 9.

$$\begin{aligned} 3 \cdot 29 = 3(20 + 9) &= (3 \cdot 20) + (3 \cdot 9) \\ &= 60 + 27 \\ &= 6 \text{ tens} + 2 \text{ tens} + 7 \text{ ones} \\ &= 8 \text{ tens} + 7 \text{ ones} \\ &= 87 \end{aligned}$$

Now we will multiply $29 \cdot 3$ in vertical form.

$$\begin{array}{r} \scriptstyle 2 \\ 29 \\ \times\ \ 3 \\ \hline 7 \end{array}$$

Multiply $3 \cdot 9 = 27$. Write the 7 in the ones column and carry the 2.

$$\begin{array}{r} \scriptstyle 2 \\ 29 \\ \times\ \ 3 \\ \hline 87 \end{array}$$

Multiply $3 \cdot 2$ tens $= 6$ tens. Add the carry: 6 tens + 2 tens = 8 tens. Write the 8 in the tens place.

Skill Practice

8. Multiply.

$$\begin{array}{r} 247 \\ \times\ \ 3 \\ \hline \end{array}$$

Example 4 **Multiplying a Many-Digit Number by a One-Digit Number**

Multiply.

$$\begin{array}{r} 368 \\ \times\ \ 5 \\ \hline \end{array}$$

Solution:

Using the distributive property, we have

$$5(300 + 60 + 8) = 1500 + 300 + 40 = 1840$$

This can be written vertically as:

368	
× 5	
40	Multiply $5 \cdot 8$.
300	Multiply $5 \cdot 60$.
+ 1500	Multiply $5 \cdot 300$.
1840	Add.

The numbers 40, 300, and 1500 are called *partial sums.* The product of 386 and 5 is found by adding the partial sums. The product is 1840.

The solution to Example 4 can also be found by using a shorter form of multiplication. We outline the procedure:

$$\begin{array}{r} \scriptstyle 4\ \ \\ 368 \\ \times\ \ \ 5 \\ \hline 0 \end{array}$$

Multiply $5 \cdot 8 = 40$. Write the 0 in the ones place and carry the 4.

Answer

8. 741

$$\begin{array}{r} {\scriptstyle 3\,4} \\ 368 \\ \times \quad 5 \\ \hline 40 \end{array}$$

Multiply 5 · 6 tens = 300. Add the carry. 300 + 4 tens = 340. Write the 4 in the tens place and carry the 3.

$$\begin{array}{r} {\scriptstyle 3\,4} \\ 368 \\ \times \quad 5 \\ \hline 1840 \end{array}$$

Multiply 5 · 3 hundreds = 1500. Add the carry. 1500 + 3 hundreds = 1800. Write the 8 in the hundreds place and the 1 in the thousands place.

Example 5 demonstrates the process to multiply two factors with many digits.

Example 5 **Multiplying a Two-Digit Number by a Two-Digit Number**

Multiply.

$$\begin{array}{r} 72 \\ \times\ 83 \\ \hline \end{array}$$

Solution:

Writing the problem vertically and computing the partial sums, we have

$$\begin{array}{rl} {\scriptstyle 1}\ \ \\ 72 & \\ \times\ 83 & \\ \hline 216 & \text{Multiply } 3 \cdot 72. \\ +\ 5760 & \text{Multiply } 80 \cdot 72. \\ \hline 5976 & \text{Add.} \end{array}$$

The product is 5976.

Skill Practice

9. Multiply.

$$\begin{array}{r} 59 \\ \times\ 26 \\ \hline \end{array}$$

Example 6 **Multiplying Two Multidigit Whole Numbers**

Multiply. 368(497)

Solution:

$$\begin{array}{rl} {\scriptstyle 2\,3} & \\ {\scriptstyle 6\,7} & \\ {\scriptstyle 4\,5}\ \ & \\ 368 & \\ \times\ 497 & \\ \hline 2576 & \text{Multiply } 7 \cdot 368. \\ 33120 & \text{Multiply } 90 \cdot 368. \\ +\ 147200 & \text{Multiply } 400 \cdot 368. \\ \hline 182{,}896 & \end{array}$$

Skill Practice

10. Multiply.

$$\begin{array}{r} 274 \\ \times\ 586 \\ \hline \end{array}$$

4. Estimating Products by Rounding

A special pattern occurs when one or more factors in a product ends in zero. Consider the following products:

$$\begin{array}{llll} 12 \cdot 20 & = 240 & 120 \cdot 20 & = 2400 \\ 12 \cdot 200 & = 2400 & 1200 \cdot 20 & = 24{,}000 \\ 12 \cdot 2000 & = 24{,}000 & 12{,}000 \cdot 20 & = 240{,}000 \end{array}$$

Answers

9. 1534 **10.** 160,564

Notice in each case the product is $12 \cdot 2 = 24$ followed by the total number of zeros from each factor. Consider the product $1200 \cdot 20$.

$$\begin{array}{r|l} 12 & 00 \\ \times\ 2 & 0 \\ \hline 24 & 000 \end{array}$$

Shift the numbers 1200 and 20 so that the zeros appear to the right of the multiplication process. Multiply $12 \cdot 2 = 24$.
Write the product 24 followed by the total number of zeros from each factor.

Skill Practice

11. Estimate the product 421(869) by rounding each factor to the nearest hundred.

Example 7 Estimating a Product

Estimate the product 795(4060) by rounding 795 to the nearest hundred and 4060 to the nearest thousand.

Solution:

795 rounds to ⟶ 800

4060 rounds to ⟶ 4000

$$\begin{array}{r|l} 8 & 00 \\ \times\ 4 & 000 \\ \hline 32 & 00000 \end{array}$$

The product is approximately 3,200,000.

Skill Practice

12. A small newspaper has 16,850 subscribers. Each subscription costs $149 per year. Estimate the revenue for the year by rounding the number of subscriptions to the nearest thousand and the cost to the nearest ten.

Example 8 Estimating a Product in an Application

For a trip from Atlanta to Los Angeles, the average cost of a plane ticket was $495. If the plane carried 218 passengers, estimate the total revenue for the airline. (*Hint*: Round each number to the hundreds place and find the product.)

Solution:

$495 rounds to ⟶

218 rounds to ⟶

$$\begin{array}{r|l} \$\ 5 & 00 \\ \times 2 & 00 \\ \hline \$10 & 0000 \end{array}$$

The airline received approximately $100,000 in revenue.

5. Applications Involving Multiplication

In English there are many different words that imply multiplication. A partial list is given in Table 1-3.

Table 1-3

Word/Phrase	Example	In Symbols
Product	The product of 4 and 7	$4 \cdot 7$
Times	8 times 4	$8 \cdot 4$
Multiply . . . by . . .	Multiply 6 by 3	$6 \cdot 3$

Multiplication may also be warranted in applications involving unit rates. In Example 8, we multiplied the cost per customer ($495) by the number of customers (218). The value $495 is a unit rate because it gives the cost per one customer (per one unit).

Answers

11. 360,000 **12.** $2,550,000

Example 9 Solving an Application Involving Multiplication

The average weekly income for production workers is \$489. How much does a production worker make in 1 year (assume 52 weeks in 1 year).

Solution:

The value \$489 per week is a unit rate. The total earnings for 1 year is given by \$489 · 52.

$$\begin{array}{r} {\scriptstyle 4\,4} \\ {\scriptstyle 1\,1} \\ 489 \\ \times\ 52 \\ \hline 978 \\ +\ 24450 \\ \hline 25{,}428 \end{array}$$

The yearly earnings total \$25,428.

TIP: This product can be estimated quickly by rounding the factors.

489	rounds to →	5 | 00
52	rounds to →	× 5 | 0
		25 | 000

The total yearly income is approximately \$25,000. Estimating gives a quick approximation of a product. Furthermore, it also checks for the reasonableness of our exact product. In this case \$25,000 is close to our exact value of \$25,428.

Skill Practice

13. Ella can type 65 words per minute. How many words can she type in 45 minutes?

6. Area of a Rectangle

Another application of multiplication of whole numbers lies in finding the area of a region. **Area** measures the amount of surface contained within the region. For example, a square that is 1 in. by 1 in. occupies an area of 1 square inch, denoted as 1 in^2. Similarly, a square that is 1 centimeter (cm) by 1 cm occupies an area of 1 square centimeter. This is denoted by 1 cm^2.

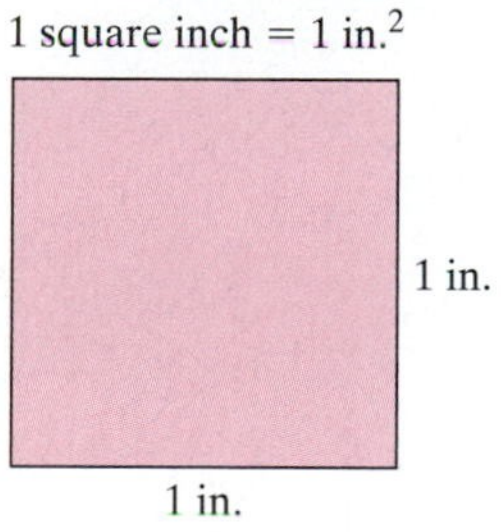

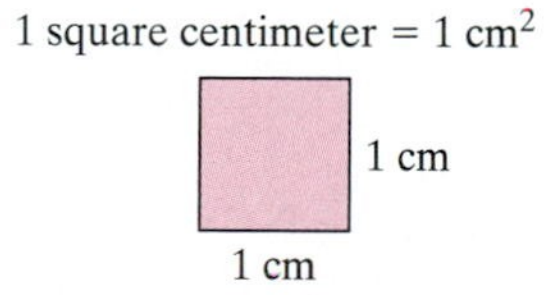

The units of square inches and square centimeters (in.^2 and cm^2) are called *square units*. To find the area of a region, measure the number of square units occupied in that region. For example, the region in Figure 1-4 occupies 6 cm^2.

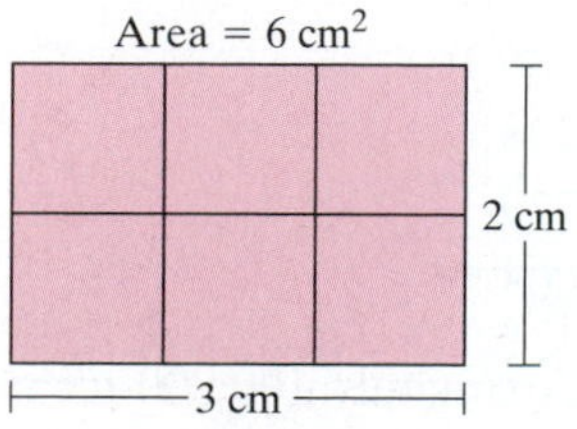

Figure 1-4

Answer

13. 2925 words

The 3-cm by 2-cm region in Figure 1-4 suggests that to find the **area of a rectangle**, multiply the length by the width. If the area is represented by A, the length is represented by l, and the width is represented by w, then we have

$$\textbf{Area of rectangle} = (\text{length}) \cdot (\text{width})$$

$$A = l \cdot w$$

The letters A, l, and w are variables because their values *vary* as they are replaced by different numbers.

Skill Practice

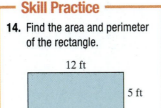

14. Find the area and perimeter of the rectangle.

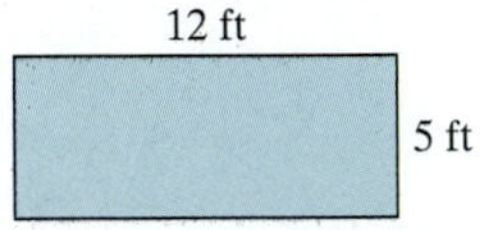

Example 10 Finding the Area of a Rectangle

Find the area and perimeter of the rectangle.

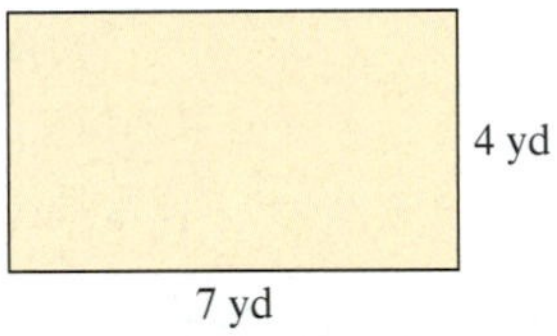

Solution:

Area:

$A = l \cdot w$

$A = (7 \text{ yd}) \cdot (4 \text{ yd})$

$= 28 \text{ yd}^2$

Recall from Section 1.3 that the perimeter of a polygon is the sum of the lengths of the sides. In a rectangle the opposite sides are equal in length.

Perimeter:

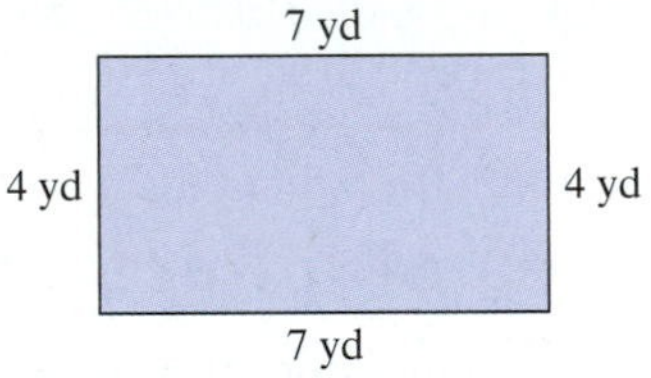

$P = 7 \text{ yd} + 4 \text{ yd} + 7 \text{ yd} + 4 \text{ yd}$

$= 22 \text{ yd}$

The area is 28 yd^2 and the perimeter is 22 yd.

Avoiding Mistakes

Notice that area is measured in square units (such as yd^2) and perimeter is measured in units of length (such as yd). It is important to apply the correct units of measurement.

Skill Practice

15. A house sits on a rectangular lot that is 192 ft by 96 ft. Approximate the area of the lot by rounding the length and width to the nearest hundred.

Example 11 Finding Area in an Application

The state of Wyoming is approximately the shape of a rectangle (Figure 1-5). Its length is 355 mi and its width is 276 mi. Approximate the total area of Wyoming by rounding the length and width to the nearest ten.

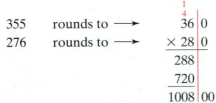

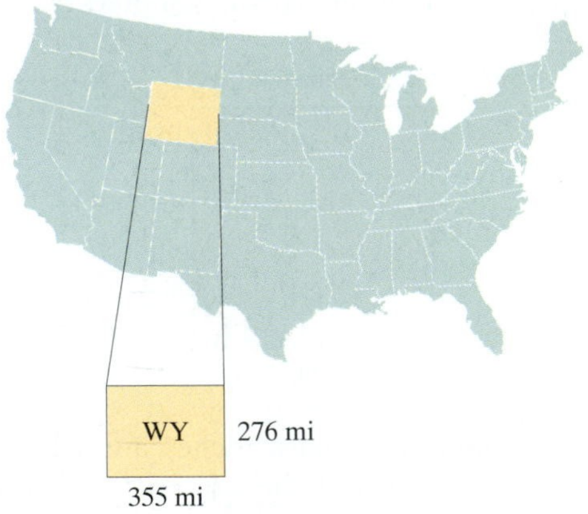

Figure 1-5

Solution:

355 rounds to → 36|0

276 rounds to → × 28|0

$$\begin{array}{r|l} \overset{1}{\overset{4}{3}}6 & 0 \\ \times\ 28 & 0 \\ \hline 288 & \\ 720 & \\ \hline 1008 & 00 \end{array}$$

The area of Wyoming is approximately $100{,}800 \text{ mi}^2$.

Answers

14. Area: 60 ft^2; perimeter: 34 ft

15. $20{,}000 \text{ ft}^2$

Section 1.5 Practice Exercises

Study Skills Exercises

1. Sometimes you may run into a problem with homework, or you may find that you are having trouble keeping up with the pace of the class. A tutor can be a good resource. Answer the following questions.

a. Does your college offer tutoring? **b.** Is it free?

c. Where would you go to sign up for a tutor?

2. Define the key terms.

a. Multiplication **b. Factor** **c. Product**

d. Area **e. Area of a rectangle**

Review Exercises

For Exercises 3–5, estimate the answer by rounding to the indicated place value.

3. 869,240 + 34,921 + 108,332; ten-thousands **4.** 907,801 − 413,560; hundred-thousands

5. 8821 − 3401; hundreds

Objective 1: Introduction to Multiplication

6. Fill in the table of multiplication facts.

×	0	1	2	3	4	5	6	7	8	9
0										
1										
2										
3										
4										
5										
6										
7										
8										
9										

For Exercises 7–10, write the repeated addition as multiplication and simplify.

7. $5 + 5 + 5 + 5 + 5 + 5$ **8.** $2 + 2 + 2 + 2 + 2 + 2 + 2 + 2 + 2$

9. $9 + 9 + 9$ **10.** $7 + 7 + 7 + 7$

For Exercises 11–14, identify the factors and the product. **(See Example 1.)**

11. $13 \times 42 = 546$ **12.** $26 \times 9 = 234$

13. $3 \cdot 5 \cdot 2 = 30$ **14.** $4 \cdot 3 \cdot 8 = 96$

15. Write the product of 5 and 12, using three different notations. (Answers may vary.)

16. Write the product of 23 and 14, using three different notations. (Answers may vary.)

Objective 2: Properties of Multiplication

For Exercises 17–22, match the property with the statement.

17. $8 \cdot 1 = 8$

18. $6 \cdot 13 = 13 \cdot 6$

19. $2(6 + 12) = 2 \cdot 6 + 2 \cdot 12$

20. $5 \cdot (3 \cdot 2) = (5 \cdot 3) \cdot 2$

21. $0 \cdot 4 = 0$

22. $7(14) = 14(7)$

a. Commutative property of multiplication

b. Associative property of multiplication

c. Multiplication property of 0

d. Multiplication property of 1

e. Distributive property of multiplication over addition

For Exercises 23–28, rewrite the expression, using the indicated property. **(See Examples 2–3.)**

23. $14 \cdot 8$; commutative property of multiplication

24. $3 \cdot 9$; commutative property of multiplication

25. $6 \cdot (2 \cdot 10)$; associative property of multiplication

26. $(4 \cdot 15) \cdot 5$; associative property of multiplication

27. $5(7 + 4)$; distributive property of multiplication over addition

28. $3(2 + 6)$; distributive property of multiplication over addition

Objective 3: Multiplying Many-Digit Whole Numbers

For Exercises 29–60, multiply. **(See Examples 4–6.)**

29. $\begin{array}{r} 24 \\ \times\ 6 \\ \hline \end{array}$

30. $\begin{array}{r} 18 \\ \times\ 5 \\ \hline \end{array}$

31. $\begin{array}{r} 26 \\ \times\ 2 \\ \hline \end{array}$

32. $\begin{array}{r} 71 \\ \times\ 3 \\ \hline \end{array}$

33. $\begin{array}{r} 131 \\ \times\ 5 \\ \hline \end{array}$

34. $\begin{array}{r} 725 \\ \times\ 3 \\ \hline \end{array}$

35. $\begin{array}{r} 344 \\ \times\ 4 \\ \hline \end{array}$

36. $\begin{array}{r} 105 \\ \times\ 9 \\ \hline \end{array}$

37. $\begin{array}{r} 1410 \\ \times\ 8 \\ \hline \end{array}$

38. $\begin{array}{r} 2016 \\ \times\ 6 \\ \hline \end{array}$

39. $\begin{array}{r} 3312 \\ \times\ 7 \\ \hline \end{array}$

40. $\begin{array}{r} 4801 \\ \times\ 5 \\ \hline \end{array}$

41. $\begin{array}{r} 42{,}014 \\ \times\ 9 \\ \hline \end{array}$

42. $\begin{array}{r} 51{,}006 \\ \times\ 8 \\ \hline \end{array}$

43. $\begin{array}{r} 32 \\ \times\ 14 \\ \hline \end{array}$

44. $\begin{array}{r} 41 \\ \times\ 21 \\ \hline \end{array}$

45. $68 \cdot 24$

46. $55 \cdot 41$

47. $72 \cdot 12$

48. $13 \cdot 46$

49. $(143)(17)$

50. $(722)(28)$

51. $(349)(19)$

52. $(512)(31)$

53. $\begin{array}{r} 151 \\ \times\ 127 \\ \hline \end{array}$ **54.** $\begin{array}{r} 703 \\ \times\ 146 \\ \hline \end{array}$ **55.** $\begin{array}{r} 222 \\ \times\ 841 \\ \hline \end{array}$ **56.** $\begin{array}{r} 387 \\ \times\ 506 \\ \hline \end{array}$

57. $\begin{array}{r} 3532 \\ \times\ 6014 \\ \hline \end{array}$ **58.** $\begin{array}{r} 2810 \\ \times\ 1039 \\ \hline \end{array}$ **59.** $\begin{array}{r} 4122 \\ \times\ 982 \\ \hline \end{array}$ **60.** $\begin{array}{r} 7026 \\ \times\ 528 \\ \hline \end{array}$

Objective 4: Estimating Products by Rounding

For Exercises 61–68, multiply the numbers, using the method found on page 40. **(See Example 7.)**

61. $\begin{array}{r} 600 \\ \times\ 40 \\ \hline \end{array}$ **62.** $\begin{array}{r} 900 \\ \times\ 50 \\ \hline \end{array}$ **63.** $\begin{array}{r} 3000 \\ \times\ 700 \\ \hline \end{array}$ **64.** $\begin{array}{r} 4000 \\ \times\ 400 \\ \hline \end{array}$

65. $\begin{array}{r} 8000 \\ \times\ 9000 \\ \hline \end{array}$ **66.** $\begin{array}{r} 1000 \\ \times\ 2000 \\ \hline \end{array}$ **67.** $\begin{array}{r} 90{,}000 \\ \times\ 400 \\ \hline \end{array}$ **68.** $\begin{array}{r} 50{,}000 \\ \times\ 6000 \\ \hline \end{array}$

For Exercises 69–72, estimate the product by first rounding the number to the indicated place value.

69. 11,784 · 5201; thousands place

70. 45,046 · 7812; thousands place

71. 82,941 · 29,740; ten-thousands place

72. 630,229 · 71,907; ten-thousands place

73. Suppose a hotel room costs \$189 per night. Round this number to the nearest hundred to estimate the cost for a five-night stay. **(See Example 8.)**

74. The science department of Comstock High School must purchase a set of calculators for a class. If the cost of one calculator is \$129, estimate the cost of 28 calculators by rounding the numbers to the tens place.

75. The average price for a ticket to see Kenny Chesney is \$137. If a concert stadium seats 10,256 fans, estimate the amount of money received during that performance by rounding the number of seats to the nearest ten-thousand.

76. A breakfast buffet at a local restaurant serves 48 people. Estimate the maximum revenue for one week (7 days) if the price of a breakfast is \$12.

Objective 5: Applications Involving Multiplication

77. The 4-gigabyte (4-GB) iPod nano is advertised to store approximately 1000 songs. Assuming the average length of a song is 4 minutes (min), how many minutes of music can be stored on the iPod nano? **(See Example 9.)**

78. One CD can hold 700 megabytes (MB) of data. How many megabytes can 15 CDs hold?

79. It costs about \$45 for a cat to have a medical exam. If a humane society has 37 cats, find the cost of medical exams for its cats.

80. The Toyota Prius gets 55 miles per gal (mpg) on the highway. How many miles can it go on 20 gal of gas?

81. A can of Coke contains 12 fluid ounces (fl oz). Find the number of ounces in a case of Coke containing 12 cans.

82. A 3 credit-hour class at a college meets 3 hours (hr) per week. If a semester is 16 weeks long, for how many hours will the class meet during the semester?

83. PaperWorld shipped 115 cases of copy paper to a business. There are 5 reams of paper in each case and 500 sheets of paper in each ream. Find the number of sheets of paper delivered to the business.

84. A dietary supplement bar has 14 grams (g) of protein. If Kathleen eats 2 bars a day for 6 days, how many grams of protein will she get from this supplement?

85. Tylee's car gets 31 miles per gallon (mpg) on the highway. How many miles can he travel if he has a full tank of gas (12 gal)?

86. Sherica manages a small business called Pizza Express. She has 23 employees who work an average of 32 hr per week. How many hours of work does Sherica have to schedule each week?

Objective 6: Area of a Rectangle

For Exercises 87–90, find the area. **(See Example 10.)**

87.

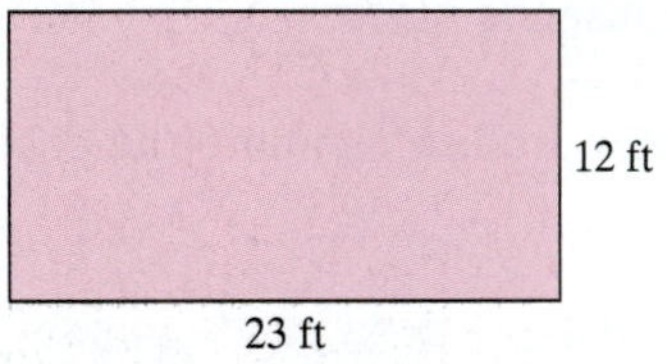

88.

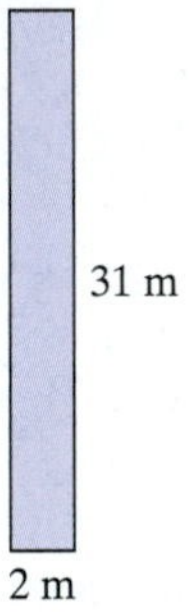

89.

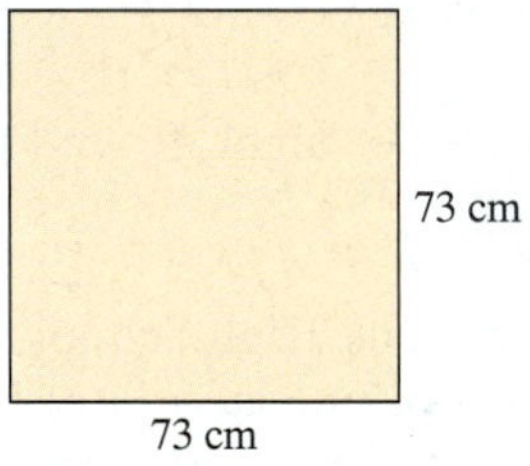

90.

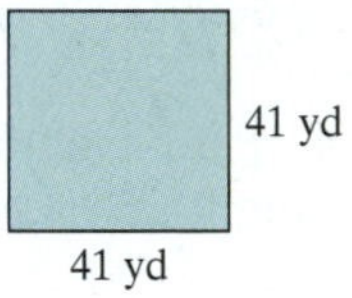

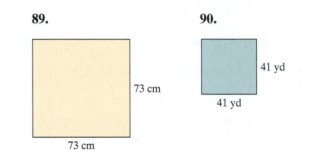

91. The state of Colorado is approximately the shape of a rectangle. Its length is 388 mi and its width is 269 mi. Approximate the total area of Colorado by rounding the length and width to the nearest ten. **(See Example 11.)**

92. A parcel of land has a width of 132 yd and a length of 149 yd. Approximate the total area by rounding each dimension to the nearest ten.

93. The front of a building has windows that are 44 in. by 58 in.

a. Approximate the area of one window.

b. If the building has three floors and each floor has 14 windows, how many windows are there?

c. What is the approximate total area of all of the windows?

94. The length of a carport is 51 ft and its width is 29 ft. Approximate the area of the carport.

95. Mr. Slackman wants to paint his garage door that is 8 ft by 16 ft. To decide how much paint to buy, he must find the area of the door. What is the area of the door?

96. To carpet a rectangular room, Erika must find the area of the floor. If the dimensions of the room are 10 yd by 15 yd, how much carpeting does she need?

Division of Whole Numbers

Section 1.6

Objectives

1. Introduction to Division
2. Properties of Division
3. Long Division
4. Dividing by a Many-Digit Divisor
5. Translations and Applications Involving Division

1. Introduction to Division

Suppose 12 pieces of pizza are to be divided evenly among 4 children (Figure 1-6). The number of pieces that each child would receive is given by $12 \div 4$, read "12 divided by 4."

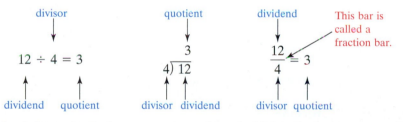

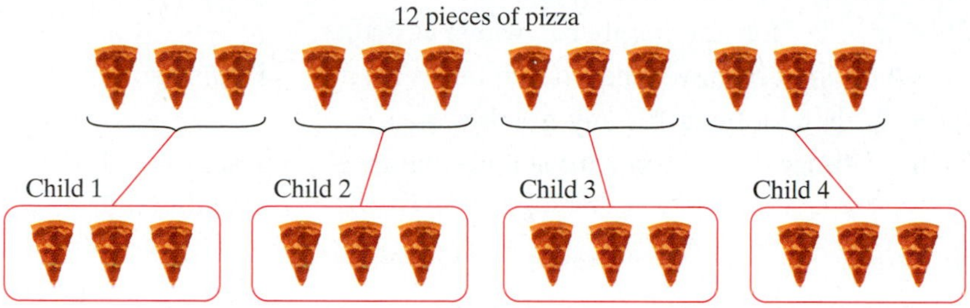

Figure 1-6

The process of separating 12 pieces of pizza evenly among 4 children is called **division**. The statement $12 \div 4 = 3$ indicates that each child receives 3 pieces of pizza. The number 12 is called the **dividend**. It represents the number to be divided. The number 4 is called the **divisor**, and it represents the number of groups. The result of the division (in this case 3) is called the **quotient**. It represents the number of items in each group.

Division can be represented in several ways. For example, the following are all equivalent statements.

$12 \div 4 = 3$ (12 is the dividend, 4 is the divisor, 3 is the quotient)

$4\overline{)12}$ with quotient 3 above (4 is the divisor, 12 is the dividend, 3 is the quotient)

$\frac{12}{4} = 3$ (12 is the dividend, 4 is the divisor, 3 is the quotient). This bar is called a fraction bar.

Recall that subtraction is the reverse operation of addition. In the same way, division is the reverse operation of multiplication. For example, we say $12 \div 4 = 3$ because $3 \cdot 4 = 12$.

Example 1 Identifying the Dividend, Divisor, and Quotient

Simplify each expression. Then identify the dividend, divisor, and quotient.

a. $48 \div 6$ **b.** $9\overline{)36}$ **c.** $\frac{63}{7}$

Solution:

a. $48 \div 6 = 8$ because $8 \cdot 6 = 48$

The dividend is 48, the divisor is 6, and the quotient is 8.

b. $\begin{array}{r} 4 \\ 9\overline{)36} \end{array}$ because $4 \cdot 9 = 36$

The dividend is 36, the divisor is 9, and the quotient is 4.

c. $\frac{63}{7} = 9$ because $9 \cdot 7 = 63$

The dividend is 63, the divisor is 7, and the quotient is 9.

Skill Practice

Identify the dividend, divisor, and quotient.

1. $56 \div 7$

2. $4\overline{)20}$

3. $\frac{18}{2}$

Answers

1. Dividend: 56; divisor: 7; quotient: 8
2. Dividend: 20; divisor: 4; quotient: 5
3. Dividend: 18; divisor: 2; quotient: 9

2. Properties of Division

PROPERTY Properties of Division

1. $a \div a = 1$ for any number a, where $a \neq 0$.
   Any nonzero number divided by itself is 1. Example: $9 \div 9 = 1$
2. $a \div 1 = a$ for any number a.
   Any number divided by 1 is the number itself. Example: $3 \div 1 = 3$
3. $0 \div a = 0$ for any number a, where $a \neq 0$.
   Zero divided by any nonzero number is zero. Example: $0 \div 5 = 0$
4. $a \div 0$ is undefined for any number a.
   Any number divided by zero is undefined. Example: $9 \div 0$ is undefined.

Example 2 illustrates the important properties of division.

Skill Practice

Divide.

4. $3\overline{)3}$
5. $5 \div 5$
6. $\frac{4}{1}$
7. $8 \div 0$
8. $\frac{0}{7}$
9. $3\overline{)0}$

Example 2 Dividing Whole Numbers

Divide.

a. $8 \div 8$ b. $\frac{6}{6}$ c. $5 \div 1$ d. $1\overline{)7}$

e. $0 \div 6$ f. $\frac{0}{4}$ g. $6 \div 0$ h. $\frac{10}{0}$

Solution:

a. $8 \div 8 = 1$ because $1 \cdot 8 = 8$

b. $\frac{6}{6} = 1$ because $1 \cdot 6 = 6$

c. $5 \div 1 = 5$ because $5 \cdot 1 = 5$

d. $\begin{array}{r} 7 \\ 1\overline{)7} \end{array}$ because $7 \cdot 1 = 7$

e. $0 \div 6 = 0$ because $0 \cdot 6 = 0$

f. $\frac{0}{4} = 0$ because $0 \cdot 4 = 0$

g. $6 \div 0$ is *undefined* because there is no number that when multiplied by 0 will produce a product of 6.

h. $\frac{10}{0}$ is *undefined* because there is no number that when multiplied by 0 will produce a product of 10.

Avoiding Mistakes

There is no mathematical symbol to describe the result when dividing by 0. We must write out the word *undefined* to denote that we cannot divide by 0.

Concept Connections

10. Which expression is undefined?
    $\frac{5}{0}$ or $\frac{0}{5}$
11. Which expression is equal to zero?
    $0\overline{)4}$ or $4\overline{)0}$

You should also note that unlike addition and multiplication, division is neither commutative nor associative. In other words, reversing the order of the dividend and divisor may produce a different quotient. Similarly, changing the manner in which numbers are grouped with division may affect the outcome. See Exercises 31 and 32.

Answers

4. 1 5. 1 6. 4 7. Undefined
8. 0 9. 0 10. $\frac{5}{0}$ 11. $4\overline{)0}$

3. Long Division

To divide larger numbers we use a process called **long division**. This process uses a series of estimates to find the quotient. We illustrate long division in Example 3.

Example 3 Using Long Division

Divide. $7\overline{)161}$

Solution:

Estimate $7\overline{)161}$ by first estimating $7\overline{)16}$ and writing the result above the tens place of the dividend. Since $7 \cdot 2 = 14$, there are at least 2 sevens in 16.

$$\begin{array}{r} 2 \\ 7\overline{)161} \\ -140 \\ \hline 21 \end{array}$$

The 2 in the tens place represents 20 in the quotient.
← Multiply $7 \cdot 20$ and write the result under the dividend.
Subtract 140. We see that our estimate leaves 21.

Repeat the process. Now divide $7\overline{)21}$ and write the result in the ones place of the quotient.

$$\begin{array}{r} 23 \\ 7\overline{)161} \\ -140 \\ \hline 21 \\ -21 \\ \hline 0 \end{array}$$

−21 ← Multiply $7 \cdot 3$.
0 Subtract.

The quotient is 23.

Check:
$$\begin{array}{r} 23 \\ \times\ 7 \\ \hline 161 \end{array}\ ✔$$

Skill Practice

12. Divide.
$8\overline{)136}$

We can streamline the process of long division by "bringing down" digits of the dividend one at a time.

Example 4 Using Long Division

Divide. $6138 \div 9$

Solution:

$$\begin{array}{r} 682 \\ 9\overline{)6138} \\ -54\downarrow \\ \hline 73 \\ -72\downarrow \\ \hline 18 \\ -18 \\ \hline 0 \end{array}$$

$61 \div 9$ is estimated at 6. Multiply $9 \cdot 6 = 54$ and subtract.

$73 \div 9$ is estimated at 8. Multiply $9 \cdot 8 = 72$ and subtract.

$18 \div 9 = 2$. Multiply $9 \cdot 2 = 18$ and subtract.

The quotient is 682.

Check:
$$\begin{array}{r} {}^{7\,1} \\ 682 \\ \times\ \ 9 \\ \hline 6138 \end{array}\ ✔$$

Skill Practice

13. Divide.
$2891 \div 7$

Answers

12. 17 **13.** 413

In many instances, quotients do not come out evenly. For example, suppose we had 13 pieces of pizza to distribute among 4 children (Figure 1-7).

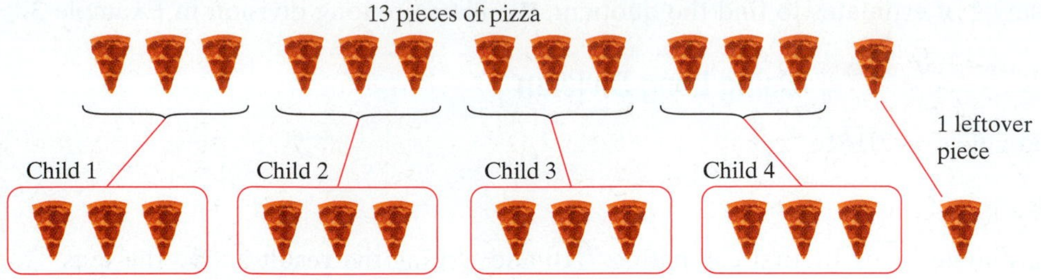

Figure 1-7

The mathematical term given to the "leftover" piece is called the **remainder**. The division process may be written as

$$\begin{array}{r} 3\text{ R}1 \\ 4\overline{)13} \\ \underline{-12} \\ 1 \end{array}$$

The remainder is written next to the 3.

The **whole part of the quotient** is 3, and the remainder is 1. Notice that the remainder is written next to the whole part of the quotient.

We can check a division problem that has a remainder. To do so, multiply the divisor by the whole part of the quotient and then add the remainder. The result must equal the dividend. That is,

$$(\text{Divisor})(\text{whole part of quotient}) + \text{remainder} = \text{dividend}$$

Thus,

$$(4)(3) + 1 \stackrel{?}{=} 13$$
$$12 + 1 \stackrel{?}{=} 13$$
$$13 = 13 \checkmark$$

Skill Practice

14. Divide.
5107 ÷ 5

Example 5 Using Long Division

Divide. $1253 \div 6$

Solution:

$$\begin{array}{r} 208\text{ R}5 \\ 6\overline{)1253} \\ \underline{-12} \\ 05 \\ \underline{-00} \\ 53 \\ \underline{-48} \\ 5 \end{array}$$

- $6 \cdot 2 = 12$ and subtract.
- Bring down the 5.
- Note that 6 does not divide into 5, so we put a 0 in the quotient.
- Bring down the 3.
- $6 \cdot 8 = 48$ and subtract.
- The remainder is 5.

To check, verify that $(6)(208) + 5 = 1253$. ✔

Answer

14. 1021 R2

4. Dividing by a Many-Digit Divisor

When the divisor has more than one digit, we still use a series of estimations to find the quotient.

Example 6 Dividing by a Two-Digit Number

Divide. $32\overline{)1259}$

Solution:

To estimate the leading digit of the quotient, estimate the number of times 30 will go into 125. Since $30 \cdot 4 = 120$, our estimate is 4.

$$\begin{array}{r} 4 \\ 32\overline{)1259} \\ -128 \end{array}$$

$32 \cdot 4 = 128$ is too big. We cannot subtract 128 from 125. Revise the estimate in the quotient to 3.

$$\begin{array}{r} 3 \\ 32\overline{)1259} \\ -96 \\ \hline 299 \end{array}$$

$32 \cdot 3 = 96$ and subtract.
Bring down the 9.

Now estimate the number of times 30 will go into 299. Because $30 \cdot 9 = 270$, our estimate is 9.

$$\begin{array}{r} 39 \text{ R}11 \\ 32\overline{)1259} \\ -96 \\ \hline 299 \\ -288 \\ \hline 11 \end{array}$$

$32 \cdot 9 = 288$ and subtract.
The remainder is 11.

To check, verify that $(32)(39) + 11 = 1259$. ✔

Skill Practice

15. $63\overline{)4516}$

Example 7 Dividing by a Many-Digit Number

Divide. $\dfrac{82{,}705}{602}$

Solution:

$$\begin{array}{r} 137 \text{ R}231 \\ 602\overline{)82{,}705} \\ -602 \\ \hline 2250 \\ -1806 \\ \hline 4445 \\ -4214 \\ \hline 231 \end{array}$$

$602 \cdot 1 = 602$ and subtract.
Bring down the 0.
$602 \cdot 3 = 1806$ and subtract.
Bring down the 5.
$602 \cdot 7 = 4214$ and subtract.
The remainder is 231.

To check, verify that $(602)(137) + 231 = 82{,}705$. ✔

Skill Practice

16. $304\overline{)62{,}405}$

Answers

15. 71 R43 **16.** 205 R85

5. Translations and Applications Involving Division

Several words and phrases imply division. A partial list is given in Table 1-4.

Table 1-4

Word/Phrase	Example	In Symbols
Divide	Divide 12 by 3	$12 \div 3$ or $\frac{12}{3}$ or $3\overline{)12}$
Quotient	The quotient of 20 and 2	$20 \div 2$ or $\frac{20}{2}$ or $2\overline{)20}$
Per	110 mi per 2 hr	$110 \div 2$ or $\frac{110}{2}$ or $2\overline{)110}$
Divides into	4 divides into 28	$28 \div 4$ or $\frac{28}{4}$ or $4\overline{)28}$
Divided, or shared equally among	64 shared equally among 4	$64 \div 4$ or $\frac{64}{4}$ or $4\overline{)64}$

Skill Practice

17. Four people play Hearts with a standard 52-card deck of cards. If the cards are equally distributed, how many cards does each player get?

Example 8 Solving an Application Involving Division

A painting business employs 3 painters. The business collects $1950 for painting a house. If all painters are paid equally, how much does each person make?

Solution:

This is an example where $1950 is shared equally among 3 people. Therefore, we divide.

$$\begin{array}{r} 650 \\ 3\overline{)1950} \\ -18 \\ \hline 15 \\ -15 \\ \hline 00 \\ -0 \\ \hline 0 \end{array}$$

$3 \cdot 6 = 18$ and subtract.
Bring down the 5.
$3 \cdot 5 = 15$ and subtract.
Bring down the 0.
$3 \cdot 0 = 0$ and subtract.
The remainder is 0.

Each painter makes $650.

Answer

17. 13 cards

Example 9 Solving an Application Involving Division

The graph in Figure 1-8 depicts the number of calories burned per hour for selected activities.

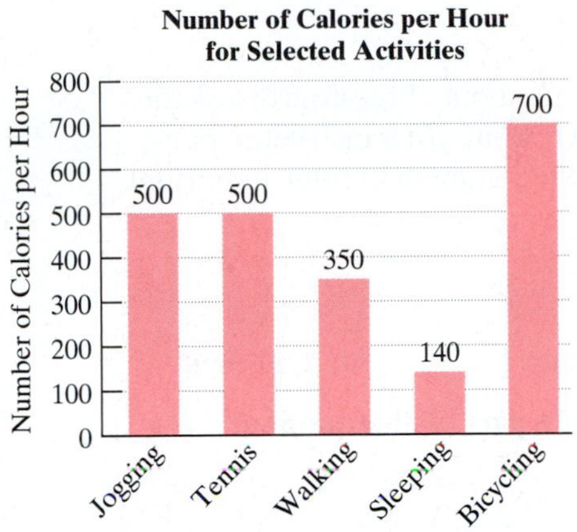

Figure 1-8

a. Janie wants to burn 3500 calories per week exercising. For how many hours must she jog?

b. For how many hours must Janie bicycle to burn 3500 calories?

Solution:

a. The total number of calories must be divided into 500-calorie increments. Thus, the number of hours required is given by $3500 \div 500$.

$$\begin{array}{r} 7 \\ 500\overline{)3500} \\ \underline{-3500} \\ 0 \end{array}$$

Janie requires 7 hr of jogging to burn 3500 calories.

b. 3500 calories must be divided into 700-calorie increments. The number of hours required is given by $3500 \div 700$.

$$\begin{array}{r} 5 \\ 700\overline{)3500} \\ \underline{-3500} \\ 0 \end{array}$$

Janie requires 5 hr of bicycling to burn 3500 calories.

Skill Practice

18. The cost for four different types of pastry at a French bakery is shown in the graph.

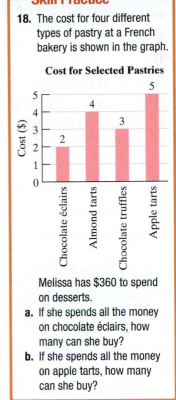

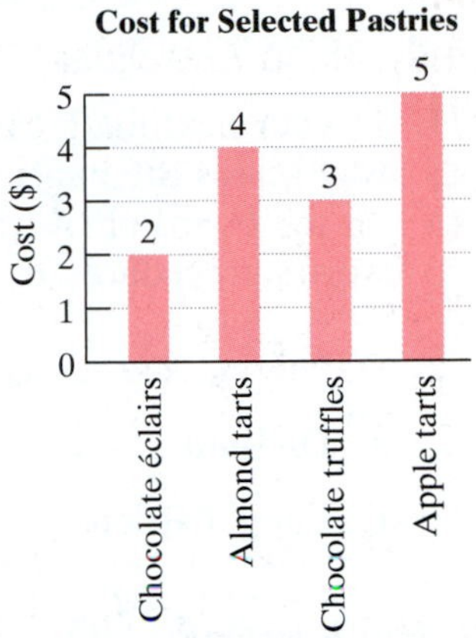

Melissa has $360 to spend on desserts.

a. If she spends all the money on chocolate éclairs, how many can she buy?

b. If she spends all the money on apple tarts, how many can she buy?

Answer

19. a. 180 chocolate éclairs
b. 72 apple tarts

Section 1.6 Practice Exercises

Study Skills Exercises

1. In your next math class, take notes by drawing a vertical line about three-fourths of the way across the paper, as shown. On the left side, write down what your instructor puts on the board or overhead. On the right side, make your own comments about important words, procedures, or questions that you have.

2. Define the key terms.

a. Division **b. Dividend** **c. Divisor** **d. Quotient**

e. Long division **f. Remainder** **g. Whole part of the quotient**

Review Exercises

For Exercises 3–10, add, subtract, or multiply as indicated.

3. $48 \cdot 103$ **4.** $678 - 83$ **5.** $1008 + 245$ **6.** $14(220)$

7. $5230 \cdot 127$ **8.** $789(25)$ **9.** $4890 - 3988$ **10.** $38{,}002 + 3902$

Objective 1: Introduction to Division

For Exercises 11–16, simplify each expression. Then identify the dividend, divisor, and quotient. **(See Example 1.)**

11. $72 \div 8$ **12.** $32 \div 4$ **13.** $8\overline{)64}$

14. $5\overline{)35}$ **15.** $\frac{45}{9}$ **16.** $\frac{20}{5}$

Objective 2: Properties of Division

17. In your own words, explain the difference between dividing a number by zero and dividing zero by a number.

18. Explain what happens when a number is either divided or multiplied by 1.

For Exercises 19–30, use the properties of division to simplify the expression, if possible. **(See Example 2.)**

19. $15 \div 1$ **20.** $21\overline{)21}$ **21.** $0 \div 10$ **22.** $\frac{0}{3}$

23. $0\overline{)9}$ **24.** $4 \div 0$ **25.** $\frac{20}{20}$ **26.** $1\overline{)9}$

27. $\frac{16}{0}$ **28.** $\frac{5}{1}$ **29.** $8\overline{)0}$ **30.** $13 \div 13$

31. Show that $6 \div 3 = 2$ but $3 \div 6 \neq 2$ by using multiplication to check.

32. Show that division is not associative, using the numbers 36, 12, and 3.

Objective 3: Long Division

33. Explain the process for checking a division problem when there is no remainder.

34. Show how checking by multiplication can help us remember that $0 \div 5 = 0$ and that $5 \div 0$ is undefined.

For Exercises 35–46, divide and check by multiplying. **(See Examples 3 and 4.)**

35. $78 \div 6$
Check: $6 \cdot \square = 78$

36. $364 \div 7$
Check: $7 \cdot \square = 364$

37. $5\overline{)205}$
Check: $5 \cdot \square = 205$

38. $8\overline{)152}$
Check: $8 \cdot \square = 152$

39. $\dfrac{972}{2}$

40. $\dfrac{582}{6}$

41. $1227 \div 3$

42. $236 \div 4$

43. $5\overline{)1015}$

44. $5\overline{)2035}$

45. $\dfrac{4932}{6}$

46. $\dfrac{3619}{7}$

For Exercises 47–54, check the following division problems. If it does not check, find the correct answer.

47. $\begin{array}{r} 56 \\ 4\overline{)224} \end{array}$

48. $\begin{array}{r} 82 \\ 7\overline{)574} \end{array}$

49. $761 \div 3 = 253$

50. $604 \div 5 = 120$

51. $\dfrac{1021}{9} = 113 \text{ R}4$

52. $\dfrac{1311}{6} = 218 \text{ R}3$

53. $\begin{array}{r} 25 \text{ R}6 \\ 8\overline{)203} \end{array}$

54. $\begin{array}{r} 117 \text{ R}5 \\ 7\overline{)821} \end{array}$

For Exercises 55–70, divide and check the answer. **(See Example 5.)**

55. $61 \div 8$

56. $89 \div 3$

57. $9\overline{)92}$

58. $5\overline{)74}$

59. $\dfrac{55}{2}$

60. $\dfrac{49}{3}$

61. $593 \div 3$

62. $801 \div 4$

63. $\dfrac{382}{9}$

64. $\dfrac{428}{8}$

65. $3115 \div 2$

66. $4715 \div 6$

67. $6014 \div 8$

68. $9013 \div 7$

69. $6\overline{)5012}$

70. $2\overline{)1101}$

Objective 4: Dividing by a Many-Digit Divisor

For Exercises 71–86, divide. **(See Examples 6 and 7.)**

71. $9110 \div 19$

72. $3505 \div 13$

73. $24\overline{)1051}$

74. $41\overline{)8104}$

75. $\dfrac{8008}{26}$

76. $\dfrac{9180}{15}$

77. $68{,}012 \div 54$

78. $92{,}013 \div 35$

79. $\dfrac{1650}{75}$

80. $\dfrac{3649}{89}$

81. $520\overline{)18{,}201}$

82. $298\overline{)6278}$

83. $69{,}712 \div 304$

84. $51{,}107 \div 221$

85. $114\overline{)34{,}428}$

86. $421\overline{)87{,}989}$

Objective 5: Translations and Applications Involving Division

For Exercises 87–92, for each English sentence, write a mathematical expression and simplify.

87. Find the quotient of 497 and 71.

88. Find the quotient of 1890 and 45.

89. Divide 877 by 14.

90. Divide 722 by 53.

91. Divide 6 into 42.

92. Divide 9 into 108.

93. There are 392 students signed up for Anatomy 101. If each classroom can hold 28 students, find the number of classrooms needed. (See Example 8.)

94. A wedding reception is planned to take place in the fellowship hall of a church. The bride anticipates 120 guests, and each table will seat 8 people. How many tables should be set up for the reception to accommodate all the guests?

95. A case of tomato sauce contains 32 cans. If a grocer has 168 cans, how many cases can he fill completely? How many cans will be left over?

96. Austin has $425 to spend on dining room chairs. If each chair costs $52, does he have enough to purchase 8 chairs? If so, will he have any money left over?

97. At one time, Seminole Community College had 3000 students who registered for Beginning Algebra. If the average class size is 25 students, how many Beginning Algebra classes will the college have to offer?

98. Eight people are to share equally in an inheritance of $84,480. How much money will each person receive?

99. The Honda Hybrid gets 45 mpg in stop-and-go traffic. How many gallons will it use in 405 mi of stop-and-go driving?

100. A couple traveled at an average speed of 52 mph for a cross-country trip. If the couple drove 1352 mi, how many hours was the trip?

101. Suppose Genny can type 1234 words in 22 min. Round each number to estimate her rate in words per minute.

102. On a trip to California from Illinois, Lavu drove 2780 mi. The gas tank in his car allows him to travel 405 mi. Round each number to the hundreds place to estimate the number of tanks of gas needed for the trip.

103. A group of 18 people goes to a concert. Ticket prices are given in the graph. If the group has $450, can they all attend the concert? If so, which type of seats can they buy? (See Example 9.)

104. The graph gives the average annual income for four professions: teacher, professor, CEO, and programmer. Find the monthly income for each of the four professions.

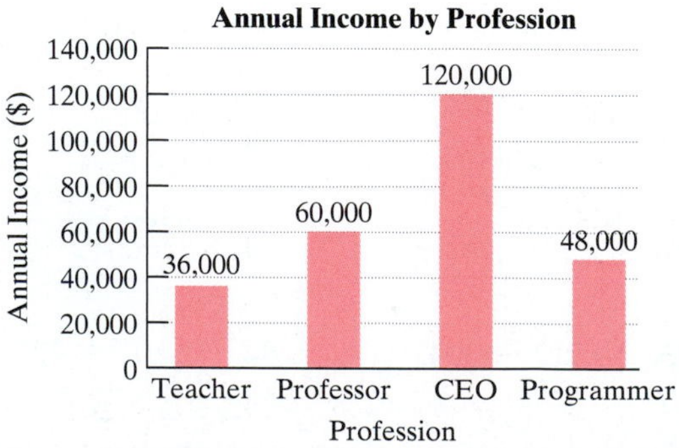

Calculator Connections

Topic: Multiplying and Dividing Whole Numbers

To multiply and divide numbers on a calculator, use the [×] and [÷] keys, respectively.

Expression	Keystrokes	Result
38,319 × 1561	38319 [×] 1561 [ENTER]	59815959
2,449,216 ÷ 6248	2449216 [÷] 6248 [ENTER]	392

Calculator Exercises

For Exercises 105–108, solve the problem. Use a calculator to perform the calculations.

105. The United States consumes approximately 21,000,000 barrels (bbl) of oil per day. (*Source:* U.S. Energy Information Administration) How much does it consume in 1 year?

106. The average time to commute to work for people living in Washington state is 26 min (round trip 52 min). (*Source:* U.S. Census Bureau) How much time does a person spend commuting to and from work in 1 year if the person works 5 days a week for 50 weeks per year?

107. The budget for the U.S. federal government for 2008 was approximately $2532 billion dollars. (*Source:* U.S. Department of the Treasury) How much could the government spend each month and still stay within its budget?

108. At a weigh station, a truck carrying 96 crates weighs in at 34,080 lb. If the truck weighs 9600 lb when empty, how much does each crate weigh?

Problem Recognition Exercises

Operations on Whole Numbers

For Exercises 1–18, perform the indicated operations.

1. a. $\begin{array}{r} 96 \\ +\ 24 \\ \hline \end{array}$ **b.** $\begin{array}{r} 96 \\ -\ 24 \\ \hline \end{array}$ **c.** $\begin{array}{r} 96 \\ \times\ 24 \\ \hline \end{array}$ **d.** $24\overline{)96}$

2. a. $\begin{array}{r} 550 \\ +\ 25 \\ \hline \end{array}$ **b.** $\begin{array}{r} 550 \\ -\ 25 \\ \hline \end{array}$ **c.** $\begin{array}{r} 550 \\ \times\ 25 \\ \hline \end{array}$ **d.** $25\overline{)550}$

3. a. $\begin{array}{r} 612 \\ +\ 334 \\ \hline \end{array}$ **b.** $\begin{array}{r} 946 \\ -\ 334 \\ \hline \end{array}$

4. a. $\begin{array}{r} 612 \\ -\ 334 \\ \hline \end{array}$ **b.** $\begin{array}{r} 278 \\ +\ 334 \\ \hline \end{array}$

5. a. $\begin{array}{r} 5500 \\ -\ 4299 \\ \hline \end{array}$ **b.** $\begin{array}{r} 1201 \\ +\ 4299 \\ \hline \end{array}$

6. a. $\begin{array}{r} 22{,}718 \\ +\ 12{,}137 \\ \hline \end{array}$ **b.** $\begin{array}{r} 34{,}855 \\ -\ 12{,}137 \\ \hline \end{array}$

7. a. $50 \cdot 400$ **b.** $20{,}000 \div 50$

8. a. $548 \cdot 63$ **b.** $34{,}524 \div 63$

9. a. $5060 \div 22$ **b.** $230 \cdot 22$

10. a. $1875 \div 125$ **b.** $125 \cdot 15$

11. a. $4\overline{)1312}$ **b.** $328\overline{)1312}$

12. a. $547\overline{)4376}$ **b.** $8\overline{)4376}$

13. a. $418 \cdot 10$ **b.** $418 \cdot 100$ **c.** $418 \cdot 1000$ **d.** $418 \cdot 10{,}000$

14. a. $350{,}000 \div 10$ **b.** $350{,}000 \div 100$ **c.** $350{,}000 \div 1000$ **d.** $350{,}000 \div 10{,}000$

15. $159 + 224 + 123$

16. $5064 \div 22$

17. $843 \cdot 27$

18. $7000 - 439$

Section 1.7 Exponents, Algebraic Expressions, and the Order of Operations

Objectives

1. Exponents
2. Square Roots
3. Order of Operations
4. Algebraic Expressions

1. Exponents

Thus far in the text we have learned to add, subtract, multiply, and divide whole numbers. We now present the concept of an **exponent** to represent repeated multiplication. For example, the product

$$3 \cdot 3 \cdot 3 \cdot 3 \cdot 3 \quad \text{can be written as} \quad 3^5$$

(exponent: 5; base: 3)

The expression 3^5 is written in exponential form. The exponent, or **power**, is 5 and represents the number of times the **base**, 3, is used as a factor. The expression 3^5 is read as "three to the fifth power." Other expressions in exponential form are shown next.

5^2	is read as	"five squared" or "five to the second power"
5^3	is read as	"five cubed" or "five to the third power"
5^4	is read as	"five to the fourth power"
5^5	is read as	"five to the fifth power"

TIP: The expression $5^1 = 5$. Any number without an exponent explicitly written has a power of 1.

Exponential form is a shortcut notation for repeated multiplication. However, to simplify an expression in exponential form, we often write out the individual factors.

Skill Practice

Evaluate.

1. 8^2 **2.** 4^3 **3.** 2^5

Example 1 Evaluating Exponential Expressions

Evaluate.

a. 6^2 **b.** 5^3 **c.** 2^4

Solution:

a. $6^2 = 6 \cdot 6$ The exponent, 2, indicates the number of times the base, 6, is used as a factor.

$= 36$

b. $5^3 = 5 \cdot 5 \cdot 5$ When three factors are multiplied, we can group the first two factors and perform the multiplication.

$= (5 \cdot 5) \cdot 5$

$= (25) \cdot 5$ Then multiply the product of the first two factors by the last factor.

$= 125$

c. $2^4 = 2 \cdot 2 \cdot 2 \cdot 2$

$= (2 \cdot 2) \cdot 2 \cdot 2$ Group the first two factors.

$= 4 \cdot 2 \cdot 2$ Multiply the first two factors.

$= (4 \cdot 2) \cdot 2$ Multiply the product by the next factor to the right.

$= 8 \cdot 2$

$= 16$

Answers

1. 64 **2.** 64 **3.** 32

One important application of exponents lies in recognizing **powers of 10**, that is, 10 raised to a whole-number power. For example, consider the following expressions.

$$10^1 = 10$$
$$10^2 = 10 \cdot 10 = 100$$
$$10^3 = 10 \cdot 10 \cdot 10 = 1000$$
$$10^4 = 10 \cdot 10 \cdot 10 \cdot 10 = 10{,}000$$
$$10^5 = 10 \cdot 10 \cdot 10 \cdot 10 \cdot 10 = 100{,}000$$
$$10^6 = 10 \cdot 10 \cdot 10 \cdot 10 \cdot 10 \cdot 10 = 1{,}000{,}000$$

From these examples, we see that a power of 10 results in a 1 followed by several zeros. The number of zeros is the same as the exponent on the base of 10.

2. Square Roots

To square a number means that we multiply the base times itself. For example, $5^2 = 5 \cdot 5 = 25$.

To find a positive **square root** of a number means that we reverse the process of squaring. For example, finding the square root of 25 is equivalent to asking, "What positive number, when squared, equals 25?" The symbol $\sqrt{\ }$, (called a *radical sign*) is used to denote the positive square root of a number. Therefore, $\sqrt{25}$ is the positive number, that when squared, equals 25. Thus, $\sqrt{25} = 5$ because $(5)^2 = 25$.

Example 2 Evaluating Square Roots

Find the square roots.

a. $\sqrt{9}$ **b.** $\sqrt{64}$ **c.** $\sqrt{1}$ **d.** $\sqrt{0}$

Solution:

a. $\sqrt{9} = 3$ because $(3)^2 = 3 \cdot 3 = 9$

b. $\sqrt{64} = 8$ because $(8)^2 = 8 \cdot 8 = 64$

c. $\sqrt{1} = 1$ because $(1)^2 = 1 \cdot 1 = 1$

d. $\sqrt{0} = 0$ because $(0)^2 = 0 \cdot 0 = 0$

Skill Practice

Find the square roots.

4. $\sqrt{4}$

5. $\sqrt{100}$

6. $\sqrt{400}$

7. $\sqrt{121}$

TIP: To simplify square roots, it is advisable to become familiar with the following squares and square roots.

$0^2 = 0 \longrightarrow \sqrt{0} = 0$ $7^2 = 49 \longrightarrow \sqrt{49} = 7$

$1^2 = 1 \longrightarrow \sqrt{1} = 1$ $8^2 = 64 \longrightarrow \sqrt{64} = 8$

$2^2 = 4 \longrightarrow \sqrt{4} = 2$ $9^2 = 81 \longrightarrow \sqrt{81} = 9$

$3^2 = 9 \longrightarrow \sqrt{9} = 3$ $10^2 = 100 \longrightarrow \sqrt{100} = 10$

$4^2 = 16 \longrightarrow \sqrt{16} = 4$ $11^2 = 121 \longrightarrow \sqrt{121} = 11$

$5^2 = 25 \longrightarrow \sqrt{25} = 5$ $12^2 = 144 \longrightarrow \sqrt{144} = 12$

$6^2 = 36 \longrightarrow \sqrt{36} = 6$ $13^2 = 169 \longrightarrow \sqrt{169} = 13$

Answers

4. 2 **5.** 10 **6.** 20 **7.** 11

3. Order of Operations

A numerical expression may contain more than one operation. For example, the following expression contains both multiplication and subtraction.

$$18 - 5(2)$$

The order in which the multiplication and subtraction are performed will affect the overall outcome.

Multiplying first yields	**Subtracting first yields**
$18 - 5(2) = 18 - 10$	$18 - 5(2) = 13(2)$
$= 8$ (correct)	$= 26$ (incorrect)

To avoid confusion, mathematicians have outlined the proper order of operations. In particular, multiplication is performed before addition or subtraction. The guidelines for the order of operations are given next. These rules must be followed in all cases.

Animation

PROCEDURE Order of Operations

Step 1 First perform all operations inside parentheses or other grouping symbols.

Step 2 Simplify any expressions containing exponents or square roots.

Step 3 Perform multiplication or division in the order that they appear from left to right.

Step 4 Perform addition or subtraction in the order that they appear from left to right.

Skill Practice

Simplify.

8. $18 + 6 \div 2 - 4$

9. $(20 - 4) \div 2 + 1$

10. $2^3 - \sqrt{16 + 9}$

Example 3 Using the Order of Operations

Simplify.

a. $15 - 10 \div 2 + 3$ **b.** $(5 - 2) \cdot 7 - 1$ **c.** $\sqrt{64 + 36} - 2^3$

Solution:

a. $15 - \underbrace{10 \div 2} + 3$

$= \underbrace{15 - 5} + 3$ Perform the division $10 \div 2$ first.

$= 10 + 3$ Perform addition and subtraction from left to right.

$= 13$ Add.

b. $(\underbrace{5 - 2}) \cdot 7 - 1$

$= \underbrace{(3) \cdot 7} - 1$ Perform the operation inside parentheses first.

$= 21 - 1$ Perform multiplication before subtraction.

$= 20$ Subtract.

Answers

8. 17 **9.** 9 **10.** 3

c. $\sqrt{64 + 36} - 2^3$

$= \sqrt{100} - 2^3$ The radical sign is a grouping symbol. Perform the operation within the radical first.

$= 10 - 8$ Simplify any expressions with exponents or square roots. Note that $\sqrt{100} = 10$, and $2^3 = 2 \cdot 2 \cdot 2 = 8$.

$= 2$ Subtract.

Example 4 Using the Order of Operations

Simplify.

a. $300 \div (7 - 2)^2 \cdot 2^2$ **b.** $36 + (7^2 - 3)$ **c.** $\dfrac{3^2 + 6 \cdot 1}{10 - 7}$

Solution:

a. $300 \div (7 - 2)^2 \cdot 2^2$ Perform the operation within parentheses first.

$= 300 \div (5)^2 \cdot 2^2$ Simplify exponents: $5^2 = 5 \cdot 5 = 25$ and $2^2 = 2 \cdot 2 = 4$.

$= 300 \div 25 \cdot 4$ From left to right, division appears before multiplication.

$= 12 \cdot 4$ Multiply.

$= 48$

b. $36 + (7^2 - 3)$

$= 36 + (49 - 3)$ Perform the operations within parentheses first. The guidelines indicate that we simplify the expression with the exponent before we subtract: $7^2 = 49$.

$= 36 + 46$ Add.

$= 82$

c. $\dfrac{3^2 + 6 \cdot 1}{10 - 7}$

$= \dfrac{9 + 6}{3}$ Simplify the expressions above and below the fraction bar by using the order of operations.

$= \dfrac{15}{3}$ Simplify.

$= 5$ Divide.

Skill Practice

Simplify.

11. $40 \div (3 - 1)^2 \cdot 5^2$

12. $42 - (50 - 6^2)$

13. $\dfrac{6^2 - 3 \cdot 4}{8 \cdot 3}$

TIP: A division bar within an expression acts as a grouping symbol. In Example 4(c), we must simplify the expressions above and below the division bar first before dividing.

Sometimes an expression will have parentheses within parentheses. These are called *nested parentheses*. The grouping symbols ( ), [ ], or { } are all used as parentheses. The different shapes make it easier to match up the pairs of parentheses. For example,

$$\{300 - 4[4 + (5 + 2)^2] + 8\} - 31$$

When nested parentheses are present, simplify the innermost set first. Then work your way out.

Answers

11. 250 **12.** 28 **13.** 1

Skill Practice

14. Simplify.
$4^2 - 2[12 - (3 + 6)]$

Example 5 Using the Order of Operations

Simplify. $\{300 - 4[4 + (5 + 2)^2] + 8\} - 31$

Solution:

$\{300 - 4[4 + (5 + 2)^2] + 8\} - 31$	Simplify within the innermost parentheses first ( ).
$= \{300 - 4[4 + (7)^2] + 8\} - 31$	Simplify the exponent.
$= \{300 - 4[4 + 49] + 8\} - 31$	Simplify within the next innermost parentheses [ ].
$= \{300 - 4[53] + 8\} - 31$	Multiply before adding.
$= \{300 - 212 + 8\} - 31$	Subtract and add in order from left to right within the parentheses { }.
$= \{88 + 8\} - 31$	Simplify within the parentheses { }.
$= 96 - 31$	Simplify.
$= 65$	

4. Algebraic Expressions

In Section 1.5, we introduced the formula $A = l \cdot w$. This represents the area of a rectangle in terms of its length and width. The letters A, l, and w are called variables. **Variables** are used to represent quantities that are subject to change. Quantities that do not change are called **constants**. Variables and constants are used to build algebraic expressions. The following are all examples of algebraic expressions.

$$n + 30, \quad x - y, \quad 3w, \quad \frac{a}{4}$$

The value of an algebraic expression depends on the values of the variables within the expression. In Examples 6 and 7, we practice evaluating expressions for given values of the variables.

Skill Practice

15. Evaluate $x + 7y$ for $x = 8$ and $y = 4$.

Example 6 Evaluating an Algebraic Expression

Evaluate the expression for the given values of the variables.

$$5a + b \quad \text{for } a = 6 \text{ and } b = 10$$

Solution:

$5a + b$	
$= 5(\ ) + (\ )$	When we substitute a number for a variable, use parentheses in place of the variable.
$= 5(6) + (10)$	Substitute 6 for a and 10 for b, by placing the values within the parentheses.
$= 30 + 10$	Apply the order of operations. Multiplication is performed before addition.
$= 40$	

Answers

14. 10 **15.** 36

Example 7 Evaluating an Algebraic Expression

Evaluate the expression for the given values of the variables.

$$(x - y + z)^2 \quad \text{for } x = 12, y = 9, \text{ and } z = 4$$

Solution:

$(x - y + z)^2$

$= [(\ ) - (\ ) + (\ )]^2$ Use parentheses in place of the variables.

$= [(12) - (9) + (4)]^2$ Substitute 12 for x, 9 for y, and 4 for z.

$= [3 + 4]^2$ Subtract and add within the grouping symbols from left to right.

$= (7)^2$

$= 49$ The value $7^2 = 7 \cdot 7 = 49$.

Skill Practice

16. Evaluate $(m - n)^2 + p$ for $m = 11$, $n = 5$, and $p = 2$.

Answer

16. 38

Section 1.7 Practice Exercises

Study Skills Exercises

1. Look over the notes that you took today. Do you understand what you wrote? If there were any rules, definitions, or formulas, highlight them so that they can be easily found when studying for the test. You may want to begin by highlighting the order of operations.

2. Define the key terms.

a. **Exponent** **b.** **Power** **c.** **Base** **d.** **Power of 10**

e. **Square root** **f.** **Variable** **g.** **Constant**

Review Exercises

For Exercises 3–8, write true or false for each statement.

3. Addition is commutative; for example, $5 + 3 = 3 + 5$.

4. Subtraction is commutative; for example, $5 - 3 = 3 - 5$.

5. $6 \cdot 0 = 6$ **6.** $0 \div 8 = 0$ **7.** $0 \cdot 8 = 0$ **8.** $5 \div 0$ is undefined

Objective 1: Exponents

9. Write an exponential expression with 9 as the base and 4 as the exponent.

10. Write an exponential expression with 3 as the base and 8 as the exponent.

For Exercises 11–14, write the repeated multiplication in exponential form. Do not simplify.

11. $3 \cdot 3 \cdot 3 \cdot 3 \cdot 3 \cdot 3$ **12.** $7 \cdot 7 \cdot 7 \cdot 7$ **13.** $4 \cdot 4 \cdot 4 \cdot 4 \cdot 2 \cdot 2 \cdot 2$ **14.** $5 \cdot 5 \cdot 5 \cdot 10 \cdot 10 \cdot 10$

For Exercises 15–18, expand the exponential expression as a repeated multiplication. Do not simplify.

15. 8^4 **16.** 2^6 **17.** 4^8 **18.** 6^2

For Exercises 19–30, evaluate the exponential expressions. **(See Example 1.)**

19. 2^3 **20.** 4^2 **21.** 3^2 **22.** 5^2

23. 3^3 **24.** 11^2 **25.** 5^3 **26.** 10^3

27. 2^5 **28.** 6^3 **29.** 3^4 **30.** 5^4

31. Evaluate 1^2, 1^3, 1^4, and 1^5. Explain what happens when 1 is raised to any power.

For Exercises 32–35, evaluate the powers of 10.

32. 10^2 **33.** 10^3 **34.** 10^4 **35.** 10^5

36. Explain how to get 10^9 *without* performing the repeated multiplication. **(See Exercises 32–35.)**

Objective 2: Square Roots

For Exercises 37–44, evaluate the square roots. **(See Example 2.)**

37. $\sqrt{4}$ **38.** $\sqrt{9}$ **39.** $\sqrt{36}$ **40.** $\sqrt{81}$

41. $\sqrt{100}$ **42.** $\sqrt{49}$ **43.** $\sqrt{0}$ **44.** $\sqrt{16}$

Objective 3: Order of Operations

45. Does the order of operations indicate that addition is always performed before subtraction? Explain.

46. Does the order of operations indicate that multiplication is always performed before division? Explain.

For Exercises 47–87, simplify using the order of operations. **(See Examples 3–4.)**

Animation

47. $6 + 10 \cdot 2$ **48.** $4 + 3 \cdot 7$ **49.** $10 - 3^2$ **50.** $11 - 2^2$

51. $(10 - 3)^2$ **52.** $(11 - 2)^2$ **53.** $36 \div 2 \div 6$ **54.** $48 \div 4 \div 2$

55. $15 - (5 + 8)$ **56.** $41 - (13 + 8)$ **57.** $(13 - 2) \cdot 5 - 2$ **58.** $(8 + 4) \cdot 6 + 8$

59. $4 + 12 \div 3$ **60.** $9 + 15 \div \sqrt{25}$ **61.** $30 \div 2 \cdot \sqrt{9}$ **62.** $55 \div 11 \cdot 5$

63. $7^2 - 5^2$ **64.** $3^3 - 2^3$ **65.** $(7 - 5)^2$ **66.** $(3 - 2)^3$

67. $100 \div 5 \cdot 2$ **68.** $60 \div 3 \cdot 2$ **69.** $20 - 5(11 - 8)$ **70.** $38 - 6(10 - 5)$

71. $\sqrt{36 + 64} + 2(9 - 1)$ **72.** $\sqrt{16 + 9} + 3(8 - 3)$ **73.** $\dfrac{36}{2^2 + 5}$ **74.** $\dfrac{42}{3^2 - 2}$

75. $80 - 20 \div 4 \cdot 6$ **76.** $300 - 48 \div 8 \cdot 40$ **77.** $\dfrac{42 - 26}{4^2 - 8}$ **78.** $\dfrac{22 + 14}{2^2 \cdot 3}$

79. $(18 - 5) - (23 - \sqrt{100})$ **80.** $(\sqrt{36} + 11) - (31 - 16)$ **81.** $80 \div (9^2 - 7 \cdot 11)^2$

82. $108 \div (3^3 - 6 \cdot 4)^2$ **83.** $22 - 4(\sqrt{25} - 3)^2$ **84.** $17 + 3(7 - \sqrt{9})^2$

85. $96 - 3(42 \div 7 \cdot 6 - 5)$ **86.** $50 - 2(36 \div 12 \cdot 2 - 4)$ **87.** $16 + 5(20 \div 4 \cdot 8 - 3)$

For Exercises 88–93, simplify the expressions with nested parentheses. **(See Example 5.)**

88. $3[4 + (6 - 3)^2] - 15$

89. $2[5(4 - 1) + 3] \div 6$

90. $8^2 - 5[12 - (8 - 6)]$

91. $3^3 - 2[15 - (2 + 1)^2]$

92. $5\{21 - [3^2 - (4 - 2)]\}$

93. $4\{18 - [(10 - 8) + 2^3]\}$

Objective 4: Algebraic Expressions

For Exercises 94–101, evaluate the expressions for the given values of the variables. $x = 12, y = 4, z = 25$, and $w = 9$. **(See Examples 6 – 7.)**

94. $10y - z$

95. $8w - 4x$

96. $3x + 6y + 9w$

97. $9y - 4w + 3z$

98. $(z - x - y)^2$

99. $(y + z - w)^2$

100. $\sqrt{z}$

101. $\sqrt{w}$

Calculator Connections

Topic: Evaluating Expressions with Exponents

Many calculators use the [x^2] key to square a number. To raise a number to a higher power, use the [^] key (or on some calculators, the [x^y] key or [y^x] key).

Expression	Keystrokes	Result
26^2	26 [x^2] [ENTER]	676
	On some calculators, you do not need to press [ENTER].	
3^4	3 [^] 4 [ENTER]	81
or	3 [y^x] 4 [=]	81

Calculator Exercises

For Exercises 102–107, use a calculator to perform the indicated operations.

102. 156^2

103. 418^2

104. 12^5

105. 35^4

106. 43^3

107. 71^3

For Exercises 108–113, simplify the expressions by using the order of operations. For each step use the calculator to simplify the given operation.

108. $8126 - 54{,}978 \div 561$

109. $92{,}168 + 6954 \times 29$

110. $(3548 - 3291)^2$

111. $(7500 \div 625)^3$

112. $\dfrac{89{,}880}{384 + 2184}$ *Hint:* This expression has implied grouping symbols. $\dfrac{89{,}880}{(384 + 2184)}$

113. $\dfrac{54{,}137}{3393 - 2134}$ *Hint:* This expression has implied grouping symbols. $\dfrac{54{,}137}{(3393 - 2134)}$

Section 1.8 Mixed Applications and Computing Mean

Objectives

1. Applications Involving Multiple Operations
2. Computing a Mean (Average)

1. Applications Involving Multiple Operations

Sometimes more than one operation is needed to solve an application problem.

Example 1 Solving a Consumer Application

Jorge bought a car for \$18,340. He paid \$2500 down and then paid the rest in equal monthly payments over a 4-year period. Find the amount of Jorge's monthly payment (not including interest).

Solution:

Familiarize and draw a picture.

Given: total price: \$18,340
down payment: \$2500
payment plan: 4 years (48 months)

Find: monthly payment

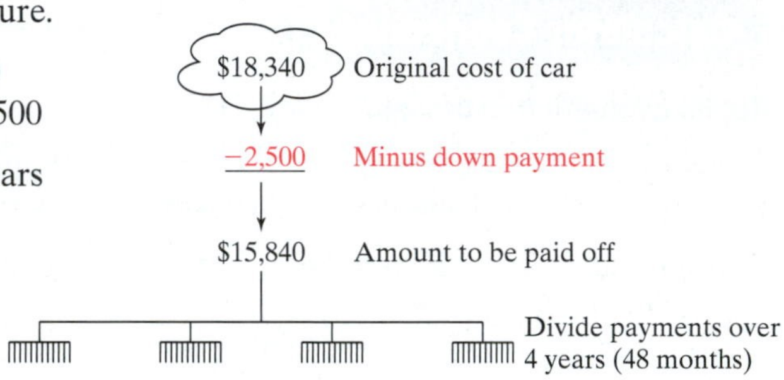

Operations:

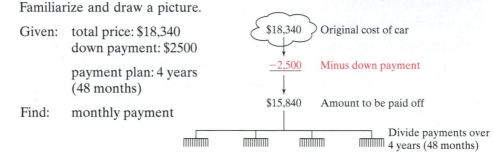

1. The amount of the loan to be paid off is equal to the original cost of the car minus the down payment. We use subtraction:

$$\begin{array}{r} \$18{,}340 \\ -\ 2{,}500 \\ \hline \$15{,}840 \end{array}$$

2. This money is distributed in equal payments over a 4-year period. Because there are 12 months in 1 year, there are $4 \cdot 12 = 48$ months in a 4-year period. To distribute \$15,840 among 48 equal payments, we divide.

$$\begin{array}{r} 330 \\ 48\overline{)15{,}840} \\ -144 \\ \hline 144 \\ -144 \\ \hline 00 \end{array}$$

Jorge's monthly payments will be \$330.

TIP: The solution to Example 1 can be checked by multiplication. Forty-eight payments of \$330 each amount to $48(\$330) = \$15{,}840$. This added to the down payment totals \$18,340 as desired.

Skill Practice

1. Danielle buys a new entertainment center with a new plasma television for \$4240. She pays \$1000 down, and the rest is paid off in equal monthly payments over 2 years. Find Danielle's monthly payment.

Answer

1. \$135 per month

Example 2 Solving a Travel Application

Linda must drive from Clayton to Oakley. She can travel directly from Clayton to Oakley on a mountain road, but will only average 40 mph. On the route through Pearson, she travels on highways and can average 60 mph. Which route will take less time?

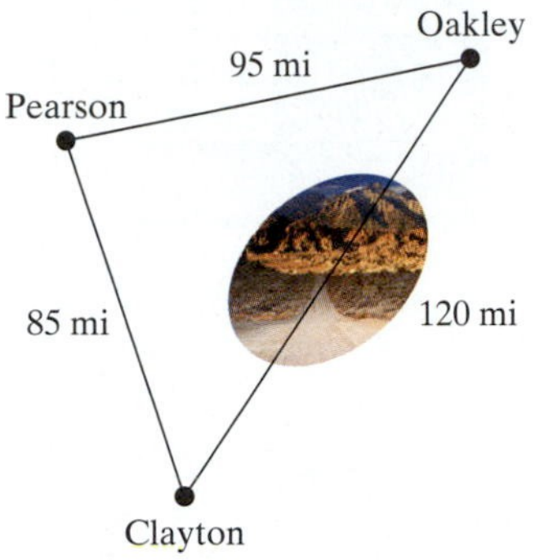

Solution:

Read and familiarize: A map is presented in the problem.

Given: The distance for each route and the speed traveled along each route

Find: Find the time required for each route. Then compare the times to determine which will take less time.

Operations:

1. First note that the total distance of the route through Pearson is found by using addition.

$$85 \text{ mi} + 95 \text{ mi} = 180 \text{ mi}$$

2. The speed of the vehicle gives us the distance traveled per hour. Therefore, the time of travel equals the total distance divided by the speed.

   From Clayton to Oakley through the mountains, we divide 120 mi by 40-mph increments to determine the number of hours.

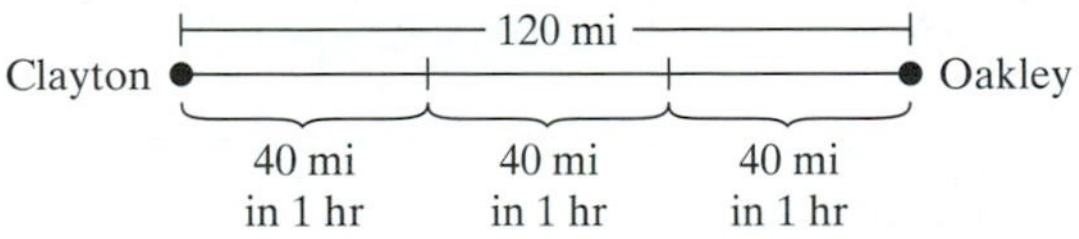

$$\text{Time} = \frac{120 \text{ mi}}{40 \text{ mph}} = 3 \text{ hr}$$

   From Clayton to Oakley through Pearson, we divide 180 mi by 60-mph increments to determine the number of hours.

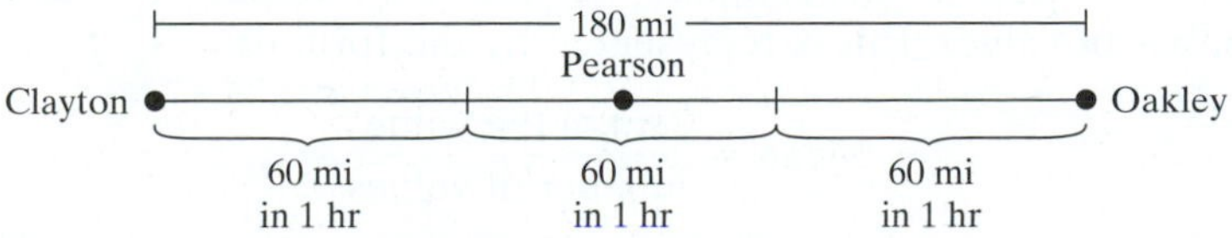

$$\text{Time} = \frac{180 \text{ mi}}{60 \text{ mph}} = 3 \text{ hr}$$

Therefore, each route takes the same amount of time, 3 hr.

Skill Practice

2. Taylor makes $18 per hour for the first 40 hr worked each week. His overtime rate is $27 per hour for hours exceeding the normal 40-hr workweek. If his total salary for one week is $963, determine the number of hours of overtime worked.

Answer

2. 9 hr overtime

Skill Practice

3. Alain wants to put molding around the base of the room shown in the figure. No molding is needed where the door, closet, and bathroom are located. Find the total cost if molding is $2 per foot.

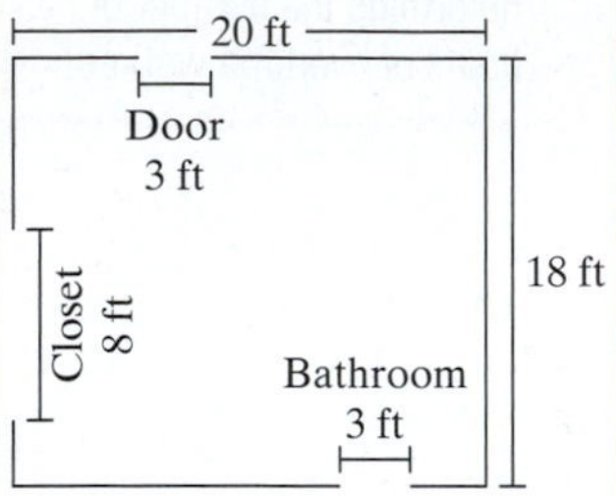

Example 3 Solving a Construction Application

A rancher must fence the corral shown in Figure 1-9. However, no fencing is required on the side adjacent to the barn. If fencing costs $4 per foot, what is the total cost?

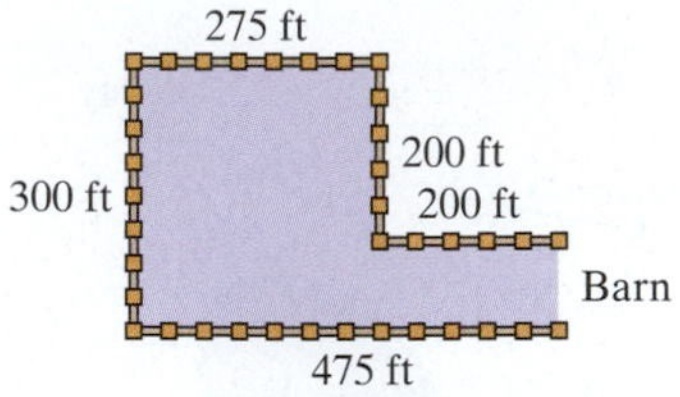

Figure 1-9

Solution:

Read and familiarize: A figure is provided.

Strategy

With some application problems, it helps to work backward from your final goal. In this case, our final goal is to find the total cost. However, to find the total cost, we must first find the total distance to be fenced. To find the total distance, we add the lengths of the sides that are being fenced.

$$\begin{array}{r} \overset{1\,1}{275}\text{ ft} \\ 200\text{ ft} \\ 200\text{ ft} \\ 475\text{ ft} \\ +\ 300\text{ ft} \\ \hline 1450\text{ ft} \end{array}$$

Therefore,

$$\begin{pmatrix}\text{Total cost}\\ \text{of fencing}\end{pmatrix} = \begin{pmatrix}\text{total}\\ \text{distance}\\ \text{in feet}\end{pmatrix}\begin{pmatrix}\text{cost}\\ \text{per foot}\end{pmatrix}$$

$$= (1450\text{ ft})(\$4\text{ per ft})$$

$$= \$5800$$

The total cost of fencing is $5800.

2. Computing a Mean (Average)

The order of operations must be used when we compute an average. The technical term for the average of a list of numbers is the **mean** of the numbers. To find the mean of a set of numbers, first compute the sum of the values. Then divide the sum by the number of values. This is represented by the formula

$$\text{Mean} = \frac{\text{sum of the values}}{\text{number of values}}$$

Answer

3. $124

Example 4 Computing a Mean (Average)

Ashley took 6 tests in Chemistry. Find her mean (average) score.

$$89, 91, 72, 86, 94, 96$$

Solution:

$$\text{Average score} = \frac{89 + 91 + 72 + 86 + 94 + 96}{6}$$

$$= \frac{528}{6} \quad \text{Add the values in the list first.}$$

$$= 88 \quad \text{Divide.} \quad \begin{array}{r} 88 \\ 6\overline{)528} \\ -48 \\ \hline 48 \\ -48 \\ \hline 0 \end{array}$$

Ashley's mean (average) score is 88.

Skill Practice

4. The ages (in years) of 5 students in an algebra class are given here. Find the mean age.

22, 18, 22, 32, 46

TIP: The division bar in $\frac{89 + 91 + 72 + 86 + 94 + 96}{6}$ is also a grouping symbol and implies parentheses:

$$\frac{(89 + 91 + 72 + 86 + 94 + 96)}{6}$$

Answer

4. 28 years

Section 1.8 Practice Exercises

Study Skills Exercises

1. Check yourself.

Yes _____ No _____ Did you have sufficient time to study for the test on this chapter? If not, what could you have done to create more time for studying?

Yes _____ No _____ Did you work all the assigned homework problems in this chapter?

Yes _____ No _____ If you encountered difficulty in this chapter, did you see your instructor or tutor for help?

Yes _____ No _____ Have you taken advantage of the textbook supplements such as the *Student Solution Manual*?

2. Define the key term, **mean**.

Review Exercises

For Exercises 3–13, translate the English phrase into a mathematical statement and simplify.

3. 71 increased by 14

4. 16 more than 42

5. Twice 14

6. The difference of 93 and 79

7. Subtract 32 from 102

8. Divide 12 into 60

9. The product of 10 and 13

10. The total of 12, 14, and 15

11. The quotient of 24 and 6

12. 41 less than 78

13. The sum of 5, 13, and 25

Objective 1: Applications Involving Multiple Operations

14. Jackson purchased a car for \$16,540. He paid \$2500 down and paid the rest in equal monthly payments over a 36-month period. How much were his monthly payments?

15. Lucio purchased a refrigerator for \$1170. He paid \$150 at the time of purchase and then paid off the rest in equal monthly payments over 1 year. How much was his monthly payment? **(See Example 1.)**

16. Monika must drive from Watertown to Utica. She can travel directly from Watertown to Utica on a small county road, but will only average 40 mph. On the route through Syracuse, she travels on highways and can average 60 mph. Which route will take less time?

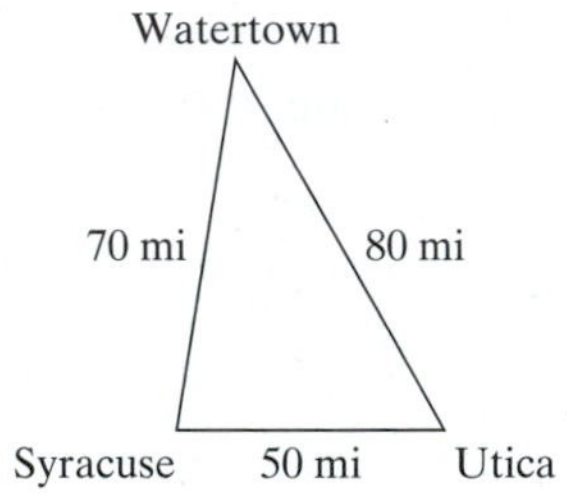

Figure for Exercise 16

17. Rex has a choice of two routes to drive from Oklahoma City to Fort Smith. On the interstate, the distance is 220 mi and he can drive 55 mph. If he takes the back roads, he can only travel 40 mph, but the distance is 200 mi. Which route will take less time? **(See Example 2.)**

18. If you wanted to line the outside of a garden with a decorative border, would you need to know the area of the garden or the perimeter of the garden?

19. If you wanted to know how much sod to lay down within a rectangular backyard, would you need to know the area of the yard or the perimeter of the yard?

20. Alexis wants to buy molding for a room that is 12 ft by 11 ft. No molding is needed for the doorway, which measures 3 ft. See the figure. If molding costs \$2 per foot, how much money will it cost?

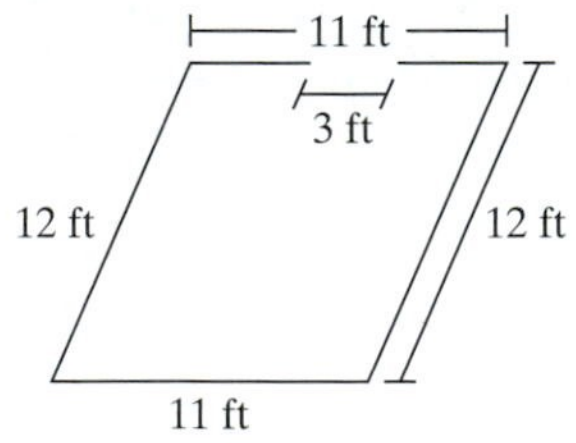

21. A homeowner wants to fence her rectangular backyard. The yard is 75 ft by 90 ft. If fencing costs \$5 per foot, how much will it cost to fence the yard? **(See Example 3.)**

22. What is the cost to carpet the room whose dimensions are shown in the figure? Assume that carpeting costs \$34 per square yard and that there is no waste.

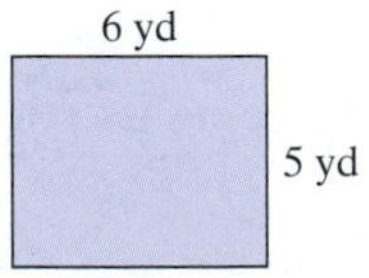

23. What is the cost to tile the room whose dimensions are shown in the figure? Assume that tile costs \$3 per square foot.

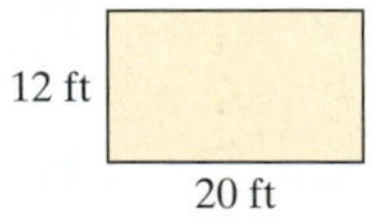

24. The balance in Gina's checking account is \$278. If she writes checks for \$82, \$59, and \$101, how much will be left over?

25. The balance in Jose's checking account is \$3455. If he write checks for \$587, \$36, and \$156, how much will be left over?

26. A community college bought 72 new computers and 6 new printers for a computer lab. If computers were purchased for $2118 each and the printers for $256 each, what was the total bill (not including tax)?

27. Tickets to the San Diego Zoo in California cost $22 for children aged 3–11 and $33 for adults. How much money is required to buy tickets for a class of 33 children and 6 adult chaperones?

28. A discount music store buys used CDs from its customers for $3. Furthermore, a customer can buy any used CD in the store for $8. Latayne sells 16 CDs.

a. How much money does she receive by selling the 16 CDs?

b. How many CDs can she then purchase with the money?

29. Shevona earns $8 per hour and works a 40-hr workweek. At the end of the week, she cashes her paycheck and then buys two tickets to a Beyoncé concert.

a. How much is her paycheck?

b. If the concert tickets cost $64 each, how much money does she have left over from her paycheck after buying the tickets?

30. During his 13-year career with the Chicago Bulls, Michael Jordan scored 12,192 field goals (worth 2 points each). He scored 581 three-point shots and 7327 free-throws (worth 1 point each). How many total points did he score during his career with the Bulls?

31. A matte is to be cut and placed over five small square pictures before framing. Each picture is 5 in. wide, and the matte frame is 37 in. wide, as shown in the figure. If the pictures are to be equally spaced (including the space on the left and right edges), how wide is the matte between them?

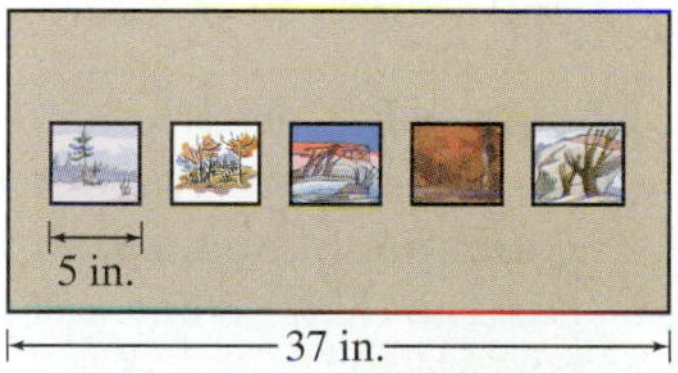

32. Mortimer the cat was prescribed a suspension of methimazole for hyperthyroidism. This suspension comes in a 60 milliliter bottle with instructions to give 1 milliliter twice a day. The label also shows there is one refill, but it must be called in 2 days ahead. Mortimer had his first two doses on September 1.

a. For how many days will one bottle last?

b. On what day, at the latest, should his owner order a refill to avoid running out of medicine?

33. Recently, the American Medical Association reported that there were 630,300 male doctors and 205,900 female doctors in the United States.

a. What is the difference between the number of male doctors and the number of female doctors?

b. What is the total number of doctors?

34. On a map, 1 in. represents 60 mi.

a. If Las Vegas and Salt Lake City are approximately 6 in. apart on the map, what is the actual distance between the cities?

b. If Madison, Wisconsin, and Dallas, Texas, are approximately 840 mi apart, how many inches would this represent on the map?

35. On a map, each inch represents 40 mi.

a. If Wichita, Kansas, and Des Moines, Iowa, are approximately 8 in. apart on the map, what is the actual distance between the cities?

b. If Seattle, Washington, and Sacramento, California, are approximately 600 mi apart, how many inches would this represent on the map?

36. A textbook company ships books in boxes containing a maximum of 12 books. If a bookstore orders 1250 books, how many boxes can be filled completely? How many books will be left over?

37. A farmer sells eggs in containers holding a dozen eggs. If he has 4257 eggs, how many containers will be filled completely? How many eggs will be left over?

38. Marc pays for an $84 dinner with $20 bills.

a. How many bills must he use?

b. How much change will he receive?

39. Byron buys three CDs for a total of $54 and pays with $10 bills.

a. How many bills must he use?

b. How much change will he receive?

40. Ling has three jobs. He works for a lawn maintenance service 4 days a week. He also tutors math and works as a waiter on weekends. His hourly wage and the number of hours for each job are given for a 1-week period. How much money did Ling earn for the week?

	Hourly Wage	Number of Hours
Tutor	$30/hr	4
Waiter	10/hr	16
Lawn maintenance	8/hr	30

41. An electrician, a plumber, a mason, and a carpenter work at a construction site. The hourly wage and the number of hours each person worked are summarized in the table. What was the total amount paid for all four workers?

	Hourly Wage	Number of Hours
Electrician	$36/hr	18
Plumber	28/hr	15
Mason	26/hr	24
Carpenter	22/hr	48

Objective 2: Computing a Mean (Average)

For Exercises 42–44, find the mean (average) of each set of numbers. **(See Example 4.)**

42. 19, 21, 18, 21, 16

43. 105, 114, 123, 101, 100, 111

44. 1480, 1102, 1032, 1002

45. Neelah took six quizzes and received the following scores: 19, 20, 18, 19, 18, 14. Find her quiz average.

46. Shawn's scores on his last four tests were 83, 95, 87, and 91. What is his average for these tests?

47. At a certain grocery store, Jessie notices that the price of bananas varies from week to week. During a 3-week period she buys bananas for 89¢ per pound, 79¢ per pound, and 66¢ per pound. What does Jessie pay on average per pound?

48. On a trip, Stephen had his car washed four times and paid $7, $10, $8, and $7. What was the average amount spent per wash?

49. The monthly rainfall for Seattle, Washington, is given in the table. All values are in millimeters (mm).

	Jan.	Feb.	Mar.	Apr.	May	Jun.	Jul.	Aug.	Sep.	Oct.	Nov.	Dec.
Rainfall	122	94	80	52	47	40	15	21	44	90	118	123

Find the average monthly rainfall for the months of November, December, and January.

50. The monthly snowfall for Alpena, Michigan, is given in the table. All values are in inches.

	Jan.	Feb.	Mar.	Apr.	May	Jun.	Jul.	Aug.	Sep.	Oct.	Nov.	Dec.
Snowfall	22	16	13	5	1	0	0	0	0	1	9	20

Find the average monthly snowfall for the months of November, December, January, February, and March.

Chapter 1 Summary

Section 1.2 Introduction to Whole Numbers

Key Concepts

The place value for each **digit** of a number is shown in the chart.

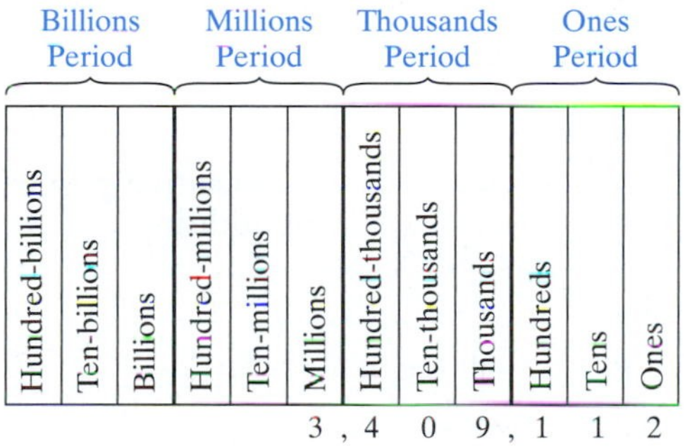

Numbers can be written in different forms, for example:

Standard Form: 3,409,112

Expanded Form: 3 millions + 4 hundred-thousands + 9 thousands + 1 hundred + 1 ten + 2 ones

Words: three million, four hundred nine thousand, one hundred twelve

The order of whole numbers can be visualized by placement on a number line.

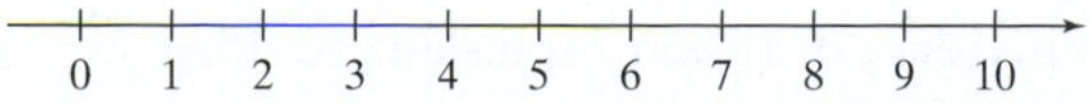

Examples

Example 1

The digit 9 in the number 24,891,321 is in the ten-thousands place.

Example 2

The standard form of the number forty-one million, three thousand, fifty-six is 41,003,056.

Example 3

The expanded form of the number 76,903 is 7 ten-thousands + 6 thousands + 9 hundreds + 3 ones.

Example 4

In words the number 2504 is two thousand, five hundred four.

Example 5

To show that $8 > 4$, note the placement on the number line: 8 is to the right of 4.

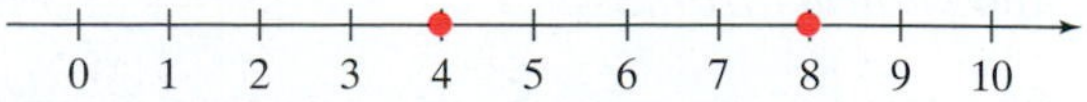

Section 1.3 Addition and Subtraction of Whole Numbers and Perimeter

Key Concepts

The **sum** is the result of adding numbers called **addends**.

Example 1

For $2 + 7 = 9$, the addends are 2 and 7, and the sum is 9.

Addition is performed with and without carrying (or regrouping).

Example 2

$$\begin{array}{r} 23 \\ +\ 41 \\ \hline 64 \end{array} \qquad \begin{array}{r} {\scriptstyle 1\,1} \\ 189 \\ +\ 76 \\ \hline 265 \end{array}$$

Addition Property of Zero:

The sum of any number and zero is that number.

Example 3

$16 + 0 = 16$ Addition property of zero

Commutative Property of Addition:

Changing the order of the addends does not affect the sum.

$3 + 12 = 12 + 3$ Commutative property of addition

Associative Property of Addition:

The manner in which the addends are grouped does not affect the sum.

$2 + (9 + 3) = (2 + 9) + 3$ Associative property of addition

There are several words and phrases that indicate addition, such as *sum, added to, increased by, more than, plus,* and *total of.*

Example 4

18 added to 4 translates to $4 + 18$.

The **perimeter** of a **polygon** is the distance around the outside of the figure. To find perimeter, take the sum of the lengths of all sides of the figure.

Example 5

The perimeter is found by adding the lengths of all sides.

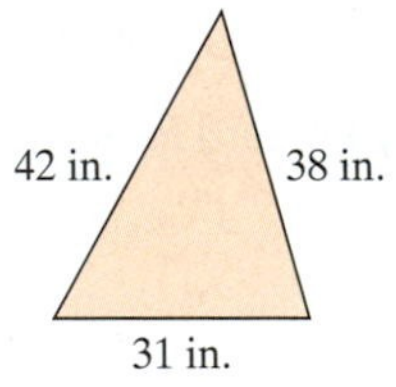

$$\begin{aligned} \text{Perimeter} &= 42 \text{ in.} + 38 \text{ in.} + 31 \text{ in.} \\ &= 111 \text{ in.} \end{aligned}$$

The **difference** is the result of subtracting the **subtrahend** from the **minuend**.

Example 6

For $19 - 13 = 6$, the minuend is 19, the subtrahend is 13, and the difference is 6.

Subtract numbers with and without borrowing.

Example 7

$$\begin{array}{r} 398 \\ -\ 227 \\ \hline 171 \end{array} \qquad \begin{array}{r} {\scriptstyle 9} \\ {\scriptstyle 1\ \not{1}\!\not{0}\ 14} \\ \not{2}\,\not{0}\,4 \\ -\ 88 \\ \hline 116 \end{array}$$

There are several words and phrases that indicate subtraction, such as *minus, difference, decreased by, less than,* and *subtract from.*

Example 8

The difference of 15 and 7 translates to $15 - 7$.

Section 1.4 Rounding and Estimating

Key Concepts

To **round a number**, follow these steps.

Step 1 Identify the digit one position to the right of the given place value.
Step 2 If the digit in step 1 is a 5 or greater, then add 1 to the digit in the given place value. If the digit in step 1 is less than 5, leave the given place value unchanged.
Step 3 Replace each digit to the right of the given place value by 0.

Use rounding to estimate sums and differences.

Examples

Example 1

Round each number to the indicated place.

a. 4942; hundreds place ⟶ 4900
b. 3712; thousands place ⟶ 4000
c. 135; tens place ⟶ 140
d. 199; tens place ⟶ 200

Example 2

Round to the thousands place to estimate the sum: $3929 + 2528 + 5452$.

$4000 + 3000 + 5000 = 12{,}000$

The sum is approximately 12,000.

Section 1.5 Multiplication of Whole Numbers and Area

Key Concepts

Multiplication is repeated addition.

The **product** is the result of multiplying **factors**.

Properties of Multiplication

1. Commutative Property of Multiplication: Changing the order of the factors does not affect the product.
2. Associative Property of Multiplication: The manner in which the factors are grouped does not affect the product.
3. Multipliction Property of 0: The product of any number and 0 is 0.
4. Multiplication Property of 1: The product of any number and 1 is that number.
5. Distributive Property of Multiplication over Addition

Examples

Example 1

$16 + 16 + 16 + 16 = 4 \cdot 16 = 64$

Example 2

For $3 \cdot 13 \cdot 2 = 78$ the factors are 3, 13, and 2, and the product is 78.

Example 3

1. $4 \cdot 7 = 7 \cdot 4$
2. $6 \cdot (5 \cdot 7) = (6 \cdot 5) \cdot 7$
3. $43 \cdot 0 = 0$
4. $290 \cdot 1 = 290$
5. $5 \cdot (4 + 8) = (5 \cdot 4) + (5 \cdot 8)$

Multiply whole numbers.

Example 4

$3 \cdot 14 = 42 \qquad 7(4) = 28$

$$\begin{array}{r} 312 \\ \times\ 23 \\ \hline 936 \\ 6240 \\ \hline 7176 \end{array}$$

The **area of a rectangle** with length l and width w is given by $A = l \cdot w$.

Example 5

Find the area of the rectangle.

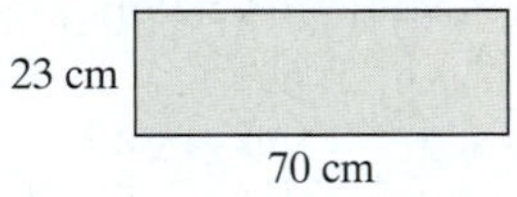

$A = (23\text{ cm}) \cdot (70\text{ cm}) = 1610\text{ cm}^2$

Section 1.6 Division of Whole Numbers

Key Concepts

A **quotient** is the result of dividing the **dividend** by the **divisor**.

Examples

Example 1

For $36 \div 4 = 9$, the dividend is 36, the divisor is 4, and the quotient is 9.

Properties of Division:

1. Any nonzero number divided by itself is 1.
2. Any number divided by 1 is the number itself.
3. Zero divided by any nonzero number is zero.
4. A number divided by zero is undefined.

Example 2

1. $13 \div 13 = 1$
2. $1\overline{)37}$ = 37
3. $\frac{0}{2} = 0$

Example 3

$\frac{2}{0}$ is undefined.

Long division, with and without a **remainder**

Example 4

$$\begin{array}{r} 263 \\ 3\overline{)789} \\ -6 \\ \hline 18 \\ -18 \\ \hline 09 \\ -9 \\ \hline 0 \end{array} \qquad \begin{array}{r} 41 \text{ R } 12 \\ 21\overline{)873} \\ -84 \\ \hline 33 \\ -21 \\ \hline 12 \end{array}$$

Section 1.7 Exponents, Algebraic Expressions, and the Order of Operations

Key Concepts

A number raised to an **exponent** represents repeated multiplication.

For 6^3, 6 is the **base** and 3 is the exponent or **power**.

The **square root** of 16 is 4 because $4^2 = 16$. That is, $\sqrt{16} = 4$.

Order of Operations

1. First perform all operations inside parentheses or other grouping symbols.
2. Simplify any expressions containing exponents or square roots.
3. Perform multiplication or division in the order that they appear from left to right.
4. Perform addition or subtraction in the order that they appear from left to right.

Powers of 10

$10^1 = 10$

$10^2 = 100$

$10^3 = 1000$ and so on.

Variables are used to represent quantities that are subject to change. Quantities that do not change are called **constants**. Variables and constants are used to build **algebraic expressions**.

Examples

Example 1

$9^4 = 9 \cdot 9 \cdot 9 \cdot 9 = 6561$

Example 2

$\sqrt{49} = 7$

Example 3

$$\begin{aligned} &32 \div \sqrt{16} + (9 - 6)^2 \\ &= 32 \div \sqrt{16} + (3)^2 \\ &= 32 \div 4 + 9 \\ &= 8 + 9 \\ &= 17 \end{aligned}$$

Example 4

$10^5 = 100{,}000$ 1 followed by 5 zeros

Example 5

Evaluate the expression $3x + y^2$ for $x = 10$ and $y = 5$.

$$\begin{aligned} 3x + y^2 &= 3(\quad) + (\quad)^2 \\ &= 3(10) + (5)^2 \\ &= 3(10) + 25 \\ &= 30 + 25 \\ &= 55 \end{aligned}$$

Section 1.8 Mixed Applications and Computing Mean

Key Concepts

Many applications require several steps and several mathematical operations.

Examples

Example 1

Nolan received a doctor's bill for \$984. His insurance will pay \$200, and the balance can be paid in 4 equal monthly payments. How much will each payment be?

Solution:

To find the amount not paid by insurance, subtract \$200 from the total bill.

$$984 - 200 = 784$$

To find Nolan's 4 equal payments, divide the amount not covered by insurance by 4.

$$784 \div 4 = 196$$

Nolan must make 4 payments of \$196 each.

The **mean** is the average of a set of numbers. To find the mean, add all the values and divide by the number of values.

Example 2

Find the mean (average) of Michael's scores from his homework assignments.

40, 41, 48, 38, 42, 43

Solution:

$$\frac{40 + 41 + 48 + 38 + 42 + 43}{6} = \frac{252}{6} = 42$$

The average is 42.

Chapter 1 Review Exercises

Section 1.2

For Exercises 1–2, determine the place value for each underlined digit.

1. $\underline{1}0{,}024$ **2.** $\underline{8}21{,}811$

For Exercises 3–4, convert the numbers to standard form.

3. 9 ten-thousands + 2 thousands + 4 tens + 6 ones

4. 5 hundred-thousands + 3 thousands + 1 hundred + 6 tens

For Exercises 5–6, convert the numbers to expanded form.

5. 3,400,820 **6.** 30,554

For Exercises 7–8, write the numbers in words.

7. 245 **8.** 30,861

For Exercises 9–10, write the numbers in standard form.

9. Three thousand, six-hundred two

10. Eight hundred thousand, thirty-nine

For Exercises 11–12, place the numbers on the number line.

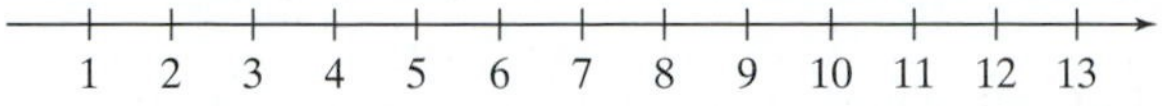

11. 2

12. 7

For Exercises 13–14, determine if the inequality is true or false.

13. $3 < 10$

14. $10 > 12$

Section 1.3

For Exercises 15–16, identify the addends and the sum.

15. $105 + 119 = 224$

16. $\begin{array}{r} 53 \\ +\ 21 \\ \hline 74 \end{array}$

For Exercises 17–20, add.

17. $18 + 24 + 29$

18. $27 + 9 + 18$

19. $\begin{array}{r} 8403 \\ +\ 9007 \\ \hline \end{array}$

20. $\begin{array}{r} 68{,}421 \\ +\ \ 2{,}221 \\ \hline \end{array}$

21. For each of the mathematical statements, identify the property used. Choose from the commutative property or the associative property.

a. $6 + (8 + 2) = (8 + 2) + 6$

b. $6 + (8 + 2) = (6 + 8) + 2$

c. $6 + (8 + 2) = 6 + (2 + 8)$

For Exercises 22–23, identify the minuend, subtrahend, and difference.

22. $14 - 8 = 6$

23. $\begin{array}{r} 102 \\ -\ 78 \\ \hline 24 \end{array}$

For Exercises 24–25, subtract and check your answer by addition.

24. $\begin{array}{r} 37 \\ -\ 11 \\ \hline \end{array}$ Check: $\square + 11 = 37$

25. $\begin{array}{r} 61 \\ -\ 41 \\ \hline \end{array}$ Check: $\square + 41 = 61$

For Exercises 26–29, subtract.

26. $\begin{array}{r} 2005 \\ -\ 1884 \\ \hline \end{array}$

27. $1389 - 299$

28. $86{,}000 - 54{,}981$

29. $67{,}000 - 32{,}812$

For Exercises 30–37, translate the English phrase to a mathematical statement and simplify.

30. The sum of 403 and 79

31. 92 added to 44

32. 38 minus 31

33. 111 decreased by 15

34. 7 more than 36

35. 23 increased by 6

36. Subtract 42 from 251.

37. The difference of 90 and 52

38. The table gives the number of cars sold by three dealerships during one week.

	Honda	Ford	Toyota
Bob's Discount Auto	23	21	34
AA Auto	31	25	40
Car World	33	20	22

a. What is the total number of cars sold by AA Auto?

b. What is the total number of Fords sold by these three dealerships?

Use the bar graph to answer exercises 39–41. The graph represents the distribution of the U.S. population by age group for a recent year.

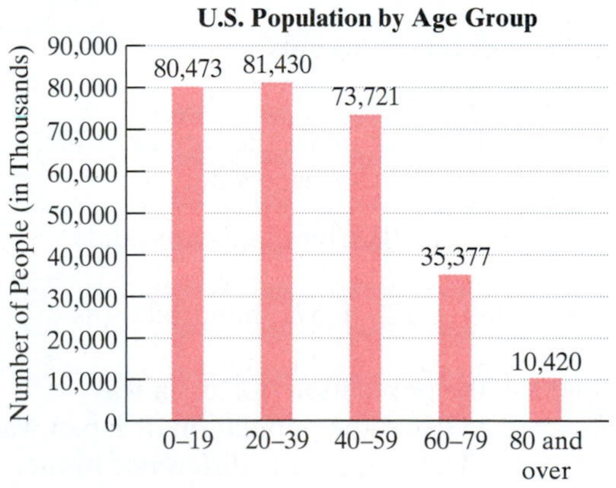

Source: U.S. Census Bureau

39. Determine the number of seniors (aged 60 and over).

40. Compute the difference in the number of people in the 20–39 age group and the number in the 40–59 age group.

41. How many more people are in the 0–19 age group than in the 60–79 age group?

42. For a recent year, 95,192,000 tons of watermelon and 23,299,000 tons of cantaloupe were grown. Determine the difference between the amount of watermelon grown and the amount of cantaloupe grown.

43. Tiger Woods earned $57,940,144 from the PGA tour as of 2006. If Phil Mickelson earned $36,167,360, find the difference in their winnings.

44. Find the perimeter of the figure.

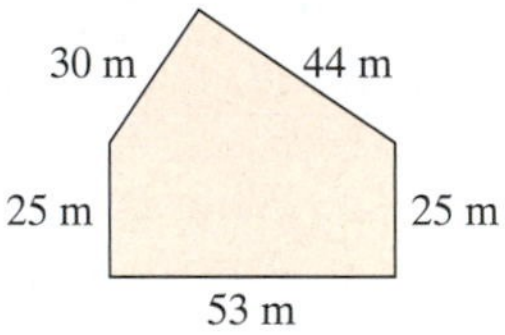

Section 1.4

For Exercises 45–46, round each number to the given place value.

45. 5,234,446; millions

46. 9,332,945; ten-thousands

For Exercises 47–48, estimate the sum or difference by rounding to the indicated place value.

47. 894,004 − 123,883; hundred-thousands

48. 330 + 489 + 123 + 571; hundreds

49. In 2004, the population of Russia was 144,112,353, and the population of Japan was 127,295,333. Estimate the difference in their populations by rounding to the nearest million.

50. The state of Missouri has two dams: Fort Peck with a volume of 96,050 cubic meters (m^3) and Oahe with a volume of 66,517 m^3. Round the numbers to the nearest thousand to estimate the total volume of these two dams.

Section 1.5

51. Identify the factors and the product. $33 \cdot 40 = 1320$

52. Indicate whether the statement is equal to the product of 8 and 13.

a. 8(13) **b.** $(8) \cdot 13$ **c.** (8) + (13)

For Exercises 53–57, for each property listed, choose an expression from the right column that demonstrates the property.

53. Associative property of multiplication — **a.** $3(4) = 4(3)$

54. Distributive property of multiplication over addition — **b.** $19 \cdot 1 = 19$

55. Multiplication property of 0 — **c.** $(1 \cdot 8) \cdot 3 = 1 \cdot (8 \cdot 3)$

56. Commutative property of multiplication — **d.** $0 \cdot 29 = 0$

57. Multiplication property of 1 — **e.** $4(3 + 1) = 4 \cdot 3 + 4 \cdot 1$

For Exercises 58–60, multiply.

58. $\begin{array}{r} 142 \\ \times\ 43 \\ \hline \end{array}$ **59.** (1024)(51) **60.** $\begin{array}{r} 6000 \\ \times\ 500 \\ \hline \end{array}$

61. A discussion group needs to purchase books that are accompanied by a workbook. The price of the book is $26, and the workbook costs an additional $13. If there are 11 members in the group, how much will it cost the group to purchase both the text and workbook for each student?

62. Orcas, or killer whales, eat 551 pounds (lb) of food a day. If Sea World has two adult killer whales, how much food will they eat in 1 week?

Section 1.6

For Exercises 63–64, perform the division. Then identify the divisor, dividend, and quotient.

63. $42 \div 6$ **64.** $4\overline{)52}$

For Exercises 65–68, use the properties of division to simplify the expression, if possible.

65. $3 \div 1$ **66.** $3 \div 3$

67. $3 \div 0$ **68.** $0 \div 3$

69. Explain how you check a division problem if there is no remainder.

70. Explain how you check a division problem if there is a remainder.

For Exercises 71–73, divide and check the answer.

71. $348 \div 6$ **72.** $11\overline{)458}$ **73.** $\dfrac{1043}{20}$

For Exercises 74–75, write the English phrase as a mathematical expression and simplify.

74. The quotient of 72 and 4

75. 108 divided by 9

76. Quinita has 105 photographs that she wants to divide equally among herself and three siblings. How many photos will each person receive? How many photos will be left over?

77. Ashley has \$60 to spend on souvenirs at a surf shop. The prices of several souvenirs are given in the graph.

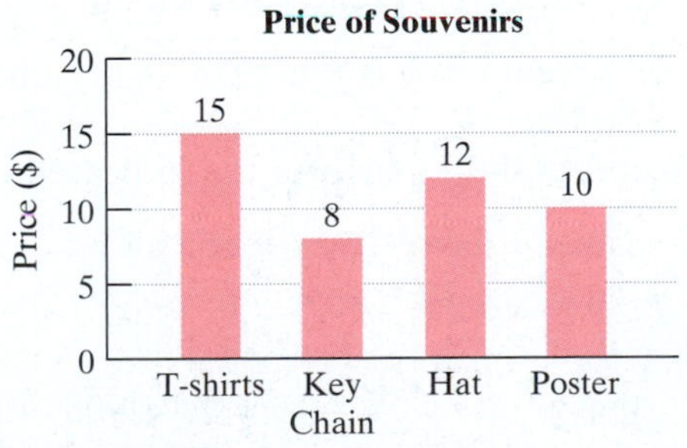

a. How many souvenirs can Ashley buy if she chooses all T-shirts?

b. How many souvenirs can Ashley buy if she chooses all hats?

Section 1.7

For Exercises 78–79, write the repeated multiplication in exponential form. Do not simplify.

78. $8 \cdot 8 \cdot 8 \cdot 8 \cdot 8$

79. $2 \cdot 2 \cdot 2 \cdot 2 \cdot 5 \cdot 5 \cdot 5$

For Exercises 80–83, evaluate the exponential expressions.

80. 5^3 **81.** 4^4 **82.** 1^7 **83.** 10^6

For Exercises 84–85, evaluate the square roots.

84. $\sqrt{64}$ **85.** $\sqrt{144}$

For Exercises 86–92, evaluate the expression using the order of operations.

86. $14 \div 7 \cdot 4 - 1$ **87.** $10^2 - 5^2$

88. $90 - 4 + 6 \div 3 \cdot 2$

89. $2 + 3 \cdot 12 \div 2 - \sqrt{25}$

90. $6^2 - [4^2 + (9 - 7)^3]$

91. $26 - 2(10 - 1) + (3 + 4 \cdot 11)$

92. $\dfrac{5 \cdot 3^2}{7 + 8}$

For Exercises 93–96, evaluate the expressions for $a = 20$, $b = 10$, and $c = 6$.

93. $a + b + 2c$ **94.** $5a - b^2$

95. $\sqrt{b + c}$ **96.** $(a - b)^2$

Section 1.8

97. Doris drives her son to extracurricular activities each week. She drives 5 mi round-trip to baseball practice 3 times a week and 6 mi round-trip to piano lessons once a week.

a. How many miles does she drive in 1 week to get her child to his activities?

b. Approximately how many miles does she travel during a school year consisting of 10 months (there are approximately 4 weeks per month)?

98. At one point in his baseball career, Alex Rodriquez signed a contract for \$252,000,000 for a 9-year period. Suppose federal taxes amount to \$75,600,000 for the contract. After taxes, how much did Alex receive per year?

99. Aletha wants to buy plants for a rectangular garden in her backyard that measures 12 ft by 8 ft. She wants to divide the garden into 2-square-foot (2 ft^2) areas, one for each plant.

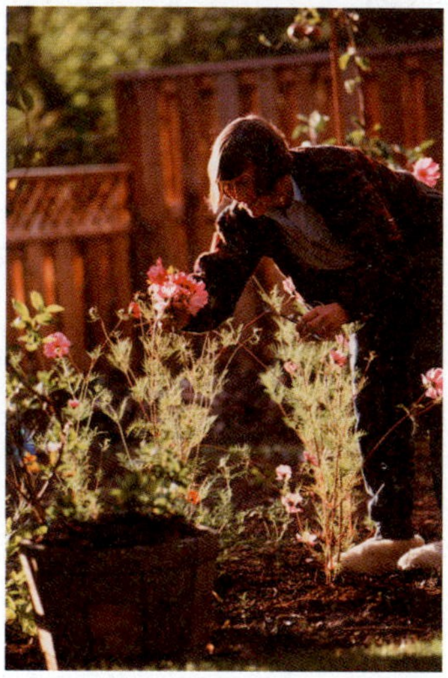

a. How many plants should Aletha buy?

b. If the plants cost \$3 each, how much will it cost Aletha for the plants?

c. If she puts a fence around the perimeter of the garden that costs \$2 per foot, how much will it cost for the fence?

d. What will be Aletha's total cost for this garden?

100. Find the mean (average) for the set of numbers 7, 6, 12, 5, 7, 6, 13.

101. Carolyn's electric bills for the past 5 months have been \$80, \$78, \$101, \$92, and \$94. Find her average monthly charge.

102. The table shows the number of homes sold by a realty company in the last 6 months. Determine the average number of houses sold per month for these 6 months.

Month	Number of Houses
May	6
June	9
July	11
August	13
September	5
October	4

Chapter 1 Test

1. Determine the place value for the underlined digit.

a. 492 **b.** 23,441 **c.** 2,340,711 **d.** 340,592

2. Fill in the table with either the word name for the number or the number in standard form.

	Population	
State / Province	**Standard Form**	**Word Name**
a. Kentucky		Four million, sixty-five thousand
b. Texas	21,325,000	
c. Pennsylvania	12,287,000	
d. New Brunswick, Canada		Seven hundred twenty-nine thousand
e. Ontario, Canada	11,410,000	

3. Translate the phrase by writing the numbers in standard form and inserting the appropriate inequality. Choose from < or >.

a. Fourteen is greater than six.

b. Seventy-two is less than eighty-one.

For Exercises 4–17, perform the indicated operation.

4. $\begin{array}{r} 51 \\ +\ 78 \\ \hline \end{array}$

5. $\begin{array}{r} 82 \\ \times\ 4 \\ \hline \end{array}$

6. $\begin{array}{r} 154 \\ -\ 41 \\ \hline \end{array}$

7. $4\overline{)908}$

8. $58 \cdot 49$

9. $149 + 298$

10. $324 \div 15$

11. $3002 - 2456$

12. $10{,}984 - 2881$

13. $\dfrac{840}{42}$

14. (500,000)(3000)

15. $34 + 89 + 191 + 22$

16. 403(0)

17. $0\overline{)16}$

18. For each of the mathematical statements, identify the property used. Choose from the commutative property of multiplication and the associative property of multiplication. Explain your answer.

a. $(11 \cdot 6) \cdot 3 = 11 \cdot (6 \cdot 3)$

b. $(11 \cdot 6) \cdot 3 = 3 \cdot (11 \cdot 6)$

19. Round each number to the indicated place value.

a. 4850; hundreds

b. 12,493; thousands

c. 7,963,126; hundred-thousands

20. The attendance to the Van Gogh and Gauguin exhibit in Chicago was 690,951. The exhibit moved to Amsterdam, and the attendance was 739,117. Round the numbers to the ten-thousands place to estimate the total attendance for this exhibit.

For Exercises 21–24, simplify, using the order of operations.

21. $8^2 \div 2^4$

22. $26 \cdot \sqrt{4} - 4(8 - 1)$

23. $36 \div 3(14 - 10)$

24. $65 - 2(5 \cdot 3 - 11)^2$

For Exercises 25–26, evaluate the expressions for $x = 5$ and $y = 16$.

25. $x^2 + 2y$

26. $x + \sqrt{y}$

27. Brittany and Jennifer are taking an online course in business management. Brittany has taken 6 quizzes worth 30 points each and received the following scores: 29, 28, 24, 27, 30, and 30. Jennifer has only taken 5 quizzes so far, and her scores are 30, 30, 29, 28, and 28. At this point in the course, which student has a higher average?

28. The use of the cell phone has grown every year for the past 13 years. See the graph.

a. Find the change in the number of phones used from 2003 to 2004.

b. Of the years presented in the graph, between which two years was the increase the greatest?

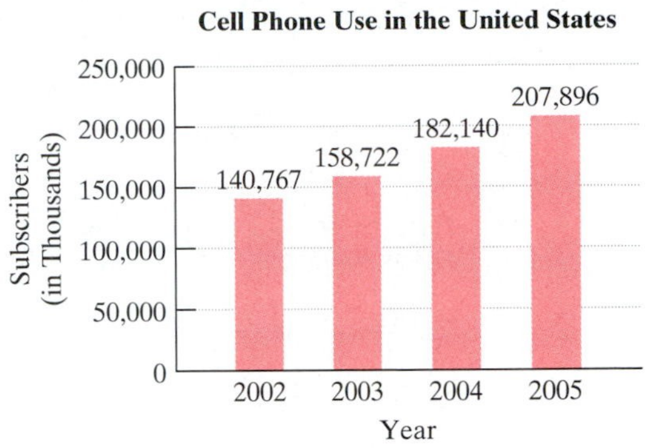

29. The table gives the number of calls to three fire departments during a selected number of weeks. Find the number of calls per week for each department to determine which department is the busiest.

	Number of Calls	Time Period (Number of Weeks)
North Side Fire Department	80	16
South Side Fire Department	72	18
East Side Fire Department	84	28

30. Find the perimeter of the figure.

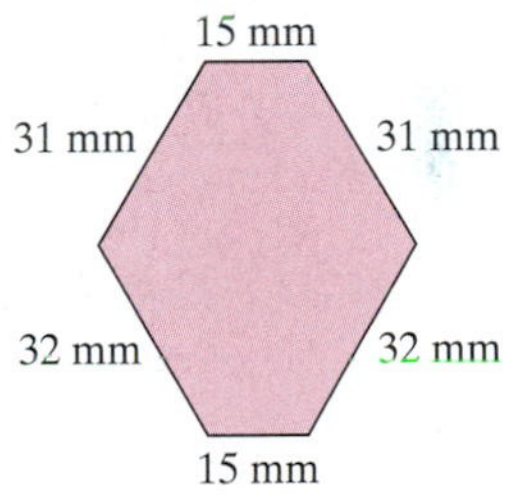

31. Find the perimeter and the area of the rectangle.

32. Round to the nearest hundred to estimate the area of the rectangle.

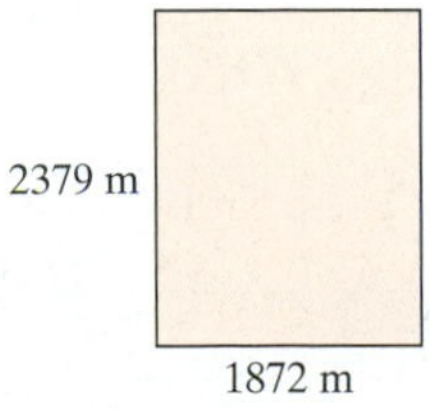

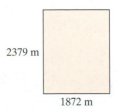

Integers and Algebraic Expressions

2

CHAPTER OUTLINE

Chapter 2

In this chapter, we begin our study of algebra by learning how to add, subtract, multiply, and divide positive and negative numbers. These are necessary skills to continue in algebra.

Are You Prepared?

To help you prepare for this chapter, try the following problems to review the order of operations. As you simplify the expressions, fill out the boxes labeled A through M. Then fill in the remaining part of the grid so that every row, every column, and every 2×3 box contains the digits 1 through 6.

A. $12 - 10 - 1 + 4$

B. $22 - 3 \cdot 6 - 1$

C. $24 \div 8 \cdot 2$

D. 2^2

E. $32 \div 4 \div 2$

F. $9^2 - 4(30 - 2 \cdot 5)$

G. $13 - 8 \div 2 \cdot 3$

H. $\sqrt{16 - 3 \cdot 4}$

I. $\sqrt{10^2 - 8^2}$

J. $50 \div 2 \div 5$

K. $18 \div 9 \cdot 3$

L. $\dfrac{50 - 40}{5 - 3}$

M. $\sqrt{5^2 - 3^2}$

	A	2	B	6	
C		D			1
E				F	6
5	G				H
3			I		J
	K	L	1	M	

Section 2.1 Integers, Absolute Value, and Opposite

Objectives

1. Integers
2. Absolute Value
3. Opposite

1. Integers

The numbers 1, 2, 3, . . . are **positive numbers** because they lie to the right of zero on the number line (Figure 2-1).

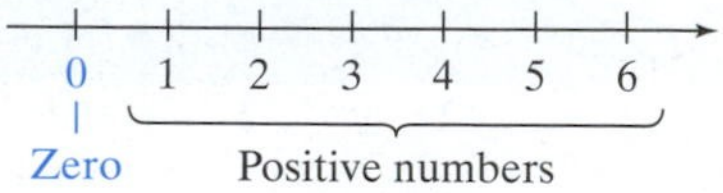

Figure 2-1

In some applications of mathematics we need to use *negative* numbers. For example:

- On a winter day in Buffalo, the low temperature was 3 degrees below zero: −3°
- Tiger Woods' golf score in the U.S. Open was 7 below par: −7
- Carmen is \$128 overdrawn on her checking account. Her balance is: −\$128

The values −3°, −7, and −\$128 are negative numbers. **Negative numbers** lie to the *left* of zero on a number line (Figure 2-2). The number 0 is neither negative nor positive.

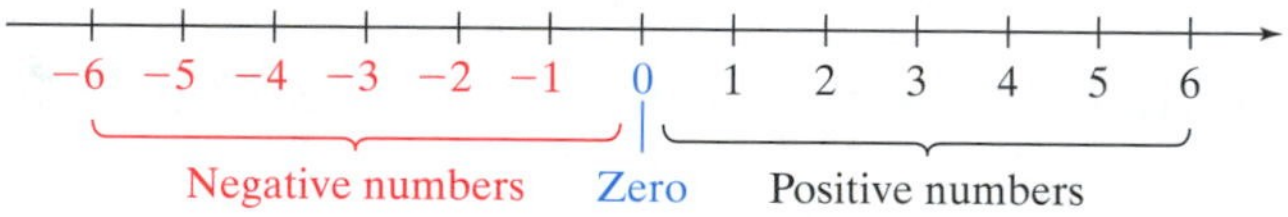

Figure 2-2

The numbers . . . −3, −2, −1, 0, 1, 2, 3, . . . and so on are called **integers**.

Skill Practice

Write an integer that denotes each number.

1. The average temperature at the South Pole in July is 65°C below zero.
2. Sylvia's checking account is overdrawn by \$156.
3. The price of a new car is \$2000 more than it was one year ago.

Example 1 Writing Integers

Write an integer that denotes each number.

a. Liquid nitrogen freezes at 346°F below zero.

b. The shoreline of the Dead Sea on the border of Israel and Jordan is the lowest land area on Earth. It is 1300 ft below sea level.

c. Jenna's 10-year-old daughter weighs 14 lb more than the average child her age.

Solution:

a. −346°F

b. −1300 ft

c. 14 lb

Answers

1. −65°C 2. −\$156
3. \$2000

Example 2 Locating Integers on the Number Line

Locate each number on the number line.

a. -4 **b.** -7 **c.** 0 **d.** 2

Solution:

On the number line, negative numbers lie to the left of 0, and positive numbers lie to the right of 0.

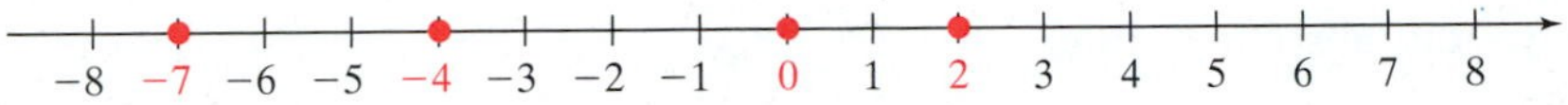

Skill Practice

Locate each number on the number line.

4. -5 **5.** -1 **6.** 4

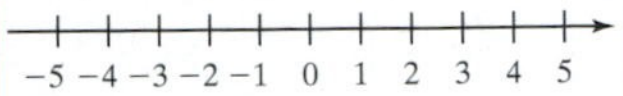

As with whole numbers, the order between two integers can be determined using the number line.

- A number a is less than b (denoted $a < b$) if a lies to the left of b on the number line (Figure 2-3).
- A number a is greater than b (denoted $a > b$) if a lies to the right of b on the number line (Figure 2-4).

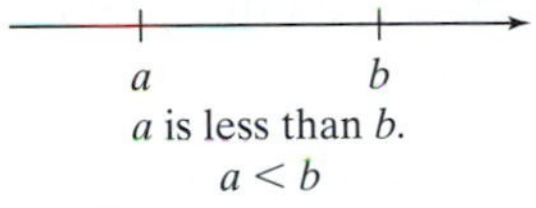

Figure 2-3

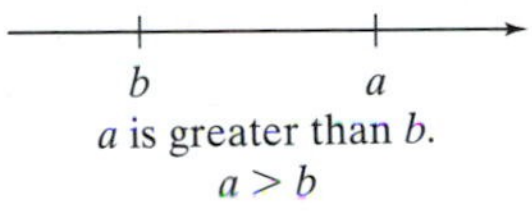

Figure 2-4

Example 3 Determining Order Between Two Integers

Use the number line from Example 2 to fill in the blank with $<$ or $>$ to make a true statement.

a. -7 ☐ -4 **b.** 0 ☐ -4 **c.** 2 ☐ -7

Solution:

a. -7 [$<$] -4 -7 lies to the *left* of -4 on the number line. Therefore, $-7 < -4$.

b. 0 [$>$] -4 0 lies to the *right* of -4 on the number line. Therefore, $0 > -4$.

c. 2 [$>$] -7 2 lies to the *right* of -7 on the number line. Therefore, $2 > -7$.

Skill Practice

Fill in the blank with $<$ or $>$.

7. -3 ☐ -8

8. -3 ☐ 8

9. 0 ☐ -11

2. Absolute Value

On the number line, pairs of numbers like 4 and -4 are the same distance from zero (Figure 2-5). The distance between a number and zero on the number line is called its **absolute value**.

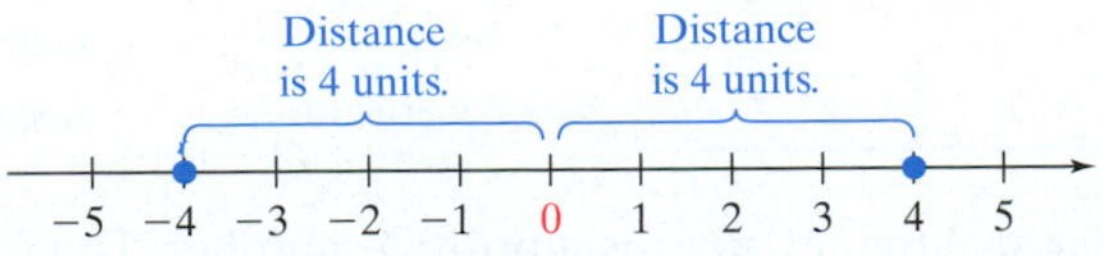

Figure 2-5

Answers

4–6

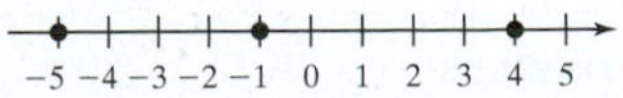

7. $>$ **8.** $<$ **9.** $>$

DEFINITION Absolute Value

The absolute value of a number a is denoted $|a|$. The value of $|a|$ is the distance between a and 0 on the number line.

From the number line, we see that $|-4| = 4$ and $|4| = 4$.

Skill Practice

Determine the absolute value.

10. $|-8|$

11. $|1|$

12. $|-16|$

Example 4 Finding Absolute Value

Determine the absolute value.

a. $|-5|$ **b.** $|3|$ **c.** $|0|$

Solution:

a. $|-5| = 5$ The number -5 is 5 units from 0 on the number line.

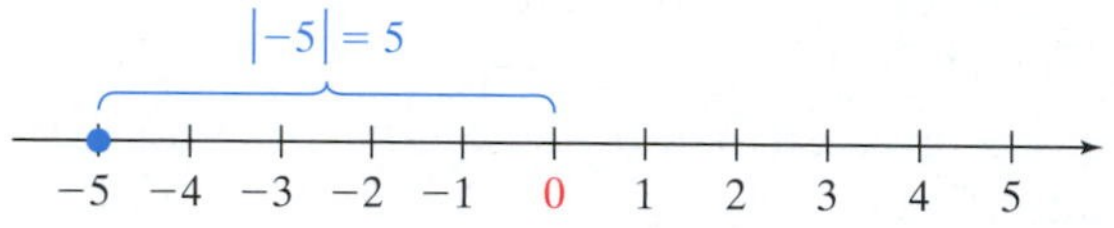

b. $|3| = 3$ The number 3 is 3 units from 0 on the number line.

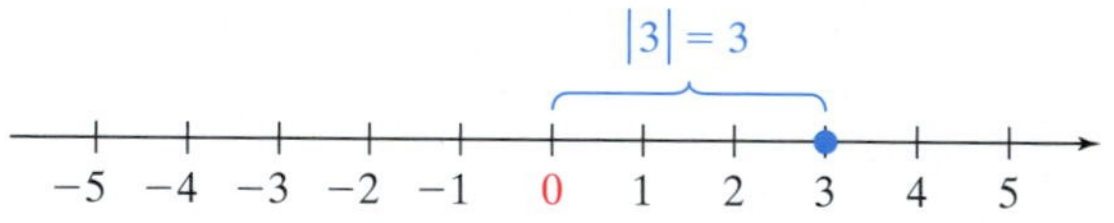

c. $|0| = 0$ The number 0 is 0 units from 0 on the number line.

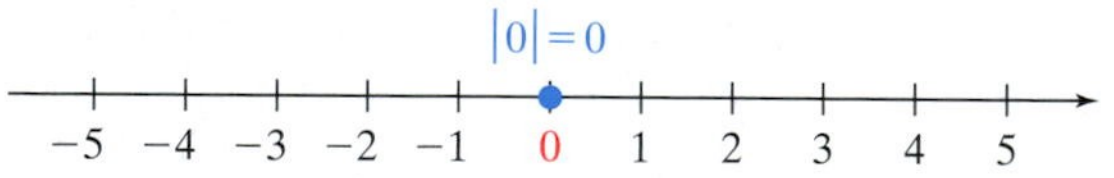

TIP: The absolute value of a nonzero number is always positive. The absolute value of zero is 0.

3. Opposite

Two numbers that are the same distance from zero on the number line, but on opposite sides of zero, are called **opposites**. For example, the numbers -2 and 2 are opposites (see Figure 2-6).

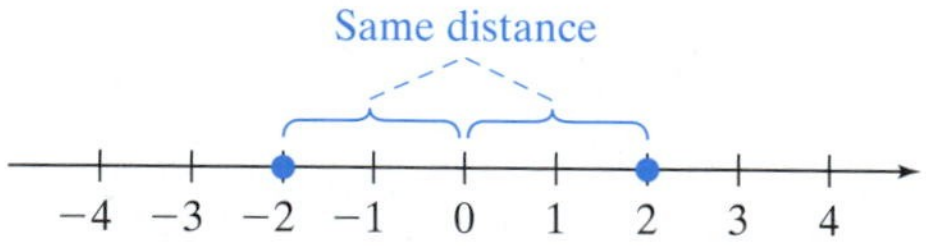

Figure 2-6

The opposite of a number, a, is denoted $-(a)$.

Original number a	Opposite $-(a)$	Simplified Form
5	$-(5)$	-5
-7	$-(-7)$	7

The opposite of a positive number is a negative number.

The opposite of a negative number is a positive number.

The opposite of a negative number is a positive number. Thus, for a positive value, a $(a > 0)$, we have

$$-(-a) = a$$ This is sometimes called the *double negative property*.

Answers

10. 8 **11.** 1 **12.** 16

Example 5 Finding the Opposite of an Integer

Find the opposite.

a. 4 **b.** -99

Solution:

a. If a number is positive, its opposite is negative. The opposite of 4 is -4.

b. If a number is negative, its opposite is positive. The opposite of -99 is 99.

TIP: To find the opposite of a number, change the sign.

Skill Practice

Find the opposite.

13. -108

14. 54

Example 6 Simplifying Expressions

Simplify.

a. $-(-9)$ **b.** $-|-12|$ **c.** $-|7|$

Solution:

a. $-(-9) = 9$ This represents the opposite of -9, which is 9.

b. $-|-12| = -12$ This represents the opposite of $|-12|$. Since $|-12|$ is equal to 12, the opposite is -12.

c. $-|7| = -7$ This represents the opposite of $|7|$. Since $|7|$ is equal to 7, the opposite is -7.

Skill Practice

Simplify.

15. $-(-34)$

16. $-|-20|$ **17.** $-|4|$

Avoiding Mistakes

In Example 6(b) two operations are performed. First take the absolute value of -12. Then determine the opposite of the result.

Answers

13. 108 **14.** -54
15. 34 **16.** -20 **17.** -4

Section 2.1 Practice Exercises

Boost *your* GRADE at ALEKS.com!

- Practice Problems
- Self-Tests
- NetTutor
- e-Professors
- Videos

Study Skills Exercises

1. When working with signed numbers, keep a simple example in your mind, such as temperature. We understand that 10 degrees below zero is colder than 2 degrees below zero, so the inequality $-10 < -2$ makes sense. Write down another example involving signed numbers that you can easily remember.

2. Define the key terms.

a. **Absolute value** **b.** **Integers** **c.** **Negative numbers**

d. **Opposite** **e.** **Positive numbers**

Objective 1: Integers

For Exercises 3–12, write an integer that represents each numerical value. (See Example 1.)

3. Death Valley, California, is 86 m below sea level.

Figure for Exercise 3

4. In a card game, Jack lost $45.

5. Playing *Wheel of Fortune*, Sally won \$3800.

6. Jim's golf score is 5 over par.

7. Rena lost \$500 in the stock market in 1 month.

8. LaTonya earned \$23 in interest on her saving account.

9. Patrick lost 14 lb on a diet.

10. A plane descended 2000 ft.

11. The number of Internet users rose by about 1,400,000.

12. A small business experienced a loss of \$20,000 last year.

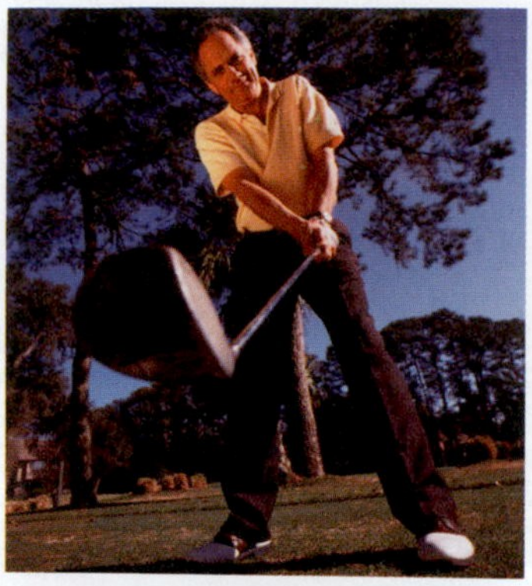

Figure for Exercise 6

Figure for Exercise 11

For Exercises 13–14, graph the numbers on the number line. **(See Example 2.)**

13. $-6, 0, -1, 2$

−6 −5 −4 −3 −2 −1 0 1 2 3 4 5 6

14. $-2, 4, -5, 1$

−6 −5 −4 −3 −2 −1 0 1 2 3 4 5 6

15. Which number is closer to -4 on the number line? -2 or -7

16. Which number is closer to 2 on the number line? -5 or 8

For Exercises 17–24, fill in the blank with $<$ or $>$ to make a true statement. **(See Example 3.)**

17. 0 □ −3 **18.** −1 □ 0 **19.** −8 □ −9 **20.** −5 □ −2

21. 8 □ 9 **22.** 5 □ 2 **23.** −226 □ 198 **24.** 408 □ −416

Objective 2: Absolute Value

For Exercises 25–32, determine the absolute value. **(See Example 4.)**

25. $|-2|$ **26.** $|-9|$ **27.** $|2|$ **28.** $|9|$

29. $|-427|$ **30.** $|-615|$ **31.** $|100{,}000|$ **32.** $|64{,}000|$

33. a. Which is great… 12 or −8?
b. Which is greater, …| or |−8|?

34. a. Which is greater, −14 or −20?
b. Which is greater, |−14| or |−20|?

35. a. Which is greater, 5 …
b. Which is greater, |5| o…

36. a. Which is greater, 3 or 4?
b. Which is greater, |3| or |4|?

37. Which is greater, −5 or |−5…

38. Which is greater −9 or |−9|?

39. Which is greater 10 or |10|?

40. Which is greater 256 or |256|?

Objective 3: Opposite

For Exercises 41–48 find the opposite. **(See…**

41. 5 **42.** 31 **43.** −12 **44.** −25

45. 0 **46.** 1 **47.** −1 **48.** −612

For Exercises 49–60, simplify the expression. **(See Example 6.)**

49. $-(-15)$ **50.** $-(-4)$ **51.** $-|-15|$ **52.** $-|-4|$

53. $-|15|$ **54.** $-|4|$ **55.** $|-15|$ **56.** $|-4|$

57. $-(-36)$ **58.** $-(-19)$ **59.** $-|-107|$ **60.** $-|-26|$

Mixed Exercises

For Exercises 61–64, simplify the expression.

61. **a.** $|-6|$ **b.** $-(-6)$ **c.** $-|6|$ **d.** $|6|$ **e.** $-|-6|$

62. **a.** $-(-12)$ **b.** $|12|$ **c.** $|-12|$ **d.** $-|-12|$ **e.** $-|12|$

63. **a.** $-|8|$ **b.** $|8|$ **c.** $-|-8|$ **d.** $-(-8)$ **e.** $|-8|$

64. **a.** $-|-1|$ **b.** $-(-1)$ **c.** $|1|$ **d.** $|-1|$ **e.** $-|1|$

For Exercises 65–74, write in symbols, do not simplify.

65. The opposite of 6

66. The opposite of 23

67. The opposite of negative 2

68. The opposite of negative 9

69. The absolute value of 7

70. The absolute value of 11

71. The absolute value of negative 3

72. The absolute value of negative 10

73. The opposite of the absolute value of 14

74. The opposite of the absolute value of 42

For Exercises 75–84, fill in the blank with <, >, or =.

75. $|-12|$ □ $|12|$ **76.** $-(-4)$ □ $-|-4|$ **77.** $|-22|$ □ $-(22)$ **78.** -8 □ -10

79. -44 □ -54 **80.** $-|0|$ □ $-|1|$ **81.** $|-55|$ □ $-(-65)$ **82.** $-(82)$ □ $|46|$

83. $-|32|$ □ $|0|$ **84.** $-|22|$ □ 0

For Exercises 85–91, refer to the contour map for wind chill temperatures for a day in January. Give an *estimate* of the wind chill for the given city. For example, the wind chill in Phoenix is between 30°F and 40°F, but closer to 30°F. We might estimate the wind chill in Phoenix to be 33°F.

85. Portland **86.** Atlanta

87. Bismarck **88.** Denver

89. Eugene **90.** Orlando

91. Dallas

Wind Chill Temperatures, °F

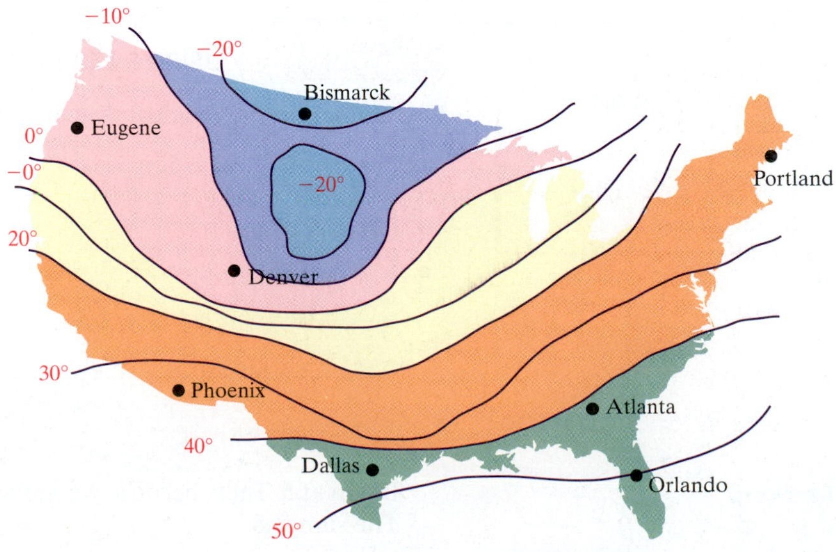

Late spring and summer rainfall in the Lake Okeechobee region in Florida is important to replenish the water supply for large areas of south Florida. Each bar in the graph indicates the number of inches of rainfall above or below average for the given month. Use the graph for Exercises 92–94.

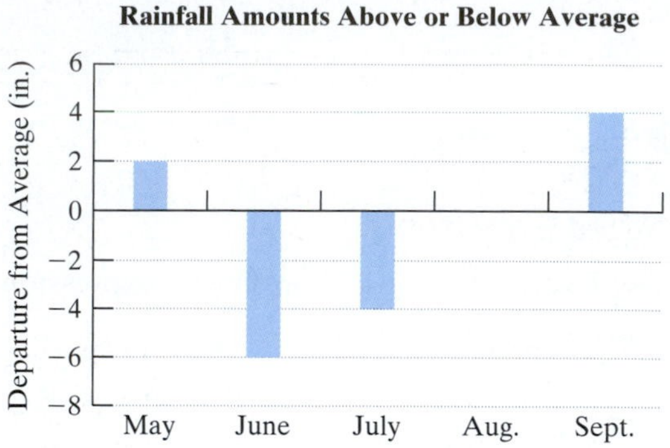

92. Which month had the greatest amount of rainfall *below* average? What was the departure from average?

93. Which month had the greatest amount of rainfall *above* average?

94. Which month had the average amount of rainfall?

Expanding Your Skills

For Exercises 95–96, rank the numbers from least to greatest.

95. $-|-46|, -(-24), -60, 5^2, |-12|$

96. $-15, -(-18), -|20|, 4^2, |-3|^2$

97. If a represents a negative number, then what is the sign of $-a$?

98. If b represents a negative number, then what is the sign of $|b|$?

99. If c represents a negative number, then what is the sign of $-|c|$?

100. If d represents a negative number, then what is the sign of $-(-d)$?

Section 2.2 Addition of Integers

Objectives

1. Addition of Integers by Using a Number Line
2. Addition of Integers
3. Translations and Applications of Addition

1. Addition of Integers by Using a Number Line

Addition of integers can be visualized on a number line. To do so, we locate the first addend on the number line. Then to add a positive number, we move to the right on the number line. To add a negative number, we move to the left on the number line. This is demonstrated in Example 1.

Skill Practice

Use a number line to add.

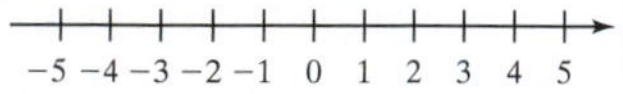

1. $3 + 2$
2. $-3 + 2$

Example 1 Using a Number Line to Add Integers

Use a number line to add.

a. $5 + 3$ **b.** $-5 + 3$

Solution:

a. $5 + 3 = 8$

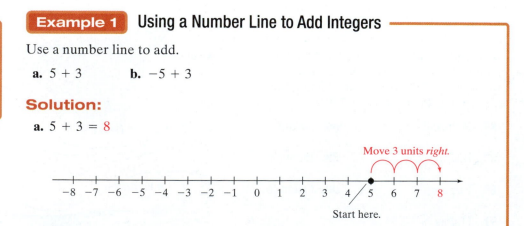

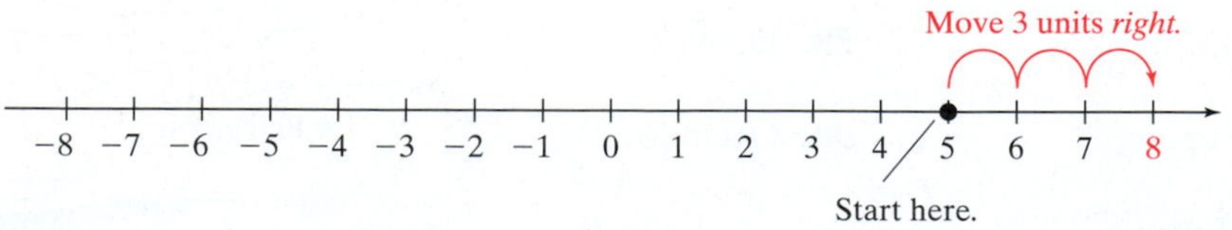

Begin at 5. Then, because we are adding *positive* 3, move to the *right* 3 units. The sum is 8.

Answers

1. 5 2. −1

b. $-5 + 3 = -2$

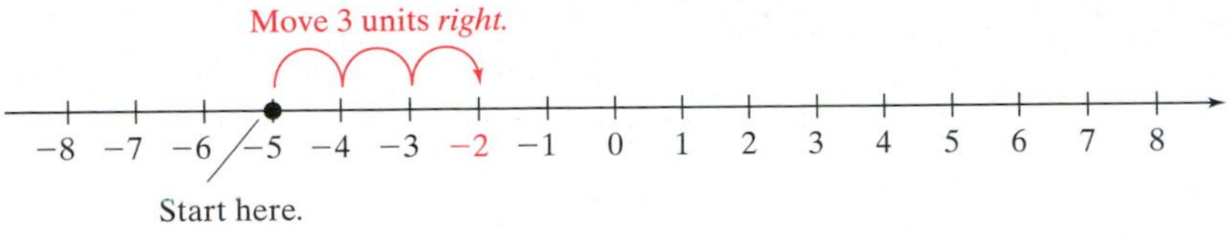

Begin at −5. Then, because we are adding *positive* 3, move to the *right* 3 units. The sum is −2.

Example 2 Using a Number Line to Add Integers

Use a number line to add.

a. $5 + (-3)$ **b.** $-5 + (-3)$

Solution:

a. $5 + (-3) = 2$

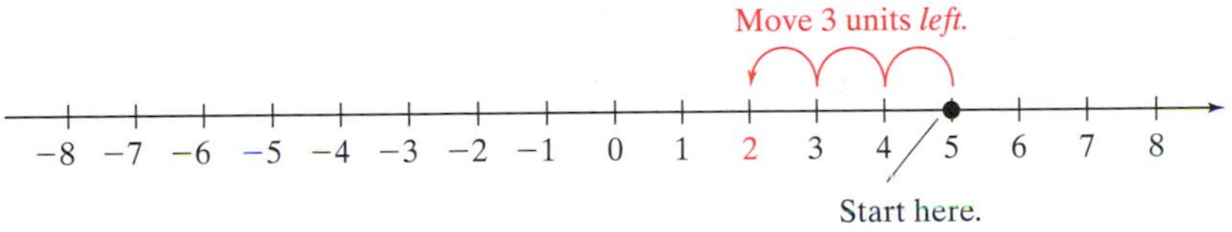

Begin at 5. Then, because we are adding *negative* 3, move to the *left* 3 units. The sum is 2.

b. $-5 + (-3) = -8$

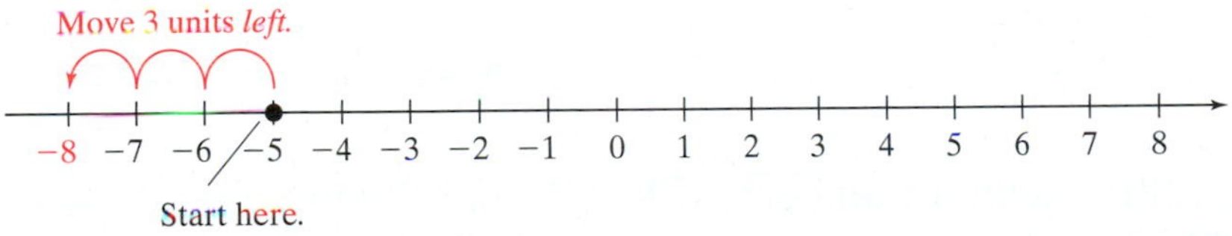

Begin at −5. Then, because we are adding *negative* 3, move to the *left* 3 units. The sum is −8.

Skill Practice

Use a number line to add.

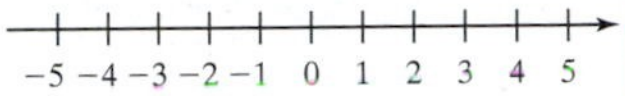

3. $3 + (-2)$

4. $-3 + (-2)$

TIP: In Example 2, parentheses are inserted for clarity. The parentheses separate the number −3 from the symbol for addition, +.

$5 + (-3)$ and $-5 + (-3)$

2. Addition of Integers

It is inconvenient to draw a number line each time we want to add signed numbers. Therefore, we offer two rules for adding integers. The first rule is used when the addends have the *same* sign (that is, if the numbers are both positive or both negative).

PROCEDURE Adding Numbers with the Same Sign

To add two numbers with the same sign, add their absolute values and apply the common sign.

Answers

3. 1 **4.** −5

Skill Practice

Add.

5. $-6 + (-8)$

6. $-84 + (-27)$

7. $14 + 31$

TIP: Parentheses are used to show that the absolute values are added *before* applying the common sign.

Example 3 Adding Integers with the Same Sign

Add.

a. $-2 + (-4)$ **b.** $-12 + (-37)$ **c.** $10 + 66$

Solution:

a. $-2 + (-4)$ — First find the absolute value of each addend. $|-2| = 2$ and $|-4| = 4$.

$= -(2 + 4)$ — Add their absolute values and apply the common sign (in this case, the common sign is negative).

Common sign is negative.

$= -6$ — The sum is -6.

b. $-12 + (-37)$ — First find the absolute value of each addend. $|-12| = 12$ and $|-37| = 37$.

$= -(12 + 37)$ — Add their absolute values and apply the common sign (in this case, the common sign is negative).

Common sign is negative.

$= -49$ — The sum is -49.

c. $10 + 66$ — First find the absolute value of each addend. $|10| = 10$ and $|66| = 66$.

$= +(10 + 66)$ — Add their absolute values and apply the common sign (in this case, the common sign is positive).

Common sign is positive.

$= 76$ — The sum is 76.

The next rule helps us add two numbers with different signs.

Concept Connections

State the sign of the sum.

8. $-9 + 11$

9. $-9 + 7$

PROCEDURE Adding Numbers with Different Signs

To add two numbers with different signs, subtract the smaller absolute value from the larger absolute value. Then apply the sign of the number having the larger absolute value.

Skill Practice

Add.

10. $5 + (-8)$

11. $-12 + 37$

12. $-4 + 4$

Example 4 Adding Integers with Different Signs

Add.

a. $2 + (-7)$ **b.** $-6 + 24$ **c.** $-8 + 8$

Solution:

a. $2 + (-7)$ — First find the absolute value of each addend. $|2| = 2$ and $|-7| = 7$

Note: The absolute value of -7 is greater than the absolute value of 2. Therefore, the sum is negative.

$= -(7 - 2)$ — Next, subtract the smaller absolute value from the larger absolute value.

Apply the sign of the number with the larger absolute value.

$= -5$

Answers

5. -14 **6.** -111 **7.** 45
8. Positive **9.** Negative **10.** -3
11. 25 **12.** 0

b. $-6 + 24$ First find the absolute value of each addend.

$|-6| = 6$ and $|24| = 24$

Note: The absolute value of 24 is greater than the absolute value of -6. Therefore, the sum is positive.

$= +(24 - 6)$ Next, subtract the smaller absolute value from the larger absolute value.

Apply the sign of the number with the larger absolute value.

$= 18$

c. $-8 + 8$ First find the absolute value of each addend.

$|-8| = 8$ and $|8| = 8$

$= (8 - 8)$ The absolute values are equal. Therefore, their difference is 0. The number zero is neither positive nor negative.

$= 0$

TIP: Parentheses are used to show that the absolute values are subtracted *before* applying the appropriate sign.

Example 4(c) illustrates that the sum of a number and its opposite is zero. For example:

$$-8 + 8 = 0 \qquad -12 + 12 = 0 \qquad 6 + (-6) = 0$$

PROPERTY Adding Opposites

For any number a,

$$a + (-a) = 0 \quad \text{and} \quad -a + a = 0$$

That is, the sum of any number and its opposite is zero. (This is also called the *additive inverse property*.)

Example 5 Adding Several Integers

Simplify. $-30 + (-12) + 4 + (-10) + 6$

Solution:

$-30 + (-12) + 4 + (-10) + 6$ Apply the order of operations by adding from left to right.

$= -42 + 4 + (-10) + 6$

$= -38 + (-10) + 6$

$= -48 + 6$

$= -42$

Skill Practice

Simplify.

13. $-24 + (-16) + 8 + 2 + (-20)$

TIP: When several numbers are added, we can reorder and regroup the addends using the commutative property and associative property of addition. In particular, we can group all the positive addends together, and we can group all the negative addends together. This makes the arithmetic easier. For example,

$$-30 + (-12) + 4 + (-10) + 6 = \underbrace{4 + 6}_{\text{positive addends}} + \underbrace{(-30) + (-12) + (-10)}_{\text{negative addends}}$$

$$= 10 + (-52)$$

$$= -42$$

Answer

13. -50

3. Translations and Applications of Addition

Skill Practice

Translate to a mathematical expression and simplify.

14. -2 more than the total of -8, -10, and 5

Example 6 Translating an English Phrase to a Mathematical Expression

Translate to a mathematical expression and simplify.

-6 added to the sum of 2 and -11

Solution:

$[2 + (-11)] + (-6)$ Translate: -6 added to the sum of 2 and -11. Notice that the sum of 2 and -11 is written first, and then -6 is added to that value.

$= -9 + (-6)$ Find the sum of 2 and -11 first. $2 + (-11) = -9$.

$= -15$

Skill Practice

15. Jonas played 9 holes of golf and received the following scores. Positive scores indicate that Jonas was above par. Negative scores indicate that he was below par. Find Jonas' total after 9 holes. Is his score above or below par?

$+2, +1, -1, 0, 0, -1, +3, +4, -1$

Example 7 Applying Addition of Integers to Determine Rainfall Amount

Late spring and summer rainfall in the Lake Okeechobee region in Florida is important to replenish the water supply for large areas of south Florida. The graph indicates the number of inches of rainfall above or below average for given months.

Find the total departure from average rainfall for these months. Do the results show that the region received above average rainfall for the summer or below average?

Solution:

From the graph, we have the following departures from the average rainfall.

Month	Amount (in.)
May	2
June	-6
July	-4
Aug.	0
Sept.	4

Rainfall Amounts Above or Below Average

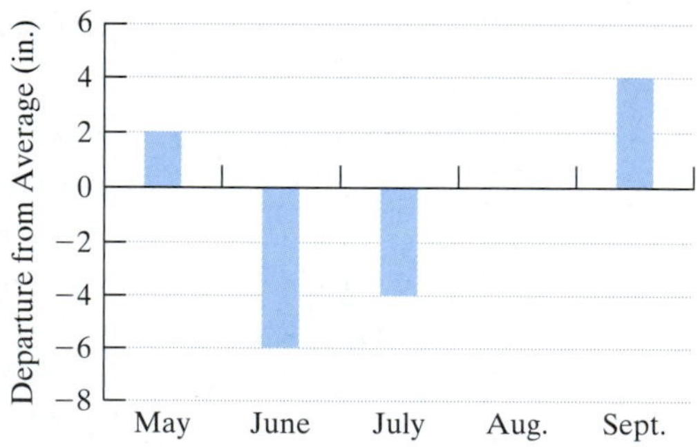

To find the total, we can group the addends conveniently.

Total: $\underbrace{(-6) + (-4)} + \underbrace{2 + 0 + 4}$

$= -10 + 6$

$= -4$ The total departure from average is -4 in. The region received below average rainfall.

Answers

14. $[-8 + (-10) + 5] + (-2)$; -15

15. The total score is 7. Jonas' score is above par.

Section 2.2 Practice Exercises

Boost *your* GRADE at ALEKS.com!

ALEKS version 3.0

- Practice Problems
- Self-Tests
- NetTutor
- e-Professors
- Videos

Study Skill Exercise

1. Instructors vary in what they emphasize on tests. For example, test material may come from the textbook, notes, handouts, homework, etc. What does your instructor emphasize?

Review Exercises

For Exercises 2–8, place the correct symbol ($>$, $<$, or $=$) between the two numbers.

2. $-6 \square -5$ **3.** $-33 \square -44$ **4.** $|-4| \square -|4|$ **5.** $|6| \square |-6|$

6. $0 \square -6$ **7.** $-|-10| \square 10$ **8.** $-(-2) \square 2$

Objective 1: Addition of Integers by Using a Number Line

For Exercises 9–20, refer to the number line to add the integers. **(See Examples 1–2.)**

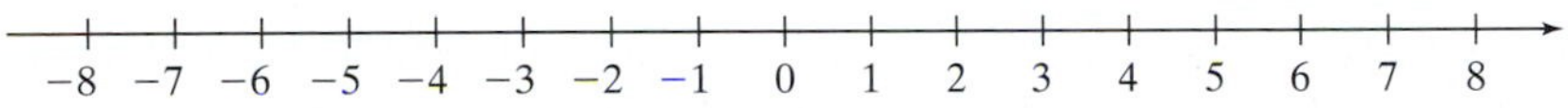

9. $-3 + 5$ **10.** $-6 + 3$ **11.** $2 + (-4)$ **12.** $5 + (-1)$

13. $-4 + (-4)$ **14.** $-2 + (-5)$ **15.** $-3 + 9$ **16.** $-1 + 5$

17. $0 + (-7)$ **18.** $(-5) + 0$ **19.** $-1 + (-3)$ **20.** $-4 + (-3)$

Objective 2: Addition of Integers

21. Explain the process to add two numbers with the same sign.

For Exercises 22–29, add the numbers with the same sign. **(See Example 3.)**

22. $23 + 12$ **23.** $12 + 3$ **24.** $-8 + (-3)$ **25.** $-10 + (-6)$

26. $-7 + (-9)$ **27.** $-100 + (-24)$ **28.** $23 + 50$ **29.** $44 + 45$

30. Explain the process to add two numbers with different signs.

For Exercises 31–42, add the numbers with different signs. **(See Example 4.)**

31. $7 + (-10)$ **32.** $-8 + 2$ **33.** $12 + (-7)$ **34.** $-3 + 9$

35. $-90 + 66$ **36.** $-23 + 49$ **37.** $78 + (-33)$ **38.** $10 + (-23)$

39. $2 + (-2)$ **40.** $-6 + 6$ **41.** $-13 + 13$ **42.** $45 + (-45)$

Mixed Exercises

Animation

For Exercises 43–66, simplify. **(See Example 5.)**

43. $12 + (-3)$ **44.** $-33 + (-1)$ **45.** $-23 + (-3)$

46. $-5 + 15$ **47.** $4 + (-45)$ **48.** $-13 + (-12)$

49. $(-103) + (-47)$

50. $119 + (-59)$

51. $0 + (-17)$

52. $-29 + 0$

53. $-19 + (-22)$

54. $-300 + (-24)$

55. $6 + (-12) + 8$

56. $20 + (-12) + (-5)$

57. $-33 + (-15) + 18$

58. $3 + 5 + (-1)$

59. $7 + (-3) + 6$

60. $12 + (-6) + (-9)$

61. $-10 + (-3) + 5$

62. $-23 + (-4) + (-12) + (-5)$

63. $-18 + (-5) + 23$

64. $14 + (-15) + 20 + (-42)$

65. $4 + (-12) + (-30) + 16 + 10$

66. $24 + (-5) + (-19)$

Objective 3: Translations and Applications of Addition

For Exercises 67–72, translate to a mathematical expression. Then simplify the expression. **(See Example 6.)**

67. The sum of -23 and 49

68. The sum of 89 and -11

69. The total of 3, -10, and 5

70. The total of -2, -4, 14, and 20

71. -5 added to the sum of -8 and 6

72. -15 added to the sum of -25 and 7

73. The graph gives the number of inches below or above the average snowfall for the given months for Marquette, Michigan. Find the total departure from average. Is the snowfall for Marquette above or below average? **(See Example 7.)**

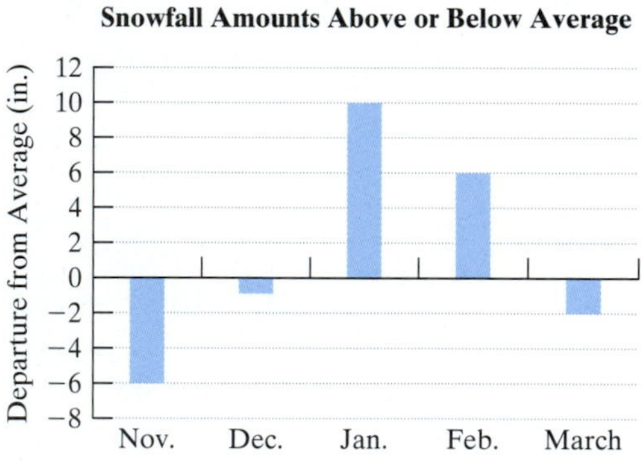

74. The graph gives the number of inches below or above the average rainfall for the given months for Hilo, Hawaii. Find the total departure from average. Is the total rainfall for these months above or below average?

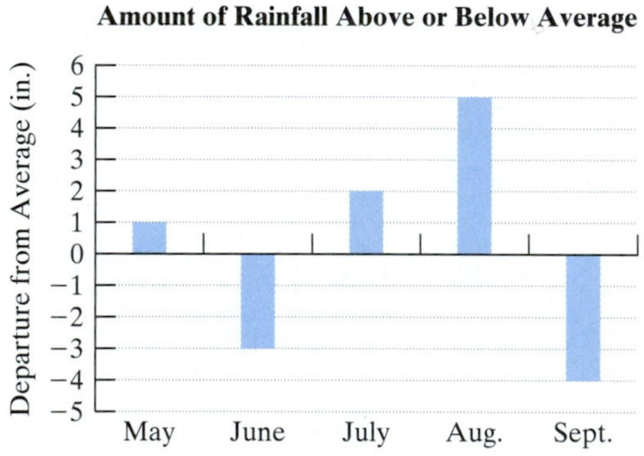

For Exercises 75–76, refer to the table. The table gives the scores for the top two finishers at a recent PGA Open golf tournament.

75. Compute Tiger Woods' total score.

76. Compute Woody Austin's total score.

	Round 1	Round 2	Round 3	Round 4
Tiger Woods	1	−7	−1	−1
Woody Austin	−2	0	−1	−3

77. At 6:00 A.M. the temperature was $-4°F$. By noon, the temperature had risen by 12°F. What was the temperature at noon?

78. At midnight the temperature was $-14°F$. By noon, the temperature had risen 10°F. What was the temperature at noon?

79. Jorge's checking account is overdrawn. His beginning balance was −$56. If he deposits his paycheck for $389, what is his new balance?

80. Ellen's checking account balance is $23. If she writes a check for $40, what is her new balance?

81. A contestant on *Jeopardy* scored the following amounts for several questions he answered. Determine his total score.

−$200,   −$400, $1000,   −$400, $600

82. The number of yards gained or lost by a running back in a football game are given. Find the total number of yards.

3,   2,   −8,   5,   −2,   4,   21

83. Christie Kerr won the U.S. Open Women's Golf Championship for a recent year. The table gives her scores for the first 9 holes in the first round. Find the sum of the scores.

Hole	1	2	3	4	5	6	7	8	9
Score	0	2	−1	−1	0	−1	1	0	0

84. Se Ri Pak tied for fourth place in the U.S. Open Women's Golf Championship for a recent year. The table gives her scores for the first 9 holes in the first round. Find the sum of the scores.

Hole	1	2	3	4	5	6	7	8	9
Score	1	1	0	0	−1	−1	0	0	2

Expanding Your Skills

85. Find two integers whose sum is −10. Answers may vary.

86. Find two integers whose sum is −14. Answers may vary.

87. Find two integers whose sum is −2. Answers may vary.

88. Find two integers whose sum is 0. Answers may vary.

Calculator Connections

Topic: Adding Integers on a Calculator

To enter negative numbers on a calculator, use the [(−)] key or the [+○-] key. To use the [(−)] key, enter the number the same way that it is written. That is, enter the negative sign first and then the number, such as: [(−)] 5. If your calculator has the [+○-] key, type the number first, followed by the [+○-] key. Thus, −5 is entered as: 5 [+○-]. Try entering the expressions below to determine which method your calculator uses.

Expression	**Keystrokes**		**Result**
−10 + (−3)	[(−)] 10 [+] [(−)] 3 [ENTER]	or 10 [+○-] [+] 3 [+○-] [=]	−13
−4 + 6	[(−)] 4 [+] 6 [ENTER]	or 4 [+○-] [+] 6 [=]	2

Calculator Exercises

For Exercises 89–94, add using a calculator.

89. 302 + (−422)

90. −900 + 334

91. −23,991 + (−4423)

92. −1034 + (−23,291)

93. 23 + (−125) + 912 + (−99)

94. 891 + 12 + (−223) + (−341)

Section 2.3 Subtraction of Integers

Objectives

1. Subtraction of Integers
2. Translations and Applications of Subtraction

1. Subtraction of Integers

In Section 2.2, we learned the rules for adding integers. Subtraction of integers is defined in terms of the addition process. For example, consider the following subtraction problem. The corresponding addition problem produces the same result.

$$6 - 4 = 2 \quad \Leftrightarrow \quad 6 + (-4) = 2$$

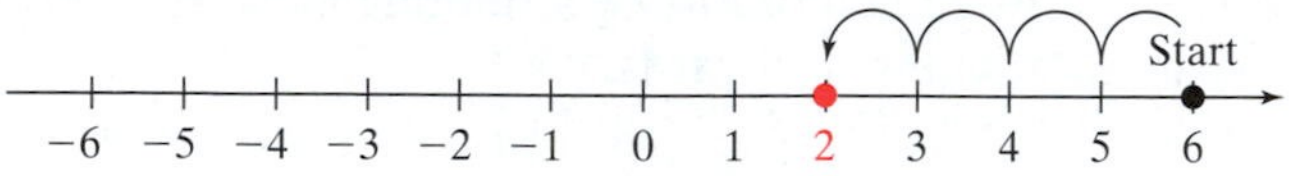

In each case, we start at 6 on the number line and move to the *left* 4 units. Adding the *opposite* of 4 produces the same result as subtracting 4. This is true in general. To subtract two integers, add the opposite of the second number to the first number.

PROCEDURE Subtracting Signed Numbers

For two numbers a and b, $a - b = a + (-b)$.

Therefore, to perform subtraction, follow these steps:

Step 1 Leave the first number (the minuend) unchanged.
Step 2 Change the subtraction sign to an addition sign.
Step 3 Add the opposite of the second number (subtrahend).

For example:

$$\left.\begin{aligned} 10 - 4 &= 10 + (-4) = 6 \\ -10 - 4 &= -10 + (-4) = -14 \end{aligned}\right\} \quad \text{Subtracting 4 is the same as adding } -4.$$

$$\left.\begin{aligned} 10 - (-4) &= 10 + (4) = 14 \\ -10 - (-4) &= -10 + (4) = -6 \end{aligned}\right\} \quad \text{Subtracting } -4 \text{ is the same as adding 4.}$$

Concept Connections

Fill in the blank to change subtraction to addition of the opposite.

1. $9 - 3 = 9 + \square$
2. $-9 - 3 = -9 + \square$
3. $9 - (-3) = 9 + \square$
4. $-9 - (-3) = -9 + \square$

Example 1 Subtracting Integers

Subtract. **a.** $15 - 20$ **b.** $-7 - 12$ **c.** $40 - (-8)$

Solution:

a. $15 - \textcircled{20} = 15 + (-20) = -5$ (Add the opposite of 20. Change subtraction to addition.)

Rewrite the subtraction in terms of addition. Subtracting 20 is the same as adding −20.

b. $-7 - 12 = -7 + (-12)$
$= -19$

Rewrite the subtraction in terms of addition. Subtracting 12 is the same as adding −12.

c. $40 - (-8) = 40 + (8)$
$= 48$

Rewrite the subtraction in terms of addition. Subtracting −8 is the same as adding 8.

Skill Practice

Subtract.

5. $12 - 19$
6. $-8 - 14$
7. $30 - (-3)$

Answers

1. −3 2. −3 3. 3 4. 3
5. −7 6. −22 7. 33

TIP: After subtraction is written in terms of addition, the rules of addition are applied.

Example 2 Adding and Subtracting Several Integers

Simplify. $-4 - 6 + (-3) - 5 + 8$

Solution:

$-4 - 6 + (-3) - 5 + 8$

$= \underbrace{-4 + (-6)} + (-3) + (-5) + 8$ Rewrite all subtractions in terms of addition.

$= \underbrace{-10 + (-3)} + (-5) + 8$ Add from left to right.

$= \underbrace{-13 + (-5)} + 8$

$= \underbrace{-18 + 8}$

$= -10$

Skill Practice

Simplify.

8. $-8 - 10 + (-6) - (-1) + 4$

2. Translations and Applications of Subtraction

Recall from Section 1.3 that several key words imply subtraction.

Word/Phrase	Example	In Symbols
a minus *b*	-15 minus 10	$-15 - 10$
The difference of *a* and *b*	The difference of 10 and -2	$10 - (-2)$
a decreased by *b*	9 decreased by 1	$9 - 1$
a less than *b*	-12 less than 5	$5 - (-12)$
Subtract *a* from *b*	Subtract -3 from 8	$8 - (-3)$
b subtracted from *a*	-2 subtracted from -10	$-10 - (-2)$

Example 3 Translating to a Mathematical Expression

Translate to a mathematical expression. Then simplify.

a. The difference of -52 and 10.

b. -35 decreased by -6.

Solution:

a. the difference of

$-52 - 10$ Translate: The difference of -52 and 10.

$= -52 + (-10)$ Rewrite subtraction in terms of addition.

$= -62$ Add.

b. decreased by

$-35 - (-6)$ Translate: -35 decreased by -6.

$= -35 + (6)$ Rewrite subtraction in terms of addition.

$= -29$ Add.

Skill Practice

Translate to a mathematical expression. Then simplify.

9. The difference of -16 and 4

10. -8 decreased by -9

Avoiding Mistakes

Subtraction is not commutative. The order of the numbers being subtracted is important.

Answers

8. -19 **9.** $-16 - 4; -20$
10. $-8 - (-9); 1$

Skill Practice

Translate to a mathematical expression. Then simplify.

11. 6 less than 2

12. Subtract −4 from −1.

Example 4 Translating to a Mathematical Expression

Translate each English phrase to a mathematical expression. Then simplify.

a. 12 less than −8 **b.** Subtract 27 from 5.

Solution:

a. To translate "12 less than −8," we must *start* with −8 and subtract 12.

$-8 - 12$	Translate: 12 less than −8.
$= -8 + (-12)$	Rewrite subtraction in terms of addition.
$= -20$	Add.

b. To translate "subtract 27 from 5," we must *start* with 5 and subtract 27.

$5 - 27$	Translate: Subtract 27 from 5.
$= 5 + (-27)$	Rewrite subtraction in terms of addition.
$= -22$	Add.

Skill Practice

13. The highest point in California is Mt. Whitney at 14,494 ft above sea level. The lowest point in California is Death Valley, which has an "altitude" of −282 ft (282 ft below sea level). Find the difference in the elevations of the highest point and lowest point in California.

Example 5 Applying Subtraction of Integers

A helicopter is hovering at a height of 200 ft above the ocean. A submarine is directly below the helicopter 125 ft below sea level. Find the difference in elevation between the helicopter and the submarine.

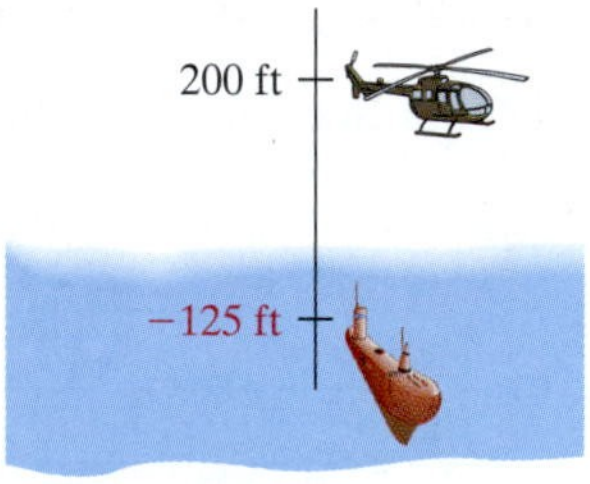

Solution:

$$\begin{pmatrix}\text{Difference between}\\ \text{elevation of helicopter}\\ \text{and submarine}\end{pmatrix} = \begin{pmatrix}\text{Elevation of}\\ \text{helicopter}\end{pmatrix} - \begin{pmatrix}\text{"Elevation" of}\\ \text{submarine}\end{pmatrix}$$

$= 200 \text{ ft} - (-125 \text{ ft})$

$= 200 \text{ ft} + (125 \text{ ft})$ Rewrite as addition.

$= 325 \text{ ft}$

The helicopter and submarine are 325 ft apart.

Answers

11. $2 - 6$; -4 **12.** $-1 - (-4)$; 3

13. 14,776 ft

Section 2.3 Practice Exercises

Boost *your* GRADE at ALEKS.com!

ALEKS version 3.0

- Practice Problems
- Self-Tests
- NetTutor
- e-Professors
- Videos

Study Skills Exercise

1. Which activities might you try when working in a study group to help you learn and understand the material?

☐ Quiz one another by asking one another questions.

☐ Practice teaching one another.

☐ Share and compare class notes.

☐ Support and encourage one another.

☐ Work together on exercises and sample problems.

Review Exercises

For Exercises 2–7, simplify.

2. $34 + (-13)$

3. $-34 + (-13)$

4. $-34 + 13$

5. $-|-26|$

6. $-(-32)$

7. $-9 + (-8) + 5 + (-3) + 7$

Objective 1: Subtraction of Integers

8. Explain the process to subtract integers.

For Exercises 9–16, rewrite the subtraction problem as an equivalent addition problem. Then simplify. **(See Example 1.)**

9. $2 - 9 =$ ____ + ____ = ____

10. $5 - 11 =$ ____ + ____ = ____

11. $4 - (-3) =$ ____ + ____ = ____

12. $12 - (-8) =$ ____ + ____ = ____

13. $-3 - 15 =$ ____ + ____ = ____

14. $-7 - 21 =$ ____ + ____ = ____

15. $-11 - (-13) =$ ____ + ____ = ____

16. $-23 - (-9) =$ ____ + ____ = ____

For Exercises 17–46, simplify. **(See Examples 1–2.)**

17. $35 - (-17)$

18. $23 - (-12)$

19. $-24 - 9$

20. $-5 - 15$

21. $50 - 62$

22. $38 - 46$

23. $-17 - (-25)$

24. $-2 - (-66)$

25. $-8 - (-8)$

26. $-14 - (-14)$

27. $120 - (-41)$

28. $91 - (-62)$

29. $-15 - 19$

30. $-82 - 44$

31. $3 - 25$

32. $6 - 33$

33. $-13 - 13$

34. $-43 - 43$

35. $24 - 25$

36. $43 - 98$

37. $-6 - (-38)$

38. $-75 - (-21)$

39. $-48 - (-33)$

40. $-29 - (-32)$

41. $2 + 5 - (-3) - 10$

42. $4 - 8 + 12 - (-1)$

43. $-5 + 6 + (-7) - 4 - (-9)$

44. $-2 - 1 + (-11) + 6 - (-8)$

45. $25 - 13 - (-40)$

46. $-35 + 15 - (-28)$

Objective 2: Translations and Applications of Subtraction

47. State at least two words or phrases that would indicate subtraction.

48. Is subtraction commutative. For Example, does $3 - 7 = 7 - 3$?

For Exercises 49–60, translate each English phrase to a mathematical expression. Then simplify. **(See Examples 3–4.)**

49. 14 minus 23

50. 27 minus 40

51. The difference of 105 and 110

52. The difference of 70 and 98

53. 320 decreased by -20

54. 150 decreased by 75

55. Subtract 12 from 5.

56. Subtract 10 from 16.

57. 21 less than −34

58. 22 less than −90

59. Subtract 24 from −35

60. Subtract 189 from 175

61. The liquid hydrogen in the space shuttle's main engine is −423°F. The temperature in the engine's combustion chamber reaches 6000°F. Find the difference between the temperature in the combustion chamber and the temperature of the liquid hydrogen. **(See Example 5.)**

62. Temperatures on the moon range from −184°C during its night to 214°C during its day. Find the difference between the highest temperature on the moon and the lowest temperature.

63. Ivan owes $320 on his credit card; that is, his balance is −$320. If he charges $55 for a night out, what is his new balance?

64. If Justin's balance on his credit card was −$210 and he made the minimum payment of $25, what is his new balance?

65. The Campus Food Court reports its total profit or loss each day. During a 1-week period, the following profits or losses were reported. If the Campus Food Court's balance was $17,476 at the beginning of the week, what is the balance at the end of the reported week?

Monday	$1786
Tuesday	−$2342
Wednesday	−$754
Thursday	$321
Friday	$1597

66. Jeff's balance in his checking account was $2036 at the beginning of the week. During the week, he wrote two checks, made two deposits, and made one ATM withdrawal. What is his ending balance?

Check	−$150
Check	−$25
Paycheck (deposit)	$480
ATM	−$200
Cash (deposit)	$80

For Exercises 67–70, refer to the graph indicating the change in value of the Dow Jones Industrial Average for a given week.

67. What is the difference in the change in value between Tuesday and Wednesday?

68. What is the difference in the change in value between Thursday and Friday?

69. What is the total change for the week?

70. Based on the values in the graph, did the Dow gain or lose points for the given week?

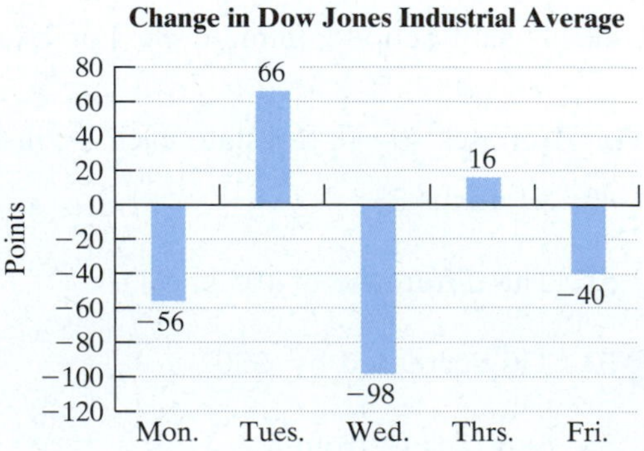

For Exercises 71–72, find the range. The *range* of a set of numbers is the difference between the highest value and the lowest value. That is, range = highest − lowest.

71. Low temperatures for 1 week in Anchorage (°C): −4°, −8°, 0°, 3°, −8°, −1°, 2°

72. Low temperatures for 1 week in Fargo (°C): −6°, −2°, −10°, −4°, −12°, −1°, −3°

73. Find two integers whose difference is −6. Answers may vary.

74. Find two integers whose difference is −20. Answers may vary.

For Exercises 75–76, write the next three numbers in the sequence.

75. 5, 1, −3, −7, ____, ____, ____

76. −13, −18, −23, −28, ____, ____, ____

Expanding Your Skills

For Exercises 77–84, assume $a > 0$ (this means that a is positive) and $b < 0$ (this means that b is negative). Find the sign of each expression.

77. $a - b$

78. $b - a$

79. $|a| + |b|$

80. $|a + b|$

81. $-|a|$

82. $-|b|$

83. $-(a)$

84. $-(b)$

Calculator Connections

Topic: Subtracting Integers on a Calculator

The [−] key is used for subtraction. This should not be confused with the [(−)] key or [+○-] key, which is used to enter a negative number.

Expression	Keystrokes	Result
−7 − 4	[(−)] 7 [−] 4 [ENTER] or 7 [+○-] [−] 4 [=]	−11

Calculator Exercises

For Exercises 85–90, subtract the integers using a calculator.

85. −190 − 223

86. −288 − 145

87. −23,624 − (−9001)

88. −14,593 − (−34,499)

89. 892,904 − (−23,546)

90. 104,839 − (−24,938)

91. The height of Mount Everest is 29,029 ft. The lowest point on the surface of the Earth is −35,798 ft (that is, 35,798 ft below sea level) occurring at the Mariana Trench on the Pacific Ocean floor. What is the difference in altitude between the height of Mt. Everest and the Mariana Trench?

92. Mt. Rainier is 4392 m at its highest point. Death Valley, California, is 86 m below sea level (−86 m) at the basin, Badwater. What is the difference between the altitude of Mt. Rainier and the altitude at Badwater?

Section 2.4 Multiplication and Division of Integers

Objectives

1. Multiplication of Integers
2. Multiplying Many Factors
3. Exponential Expressions
4. Division of Integers

1. Multiplication of Integers

We know from our knowledge of arithmetic that the product of two positive numbers is a positive number. This can be shown by using repeated addition.

For example: $3(4) = 4 + 4 + 4 = 12$

Now consider a product of numbers with different signs.

For example: $3(-4) = -4 + (-4) + (-4) = -12$ (3 times -4)

This example suggests that the product of a positive number and a negative number is *negative*.

Now what if we have a product of two negative numbers? To determine the sign, consider the following pattern of products.

$$\begin{aligned} 3 \cdot -4 &= -12 \\ 2 \cdot -4 &= -8 \\ 1 \cdot -4 &= -4 \\ 0 \cdot -4 &= 0 \\ -1 \cdot -4 &= 4 \\ -2 \cdot -4 &= 8 \\ -3 \cdot -4 &= 12 \end{aligned}$$

The pattern increases by 4 with each row.

The product of two negative numbers is *positive.*

From the first four rows, we see that the product increases by 4 for each row. For the pattern to continue, it follows that the product of two negative numbers must be *positive*.

Concept Connections

1. Write $4(-5)$ as repeated addition.

Avoiding Mistakes

Try not to confuse the rule for multiplying two negative numbers with the rule for adding two negative numbers.

- The product of two negative numbers is positive.
- The sum of two negative numbers is negative.

PROCEDURE Multipliying Signed Numbers

1. The product of two numbers with the *same* sign is positive.

   Examples: $(5)(6) = 30$
   $(-5)(-6) = 30$

2. The product of two numbers with *different* signs is negative.

   Examples: $4(-10) = -40$
   $-4(10) = -40$

3. The product of any number and zero is zero.

   Examples: $3(0) = 0$
   $0(-6) = 0$

Answer

1. $-5 + (-5) + (-5) + (-5)$

Example 1 Multiplying Integers

Multiply.

a. $-8(-7)$ **b.** $-5 \cdot 10$ **c.** $(18)(-2)$ **d.** $16 \cdot 2$ **e.** $-3 \cdot 0$

Solution:

a. $-8(-7) = 56$ Same signs. Product is positive.

b. $-5 \cdot 10 = -50$ Different signs. Product is negative.

c. $(18)(-2) = -36$ Different signs. Product is negative.

d. $16 \cdot 2 = 32$ Same signs. Product is positive.

e. $-3 \cdot 0 = 0$ The product of any number and zero is zero.

Skill Practice

Multiply.

2. $-2(-6)$
3. $3 \cdot (-10)$
4. $-14(3)$
5. $8 \cdot 4$
6. $-5 \cdot 0$

Recall that the terms *product*, *multiply*, and *times* imply multiplication.

Example 2 Translating to a Mathematical Expression

Translate each phrase to a mathematical expression. Then simplify.

a. The product of 7 and -8 **b.** -3 times -11

Solution:

a. $7(-8)$ Translate: The product of 7 and -8.

$= -56$ Different signs. Product is negative.

b. $(-3)(-11)$ Translate: -3 times -11.

$= 33$ Same signs. Product is positive.

Skill Practice

Translate each phrase to a mathematical expression. Then simplify.

7. The product of -4 and -12.
8. Eight times -5.

2. Multiplying Many Factors

In each of the following products, we can apply the order of operations and multiply from left to right.

two negative factors

$(-2)(-2)$
$= 4$

Product is positive.

three negative factors

$(-2)(-2)(-2)$
$= 4(-2)$
$= -8$

Product is negative.

four negative factors

$(-2)(-2)(-2)(-2)$
$= 4(-2)(-2)$
$= (-8)(-2)$
$= 16$

Product is positive.

five negative factors

$(-2)(-2)(-2)(-2)(-2)$
$= 4(-2)(-2)(-2)$
$= (-8)(-2)(-2)$
$= 16(-2)$
$= -32$

Product is negative.

These products indicate the following rules.

- The product of an *even* number of negative factors is *positive*.
- The product of an *odd* number of negative factors is *negative*.

Concept Connections

Multiply.

9. $(-1)(-1)$
10. $(-1)(-1)(-1)$
11. $(-1)(-1)(-1)(-1)$
12. $(-1)(-1)(-1)(-1)(-1)$

Answers

2. 12 **3.** -30 **4.** -42
5. 32 **6.** 0 **7.** $(-4)(-12)$; 48
8. $8(-5)$; -40 **9.** 1
10. -1 **11.** 1 **12.** -1

Skill Practice

Multiply.

13. $(-3)(-4)(-8)(-1)$

14. $(-1)(-4)(-6)(5)$

Example 3 Multiplying Several Factors

Multiply.

a. $(-2)(-5)(-7)$ **b.** $(-4)(2)(-1)(5)$

Solution:

a. $(-2)(-5)(-7)$	This product has an odd number of negative factors.
$= -70$	The product is negative.
b. $(-4)(2)(-1)(5)$	This product has an even number of negative factors.
$= 40$	The product is positive.

3. Exponential Expressions

Be particularly careful when evaluating exponential expressions involving negative numbers. An exponential expression with a negative base is written with parentheses around the base, such as $(-3)^4$.

To evaluate $(-3)^4$, the base -3 is multiplied 4 times:

$$(-3)^4 = (-3)(-3)(-3)(-3) = 81$$

If parentheses are *not* used, the expression -3^4 has a different meaning:

- The expression -3^4 has a base of 3 (not -3) and can be interpreted as $-1 \cdot 3^4$. Hence,

$$-3^4 = -1 \cdot (3)(3)(3)(3) = -81$$

- The expression -3^4 can also be interpreted as "the opposite of 3^4." Hence,

$$-3^4 = -(3 \cdot 3 \cdot 3 \cdot 3) = -81$$

Skill Practice

Simplify.

15. $(-5)^2$ **16.** -5^2

17. $(-2)^3$ **18.** -2^3

Example 4 Simplifying Exponential Expressions

Simplify.

a. $(-4)^2$ **b.** -4^2 **c.** $(-5)^3$ **d.** -5^3

Solution:

a. $(-4)^2 = (-4)(-4)$	The base is -4.
$= 16$	Multiply.
b. $-4^2 = -(4)(4)$	The base is 4. This is equal to $-1 \cdot 4^2 = -1 \cdot (4)(4)$.
$= -16$	Multiply.
c. $(-5)^3 = (-5)(-5)(-5)$	The base is -5.
$= -125$	Multiply.
d. $-5^3 = -(5)(5)(5)$	The base is 5. This is equal to $-1 \cdot 5^3 = -1 \cdot (5)(5)(5)$.
$= -125$	Multiply.

Avoiding Mistakes

In Example 4(b) the base is positive because the negative sign is not enclosed in parentheses with the base.

Answers

13. 96 **14.** -120

15. 25 **16.** -25

17. -8 **18.** -8

4. Division of Integers

From Section 1.6, we learned that when we divide two numbers, we can check our result by using multiplication. Because multiplication and division are related in this way, it seems reasonable that the same sign rules that apply to multiplication also apply to division. For example:

$$24 \div 6 = 4 \qquad \underline{\text{Check:}} \qquad 6 \cdot 4 = 24 \checkmark$$
$$-24 \div 6 = -4 \qquad \underline{\text{Check:}} \qquad 6 \cdot (-4) = -24 \checkmark$$
$$24 \div (-6) = -4 \qquad \underline{\text{Check:}} \qquad -6 \cdot (-4) = 24 \checkmark$$
$$-24 \div (-6) = 4 \qquad \underline{\text{Check:}} \qquad -6 \cdot 4 = -24 \checkmark$$

We summarize the rules for dividing signed numbers along with the properties of division learned in Section 1.6.

PROCEDURE Dividing Signed Numbers

1. The quotient of two numbers with the *same* sign is positive.

   Examples: $20 \div 5 = 4$ and $-20 \div (-5) = 4$

2. The quotient of two numbers with different signs is negative.

   Examples: $16 \div (-8) = -2$ and $-16 \div 8 = -2$

3. Zero divided by any nonzero number is zero.

   Examples: $0 \div 12 = 0$ and $0 \div (-3) = 0$

4. Any nonzero number divided by zero is undefined.

   Examples: $-5 \div 0$ is undefined

Example 5 Dividing Integers

Divide.

a. $50 \div (-5)$ **b.** $\dfrac{-42}{-7}$ **c.** $\dfrac{-39}{3}$ **d.** $0 \div (-7)$ **e.** $\dfrac{-8}{0}$

Solution:

a. $50 \div (-5) = -10$ Different signs. The quotient is negative.

b. $\dfrac{-42}{-7} = 6$ Same signs. The quotient is positive.

c. $\dfrac{-39}{3} = -13$ Different signs. The quotient is negative.

d. $0 \div (-7) = 0$ Zero divided by any nonzero number is 0.

e. $\dfrac{-8}{0}$ is undefined. Any nonzero number divided by 0 is undefined.

Skill Practice

Divide.

19. $-40 \div 10$

20. $\dfrac{-36}{-12}$

21. $\dfrac{18}{-2}$

22. $0 \div (-12)$

23. $\dfrac{-2}{0}$

TIP: In Example 5(e), $\frac{-8}{0}$ is undefined because there is no number that when multiplied by 0 will equal −8.

Answers

19. −4 **20.** 3
21. −9 **22.** 0
23. Undefined

Skill Practice

Translate the phrase into a mathematical expression. Then simplify.

24. The quotient of -40 and -4
25. 60 divided by -3
26. -8 divided into -24

Example 6 Translating to a Mathematical Expression

Translate the phrase into a mathematical expression. Then simplify.

a. The quotient of 26 and -13 **b.** -45 divided by 5
c. -4 divided into -24

Solution:

a. The word quotient implies division. The quotient of 26 and -13 translates to $26 \div (-13)$.

$$26 \div (-13) = -2 \quad \text{Simplify.}$$

b. -45 divided by 5 translates to $-45 \div 5$.

$$-45 \div 5 = -9 \quad \text{Simplify.}$$

c. -4 divided into -24 translates to $-24 \div (-4)$.

$$-24 \div (-4) = 6 \quad \text{Simplify.}$$

Skill Practice

27. A severe cold front blew through Atlanta and the temperature change over a 6-hr period was -24°F. Find the average hourly change in temperature.

Example 7 Applying Division of Integers

Between midnight and 6:00 A.M., the change in temperature was -18°F. Find the average hourly change in temperature.

Solution:

In this example, we have a change of -18°F in temperature to distribute evenly over a 6-hr period (from midnight to 6:00 A.M. is 6 hr). This implies division.

$$-18 \div 6 = -3 \quad \text{Divide } -18\text{°F by 6 hr.}$$

The temperature changed by -3°F per hour.

Answers

24. $-50 \div (-5)$; 10
25. $60 \div (-3)$; -20
26. $-24 \div (-8)$; 3 **27.** -4°F

Section 2.4 Practice Exercises

Study Skills Exercise

1. Students often learn a rule about signs that states "two negatives make a positive." This rule is incomplete and therefore not always true. Note the following combinations of two negatives:

$-2 + (-4)$	the sum of two negatives
$-(-5)$	the opposite of a negative
$-\lvert -10 \rvert$	the opposite of an absolute value
$(-3)(-6)$	the product of two negatives

Simplify the expressions to determine which are negative and which are positive. Then write the rule for multiplying two numbers with the same sign.

$-2 + (-4)$ ______________

$-(-5)$ ______________

$-|-10|$ ______________

$(-3)(-6)$ ______________

When multiplying two numbers with the same sign, the product is ______________.

Review Exercises

2. Simplify.

a. $|-5|$ **b.** $|5|$ **c.** $-|5|$ **d.** $-|-5|$ **e.** $-(-5)$

For Exercises 3–8, add or subtract as indicated.

3. $14 - (-5)$ **4.** $-24 - 50$ **5.** $-33 + (-11)$

6. $-7 - (-23)$ **7.** $23 - 12 + (-4) - (-10)$ **8.** $9 + (-12) - 17 - 4 - (-15)$

Objective 1: Multiplication of Integers

For Exercises 9–24, multiply the integers. **(See Example 1.)**

9. $-3(5)$ **10.** $-2(13)$ **11.** $(-5)(-8)$ **12.** $(-12)(-2)$

13. $7(-3)$ **14.** $5(-12)$ **15.** $-12(-4)$ **16.** $-6(-11)$

17. $-15(3)$ **18.** $-3(25)$ **19.** $9(-8)$ **20.** $8(-3)$

21. $-14 \cdot 0$ **22.** $-8 \cdot 0$ **23.** $-95(-1)$ **24.** $-144(-1)$

For Exercises 25–30, translate to a mathematical expression. Then simplify. **(See Example 2.)**

25. Multiply -3 and -1 **26.** Multiply -12 and -4 **27.** The product of -5 and 3

28. The product of 9 and -2 **29.** 3 times -5 **30.** -3 times 6

Objective 2: Multiplying Many Factors

For Exercises 31–40, multiply. **(See Example 3.)**

31. $(-5)(-2)(-4)(-10)$ **32.** $(-3)(-5)(-2)(-4)$ **33.** $(-11)(-4)(-2)$

34. $(-20)(-3)(-1)$ **35.** $(24)(-2)(0)(-3)$ **36.** $(3)(0)(-13)(22)$

37. $(-1)(-1)(-1)(-1)(-1)(-1)$ **38.** $(-1)(-1)(-1)(-1)(-1)(-1)(-1)$

39. $(-2)(2)(2)(-2)(2)$ **40.** $(2)(-2)(2)(2)$

Objective 3: Exponential Expressions

For Exercises 41–56, simplify. **(See Example 4.)**

41. -10^2 **42.** -8^2 **43.** $(-10)^2$ **44.** $(-8)^2$

45. -10^3 **46.** -8^3 **47.** $(-10)^3$ **48.** $(-8)^3$

49. -5^4 **50.** -4^4 **51.** $(-5)^4$ **52.** $(-4)^4$

53. $(-1)^2$ **54.** $(-1)^3$ **55.** -1^4 **56.** -1^5

Objective 4: Division of Integers

For Exercises 57–72, divide the real numbers, if possible. **(See Example 5.)**

57. $60 \div (-3)$ **58.** $46 \div (-2)$ **59.** $\frac{-56}{-8}$ **60.** $\frac{-48}{-3}$

61. $\frac{-15}{5}$ **62.** $\frac{30}{-6}$ **63.** $-84 \div (-4)$ **64.** $-48 \div (-6)$

65. $\frac{-13}{0}$ **66.** $\frac{-41}{0}$ **67.** $\frac{0}{-18}$ **68.** $\frac{0}{-6}$

69. $(-20) \div (-5)$ **70.** $(-10) \div (-2)$ **71.** $\frac{204}{-6}$ **72.** $\frac{300}{-2}$

For Exercises 73–78, translate the English phrase to a mathematical expression. Then simplify. **(See Example 6.)**

73. The quotient of -100 and 20 **74.** The quotient of 46 and -23 **75.** -64 divided by -32

76. -108 divided by -4 **77.** 13 divided into -52 **78.** -15 divided into -45

79. During a 10-min period, a SCUBA diver's depth changed by -60 ft. Find the average change in depth per minute. **(See Example 7.)**

80. When a severe winter storm moved through Albany, New York, the change in temperature between 4:00 P.M. and 7:00 P.M. was -27°F. What was the average hourly change in temperature?

81. One of the most famous blizzards in the United States was the blizzard of 1888. In one part of Nebraska, the temperature plunged from 40°F to -25°F in 5 hr. What was the average change in temperature during this time?

82. A submarine descended from a depth of -528 m to -1804 m in a 2-hr period. What was the average change in depth per hour during this time?

83. Travis wrote five checks to the employees of his business, each for \$225. If the original balance in his checking account was \$890, what is his new balance?

84. Jennifer's checking account had \$320 when she wrote two checks for \$150, and one check for \$82. What is her new balance?

85. During a severe drought in Georgia, the water level in Lake Lanier dropped. During a 1-month period from June to July, the lake's water level changed by -3 ft. If this continued for 6 months, by how much did the water level change?

86. During a drought, the change in elevation for a retention pond was -9 in. over a 1-month period. At this rate, what will the change in elevation be after 5 months?

Mixed Exercises

Animation

For Exercises 87–102, perform the indicated operation.

87. $18(-6)$ **88.** $24(-2)$ **89.** $18 \div (-6)$ **90.** $24 \div (-2)$

91. $(-9)(-12)$ **92.** $-36 \div (-12)$ **93.** $-90 \div (-6)$ **94.** $(-5)(-4)$

95. $\dfrac{0}{-2}$ **96.** $-24 \div 0$ **97.** $-90 \div 0$ **98.** $\dfrac{0}{-5}$

99. $(-2)(-5)(4)$ **100.** $(10)(-2)(-3)(-5)$ **101.** $(-7)^2$ **102.** -7^2

103. **a.** What number must be multiplied by -5 to obtain -35?

b. What number must be multiplied by -5 to obtain 35?

104. **a.** What number must be multiplied by -4 to obtain -36?

b. What number must be multiplied by -4 to obtain 36?

Expanding Your Skills

The electrical charge of an atom is determined by the number of protons and number of electrons an atom has. Each proton gives a positive charge $(+1)$ and each electron gives a negative charge (-1). An atom that has a total charge of 0 is said to be electrically neutral. An atom that is not electrically neutral is called an ion. For Exercises 105–108, determine the total charge for an atom with the given number of protons and electrons.

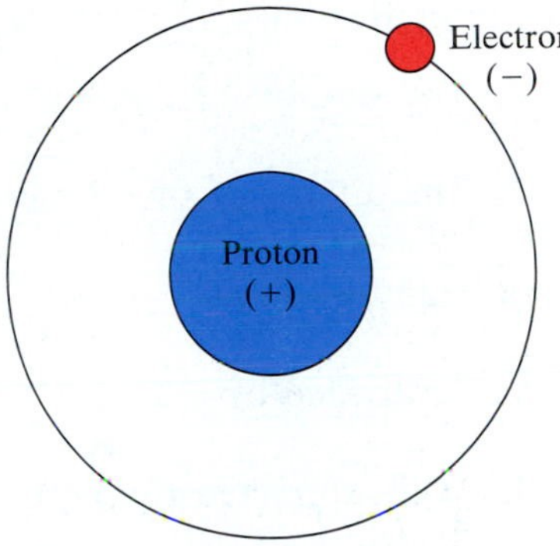

The Hydrogen Atom

105. 1 proton, 0 electrons **106.** 17 protons, 18 electrons

107. 8 protons, 10 electrons **108.** 20 protons, 18 electrons

For Exercises 109–114, assume $a > 0$ (this means that a is positive) and $b < 0$ (this means that b is negative). Find the sign of each expression.

109. $a \cdot b$ **110.** $b \div a$ **111.** $|a| \div b$

112. $a \cdot |b|$ **113.** $-a \div b$ **114.** $a(-b)$

Calculator Connections

Topic: Multiplying and Dividing Integers on a Calculator

Knowing the sign of the product or quotient can make using the calculator easier. For example, look at $\frac{-78}{-26}$. Note the keystrokes if we enter this into the calculator as written:

Expression	Keystrokes	Result
$\dfrac{-78}{-26}$	78 [+○-] [÷] 26 [+○-] [=]	3
	or [(−)] 78 [÷] [(−)] 26 [=]	3

But since we know that the quotient of two negative numbers is positive, we can simply enter:

Keystrokes	Result
78 [÷] 26 [=]	3

Calculator Exercises

For Exercises 115–118, use a calculator to perform the indicated operations.

115. $(-413)(871)$ **116.** $-6125 \cdot (-97)$

117. $\dfrac{-576{,}828}{-10{,}682}$ **118.** $5{,}945{,}308 \div (-9452)$

Problem Recognition Exercises

Operations on Integers

1. Perform the indicated operations

a. $(-24)(-2)$

b. $(-24) - (-2)$

c. $(-24) + (-2)$

d. $(-24) \div (-2)$

2. Perform the indicated operations.

a. $12(-3)$

b. $12 - (-3)$

c. $12 + (-3)$

d. $12 \div (-3)$

For Exercises 3–14, translate each phrase to a mathematical expression. Then simplify.

3. The sum of -5 and -3

4. The product of 9 and -5

5. The difference of -3 and -7

6. The quotient of 28 and -4

7. -23 times -2

8. 18 subtracted from -4

9. 42 divided by -2

10. -13 added to -18

11. -12 subtracted from 10

12. -7 divided into -21

13. The product of -6 and -9

14. The total of -7, 4, 8, -16, and -5

For Exercises 15–37, perform the indicated operations.

15. a. $15 - (-5)$ **b.** $15(-5)$ **c.** $15 + (-5)$ **d.** $15 \div (-5)$

16. a. $-36(-2)$ **b.** $-36 - (-2)$ **c.** $\dfrac{-36}{-2}$ **d.** $-36 + (-2)$

17. a. $20(-4)$ **b.** $-20(-4)$ **c.** $-20(4)$ **d.** $20(4)$

18. a. $-5 - 9 - 2$ **b.** $-5(-9)(-2)$

19. a. $10 + (-3) + (-12)$ **b.** $10 - (-3) - (-12)$

20. a. $(-1)(-2)(-3)(-4)$ **b.** $(-1)(-2)(3)(4)$ **c.** $(-1)(-2)(-3)(4)$ **d.** $(-1)(2)(3)(4)$

21. a. $|-50|$ **b.** $-(-50)$ **c.** $|50|$ **d.** $-|-50|$

22. $\dfrac{0}{-8}$

23. $-55 \div 0$

24. $-615 - (-705)$

25. $-184 - 409$

26. $420 \div (-14)$

27. $-3600 \div (-90)$

28. $-44 - (-44)$

29. $-37 - (-37)$

30. $(-9)^2$

31. $(-2)^5$

32. -9^2

33. -2^5

34. $\dfrac{-46}{0}$

35. $0 \div (-16)$

36. $-15{,}042 + 4893$

37. $-84{,}506 + (-542)$

Order of Operations and Algebraic Expressions

Section 2.5

Objectives

1. Order of Operations
2. Translations Involving Variables
3. Evaluating Algebraic Expressions

1. Order of Operations

The order of operations was first introduced in Section 1.7. The order of operations also applies when simplifying expressions with integers.

PROCEDURE Applying the Order of Operations

1. First perform all operations inside parentheses and other grouping symbols.
2. Simplify expressions containing exponents, square roots, or absolute values.
3. Perform multiplication or division in the order that they appear from left to right.
4. Perform addition or subtraction in the order that they appear from left to right.

Example 1 Applying the Order of Operations

Simplify. $-12 - 6(7 - 5)$

Solution:

$-12 - 6(7 - 5)$

$= -12 - 6(2)$ Simplify within parentheses first.

$= -12 - 12$ Multiply before subtracting.

$= -24$ Subtract. *Note:* $-12 - 12 = -12 + (-12) = -24$.

Skill Practice

Simplify.

1. $8 - 2(3 - 10)$

Example 2 Applying the Order of Operations

Simplify. $6 + 48 \div 4 \cdot (-2)$

Solution:

$6 + \underbrace{48 \div 4} \cdot (-2)$ Perform division and multiplication from left to right before addition.

$= 6 + \underbrace{12 \cdot (-2)}$ Perform multiplication before addition.

$= 6 + (-24)$ Add.

$= -18$

Skill Practice

Simplify.

2. $-10 + 24 \div 2 \cdot (-3)$

Answers

1. 22 2. -46

Skill Practice

Simplify.

3. $8^2 - 2^3 \div (-7 + 5)$

Example 3 Applying the Order of Operations

Simplify. $3^2 - 10^2 \div (-1 - 4)$

Solution:

$3^2 - 10^2 \div (-1 - 4)$

$= 3^2 - 10^2 \div (-5)$ Simplify within parentheses. *Note:* $-1 - 4 = -1 + (-4) = -5$.

$= 9 - 100 \div (-5)$ Simplify exponents. *Note:* $3^2 = 3 \cdot 3 = 9$ and $10^2 = 10 \cdot 10 = 100$.

$= 9 - (-20)$ Perform division before subtraction. *Note:* $100 \div (-5) = -20$.

$= 29$ Subtract. *Note:* $9 - (-20) = 9 + (20) = 29$.

Animation

Skill Practice

Simplify.

4. $\dfrac{|50 + (-10)|}{-2^2}$

5. $1 - 3[5 - (8 - 1)]$

Example 4 Applying the Order of Operations

Simplify.

a. $\dfrac{|-8 + 24|}{5 - 3^2}$ **b.** $5 - 2[8 + (-7 - 5)]$

Solution:

a. $\dfrac{|-8 + 24|}{5 - 3^2}$ Simplify the expressions above and below the division bar by first adding within the absolute value and simplifying exponents.

$= \dfrac{|16|}{5 - 9}$

$= \dfrac{16}{-4}$ Simplify the absolute value and subtract $5 - 9$.

$= -4$ Divide.

b. $5 - 2[8 + (-7 - 5)]$

$= 5 - 2[8 + (-12)]$ First simplify the expression within the innermost parentheses.

$= 5 - 2[-4]$ Continue simplifying within parentheses.

$= 5 - (-8)$ Perform multiplication before subtraction.

$= 13$ Subtract. *Note:* $5 - (-8) = 5 + 8 = 13$

2. Translations Involving Variables

Recall that **variables** are used to represent quantities that are subject to change. For this reason, we can use variables and algebraic expressions to represent one or more unknowns in a word problem.

Answers

3. 68 **4.** −10 **5.** 7

Example 5 Using Algebraic Expressions in Applications

a. At a discount CD store, each CD costs \$8. Suppose n is the number of CDs that a customer buys. Write an expression that represents the cost for n CDs.

b. The length of a rectangle is 5 in. longer than the width w. Write an expression that represents the length of the rectangle.

Solution:

a. The cost of 1 CD is 8(1) dollars.
The cost of 2 CDs is 8(2) dollars.
The cost of 3 CDs is 8(3) dollars.
The cost of n CDs is $8(n)$ dollars or simply $8n$ dollars.

From this pattern, we see that the total cost is the cost per CD times the number of CDs.

b. The length of a rectangle is 5 in. more than the width. The phrase "more than" implies addition. Thus, the length (in inches) is represented by

$$\text{length} = w + 5$$

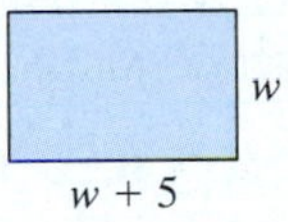

Skill Practice

6. Smoked turkey costs \$7 per pound. Write an expression that represents the cost of p pounds of turkey.

7. The width of a basketball court is 44 ft shorter than its length L. Write an expression that represents the width.

Example 6 Translating to an Algebraic Expression

Write each phrase as an algebraic expression.

a. The product of −6 and x

b. p subtracted from 7

c. The quotient of c and d

d. Twice the sum of y and 4

Solution:

a. The product of −6 and x: $-6x$ — "Product" implies multiplication.

b. p subtracted from 7: $7 - p$ — To subtract p from 7, we must "start" with 7 and then perform the subtraction.

c. The quotient of c and d: $\frac{c}{d}$ — "Quotient" implies division.

d. Twice the sum of y and 4: $2(y + 4)$ — The word "twice the sum" implies that we multiply the *sum* by 2. The sum must be enclosed in parentheses so that the entire quantity is doubled.

Skill Practice

Write each phrase as an algebraic expression.

8. The quotient of w and −4

9. 11 subtracted from x

10. The product of −9 and p

11. Four times the sum of m and n

3. Evaluating Algebraic Expressions

The value of an algebraic expression depends on the values of the variables within the expression.

Answers

6. $7p$ **7.** $L - 44$ **8.** $\frac{w}{-4}$
9. $x - 11$ **10.** $-9p$ **11.** $4(m + n)$

Skill Practice

12. Evaluate $4a + b$ for $a = 8$ and $b = -52$.

Example 7 Evaluating an Algebraic Expression

Evaluate the expression for the given values of the variables.

$$3x - y \quad \text{for } x = 7 \text{ and } y = -3$$

Solution:

$3x - y$

$= 3(\quad) - (\quad)$ When substituting a number for a variable, use parentheses in place of the variable.

$= 3(7) - (-3)$ Substitute 7 for x and -3 for y.

$= 21 + 3$ Apply the order of operations. Multiplication is performed before subtraction.

$= 24$

Skill Practice

Evaluate the expressions for the given values of the variable.

13. Evaluate $-|-y|$ for $y = -12$.
14. Evaluate p^2 for $p = -5$.
15. Evaluate $-w^2$ for $w = -10$.

Example 8 Evaluating an Algebraic Expression

Evaluate the expressions for the given values of the variables.

a. $-|-z|$ for $z = -6$ **b.** x^2 for $x = -4$ **c.** $-y^2$ for $y = -6$

Solution:

a. $-|-z|$

$= -|-(\quad)|$ Replace the variable with empty parentheses.

$= -|-(-6)|$ Substitute the value -6 for z. Apply the order of operations. Inside the absolute value bars, we take the opposite of -6, which is 6.

$= -|6|$

$= -6$ The opposite of $|6|$ is -6.

b. x^2

$= (\quad)^2$ Replace the variable with empty parentheses.

$= (-4)^2$ Substitute -4 for x.

$= (-4)(-4)$ The expression $(-4)^2$ has a base of -4. Therefore, $(-4)^2 = (-4)(-4) = 16$.

$= 16$

c. $-y^2$

$= -(\quad)^2$ Replace the variable with empty parentheses.

$= -(-6)^2$ Substitute -6 for y.

$= -(36)$ Simplify the expression with exponents first. The expression $(-6)^2 = (-6)(-6) = 36$.

$= -36$

Answers

12. -20 13. -12
14. 25 15. -100

Example 9 **Evaluating an Algebraic Expression**

Evaluate the expression for the given values of the variables.

$$-5|x - y + z|, \quad \text{for } x = -9, y = -4, \text{ and } z = 2$$

Solution:

$-5|x - y + z|$

$= -5|(\ \ ) - (\ \ ) + (\ \ )|$ Replace the variables with empty parentheses.

$= -5|(-9) - (-4) + (2)|$ Substitute -9 for x, -4 for y, and 2 for z.

$= -5|-9 + 4 + 2|$ Simplify inside absolute value bars first. Rewrite subtraction in terms of addition.

$= -5|-3|$ Add within the absolute value bars.

$= -5 \cdot 3$ Evaluate the absolute value before multiplying.

$= -15$

TIP: Absolute value bars act as grouping symbols. Therefore, you must perform the operations within the absolute value first.

Skill Practice

16. Evaluate the expression for the given value of the variable.

$3 - |a + b + 4|$ for $a = -5$ and $b = -12$

Answer

16. -10

Section 2.5 Practice Exercises

Boost *your* GRADE at ALEKS.com!

ALEKS version 3.0

- Practice Problems
- Self-Tests
- NetTutor
- e-Professors
- Videos

Study Skills Exercises

1. When you take a test, go through the test and do all the problems that you know first. Then go back and work on the problems that were more difficult. Give yourself a time limit for how much time you spend on each problem (maybe 3 to 5 minutes the first time through). Circle the importance of each statement.

	Not important	Somewhat important	Very important
a. Read through the entire test first.	1	2	3
b. If time allows, go back and check each problem.	1	2	3
c. Write out all steps instead of doing the work in your head.	1	2	3

2. Define the key term **variable**.

Review Exercises

For Exercises 3–8, perform the indicated operation.

3. $-100 \div (-4)$

4. $-100 - (-4)$

5. $-100(-4)$

6. $-100 + (-4)$

7. $(-12)^2$

8. -12^2

Objective 1: Order of Operations

For Exercises 9–54, simplify using the order of operations. **(See Examples 1–4.)**

9. $-1 - 5 - 8 - 3$
10. $-2 - 6 - 3 - 10$
11. $(-1)(-5)(-8)(-3)$
12. $(-2)(-6)(-3)(-10)$
13. $5 + 2(3 - 5)$
14. $6 - 4(8 - 10)$
15. $-2(3 - 6) + 10$
16. $-4(1 - 3) - 8$
17. $-8 - 6^2$
18. $-10 - 5^2$
19. $120 \div (-4)(5)$
20. $36 \div (-2)(3)$
21. $40 - 32 \div (-4)(2)$
22. $48 - 36 \div (6)(-2)$
23. $100 - 2(3 - 8)$
24. $55 - 3(2 - 6)$
25. $|-10 + 13| - |-6|$
26. $|4 - 9| - |-10|$
27. $\sqrt{100 - 36} - 3\sqrt{16}$
28. $\sqrt{36 - 11} + 2\sqrt{9}$
29. $5^2 - (-3)^2$
30. $6^2 - (-4)^2$
31. $-3 + (5 - 9)^2$
32. $-5 + (3 - 10)^2$
33. $12 + (14 - 16)^2 \div (-4)$
34. $-7 + (1 - 5)^2 \div 4$
35. $-48 \div 12 \div (-2)$
36. $-100 \div (-5) \div (-5)$
37. $90 \div (-3) \cdot (-1) \div (-6)$
38. $64 \div (-4) \cdot 2 \div (-16)$
39. $[7^2 - 9^2] \div (-5 + 1)$
40. $[(-8)^2 - 5^2] \div (-4 + 1)$
41. $2 + 2^2 - 10 - 12$
42. $14 - 4^2 + 2 - 10$
43. $\dfrac{3^2 - 27}{-9 + 6}$
44. $\dfrac{8 + (-2)^2}{-5 + (-1)}$
45. $\dfrac{13 - (2)(4)}{-1 - 2^2}$
46. $\dfrac{10 - (-3)(5)}{-9 - 4^2}$
47. $\dfrac{|-23 + 7|}{5^2 - (-3)^2}$
48. $\dfrac{|10 - 50|}{6^2 - (-4)^2}$
49. $21 - [4 - (5 - 8)]$
50. $15 - [10 - (20 - 25)]$
51. $-17 - 2[18 \div (-3)]$
52. $-8 - 5(-45 \div 15)$
53. $4 + 2[9 + (-4 + 12)]$
54. $-13 + 3[11 + (-15 + 10)]$

Objective 2: Translations Involving Variables

55. Carolyn sells homemade candles. Write an expression for her total revenue if she sells x candles for \$15 each. **(See Example 5.)**

56. Maria needs to buy 12 wine glasses. Write an expression of the cost of 12 glasses at p dollars each.

57. Jonathan is 4 in. taller than his brother. Write an expression for Jonathan's height if his brother is t inches tall. **(See Example 5.)**

58. It takes Perry 1 hr longer than David to mow the lawn. If it takes David h hours to mow the lawn, write an expression for the amount of time it takes Perry to mow the lawn.

59. A sedan travels 6 mph slower than a sports car. Write an expression for the speed of the sedan if the sports car travels v mph.

60. Bill's daughter is 30 years younger than he is. Write an expression for his daughter's age if Bill is A years old.

61. The price of gas has doubled over the last 3 years. If gas cost g dollars per gallon 3 years ago, write an expression of the current price per gallon.

62. Suppose that the amount of rain that fell on Monday was twice the amount that fell on Sunday. Write an expression for the amount of rain on Monday, if Sunday's amount was t inches.

For Exercises 63–74, write each phrase as an algebraic expression. **(See Example 6.)**

63. The product of -12 and n

64. The product of -3 and z

65. x subtracted from -9

66. p subtracted from -18

67. The quotient of t and -2

68. The quotient of -10 and w

69. -14 added to y

70. -150 added to c

71. Twice the sum of c and d

72. Twice the sum of a and b

73. The difference of x and -8

74. The difference of m and -5

Objective 3: Evaluating Algebraic Expressions

For Exercises 75–94, evaluate the expressions for the given values of the variables. **(See Examples 7–9.)**

75. $x + 9z$ for $x = -10$ and $z = -3$

76. $a + 7b$ for $a = -3$ and $b = -6$

77. $x + 5y + z$ for $x = -10, y = 5$, and $z = 2$

78. $9p + 4t + w$ for $p = 2, t = 6$, and $w = -50$

79. $a - b + 3c$ for $a = -7, b = -2$, and $c = 4$

80. $w + 2y - z$ for $w = -9, y = 10$, and $z = -3$

81. $-3mn$ for $m = -8$ and $n = -2$

82. $-5pq$ for $p = -4$ and $q = -2$

83. $|-y|$ for $y = -9$

84. $|-z|$ for $z = -18$

85. $-|-w|$ for $w = -4$

86. $-|-m|$ for $m = -15$

87. x^2 for $x = -3$

88. n^2 for $n = -9$

89. $-x^2$ for $x = -3$

90. $-n^2$ for $n = -9$

91. $-4|x + 3y|$ for $x = 5$ and $y = -6$

92. $-2|4a - b|$ for $a = -8$ and $b = -2$

93. $6 - |m - n^2|$ for $m = -2$ and $n = 3$

94. $4 - |c^2 - d^2|$ for $c = 3$ and $d = -5$

Expanding Your Skills

95. Find the average temperature: $-8°, -11°, -4°, 1°, 9°, 4°, -5°$

96. Find the average temperature: $15°, 12°, 10°, 3°, 0°, -2°, -3°$

97. Find the average score: $-8, -8, -6, -5, -2, -3, 3, 3, 0, -4$

98. Find the average score: $-6, -2, 5, 1, 0, -3, 4, 2, -7, -4$

Group Activity

Checking Weather Predictions

Materials: A computer with online access

Estimated time: 2–3 minutes each day for 10 days

Group Size: 3

1. Go to a website such as http://www.weather.com/ to find the predicted high and low temperatures for a 10-day period for a city of your choice.

2. Record the predicted high and low temperatures for each of the 10 days. Record these values in the second column of each table.

Day	Predicted High	Actual High	Difference (error)
1			
2			
3			
4			
5			
6			
7			
8			
9			
10			

Day	Predicted Low	Actual Low	Difference (error)
1			
2			
3			
4			
5			
6			
7			
8			
9			
10			

3. For the next 10 days, record the actual high and low temperatures for your chosen city for that day. Record these values in the third column of each table.

4. For each day, compute the difference between the predicted and actual temperature and record the results in the fourth column of each table. We will call this difference the *error*.

$$\text{error} = (\text{predicted temperature}) - (\text{actual temperature})$$

5. If the error is *negative*, does this mean that the weather service overestimated or underestimated the temperature?

6. If the error is *positive*, does this mean that the weather service overestimated or underestimated the temperature?

Chapter 2 Summary

Section 2.1 Integers, Absolute Value, and Opposite

Key Concepts

The numbers . . . −3, −2, −1, 0, 1, 2, 3, . . . and so on are called **integers**. The negative integers lie to the left of zero on the number line.

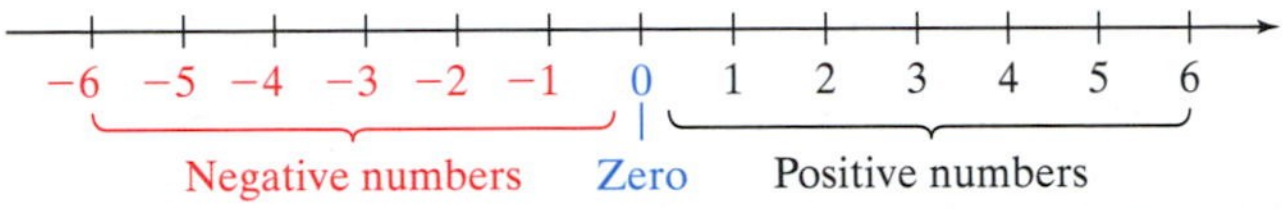

The **absolute value** of a number a is denoted $|a|$. The value of $|a|$ is the distance between a and 0 on the number line.

Two numbers that are the same distance from zero on the number line, but on opposite sides of zero are called **opposites**.

The double negative property states that the opposite of a negative number is a positive number. That is, $-(-a) = a$, for $a > 0$.

Examples

Example 1

The temperature 5° below zero can be represented by a negative number: −5°.

Example 2

a. $|5| = 5$ b. $|-13| = 13$ c. $|0| = 0$

Example 3

The opposite of 12 is $-(12) = -12$.

Example 4

The opposite of −23 is $-(-23) = 23$.

Section 2.2 Addition of Integers

Key Concepts

To add integers using a number line, locate the first number on the number line. Then to add a positive number, move to the right on the number line. To add a negative number, move to the left on the number line.

Integers can be added using the following rules:

Adding Numbers with the Same Sign

To add two numbers with the same sign, add their absolute values and apply the common sign.

Adding Numbers with Different Signs

To add two numbers with different signs, subtract the smaller absolute value from the larger absolute value. Then apply the sign of the number having the larger absolute value.

An English phrase can be translated to a mathematical expression involving addition of integers.

Examples

Example 1

Add $-2 + (-4)$ using the number line.

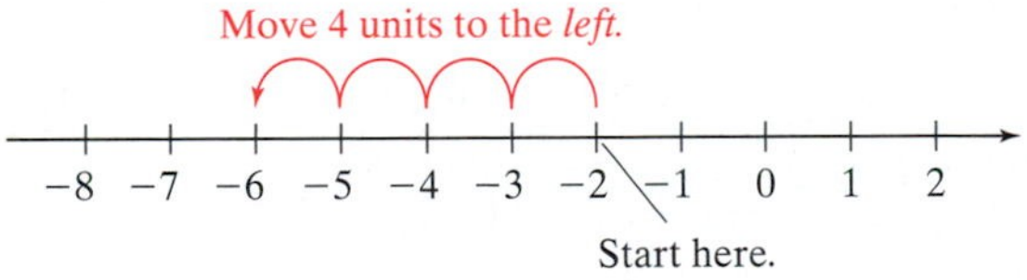

$-2 + (-4) = -6$

Example 2

a. $5 + 2 = 7$
b. $-5 + (-2) = -7$

Example 3

a. $6 + (-5) = 1$
b. $(-6) + 5 = -1$

Example 4

−3 added to the sum of −8 and 6 translates to $(-8 + 6) + (-3)$.

Section 2.3 Subtraction of Integers

Key Concepts

Subtraction of Signed Numbers

For two numbers a and b,

$$a - b = a + (-b)$$

To perform subtraction, follow these steps:

1. Leave the first number (the minuend) unchanged.
2. Change the subtraction sign to an addition sign.
3. Add the opposite of the second number (the subtrahend).

An English phrase can be translated to a mathematical expression involving subtraction of integers.

Examples

Example 1

a. $3 - 9 = 3 + (-9) = -6$

b. $-3 - 9 = -3 + (-9) = -12$

c. $3 - (-9) = 3 + (9) = 12$

d. $-3 - (-9) = -3 + (9) = 6$

Example 2

2 decreased by -10 translates to $2 - (-10)$.

Section 2.4 Multiplication and Division of Integers

Key Concepts

Multiplication of Signed Numbers

1. The product of two numbers with the same sign is positive.
2. The product of two numbers with different signs is negative.
3. The product of any number and zero is zero.

The product of an *even* number of negative factors is *positive*.
The product of an *odd* number of negative factors is *negative*.

When evaluating an exponential expression, attention must be given when parentheses are used.
That is, $(-2)^4 = (-2)(-2)(-2)(-2) = 16$,
while $-2^4 = -1 \cdot (2)(2)(2)(2) = -16$.

Division of Signed Numbers

1. The quotient of two numbers with the same sign is positive.
2. The quotient of two numbers with different signs is negative.
3. Division by zero is undefined.
4. Zero divided by a nonzero number is 0.

Examples

Example 1

a. $-8(-3) = 24$
b. $8(-3) = -24$
c. $-8(0) = 0$

Example 2

a. $(-5)(-4)(-1)(-3) = 60$
b. $(-2)(-1)(-6)(-3)(-2) = -72$

Example 3

a. $(-3)^2 = (-3)(-3) = 9$
b. $-3^2 = -(3)(3) = -9$

Example 4

a. $-36 \div (-9) = 4$

b. $\dfrac{42}{-6} = -7$

Example 5

a. $-15 \div 0$ is undefined.

b. $0 \div (-3) = 0$

Section 2.5 Order of Operations and Algebraic Expressions

Key Concepts

Order of Operations

1. First perform all operations inside parentheses and other grouping symbols.
2. Simplify expressions containing exponents, square roots, or absolute values.
3. Perform multiplication or division in the order that they appear from left to right.
4. Perform addition or subtraction in the order that they appear from left to right.

Evaluating an Algebraic Expression

To evaluate an expression, first replace the variable with parentheses. Then insert the values and simplify using the order of operations.

Examples

Example 1

$$\begin{aligned} -15 - 2(8 - 11)^2 &= -15 - 2(-3)^2 \\ &= -15 - 2 \cdot 9 \\ &= -15 - 18 \\ &= -33 \end{aligned}$$

Example 2

Evaluate $4x - 5y$ for $x = -2$ and $y = 3$.

$$\begin{aligned} 4x - 5y &= 4(\quad) - 5(\quad) \\ &= 4(-2) - 5(3) \\ &= -8 - 15 \\ &= -23 \end{aligned}$$

Chapter 2 Review Exercises

Section 2.1

For Exercises 1–2, write an integer that represents each numerical value.

1. The plane descended 4250 ft.

2. The company's profit fell by \$3,000,000.

For Exercises 3–6, graph the numbers on the number line.

3. -2 **4.** -5 **5.** 0 **6.** 3

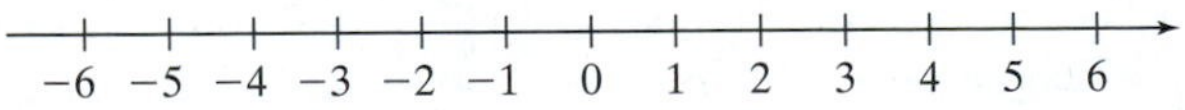

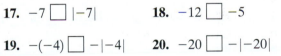

For Exercises 7–8, determine the opposite and the absolute value for each number.

7. -4 **8.** 6

For Exercises 9–16, simplify.

9. $|-3|$ **10.** $|-1000|$ **11.** $|74|$

12. $|0|$ **13.** $-(-9)$ **14.** $-(-28)$

15. $-|-20|$ **16.** $-|-45|$

For Exercises 17–20, fill in the blank with $<$, $>$, or $=$, to make a true statement.

17. $-7 \;\square\; |-7|$ **18.** $-12 \;\square\; -5$

19. $-(-4) \;\square\; -|-4|$ **20.** $-20 \;\square\; -|-20|$

Section 2.2

For Exercises 21–24, add the integers using the number line.

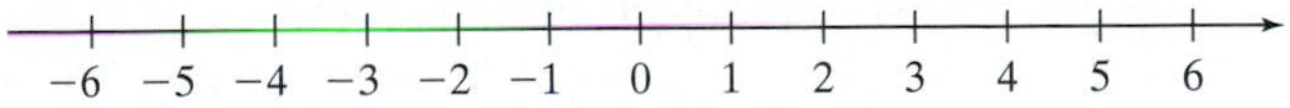

21. $6 + (-2)$ **22.** $-3 + 6$

23. $-3 + (-2)$ **24.** $-3 + 0$

25. State the rule for adding two numbers with the same sign.

26. State the rule for adding two numbers with different signs.

For Exercises 27–32, add the integers.

27. $35 + (-22)$ **28.** $-105 + 90$

29. $-29 + (-41)$

30. $-98 + (-42)$

31. $-3 + (-10) + 12 + 14 + (-10)$

32. $9 + (-15) + 2 + (-7) + (-4)$

For Exercises 33–38, translate each phrase to a mathematical expression. Then simplify the expression.

33. The sum of 23 and -35

34. 57 plus -10

35. The total of -5, -13, and 20

36. -42 increased by 12

37. 3 more than -12

38. -89 plus -22

39. The graph gives the number of inches below or above the average snowfall for the given months for Caribou, Maine. Find the total departure from average. Is the snowfall for Caribou above or below average?

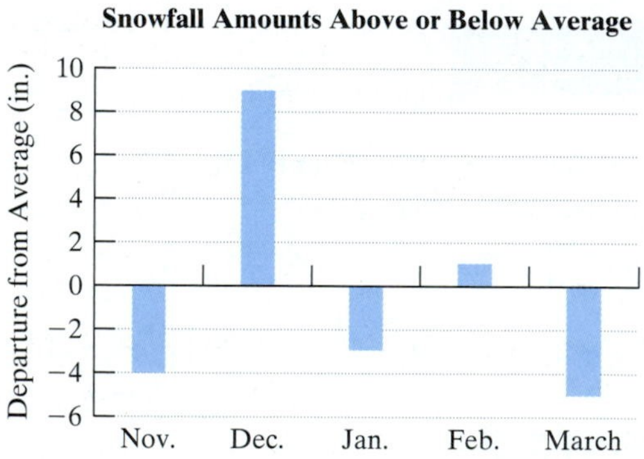

40. The table gives the scores for golfer Ernie Els at a recent PGA Open golf tournament. Find Ernie's total score after all four rounds.

	Round 1	Round 2	Round 3	Round 4
Ernie Els	2	−2	−1	−4

Section 2.3

41. State the steps for subtracting two numbers.

For Exercises 42–49, simplify.

42. $4 - (-23)$

43. $19 - 44$

44. $-2 - (-24)$

45. $-289 - 130$

46. $2 - 7 - 3$

47. $-45 - (-77) + 8$

48. $-16 - 4 - (-3)$

49. $99 - (-7) - 6$

50. Translate the phrase to a mathematical expression. Then simplify.

a. The difference of 8 and 10

b. 8 subtracted from 10

For Exercises 51–52, translate the mathematical statement to an English phrase. Answers will vary.

51. $-2 - 14$

52. $-25 - (-7)$

53. The temperature in Fargo, North Dakota, rose from $-6°F$ to $-1°F$. By how many degrees did the temperature rise?

54. Sam's balance in his checking account was $-\$40$, so he deposited \$132. What is his new balance?

55. Find the average of the golf scores: $-3, 4, 0, 9, -2, -1, 0, 5, -3$ (These scores are the number of holes above or below par.)

56. A missile was launched from a submarine from a depth of -1050 ft below sea level. If the maximum height of the missile is 2400 ft, find the vertical distance between its greatest height and its depth at launch.

Section 2.4

For Exercises 57–72, simplify.

57. $6(-3)$

58. $\frac{-12}{4}$

59. $\frac{-900}{-60}$

60. $(-7)(-8)$

61. $-36 \div 9$

62. $60 \div (-5)$

63. $(-12)(-4)(-1)(-2)$

64. $(-1)(-8)(2)(1)(-2)$

65. $-15 \div 0$

66. $\frac{0}{-5}$

67. -5^3

68. $(-5)^3$

69. $(-6)^2$

70. -6^2

71. $(-1)^{10}$

72. $(-1)^{21}$

73. What is the sign of the product of three negative factors?

74. What is the sign of the product of four negative factors?

For Exercises 75–76, translate the English phrase to a mathematical expression. Then simplify.

75. The quotient of -45 and -15

76. The product of -4 and 19

77. Between 8:00 P.M. and midnight, the change in temperature was $-12°$F. Find the average hourly change in temperature.

78. Suzie wrote four checks to a vendor, each for \$160. If the original balance in her checking account was \$550, what is her new balance?

Section 2.5

For Exercises 79–88, simplify using the order of operations.

79. $50 - 3(6 - 2)$

80. $48 - 8 \div (-2) + 5$

81. $28 \div (-7) \cdot 3 - (-1)$

82. $(-4)^2 \div 8 - (-6)$

83. $[10 - (-3)^2] \cdot (-11) + 4$

84. $[-9 - (-7)]^2 \cdot 3 \div (-6)$

85. $\dfrac{100 - 4^2}{(-7)(6)}$

86. $\dfrac{18 - 3(-2)}{4^2 - 8}$

87. $5 - 2[-3 + (2 - 5)]$

88. $-10 + 3[4 - (-2 + 7)]$

89. Michael is 8 years older than his sister. Write an expression for Michael's age if his sister is a years old.

90. At a movie theater, drinks are \$3 each. Write an expression for the cost of n drinks.

For Exercises 91–96, write each phrase as an algebraic expression.

91. The product of -5 and x

92. The difference of p and 12

93. Two more than the sum of a and b

94. The quotient of w and 4

95. -8 subtracted from y

96. Twice the sum of 5 and z

For Exercises 97–104, evaluate the expression for the given values of the variable.

97. $3x - 2y$ for $x = -5$ and $y = 4$

98. $5(a - 4b)$ for $a = -3$ and $b = 2$

99. $-2(x + y)^2$ for $x = 6$ and $y = -9$

100. $-3w^2 - 2z$ for $w = -4$ and $z = -9$

101. $-|x|$ for $x = -2$

102. $-|-x|$ for $x = -5$

103. $-(-x)$ for $x = -10$

104. $-(-x)$ for $x = 5$

Chapter 2 Test

For Exercises 1–2, write an integer that represents the numerical value.

1. Dwayne lost \$220 during in his last trip to Las Vegas.

2. Garth Brooks has 26 more platinum albums than Elvis Presley.

For Exercises 3–8, fill in the blank with $>$, $<$, or $=$ to make the statement true.

3. $-5 \square -2$ **4.** $|-5| \square |-2|$

5. $0 \square -(-2)$ **6.** $-|-12| \square -12$

7. $-|-9| \square 9$ **8.** $-5^2 \square (-5)^2$

9. Determine the absolute value of -10.

10. Determine the opposite of -10.

For Exercises 11–22, perform the indicated operations.

11. $9 + (-14)$ **12.** $-23 + (-5)$

13. $-4 - (-13)$ **14.** $-30 - 11$

15. $-15 + 21$ **16.** $5 - 28$

17. $6(-12)$ **18.** $(-11)(-8)$

19. $\dfrac{-24}{-12}$ **20.** $\dfrac{54}{-3}$

21. $\dfrac{-44}{0}$ **22.** $(-91)(0)$

For Exercises 23–28, translate to a mathematical expression. Then simplify the expression.

23. The product of -3 and -7

24. 8 more than -13

25. Subtract -4 from 18.

26. The quotient of 6 and -2

27. -8 increased by 5

28. The total of $-3, 15, -6$, and -1

29. The graph gives the number of inches below or above the average rainfall for the given months for Atlanta, Georgia. Find the total departure from average. Is the total rainfall for these months above or below average?

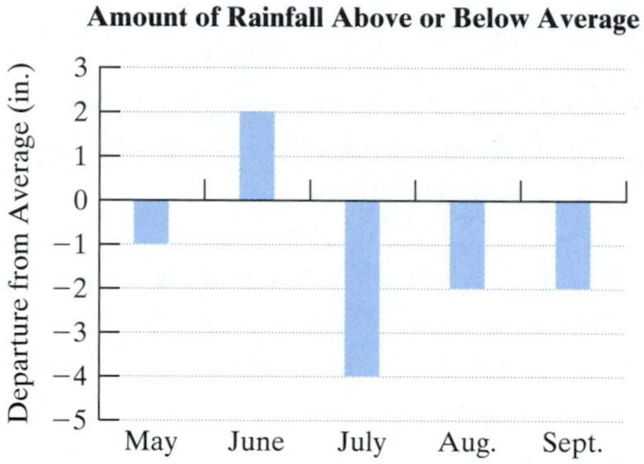

30. The "Great White Hurricane" was a severe winter blizzard that dumped 50 in. of snow in Connecticut and Massachusetts. In one part of Connecticut, between 3:00 P.M. and 8:00 P.M., the change in temperature was $-35°$F. Find the average hourly change in temperature.

31. Simplify the expressions.

a. $(-8)^2$ **b.** -8^2

c. $(-4)^3$ **d.** -4^3

For Exercises 32–37, simplify the expressions.

32. $-14 + 22 - (-5) + (-10)$

33. $(-3)(-1)(-4)(-1)(-5)$

34. $16 - 2[5 - (1 - 4)]$

35. $-20 \div (-2)^2 + (-14)$

36. $12 \cdot (-6) + [20 - (-12)] - 15$

37. $\dfrac{24 - 2|3 - 9|}{8 - 2^2}$

38. A high school student sells magazine subscriptions at \$18 each. Write an expression that represents the total value sold for m magazines.

For Exercises 39–40, evaluate the expressions for the given values of the variables.

39. $-x^2 + y^2$ for $x = 4$ and $y = -1$

40. $-4m - 3n$ for $m = -6$ and $n = 4$

Chapters 1–2 Cumulative Review Exercises

1. For the number $6{,}8\underline{7}3{,}129$ identify the place value of the underlined digit.

2. Write the following inequality in words: $130 < 244$

3. Approximate the perimeter of the triangle by first rounding the numbers to the hundreds place.

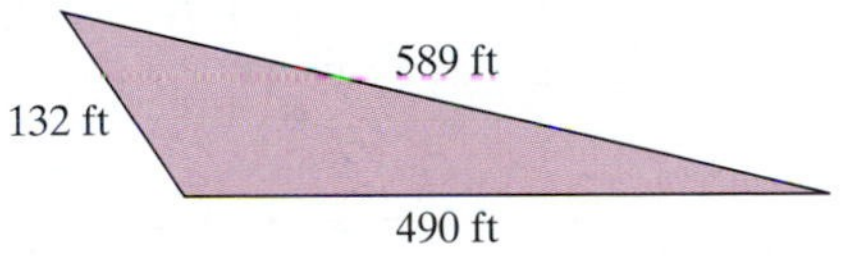

For Exercises 4–7, add or subtract as indicated.

4. $73 + 41$

5. $71 + (-4) + 81 + (-106)$

6. $284 - 171 - (-84) - 393$

7. $\begin{array}{r} 1001 \\ -235 \\ \hline \end{array}$

For Exercises 8–13, multiply or divide as indicated.

8. $31 \cdot (-8)$

9. $-386 \div (-2)$

10. $737 \div 7$

11. $\begin{array}{r} 409 \\ \times\ 228 \\ \hline \end{array}$

12. $\dfrac{0}{-61}$

13. $0\overline{)341}$

14. Find the area of the rectangle.

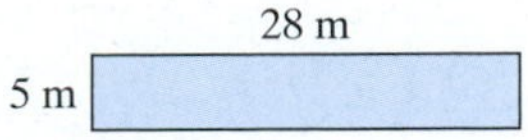

15. Simplify the expressions.

a. $-|-4|$

b. $-(-4)$

c. -4^2

d. $(-4)^2$

16. Simplify the expression using the order of operations: $-14 - 2(9 - 5^2)$

For Exercises 17–18, evaluate the expressions for $x = -4$ and $y = 1$.

17. $x^2 - x + y$

18. $|x - y|$

19. Torie is taking a natural herb in capsule form. She purchased four bottles containing 30 capsules each. The directions state that she can take either 2 or 3 capsules a day.

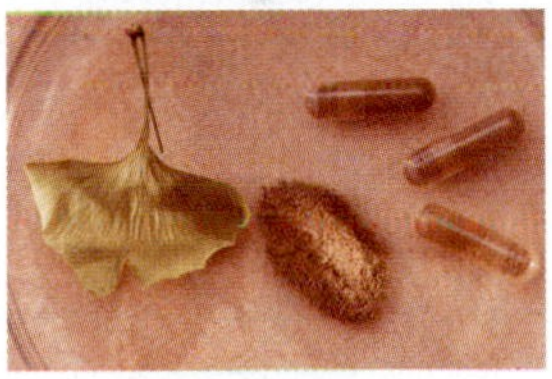

a. How many days will the capsules last if Torie takes 3 a day?

b. How many days will the capsules last if she takes only 2 a day?

20. The low temperatures for Fairbanks, Alaska, are given for a 1-week period. Find the average low temperature.

$-8°F$, $-11°F$, $3°F$, $6°F$, $0°F$, $-8°F$, $-10°F$

Solving Equations

3

CHAPTER OUTLINE

Chapter 3

In this chapter, we continue our study of algebra by learning how to solve linear equations. An *equation* is a statement indicating that two expressions are equal. A *solution* to an equation is a value of the variable that makes the right-hand side of an equation equal to the left-hand side of the equation.

Are You Prepared?

To help you prepare for this chapter, try the following exercises. Substitute the given value into the equation in place of the variable. Then determine if this value is a solution to the equation. That is, determine if the right-hand side equals the left-hand side.

1. $x + 12 = 16$ Is 4 a solution to this equation? yes or no

2. $y - 7 = 18$ Is 11 a solution to this equation? yes or no

3. $\dfrac{w}{-3} = 8$ Is −24 a solution to this equation? yes or no

4. $18 = -3 + z$ Is 15 a solution to this equation? yes or no

5. $120 = -5p$ Is −24 a solution to this equation? yes or no

6. $3x - 7 = 29$ Is 12 a solution to this equation? yes or no

7. $44 = 4 - 5x$ Is 8 a solution to this equation? yes or no

8. $11x - 6 = 12x + 2$ Is −8 a solution to this equation? yes or no

Section 3.1 Simplifying Expressions and Combining *Like* Terms

Objectives

1. Identifying *Like* Terms
2. Commutative, Associative, and Distributive Properties
3. Combining *Like* Terms
4. Simplifying Expressions

1. Identifying *Like* Terms

A **term** is a number or the product or quotient of numbers and variables. An algebraic expression is the sum of one or more terms. For example, the expression

$$-8x^3 + xy - 40 \quad \text{can be written as} \quad -8x^3 + xy + (-40)$$

This expression consists of the terms $-8x^3$, xy, and -40. The terms $-8x^3$ and xy are called **variable terms** because the value of the term will change when different numbers are substituted for the variables. The term -40 is called a **constant term** because its value will never change.

It is important to distinguish between a term and the factors within a term. For example, the quantity xy is one term, and the values x and y are factors within the term. The **coefficient** of the term is the numerical factor of the term.

TIP: Variables without a coefficient explicitly written have a coefficient of 1. Thus, the term xy is equal to $1xy$. The 1 is understood.

Term	Coefficient of the Term
$-8x^3$	-8
xy or $1xy$	1
-40	-40

Terms are said to be ***like* terms** if they each have the same variables, and the corresponding variables are raised to the same powers. For example:

Like Terms	Unlike Terms	
$-4x$ and $6x$	$-4x$ and $6y$	(different variables)
$18ab$ and $4ba$	$18ab$ and $4a$	(different variables)
$7m^2n^5$ and $3m^2n^5$	$7m^2n^5$ and $3mn^5$	(different powers of m)
$5p$ and $-3p$	$5p$ and 3	(different variables)
8 and 10	8 and $10x$	(different variables)

Skill Practice

For Exercises 1–2, use the expression:

$$-48y^5 + 8y^2 - y - 8$$

1. List the terms of the expression.
2. List the coefficient of each term.
3. Are the terms $5x$ and x *like* terms?
4. Are the terms $-6a^3b^2$ and a^3b^2 *like* terms?

Example 1 Identifying Terms, Coefficients, and *Like* Terms

a. List the terms of the expression: $14x^3 - 6x^2 + 3x + 5$

b. Identify the coefficient of each term: $14x^3 - 6x^2 + 3x + 5$

c. Which two terms are *like* terms? $-6x$, 5, $-3y$, and $4x$

d. Are the terms $3x^2y$ and $5xy^2$ *like* terms?

Solution:

a. The expression $14x^3 - 6x^2 + 3x + 5$ can be written as $14x^3 + (-6x^2) + 3x + 5$

Therefore, the terms are $14x^3$, $-6x^2$, $3x$, and 5.

b. The coefficients are 14, -6, 3, and 5.

c. The terms $-6x$ and $4x$ are *like* terms.

d. $3x^2y$ and $5xy^2$ are not *like* terms because the exponents on the x variables are different, and the exponents on the y variables are different.

Answers

1. $-48y^5, 8y^2, -y, -8$
2. $-48, 8, -1, -8$
3. Yes
4. Yes

2. Commutative, Associative, and Distributive Properties

Several important properties of whole numbers were introduced in Sections 1.3 and 1.5 involving addition and multiplication of whole numbers. These properties also hold for integers and algebraic expressions and will be used to simplify expressions (Tables 3-1 through 3-3).

Table 3-1 Commutative Properties

Property	In Symbols/Examples	Comments/Notes
Commutative property of addition	$a + b = b + a$ ex: $-4 + 7 = 7 + (-4)$ $x + 3 = 3 + x$	The order in which two numbers are added does not affect the sum.
Commutative property of multiplication	$a \cdot b = b \cdot a$ ex: $-5 \cdot 9 = 9 \cdot (-5)$ $8y = y \cdot 8$	The order in which two numbers are multiplied does not affect the product.

Avoiding Mistakes

In an expression such as $x + 3$, the two terms x and 3 cannot be combined into one term because they are not *like* terms. Addition of *like* terms will be discussed in Examples 6 and 7.

Table 3-2 Associative Properties

Property	In Symbols/Examples	Comments/Notes
Associative property of addition	$(a + b) + c = a + (b + c)$ ex: $(5 + 8) + 1 = 5 + (8 + 1)$ $(t + n) + 3 = t + (n + 3)$	The manner in which three numbers are grouped under addition does not affect the sum.
Associative property of multiplication	$(a \cdot b) \cdot c = a \cdot (b \cdot c)$ ex: $(-2 \cdot 3) \cdot 6 = -2 \cdot (3 \cdot 6)$ $3 \cdot (m \cdot n) = (3 \cdot m) \cdot n$	The manner in which three numbers are grouped under multiplication does not affect the product.

Table 3-3 Distributive Property of Multiplication over Addition

Property	In Symbols/Examples	Comments/Notes
Distributive property of multiplication over addition*	$a \cdot (b + c) = a \cdot b + a \cdot c$ ex: $-3(2 + x) = -3(2) + (-3)(x)$ $= -6 + (-3)x$ $= -6 - 3x$	The factor outside the parentheses is multiplied by each term in the sum.

*Note that the distributive property of multiplication over addition is sometimes referred to as the *distributive property*.

Example 2 demonstrates the use of the commutative properties.

Skill Practice

Apply the commutative property of addition or multiplication to rewrite the expression.

5. $x + 3$ **6.** $z(2)$

7. $-12 + p$ **8.** ab

Example 2 Applying the Commutative Properties

Apply the commutative property of addition or multiplication to rewrite the expression.

a. $6 + p$ **b.** $-5 + n$ **c.** $y(7)$ **d.** xy

Solution:

a. $6 + p = p + 6$ Commutative property of addition

b. $-5 + n = n + (-5)$ or $n - 5$ Commutative property of addition

c. $y(7) = 7y$ Commutative property of multiplication

d. $xy = yx$ Commutative property of multiplication

Recall from Section 1.3 that subtraction is not a commutative operation. However, if we rewrite the difference of two numbers $a - b$ as $a + (-b)$, then we can apply the commutative property of addition. For example:

$x - 9 = x + (-9)$ Rewrite as addition of the opposite.

$= -9 + x$ Apply the commutative property of addition.

Example 3 demonstrates the associative properties of addition and multiplication.

Skill Practice

Use the associative property of addition or multiplication to rewrite the expression. Then simplify the expression.

9. $6 + (14 + x)$

10. $2(8w)$

Example 3 Applying the Associative Properties

Use the associative property of addition or multiplication to rewrite each expression. Then simplify the expression.

a. $12 + (45 + y)$ **b.** $5(7w)$

Solution:

a. $12 + (45 + y) = (12 + 45) + y$ Apply the associative property of addition.

$= 57 + y$ Simplify.

b. $5(7w) = (5 \cdot 7)w$ Apply the associative property of multiplication.

$= 35w$ Simplify.

Note that in most cases, a detailed application of the associative properties will not be given. Instead the process will be written in one step, such as:

$$12 + (45 + y) = 57 + y, \qquad 5(7w) = 35w$$

Example 4 demonstrates the use of the distributive property.

Answers

5. $3 + x$ **6.** $2z$

7. $p + (-12)$ or $p - 12$ **8.** ba

9. $(6 + 14) + x$; $20 + x$

10. $(2 \cdot 8)w$; $16w$

Example 4 **Applying the Distributive Property**

Apply the distributive property.

a. $3(x + 4)$ **b.** $2(3y - 5z + 1)$

Solution:

a. $3(x + 4) = 3(x) + 3(4)$ Apply the distributive property.

$= 3x + 12$ Simplify.

b. $2(3y - 5z + 1) = 2[3y + (-5z) + 1]$ First write the subtraction as addition of the opposite.

$= 2[3y + (-5z) + 1]$ Apply the distributive property.

$= 2(3y) + 2(-5z) + 2(1)$

$= 6y + (-10z) + 2$ Simplify.

$= 6y - 10z + 2$

TIP: In Example 4(b), we rewrote the expression by writing the subtraction as addition of the opposite. Often this step is not shown and fewer steps are shown overall. For example:

$$2(3y - 5z + 1) = 2(3y) + 2(-5z) + 2(1)$$
$$= 6y - 10z + 2$$

Skill Practice

Apply the distributive property.

11. $4(2 + m)$

12. $6(5p - 3q + 1)$

Avoiding Mistakes

Note that $6y - 10z + 2$ cannot be simplified further because $6y$, $-10z$, and 2 are not *like* terms.

Example 5 **Applying the Distributive Property**

Apply the distributive property.

a. $-8(2 - 5y)$ **b.** $-(-4a + b + 3c)$

Solution:

a. $-8(2 - 5y)$

$= -8[2 + (-5y)]$ Write the subtraction as addition of the opposite.

$= -8[2 + (-5y)]$ Apply the distributive property.

$= -8(2) + (-8)(-5y)$

$= -16 + 40y$ Simplify.

b. $-(-4a + b + 3c)$ The negative sign preceding the parentheses indicates that we take the opposite of the expression within parentheses. This is equivalent to multiplying the expression within parentheses by -1.

$= -1 \cdot (-4a + b + 3c)$

$= -1(-4a) + (-1)(b) + (-1)(3c)$ Apply the distributive property.

$= 4a - b - 3c$ Simplify.

Skill Practice

Apply the distributive property.

13. $-4(6 - 10x)$

14. $-(2x - 3y + 4z)$

TIP: Notice that a negative factor outside the parentheses changes the signs of all terms to which it is multiplied.

$$-1 \cdot (-4a + b + 3c)$$
$$= +4a - b - 3c$$

Answers

11. $8 + 4m$ **12.** $30p - 18q + 6$
13. $-24 + 40x$ **14.** $-2x + 3y - 4z$

3. Combining *Like* Terms

Two terms may be combined if they are *like* terms. To add or subtract *like* terms, we use the distributive property as shown in Example 6.

Skill Practice

Add or subtract as indicated.

15. $-4w + 11w$

16. $6a^3b^2 - 11a^3b^2$

17. $z - 4z + 22z$

Example 6 Using the Distributive Property to Add and Subtract *Like* Terms

Add or subtract as indicated.

a. $8y + 6y$ **b.** $7x^2y - 10x^2y$ **c.** $-15w + 4w - w$

Solution:

a. $8y + 6y = (8 + 6)y$ — Apply the distributive property.

$= 14y$ — Simplify.

b. $7x^2y - 10x^2y = (7 - 10)x^2y$ — Apply the distributive property.

$= -3x^2y$ — Simplify.

c. $-15w + 4w - w = -15w + 4w - 1w$ — First note that $w = 1w$.

$= (-15 + 4 - 1)w$ — Apply the distributive property.

$= (-12)w$ — Simplify within parentheses.

$= -12w$

Although the distributive property is used to add and subtract *like* terms, it is tedious to write each step. Observe that adding or subtracting *like* terms is a matter of adding or subtracting the coefficients and leaving the variable factors unchanged. This can be shown in one step.

$$8y + 6y = 14y \quad \text{and} \quad -15w + 4w - 1w = -12w$$

This shortcut will be used throughout the text.

Skill Practice

Simplify.

18. $4a - 10b - a + 16b + 9$

Example 7 Adding and Subtracting *Like* Terms

Simplify by combining *like* terms. $-3x + 8y + 4x - 19 - 10y$

Solution:

$-3x + 8y + 4x - 19 - 10y$

$= -3x + 4x + 8y - 10y - 19$ — Arrange *like* terms together.

$= 1x - 2y - 19$ — Combine *like* terms.

$= x - 2y - 19$ — Note that $1x = x$. Also note that the remaining terms cannot be combined further because they are not *like*. The variable factors are different.

TIP: The commutative and associative properties enable us to arrange *like* terms together.

Answers

15. $7w$ **16.** $-5a^3b^2$

17. $19z$ **18.** $3a + 6b + 9$

4. Simplifying Expressions

For expressions containing parentheses, it is necessary to apply the distributive property before combining *like* terms. This is demonstrated in Examples 8 and 9. Notice that when we apply the distributive property, the parentheses are dropped. This is often called *clearing parentheses*.

Example 8 Clearing Parentheses and Combining *Like* Terms

Simplify by clearing parentheses and combining *like* terms. $6 - 3(2y + 9)$

Solution:

$6 - 3(2y + 9)$ The order of operations indicates that we must perform multiplication before subtraction.

It is also important to understand that a factor of -3 (not 3) will be multiplied to all terms within the parentheses. To see why, we can rewrite the subtraction in terms of addition of the opposite.

$6 - 3(2y + 9) = 6 + (-3)(2y + 9)$	Rewrite subtraction as addition of the opposite.
$= 6 + (-3)(2y) + (-3)(9)$	Apply the distributive property.
$= 6 + (-6y) + (-27)$	Simplify.
$= -6y + 6 + (-27)$	Arrange *like* terms together.
$= -6y + (-21)$ or $-6y - 21$	Combine *like* terms.

Skill Practice

Simplify.

19. $8 - 6(w + 4)$

Avoiding Mistakes

Multiplication is performed before subtraction. It is incorrect to subtract $6 - 3$ first.

Example 9 Clearing Parentheses and Combining *Like* Terms

Simplify by clearing parentheses and combining *like* terms.

$$-8(x - 4) - 5(x + 7)$$

Solution:

$-8(x - 4) - 5(x + 7)$	
$= -8[x + (-4)] + (-5)(x + 7)$	Rewrite subtraction as addition of the opposite.
$= -8[x + (-4)] + (-5)(x + 7)$	Apply the distributive property.
$= -8(x) + (-8)(-4) + (-5)(x) + (-5)(7)$	
$= -8x + 32 - 5x - 35$	Simplify.
$= -8x - 5x + 32 - 35$	Arrange *like* terms together.
$= -13x - 3$	Combine *like* terms.

Skill Practice

Simplify.

20. $-5(10 - m) - 2(m + 1)$

Answers

19. $-6w - 16$ **20.** $3m - 52$

Section 3.1 Practice Exercises

Study Skills Exercises

1. After you get a test back, it is a good idea to correct the test so that you do not make the same errors again. One recommended approach is to use a clean sheet of paper and divide the paper down the middle vertically, as shown. For each problem that you missed on the test, rework the problem correctly on the left-hand side of the paper. Then write a written explanation on the right-hand side of the paper. Take the time this week to make corrections from your last test.

Perform the correct math here.	Explain the process here.
↓	↓
$2 + 4(5)$ $= 2 + 20$ $= 22$	Do multiplication before addition.

2. Define the key terms.

   a. **Coefficient** **b.** **Constant term** **c.** ***Like* terms**

   d. **Term** **e.** **Variable term**

Review Exercises

For Exercises 3–8, simplify.

3. $-4 - 2(9 - 3)$ **4.** $6 - 4^2 \div (-2)$ **5.** $|-9 + 3|$

6. $|-9| + |3|$ **7.** -12^2 **8.** $(-12)^2$

Objective 1: Identifying *Like* Terms

For Exercises 9–12, for each expression, list the terms and identify each term as a variable term or a constant term. **(See Example 1.)**

9. $2a + 5b^2 + 6$ **10.** $-5x - 4 + 7y$

11. $8 + 9a$ **12.** $12 - 8k$

For Exercises 13–16, identify the coefficients for each term. **(See Example 1.)**

13. $6p - 4q$ **14.** $-5a^3 - 2a$ **15.** $-h - 12$ **16.** $8x - 9$

For Exercises 17–24, determine if the two terms are *like* terms or unlike terms. If they are unlike terms, explain why. **(See Example 1.)**

17. $3a, -2a$ **18.** $8b, 12b$ **19.** $4xy, 4y$ **20.** $-9hk, -9h$

21. $14y^2, 14y$ **22.** $25x, 25x^2$ **23.** $17, 17y$ **24.** $-22t, -22$

Objective 2: Commutative, Associative, and Distributive Properties

For Exercises 25–32, apply the commutative property of addition or multiplication to rewrite each expression. **(See Example 2.)**

25. $5 + w$ **26.** $t + 2$ **27.** $r(2)$ **28.** $a(-4)$

29. $t(-s)$ **30.** $d(-c)$ **31.** $-p + 7$ **32.** $-q + 8$

For Exercises 33–40, apply the associative property of addition or multiplication to rewrite each expression. Then simplify the expression. **(See Example 3.)**

33. $3 + (8 + t)$ **34.** $7 + (5 + p)$ **35.** $-2(6b)$ **36.** $-3(2c)$

37. $3(6x)$ **38.** $9(5k)$ **39.** $-9 + (-12 + h)$ **40.** $-11 + (-4 + s)$

For Exercises 41–52, apply the distributive property. **(See Examples 4–5.)**

41. $4(x + 8)$ **42.** $5(3 + w)$ **43.** $4(a + 4b - c)$ **44.** $2(3q - r + s)$

45. $-2(p + 4)$ **46.** $-6(k + 2)$ **47.** $-(3x + 9 - 5y)$ **48.** $-(a - 8b + 4c)$

49. $-4(3 - n^2)$ **50.** $-2(13 - t^2)$ **51.** $-3(-5q - 2s - 3t)$ **52.** $-2(-10p - 12q + 3)$

For Exercises 53–60, apply the appropriate property to simplify the expression.

53. $6(2x)$ **54.** $-3(12k)$ **55.** $6(2 + x)$ **56.** $-3(12 + k)$

57. $-8 + (4 - p)$ **58.** $3 + (25 - m)$ **59.** $-8(4 - p)$ **60.** $-3(25 - m)$

Objective 3: Combining *Like* Terms

For Exercises 61–72, combine the *like* terms. **(See Examples 6–7.)**

61. $6r + 8r$ **62.** $4x + 21x$ **63.** $-4h + 12h - h$

64. $9p - 13p + p$ **65.** $4a^2b - 6a^2b$ **66.** $13xy^2 + 8xy^2$

67. $10x - 12y - 4x - 3y + 9$ **68.** $14a - 5b + 3a - b - 3$ **69.** $-8 - 6k - 9k + 12k + 4$

70. $-5 - 11p + 23p - p + 4$ **71.** $-8uv + 6u + 12uv$ **72.** $9pq - 9p + 13pq$

Objective 4: Simplifying Expressions

For Exercises 73–94, clear parentheses and combine *like* terms. **(See Examples 8–9.)**

73. $5(t - 6) + 2$ **74.** $7(a - 4) + 8$ **75.** $-3(2x + 1) - 13$

76. $-2(4b + 3) - 10$ **77.** $4 + 6(y - 3)$ **78.** $11 + 2(p - 8)$

79. $21 - 7(3 - q)$ **80.** $10 - 5(2 - 5m)$ **81.** $-3 - (2n + 1)$

82. $-13 - (6s + 5)$ **83.** $10(x + 5) - 3(2x + 9)$ **84.** $6(y - 9) - 5(2y - 5)$

85. $-(12z + 1) + 2(7z - 5)$ **86.** $-(8w + 5) + 3(w - 15)$ **87.** $3(w + 3) - (4w + y) - 3y$

88. $2(s + 6) - (8s - t) + 6t$ **89.** $20a - 4(b + 3a) - 5b$ **90.** $16p - 3(2p - q) + 7q$

91. $6 - (3m - n) - 2(m + 8) + 5n$ **92.** $12 - (5u + v) - 4(u - 6) + 2v$

93. $15 + 2(w - 4) - (2w - 5z^2) + 7z^2$ **94.** $7 + 3(2a - 5) - (6a - 8b^2) - 2b^2$

Mixed Exercises

95. Demonstrate the commutative property of addition by evaluating the expressions for $x = -3$ and $y = 9$.

a. $x + y$ **b.** $y + x$

96. Demonstrate the commutative property of addition by evaluating the expressions for $m = -12$ and $n = -5$.

a. $m + n$ **b.** $n + m$

97. Demonstrate the associative property of addition by evaluating the expressions for $x = -7$, $y = 2$, and $z = 10$.

a. $(x + y) + z$ **b.** $x + (y + z)$

98. Demonstrate the associative property of addition by evaluating the expressions for $a = -4$, $b = -6$, and $c = 18$.

a. $(a + b) + c$ **b.** $a + (b + c)$

99. Demonstrate the commutative property of multiplication by evaluating the expressions for $x = -9$ and $y = 5$.

a. $x \cdot y$ **b.** $y \cdot x$

100. Demonstrate the commutative property of multiplication by evaluating the expressions for $c = 12$ and $d = -4$.

a. $c \cdot d$ **b.** $d \cdot c$

101. Demonstrate the associative property of multiplication by evaluating the expressions for $x = -2$, $y = 6$, and $z = -3$.

a. $(x \cdot y) \cdot z$ **b.** $x \cdot (y \cdot z)$

102. Demonstrate the associative property of multiplication by evaluating the expressions for $b = -4$, $c = 2$, and $d = -5$.

a. $(b \cdot c) \cdot d$ **b.** $b \cdot (c \cdot d)$

Section 3.2 Addition and Subtraction Properties of Equality

Objectives

1. Definition of a Linear Equation in One Variable
2. Addition and Subtraction Properties of Equality

1. Definition of a Linear Equation in One Variable

An **equation** is a statement that indicates that two quantities are equal. The following are equations.

$$x = 7 \qquad z + 3 = 8 \qquad -6p = 18$$

All equations have an equal sign. Furthermore, notice that the equal sign separates the equation into two parts, the left-hand side and the right-hand side. A **solution to an equation** is a value of the variable that makes the equation a true statement. Substituting a solution to an equation for the variable makes the right-hand side equal to the left-hand side.

Equation	Solution	Check	
$x = 7$	7	$x = 7$ $7 \stackrel{?}{=} 7$ ✓	Substitute 7 for x. The right-hand side equals the left-hand side.
$z + 3 = 8$	5	$z + 3 = 8$ $5 + 3 \stackrel{?}{=} 8$ $8 \stackrel{?}{=} 8$ ✓	Substitute 5 for z. The right-hand side equals the left-hand side.
$-6p = 18$	-3	$-6p = 18$ $-6(-3) \stackrel{?}{=} 18$ $18 \stackrel{?}{=} 18$ ✓	Substitute -3 for p. The right-hand side equals the left-hand side.

Avoiding Mistakes

It is important to distinguish between an equation and an expression. An equation has an equal sign, whereas an expression does not. For example:

$2x + 4 = 16$ equation

$7x - 9$ expression

Concept Connections

Identify as an expression or an equation.

1. $-2w = 6$
2. $x + 8$

Example 1 Determining Whether a Number Is a Solution to an Equation

Determine whether the given number is a solution to the equation.

a. $2x - 9 = 3;\quad 6$ **b.** $20 = 8p - 4;\quad -2$

Solution:

a. $2x - 9 = 3$

$2(6) - 9 \stackrel{?}{=} 3$ Substitute 6 for x.

$12 - 9 \stackrel{?}{=} 3$ Simplify.

$3 \stackrel{?}{=} 3$ ✓ The right-hand side equals the left-hand side. Thus, 6 is a solution to the equation $2x - 9 = 3$.

b. $20 = 8p - 4$

$20 \stackrel{?}{=} 8(-2) - 4$ Substitute -2 for p.

$20 \stackrel{?}{=} -16 - 4$ Simplify.

$20 \neq -20$ The right-hand side does not equal the left-hand side. Thus, -2 is *not* a solution to the equation $20 = 8p - 4$.

Skill Practice

Determine whether the given number is a solution to the equation.

3. $2 + 3x = 23;\ 7$
4. $9 = -4x + 1;\ 2$

In the study of algebra, you will encounter a variety of equations. In this chapter, we will focus on a specific type of equation called a linear equation in one variable.

DEFINITION Linear Equation in One Variable

Let a and b be numbers such that $a \neq 0$. A **linear equation in one variable** is an equation that can be written in the form

$$ax + b = 0$$

Note: A linear equation in one variable contains only one variable and the exponent on the variable is 1.

2. Addition and Subtraction Properties of Equality

Given the equation $x = 3$, we can easily determine that the solution is 3. The solution to the equation $2x + 14 = 20$ is also 3. These two equations are called **equivalent equations** because they have the same solution. However, while the solution to $x = 3$ is obvious, the solution to $2x + 14 = 20$ is not. Our goal in this chapter is to learn how to *solve* equations.

Answers

1. Equation
2. Expression
3. Yes
4. No

To solve an equation we use algebraic principles to write an equation like $2x + 14 = 20$ in an equivalent but simpler form, such as $x = 3$. The addition and subtraction properties of equality are the first tools we will use to solve an equation.

PROPERTY Addition and Subtraction Properties of Equality

Let a, b, and c represent algebraic expressions.

1. The **addition property of equality:** If $a = b$, then, $a + c = b + c$
2. The **subtraction property of equality:** If $a = b$, then, $a - c = b - c$

The addition and subtraction properties of equality indicate that adding or subtracting the same quantity to each side of an equation results in an equivalent equation. This is true because if two equal quantities are increased (or decreased) by the same amount, then the resulting quantities will also be equal (Figure 3-1).

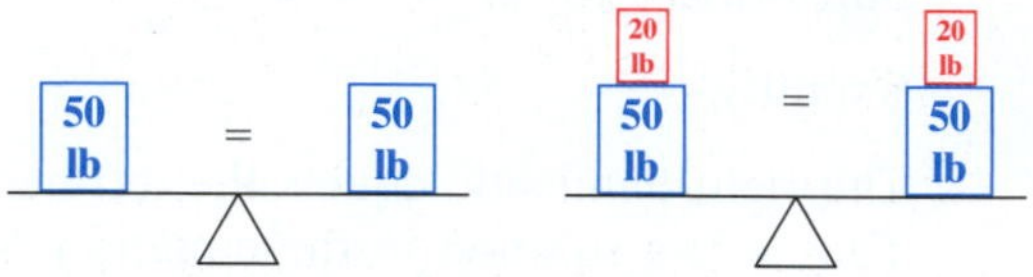

Figure 3-1

Skill Practice

Solve the equation and check the solution.

5. $y - 4 = 12$
6. $-18 = w + 5$
7. $-9 + z = 2$

Example 2 Applying the Addition Property of Equality

Solve the equations and check the solutions.

a. $x - 6 = 18$ **b.** $-12 = x - 7$ **c.** $-4 + y = 5$

Solution:

To solve an equation, the goal is to isolate the variable on one side of the equation. That is, we want to create an equivalent equation of the form $x = number$. To accomplish this, we can use the fact that the sum of a number and its opposite is zero.

a. $x - 6 = 18$

$x - 6 + 6 = 18 + 6$ To isolate x, add 6 to both sides, because $-6 + 6 = 0$.

$x + 0 = 24$ Simplify.

$x = 24$ The variable is isolated (by itself) on the left-hand side of the equation. The solution is 24.

Check: $x - 6 = 18$ Original equation

$(24) - 6 \stackrel{?}{=} 18$ Substitute 24 for x.

$18 \stackrel{?}{=} 18$ ✓ True

TIP: Notice that the variable may be isolated on *either* side of the equal sign. In Example 2(a), the variable appears on the left. In Example 2(b), the variable appears on the right.

b. $-12 = x - 7$

$-12 + 7 = x - 7 + 7$ To isolate x, add 7 to both sides, because $-7 + 7 = 0$.

$-5 = x + 0$ Simplify.

$-5 = x$ The variable is isolated on the right-hand side of the equation. The solution is -5.

The equation $-5 = x$ is equivalent to $x = -5$.

Answers

5. 16 6. −23 7. 11

Check:	$-12 = x - 7$	Original equation
	$-12 \stackrel{?}{=} (-5) - 7$	Substitute -5 for x.
	$-12 \stackrel{?}{=} -12$ ✓	True

Avoiding Mistakes

To check the solution to an equation, always substitute the solution into the *original* equation.

c. $-4 + y = 5$

$-4 + 4 + y = 5 + 4$	To isolate y, add 4 to both sides, because $-4 + 4 = 0$.
$0 + y = 9$	Simplify.
$y = 9$	The solution is 9.

Check:	$-4 + y = 5$	Original equation
	$-4 + (9) \stackrel{?}{=} 5$	Substitute 9 for y.
	$5 \stackrel{?}{=} 5$ ✓	True

In Example 3, we apply the subtraction property of equality. This indicates that we can subtract the same quantity from both sides of the equation to obtain an equivalent equation.

Example 3 Applying the Subtraction Property of Equality

Solve the equations and check.

a. $z + 11 = 14$ **b.** $-8 = 2 + q$

Solution:

a. $z + 11 = 14$

$z + 11 - 11 = 14 - 11$	To isolate z, subtract 11 from both sides, because $11 - 11 = 0$.
$z + 0 = 3$	Simplify.
$z = 3$	The solution is 3.

Check:	$z + 11 = 14$	Original equation
	$(3) + 11 \stackrel{?}{=} 14$	Substitute 3 for z.
	$14 \stackrel{?}{=} 14$ ✓	True

b. $-8 = 2 + q$

$-8 - 2 = 2 - 2 + q$	To isolate q, subtract 2 from both sides, because $2 - 2 = 0$.
$-10 = 0 + q$	Simplify.
$-10 = q$	The solution is -10.

Check:	$-8 = 2 + q$	Original equation
	$-8 \stackrel{?}{=} 2 + (-10)$	Substitute -10 for q.
	$-8 \stackrel{?}{=} -8$ ✓	True

Skill Practice

Solve the equation and check the solution.

8. $m + 8 = 21$

9. $-16 = 1 + z$

Answers

8. 13 **9.** -17

The equations in Example 4 require that we simplify the expressions on both sides of the equation, then apply the addition or subtraction property as needed. As you read through each example, remember that you want to isolate the variable.

Skill Practice

Solve the equations.

10. $15 + 6y - 5y = 4 - 9$

11. $-1 = 3(x + 4) - 2x$

Example 4 Applying the Addition and Subtraction Properties of Equality

Solve the equations.

a. $-8 + 3x - 2x = 12 - 9$ **b.** $-5 = 2(y + 1) - y$

Solution:

a. $-8 + 3x - 2x = 12 - 9$

$-8 + x = 3$ Combine like terms.

$-8 + 8 + x = 3 + 8$ To isolate x, add 8 to both sides, because $-8 + 8 = 0$.

$x = 11$ The solution is 11.

Check: $-8 + 3x - 2x = 12 - 9$

$-8 + 3(11) - 2(11) \stackrel{?}{=} 12 - 9$

$-8 + 33 - 22 \stackrel{?}{=} 3$

$25 - 22 \stackrel{?}{=} 3$

$3 \stackrel{?}{=} 3$ ✓ True

b. $-5 = 2(y + 1) - y$

$-5 = 2y + 2 - y$ Apply the distributive property to clear parentheses.

$-5 = y + 2$ Combine like terms.

$-5 - 2 = y + 2 - 2$ To isolate y, subtract 2 from both sides, because $2 - 2 = 0$.

$-7 = y$ The solution is -7.

Check: $-5 = 2(y + 1) - y$

$-5 \stackrel{?}{=} 2(-7 + 1) - (-7)$

$-5 \stackrel{?}{=} 2(-6) + 7$

$-5 \stackrel{?}{=} -12 + 7$

$-5 \stackrel{?}{=} -5$ ✓ True

Answers

10. -20 **11.** -13

Section 3.2 Practice Exercises

Boost *your* GRADE at ALEKS.com!

ALEKS version 3.0

- Practice Problems
- Self-Tests
- NetTutor
- e-Professors
- Videos

Study Skills Exercises

1. Does your school have a learning resource center or a tutoring center? If so, do you remember the location and hours of operation? Write them here.

Location of learning resource center or tutoring center: ________________

Hours of operation: ________________

2. Define the key terms.

a. Addition property of equality

b. Equation

c. Equivalent equations

d. Linear equation in one variable

e. Solution to an equation

f. Subtraction property of equality

Review Exercises

For Exercises 3–8, simplify the expression.

3. $-10a + 3b - 3a + 13b$

4. $4 - 23y^2 + 11 - 16y^2$

5. $-(-8h + 2k - 13)$

6. $3(-4m + 3) - 12$

7. $5z - 8(z - 3) - 20$

8. $-(7p - 12) - 10(1 - p) + 6$

Objective 1: Definition of a Linear Equation in One Variable

For Exercises 9–16, determine whether the given number is a solution to the equation. **(See Example 1.)**

9. $5x + 3 = -2;\quad -1$

10. $3y - 2 = 4;\quad 2$

11. $-z + 8 = 20;\quad 12$

12. $-7 - w = -10;\quad -3$

13. $13 = 13 + 6t;\quad 0$

14. $25 = 25 - 2x;\quad 0$

15. $15 = -2q + 9;\quad 3$

16. $39 = -7p + 4;\quad 5$

For Exercises 17–22, identify as an expression or an equation.

17. $8x - 9 = 7$

18. $24 - 5x = 12 + x$

19. $8x - 9 - 7$

20. $24 - 5x + 12 + x$

21. $2(x - 4) - x = 1$

22. $2(x - 4) - x + 1$

Objective 2: Addition and Subtraction Properties of Equality

For Exercises 23–28, fill in the blank with the appropriate number.

23. $13 + (-13) =$ ______

24. $6 +$ ______ $= 0$

25. ______ $+ (-7) = 0$

26. $1 + (-1) =$ ______

27. $0 = -3 +$ ______

28. $0 = 9 +$ ______

For Exercises 29–40, solve the equation using the addition property of equality. **(See Example 2.)**

29. $x - 23 = 14$

30. $y - 12 = 30$

31. $-4 + k = 12$

32. $-16 + m = 4$

33. $-18 = n - 3$

34. $-9 = t - 6$

35. $9 = -7 + t$

36. $-6 = -10 + z$

37. $k - 44 = -122$

38. $a - 465 = -206$

39. $13 = -21 + w$

40. $2 = -17 + p$

For Exercises 41–46, fill in the blank with the appropriate number.

41. $52 -$ ______ $= 0$

42. $2 - 2 =$ ______

43. $18 - 18 =$ ______

44. ______ $- 15 = 0$

45. ______ $- 100 = 0$

46. $21 -$ ______ $= 0$

For Exercises 47–58, solve the equation using the subtraction property of equality. **(See Example 3.)**

47. $x + 34 = 6$

48. $y + 12 = 4$

49. $17 + b = 20$

50. $5 + c = 14$

51. $-32 = t + 14$

52. $-23 = k + 11$

53. $82 = 21 + m$

54. $16 = 88 + n$

55. $z + 145 = 90$

56. $c + 80 = -15$

57. $52 = 10 + k$

58. $43 = 12 + p$

Mixed Exercises

For Exercises 59–78, solve the equation using the appropriate property. **(See Example 4.)**

59. $1 + p = 0$

60. $r - 12 = 13$

61. $-34 + t = -40$

62. $7 + q = 4$

63. $-11 = w - 23$

64. $-9 = p - 10$

65. $16 = x + 21$

66. $-4 = y + 18$

67. $5h - 4h + 4 = 3$

68. $10x - 9x - 11 = 15$

69. $-9 - 4x + 5x = 3 - 1$

70. $11 - 7x + 8x = -4 - 2$

71. $3(x + 2) - 2x = 5$

72. $4(x - 1) - 3x = 2$

73. $3(r - 2) - 2r = -2 + 6$

74. $4(k + 2) - 3k = -6 + 9$

75. $9 + (-2) = 4 + t$

76. $-13 + 15 = p + 5$

77. $-2 = 2(a - 15) - a$

78. $-1 = 6(t - 4) - 5t$

Multiplication and Division Properties of Equality

Section 3.3

Objectives

1. Multiplication and Division Properties of Equality
2. Comparing the Properties of Equality

1. Multiplication and Division Properties of Equality

Adding or subtracting the same quantity on both sides of an equation results in an equivalent equation. The same is true when we multiply or divide both sides of an equation by the same nonzero quantity.

PROPERTY Multiplication and Division Properties of Equality

Let a, b, and c represent algebraic expressions.

1. The **multiplication property of equality:** If $a = b$, then, $a \cdot c = b \cdot c$
2. The **division property of equality:** If $a = b$ then, $\frac{a}{c} = \frac{b}{c}$ (provided $c \neq 0$)

To understand the multiplication property of equality, suppose we start with a true equation such as $10 = 10$. If both sides of the equation are multiplied by a constant such as 3, the result is also a true statement (Figure 3-2).

$$10 = 10$$
$$3 \cdot 10 = 3 \cdot 10$$
$$30 = 30$$

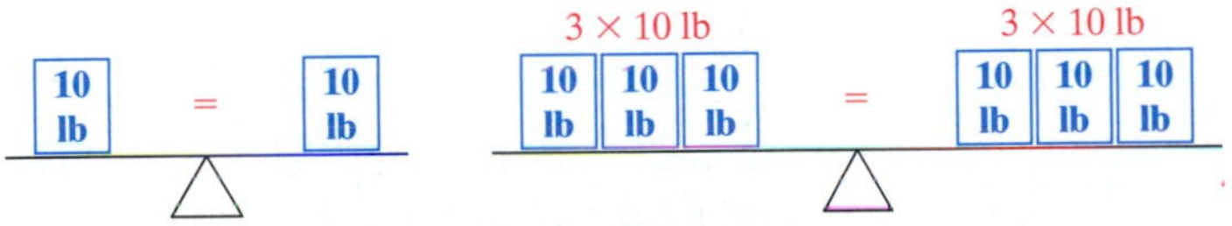

Figure 3-2

To solve an equation in the variable x, the goal is to write the equation in the form $x =$ number. In particular, notice that we want the coefficient of x to be 1. That is, we want to write the equation as $1 \cdot x =$ number. Therefore, to solve an equation such as $3x = 12$, we can *divide* both sides of the equation by the 3. We do this because $\frac{3}{3} = 1$, and that leaves $1x$ on the left-hand side of the equation.

$3x = 12$

$\frac{3x}{3} = \frac{12}{3}$ Divide both sides by 3.

$1 \cdot x = 4$ The coefficient of the x term is now 1.

$x = 4$ Simplify.

TIP: Recall that the quotient of a nonzero real number and itself is 1. For example:

$$\frac{3}{3} = 1 \quad \text{and} \quad \frac{-5}{-5} = 1$$

Skill Practice

Solve the equations.

1. $4x = 32$
2. $18 = -2w$
3. $19 = -m$

Example 1 Applying the Division Property of Equality

Solve the equations.

a. $10x = 50$ **b.** $28 = -4p$ **c.** $-y = 34$

Solution:

a. $10x = 50$

$\frac{10x}{10} = \frac{50}{10}$ To obtain a coefficient of 1 for the x term, divide both sides by 10, because $\frac{10}{10} = 1$.

$1x = 5$ Simplify.

$x = 5$ The solution is 5.

Check: $10x = 50$ Original equation

$10(5) \stackrel{?}{=} 50$ Substitute 5 for x.

$50 \stackrel{?}{=} 50$ ✓ True

b. $28 = -4p$

$\frac{28}{-4} = \frac{-4p}{-4}$ To obtain a coefficient of 1 for the p term, divide both sides by -4, because $\frac{-4}{-4} = 1$.

$-7 = 1p$ Simplify.

$-7 = p$ The solution is -7 and checks in the original equation.

Avoiding Mistakes

In Example 1(b), the operation between -4 and p is multiplication. We must divide by -4 (rather than 4) so that the resulting coefficient on p is positive 1.

c. $-y = 34$

$-1y = 34$ Note that $-y$ is the same as $-1 \cdot y$. To isolate y, we need a coefficient of *positive* 1.

$\frac{-1y}{-1} = \frac{34}{-1}$ To obtain a coefficient of 1 for the y term, divide both sides by -1.

$1y = -34$ Simplify.

$y = -34$ The solution is -34 and checks in the original equation.

TIP: In Example 1(c), we could have also multiplied both sides by -1 to obtain a coefficient of 1 for x.

$(-1)(-y) = (-1)34$

$y = -34$

The multiplication property of equality indicates that multiplying both sides of an equation by the same nonzero quantity results in an equivalent equation. For example, consider the equation $\frac{x}{2} = 6$. The variable x is being divided by 2. To solve for x, we need to reverse this process. Therefore, we will *multiply* by 2.

$\frac{x}{2} = 6$

$2 \cdot \frac{x}{2} = 2 \cdot 6$ Multiply both sides by 2.

$\frac{2}{2} \cdot x = 12$ The expression $2 \cdot \frac{x}{2}$ can be written as $\frac{2}{2} \cdot x$. This process is called *regrouping factors.*

$1 \cdot x = 12$ The x coefficient is 1, because $\frac{2}{2}$ equals 1.

$x = 12$ The solution is 12.

Answers

1. 8 2. -9 3. -19

Example 2 **Applying the Multiplication Property of Equality**

Solve the equations.

a. $\frac{x}{4} = -5$ **b.** $2 = \frac{t}{-8}$

Solution:

a. $\frac{x}{4} = -5$

$4 \cdot \frac{x}{4} = 4 \cdot (-5)$ To isolate x, multiply both sides by 4.

$\frac{4}{4} \cdot x = -20$ Regroup factors.

$1x = -20$ The x coefficient is now 1, because $\frac{4}{4}$ equals 1.

$x = -20$ The solution is -20. Check: $\frac{x}{4} = -5$

$\frac{(-20)}{4} \stackrel{?}{=} -5$

$-5 \stackrel{?}{=} -5$ ✓ True

b. $2 = \frac{t}{-8}$

$-8 \cdot (2) = -8 \cdot \frac{t}{-8}$ To isolate t, multiply both sides by -8.

$-16 = \frac{-8}{-8} \cdot t$ Regroup factors.

$-16 = 1 \cdot t$ The coefficient on t is now 1.

$-16 = t$ The solution is -16 and checks in the original equation.

Skill Practice

Solve the equations.

4. $\frac{y}{6} = -3$

5. $5 = \frac{w}{-10}$

2. Comparing the Properties of Equality

It is important to determine which property of equality should be used to solve an equation. For example, compare equations:

$$4 + x = 12 \quad \text{and} \quad 4x = 12$$

In the first equation, the operation between 4 and x is addition. Therefore, we want to reverse the process by *subtracting* 4 from both sides. In the second equation, the operation between 4 and x is multiplication. To isolate x, we reverse the process by *dividing* by 4.

$4 + x = 12$	$4x = 12$
$4 - 4 + x = 12 - 4$	$\frac{4x}{4} = \frac{12}{4}$
$x = 8$	$x = 3$

In Example 3, we practice distinguishing which property of equality to use.

Answers

4. -18 **5.** -50

Skill Practice

Solve the equations.

6. $\frac{t}{5} = -8$
7. $x + 46 = 12$
8. $10 = 5p - 7p$
9. $-5 + 20 = -3 + 6w - 5(w + 1)$

Example 3 Solving Linear Equations

Solve the equations.

a. $\frac{m}{12} = -3$ **b.** $x + 18 = 2$

c. $20 + 8 = -10t + 3t$ **d.** $-4 + 10 = 4t - 3(t + 2) - 1$

Solution:

a. $\frac{m}{12} = -3$ The operation between m and 12 is division. To obtain a coefficient of 1 for the m term, *multiply* both sides by 12.

$12 \cdot \frac{m}{12} = 12(-3)$ Multiply both sides by 12.

$\frac{12}{12} \cdot m = 12(-3)$ Regroup.

$m = -36$ Simplify both sides. The solution -36 checks in the original equation.

b. $x + 18 = 2$ The operation between x and 18 is addition. To isolate x, *subtract* 18 from both sides.

$x + 18 - 18 = 2 - 18$ Subtract 18 on both sides.

$x = -16$ Simplify. The solution is -16 and checks in the original equation.

c. $20 + 8 = -10t + 3t$ Begin by simplifying both sides of the equation.

$28 = -7t$ The relationship between t and -7 is multiplication. To obtain a coefficient of 1 on the t-term, we *divide* both sides by -7.

$\frac{28}{-7} = \frac{-7t}{-7}$ Divide both sides by -7.

$-4 = t$ Simplify. The solution is -4 and checks in the original equation.

d. $-4 + 10 = 4t - 3(t + 2) - 1$ Begin by simplifying both sides of the equation.

$6 = 4t - 3t - 6 - 1$ Apply the distributive property on the right-hand side.

$6 = t - 7$ Combine *like* terms on the right-hand side.

$6 + 7 = -7 + 7 + t$ To isolate the t term, *add* 7 to both sides. This is because $-7 + 7 = 0$.

$13 = t$ The solution is 13 and checks in the original equation.

Answers

6. -40 7. -34
8. -5 9. 23

Section 3.3 Practice Exercises

Study Skills Exercises

1. One way to know that you really understand a concept is to explain it to someone else. In your own words, explain the circumstances in which you would apply the multiplication property of equality or the division property of equality.

2. Define the key terms. **a. Division property of equality** **b. Multiplication property of equality**

Review Exercises

For Exercises 3–6, simplify the expression.

3. $-3x + 5y + 9x - y$ **4.** $-4ab - 2b - 3ab + b$ **5.** $3(m - 2n) - (m + 4n)$ **6.** $2(5w - z) - (3w + 4z)$

For Exercises 7–10, solve the equation.

7. $p - 12 = 33$ **8.** $-8 = 10 + k$ **9.** $24 + z = -12$ **10.** $-4 + w = 22$

Objective 1: Multiplication and Division Properties of Equality

For Exercises 11–34, solve the equation using the multiplication or division properties of equality. **(See Examples 1–2.)**

11. $14b = 42$ **12.** $6p = 12$ **13.** $-8k = 56$ **14.** $-5y = 25$

15. $-16 = -8z$ **16.** $-120 = -10p$ **17.** $-t = -13$ **18.** $-h = -17$

19. $5 = -x$ **20.** $30 = -a$ **21.** $\frac{b}{7} = -3$ **22.** $\frac{a}{4} = -12$

23. $\frac{u}{-2} = -15$ **24.** $\frac{v}{-10} = -4$ **25.** $-4 = \frac{p}{7}$ **26.** $-9 = \frac{w}{6}$

27. $-28 = -7t$ **28.** $-33 = -3r$ **29.** $5m = 0$ **30.** $6q = 0$

31. $\frac{x}{7} = 0$ **32.** $\frac{t}{8} = 0$ **33.** $-6 = -z$ **34.** $-9 = -n$

Objective 2: Comparing the Properties of Equality

35. In your own words, explain how to determine when to use the addition property of equality.

36. In your own words, explain how to determine when to use the subtraction property of equality.

37. Explain how to determine when to use the division property of equality.

38. Explain how to determine when to use the multiplication property of equality.

Mixed Exercises

For Exercises 39–74, solve the equation. **(See Example 3.)**

39. $4 + x = -12$ **40.** $6 + z = -18$ **41.** $4y = -12$ **42.** $6p = -18$

43. $q - 4 = -12$
44. $p - 6 = -18$
45. $\frac{h}{4} = -12$
46. $\frac{w}{6} = -18$

47. $-18 = -9a$
48. $-40 = -8x$
49. $7 = r - 23$
50. $11 = s - 4$

51. $5 = \frac{y}{-3}$
52. $1 = \frac{h}{-5}$
53. $-52 = 5 + y$
54. $-47 = 12 + z$

55. $-4a = 0$
56. $-7b = 0$
57. $100 = 5k$
58. $95 = 19h$

59. $31 = -p$
60. $11 = -q$
61. $-3x + 7 + 4x = 12$
62. $6x + 7 - 5x = 10$

63. $5(x - 2) - 4x = 3$
64. $3(y - 6) - 2y = 8$

65. $3p + 4p = 25 - 4$
66. $2q + 3q = 54 - 9$

67. $-5 + 7 = 5x - 4(x - 1)$
68. $-3 + 11 = -2z + 3(z - 2)$

69. $-10 - 4 = 6m - 5(3 + m)$
70. $-15 - 5 = 9n - 8(2 + n)$

71. $5x - 2x = -15$
72. $13y - 10y = -18$

73. $-2(a + 3) - 6a + 6 = 8$
74. $-(b - 11) - 3b - 11 = -16$

Section 3.4 Solving Equations with Multiple Steps

Objectives

1. Solving Equations with Multiple Steps
2. General Procedure to Solve a Linear Equation

1. Solving Equations with Multiple Steps

In Sections 3.2 and 3.3 we studied a one-step process to solve linear equations. We used the addition, subtraction, multiplication, and division properties of equality. In this section, we combine these properties to solve equations that require multiple steps. This is shown in Example 1.

Skill Practice

Solve.

1. $3x + 7 = 25$

Example 1 Solving a Linear Equation

Solve. $2x - 3 = 15$

Solution:

Remember that our goal is to isolate x. Therefore, in this equation, we will first isolate the *term* containing x. This can be done by adding 3 to both sides.

$2x - 3 + 3 = 15 + 3$	Add 3 to both sides, because $-3 + 3 = 0$.
$2x = 18$	The term containing x is now isolated (by itself). The resulting equation now requires only one step to solve.
$\frac{2x}{2} = \frac{18}{2}$	Divide both sides by 2 to make the x coefficient equal to 1.
$x = 9$	Simplify. The solution is 9.

Check:	$2x - 3 = 15$	Original equation
	$2(9) - 3 \stackrel{?}{=} 15$	Substitute 9 for x.
	$18 - 3 \stackrel{?}{=} 15$ ✓	True

Answer

1. 6

As Example 1 shows, we will generally apply the addition (or subtraction) property of equality to isolate the variable term first. Then we will apply the multiplication (or division) property of equality to obtain a coefficient of 1 on the variable term.

Example 2 Solving a Linear Equation

Solve. $22 = -3c + 10$

Solution:

We first isolate the term containing the variable by subtracting 10 from both sides.

$22 - 10 = -3c + 10 - 10$ Subtract 10 from both sides because $10 - 10 = 0$.

$12 = -3c$ The term containing c is now isolated.

$\dfrac{12}{-3} = \dfrac{-3c}{-3}$ Divide both sides by -3 to make the c coefficient equal to 1.

$-4 = c$ Simplify. The solution is -4.

Check: $22 = -3c + 10$ Original equation

$22 = -3(-4) + 10$ Substitute -4 for c.

$22 = 12 + 10$ ✓ True

Skill Practice

Solve.

2. $-12 = -5t + 13$

Example 3 Solving a Linear Equation

Solve. $14 = \dfrac{y}{2} + 8$

Solution:

$14 = \dfrac{y}{2} + 8$

$14 - 8 = \dfrac{y}{2} + 8 - 8$ Subtract 8 from both sides. This will isolate the term containing the variable, y.

$6 = \dfrac{y}{2}$ Simplify.

$2 \cdot 6 = 2 \cdot \dfrac{y}{2}$ Multiply both sides by 2 to make the y coefficient equal to 1.

$12 = y$ Simplify. The solution is 12.

Check: $14 = \dfrac{y}{2} + 8$

$14 \stackrel{?}{=} \dfrac{(12)}{2} + 8$ Substitute 12 for y.

$14 \stackrel{?}{=} 6 + 8$ ✓ True

Skill Practice

Solve.

3. $-3 = \dfrac{x}{3} + 9$

Answers

2. 5 **3.** -36

In Example 4, the variable x appears on both sides of the equation. In this case, apply the addition or subtraction properties of equality to collect the variable terms on one side of the equation and the constant terms on the other side.

Skill Practice

Solve.

4. $8y - 3 = 6y + 17$

Example 4 Solving a Linear Equation with Variables on Both Sides

Solve. $4x + 5 = -2x - 13$

Solution:

To isolate x, we must first "move" all x terms to one side of the equation. For example, suppose we add $2x$ to both sides. This would "remove" the x term from the right-hand side because $-2x + 2x = 0$. The term $2x$ is then combined with $4x$ on the left-hand side.

$$4x + 5 = -2x - 13$$

$$4x + 2x + 5 = -2x + 2x - 13$$ Add $2x$ to both sides.

$$6x + 5 = -13$$ Simplify. Next, we want to isolate the *term* containing x.

$$6x + 5 - 5 = -13 - 5$$ Subtract 5 from both sides to isolate the x term.

$$6x = -18$$ Simplify.

$$\frac{6x}{6} = \frac{-18}{6}$$ Divide both sides by 6 to obtain an x coefficient of 1.

$$x = -3$$ The solution is -3 and checks in the original equation.

TIP: It should be noted that the variable may be isolated on either side of the equation. In Example 4 for instance, we could have isolated the x term on the right-hand side of the equation.

$$4x + 5 = -2x - 13$$

$$4x - 4x + 5 = -2x - 4x - 13$$ Subtract $4x$ from both sides. This "removes" the x term from the left-hand side.

$$5 = -6x - 13$$

$$5 + 13 = -6x - 13 + 13$$ Add 13 to both sides to isolate the x term.

$$18 = -6x$$ Simplify.

$$\frac{18}{-6} = \frac{-6x}{-6}$$ Divide both sides by -6.

$$-3 = x$$ This is the same solution as in Example 4.

Answer

4. 10

2. General Procedure to Solve a Linear Equation

In Examples 1–4, we used multiple steps to solve equations. We also learned how to collect the variable terms on one side of the equation so that the variable could be isolated. The following procedure summarizes the steps to solve a linear equation.

PROCEDURE Solving a Linear Equation in One Variable

Step 1 Simplify both sides of the equation.
- Clear parentheses if necessary.
- Combine *like* terms if necessary.

Step 2 Use the addition or subtraction property of equality to collect the variable terms on one side of the equation.

Step 3 Use the addition or subtraction property of equality to collect the constant terms on the *other* side of the equation.

Step 4 Use the multiplication or division property of equality to make the coefficient of the variable term equal to 1.

Step 5 Check the answer in the original equation.

Example 5 Solving a Linear Equation

Solve. $2(y + 10) = 8 - 4y$

Solution:

$2(y + 10) = 8 - 4y$

$2y + 20 = 8 - 4y$ — **Step 1:** Simplify both sides of the equation. Clear parentheses.

$2y + 4y + 20 = 8 - 4y + 4y$ — **Step 2:** Add $4y$ to both sides to collect the variable terms on the left.

$6y + 20 = 8$ — Simplify.

$6y + 20 - 20 = 8 - 20$ — **Step 3:** Subtract 20 from both sides to collect the constants on the right.

$6y = -12$ — Simplify.

$\dfrac{6y}{6} = \dfrac{-12}{6}$ — **Step 4:** Divide both sides by 6 to obtain a coefficient of 1 on the y term

$y = -2$ — The solution is -2.

Check: $2(y + 10) = 8 - 4y$ — **Step 5:** Check the solution in the original equation.

$2(-2 + 10) \stackrel{?}{=} 8 - 4(-2)$ — Substitute -2 for y.

$2(8) \stackrel{?}{=} 8 - (-8)$

$16 \stackrel{?}{=} 16$ ✓ — The solution checks.

Skill Practice

Solve.

5. $6(z + 4) = -16 - 4z$

Answer

5. -4

Skill Practice

Solve.

6. $-3y - y - 4 = -5(y - 8)$

Example 6 Solving a Linear Equation

Solve. $2x + 3x + 2 = -4(3 - x)$

Solution:

$2x + 3x + 2 = -4(3 - x)$

$5x + 2 = -12 + 4x$ — **Step 1:** Simplify both sides of the equation. On the left, combine *like* terms. On the right, clear parentheses.

$5x - 4x + 2 = -12 + 4x - 4x$ — **Step 2:** Subtract $4x$ from both sides to collect the variable terms on the left.

$x + 2 = -12$ — Simplify.

$x + 2 - 2 = -12 - 2$ — **Step 3:** Subtract 2 from both sides to collect the constants on the right.

$x = -14$ — **Step 4:** The x coefficient is already 1. The solution is -14.

Check: $2x + 3x + 2 = -4(3 - x)$ — **Step 5:** Check in the original equation.

$2(-14) + 3(-14) + 2 \stackrel{?}{=} -4[3 - (-14)]$ — Substitute -14 for x.

$-28 - 42 + 2 \stackrel{?}{=} -4(17)$

$-70 + 2 \stackrel{?}{=} -68$

$-68 \stackrel{?}{=} -68$ ✓ — The solution checks.

Answer

6. 44

Section 3.4 Practice Exercises

Study Skills Exercise

1. A good way to determine what will be on a test is to look at both your notes and the exercises assigned by your instructor. List five types of problems that you think will be on the test for this chapter.

Review Exercises

For Exercises 2–10, solve the equation.

2. $4c = 12$

3. $\frac{t}{3} = -4$

4. $9 + (-11) = 4x - 3x + 5$

5. $-2x + 3x - 6 = -8 + 3$

6. $4 - 5(y + 3) + 6y = 7$

7. $-8 = 3 - 2(z - 4) + 3z$

8. $8 + h = 0$

9. $-8h = 0$

10. $9p - 6p = -8 + 11$

Objective 1: Solving Equations with Multiple Steps

For Exercises 11–26, solve the equation. **(See Examples 1–3.)**

11. $3m - 2 = 16$

12. $2n - 5 = 3$

13. $8c - 12 = 36$

14. $5t - 1 = -11$

15. $1 = -4z + 21$

16. $-4 = -3p + 14$

17. $9 = 12x - 15$

18. $7 = 5y - 8$

19. $\frac{b}{3} - 12 = -9$

20. $\frac{c}{5} + 2 = 4$

21. $-9 = \frac{w}{2} - 3$

22. $-16 = \frac{t}{4} - 14$

23. $\frac{m}{-3} + 40 = 60$

24. $\frac{c}{-4} - 3 = 5$

25. $-y - 7 = 14$

26. $-p + 8 = 20$

For Exercises 27–38, solve the equation. **(See Example 4.)**

27. $8 + 4b = 2 + 2b$

28. $2w + 10 = 5w - 5$

29. $7 - 5t = 3t - 9$

30. $4 - 2p = -3 + 5p$

31. $4 - 3d = 5d - 4$

32. $-3k + 14 = -4 + 3k$

33. $12p = 3p + 36$

34. $2x + 10 = 4x$

35. $4 + 2a - 7 = 3a + a + 3$

36. $4b + 2b - 7 = 2 + 4b + 5$

37. $-8w + 8 + 3w = 2 - 6w + 2$

38. $-12 + 5m + 10 = -2m - 10 - m$

Objective 2: General Procedure to Solve a Linear Equation

For Exercises 39–56, solve the equation. **(See Examples 5–6.)**

39. $5(z + 7) = 9 + 3z$

40. $6(w + 2) = 20 + 2w$

41. $2(1 - m) = 5 - 3m$

42. $3(2 - g) = 12 - g$

43. $3n - 4(n - 1) = 16$

44. $4p - 3(p + 2) = 18$

45. $4x + 2x - 9 = -3(5 - x)$

46. $-3x + x - 8 = -2(6 - x)$

47. $-4 + 2x + 1 = 3(x - 1)$

48. $-2 + 5x + 8 = 6(x + 2) - 6$

49. $9q - 5(q - 3) = 5q$

50. $6h - 2(h + 6) = 10h$

51. $-4(k - 2) + 14 = 3k - 20$

52. $-3(x + 4) - 9 = -2x + 12$

53. $3z + 9 = 3(5z - 1)$

54. $4y + 4 = 8(y - 2)$

55. $6(u - 1) + 5u + 1 = 5(u + 6) - u$

56. $2(2v + 3) + 8v = 6(v - 1) + 3v$

Problem Recognition Exercises

Comparing Expressions and Equations

For Exercises 1–6, identify the problem as an expression or as an equation.

1. $-5 + 4x - 6x = 7$

2. $-8(4 - 7x) + 4$

3. $4 - 6(2x - 3) + 1$

4. $10 - x = 2x + 19$

5. $6 - 3(x + 4) = 6$

6. $9 - 6(x + 1)$

For Exercises 7–30, solve the equation or simplify the expression.

7. $5t = 20$

8. $6x = 36$

9. $5t - 20t$

10. $6x - 36x$

11. $5(w - 3)$

12. $6(x - 2)$

13. $5(w - 3) = 20$ **14.** $6(x - 2) = 36$ **15.** $5 + t = 20$

16. $6 + x = 36$ **17.** $5 + t + 20$ **18.** $6 + x + 36$

19. $5 + 3p - 2 = 0$ **20.** $16 - 2k + 2 = 0$ **21.** $23u + 2 = -12u + 72$

22. $75w - 27 = 14w + 156$ **23.** $23u + 2 - 12u + 72$ **24.** $75w - 27 - 14w + 156$

25. $-2(x - 3) + 14 + 10 - (x + 4)$ **26.** $26 - (3x + 12) + 11 - 4(x + 1)$ **27.** $-2(x - 3) + 14 = 10 - (x + 4)$

28. $26 - (3x + 12) = 11 - 4(x + 1)$ **29.** $2 - 3(y + 1) = -4y + 7$ **30.** $5(t + 5) - 3t = t - 9$

Section 3.5 Applications and Problem Solving

Objectives

1. Problem-Solving Flowchart
2. Translating Verbal Statements into Equations
3. Applications of Linear Equations

1. Problem-Solving Flowchart

Linear equations can be used to solve many real-world applications. In Section 1.8, we introduced guidelines for problem solving. In this section, we expand on these guidelines to write equations to solve applications. Consider the problem-solving flowchart.

Problem-Solving Flowchart for Word Problems

Step 1	Read the problem completely.	• Familiarize yourself with the problem. Estimate the answer if possible.
Step 2	Assign labels to unknown quantities.	• Identify the unknown quantity or quantities. Let a variable represent one of the unknowns. Draw a picture and write down relevant formulas.
Step 3	Write an equation in words.	• Verbalize what quantities must be equal.
Step 4	Write a mathematical equation.	• Replace the verbal equation with a mathematical equation using x or some other variable.
Step 5	Solve the equation.	• Solve for the variable using the steps for solving linear equations.
Step 6	Interpret the results and write the final answer in words.	• Once you have obtained a numerical value for the variable, recall what it represents in the context of the problem. Can this value be used to determine other unknowns in the problem? Write an answer to the word problem *in words*.

Avoiding Mistakes

It is always a good idea to check your answer in the context of the problem. This will help you determine if your answer is reasonable.

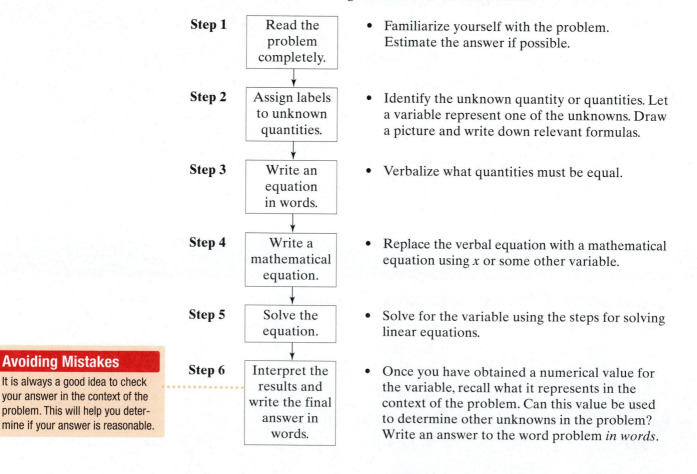

2. Translating Verbal Statements into Equations

We begin solving word problems with practice translating between an English sentence and an algebraic equation. First, spend a minute to recall some of the key words that represent addition, subtraction, multiplication, and division. See Table 3-4.

Table 3-4

Addition: $a + b$	**Subtraction:** $a - b$
The *sum* of a and b a plus b b added to a b more than a a increased by b The total of a and b	The *difference* of a and b a minus b b subtracted from a a decreased by b b less than a
Multiplication: $a \cdot b$	**Division:** $a \div b$
The *product* of a and b a times b a multiplied by b	The *quotient* of a and b a divided by b b divided into a The ratio of a and b a over b a per b

Example 1 Translating Sentences to Mathematical Equations

A number decreased by 7 is 12. Find the number.

Solution:

Step 1: Read the problem completely.

Let x represent the number. **Step 2:** Label the variable.

A number decreased by 7 is 12. **Step 3:** Write the equation in words

$x \quad - \quad 7 = 12$ **Step 4:** Translate to a mathematical equation.

$x - 7 = 12$ **Step 5:** Solve the equation.

$x - 7 + 7 = 12 + 7$ Add 7 to both sides.

$x = 19$

The number is 19. **Step 6:** Interpret the answer in words.

Skill Practice

1. A number minus 6 is -22. Find the number.

Avoiding Mistakes

To check the answer to Example 1, we see that 19 decreased by 7 is 12.

Answer

1. The number is -16.

Skill Practice

2. 4 subtracted from 8 times a number is 36. Find the number.

Example 2 Translating Sentences to Mathematical Equations

Seven less than 3 times a number results in 11. Find the number.

Solution:

	Step 1: Read the problem completely.
Let x represent the number.	**Step 2:** Label the variable.
Seven less than 3 times a number results in 11.	**Step 3:** Write the equation in words
$\underbrace{3x}_{\text{three times a number}} \overbrace{- 7}^{\text{7 less than}} \overbrace{= \quad 11}^{\text{results in 11}}$	**Step 4:** Translate to a mathematical equation.
$3x - 7 + 7 = 11 + 7$	**Step 5:** Solve the equation. Add 7 to both sides.
$3x = 18$	Simplify.
$\dfrac{3x}{3} = \dfrac{18}{3}$	Divide both sides by 3.
$x = 6$	
The number is 6.	**Step 6:** Interpret the answer in words.

Avoiding Mistakes

To check the answer to Example 2, we have:

The value 3 times 6 is 18, and 7 less than 18 is 11.

Skill Practice

3. 8 added to twice a number is 22. Find the number.

Example 3 Translating Sentences to Mathematical Equations

Two times the sum of a number and 8 equals 38.

Solution:

	Step 1: Read the problem completely.
Let x represent the number.	**Step 2:** Label the variable.
Two times the sum of a number and 8 equals 38.	**Step 3:** Write the equation in words.
$\overbrace{2 \cdot}^{\text{two times}} \underbrace{(x + 8)}_{\text{the sum of a number and 8}} \overbrace{= 38}^{\text{equals 38}}$	**Step 4:** Translate to a mathematical equation.
$2(x + 8) = 38$	**Step 5:** Solve the equation.
$2x + 16 = 38$	Clear parentheses.
$2x + 16 - 16 = 38 - 16$	Subtract 16 from both sides.
$2x = 22$	Simplify.
$\dfrac{2x}{2} = \dfrac{22}{2}$	Divide both sides by 2.
$x = 11$	
The number is 11.	**Step 6:** Interpret the answer in words.

Avoiding Mistakes

The sum $(x + 8)$ must be enclosed in parentheses so that the entire sum is multiplied by 2.

Avoiding Mistakes

To check the answer to Example 3, we have:

The sum of 11 and 8 is 19. Twice this amount gives 38 as expected.

Answers

2. The number is 5.
3. The number is 7.

3. Applications of Linear Equations

In Example 4, we practice representing quantities within a word problem by using variables.

Example 4 Representing Quantities Algebraically

a. Kathleen works twice as many hours in one week as Kevin. If Kevin works for h hours, write an expression representing the number of hours that Kathleen works.

b. At a carnival, rides cost \$3 each. If Alicia takes n rides during the day, write an expression for the total cost.

c. Josie made \$430 less during one week than her friend Annie made. If Annie made D dollars, write an expression for the amount that Josie made.

Solution:

a. Let h represent the number of hours that Kevin works.

Kathleen works twice as many hours as Kevin.

$2h$ is the number of hours that Kathleen works.

b. Let n represent the number of rides Alicia takes during the day.

Rides cost \$3 each.

$3n$ is the total cost.

c. Let D represent the amount of money that Annie made during the week.

Josie made \$430 less than Annie.

$D - 430$ represents the amount that Josie made.

In Examples 5 and 6, we practice solving application problems using linear equations.

Skill Practice

4. Tasha ate three times as many M&Ms as her friend Kate. If Kate ate x M&Ms, write an expression for the number that Tasha ate.
5. Casey bought seven books from a sale rack. If the books cost d dollars each, write an expression for the total cost.
6. One week Kim worked 8 hr more than her friend Tom. If Tom worked x hours, write an expression for the number of hours that Kim worked.

Answers

4. $3x$ **5.** $7d$ **6.** $x + 8$

Skill Practice

7. A piece of cable 92 ft long is to be cut into two pieces. One piece must be three times longer than the other. How long should each piece be?

Example 5 Applying a Linear Equation to Carpentry

A carpenter must cut a 10-ft board into two pieces to build a brace for a picnic table. If one piece is to be four times longer than the other piece, how long should each piece be?

Solution:

Step 1: Read the problem completely.

We can let x represent the length of either piece. However, if we choose x to be the length of the shorter piece, then the longer piece has to be $4x$ (4 times as long).

Step 2: Label the variables. Draw a picture.

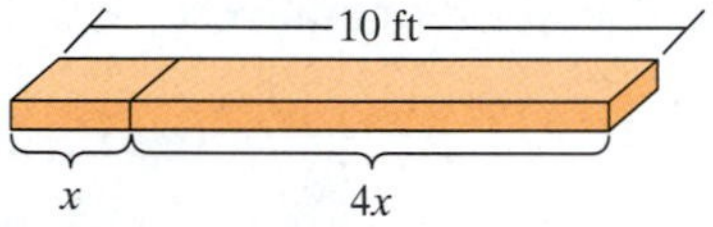

Let x = the length of the shorter piece.

Then $4x$ = the length of the longer piece.

$$\begin{pmatrix}\text{length of}\\ \text{one piece}\end{pmatrix} + \begin{pmatrix}\text{length of the}\\ \text{other piece}\end{pmatrix} = \begin{pmatrix}\text{total}\\ \text{length}\end{pmatrix}$$

Step 3: Write an equation in words.

$$x \quad + \quad 4x \quad = \quad 10$$

Step 4: Write a mathematical equation.

$$x + 4x = 10$$

Step 5: Solve the equation.

$$5x = 10$$

Combine *like* terms.

$$\frac{5x}{5} = \frac{10}{5}$$

Divide both sides by 5.

$$x = 2$$

Step 6: Interpret the results in words.

Recall that x represents the length of the shorter piece. Therefore, the shorter piece is 2 ft. The longer piece is given by $4x$ or $4(2 \text{ ft}) = 8$ ft.

The pieces are 2 ft and 8 ft.

Avoiding Mistakes

In Example 5, the two pieces should total 10 ft. We have,

2 ft + 8 ft = 10 ft as desired.

Skill Practice

8. A kit of cordless 18-volt tools made by Craftsman cost \$310 less than a similar kit made by DeWalt. The combined cost for both models is \$690. Find the cost for each model. (*Source: Consumer Reports*)

Example 6 Applying a Linear Equation

One model of a Panasonic high-definition plasma TV sold for \$1500 more than a certain model made by Planar. The combined cost for these two models is \$8500. Find the cost for each model.

(*Source: Consumer Reports*)

Answers

7. One piece should be 23 ft and the other should be 69 ft long.
8. The Craftsman model costs \$190 and the DeWalt model costs \$500.

Solution:

	Step 1: Read the problem completely.
Let x represent the cost of the Planar TV. Then $x + 1500$ represents the cost of the Panasonic.	**Step 2:** Label the variables
$\begin{pmatrix}\text{cost of}\\ \text{Planar}\end{pmatrix} + \begin{pmatrix}\text{cost of}\\ \text{Panasonic}\end{pmatrix} = \begin{pmatrix}\text{total}\\ \text{cost}\end{pmatrix}$	**Step 3:** Write an equation in words.
$x \quad + \quad x + 1500 \quad = \quad 8500$	**Step 4:** Write a mathematical equation.
$x + x + 1500 = 8500$	**Step 5:** Solve the equation.
$2x + 1500 = 8500$	Combine *like* terms.
$2x + 1500 - 1500 = 8500 - 1500$	Subtract 1500 from both sides.
$2x = 7000$	
$\dfrac{2x}{2} = \dfrac{7000}{2}$	Divide both sides by 2.
$x = 3500$	
	Step 6: Interpret the results in words.

Since $x = 3500$, the Planar model TV costs \$3500.

The Panasonic model is represented by $x + 1500 = \$3500 + \$1500 = \$5000$.

TIP: In Example 6, we could have let x represent *either* the cost of the Planar model TV or the Panasonic model.

Suppose we had let x represent the cost of the Panasonic model.
Then $x - 1500$ is the cost of the Planar model (the Planar model is *less* expensive).

$$\begin{pmatrix}\text{cost of}\\ \text{Planar}\end{pmatrix} + \begin{pmatrix}\text{cost of}\\ \text{Panasonic}\end{pmatrix} = \begin{pmatrix}\text{total}\\ \text{cost}\end{pmatrix}$$

$$x - 1500 + \quad x \quad = 8500$$

$$2x - 1500 = 8500$$

$$2x - 1500 + 1500 = 8500 + 1500$$

$$2x = 10{,}000$$

$$x = 5000$$

Therefore, the Panasonic model costs \$5000 as expected.
The Planar model costs $x - 1500$ or $\$5000 - \$1500 = \$3500$.

Section 3.5 Practice Exercises

Study Skills Exercise

1. In solving an application it is very important first to read and understand what is being asked in the problem. One way to do this is to read the problem several times. Another is to read it out loud so you can hear yourself. Another is to rewrite the problem in your own words. Which of these methods do you think will help you in understanding an application?

Review Exercises

2. Use substitution to determine if -4 is a solution to the equation $-3x + 9 = 21$.

For Exercises 3–8, solve the equation.

3. $3t - 15 = -24$

4. $-6x + 4 = 16$

5. $\frac{b}{5} - 5 = -14$

6. $\frac{w}{8} - 3 = 3$

7. $2x + 22 = 6x - 2$

8. $-5y - 34 = -3y + 12$

Objective 2: Translating Verbal Statements into Equations

For Exercises 9–40,

a. write an equation that represents the given statement.

b. solve the problem. **(See Examples 1–3.)**

9. The sum of a number and 6 is 19. Find the number.

10. The sum of 12 and a number is 49. Find the number.

11. The difference of a number and 14 is 20. Find the number.

12. The difference of a number and 10 is 18. Find the number.

13. The quotient of a number and 3 is -8. Find the number.

14. The quotient of a number and -2 is 10. Find the number.

15. The product of a number and -6 is -60. Find the number.

16. The product of a number and -5 is -20. Find the number.

17. The difference of -2 and a number is -14. Find the number.

18. A number subtracted from -30 results in 42. Find the number.

19. 13 increased by a number results in -100. Find the number.

20. The total of 30 and a number is 13. Find the number.

21. Sixty is -5 times a number. Find the number.

22. Sixty-four is -4 times a number. Find the number.

23. Nine more than 3 times a number is 15. Find the number.

24. Eight more than twice a number is 20. Find the number.

25. Five times a number when reduced by 12 equals -27. Find the number.

26. Negative four times a number when reduced by 6 equals 14. Find the number.

27. Five less than the quotient of a number and 4 is equal to -12. Find the number.

28. Ten less than the quotient of a number and -6 is -2. Find the number.

29. Eight decreased by the product of a number and 3 is equal to 5. Find the number.

30. Four decreased by the product of a number and 7 is equal to 11. Find the number.

31. Three times the sum of a number and 4 results in -24. Find the number.

32. Twice the sum of 9 and a number is 30. Find the number.

33. Negative four times the difference of 3 and a number is -20. Find the number.

34. Negative five times the difference of 8 and a number is -55. Find the number.

35. The product of -12 and a number is the same as the sum of the number and 26. Find the number.

36. The difference of a number and 16 is the same as the product of the number and -3. Find the number.

37. Ten times the total of a number and 5 is 80. Find the number.

38. Three times the difference of a number and 5 is 15. Find the number.

39. The product of 3 and a number is the same as 10 less than twice the number.

40. Six less than a number is the same as 3 more than twice the number.

Objective 3: Applications of Linear Equations

41. A metal rod is cut into two pieces. One piece is five times as long as the other. If x represents the length of the shorter piece, write an expression for the length of the longer piece. **(See Example 4a.)**

42. Jackson scored three times as many points in a basketball game as Tony did. If p represents the number of points that Tony scored, write an expression for the number of points that Jackson scored.

43. Tickets to a college baseball game cost \$6 each. If Stan buys n tickets, write an expression for the total cost. **(See Example 4b.)**

44. At a concession stand, drinks cost \$2 each. If Frank buys x drinks, write an expression for the total cost.

45. Bill's daughter is 30 years younger than he is. Write an expression for his daughter's age if Bill is A years old. **(See Example 4c.)**

46. Carlita spent \$88 less on tuition and fees than her friend Carlo did. If Carlo spent d dollars, write an expression for the amount that Carlita spent.

47. The number of prisoners at the Fort Dix Federal Correctional Facility is 1481 more than the number at Big Spring in Texas. If p represents the number of prisoners at Big Spring, write an expression for the number at Fort Dix.

48. Race car driver A. J. Foyt won 15 more races than Mario Andretti. If Mario Andretti won r races, write an expression for the number of races won by A. J. Foyt.

49. Carol sells homemade candles. Write an expression for her total revenue if she sells five candles for r dollars each.

50. Sandy bought eight tomatoes for c cents each. Write an expression for the total cost.

51. The cost to rent space in a shopping center doubled over a 10-year period. If c represents the original cost 10 years ago, write an expression for the cost now.

52. Jacob scored three times as many points in his basketball game on Monday as he did in Wednesday's game. If p represents the number of points scored on Monday, write an expression for the number scored on Wednesday.

For Exercises 53–64, use the problem-solving flowchart to set up and solve an equation to solve each problem.

53. Felicia has an 8-ft piece of ribbon. She wants to cut the ribbon into two pieces so that one piece is three times the length of the other. Find the length of each piece. **(See Example 5.)**

54. Richard and Linda enjoy visiting Hilton Head Island, South Carolina. The distance from their home to Hilton Head is 954 mi, so the drive takes them 2 days. Richard and Linda travel twice as far the first day as they do the second. How many miles do they travel each day?

55. A motorist drives from Minneapolis, Minnesota, to Madison, Wisconsin, and then on to Chicago, Illinois. The total distance is 360 mi. The distance between Minneapolis and Madison is 120 mi more than the distance from Madison to Chicago. Find the distance between Minneapolis and Madison.

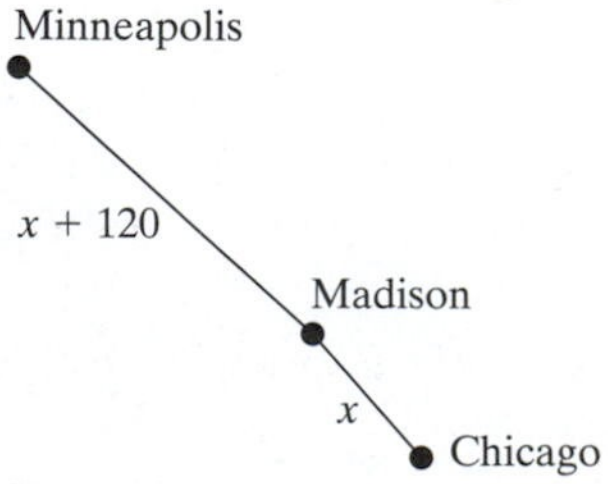

56. Two hikers on the Appalachian Trail hike from Hawk Mountain to Gooch Mountain. The next day, they hike from Gooch Mountain to Woods Hole. The total distance is 19 mi. If the distance between Gooch Mountain and Woods Hole is 5 mi more than the distance from Hawk Mountain to Gooch Mountain, find the distance they hiked each day.

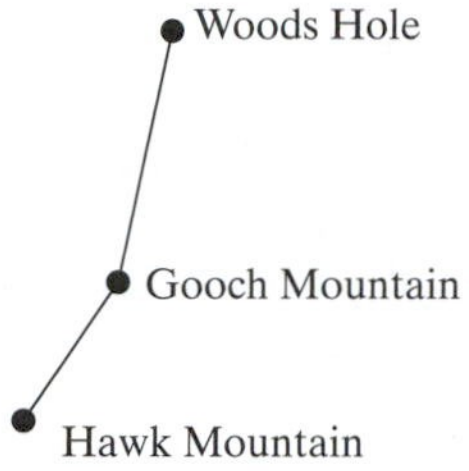

57. The Beatles had 9 more number one albums than Elvis Presley. Together, they had a total of 29 number one albums. How many number one albums did each have?
(See Example 6.)

58. A two-piece set of luggage costs \$150. If sold individually, the larger bag costs \$40 more than the smaller bag. What are the individual costs for each bag?

59. In the 2008 Super Bowl, the New England Patriots scored 3 points less than the New York Giants. A total of 31 points was scored in the game. How many points did each team score?

60. The most goals scored in a regular season NHL hockey game occurred during a game between the Edmonton Oilers and the Chicago Black Hawks. The total number of goals scored was 21. The Oilers scored 3 more goals than the Black Hawks. How many goals did each team score?

61. An apartment complex charges a refundable security deposit when renting an apartment. The deposit is \$350 less than the monthly rent. If Charlene paid a total of \$950 for her first month's rent and security deposit, how much is her monthly rent? How much is the security deposit?

62. Becca paid a total of \$695 for tuition and lab fees for fall semester. Tuition for her courses cost \$605 more than lab fees. How much did her tuition cost? What was the cost of the lab fees?

63. Stefan is paid a salary of \$480 a week at his job. When he works overtime, he receives \$18 an hour. If his weekly paycheck came to \$588, how many hours of overtime did he put in that week?

64. Mercedes is paid a salary of \$480 a week at her job. She worked 8 hr of overtime during the holidays and her weekly paycheck came to \$672. What is her overtime pay per hour?

Group Activity

Constructing Linear Equations

Estimated time: 20 minutes

Group Size: 4

1. Construct a linear equation in which the solution is your age. The equation should require at least two operations. For example:

   "Solve the equation to guess my age": $2(x + 4) = 106$

2. Pass your paper to the student sitting to your right. That student will solve the equation and verify that it is correct.

3. Next, each of you will construct another linear equation. The solution to this equation should be the number of pets that you have.

   "Solve the equation to guess the number of pets I have":

   Equation ________________________________

4. Pass your paper to the right. That student will solve the equation and verify that the solution is correct.

5. Each of you will construct another equation. This time, the solution should be the number of credit hours you are taking.

   "Solve the equation to guess the number of credit hours I am taking."

   Equation ________________________________

6. Pass the papers to the right, solve the equation, and check the answer.

7. Construct an equation in which the solution is the number of sisters and brothers you have.

   "Solve the equation to guess the number of sisters and brothers I have."

   Equation ________________________________

8. Pass the papers to the right again, solve the equation, and check the answer. The papers should now be back in the hands of their original owners.

Chapter 3 Summary

Section 3.1 Simplifying Expressions and Combining *Like* Terms

Key Concepts

An algebraic expression is the sum of one or more terms. A **term** is a constant or the product of a constant and one or more variables. If a term contains a variable it is called a **variable term**. A number multiplied by the variable is called a **coefficient**. A term with no variable is called a **constant term**.

Terms that have exactly the same variable factors with the same exponents are called ***like* terms**.

Properties

1. Commutative property of addition: $a + b = b + a$
2. Commutative property of multiplication: $a \cdot b = b \cdot a$
3. Associative property of addition: $(a + b) + c = a + (b + c)$
4. Associative property of multiplication: $(a \cdot b) \cdot c = a \cdot (b \cdot c)$
5. Distributive property of multiplication over addition: $a(b + c) = a \cdot b + a \cdot c$

Like terms can be combined by applying the distributive property.

To simplify an expression, first clear parentheses using the distributive property. Arrange *like* terms together. Then combine *like* terms.

Examples

Example 1

In the expression $12x + 3$,
$12x$ is a variable term.
12 is the coefficient of the term $12x$.
The term 3 is a constant term.

Example 2

$5h$ and $-2h$ are *like* terms because the variable factor, h, is the same.

$6t$ and $6v$ are not *like* terms because the variable factors, t and v, are not the same.

Example 3

1. $5 + (-8) = -8 + 5$
2. $(3)(-9) = (-9)(3)$
3. $(-7 + 5) + 11 = -7 + (5 + 11)$
4. $(-3 \cdot 10) \cdot 2 = -3 \cdot (10 \cdot 2)$
5. $-3(4x + 12) = -3(4x) + (-3)(12)$
   $= -12x - 36$

Example 4

$$3x + 15x - 7x = (3 + 15 - 7)x$$
$$= 11x$$

Example 5

Simplify: $3(k - 4) - (6k + 10) + 14$

$3(k - 4) - (6k + 10) + 14$

$= 3k - 12 - 6k - 10 + 14$ Clear parentheses.

$= 3k - 6k - 12 - 10 + 14$

$= -3k - 8$

Section 3.2 Addition and Subtraction Properties of Equality

Key Concepts

An **equation** is a statement that indicates that two quantities are equal.

A **solution** to an equation is a value of the variable that makes the equation a true statement.

Definition of a Linear Equation in One Variable

Let a and b be numbers such that $a \neq 0$. A **linear equation in one variable** is an equation that can be written in the form.

$$ax + b = 0$$

Two equations that have the same solution are called **equivalent equations**.

The Addition and Subtraction Properties of Equality

Let a, b, and c represent algebraic expressions.

1. The **addition property of equality:**
   If $a = b$, then $a + c = b + c$
2. The **subtraction property of equality:**
   If $a = b$, then $a - c = b - c$

Examples

Example 1

$3x + 4 = 6$ is an equation, compared to $3x + 4$, which is an expression.

Example 2

The number -4 is a solution to the equation $5x + 7 = -13$ because when we substitute -4 for x we get a true statement.

$$5(-4) + 7 \stackrel{?}{=} -13$$
$$-20 + 7 \stackrel{?}{=} -13$$
$$-13 \stackrel{?}{=} -13 \checkmark$$

Example 3

The equation $5x + 7 = -13$ is equivalent to the equation $x = -4$ because they both have the same solution, -4.

Example 4

To solve the equation $t - 12 = -3$, use the addition property of equality.

$$t - 12 = -3$$
$$t - 12 + 12 = -3 + 12$$
$$t = 9$$ The solution is 9.

Example 5

To solve the equation $-1 = p + 2$, use the subtraction property of equality.

$$-1 = p + 2$$
$$-1 - 2 = p + 2 - 2$$
$$-3 = p$$ The solution is -3.

Section 3.3 Multiplication and Division Properties of Equality

Key Concepts

The Multiplication and Division Properties of Equality

Let a, b, and c represent algebraic expressions.

1. The **multiplication property of equality:** If $a = b$, then $a \cdot c = b \cdot c$
2. The **division property of equality:** If $a = b$, then $\frac{a}{c} = \frac{b}{c}$ (provided $c \neq 0$)

To determine which property to use to solve an equation, first identify the operation on the variable. Then use the property of equality that *reverses* the operation.

Examples

Example 1

Solve $\frac{w}{2} = -11$

$2\left(\frac{w}{2}\right) = 2(-11)$ Multiply both sides by 2.

$w = -22$ The solution is -22.

Example 2

Solve $3a = -18$

$\frac{3a}{3} = \frac{-18}{3}$ Divide both sides by 3.

$a = -6$ The solution is -6.

Example 3

Solve. $4x = 20$ and $4 + x = 20$

$\frac{4x}{4} = \frac{20}{4}$ $\qquad$ $4 - 4 + x = 20 - 4$

$x = 5$ $\qquad$ $x = 16$

The solution is 5. $\qquad$ The solution is 16.

Section 3.4 Solving Equations with Multiple Steps

Key Concepts

Steps to Solve a Linear Equation in One Variable

1. Simplify both sides of the equation.
   - Clear parentheses if necessary.
   - Combine *like* terms if necessary.
2. Use the addition or subtraction property of equality to collect the variable terms on one side of the equation.
3. Use the addition or subtraction property of equality to collect the constant terms on the *other* side of the equation.
4. Use the multiplication or division property of equality to make the coefficient of the variable term equal to 1.
5. Check the answer in the original equation.

Examples

Example 1

Solve: $4(x - 3) - 6 = -2x$

$4x - 12 - 6 = -2x$ **Step 1**

$4x - 18 = -2x$

$4x + 2x - 18 = -2x + 2x$ **Step 2**

$6x - 18 = 0$

$6x - 18 + 18 = 0 + 18$ **Step 3**

$6x = 18$

$\frac{6x}{6} = \frac{18}{6}$ **Step 4**

$x = 3$ The solution is 3.

<u>Check</u>: $4(x - 3) - 6 = -2x$ **Step 5**

$4(3 - 3) - 6 \stackrel{?}{=} -2(3)$

$4(0) - 6 \stackrel{?}{=} -6$

$-6 \stackrel{?}{=} -6$ ✓ True

Section 3.5 Applications and Problem Solving

Key Concepts

Problem-Solving Flowchart for Word Problems

Step 1 Read the problem completely.
↓
Step 2 Assign labels to unknown quantities.
↓
Step 3 Write an equation in words.
↓
Step 4 Write a mathematical equation.
↓
Step 5 Solve the equation.
↓
Step 6 Interpret the results and write the final answer in words.

Examples

Example 1

Subtract 5 times a number from 14. The result is −6. Find the number. **Step 1**

Let n represent the unknown number. **Step 2**

from 14 — subtract $5n$ — result is −6 **Step 3**

$14 - 5n = -6$ **Step 4**

$14 - 14 - 5n = -6 - 14$ **Step 5**

$-5n = -20$

$\frac{-5n}{-5} = \frac{-20}{-5}$

$n = 4$

The number is 4. **Step 6**

Example 2

An electrician needs to cut a 20-ft wire into two pieces so that one piece is four times as long as the other. How long should each piece be? **Step 1**

Let x be the length of the shorter piece. Then $4x$ is the length of the longer piece. **Step 2**

The two pieces added together will be 20 ft. **Step 3**

$x + 4x = 20$ **Step 4**

$5x = 20$ **Step 5**

$\frac{5x}{5} = \frac{20}{5}$

$x = 4$ **Step 6**

One piece of wire is 4 ft and the other is 4(4 ft), which is 16 ft.

Chapter 3 Review Exercises

Section 3.1

For Exercises 1–2, list the terms of the expression and identify the term as a variable term or a constant term. Then identify the coefficient.

1. $3a^2 - 5a + 12$

2. $-6xy - y + 2x$

For Exercises 3–6, determine if the two terms are *like* terms or unlike terms.

3. $5t^2, 5t$

4. $4h, -2h$

5. $21, -5$

6. $-8, -8k$

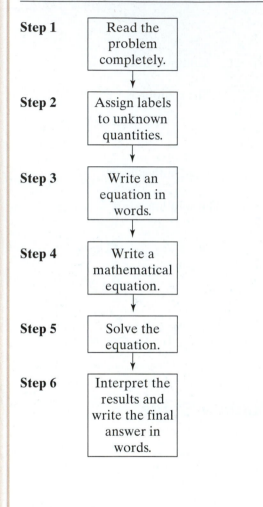

For Exercises 7–8, apply the commutative property of addition or multiplication to rewrite the expression.

7. $t - 5$ **8.** $h \cdot 3$

For Exercises 9–10, apply the associative property of addition or multiplication to rewrite the expression. Then simplify the expression.

9. $-4(2 \cdot p)$ **10.** $(m + 10) - 12$

For Exercises 11–14, apply the distributive property.

11. $3(2b + 5)$ **12.** $5(4x + 6y - 3z)$

13. $-(4c - 6d)$ **14.** $-(-4k + 8w - 12)$

For Exercises 15–22, combine *like* terms. Clear parentheses if necessary.

15. $-5x - x + 8x$

16. $-3y - 7y + y$

17. $6y + 8x - 2y - 2x + 10$

18. $12a - 5 + 9b - 5a + 14$

19. $5 - 3(x - 7) + x$

20. $6 - 4(z + 2) + 2z$

21. $4(u - 3v) - 5(u + v)$

22. $-5(p + 4) + 6(p + 1) - 2$

Section 3.2

For Exercises 23–24, determine if -3 is a solution to the equation.

23. $5x + 10 = -5$ **24.** $-3(x - 1) = -9 + x$

For Exercises 25–32, solve the equation using either the addition property of equality or the subtraction property of equality.

25. $r + 23 = -12$ **26.** $k - 3 = -15$

27. $10 = p - 4$ **28.** $21 = q + 3$

29. $5a + 7 - 4a = 20$ **30.** $-7t - 4 + 8t = 11$

31. $-4(m - 3) + 7 + 5m = 21$

32. $-2(w - 3) + 3w = -8$

Section 3.3

For Exercises 33–38, solve the equation using either the multiplication property of equality or the division property of equality.

33. $4d = -28$ **34.** $-3c = -12$

35. $\frac{t}{-2} = -13$ **36.** $\frac{p}{5} = 7$

37. $-42 = -7p$ **38.** $-12 = \frac{m}{4}$

Section 3.4

For Exercises 39–52, solve the equation.

39. $9x + 7 = -2$ **40.** $8y + 3 = 27$

41. $45 = 6m - 3$ **42.** $-25 = 2n - 1$

43. $\frac{p}{8} + 1 = 5$ **44.** $\frac{x}{-5} - 2 = -3$

45. $5x + 12 = 4x - 16$ **46.** $-4t - 2 = -3t + 5$

47. $-8 + 4y = 7y + 4$ **48.** $15 - 2c = 5c + 1$

49. $6(w - 2) + 15 = 3w$ **50.** $-4(h - 5) + h = 7h$

51. $-(5a + 3) - 3(a - 2) = 24 - a$

52. $-(4b - 7) = 2(b + 3) - 4b + 13$

Section 3.5

For Exercises 53–58,

a. write an equation that represents the statement.

b. solve the problem.

53. Four subtracted from a number is 13. Find the number.

54. The quotient of a number and -7 is -6. Find the number.

55. Three more than -4 times a number is -17. Find the number.

56. Seven less than 3 times a number is -22. Find the number.

57. Twice the sum of a number and 10 is 16. Find the number.

58. Three times the difference of 4 and a number is -9. Find the number.

59. A rack of discount CDs are on sale for \$9 each. If Mario buys n CDs, write an expression for the total cost.

60. Henri bought four sandwiches for x dollars each. Write an expression that represents the total cost.

61. It took Winston 2 hr longer to finish his psychology paper than it did for Gus to finish. If Gus finished in x hours, write an expression for the amount of time it took Winston.

62. Gerard is 6 in. taller than his younger brother Dwayne. If Dwayne is h inches tall, write an expression for Gerard's height.

63. Monique and Michael drove from Ormond Beach, Florida, to Knoxville, Tennessee. Michael drove three times as far as Monique drove. If the total distance is 480 mi, how far did each person drive?

64. Joel ate twice as much pizza as Angela. Together, they finished off a 12-slice pizza. How many pieces did each person have?

65. As of a recent year, Tom Cruise starred in 5 fewer films than Tom Hanks. If together they starred in 65 films, how many films did Tom Hanks and Tom Cruise star in individually?

66. Raul signed up for his classes for the spring semester. His load was 4 credit-hours less in the spring than in the fall. If he took a total of 28 hours in the two semesters combined, how many hours did he take in the fall? How many hours did he take in the spring?

Chapter 3 Test

For Exercises 1–5, state the property demonstrated. Choose from:

a. commutative property of addition

b. commutative property of multiplication

c. associative property of addition

d. associative property of multiplication

e. distributive property of multiplication over addition.

1. $-5(9x) = (-5 \cdot 9)x$ **2.** $-5x + 9 = 9 + (-5x)$

3. $-3 + (u + v) = (-3 + u) + v$

4. $-4(b + 2) = -4b - 8$ **5.** $g(-6) = -6g$

For Exercises 6–11, simplify the expressions.

6. $-5x - 3x + x$

7. $-2a - 3b + 8a + 4b - a$

8. $4(a + 9) - 12$ **9.** $-3(6b) + 5b + 8$

10. $14y - 2(y - 9) + 21$

11. $2 - (5 - w) + 3(-2w)$

12. Explain the difference between an expression and an equation.

For Exercises 13–16, identify as either an expression or an equation.

13. $4x + 5$

14. $4x + 5 = 2$

15. $2(q - 3) = 6$

16. $2(q - 3) + 6$

For Exercises 17–36, solve the equation.

17. $a - 9 = 12$

18. $t - 6 = 12$

19. $7 = 10 + x$

20. $19 = 12 + y$

21. $-4p = 28$

22. $-3c = 30$

23. $-7 = \dfrac{d}{3}$

24. $8 = \dfrac{m}{4}$

25. $-6x = 12$

26. $-6 + x = 12$

27. $\dfrac{x}{-6} = 12$

28. $-6x = 12$

29. $4x - 5 = 23$

30. $-9x - 6 = 21$

31. $\dfrac{x}{7} + 1 = -11$

32. $\dfrac{z}{-2} - 3 = 4$

33. $5h - 2 = -h + 22$

34. $6p + 3 = 15 + 2p$

35. $-2(q - 5) = 6q + 10$

36. $-(4k - 2) - k = 2(k - 6)$

37. The product of -2 and a number is the same as the total of 15 and the number. Find the number.

38. Two times the sum of a number and 8 is 10. Find the number.

39. A high school student sells magazine subscriptions for \$15 each. Write an expression that represents the total cost of m magazines.

40. Alberto is 5 years older than Juan. If Juan's age is represented by a, write an expression for Alberto's age.

41. Monica and Phil each have part-time jobs. Monica makes twice as much money in a week as Phil. If the total of their weekly earnings is \$756, how much does each person make?

42. A computer with a monitor costs \$899. If the computer costs \$241 more than the monitor, what is the price of the computer? What is the price of the monitor?

Chapters 1–3 Cumulative Review Exercises

1. Identify the place value of the underlined digit.

a. $34,\underline{9}11$ **b.** $20\underline{9},001$ **c.** $5,\underline{9}01,888$

For Exercises 2–4, round the number to the indicated place value.

2. 45,921; thousands

3. 1,285,000; ten-thousands

4. 25,449; hundreds

5. Divide 39,190 by 46. Identify the dividend, divisor, quotient, and remainder.

6. Sarah cleans three apartments in a weekend. The apartments have five, six, and four rooms, respectively. If she earns \$300 for the weekend, how much does she make per room?

7. The data represent the number of turkey subs sold between 11:00 A.M. and 12:00 P.M. at

Subs-R-Us over a 14-day period. Find the mean number of subs sold per day.

15	12	8	4	6	12	10
20	7	5	8	9	11	13

8. Find the perimeter of the lot.

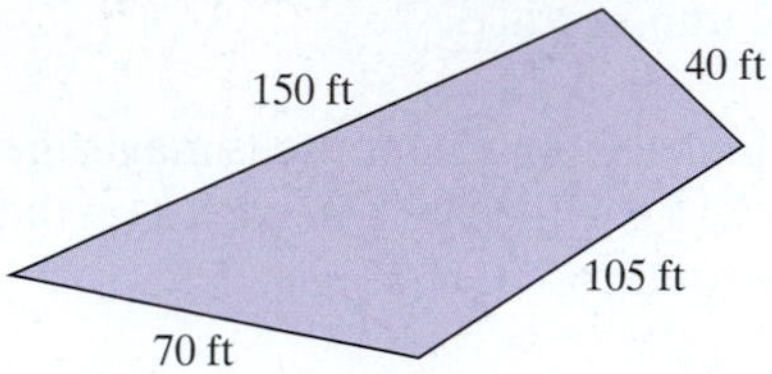

For Exercises 9–14, perform the indicated operations.

9. $18 - (-5) + (-3)$

10. $-4 - (-7) - 8$

11. $4(5 - 11) + (-1)$

12. $5 - 23 + 12 - 3$

13. $-36 \div (0)$

14. $16 \div (-4) \cdot 3$

For Exercises 15–16, simplify the expression.

15. $-2x + 14 + 15 - 3x$

16. $3y - (5y + 6) - 12$

For Exercises 17–18, solve the equation.

17. $4p + 5 = -11$

18. $9(t - 1) - 7t + 2 = t - 15$

19. Evaluate the expression $a^2 - b$ for $a = -3$ and $b = 5$.

20. In a recent year, the recording artist Kanye West received 2 more Grammy nominations than Alicia Keys. Together they received 18 nominations. How many nominations did each receive?

Fractions and Mixed Numbers

4

CHAPTER OUTLINE

Chapter 4

In this chapter, we study the concept of a fraction and a mixed number. We begin by learning how to simplify fractions to lowest terms. We follow with addition, subtraction, multiplication, and division of fractions and mixed numbers. At the end of the chapter, we solve equations with fractions.

Are You Prepared?

To review for this chapter, take a minute to practice the technique to solve linear equations. Use the solutions to complete the crossword. If you have trouble, see Section 3.4.

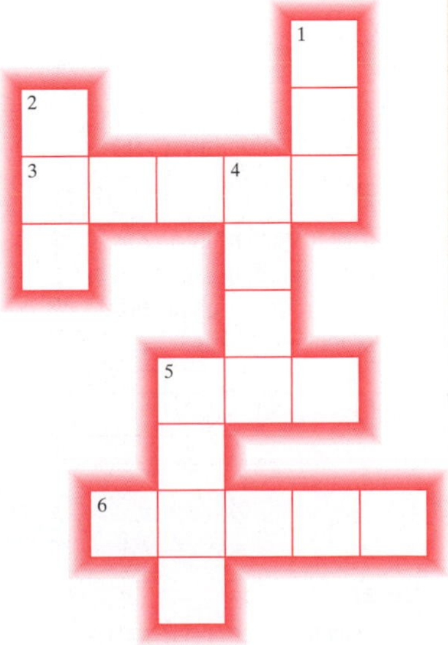

Across

3. $x + 489 = 23{,}949$

5. $\dfrac{x}{3} = 81$

6. $12{,}049 = -58{,}642 + y$

Down

1. $10x = 1000$

2. $2x - 4 = 254$

4. $-18{,}762 = -3z - 540$

5. $3x - 120 = 9920 - 2x$

Section 4.1 Introduction to Fractions and Mixed Numbers

Objectives

1. Definition of a Fraction
2. Proper and Improper Fractions
3. Mixed Numbers
4. Fractions and the Number Line

1. Definition of a Fraction

In Chapter 1, we studied operations on whole numbers. In this chapter, we work with numbers that represent part of a whole. When a whole unit is divided into equal parts, we call the parts **fractions** of a whole. For example, the pizza in Figure 4-1 is divided into 5 equal parts. One-fifth $\left(\frac{1}{5}\right)$ of the pizza has been eaten, and four-fifths $\left(\frac{4}{5}\right)$ of the pizza remains.

Figure 4-1

A fraction is written in the form $\frac{a}{b}$, where a and b are whole numbers and $b \neq 0$. In the fraction $\frac{5}{8}$, the "top" number, 5, is called the **numerator**. The "bottom" number, 8, is called the **denominator**.

$$\text{numerator} \longrightarrow 5 \qquad 2x^2 \longleftarrow \text{numerator}$$
$$\text{denominator} \longrightarrow 8 \qquad 3y \longleftarrow \text{denominator}$$

A fraction whose numerator is an integer and whose denominator is a nonzero integer is also called a **rational number**.

The denominator of a fraction denotes the number of equal pieces into which a whole unit is divided. The numerator denotes the number of pieces being considered. For example, the garden in Figure 4-2 is divided into 10 equal parts. Three sections contain tomato plants. Therefore, $\frac{3}{10}$ of the garden contains tomato plants.

Avoiding Mistakes

The fraction $\frac{3}{10}$ can also be written as 3/10. However, we discourage the use of the "slanted" fraction bar. In later applications of algebra, the slanted fraction bar can cause confusion.

Figure 4-2

Skill Practice

1. Write a fraction for the shaded portion and a fraction for the unshaded portion.

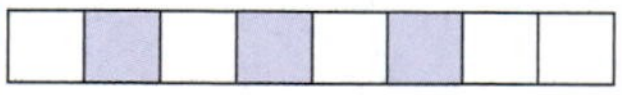

Example 1 Writing Fractions

Write a fraction for the shaded portion and a fraction for the unshaded portion of the figure.

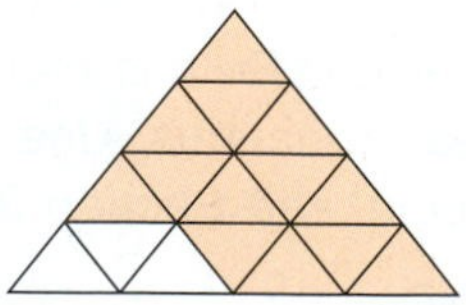

Solution:

Shaded portion: $\frac{13}{16}$ — 13 pieces are shaded. The triangle is divided into 16 equal pieces.

Unshaded portion: $\frac{3}{16}$ — 3 pieces are not shaded. The triangle is divided into 16 equal pieces.

Answer

1. Shaded portion: $\frac{3}{8}$; unshaded portion: $\frac{5}{8}$

In Section 1.6, we learned that fractions represent division. For example, the fraction $\frac{5}{1} = 5 \div 1 = 5$. Interpreting fractions as division leads to the following important properties.

PROPERTY Properties of Fractions

Suppose that a and b represent nonzero numbers.

1. $\dfrac{a}{1} = a$ Example: $\dfrac{-8}{1} = -8$
2. $\dfrac{0}{a} = 0$ Example: $\dfrac{0}{-7} = 0$
3. $\dfrac{a}{0}$ is undefined Example: $\dfrac{11}{0}$ is undefined
4. $\dfrac{a}{a} = 1$ Example: $\dfrac{-3}{-3} = 1$
5. $\dfrac{-a}{-b} = \dfrac{a}{b}$ Example: $\dfrac{-2}{-5} = \dfrac{2}{5}$
6. $-\dfrac{a}{b} = \dfrac{-a}{b} = \dfrac{a}{-b}$ Example: $-\dfrac{2}{3} = \dfrac{-2}{3} = \dfrac{2}{-3}$

Concept Connections

2. Which of the expressions is equivalent to $\frac{-6}{1}$?

a. $\dfrac{6}{-1}$ b. $-\dfrac{6}{1}$

c. -6 d. $\dfrac{-6}{-1}$

Property 6 tells us that a negative fraction can be written with the negative sign in the numerator, in the denominator, or out in front. This is because a quotient of two numbers with opposite signs is negative.

2. Proper and Improper Fractions

A positive fraction whose numerator is less than its denominator (or the opposite of such a fraction) is called a **proper fraction**. Furthermore, a positive proper fraction represents a number less than 1 whole unit. The following are proper fractions.

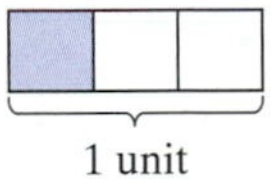

$\frac{1}{3}$

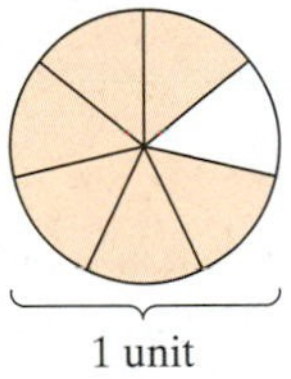

$\frac{6}{7}$

A positive fraction whose numerator is greater than or equal to its denominator (or the opposite of such a fraction) is called an **improper fraction**. For example:

numerator greater than denominator → $\dfrac{4}{3}$ and $\dfrac{7}{7}$ ← numerator equal to denominator

A positive improper fraction represents a quantity greater than 1 whole unit or equal to 1 whole unit.

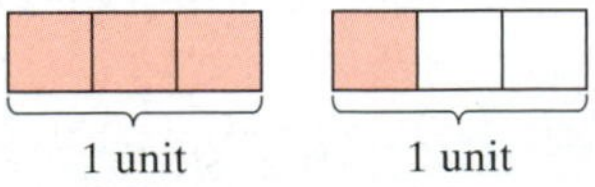

$\frac{4}{3}$

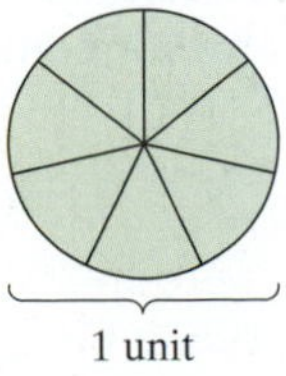

$\frac{7}{7}$

Answer

2. a, b, c

Skill Practice

Identify each fraction as proper or improper.

3. $\frac{10}{10}$ **4.** $\frac{7}{9}$ **5.** $\frac{9}{7}$

Example 2 Categorizing Fractions

Identify each fraction as proper or improper.

a. $\frac{12}{5}$ **b.** $\frac{5}{12}$ **c.** $\frac{12}{12}$

Solution:

a. $\frac{12}{5}$ Improper fraction (numerator is greater than denominator)

b. $\frac{5}{12}$ Proper fraction (numerator is less than denominator)

c. $\frac{12}{12}$ Improper fraction (numerator is equal to denominator)

Skill Practice

6. Write an improper fraction representing the shaded area.

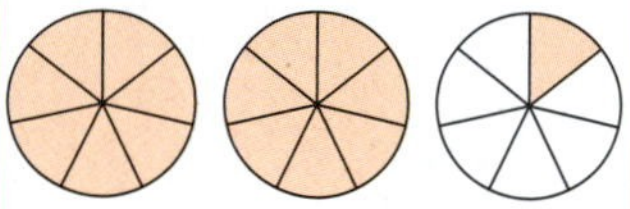

Example 3 Writing Improper Fractions

Write an improper fraction to represent the fractional part of an inch for the screw shown in the figure.

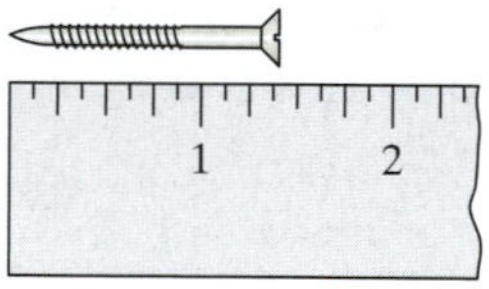

Avoiding Mistakes

Each whole unit is divided into 8 pieces. Therefore the screw is $\frac{11}{8}$ in., not $\frac{11}{16}$ in.

Solution:

Each 1-in. unit is divided into 8 parts, and the screw extends for 11 parts. Therefore, the screw is $\frac{11}{8}$ in.

3. Mixed Numbers

Sometimes a mixed number is used instead of an improper fraction to denote a quantity greater than one whole. For example, suppose a typist typed $\frac{9}{4}$ pages of a report. We would be more likely to say that the typist typed $2\frac{1}{4}$ pages (read as "two and one-fourth pages"). The number $2\frac{1}{4}$ is called a *mixed number* and represents 2 wholes plus $\frac{1}{4}$ of a whole.

$\frac{9}{4} = 2\frac{1}{4}$

In general, a **mixed number** is a sum of a whole number and a fractional part of a whole. However, by convention the plus sign is left out. For example,

$$3\frac{1}{2} \quad \text{means} \quad 3 + \frac{1}{2}$$

Answers

3. Improper **4.** Proper

5. Improper **6.** $\frac{15}{7}$

A negative mixed number implies that both the whole number part and the fraction part are negative. Therefore, we interpret the mixed number $-3\frac{1}{2}$ as

$$-3\frac{1}{2} = -\left(3 + \frac{1}{2}\right) = -3 + \left(-\frac{1}{2}\right)$$

Suppose we want to change a mixed number to an improper fraction. From Figure 4-3, we see that the mixed number $3\frac{1}{2}$ is the same as $\frac{7}{2}$.

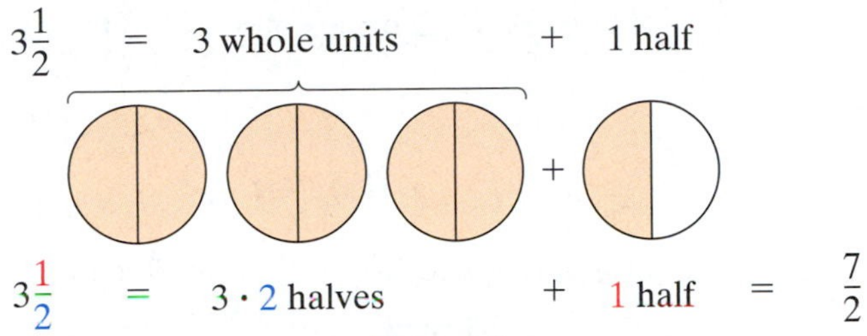

Figure 4-3

The process to convert a mixed number to an improper fraction can be summarized as follows.

> **PROCEDURE Changing a Mixed Number to an Improper Fraction**
>
> **Step 1** Multiply the whole number by the denominator.
>
> **Step 2** Add the result to the numerator.
>
> **Step 3** Write the result from step 2 over the denominator.

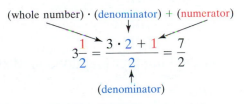

(whole number) · (denominator) + (numerator)

$$3\frac{1}{2} = \frac{3 \cdot 2 + 1}{2} = \frac{7}{2}$$

(denominator)

Example 4 Converting Mixed Numbers to Improper Fractions

Convert the mixed number to an improper fraction.

a. $7\frac{1}{4}$ **b.** $-8\frac{2}{5}$

Solution:

a.
$$7\frac{1}{4} = \frac{7 \cdot 4 + 1}{4} = \frac{28 + 1}{4} = \frac{29}{4}$$

b.
$$-8\frac{2}{5} = -\left(8\frac{2}{5}\right)$$

The entire mixed number is negative.

$$= -\left(\frac{8 \cdot 5 + 2}{5}\right) = -\left(\frac{40 + 2}{5}\right) = -\frac{42}{5}$$

Skill Practice

Convert the mixed number to an improper fraction.

7. $10\frac{5}{8}$ **8.** $-15\frac{1}{2}$

Avoiding Mistakes

The negative sign in the mixed number applies to both the whole number and the fraction.

Answers

7. $\frac{85}{8}$ **8.** $-\frac{31}{2}$

Now suppose we want to convert an improper fraction to a mixed number. In Figure 4-4, the improper fraction $\frac{13}{5}$ represents 13 slices of pizza where each slice is $\frac{1}{5}$ of a whole pizza. If we divide the 13 pieces into groups of 5, we make 2 whole pizzas with 3 pieces left over. Thus,

$$\frac{13}{5} = 2\frac{3}{5}$$

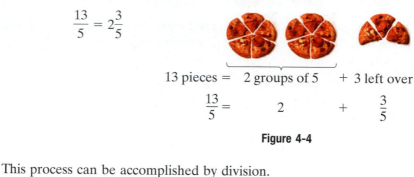

13 pieces = 2 groups of 5 + 3 left over

$$\frac{13}{5} = 2 + \frac{3}{5}$$

Figure 4-4

Avoiding Mistakes

When writing a mixed number, the + sign between the whole number and fraction should not be written.

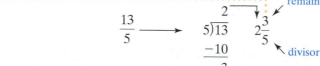

This process can be accomplished by division.

$$\frac{13}{5} \longrightarrow \begin{array}{r} 2 \\ 5\overline{)13} \\ \underline{-10} \\ 3 \end{array} \qquad 2\frac{3}{5}$$

(3: remainder; 5: divisor)

PROCEDURE Changing an Improper Fraction to a Mixed Number

Step 1 Divide the numerator by the denominator to obtain the quotient and remainder.

Step 2 The mixed number is then given by

$$\text{Quotient} + \frac{\text{remainder}}{\text{divisor}}$$

Skill Practice

Convert the improper fraction to a mixed number.

9. $\frac{14}{5}$ **10.** $-\frac{95}{22}$

Example 5 Converting Improper Fractions to Mixed Numbers

Convert to a mixed number.

a. $\frac{25}{6}$ **b.** $-\frac{39}{4}$

Solution:

a. $\frac{25}{6} \longrightarrow \begin{array}{r} 4 \\ 6\overline{)25} \\ \underline{-24} \\ 1 \end{array} \qquad 4\frac{1}{6}$

(4: quotient; 1: remainder; 6: divisor)

b. $-\frac{39}{4} = -\left(\frac{39}{4}\right)$ First perform division.

$$\begin{array}{r} 9 \\ 4\overline{)39} \\ \underline{-36} \\ 3 \end{array} \qquad 9\frac{3}{4}$$

(9: quotient; 3: remainder; 4: divisor)

$= -9\frac{3}{4}$ Then take the opposite of the result.

The process to convert an improper fraction to a mixed number indicates that the result of a division problem can be written as a mixed number.

Answers

9. $2\frac{4}{5}$ **10.** $-4\frac{7}{22}$

Example 6 Writing a Quotient as a Mixed Number

Divide. Write the quotient as a mixed number.

$$25\overline{)529}$$

Solution:

$$\begin{array}{r} 21 \\ 25\overline{)529} \\ \underline{-50} \\ 29 \\ \underline{-25} \\ 4 \end{array} \qquad 21\frac{4}{25}$$

remainder (4); divisor (25)

4. Fractions and the Number Line

Fractions can be visualized on a number line. For example, to graph the fraction $\frac{3}{4}$, divide the distance between 0 and 1 into 4 equal parts. To plot the number $\frac{3}{4}$, start at 0 and count over 3 parts.

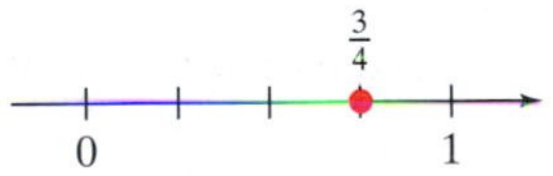

Example 7 Plotting Fractions on a Number Line

Plot the point on the number line corresponding to each fraction.

a. $\frac{1}{2}$ **b.** $\frac{5}{6}$ **c.** $-\frac{21}{5}$

Solution:

a. $\frac{1}{2}$ Divide the distance between 0 and 1 into 2 equal parts.

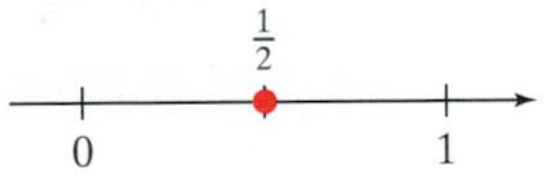

b. $\frac{5}{6}$ Divide the distance between 0 and 1 into 6 equal parts.

c. $-\frac{21}{5} = -4\frac{1}{5}$ Write $-\frac{21}{5}$ as a mixed number.

The value $-4\frac{1}{5}$ is located one-fifth of the way between -4 and -5 on the number line. Divide the distance between -4 and -5 into 5 equal parts. Plot the point one-fifth of the way from -4 to -5.

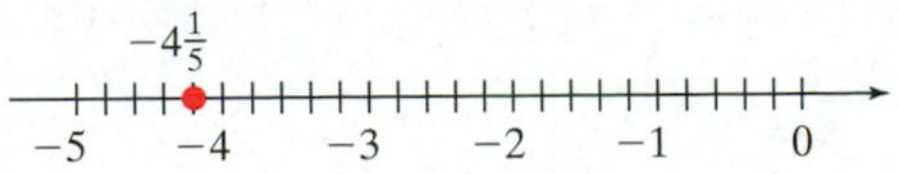

Skill Practice

Divide and write the quotient as a mixed number.

11. $5967 \div 41$

Skill Practice

Plot the numbers on a number line.

12. $\frac{4}{5}$ **13.** $\frac{1}{3}$ **14.** $-\frac{13}{4}$

Answers

11. $145\frac{22}{41}$

12.

13.

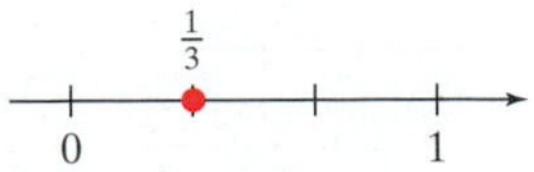

14.

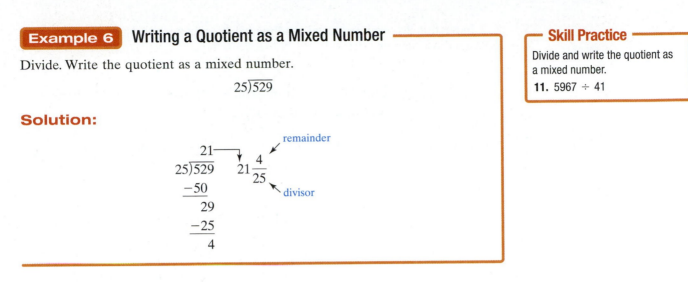

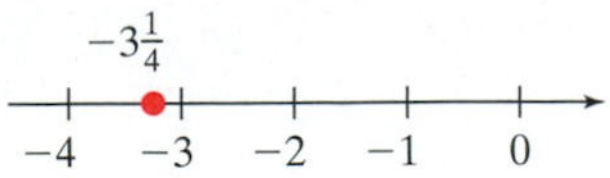

In Section 2.1, we introduced absolute value and opposite. Now we apply these concepts to fractions.

Skill Practice

Simplify.

15. $\left|-\dfrac{12}{5}\right|$

16. $\left|\dfrac{3}{7}\right|$

17. $-\left|-\dfrac{3}{4}\right|$

18. $-\left(-\dfrac{3}{4}\right)$

Example 8 Determining the Absolute Value and Opposite of a Fraction

Simplify. **a.** $\left|-\dfrac{2}{7}\right|$ **b.** $\left|\dfrac{1}{5}\right|$ **c.** $-\left|-\dfrac{4}{9}\right|$ **d.** $-\left(-\dfrac{4}{9}\right)$

Solution:

a. $\left|-\dfrac{2}{7}\right| = \dfrac{2}{7}$ The distance between $-\frac{2}{7}$ and 0 on the number line is $\frac{2}{7}$.

b. $\left|\dfrac{1}{5}\right| = \dfrac{1}{5}$ The distance between $\frac{1}{5}$ and 0 on the number line is $\frac{1}{5}$.

c. $-\left|-\dfrac{4}{9}\right| = -\left(\dfrac{4}{9}\right)$ Take the absolute value of $-\frac{4}{9}$ first. This gives $\frac{4}{9}$. Then take the opposite of $\frac{4}{9}$, which is $-\frac{4}{9}$.

$= -\dfrac{4}{9}$

d. $-\left(-\dfrac{4}{9}\right) = \dfrac{4}{9}$ Take the opposite of $-\frac{4}{9}$, which is $\frac{4}{9}$.

Answers

15. $\dfrac{12}{5}$ 16. $\dfrac{3}{7}$

17. $-\dfrac{3}{4}$ 18. $\dfrac{3}{4}$

Section 4.1 Practice Exercises

Boost *your* GRADE at ALEKS.com! ALEKS version 3.0

- Practice Problems
- Self-Tests
- NetTutor
- e-Professors
- Videos

Study Skills Exercise

1. Define the key terms.

a. **Fraction** **b.** **Numerator** **c.** **Denominator** **d.** **Rational Number**

e. **Proper fraction** **f.** **Improper fraction** **g.** **Mixed number**

Objective 1: Definition of a Fraction

For Exercises 2–5, identify the numerator and the denominator for each fraction.

2. $\dfrac{2}{3}$

3. $\dfrac{8}{9}$

4. $\dfrac{12x}{11y^2}$

5. $\dfrac{7p}{9q}$

For Exercises 6–11, write a fraction that represents the shaded area. (See Example 1.)

6. 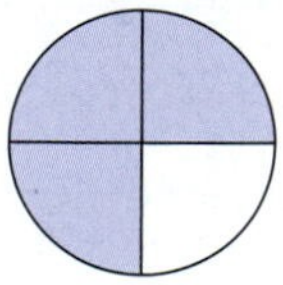

7. 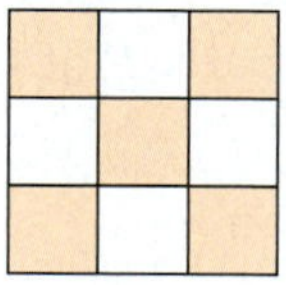

8. 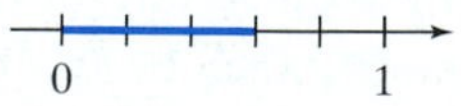

9.

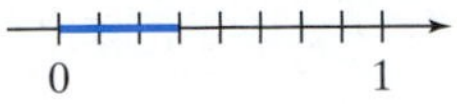

10.

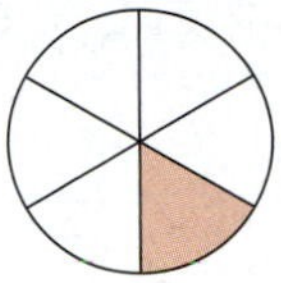

11.

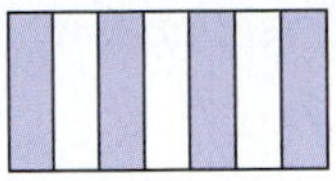

12. Write a fraction to represent the portion of gas in a gas tank represented by the gauge.

13. The scoreboard for a recent men's championship swim meet in Melbourne, Australia, shows the final standings in the event. What fraction of the finalists are from the USA?

Name	Country	Time
Maginni, Filippo	ITA	48.43
Hayden, Brent	CAN	48.43
Sullivan, Eamon	AUS	48.47
Cielo Filho, Cesar	BRA	48.51
Lezak, Jason	USA	48.52
Van Den Hoogenband, Pieter	NED	48.63
Schoeman, Roland Mark	RSA	48.72
Neethling, Ryk	RSA	48.81

14. Refer to the scoreboard from Exercise 13. What fraction of the finalists are from the Republic of South Africa (RSA)?

15. The graph categorizes a sample of people by blood type. What fraction of the sample represents people with type O blood?

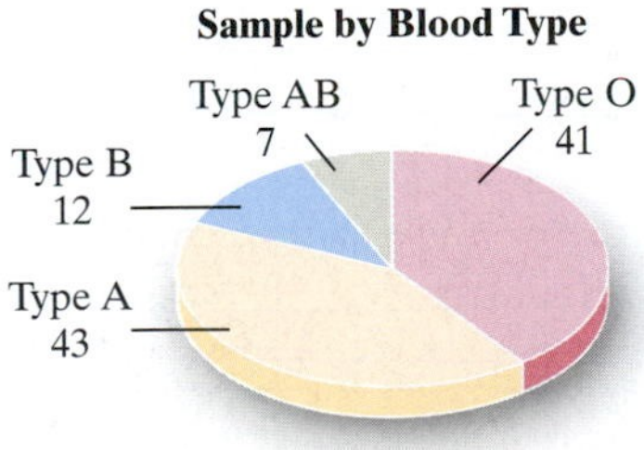

16. Refer to the graph from Exercise 15. What fraction of the sample represents people with type A blood?

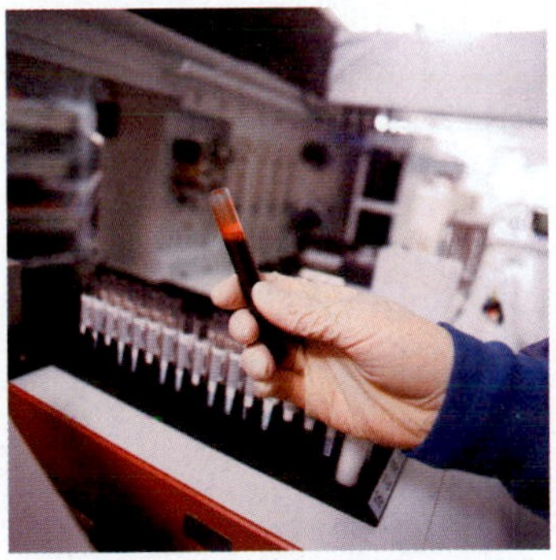

17. A class has 21 children—11 girls and 10 boys. What fraction of the class is made up of boys?

18. In a neighborhood in Ft. Lauderdale, Florida, 10 houses are for sale and 53 are not for sale. Write a fraction representing the portion of houses that are for sale.

For Exercises 19–28, simplify if possible.

19. $\frac{-13}{1}$ **20.** $\frac{-14}{1}$ **21.** $\frac{2}{2}$ **22.** $\frac{8}{8}$

23. $\frac{0}{-3}$ **24.** $\frac{0}{7}$ **25.** $\frac{-3}{0}$ **26.** $\frac{-11}{0}$

27. $\frac{-9}{-10}$ **28.** $\frac{-13}{-6}$

29. Which expressions are equivalent to $\frac{-9}{10}$?

a. $\frac{9}{10}$ **b.** $-\frac{9}{10}$ **c.** $\frac{9}{-10}$ **d.** $\frac{-9}{-10}$

30. Which expressions are equivalent to $-\frac{4}{5}$?

a. $\frac{-4}{5}$ **b.** $\frac{-4}{-5}$ **c.** $\frac{4}{-5}$ **d.** $-\frac{5}{4}$

31. Which expressions are equivalent to $\frac{-4}{1}$?

a. $\frac{4}{-1}$ **b.** -4 **c.** $-\frac{4}{1}$ **d.** $\frac{1}{-4}$

32. Which expressions are equivalent to $\frac{-9}{1}$?

a. $\frac{-9}{-1}$ **b.** $\frac{9}{-1}$ **c.** -9 **d.** $-\frac{9}{1}$

Objective 2: Proper and Improper Fractions

For Exercises 33–38, label the fraction as proper or improper. **(See Example 2.)**

33. $\frac{7}{8}$ **34.** $\frac{2}{3}$ **35.** $\frac{10}{10}$

36. $\frac{3}{3}$ **37.** $\frac{7}{2}$ **38.** $\frac{21}{20}$

For Exercises 39–42, write an improper fraction for the shaded portion of each group of figures. **(See Example 3.)**

39.

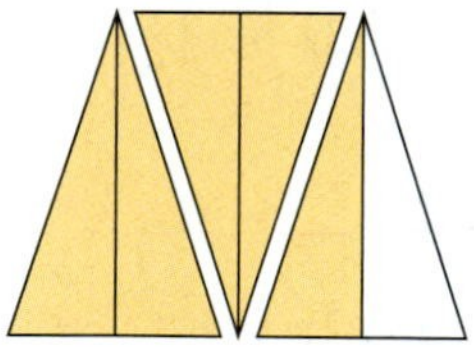

40.

41.

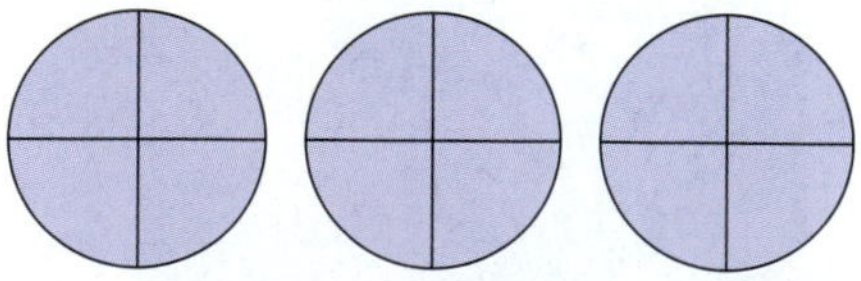

42.

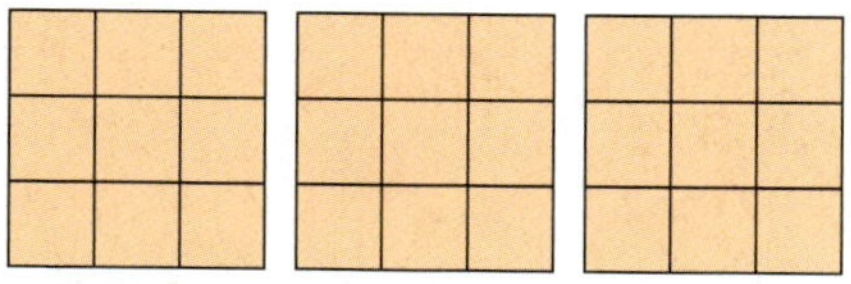

Objective 3: Mixed Numbers

For Exercises 43–44, write an improper fraction and a mixed number for the shaded portion of each group of figures.

43.

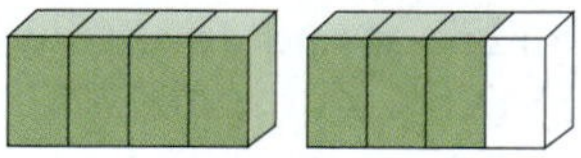

44.

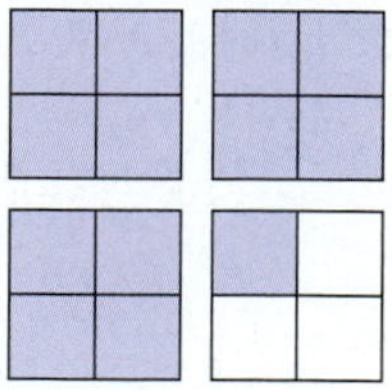

45. Write an improper fraction and a mixed number to represent the length of the nail.

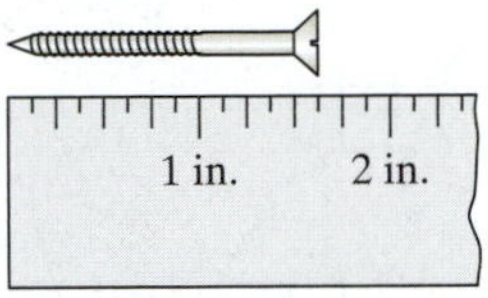

46. Write an improper fraction and a mixed number that represent the number of cups of sugar needed for a batch of cookies, as indicated in the figure.

For Exercises 47–58, convert the mixed number to an improper fraction. **(See Example 4.)**

47. $1\frac{3}{4}$ **48.** $6\frac{1}{3}$ **49.** $-4\frac{2}{9}$ **50.** $-3\frac{1}{5}$

51. $-3\frac{3}{7}$ **52.** $-8\frac{2}{3}$ **53.** $6\frac{3}{4}$ **54.** $10\frac{3}{5}$

55. $11\frac{5}{12}$ **56.** $12\frac{1}{6}$ **57.** $-21\frac{3}{8}$ **58.** $-15\frac{1}{2}$

59. How many thirds are in 10?

60. How many sixths are in 2?

61. How many eighths are in $2\frac{3}{8}$?

62. How many fifths are in $2\frac{3}{5}$?

63. How many fourths are in $1\frac{3}{4}$?

64. How many thirds are in $5\frac{2}{3}$?

For Exercises 65–76, convert the improper fraction to a mixed number. **(See Example 5.)**

65. $\frac{37}{8}$ **66.** $\frac{13}{7}$ **67.** $-\frac{39}{5}$ **68.** $-\frac{19}{4}$

69. $-\frac{27}{10}$ **70.** $-\frac{43}{18}$ **71.** $\frac{52}{9}$ **72.** $\frac{67}{12}$

73. $\frac{133}{11}$ **74.** $\frac{51}{10}$ **75.** $-\frac{23}{6}$ **76.** $-\frac{115}{7}$

For Exercises 77–84, divide. Write the quotient as a mixed number. **(See Example 6.)**

77. $7\overline{)309}$ **78.** $4\overline{)921}$ **79.** $5281 \div 5$ **80.** $7213 \div 8$

81. $8913 \div 11$ **82.** $4257 \div 23$ **83.** $15\overline{)187}$ **84.** $34\overline{)695}$

Objective 4: Fractions and the Number Line

For Exercises 85–94, plot the fraction on the number line. **(See Example 7.)**

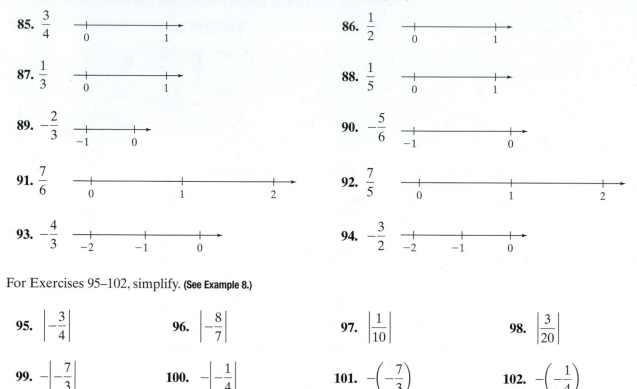

85. $\frac{3}{4}$

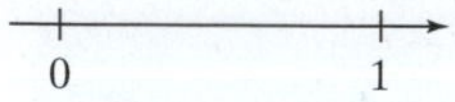

86. $\frac{1}{2}$

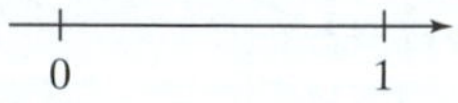

87. $\frac{1}{3}$

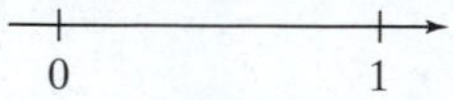

88. $\frac{1}{5}$

89. $-\frac{2}{3}$

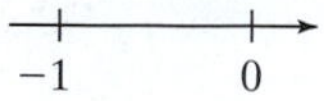

90. $-\frac{5}{6}$

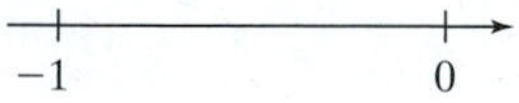

91. $\frac{7}{6}$

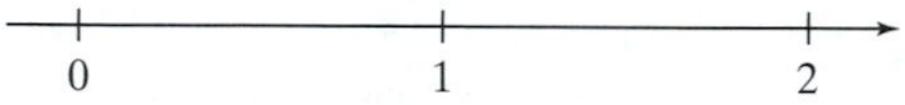

92. $\frac{7}{5}$

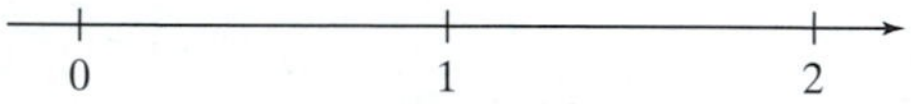

93. $-\frac{4}{3}$

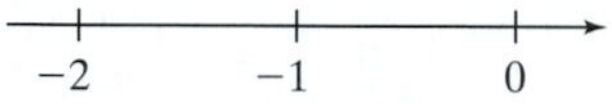

94. $-\frac{3}{2}$

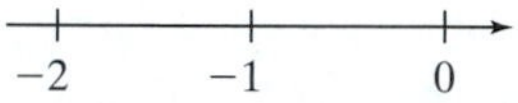

For Exercises 95–102, simplify. **(See Example 8.)**

95. $\left|-\frac{3}{4}\right|$

96. $\left|-\frac{8}{7}\right|$

97. $\left|\frac{1}{10}\right|$

98. $\left|\frac{3}{20}\right|$

99. $-\left|-\frac{7}{3}\right|$

100. $-\left|-\frac{1}{4}\right|$

101. $-\left(-\frac{7}{3}\right)$

102. $-\left(-\frac{1}{4}\right)$

Expanding Your Skills

103. True or false? Whole numbers can be written both as proper and improper fractions.

104. True or false? Suppose m and n are nonzero numbers, where $m > n$. Then $\frac{m}{n}$ is an improper fraction.

105. True or false? Suppose m and n are nonzero numbers, where $m > n$. Then $\frac{n}{m}$ is a proper fraction.

106. True or false? Suppose m and n are nonzero numbers, where $m > n$. Then $\frac{n}{3m}$ is a proper fraction.

Section 4.2 Simplifying Fractions

Objectives

1. Factorizations and Divisibility
2. Prime Factorization
3. Equivalent Fractions
4. Simplifying Fractions to Lowest Terms
5. Applications of Simplifying Fractions

1. Factorizations and Divisibility

Recall from Section 1.5 that two numbers multiplied to form a product are called factors. For example, $2 \cdot 3 = 6$ indicates that 2 and 3 are factors of 6. Likewise, because $1 \cdot 6 = 6$, the numbers 1 and 6 are factors of 6. In general, a **factor** of a number n is a nonzero whole number that divides evenly into n.

The products $2 \cdot 3$ and $1 \cdot 6$ are called factorizations of 6. In general, a **factorization** of a number n is a product of factors that equals n.

Example 1 Finding Factorizations of a Number

Find four different factorizations of 12.

Solution:

$$12 = \begin{cases} 1 \cdot 12 \\ 2 \cdot 6 \\ 3 \cdot 4 \\ 2 \cdot 2 \cdot 3 \end{cases}$$

TIP: Notice that a factorization may include more than two factors.

Skill Practice

1. Find four different factorizations of 18.

A factor of a number must divide evenly into the number. There are several rules by which we can quickly determine whether a number is divisible by 2, 3, 5, or 10. These are called divisibility rules.

PROCEDURE Divisibility Rules for 2, 3, 5, and 10

- *Divisibility by 2.* A whole number is divisible by 2 if it is an even number. That is, the ones-place digit is 0, 2, 4, 6, or 8.
  Examples: 26 and 384
- *Divisibility by 3.* A whole number is divisible by 3 if the sum of its digits is divisible by 3.
  Example: 312 (sum of digits is $3 + 1 + 2 = 6$, which is divisible by 3)
- *Divisibility by 5.* A whole number is divisible by 5 if its ones-place digit is 5 or 0.
  Examples: 45 and 260
- *Divisibility by 10.* A whole number is divisible by 10 if its ones-place digit is 0.
  Examples: 30 and 170

The divisibility rules for other numbers are harder to remember. In these cases, it is often easier simply to perform division to test for divisibility.

Example 2 Applying the Divisibility Rules

Determine whether the given number is divisible by 2, 3, 5, or 10.

a. 720 **b.** 82

TIP: When in doubt about divisibility, you can check by division. When we divide 82 by 2, the remainder is zero. This means that 2 divides evenly into 82.

Solution:

		Test for Divisibility	
a. 720	By 2:	Yes.	The number 720 is even.
	By 3:	Yes.	The sum $7 + 2 + 0 = 9$ is divisible by 3.
	By 5:	Yes.	The ones-place digit is 0.
	By 10:	Yes.	The ones-place digit is 0.
b. 82	By 2:	Yes.	The number 82 is even.
	By 3:	No.	The sum $8 + 2 = 10$ is not divisible by 3.
	By 5:	No.	The ones-place digit is not 5 or 0.
	By 10:	No.	The ones-place digit is not 0.

Skill Practice

Determine whether the given number is divisible by 2, 3, 5, or 10.

2. 75 **3.** 2100

Answers

1. For example.
$1 \cdot 18$
$2 \cdot 9$
$3 \cdot 6$
$2 \cdot 3 \cdot 3$
2. Divisible by 3 and 5
3. Divisible by 2, 3, 5, and 10

2. Prime Factorization

Two important classifications of whole numbers are prime numbers and composite numbers.

DEFINITION Prime and Composite Numbers

- A **prime number** is a whole number greater than 1 that has only two factors (itself and 1).
- A **composite number** is a whole number greater than 1 that is not prime. That is, a composite number will have at least one factor other than 1 and the number itself.

Note: The whole numbers 0 and 1 are neither prime nor composite.

Skill Practice

Determine whether the number is prime, composite, or neither.

4. 39 **5.** 0 **6.** 41

Example 3 Identifying Prime and Composite Numbers

Determine whether the number is prime, composite, or neither.

a. 19 **b.** 51 **c.** 1

Solution:

a. The number 19 is prime because its only factors are 1 and 19.

b. The number 51 is composite because $3 \cdot 17 = 51$. That is, 51 has factors other than 1 and 51.

c. The number 1 is neither prime nor composite by definition.

TIP: The number 2 is the only even prime number.

Prime numbers are used in a variety of ways in mathematics. It is advisable to become familiar with the first several prime numbers: 2, 3, 5, 7, 11, 13, 17, 19, 23, 29, . . .

In Example 1 we found four factorizations of 12.

$$1 \cdot 12$$
$$2 \cdot 6$$
$$3 \cdot 4$$
$$2 \cdot 2 \cdot 3$$

The last factorization $2 \cdot 2 \cdot 3$ consists of only prime-number factors. Therefore, we say $2 \cdot 2 \cdot 3$ is the prime factorization of 12.

Concept Connections

7. Is the product $2 \cdot 3 \cdot 10$ the prime factorization of 60? Explain.

DEFINITION Prime Factorization

The **prime factorization** of a number is the factorization in which every factor is a prime number.

Note: The order in which the factors are written does not affect the product.

Prime factorizations of numbers will be particularly helpful when we add, subtract, multiply, divide, and simplify fractions.

Answers

4. Composite **5.** Neither **6.** Prime

7. No. The factor 10 is not a prime number. The prime factorization of 60 is $2 \cdot 2 \cdot 3 \cdot 5$.

Example 4 Determining the Prime Factorization of a Number

Find the prime factorization of 220.

Solution:

One method to factor a whole number is to make a factor tree. Begin by determining *any* two numbers that when multiplied equal 220. Then continue factoring each factor until the branches "end" in prime numbers.

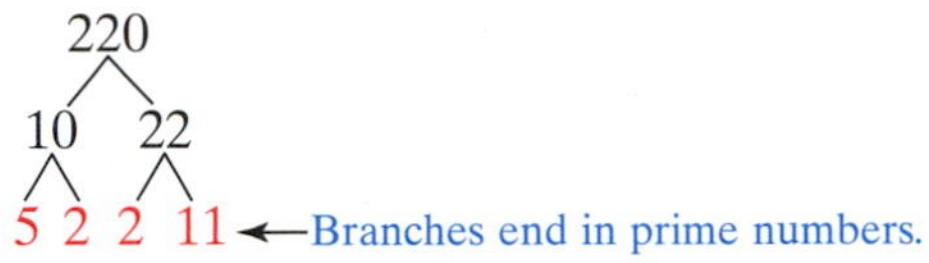

TIP: The prime factorization from Example 4 can also be expressed by using exponents as $2^2 \cdot 5 \cdot 11$.

Therefore, the prime factorization of 220 is $2 \cdot 2 \cdot 5 \cdot 11$.

Skill Practice

8. Find the prime factorization of 90.

TIP: In creating a factor tree, you can begin with any two factors of the number. The result will be the same. In Example 4, we could have started with the factors of 4 and 55.

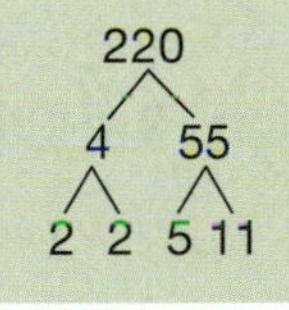

The prime factorization is $2 \cdot 2 \cdot 5 \cdot 11$ as expected.

Another technique to find the prime factorization of a number is to divide the number by the smallest known prime factor of the number. Then divide the quotient by its smallest prime factor. Continue dividing in this fashion until the quotient is a prime number. The prime factorization is the product of divisors and the final quotient. This is demonstrated in Example 5.

Example 5 Determining Prime Factorizations

Find the prime factorization.

a. 198 **b.** 153

Solution:

a.

2 is the smallest prime factor of 198.	⟶	2)198
3 is the smallest prime factor of 99.	⟶	3)99
3 is the smallest prime factor of 33	⟶	3)33
The last quotient is prime.	⟶	11

The prime factorization of 198 is $2 \cdot 3 \cdot 3 \cdot 11$ or $2 \cdot 3^2 \cdot 11$.

b.
3)153
3)51
17

The prime factorization of 153 is $3 \cdot 3 \cdot 17$ or $3^2 \cdot 17$.

Skill Practice

Find the prime factorization of the given number.

9. 168 **10.** 990

Answers

8. $2 \cdot 3 \cdot 3 \cdot 5$ or $2 \cdot 3^2 \cdot 5$
9. $2 \cdot 2 \cdot 2 \cdot 3 \cdot 7$ or $2^3 \cdot 3 \cdot 7$
10. $2 \cdot 3 \cdot 3 \cdot 5 \cdot 11$ or $2 \cdot 3^2 \cdot 5 \cdot 11$

3. Equivalent Fractions

The fractions $\frac{3}{6}$, $\frac{2}{4}$, and $\frac{1}{2}$ all represent the same portion of a whole. See Figure 4-5. Therefore, we say that the fractions are *equivalent*.

Animation

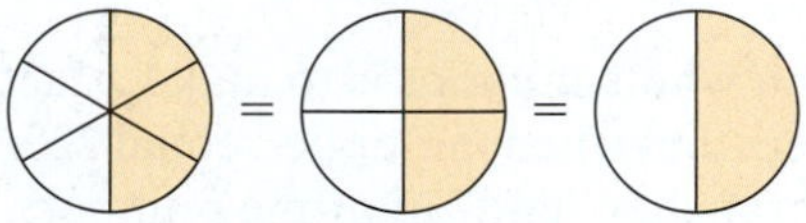

$$\frac{3}{6} = \frac{2}{4} = \frac{1}{2}$$

Figure 4-5

> **Avoiding Mistakes**
> The test to determine whether two fractions are equivalent is not the same process as multiplying fractions. Multiplying fractions is covered in Section 4.3.

One method to show that two fractions are equivalent is to calculate their cross products. For example, to show that $\frac{3}{6} = \frac{2}{4}$, we have

$$\frac{3}{6} \times \frac{2}{4}$$

$$3 \cdot 4 \stackrel{?}{=} 6 \cdot 2$$

$$12 = 12 \qquad \text{Yes. The fractions are equivalent.}$$

> **Skill Practice**
> Fill in the blank $\square$ with $=$ or $\neq$.
> **11.** $\frac{13}{24} \square \frac{6}{11}$ **12.** $\frac{9}{4} \square \frac{54}{24}$

Example 6 Determining Whether Two Fractions Are Equivalent

Fill in the blank $\square$ with $=$ or $\neq$.

a. $\frac{18}{39} \square \frac{6}{13}$ **b.** $\frac{5}{7} \square \frac{7}{9}$

Solution:

a. $\frac{18}{39} \stackrel{?}{\times} \frac{6}{13}$

$$18 \cdot 13 \stackrel{?}{=} 39 \cdot 6$$

$$234 = 234$$

Therefore, $\frac{18}{39} \boxed{=} \frac{6}{13}$.

b. $\frac{5}{7} \stackrel{?}{\times} \frac{7}{9}$

$$5 \cdot 9 \stackrel{?}{=} 7 \cdot 7$$

$$45 \neq 49$$

Therefore, $\frac{5}{7} \boxed{\neq} \frac{7}{9}$.

4. Simplifying Fractions to Lowest Terms

In Figure 4-5, we see that $\frac{3}{6}$, $\frac{2}{4}$, and $\frac{1}{2}$ all represent equal quantities. However, the fraction $\frac{1}{2}$ is said to be in **lowest terms** because the numerator and denominator share no common factors other than 1.

To simplify a fraction to lowest terms, we apply the following important principle.

> **PROPERTY Fundamental Principle of Fractions**
> Suppose that a number, c, is a common factor in the numerator and denominator of a fraction. Then
> $$\frac{a \cdot c}{b \cdot c} = \frac{a}{b} \cdot \frac{c}{c} = \frac{a}{b} \cdot 1 = \frac{a}{b} \qquad \text{provided } b \neq 0.$$

Answers
11. $\neq$ **12.** $=$

To simplify a fraction, we begin by factoring the numerator and denominator into prime factors. This will help identify the common factors.

Example 7 Simplifying a Fraction to Lowest Terms

Simplify to lowest terms.

a. $\frac{6}{10}$ **b.** $-\frac{170}{102}$ **c.** $\frac{20}{24}$

Solution:

a. $\frac{6}{10} = \frac{3 \cdot 2}{5 \cdot 2}$ Factor the numerator and denominator. Notice that 2 is a common factor.

$= \frac{3}{5} \cdot \frac{2}{2}$ Apply the fundamental principle of fractions.

$= \frac{3}{5} \cdot 1$ Any nonzero number divided by itself is 1.

$= \frac{3}{5}$

b. $-\frac{170}{102} = -\frac{5 \cdot 2 \cdot 17}{3 \cdot 2 \cdot 17}$ Factor the numerator and denominator.

$= -\frac{5}{3} \cdot \frac{2}{2} \cdot \frac{17}{17}$ Apply the fundamental principle of fractions.

$= -\frac{5}{3} \cdot 1 \cdot 1$ Any nonzero number divided by itself is 1.

$= -\frac{5}{3}$

c. $\frac{20}{24} = \frac{5 \cdot 2 \cdot 2}{3 \cdot 2 \cdot 2 \cdot 2}$ Factor the numerator and denominator.

$= \frac{5}{3 \cdot 2} \cdot \frac{2}{2} \cdot \frac{2}{2}$ Apply the fundamental principle of fractions.

$= \frac{5}{6} \cdot 1 \cdot 1$

$= \frac{5}{6}$

Skill Practice

Simplify to lowest terms.

13. $\frac{15}{35}$ **14.** $-\frac{26}{195}$

15. $\frac{150}{105}$

TIP: To check that you have simplified a fraction correctly, verify that the cross products are equal.

$\frac{6}{10} \quad \frac{3}{5}$

$6 \cdot 5 \stackrel{?}{=} 10 \cdot 3$

$30 = 30$ ✓

In Example 7, we show numerous steps to simplify fractions to lowest terms. However, the process is often made easier. For instance, we sometimes divide common factors, and replace them with the new common factor of 1.

$$\frac{20}{24} = \frac{5 \cdot \overset{1}{\cancel{2}} \cdot \overset{1}{\cancel{2}}}{3 \cdot 2 \cdot \underset{1}{\cancel{2}} \cdot \underset{1}{\cancel{2}}} = \frac{5}{6}$$

Answers

13. $\frac{3}{7}$ **14.** $-\frac{2}{15}$ **15.** $\frac{10}{7}$

The largest number that divides evenly into the numerator and denominator is called their **greatest common factor**. By identifying the greatest common factor you can simplify the process even more. For example, the greatest common factor of 20 and 24 is 4.

$$\frac{20}{24} = \frac{5 \cdot \overset{1}{\cancel{4}}}{6 \cdot \underset{1}{\cancel{4}}} = \frac{5}{6}$$

Notice that "dividing out" the common factor of 4 has the same effect as dividing the numerator and denominator by 4. This is often done mentally.

$$\frac{\overset{5}{\cancel{20}}}{\underset{6}{\cancel{24}}} = \frac{5}{6} \begin{array}{l} \longleftarrow \text{20 divided by 4 equals 5.} \\ \longleftarrow \text{24 divided by 4 equals 6.} \end{array}$$

TIP: Simplifying a fraction is also called reducing a fraction to lowest terms. For example, the simplified (or reduced) form of $\frac{20}{24}$ is $\frac{5}{6}$.

Skill Practice

Simplify the fraction. Write the answer as a fraction or whole number.

16. $\frac{39}{3}$ **17.** $\frac{15}{90}$

Example 8 Simplifying Fractions to Lowest Terms

Simplify the fraction. Write the answer as a fraction or whole number.

a. $\frac{75}{25}$ **b.** $\frac{12}{60}$

Solution:

a. $\frac{75}{25} = \frac{3 \cdot \overset{1}{\cancel{25}}}{1 \cdot \underset{1}{\cancel{25}}} = \frac{3}{1} = 3$ The greatest common factor in the numerator and denominator is 25.

Or alternatively: $\frac{\overset{3}{\cancel{75}}}{\underset{1}{\cancel{25}}} = \frac{3}{1} \begin{array}{l} \longleftarrow \text{75 divided by 25 equals 3.} \\ \longleftarrow \text{25 divided by 25 equals 1.} \end{array}$

$= 3$

TIP: Recall that any fraction of the form $\frac{n}{1} = n$. Therefore, $\frac{3}{1} = 3$.

b. $\frac{12}{60} = \frac{1 \cdot \overset{1}{\cancel{12}}}{5 \cdot \underset{1}{\cancel{12}}} = \frac{1}{5}$ The greatest common factor in the numerator and denominator is 12.

Or alternatively: $= \frac{\overset{1}{\cancel{12}}}{\underset{5}{\cancel{60}}} = \frac{1}{5} \begin{array}{l} \longleftarrow \text{12 divided by 12 equals 1.} \\ \longleftarrow \text{60 divided by 12 equals 5.} \end{array}$

Avoiding Mistakes

Do not forget to write the "1" in the numerator of the fraction $\frac{1}{5}$.

Avoiding Mistakes

Suppose that you do not recognize the *greatest* common factor in the numerator and denominator. You can still divide by *any* common factor. However, you will have to repeat this process more than once to simplify the fraction completely. For instance, consider the fraction from Example 8(b).

$\frac{\overset{2}{\cancel{12}}}{\underset{10}{\cancel{60}}} = \frac{2}{10}$ Dividing by the common factor of 6 leaves a fraction that can be simplified further.

$= \frac{\overset{1}{\cancel{2}}}{\underset{5}{\cancel{10}}} = \frac{1}{5}$ Divide again, this time by 2. The fraction is now simplified completely because the greatest common factor in the numerator and denominator is 1.

Answers

16. 13 **17.** $\frac{1}{6}$

Example 9 Simplifying Fractions by 10, 100, and 1000

Simplify each fraction to lowest terms by first reducing by 10, 100, or 1000. Write the answer as a fraction.

a. $\dfrac{170}{30}$ **b.** $\dfrac{2500}{75{,}000}$

Solution:

a. $\dfrac{170}{30} = \dfrac{17\cancel{0}}{3\cancel{0}}$ Both 170 and 30 are divisible by 10. "Strike through" one zero. This is equivalent to dividing by 10.

$= \dfrac{17}{3}$ The fraction $\frac{17}{3}$ is simplified completely.

b. $\dfrac{2500}{75{,}000} = \dfrac{25\cancel{00}}{75{,}0\cancel{00}}$ Both 2500 and 75,000 are divisible by 100. "Strike through" two zeros. This is equivalent to dividing by 100.

$= \dfrac{\overset{1}{\cancel{25}}}{\underset{30}{\cancel{750}}}$ Simplify further. Both 25 and 750 have a common factor of 25.

$= \dfrac{1}{30}$

Skill Practice

Simplify to lowest terms by first reducing by 10, 100, or 1000.

18. $\dfrac{630}{190}$ **19.** $\dfrac{1300}{52{,}000}$

Avoiding Mistakes

The "strike through" method only works for the digit 0 at the *end* of the numerator and denominator.

Concept Connections

20. How many zeros may be eliminated from the numerator and denominator of the fraction $\frac{430{,}000}{154{,}000{,}000}$?

The process to simplify a fraction is the same for fractions that contain variables. This is shown in Example 10.

Example 10 Simplifying a Fraction Containing Variables

Simplify. **a.** $\dfrac{10xy}{6x}$ **b.** $\dfrac{2a^3}{4a^4}$

Solution:

a. $\dfrac{10xy}{6x} = \dfrac{2 \cdot 5 \cdot x \cdot y}{2 \cdot 3 \cdot x}$ Factor the numerator and denominator. Common factors are shown in red.

$= \dfrac{\overset{1}{\cancel{2}} \cdot 5 \cdot \overset{1}{\cancel{x}} \cdot y}{\underset{1}{\cancel{2}} \cdot 3 \cdot \underset{1}{\cancel{x}}}$ Simplify.

$= \dfrac{5y}{3}$

b. $\dfrac{2a^3}{4a^4} = \dfrac{2 \cdot a \cdot a \cdot a}{2 \cdot 2 \cdot a \cdot a \cdot a \cdot a}$ Factor the numerator and denominator. Common factors are shown in red.

$= \dfrac{\overset{1}{\cancel{2}} \cdot \overset{1}{\cancel{a}} \cdot \overset{1}{\cancel{a}} \cdot \overset{1}{\cancel{a}}}{2 \cdot \underset{1}{\cancel{2}} \cdot \underset{1}{\cancel{a}} \cdot \underset{1}{\cancel{a}} \cdot \underset{1}{\cancel{a}} \cdot a}$ Simplify.

$= \dfrac{1}{2a}$

Skill Practice

Simplify.

21. $\dfrac{18d}{15cd}$ **22.** $\dfrac{9w^2}{36w^3}$

Avoiding Mistakes

Since division by 0 is undefined, we know that the value of the variable cannot make the denominator equal 0. In Example 10, $x \neq 0$ and $a \neq 0$.

Answers

18. $\dfrac{63}{19}$ **19.** $\dfrac{1}{40}$

20. Four zeros; the numerator and denominator are both divisible by 10,000.

21. $\dfrac{6}{5c}$ **22.** $\dfrac{1}{4w}$

5. Applications of Simplifying Fractions

Skill Practice

23. Joanne planted 77 seeds in her garden and 55 sprouted. Geoff planted 140 seeds and 80 sprouted.
- **a.** What fractional part of the seeds sprouted for Joanne and what part sprouted for Geoff?
- **b.** For which person did a greater portion of seeds sprout?

Example 11 Simplifying Fractions in an Application

Madeleine got 28 out of 35 problems correct on an algebra exam. David got 27 out of 45 questions correct on a different algebra exam.

a. What fractional part of the exam did each student answer correctly?

b. Which student performed better?

Solution:

a. Fractional part correct for Madeleine:

$$\frac{28}{35} \quad \text{or equivalently} \quad \frac{28}{35} = \frac{4 \cdot \overset{1}{\cancel{7}}}{5 \cdot \underset{1}{\cancel{7}}} = \frac{4}{5}$$

Fractional part correct for David:

$$\frac{27}{45} \quad \text{or equivalently} \quad \frac{27}{45} = \frac{3 \cdot \overset{1}{\cancel{9}}}{5 \cdot \underset{1}{\cancel{9}}} = \frac{3}{5}$$

b. From the simplified form of each fraction, we see that Madeleine performed better because $\frac{4}{5} > \frac{3}{5}$. That is, 4 parts out of 5 is greater than 3 parts out of 5. This is also easily verified on a number line.

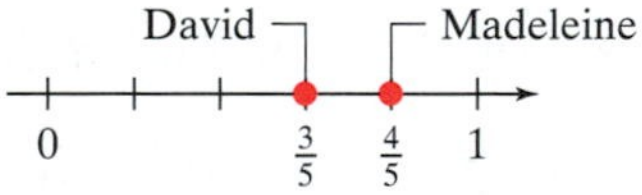

Answer

23. a. Joanne: $\frac{5}{7}$; Geoff: $\frac{4}{7}$
b. Joanne had a greater portion of seeds sprout.

Section 4.2 Practice Exercises

Boost *your* GRADE at ALEKS.com!

ALEKS® version 3.0

- Practice Problems
- Self-Tests
- NetTutor
- e-Professors
- Videos

Study Skills Exercise

1. Define the key terms.

a. **Factor** **b.** **Factorization** **c.** **Prime number** **d.** **Composite number**

e. **Prime factorization** **f.** **Lowest terms** **g.** **Greatest common factor**

Review Exercises

For Exercises 2–3, write two fractions, one representing the shaded area and one representing the unshaded area.

2. 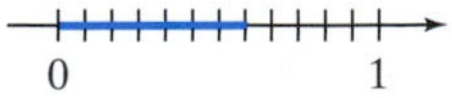

3. 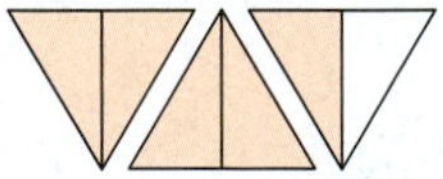

4. Write a fraction with numerator 6 and denominator 5. Is this fraction proper or improper?

5. Write the fraction $\frac{23}{5}$ as a mixed number.

6. Write the mixed number $6\frac{2}{7}$ as a fraction.

Objective 1: Factorizations and Divisibility

For Exercises 7–10, find two different factorizations of each number. (Answers may vary.) **(See Example 1.)**

7. 8 **8.** 20 **9.** 24 **10.** 14

11. State the divisibility rule for dividing by 2.

12. State the divisibility rule for dividing by 10.

13. State the divisibility rule for dividing by 3.

14. State the divisibility rule for dividing by 5.

For Exercises 15–22, determine if the number is divisible by **a.** 2 **b.** 3 **c.** 5 **d.** 10 **(See Example 2.)**

15. 45 **16.** 100 **17.** 137 **18.** 241

19. 108 **20.** 1040 **21.** 3140 **22.** 2115

23. Ms. Berglund has 28 students in her class. Can she distribute a package of 84 candies evenly to her students?

24. Mr. Blankenship has 22 students in an algebra class. He has 110 sheets of graph paper. Can he distribute the graph paper evenly among his students?

Objective 2: Prime Factorization

For Exercises 25–32, determine whether the number is prime, composite, or neither. **(See Example 3.)**

25. 7 **26.** 17 **27.** 10 **28.** 21

29. 1 **30.** 0 **31.** 97 **32.** 57

33. One method for finding prime numbers is the *sieve of Eratosthenes*. The natural numbers from 2 to 50 are shown in the table. Start at the number 2 (the smallest prime number). Leave the number 2 and cross out every second number after the number 2. This will eliminate all numbers that are multiples of 2. Then go back to the beginning of the chart and leave the number 3, but cross out every third number after the number 3 (thus eliminating the multiples of 3). Begin at the next open number and continue this process. The numbers that remain are prime numbers. Use this process to find the prime numbers less than 50.

	2	3	4	5	6	7	8	9	10
11	12	13	14	15	16	17	18	19	20
21	22	23	24	25	26	27	28	29	30
31	32	33	34	35	36	37	38	39	40
41	42	43	44	45	46	47	48	49	50

34. True or false? The square of any prime number is also a prime number.

35. True or false? All odd numbers are prime.

36. True or false? All even numbers are composite.

For Exercises 37–40, determine whether the factorization represents the prime factorization. If not, explain why.

37. $36 = 2 \cdot 2 \cdot 9$ **38.** $48 = 2 \cdot 3 \cdot 8$ **39.** $210 = 5 \cdot 2 \cdot 7 \cdot 3$ **40.** $126 = 3 \cdot 7 \cdot 3 \cdot 2$

For Exercises 41–48, find the prime factorization. **(See Examples 4 and 5.)**

41. 70 **42.** 495 **43.** 260 **44.** 175

45. 147 **46.** 231 **47.** 616 **48.** 364

Objective 3: Equivalent Fractions

For Exercises 49–50, shade the second figure so that it expresses a fraction equivalent to the first figure.

49. 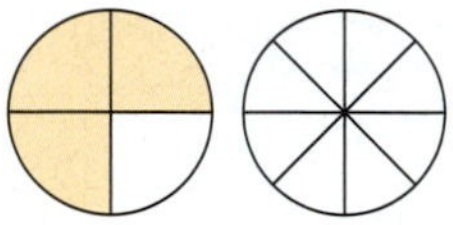

50. 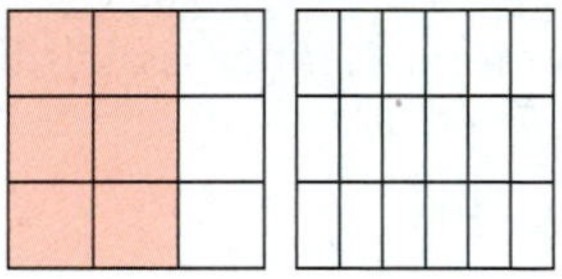

51. True or false? The fractions $\frac{4}{5}$ and $\frac{5}{4}$ are equivalent.

52. True or false? The fractions $\frac{3}{1}$ and $\frac{1}{3}$ are equivalent.

For Exercises 53–60, determine if the fractions are equivalent. Then fill in the blank with either = or ≠. **(See Example 6.)**

53. $\frac{2}{3} \square \frac{3}{5}$ **54.** $\frac{1}{4} \square \frac{2}{9}$ **55.** $\frac{1}{2} \square \frac{3}{6}$ **56.** $\frac{6}{16} \square \frac{3}{8}$

57. $\frac{12}{16} \square \frac{3}{4}$ **58.** $\frac{4}{5} \square \frac{12}{15}$ **59.** $\frac{8}{9} \square \frac{20}{27}$ **60.** $\frac{5}{6} \square \frac{12}{18}$

Objective 4: Simplifying Fractions to Lowest Terms

For Exercises 61–80, simplify the fraction to lowest terms. Write the answer as a fraction or a whole number. **(See Examples 7–8.)**

61. $\frac{12}{24}$ **62.** $\frac{15}{18}$ **63.** $\frac{6}{18}$ **64.** $\frac{21}{24}$

65. $-\frac{36}{20}$ **66.** $-\frac{49}{42}$ **67.** $-\frac{15}{12}$ **68.** $-\frac{30}{25}$

69. $\frac{9}{9}$ **70.** $\frac{2}{2}$ **71.** $\frac{105}{140}$ **72.** $\frac{84}{126}$

73. $-\frac{33}{11}$ **74.** $-\frac{65}{5}$ **75.** $\frac{77}{110}$ **76.** $\frac{85}{153}$

77. $\frac{385}{195}$ **78.** $\frac{39}{130}$ **79.** $\frac{34}{85}$ **80.** $\frac{69}{92}$

For Exercises 81–88, simplify to lowest terms by first reducing the powers of 10. **(See Example 9.)**

81. $\frac{120}{160}$ **82.** $\frac{720}{800}$ **83.** $-\frac{3000}{1800}$ **84.** $-\frac{2000}{1500}$

85. $\frac{42{,}000}{22{,}000}$ **86.** $\frac{50{,}000}{65{,}000}$ **87.** $\frac{5100}{30{,}000}$ **88.** $\frac{9800}{28{,}000}$

For Exercises 89–96, simplify the expression. **(See Example 10.)**

89. $\frac{16ab}{10a}$ **90.** $\frac{25mn}{10n}$ **91.** $\frac{14xyz}{7z}$ **92.** $\frac{18pqr}{6q}$

93. $\frac{5x^4}{15x^3}$ **94.** $\frac{4y^3}{20y}$ **95.** $-\frac{6ac^2}{12ac^4}$ **96.** $-\frac{3m^2n}{9m^2n^4}$

Objective 5: Applications of Simplifying Fractions

97. André tossed a coin 48 times and heads came up 20 times. What fractional part of the tosses came up heads? What fractional part came up tails?

98. At Pizza Company, Lee made 70 pizzas one day. There were 105 pizzas sold that day. What fraction of the pizzas did Lee make?

99. **a.** What fraction of the alphabet is made up of vowels? (Include the letter y as a vowel, not a consonant.)

b. What fraction of the alphabet is made up of consonants?

100. Of the 88 constellations that can be seen in the night sky, 12 are associated with astrological horoscopes. The names of as many as 36 constellations are associated with animals or mythical creatures.

a. Of the 88 constellations, what fraction is associated with horoscopes?

b. What fraction of the constellations have names associated with animals or mythical creatures?

101. Jonathan and Jared both sold candy bars for a fundraiser. Jonathan sold 25 of his 35 candy bars, and Jared sold 24 of his 28 candy bars. **(See Example 11.)**

a. What fractional part of his total number of candy bars did each person sell?

b. Which person sold the greater fractional part?

102. Lisa and Lynette are taking online courses. Lisa has completed 14 out of 16 assignments in her course while Lynette has completed 15 out of 24 assignments.

a. What fractional part of her total number of assignments did each woman complete?

b. Which woman has completed more of her course?

103. Raymond read 720 pages of a 792-page book. His roommate, Travis, read 540 pages from a 660-page book.

a. What fractional part of the book did each person read?

b. Which of the roommates read a greater fraction of his book?

104. Mr. Bishop and Ms. Waymire both gave exams today. By mid-afternoon, Mr. Bishop had finished grading 16 out of 36 exams, and Ms. Waymire had finished grading 15 out of 27 exams.

a. What fractional part of her total has Ms. Waymire completed?

b. What fractional part of his total has Mr. Bishop completed?

105. For a recent year, the population of the United States was reported to be 296,000,000. During the same year, the population of California was 36,458,000.

a. Round the U.S. population to the nearest hundred million.

b. Round the population of California to the nearest million.

c. Using the results from parts (a) and (b), write a simplified fraction showing the portion of the U.S. population represented by California.

106. For a recent year, the population of the United States was reported to be 296,000,000. During the same year, the population of Ethiopia was 75,067,000.

a. Round the U.S. population to the nearest hundred million.

b. Round the population of Ethiopia to the nearest million.

c. Using the results from parts (a) and (b), write a simplified fraction comparing the population of the United States to the population of Ethiopia.

d. Based on the result from part (c), how many times greater is the U.S. population than the population of Ethiopia?

Expanding Your Skills

107. Write three fractions equivalent to $\frac{3}{4}$.

108. Write three fractions equivalent to $\frac{1}{3}$.

109. Write three fractions equivalent to $-\frac{12}{18}$.

110. Write three fractions equivalent to $-\frac{80}{100}$.

Calculator Connections

Topic: Simplifying Fractions on a Calculator

Some calculators have a fraction key, a b/c. To enter a fraction, follow this example.

Expression: $\frac{3}{4}$

Keystrokes: 3 [a b/c] 4 [=]

Result: 3⌋4

numerator — denominator

To simplify a fraction to lowest terms, follow this example.

Expression: $\frac{22}{10}$

Keystrokes: 22 [a b/c] 10 [=]

Result: 2_1⌋5 $= 2\frac{1}{5}$

whole number — fraction

To convert to an improper fraction, press [2nd] [d/e] 11⌋5 $= \frac{11}{5}$

Calculator Exercises

For Exercises 111–118, use a calculator to simplify the fractions. Write the answer as a proper or improper fraction.

111. $\frac{792}{891}$

112. $\frac{728}{784}$

113. $\frac{779}{969}$

114. $\frac{462}{220}$

115. $\frac{493}{510}$

116. $\frac{871}{469}$

117. $\frac{969}{646}$

118. $\frac{713}{437}$

Multiplication and Division of Fractions

Section 4.3

Objectives

1. Multiplication of Fractions
2. Area of a Triangle
3. Reciprocal
4. Division of Fractions
5. Applications of Multiplication and Division of Fractions

1. Multiplication of Fractions

Suppose Elija takes $\frac{1}{3}$ of a cake and then gives $\frac{1}{2}$ of this portion to his friend Max. Max gets $\frac{1}{2}$ of $\frac{1}{3}$ of the cake. This is equivalent to the expression $\frac{1}{2} \cdot \frac{1}{3}$. See Figure 4-6.

Elija takes $\frac{1}{3}$
Max gets $\frac{1}{2}$ of $\frac{1}{3} = \frac{1}{6}$

Figure 4-6

From the illustration, the product $\frac{1}{2} \cdot \frac{1}{3} = \frac{1}{6}$. Notice that the product $\frac{1}{6}$ is found by multiplying the numerators and multiplying the denominators. This is true in general to multiply fractions.

PROCEDURE Multiplying Fractions

To multiply fractions, write the product of the numerators over the product of the denominators. Then simplify the resulting fraction, if possible.

$$\frac{a}{b} \cdot \frac{c}{d} = \frac{a \cdot c}{b \cdot d} \quad \text{provided } b \text{ and } d \text{ are not equal to 0.}$$

Concept Connections

1. What fraction is $\frac{1}{2}$ of $\frac{1}{4}$ of a whole?

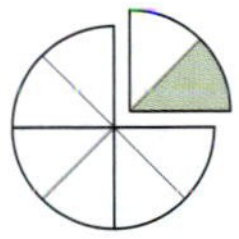

Example 1 Multiplying Fractions

Multiply.

a. $\frac{2}{5} \cdot \frac{4}{7}$ **b.** $-\frac{8}{3} \cdot 5$

Solution:

a. $\frac{2}{5} \cdot \frac{4}{7} = \frac{2 \cdot 4}{5 \cdot 7} = \frac{8}{35}$ ← Multiply the numerators. ← Multiply the denominators.

Notice that the product $\frac{8}{35}$ is simplified completely because there are no common factors shared by 8 and 35.

b. $-\frac{8}{3} \cdot 5 = -\frac{8}{3} \cdot \frac{5}{1}$ First write the whole number as a fraction.

$= -\frac{8 \cdot 5}{3 \cdot 1}$ Multiply the numerators. Multiply the denominators. The product of two numbers of different signs is negative.

$= -\frac{40}{3}$ The product cannot be simplified because there are no common factors shared by 40 and 3.

Skill Practice

Multiply. Write the answer as a fraction.

2. $\frac{2}{3} \cdot \frac{5}{9}$ **3.** $-\frac{7}{12} \cdot 11$

Answers

1. $\frac{1}{8}$ **2.** $\frac{10}{27}$ **3.** $-\frac{77}{12}$

Example 2 illustrates a case where the product of fractions must be simplified.

Skill Practice

Multiply and simplify.

4. $\frac{7}{20} \cdot \frac{4}{3}$

Example 2 Multiplying and Simplifying Fractions

Multiply the fractions and simplify if possible. $\frac{4}{30} \cdot \frac{5}{14}$

Solution:

$$\frac{4}{30} \cdot \frac{5}{14} = \frac{4 \cdot 5}{30 \cdot 14}$$ Multiply the numerators. Multiply the denominators.

$$= \frac{20}{420}$$ Simplify by first dividing 20 and 420 by 10.

$$= \frac{\overset{1}{\cancel{2}}}{\underset{21}{\cancel{42}}}$$ Simplify further by dividing 2 and 42 by 2.

$$= \frac{1}{21}$$

It is often easier to simplify *before* multiplying. Consider the product from Example 2.

$$\frac{4}{30} \cdot \frac{5}{14} = \frac{\overset{2}{\cancel{4}}}{\underset{6}{\cancel{30}}} \cdot \frac{\overset{1}{\cancel{5}}}{\underset{7}{\cancel{14}}}$$ 4 and 14 share a common factor of 2. 30 and 5 share a common factor of 5.

$$= \frac{\overset{1}{\cancel{2}}}{\underset{3}{\cancel{6}}} \cdot \frac{1}{7}$$ 2 and 6 share a common factor of 2.

$$= \frac{1}{21}$$

Skill Practice

Multiply and simplify.

5. $\left(-\frac{6}{25}\right)\left(-\frac{15}{18}\right)$

Example 3 Multiplying and Simplifying Fractions

Multiply and simplify. $\left(-\frac{10}{18}\right)\left(-\frac{21}{55}\right)$

Solution:

$$\left(-\frac{10}{18}\right)\left(-\frac{21}{55}\right) = +\left(\frac{10}{18} \cdot \frac{21}{55}\right)$$ First note that the product will be positive. The product of two numbers with the same sign is positive.

$$\frac{10}{18} \cdot \frac{21}{55} = \frac{\overset{2}{\cancel{10}}}{\underset{6}{\cancel{18}}} \cdot \frac{\overset{7}{\cancel{21}}}{\underset{11}{\cancel{55}}}$$ 10 and 55 share a common factor of 5. 18 and 21 share a common factor of 3.

$$= \frac{\overset{1}{\cancel{2}}}{\underset{3}{\cancel{6}}} \cdot \frac{7}{11}$$ We can simplify further because 2 and 6 share a common factor of 2.

$$= \frac{7}{33}$$

Answers

4. $\frac{7}{15}$ 5. $\frac{1}{5}$

Example 4 **Multiplying Fractions Containing Variables**

Multiply and simplify.

a. $\dfrac{5x}{7} \cdot \dfrac{2}{15x}$ **b.** $\dfrac{2a^2}{3b} \cdot \dfrac{b^3}{a}$

> **Avoiding Mistakes**
>
> In Example 4, $x \neq 0$, $a \neq 0$, and $b \neq 0$. The denominator of a fraction cannot equal 0.

Solution:

a. $\dfrac{5x}{7} \cdot \dfrac{2}{15x} = \dfrac{5 \cdot x \cdot 2}{7 \cdot 3 \cdot 5 \cdot x}$ Multiply fractions and factor.

$= \dfrac{\overset{1}{\cancel{5}} \cdot \overset{1}{\cancel{x}} \cdot 2}{7 \cdot 3 \cdot \underset{1}{\cancel{5}} \cdot \underset{1}{\cancel{x}}}$ Simplify. The common factors are in red.

$= \dfrac{2}{21}$

b. $\dfrac{2a^2}{3b} \cdot \dfrac{b^3}{a} = \dfrac{2 \cdot a \cdot a \cdot b \cdot b \cdot b}{3 \cdot b \cdot a}$ Multiply fractions and factor.

$= \dfrac{2 \cdot a \cdot \overset{1}{\cancel{a}} \cdot \overset{1}{\cancel{b}} \cdot b \cdot b}{3 \cdot \underset{1}{\cancel{b}} \cdot \underset{1}{\cancel{a}}}$ Simplify. The common factors are in red.

$= \dfrac{2ab^2}{3}$

Skill Practice

Multiply and simplify.

6. $\dfrac{3w}{11} \cdot \dfrac{5}{6w}$

7. $\dfrac{5x^3}{6y} \cdot \dfrac{y^2}{x}$

2. Area of a Triangle

Recall that the area of a rectangle with length l and width w is given by

$$A = l \cdot w$$

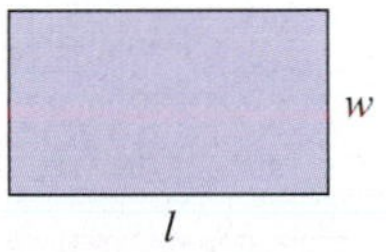

FORMULA Area of a Triangle

The formula for the area of a triangle is given by $A = \frac{1}{2}bh$, read "one-half base times height."

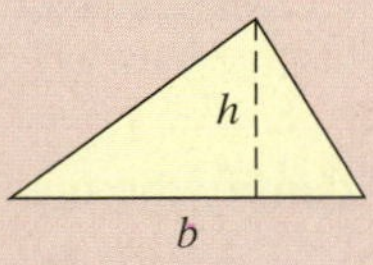

The value of b is the measure of the base of the triangle. The value of h is the measure of the height of the triangle. The base b can be chosen as the length of any of the sides of the triangle. However, once you have chosen the base, the height must be measured as the shortest distance from the base to the opposite vertex (or point) of the triangle.

Figure 4-7 shows the same triangle with different choices for the base. Figure 4-8 shows a situation in which the height must be drawn "outside" the triangle. In such a case, notice that the height is drawn down to an imaginary extension of the base line.

Animation

Answers

6. $\dfrac{5}{22}$ **7.** $\dfrac{5x^2y}{6}$

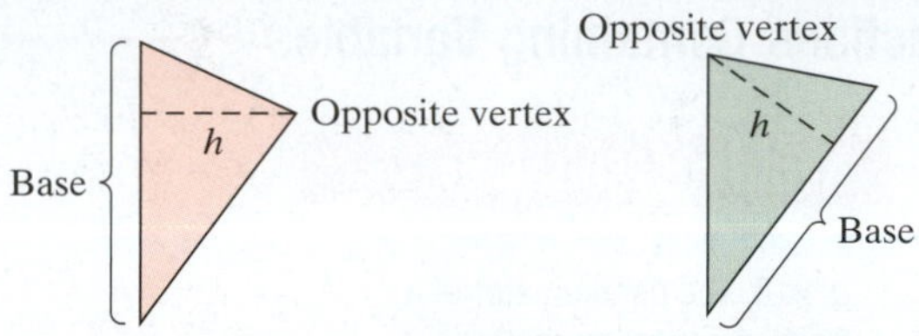

Figure 4-7

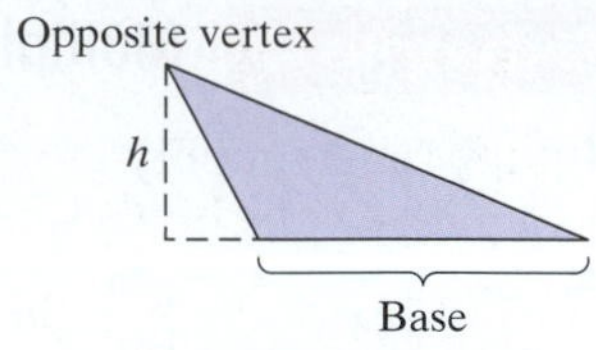

Figure 4-8

Skill Practice

Find the area of the triangle.

8.

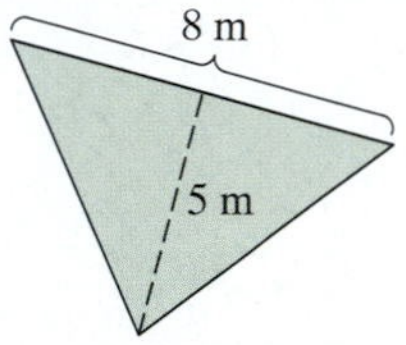

Example 5 Finding the Area of a Triangle

Find the area of the triangle.

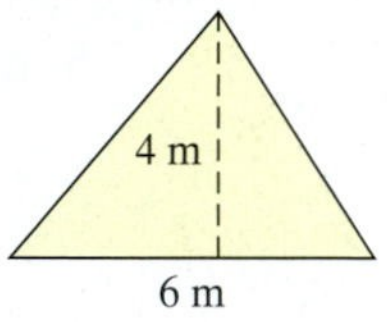

Solution:

$b = 6 \text{ m}$ and $h = 4 \text{ m}$ — Identify the measure of the base and the height.

$A = \frac{1}{2}bh$

$= \frac{1}{2}(6 \text{ m})(4 \text{ m})$ — Apply the formula for the area of a triangle.

$= \frac{1}{2}\left(\frac{6}{1} \text{ m}\right)\left(\frac{4}{1} \text{ m}\right)$ — Write the whole numbers as fractions.

$= \frac{1}{\cancel{2}_1}\left(\frac{\cancel{6}^3}{1} \text{ m}\right)\left(\frac{4}{1} \text{ m}\right)$ — Simplify.

$= \frac{12}{1} \text{ m}^2$ — Multiply numerators. Multiply denominators.

$= 12 \text{ m}^2$ — The area of the triangle is 12 square meters (m^2).

Skill Practice

Find the area of the triangle.

9.

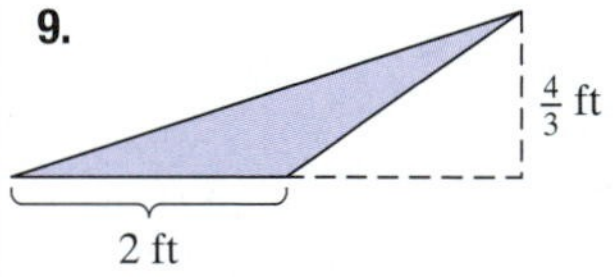

Example 6 Finding the Area of a Triangle

Find the area of the triangle.

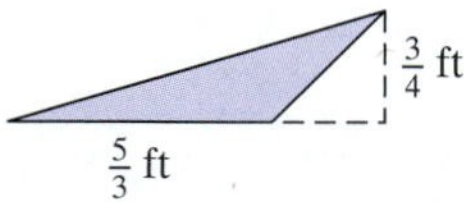

Solution:

$b = \frac{5}{3} \text{ ft}$ and $h = \frac{3}{4} \text{ ft}$ — Identify the measure of the base and the height.

$A = \frac{1}{2}bh$

$= \frac{1}{2}\left(\frac{5}{3} \text{ ft}\right)\left(\frac{3}{4} \text{ ft}\right)$ — Apply the formula for the area of a triangle.

$= \frac{1}{2}\left(\frac{5}{\cancel{3}_1} \text{ ft}\right)\left(\frac{\cancel{3}^1}{4} \text{ ft}\right)$ — Simplify.

$= \frac{5}{8} \text{ ft}^2$ — The area of the triangle is $\frac{5}{8}$ square feet (ft^2).

Answers

8. 20 m^2 **9.** $\frac{4}{3} \text{ ft}^2$ or $1\frac{1}{3} \text{ ft}^2$

3. Reciprocal

Two numbers whose product is 1 are *reciprocals* of each other. For example, consider the product of $\frac{3}{8}$ and $\frac{8}{3}$.

$$\frac{3}{8} \cdot \frac{8}{3} = \frac{\overset{1}{\cancel{3}}}{\underset{1}{\cancel{8}}} \cdot \frac{\overset{1}{\cancel{8}}}{\underset{1}{\cancel{3}}} = 1$$

Because the product equals 1, we say that $\frac{3}{8}$ is the reciprocal of $\frac{8}{3}$ and vice versa.

To divide fractions, first we need to learn how to find the reciprocal of a fraction.

Concept Connections

Fill in the blank.

10. The product of a number and its reciprocal is ______.

PROCEDURE Finding the Reciprocal of a Fraction

To find the **reciprocal** of a nonzero fraction, interchange the numerator and denominator of the fraction. If a and b are nonzero numbers, then the reciprocal of $\frac{a}{b}$ is $\frac{b}{a}$. This is because $\frac{a}{b} \cdot \frac{b}{a} = 1$.

Example 7 Finding Reciprocals

Find the reciprocal.

a. $\frac{2}{5}$ **b.** $\frac{1}{9}$ **c.** 5 **d.** 0 **e.** $-\frac{3}{7}$

Solution:

a. The reciprocal of $\frac{2}{5}$ is $\frac{5}{2}$.

b. The reciprocal of $\frac{1}{9}$ is $\frac{9}{1}$, or 9.

c. First write the whole number 5 as the improper fraction $\frac{5}{1}$.
The reciprocal of $\frac{5}{1}$ is $\frac{1}{5}$.

d. The number 0 has no reciprocal because $\frac{1}{0}$ is undefined.

e. The reciprocal of $-\frac{3}{7}$ is $-\frac{7}{3}$. This is because $-\frac{3}{7} \cdot \left(-\frac{7}{3}\right) = 1$.

Skill Practice

Find the reciprocal.

11. $\frac{7}{10}$ 12. $\frac{1}{4}$

13. 7 14. 1

15. $-\frac{9}{8}$

TIP: From Example 7(e) we see that a negative number will have a negative reciprocal.

4. Division of Fractions

To understand the division of fractions, we compare it to the division of whole numbers. The statement $6 \div 2$ asks, "How many groups of 2 can be found among 6 wholes?" The answer is 3.

$6 \div 2 = 3$ 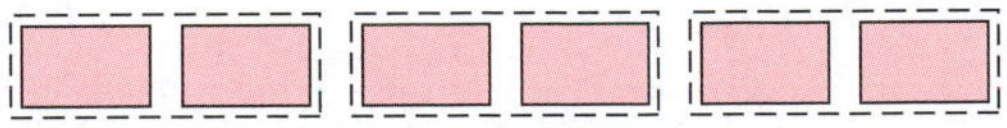

In fractional form, the statement $6 \div 2 = 3$ can be written as $\frac{6}{2} = 3$. This result can also be found by multiplying.

$$6 \cdot \frac{1}{2} = \frac{6}{1} \cdot \frac{1}{2} = \frac{6}{2} = 3$$

Answers

10. 1 **11.** $\frac{10}{7}$ **12.** 4

13. $\frac{1}{7}$ **14.** 1 **15.** $-\frac{8}{9}$

That is, to divide by 2 is equivalent to multiplying by the reciprocal $\frac{1}{2}$.

In general, to divide two nonzero numbers we can multiply the dividend by the reciprocal of the divisor. This is how we divide by a fraction.

Concept Connections

Fill in the box to make the right- and left-hand sides equal.

16. $\frac{4}{7} \div \frac{3}{5} = \frac{4}{7} \cdot \square$

17. $\frac{2}{3} \div \frac{1}{5} = \square \cdot \frac{5}{1}$

PROCEDURE Dividing Fractions

To divide two fractions, multiply the dividend (the "first" fraction) by the reciprocal of the divisor (the "second" fraction).

The process to divide fractions can be written symbolically as

Change division to multiplication.

$$\frac{a}{b} \div \frac{c}{d} = \frac{a}{b} \cdot \frac{d}{c} \quad \text{provided } b, c, \text{ and } d \text{ are not } 0.$$

Take the reciprocal of the divisor.

Skill Practice

Divide and simplify.

18. $\frac{1}{4} \div \frac{2}{5}$

19. $\frac{3}{8} \div \left(-\frac{9}{10}\right)$

Example 8 Dividing Fractions

Divide and simplify, if possible.

a. $\frac{2}{5} \div \frac{7}{4}$ **b.** $\frac{2}{27} \div \left(-\frac{8}{15}\right)$

Solution:

a. $\frac{2}{5} \div \frac{7}{4} = \frac{2}{5} \cdot \frac{4}{7}$ Multiply by the reciprocal of the divisor ("second" fraction).

$= \frac{2 \cdot 4}{5 \cdot 7}$ Multiply numerators. Multiply denominators.

$= \frac{8}{35}$

Avoiding Mistakes

Do not try to simplify until after taking the reciprocal of the divisor. In Example 8(a) it would be incorrect to "cancel" the 2 and the 4 in the expression $\frac{2}{5} \div \frac{7}{4}$.

b. $\frac{2}{27} \div \left(-\frac{8}{15}\right) = \frac{2}{27} \cdot \left(-\frac{15}{8}\right)$ Multiply by the reciprocal of the divisor.

$= -\left(\frac{2}{27} \cdot \frac{15}{8}\right)$ The product will be negative.

$= -\left(\frac{\overset{1}{\cancel{2}}}{\underset{9}{\cancel{27}}} \cdot \frac{\overset{5}{\cancel{15}}}{\underset{4}{\cancel{8}}}\right)$ Simplify.

$= -\frac{5}{36}$ Multiply.

Answers

16. $\frac{5}{3}$ 17. $\frac{2}{3}$

18. $\frac{5}{8}$ 19. $-\frac{5}{12}$

Example 9 Dividing Fractions

Divide and simplify. Write the answer as a fraction.

a. $\frac{35}{14} \div 7$ **b.** $-12 \div \left(-\frac{8}{3}\right)$

Solution:

a. $\frac{35}{14} \div 7 = \frac{35}{14} \div \frac{7}{1}$ Write the whole number 7 as an improper fraction *before* multiplying by the reciprocal.

$= \frac{35}{14} \cdot \frac{1}{7}$ Multiply by the reciprocal of the divisor.

$= \frac{\overset{5}{\cancel{35}}}{14} \cdot \frac{1}{\underset{1}{\cancel{7}}}$ Simplify.

$= \frac{5}{14}$ Multiply.

b. $-12 \div \left(-\frac{8}{3}\right) = +\left(12 \div \frac{8}{3}\right)$ First note that the quotient will be positive. The quotient of two numbers with the same sign is positive.

$12 \div \frac{8}{3} = \frac{12}{1} \div \frac{8}{3}$ Write the whole number 12 as an improper fraction.

$= \frac{12}{1} \cdot \frac{3}{8}$ Multiply by the reciprocal of the divisor.

$= \frac{\overset{3}{\cancel{12}}}{1} \cdot \frac{3}{\underset{2}{\cancel{8}}}$ Simplify.

$= \frac{9}{2}$ Multiply.

Skill Practice

Divide and simplify. Write the quotient as a fraction.

20. $\frac{15}{4} \div 10$

21. $-20 \div \left(-\frac{12}{5}\right)$

Example 10 Dividing Fractions Containing Variables

Divide and simplify. $-\frac{10x^2}{y^2} \div \frac{5}{y}$

Solution:

$-\frac{10x^2}{y^2} \div \frac{5}{y} = -\frac{10x^2}{y^2} \cdot \frac{y}{5}$ Multiply by the reciprocal of the divisor.

$= -\frac{2 \cdot 5 \cdot x \cdot x \cdot y}{y \cdot y \cdot 5}$ Multiply fractions and factor.

$= -\frac{2 \cdot \overset{1}{\cancel{5}} \cdot x \cdot x \cdot \overset{1}{\cancel{y}}}{y \cdot \underset{1}{\cancel{y}} \cdot \underset{1}{\cancel{5}}}$ Simplify. The common factors are in red.

$= -\frac{2x^2}{y}$

Skill Practice

Divide and simplify.

22. $-\frac{8y^3}{7} \div \frac{4y}{5}$

Answers

20. $\frac{3}{8}$ **21.** $\frac{25}{3}$ **22.** $-\frac{10y^2}{7}$

5. Applications of Multiplication and Division of Fractions

Sometimes it is difficult to determine whether multiplication or division is appropriate to solve an application problem. Division is generally used for a problem that requires you to separate or "split up" a quantity into pieces. Multiplication is generally used if it is necessary to take a fractional part of a quantity.

Skill Practice

23. A cookie recipe requires $\frac{2}{5}$ package of chocolate chips for each batch of cookies. If a restaurant has 20 packages of chocolate chips, how many batches of cookies can it make?

Example 11 Using Division in an Application

A road crew must mow the grassy median along a stretch of highway I-95. If they can mow $\frac{5}{8}$ mile (mi) in 1 hr, how long will it take them to mow a 15-mi stretch?

Solution:

Read and familiarize.

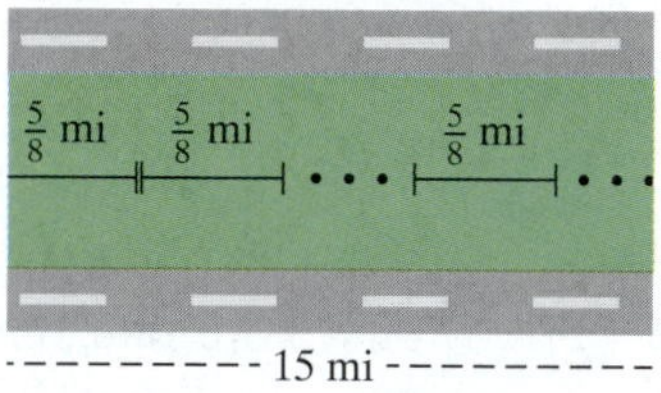

Strategy/operation: From the figure, we must separate or "split up" a 15-mi stretch of highway into pieces that are $\frac{5}{8}$ mi in length. Therefore, we must divide 15 by $\frac{5}{8}$.

$$15 \div \frac{5}{8} = \frac{15}{1} \cdot \frac{8}{5}$$ Write the whole number as a fraction. Multiply by the reciprocal of the divisor.

$$= \frac{\overset{3}{\cancel{15}}}{1} \cdot \frac{8}{\underset{1}{\cancel{5}}}$$

$$= 24$$

The 15-mi stretch of highway will take 24 hr to mow.

Skill Practice

24. A $\frac{25}{2}$-yd ditch will be dug to put in a new water line. If piping comes in segments of $\frac{5}{4}$ yd, how many segments are needed to line the ditch?

Example 12 Using Division in an Application

A $\frac{9}{4}$-ft length of wire must be cut into pieces of equal length that are $\frac{3}{8}$ ft long. How many pieces can be cut?

Solution:

Read and familiarize.
Operation: Here we divide the total length of wire into pieces of equal length.

$$\frac{9}{4} \div \frac{3}{8} = \frac{9}{4} \cdot \frac{8}{3}$$ Multiply by the reciprocal of the divisor.

$$= \frac{\overset{3}{\cancel{9}}}{\underset{1}{\cancel{4}}} \cdot \frac{\overset{2}{\cancel{8}}}{\underset{1}{\cancel{3}}}$$ Simplify.

$$= 6$$

Six pieces of wire can be cut.

Answers

23. 50 batches
24. 10 segments of piping

Example 13 Using Multiplication in an Application

Carson estimates that his total cost for college for 1 year is \$12,600. He has financial aid to pay $\frac{2}{3}$ of the cost.

a. How much money will be paid by financial aid?

b. How much money will Carson have to pay?

c. If Carson's parents help him by paying $\frac{1}{2}$ of the amount not paid by financial aid, how much money will be paid by Carson's parents?

Solution:

a. Carson's financial aid will pay $\frac{2}{3}$ of \$12,600. Because we are looking for a fraction of a quantity, we multiply.

$$\frac{2}{3} \cdot 12{,}600 = \frac{2}{3} \cdot \frac{12{,}600}{1}$$

$$= \frac{2}{\cancel{3}_{1}} \cdot \frac{\cancel{12{,}600}^{4200}}{1}$$

$$= 8400$$

Financial aid will pay \$8400.

b. Carson will have to pay the remaining portion of the cost. This can be found by subtraction.

$$\$12{,}600 - \$8400 = \$4200$$

Carson will have to pay \$4200.

TIP: The answer to Example 13(b) could also have been found by noting that financial aid paid $\frac{2}{3}$ of the cost. This means that Carson must pay $\frac{1}{3}$ of the cost, or

$$\frac{1}{3} \cdot \frac{\$12{,}600}{1} = \frac{1}{\cancel{3}_{1}} \cdot \frac{\cancel{\$12{,}600}^{4200}}{1}$$

$$= \$4200$$

c. Carson's parents will pay $\frac{1}{2}$ of \$4200.

$$\frac{1}{\cancel{2}_{1}} \cdot \frac{\cancel{4200}^{2100}}{1}$$

Carson's parents will pay \$2100.

Skill Practice

25. A new school will cost \$20,000,000 to build, and the state will pay $\frac{3}{5}$ of the cost.

a. How much will the state pay?

b. How much will the state not pay?

c. The county school district issues bonds to pay $\frac{4}{5}$ of the money not covered by the state. How much money will be covered by bonds?

Answer

25. a. \$12,000,000 **b.** \$8,000,000 **c.** \$6,400,000

Section 4.3 Practice Exercises

Study Skills Exercise

1. Define the key term **reciprocal**.

Review Exercises

2. The number 126 is divisible by which of the following?

a. 2 **b.** 3 **c.** 5 **d.** 10

3. Identify the numerator and denominator. Then simplify the fraction. $\dfrac{2100}{7000}$

4. Simplify. $\dfrac{12x^2}{15x}$

5. Convert $2\dfrac{7}{8}$ to an improper fraction.

6. Convert $\dfrac{32}{9}$ to a mixed number.

Objective 1: Multiplication of Fractions

7. Shade the portion of the figure that represents $\frac{1}{4}$ of $\frac{1}{4}$.

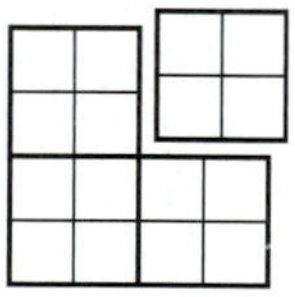

8. Shade the portion of the figure that represents $\frac{1}{3}$ of $\frac{1}{4}$.

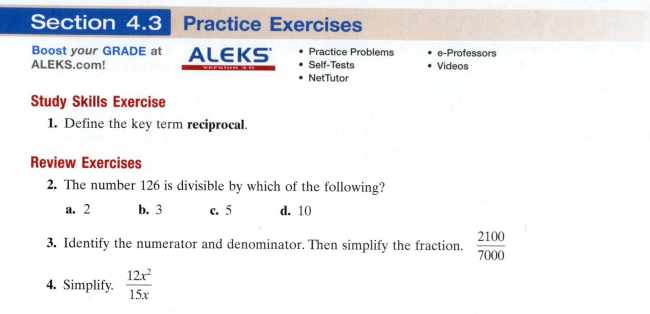

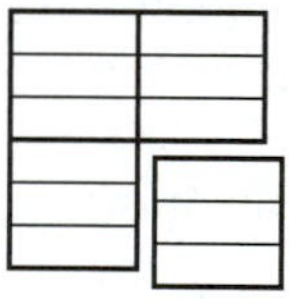

For Exercises 9–36, multiply the fractions and simplify to lowest terms. Write the answer as an improper fraction when necessary. **(See Examples 1–4.)**

9. $\dfrac{1}{2} \cdot \dfrac{3}{8}$

10. $\dfrac{2}{3} \cdot \dfrac{1}{3}$

11. $\left(-\dfrac{12}{7}\right)\left(-\dfrac{2}{5}\right)$

12. $\left(-\dfrac{9}{10}\right)\left(-\dfrac{7}{4}\right)$

13. $8 \cdot \left(\dfrac{1}{11}\right)$

14. $3 \cdot \left(\dfrac{2}{7}\right)$

15. $-\dfrac{4}{5} \cdot 6$

16. $-\dfrac{5}{8} \cdot 5$

17. $\dfrac{2}{9} \cdot \dfrac{3}{5}$

18. $\dfrac{1}{8} \cdot \dfrac{4}{7}$

19. $\dfrac{5}{6} \cdot \dfrac{3}{4}$

20. $\dfrac{7}{12} \cdot \dfrac{18}{5}$

21. $\dfrac{21}{5} \cdot \dfrac{25}{12}$

22. $\dfrac{16}{25} \cdot \dfrac{15}{32}$

23. $\dfrac{24}{15} \cdot \left(-\dfrac{5}{3}\right)$

24. $\dfrac{49}{24} \cdot \left(-\dfrac{6}{7}\right)$

25. $\left(\dfrac{6}{11}\right)\left(\dfrac{22}{15}\right)$

26. $\left(\dfrac{12}{45}\right)\left(\dfrac{5}{4}\right)$

27. $-12 \cdot \left(-\dfrac{15}{42}\right)$

28. $-4 \cdot \left(-\dfrac{8}{92}\right)$

29. $\dfrac{3y}{10} \cdot \dfrac{5}{y}$

30. $\dfrac{7z}{12} \cdot \dfrac{4}{z}$

31. $-\dfrac{4ab}{5} \cdot \dfrac{1}{8b}$

32. $-\dfrac{6cd}{7} \cdot \dfrac{1}{18d}$

33. $\dfrac{5x}{4y^2} \cdot \dfrac{y}{25x}$

34. $\dfrac{14w}{3z^4} \cdot \dfrac{z^2}{28w}$

35. $\left(-\dfrac{12m^3}{n}\right)\left(-\dfrac{n}{3m}\right)$

36. $\left(-\dfrac{15p^4}{t^2}\right)\left(-\dfrac{t^3}{3p}\right)$

Objective 2: Area of a Triangle

For Exercises 37–40, label the height with h and the base with b, as shown in the figure.

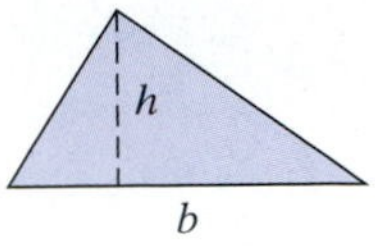

37. 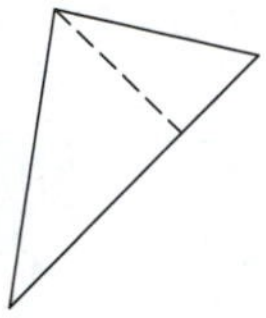

38. 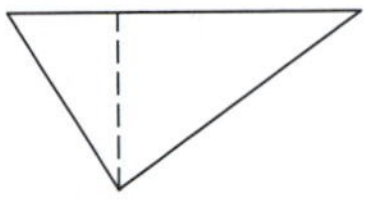

39. 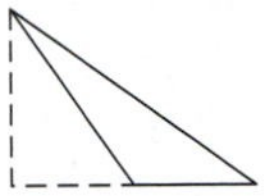

40. 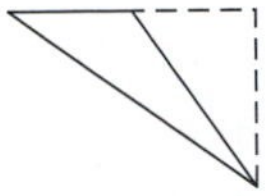

For Exercises 41–46, find the area of the triangle. **(See Examples 5–6.)**

41. 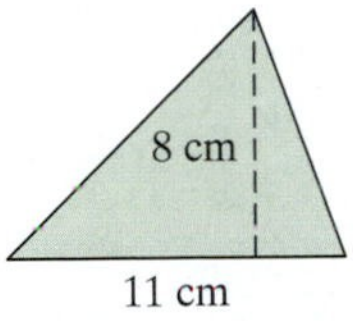

42. 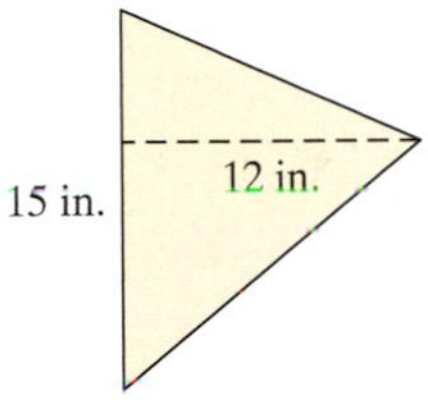

43. 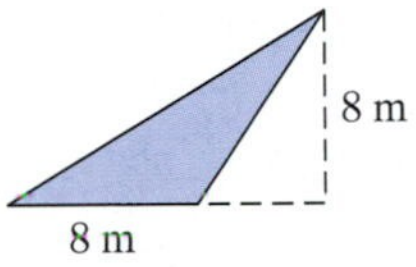

44. 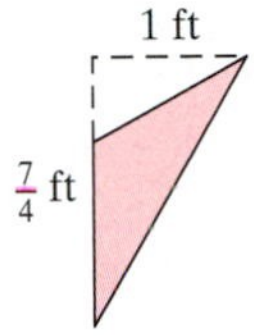

45. 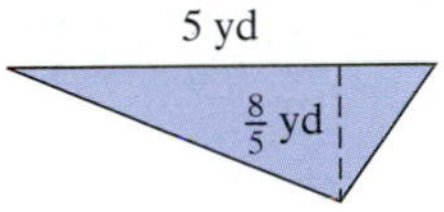

46. 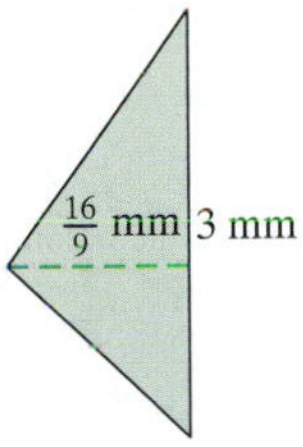

Objective 3: Reciprocal

For Exercises 47–54, find the reciprocal of the number, if it exists. **(See Example 7.)**

47. $\frac{7}{8}$ **48.** $\frac{5}{6}$ **49.** $-\frac{10}{9}$ **50.** $-\frac{14}{5}$

51. -4 **52.** -9 **53.** 0 **54.** $\frac{0}{4}$

Objective 4: Division of Fractions

For Exercises 55–58, fill in the blank.

55. Dividing by 3 is the same as multiplying by____.

56. Dividing by 5 is the same as multiplying by____.

57. Dividing by -8 is the same as ________ by $-\frac{1}{8}$.

58. Dividing by -12 is the same as ________ by $-\frac{1}{12}$.

For Exercises 59–78, divide and simplify the answer to lowest terms. Write the answer as a fraction or whole number. **(See Examples 8–10.)**

59. $\frac{2}{15} \div \frac{5}{12}$ **60.** $\frac{11}{3} \div \frac{6}{5}$ **61.** $\left(-\frac{7}{13}\right) \div \left(-\frac{2}{5}\right)$ **62.** $\left(-\frac{8}{7}\right) \div \left(-\frac{3}{10}\right)$

63. $\frac{14}{3} \div \frac{6}{5}$ **64.** $\frac{11}{2} \div \frac{3}{4}$ **65.** $\frac{15}{2} \div \left(-\frac{3}{2}\right)$ **66.** $\frac{9}{10} \div \left(-\frac{9}{2}\right)$

67. $\frac{3}{4} \div \frac{3}{4}$ **68.** $\frac{6}{5} \div \frac{6}{5}$ **69.** $-7 \div \frac{2}{3}$ **70.** $-4 \div \frac{3}{5}$

71. $\frac{12}{5} \div 4$ **72.** $\frac{20}{6} \div 2$ **73.** $-\frac{9}{100} \div \frac{13}{1000}$ **74.** $-\frac{1000}{17} \div \frac{10}{3}$

75. $\frac{4xy}{3} \div \frac{14x}{9}$ **76.** $\frac{3ab}{7} \div \frac{15a}{14}$ **77.** $-\frac{20c^3}{d^2} \div \frac{5c}{d^3}$ **78.** $-\frac{24w^2}{z} \div \frac{3w}{z^3}$

Mixed Exercises

For Exercises 79–94, multiply or divide as indicated. Write the answer as a fraction or whole number.

79. $\frac{7}{8} \div \frac{1}{4}$ **80.** $\frac{7}{12} \div \frac{5}{3}$ **81.** $\frac{5}{8} \cdot \frac{2}{9}$ **82.** $\frac{1}{16} \cdot \frac{4}{3}$

83. $6 \cdot \left(-\frac{4}{3}\right)$ **84.** $-12 \cdot \frac{5}{6}$ **85.** $\left(-\frac{16}{5}\right) \div (-8)$ **86.** $\left(-\frac{42}{11}\right) \div (-7)$

87. $\frac{1}{8} \cdot 16$ **88.** $\frac{2}{3} \cdot 9$ **89.** $8 \div \frac{16}{3}$ **90.** $5 \div \frac{15}{4}$

91. $\frac{13x^2}{y^2} \div \left(-\frac{26x}{y^3}\right)$ **92.** $\frac{11z}{w^3} \div \left(-\frac{33}{w}\right)$ **93.** $\left(-\frac{ad}{3}\right) \div \left(-\frac{ad^2}{6}\right)$ **94.** $\left(-\frac{c^2}{5}\right) \div \left(-\frac{c^3}{25}\right)$

Objective 5: Applications of Multiplication and Division of Fractions

95. During the month of December, a department store wraps packages free of charge. Each package requires $\frac{2}{3}$ yd of ribbon. If Li used up a 36-yd roll of ribbon, how many packages were wrapped? **(See Example 11.)**

96. A developer sells lots of land in increments of $\frac{3}{4}$ acre. If the developer has 60 acres, how many lots can be sold?

97. If one cup is $\frac{1}{16}$ gal, how many cups of orange juice can be filled from $\frac{3}{2}$ gal? **(See Example 12.)**

98. If 1 centimeter (cm) is $\frac{1}{100}$ meter (m), how many centimeters are in a $\frac{5}{4}$-m piece of rope?

99. Dorci buys 16 sheets of plywood, each $\frac{3}{4}$ in. thick, to cover her windows in the event of a hurricane. She stacks the wood in the garage. How high will the stack be?

100. Davey built a bookshelf 36 in. long. Can the shelf hold a set of encyclopedias if there are 24 books and each book averages $\frac{5}{4}$ in. thick? Explain your answer.

101. A radio station allows 18 minutes (min) of advertising each hour. How many 40-second ($\frac{2}{3}$-min) commercials can be run in

a. 1 hr **b.** 1 day

102. A television station has 20 min of advertising each hour. How many 30-second ($\frac{1}{2}$-min) commercials can be run in

a. 1 hr **b.** 1 day

103. Ricardo wants to buy a new house for \$240,000. The bank requires $\frac{1}{10}$ of the cost of the house as a down payment. As a gift, Ricardo's mother will pay $\frac{2}{3}$ of the down payment. **(See Example 13.)**

a. How much money will Ricardo's mother pay toward the down payment?

b. How much money will Ricardo have to pay toward the down payment?

c. How much is left over for Ricardo to finance?

104. Althea wants to buy a Toyota Camry for a total cost of $18,000. The dealer requires $\frac{1}{12}$ of the money as a down payment. Althea's parents have agreed to pay one-half of the down payment for her.

a. How much money will Althea's parents pay toward the down payment?

b. How much will Althea pay toward the down payment?

c. How much will Althea have to finance?

105. Frankie's lawn measures 40 yd by 36 yd. In the morning he mowed $\frac{2}{3}$ of the lawn. How many square yards of lawn did he already mow? How much is left to be mowed?

106. Bob laid brick to make a rectangular patio in the back of his house. The patio measures 20 yd by 12 yd. On Saturday, he put down bricks for $\frac{3}{8}$ of the patio area. How many square yards is this?

107. In a certain sample of individuals, $\frac{2}{5}$ are known to have blood type O. Of the individuals with blood type O, $\frac{1}{4}$ are Rh-negative. What fraction of the individuals in the sample have O negative blood?

108. Jim has half a pizza left over from dinner. If he eats $\frac{1}{4}$ of this for breakfast, what fractional part did he eat for breakfast?

109. A lab technician has $\frac{7}{4}$ liters (L) of alcohol. If she needs samples of $\frac{1}{8}$ L, how many samples can she prepare?

110. Troy has a $\frac{7}{8}$-in. nail that he must hammer into a board. Each strike of the hammer moves the nail $\frac{1}{16}$ in. into the board. How many strikes of the hammer must he make?

111. In the summer at the South Pole, heavy equipment is used 24 hr a day. For every gallon of fuel actually used at the South Pole, it takes $3\frac{1}{2}$ gal ($\frac{7}{2}$ gal) to get it there.

a. If in one day 130 gal of fuel was used, how many gallons of fuel did it take to transport the 130 gal?

b. How much fuel was used in all?

112. The Bishop Gaming Center hosts a football pool. There is $1200 in prize money. The first-place winner receives $\frac{2}{3}$ of the prize money. The second-place winner receives $\frac{1}{4}$ of the prize money, and the third-place winner receives $\frac{1}{12}$ of the prize money. How much money does each person get?

113. How many eighths are in $\frac{9}{4}$?

114. How many sixths are in $\frac{4}{3}$?

115. Find $\frac{2}{5}$ of $\frac{1}{5}$.

116. Find $\frac{2}{3}$ of $\frac{1}{3}$.

Expanding Your Skills

117. The rectangle shown here has an area of 30 ft^2. Find the length.

118. The rectangle shown here has an area of 8 m^2. Find the width.

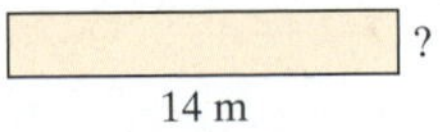

119. Find the next number in the sequence: $\frac{1}{2}, \frac{1}{4}, \frac{1}{8}, \frac{1}{16},$ _____

120. Find the next number in the sequence: $\frac{2}{3}, \frac{2}{9}, \frac{2}{27},$ _____

121. Which is greater, $\frac{1}{2}$ of $\frac{1}{8}$ or $\frac{1}{8}$ of $\frac{1}{2}$?

122. Which is greater, $\frac{2}{3}$ of $\frac{1}{4}$ or $\frac{1}{4}$ of $\frac{2}{3}$?

Section 4.4 Least Common Multiple and Equivalent Fractions

Objectives

1. Least Common Multiple
2. Applications of the Least Common Multiple
3. Writing Equivalent Fractions
4. Ordering Fractions

1. Least Common Multiple

In Section 4.5, we will learn how to add and subtract fractions. To add or subtract fractions with different denominators, we must learn how to convert unlike fractions into like fractions. An essential concept in this process is the idea of a least common multiple of two or more numbers.

When we multiply a number by the whole numbers 1, 2, 3, and so on, we form the **multiples** of the number. For example, some of the multiples of 6 and 9 are shown below.

Multiples of 6	Multiples of 9
$6 \cdot 1 = 6$	$9 \cdot 1 = 9$
$6 \cdot 2 = 12$	$9 \cdot 2 = 18$
$6 \cdot 3 = 18$	$9 \cdot 3 = 27$
$6 \cdot 4 = 24$	$9 \cdot 4 = 36$
$6 \cdot 5 = 30$	$9 \cdot 5 = 45$
$6 \cdot 6 = 36$	$9 \cdot 6 = 54$
$6 \cdot 7 = 42$	$9 \cdot 7 = 63$
$6 \cdot 8 = 48$	$9 \cdot 8 = 72$
$6 \cdot 9 = 54$	$9 \cdot 9 = 81$

In red, we have indicated several multiples that are common to both 6 and 9.

Concept Connections

1. Explain the difference between a multiple of a number and a factor of a number.

The **least common multiple (LCM)** of two given numbers is the smallest whole number that is a multiple of each given number. For example, the LCM of 6 and 9 is 18.

Multiples of 6: 6, 12, 18, 24, 30, 36, 42, . . .

Multiples of 9: 9, 18, 27, 36, 45, 54, 63, . . .

TIP: There are infinitely many numbers that are common multiples of both 6 and 9. These include 18, 36, 54, 72, and so on. However, 18 is the smallest, and is therefore the *least* common multiple.

If one number is a multiple of another number, then the LCM is the larger of the two numbers. For example, the LCM of 4 and 8 is 8.

Multiples of 4: 4, 8, 12, 16, . . .

Multiples of 8: 8, 16, 24, 32, . . .

Skill Practice

Find the LCM by listing several multiples of each number.

2. 15 and 25 **3.** 4, 6, and 10

Example 1 Finding the LCM by Listing Multiples

Find the LCM of the given numbers by listing several multiples of each number.

a. 15 and 12 **b.** 10, 15, and 8

Solution:

a. Multiples of 15: 15, 30, 45, 60
Multiples of 12: 12, 24, 36, 48, 60

The LCM of 15 and 12 is 60.

Answers

1. A multiple of a number is the product of the number and a whole number 1 or greater. A factor of a number is a value that divides evenly into the number.
2. 75 3. 60

b. Multiples of 10: 10, 20, 30, 40, 50, 60, 70, 80, 90, 100, 110, 120
Multiples of 15: 15, 30, 45, 60, 75, 90, 105, 120
Multiples of 8: 8, 16, 24, 32, 40, 48, 56, 64, 72, 80, 88, 96, 104, 112, 120

The LCM of 10, 15, and 8 is 120.

In Example 1 we used the method of listing multiples to find the LCM of two or more numbers. As you can see, the solution to Example 1(b) required several long lists of multiples. Here we offer another method to find the LCM of two given numbers by using their prime factors.

PROCEDURE Using Prime Factors to Find the LCM of Two Numbers

Step 1 Write each number as a product of prime factors.

Step 2 The LCM is the product of unique prime factors from both numbers. Use repeated factors the maximum number of times they appear in either factorization.

This process is demonstrated in Example 2.

Example 2 Finding the LCM by Using Prime Factors

Find the LCM.

a. 14 and 12 **b.** 50 and 24 **c.** 45, 54, and 50

Solution:

a. Find the prime factorization for 14 and 12.

	2's	3's	7's
14 =	2 ·		(7)
12 =	(2 · 2 ·)	(3)	

For the factors of 2, 3, and 7, we circle the greatest number of times each occurs. The LCM is the product.

LCM = $2 \cdot 2 \cdot 3 \cdot 7 = 84$

b. Find the prime factorization for 50 and 24.

	2's	3's	5's
50 =	2 ·		(5 · 5)
24 =	(2 · 2 · 2 ·)	(3)	

The factor 5 is repeated twice. The factor 2 is repeated 3 times. The factor 3 is used only once.

LCM = $2 \cdot 2 \cdot 2 \cdot 3 \cdot 5 \cdot 5 = 600$

(The LCM can also be written as $2^3 \cdot 3 \cdot 5^2$.)

c. Find the prime factorization for 45, 54, and 50.

	2's	3's	5's
45 =		3 · 3 ·	5
54 =	2 ·	(3 · 3 · 3)	
50 =	(2 ·)		(5 · 5)

LCM = $2 \cdot 3 \cdot 3 \cdot 3 \cdot 5 \cdot 5 = 1350$

(The LCM can also be written as $2 \cdot 3^3 \cdot 5^2$.)

Skill Practice

Find the LCM by using prime factors.

4. 9 and 24
5. 16 and 9
6. 36, 42, and 30

TIP: The product $2 \cdot 2 \cdot 3 \cdot 7$ can also be written as $2^2 \cdot 3 \cdot 7$.

Answers

4. 72 **5.** 144 **6.** 1260

2. Applications of the Least Common Multiple

Skill Practice

7. Three runners run on an oval track. One runner takes 60 sec to complete the loop. The second runner requires 75 sec, and the third runner requires 90 sec. Suppose the runners begin "lined up" at the same point on the track. Find the minimum amount of time required for all three runners to be lined up again.

Example 3 Using the LCM in an Application

A tile wall is to be made from 6-in., 8-in., and 12-in. square tiles. A design is made by alternating rows with different-size tiles. The first row uses only 6-in. tiles, the second row uses only 8-in. tiles, and the third row uses only 12-in. tiles. Neglecting the grout seams, what is the shortest length of wall space that can be covered using only whole tiles?

Solution:

The length of the first row must be a multiple of 6 in., the length of the second row must be a multiple of 8 in., and the length of the third row must be a multiple of 12 in. Therefore, the shortest-length wall that can be covered is given by the LCM of 6, 8, and 12.

$$6 = 2 \cdot 3$$
$$8 = 2 \cdot 2 \cdot 2$$
$$12 = 2 \cdot 2 \cdot 3$$

The LCM is $2 \cdot 2 \cdot 2 \cdot 3 = 24$. The shortest-length wall is 24 in.

Animation

This means that four 6-in. tiles can be placed on the first row, three 8-in. tiles can be placed on the second row, and two 12-in. tiles can be placed in the third row. See Figure 4-9.

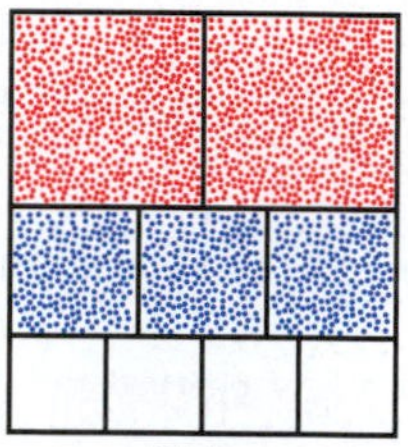

Figure 4-9

3. Writing Equivalent Fractions

A fractional amount of a whole may be represented by many fractions. For example, the fractions $\frac{1}{2}, \frac{2}{4}, \frac{3}{6}$, and $\frac{4}{8}$ all represent the same portion of a whole. See Figure 4-10.

Animation

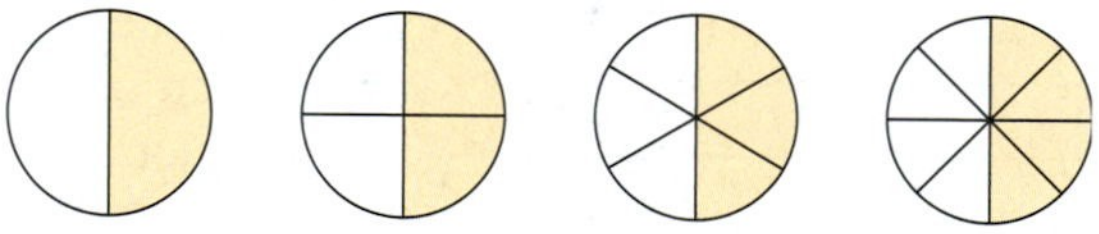

$$\frac{1}{2} = \frac{2}{4} = \frac{3}{6} = \frac{4}{8}$$

Figure 4-10

Expressing a fraction in an equivalent form is important for several reasons. We need this skill to order fractions and to add and subtract fractions.

Writing a fraction as an equivalent fraction is an application of the fundamental principle of fractions given in Section 4.2.

Skill Practice

Write the fraction with the indicated denominator.

8. $\frac{2}{3} = \frac{}{15}$

Example 4 Writing Equivalent Fractions

Write the fraction with the indicated denominator. $\frac{2}{9} = \frac{}{36}$

Solution:

$\frac{2}{9} = \frac{}{36}$ What number must we multiply 9 by to get 36?

$\frac{2 \cdot 4}{9 \cdot 4} = \frac{8}{36}$ Multiply numerator and denominator by 4.

Therefore, $\frac{2}{9}$ is equivalent to $\frac{8}{36}$.

Answers

7. After 900 sec (15 min) the runners will again be "lined up."
8. $\frac{10}{15}$

TIP: In Example 4, we multiplied numerator and denominator of the fraction by 4. This is the same as multiplying the fraction by a convenient form of 1.

$$\frac{2}{9} = \frac{2}{9} \cdot 1 = \frac{2}{9} \cdot \frac{4}{4} = \frac{2 \cdot 4}{9 \cdot 4} = \frac{8}{36}$$

This is the same as multiplying numerator and denominator by 4.

Example 5 Writing Equivalent Fractions

Write the fraction with the indicated denominator. $\frac{11}{8} = \frac{\quad}{56}$

Solution:

$\frac{11}{8} = \frac{\quad}{56}$ — What number must we multiply 8 by to get 56?

$\frac{11 \cdot 7}{8 \cdot 7} = \frac{77}{56}$ — Multiply numerator and denominator by 7.

Therefore, $\frac{11}{8}$ is equivalent to $\frac{77}{56}$.

Skill Practice

Write the fraction with the indicated denominator.

9. $\frac{5}{6} = \frac{\quad}{54}$

Example 6 Writing Equivalent Fractions

Write the fractions with the indicated denominator.

a. $-\frac{5}{6} = -\frac{\quad}{30}$ **b.** $\frac{9}{-4} = \frac{\quad}{8}$

Solution:

a. $-\frac{5}{6} = -\frac{\quad}{30}$

$-\frac{5 \cdot 5}{6 \cdot 5} = -\frac{25}{30}$ We must multiply by 5 to get a denominator of 30. Multiply numerator and denominator by 5.

Therefore, $-\frac{5}{6}$ is equivalent to $-\frac{25}{30}$.

b. $\frac{9 \cdot (-2)}{-4 \cdot (-2)} = \frac{-18}{8}$ We must multiply by −2 to get a denominator of 8. Multiply numerator and denominator by −2.

Therefore, $\frac{9}{-4}$ is equivalent to $\frac{-18}{8}$.

TIP: Recall that $\frac{-18}{8} = \frac{18}{-8} = -\frac{18}{8}$. All are equivalent to $\frac{9}{-4}$.

Skill Practice

Write the fractions with the indicated denominator.

10. $-\frac{10}{3} = -\frac{\quad}{12}$

11. $\frac{3}{-8} = \frac{\quad}{16}$

Answers

9. $\frac{45}{54}$ **10.** $-\frac{40}{12}$ **11.** $\frac{-6}{16}$

Skill Practice

Write the fractions with the indicated denominator.

12. $\dfrac{7}{9} = \dfrac{\quad}{18y}$

13. $\dfrac{10}{w} = \dfrac{\quad}{w^2}$

Example 7 Writing Equivalent Fractions

Write the fractions with the indicated denominator.

a. $\dfrac{2}{3} = \dfrac{\quad}{9x}$ **b.** $\dfrac{4}{y} = \dfrac{\quad}{y^2}$

Solution:

a. $\dfrac{2}{3} = \dfrac{\quad}{9x}$

What must we multiply 3 by to get $9x$?

$\dfrac{2 \cdot 3x}{3 \cdot 3x} = \dfrac{6x}{9x}$

Multiply numerator and denominator by $3x$.

Therefore, $\dfrac{2}{3}$ is equivalent to $\dfrac{6x}{9x}$.

b. $\dfrac{4}{y} = \dfrac{\quad}{y^2}$

What must we multiply y by to get y^2?

$\dfrac{4 \cdot y}{y \cdot y} = \dfrac{4y}{y^2}$

Multiply numerator and denominator by y.

Therefore, $\dfrac{4}{y}$ is equivalent to $\dfrac{4y}{y^2}$.

4. Ordering Fractions

Suppose we want to determine which of two fractions is larger. Comparing fractions with the same denominator, such as $\frac{3}{5}$ and $\frac{2}{5}$ is relatively easy. Clearly 3 parts out of 5 is greater than 2 parts out of 5.

Animation

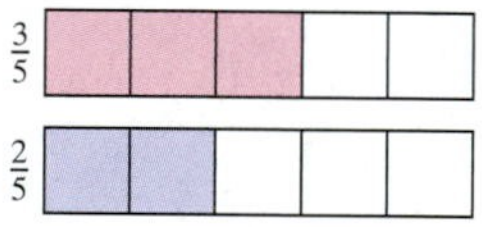

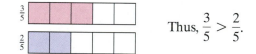

Thus, $\dfrac{3}{5} > \dfrac{2}{5}$.

So how would we compare the relative size of two fractions with *different* denominators such as $\frac{3}{5}$ and $\frac{4}{7}$? Our first step is to write the fractions as equivalent fractions with the same denominator, called a common denominator. The **least common denominator (LCD)** of two fractions is the LCM of the denominators of the fractions. The LCD of $\frac{3}{5}$ and $\frac{4}{7}$ is 35, because this is the least common multiple of 5 and 7. In Example 8, we convert the fractions $\frac{3}{5}$ and $\frac{4}{7}$ to equivalent fractions having 35 as the denominator.

Skill Practice

Fill in the blank with <, >, or =.

14. $\dfrac{3}{8} \square \dfrac{4}{9}$

Example 8 Comparing Two Fractions

Fill in the blank with <, >, or =. $\dfrac{3}{5} \square \dfrac{4}{7}$

Solution:

The fractions have different denominators and cannot be compared by inspection. The LCD is 35. We need to convert each fraction to an equivalent fraction with a denominator of 35.

$\dfrac{3}{5} = \dfrac{3 \cdot 7}{5 \cdot 7} = \dfrac{21}{35}$ Multiply numerator and denominator by 7 because $5 \cdot 7 = 35$.

$\dfrac{4}{7} = \dfrac{4 \cdot 5}{7 \cdot 5} = \dfrac{20}{35}$ Multiply numerator and denominator by 5 because $7 \cdot 5 = 35$.

Because $\dfrac{21}{35} > \dfrac{20}{35}$, then $\dfrac{3}{5} \boxed{>} \dfrac{4}{7}$.

Answers

12. $\dfrac{14y}{18y}$ **13.** $\dfrac{10w}{w^2}$ **14.** <

The relationship between $\frac{3}{5}$ and $\frac{4}{7}$ is shown in Figure 4-11. The position of the two fractions is also illustrated on the number line. See Figure 4-12.

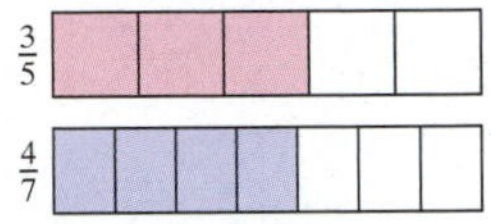

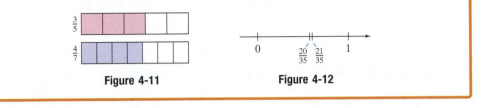

Figure 4-11

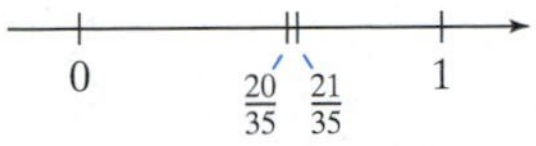

Figure 4-12

Example 9 Ranking Fractions in Order from Least to Greatest

Rank the fractions from least to greatest. $-\frac{9}{20}, -\frac{7}{15}, -\frac{4}{9}$

Solution:

We want to convert each fraction to an equivalent fraction with a common denominator. The least common denominator is the LCM of 20, 15, and 9.

$$\left.\begin{aligned} 20 &= 2 \cdot 2 \cdot 5 \\ 15 &= 3 \cdot 5 \\ 9 &= 3 \cdot 3 \end{aligned}\right\}$$ The least common denominator is $2 \cdot 2 \cdot 3 \cdot 3 \cdot 5 = 180$.

Now convert each fraction to an equivalent fraction with a denominator of 180.

$-\frac{9}{20} = -\frac{9 \cdot 9}{20 \cdot 9} = -\frac{81}{180}$ Multiply numerator and denominator by 9 because $20 \cdot 9 = 180$.

$-\frac{7}{15} = -\frac{7 \cdot 12}{15 \cdot 12} = -\frac{84}{180}$ Multiply numerator and denominator by 12 because $15 \cdot 12 = 180$.

$-\frac{4}{9} = -\frac{4 \cdot 20}{9 \cdot 20} = -\frac{80}{180}$ Multiply numerator and denominator by 20 because $9 \cdot 20 = 180$.

The relative position of these fractions is shown on the number line.

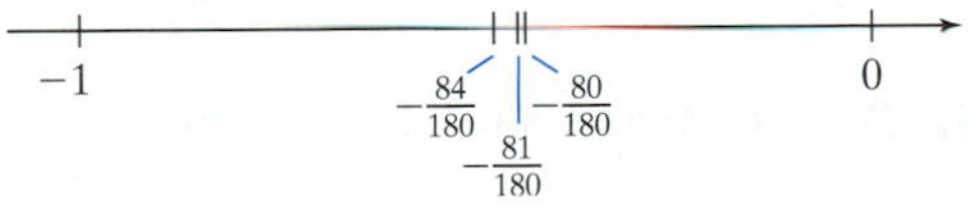

Ranking the fractions from least to greatest we have $-\frac{84}{180}, -\frac{81}{180}$, and $-\frac{80}{180}$. This is equivalent to $-\frac{7}{15}, -\frac{9}{20}$, and $-\frac{4}{9}$.

Skill Practice

Rank the fractions from least to greatest.

15. $-\frac{5}{9}, -\frac{8}{15}$, and $-\frac{3}{5}$

Answer

15. $-\frac{3}{5}, -\frac{5}{9}$, and $-\frac{8}{15}$

Section 4.4 Practice Exercises

Boost *your* GRADE at ALEKS.com!

ALEKS version 3.0

- Practice Problems
- Self-Tests
- NetTutor
- e-Professors
- Videos

Study Skills Exercise

1. Define the key terms.

a. **Multiple** **b.** **Least common multiple (LCM)** **c.** **Least common denominator (LCD)**

Review Exercises

For Exercises 2–3, simplify the fraction.

2. $-\frac{104}{36}$

3. $\frac{30xy}{20x^3}$

For Exercises 4–6, multiply or divide as indicated. Write the answers as fractions.

4. $\left(-\frac{22}{5}\right)\left(-\frac{15}{4}\right)$

5. $\frac{60}{7} \div (-6)$

6. $\frac{4w}{9x^2} \div \frac{2w}{3x}$

7. Convert $-5\frac{3}{4}$ to an improper fraction.

8. Convert $-\frac{16}{7}$ to a mixed number.

Objective 1: Least Common Multiple

9. **a.** Circle the multiples of 24: 4, 8, 48, 72, 12, 240

b. Circle the factors of 24: 4, 8, 48, 72, 12, 240

10. **a.** Circle the multiples of 30: 15, 90, 120, 3, 5, 60

b. Circle the factors of 30: 15, 90, 120, 3, 5, 60

11. **a.** Circle the multiples of 36: 72, 6, 360, 12, 9, 108

b. Circle the factors of 36: 72, 6, 360, 12, 9, 108

12. **a.** Circle the multiples of 28: 7, 4, 2, 56, 140, 280

b. Circle the factors of 28: 7, 4, 2, 56, 140, 280

For Exercises 13–32, find the LCM. **(See Examples 1–2.)**

13. 10 and 25

14. 21 and 14

15. 16 and 12

16. 20 and 12

17. 18 and 24

18. 9 and 30

19. 12 and 15

20. 27 and 45

21. 42 and 70

22. 6 and 21

23. 8, 10, and 12

24. 4, 6, and 14

25. 12, 15, and 20

26. 20, 30, and 40

27. 16, 24, and 30

28. 20, 42, and 35

29. 6, 12, 18, and 20

30. 21, 35, 50, and 75

31. 5, 15, 18, and 20

32. 28, 10, 21, and 35

Objective 2: Applications of the Least Common Multiple

33. A tile floor is to be made from 10-in., 12-in., and 15-in. square tiles. A design is made by alternating rows with different-size tiles. The first row uses only 10-in. tiles, the second row uses only 12-in. tiles, and the third row uses only 15-in. tiles. Neglecting the grout seams, what is the shortest length of floor space that can be covered evenly by each row? **(See Example 3.)**

34. A patient admitted to the hospital was prescribed a pain medication to be given every 4 hr and an antibiotic to be given every 5 hr. Bandages applied to the patient's external injuries needed changing every 12 hr. The nurse changed the bandages and gave the patient both medications at 6:00 A.M. Monday morning.

a. How many hours will pass before the patient is given both medications and has his bandages changed at the same time?

b. What day and time will this be?

35. Four satellites revolve around the earth once every 6, 8, 10, and 15 hr, respectively. If the satellites are initially "lined up," how many hours must pass before they will again be lined up?

36. Mercury, Venus, and Earth revolve around the Sun approximately once every 3 months, 7 months, and 12 months, respectively (see the figure). If the planets begin "lined up," what is the minimum number of months required for them to be aligned again? (Assume that the planets lie roughly in the same plane.)

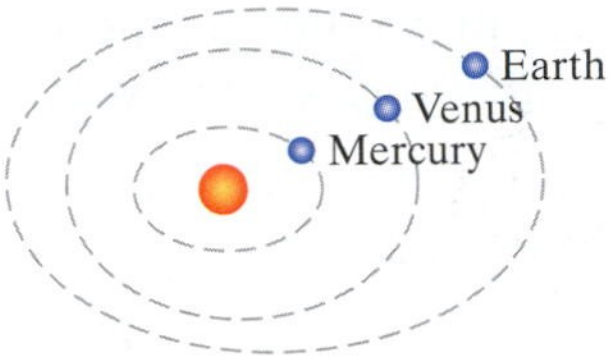

Objective 3: Writing Equivalent Fractions

For Exercises 37–66, rewrite each fraction with the indicated denominators. **(See Examples 4-7.)**

37. $\frac{2}{3} = \frac{}{21}$ **38.** $\frac{7}{4} = \frac{}{32}$ **39.** $\frac{5}{8} = \frac{}{16}$ **40.** $\frac{2}{9} = \frac{}{27}$

41. $-\frac{3}{4} = -\frac{}{16}$ **42.** $-\frac{3}{10} = -\frac{}{50}$ **43.** $\frac{4}{-5} = \frac{}{15}$ **44.** $\frac{3}{-7} = \frac{}{70}$

45. $\frac{7}{6} = \frac{}{42}$ **46.** $\frac{10}{3} = \frac{}{18}$ **47.** $\frac{11}{9} = \frac{}{99}$ **48.** $\frac{7}{5} = \frac{}{35}$

49. $5 = \frac{}{4}$ (*Hint*: $5 = \frac{5}{1}$) **50.** $3 = \frac{}{12}$ (*Hint*: $3 = \frac{3}{1}$)

51. $\frac{11}{4} = \frac{}{4000}$ **52.** $\frac{18}{7} = \frac{}{700}$ **53.** $-\frac{11}{3} = -\frac{}{15}$ **54.** $-\frac{1}{6} = -\frac{}{60}$

55. $\frac{-5}{8} = \frac{}{24}$ **56.** $\frac{-20}{7} = \frac{}{35}$ **57.** $\frac{4y}{7} = \frac{}{28}$ **58.** $\frac{3v}{13} = \frac{}{26}$

59. $\frac{3}{8} = \frac{}{8y}$ **60.** $\frac{7}{13} = \frac{}{13u}$ **61.** $\frac{3}{5} = \frac{}{25p}$ **62.** $\frac{4}{9} = \frac{}{18v}$

63. $\frac{2}{x} = \frac{}{x^2}$ **64.** $\frac{6}{w} = \frac{}{w^2}$ **65.** $\frac{8}{ab} = \frac{}{ab^3}$ **66.** $\frac{9}{cd} = \frac{}{c^3d}$

Objective 4: Ordering Fractions

For Exercises 67–74, fill in the blanks with <, >, or =. **(See Example 8.)**

67. $\frac{7}{8} \square \frac{3}{4}$ **68.** $\frac{7}{15} \square \frac{11}{20}$ **69.** $\frac{13}{10} \square \frac{22}{15}$ **70.** $\frac{15}{4} \square \frac{21}{6}$

71. $-\frac{3}{12} \square -\frac{2}{8}$ **72.** $-\frac{4}{20} \square -\frac{6}{30}$ **73.** $-\frac{5}{18} \square -\frac{8}{27}$ **74.** $-\frac{9}{24} \square -\frac{8}{21}$

75. Which of the following fractions has the greatest value? $\frac{2}{3}, \frac{7}{8}, \frac{5}{6}, \frac{1}{2}$

76. Which of the following fractions has the least value? $\frac{1}{6}, \frac{1}{4}, \frac{2}{15}, \frac{2}{9}$

For Exercises 77–82, rank the fractions from least to greatest. **(See Example 9.)**

77. $\frac{7}{8}, \frac{2}{3}, \frac{3}{4}$

78. $\frac{5}{12}, \frac{3}{8}, \frac{2}{3}$

79. $-\frac{5}{16}, -\frac{3}{8}, -\frac{1}{4}$

80. $-\frac{2}{5}, -\frac{3}{10}, -\frac{5}{6}$

81. $-\frac{4}{3}, -\frac{13}{12}, \frac{17}{15}$

82. $-\frac{5}{7}, \frac{11}{21}, -\frac{18}{35}$

83. A patient had three cuts that needed stitches. A nurse recorded the lengths of the cuts. Where did the patient have the longest cut? Where did the patient have the shortest cut?

upper right arm 3/4 in.
Right hand $\frac{11}{16}$ in.
above left eye $\frac{7}{8}$ in.

84. Three screws have lengths equal to $\frac{3}{4}$ in., $\frac{5}{8}$ in., and $\frac{11}{16}$ in. Which screw is the longest? Which is the shortest?

85. For a party, Aman had $\frac{3}{4}$ lb of cheddar cheese, $\frac{7}{8}$ lb of Swiss cheese, and $\frac{4}{5}$ lb of pepper jack cheese. Which type of cheese is in the least amount? Which type is in the greatest amount?

86. Susan buys $\frac{2}{3}$ lb of smoked turkey, $\frac{3}{5}$ lb of ham, and $\frac{5}{8}$ lb of roast beef. Which type of meat did she buy in the greatest amount? Which type did she buy in the least amount?

Expanding Your Skills

87. Which of the following fractions is between $\frac{1}{4}$ and $\frac{5}{6}$? Identify all that apply.

a. $\frac{5}{12}$ **b.** $\frac{2}{3}$ **c.** $\frac{1}{8}$

88. Which of the following fractions is between $\frac{1}{3}$ and $\frac{11}{15}$? Identify all that apply.

a. $\frac{2}{3}$ **b.** $\frac{4}{5}$ **c.** $\frac{2}{5}$

The LCM of two or more numbers can also be found with a method called "division of primes." For example, consider the numbers 32, 48, and 30. To find the LCM, first divide by any prime number that divides evenly into any of the numbers. Then divide and write the quotient as shown.

2)32 48 30
16 24 15

Repeat this process and bring down any number that is not divisible by the chosen prime.

2)32 48 30
2)16 24 15
)8 12 15 ← Bring down the 15.

Continue until all quotients are 1. The LCM is the product of the prime factors on the left.

2)32 48 30
2)16 24 15
2)8 12 15
2)4 6 15
2)2 3 15
3)1 3 15 ← At this point, the prime number 2 does not divide evenly into any of the quotients. We try the next-greater prime number, 3.
5)1 1 5
1 1 1

The LCM is $2 \cdot 2 \cdot 2 \cdot 2 \cdot 2 \cdot 3 \cdot 5 = 480$.

For Exercises 89–92, use "division of primes" to determine the LCM of the given numbers.

89. 16, 24, and 28

90. 15, 25, and 35

91. 20, 18, and 27

92. 9, 15, and 42

Addition and Subtraction of Fractions

Section 4.5

Objectives

1. Addition and Subtraction of Like Fractions
2. Addition and Subtraction of Unlike Fractions
3. Applications of Addition and Subtraction of Fractions

1. Addition and Subtraction of Like Fractions

In Section 4.3 we learned how to multiply and divide fractions. The main focus of this section is to add and subtract fractions. The operation of addition can be thought of as combining like groups of objects. For example:

3 apples + 1 apple = 4 apples

three-fifths + one-fifth = four-fifths

$$\frac{3}{5} + \frac{1}{5} = \frac{4}{5}$$

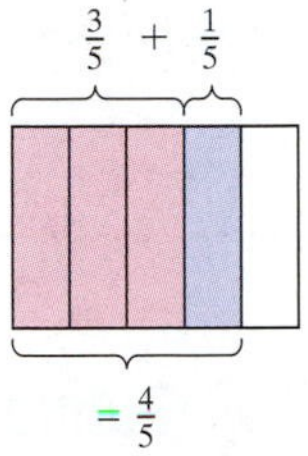

The fractions $\frac{3}{5}$ and $\frac{1}{5}$ are **like fractions** because their denominators are the same. That is, the fractions have a **common denominator**.

The following property leads to the procedure to add and subtract like fractions.

PROPERTY Addition and Subtraction of Like Fractions

Suppose a, b, and c represent numbers, and $c \neq 0$, then

$$\frac{a}{c} + \frac{b}{c} = \frac{a+b}{c} \quad \text{and} \quad \frac{a}{c} - \frac{b}{c} = \frac{a-b}{c}$$

PROCEDURE Adding and Subtracting Like Fractions

Step 1 Add or subtract the numerators.
Step 2 Write the result over the common denominator.
Step 3 Simplify to lowest terms, if possible.

Example 1 Adding and Subtracting Like Fractions

Add. Write the answer as a fraction or whole number.

a. $\frac{1}{4} + \frac{5}{4}$ **b.** $\frac{2}{15} + \frac{1}{15} - \frac{13}{15}$

Solution:

a. $\frac{1}{4} + \frac{5}{4} = \frac{1+5}{4}$ Add the numerators.

$= \frac{6}{4}$ Write the sum over the common denominator.

$= \frac{\overset{3}{\cancel{6}}}{\underset{2}{\cancel{4}}}$ Simplify to lowest terms.

$= \frac{3}{2}$

Avoiding Mistakes

Notice that when adding fractions, we do not add the denominators. We add *only* the numerators.

Skill Practice

Add. Write the answer as a fraction or whole number.

1. $\frac{2}{9} + \frac{4}{9}$
2. $\frac{7}{12} - \frac{5}{12} - \frac{11}{12}$

Answers

1. $\frac{2}{3}$ 2. $-\frac{3}{4}$

b. $\dfrac{2}{15}+\dfrac{1}{15}-\dfrac{13}{15}=\dfrac{2+1-13}{15}$ Add and subtract terms in the numerator. Write the result over the common denominator.

$=\dfrac{-10}{15}$ Simplify. The answer will be a negative fraction.

$=-\dfrac{\overset{2}{\cancel{10}}}{\underset{3}{\cancel{15}}}$ Simplify to lowest terms.

$=-\dfrac{2}{3}$

Skill Practice

Simplify.

3. $\dfrac{4a}{7b}-\dfrac{19}{7b}$

Example 2 Subtracting Fractions

Subtract. $\dfrac{3x}{5y}-\dfrac{2}{5y}$

Solution:

The fractions $\frac{3x}{5y}$ and $\frac{2}{5y}$ are like fractions because they have the same denominator.

$\dfrac{3x}{5y}-\dfrac{2}{5y}=\dfrac{3x-2}{5y}$ Subtract the numerators. Write the result over the common denominator.

$=\dfrac{3x-2}{5y}$ Notice that the numerator cannot be simplified further because $3x$ and 2 are not like terms.

2. Addition and Subtraction of Unlike Fractions

Concept Connections

Use the figure to add the fractions.

4. $\dfrac{1}{3}+\dfrac{2}{5}$

Adding fractions can be visualized by using a diagram. For example, the sum $\frac{1}{2}+\frac{1}{3}$ is illustrated in Figure 4-13.

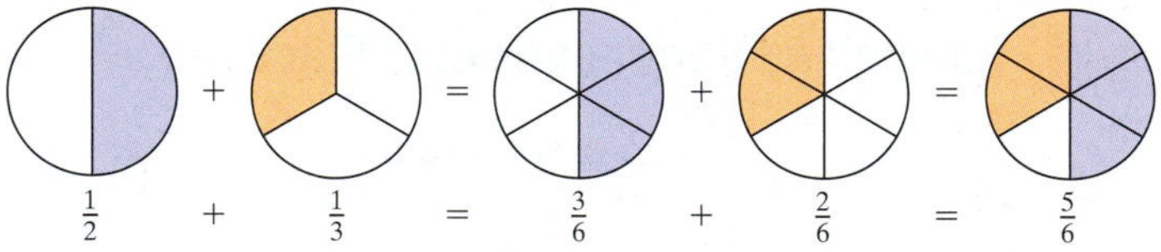

Figure 4-13

In Example 3, we use the concept of the LCD to help us add and subtract unlike fractions. The first step in adding or subtracting unlike fractions is to identify the LCD. Then we change the unlike fractions to like fractions having the LCD as the denominator.

Example 3 Adding Unlike Fractions

Add. $\dfrac{1}{6}+\dfrac{3}{4}$

Solution:

The LCD of $\frac{1}{6}$ and $\frac{3}{4}$ is 12. We can convert each individual fraction to an equivalent fraction with 12 as the denominator.

Answers

3. $\dfrac{4a-19}{7b}$

4.

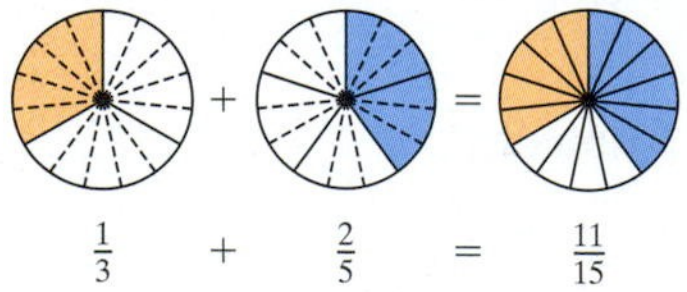

$$\frac{1}{6} = \frac{1 \cdot 2}{6 \cdot 2} = \frac{2}{12}$$ Multiply numerator and denominator by 2 because $6 \cdot 2 = 12$.

$$\frac{3}{4} = \frac{3 \cdot 3}{4 \cdot 3} = \frac{9}{12}$$ Multiply numerator and denominator by 3 because $4 \cdot 3 = 12$.

Thus, $\frac{1}{6} + \frac{3}{4}$ becomes $\frac{2}{12} + \frac{9}{12} = \frac{11}{12}$.

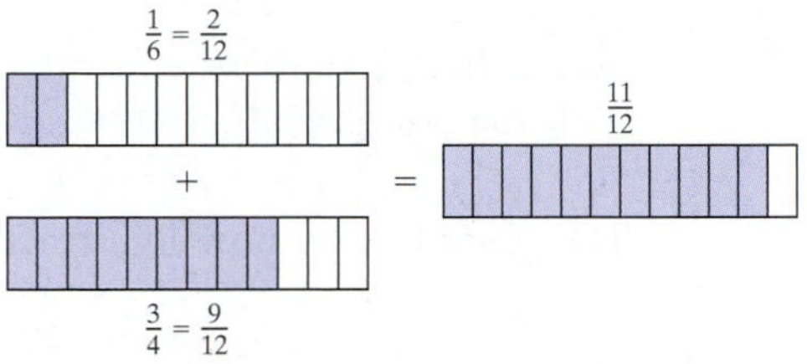

Skill Practice

Add.

5. $\frac{1}{10} + \frac{1}{15}$

The general procedure to add or subtract unlike fractions is outlined as follows.

PROCEDURE Adding and Subtracting Unlike Fractions

Step 1 Identify the LCD.

Step 2 Write each individual fraction as an equivalent fraction with the LCD.

Step 3 Add or subtract the numerators and write the result over the common denominator.

Step 4 Simplify to lowest terms, if possible.

Example 4 Subtracting Unlike Fractions

Subtract. $\frac{7}{10} - \frac{1}{5}$

Solution:

$\frac{7}{10} - \frac{1}{5}$ The LCD is 10. We must convert $\frac{1}{5}$ to an equivalent fraction with 10 as the denominator.

$= \frac{7}{10} - \frac{1 \cdot 2}{5 \cdot 2}$ Multiply numerator and denominator by 2 because $5 \cdot 2 = 10$.

$= \frac{7}{10} - \frac{2}{10}$ The fractions are now like fractions.

$= \frac{7 - 2}{10}$ Subtract the numerators.

$= \frac{5}{10}$

$= \frac{\cancel{5}^{1}}{\cancel{10}_{2}}$ Simplify to lowest terms.

$= \frac{1}{2}$

Skill Practice

Subtract. Write the answer as a fraction.

6. $\frac{9}{5} - \frac{7}{15}$

Avoiding Mistakes

When adding or subtracting fractions, we do not add or subtract the denominators.

Answers

5. $\frac{1}{6}$ **6.** $\frac{4}{3}$

Skill Practice

Subtract.

7. $-\frac{7}{12} - \frac{1}{8}$

Example 5 Subtracting Unlike Fractions

Subtract. $-\frac{4}{15} - \frac{1}{10}$

Solution:

$-\frac{4}{15} - \frac{1}{10}$ The LCD is 30.

$= -\frac{4 \cdot 2}{15 \cdot 2} - \frac{1 \cdot 3}{10 \cdot 3}$ Write the fractions as equivalent fractions with the denominators equal to the LCD.

$= -\frac{8}{30} - \frac{3}{30}$ The fractions are now like fractions.

$= \frac{-8 - 3}{30}$ The fraction $-\frac{8}{30}$ can be written as $\frac{-8}{30}$.

$= \frac{-11}{30}$ Subtract the numerators.

$= -\frac{11}{30}$ The fraction $-\frac{11}{30}$ is in lowest terms because the only common factor of 11 and 30 is 1.

Skill Practice

Add. Write the answer as a fraction.

8. $-3 + \frac{4}{5}$

Example 6 Adding Unlike Fractions

Add. Write the answer as a fraction. $-5 + \frac{3}{4}$

Solution:

$-5 + \frac{3}{4} = -\frac{5}{1} + \frac{3}{4}$ Write the whole number as a fraction.

$= -\frac{5 \cdot 4}{1 \cdot 4} + \frac{3}{4}$ The LCD is 4.

$= -\frac{20}{4} + \frac{3}{4}$ The fractions are now like fractions.

$= \frac{-20 + 3}{4}$ Add the terms in the numerator.

$= \frac{-17}{4}$ Simplify.

$= -\frac{17}{4}$

Example 7 Adding and Subtracting Unlike Fractions

Add and subtract as indicated. $-\frac{7}{12} - \frac{2}{15} + \frac{7}{24}$

Answers

7. $-\frac{17}{24}$ 8. $-\frac{11}{5}$

Solution:

$$-\frac{7}{12}-\frac{2}{15}+\frac{7}{24}$$ To find the LCD, factor each denominator.

$$=-\frac{7}{2\cdot2\cdot3}-\frac{2}{3\cdot5}+\frac{7}{2\cdot2\cdot2\cdot3}$$

$$\left.\begin{array}{l}12=2\cdot2\cdot3\\15=3\cdot5\\24=2\cdot2\cdot2\cdot3\end{array}\right\}$$ The LCD is $2\cdot2\cdot2\cdot3\cdot5=120$.

We want to convert each fraction to an equivalent fraction having a denominator of $2\cdot2\cdot2\cdot3\cdot5=120$. Multiply numerator and denominator of each fraction by the factors missing from the denominator.

$$-\frac{7\cdot(2\cdot5)}{2\cdot2\cdot3\cdot(2\cdot5)}-\frac{2\cdot(2\cdot2\cdot2)}{3\cdot5\cdot(2\cdot2\cdot2)}+\frac{7\cdot(5)}{2\cdot2\cdot2\cdot3\cdot(5)}$$

$$=-\frac{70}{120}-\frac{16}{120}+\frac{35}{120}$$ The fractions are now like fractions.

$$=\frac{-70-16+35}{120}$$ Add and subtract terms in the numerator.

$$=\frac{-51}{120}$$

$$=-\frac{\overset{17}{\cancel{51}}}{\underset{40}{\cancel{120}}}$$ Simplify to lowest terms. The numerator and denominator share a common factor of 3.

$$=-\frac{17}{40}$$

Skill Practice

Subtract.

9. $-\frac{7}{18}-\frac{4}{15}+\frac{7}{30}$

Example 8 Adding and Subtracting Fractions with Variables

Add or subtract as indicated. **a.** $\frac{4}{5}+\frac{3}{x}$ **b.** $\frac{7}{x}-\frac{3}{x^2}$

Solution:

a. $\frac{4}{5}+\frac{3}{x}=\frac{4\cdot x}{5\cdot x}+\frac{3\cdot5}{x\cdot5}$ The LCD is $5x$. Multiply by the missing factor from each denominator.

$=\frac{4x}{5x}+\frac{15}{5x}$ Simplify each fraction.

$=\frac{4x+15}{5x}$ The numerator cannot be simplified further because $4x$ and 15 are not like terms.

b. $\frac{7}{x}-\frac{3}{x^2}=\frac{7\cdot x}{x\cdot x}-\frac{3}{x^2}$ The LCD is x^2.

$=\frac{7x}{x^2}-\frac{3}{x^2}$ Simplify each fraction.

$=\frac{7x-3}{x^2}$

Skill Practice

Add or subtract as indicated.

10. $\frac{7}{8}+\frac{5}{z}$ 11. $\frac{11}{t^2}-\frac{4}{t}$

Avoiding Mistakes

The fraction $\frac{4x+15}{5x}$ cannot be simplified because the numerator is a sum and not a product. Only *factors* can be "divided out" when simplifying a fraction.

Answers

9. $-\frac{19}{45}$ 10. $\frac{7z+40}{8z}$

11. $\frac{11-4t}{t^2}$

3. Applications of Addition and Subtraction of Fractions

Skill Practice

12. On Monday, $\frac{2}{3}$ in. of rain fell on a certain town. On Tuesday, $\frac{1}{5}$ in. of rain fell. How much more rain fell on Monday than on Tuesday?

Example 9 Applying Operations on Unlike Fractions

A new Kelly Safari SUV tire has $\frac{7}{16}$-in. tread. After being driven 50,000 mi, the tread depth has worn down to $\frac{7}{32}$ in. By how much has the tread depth worn away?

Solution:

In this case, we are looking for the difference in the tread depth.

$$\text{Difference in tread depth} = \left(\text{original tread depth}\right) - \left(\text{final tread depth}\right)$$

$$= \frac{7}{16} - \frac{7}{32} \quad \text{The LCD is 32.}$$

$$= \frac{7 \cdot 2}{16 \cdot 2} - \frac{7}{32} \quad \text{Multiply numerator and denominator by 2 because } 16 \cdot 2 = 32.$$

$$= \frac{14}{32} - \frac{7}{32} \quad \text{The fractions are now like.}$$

$$= \frac{7}{32} \quad \text{Subtract.}$$

The tire lost $\frac{7}{32}$ in. in tread depth after 50,000 mi of driving.

TIP: You can check your result by adding the final tread depth to the difference in tread depth to get the original tread depth.

$$\frac{7}{32} + \frac{7}{32} = \frac{14}{32} = \frac{7}{16}$$

Skill Practice

13. Twelve members of a college hiking club hiked the perimeter of a canyon. How far did they hike?

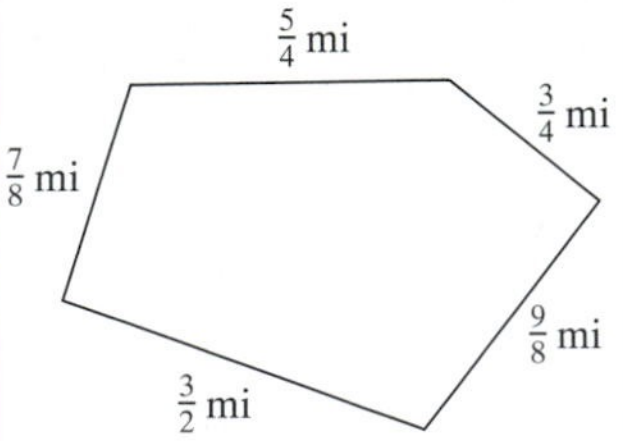

Example 10 Finding Perimeter

A parcel of land has the following dimensions. Find the perimeter.

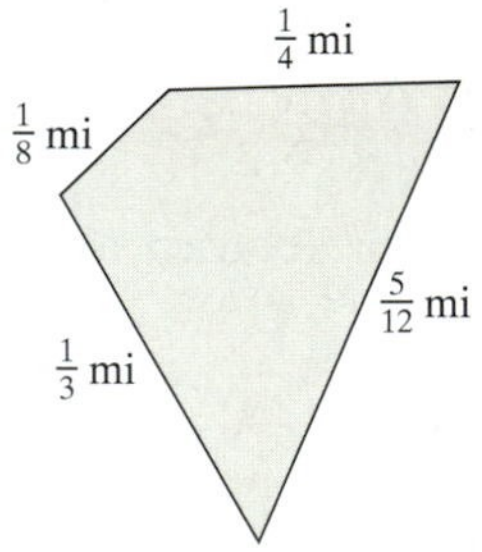

Solution:

To find the perimeter, add the lengths of the sides.

$$\frac{1}{8} + \frac{1}{4} + \frac{5}{12} + \frac{1}{3} \quad \text{The LCD is 24.}$$

$$= \frac{1 \cdot 3}{8 \cdot 3} + \frac{1 \cdot 6}{4 \cdot 6} + \frac{5 \cdot 2}{12 \cdot 2} + \frac{1 \cdot 8}{3 \cdot 8} \quad \text{Convert to like fractions.}$$

$$= \frac{3}{24} + \frac{6}{24} + \frac{10}{24} + \frac{8}{24}$$

$$= \frac{27}{24} \quad \text{Add the fractions.}$$

$$= \frac{9}{8} \quad \text{Simplify to lowest terms.}$$

The perimeter is $\frac{9}{8}$ mi or $1\frac{1}{8}$ mi.

Answers

12. $\frac{7}{15}$ in.

13. They hiked $\frac{11}{2}$ mi. or equivalently $5\frac{1}{2}$ mi.

Section 4.5 Practice Exercises

Study Skills Exercise

1. Define the key terms.

a. Like fractions **b. Common denominator**

Review Exercises

For Exercises 2–7, write the fraction as an equivalent fraction with the indicated denominator.

2. $\dfrac{3}{5} = \dfrac{}{15}$ **3.** $-\dfrac{6}{7} = -\dfrac{}{14}$ **4.** $\dfrac{3}{1} = \dfrac{}{10}$

5. $\dfrac{5}{1} = \dfrac{}{5}$ **6.** $\dfrac{3}{4x} = \dfrac{}{12x^2}$ **7.** $\dfrac{4}{t} = \dfrac{}{t^3}$

Objective 1: Addition and Subtraction of Like Fractions

For Exercises 8–9, shade in the portion of the third figure that represents the addition of the first two figures.

8.

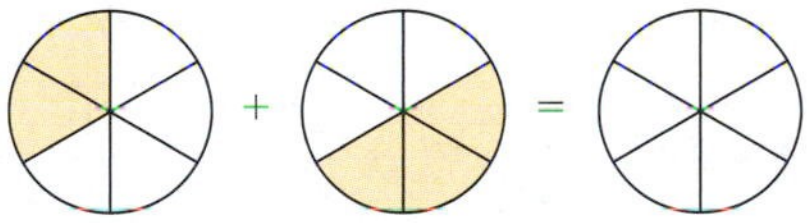

9.

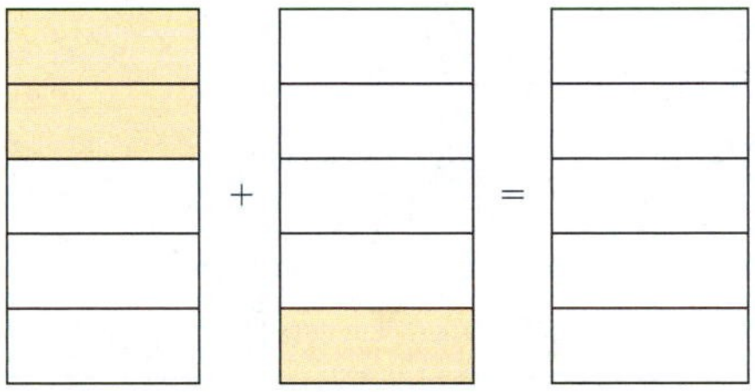

For Exercises 10–11, shade in the portion of the third figure that represents the subtraction of the first two figures.

10.

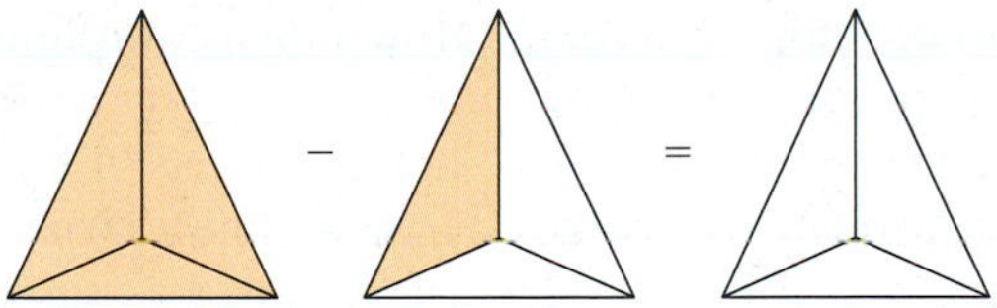

11.

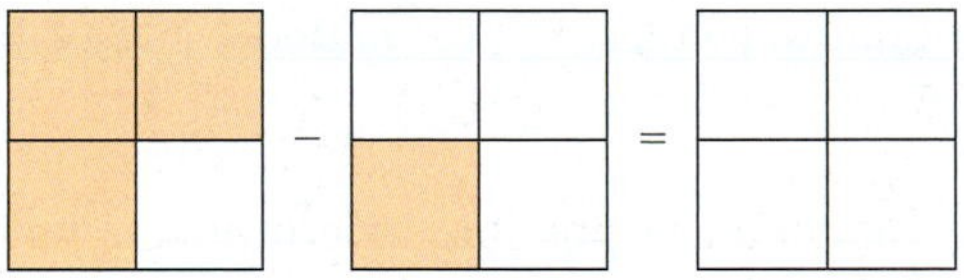

12. Explain the difference between evaluating the two expressions $\frac{2}{5} \cdot \frac{7}{5}$ and $\frac{2}{5} + \frac{7}{5}$.

For Exercises 13–28, add or subtract as indicated. Write the answer as a fraction or whole number. **(See Examples 1–2.)**

13. $\dfrac{7}{8} + \dfrac{5}{8}$ **14.** $\dfrac{1}{21} + \dfrac{13}{21}$ **15.** $\dfrac{23}{12} - \dfrac{15}{12}$ **16.** $\dfrac{13}{6} - \dfrac{5}{6}$

17. $\dfrac{18}{14} + \dfrac{11}{14} + \dfrac{6}{14}$ **18.** $\dfrac{7}{18} + \dfrac{22}{18} + \dfrac{10}{18}$ **19.** $\dfrac{14}{15} + \dfrac{2}{15} - \dfrac{4}{15}$ **20.** $\dfrac{19}{6} - \dfrac{11}{6} + \dfrac{5}{6}$

21. $-\dfrac{7}{2} + \dfrac{3}{2} - \dfrac{1}{2}$ **22.** $-\dfrac{8}{3} - \dfrac{2}{3} + \dfrac{1}{3}$ **23.** $\dfrac{5}{12} - \dfrac{19}{12} - \dfrac{7}{12}$ **24.** $\dfrac{5}{18} - \dfrac{7}{18} - \dfrac{13}{18}$

25. $\dfrac{3y}{2w} + \dfrac{5}{2w}$ **26.** $\dfrac{11a}{4b} + \dfrac{7}{4b}$ **27.** $\dfrac{x}{5y} - \dfrac{3x}{5y}$ **28.** $\dfrac{a}{9x} - \dfrac{5a}{9x}$

Objective 2: Addition and Subtraction of Unlike Fractions

For Exercises 29–64, add or subtract. Write the answer as a fraction simplified to lowest terms. **(See Examples 3–8.)**

29. $\frac{7}{8} + \frac{5}{16}$

30. $\frac{2}{9} + \frac{1}{18}$

31. $\frac{1}{15} + \frac{1}{10}$

32. $\frac{5}{6} + \frac{3}{8}$

33. $\frac{5}{6} + \frac{8}{7}$

34. $\frac{2}{11} + \frac{4}{5}$

35. $\frac{7}{8} - \frac{1}{2}$

36. $\frac{9}{10} - \frac{4}{5}$

37. $\frac{13}{12} - \frac{3}{4}$

38. $\frac{29}{30} - \frac{7}{10}$

39. $-\frac{10}{9} - \frac{5}{12}$

40. $-\frac{7}{6} - \frac{1}{15}$

41. $\frac{9}{8} - 2$

42. $\frac{11}{9} - 3$

43. $4 - \frac{4}{3}$

44. $2 - \frac{3}{8}$

45. $-\frac{16}{7} + 2$

46. $-\frac{15}{4} + 3$

47. $-\frac{1}{10} - \left(-\frac{9}{100}\right)$

48. $-\frac{3}{100} - \left(-\frac{21}{1000}\right)$

49. $\frac{3}{10} + \frac{9}{100} + \frac{1}{1000}$

50. $\frac{1}{10} + \frac{3}{100} + \frac{7}{1000}$

51. $\frac{5}{3} - \frac{7}{6} + \frac{5}{8}$

52. $\frac{7}{12} - \frac{2}{15} + \frac{5}{18}$

53. $-\frac{7}{10} - \frac{1}{20} - \left(-\frac{5}{8}\right)$

54. $\frac{1}{12} - \frac{3}{5} - \left(-\frac{3}{10}\right)$

55. $\frac{1}{2} + \frac{1}{4} - \frac{1}{8} - \frac{1}{16}$

56. $\frac{1}{3} - \frac{1}{9} + \frac{1}{27} - \frac{1}{81}$

57. $\frac{3}{4} + \frac{2}{x}$

58. $\frac{9}{5} + \frac{3}{y}$

59. $\frac{10}{x} + \frac{7}{y}$

60. $\frac{7}{a} + \frac{8}{b}$

61. $\frac{10}{x} - \frac{2}{x^2}$

62. $\frac{11}{z} - \frac{8}{z^2}$

63. $\frac{5}{3x} - \frac{2}{3}$

64. $\frac{13}{5t} - \frac{4}{5}$

Objective 3: Applications of Addition and Subtraction of Fractions

65. When doing her laundry, Inez added $\frac{3}{4}$ cup of bleach to $\frac{3}{8}$ cup of liquid detergent. How much total liquid is added to her wash?

66. What is the smallest possible length of screw needed to pass through two pieces of wood, one that is $\frac{7}{8}$ in. thick and one that is $\frac{1}{2}$ in. thick?

67. Before a storm, a rain gauge has $\frac{1}{8}$ in. of water. After the storm, the gauge has $\frac{9}{32}$ in. How many inches of rain did the storm deliver? **(See Example 9.)**

68. In one week it rained $\frac{5}{16}$ in. If a garden needs $\frac{9}{8}$ in. of water per week, how much more water does it need?

69. The information in the graph shows the distribution of a college student body by class.

a. What fraction of the student body consists of upper classmen (juniors and seniors)?

b. What fraction of the student body consists of freshmen and sophomores?

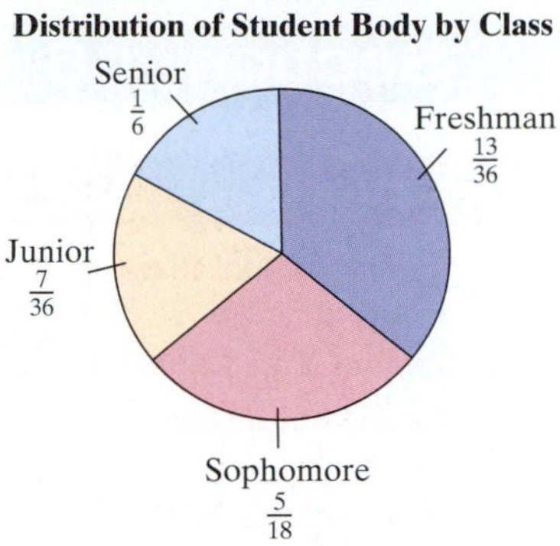

70. A group of college students took part in a survey. One of the survey questions read:

"Do you think the government should spend more money on research to produce alternative forms of fuel?"

The results of the survey are shown in the figure.

a. What fraction of the survey participants chose to strongly agree or agree?

b. What fraction of the survey participants chose to strongly disagree or disagree?

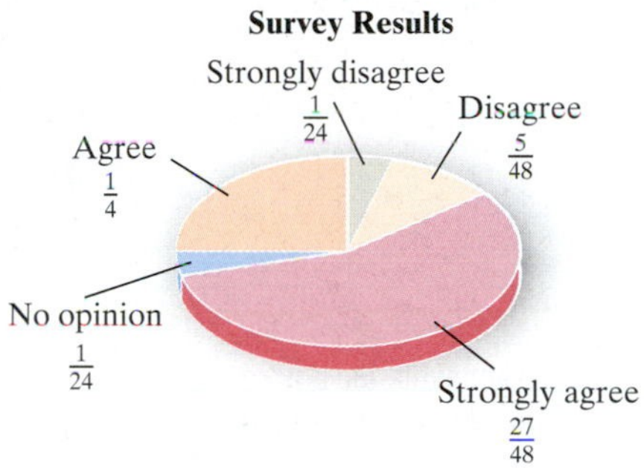

For Exercises 71–72, find the perimeter. **(See Example 10.)**

71.

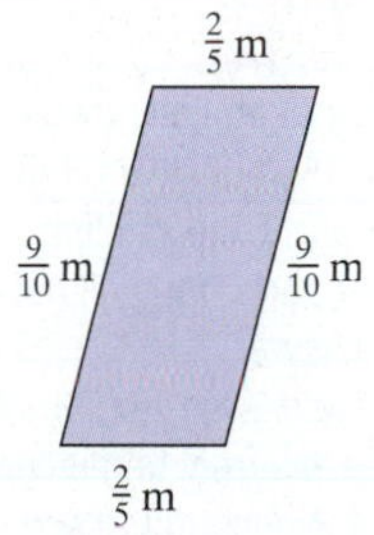

72.

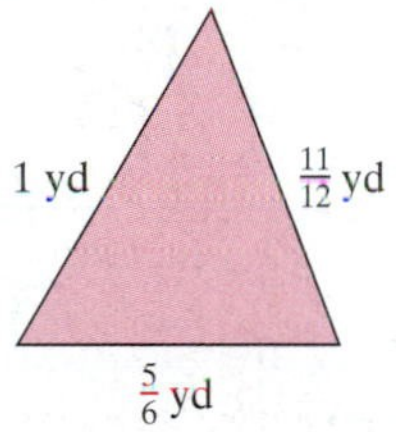

For Exercises 73–74, find the missing dimensions. Then calculate the perimeter.

73.

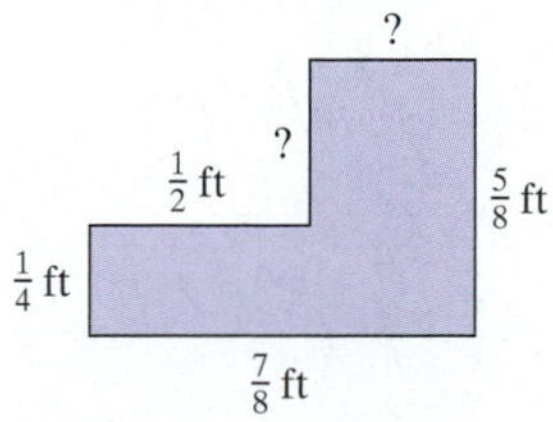

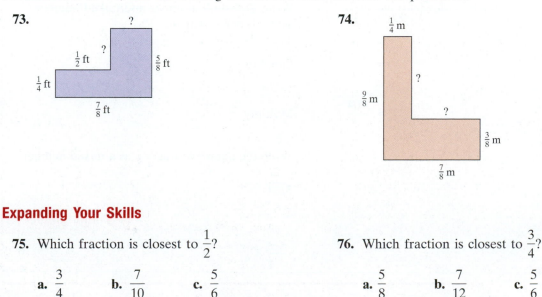

74.

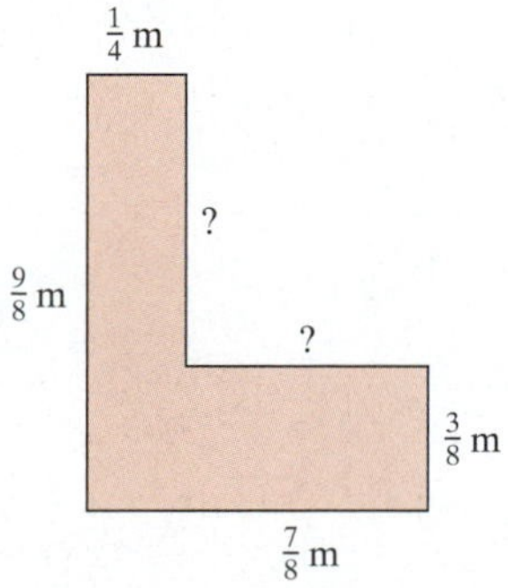

Expanding Your Skills

75. Which fraction is closest to $\frac{1}{2}$?

a. $\frac{3}{4}$ **b.** $\frac{7}{10}$ **c.** $\frac{5}{6}$

76. Which fraction is closest to $\frac{3}{4}$?

a. $\frac{5}{8}$ **b.** $\frac{7}{12}$ **c.** $\frac{5}{6}$

Section 4.6 Estimation and Operations on Mixed Numbers

Objectives

1. Multiplication and Division of Mixed Numbers
2. Addition of Mixed Numbers
3. Subtraction of Mixed Numbers
4. Addition and Subtraction of Negative Mixed Numbers
5. Applications of Mixed Numbers

1. Multiplication and Division of Mixed Numbers

To multiply mixed numbers, we follow these steps.

PROCEDURE Multiplying Mixed Numbers

Step 1 Change each mixed number to an improper fraction.

Step 2 Multiply the improper fractions and simplify to lowest terms, if possible (see Section 4.3).

Answers greater than or equal to 1 may be written as an improper fraction or as a mixed number, depending on the directions of the problem.

Example 1 demonstrates this process.

Skill Practice

Multiply and write the answer as a mixed number or whole number.

1. $16\frac{1}{2} \cdot 3\frac{7}{11}$

2. $-10 \cdot \left(7\frac{1}{6}\right)$

Example 1 Multiplying Mixed Numbers

Multiply and write the answer as a mixed number or integer.

a. $7\frac{1}{2} \cdot 4\frac{2}{3}$ **b.** $-12 \cdot \left(1\frac{7}{9}\right)$

Solution:

a. $7\frac{1}{2} \cdot 4\frac{2}{3} = \frac{15}{2} \cdot \frac{14}{3}$ Write each mixed number as an improper fraction.

$= \frac{\overset{5}{\cancel{15}}}{\underset{1}{\cancel{2}}} \cdot \frac{\overset{7}{\cancel{14}}}{\underset{1}{\cancel{3}}}$ Simplify.

$= \frac{35}{1}$ Multiply.

$= 35$

b. $-12 \cdot \left(1\frac{7}{9}\right) = -\frac{12}{1} \cdot \frac{16}{9}$ Write the whole number and mixed number as improper fractions.

$= -\frac{\overset{4}{\cancel{12}}}{1} \cdot \frac{16}{\underset{3}{\cancel{9}}}$ Simplify.

$= -\frac{64}{3}$ Multiply.

$= -21\frac{1}{3}$ Write the improper fraction as a mixed number.

$$\begin{array}{r} 21 \\ 3\overline{)64} \\ \underline{-6} \\ 4 \\ \underline{-3} \\ 1 \end{array}$$

Avoiding Mistakes

Do not try to multiply mixed numbers by multiplying the whole-number parts and multiplying the fractional parts. You will not get the correct answer.

For the expression $7\frac{1}{2} \cdot 4\frac{2}{3}$, it would be incorrect to multiply (7)(4) and $\frac{1}{2} \cdot \frac{2}{3}$. Notice that these values do not equal 35.

Answers

1. 60 **2.** $-71\frac{2}{3}$

To divide mixed numbers, we use the following steps.

PROCEDURE Dividing Mixed Numbers

Step 1 Change each mixed number to an improper fraction.

Step 2 Divide the improper fractions and simplify to lowest terms, if possible. Recall that to divide fractions, we multiply the dividend by the reciprocal of the divisor (see Section 4.3).

Answers greater than or equal to 1 may be written as an improper fraction or as a mixed number, depending on the directions of the problem.

Example 2 Dividing Mixed Numbers

Divide and write the answer as a mixed number or whole number.

$$7\frac{1}{2} \div 4\frac{2}{3}$$

Solution:

$7\frac{1}{2} \div 4\frac{2}{3} = \frac{15}{2} \div \frac{14}{3}$ Write the mixed numbers as improper fractions.

$= \frac{15}{2} \cdot \frac{3}{14}$ Multiply by the reciprocal of the divisor.

$= \frac{45}{28}$ Multiply.

$= 1\frac{17}{28}$ Write the improper fraction as a mixed number.

Avoiding Mistakes

Be sure to take the reciprocal of the divisor *after* the mixed number is changed to an improper fraction.

Skill Practice

Divide and write the answer as a mixed number or whole number.

3. $10\frac{1}{3} \div 2\frac{5}{6}$

Example 3 Dividing Mixed Numbers

Divide and write the answers as mixed numbers.

a. $-6 \div \left(-5\frac{1}{7}\right)$ **b.** $13\frac{5}{6} \div (-7)$

Solution:

a. $-6 \div \left(-5\frac{1}{7}\right) = -\frac{6}{1} \div \left(-\frac{36}{7}\right)$ Write the whole number and mixed number as improper fractions.

$= -\frac{6}{1} \cdot \left(-\frac{7}{36}\right)$ Multiply by the reciprocal of the divisor.

$= -\frac{\overset{1}{\cancel{6}}}{1} \cdot \left(-\frac{7}{\underset{6}{\cancel{36}}}\right)$ Simplify.

$= \frac{7}{6}$ Multiply.

$= 1\frac{1}{6}$ Write the improper fraction as a mixed number.

Skill Practice

Divide and write the answers as mixed numbers.

4. $-8 \div \left(-4\frac{4}{5}\right)$

5. $12\frac{4}{9} \div (-8)$

Answers

3. $3\frac{11}{17}$ **4.** $1\frac{2}{3}$ **5.** $-1\frac{5}{9}$

b. $13\frac{5}{6} \div (-7) = \frac{83}{6} \div \left(-\frac{7}{1}\right)$ Write the whole number and mixed number as improper fractions.

$= \frac{83}{6} \cdot \left(-\frac{1}{7}\right)$ Multiply by the reciprocal of the divisor.

$= -\frac{83}{42}$ Multiply. The product is negative.

$= -1\frac{41}{42}$ Write the improper fraction as a mixed number.

2. Addition of Mixed Numbers

Now we will learn to add and subtract mixed numbers. To find the sum of two or more mixed numbers, add the whole-number parts and add the fractional parts.

Skill Practice

Add.

6. $7\frac{2}{15} + 2\frac{1}{15}$

Example 4 Adding Mixed Numbers

Add. $1\frac{5}{9} + 2\frac{1}{9}$

Solution:

$$\begin{array}{r} 1\frac{5}{9} \\ +2\frac{1}{9} \\ \hline 3\frac{6}{9} \end{array}$$

Add the whole numbers. Add the fractional parts.

The sum is $3\frac{6}{9}$ which simplifies to $3\frac{2}{3}$.

TIP: To understand why mixed numbers can be added in this way, recall that $1\frac{5}{9} = 1 + \frac{5}{9}$ and $2\frac{1}{9} = 2 + \frac{1}{9}$. Therefore,

$$1\frac{5}{9} + 2\frac{1}{9} = 1 + \frac{5}{9} + 2 + \frac{1}{9}$$
$$= 3 + \frac{6}{9}$$
$$= 3\frac{6}{9}$$
$$= 3\frac{2}{3}$$

When we perform operations on mixed numbers, it is often desirable to estimate the answer first. When rounding a mixed number, we offer the following convention.

1. If the fractional part of a mixed number is greater than or equal to $\frac{1}{2}$ (that is, if the numerator is half the denominator or greater), round to the next-greater whole number.
2. If the fractional part of the mixed number is less than $\frac{1}{2}$ (that is, if the numerator is less than half the denominator), the mixed number rounds down to the whole number.

Example 5 Adding Mixed Numbers

Estimate the sum and then find the actual sum.

$$42\frac{1}{12} + 17\frac{7}{8}$$

Answer

6. $9\frac{1}{5}$

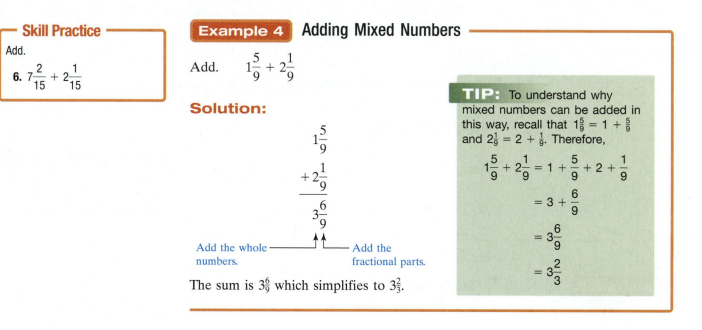

Solution:

To estimate the sum, first round the addends.

$$\begin{array}{rll} 42\frac{1}{12} & \text{rounds to} & 42 \\ +\ 17\frac{7}{8} & \text{rounds to} & +\ 18 \\ \hline & & 60 \end{array}$$

The estimated value is 60.

To find the actual sum, we must first write the fractional parts as like fractions. The LCD is 24.

$$\begin{array}{rcrcr} 42\frac{1}{12} & = & 42\frac{1 \cdot 2}{12 \cdot 2} & = & 42\frac{2}{24} \\ +\ 17\frac{7}{8} & = & +\ 17\frac{7 \cdot 3}{8 \cdot 3} & = & +\ 17\frac{21}{24} \\ \hline & & & & 59\frac{23}{24} \end{array}$$

The actual sum is $59\frac{23}{24}$. This is close to our estimate of 60.

Skill Practice

Estimate the sum and then find the actual sum.

7. $6\frac{1}{11} + 3\frac{1}{2}$

Example 6 Adding Mixed Numbers with Carrying

Estimate the sum and then find the actual sum.

$$7\frac{5}{6} + 3\frac{3}{5}$$

Solution:

$$\begin{array}{rll} 7\frac{5}{6} & \text{rounds to} & 8 \\ +\ 3\frac{3}{5} & \text{rounds to} & +\ 4 \\ \hline & & 12 \end{array}$$

The estimated value is 12.

To find the actual sum, first write the fractional parts as like fractions. The LCD is 30.

$$\begin{array}{rcrcr} 7\frac{5}{6} & = & 7\frac{5 \cdot 5}{6 \cdot 5} & = & 7\frac{25}{30} \\ +\ 3\frac{3}{5} & = & +\ 3\frac{3 \cdot 6}{5 \cdot 6} & = & +\ 3\frac{18}{30} \\ \hline & & & & 10\frac{43}{30} \end{array}$$

Notice that the number $\frac{43}{30}$ is an improper fraction. By convention, a mixed number is written as a whole number and a *proper* fraction. We have $\frac{43}{30} = 1\frac{13}{30}$. Therefore,

$$10\frac{43}{30} = 10 + 1\frac{13}{30} = 11\frac{13}{30}$$

The sum is $11\frac{13}{30}$. This is close to our estimate of 12.

Skill Practice

Estimate the sum and then find the actual sum.

8. $5\frac{2}{5} + 7\frac{8}{9}$

Concept Connections

9. Explain how you would rewrite $2\frac{9}{8}$ as a mixed number containing a proper fraction.

Answers

7. Estimate: 10; actual sum: $9\frac{13}{22}$

8. Estimate: 13; actual sum: $13\frac{13}{45}$

9. Write the improper fraction $\frac{9}{8}$ as $1\frac{1}{8}$, and add the result to 2. The result is $3\frac{1}{8}$.

We have shown how to add mixed numbers by writing the numbers in columns. Another approach to add or subtract mixed numbers is to write the numbers first as improper fractions. Then add or subtract the fractions, as you learned in Section 4.5. To demonstrate this process, we add the mixed numbers from Example 6.

Skill Practice

Add the mixed numbers by first converting the addends to improper fractions. Write the answer as a mixed number.

10. $12\frac{1}{3} + 4\frac{3}{4}$

Example 7 Adding Mixed Numbers by Using Improper Fractions

Add. $7\frac{5}{6} + 3\frac{3}{5}$

Solution:

$$7\frac{5}{6} + 3\frac{3}{5} = \frac{47}{6} + \frac{18}{5}$$ Write each mixed number as an improper fraction.

$$= \frac{47 \cdot 5}{6 \cdot 5} + \frac{18 \cdot 6}{5 \cdot 6}$$ Convert the fractions to like fractions. The LCD is 30.

$$= \frac{235}{30} + \frac{108}{30}$$ The fractions are now like fractions.

$$= \frac{343}{30}$$ Add the like fractions.

$$= 11\frac{13}{30}$$ Convert the improper fraction to a mixed number.

$$\begin{array}{r} 11 \\ 30\overline{)343} \\ -30 \\ \hline 43 \\ -30 \\ \hline 13 \end{array} \quad 11\tfrac{13}{30}$$

The mixed number $11\frac{13}{30}$ is the same as the value obtained in Example 6.

3. Subtraction of Mixed Numbers

To subtract mixed numbers, we subtract the fractional parts and subtract the whole-number parts.

Skill Practice

Subtract.

11. $6\frac{3}{4} - 2\frac{1}{3}$

Example 8 Subtracting Mixed Numbers

Subtract. $15\frac{2}{3} - 4\frac{1}{6}$

Answers

10. $17\frac{1}{12}$ **11.** $4\frac{5}{12}$

Solution:

To subtract the fractional parts, we need a common denominator. The LCD is 6.

$$\begin{array}{rcrcr} 15\frac{2}{3} & = & 15\frac{2\cdot 2}{3\cdot 2} & = & 15\frac{4}{6} \\ -\,4\frac{1}{6} & = & -\,4\frac{1}{6} & = & -\,4\frac{1}{6} \\ \hline & & & & 11\frac{3}{6} \end{array}$$

Subtract the whole numbers. → ← Subtract the fractional parts.

The difference is $11\frac{3}{6}$, which simplifies to $11\frac{1}{2}$.

Borrowing is sometimes necessary when subtracting mixed numbers.

Example 9 Subtracting Mixed Numbers with Borrowing

Subtract.

a. $17\frac{2}{7} - 11\frac{5}{7}$ **b.** $14\frac{2}{9} - 9\frac{3}{5}$

Solution:

a. We will subtract $\frac{5}{7}$ from $\frac{2}{7}$ by borrowing 1 from the whole number 17. The borrowed 1 is written as $\frac{7}{7}$ because the common denominator is 7.

$$\begin{array}{rcrcr} 17\frac{2}{7} & = & \overset{16}{\cancel{17}}\frac{2}{7} + \frac{7}{7} & = & 16\frac{9}{7} \\ -\,11\frac{5}{7} & = & -\,11\frac{5}{7} & = & -\,11\frac{5}{7} \\ \hline & & & & 5\frac{4}{7} \end{array}$$

The difference is $5\frac{4}{7}$.

b. To subtract the fractional parts, we need a common denominator. The LCD is 45.

$$\begin{array}{rcrcr} 14\frac{2}{9} & = & 14\frac{2\cdot 5}{9\cdot 5} & = & 14\frac{10}{45} \\ -\,9\frac{3}{5} & = & -\,9\frac{3\cdot 9}{5\cdot 9} & = & -\,9\frac{27}{45} \end{array}$$

We will subtract $\frac{27}{45}$ from $\frac{10}{45}$ by borrowing. Therefore, borrow 1 (or equivalently $\frac{45}{45}$) from 14.

$$\begin{array}{rcr} = \overset{13}{\cancel{14}}\frac{10}{45} + \frac{45}{45} & = & 13\frac{55}{45} \\ = -\,9\frac{27}{45} & = & -\,9\frac{27}{45} \\ \hline & & 4\frac{28}{45} \end{array}$$

The difference is $4\frac{28}{45}$.

Skill Practice

Subtract.

12. $24\frac{2}{7} - 8\frac{5}{7}$

13. $9\frac{2}{3} - 8\frac{3}{4}$

Answers

12. $15\frac{4}{7}$ **13.** $\frac{11}{12}$

Skill Practice

Subtract.

14. $10 - 3\frac{1}{6}$

Example 10 Subtracting Mixed Numbers with Borrowing

Subtract. $4 - 2\frac{5}{8}$

Solution:

$$\begin{array}{r} 4 \\ -\,2\frac{5}{8} \\ \hline \end{array}$$

In this case, we have no fractional part from which to subtract.

$$\begin{array}{r} \overset{3}{\cancel{4}}\frac{8}{8} \\ -\,2\frac{5}{8} \\ \hline 1\frac{3}{8} \end{array}$$

We can borrow 1 or equivalently $\frac{8}{8}$ from the whole number 4.

TIP: The borrowed 1 is written as $\frac{8}{8}$ because the common denominator is 8.

The difference is $1\frac{3}{8}$.

TIP: The subtraction problem $4 - 2\frac{5}{8} = 1\frac{3}{8}$ can be checked by adding:

$$1\frac{3}{8} + 2\frac{5}{8} = 3\frac{8}{8} = 3 + 1 = 4 \checkmark$$

In Example 11, we show the alternative approach to subtract mixed numbers by first writing each mixed number as an improper fraction.

Skill Practice

Subtract by first converting to improper fractions. Write the answer as a mixed number.

15. $8\frac{2}{9} - 3\frac{5}{6}$

Example 11 Subtracting Mixed Numbers by Using Improper Fractions

Subtract by first converting to improper fractions. Write the answer as a mixed number.

$$10\frac{2}{5} - 4\frac{3}{4}$$

Solution:

$10\frac{2}{5} - 4\frac{3}{4} = \frac{52}{5} - \frac{19}{4}$ Write each mixed number as an improper fraction.

$= \frac{52 \cdot 4}{5 \cdot 4} - \frac{19 \cdot 5}{4 \cdot 5}$ Convert the fractions to like fractions. The LCD is 20.

$= \frac{208}{20} - \frac{95}{20}$ Subtract the like fractions.

$= \frac{113}{20}$

$= 5\frac{13}{20}$ Write the result as a mixed number.

$$\begin{array}{r} 5 \\ 20\overline{)113} \\ -100 \\ \hline 13 \end{array}$$

Answers

14. $6\frac{5}{6}$ **15.** $4\frac{7}{18}$

As you can see from Examples 7 and 11, when we convert mixed numbers to improper fractions, the numerators of the fractions become larger numbers. Thus, we must add (or subtract) larger numerators than if we had used the method involving columns. This is one drawback. However, an advantage of converting to improper fractions first is that there is no need for carrying or borrowing.

4. Addition and Subtraction of Negative Mixed Numbers

In Examples 4–11, we have shown two methods for adding and subtracting mixed numbers. The method of changing mixed numbers to improper fractions is preferable when adding or subtracting negative mixed numbers.

Example 12 Adding and Subtracting Signed Mixed Numbers

Add or subtract as indicated. Write the answer as a fraction.

a. $-3\frac{1}{2} - \left(-6\frac{2}{3}\right)$ **b.** $-4\frac{2}{3} - 1\frac{1}{6} + 2\frac{3}{4}$

Solution:

a. $-3\frac{1}{2} - \left(-6\frac{2}{3}\right) = -3\frac{1}{2} + 6\frac{2}{3}$ Rewrite subtraction in terms of addition.

$= -\frac{7}{2} + \frac{20}{3}$ Change the mixed numbers to improper fractions.

$= -\frac{7 \cdot 3}{2 \cdot 3} + \frac{20 \cdot 2}{3 \cdot 2}$ The LCD is 6.

$= -\frac{21}{6} + \frac{40}{6}$ Simplify.

$= \frac{-21 + 40}{6}$ Add the numerators. Note that the fraction $-\frac{21}{6}$ can be written as $\frac{-21}{6}$.

$= \frac{19}{6}$ Simplify.

b. $-4\frac{2}{3} - 1\frac{1}{6} + 2\frac{3}{4} = -\frac{14}{3} - \frac{7}{6} + \frac{11}{4}$ Change the mixed numbers to improper fractions.

$= -\frac{14 \cdot 4}{3 \cdot 4} - \frac{7 \cdot 2}{6 \cdot 2} + \frac{11 \cdot 3}{4 \cdot 3}$ The LCD is 12.

$= -\frac{56}{12} - \frac{14}{12} + \frac{33}{12}$ Change each fraction to an equivalent fraction with denominator 12.

$= \frac{-56 - 14 + 33}{12}$ Add and subtract the numerators.

$= \frac{-37}{12}$ or $-\frac{37}{12}$ Simplify.

Skill Practice

Add or subtract as indicated. Write the answers as fractions.

16. $-5\frac{3}{4} - \left(-1\frac{1}{2}\right)$

17. $-2\frac{5}{8} + 4\frac{3}{4} - 3\frac{1}{2}$

Answers

16. $-\frac{17}{4}$ **17.** $-\frac{11}{8}$

5. Applications of Mixed Numbers

Skill Practice

18. On December 1, the snow base at the Bear Mountain Ski Resort was $4\frac{1}{3}$ ft. By January 1, the base was $6\frac{1}{2}$ ft. By how much did the base amount of snow increase?

Example 13 Subtracting Mixed Numbers in an Application

The average height of a 3-year-old girl is $38\frac{1}{3}$ in. The average height of a 4-year-old girl is $41\frac{3}{4}$ in. On average, by how much does a girl grow between the ages of 3 and 4?

Solution:

We use subtraction to find the difference in heights.

$$\begin{array}{rcrcr} 41\frac{3}{4} & = & 41\frac{3 \cdot 3}{4 \cdot 3} & = & 41\frac{9}{12} \\ -\ 38\frac{1}{3} & = & -\ 38\frac{1 \cdot 4}{3 \cdot 4} & = & -\ 38\frac{4}{12} \\ \hline & & & & 3\frac{5}{12} \end{array}$$

The average amount of growth is $3\frac{5}{12}$ in.

Skill Practice

19. A department store wraps packages for \$2 each. Ribbon $2\frac{5}{8}$ ft long is used to wrap each package. How many packages can be wrapped from a roll of ribbon 168 ft long?

Example 14 Dividing Mixed Numbers in an Application

A construction site brings in $6\frac{2}{3}$ tons of soil. Each truck holds $\frac{2}{3}$ ton. How many truckloads are necessary?

Solution:

The $6\frac{2}{3}$ tons of soil must be distributed in $\frac{2}{3}$-ton increments. This will require division.

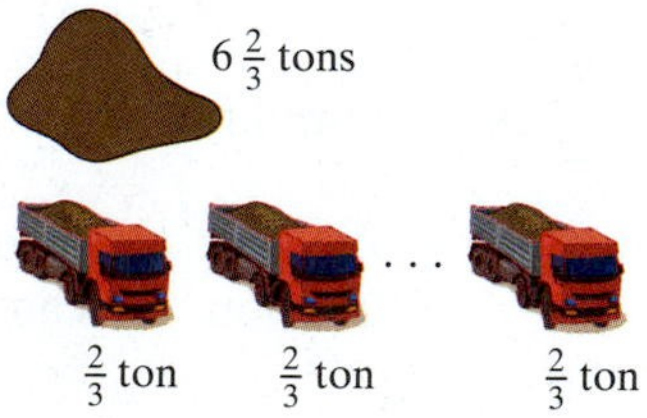

$6\frac{2}{3} \div \frac{2}{3} = \frac{20}{3} \div \frac{2}{3}$ Write the mixed number as an improper fraction.

$= \frac{20}{3} \cdot \frac{3}{2}$ Multiply by the reciprocal of the divisor.

$= \frac{\overset{10}{\cancel{20}}}{\underset{1}{\cancel{3}}} \cdot \frac{\overset{1}{\cancel{3}}}{\underset{1}{\cancel{2}}}$ Simplify.

$= \frac{10}{1}$ Multiply.

$= 10$

A total of 10 truckloads of soil will be required.

Answers

18. $2\frac{1}{6}$ ft **19.** 64 packages

Section 4.6 Practice Exercises

Review Exercises

For Exercises 1–8, perform the indicated operations. Write the answers as fractions.

1. $\frac{9}{5} + 3$ **2.** $\frac{3}{16} + \frac{7}{12}$ **3.** $\frac{25}{8} - \frac{23}{24}$ **4.** $-\frac{20}{9} \div \left(-\frac{10}{3}\right)$

5. $-\frac{42}{11} \div \left(-\frac{7}{2}\right)$ **6.** $\frac{52}{18} \div (-13)$ **7.** $\frac{125}{32} - \frac{51}{32} - \frac{58}{32}$ **8.** $\frac{17}{10} - \frac{23}{100} + \frac{321}{1000}$

For Exercises 9–10, write the mixed number as an improper fraction.

9. $3\frac{2}{5}$ **10.** $2\frac{7}{10}$

For Exercises 11–12, write the improper fraction as a mixed number.

11. $-\frac{77}{6}$ **12.** $-\frac{57}{11}$

Objective 1: Multiplication and Division of Mixed Numbers

For Exercises 13–32, multiply or divide the mixed numbers. Write the answer as a mixed number or whole number. **(See Examples 1–3.)**

13. $\left(2\frac{2}{5}\right)\left(3\frac{1}{12}\right)$ **14.** $\left(5\frac{1}{5}\right)\left(3\frac{3}{4}\right)$ **15.** $-2\frac{1}{3} \cdot \left(-6\frac{3}{5}\right)$ **16.** $-6\frac{1}{8} \cdot \left(-2\frac{3}{4}\right)$

17. $(-9) \cdot 4\frac{2}{9}$ **18.** $(-6) \cdot 3\frac{1}{3}$ **19.** $\left(5\frac{3}{16}\right)\left(5\frac{1}{3}\right)$ **20.** $\left(8\frac{2}{3}\right)\left(2\frac{1}{13}\right)$

21. $\left(7\frac{1}{4}\right) \cdot 10$ **22.** $\left(2\frac{2}{3}\right) \cdot 3$ **23.** $4\frac{1}{2} \div 2\frac{1}{4}$ **24.** $5\frac{5}{6} \div 2\frac{1}{3}$

25. $5\frac{8}{9} \div \left(-1\frac{1}{3}\right)$ **26.** $12\frac{4}{5} \div \left(-2\frac{3}{5}\right)$ **27.** $-2\frac{1}{2} \div \left(-1\frac{1}{16}\right)$ **28.** $-7\frac{3}{5} \div \left(-1\frac{7}{12}\right)$

29. $2 \div 3\frac{1}{3}$ **30.** $6 \div 4\frac{2}{5}$ **31.** $8\frac{1}{4} \div (-3)$ **32.** $6\frac{2}{5} \div (-2)$

Objective 2: Addition of Mixed Numbers

For Exercises 33–40, add the mixed numbers. **(See Examples 4–5.)**

33. $2\frac{1}{11} + 5\frac{3}{11}$ **34.** $5\frac{2}{7} + 4\frac{3}{7}$ **35.** $12\frac{1}{14} + 3\frac{5}{14}$ **36.** $1\frac{3}{20} + 17\frac{7}{20}$

37. $\begin{array}{r} 4\frac{5}{16} \\ +\ 11\frac{1}{4} \\ \hline \end{array}$

38. $\begin{array}{r} 21\frac{2}{9} \\ +\ 10\frac{1}{3} \\ \hline \end{array}$

39. $\begin{array}{r} 6\frac{2}{3} \\ +\ 4\frac{1}{5} \\ \hline \end{array}$

40. $\begin{array}{r} 7\frac{1}{6} \\ +\ 3\frac{5}{8} \\ \hline \end{array}$

For Exercises 41–44, round the mixed number to the nearest whole number.

41. $5\frac{1}{3}$

42. $2\frac{7}{8}$

43. $1\frac{3}{5}$

44. $6\frac{3}{7}$

For Exercises 45–48, write the mixed number in proper form (that is, as a whole number with a proper fraction that is simplified to lowest terms).

45. $2\frac{6}{5}$

46. $4\frac{8}{7}$

47. $7\frac{5}{3}$

48. $1\frac{9}{5}$

For Exercises 49–54, round the numbers to estimate the answer. Then find the exact sum. In Exercise 49, the estimate is done for you. **(See Examples 6–7.)**

49. Estimate: $\begin{array}{r} 7 \\ +\ 8 \\ \hline 15 \end{array}$ Exact: $\begin{array}{r} 6\frac{3}{4} \\ +\ 7\frac{3}{4} \\ \hline \end{array}$

50. Estimate: $\begin{array}{r} \\ + \quad \\ \hline \end{array}$ Exact: $\begin{array}{r} 8\frac{3}{5} \\ +\ 13\frac{4}{5} \\ \hline \end{array}$

51. Estimate: $\begin{array}{r} \\ + \quad \\ \hline \end{array}$ Exact: $\begin{array}{r} 14\frac{7}{8} \\ +\ 8\frac{1}{4} \\ \hline \end{array}$

52. Estimate: $\begin{array}{r} \\ + \quad \\ \hline \end{array}$ Exact: $\begin{array}{r} 21\frac{3}{5} \\ +\ 24\frac{9}{10} \\ \hline \end{array}$

53. Estimate: $\begin{array}{r} \\ + \quad \\ \hline \end{array}$ Exact: $\begin{array}{r} 3\frac{7}{16} \\ +\ 15\frac{11}{12} \\ \hline \end{array}$

54. Estimate: $\begin{array}{r} \\ + \quad \\ \hline \end{array}$ Exact: $\begin{array}{r} 7\frac{7}{9} \\ +\ 8\frac{5}{6} \\ \hline \end{array}$

For Exercises 55–62, add the mixed numbers. Write the answer as a mixed number or whole number. **(See Examples 6–7.)**

55. $3\frac{3}{4} + 5\frac{2}{3}$

56. $6\frac{5}{7} + 10\frac{3}{5}$

57. $11\frac{5}{8} + \frac{7}{6}$

58. $9\frac{5}{6} + \frac{3}{4}$

59. $3 + 6\frac{7}{8}$

60. $5 + 11\frac{1}{13}$

61. $124\frac{2}{3} + 46\frac{5}{6}$

62. $345\frac{3}{5} + 84\frac{7}{10}$

Objective 3: Subtraction of Mixed Numbers

For Exercises 63–66, subtract the mixed numbers. **(See Example 8.)**

63. $\begin{array}{r} 21\frac{9}{10} \\ -\ 10\frac{3}{10} \\ \hline \end{array}$

64. $\begin{array}{r} 19\frac{2}{3} \\ -\ 4\frac{1}{3} \\ \hline \end{array}$

65. $\begin{array}{r} 18\frac{5}{6} \\ -\ 6\frac{2}{3} \\ \hline \end{array}$

66. $\begin{array}{r} 21\frac{17}{20} \\ -\ 20\frac{1}{10} \\ \hline \end{array}$

For Exercises 67–72, round the numbers to estimate the answer. Then find the exact difference. In Exercise 67, the estimate is done for you. **(See Examples 9–11.)**

	Estimate	Exact		Estimate	Exact		Estimate	Exact
67.	25	$25\frac{1}{4}$	**68.**		$36\frac{1}{5}$	**69.**		$17\frac{1}{6}$
	$-\ 14$	$-\ 13\frac{3}{4}$		$\underline{\hspace{1cm}}$	$-\ 12\frac{3}{5}$		$\underline{\hspace{1cm}}$	$-\ 15\frac{5}{12}$
	$\overline{11}$	$\overline{\hspace{1cm}}$			$\overline{\hspace{1cm}}$			$\overline{\hspace{1cm}}$

	Estimate	Exact		Estimate	Exact		Estimate	Exact
70.		$22\frac{5}{18}$	**71.**		$46\frac{3}{7}$	**72.**		$23\frac{1}{2}$
	$\underline{\hspace{1cm}}$	$-\ 10\frac{7}{9}$		$\underline{\hspace{1cm}}$	$-\ 38\frac{1}{2}$		$\underline{\hspace{1cm}}$	$-\ 18\frac{10}{13}$
		$\overline{\hspace{1cm}}$			$\overline{\hspace{1cm}}$			$\overline{\hspace{1cm}}$

For Exercises 73–80, subtract the mixed numbers. Write the answers as fractions or mixed numbers. **(See Examples 9–11.)**

73. $6 - 2\frac{5}{6}$

74. $9 - 4\frac{1}{2}$

75. $12 - 9\frac{2}{9}$

76. $10 - 9\frac{1}{3}$

77. $3\frac{7}{8} - 3\frac{3}{16}$

78. $3\frac{1}{6} - 1\frac{23}{24}$

79. $12\frac{1}{5} - 11\frac{2}{7}$

80. $10\frac{1}{8} - 2\frac{17}{18}$

Objective 4: Addition and Subtraction of Negative Mixed Numbers

For Exercises 81–88, add or subtract the mixed numbers. Write the answer as a mixed number. **(See Example 12.)**

81. $-4\frac{2}{7} - 3\frac{1}{2}$

82. $-9\frac{1}{5} - 2\frac{1}{10}$

83. $-4\frac{3}{8} + 7\frac{1}{4}$

84. $-5\frac{2}{11} + 8\frac{1}{2}$

85. $3\frac{1}{2} - \left(-2\frac{1}{4}\right)$

86. $6\frac{1}{6} - \left(-4\frac{1}{3}\right)$

87. $5 + \left(-7\frac{5}{6}\right)$

88. $6 + \left(-10\frac{3}{4}\right)$

Mixed Exercises

For Exercises 89–96, perform the indicated operations. Write the answers as fractions or integers.

89. $\left(3\frac{1}{5}\right)\left(-2\frac{1}{4}\right)$

90. $(-10)\left(-3\frac{2}{5}\right)$

91. $4\frac{1}{8} - 3\frac{5}{6}$

92. $-5\frac{5}{6} - 3\frac{1}{3}$

93. $-8\frac{1}{3} \div \left(-2\frac{1}{6}\right)$

94. $10\frac{2}{3} \div (-8)$

95. $-6\frac{3}{10} + 4\frac{5}{6} - \left(-3\frac{1}{2}\right)$

96. $4\frac{4}{5} - \left(-2\frac{1}{4}\right) - 1\frac{3}{10}$

Objective 5: Applications of Mixed Numbers

For Exercises 97–100, use the table to find the lengths of several common birds.

Bird	Length
Cuban Bee Hummingbird	$2\frac{1}{4}$ in.
Sedge Wren	$3\frac{1}{2}$ in.
Great Carolina Wren	$5\frac{1}{2}$ in.
Belted Kingfisher	$11\frac{1}{4}$ in.

97. How much longer is the Belted Kingfisher than the Sedge Wren? **(See Example 13.)**

98. How much longer is the Great Carolina Wren than the Cuban Bee Hummingbird?

99. Estimate or measure the length of your index finger. Which is longer, your index finger or a Cuban Bee Hummingbird?

100. For a recent year, the smallest living dog in the United States was Brandy, a female Chihuahua, who measures 6 in. in length. How much longer is a Belted Kingfisher than Brandy?

101. Tabitha charges \$8 per hour for baby sitting. If she works for $4\frac{3}{4}$ hr, how much should she be paid?

102. Kurt bought $2\frac{2}{3}$ acres of land. If land costs \$10,500 per acre, how much will the land cost him?

103. According to the U.S. Census Bureau's Valentine's Day Press Release, the average American consumes $25\frac{7}{10}$ lb of chocolate in a year. Over the course of 25 years, how many pounds of chocolate would the average American consume?

104. Florida voters approved an amendment to the state constitution to set Florida's minimum wage at $\$6\frac{2}{3}$ per hour. Kayla earns minimum wage while working at the campus bookstore. If she works for $15\frac{3}{4}$ hr, how much should she be paid?

105. A student has three part-time jobs. She tutors, delivers newspapers, and takes notes for a blind student. During a typical week she works $8\frac{2}{3}$ hr delivering newspapers, $4\frac{1}{2}$ hr tutoring, and $3\frac{3}{4}$ hr note-taking. What is the total number of hours worked in a typical week?

106. A contractor ordered three loads of gravel. The orders were for $2\frac{1}{2}$ tons, $3\frac{1}{8}$ tons, and $4\frac{1}{3}$ tons. What is the total amount of gravel ordered?

107. The age of a small kitten can be approximated by the following rule. The kitten's age is given as 1 week for every quarter pound of weight. **(See Example 14.)**

a. Approximately how old is a $1\frac{3}{4}$-lb kitten?

b. Approximately how old is a $2\frac{1}{8}$-lb kitten?

108. Richard's estate is to be split equally among his three children. If his estate is worth $\$1\frac{3}{4}$ million, how much will each child inherit?

109. Lucy earns \$14 per hour and Ricky earns \$10 per hour. Suppose Lucy worked $35\frac{1}{2}$ hr last week and Ricky worked $42\frac{1}{2}$ hr.

a. Who earned more money and by how much?

b. How much did they earn altogether?

110. A roll of wallpaper covers an area of 28 ft^2. If the roll is $1\frac{17}{24}$ ft wide, how long is the roll?

Figure for Exercise 110

111. A plumber fits together two pipes. Find the length of the larger piece.

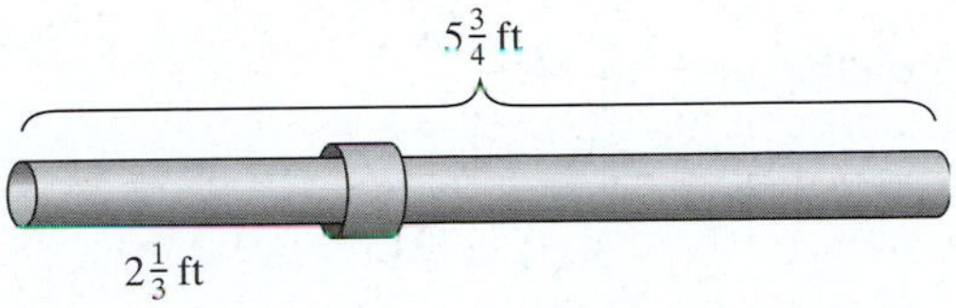

112. Find the thickness of the carpeting and pad.

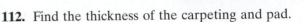

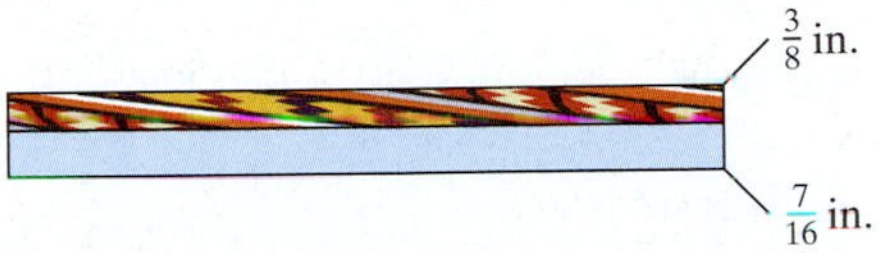

113. When using the word processor, Microsoft Word, the default margins are $1\frac{1}{4}$ in. for the left and right margins. If an $8\frac{1}{2}$ in. by 11 in. piece of paper is used, what is the width of the printing area?

114. A water gauge in a pond measured $25\frac{7}{8}$ in. on Monday. After 2 days of rain and runoff, the gauge read $32\frac{1}{2}$ in. By how much did the water level rise?

115. A patient admitted to the hospital was dehydrated. In addition to intravenous (IV) fluids, the doctor told the patient that she must drink at least 4 L of an electrolyte solution within the next 12 hr. A nurse recorded the amounts the patient drank in the patient's chart.

a. How many L of electrolyte solution did the patient drink?

b. How much more would the patient need to drink to reach 4 L?

Time	Amount
7 A.M.–10 A.M.	$1\frac{1}{4}$ L
10 A.M.–1 P.M.	$\frac{7}{8}$ L
1 P.M.–4 P.M.	$\frac{3}{4}$ L
4 P.M.–7 P.M.	$\frac{1}{2}$ L

116. Benjamin loves to hike in the White Mountains of New Hampshire. It takes him $4\frac{1}{2}$ hr to hike from the Pinkham Notch Visitor Center to the summit of Mt. Washington. If the round trip usually takes 8 hr, how long does it take for the return trip?

Expanding Your Skills

For Exercises 117–120, fill in the blank to complete the pattern.

117. $1, 1\frac{1}{3}, 1\frac{2}{3}, 2, 2\frac{1}{3}, \square$

118. $\frac{1}{4}, 1, 1\frac{3}{4}, 2\frac{1}{2}, 3\frac{1}{4}, \square$

119. $\frac{5}{6}, 1\frac{1}{6}, 1\frac{1}{2}, 1\frac{5}{6}, \square$

120. $\frac{1}{2}, 1\frac{1}{4}, 2, 2\frac{3}{4}, 3\frac{1}{2}, \square$

Calculator Connections

Topic: Multiplying and Dividing Fractions and Mixed Numbers on a Scientific Calculator

Expression	Keystrokes	Result
$\frac{8}{15} \cdot \frac{25}{28}$	8 [a b/c] 15 × 25 [a b/c] 28 [=]	10⌟21 $= \frac{10}{21}$
$2\frac{3}{4} \div \frac{1}{6}$	2 [a b/c] 3 [a b/c] 4 [÷] 1 [a b/c] 6 [=]	16_1⌟2 $= 16\frac{1}{2}$

(Under "2 [a b/c] 3 [a b/c] 4": This is how you enter the mixed number $2\frac{3}{4}$.)

To convert the result to an improper fraction, press [2nd] [d/e] — 33⌟2 $= \frac{33}{2}$

Topic: Adding and Subtracting Fractions and Mixed Numbers on a Calculator

Expression	Keystrokes	Result
$\frac{7}{18} + \frac{1}{3}$	7 [a b/c] 18 + 1 [a b/c] 3 [=]	13⌟18 $= \frac{13}{18}$
$7\frac{5}{8} - 4\frac{2}{3}$	7 [a b/c] 5 [a b/c] 8 [−] 4 [a b/c] 2 [a b/c] 3 [=]	2_23⌟24 $= 2\frac{23}{24}$

(Under "7 [a b/c] 5 [a b/c] 8": $7\frac{5}{8}$; under "4 [a b/c] 2 [a b/c] 3": $4\frac{2}{3}$)

To convert the result to an improper fraction, press [2nd] [d/e] — 71⌟24 $= \frac{71}{24}$

Calculator Exercises

For Exercises 121–132, use a calculator to perform the indicated operations and simplify. Write the answer as a mixed number.

121. $12\frac{2}{3} \cdot 25\frac{1}{8}$

122. $38\frac{1}{3} \div 12\frac{1}{2}$

123. $56\frac{5}{6} \div 3\frac{1}{6}$

124. $25\frac{1}{5} \cdot 18\frac{1}{2}$

125. $\frac{23}{42} + \frac{17}{24}$

126. $\frac{14}{75} + \frac{9}{50}$

127. $\frac{31}{44} - \frac{14}{33}$

128. $\frac{29}{68} - \frac{7}{92}$

129. $32\frac{7}{18} + 14\frac{2}{27}$

130. $21\frac{3}{28} + 4\frac{31}{42}$

131. $7\frac{11}{21} - 2\frac{10}{33}$

132. $5\frac{14}{17} - 2\frac{47}{68}$

Problem Recognition Exercises

Operations on Fractions and Mixed Numbers

For Exercises 1–10, perform the indicated operations. Check the reasonableness of your answers by estimating.

1. a. $-\frac{7}{5} + \frac{2}{5}$ **b.** $-\frac{7}{5} \cdot \frac{2}{5}$ **c.** $-\frac{7}{5} \div \frac{2}{5}$ **d.** $-\frac{7}{5} - \frac{2}{5}$

2. a. $\frac{4}{3} \cdot \frac{5}{6}$ **b.** $\frac{4}{3} \div \frac{5}{6}$ **c.** $\frac{4}{3} + \frac{5}{6}$ **d.** $\frac{4}{3} - \frac{5}{6}$

3. a. $2\frac{3}{4} + \left(-1\frac{1}{2}\right)$ **b.** $2\frac{3}{4} - \left(-1\frac{1}{2}\right)$ **c.** $2\frac{3}{4} \div \left(-1\frac{1}{2}\right)$ **d.** $2\frac{3}{4} \cdot \left(-1\frac{1}{2}\right)$

4. a. $\left(4\frac{1}{3}\right) \cdot \left(2\frac{5}{6}\right)$ **b.** $\left(4\frac{1}{3}\right) \div \left(2\frac{5}{6}\right)$ **c.** $\left(4\frac{1}{3}\right) - \left(2\frac{5}{6}\right)$ **d.** $\left(4\frac{1}{3}\right) + \left(2\frac{5}{6}\right)$

5. a. $-4 - \frac{3}{8}$ **b.** $-4 \cdot \frac{3}{8}$ **c.** $-4 \div \frac{3}{8}$ **d.** $-4 + \frac{3}{8}$

6. a. $3\frac{2}{3} \div 2$ **b.** $3\frac{2}{3} - 2$ **c.** $3\frac{2}{3} + 2$ **d.** $3\frac{2}{3} \cdot 2$

7. a. $-4\frac{1}{5} - \left(-\frac{2}{3}\right)$ **b.** $-4\frac{1}{5} + \left(-\frac{2}{3}\right)$ **c.** $\left(-4\frac{1}{5}\right) \cdot \left(-\frac{2}{3}\right)$ **d.** $\left(-4\frac{1}{5}\right) \div \left(-\frac{2}{3}\right)$

8. a. $\frac{25}{9} \div 2$ **b.** $\frac{25}{9} \cdot 2$ **c.** $\frac{25}{9} - 2$ **d.** $\frac{25}{9} + 2$

9. a. $-1\frac{4}{5} \cdot \frac{5}{9}$ **b.** $-1\frac{4}{5} + \frac{5}{9}$ **c.** $-1\frac{4}{5} \div \frac{5}{9}$ **d.** $-1\frac{4}{5} - \frac{5}{9}$

10. a. $8 \cdot \frac{1}{8}$ **b.** $\frac{1}{9} \cdot 9$ **c.** $-\frac{3}{7} \cdot \left(-\frac{7}{3}\right)$ **d.** $-\frac{5}{13} \cdot \left(-\frac{13}{5}\right)$

Order of Operations and Complex Fractions

Section 4.7

Objectives

1. Order of Operations
2. Complex Fractions
3. Simplifying Algebraic Expressions

1. Order of Operations

We will begin this section with a review of expressions containing exponents. Recall that to square a number, multiply the number times itself. For example: $5^2 = 5 \cdot 5 = 25$. The same process is used to square a fraction.

$$\left(\frac{3}{7}\right)^2 = \frac{3}{7} \cdot \frac{3}{7} = \frac{9}{49}$$

Skill Practice

Simplify.

1. $\left(\frac{5}{4}\right)^2$ 2. $\left(-\frac{5}{4}\right)^2$

3. $-\left(\frac{5}{4}\right)^2$ 4. $\left(-\frac{1}{3}\right)^3$

Example 1 Simplifying Expressions with Exponents

Simplify. **a.** $\left(\frac{2}{5}\right)^2$ **b.** $\left(-\frac{2}{5}\right)^2$ **c.** $-\left(\frac{2}{5}\right)^2$ **d.** $\left(-\frac{1}{4}\right)^3$

Solution:

a. $\left(\frac{2}{5}\right)^2 = \frac{2}{5} \cdot \frac{2}{5} = \frac{4}{25}$

b. $\left(-\frac{2}{5}\right)^2 = \left(-\frac{2}{5}\right)\left(-\frac{2}{5}\right) = \frac{4}{25}$ The base is negative. The product of two negatives is positive.

c. $-\left(\frac{2}{5}\right)^2 = -\left(\frac{2}{5}\right)\left(\frac{2}{5}\right) = -\frac{4}{25}$ First square the base. Then take the opposite.

d. $\left(-\frac{1}{4}\right)^3 = \left(-\frac{1}{4}\right)\left(-\frac{1}{4}\right)\left(-\frac{1}{4}\right) = -\frac{1}{64}$ The base is negative. The product of *three* negatives is negative.

In Section 1.7 we learned to recognize powers of 10. These are $10^1 = 10$, $10^2 = 100$, and so on. In this section, we learn to recognize the **powers of one-tenth**. That is, $\frac{1}{10}$ raised to a whole number power.

$$\left(\frac{1}{10}\right)^1 = \frac{1}{10}$$
$$\left(\frac{1}{10}\right)^2 = \frac{1}{10} \cdot \frac{1}{10} = \frac{1}{100}$$
$$\left(\frac{1}{10}\right)^3 = \frac{1}{10} \cdot \frac{1}{10} \cdot \frac{1}{10} = \frac{1}{1000}$$
$$\left(\frac{1}{10}\right)^4 = \frac{1}{10} \cdot \frac{1}{10} \cdot \frac{1}{10} \cdot \frac{1}{10} = \frac{1}{10{,}000}$$
$$\left(\frac{1}{10}\right)^5 = \frac{1}{10} \cdot \frac{1}{10} \cdot \frac{1}{10} \cdot \frac{1}{10} \cdot \frac{1}{10} = \frac{1}{100{,}000}$$
$$\left(\frac{1}{10}\right)^6 = \frac{1}{10} \cdot \frac{1}{10} \cdot \frac{1}{10} \cdot \frac{1}{10} \cdot \frac{1}{10} \cdot \frac{1}{10} = \frac{1}{1{,}000{,}000}$$

TIP: The instruction "simplify" means to perform all indicated operations. Fractional answers should always be written in lowest terms.

From these examples, we see that a power of one-tenth results in a fraction with a 1 in the numerator. The denominator has a 1 followed by the same number of zeros as the exponent.

Skill Practice

Simplify.

5. $\left(-\frac{7}{8} \cdot \frac{4}{35}\right)^3$

Example 2 Applying the Order of Operations

Simplify. $\left(-\frac{2}{15} \cdot \frac{3}{4}\right)^2$

Solution:

$\left(-\frac{2}{15} \cdot \frac{3}{4}\right)^2 = \left(-\frac{\overset{1}{\cancel{2}}}{\underset{5}{\cancel{15}}} \cdot \frac{\overset{1}{\cancel{3}}}{\underset{2}{\cancel{4}}}\right)^2 = \left(-\frac{1}{10}\right)^2$ Multiply fractions within parentheses.

$= \left(-\frac{1}{10}\right)\left(-\frac{1}{10}\right)$ Square the fraction $-\frac{1}{10}$.

$= \frac{1}{100}$

Answers

1. $\frac{25}{16}$ 2. $\frac{25}{16}$ 3. $-\frac{25}{16}$

4. $-\frac{1}{27}$ 5. $-\frac{1}{1000}$

Example 3 Applying the Order of Operations

Simplify. Write the answer as a fraction. $\frac{1}{2} + \left(3\frac{3}{5}\right) \cdot \left(-\frac{10}{9}\right)$

Solution:

$\frac{1}{2} + \left(3\frac{3}{5}\right) \cdot \left(-\frac{10}{9}\right)$	We must perform multiplication before addition.
$= \frac{1}{2} + \left(\frac{18}{5}\right) \cdot \left(-\frac{10}{9}\right)$	Write the mixed number $3\frac{3}{5}$ as $\frac{18}{5}$.
$= \frac{1}{2} + \left(\frac{\overset{2}{\cancel{18}}}{\underset{1}{\cancel{5}}}\right) \cdot \left(-\frac{\overset{2}{\cancel{10}}}{\underset{1}{\cancel{9}}}\right)$	Multiply fractions. The product will be negative.
$= \frac{1}{2} - \frac{4}{1}$	
$= \frac{1}{2} - \frac{4 \cdot 2}{1 \cdot 2}$	Write the whole number over 1 and obtain a common denominator.
$= \frac{1}{2} - \frac{8}{2}$	
$= \frac{1 - 8}{2}$	Subtract the fractions.
$= -\frac{7}{2}$	Simplify.

Skill Practice

Simplify. Write the answer as a fraction.

6. $\frac{2}{3} + \left(3\frac{3}{4}\right) \cdot \left(-\frac{8}{5}\right)$

Example 4 Evaluating an Algebraic Expression

Evaluate the expression. $2 \div y \cdot z$ for $y = -\frac{14}{3}$ and $z = -\frac{1}{3}$

Solution:

$2 \div y \cdot z$	
$= 2 \div \left(-\frac{14}{3}\right) \cdot \left(-\frac{1}{3}\right)$	Substitute $-\frac{14}{3}$ for y and $-\frac{1}{3}$ for z.
$= \frac{2}{1} \cdot \left(-\frac{3}{14}\right) \cdot \left(-\frac{1}{3}\right)$	Write the whole number over 1. Multiply by the reciprocal of $-\frac{14}{3}$.
$= \frac{\overset{1}{\cancel{2}}}{1} \cdot \left(-\frac{\overset{1}{\cancel{3}}}{\underset{7}{\cancel{14}}}\right) \cdot \left(-\frac{1}{\underset{1}{\cancel{3}}}\right)$	Simplify by dividing out common factors. The product is positive.
$= \frac{1}{7}$	

Skill Practice

Evaluate the expression for $a = -\frac{7}{15}$ and $b = \frac{21}{10}$.

7. $3 \cdot a \div b$

Avoiding Mistakes

Do not forget to write the "1" in the numerator of the fraction.

Answers

6. $-\frac{16}{3}$ **7.** $-\frac{2}{3}$

2. Complex Fractions

A **complex fraction** is a fraction in which the numerator and denominator contains one or more terms with fractions. We will simplify complex fractions in Examples 5 and 6.

Skill Practice

Simplify.

8. $\dfrac{\frac{5}{6}}{\frac{y}{21}}$

Example 5 Simplifying a Complex Fraction

Simplify. $\dfrac{\frac{3}{5}}{\frac{m}{10}}$

Solution:

a. $\dfrac{\frac{3}{5}}{\frac{m}{10}}$ ← This fraction bar denotes division.

$= \dfrac{3}{5} \div \dfrac{m}{10}$ Write the expression as a division of fractions.

$= \dfrac{3}{5} \cdot \dfrac{10}{m}$ Multiply by the reciprocal of $\dfrac{m}{10}$.

$= \dfrac{3}{\cancel{5}_1} \cdot \dfrac{\cancel{10}^2}{m}$ Reduce common factors.

$= \dfrac{6}{m}$ Simplify.

Skill Practice

Simplify.

9. $\dfrac{\frac{3}{10} - \frac{1}{2}}{2 + \frac{1}{5}}$

Example 6 Simplifying a Complex Fraction

Simplify. $\dfrac{\frac{2}{5} - \frac{1}{3}}{1 - \frac{3}{5}}$

Solution:

$\dfrac{\frac{2}{5} - \frac{1}{3}}{1 - \frac{3}{5}}$ This expression can be simplified using the order of operations. Subtract the fractions in the numerator and denominator separately. Then divide the results.

$= \dfrac{\frac{2 \cdot 3}{5 \cdot 3} - \frac{1 \cdot 5}{3 \cdot 5}}{\frac{1 \cdot 5}{1 \cdot 5} - \frac{3}{5}}$ The LCD in the numerator is 15. In the denominator, the LCD is 5.

$= \dfrac{\frac{6}{15} - \frac{5}{15}}{\frac{5}{5} - \frac{3}{5}}$

Answers

8. $\dfrac{35}{2y}$ 9. $-\dfrac{1}{11}$

$$= \frac{\frac{1}{15}}{\frac{2}{5}}$$ ← This fraction bar represents division.

$$= \frac{1}{\cancel{15}_3} \cdot \frac{\cancel{5}^1}{2}$$ Multiply by the reciprocal of $\frac{2}{5}$ and reduce common factors.

$$= \frac{1}{6}$$

Avoiding Mistakes

Perform all operations in the numerator and denominator before performing division.

As you can see from Example 6, simplifying a complex fraction can be a tedious process. For this reason, we offer an alternative approach, as demonstrated in Example 7.

Example 7 **Simplifying a Complex Fraction by Multiplying by the LCD**

Simplify. $\dfrac{\frac{2}{5} - \frac{1}{3}}{1 - \frac{3}{5}}$

Solution:

$$\frac{\frac{2}{5} - \frac{1}{3}}{1 - \frac{3}{5}}$$ First identify the LCD of all four terms in the expression. The LCD of $\frac{2}{5}$, $\frac{1}{3}$, 1, and $\frac{3}{5}$ is 15.

$$= \frac{15 \cdot \left(\frac{2}{5} - \frac{1}{3}\right)}{15 \cdot \left(1 - \frac{3}{5}\right)}$$ Group the terms in the numerator and denominator within parentheses. Then multiply each term in the numerator and denominator by the LCD, 15.

$$= \frac{\left(\overset{3}{\cancel{15}} \cdot \frac{2}{\cancel{5}_1}\right) - \left(\overset{5}{\cancel{15}} \cdot \frac{1}{\cancel{3}_1}\right)}{(15 \cdot 1) - \left(\overset{3}{\cancel{15}} \cdot \frac{3}{\cancel{5}_1}\right)}$$ Apply the distributive property to multiply each term by 15. Then simplify each product.

$$= \frac{6 - 5}{15 - 9} = \frac{1}{6}$$ Simplify the fraction.

Skill Practice

Simplify the complex fraction by multiplying numerator and denominator by the LCD of all four terms in the fraction.

10. $\dfrac{\frac{3}{10} - \frac{1}{2}}{2 + \frac{1}{5}}$

3. Simplifying Algebraic Expressions

In Section 3.1, we simplified algebraic expressions by combining like terms. For example,

$$2x - 3x + 8x = (2 - 3 + 8)x \quad \text{Apply the distributive property.}$$
$$= 7x$$

In a similar way, we can combine like terms when the coefficients are fractions. This is demonstrated in Example 8.

TIP: Recall that like terms can also be combined by combining the coefficients and keeping the variable factor unchanged.

$$2x - 3x + 8x = 7x$$

Answer

10. $-\frac{1}{11}$

Skill Practice

Simplify.

11. $\frac{7}{4}y + \frac{2}{3}y$

Example 8 Simplifying an Algebraic Expression

Simplify. $\frac{2}{5}x - \frac{1}{3}x$

Solution:

$\frac{2}{5}x - \frac{1}{3}x$ — The two terms are like terms.

$= \left(\frac{2}{5} - \frac{1}{3}\right)x$ — Combine like terms by applying the distributive property.

$= \left(\frac{2 \cdot 3}{5 \cdot 3} - \frac{1 \cdot 5}{3 \cdot 5}\right)x$ — The LCD is 15.

$= \left(\frac{6}{15} - \frac{5}{15}\right)x$ — Simplify.

$= \frac{1}{15}x$

TIP: $\frac{1}{15}x$ can also be written as $\frac{x}{15}$.

$$\frac{1}{15}x = \frac{1}{15} \cdot x = \frac{1}{15} \cdot \frac{x}{1} = \frac{x}{15}$$

Answer

11. $\frac{29}{12}y$

Section 4.7 Practice Exercises

Boost *your* GRADE at ALEKS.com!

ALEKS version 3.0

- Practice Problems
- Self-Tests
- NetTutor
- e-Professors
- Videos

Study Skills Exercise

1. Define the key terms.

   a. **Powers of one-tenth** b. **Complex fraction**

Review Exercises

For Exercises 2–8, simplify.

2. $-\frac{2}{5} + \frac{7}{6}$
3. $-\frac{2}{5} - \frac{7}{6}$
4. $\left(-\frac{2}{5}\right)\left(\frac{7}{6}\right)$
5. $-\frac{2}{5} \div \frac{7}{6}$
6. $3\frac{1}{4} + \left(-2\frac{5}{6}\right)$
7. $\left(3\frac{1}{4}\right) \cdot \left(-2\frac{5}{6}\right)$
8. $3\frac{1}{4} \div \left(-2\frac{5}{6}\right)$

Objective 1: Order of Operations

For Exercises 9–44, simplify. **(See Examples 1–3.)**

9. $\left(\frac{1}{9}\right)^2$
10. $\left(\frac{1}{4}\right)^2$
11. $\left(-\frac{1}{9}\right)^2$
12. $\left(-\frac{1}{4}\right)^2$
13. $-\left(\frac{3}{2}\right)^3$
14. $-\left(\frac{4}{3}\right)^3$
15. $\left(-\frac{3}{2}\right)^3$
16. $\left(-\frac{4}{3}\right)^3$

17. $\left(\frac{1}{10}\right)^3$ **18.** $\left(\frac{1}{10}\right)^4$ **19.** $\left(-\frac{1}{10}\right)^6$ **20.** $\left(-\frac{1}{10}\right)^5$

21. $\left(-\frac{10}{3} \cdot \frac{3}{100}\right)^3$ **22.** $\left(-\frac{1}{6} \cdot \frac{3}{5}\right)^2$ **23.** $-\left(4 \cdot \frac{3}{4}\right)^3$ **24.** $-\left(5 \cdot \frac{2}{5}\right)^2$

25. $\frac{1}{6} + \left(2\frac{1}{3}\right) \cdot \left(1\frac{3}{4}\right)$ **26.** $\frac{7}{9} + \left(2\frac{1}{6}\right) \cdot \left(3\frac{1}{3}\right)$ **27.** $6 - 5\frac{1}{7} \div \left(-\frac{1}{7}\right)$ **28.** $11 - 6\frac{1}{3} \div \left(-1\frac{1}{6}\right)$

29. $-\frac{1}{3} \cdot \left|-\frac{21}{4} \cdot \frac{8}{7}\right|$ **30.** $-\frac{1}{6} \cdot \left|-\frac{24}{5} \cdot \frac{30}{8}\right|$ **31.** $\frac{16}{9} \cdot \left(\frac{1}{2}\right)^3$ **32.** $\frac{28}{6} \cdot \left(\frac{3}{2}\right)^2$

33. $\frac{54}{21} \div \frac{2}{3} \cdot \frac{7}{9}$ **34.** $\frac{48}{56} \div \frac{3}{8} \cdot \frac{7}{8}$ **35.** $7\frac{1}{8} \div \left(-1\frac{1}{3}\right) \div \left(-2\frac{1}{4}\right)$ **36.** $\left(-3\frac{1}{8}\right) \div 5\frac{5}{7} \div 1\frac{5}{16}$

37. $\frac{5}{4} \div \frac{3}{2} - \left(-\frac{5}{6}\right)$ **38.** $\frac{1}{7} \div \frac{2}{21} - \left(-\frac{5}{2}\right)$ **39.** $\left(\frac{1}{3} - \frac{1}{2}\right)^2$ **40.** $\left(-\frac{2}{3} + \frac{1}{6}\right)^2$

41. $\left(\frac{1}{4}\right)^2 \div \left(\frac{5}{6} - \frac{2}{3}\right) + \frac{7}{12}$ **42.** $\left(\frac{1}{2} + \frac{1}{3}\right) \cdot \left(\frac{2}{5}\right)^2 + \frac{3}{10}$ **43.** $\left(5 - 1\frac{7}{8}\right) \div \left(3 - \frac{13}{16}\right)$ **44.** $\left(4 + 2\frac{1}{9}\right) \div \left(2 - 1\frac{11}{36}\right)$

For Exercises 45–52, evaluate the expression for the given values of the variables. **(See Example 4.)**

45. $-3 \cdot a \div b$ for $a = -\frac{5}{6}$ and $b = \frac{3}{10}$ **46.** $4 \div w \cdot z$ for $w = \frac{2}{7}$ and $z = -\frac{1}{5}$

47. xy^2 for $x = 2\frac{1}{3}$ and $y = \frac{3}{2}$ **48.** c^3d for $c = -1\frac{1}{2}$ and $d = \frac{1}{3}$

49. $4x + 6y$ for $x = \frac{1}{2}$ and $y = -\frac{3}{2}$ **50.** $2m - 3n$ for $m = -\frac{3}{4}$ and $n = \frac{1}{6}$

51. $(4 - w)(3 + z)$ for $w = 2\frac{1}{3}$ and $z = 1\frac{2}{3}$ **52.** $(2 + a)(7 - b)$ for $a = -1\frac{1}{2}$ and $b = 5\frac{1}{4}$

Objective 2: Complex Fractions

For Exercises 53–68, simplify the complex fractions. **(See Examples 5–7.)**

53. $\dfrac{\frac{5}{8}}{\frac{3}{4}}$ **54.** $\dfrac{\frac{8}{9}}{\frac{7}{12}}$ **55.** $\dfrac{-\frac{21}{10}}{\frac{6}{5}}$ **56.** $\dfrac{\frac{20}{3}}{-\frac{5}{8}}$

57. $\dfrac{\frac{3}{7}}{\frac{12}{x}}$ **58.** $\dfrac{\frac{5}{p}}{\frac{30}{11}}$ **59.** $\dfrac{\frac{-15}{w}}{\frac{25}{w}}$ **60.** $\dfrac{-\frac{18}{t}}{\frac{15}{t}}$

61. $\dfrac{\frac{4}{3} - \frac{1}{6}}{1 - \frac{1}{3}}$ **62.** $\dfrac{\frac{6}{5} + \frac{3}{10}}{2 + \frac{1}{2}}$ **63.** $\dfrac{\frac{1}{2} + 3}{\frac{9}{8} + \frac{1}{4}}$ **64.** $\dfrac{\frac{5}{2} + 1}{\frac{3}{4} + \frac{1}{3}}$

65. $\dfrac{\frac{5}{7} - \frac{1}{14}}{\frac{1}{2} - \frac{3}{7}}$

66. $\dfrac{-\frac{1}{4} + \frac{1}{6}}{-\frac{4}{3} - \frac{5}{6}}$

67. $\dfrac{-\frac{7}{4} + \frac{3}{2}}{-\frac{7}{8} - \frac{1}{4}}$

68. $\dfrac{\frac{8}{3} - \frac{5}{6}}{\frac{1}{4} - \frac{4}{3}}$

Objective 3: Simplifying Algebraic Expressions

For Exercises 69–76, simplify the expressions. **(See Example 8.)**

69. $\frac{1}{2}y + \frac{3}{2}y$

70. $-\frac{4}{5}p + \frac{2}{5}p$

71. $\frac{3}{4}a - \frac{1}{8}a$

72. $\frac{1}{3}b + \frac{2}{9}b$

73. $\frac{4}{5}x - \frac{3}{10}x + \frac{1}{15}x$

74. $-\frac{3}{2}y - \frac{4}{3}y + \frac{3}{4}y$

75. $\frac{3}{2}y + \frac{1}{4}z - \frac{1}{6}y - 2z$

76. $-\frac{5}{8}a + 3b + \frac{1}{4}a - \frac{7}{2}b$

Expanding Your Skills

77. Evaluate

a. $\left(\frac{1}{6}\right)^2$

b. $\sqrt{\frac{1}{36}}$

78. Evaluate

a. $\left(\frac{2}{7}\right)^2$

b. $\sqrt{\frac{4}{49}}$

For Exercises 79–82, evaluate the square roots.

79. $\sqrt{\frac{1}{25}}$

80. $\sqrt{\frac{1}{100}}$

81. $\sqrt{\frac{64}{81}}$

82. $\sqrt{\frac{9}{4}}$

Section 4.8 Solving Equations Containing Fractions

Objectives

1. Solving Equations Containing Fractions
2. Solving Equations by Clearing Fractions

1. Solving Equations Containing Fractions

In this section we will solve linear equations that contain fractions. To begin, review the addition, subtraction, multiplication, and division properties of equality.

Property	Example
Addition Property of Equality If $a = b$, then $a + c = b + c$	Solve. $x - 4 = 6$ $x - 4 + 4 = 6 + 4$ $x = 10$
Subtraction Property of Equality If $a = b$, then $a - c = b - c$	Solve. $x + 3 = 5$ $x + 3 - 3 = 5 - 3$ $x = 2$
Multiplication Property of Equality If $a = b$, then $a \cdot c = b \cdot c$	Solve. $\frac{x}{5} = 3$ $\frac{x}{5} \cdot 5 = 3 \cdot 5$ $x = 15$
Division Property of Equality If $a = b$, then $\frac{a}{c} = \frac{b}{c}$ (provided that $c \neq 0$)	Solve. $9x = 27$ $\frac{9x}{9} = \frac{27}{9}$ $x = 3$

We will use the same properties to solve equations containing fractions. In Examples 1 and 2, we use the addition and subtraction properties of equality.

Example 1 Using the Addition Property of Equality

Solve. $x - \frac{3}{10} = \frac{9}{20}$

Solution:

$$x - \frac{3}{10} = \frac{9}{20}$$

$$x - \frac{3}{10} + \frac{3}{10} = \frac{9}{20} + \frac{3}{10}$$ Add $\frac{3}{10}$ to both sides to isolate x.

$$x = \frac{9}{20} + \frac{3 \cdot 2}{10 \cdot 2}$$ To add the fractions on the right, the LCD is 20.

$$x = \frac{9}{20} + \frac{6}{20}$$

$$x = \frac{15}{20}$$ Now simplify the fraction.

$$x = \frac{3}{4}$$ The solution is $\frac{3}{4}$.

Check: $x - \frac{3}{10} = \frac{9}{20}$

$\frac{3}{4} - \frac{3}{10} \stackrel{?}{=} \frac{9}{20}$ Substitute $\frac{3}{4}$ for x.

$\frac{3 \cdot 5}{4 \cdot 5} - \frac{3 \cdot 2}{10 \cdot 2} \stackrel{?}{=} \frac{9}{20}$

$\frac{15}{20} - \frac{6}{20} \stackrel{?}{=} \frac{9}{20}$ ✓ True

Skill Practice

Solve.

1. $w - \frac{1}{3} = \frac{4}{15}$

Example 2 Using the Subtraction Property of Equality

Solve. $-\frac{5}{4} = \frac{1}{2} + t$

Solution:

$$-\frac{5}{4} = \frac{1}{2} + t$$

$$-\frac{5}{4} - \frac{1}{2} = \frac{1}{2} - \frac{1}{2} + t$$ Subtract $\frac{1}{2}$ on both sides to isolate t.

$$-\frac{5}{4} - \frac{1 \cdot 2}{2 \cdot 2} = t$$ To subtract the fractions on the left, the LCD is 4.

$$-\frac{5}{4} - \frac{2}{4} = t$$

$$-\frac{7}{4} = t$$ The solution is $-\frac{7}{4}$ and checks in the original equation.

Skill Practice

Solve.

2. $-\frac{7}{9} = \frac{1}{3} + y$

Answers

1. $\frac{3}{5}$ **2.** $-\frac{10}{9}$

Recall that the product of a number and its reciprocal is 1. For example:

$$\frac{2}{7} \cdot \frac{7}{2} = 1, \qquad -\frac{3}{8} \cdot \left(-\frac{8}{3}\right) = 1, \qquad 5 \cdot \frac{1}{5} = 1$$

We will use this fact and the multiplication property of equality to solve the equations in Examples 3–5.

Skill Practice

Solve.

3. $\frac{3}{4}x = \frac{5}{6}$

Example 3 Using the Multiplication Property of Equality

Solve. $\frac{4}{5}x = \frac{2}{7}$

Solution:

$\frac{4}{5}x = \frac{2}{7}$ In this equation, we will apply the multiplication property of equality. Multiply both sides of the equation by the reciprocal of $\frac{4}{5}$. Do this because $\frac{5}{4} \cdot \frac{4}{5}x$ is equal to $1x$. This isolates the variable x.

$\frac{5}{4} \cdot \frac{4}{5}x = \frac{5}{4} \cdot \frac{2}{7}$ Multiply both sides of the equation by the reciprocal of $\frac{4}{5}$.

$1x = \frac{10}{28}$

$x = \frac{\overset{5}{\cancel{10}}}{\underset{14}{\cancel{28}}}$ Simplify.

$x = \frac{5}{14}$ The solution is $\frac{5}{14}$ and checks in the original equation.

Skill Practice

Solve.

4. $7 = -\frac{1}{4}y$

Example 4 Using the Multiplication Property of Equality

Solve. $8 = -\frac{1}{6}y$

Solution:

$8 = -\frac{1}{6}y$

$-6 \cdot (8) = -6 \cdot \left(-\frac{1}{6}y\right)$ Multiply both sides of the equation by the reciprocal of $-\frac{1}{6}$. Do this because $-6 \cdot (-\frac{1}{6}y)$ is equal to $1y$. This isolates the variable y.

$-48 = 1y$ The solution is -48 and checks in the original equation.

$-48 = y$

Answers

3. $\frac{10}{9}$ **4.** -28

Example 5 Using the Multiplication Property of Equality

Solve. $-\frac{2}{9} = -4w$

Solution:

$$-\frac{2}{9} = -4w$$

$$-\frac{1}{4}\left(-\frac{2}{9}\right) = -\frac{1}{4}(-4w)$$ Multiply both sides of the equation by the reciprocal of −4. Do this because $-\frac{1}{4} \cdot (-4w)$ is equal to $1w$. This isolates the variable w.

$$-\frac{1}{\cancel{4}_2}\left(-\frac{\cancel{2}^1}{9}\right) = 1w$$

$$\frac{1}{18} = w$$ The solution is $\frac{1}{18}$ and checks in the original equation.

Skill Practice

Solve.

5. $-\frac{2}{5} = -4x$

In Example 6, we solve an equation in which we must apply both the addition property of equality and the multiplication property of equality.

Example 6 Using Multiple Steps to Solve an Equation with Fractions

Solve. $\frac{3}{4}x - \frac{1}{3} = \frac{5}{6}$

Solution:

$$\frac{3}{4}x - \frac{1}{3} = \frac{5}{6}$$

$$\frac{3}{4}x - \frac{1}{3} + \frac{1}{3} = \frac{5}{6} + \frac{1}{3}$$ To isolate the x-term, add $\frac{1}{3}$ to both sides.

$$\frac{3}{4}x = \frac{5}{6} + \frac{1 \cdot 2}{3 \cdot 2}$$ To add the fractions on the right, the LCD is 6.

$$\frac{3}{4}x = \frac{5}{6} + \frac{2}{6}$$

$$\frac{3}{4}x = \frac{7}{6}$$

$$\frac{4}{3} \cdot \frac{3}{4}x = \frac{4}{3} \cdot \frac{7}{6}$$ Multiply both sides of the equation by the reciprocal of $\frac{3}{4}$. Do this because $\frac{4}{3} \cdot (\frac{3}{4}x)$ is equal to $1x$. This isolates the variable x.

$$x = \frac{\cancel{4}^2}{3} \cdot \frac{7}{\cancel{6}_3}$$ Simplify the product.

$$x = \frac{14}{9}$$ The solution is $\frac{14}{9}$ and checks in the original equation.

Skill Practice

Solve.

6. $\frac{2}{3}x - \frac{1}{4} = \frac{3}{2}$

Answers

5. $\frac{1}{10}$ 6. $\frac{21}{8}$

2. Solving Equations by Clearing Fractions

As you probably noticed, Example 6 required tedious manipulation of fractions to isolate the variable. Therefore, we will now show you an alternative technique that eliminates the fractions immediately. This technique is called **clearing fractions**. Its basis is to multiply both sides of the equation by the LCD of all terms in the equation. This is demonstrated in Examples 7–8.

Skill Practice

Solve.

7. $\frac{x}{10} + \frac{1}{2} = \frac{x}{5}$

Example 7 Solving an Equation by First Clearing Fractions

Solve. $\frac{y}{8} + \frac{3}{2} = \frac{y}{4}$

Solution:

$\frac{y}{8} + \frac{3}{2} = \frac{y}{4}$	The LCD of $\frac{y}{8}, \frac{3}{2}$, and $\frac{y}{4}$ is 8.
$8 \cdot \left(\frac{y}{8} + \frac{3}{2}\right) = 8 \cdot \left(\frac{y}{4}\right)$	Apply the multiplication property of equality. Multiply both sides of the equation by 8.
$8 \cdot \left(\frac{y}{8}\right) + 8 \cdot \left(\frac{3}{2}\right) = 8 \cdot \left(\frac{y}{4}\right)$	Use the distributive property to multiply each term by 8.
$\overset{1}{\cancel{8}} \cdot \left(\frac{y}{\cancel{8}_1}\right) + \overset{4}{\cancel{8}} \cdot \left(\frac{3}{\cancel{2}_1}\right) = \overset{2}{\cancel{8}} \cdot \left(\frac{y}{\cancel{4}_1}\right)$	Simplify each term.
$y + 12 = 2y$	The fractions have been "cleared." Now isolate the variable.
$y - y + 12 = 2y - y$	Subtract y from both sides to collect the variable terms on one side.
$12 = y$	Simplify.

The solution is 12 and checks in the original equation.

Clearing fractions is a technique that "removes" fractions from an equation and produces a simpler equation. Because this technique is so powerful, we will add it to step 1 of our procedure box for solving a linear equation in one variable.

PROCEDURE Solving a Linear Equation in One Variable

Step 1 Simplify both sides of the equation.
- Clear parentheses if necessary.
- Combine *like* terms if necessary.
- Consider clearing fractions if necessary.

Step 2 Use the addition or subtraction property of equality to collect the variable terms on one side of the equation.

Step 3 Use the addition or subtraction property of equality to collect the constant terms on the *other* side of the equation.

Step 4 Use the multiplication or division property of equality to make the coefficient of the variable term equal to 1.

Step 5 Check the answer in the original equation.

Answer

7. 5

Example 8 **Solving an Equation by First Clearing Fractions**

Solve. $-\frac{4}{7}x - 2 = \frac{3}{14}$

Solution:

$-\frac{4}{7}x - 2 = \frac{3}{14}$ The LCD of $-\frac{4}{7}x$, -2, and $\frac{3}{14}$ is 14.

$14 \cdot \left(-\frac{4}{7}x - 2\right) = 14 \cdot \left(\frac{3}{14}\right)$ Multiply both sides by 14.

$14 \cdot \left(-\frac{4}{7}x\right) + 14 \cdot (-2) = 14 \cdot \left(\frac{3}{14}\right)$ Use the distributive property to multiply each term by 14.

$\overset{2}{\cancel{14}} \cdot \left(-\frac{4}{\underset{1}{\cancel{7}}}x\right) + 14 \cdot (-2) = \overset{1}{\cancel{14}} \cdot \left(\frac{3}{\underset{1}{\cancel{14}}}\right)$ Simplify each term.

$-8x - 28 = 3$ The fractions have been "cleared."

$-8x - 28 + 28 = 3 + 28$ Add 28 to both sides to isolate the x term.

$-8x = 31$

$\frac{-8x}{-8} = \frac{31}{-8}$ Apply the division property of equality to isolate x.

$x = -\frac{31}{8}$ The solution is $-\frac{31}{8}$ and checks in the original equation.

Skill Practice

Solve.

8. $-\frac{3}{8}x - 3 = \frac{1}{4}$

Avoiding Mistakes

Be sure to multiply each term by 14. This includes the constant term, -2.

Answer

8. $-\frac{26}{3}$

Section 4.8 Practice Exercises

Boost *your* GRADE at ALEKS.com! ALEKS version 3.0

- Practice Problems
- Self-Tests
- NetTutor
- e-Professors
- Videos

Study Skills Exercises

1. When you solve equations involving several steps, it is recommended that you write an explanation for each step. In Example 8, an equation is solved with each step shown. Your job is to write an explanation for each step for the following equation.

Explanation

$\frac{3}{5}x - \frac{7}{10} = \frac{1}{2}$

$10\left(\frac{3}{5}x - \frac{7}{10}\right) = 10\left(\frac{1}{2}\right)$

$6x - 7 = 5$

$6x - 7 + 7 = 5 + 7$

$6x = 12$

$\frac{6x}{6} = \frac{12}{6}$

$x = 2$ The solution is 2.

2. Define the key terms.

a. **Addition property of equality**
b. **Subtraction property of equality**
c. **Multiplication property of equality**
d. **Division property of equality**
e. **Clearing fractions**

Review Exercises

For Exercises 3–8, simplify.

3. $\left(\frac{2}{3} - \frac{3}{2}\right)^2$

4. $\frac{3}{5} + \frac{9}{4} - 3\frac{1}{2}$

5. $\left(\frac{2}{3}\right)^2 - \left(\frac{3}{2}\right)^2$

6. $-\frac{3}{5} \div \frac{9}{4} \cdot \left(3\frac{1}{2}\right)$

7. $-5\frac{2}{3} - 4\frac{3}{8}$

8. $\left(3 - 1\frac{3}{4}\right)^2$

Objective 1: Solving Equations Containing Fractions

For Exercises 9–42, solve the equations. Write the answers as fractions or whole numbers. **(See Examples 1–6.)**

9. $p - \frac{5}{6} = \frac{1}{3}$

10. $q - \frac{3}{4} = \frac{3}{2}$

11. $-\frac{7}{10} = \frac{3}{5} + a$

12. $-\frac{3}{8} = \frac{1}{4} + b$

13. $\frac{2}{3} = y - \frac{5}{12}$

14. $\frac{7}{11} = z + \frac{3}{11}$

15. $t + \frac{3}{8} = 2$

16. $r - \frac{4}{7} = -1$

17. $\frac{1}{6} = -\frac{11}{6} + m$

18. $n + \frac{1}{2} = -\frac{2}{3}$

19. $\frac{3}{5}y = \frac{7}{10}$

20. $\frac{7}{2}x = \frac{5}{4}$

21. $\frac{5}{4}k = -\frac{1}{2}$

22. $-\frac{11}{12}h = -\frac{1}{6}$

23. $6 = -\frac{1}{4}x$

24. $3 = -\frac{1}{5}y$

25. $\frac{2}{3}m = 14$

26. $\frac{5}{9}n = 40$

27. $\frac{b}{7} = -3$

28. $\frac{a}{4} = 12$

29. $-\frac{u}{2} = -15$

30. $-\frac{v}{10} = -4$

31. $0 = \frac{3}{8}m$

32. $0 = \frac{1}{10}n$

33. $6x = \frac{12}{5}$

34. $7t = \frac{14}{3}$

35. $-\frac{5}{9} = -10x$

36. $-\frac{4}{3} = -6y$

37. $\frac{2}{5}x - \frac{1}{4} = \frac{3}{2}$

38. $\frac{5}{9}y - \frac{1}{3} = \frac{5}{6}$

39. $-\frac{4}{7} = \frac{1}{2} + \frac{3}{14}w$

40. $-\frac{1}{8} = \frac{3}{4} + \frac{5}{2}z$

41. $3p + \frac{1}{2} = \frac{5}{4}$

42. $2t - \frac{3}{8} = \frac{9}{16}$

Objective 2: Solving Equations by Clearing Fractions

For Exercises 43–54, solve the equations by first clearing fractions. **(See Examples 7–8.)**

43. $\frac{x}{5} + \frac{1}{2} = \frac{7}{10}$

44. $\frac{p}{4} + \frac{1}{2} = \frac{5}{8}$

45. $-\frac{5}{7}y - 1 = \frac{3}{2}$

46. $-\frac{2}{3}w - 3 = -\frac{1}{2}$

47. $\frac{2}{3} = \frac{5}{9} + \frac{1}{6}t$

48. $\frac{4}{5} = \frac{9}{15} + \frac{1}{3}n$

49. $\frac{m}{3} + \frac{m}{6} = \frac{5}{9}$

50. $\frac{4}{5} = \frac{n}{15} - \frac{n}{3}$

51. $\frac{x}{3} + \frac{7}{6} = \frac{x}{2}$

52. $\frac{p}{4} = \frac{p}{8} + \frac{1}{2}$

53. $\frac{3}{2}y + 3 = 2y + \frac{1}{2}$

54. $\frac{1}{4}x - 1 = 2 - \frac{1}{2}x$

Mixed Exercises

For Exercises 55–72, solve the equations.

55. $\frac{h}{4} = -12$

56. $\frac{w}{6} = -18$

57. $\frac{2}{3} + t = 1$

58. $\frac{3}{4} + q = 1$

59. $-\frac{3}{7}x = \frac{9}{10}$

60. $-\frac{2}{11}y = \frac{4}{15}$

61. $4c = -\frac{1}{3}$

62. $\frac{1}{3}b = -4$

63. $-p = -\frac{7}{10}$

64. $-8h = 0$

65. $-9 = \frac{w}{2} - 3$

66. $-16 = \frac{t}{4} - 14$

67. $2x - \frac{1}{2} = \frac{1}{6}$

68. $3z - \frac{3}{4} = \frac{1}{2}$

69. $\frac{5}{4}x = \frac{5}{6}x + \frac{2}{3}$

70. $\frac{3}{4}y = \frac{3}{2}y + \frac{1}{5}$

71. $-4 - \frac{3}{2}d = \frac{2}{5}$

72. $-2 - \frac{5}{4}z = \frac{5}{8}$

Expanding Your Skills

For Exercises 73–76, solve the equations by clearing fractions.

73. $p - 1 + \frac{1}{4}p = 2 + \frac{3}{4}p$

74. $\frac{4}{3} + \frac{2}{3}q = -\frac{5}{3} - q - \frac{1}{3}$

75. $\frac{5}{3}x - \frac{4}{5} = \frac{2}{3}x + 1$

76. $\frac{3}{4}y + \frac{9}{7} = -\frac{1}{4}y + 2$

Problem Recognition Exercises

Comparing Expressions and Equations

For Exercises 1–24, first identify the problem as an expression or as an equation. Then simplify the expression or solve the equation. Two examples are given for you.

Example: $\frac{1}{3} + \frac{1}{6} - \frac{5}{6}$	Example: $\frac{1}{3}x + \frac{1}{6} = \frac{5}{6}$
This is an expression. Combine like terms. $\frac{1}{3} + \frac{1}{6} - \frac{5}{6}$ $= \frac{2 \cdot 1}{2 \cdot 3} + \frac{1}{6} - \frac{5}{6}$ $= \frac{2}{6} + \frac{1}{6} - \frac{5}{6}$ $= \frac{-2}{6}$ $= -\frac{1}{3}$	This is an equation. Solve the equation. $\frac{1}{3}x + \frac{1}{6} = \frac{5}{6}$ $6 \cdot \left(\frac{1}{3}x + \frac{1}{6}\right) = 6 \cdot \left(\frac{5}{6}\right)$ $2x + 1 = 5$ $2x + 1 - 1 = 5 - 1$ $2x = 4$ $\frac{2x}{2} = \frac{4}{2}$ $x = 2$ The solution is 2.

1. $\frac{5}{3}x = \frac{1}{6}$

2. $\frac{7}{8}y = \frac{3}{4}$

3. $\frac{5}{3} - \frac{1}{6}$

4. $\frac{7}{8} - \frac{3}{4}$

5. $1 + \frac{2}{5}$

6. $1 + \frac{4}{5}$

7. $z + \frac{2}{5} = 0$

8. $w + \frac{4}{5} = 0$

9. $\frac{2}{9}x - \frac{1}{3} = \frac{5}{9}$

10. $\frac{3}{10}x - \frac{2}{5} = \frac{1}{10}$

11. $\frac{2}{9} - \frac{1}{3} + \frac{5}{9}$

12. $\frac{3}{10} - \frac{2}{5} + \frac{1}{10}$

13. $3(x - 4) + 2x = 12 + x$

14. $5(x + 4) - 2x = 10 - x$

15. $3(x - 4) + 2x - 12 + x$

16. $5(x + 4) - 2x + 10 - x$

17. $\frac{3}{4}x = 2$

18. $\frac{5}{7}y = 5$

19. $\frac{3}{4} \cdot 2$

20. $\frac{5}{7} \cdot 5$

21. $\frac{4}{7} = 2c$

22. $\frac{9}{4} = 3d$

23. $\frac{4}{7}c + 2c$

24. $\frac{9}{4}d + 3d$

Group Activity

Card Games with Fractions

Materials: A deck of fraction cards for each group. These can be made from index cards where one side of the card is blank, and the other side has a fraction written on it. The deck should consist of several cards of each of the following fractions.

$$\frac{1}{4}, \frac{1}{2}, \frac{3}{4}, \frac{1}{6}, \frac{1}{3}, \frac{2}{3}, \frac{5}{6}, \frac{1}{8}, \frac{3}{8}, \frac{5}{8}, \frac{7}{8}, \frac{1}{5}, \frac{2}{5}, \frac{3}{5}, \frac{4}{5}, \frac{1}{10}, \frac{3}{10}, \frac{4}{9}, \frac{2}{9}, \frac{3}{7}$$

Estimated time: Instructor discretion

In this activity, we outline three different games for students to play in their groups as a fun way to reinforce skills of adding fractions, recognizing equivalent fractions, and ordering fractions.

Game 1 "Blackjack"

Group Size: 3

1. In this game, one student in the group will be the dealer, and the other two will be players. The dealer will deal each player one card face down and one card face up. Then the players individually may elect to have more cards given to them (face up). The goal is to have the sum of the fractions get as close to "2" without going over.

2. Once the players have taken all the cards that they want, they will display their cards face up for the group to see. The player who has a sum closest to "2" without going over wins. The dealer will resolve any "disputes."

3. The members of the group should rotate after several games so that each person has the opportunity to be a player and to be the dealer.

Game 2 "War"

Group Size: 2

1. In this game, each player should start with half of the deck of cards. The players should shuffle the cards and then stack them neatly face down on the table. Then each player will select the top card from the deck, turn it over and place it on the table. The player who has the fraction with the greatest value "wins" that round and takes both cards.

2. Continue overturning cards and deciding who "wins" each round until all of the cards have been overturned. Then the players will count the number of cards they each collected. The player with the most cards wins.

Game 3 "Bingo"

Group Size: The whole class

1. Each student gets five fraction cards. The instructor will call out fractions that are not in lowest terms. The students must identify whether the fraction that was called is the same as one of the fractions on their cards. For example, if the instructor calls out "three-ninths," then students with the fraction card $\frac{1}{3}$ would have a match.

2. The student who first matches all five cards wins.

Chapter 4 Summary

Section 4.1 Introduction to Fractions and Mixed Numbers

Key Concepts

A **fraction** represents a part of a whole unit. For example, $\frac{1}{3}$ represents one part of a whole unit that is divided into 3 equal pieces. A fraction whose numerator is an integer and whose denominator is a nonzero integer is also called a **rational number**.

In the fraction $\frac{1}{3}$, the "top" number, 1, is the **numerator**, and the "bottom" number, 3, is the **denominator**.

A positive fraction in which the numerator is less than the denominator (or the opposite of such a fraction) is called a **proper fraction**. A positive fraction in which the numerator is greater than or equal to the denominator (or the opposite of such a fraction) is called an **improper fraction**.

The fraction $-\dfrac{a}{b}$ is equivalent to $\dfrac{-a}{b}$ or $\dfrac{a}{-b}$.

An improper fraction can be written as a **mixed number** by dividing the numerator by the denominator. Write the quotient as a whole number, and write the remainder over the divisor.

A mixed number can be written as an improper fraction by multiplying the whole number by the denominator and adding the numerator. Then write that total over the denominator.

In a negative mixed number, both the whole number part and the fraction are negative.

Fractions can be represented on a number line. For example,

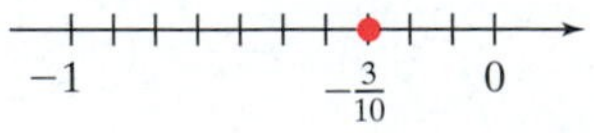

Examples

Example 1

$\frac{1}{3}$ of the pie is shaded.

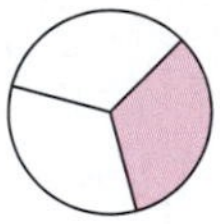

Example 2

For the fraction $\frac{7}{9}$, the numerator is 7 and the denominator is 9.

Example 3

$\frac{5}{3}$ is an improper fraction, $\frac{3}{5}$ is a proper fraction, and $\frac{3}{3}$ is an improper fraction.

Example 4

$$-\frac{5}{7} = \frac{-5}{7} = \frac{5}{-7}$$

Example 5

$\dfrac{10}{3}$ can be written as $3\dfrac{1}{3}$ because

$$\begin{array}{r} 3 \\ 3\overline{)10} \\ \underline{-9} \\ 1 \end{array}$$

Example 6

$2\dfrac{4}{5}$ can be written as $\dfrac{14}{5}$ because $\dfrac{2 \cdot 5 + 4}{5} = \dfrac{14}{5}$

Example 7

$$-5\frac{2}{3} = -\left(5 + \frac{2}{3}\right) = -5 - \frac{2}{3}$$

Example 8

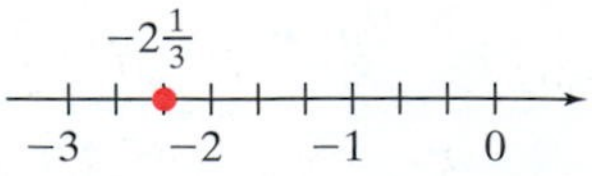

Section 4.2 Simplifying Fractions

Key Concepts

A **factorization** of a number is a product of factors that equals the number.

Divisibility Rules for 2, 3, 5, and 10

A whole number is divisible by
- 2 if the ones-place digit is 0, 2, 4, 6, or 8.
- 3 if the sum of the digits is divisible by 3.
- 5 if the ones-place digit is 0 or 5.
- 10 if the ones-place digit is 0.

A **prime number** is a whole number greater than 1 that has exactly two factors, 1 and itself.

Composite numbers are whole numbers that have more than two factors. The numbers 0 and 1 are neither prime nor composite.

Prime Factorization

The **prime factorization** of a number is the factorization in which every factor is a prime number.

A factor of a number n is any number that divides evenly into n. For example, the factors of 80 are 1, 2, 4, 5, 8, 10, 16, 20, 40, and 80.

Equivalent fractions are fractions that represent the same portion of a whole unit.

To determine if two fractions are equivalent, calculate the cross products. If the cross products are equal, then the fractions are equivalent.

Examples

Example 1

$4 \cdot 4$ and $8 \cdot 2$ are two factorizations of 16.

Example 2

382 is divisible by 2.
640 is divisible by 2, 5, and 10.
735 is divisible by 3 and 5.

Example 3

9 is a composite number.
2 is a prime number.
1 is neither prime nor composite.

Example 4

$$2\overline{)378}$$
$$3\overline{)189}$$
$$3\overline{)63}$$
$$3\overline{)21}$$
$$7$$

The prime factorization of 378 is $2 \cdot 3 \cdot 3 \cdot 3 \cdot 7$ or $2 \cdot 3^3 \cdot 7$

Example 5

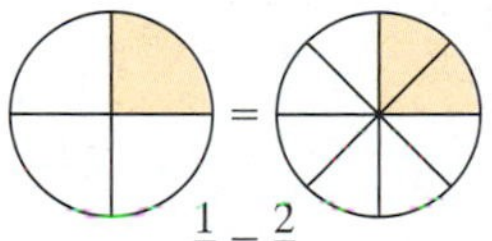

$\frac{1}{4} = \frac{2}{8}$

Example 6

a. Compare $\frac{5}{3}$ and $\frac{6}{4}$.

$\frac{5}{3} \stackrel{?}{=} \frac{6}{4}$

$20 \neq 18$

The fractions are not equivalent.

b. Compare $\frac{4}{5}$ and $\frac{8}{10}$.

$\frac{4}{5} \stackrel{?}{=} \frac{8}{10}$

$40 = 40$

The fractions are equivalent.

To simplify fractions to **lowest terms**, use the fundamental principle of fractions:

Given $\frac{a}{b}$ and the nonzero number c. Then

$$\frac{a \cdot c}{b \cdot c} = \frac{a}{b} \cdot \frac{c}{c} = \frac{a}{b} \cdot 1 = \frac{a}{b}$$

Example 7

$$\frac{25}{15} = \frac{5 \cdot 5}{3 \cdot 5} = \frac{5}{3} \cdot \frac{5}{5} = \frac{5}{3} \cdot 1 = \frac{5}{3}$$

$$\frac{3x^2}{5x^3} = \frac{3 \cdot \overset{1}{\cancel{x}} \cdot \overset{1}{\cancel{x}}}{5 \cdot \underset{1}{\cancel{x}} \cdot \underset{1}{\cancel{x}} \cdot x} = \frac{3}{5x}$$

To simplify fractions with common powers of 10, "strike through" the common zeros first.

Example 8

$$\frac{3{,}\cancel{000}}{12{,}\cancel{000}} = \frac{3}{12} = \frac{\overset{1}{\cancel{3}}}{\underset{4}{\cancel{12}}} = \frac{1}{4}$$

Section 4.3 Multiplication and Division of Fractions

Key Concepts

Multiplication of Fractions

To multiply fractions, write the product of the numerators over the product of the denominators. Then simplify the resulting fraction, if possible.

Examples

Example 1

$$\frac{4}{7} \cdot \frac{6}{5} = \frac{24}{35}$$

$$\left(\frac{15}{16}\right)\left(\frac{4}{5}\right) = \frac{\overset{3}{\cancel{15}}}{\underset{4}{\cancel{16}}} \cdot \frac{\overset{1}{\cancel{4}}}{\underset{1}{\cancel{5}}} = \frac{3}{4}$$

When multiplying a whole number and a fraction, first write the whole number as a fraction by writing the whole number over 1.

Example 2

$$-8 \cdot \left(\frac{5}{6}\right) = -\frac{\overset{4}{\cancel{8}}}{1} \cdot \frac{5}{\underset{3}{\cancel{6}}} = -\frac{20}{3}$$

The formula for the area of a triangle is given by $A = \frac{1}{2}bh$.

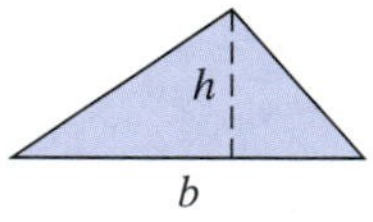

Area is expressed in square units such as ft^2, in.^2, yd^2, m^2, and cm^2.

Example 3

The area of the triangle is

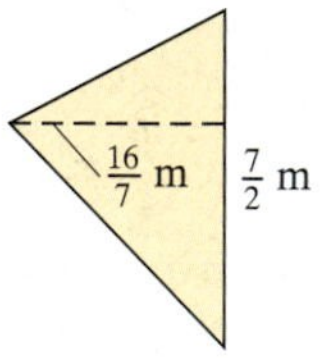

$$A = \frac{1}{2}bh$$

$$= \frac{1}{2}\left(\frac{7}{2}\text{m}\right)\left(\frac{16}{7}\text{m}\right)$$

$$= \frac{\overset{4}{\cancel{112}}}{\underset{1}{\cancel{28}}}\text{m}^2$$

$$= 4\ \text{m}^2$$

The area is $4\ \text{m}^2$.

The **reciprocal** of $\frac{a}{b}$ is $\frac{b}{a}$ for $a, b \neq 0$. The product of a fraction and its reciprocal is 1. For example, $\frac{6}{11} \cdot \frac{11}{6} = 1$.

Example 4

The reciprocal of $-\frac{5}{8}$ is $-\frac{8}{5}$.
The reciprocal of 4 is $\frac{1}{4}$.

The number 0 does not have a reciprocal because $\frac{1}{0}$ is undefined.

Dividing Fractions

To divide two fractions, multiply the dividend (the "first" fraction) by the reciprocal of the divisor (the "second" fraction).

Example 5

$$\frac{18}{25} \div \frac{30}{35} = \frac{\overset{3}{\cancel{18}}}{\underset{5}{\cancel{25}}} \cdot \frac{\overset{7}{\cancel{35}}}{\underset{5}{\cancel{30}}} = \frac{21}{25}$$

When dividing by a whole number, first write the whole number as a fraction by writing the whole number over 1. Then multiply by its reciprocal.

Example 6

$$-\frac{9}{8} \div (-4) = -\frac{9}{8} \div \left(-\frac{4}{1}\right) = -\frac{9}{8} \cdot \left(-\frac{1}{4}\right) = \frac{9}{32}$$

Section 4.4 Least Common Multiple and Equivalent Fractions

Key Concepts

The numbers obtained by multiplying a number n by the whole numbers 1, 2, 3, and so on are called **multiples** of n.

The **least common multiple (LCM)** of two given numbers is the smallest whole number that is a multiple of each given number.

Examples

Example 1

The numbers 5, 10, 15, 20, 25, 30, 35, and 40 are several multiples of 5.

Example 2

Find the LCM of 8 and 10.
Some multiples of 8 are 8, 16, 24, 32, 40.
Some multiples of 10 are 10, 20, 30, 40.

40 is the least common multiple.

Using Prime Factors to Find the LCM of Two Numbers

1. Write each number as a product of prime factors.
2. The LCM is the product of unique prime factors from both numbers. Use repeated factors the maximum number of times they appear in either factorization.

Example 3

Find the LCM for the numbers 24 and 16.

$24 = 2 \cdot 2 \cdot 2 \cdot 3$

$16 = 2 \cdot 2 \cdot 2 \cdot 2$

$\text{LCM} = 2 \cdot 2 \cdot 2 \cdot 2 \cdot 3 = 48$

Writing Equivalent Fractions

Use the fundamental principle of fractions to convert a fraction to an equivalent fraction with a given denominator.

Example 4

Write the fraction with the indicated denominator.

$$\frac{3}{4} = \frac{}{36x}$$

$$\frac{3 \cdot 9x}{4 \cdot 9x} = \frac{27x}{36x}$$

The fraction $\frac{27x}{36x}$ is equivalent to $\frac{3}{4}$.

Ordering Fractions

Write the fractions with a common denominator. Then compare the numerators.

The **least common denominator (LCD)** of two fractions is the LCM of their denominators.

Example 5

Fill in the blank with the appropriate symbol, $<$ or $>$.

$-\frac{5}{9} \square -\frac{7}{12}$ The LCD is 36.

$$-\frac{5 \cdot 4}{9 \cdot 4} \square -\frac{7 \cdot 3}{12 \cdot 3}$$

$$-\frac{20}{36} \boxed{>} -\frac{21}{36}$$

Section 4.5 Addition and Subtraction of Fractions

Key Concepts

Adding or Subtracting Like Fractions

1. Add or subtract the numerators.
2. Write the sum or difference over the common denominator.
3. Simplify the fraction to lowest terms if possible.

Examples

Example 1

$$\frac{5}{8} + \frac{7}{8} = \frac{12}{8} = \frac{\overset{3}{\cancel{12}}}{\underset{2}{\cancel{8}}} = \frac{3}{2}$$

Example 2

A nail that is $\frac{13}{8}$ in. long is driven through a board that is $\frac{11}{8}$ in. thick. How much of the nail extends beyond the board?

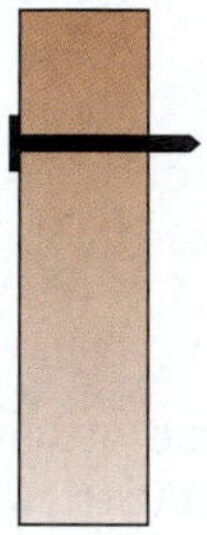

$$\frac{13}{8} - \frac{11}{8} = \frac{2}{8} = \frac{1}{4}$$

The nail will extend $\frac{1}{4}$ in.

To add or subtract unlike fractions, first we must write each fraction as an equivalent fraction with a common denominator.

Adding or Subtracting Unlike Fractions

1. Identify the LCD.
2. Write each individual fraction as an equivalent fraction with the LCD.
3. Add or subtract the numerators and write the result over the common denominator.
4. Simplify to lowest terms, if possible.

Example 3

Simplify. $\frac{7}{5} - \frac{3}{10} + \frac{13}{15}$

$$\frac{7 \cdot 6}{5 \cdot 6} - \frac{3 \cdot 3}{10 \cdot 3} + \frac{13 \cdot 2}{15 \cdot 2}$$

The LCD is 30.

$$= \frac{42}{30} - \frac{9}{30} + \frac{26}{30}$$

$$= \frac{42 - 9 + 26}{30}$$

$$= \frac{59}{30}$$

Section 4.6 Estimation and Operations on Mixed Numbers

Key Concepts

Multiplication of Mixed Numbers

Step 1 Change each mixed number to an improper fraction.

Step 2 Multiply the improper fractions and simplify to lowest terms, if possible.

Division of Mixed Numbers

Step 1 Change each mixed number to an improper fraction.

Step 2 Divide the improper fractions and simplify to lowest terms, if possible. Recall that to divide fractions, multiply the dividend by the reciprocal of the divisor.

Addition of Mixed Numbers

To find the sum of two or more mixed numbers, add the whole-number parts and add the fractional parts.

Subtraction of Mixed Numbers

To subtract mixed numbers, subtract the fractional parts and subtract the whole-number parts.
When the fractional part in the subtrahend is larger than the fractional part in the minuend, we borrow from the whole number part of the minuend.

We can also add or subtract mixed numbers by writing the numbers as improper fractions. Then add or subtract the fractions.

Examples

Example 1

$$4\frac{4}{5}\cdot 2\frac{1}{2} = \frac{\overset{12}{\cancel{24}}}{\underset{1}{\cancel{5}}}\cdot\frac{\overset{1}{\cancel{5}}}{\underset{1}{\cancel{2}}} = \frac{12}{1} = 12$$

Example 2

$$6\frac{2}{3} \div 2\frac{7}{9} = \frac{20}{3} \div \frac{25}{9} = \frac{\overset{4}{\cancel{20}}}{\underset{1}{\cancel{3}}}\cdot\frac{\overset{3}{\cancel{9}}}{\underset{5}{\cancel{25}}} = \frac{12}{5} = 2\frac{2}{5}$$

Example 3

$$\begin{array}{rcl} 3\frac{5}{8} & = & 3\frac{10}{16} \\ +\ 1\frac{1}{16} & = & 1\frac{1}{16} \\ \hline & & 4\frac{11}{16} \end{array}$$

Example 4

$$\begin{array}{rcl} 2\frac{9}{10} & = & 2\frac{27}{30} \\ +\ 6\frac{5}{6} & = & 6\frac{25}{30} \\ \hline & & 8\frac{52}{30} = 8 + 1\frac{22}{30} \\ & & = 9\frac{11}{15} \end{array}$$

Example 5

$$\begin{array}{rcl} 5\frac{3}{4} & = & 5\frac{9}{12} \\ -\ 2\frac{2}{3} & = & 2\frac{8}{12} \\ \hline & & 3\frac{1}{12} \end{array}$$

Example 6

$$\begin{array}{rcccl} 7\frac{1}{2} & = & \overset{6}{\cancel{7}}\frac{\overset{5+10}{5}}{10} & = & 6\frac{15}{10} \\ -\ 3\frac{4}{5} & = & 3\frac{8}{10} & = & 3\frac{8}{10} \\ \hline & & & & 3\frac{7}{10} \end{array}$$

Example 7

$$-4\frac{7}{8} + 2\frac{1}{16} - 3\frac{1}{4} = -\frac{39}{8} + \frac{33}{16} - \frac{13}{4}$$

$$= -\frac{39\cdot 2}{8\cdot 2} + \frac{33}{16} - \frac{13\cdot 4}{4\cdot 4}$$

$$= -\frac{78}{16} + \frac{33}{16} - \frac{52}{16}$$

$$= \frac{-78 + 33 - 52}{16} = \frac{-97}{16} = -6\frac{1}{16}$$

Section 4.7 Order of Operations and Complex Fractions

Key Concepts

To simplify an expression with more than one operation, apply the order of operations.

Examples

Example 1

Simplify. $\left(\frac{2}{5}\cdot\frac{10}{7}\right)^2+\frac{6}{7}$

$$=\left(\frac{2}{\cancel{5}_1}\cdot\frac{\cancel{10}^2}{7}\right)^2+\frac{6}{7}=\left(\frac{4}{7}\right)^2+\frac{6}{7}$$

$$=\frac{16}{49}+\frac{6}{7}=\frac{16}{49}+\frac{6\cdot 7}{7\cdot 7}=\frac{16}{49}+\frac{42}{49}$$

$$=\frac{58}{49}\quad\text{or}\quad 1\frac{9}{49}$$

A complex fraction is a fraction with one or more fractions in the numerator or denominator.

Example 2

Simplify. $\dfrac{\frac{5}{m}}{\frac{15}{4}}$

$$=\frac{5}{m}\cdot\frac{4}{15}\qquad\text{Multiply by the reciprocal.}$$

$$=\frac{\cancel{5}^1}{m}\cdot\frac{4}{\cancel{15}_3}=\frac{4}{3m}$$

To add or subtract like terms, apply the distributive property.

Example 3

Simplify. $\frac{3}{4}x-\frac{1}{8}x$

$$=\left(\frac{3}{4}-\frac{1}{8}\right)x=\left(\frac{3\cdot 2}{4\cdot 2}-\frac{1}{8}\right)x$$

$$=\left(\frac{6}{8}-\frac{1}{8}\right)x=\frac{5}{8}x$$

Section 4.8 Solving Equations Containing Fractions

Key Concepts

To solve equations containing fractions, we use the addition, subtraction, multiplication, and division properties of equality.

Example 1

Solve. $\frac{2}{3}x - \frac{1}{4} = \frac{1}{2}$

$\frac{2}{3}x - \frac{1}{4} + \frac{1}{4} = \frac{1}{2} + \frac{1}{4}$ — Add $\frac{1}{4}$ to both sides.

$\frac{2}{3}x = \frac{1 \cdot 2}{2 \cdot 2} + \frac{1}{4}$ — On the right-hand side, the LCD is 4.

$\frac{2}{3}x = \frac{2}{4} + \frac{1}{4}$

$\frac{2}{3}x = \frac{3}{4}$

$\frac{3}{2} \cdot \frac{2}{3}x = \frac{3}{2} \cdot \frac{3}{4}$ — Multiply by the reciprocal of $\frac{2}{3}$.

$x = \frac{9}{8}$ — The solution is $\frac{9}{8}$.

Examples

Another technique to solve equations with fractions is to multiply both sides of the equation by the LCD of all terms in the equation. This "clears" the fractions within the equation.

Example 2

Solve. $\frac{1}{9}x - 2 = \frac{5}{3}$ — The LCD is 9.

$9 \cdot \left(\frac{1}{9}x - 2\right) = 9 \cdot \left(\frac{5}{3}\right)$

$\overset{1}{\cancel{9}} \cdot \left(\frac{1}{\underset{1}{\cancel{9}}}x\right) - 9 \cdot (2) = \overset{3}{\cancel{9}} \cdot \left(\frac{5}{\underset{1}{\cancel{3}}}\right)$

$x - 18 = 15$

$x - 18 + 18 = 15 + 18$

$x = 33$ — The solution is 33.

Chapter 4 Review Exercises

Section 4.1

For Exercises 1–2, write a fraction that represents the shaded area.

1. 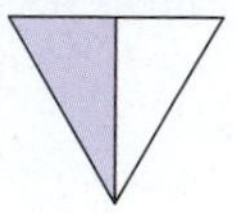

2. 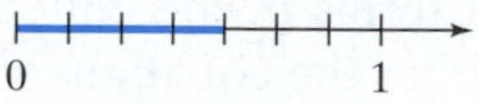

3. a. Write a fraction that has denominator 3 and numerator 5.

b. Label this fraction as proper or improper.

4. a. Write a fraction that has numerator 1 and denominator 6.

b. Label this fraction as proper or improper.

5. Simplify.

a. $\left|-\frac{3}{8}\right|$ **b.** $\left|\frac{2}{3}\right|$ **c.** $-\left(-\frac{4}{9}\right)$

For Exercises 6–7, write a fraction and a mixed number that represent the shaded area.

6. 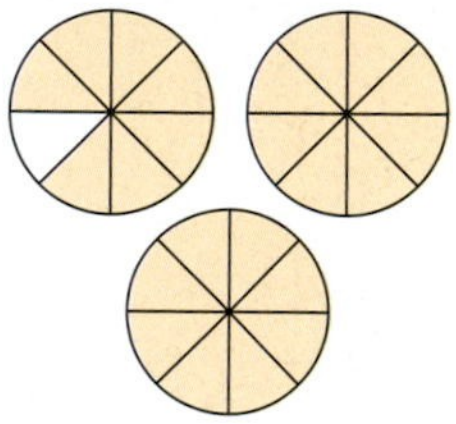

7. 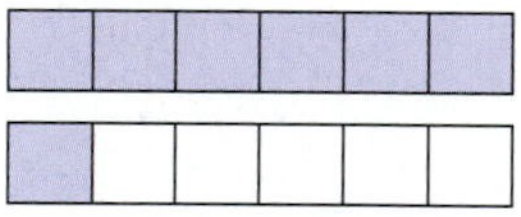

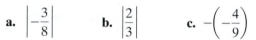

For Exercises 8–9, convert the mixed number to a fraction.

8. $6\frac{1}{7}$ **9.** $11\frac{2}{5}$

For Exercises 10–11, convert the improper fraction to a mixed number.

10. $\frac{47}{9}$ **11.** $\frac{23}{21}$

For Exercises 12–15, locate the numbers on the number line.

12. $-\frac{10}{5}$ **13.** $-\frac{7}{8}$ **14.** $\frac{13}{8}$ **15.** $-1\frac{3}{8}$

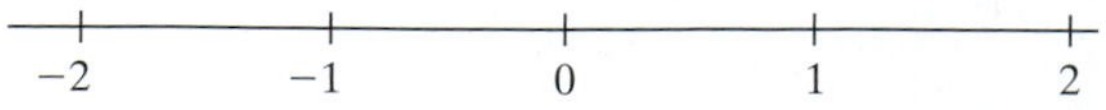

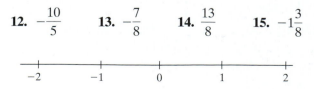

For Exercises 16–17, divide. Write the answer as a mixed number.

16. $7\overline{)941}$ **17.** $26\overline{)1582}$

Section 4.2

For Exercises 18–19, refer to this list of numbers: 21, 43, 51, 55, 58, 124, 140, 260, 1200.

18. List all the numbers that are divisible by 3.

19. List all the numbers that are divisible by 5.

20. Identify the prime numbers in the following list. 2, 39, 53, 54, 81, 99, 112, 113

21. Identify the composite numbers in the following list. 1, 12, 27, 51, 63, 97, 130

For Exercises 22–23, find the prime factorization.

22. 330 **23.** 900

For Exercises 24–25, determine if the fractions are equivalent. Fill in the blank with = or ≠.

24. $\frac{3}{6} \square \frac{5}{9}$ **25.** $\frac{15}{21} \square \frac{10}{14}$

For Exercises 26–33, simplify the fraction to lowest terms. Write the answer as a fraction.

26. $\frac{5}{20}$ **27.** $\frac{7}{35}$

28. $-\frac{24}{16}$ **29.** $-\frac{63}{27}$

30. $\frac{120}{1500}$ **31.** $\frac{140}{20{,}000}$

32. $\frac{4ac}{10c^2}$ **33.** $\frac{24t^3}{30t}$

34. On his final exam, Gareth got 42 out of 45 questions correct. What fraction of the test represents correct answers? What fraction represents incorrect answers?

35. Isaac proofread 6 pages of his 10-page term paper. Yulisa proofread 6 pages of her 15-page term paper.

a. What fraction of his paper did Isaac proofread?

b. What fraction of her paper did Yulisa proofread?

Section 4.3

For Exercises 36–41, multiply the fractions and simplify to lowest terms. Write the answer as a fraction or whole number.

36. $-\frac{2}{5} \cdot \frac{15}{14}$ **37.** $-\frac{4}{3} \cdot \frac{9}{8}$

38. $-14 \cdot \left(-\frac{9}{2}\right)$ **39.** $-33 \cdot \left(-\frac{5}{11}\right)$

40. $\frac{2x}{5} \cdot \frac{10}{x^2}$ **41.** $\frac{3y^3}{14} \cdot \frac{7}{y}$

42. Write the formula for the area of a triangle.

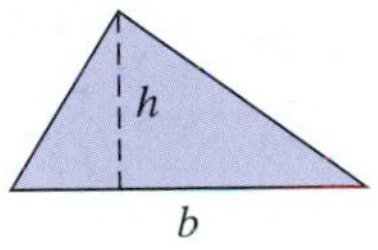

43. Find the area of the shaded region.

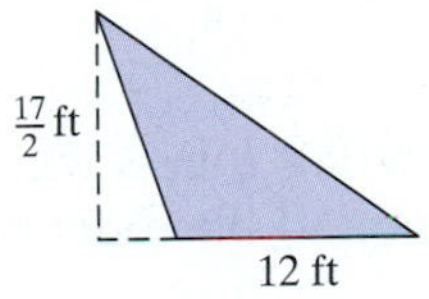

For Exercises 44–45, multiply.

44. $-\frac{3}{4} \cdot \left(-\frac{4}{3}\right)$ **45.** $\frac{1}{12} \cdot 12$

For Exercises 46–47, find the reciprocal of the number, if it exists.

46. $\frac{7}{2}$ **47.** -7

For Exercises 48–53, divide and simplify the answer to lowest terms. Write the answer as a fraction or whole number.

48. $\frac{28}{15} \div \frac{21}{20}$ **49.** $\frac{7}{9} \div \frac{35}{63}$

50. $-\frac{6}{7} \div 18$ **51.** $12 \div \left(-\frac{6}{7}\right)$

52. $\frac{4a^2}{7} \div \frac{a}{14}$ **53.** $\frac{11}{3y} \div \frac{22}{9y^3}$

54. How many $\frac{2}{3}$-lb bags of candy can be filled from a 24-lb sack of candy?

55. Chuck is an elementary school teacher and needs 22 pieces of wood, $\frac{3}{8}$ ft long, for a class project. If he has a 9-ft board from which to cut the pieces, will he have enough $\frac{3}{8}$-ft pieces for his class? Explain.

For Exercises 56–57, refer to the graph. The graph represents the distribution of the students at a college by race/ethnicity.

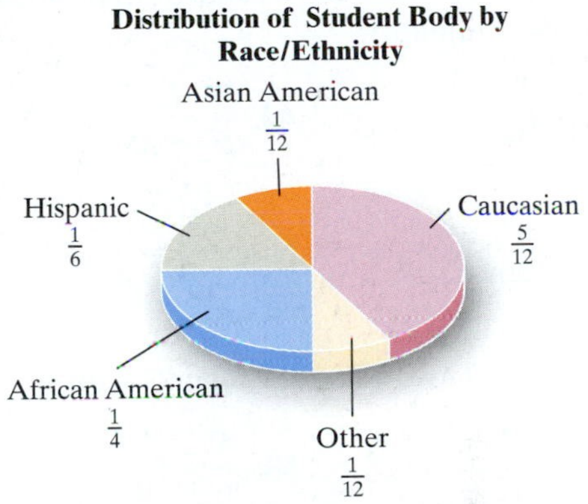

56. If the college has 3600 students, how many are African American?

57. If the college has 3600 students, how many are Asian American?

58. Amelia worked only $\frac{4}{5}$ of her normal 40-hr workweek. If she makes \$18 per hour, how much money did she earn for the week?

Section 4.4

59. Find the prime factorization.

a. 100 **b.** 65 **c.** 70

For Exercises 60–61, find the LCM by using any method.

60. 105 and 28 **61.** 16, 24, and 32

62. Sharon and Tony signed up at a gym on the same day. Sharon will be able to go to the gym every third day and Tony will go to the gym every fourth day. In how many days will they meet again at the gym?

For Exercises 63–66, rewrite each fraction with the indicated denominator.

63. $\dfrac{5}{16} = \dfrac{\quad}{48}$ **64.** $\dfrac{9}{5} = \dfrac{\quad}{35}$

65. $\dfrac{7}{12} = \dfrac{\quad}{60y}$ **66.** $\dfrac{-7}{x} = \dfrac{\quad}{4x}$

For Exercises 67–69, fill in the blanks with <, >, or =.

67. $\dfrac{11}{24} \square \dfrac{7}{12}$ **68.** $\dfrac{5}{6} \square \dfrac{7}{9}$

69. $-\dfrac{5}{6} \square -\dfrac{15}{18}$

70. Rank the numbers from least to greatest.
$-\dfrac{7}{10}, -\dfrac{72}{105}, -\dfrac{8}{15}, -\dfrac{27}{35}$

Section 4.5

For Exercises 71–80, add or subtract. Write the answer as a fraction simplified to lowest terms.

71. $\dfrac{5}{6} + \dfrac{4}{6}$ **72.** $\dfrac{4}{15} + \dfrac{6}{15}$

73. $\dfrac{9}{10} - \dfrac{61}{100}$ **74.** $\dfrac{11}{25} - \dfrac{2}{5}$

75. $-\dfrac{25}{11} - 2$ **76.** $-4 - \dfrac{37}{20}$

77. $\dfrac{2}{15} - \left(-\dfrac{5}{8}\right) - \dfrac{1}{3}$ **78.** $\dfrac{11}{14} - \dfrac{4}{7} - \left(-\dfrac{3}{2}\right)$

79. $\dfrac{7}{5w} + \dfrac{2}{w}$ **80.** $\dfrac{11}{a} + \dfrac{4}{b}$

For Exercises 81–82, find (a) the perimeter and (b) the area.

81.

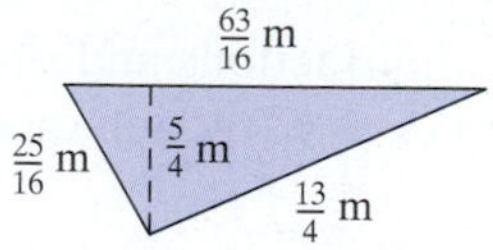

82.

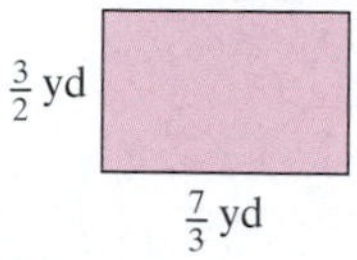

Section 4.6

For Exercises 83–88, multiply or divide as indicated.

83. $\left(3\dfrac{2}{3}\right)\left(6\dfrac{2}{5}\right)$ **84.** $\left(11\dfrac{1}{3}\right)\left(2\dfrac{3}{34}\right)$

85. $-3\dfrac{5}{11} \div \left(-3\dfrac{4}{5}\right)$ **86.** $-7 \div \left(-1\dfrac{5}{9}\right)$

87. $-4\dfrac{6}{11} \div 2$ **88.** $10\dfrac{1}{5} \div (-17)$

For Exercises 89–90, round the numbers to estimate the answer. Then find the exact sum or difference.

89. $65\dfrac{1}{8} - 14\dfrac{9}{10}$

Estimate: ______

Exact: ________

90. $43\dfrac{13}{15} - 20\dfrac{23}{25}$

Estimate: ______

Exact: ________

For Exercises 91–100, add or subtract the mixed numbers.

91. $9\dfrac{8}{9} + 1\dfrac{2}{7}$ **92.** $10\dfrac{1}{2} + 3\dfrac{15}{16}$ **93.** $7\dfrac{5}{24} - 4\dfrac{7}{12}$

94. $5\dfrac{1}{6} - 3\dfrac{1}{4}$ **95.** $6 - 2\dfrac{3}{5}$ **96.** $8 - 4\dfrac{11}{14}$

97. $42\frac{1}{8} - \left(-21\frac{13}{16}\right)$

98. $38\frac{9}{10} - \left(-11\frac{3}{5}\right)$

99. $-4\frac{2}{3} + 1\frac{5}{6}$

100. $6\frac{3}{8} + \left(-10\frac{1}{4}\right)$

101. Corry drove for $4\frac{1}{2}$ hr in the morning and $3\frac{2}{3}$ hr in the afternoon. Find the total number of hours he drove.

102. Denise owned $2\frac{1}{8}$ acres of land. If she sells $1\frac{1}{4}$ acres, how much will she have left

103. It takes $1\frac{1}{4}$ gal of paint for Neva to paint her living room. If her great room is $2\frac{1}{2}$ times larger than the living room, how many gallons will it take to paint the great room?

104. A roll of ribbon contains $12\frac{1}{2}$ yd. How many pieces of length $1\frac{1}{4}$ yd can be cut from this roll?

Section 4.7

For Exercises 105–116, simplify.

105. $\left(\frac{3}{8}\right)^2$

106. $\left(-\frac{3}{8}\right)^2$

107. $\left(-\frac{3}{8} \cdot \frac{4}{15}\right)^5$

108. $\left(\frac{1}{25} \cdot \frac{15}{6}\right)^4$

109. $-\frac{2}{5} - \left(1\frac{2}{3}\right) \cdot \frac{3}{2}$

110. $\frac{7}{5} - \left(-2\frac{1}{3}\right) \div \frac{7}{2}$

111. $\left(\frac{2}{3} - \frac{5}{6}\right)^2 + \frac{5}{36}$

112. $\left(-\frac{1}{4} - \frac{1}{2}\right)^2 - \frac{1}{8}$

113. $\dfrac{\frac{8}{5}}{\frac{4}{7}}$

114. $\dfrac{\frac{14}{9}}{\frac{7}{x}}$

115. $\dfrac{-\frac{2}{3} - \frac{5}{6}}{3 + \frac{1}{2}}$

116. $\dfrac{\frac{3}{5} - 1}{\frac{1}{2} - \frac{3}{10}}$

For Exercises 117–120, evaluate the expressions for the given values of the variables.

117. $x \div y \div z$ for $x = \frac{2}{3}$, $y = \frac{5}{6}$, and $z = -\frac{3}{5}$

118. a^2b for $a = -\frac{3}{5}$ and $b = 1\frac{2}{3}$

119. $t^2 + v^2$ for $t = \frac{1}{2}$ and $v = -\frac{1}{4}$

120. $2(w + z)$ for $w = 3\frac{1}{3}$ and $z = 2\frac{1}{2}$

For Exercises 121–124, simplify the expressions.

121. $-\frac{3}{4}x - \frac{2}{3}x$

122. $\frac{1}{5}y - \frac{3}{2}y$

123. $-\frac{4}{3}a + \frac{1}{2}c + 2a - \frac{1}{3}c$

124. $\frac{4}{5}w + \frac{2}{3}y + \frac{1}{10}w + y$

Section 4.8

For Exercises 125–134, solve the equations.

125. $x - \frac{3}{5} = \frac{2}{3}$

126. $y + \frac{4}{7} = \frac{3}{14}$

127. $\frac{3}{5}x = \frac{2}{3}$

128. $\frac{4}{7}y = \frac{3}{14}$

129. $-\frac{6}{5} = -2c$

130. $-\frac{9}{4} = -3d$

131. $-\frac{2}{5} + \frac{y}{10} = \frac{y}{2}$

132. $-\frac{3}{4} + \frac{w}{2} = \frac{w}{8}$

133. $2 = \frac{1}{2} - \frac{x}{10}$

134. $1 = \frac{7}{3} - \frac{t}{9}$

Chapter 4 Test

1. **a.** Write a fraction that represents the shaded portion of the figure.

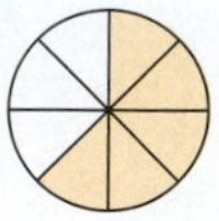

b. Is the fraction proper or improper?

2. **a.** Write a fraction that represents the total shaded portion of the three figures.

b. Is the fraction proper or improper?

3. **a.** Write $\frac{11}{3}$ as a mixed number.

b. Write $3\frac{7}{9}$ as an improper fraction.

For Exercises 4–5, plot the fraction on the number line.

4. $\dfrac{13}{5}$

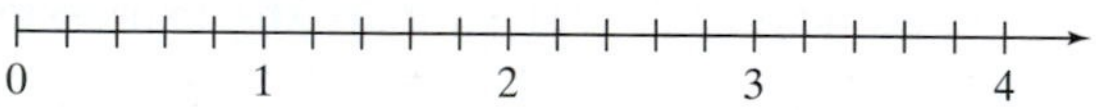

5. $-2\dfrac{3}{5}$

For Exercises 6–8, simplify.

6. $\left|-\dfrac{2}{11}\right|$

7. $-\left|-\dfrac{2}{11}\right|$

8. $-\left(-\dfrac{2}{11}\right)$

9. Label the following numbers as prime, composite, or neither.

a. 15 **b.** 0

c. 53 **d.** 1

e. 29 **f.** 39

10. Write the prime factorization of 45.

11. **a.** What is the divisibility rule for 3?

b. Is 1,981,011 divisible by 3?

12. Determine whether 1155 is divisible by

a. 2 **b.** 3

c. 5 **d.** 10

For Exercises 13–14, determine if the fractions are equivalent. Then fill in the blank with either = or ≠ .

13. $\dfrac{15}{12} \square \dfrac{5}{4}$

14. $-\dfrac{2}{5} \square -\dfrac{4}{25}$

For Exercises 15–16, simplify the fractions to lowest terms.

15. $\dfrac{150}{105}$

16. $\dfrac{100a}{350ab}$

17. Christine and Brad are putting their photographs in scrapbooks. Christine has placed 15 of her 25 photos and Brad has placed 16 of his 20 photos.

a. What fractional part of the total photos has each person placed?

b. Which person has a greater fractional part completed?

For Exercises 18–23, multiply or divide as indicated. Simplify the fraction to lowest terms.

18. $\dfrac{2}{9} \cdot \dfrac{57}{46}$

19. $\left(-\dfrac{75}{24}\right)(-4)$

20. $\dfrac{28}{24} \div \dfrac{21}{8}$

21. $-\dfrac{105}{42} \div 5$

22. $\dfrac{4}{3y} \cdot \dfrac{y^2}{2}$

23. $-\dfrac{5ab}{c} \div \dfrac{a}{c^2}$

24. Find the area of the triangle.

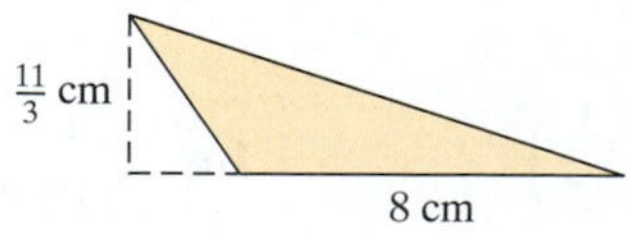

25. Which is greater, $20 \cdot \frac{1}{4}$ or $20 \div \frac{1}{4}$?

26. How many "quarter-pounders" can be made from 12 lb of ground beef?

27. A zoning requirement indicates that a house built on less than 1 acre of land may take up no more than one-half of the land. If Liz and George purchased a $\frac{4}{5}$-acre lot of land, what is the maximum land area that they can use to build the house?

28. **a.** List the first four multiples of 24.

b. List all factors of 24.

c. Write the prime factorization of 24.

29. Find the LCM for the numbers 16, 24, and 30.

For Exercises 30–31, write each fraction with the indicated denominator.

30. $\frac{5}{9} = \frac{\quad}{63}$

31. $\frac{11}{21} = \frac{\quad}{42w}$

32. Rank the fractions from least to greatest.

$-\frac{5}{3}, -\frac{11}{21}, -\frac{4}{7}$

33. Explain the difference between evaluating these two expressions:

$\frac{5}{11} - \frac{3}{11}$ and $\frac{5}{11} \cdot \frac{3}{11}$

For Exercises 34–43, perform the indicated operations. Write the answer as a fraction or mixed number.

34. $\frac{3}{8} + \frac{3}{16}$

35. $\frac{7}{3} - 2$

36. $\frac{1}{4} - \frac{7}{12}$

37. $\frac{12}{y} - \frac{6}{y^2}$

38. $-7\frac{2}{3} \div 4\frac{1}{6}$

39. $-4\frac{4}{17} \cdot \left(-2\frac{4}{15}\right)$

40. $6\frac{3}{4} + 10\frac{5}{8}$

41. $12 - 9\frac{10}{11}$

42. $-3 - 4\frac{4}{9}$

43. $-2\frac{1}{5} - \left(-6\frac{1}{10}\right)$

44. A fudge recipe calls for $1\frac{1}{2}$ lb of chocolate. How many pounds are required for $\frac{2}{3}$ of the recipe?

45. Find the area and perimeter of this parking area.

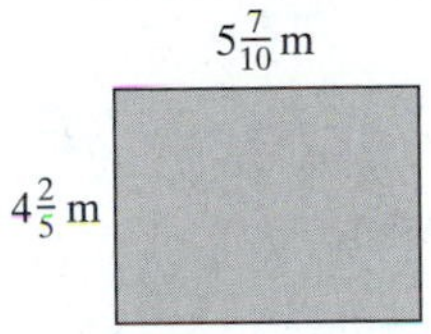

For Exercises 46–50, simplify.

46. $\left(-\frac{6}{7}\right)^2$

47. $\left(\frac{1}{2} - \frac{3}{5}\right)^3$

48. $\frac{2}{5} - 3\frac{2}{3} \div \frac{11}{2}$

49. $\dfrac{\frac{24}{35}}{\frac{8}{15}}$

50. $\dfrac{\frac{2}{5} - \frac{1}{2}}{-\frac{3}{10} + 2}$

51. Evaluate the expression $x \div z + y$ for $x = -\frac{2}{3}$, $y = \frac{7}{4}$, and $z = 2\frac{2}{3}$.

52. Simplify the expression. $-\frac{4}{5}m - \frac{2}{3}m + 2m$

For Exercises 53–58, solve the equation.

53. $-\frac{5}{9} + k = \frac{2}{3}$

54. $-\frac{5}{9}k = \frac{2}{3}$

55. $\frac{6}{11} = -3t$

56. $-2 = -\frac{1}{8}p$

57. $\frac{3}{2} = -\frac{2}{3}x - \frac{5}{6}$

58. $\frac{3}{14}y - 1 = \frac{3}{7}$

Chapters 1–4 Cumulative Review Exercises

1. For the number 6,8$\underline{7}$3,129 identify the place value of the underlined digit.

2. Write the following inequality in words: $130 < 244$

3. Find the prime factorization of 360.

For Exercises 4–10, simplify.

4. $71 + (-4) + 81 + (-106)$

5. $-\dfrac{15}{16} \cdot \dfrac{2}{5}$

6. $368 \div (-4)$

7. $\dfrac{0}{-61}$

8. $-\dfrac{13}{8} + \dfrac{7}{4}$

9. $4 - \dfrac{18}{5}$

10. $2\dfrac{3}{5} \div 1\dfrac{7}{10}$

11. Simplify the expression. $\left(8\dfrac{1}{4} \div 2\dfrac{3}{4}\right)^2 \cdot \dfrac{5}{18} + \dfrac{5}{6}$

12. Simplify the fraction to lowest terms: $\dfrac{180}{900}$

13. Find the area of the rectangle.

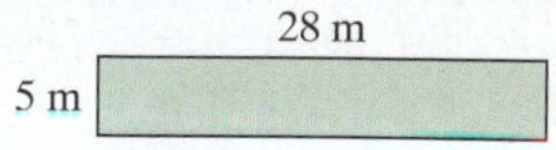

14. Simplify the expressions.

a. $-|-4|$ **b.** $-(-4)$ **c.** -4^2 **d.** $(-4)^2$

15. Simplify the expression. $-14 - 2(9 - 5^2)$

16. Combine *like* terms.
$-6x - 4y - 9x + y + 4x - 5$

17. Simplify. $-4(x - 5) - (3x + 2)$

18. Solve the equation. $-2(x - 3) + 4x = 3(x - 6)$

19. Solve the equation. $\dfrac{3}{4} = \dfrac{1}{2} + x$

20. Solve the equation. $\dfrac{x}{10} - 1 = \dfrac{2}{5}$

Decimals

5

CHAPTER OUTLINE

Chapter 5

This chapter is devoted to the study of decimal numbers and their important applications in day-to-day life. We begin with a discussion of place value and then we perform operations on signed decimal numbers. The chapter closes with applications and equations involving decimals.

Are You Prepared?

To prepare for your work with decimal numbers, take a minute to review place value, operations on integers, expressions, and equations. Write each answer in words and complete the crossword puzzle.

Across

2. Are $3x$ and 3 like terms?
3. Identify the place value of the underlined digit. 15,$\underline{4}$27
5. Solve. $-5x + 4 = -21$

Down

1. Simplify. $(27 - 17)^2 - 80$
2. Simplify. $(-3)^2$
4. Simplify. $x - 3(x - 4) + 2x - 5$

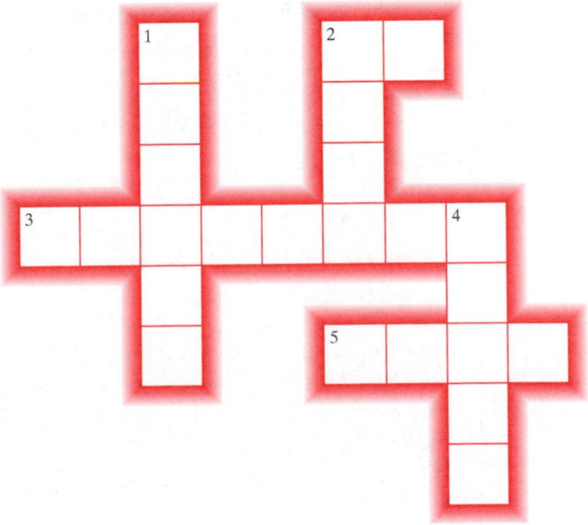

Section 5.1 Decimal Notation and Rounding

Objectives

1. Decimal Notation
2. Writing Decimals as Mixed Numbers or Fractions
3. Ordering Decimal Numbers
4. Rounding Decimals

1. Decimal Notation

In Chapter 4, we studied fraction notation to denote equal parts of a whole. In this chapter, we introduce decimal notation to denote parts of a whole. We first introduce the concept of a decimal fraction. A **decimal fraction** is a fraction whose denominator is a power of 10. The following are examples of decimal fractions.

$$\frac{3}{10} \text{ is read as "three-tenths"}$$

$$\frac{7}{100} \text{ is read as "seven-hundredths"}$$

$$-\frac{9}{1000} \text{ is read as "negative nine-thousandths"}$$

We now want to write these fractions in **decimal notation**. This means that we will write the numbers by using place values, as we did with whole numbers. The place value chart from Section 1.2 can be extended as shown in Figure 5-1.

Ten-thousands	Thousands	Hundreds	Tens	Ones	.	Ten*ths*	Hundred*ths*	Thousand*ths*	Ten-thousand*ths*
10,000	1000	100	10	1		$\frac{1}{10}$	$\frac{1}{100}$	$\frac{1}{1000}$	$\frac{1}{10,000}$

Whole-number place values — Decimal point — Fractional place values

Figure 5-1

From Figure 5-1, we see that the decimal point separates the whole-number part from the fractional part. The place values for decimal fractions are located to the right of the decimal point. Their place value names are similar to those for whole numbers, but end in *ths*. Notice the correspondence between the tens place and the ten*ths* place. Similarly notice the hundreds place and the hundred*ths* place. Each place value on the left has a corresponding place value on the right, with the exception of the ones place. There is no "one*ths*" place.

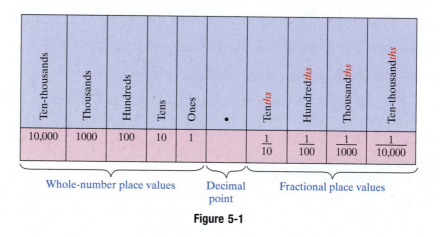

Example 1 Identifying Place Values

Identify the place value of each underlined digit.

a. $30{,}804.0\underline{9}$ **b.** $-0.8469\underline{2}0$ **c.** $2\underline{9}3.604$

Solution:

a. $30{,}804.0\underline{9}$ The digit 9 is in the hundredths place.

b. $-0.8469\underline{2}0$ The digit 2 is in the hundred-thousandths place.

c. $2\underline{9}3.604$ The digit 9 is in the tens place.

Skill Practice

Identify the place value of each underlined digit.

1. $24.38\underline{6}$
2. $-218.02168\underline{4}$
3. $1\underline{3}16.42$

For a whole number, the decimal point is understood to be after the ones place, and is usually not written. For example:

$$42. = 42$$

Concept Connections

4. Where is the decimal point located for the number 65?

Using Figure 5-1, we can write the numbers $\frac{3}{10}$, $\frac{7}{100}$, and $-\frac{9}{1000}$ in decimal notation.

Fraction	Word name	Decimal notation
$\frac{3}{10}$	Three-tenths	0.3 (↑ tenths place)
$\frac{7}{100}$	Seven-hundredths	0.07 (↑ hundredths place)
$-\frac{9}{1000}$	Negative nine-thousandths	−0.009 (↑ thousandths place)

TIP: The 0 to the left of the decimal point is a placeholder so that the position of the decimal point can be easily identified. It does not contribute to the value of the number. Thus, 0.3 and .3 are equal.

Now consider the number $15\frac{7}{10}$. This value represents 1 ten + 5 ones + 7 tenths. In decimal form we have 15.7.

15.7

1 ten — 5 ones — 7 tenths

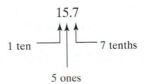

The decimal point is interpreted as the word *and*. Thus, 15.7 is read as "fifteen *and* seven tenths." The number 356.29 can be represented as

$$356 + 2 \text{ tenths} + 9 \text{ hundredths} = 356 + \frac{2}{10} + \frac{9}{100}$$

$$= 356 + \frac{20}{100} + \frac{9}{100}$$ We can use the LCD of 100 to add the fractions.

$$= 356\frac{29}{100}$$

We can read the number 356.29 as "three hundred fifty-six *and* twenty-nine hundredths."

This discussion leads to a quicker method to read decimal numbers.

Answers

1. Thousandths
2. Millionths
3. Hundreds
4. The decimal point is located after the ones-place digit: 65.

PROCEDURE Reading a Decimal Number

Step 1 The part of the number to the left of the decimal point is read as a whole number. *Note:* If there is no whole-number part, skip to step 3.

Step 2 The decimal point is read *and*.

Step 3 The part of the number to the right of the decimal point is read as a whole number but is followed by the name of the place position of the digit farthest to the right.

Skill Practice

Write a word name for each number.

5. 1004.6 **6.** 3.042

7. -0.0063

Example 2 Reading Decimal Numbers

Write the word name for each number.

a. 1028.4 **b.** 2.0736 **c.** -0.478

Solution:

a. 1028.4 is written as "one thousand, twenty-eight and four-tenths."

b. 2.0736 is written as "two and seven hundred thirty-six ten-thousandths."

c. -0.478 is written as "negative four hundred seventy-eight thousandths."

Skill Practice

Write the word name as a numeral.

8. Two hundred and two hundredths

9. Negative seventy-nine and sixteen thousandths

Example 3 Writing a Numeral from a Word Name

Write the word name as a numeral.

a. Four hundred eight and fifteen ten-thousandths

b. Negative five thousand eight hundred and twenty-three hundredths

Solution:

a. Four hundred eight and fifteen ten-thousandths: 408.0015

b. Negative five thousand eight hundred and twenty-three hundredths: -5800.23

2. Writing Decimals as Mixed Numbers or Fractions

A fractional part of a whole may be written as a fraction or as a decimal. To convert a decimal to an equivalent fraction, it is helpful to think of the decimal in words. For example:

Decimal	Word Name	Fraction	
0.3	Three tenths	$\frac{3}{10}$	
0.67	Sixty-seven hundredths	$\frac{67}{100}$	
0.048	Forty-eight thousandths	$\frac{48}{1000} = \frac{6}{125}$	(simplified)
6.8	Six and eight-tenths	$6\frac{8}{10} = 6\frac{4}{5}$	(simplified)

From the list, we notice several patterns that can be summarized as follows.

Answers

5. One thousand, four and six-tenths

6. Three and forty-two thousandths

7. Negative sixty-three ten-thousandths

8. 200.02 **9.** -79.016

PROCEDURE Converting a Decimal to a Mixed Number or Proper Fraction

Step 1 The digits to the right of the decimal point are written as the numerator of the fraction.

Step 2 The place value of the digit farthest to the right of the decimal point determines the denominator.

Step 3 The whole-number part of the number is left unchanged.

Step 4 Once the number is converted to a fraction or mixed number, simplify the fraction to lowest terms, if possible.

Example 4 Writing Decimals as Proper Fractions or Mixed Numbers

Write the decimals as proper fractions or mixed numbers and simplify.

a. 0.847 **b.** −0.0025 **c.** 4.16

Solution:

a. $0.847 = \frac{847}{1000}$

thousandths place

b. $-0.0025 = -\frac{25}{10{,}000} = -\frac{\overset{1}{\cancel{25}}}{\underset{400}{\cancel{10{,}000}}} = -\frac{1}{400}$

ten-thousandths place

c. $4.16 = 4\frac{16}{100} = 4\frac{\overset{4}{\cancel{16}}}{\underset{25}{\cancel{100}}} = 4\frac{4}{25}$

hundredths place

Skill Practice

Write the decimals as proper fractions or mixed numbers.

10. 0.034 **11.** −0.00086

12. 3.184

A decimal number greater than 1 can be written as a mixed number or as an improper fraction. The number 4.16 from Example 4(c) can be expressed as follows.

$$4.16 = 4\frac{16}{100} = 4\frac{4}{25} \quad \text{or} \quad \frac{104}{25}$$

Concept Connections

13. Which is a correct representation of 3.17?

$3\frac{17}{100}$ or $\frac{317}{100}$

A quick way to obtain an improper fraction for a decimal number greater than 1 is outlined here.

PROCEDURE Writing a Decimal Number Greater Than 1 as an Improper Fraction

Step 1 The denominator is determined by the place position of the digit farthest to the right of the decimal point.

Step 2 The numerator is obtained by removing the decimal point of the original number. The resulting whole number is then written over the denominator.

Step 3 Simplify the improper fraction to lowest terms, if possible.

Answers

10. $\frac{17}{500}$ **11.** $-\frac{43}{50{,}000}$ **12.** $3\frac{23}{125}$

13. They are both correct representations.

For example:

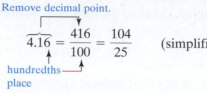

Remove decimal point.

$$4.16 = \frac{416}{100} = \frac{104}{25} \quad \text{(simplified)}$$

hundredths place

Skill Practice

Write the decimals as improper fractions and simplify.

14. 6.38 **15.** −15.1

Example 5 Writing Decimals as Improper Fractions

Write the decimals as improper fractions and simplify.

a. 40.2 **b.** −2.113

Solution:

a. $40.2 = \dfrac{402}{10} = \dfrac{\overset{201}{\cancel{402}}}{\underset{5}{\cancel{10}}} = \dfrac{201}{5}$

b. $-2.113 = -\dfrac{2113}{1000}$ Note that the fraction is already in lowest terms.

3. Ordering Decimal Numbers

It is often necessary to compare the values of two decimal numbers.

PROCEDURE Comparing Two Positive Decimal Numbers

Step 1 Starting at the left (and moving toward the right), compare the digits in each corresponding place position.

Step 2 As we move from left to right, the first instance in which the digits differ determines the order of the numbers. The number having the greater digit is greater overall.

Skill Practice

Fill in the blank with < or >.

16. 4.163 ☐ 4.159

17. 218.38 ☐ 218.41

Example 6 Ordering Decimals

Fill in the blank with < or >.

a. 0.68 ☐ 0.7 **b.** 3.462 ☐ 3.4619

Solution:

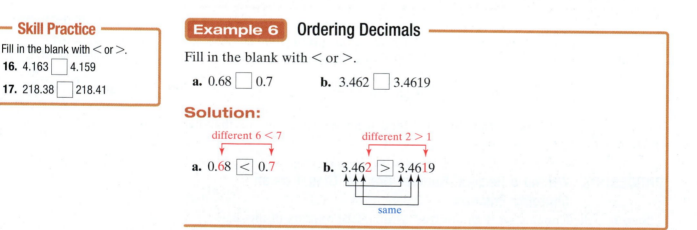

a. 0.68 $\boxed{<}$ 0.7 (different 6 < 7)

b. 3.462 $\boxed{>}$ 3.4619 (different 2 > 1; 3, 4, 6 same)

Answers

14. $\dfrac{319}{50}$ **15.** $-\dfrac{151}{10}$

16. > **17.** <

When ordering two negative numbers, you must be careful to visualize their relative positions on the number line. For example, $1.4 < 1.5$ but $-1.4 > -1.5$. See Figure 5-2.

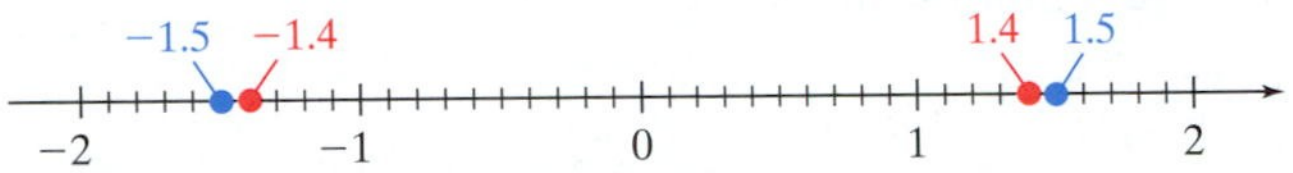

Figure 5-2

Also note that when ordering two decimal numbers, sometimes it is necessary to insert additional zeros after the rightmost digit in a number. This does not change the value of the number. For example:

$$0.7 = 0.70 \quad \text{because} \quad \frac{7}{10} = \frac{70}{100}$$

Example 7 **Ordering Negative Decimals**

Fill in the blank with $<$ or $>$.

a. $-0.2 \;\square\; -0.05$ **b.** $-0.04591 \;\square\; -0.0459$

Solution:

a. Insert a zero in the hundredths place for the number on the left. The digits in the tenths place are different.

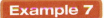

different

$-0.20 \;\boxed{<}\; -0.05$

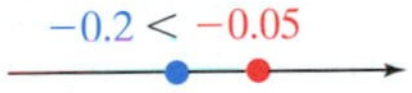

b. Insert a zero in the hundred-thousandths place for the number on the right. The digits in the hundred-thousandths place are different.

different

$-0.04591 \;\boxed{<}\; -0.04590$

Skill Practice

Fill in the blank with $<$ or $>$.

18. $-0.32 \;\square\; -0.062$

19. $-0.873 \;\square\; -0.8731$

4. Rounding Decimals

The process to round the decimal part of a number is nearly the same as rounding whole numbers (see Section 1.4). The main difference is that the digits to the right of the rounding place position are dropped instead of being replaced by zeros.

PROCEDURE Rounding Decimals to a Place Value to the Right of the Decimal Point

Step 1 Identify the digit one position to the right of the given place value.

Step 2 If the digit in step 1 is 5 or greater, add 1 to the digit in the given place value. Then discard the digits to its right.

Step 3 If the digit in step 1 is less than 5, discard it and any digits to its right.

Answers

18. $<$ **19.** $>$

Skill Practice

Round 187.26498 to the indicated place value.

20. Hundredths

21. Ten-thousandths

Example 8 Rounding Decimal Numbers

Round 14.795 to the indicated place value.

a. Tenths **b.** Hundredths

Solution:

a. $14.\overset{+1}{7}95 \approx 14.8$ (remaining digits discarded)

tenths place; 9: This digit is 5 or greater. Add 1 to the tenths place.

b. $14.7\overset{+1}{9}5 \approx 14.80$ (remaining digit discarded)

hundredths place; 5: This digit is 5 or greater. Add 1 to the hundredths place.

- Since the hundredths place digit is 9, adding 1 requires us to carry 1 to the tenths place digit.

In Example 8(b) the 0 in 14.80 indicates that the number was rounded to the hundredths place. It would be incorrect to drop the zero. Even though 14.8 has the same numerical value as 14.80, it implies a different level of accuracy. For example, when measurements are taken using some instrument such as a ruler or scale, the measured values are not exact. The place position to which a number is rounded reflects the accuracy of the measuring device. Thus, the value 14.8 lb indicates that the scale is accurate to the nearest tenth of a pound. The value 14.80 lb indicates that the scale is accurate to the nearest hundredth of a pound.

Skill Practice

22. Round 45.372 to the hundredths place.

23. Round 134.9996 to the thousandths place.

Example 9 Rounding Decimal Numbers

a. Round 4.81542 to the thousandths place.

b. Round 52.9999 to the hundredths place.

Solution:

a. $4.81542 \approx 4.815$ (remaining digits discarded)

thousandths place; 4: This digit is less than 5. Discard it and all digits to the right.

b. $5\overset{1}{2}.\overset{1}{9}\overset{+1}{9}99$ (discard remaining digits)

hundredths place; 9: This digit is greater than 5. Add 1 to the hundredths place digit.

- Since the hundredths place digit is 9, adding 1 requires us to carry 1 to the tenths place digit.
- Since the tenths place digit is 9, adding 1 requires us to carry 1 to the ones place digit.

≈ 53.00

Answers

20. 187.26 **21.** 187.2650

22. 45.37 **23.** 135.000

Section 5.1 Practice Exercises

Study Skills Exercise

1. Define the key terms.

a. Decimal fraction **b. Decimal notation**

Objective 1: Decimal Notation

2. State the first five place positions to the right of the decimal point in order from left to right.

For Exercises 3–10, expand the powers of 10 or $\frac{1}{10}$.

3. 10^2 **4.** 10^3 **5.** 10^4 **6.** 10^5

7. $\left(\frac{1}{10}\right)^2$ **8.** $\left(\frac{1}{10}\right)^3$ **9.** $\left(\frac{1}{10}\right)^4$ **10.** $\left(\frac{1}{10}\right)^5$

For Exercises 11–22, identify the place value of each underlined digit. **(See Example 1.)**

11. $3.\underline{9}83$ **12.** $34.8\underline{2}$ **13.** $440.3\underline{9}$ **14.** $2\underline{4}8.94$

15. $4\underline{8}9.02$ **16.** $4.092\underline{8}4$ **17.** $-9.283\underline{4}5$ **18.** $-0.32\underline{1}$

19. $0.48\underline{9}$ **20.** $5\underline{8}.211$ **21.** $-9\underline{3}.834$ **22.** $-5.00000\underline{1}$

For Exercises 23–30, write the word name for each decimal fraction.

23. $\frac{9}{10}$ **24.** $\frac{7}{10}$ **25.** $\frac{23}{100}$ **26.** $\frac{19}{100}$

27. $-\frac{33}{1000}$ **28.** $-\frac{51}{1000}$ **29.** $\frac{407}{10,000}$ **30.** $\frac{20}{10,000}$

For Exercises 31–38, write the word name for the decimal. **(See Example 2.)**

31. 3.24 **32.** 4.26 **33.** -5.9 **34.** -3.4

35. 52.3 **36.** 21.5 **37.** 6.219 **38.** 7.338

For Exercises 39–44, write the word name as a numeral. **(See Example 3.)**

39. Negative eight thousand, four hundred seventy-two and fourteen thousandths

40. Negative sixty thousand, twenty-five and four hundred one ten-thousandths

41. Seven hundred and seven hundredths

42. Nine thousand and nine thousandths

43. Negative two million, four hundred sixty-nine thousand and five hundred six thousandths

44. Negative eighty-two million, six hundred fourteen and ninety-seven ten-thousandths

Objective 2: Writing Decimals as Mixed Numbers or Fractions

For Exercises 45–56, write the decimal as a proper fraction or as a mixed number and simplify. **(See Example 4.)**

45. 3.7 **46.** 1.9 **47.** 2.8 **48.** 4.2

49. 0.25 **50.** 0.75 **51.** −0.55 **52.** −0.45

53. 20.812 **54.** 32.905 **55.** −15.0005 **56.** −4.0015

For Exercises 57–64, write the decimal as an improper fraction and simplify. **(See Example 5.)**

57. 8.4 **58.** 2.5 **59.** 3.14 **60.** 5.65

61. −23.5 **62.** −14.6 **63.** 11.91 **64.** 21.33

Objective 3: Ordering Decimal Numbers

For Exercises 65–72, fill in the blank with $<$ or $>$. **(See Examples 6–7.)**

65. 6.312 ☐ 6.321 **66.** 8.503 ☐ 8.530 **67.** 11.21 ☐ 11.2099 **68.** 10.51 ☐ 10.5098

69. −0.762 ☐ −0.76 **70.** −0.1291 ☐ −0.129 **71.** −51.72 ☐ −51.721 **72.** −49.06 ☐ −49.062

73. Which number is between 3.12 and 3.13? Circle all that apply.

a. 3.127 **b.** 3.129 **c.** 3.134 **d.** 3.139

74. Which number is between 42.73 and 42.86? Circle all that apply.

a. 42.81 **b.** 42.64 **c.** 42.79 **d.** 42.85

75. The batting averages for five legends are given in the table. Rank the players' batting averages from lowest to highest. (*Source:* Baseball Almanac)

Player	Average
Joe Jackson	0.3558
Ty Cobb	0.3664
Lefty O'Doul	0.3493
Ted Williams	0.3444
Rogers Hornsby	0.3585

76. The average speed, in miles per hour (mph), of the Daytona 500 for selected years is given in the table. Rank the speeds from slowest to fastest. (*Source:* NASCAR)

Year	Driver	Speed (mph)
1989	Darrell Waltrip	148.466
1991	Ernie Irvan	148.148
1997	Jeff Gordon	148.295
2007	Kevin Harvick	149.333

Objective 4: Rounding Decimals

77. The numbers given all have equivalent value. However, suppose they represent measured values from a scale. Explain the difference in the interpretation of these numbers.

0.25, 0.250, 0.2500, 0.25000

78. Which number properly represents 3.499999 rounded to the thousandths place?

a. 3.500 **b.** 3.5 **c.** 3.500000 **d.** 3.499

79. Which value is rounded to the nearest tenth, 7.1 or 7.10?

80. Which value is rounded to the nearest hundredth, 34.50 or 34.5?

For Exercises 81–92, round the decimals to the indicated place values. **(See Examples 8–9.)**

81. 49.943; tenths

82. 12.7483; tenths

83. 33.416; hundredths

84. 4.359; hundredths

85. −9.0955; thousandths

86. −2.9592; thousandths

87. 21.0239; tenths

88. 16.804; hundredths

89. 6.9995; thousandths

90. 21.9997; thousandths

91. 0.0079499; ten-thousandths

92. 0.00084985; ten-thousandths

93. A snail moves at a rate of about 0.00362005 miles per hour. Round the decimal value to the ten-thousandths place.

For Exercises 94–97, round the number to the indicated place value.

	Number	Hundreds	Tens	Tenths	Hundredths	Thousandths
94.	349.2395					
95.	971.0948					
96.	79.0046					
97.	21.9754					

Expanding Your Skills

98. What is the least number with three places to the right of the decimal that can be created with the digits 2, 9, and 7? Assume that the digits cannot be repeated.

99. What is the greatest number with three places to the right of the decimal that can be created from the digits 2, 9, and 7? Assume that the digits cannot be repeated.

Section 5.2 Addition and Subtraction of Decimals

Objectives

1. Addition and Subtraction of Decimals
2. Applications of Addition and Subtraction of Decimals
3. Algebraic Expressions

1. Addition and Subtraction of Decimals

In this section, we learn to add and subtract decimals. To begin, consider the sum 5.67 + 3.12.

$$\begin{array}{rl} 5.67 = & 5 + \frac{6}{10} + \frac{7}{100} \\ \underline{+\ 3.12} = & +\ 3 + \frac{1}{10} + \frac{2}{100} \\ \hline & 8 + \frac{7}{10} + \frac{9}{100} = 8.79 \end{array}$$

Notice that the decimal points and place positions are lined up to add the numbers. In this way, we can add digits with the same place values because we are effectively adding decimal fractions with like denominators. The intermediate step of using fraction notation is often skipped. We can get the same result more quickly by adding digits in like place positions.

PROCEDURE Adding and Subtracting Decimals

Step 1 Write the numbers in a column with the decimal points and corresponding place values lined up. (You may insert additional zeros as placeholders after the last digit to the right of the decimal point.)

Step 2 Add or subtract the digits in columns from right to left, as you would whole numbers. The decimal point in the answer should be lined up with the decimal points from the original numbers.

Skill Practice

Add.

1. 184.218 + 14.12

Example 1 Adding Decimals

Add. 27.486 + 6.37

Solution:

$$\begin{array}{r} 27.486 \\ +\ 6.370 \\ \hline \end{array}$$ Line up the decimal points. Insert an extra zero as a placeholder.

$$\begin{array}{r} \overset{1}{2}7.\overset{1}{4}86 \\ +\ 6.370 \\ \hline 33.856 \end{array}$$ Add digits with common place values. Line up the decimal point in the answer.

Concept Connections

2. Check your answer to problem 1 by estimation.

With operations on decimals it is important to locate the correct position of the decimal point. A quick estimate can help you determine whether your answer is reasonable. From Example 1, we have

27.486	rounds to	27
6.370	rounds to	+6
		33

The estimated value, 33, is close to the actual value of 33.856.

Answers

1. 198.338 2. ≈184 + 14 = 198, which is close to 198.338.

Example 2 Adding Decimals

Add. 3.7026 + 43 + 816.3

Solution:

3.7026	Line up the decimal points.
43.0000	Insert a decimal point and four zeros after it.
+ 816.3000	Insert three zeros.

$\overset{1\,1}{3}$.7026	Add digits with common place values.
43.0000	
+ 816.3000	
863.0026	Line up the decimal point in the answer.

The sum is 863.0026.

TIP: To check that the answer is reasonable, round each addend.

3.7026	rounds to	4
43	rounds to	$\overset{1}{4}$3
816.3	rounds to	+ 816
		863

which is close to the actual sum, 863.0026.

Skill Practice

Add.

3. 2.90741 + 15.13 + 3

Example 3 Subtracting Decimals

Subtract.

a. 0.2868 − 0.056 **b.** 139 − 28.63 **c.** 192.4 − 89.387

Solution:

a.	0.2868	Line up the decimal points.
	− 0.0560	Insert an extra zero as a placeholder.
	0.2308	Subtract digits with common place values.
		Decimal point in the answer is lined up.
b.	139.00	Insert extra zeros as placeholders.
	− 28.63	Line up the decimal points.
	13$\overset{8}{\not 9}$.$\overset{9}{\overset{\not 10}{\not 0}}$$\overset{10}{\not 0}$	Subtract digits with common place values.
	− 28.63	Borrow where necessary.
	110.37	Line up the decimal point in the answer.
c.	192.400	Insert extra zeros as placeholders.
	− 89.387	Line up the decimal points.
	1$\overset{8}{\not 9}$$\overset{12}{\not 2}$.$\overset{3}{\not 4}$$\overset{9}{\overset{\not 10}{\not 0}}$$\overset{10}{\not 0}$	Subtract digits with common place values.
	− 89.387	Borrow where necessary.
	103.013	Line up the decimal point in the answer.

Skill Practice

Subtract.

4. 3.194 − 0.512

5. 0.397 − 0.1584

6. 566.4 − 414.231

Answers

3. 21.03741 **4.** 2.682

5. 0.2386 **6.** 152.169

In Example 4, we will use the rules for adding and subtracting signed numbers.

Skill Practice

Simplify.

7. $-39.46 + 29.005$
8. $-0.345 - 6.51$
9. $-3.79 - (-6.2974)$

Example 4 Adding and Subtracting Negative Decimal Numbers

Simplify.

a. $-23.9 + 45.8$ **b.** $-0.694 - 0.482$ **c.** $2.61 - 3.79 - (-6.29)$

Solution:

a. $-23.9 + 45.8$ The sum will be *positive* because $|45.8|$ is greater than $|-23.9|$.

$= +(45.8 - 23.9)$ To add two numbers with different signs, subtract the smaller absolute value from the larger absolute value. Apply the sign from the number with the larger absolute value.

$$\begin{array}{r} \overset{4\ 18}{4\not{5}.8} \\ -23.9 \\ \hline 21.9 \end{array}$$

$= 21.9$ The result is positive.

b. $-0.694 - 0.482$

$= -0.694 + (-0.482)$ Change subtraction to addition of the opposite.

$= -(0.694 + 0.482)$ To add two numbers with the same signs, add their absolute values and apply the common sign.

$$\begin{array}{r} \overset{1}{} \\ 0.694 \\ +\ 0.482 \\ \hline 1.176 \end{array}$$

$= -1.176$ The result is negative.

c. $2.61 - 3.79 - (-6.29)$

$= \underbrace{2.61 + (-3.79)} + (6.29)$ Change subtraction to addition of the opposite. Add from left to right.

$= -1.18 + 6.29$

$$\begin{array}{r} 3.79 \\ -2.61 \\ \hline 1.18 \end{array}$$

Subtract the smaller absolute value from the larger absolute value.

$= 5.11$

$$\begin{array}{r} 6.29 \\ -1.18 \\ \hline 5.11 \end{array}$$

Subtract the smaller absolute value from the larger absolute value.

2. Applications of Addition and Subtraction of Decimals

Decimals are used often in measurements and in day-to-day applications.

Answers

7. -10.455 8. -6.855
9. 2.5074

Example 5 Applying Addition and Subtraction of Decimals in a Checkbook

Fill in the balance for each line in the checkbook register, shown in Figure 5-3. What is the ending balance?

Check No.	Description	Debit	Credit	Balance
				$684.60
2409	Doctor	$ 75.50		
2410	Mechanic	215.19		
2411	Home Depot	94.56		
	Paycheck		$981.46	
2412	Veterinarian	49.90		

Figure 5-3

Solution:

We begin with $684.60 in the checking account. For each debit, we subtract. For each credit, we add.

Check No.	Description	Debit	Credit	Balance	
				$ 684.60	
2409	Doctor	$ 75.50		609.10	= $684.60 − $75.50
2410	Mechanic	215.19		393.91	= $609.10 − $215.19
2411	Home Depot	94.56		299.35	= $393.91 − $94.56
	Paycheck		$981.46	1280.81	= $299.35 + $981.46
2412	Veterinarian	49.90		1230.91	= $1280.81 − $49.90

The ending balance is $1230.91.

Example 6 Applying Decimals to Perimeter

a. Find the length of the side labeled x.

b. Find the length of the side labeled y.

c. Find the perimeter of the figure.

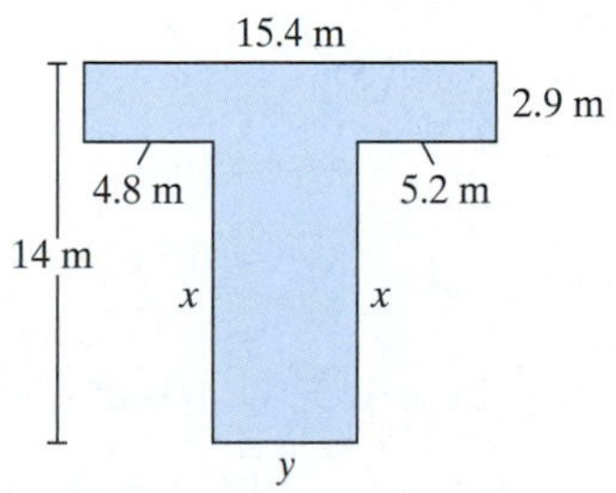

Solution:

a. If we extend the line segment labeled x with the dashed line as shown below, we see that the sum of side x and the dashed line must equal 14 m. Therefore, subtract $14 - 2.9$ to find the length of side x.

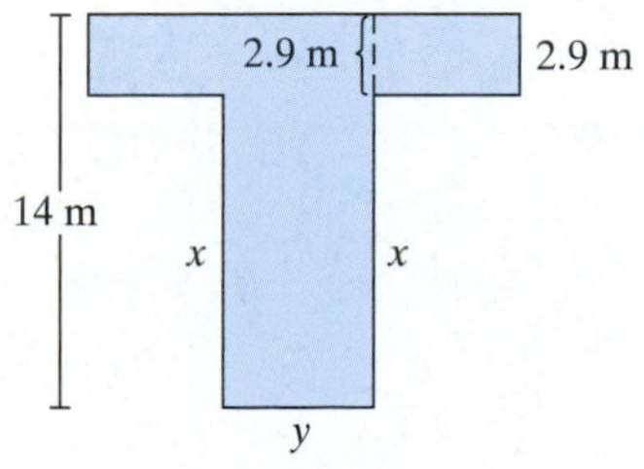

Length of side x:

$$\begin{array}{r} \overset{3\ 10}{1\not{4}.\not{0}} \\ -\ 2.9 \\ \hline 11.1 \end{array}$$

Side x is 11.1 m long.

Skill Practice

10. Fill in the balance for each line in the checkbook register.

Check	Debit	Credit	Balance
			$437.80
1426	$82.50		
Pay		$514.02	
1427	26.04		

Answer

10.

Check	Debit	Credit	Balance
			$ 437.80
1426	$82.50		355.30
Pay		$514.02	869.32
1427	26.04		843.28

Skill Practice

11. Consider the figure.

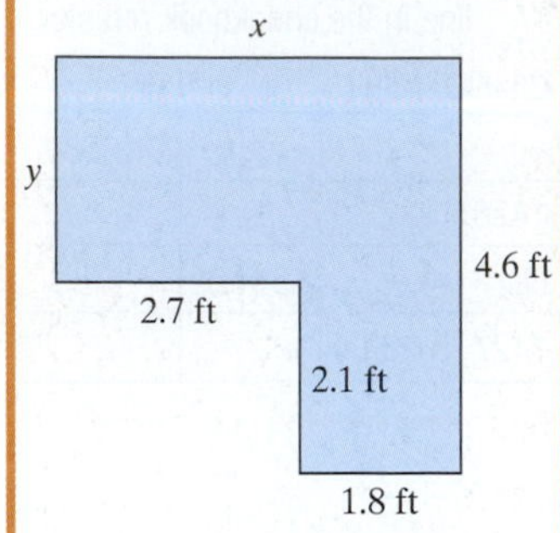

a. Find the length of side x.
b. Find the length of side y.
c. Find the perimeter.

b. The dashed line in the figure below has the same length as side y. We also know that $4.8 + 5.2 + y$ must equal 15.4. Since $4.8 + 5.2 = 10.0$,

$$y = 15.4 - 10.0$$
$$= 5.4$$

The length of side y is 5.4 m.

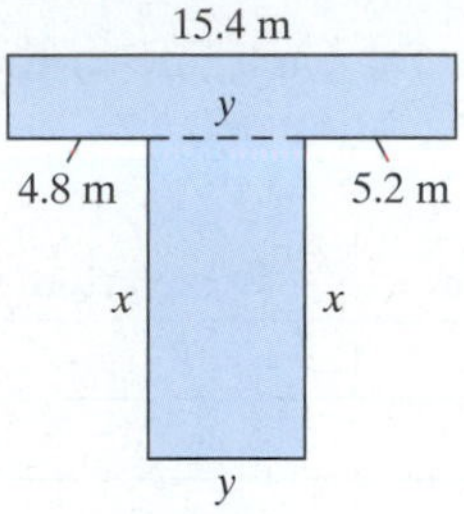

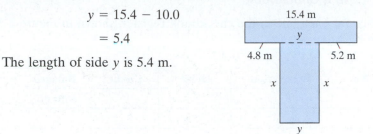

c. Now that we have the lengths of all sides, add them to get the perimeter.

$$\begin{array}{r} \overset{2\,3}{15.4} \\ 2.9 \\ 5.2 \\ 11.1 \\ 5.4 \\ 11.1 \\ 4.8 \\ +\ 2.9 \\ \hline 58.8 \end{array}$$

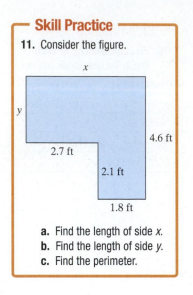

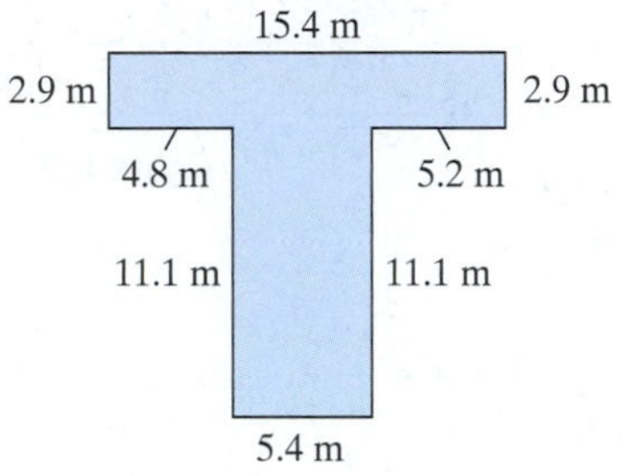

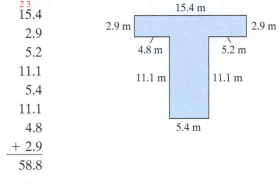

The perimeter is 58.8 m.

3. Algebraic Expressions

In Example 7, we practice combining *like* terms. In this case, the terms have decimal coefficients.

Skill Practice

Simplify by combining *like* terms.

12. $-9.15t - 2.34t$

13. $0.31w + 0.46z + 0.45w - 0.2z$

Example 7 Combining *Like* Terms

Simplify by combining *like* terms.

a. $-2.3x + 8.6x$ **b.** $0.042x + 0.539y + 0.65x - 0.21y$

Solution:

a. $-2.3x + 8.6x = (-2.3 + 8.6)x$ Apply the distributive property.

$= (6.3)x$ $\begin{array}{r} 8.6 \\ -\ 2.3 \\ \hline 6.3 \end{array}$ Subtract the smaller absolute value from the larger.

$= 6.3x$

b. $0.042x + 0.539y + 0.65x - 0.21y$

$= 0.042x + 0.65x + 0.539y - 0.21y$ Group *like* terms together.

$= 0.692x - 0.329y$ Combine *like* terms by adding or subtracting the coefficients.

TIP: To combine *like* terms, we can simply add or subtract the coefficients and leave the variable unchanged.

$0.042x + 0.65x = 0.692x$

Answers

11. a. Side x is 4.5 ft.
b. Side y is 2.5 ft.
c. The perimeter is 18.2 ft.

12. $-11.49t$ **13.** $0.76w + 0.26z$

Section 5.2 Practice Exercises

Review Exercises

1. Which number is equal to 5.03? Circle all that apply.

a. 5.030 **b.** 5.30 **c.** 5.0300 **d.** 5.3

2. Which number is equal to $\frac{7}{100}$? Circle all that apply.

a. 0.7 **b.** 0.07 **c.** 0.070 **d.** 0.007

For Exercises 3–8, round the decimals to the indicated place values.

3. 23.489; tenths

4. 42.314; hundredths

5. 8.6025; thousandths

6. 0.981; tenths

7. 2.82998; ten-thousandths

8. 2.78999; thousandths

Objective 1: Addition and Subtraction of Decimals

For Exercises 9–14, add the decimal numbers. Then round the numbers and find the sum to determine if your answer is reasonable. The first estimate is done for you. **(See Examples 1–2.)**

Expression	Estimate	Expression	Estimate
9. 44.6 + 18.6	45 + 19 = 64	**10.** 28.2 + 23.2	
11. 5.306 + 3.645		**12.** 3.451 + 7.339	
13. 12.9 + 3.091		**14.** 4.125 + 5.9	

For Exercises 15–26, add the decimals. **(See Examples 1–2.)**

15. 78.9 + 0.9005

16. 44.2 + 0.7802

17. 23 + 8.0148

18. 7.9302 + 34

19. 34 + 23.0032 + 5.6

20. 23 + 8.01 + 1.0067

21. 68.394 + 32.02

22. 2.904 + 34.229

23. $\begin{array}{r} 103.94 \\ +\ 24.5 \\ \hline \end{array}$

24. $\begin{array}{r} 93.2 \\ +\ 43.336 \\ \hline \end{array}$

25. $\begin{array}{r} 54.2 \\ 23.993 \\ +\ 3.87 \\ \hline \end{array}$

26. $\begin{array}{r} 13.9001 \\ 72.4 \\ +\ 34.13 \\ \hline \end{array}$

For Exercises 27–32, subtract the decimal numbers. Then round the numbers and find the difference to determine if your answer is reasonable. The first estimate is done for you. **(See Example 3.)**

Expression	Estimate	Expression	Estimate
27. 35.36 − 21.12	35 − 21 = 14	**28.** 53.9 − 22.4	
29. 7.24 − 3.56		**30.** 23.3 − 20.8	
31. 45.02 − 32.7		**32.** 66.15 − 42.9	

For Exercises 33–44, subtract the decimals. **(See Example 3.)**

33. 14.5 − 8.823

34. 33.2 − 21.932

35. 2 − 0.123

36. 4 − 0.42

37. 55.9 − 34.2354

38. 49.1 − 24.481

39. 18.003 − 3.238

40. 21.03 − 16.446

41. 183.01 − 23.452

42. 164.23 − 44.3893

43. 1.001 − 0.0998

44. 2.0007 − 0.0689

Mixed Exercises: Addition and Subtraction of Negative Decimals

For Exercises 45–60, add or subtract as indicated. **(See Example 4.)**

45. −506.34 + 83.4

46. −89.041 + 76.43

47. −0.489 − 0.87

48. −0.78 − 0.439

49. 47.82 − (−3.159)

50. 4.2 − (−9.8458)

51. 55.3 − 68.4 − (−9.83)

52. 3.45 − 8.7 − (−10.14)

53. 5 − 9.432

54. 7 − 11.468

55. −6.8 − (−8.2)

56. −10.3 − (−5.1)

57. −28.3 + (−82.9)

58. −92.6 + (−103.8)

59. 2.6 − 2.06 − 2.006 + 2.0006

60. 5.84 − 5.084 − 5.0084 + 58.4

Objective 2: Applications of Addition and Subtraction of Decimals

61. Fill in the balance for each line in the checkbook register shown in the figure. What is the ending balance? **(See Example 5.)**

Check No.	Description	Debit	Credit	Balance
				$ 245.62
2409	Electric bill	$ 52.48		
2410	Groceries	72.44		
2411	Department store	108.34		
	Paycheck		$1084.90	
2412	Restaurant	23.87		
	Transfer from savings		200	

62. A section of a bank statement is shown in the figure. Find the mistake that was made by the bank.

Date	Action	Debit	Credit	Balance
				$1124.35
1/2/10	Check #4214	$749.32		375.03
1/3/10	Check #4215	37.29		337.74
1/4/10	Transfer from savings		$ 400.00	737.74
1/5/10	Paycheck		1451.21	2188.95
1/6/10	Cash withdrawal	150.00		688.95

63. A normal human red blood cell count is between 4.2 and 6.9 million cells per microliter (μL). A cancer patient undergoing chemotherapy has a red blood cell count of 2.85 million cells per microliter. How far below the lower normal limit is this?

64. A laptop computer was originally priced at $1299.99 and was discounted to $998.95. By how much was it marked down?

65. Water flows into a pool at a constant rate. The water level is recorded at several 1-hr intervals.

a. From the table, how many inches is the water level rising each hour?

b. At this rate, what will the water level be at 1:00 P.M.?

c. At this rate, what will the water level be at 3:00 P.M.?

Time	Water Level
9:00 A.M.	4.2 in.
10:00 A.M.	5.9 in.
11:00 A.M.	7.6 in.
12:00 P.M.	9.3 in.

66. The amount of time that it takes Mercury, Venus, Earth, and Mars to revolve about the Sun is given in the graph.

a. How much longer does it take Mars to complete a revolution around the Sun than the Earth?

b. How much longer does it take Venus than Mercury to revolve around the Sun?

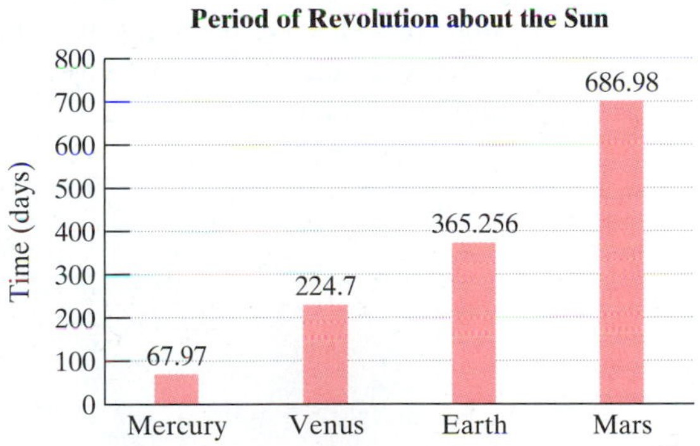

Source: National Aeronautics and Space Administration

67. The table shows the thickness of four U.S. coins. If you stacked three quarters and a dime in one pile and two nickels and two pennies in another pile, which pile would be higher?

Coin	Thickness
Quarter	1.75 mm
Dime	1.35 mm
Nickel	1.95 mm
Penny	1.55 mm

Source: U.S. Department of the Treasury

68. How much thicker is a nickel than a quarter?

For Exercises 69–72, find the lengths of the sides labeled x and y. Then find the perimeter. **(See Example 6.)**

69.

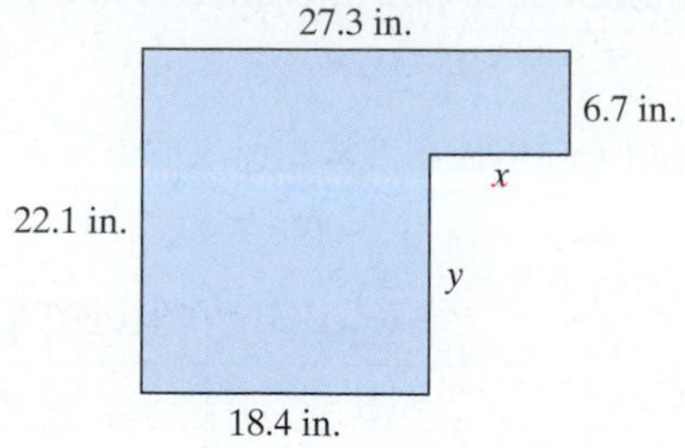

70.

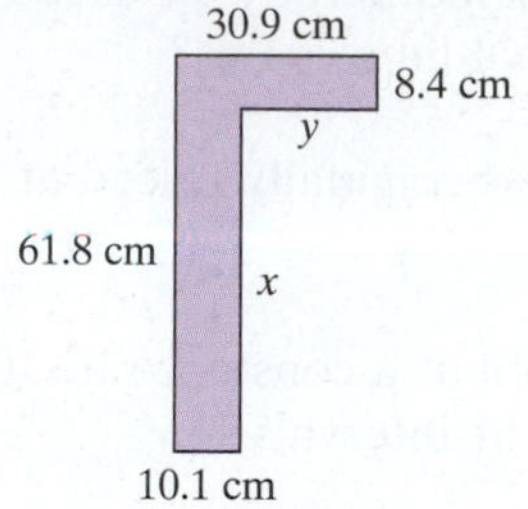

71.

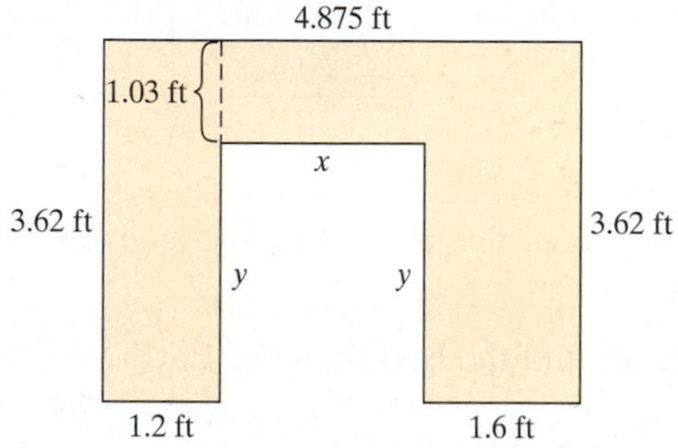

72.

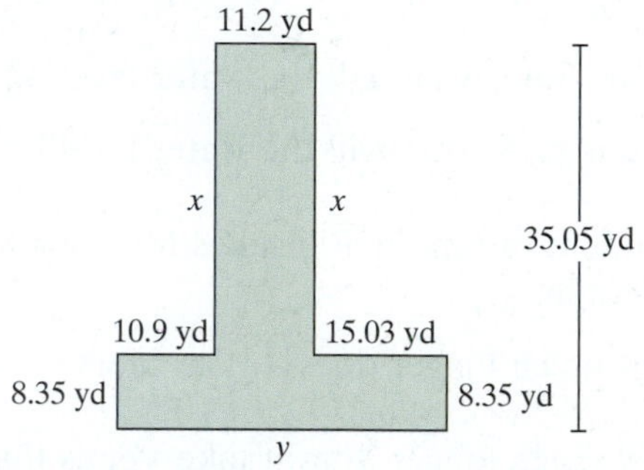

73. A city bus follows the route shown in the map. How far does it travel in one circuit?

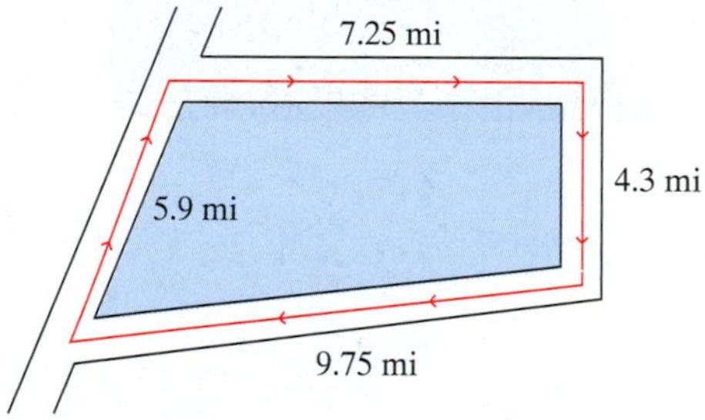

74. Santos built a new deck and needs to put a railing around the sides. He does not need railing where the deck is against the house. How much railing should he purchase?

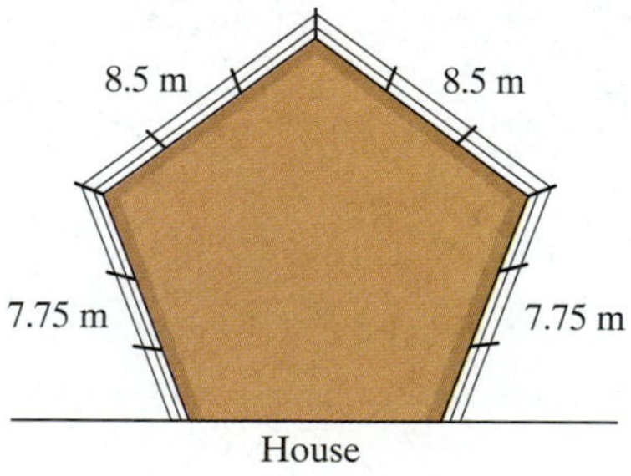

Objective 3: Algebraic Expressions

For Exercises 75–84, simplify by combining *like* terms. **(See Example 7.)**

75. $-5.83t + 9.7t$

76. $-2.158w + 10.4w$

77. $-4.5p - 8.7p$

78. $-98.46a - 12.04a$

79. $y - 0.6y$

80. $0.18x - x$

81. $0.025x + 0.83y + 0.82x - 0.31y$

82. $0.008m - 0.06n - 0.0043m + 0.092n$

83. $0.92c - 4.78d + 0.08c - 0.22d$

84. $3.62x - 80.09y - 0.62x + 10.09y$

Calculator Connections

Topic: Entering decimals on a calculator

To enter decimals on a calculator, use the [.] key.

Expression	Keystrokes	Result
984.126 + 37.11	984 [.] 126 [+] 37 [.] 11 [ENTER]	1021.236

Calculator Exercises

For Exercises 85–90, refer to the table. The table gives the closing stock prices (in dollars per share) for the first day of trading for the given month.

Stock	January	February	March	April	May
IBM	\$97.27	\$99.00	\$92.27	\$95.21	\$103.17
FedEx	\$109.77	\$111.89	\$114.16	\$106.07	\$106.11

85. By how much did the IBM stock increase between January and May?

86. By how much did the FedEx stock decrease between January and May?

87. Between which two consecutive months did the FedEx stock increase the most? What was the amount of increase?

88. Between which two consecutive months did the IBM stock increase the most? What was the amount of increase?

89. Between which two consecutive months did the FedEx stock decrease the most? What was the amount of decrease?

90. Between which two consecutive months did the IBM stock decrease the most? What was the amount of decrease?

Multiplication of Decimals and Applications with Circles

Section 5.3

Objectives

1. Multiplication of Decimals
2. Multiplication by a Power of 10 and by a Power of 0.1
3. Applications Involving Multiplication of Decimals
4. Circumference and Area of a Circle

1. Multiplication of Decimals

Multiplication of decimals is much like multiplication of whole numbers. However, we need to know where to place the decimal point in the product. Consider the product (0.3)(0.41). One way to multiply these numbers is to write them first as decimal fractions.

$$(0.3)(0.41) = \frac{3}{10} \cdot \frac{41}{100} = \frac{123}{1000} \text{ or } 0.123$$

Another method multiplies the factors vertically. First we multiply the numbers as though they were whole numbers. We temporarily disregard the decimal point in the product because it will be placed later.

$$\begin{array}{r} 0.41 \\ \times\ 0.3 \\ \hline 123 \end{array} \longleftarrow \text{decimal point not yet placed}$$

Concept Connections

1. How many decimal places will be in the product (2.72)(1.4)?
2. Explain the difference between the process to multiply 123 × 51 and the process to multiply 1.23 × 5.1.

From the first method, we know that the correct answer to this problem is 0.123. Notice that 0.123 contains the same number of decimal places as the two factors combined. That is,

$$\begin{array}{rcl} 0.41 & \longleftarrow & 2 \text{ decimal places} \\ \times\ 0.3 & \longleftarrow & 1 \text{ decimal place} \\ \hline .123 & \longleftarrow & 3 \text{ decimal places} \end{array}$$

TIP: When multiplying decimals, it is *not* necessary to line up the decimal points as we do when we add or subtract decimals. Instead, we line up the right-most digits.

The process to multiply decimals is summarized as follows.

PROCEDURE Multiplying Two Decimals

Step 1 Multiply as you would integers.

Step 2 Place the decimal point in the product so that the number of decimal places equals the combined number of decimal places of both factors.

Note: You may need to insert zeros to the left of the whole-number product to get the correct number of decimal places in the answer.

In Example 1, we multiply decimals by using this process.

Skill Practice

Multiply.

3. (19.7)(4.1)

Example 1 Multiplying Decimals

Multiply. $\begin{array}{r} 11.914 \\ \times\ 0.8 \\ \hline \end{array}$

Solution:

$$\begin{array}{r} \overset{1\,7\ \,1\,3}{11.914} \\ \times\ 0.8 \\ \hline 9.5312 \end{array} \qquad \begin{array}{r} 3 \text{ decimal places} \\ +\ 1 \text{ decimal place} \\ \hline 4 \text{ decimal places} \end{array}$$

The product is 9.5312.

Example 2 Multiplying Decimals

Multiply. Then use estimation to check the location of the decimal point.

(29.3)(2.8)

Solution:

$$\begin{array}{r} \overset{1}{\overset{7\,2}{29.3}} \\ \times\ 2.8 \\ \hline 2344 \\ 5860 \\ \hline 82.04 \end{array} \qquad \begin{array}{r} 1 \text{ decimal place} \\ +\ 1 \text{ decimal place} \\ \hline \\ \\ 2 \text{ decimal places} \end{array}$$

The product is 82.04.

Answers

1. 3
2. The actual process of vertical multiplication is the same for both cases. However, for the product 1.23 × 5.1, the decimal point must be placed so that the product has the same number of decimal places as both factors combined (in this case, 3).
3. 80.77

To check the answer, we can round the factors and estimate the product. The purpose of the estimate is primarily to determine whether we have placed the decimal point correctly. Therefore, it is usually sufficient to round each factor to the left-most nonzero digit. This is called **front-end rounding**. Thus,

$$
\begin{array}{rll}
29.3 & \text{rounds to} & 30 \\
2.8 & \text{rounds to} & \times\ 3 \\
\hline
& & 90
\end{array}
$$

The first digit for the actual product $\underline{8}2.04$ and the first digit for the estimate $\underline{9}0$ is the tens place. Therefore, we are reasonably sure that we have located the decimal point correctly. The estimate 90 is close to 82.04.

Skill Practice

4. Multiply $(1.9)(29.1)$ and check your answer using estimation.

Example 3 Multiplying Decimals

Multiply. Then use estimation to check the location of the decimal point.

$$-2.79 \times 0.0003$$

Solution:

The product will be *negative* because the factors have opposite signs.

Actual product:

$$
\begin{array}{rl}
\overset{2\ 2}{-2.79} & \text{2 decimal places} \\
\times\ 0.0003 & +\ \text{4 decimal places} \\
\hline
-.000837 & \text{6 decimal places} \\
& \text{(insert 3 zeros to the left)}
\end{array}
$$

Estimate:

$$
\begin{array}{rll}
-2.79 & \text{rounds to} & -3 \\
\times\ 0.0003 & \text{rounds to} & \times\ 0.0003 \\
\hline
& & -0.0009
\end{array}
$$

The first digit for both the actual product and the estimate is in the ten-thousandths place. We are reasonably sure the decimal point is positioned correctly.

The product is -0.000837.

Skill Practice

5. Multiply -4.6×0.00008, and check your answer using estimation.

2. Multiplication by a Power of 10 and by a Power of 0.1

Consider the number 2.7 multiplied by $10, 100, 1000 \ldots$

$$
\begin{array}{ccc}
10 & 100 & 1000 \\
\times\ 2.7 & \times\ 2.7 & \times\ 2.7 \\
\hline
70 & 700 & 7000 \\
200 & 2000 & 20000 \\
\hline
27.0 & 270.0 & 2700.0
\end{array}
$$

Multiplying 2.7 by 10 moves the decimal point 1 place to the right.
Multiplying 2.7 by 100 moves the decimal point 2 places to the right.
Multiplying 2.7 by 1000 moves the decimal point 3 places to the right.

This leads us to the following generalization.

Answers

4. $55.29; \approx 2 \cdot 30 = 60$, which is close to 55.29.
5. $-0.000368; \approx -5 \times 0.00008 = -0.0004$, which is close to -0.000368.

PROCEDURE Multiplying a Decimal by a Power of 10

Move the decimal point to the right the same number of decimal places as the number of zeros in the power of 10.

Skill Practice

Multiply.

6. 81.6×1000
7. $0.0000085 \times 10{,}000$
8. $-2.396 \times (-10{,}000{,}000)$

Example 4 Multiplying by Powers of 10

Multiply.

a. $14.78 \times 10{,}000$ **b.** 0.0064×100 **c.** $-8.271 \times (-1{,}000{,}000)$

Solution:

a. $14.78 \times 10{,}000 = 147{,}800$ Move the decimal point 4 places to the right.

b. $0.0064 \times 100 = 0.64$ Move the decimal point 2 places to the right.

c. The product will be positive because the factors have the same sign.

$-8.271 \times (-1{,}000{,}000) = 8{,}271{,}000$ Move the decimal point 6 places to the right.

Multiplying a positive decimal by 10, 100, 1000, and so on increases its value. Therefore, it makes sense to move the decimal point to the *right*. Now suppose we multiply a decimal by 0.1, 0.01, and 0.001. These numbers represent the decimal fractions $\frac{1}{10}$, $\frac{1}{100}$, and $\frac{1}{1000}$, respectively, and are easily recognized as powers of 0.1 (see Section 2.4). Taking one-tenth of a positive number or one-hundredth of a positive number makes the number smaller. To multiply by 0.1, 0.01, 0.001, and so on (powers of 0.1), move the decimal point to the *left*.

$$\begin{array}{r} 3.6 \\ \times\ 0.1 \\ \hline .36 \end{array} \qquad \begin{array}{r} 3.6 \\ \times\ 0.01 \\ \hline .036 \end{array} \qquad \begin{array}{r} 3.6 \\ \times\ 0.001 \\ \hline .0036 \end{array}$$

Concept Connections

9. Explain the difference between multiplying a positive number by 100 versus 0.01.

PROCEDURE Multiplying a Decimal by Powers of 0.1

Move the decimal point to the left the same number of places as there are decimal places in the power of 0.1.

Skill Practice

Multiply.

10. 471.034×0.01
11. $9{,}437{,}214.5 \times 0.00001$
12. -0.0004×0.001

Example 5 Multiplying by Powers of 0.1

Multiply.

a. 62.074×0.0001 **b.** 7965.3×0.1 **c.** -0.0057×0.00001

Solution:

a. $62.074 \times 0.0001 = 0.0062074$ Move the decimal point 4 places to the left. Insert extra zeros.

b. $7965.3 \times 0.1 = 796.53$ Move the decimal point 1 place to the left.

c. $-0.0057 \times 0.00001 = -0.000000057$ Move the decimal point 5 places to the left.

Answers

6. 81,600 7. 0.085 8. 23,960,000
9. Multiplying a number by 100 increases its value. Therefore, we move the decimal point to the right two places. Multiplying a number by 0.01 decreases its value. Therefore, move the decimal point to the left two places.
10. 4.71034 11. 94.372145
12. −0.0000004

Sometimes people prefer to use number names to express very large numbers. For example, we might say that the U.S. population in a recent year was approximately 280 million. To write this in decimal form, we note that 1 million = 1,000,000. In this case, we have 280 of this quantity. Thus,

$$280 \text{ million} = 280 \times 1{,}000{,}000 \text{ or } 280{,}000{,}000$$

Example 6 **Naming Large Numbers**

Write the decimal number representing each word name.

a. The distance between the Earth and Sun is approximately 92.9 million miles.

b. The number of deaths in the United States due to heart disease in 2010 was projected to be 8 hundred thousand.

c. A recent estimate claimed that collectively Americans throw away 472 billion pounds of garbage each year.

Solution:

a. $92.9 \text{ million} = 92.9 \times 1{,}000{,}000 = 92{,}900{,}000$

b. $8 \text{ hundred thousand} = 8 \times 100{,}000 = 800{,}000$

c. $472 \text{ billion} = 472 \times 1{,}000{,}000{,}000 = 472{,}000{,}000{,}000$

Skill Practice

Write a decimal number representing the word name.

13. The population in Bexar County, Texas, is approximately 1.6 million.

14. Light travels approximately 5.9 trillion miles in 1 year.

15. The legislative branch of the federal government employs approximately 31 thousand employees.

3. Applications Involving Multiplication of Decimals

Example 7 **Applying Decimal Multiplication**

Jane Marie bought eight cans of tennis balls for \$1.98 each. She paid \$1.03 in tax. What was the total bill?

Solution:

The cost of the tennis balls before tax is

$$8(\$1.98) = \$15.84 \qquad \begin{array}{r} {\scriptstyle 7\,6} \\ 1.98 \\ \times\ \ 8 \\ \hline 15.84 \end{array}$$

Adding the tax to this value, we have

$$\begin{pmatrix}\text{Total}\\ \text{cost}\end{pmatrix} = \begin{pmatrix}\text{Cost of}\\ \text{tennis balls}\end{pmatrix} + (\text{Tax})$$

$$\begin{array}{r} = \$15.84 \\ +\ 1.03 \\ \hline \$16.87 \end{array} \qquad \text{The total cost is \$16.87.}$$

Skill Practice

16. A book club ordered 12 books on www.amazon.com for \$8.99 each. The shipping cost was \$4.95. What was the total bill?

Answers

13. 1,600,000
14. 5,900,000,000,000
15. 31,000
16. The total bill was \$112.83.

Skill Practice

17. The IMAX movie screen at the Museum of Science and Discovery in Ft. Lauderdale, Florida, is 18 m by 24.4 m. What is the area of the screen?

Example 8 Finding the Area of a Rectangle

The *Mona Lisa* is perhaps the most famous painting in the world. It was painted by Leonardo da Vinci somewhere between 1503 and 1506 and now hangs in the Louvre in Paris, France. The dimensions of the painting are 30 in. by 20.875 in. What is the total area?

Solution:

Recall that the area of a rectangle is given by

$$A = l \cdot w$$

$$A = (30\text{ in.})(20.875\text{ in.}) \qquad \begin{array}{r} \scriptstyle 2\ 2\ 1 \\ 20.875 \\ \times\ 30 \\ \hline 0 \\ 626250 \\ \hline 626.250 \end{array}$$

$$= 626.25\text{ in.}^2$$

The area of the *Mona Lisa* is 626.25 in.^{2}.

4. Circumference and Area of a Circle

A **circle** is a figure consisting of all points in a flat surface located the same distance from a fixed point called the center. The **radius** of a circle is the length of any line segment from the center to a point on the circle. The **diameter** of a circle is the length of any line segment connecting two points on the circle and passing through the center. See Figure 5-4.

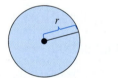

r is the length of the radius. *d* is the length of the diameter.

Figure 5-4

Notice that the length of a diameter is twice the radius. Therefore, we have

$$d = 2r \qquad \text{or equivalently} \qquad r = \frac{1}{2}d$$

Skill Practice

Find the length of a radius.

18.

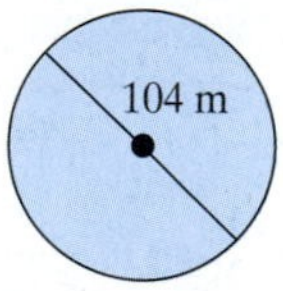

Find the length of a diameter.

19.

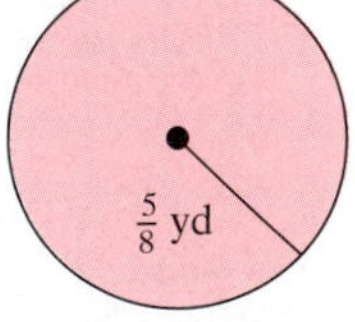

Example 9 Finding Diameter and Radius

a. Find the length of a radius. **b.** Find the length of a diameter.

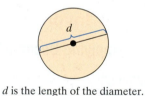

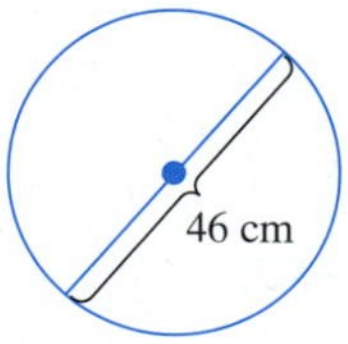

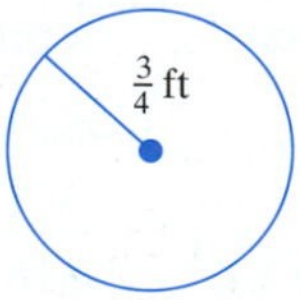

Solution:

a. $r = \frac{1}{2}d = \frac{1}{2}(46\text{ cm}) = 23\text{ cm}$

b. $d = 2r = \overset{1}{\cancel{2}}\left(\frac{3}{\underset{2}{\cancel{4}}}\text{ ft}\right) = \frac{3}{2}\text{ ft}$

Answers

17. The screen area is 439.2 m^2.

18. 52 m **19.** $\frac{5}{4}$ yd

The distance around a circle is called the **circumference**. In any circle, if the circumference C is divided by the diameter, the result is equal to the number π (read "pi"). The number π in decimal form is 3.1415926535. . . , which goes on forever without a repeating pattern. We approximate π by 3.14 or $\frac{22}{7}$ to make it easier to use in calculations. The relationship between the circumference and the diameter of a circle is $\frac{C}{d} = \pi$. This gives us the following formulas.

Concept Connections

20. What is the meaning of the circumference of a circle?

FORMULA Circumference of a Circle

The circumference C of a circle is given by

$$C = \pi d \qquad \text{or} \qquad C = 2\pi r$$

where π is approximately 3.14 or $\frac{22}{7}$.

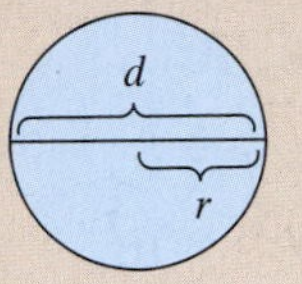

Example 10 Determining Circumference of a Circle

Determine the circumference. Use 3.14 for π.

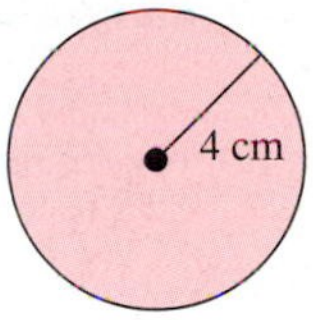

Solution:

The radius is given, $r = 4$ cm.

$C = 2\pi r$

$= 2(\pi)(4 \text{ cm})$ — Substitute 4 cm for r.

$= 8\pi$ cm

$\approx 8(3.14)$ cm — Approximate π by 3.14.

≈ 25.12 cm — The distance around the circle is approximately 25.12 cm.

Skill Practice

Find the circumference. Use 3.14 for π.

21.

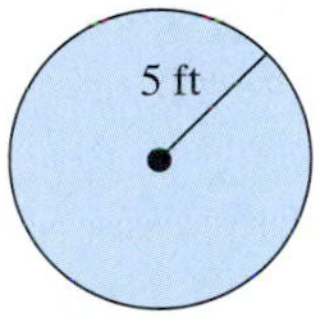

TIP: The value 3.14 is an approximation for π. Therefore, using 3.14 in a calculation results in an approximate answer.

TIP: The circumference from Example 10 can also be found from the formula $C = \pi d$. In this case, the diameter is 8 cm.

$$C = \pi d$$
$$C = \pi(8 \text{ cm})$$
$$= 8\pi \text{ cm}$$
$$\approx 8(3.14) \text{ cm}$$
$$= 25.12 \text{ cm}$$

The formula for the area of a circle also involves the number π. In Chapter 8 we will provide the geometric basis for the formula given below.

Avoiding Mistakes

To express the formula for the circumference of a circle, we can use either the radius ($C = 2\pi r$) or the diameter ($C = \pi d$).

To find the area of a circle, we will always use the radius ($A = \pi r^2$).

FORMULA Area of a Circle

The **area of a circle** is given by

$$A = \pi r^2$$

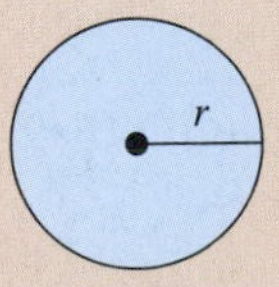

Answers

20. The circumference of a circle is the distance around the circle.

21. ≈ 31.4 ft

Skill Practice

22. Find the area of a circular clock having a radius of 6 in. Approximate the answer by using 3.14 for π. Round to the nearest whole unit.

Example 11 Determining the Area of a Circle

Determine the area of a circle that has radius 0.3 ft. Approximate the answer by using 3.14 for π. Round to two decimal places.

Solution:

$A = \pi r^2$	
$= \pi(0.3\text{ ft})^2$	Substitute $r = 0.3$ ft.
$= \pi(0.09\text{ ft}^2)$	Square the radius. $(0.3\text{ ft})^2 = (0.3\text{ ft})(0.3\text{ ft}) = 0.09\text{ ft}^2$
$= 0.09\pi\text{ ft}^2$	
$\approx 0.09(3.14)\text{ ft}^2$	Approximate π by 3.14.
$= 0.2826\text{ ft}^2$	Multiply decimals.
$\approx 0.28\text{ ft}^2$	The area is approximately 0.28 ft^2.

Avoiding Mistakes

$(0.3)^2 \neq (0.9)$. Be sure to count the number of places needed to the right of the decimal point. $(0.3)^2 = (0.3)(0.3) = 0.09$

TIP: Using the approximation 3.14 for π results in approximate answers for area and circumference. To express the exact answer, we must leave the result written in terms of π. In Example 11 the exact area is given by $0.09\pi\text{ ft}^2$.

Answer

22. $\approx 113\text{ in.}^2$

Section 5.3 Practice Exercises

Study Skills Exercises

1. To help you remember the formulas for circumference and area of a circle, list them together and note the similarities and differences in the formulas.

Circumference: __________ Area: __________

Similarities: Differences:

2. Define the key terms.

a. Front-end rounding **b. Circle** **c. Radius**

d. Diameter **e. Circumference** **f. Area of a circle**

Review Exercises

For Exercises 3–5, round to the indicated place value.

3. 49.997; tenths **4.** −0.399; hundredths **5.** −0.00298; thousandths

For Exercises 6–8, add or subtract as indicated.

6. $-2.7914 + 5.03216$ **7.** $-33.072 - (-41.03)$ **8.** $-0.0723 - 0.514$

Objecitve 1: Multiplication of Decimals

For Exercises 9–16, multiply the decimals. **(See Examples 1–3.)**

9. $\begin{array}{r} 0.8 \\ \times\ 0.5 \\ \hline \end{array}$ **10.** $\begin{array}{r} 0.6 \\ \times\ 0.5 \\ \hline \end{array}$ **11.** (0.9)(4) **12.** (0.2)(9)

13. (−60)(−0.003) **14.** (−40)(−0.005) **15.** $\begin{array}{r} 22.38 \\ \times\ 0.8 \\ \hline \end{array}$ **16.** $\begin{array}{r} 31.67 \\ \times\ 0.4 \\ \hline \end{array}$

For Exercises 17–30, multiply the decimals. Then estimate the answer by rounding. The first estimate is done for you. **(See Examples 2–3.)**

	Exact	Estimate
17.	$\begin{array}{r} 8.3 \\ \times\ 4.5 \\ \hline \end{array}$	$\begin{array}{r} 8 \\ \times\ 5 \\ \hline 40 \end{array}$
18.	$\begin{array}{r} 4.3 \\ \times\ 9.2 \\ \hline \end{array}$	
19.	$\begin{array}{r} 0.58 \\ \times\ 7.2 \\ \hline \end{array}$	

20. 0.83(6.5) **21.** 5.92(−0.8) **22.** 9.14(−0.6)

23. (−0.413)(−7) **24.** (−0.321)(−6) **25.** 35.9 × 3.2 **26.** 41.7 × 6.1

27. (562)(0.004) **28.** (984)(0.009) **29.** −0.0004 × 3.6 **30.** −0.0008 × 6.5

Objective 2: Multiplication by a Power of 10 and by a Power of 0.1

31. Multiply.

a. 5.1 × 10 **b.** 5.1 × 100 **c.** 5.1 × 1000 **d.** 5.1 × 10,000

32. If 256.8 is multiplied by 0.001, will the decimal point move to the left or to the right? By how many places?

33. Multiply.

a. 5.1 × 0.1 **b.** 5.1 × 0.01 **c.** 5.1 × 0.001 **d.** 5.1 × 0.0001

34. Multiply.

a. −6.2 × 100 **b.** −6.2 × 0.01 **c.** −6.2 × 10,000 **d.** −6.2 × 0.0001

For Exercises 35–42, multiply the numbers by the powers of 10 and 0.1. **(See Examples 4–5.)**

35. 34.9 × 100 **36.** 2.163 × 100 **37.** 96.59 × 1000 **38.** 18.22 × 1000

39. −93.3 × 0.01 **40.** −80.2 × 0.01 **41.** −54.03 × (−0.001) **42.** −23.11 × (−0.001)

For Exercises 43–48, write the decimal number representing each word name. **(See Example 6.)**

43. The number of beehives in the United States is 2.6 million. (*Source:* U.S. Department of Agriculture)

44. The people of France collectively consume 34.7 million gallons of champagne per year. (*Source*: Food and Agriculture Organization of the United Nations)

45. The most stolen make of car worldwide is Toyota. For a recent year, there were 4 hundred-thousand Toyota's stolen. (*Source:* Interpol)

46. The musical *Miss Saigon* ran for about 4 thousand performances in a 10-year period.

47. The people in the United States have spent over \$20.549 billion on DVDs.

48. Coca-Cola Classic was the greatest selling brand of soft-drinks. For a recent year, over 4.8 billion gallons were sold in the United States. (*Source:* Beverage Marketing Corporation)

Objective 3: Applications Involving Multiplication of Decimals

49. One gallon of gasoline weighs about 6.3 lb. However, when burned, it produces 20 lb of carbon dioxide (CO_2). This is because most of the weight of the CO_2 comes from the oxygen in the air.

   a. How many pounds of gasoline does a Hummer H2 carry when its tank is full (the tank holds 32 gal).

   b. How many pounds of CO_2 does a Hummer H2 produce after burning an entire tankful of gasoline?

50. Corrugated boxes for shipping cost \$2.27 each. How much will 10 boxes cost including tax of \$1.59?

51. The Athletic Department at Broward College bought 20 pizzas for \$10.95 each, 10 Greek salads for \$3.95 each, and 60 soft drinks for \$0.60 each. What was the total bill including a sales tax of \$17.67? **(See Example 7.)**

52. A hotel gift shop ordered 40 T-shirts at \$8.69 each, 10 hats at \$3.95 each, and 20 beach towels at \$4.99 each. What was the total cost of the merchandise, including the \$29.21 sales tax?

53. Firestone tires cost \$50.20 each. A set of four Lemans tires costs \$197.99. How much can a person save by buying the set of four Lemans tires compared to four Firestone tires?

54. Certain DVD titles are on sale for 2 for \$32. If they regularly sell for \$19.99, how much can a person save by buying 4 DVDs?

For Exercises 55–56, find the area. **(See Example 8.)**

55.

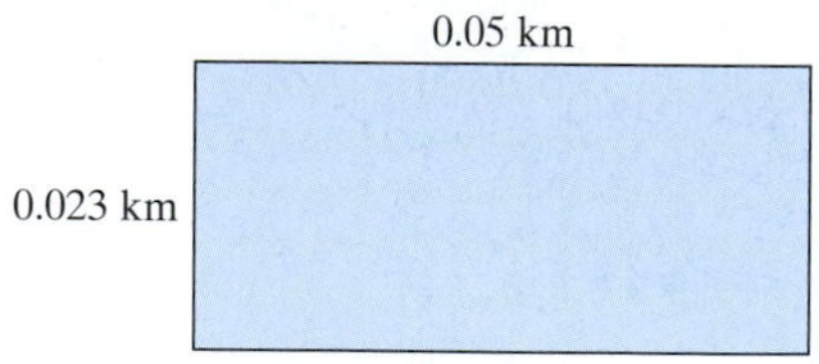

56.

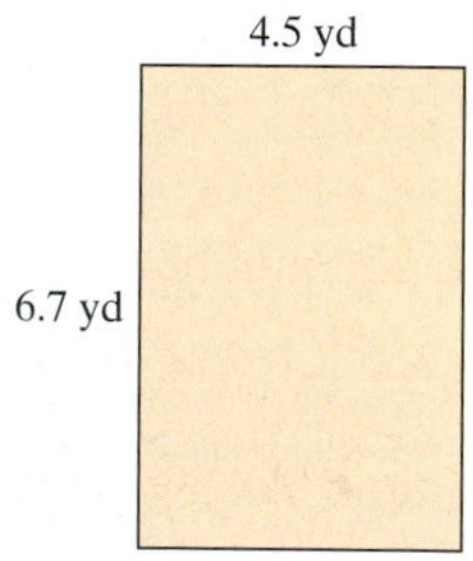

57. Blake plans to build a rectangular patio that is 15 yd by 22.2 yd. What is the total area of the patio?

58. The front page of a newspaper is 56 cm by 31.5 cm. Find the area of the page.

For Exercises 59–66, simplify the expressions.

59. $(0.4)^2$

60. $(0.7)^2$

61. $(-1.3)^2$

62. $(-2.4)^2$

63. $(0.1)^3$

64. $(0.2)^3$

65. -0.2^2

66. -0.3^2

Objective 4: Circumference and Area of a Circle

67. How does the length of a radius of a circle compare to the length of a diameter?

68. Circumference is similar to which type of measure?

a. Area **b.** Volume **c.** Perimeter **d.** Weight

For Exercises 69–70, find the length of a diameter. **(See Example 9.)**

69.

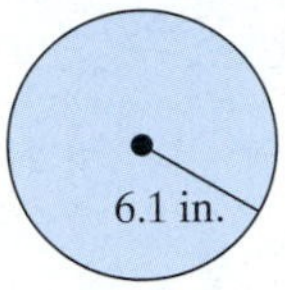

70.

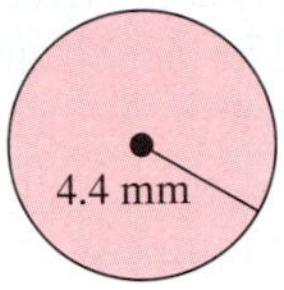

For Exercises 71–72, find the length of a radius. **(See Example 9.)**

71.

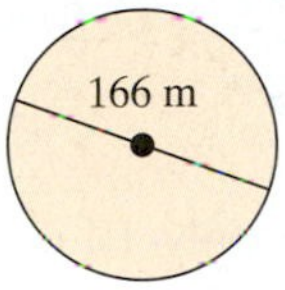

72.

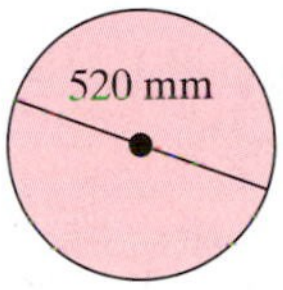

For Exercises 73–76, find the circumference of the circle. Approximate the answer by using 3.14 for π. **(See Example 10.)**

73.

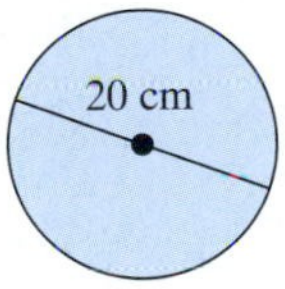

74.

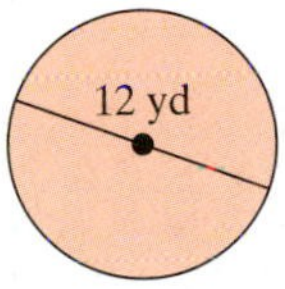

75.

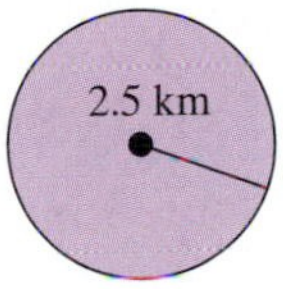

76.

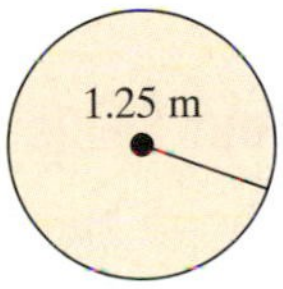

For Exercises 77–80, use 3.14 for π.

77. Find the circumference of the can of soda.

78. Find the circumference of a can of tuna.

79. Find the circumference of a compact disk.

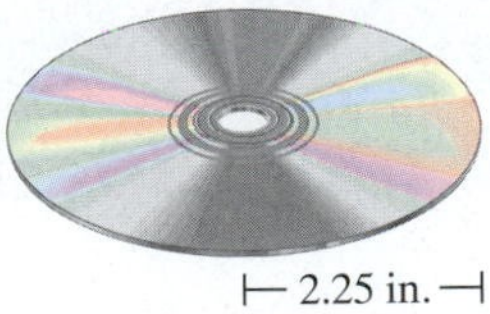

80. Find the outer circumference of a pipe with 1.75-in. radius.

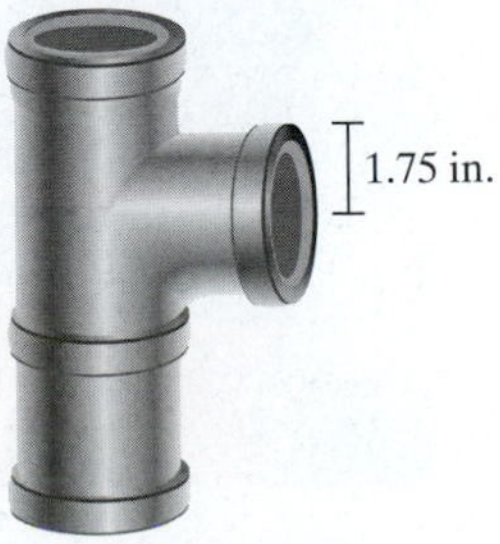

For Exercises 81–84, find the area. Approximate the answer by using 3.14 for π. Round to the nearest whole unit. **(See Example 11.)**

81.

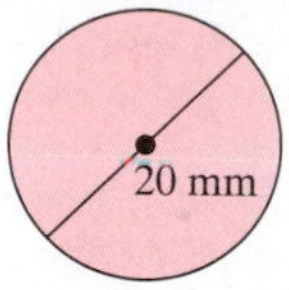

82.

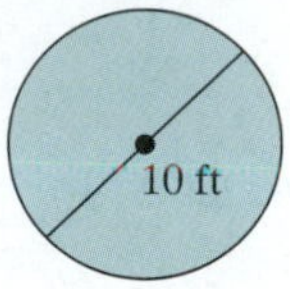

83.

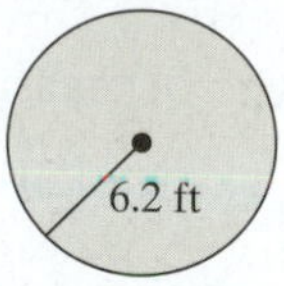

84.

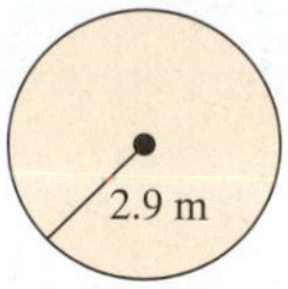

85. The Large Hadron Collider (LHC) is a particle accelerator and collider built to detect subatomic particles. The accelerator is in a huge circular tunnel that straddles the border between France and Switzerland. The diameter of the tunnel is 5.3 mi. Find the circumference of the tunnel. (*Source:* European Organization for Nuclear Research)

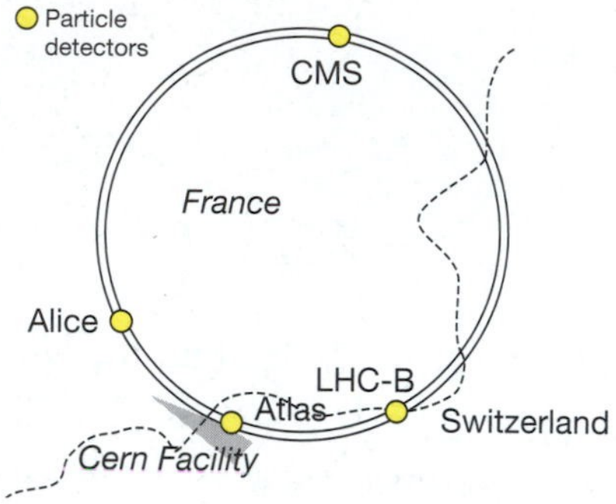

86. A ceiling fan blade rotates in a full circle. If the fan blades are 2 ft long, what is the area covered by the fan blades?

87. An outdoor torch lamp shines light a distance of 30 ft in all directions. What is the total ground area lighted?

88. Hurricane Katrina's eye was 32 mi wide. The eye of a storm of similar intensity is usually only 10 mi wide. (*Source:* Associated Press 10/8/05 "Mapping Katrina's Storm Surge")

a. What area was covered by the eye of Katrina? Round to the nearest square mile.

b. What is the usual area of the eye of a similar storm?

Expanding Your Skills

89. A hula hoop has a 20-in. diameter.

a. Find the circumference. Use 3.14 for π.

b. How many times will the hula hoop have to turn to roll down a 40-yd (1440 in.) driveway? Round to the nearest whole unit.

90. A bicycle wheel has a 26-in. diameter.

a. Find the circumference. Use 3.14 for π.

b. How many times will the wheel have to turn to go a distance of 1000 ft (12,000 in.)? Round to the nearest whole unit.

91. Latasha has a bicycle, and the wheel has a 22-in. diameter. If the wheels of the bike turned 1000 times, how far did she travel? Use 3.14 for π. Give the answer to the nearest inch and to the nearest foot (1 ft = 12 in.)

92. The exercise wheel for Al's dwarf hamster has a diameter of 6.75 in.

a. Find the circumference. Use 3.14 for π and round to the nearest inch.

b. How far does Al's hamster travel if he completes 25 revolutions? Write the answer in feet and round to the nearest foot.

Calculator Connections

Topic: Using the π Key

When finding the circumference or the area of a circle, we can use the [π] key on the calculator to lend more accuracy to our calculations. If you press the [π] key on the calculator, the display will show 3.141592654. This number is not the exact value of π (remember that in decimal form π is a nonterminating and nonrepeating decimal). However, using the [π] key provides more accuracy than by using 3.14. For example, suppose we want to find the area of a circle of radius 3 ft. Compare the values by using 3.14 for π versus using the [π] key.

Expression	**Keystrokes**	**Result**
$3.14 \cdot 3^2$	3.14 [×] 3 [x^2] [=]	28.26
$\pi \cdot 3^2$	[π] [×] 3 [x^2] [=]	28.27433388

Again, it is important to note that neither of these answers is the exact area. The only way to write the exact value is to express the answer in terms of π. The exact area is 9π ft^2.

For Exercises 93–96, find the area and circumference rounded to four decimals places. Use the [π] key on your calculator.

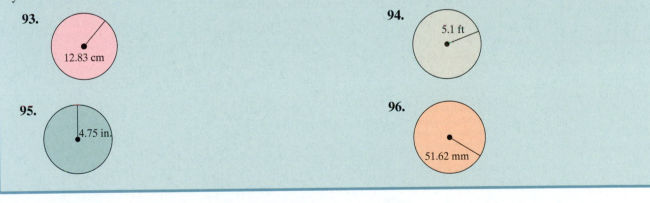

93.

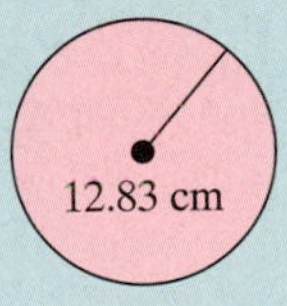

94.

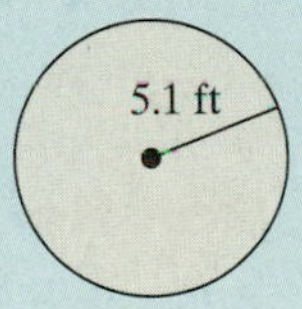

95.

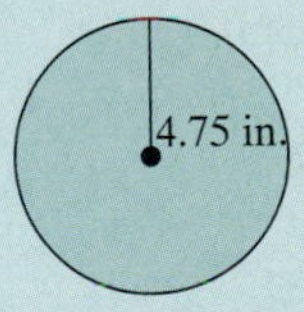

96.

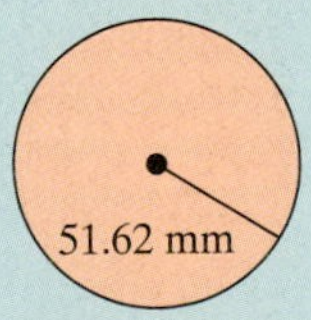

Section 5.4 Division of Decimals

Objectives

1. Division of Decimals
2. Rounding a Quotient
3. Applications of Decimal Division

1. Division of Decimals

Dividing decimals is much the same as dividing whole numbers. However, we must determine where to place the decimal point in the quotient.

First consider the quotient 3.5 ÷ 7. We can write the numbers in fractional form and then divide.

$$3.5 \div 7 = \frac{35}{10} \div \frac{7}{1} = \frac{35}{10} \cdot \frac{1}{7} = \frac{35}{70} = \frac{5}{10} = 0.5$$

Now consider the same problem by using the efficient method of long division: $7\overline{)3.5}$.

When the divisor is a whole number, we place the decimal point directly above the decimal point in the dividend. Then we divide as we would whole numbers.

$$\begin{array}{r} .5 \\ 7\overline{)3.5} \end{array}$$

decimal point placed above the decimal point in the dividend

PROCEDURE Dividing a Decimal by a Whole Number

To divide by a whole number:

Step 1 Place the decimal point in the quotient directly above the decimal point in the dividend.

Step 2 Divide as you would whole numbers.

Skill Practice

Divide. Check by using multiplication.

1. 502.96 ÷ 8

Example 1 Dividing by a Whole Number

Divide and check the answer by multiplying.

30.55 ÷ 13

Solution:

$$\begin{array}{r} . \\ 13\overline{)30.55} \end{array}$$

Locate the decimal point in the quotient.

$$\begin{array}{r} 2.35 \\ 13\overline{)30.55} \\ -26 \\ \hline 45 \\ -39 \\ \hline 65 \\ -65 \\ \hline 0 \end{array}$$

Divide as you would whole numbers.

Check by multiplying:

$$\begin{array}{r} \overset{1\,1}{2.35} \\ \times\ 13 \\ \hline 705 \\ 2350 \\ \hline 30.55 \end{array}\ \checkmark$$

Answer

1. 62.87

When dividing decimals, we do not use a remainder. Instead we insert zeros to the right of the dividend and continue dividing. This is demonstrated in Example 2.

Example 2 Dividing by an Integer

Divide and check the answer by multiplying.

$$\frac{-3.5}{-4}$$

Solution:

The quotient will be *positive* because the divisor and dividend have the same sign. We will perform the division without regard to sign, and then write the quotient as a positive number.

$$\begin{array}{r} . \\ 4\overline{)3.5} \end{array}$$

Locate the decimal point in the quotient.

$$\begin{array}{r} .8 \\ 4\overline{)3.5} \\ \underline{-32} \\ 3 \end{array}$$

Rather than using a remainder, we insert zeros in the dividend and continue dividing.

$$\begin{array}{r} .875 \\ 4\overline{)3.500} \\ \underline{-32} \\ 30 \\ \underline{-28} \\ 20 \\ \underline{-20} \\ 0 \end{array}$$

Check by multiplying:

$$\begin{array}{r} \overset{3\,2}{0.875} \\ \underline{\times\ \ 4} \\ 3.500\ \checkmark \end{array}$$

The quotient is 0.875.

Skill Practice

Divide.

2. $\dfrac{-6.8}{-5}$

Example 3 Dividing by an Integer

Divide and check the answer by multiplying. $40\overline{)5}$

Solution:

$$\begin{array}{r} . \\ 40\overline{)5.} \end{array}$$

The dividend is a whole number, and the decimal point is understood to be to its right. Insert the decimal point above it in the quotient.

$$\begin{array}{r} .125 \\ 40\overline{)5.000} \\ \underline{-40} \\ 100 \\ \underline{-80} \\ 200 \\ \underline{-200} \\ 0 \end{array}$$

Since 40 is greater than 5, we need to insert zeros to the right of the dividend.

Check by multiplying.

$$\begin{array}{r} \overset{1\,2}{0.125} \\ \underline{\times\ 40} \\ 000 \\ \underline{5000} \\ 5.000\ \checkmark \end{array}$$

The quotient is 0.125.

Skill Practice

Divide.

3. $20\overline{)3}$

Answers

2. 1.36 **3.** 0.15

Sometimes when dividing decimals, the quotient follows a repeated pattern. The result is called a **repeating decimal**.

Skill Practice

Divide.

4. $2.4 \div 9$

Example 4 Dividing Where the Quotient Is a Repeating Decimal

Divide. $1.7 \div 30$

Solution:

$$\begin{array}{r} .05666\ldots \\ 30\overline{)1.70000} \\ \underline{-150} \\ 200 \\ \underline{-180} \\ 200 \\ \underline{-180} \\ 200 \end{array}$$

Notice that as we continue to divide, we get the same values for each successive step. This causes a pattern of repeated digits in the quotient. Therefore, the quotient is a *repeating decimal.*

The quotient is $0.05666\ldots$. To denote the repeated pattern, we often use a bar over the first occurrence of the repeat cycle to the right of the decimal point. That is,

$0.05666\ldots = 0.05\overline{6}$ ← repeat bar

Avoiding Mistakes

In Example 4, notice that the repeat bar goes over only the 6. The 5 is not being repeated.

Skill Practice

Divide.

5. $11\overline{)57}$

Example 5 Dividing Where the Quotient Is a Repeating Decimal

Divide. $11\overline{)68}$

Solution:

$$\begin{array}{r} 6.1818\ldots \\ 11\overline{)68.0000} \\ \underline{-66} \\ 20 \\ \underline{-11} \\ 90 \\ \underline{-88} \\ 20 \\ \underline{-11} \\ 90 \\ \underline{-88} \\ 20 \end{array}$$

Could have stopped here → 20

Once again, we see a repeated pattern. The quotient is a repeating decimal. Notice that we could have stopped dividing when we obtained the second value of 20.

The quotient is $6.\overline{18}$.

Avoiding Mistakes

Be sure to put the repeating bar over the entire block of numbers that is being repeated. In Example 5, the bar extends over both the 1 and the 8. We have $6.\overline{18}$.

Answers

4. $0.2\overline{6}$ **5.** $5.\overline{18}$

The numbers $0.05\overline{6}$ and $6.\overline{18}$ are examples of repeating decimals. A decimal that "stops" is called a **terminating decimal**. For example, 6.18 is a terminating decimal, whereas $6.\overline{18}$ is a repeating decimal.

In Examples 1–5, we performed division where the divisor was an integer. Suppose now that we have a divisor that is *not* an integer, for example, $0.56 \div 0.7$. Because division can also be expressed in fraction notation, we have

$$0.56 \div 0.7 = \frac{0.56}{0.7}$$

If we multiply the numerator and denominator by 10, the denominator (divisor) becomes the whole number 7.

$$\frac{0.56}{0.7} = \frac{0.56 \times 10}{0.7 \times 10} = \frac{5.6}{7} \longrightarrow 7\overline{)5.6}$$

Recall that multiplying decimal numbers by 10 (or any power of 10, such as 100, 1000, etc.) moves the decimal point to the right. We use this idea to divide decimal numbers when the divisor is not a whole number.

PROCEDURE Dividing When the Divisor Is Not a Whole Number

Step 1 Move the decimal point in the divisor to the right to make it a whole number.

Step 2 Move the decimal point in the dividend to the right the same number of places as in step 1.

Step 3 Place the decimal point in the quotient directly above the decimal point in the dividend.

Step 4 Divide as you would whole numbers. Then apply the correct sign to the quotient.

Example 6 Dividing Decimals

Divide.

a. $0.56 \div (-0.7)$ **b.** $0.005\overline{)3.1}$

Solution:

a. The quotient will be *negative* because the divisor and dividend have different signs. We will perform the division without regard to sign and then apply the negative sign to the quotient.

$.7\overline{).56}$ Move the decimal point in the divisor and dividend one place to the right.

$7\overline{)5.6}$ ← Line up the decimal point in the quotient.

$$\begin{array}{r} 0.8 \\ 7\overline{)5.6} \\ -5\,6 \\ \hline 0 \end{array}$$

The quotient is -0.8.

Skill Practice

Divide.

6. $0.64 \div (-0.4)$

7. $5.4 \div 0.03$

Answers

6. -1.6 **7.** 180

b. $.005\overline{)3.100}$ Move the decimal point in the divisor and dividend three places to the right. Insert additional zeros in the dividend if necessary. Line up the decimal point in the quotient.

$$\begin{array}{r} 620.\\ 5\overline{)3100.}\\ \underline{-30}\\ 10\\ \underline{-10}\\ 00 \end{array}$$

The quotient is 620.

Skill Practice

Divide.

8. $\dfrac{-70}{-0.6}$

Example 7 Dividing Decimals

Divide. $\dfrac{-50}{-1.1}$

Solution:

The quotient will be *positive* because the divisor and dividend have the same sign. We will perform the division without regard to sign, and then write the quotient as a positive number.

$1.1\overline{)50.0}$ Move the decimal point in the divisor and dividend one place to the right. Insert an additional zero in the dividend. Line up the decimal point in the quotient.

$$\begin{array}{r} 45.45\ldots\\ 11\overline{)500.000}\\ \underline{-44}\\ 60\\ \underline{-55}\\ 50\\ \underline{-44}\\ 60\\ \underline{-55}\\ 50 \end{array}$$

The quotient is the repeating decimal, $45.\overline{45}$.

The quotient is $45.\overline{45}$.

Avoiding Mistakes

The quotient in Example 7 is a repeating decimal. The repeat cycle actually begins to the left of the decimal point. However, the repeat bar is placed on the first repeated block of digits to the *right* of the decimal point. Therefore, we write the quotient as $45.\overline{45}$.

Calculator Connections

Repeating decimals displayed on a calculator are rounded. This is because the display cannot show an infinite number of digits.

On a scientific calculator, the repeating decimal $45.\overline{45}$ might appear as

45.45454545

When we multiply a number by 10, 100, 1000, and so on, we move the decimal point to the right. However, dividing a positive number by 10, 100, or 1000 decreases its value. Therefore, we move the decimal point to the *left*.

For example, suppose 3.6 is divided by 10, 100, and 1000.

$$\begin{array}{r} .36\\ 10\overline{)3.60}\\ \underline{-30}\\ 60\\ \underline{-60}\\ 0 \end{array} \qquad \begin{array}{r} .036\\ 100\overline{)3.600}\\ \underline{-300}\\ 600\\ \underline{-600}\\ 0 \end{array} \qquad \begin{array}{r} .0036\\ 1000\overline{)3.6000}\\ \underline{-3000}\\ 6000\\ \underline{-6000}\\ 0 \end{array}$$

Answer

8. $116.\overline{6}$

PROCEDURE Dividing by a Power of 10

To divide a number by a power of 10, move the decimal point to the *left* the same number of places as there are zeros in the power of 10.

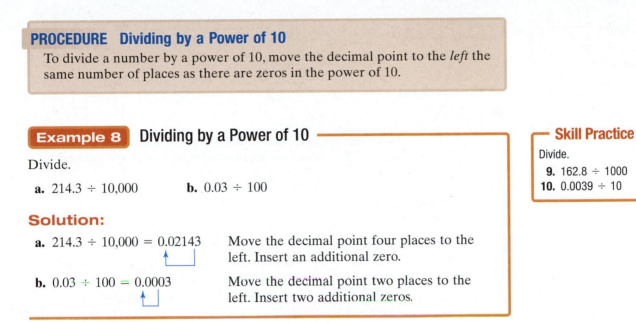

Example 8 Dividing by a Power of 10

Divide.

a. $214.3 \div 10{,}000$ **b.** $0.03 \div 100$

Solution:

a. $214.3 \div 10{,}000 = 0.02143$ Move the decimal point four places to the left. Insert an additional zero.

b. $0.03 \div 100 = 0.0003$ Move the decimal point two places to the left. Insert two additional zeros.

Skill Practice

Divide.

9. $162.8 \div 1000$

10. $0.0039 \div 10$

2. Rounding a Quotient

In Example 7, we found that $50 \div 1.1 = 45.\overline{45}$. To check this result, we could multiply $45.\overline{45} \times 1.1$ and show that the product equals 50. However, at this point we do not have the tools to multiply repeating decimals. What we can do is round the quotient and then multiply to see if the product is *close* to 50.

Example 9 Rounding a Repeating Decimal

Round $45.\overline{45}$ to the hundredths place. Then use the rounded value to estimate whether the product $45.\overline{45} \times 1.1$ is close to 50. (This will serve as a check to the division problem in Example 7.)

Solution:

To round the number $45.\overline{45}$, we must write out enough of the repeated pattern so that we can view the digit to the right of the rounding place. In this case, we must write out the number to the thousandths place.

$$45.\overline{45} = 45.454\cdots \approx 45.45$$

hundredths place

This digit is less than 5. Discard it and all others to its right.

Now multiply the rounded value by 1.1.

$$\begin{array}{r} 45.45 \\ \times\ 1.1 \\ \hline 4545 \\ 45450 \\ \hline 49.995 \end{array}$$

This value is close to 50. We are reasonably sure that we divided correctly in Example 7.

Skill Practice

Round to the indicated place value.

11. $2.3\overline{15}$; thousandths

Sometimes we may want to round a quotient to a given place value. To do so, divide until you get a digit in the quotient one place value to the right of the rounding place. At this point, you can stop dividing and round the quotient.

Answers

9. 0.1628 **10.** 0.00039

11. 2.315

Skill Practice

Round the quotient to the indicated place value.

12. 42.68 ÷ 5.1; hundredths

Example 10 Rounding a Quotient

Round the quotient to the tenths place.

$$\frac{47.3}{5.4}$$

Solution:

$5.4\overline{)47.3}$ Move the decimal point in the divisor and dividend one place to the right. Line up the decimal point in the quotient.

```
      8.75   ← hundredths place (tenths place marked)
 54)473.00
   -432
     410
    -378
     320
    -270
      50
```

To round the quotient to the tenths place, we must determine the hundredths-place digit and use it to base our decision on rounding. The hundredths-place digit is 5. Therefore, we round the quotient to the tenths place by increasing the tenths-place digit by 1 and discarding all digits to its right.

The quotient is approximately 8.8.

3. Applications of Decimal Division

Examples 11 and 12 show how decimal division can be used in applications. Remember that division is used when we need to distribute a quantity into equal parts.

Skill Practice

13. A 46.5-ft cable is to be cut into 6 pieces of equal length. How long is each piece?

Example 11 Applying Division of Decimals

A dinner costs $45.80, and the bill is to be split equally among 5 people. How much must each person pay?

Solution:

We want to distribute $45.80 equally among 5 people, so we must divide $45.80 ÷ 5.

```
     9.16
  5)45.80
   -45
     08
     -5
     30
    -30
      0
```

Each person must pay $9.16.

TIP: To check if your answer is reasonable, divide $45 among 5 people.

$$5\overline{)45} \quad 9$$

The estimated value of $9 suggests that the answer to Example 11 is correct.

Answers

12. 8.37 **13.** Each piece is 7.75 ft.

Division is also used in practical applications to express rates. In Example 12, we find the rate of speed in meters per second (m/sec) for the world record time in the men's 400-m run.

Example 12 Using Division to Find a Rate of Speed

In a recent year, the world-record time in the men's 400-m run was 43.2 sec. What is the speed in meters per second? Round to one decimal place. (*Source:* International Association of Athletics Federations)

Solution:

To find the rate of speed in meters per second, we must divide the distance in meters by the time in seconds.

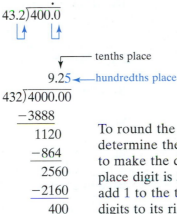

TIP: In Example 12, we had to find speed in meters per second. The units of measurement required in the answer give a hint as to the order of the division. The word *per* implies division. So to obtain meters *per* second implies 400 m ÷ 43.2 sec.

$43.2\overline{)400.0}$

tenths place

9.25 ← hundredths place

$$
\begin{array}{r}
9.25 \\
432\overline{)4000.00} \\
-3888 \\
\hline
1120 \\
-864 \\
\hline
2560 \\
-2160 \\
\hline
400
\end{array}
$$

To round the quotient to the tenths place, determine the hundredths-place digit and use it to make the decision on rounding. The hundredths-place digit is 5, which is 5 or greater. Therefore, add 1 to the tenths-place digit and discard all digits to its right.

The speed is approximately 9.3 m/sec.

Skill Practice

14. For many years, the world-record time in the women's 200-m run was held by the late Florence Griffith-Joyner. She ran the race in 21.3 sec. Find the speed in meters per second. Round to the nearest tenth.

Answer

14. The speed was 9.4 m/sec.

Section 5.4 Practice Exercises

Study Skills Exercise

1. Define the key terms.

a. Repeating decimal **b. Terminating decimal**

Review Exercises

For Exercises 2–7, perform the indicated operation.

2. 5.28×1000 **3.** $-8.003 - 2.2$ **4.** $11.8(0.32)$

5. $-102.4 + 1.239$ **6.** $16.82 - 14.8$ **7.** -5.28×0.001

For Exercises 8–10, use 3.14 for π.

8. Determine the diameter of a circle whose radius is 2.75 yd.

9. Approximate the area of a circle whose diameter is 20 ft.

10. Approximate the circumference of a circle whose radius is 10 cm.

Objective 1: Division of Decimals

For Exercises 11–18, divide. Check the answer by using multiplication. **(See Example 1.)**

11. $\dfrac{8.1}{9}$ Check: ____ $\times 9 = 8.1$

12. $\dfrac{4.8}{6}$ Check: ____ $\times 6 = 4.8$

13. $6\overline{)1.08}$ Check: ____ $\times 6 = 1.08$

14. $4\overline{)2.08}$ Check: ____ $\times 4 = 2.08$

15. $-4.24 \div (-8)$ **16.** $-5.75 \div (-25)$ **17.** $5\overline{)105.5}$ **18.** $7\overline{)221.2}$

For Exercises 19–50, divide. **(See Examples 2–7.)**

19. $5\overline{)9.8}$ **20.** $3\overline{)2.07}$ **21.** $0.28 \div 8$ **22.** $0.54 \div 8$

23. $5\overline{)84.2}$ **24.** $2\overline{)89.1}$ **25.** $50\overline{)6}$ **26.** $80\overline{)6}$

27. $-4 \div 25$ **28.** $-12 \div 60$ **29.** $16 \div 3$ **30.** $52 \div 9$

31. $-19 \div (-6)$ **32.** $-9.1 \div (-3)$ **33.** $33\overline{)71}$ **34.** $11\overline{)42}$

35. $5.03 \div 0.01$ **36.** $3.2 \div 0.001$ **37.** $0.992 \div 0.1$ **38.** $123.4 \div 0.01$

39. $\dfrac{57.12}{-1.02}$ **40.** $\dfrac{95.89}{-2.23}$ **41.** $\dfrac{-2.38}{-0.8}$ **42.** $\dfrac{-5.51}{-0.2}$

43. $0.3\overline{)62.5}$ **44.** $1.05\overline{)22.4}$ **45.** $-6.305 \div (-0.13)$ **46.** $-42.9 \div (-0.25)$

47. $1.1 \div 0.001$ **48.** $4.44 \div 0.01$ **49.** $420.6 \div 0.01$ **50.** $0.31 \div 0.1$

51. If 45.62 is divided by 100, will the decimal point move to the right or to the left? By how many places?

52. If 5689.233 is divided by 100,000, will the decimal point move to the right or to the left? By how many places?

For Exercises 53–60, divide by the powers of 10. **(See Example 8.)**

53. $3.923 \div 100$ **54.** $5.32 \div 100$ **55.** $-98.02 \div 10$ **56.** $-11.033 \div 10$

57. $-0.027 \div (-100)$ **58.** $-0.665 \div (-100)$ **59.** $1.02 \div 1000$ **60.** $8.1 \div 1000$

Objective 2: Rounding a Quotient

For Exercises 61–66, round to the indicated place value. **(See Example 9.)**

61. Round $2.\overline{4}$ to the
- **a.** Tenths place
- **b.** Hundredths place
- **c.** Thousandths place

62. Round $5.\overline{2}$ to the
- **a.** Tenths place
- **b.** Hundredths place
- **c.** Thousandths place

63. Round $1.7\overline{8}$ to the
- **a.** Tenths place
- **b.** Hundredths place
- **c.** Thousandths place

64. Round $4.2\overline{7}$ to the

a. Tenths place

b. Hundredths place

c. Thousandths place

65. Round $3.\overline{62}$ to the

a. Tenths place

b. Hundredths place

c. Thousandths place

66. Round $9.\overline{38}$ to the

a. Tenths place

b. Hundredths place

c. Thousandths place

For Exercises 67–75, divide. Round the answer to the indicated place value. Use the rounded quotient to check. **(See Example 10.)**

67. $7\overline{)1.8}$ hundredths

68. $2.1\overline{)75.3}$ hundredths

69. $-54.9 \div 3.7$ tenths

70. $-94.3 \div 21$ tenths

71. $0.24\overline{)4.96}$ thousandths

72. $2.46\overline{)27.88}$ thousandths

73. $0.9\overline{)32.1}$ hundredths

74. $0.6\overline{)81.4}$ hundredths

75. $2.13\overline{)237.1}$ tenths

Objective 3: Applications of Decimal Division

When multiplying or dividing decimals, it is important to place the decimal point correctly. For Exercises 76–79, determine whether you think the number is reasonable or unreasonable. If the number is unreasonable, move the decimal point to a position that makes more sense.

76. Steve computed the gas mileage for his Honda Civic to be 3.2 miles per gallon.

77. The sale price of a new kitchen refrigerator is $96.0.

78. Mickey makes $8.50 per hour. He estimates his weekly paycheck to be $3400.

79. Jason works in a legal office. He computes the average annual income for the attorneys in his office to be $1400 per year.

For Exercises 80–88, solve the application. Check to see if your answers are reasonable.

80. Brooke owes $39,628.68 on the mortgage for her house. If her monthly payment is $695.24, how many months does she still need to pay? How many years is this?

81. A membership at a health club costs $560 per year. The club has a payment plan in which a member can pay $50 down and the rest in 12 equal payments. How much is each payment? **(See Example 11.)**

82. It is reported that on average 42,000 tennis balls are used and 650 matches are played at the Wimbledon tennis tournament each year. On average, how many tennis balls are used per match? Round to the nearest whole unit.

83. A standard 75-watt lightbulb costs $0.75 and lasts about 800 hr. An energy efficient fluorescent bulb that gives off the same amount of light costs $5.00 and lasts about 10,000 hr.

a. How many standard lightbulbs would be needed to provide 10,000 hr of light?

b. How much would it cost using standard lightbulbs to provide 10,000 hr of light?

c. Which is more cost effective long term?

84. According to government tests, the 2009 Toyota Prius, a hybrid car, gets 45 miles per gallon of gas. If gasoline sells at \$3.20 per gallon, how many miles will a Prius travel on \$20.00 of gas?

85. In baseball, the batting average is found by dividing the number of hits by the number of times a batter was at bat. Babe Ruth had 2873 hits in 8399 times at bat. What was his batting average? Round to the thousandths place.

86. Ty Cobb was at bat 11,434 times and had 4189 hits, giving him the all-time best batting average. Find his average. Round to the thousandths place. (Refer to Exercise 85.)

87. Manny hikes 12 mi in 5.5 hr. What is his speed in miles per hour? Round to one decimal place. **(See Example 12.)**

88. Alicia rides her bike 33.2 mi in 2.5 hr. What is her speed in miles per hour? Round to one decimal place.

Expanding Your Skills

89. What number is halfway between -47.26 and -47.27?

90. What number is halfway between -22.4 and -22.5?

91. Which numbers when divided by 8.6 will produce a quotient less than 12.4? Circle all that apply.

a. 111.8 **b.** 103.2 **c.** 107.5 **d.** 105.78

92. Which numbers when divided by 5.3 will produce a quotient greater than 15.8? Circle all that apply.

a. 84.8 **b.** 84.27 **c.** 83.21 **d.** 79.5

Calculator Connections

Topic: Multiplying and Dividing Decimals on a Calculator

In some applications, the arithmetic on decimal numbers can be very tedious, and it is practical to use a calculator. To multiply or divide on a calculator, use the [×] and [÷] keys, respectively. However, be aware that for repeating decimals, the calculator cannot give an exact value. For example, the quotient of $17 \div 3$ is the repeating decimal $5.\overline{6}$. The calculator returns the rounded value 5.666666667. This is *not* the exact value. Also, when performing division, be careful to enter the dividend and divisor into the calculator in the correct order. For example:

Expression	Keystrokes	Result
$17 \div 3$	17 [÷] 3 [=]	5.666666667
$0.024\overline{)56.87}$	56.87 [÷] 0.024 [=]	2369.583333
$\frac{82.9}{3.1}$	82.9 [÷] 3.1 [=]	26.74193548

Calculator Exercises

For Exercises 93–98, multiply or divide as indicated.

93. $(2749.13)(418.2)$

94. $(139.241)(24.5)$

95. $(43.75)^2$

96. $(9.3)^5$

97. $21.5\overline{)2056.75}$

98. $14.2\overline{)4167.8}$

99. A Hummer H2 SUV uses 1260 gal of gas to travel 12,000 mi per year. A Honda Accord uses 375 gal of gas to go the same distance. Use the current cost of gasoline in your area, to determine the amount saved per year by driving a Honda Accord rather than a Hummer.

100. A Chevy Blazer gets 16.5 mpg and a Toyota Corolla averages 32 mpg. Suppose a driver drives 15,000 mi per year. Use the current cost of gasoline in your area to determine the amount saved per year by driving the Toyota rather than the Chevy.

101. As of June 2007, the U.S. capacity to generate wind power was 12,634 megawatts (MW). Texas generates approximately 3,352 MW of this power. (*Source:* American Wind Energy Association)

a. What fraction of the U.S. wind power is generated in Texas? Express this fraction as a decimal number rounded to the nearest hundredth of a megawatt.

b. Suppose there is a claim in a news article that Texas generates about one-fourth of all wind power in the United States. Is this claim accurate? Explain using your answer from part (a).

102. Population density is defined to be the number of people per square mile of land area. If California has 42,475,000 people with a land area of 155,959 square miles, what is the population density? Round to the nearest whole unit. (*Source:* U.S. Census Bureau)

103. The Earth travels approximately 584,000,000 mi around the Sun each year.

a. How many miles does the Earth travel in one day?

b. Find the speed of the Earth in miles per hour.

104. Although we say the time for the Earth to revolve about the Sun is 365 days, the actual time is 365.256 days. Multiply the fractional amount (0.256) by 4 to explain why we have a leap year every 4 years. (A leap year is a year in which February has an extra day, February 29.)

Problem Recognition Exercises

Operations on Decimals

For Exercises 1–24, perform the indicated operations.

1. **a.** $123.04 + 100$

b. 123.04×100

c. $123.04 - 100$

d. $123.04 \div 100$

e. $123.04 + 0.01$

f. 123.04×0.01

g. $123.04 \div 0.01$

h. $123.04 - 0.01$

2. **a.** $5078.3 + 1000$

b. 5078.3×1000

c. $5078.3 - 1000$

d. $5078.3 \div 1000$

e. $5078.3 + 0.001$

f. 5078.3×0.001

g. $5078.3 \div 0.001$

h. $5078.3 - 0.001$

3. **a.** $-4.8 + (-2.391)$

b. $2.391 - (-4.8)$

4. **a.** $-632.46 + (-98.0034)$

b. $98.0034 - (-632.46)$

5. **a.** $(32.9)(1.6)$

b. $(-1.6)(-32.9)$

6. **a.** $(74.23)(0.8)$

b. $(-0.8)(-74.23)$

7. **a.** 4(21.6)

b. $-0.25(21.6)$

8. **a.** 2(92.5)

b. $-0.5(92.5)$

9. **a.** $-448 \div 5.6$

b. $(5.6)(-80)$

10. **a.** $-496.8 \div 9.2$

b. $(-54)(9.2)$

11. 8(0.125)

12. 20(0.05)

13. $280 \div 0.07$

14. $6400 \div 0.001$

15. $490\overline{)98{,}000{,}000}$

16. $2000\overline{)5{,}400{,}000}$

17. $(-4500)(-300{,}000)$

18. $(-340)(-5000)$

19. $\begin{array}{r} 83.4 \\ -78.9999 \\ \hline \end{array}$

20. $\begin{array}{r} 124.7 \\ -47.9999 \\ \hline \end{array}$

21. $-3.47 - (-9.2)$

22. $-0.042 - (-0.097)$

23. $-3.47 - 9.2$

24. $-0.042 - 0.097$

Section 5.5 Fractions, Decimals, and the Order of Operations

Objectives

1. Writing Fractions as Decimals
2. Writing Decimals as Fractions
3. Decimals and the Number Line
4. Order of Operations Involving Decimals and Fractions
5. Applications of Decimals and Fractions

1. Writing Fractions as Decimals

Sometimes it is possible to convert a fraction to its equivalent decimal form by rewriting the fraction as a decimal fraction. That is, try to multiply the numerator and denominator by a number that will make the denominator a power of 10.

For example, the fraction $\frac{3}{5}$ can easily be written as an equivalent fraction with a denominator of 10.

$$\frac{3}{5} = \frac{3 \cdot 2}{5 \cdot 2} = \frac{6}{10} = 0.6$$

The fraction $\frac{3}{25}$ can easily be converted to a fraction with a denominator of 100.

$$\frac{3}{25} = \frac{3 \cdot 4}{25 \cdot 4} = \frac{12}{100} = 0.12$$

This technique is useful in some cases. However, some fractions such as $\frac{1}{3}$ cannot be converted to a fraction with a denominator that is a power of 10. This is because 3 is not a factor of any power of 10. For this reason, we recommend dividing the numerator by the denominator, as shown in Example 1.

Example 1 Writing Fractions as Decimals

Write each fraction or mixed number as a decimal.

a. $\frac{3}{5}$ **b.** $-\frac{68}{25}$ **c.** $3\frac{5}{8}$

Solution:

a. $\frac{3}{5}$ means $3 \div 5$.

$$\begin{array}{r} .6 \\ 5\overline{)3.0} \\ -30 \\ \hline 0 \end{array}$$

Divide the numerator by the denominator.

$\frac{3}{5} = 0.6$

b. $-\frac{68}{25}$ means $-(68 \div 25)$.

$$\begin{array}{r} 2.72 \\ 25\overline{)68.00} \\ -50 \\ \hline 180 \\ -175 \\ \hline 50 \\ -50 \\ \hline 0 \end{array}$$

Divide the numerator by the denominator.

$-\frac{68}{25} = -2.72$

c. $3\frac{5}{8} = 3 + \frac{5}{8} = 3 + (5 \div 8)$

$= 3 + 0.625$

$= 3.625$

$$\begin{array}{r} .625 \\ 8\overline{)5.000} \\ -48 \\ \hline 20 \\ -16 \\ \hline 40 \\ -40 \\ \hline 0 \end{array}$$

Divide the numerator by the denominator.

Skill Practice

Write each fraction or mixed number as a decimal.

1. $\frac{3}{8}$

2. $-\frac{43}{20}$

3. $12\frac{5}{16}$

The fractions in Example 1 are represented by terminating decimals. However, many fractions convert to repeating decimals.

Example 2 Converting Fractions to Repeating Decimals

Write each fraction as a decimal.

a. $\frac{4}{9}$ **b.** $-\frac{3}{22}$

Solution:

a. $\frac{4}{9}$ means $4 \div 9$.

$$\begin{array}{r} .44\ldots \\ 9\overline{)4.00} \\ -36 \\ \hline 40 \\ -36 \\ \hline 40 \end{array}$$

The quotient is a repeating decimal.

$\frac{4}{9} = 0.\overline{4}$

Skill Practice

Write each fraction as a decimal.

4. $\frac{8}{9}$ **5.** $-\frac{17}{11}$

Answers

1. 0.375 **2.** -2.15 **3.** 12.3125
4. $0.\overline{8}$ **5.** $-1.\overline{54}$

Calculator Connections

Repeating decimals displayed on a calculator must be rounded. For example, the repeating decimals from Example 2 might appear as follows.

$0.\overline{4}$	0.444444444
$-0.1\overline{36}$	−0.136363636

b. $-\dfrac{3}{22}$ means $-(3 \div 22)$

$$\begin{array}{r} .1363\ldots \\ 22\overline{)3.0000} \\ -22 \\ \hline 80 \\ -66 \\ \hline 140 \\ -132 \\ \hline 80 \end{array}$$

The quotient is a repeating decimal.

$-\dfrac{3}{22} = -0.1\overline{36}$

Avoiding Mistakes

Be sure to carry out the division far enough to see the repeating digits. In Example 2(b), the next digit in the quotient is 3. The 1 does not repeat.

$$-\frac{3}{22} = -0.1363636\ldots$$

Several fractions are used quite often. Their decimal forms are worth memorizing and are presented in Table 5-1.

Concept Connections

6. Describe the pattern between the fractions $\frac{1}{9}$ and $\frac{2}{9}$ and their decimal forms.

Table 5-1

$\frac{1}{4} = 0.25$	$\frac{2}{4} = \frac{1}{2} = 0.5$	$\frac{3}{4} = 0.75$	
$\frac{1}{9} = 0.\overline{1}$	$\frac{2}{9} = 0.\overline{2}$	$\frac{3}{9} = \frac{1}{3} = 0.\overline{3}$	$\frac{4}{9} = 0.\overline{4}$
$\frac{5}{9} = 0.\overline{5}$	$\frac{6}{9} = \frac{2}{3} = 0.\overline{6}$	$\frac{7}{9} = 0.\overline{7}$	$\frac{8}{9} = 0.\overline{8}$

Skill Practice

Convert the fraction to a decimal rounded to the indicated place value.

7. $\frac{9}{7}$; tenths

8. $-\frac{17}{37}$; hundredths

Example 3 Converting Fractions to Decimals with Rounding

Convert the fraction to a decimal rounded to the indicated place value.

a. $\dfrac{162}{7}$; tenths place **b.** $-\dfrac{21}{31}$; hundredths place

Solution:

a. $\dfrac{162}{7}$

$$\begin{array}{r} 23.14 \\ 7\overline{)162.00} \\ -14 \\ \hline 22 \\ -21 \\ \hline 1\,0 \\ -7 \\ \hline 30 \\ -28 \\ \hline 2 \end{array}$$

tenths place (3); hundredths place (4)

To round to the tenths place, we must determine the hundredths-place digit and use it to base our decision on rounding.

$23.1\not{4} \approx 23.1$

The fraction $\dfrac{162}{7}$ is approximately 23.1.

Answers

6. The numerator of the fraction is the repeated digit in the decimal form.
7. 1.3 8. −0.46

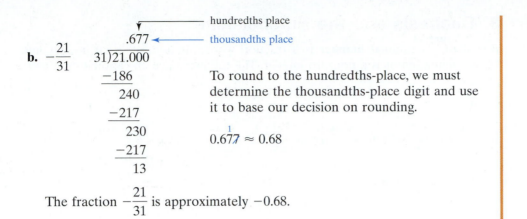

b. $-\dfrac{21}{31}$

$$\begin{array}{r} .677 \\ 31\overline{)21.000} \\ -186 \\ \hline 240 \\ -217 \\ \hline 230 \\ -217 \\ \hline 13 \end{array}$$

(hundredths place: 7 in the hundredths position; thousandths place: final 7)

To round to the hundredths-place, we must determine the thousandths-place digit and use it to base our decision on rounding.

$0.677 \approx 0.68$

The fraction $-\dfrac{21}{31}$ is approximately -0.68.

2. Writing Decimals as Fractions

In Section 5.1 we converted terminating decimals to fractions. We did this by writing the decimal as a decimal fraction and then reducing the fraction to lowest terms. For example:

$$0.46 = \frac{46}{100} = \frac{\overset{23}{\cancel{46}}}{\underset{50}{\cancel{100}}} = \frac{23}{50}$$

We do not yet have the tools to convert a repeating decimal to its equivalent fraction form. However, we can make use of our knowledge of the common fractions and their repeating decimal forms from Table 5-1.

Example 4 Writing Decimals as Fractions

Write the decimals as fractions.

a. 0.475 **b.** $0.\overline{6}$ **c.** -1.25

Solution:

a. $0.475 = \dfrac{475}{1000} = \dfrac{19 \cdot \overset{1}{\cancel{25}}}{40 \cdot \underset{1}{\cancel{25}}} = \dfrac{19}{40}$

b. From Table 5-1, the decimal $0.\overline{6} = \dfrac{2}{3}$.

c. $-1.25 = -\dfrac{125}{100} = -\dfrac{5 \cdot \overset{1}{\cancel{25}}}{4 \cdot \underset{1}{\cancel{25}}} = -\dfrac{5}{4}$

Skill Practice

Write the decimals as fractions.

9. 0.875

10. $0.\overline{7}$

11. -1.55

TIP: Recall from Section 5.1 that the place value of the farthest right digit is the denominator of the fraction.

$$0.475 = \frac{475}{1000}$$

(thousandths: the 5 in 0.475 corresponds to the denominator 1000)

Answers

9. $\dfrac{7}{8}$ **10.** $\dfrac{7}{9}$ **11.** $-\dfrac{31}{20}$

3. Decimals and the Number Line

Recall that a **rational number** is a fraction whose numerator is an integer and whose denominator is a nonzero integer. The following are all rational numbers.

$\frac{2}{3}$

$-\frac{5}{7}$ can be written as $\frac{-5}{7}$ or as $\frac{5}{-7}$.

6 can be written as $\frac{6}{1}$.

0.37 can be written as $\frac{37}{100}$.

$0.\overline{3}$ can be written as $\frac{1}{3}$.

Rational numbers consist of all numbers that can be expressed as terminating decimals or as repeating decimals.

- All numbers that can be expressed as *repeating decimals* are rational numbers.

  For example, $0.\overline{3} = 0.3333\ldots$ is a rational number.

- All numbers that can be expressed as *terminating decimals* are rational numbers.

  For example, 0.25 is a rational number.

- A number that *cannot* be expressed as a repeating or terminating decimal is not a rational number. These are called **irrational numbers**. An example of an irrational number is $\sqrt{2}$.

  For example, $\sqrt{2} \approx \underbrace{1.41421356237}_{\text{The digits never repeat and never terminate}}\ldots$

TIP: In Section 5.3 we used another irrational number, π. Recall that we approximated π with 3.14 or $\frac{22}{7}$ when calculating area and circumference of a circle.

The rational numbers and the irrational numbers together make up the set of **real numbers**. Furthermore, every real number corresponds to a point on the number line.

In Example 5, we rank the numbers from least to greatest and visualize the position of the numbers on the number line.

Example 5 Ordering Decimals and Fractions

Rank the numbers from least to greatest. Then approximate the position of the points on the number line.

$$0.\overline{45}, \quad 0.45, \quad \frac{1}{2}$$

Solution:

First note that $\frac{1}{2} = 0.5$ and that $0.\overline{45} = 0.454545\ldots$. By writing each number in decimal form, we can compare the decimals as we did in Section 5.1.

same | different

0.454545 · · · 0.450000 0.500000 This tells us that 0.5 is the greatest of the three values.

0.454545 · · · 0.450000 This tells us that $0.454545\ldots > 0.450000$.

different

Ranking the numbers from least to greatest we have: 0.45, $0.\overline{45}$, 0.5

Skill Practice

12. Rank the numbers from least to greatest. Then approximate the position of the points on the number line.
0.161, $\frac{1}{6}$, 0.16

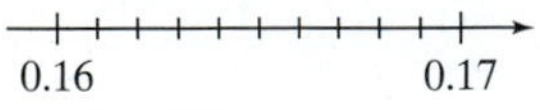

Answer

12. 0.16, 0.161, $\frac{1}{6}$

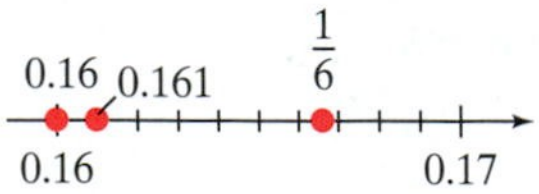

The position of these numbers can be seen on the number line. Note that we have expanded the segment of the number line between 0.4 and 0.5 to see more place values to the right of the decimal point.

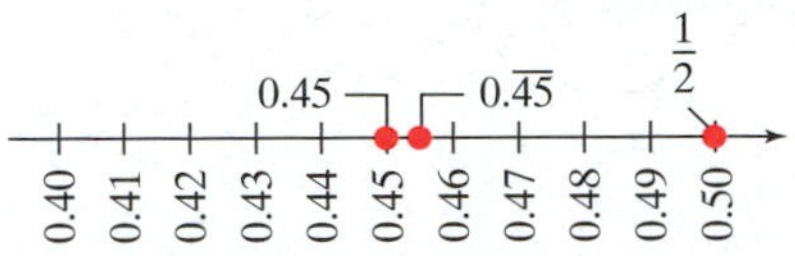

Recall that numbers that lie to the left on the number line have lesser value than numbers that lie to the right.

TIP: Be careful when ordering negative numbers. If the numbers in Example 5 had been negative, the order would be reversed. $-0.5 < -0.\overline{45} < -0.45$

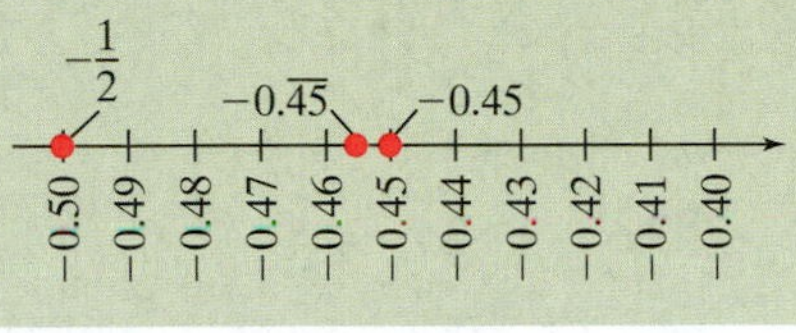

4. Order of Operations Involving Decimals and Fractions

In Example 6, we perform the order of operations with an expression involving decimal numbers.

PROCEDURE Applying the Order of Operations

Step 1 First perform all operations inside parentheses or other grouping symbols.

Step 2 Simplify expressions containing exponents, square roots, or absolute values.

Step 3 Perform multiplication or division in the order that they appear from left to right.

Step 4 Perform addition or subtraction in the order that they appear from left to right.

Example 6 Applying the Order of Operations

Simplify. $16.4 - (6.7 - 3.5)^2$

Solution:

$16.4 - (6.7 - 3.5)^2$

$= 16.4 - (3.2)^2$ Perform the subtraction within parentheses first.
$$\begin{array}{r} 6.7 \\ -\ 3.5 \\ \hline 3.2 \end{array}$$

$= 16.4 - 10.24$ Perform the operation involving the exponent.
$$\begin{array}{r} 3.2 \\ \times\ 3.2 \\ \hline 64 \\ 960 \\ \hline 10.24 \end{array}$$

$= 6.16$ Subtract.
$$\begin{array}{r} \scriptstyle 3\ 10 \\ 16.\not{4}\not{0} \\ -\ 10.24 \\ \hline 6.16 \end{array}$$

Skill Practice

Simplify.

13. $(5.8 - 4.3)^2 - 2$

Answer

13. 0.25

In Example 7, we apply the order of operations on fractions and decimals combined.

Skill Practice

Simplify.

14. $(-2.6)\cdot\frac{3}{13}\div\left(1\frac{1}{2}\right)^2$

Example 7 Applying the Order of Operations

Simplify. $(-6.4)\cdot 2\frac{5}{8}\div\left(\frac{3}{5}\right)^2$

Solution:

Approach 1

Convert all numbers to fractional form.

$$(-6.4)\cdot 2\frac{5}{8}\div\left(\frac{3}{5}\right)^2 = -\frac{64}{10}\cdot\frac{21}{8}\div\left(\frac{3}{5}\right)^2$$ Convert the decimal and mixed number to fractions.

$$= -\frac{64}{10}\cdot\frac{21}{8}\div\frac{9}{25}$$ Square the quantity $\frac{3}{5}$.

$$= -\frac{64}{10}\cdot\frac{21}{8}\cdot\frac{25}{9}$$ Multiply by the reciprocal of $\frac{9}{25}$.

$$= -\frac{\overset{\overset{4}{\cancel{8}}}{\cancel{64}}}{\underset{\underset{1}{\cancel{2}}}{\cancel{10}}}\cdot\frac{\overset{7}{\cancel{21}}}{\underset{1}{\cancel{8}}}\cdot\frac{\overset{5}{\cancel{25}}}{\underset{3}{\cancel{9}}}$$ Simplify common factors.

$$= -\frac{140}{3} \text{ or } -46\frac{2}{3} \text{ or } -46.\overline{6}$$ Multiply.

Approach 2

Convert all numbers to decimal form.

$$(-6.4)\cdot 2\frac{5}{8}\div\left(\frac{3}{5}\right)^2 = (-6.4)(2.625)\div(0.6)^2$$ The fraction $\frac{5}{8} = 0.625$ and $\frac{3}{5} = 0.6$.

$$= (-6.4)(2.625)\div 0.36$$ Square the quantity 0.6. That is, $(0.6)(0.6) = 0.36$.

Multiply $(-6.4)(2.625)$.

```
   2.625
 ×   6.4
   10500
  157500
 16.8000
```

$$= -16.8\div 0.36$$ Divide $-16.8\div 0.36$.

$$= -46.\overline{6}$$

$.36\overline{)16.80}$ (move decimal points two places right)

```
      46.6 . . .
36)1680
  -144
    240
   -216
    240
```

Answer

14. $-\frac{4}{15}$ or $-0.2\overline{6}$

When performing operations on fractions, sometimes it is desirable to write the answer as a decimal. For example, suppose that Sabina receives $\frac{1}{4}$ of a \$6245 inheritance. She would want to express this answer as a decimal. Sabina should receive:

$$\frac{1}{4}(\$6245) = (0.25)(\$6245) = \$1561.25$$

If Sabina had received $\frac{1}{3}$ of the inheritance, then the decimal form of $\frac{1}{3}$ would have to be rounded to some desired place value. This would cause *round-off error*. Any calculation performed on a rounded number will compound the error. For this reason, we recommend that fractions be kept in fractional form as long as possible as you simplify an expression. Then perform division and rounding in the last step. This is demonstrated in Example 8.

Example 8 Dividing a Fraction and Decimal

Divide $\frac{4}{7} \div 3.6$. Round the answer to the nearest hundredth.

Solution:

If we attempt to write $\frac{4}{7}$ as a decimal, we find that it is the repeating decimal $0.\overline{571428}$. Rather than rounding this number, we choose to change 3.6 to fractional form: $3.6 = \frac{36}{10}$.

$$\frac{4}{7} \div 3.6 = \frac{4}{7} \div \frac{36}{10} \quad \text{Write 3.6 as a fraction.}$$

$$= \frac{\overset{1}{\cancel{4}}}{7} \cdot \frac{10}{\underset{9}{\cancel{36}}} \quad \text{Multiply by the reciprocal of the divisor.}$$

$$= \frac{10}{63} \quad \text{Multiply and reduce to lowest terms.}$$

We must write the answer in decimal form, rounded to the nearest hundredth.

```
      .158
 63)10.00
   -63
    370
   -315
     550
    -504
      46
```

Divide. To round to the hundredths place, divide until we find the thousandths-place digit in the quotient. Use that digit to make a decision for rounding.

≈ 0.16 Round to the nearest hundredth.

Skill Practice

Divide. Round the answer to the nearest hundredth.

15. $4.1 \div \frac{12}{5}$

Answer

15. 1.71

Skill Practice

Determine the area of the triangle.

16.

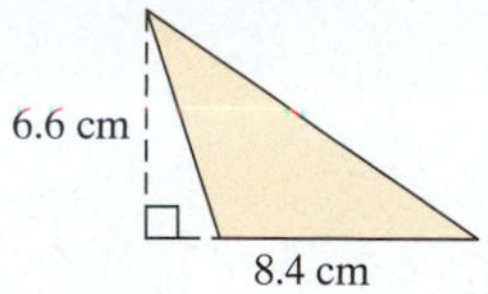

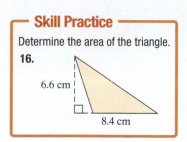

Example 9 Evaluating an Algebraic Expression

Determine the area of the triangle.

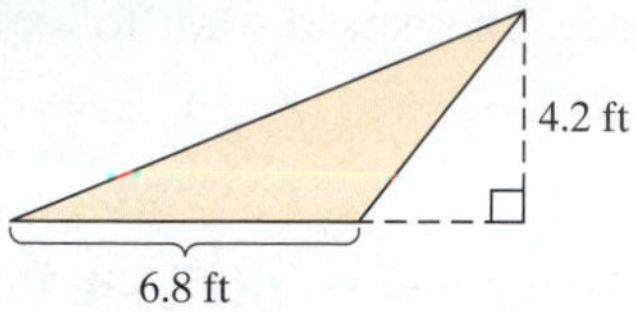

Solution:

In this triangle, the base is 6.8 ft. The height is 4.2 ft and is drawn outside the triangle.

$A = \frac{1}{2}bh$	Formula for the area of a triangle
$A = \frac{1}{2}(6.8 \text{ ft})(4.2 \text{ ft})$	Substitute $b = 6.8$ ft and $h = 4.2$ ft.
$= 0.5(6.8 \text{ ft})(4.2 \text{ ft})$	Write $\frac{1}{2}$ as the terminating decimal, 0.5.
$= 14.28 \text{ ft}^2$	Multiply from left to right.

The area is 14.28 ft^2.

5. Applications of Decimals and Fractions

Skill Practice

17. The odometer on a car read 46,125.9 mi. After a $13\frac{1}{4}$-hr trip, the odometer read 46,947.4 mi.

a. Find the total distance traveled on the trip.

b. Find the average speed in miles per hour (mph).

Example 10 Using Decimals and Fractions in a Consumer Application

Joanne filled the gas tank in her car and noted that the odometer read 22,341.9 mi. Ten days later she filled the tank again with $11\frac{1}{2}$ gal of gas. Her odometer reading at that time was 22,622.5 mi.

a. How many miles had she driven between fill-ups?

b. How many miles per gallon did she get?

Solution:

a. To find the number of miles driven, we need to subtract the initial odometer reading from the final reading.

$$\begin{array}{r} \scriptstyle 5\ 12\ 1\ \ 15 \\ 22{,}\cancel{6}\cancel{2}2.\cancel{5} \\ -\ 22{,}341.9 \\ \hline 280.6 \end{array}$$

Recall that to add or subtract decimals, line up the decimal points.

Joanne had driven 280.6 mi between fill-ups.

b. To find the number of miles per gallon (mi/gal), we divide the number of miles driven by the number of gallons.

$280.6 \div 11\frac{1}{2} = 280.6 \div 11.5$ We convert to decimal form because the fraction $11\frac{1}{2}$ is recognized as 11.5.

$= 24.4$

$$11.5\overline{)280.6} \qquad \begin{array}{r} 24.4 \\ 115\overline{)2806.0} \\ -230 \\ \hline 506 \\ -460 \\ \hline 460 \\ -460 \\ \hline 0 \end{array}$$

Joanne got 24.4 mi/gal.

Answers

16. 27.72 cm^2

17. a. 821.5 mi **b.** 62 mph

Section 5.5 Practice Exercises

Boost *your* GRADE at ALEKS.com!

ALEKS version 3.0

- Practice Problems
- Self-Tests
- NetTutor
- e-Professors
- Videos

Study Skills Exercise

1. Define the key terms.

a. Rational number **b. Irrational number** **c. Real number**

Review Exercises

2. Round $0.7\overline{84}$ to the thousandths place.

For Exercises 3–8, perform the indicated operations.

3. $(-0.15)^2$ **4.** $(-0.9)^2$ **5.** $-4.48 \div (-0.7)$

6. $-2.43 \div (0.3)$ **7.** $-4.67 - (-3.914)$ **8.** $-8.365 - 0.9$

Objective 1: Writing Fractions as Decimals

For Exercises 9–12, write each fraction as a decimal fraction, that is, a fraction whose denominator is a power of 10. Then write the number in decimal form.

9. $\frac{2}{5}$ **10.** $\frac{4}{5}$ **11.** $\frac{49}{50}$ **12.** $\frac{3}{50}$

For Exercises 13–24, write each fraction or mixed number as a decimal. **(See Example 1.)**

13. $\frac{7}{25}$ **14.** $\frac{4}{25}$ **15.** $-\frac{316}{500}$ **16.** $-\frac{19}{500}$

17. $-\frac{16}{5}$ **18.** $-\frac{68}{25}$ **19.** $-5\frac{3}{12}$ **20.** $-6\frac{5}{8}$

21. $\frac{18}{24}$ **22.** $\frac{24}{40}$ **23.** $7\frac{9}{20}$ **24.** $3\frac{11}{25}$

For Exercises 25–32, write each fraction or mixed number as a repeating decimal. **(See Example 2.)**

25. $3\frac{8}{9}$ **26.** $4\frac{7}{9}$ **27.** $\frac{19}{36}$ **28.** $\frac{7}{12}$

29. $-\frac{14}{111}$ **30.** $-\frac{58}{111}$ **31.** $\frac{25}{22}$ **32.** $\frac{45}{22}$

For Exercises 33–40, convert the fraction to a decimal and round to the indicated place value. **(See Example 3.)**

33. $\frac{15}{16}$; tenths **34.** $\frac{3}{11}$; tenths **35.** $\frac{1}{7}$; thousandths **36.** $\frac{2}{7}$; thousandths

37. $\frac{1}{13}$; hundredths **38.** $\frac{9}{13}$; hundredths **39.** $-\frac{5}{7}$; hundredths **40.** $-\frac{1}{8}$; hundredths

41. Write the fractions as decimals. Explain how to memorize the decimal form for these fractions with a denominator of 9.

a. $\frac{1}{9}$ **b.** $\frac{2}{9}$ **c.** $\frac{4}{9}$ **d.** $\frac{5}{9}$

42. Write the fractions as decimals. Explain how to memorize the decimal forms for these fractions with a denominator of 3.

a. $\frac{1}{3}$ **b.** $\frac{2}{3}$

Objective 2: Writing Decimals as Fractions

For Exercises 43–46, complete the table. **(See Example 4.)**

43.

	Decimal Form	Fraction Form
a.	0.45	
b.		$1\frac{5}{8}$ or $\frac{13}{8}$
c.	$-0.\overline{7}$	
d.		$-\frac{5}{11}$

44.

	Decimal Form	Fraction Form
a.		$\frac{2}{3}$
b.	1.6	
c.		$-\frac{152}{25}$
d.	$-0.\overline{2}$	

45.

	Decimal Form	Fraction Form
a.	$0.\overline{3}$	
b.	−2.125	
c.		$-\frac{19}{22}$
d.		$\frac{42}{25}$

46.

	Decimal Form	Fraction Form
a.	0.75	
b.		$-\frac{7}{11}$
c.	$-1.\overline{8}$	
d.		$\frac{74}{25}$

Objective 3: Decimals and the Number Line

For Exercises 47–58, identify the number as rational or irrational.

47. $-\frac{2}{5}$ **48.** $-\frac{1}{9}$ **49.** 5 **50.** 3

51. 3.5 **52.** 1.1 **53.** $\sqrt{7}$ **54.** $\sqrt{11}$

55. π **56.** 2π **57.** $0.\overline{4}$ **58.** $0.\overline{9}$

For Exercises 59–66, insert the appropriate symbol. Choose from <, >, or =.

59. $0.2 \square \frac{1}{5}$ **60.** $1.5 \square \frac{3}{2}$ **61.** $0.2 \square 0.\overline{2}$ **62.** $\frac{3}{5} \square 0.\overline{6}$

63. $\frac{1}{3} \square 0.3$ **64.** $\frac{2}{3} \square 0.66$ **65.** $-4\frac{1}{4} \square -4.\overline{25}$ **66.** $-2.12 \square -2.\overline{12}$

For Exercises 67–70, rank the numbers from least to greatest. Then approximate the position of the points on the number line. **(See Example 5.)**

67. $1.8, 1.75, 1.\overline{7}$

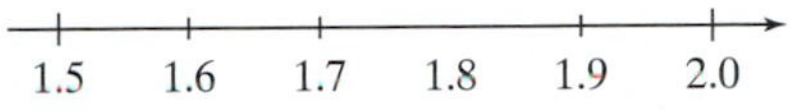

68. $5\frac{1}{6}, 5.\overline{6}, 5.0\overline{6}$

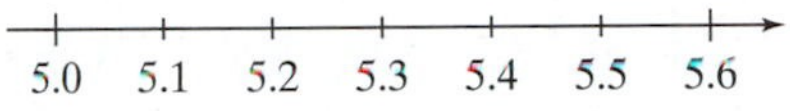

69. $-0.\overline{1}, -\frac{1}{10}, -\frac{1}{5}$

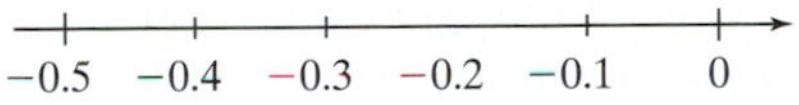

70. $-3\frac{1}{4}, -3\frac{1}{3}, -3.3$

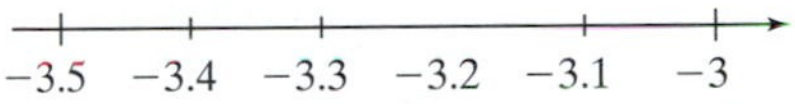

Objective 4: Order of Operations Involving Decimals and Fractions

For Exercises 71–82, simplify by using the order of operations. **(See Example 6.)**

71. $(3.7 - 1.2)^2$ **72.** $(6.8 - 4.7)^2$ **73.** $16.25 - (18.2 - 15.7)^2$ **74.** $11.38 - (10.42 - 7.52)^2$

75. $12.46 - 3.05 - 0.8^2$ **76.** $15.06 - 1.92 - 0.4^2$ **77.** $63.75 - 9.5(4)$ **78.** $6.84 + (3.6)(9)$

79. $-6.8 \div 2 \div 1.7$ **80.** $-8.4 \div 2 \div 2.1$ **81.** $2.2 - [9.34 + (1.2)^2]$ **82.** $4.2 \div 2.1 - (3.1)^2$

For Exercises 83–88, simplify by using the order of operations. Express the answer in decimal form. **(See Example 7.)**

83. $89.8 \div 1\frac{1}{3}$ **84.** $-30.12 \div \left(-1\frac{3}{5}\right)$ **85.** $-20.04 \div \left(-\frac{4}{5}\right)$

86. $(78.2 - 60.2) \div \frac{9}{13}$ **87.** $14.4\left(\frac{7}{4} - \frac{1}{8}\right)$ **88.** $6.5 + \frac{1}{8}\left(\frac{1}{5}\right)^2$

For Exercises 89–94, perform the indicated operations. Round the answer to the nearest hundredth when necessary. **(See Example 8.)**

89. $(2.3)\left(\frac{5}{9}\right)$ **90.** $(4.6)\left(\frac{7}{6}\right)$ **91.** $6.5 \div \left(-\frac{3}{5}\right)$

92. $\left(\frac{1}{12}\right)(6.24) \div (-2.1)$ **93.** $(42.81 - 30.01) \div \frac{9}{2}$ **94.** $\left(\frac{2}{7}\right)(5.1)\left(\frac{1}{10}\right)$

For Exercises 95–96, find the area of the triangle. **(See Example 9.)**

95.

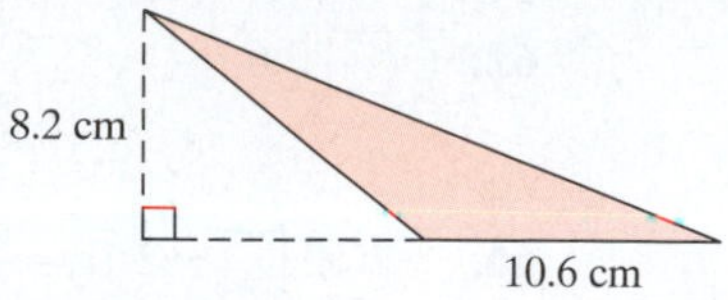

96.

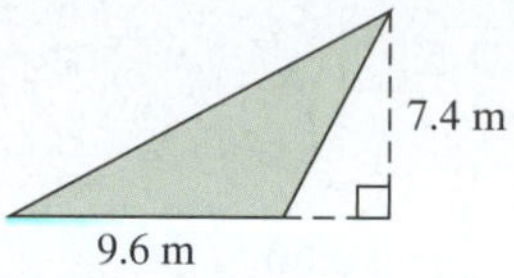

For Exercises 97–100, evaluate the expression for the given value(s) of the variables. Use 3.14 for π.

97. $\frac{1}{3}\pi r^2$ for $r = 9$

98. $\frac{4}{3}\pi r^3$ for $r = 3$

99. $-\frac{1}{2}gt^2$ for $g = 9.8$ and $t = 4$

100. $\frac{1}{2}mv^2$ for $m = 2.5$ and $v = 6$

Objective 5: Applications of Decimals and Fractions

101. Oprah and Gayle traveled for $7\frac{1}{2}$ hr between Atlanta, Georgia, and Orlando, Florida. At the beginning of the trip, the car's odometer read 21,345.6 mi. When they arrived in Orlando, the odometer read 21,816.6 mi. **(See Example 10.)**

a. How many miles had they driven on the trip?

b. Find the average speed in miles per hour (mph).

102. Jennifer earns $720.00 per week. Alex earns $\frac{3}{4}$ of what Jennifer earns per week.

a. How much does Alex earn per week?

b. If they both work 40 hr per week, how much do they each make per hour?

103. A cell phone plan has a $39.95 monthly fee and includes 450 min. For time on the phone over 450 min, the charge is $0.40 per minute. How much is Jorge charged for a month in which he talks for 597 min?

104. A night at a hotel in Dallas costs $129.95 with a nightly room tax of $20.75. The hotel also charges $1.10 per phone call made from the phone in the room. If John stays for 5 nights and makes 3 phone calls, how much is his total bill?

105. Olivia's diet allows her 60 grams (g) of fat per day. If she has $\frac{1}{4}$ of her total fat grams for breakfast and a McDonald's Quarter Pounder (20.7 g of fat) for lunch, how many grams does she have left for dinner?

106. Todd is establishing his beneficiaries for his life insurance policy. The policy is for $150,000.00 and $\frac{1}{2}$ will go to his daughter, $\frac{3}{8}$ will go to his stepson, and the rest will go to his grandson. What dollar amount will go to the grandson?

107. Hannah bought three packages of printer paper for $4.79 each. The sales tax for the merchandise was $0.86. If Hannah paid with a $20 bill, how much change should she receive?

108. Mr. Timpson bought dinner for $28.42. He left a tip of $6.00. He paid the bill with two $20 bills. How much change should he receive?

Expanding Your Skills

For Exercises 109–112, perform the indicated operations.

109. $0.\overline{3} \cdot 0.3 + 3.375$

110. $0.\overline{5} \div 0.\overline{2} - 0.75$

111. $(0.\overline{8} + 0.\overline{4}) \cdot 0.39$

112. $(0.\overline{7} - 0.\overline{6}) \cdot 5.4$

Calculator Connections

Topic: Applications Using the Order of Operations with Decimals

Calculator Exercises

113. Suppose that Deanna owns 50 shares of stock in Company A, valued at \$132.05 per share. She decides to sell these shares and use the money to buy stock in Company B, valued at \$27.80 per share. Assume there are no fees for either transaction.

a. How many full shares of Company B stock can she buy?

b. How much money will be left after she buys the Company B stock?

114. One megawatt (MW) of wind power produces enough electricity to supply approximately 275 homes. For a recent year, the state of Texas produced 3352 MW of wind power. (*Source:* American Wind Energy Association)

a. About how many homes can be supplied with electricity using wind power produced in Texas?

b. The given table outlines new proposed wind power projects in Texas. If these projects are completed, approximately how many additional homes could be supplied with electricity?

Project	MW
JD Wind IV	79.8
Buffalo Gap, Phase II	232.5
Lone Star I (3Q)	128
Sand Bluff	90
Roscoe	209
Barton Chapel	120
Stanton Wind Energy Center	120
Whirlwind Energy Center	59.8
Sweetwater V	80.5
Champion	126.5

115. Health-care providers use the *body mass index* (BMI) as one way to assess a person's risk of developing diabetes and heart disease. BMI is calculated with the following equation:

$$\text{BMI} = \frac{703w}{h^2}$$

where w is the person's weight in pounds and h is the person's height in inches. A person whose BMI is between 18.5 to 24.9 has a lower risk of developing diabetes and heart disease. A BMI of 25.0 to 29.9 indicates moderate risk. A BMI above 30.0 indicates high risk.

a. What is the risk level for a person whose height is 67.5 in. and whose weight is 195 lb?

b. What is the risk level for a person whose height is 62.5 in. and whose weight is 110 lb?

116. Marty bought a home for \$145,000. He paid \$25,000 as a down payment and then financed the rest with a 30-yr mortgage. His monthly payments are \$798.36 and go toward paying off the loan and interest on the loan.

a. How much money does Marty have to finance?

b. How many months are in a 30-yr period?

c. How much money will Marty pay over a 30-yr period to pay off the loan?

d. How much money did Marty pay in interest over the 30-yr period?

117. An inheritance for \$80,460.60 is to be divided equally among four heirs. However, before the money can be distributed, approximately one-third of the money must go to the government for taxes. How much does each person get after the taxes have been taken?

Section 5.6 Solving Equations Containing Decimals

Objectives

1. Solving Equations Containing Decimals
2. Solving Equations by Clearing Decimals
3. Applications and Problem Solving

1. Solving Equations Containing Decimals

In this section, we will practice solving linear equations. In Example 1, we will apply the addition and subtraction properties of equality.

Skill Practice

Solve.

1. $t + 2.4 = 9.68$
2. $-97.5 = -31.2 + w$

Example 1 Applying the Addition and Subtraction Properties of Equality

Solve. **a.** $x + 3.7 = 5.42$ **b.** $-35.4 = -6.1 + y$

Solution:

a.

$$x + 3.7 = 5.42$$

$$x + 3.7 - 3.7 = 5.42 - 3.7$$

The value 3.7 is added to x. To isolate x, we must reverse this process. Therefore, *subtract* 3.7 from both sides.

$$x = 1.72$$

Check: $x + 3.7 = 5.42$

$(1.72) + 3.7 \stackrel{?}{=} 5.42$

$5.42 = 5.42$ ✓

The solution is 1.72.

b.

$$-35.4 = -6.1 + y$$

$$-35.4 + 6.1 = -6.1 + 6.1 + y$$

To isolate y, add 6.1 to both sides.

$$-29.3 = y$$

Check: $-35.4 = -6.1 + y$

$-35.4 \stackrel{?}{=} -6.1 + (-29.3)$

$-35.4 = -35.4$ ✓

The solution is -29.3.

In Example 2, we will apply the multiplication and division properties of equality.

Skill Practice

Solve.

3. $\dfrac{z}{11.5} = -6.8$
4. $-8.37 = -2.7p$

Example 2 Applying the Multiplication and Division Properties of Equality

Solve. **a.** $\dfrac{t}{10.2} = -4.5$ **b.** $-25.2 = -4.2z$

Solution:

a.

$$\frac{t}{10.2} = -4.5$$

The variable t is being divided by 10.2. To isolate t, *multiply* both sides by 10.2.

$$\cancel{10.2}\left(\frac{t}{\cancel{10.2}}\right) = 10.2(-4.5)$$

$$t = -45.9$$

Check: $\dfrac{t}{10.2} = -4.5 \longrightarrow \dfrac{-45.9}{10.2} \stackrel{?}{=} -4.5$

$-4.5 = -4.5$ ✓

The solution is -45.9.

b. $-25.2 = -4.2z$

The variable z is being multiplied by -4.2. To isolate z, *divide* by -4.2. The quotient will be positive.

$$\frac{-25.2}{-4.2} = \frac{-4.2z}{-4.2}$$

Answers

1. 7.28 2. −66.3
3. −78.2 4. 3.1

$6 = z.$ Check: $-25.2 = -4.2z$

$-25.2 \stackrel{?}{=} -4.2(6)$

The solution is 6. $-25.2 = -25.2$ ✓

In Examples 3 and 4, multiple steps are required to solve the equation. As you read through Example 3, remember that we first isolate the variable term. Then we apply the multiplication or division property of equality to make the coefficient on the variable term equal to 1.

Example 3 Solving Equations Involving Multiple Steps

Solve. $2.4x - 3.85 = 8.63$

Solution:

$2.4x - 3.85 = 8.63$

$2.4x - 3.85 + 3.85 = 8.63 + 3.85$ To isolate the x term, add 3.85 to both sides.

$2.4x = 12.48$ Divide both sides by 2.4.

$\frac{2.4x}{2.4} = \frac{12.48}{2.4}$ This makes the coefficient on the x term equal to 1.

$x = 5.2$ The solution 5.2 checks in the original equation.

The solution is 5.2.

Skill Practice

Solve.

5. $5.8y - 14.4 = 55.2$

Example 4 Solving Equations Involving Multiple Steps

Solve. $0.03(x + 4) = 0.01x + 2.8$

Solution:

$0.03(x + 4) = 0.01x + 2.8$ Apply the distributive property to clear parentheses.

$0.03x + 0.12 = 0.01x + 2.8$

$0.03x - 0.01x + 0.12 = 0.01x - 0.01x + 2.8$ Subtract $0.01x$ from both sides. This places the variable terms all on one side.

$0.02x + 0.12 = 2.8$

$0.02x + 0.12 - 0.12 = 2.8 - 0.12$ Subtract 0.12 from both sides. This places the constant terms all on one side.

$0.02x = 2.68$

$\frac{0.02x}{0.02} = \frac{2.68}{0.02}$ Divide both sides by 0.02. This makes the coefficient on the x term equal to 1.

$x = 134$ The value 134 checks in the original equation.

The solution is 134.

Skill Practice

Solve.

6. $0.05(x + 6) = 0.02x - 0.18$

Answers

5. 12 **6.** −16

2. Solving Equations by Clearing Decimals

In Section 4.8 we learned that an equation containing fractions can be easier to solve if we clear the fractions. Similarly, when solving an equation with decimals, students may find it easier to clear the decimals first. To do this, we can multiply both sides of the equation by a power of 10 (10, 100, 1000, etc.). This will move the decimal point to the right in the coefficient on each term in the equation. This process is demonstrated in Example 5.

Skill Practice

Solve.

7. $0.02x - 3.42 = 1.6$

Example 5 Solving an Equation by Clearing Decimals

Solve. $0.05x - 1.45 = 2.8$

Solution:

To determine a power of 10 to use to clear fractions, identify the term with the most digits to the right of the decimal point. In this case, the terms $0.05x$ and 1.45 each have two digits to the right of the decimal point. Therefore, to clear decimals, we must multiply by 100. This will move the decimal point to the right two places.

$$0.05x - 1.45 = 2.8$$

$$100(0.05x - 1.45) = 100(2.8)$$ Multiply by 100 to clear decimals.

$$100(0.05x) - 100(1.45) = 100(2.80)$$ The decimal point will move to the right two places.

$$5x - 145 = 280$$

$$5x - 145 + 145 = 280 + 145$$ Add 145 to both sides.

$$5x = 425$$

$$\frac{5x}{5} = \frac{425}{5}$$ Divide both sides by 5.

$$x = 85$$ The value 85 checks in the original equation.

The solution is 85.

3. Applications and Problem Solving

In Examples 6–8, we will practice using linear equations to solve application problems.

Skill Practice

8. The sum of a number and 15.6 is the same as four times the number. Find the number.

Example 6 Translating to an Algebraic Expression

The sum of a number and 7.5 is three times the number. Find the number.

Solution:

Let x represent the number.

Step 1: Read the problem completely.

Step 2: Label the variable.

Step 3: Write the equation in words

the sum of → + ; 3 times the number → $3x$

$$x + 7.5 = 3x$$

a number → x ; 7.5 → 7.5

Step 4: Translate to a mathematical equation.

Answers

7. 251 **8.** 5.2

$x - x + 7.5 = 3x - x$ **Step 5:** Solve the equation.
Subtract x from both sides. This will bring all variable terms to one side.

$7.5 = 2x$

$\frac{7.5}{2} = \frac{2x}{2}$ Divide by 2.

$3.75 = x$

The number is 3.75. **Step 6:** Interpret the answer in words.

Avoiding Mistakes

Check the answer to Example 6. The sum of 3.75 and 7.5 is 11.25. The product 3(3.75) is also equal to 11.25.

Example 7 Solving an Application Involving Geometry

The perimeter of a triangle is 22.8 cm. The longest side is 8.4 cm more than the shortest side. The middle side is twice the shortest side. Find the lengths of the three sides.

Skill Practice

9. The perimeter of a triangle is 16.6 ft. The longest side is 6.1 ft more than the shortest side. The middle side is three times the shortest side. Find the lengths of the three sides.

Solution:

Step 1: Read the problem completely.

Let x represent the length of the shortest side. **Step 2:** Label the variable and draw a figure.

Then the longest side is $(x + 8.4)$.

The middle side is $2x$.

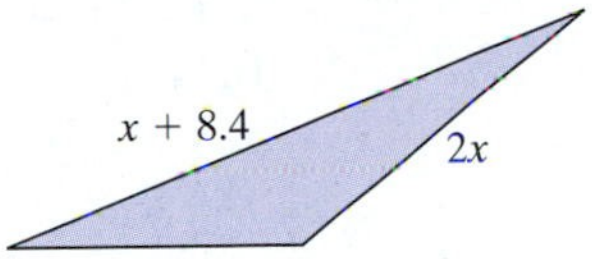

The sum of the lengths of the sides is 22.8 cm. **Step 3:** Write the equation in words

$x + (x + 8.4) + 2x = 22.8$ **Step 4:** Write a mathematical equation.

$x + x + 8.4 + 2x = 22.8$ **Step 5:** Solve the equation.

$4x + 8.4 = 22.8$ Combine *like* terms.

$4x + 8.4 - 8.4 = 22.8 - 8.4$ Subtract 8.4 from both sides.

$4x = 14.4$

$\frac{4x}{4} = \frac{14.4}{4}$ Divide by 4 on both sides.

$x = 3.6$

Step 6: Interpret the answer in words.

The shortest side is 3.6 cm.
The longest side is $(x + 8.4)$ cm $= (3.6 + 8.4)$ cm $= 12$ cm.
The middle side is $(2x)$ cm $= 2(3.6)$ cm $= 7.2$ cm.

The sides are 3.6 cm, 12 cm, and 7.2 cm.

Avoiding Mistakes

To check Example 7, notice that the sum of the lengths of the sides is 22.8 cm as expected.

$3.6 \text{ cm} + 12 \text{ cm} + 7.2 \text{ cm}$
$= 22.8 \text{ cm}$

Answer

9. 2.1 ft, 6.3 ft, and 8.2 ft

Skill Practice

10. D.J. signs up for a new credit card that earns travel miles with a certain airline. She initially earns 15,000 travel miles by signing up for the new card. Then for each dollar spent she earns 2.5 travel miles. If at the end of one year she has 38,500 travel miles, how many dollars did she charge on the credit card?

Example 8 Using a Linear Equation in a Consumer Application

Joanne has a cellular phone plan in which she pays \$39.95 per month for 450 min of air time. Additional minutes beyond 450 are charged at a rate of \$0.40 per minute. If Joanne's bill comes to \$87.95, how many minutes did she use beyond 450 min?

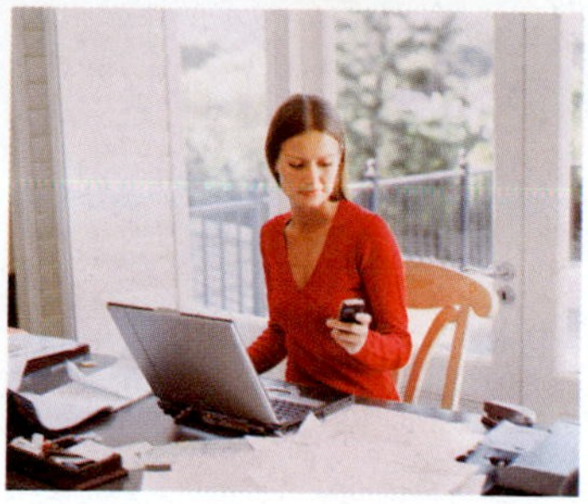

Solution:

Step 1: Read the problem.

Let x represent the number of minutes beyond 450. **Step 2:** Label the variable.

Then $0.40x$ represents the cost for x additional minutes.

$$\begin{pmatrix}\text{Monthly}\\ \text{fee}\end{pmatrix} + \begin{pmatrix}\text{Cost of}\\ \text{additional minutes}\end{pmatrix} = \begin{pmatrix}\text{Total}\\ \text{cost}\end{pmatrix}$$

Step 3: Write an equation in words.

$$39.95 \quad + \quad 0.40x \quad = \quad 87.95$$

Step 4: Write a mathematical equation.

$$39.95 + 0.40x = 87.95$$

Step 5: Solve the equation.

$$39.95 - 39.95 + 0.40x = 87.95 - 39.95$$ Subtract 39.95.

$$0.40x = 48.00$$

$$\frac{0.40x}{0.40} = \frac{48.00}{0.40}$$ Divide by 0.40.

$$x = 120$$

Joanne talked for 120 min beyond 450 min.

Answer

10. D.J. charged \$9400.

Section 5.6 Practice Exercises

Boost *your* GRADE at ALEKS.com!

ALEKS version 3.0

- Practice Problems
- Self-Tests
- NetTutor
- e-Professors
- Videos

Review Exercises

For Exercises 1–2, find the area and circumference. Use 3.14 for π.

1.

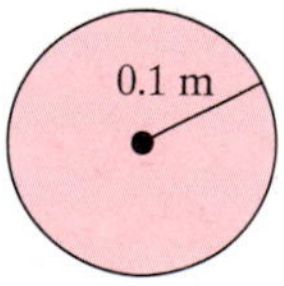

2.

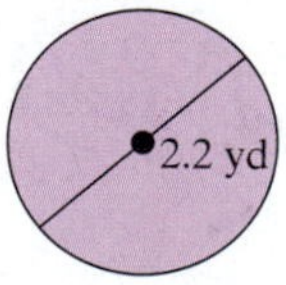

For Exercises 3–6, simplify.

3. $(2.3 - 3.8)^2$

4. $(-1.6 + 0.4)^2$

5. $\frac{1}{2}(4.8 - 9.26)$

6. $-\frac{1}{4}[62.9 + (-4.8)]$

For Exercises 7–10, simplify by clearing parentheses and combining *like* terms.

7. $-1.8x + 2.31x$

8. $-6.9y + 4.23y$

9. $-2(8.4z - 3.1) - 5.3z$

10. $-3(9.2w - 4.1) + 3.62w$

Objective 1: Solving Equations Containing Decimals

For Exercises 11–34, solve the equations. **(See Examples 1–4.)**

11. $y + 8.4 = 9.26$

12. $z + 1.9 = 12.41$

13. $t - 3.92 = -8.7$

14. $w - 12.69 = -15.4$

15. $-141.2 = -91.3 + p$

16. $-413.7 = -210.6 + m$

17. $-0.07 + n = 0.025$

18. $-0.016 + k = 0.08$

19. $\frac{x}{-4.6} = -9.3$

20. $\frac{y}{-8.1} = -1.5$

21. $6 = \frac{z}{-0.02}$

22. $7 = \frac{a}{-0.05}$

23. $19.43 = -6.7n$

24. $94.08 = -8.4q$

25. $-6.2y = -117.8$

26. $-4.1w = -73.8$

27. $8.4x + 6 = 48$

28. $9.2n + 6.4 = 43.2$

29. $-3.1x - 2 = -29.9$

30. $-5.2y - 7 = -22.6$

31. $0.04(p - 2) = 0.05p + 0.16$

32. $0.06(t - 9) = 0.07t + 0.27$

33. $-2.5x + 5.76 = 0.4(6 - 5x)$

34. $-1.5m + 14.26 = 0.2(18 - m)$

Objective 2: Solving Equations by Clearing Decimals

For Exercises 35–42, solve by first clearing decimals. **(See Example 5.)**

35. $0.04x - 1.9 = 0.1$

36. $0.03y - 2.3 = 0.7$

37. $-4.4 = -2 + 0.6x$

38. $-3.7 = -4 + 0.5x$

39. $4.2 = 3 - 0.002m$

40. $3.8 = 7 - 0.016t$

41. $6.2x - 4.1 = 5.94x - 1.5$

42. $1.32x + 5.2 = 0.12x + 0.4$

Objective 3: Applications and Problem Solving

43. Nine times a number is equal to 36 more than the number. Find the number. **(See Example 6.)**

44. Six times a number is equal to 30.5 more than the number. Find the number.

45. The difference of 13 and a number is 2.2 more than three times the number. Find the number.

46. The difference of 8 and a number is 1.7 more than two times the number. Find the number.

47. The quotient of a number and 5 is -1.88. Find the number.

48. The quotient of a number and -2.5 is 2.72. Find the number.

49. The product of 2.1 and a number is 8.36 more than the number. Find the number.

50. The product of -3.6 and a number is 48.3 more than the number. Find the number.

51. The perimeter of a triangle is 21.5 yd. The longest side is twice the shortest side. The middle side is 3.1 yd longer than the shortest side. Find the lengths of the sides. **(See Example 7.)**

52. The perimeter of a triangle is 1.32 m. The longest side is 30 times the shortest side, and the middle side is 0.04 m more than the shortest side. Find the lengths of the sides.

53. Toni, Rafa, and Henri are all servers at the Chez Joëlle Restaurant. The tips collected for the night amount to $167.80. Toni made $22.05 less in tips than Rafa. Henri made $5.90 less than Rafa. How much did each person make?

54. Bob bought a popcorn, a soda, and a hotdog at the movies for $8.25. Popcorn costs $1 more than a hotdog. A soda costs $0.25 less than a hotdog. How much is each item?

55. The U-Rent-It home supply store rents pressure cleaners for $4.95, plus $4 per hour. A painter rents a pressure cleaner and the bill comes to $18.95. For how many hours did he rent the pressure cleaner? **(See Example 8.)**

56. A cellular phone company charges $59.95 each month and this includes 500 min. For minutes used beyond the first 500, the charge is $0.25 per minute. Jim's bill came to $90.70. How many minutes over 500 min did he use?

57. Karla's credit card bill is $420.90. Part of the bill is from the previous month's balance, part is in new charges, and part is from a late fee of $39. The previous balance is $172.40 less than the new charges. How much is the previous balance and how much is in new charges?

58. Thayne's credit card bill is $879.10. This includes his charges and interest. If the new charges are $794.10 more than the interest, find the amount in charges and the amount in interest.

59. Madeline and Kim each rode 15 miles in a bicycle relay. Madeline's time was 8.25 min less than Kim's time. If the total time was 1 hr, 56.75 min, for how long did each person ride?

60. The two-night attendance for a Friday/Saturday basketball tournament at the University of Connecticut was 2570. There were 522 more people on Saturday for the finals than on Friday. How many people attended each night?

Mean, Median, and Mode

Section 5.7

Objectives

1. Mean
2. Median
3. Mode
4. Weighted Mean

1. Mean

When given a list of numerical data, it is often desirable to obtain a single number that represents the central value of the data. In this section, we discuss three such values called the mean, median, and mode. The mean (or average) of a list of data values was first presented in Section 1.8. We review the definition here.

DEFINITION Mean

The **mean** (or average) of a set of numbers is the sum of the values divided by the number of values. We can write this as a formula.

$$\text{Mean} = \frac{\text{sum of the values}}{\text{number of values}}$$

Example 1 Finding the Mean of a Data Set

A small business employs five workers. Their yearly salaries are

\$42,000 \$36,000 \$45,000 \$35,000 \$38,000

a. Find the mean yearly salary for the five employees.

b. Suppose the owner of the business makes \$218,000 per year. Find the mean salary for all six individuals (that is, include the owner's salary).

Solution:

a. Mean salary of five employees

$$= \frac{42{,}000 + 36{,}000 + 45{,}000 + 35{,}000 + 38{,}000}{5}$$

$$= \frac{196{,}000}{5} \quad \text{Add the data values.}$$

$$= 39{,}200 \quad \text{Divide.}$$

The mean salary for employees is \$39,200.

b. Mean of all six individuals

$$= \frac{42{,}000 + 36{,}000 + 45{,}000 + 35{,}000 + 38{,}000 + 218{,}000}{6}$$

$$= \frac{414{,}000}{6}$$

$$= 69{,}000$$

The mean salary with the owner's salary included is \$69,000.

Skill Practice

Housing prices for five homes in one neighborhood are given.

\$108,000 \$149,000
\$164,000 \$118,000
\$144,000

1. Find the mean price of these five houses.
2. Suppose a new home is built in the neighborhood for \$1.3 million (\$1,300,000). Find the mean price of all six homes.

Avoiding Mistakes

When computing a mean remember that the data are added first before dividing.

Answers

1. \$136,600 **2.** \$330,500

2. Median

In Example 1, you may have noticed that the mean salary was greatly affected by the unusually high value of $218,000. For this reason, you may want to use a different measure of "center" called the median. The **median** is the "middle" number in an ordered list of numbers.

PROCEDURE Finding the Median

To compute the median of a list of numbers, first arrange the numbers in order from least to greatest.

- If the number of data values in the list is *odd*, then the median is the middle number in the list.
- If the number of data values is *even*, there is no single middle number. Therefore, the median is the mean (average) of the two middle numbers in the list.

Skill Practice

3. Find the median of the five housing prices given in margin Exercise 1.
$108,000 $149,000
$164,000 $118,000
$144,000

4. Find the median of the six housing prices given in margin Exercise 2.
$108,000 $149,000
$164,000 $118,000
$144,000 $1,300,000

Example 2 Finding the Median of a Data Set

Consider the salaries of the five workers from Example 1.

$42,000 $36,000 $45,000 $35,000 $38,000

a. Find the median salary for the five workers.

b. Find the median salary including the owner's salary of $218,000.

Solution:

a. 35,000 36,000 38,000 42,000 45,000 Arrange the data in order.

Because there are five data values (an *odd* number), the median is the middle number.

The median is $38,000.

b. Now consider the scores of all six individuals (including the owner). Arrange the data in order.

Avoiding Mistakes

The data must be arranged in order before determining the median.

35,000 36,000 38,000 42,000 45,000 218,000

$$\frac{38{,}000 + 42{,}000}{2}$$ There are six data values (an *even* number). The median is the average of the two middle numbers.

$$= \frac{80{,}000}{2}$$ Add the two middle numbers.

$$= 40{,}000$$ Divide.

The median of all six salaries is $40,000.

Answers

3. $144,000 4. $146,500

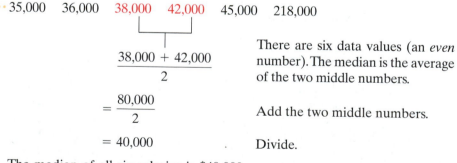

In Examples 1 and 2, the mean of all six salaries is $69,000, whereas the median is $40,000. These examples show that the median is a better representation for a central value when the data list has an unusually high (or low) value.

Example 3 **Finding the Median of a Data Set**

The average temperatures (in °C) for the South Pole are given by month. Find the median temperature. (*Source*: NOAA)

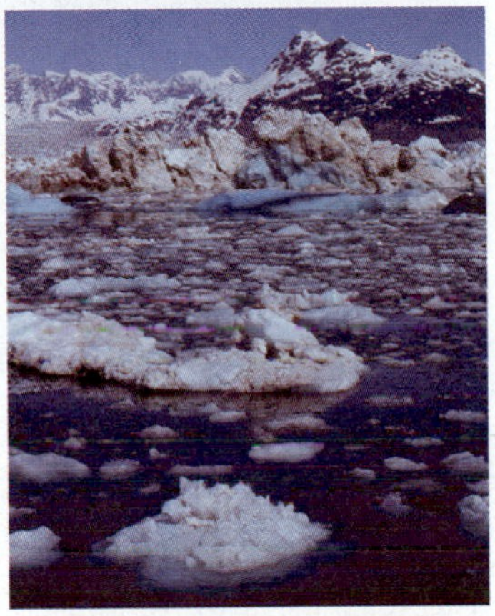

Jan.	Feb.	Mar.	April	May	June	July	Aug.	Sept.	Oct.	Nov.	Dec.
−2.9	−9.5	−19.2	−20.7	−21.7	−23.0	−25.7	−26.1	−24.6	−18.9	−9.7	−3.4

Solution:

First arrange the numbers in order from least to greatest.

−26.1 −25.7 −24.6 −23.0 −21.7 −20.7 −19.2 −18.9 −9.7 −9.5 −3.4 −2.9

$$\text{Median} = \frac{-20.7 + (-19.2)}{2} = -19.95$$

There are 12 data values (an *even* number). Therefore, the median is the average of the two middle numbers. The median temperature at the South Pole is −19.95°C.

Skill Practice

5. The gain or loss for a stock is given for an 8-day period. Find the median gain or loss.

−2.4	−2.0	1.25	0.6
−1.8	−0.4	0.6	−0.9

Note: The median may not be one of the original data values. This was true in Example 3.

3. Mode

A third representative value for a list of data is called the mode.

DEFINITION Mode

The **mode** of a set of data is the value or values that occur most often.

- If two values occur most often we say the data are **bimodal**.
- If more than two values occur most often, we say there is no mode.

Answer

5. −0.65

Skill Practice

6. The monthly rainfall amounts for Houston, Texas, are given in inches. Find the mode. (*Source:* NOAA)

4.5	3.0	3.2
3.5	5.1	6.8
4.3	4.5	5.6
5.3	4.5	3.8

Example 4 Finding the Mode of a Data Set

The student-to-teacher ratio is given for elementary schools for ten selected states. For example, California has a student-to-teacher ratio of 20.6. This means that there are approximately 20.6 students per teacher in California elementary schools. (*Source*: National Center for Education Statistics)

ME	ND	WI	NH	RI	IL	IN	MS	CA	UT
12.5	13.4	14.1	14.5	14.8	16.1	16.1	16.1	20.6	21.9

Find the mode of the student-to-teacher ratio for these states.

Solution:

The data value 16.1 appears the most often. Therefore, the mode is 16.1 students per teacher.

Skill Practice

7. Find the mode of the weights of babies (in pounds) born one day at Brackenridge Hospital in Austin, Texas.

7.2	8.1	6.9
9.3	8.3	7.7
7.9	6.4	7.5

Example 5 Finding the Mode of a Data Set

Find the mode of the list of average monthly temperatures for Albany, New York. Values are in °F.

Jan.	Feb.	March	April	May	June	July	Aug.	Sept.	Oct.	Nov.	Dec.
22	25	35	47	58	66	71	69	61	49	39	26

Solution:

No data value occurs most often. There is no mode for this set of data.

Example 6 Finding the Mode of a Data Set

The grades for a quiz in college algebra are as follows. The scores are out of a possible 10 points. Find the mode of the scores.

9	4	6	9	9	8	2	1	4	9
5	10	10	5	7	7	9	8	7	3
9	7	10	7	10	1	7	4	5	6

Answers

6. 4.5 in. 7. No mode

Solution:

Sometimes arranging the data in order makes it easier to find the repeated values.

1	1	2	3	4	4	4	5	5	5
6	6	7	7	7	7	7	7	8	8
9	9	9	9	9	9	10	10	10	10

The score of 9 occurs 6 times. The score of 7 occurs 6 times. There are two modes, 9 and 7, because these scores both occur more than any other score. We say that these data are *bimodal.*

Skill Practice

8. The ages of children participating in an after-school sports program are given. Find the mode(s).

13	15	17	15
14	15	16	16
15	16	12	13
15	14	16	15
15	16	16	13
16	13	14	18

TIP: To remember the difference between median and mode, think of the *median* of a highway that goes down the *middle*. Think of the word *mode* as sounding similar to the word *most*.

4. Weighted Mean

Sometimes data values in a list appear multiple times. In such a case, we can compute a **weighted mean**. In Example 7, we demonstrate how to use a weighted mean to compute a grade point average (GPA). To compute GPA, each grade is assigned a numerical value. For example, an "A" is worth 4 points, a "B" is worth 3 points, and so on. Then each grade for a course is "weighted" by the number of credit-hours that the course is worth.

Example 7 Using a Weighted Mean to Compute GPA

At a certain college, the grades A–F are assigned numerical values as follows.

$$A = 4.0 \quad B+ = 3.5 \quad B = 3.0 \quad C+ = 2.5$$
$$C = 2.0 \quad D+ = 1.5 \quad D = 1.0 \quad F = 0.0$$

Elmer takes the following classes with the grades as shown. Determine Elmer's GPA.

Course	Grade	Number of Credit-Hours
Prealgebra	A = 4 pts	3
Study Skills	C = 2 pts	1
First Aid	B+ = 3.5 pts	2
English I	D = 1.0 pt	4

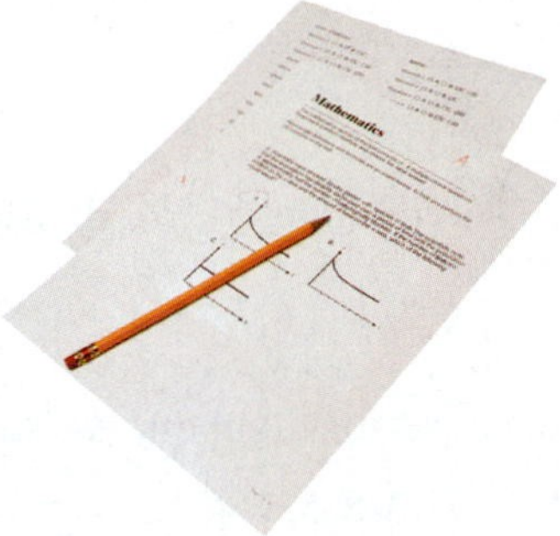

Solution:

The data in the table can be visualized as follows.

4 pts	4 pts	4 pts	2 pts	3.5 pts	3.5 pts	1 pt	1 pt	1 pt	1 pt
A	A	A	C	B+	B+	D	D	D	D
3 of these			1 of these	2 of these		4 of these			

Skill Practice

9. Clyde received the following grades for the semester. Use the numerical values assigned to grades from Example 7 to find Clyde's GPA.

Course	Grade	Credit-Hours
Math	B+	4
Science	C	3
Speech	A	3

Answers

8. There are two modes, 15 and 16.
9. Clyde's GPA is 3.2.

The number of grade points earned for each course is the product of the grade for the course and the number of credit-hours for the course. For example:

Grade points for Prealgebra: (4 pts)(3 credit-hours) = 12 points.

Course	Grade	Number of Credit-Hours (Weights)	Product Number of Grade Points
Prealgebra	A = 4 pts	3	(4 pts)(3 credit-hours) = 12 pts
Study Skills	C = 2 pts	1	(2 pts)(1 credit-hour) = 2 pts
First Aid	B+ = 3.5 pts	2	(3.5 pts)(2 credit-hours) = 7 pts
English I	D = 1.0 pt	4	(1 pt)(4 credit-hours) = 4 pt
		Total hours: 10	Total grade points: 25 pts

To determine GPA, we will add the number of grade points earned for each course and then divide by the total number of credit hours taken.

$$\text{Mean} = \frac{25}{10} = 2.5 \qquad \text{Elmer's GPA for this term is 2.5.}$$

In Example 7, notice that the value of each grade is "weighted" by the number of credit-hours. The grade of "A" for Prealgebra is weighted 3 times. The grade of "C" for the study skills course is weighted 1 time. The grade that hurt Elmer's GPA was the "D" in English. Not only did he receive a low grade, but the course was weighted heavily (4 credit-hours). In Exercise 47, we recompute Elmer's GPA with a "B" in English to see how this grade affects his GPA.

Section 5.7 Practice Exercises

Study Skills Exercise

1. Define the key terms.

a. Mean **b. Median** **c. Mode** **d. Bimodal** **e. Weighted mean**

Objective 1: Mean

For Exercises 2–7, find the mean of each set of numbers. **(See Example 1.)**

2. 4, 6, 5, 10, 4, 5, 8

3. 3, 8, 5, 7, 4, 2, 7, 4

4. 0, 5, 7, 4, 7, 2, 4, 3

5. 7, 6, 5, 10, 8, 4, 8, 6, 0

6. −10, −13, −18, −20, −15

7. −22, −14, −12, −16, −15

8. Compute the mean of your test scores for this class up to this point.

9. The flight times in hours for six flights between New York and Los Angeles are given. Find the mean flight time. Round to the nearest tenth of an hour.

5.5, 6.0, 5.8, 5.8, 6.0, 5.6

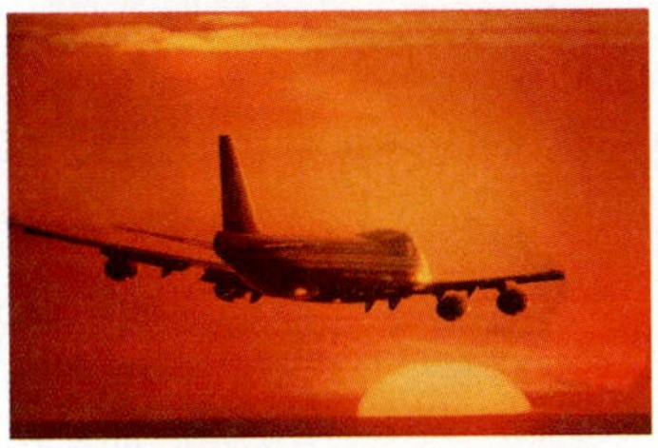

10. A nurse takes the temperature of a patient every hour and records the temperatures as follows: 98°F, 98.4°F, 98.9°F, 100.1°F, and 99.2°F. Find the patient's mean temperature.

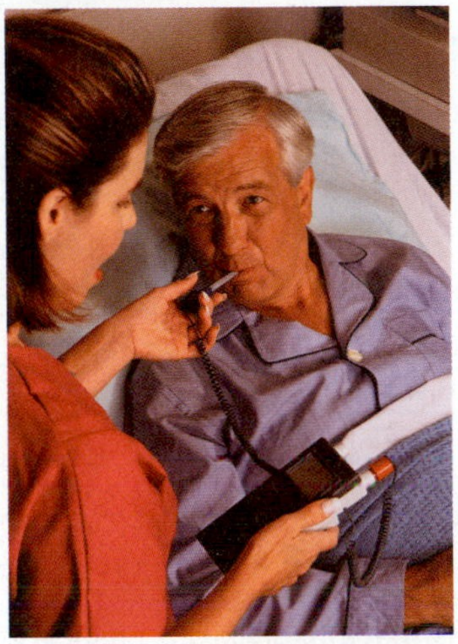

11. The number of Calories for six different chicken sandwiches and chicken salads is given in the table.

a. What is the mean number of Calories for a chicken sandwich? Round to the nearest whole unit.

b. What is the mean number of Calories for a salad with chicken? Round to the nearest whole unit.

c. What is the difference in the means?

Chicken Sandwiches	Salads with Chicken
360	310
370	325
380	350
400	390
400	440
470	500

12. The heights of the players from two NBA teams are given in the table. All heights are in inches.

a. Find the mean height for the players on the Philadelphia 76ers.

b. Find the mean height for the players on the Milwaukee Bucks.

c. What is the difference in the mean heights?

Philadelphia 76ers' Height (in.)	Milwaukee Bucks' Height (in.)
83	70
83	83
72	82
79	72
77	82
84	85
75	75
76	75
82	78
79	77

13. Zach received the following scores for his first four tests: 98%, 80%, 78%, 90%.

a. Find Zach's mean test score.

b. Zach got a 59% on his fifth test. Find the mean of all five tests.

c. How did the low score of 59% affect the overall mean of five tests?

14. The prices of four steam irons are \$50, \$30, \$25, and \$45.

a. Find the mean of these prices.

b. An iron that costs \$140 is added to the list. What is the mean of all five irons?

c. How does the expensive iron affect the mean?

Objective 2: Median

For Exercises 15–20, find the median for each set of numbers. **(See Examples 2–3.)**

15. 16, 14, 22, 13, 20, 19, 17

16. 32, 35, 22, 36, 30, 31, 38

17. 109, 118, 111, 110, 123, 100

18. 134, 132, 120, 135, 140, 118

19. −58, −55, −50, −40, −40, −55

20. −82, −90, −99, −82, −88, −87

21. The infant mortality rates for five countries are given in the table. Find the median.

Country	Infant Mortality Rate (Deaths per 1000)
Sweden	3.93
Japan	4.10
Finland	3.82
Andorra	4.09
Singapore	3.87

22. The snowfall amounts for five winter months in Burlington, Vermont, are given in the table. Find the median.

Month	Snowfall (in.)
November	6.6
December	18.1
January	18.8
February	16.8
March	12.4

23. Jonas Slackman played eight golf tournaments, each with 72-holes of golf. His scores for the tournaments are given. Find the median score.

−3, −5, 1, 4, −8, 2, 8, −1

24. Andrew Strauss recorded the daily low temperature (in °C) at his home in Virginia for 8 days in January. Find the median temperature.

5, 6, −5, 1, −4, −11, −8, −5

25. The number of passengers (in millions) on nine leading airlines for a recent year is listed. Find the median number of passengers. (*Source:* International Airline Transport Association)

48.3, 42.4, 91.6, 86.8, 46.5, 71.2, 45.4, 56.4, 51.7

26. For a recent year the number of albums sold (in millions) is listed for the 10 best sellers. Find the median number of albums sold.

2.7, 3.0, 4.8, 7.4, 3.4, 2.6, 3.0, 3.0, 3.9, 3.2

Objective 3: Mode

For Exercises 27–32, find the mode(s) for each set of numbers. **(See Examples 4–5.)**

27. 4, 5, 3, 8, 4, 9, 4, 2, 1, 4

28. 12, 14, 13, 17, 19, 18, 19, 17, 17

29. −28, −21, −24, −23, −24, −30, −21

30. −45, −42, −40, −41, −49, −49, −42

31. 90, 89, 91, 77, 88

32. 132, 253, 553, 255, 552, 234

33. The table gives the price of seven "smart" cell phones. Find the mode.

Brand and Model	Price ($)
Samsung	600
Kyocera	400
Sony Ericsson	800
PalmOne	450
Motorola	300
Siemens	600

34. The table gives the number of hazardous waste sites for selected states. Find the mode.

State	Number of Sites
Florida	51
New Jersey	112
Michigan	67
Wisconsin	39
California	96
Pennsylvania	94
Illinois	39
New York	90

35. The unemployment rates for nine countries are given. Find the mode. **(See Example 6.)**

6.3%, 7.0%, 5.8%, 9.1%, 5.2%, 8.8%, 8.4%, 5.8%, 5.2%

36. The list gives the number of children who were absent from class for an 11-day period. Find the mode.

4, 1, 6, 2, 4, 4, 4, 2, 2, 3, 2

Mixed Exercises

37. Six test scores for Jonathan's history class are listed. Find the mean and median. Round to the nearest tenth if necessary. Did the mean or median give a better overall score for Jonathan's performance?

92%, 98%, 43%, 98%, 97%, 85%

38. Nora's math test results are listed. Find the mean and median. Round to the nearest tenth if necessary. Did the mean or median give a better overall score for Nora's performance?

52%, 85%, 89%, 90%, 83%, 89%

39. Listed below are monthly costs for seven health insurance companies for a self-employed person, 55 years of age, and in good health. Find the mean, median, and mode (if one exists). Round to the nearest dollar. (*Source:* eHealth Insurance Company, 2007)

\$312, \$225, \$221, \$256, \$308, \$280, \$147

40. The salaries for seven Associate Professors at the University of Michigan are listed. These are salaries for 9-month contracts for a recent year. Find the mean, median, and mode (if one exists). Round to the nearest dollar. (*Source:* University of Michigan, University Library Volume 2006, Issue 1)

\$104,000, \$107,000, \$67,750, \$82,500, \$73,500, \$88,300, \$104,000

41. The prices of 10 single-family, three-bedroom homes for sale in Santa Rosa, California, are listed for a recent year. Find the mean, median, and mode (if one exists).

\$850,000, \$835,000, \$839,000, \$829,000, \$850,000, \$850,000, \$850,000, \$847,000, \$1,850,000, \$825,000

42. The prices of 10 single-family, three-bedroom homes for sale in Boston, Massachusetts, are listed for a recent year. Find the mean, median, and mode (if one exists).

\$300,000, \$2,495,000, \$2,120,000, \$220,000, \$194,000, \$391,000, \$315,000, \$330,000, \$435,000, \$250,000

Objective 4: Weighted Mean

For Exercises 43–46, use the following numerical values assigned to grades to compute GPA. Round each GPA to the hundredths place. **(See Example 7.)**

A = 4.0	B+ = 3.5	B = 3.0	C+ = 2.5
C = 2.0	D+ = 1.5	D = 1.0	F = 0.0

43. Compute the GPA for the following grades. Round to the nearest hundredth.

Course	Grade	Number of Credit-Hours (Weights)
Intermediate Algebra	B	4
Theater	C	1
Music Appreciation	A	3
World History	D	5

44. Compute the GPA for the following grades. Round to the nearest hundredth.

Course	Grade	Number of Credit-Hours (Weights)
General Psychology	B+	3
Beginning Algebra	A	4
Student Success	A	1
Freshman English	B	3

45. Compute the GPA for the following grades. Round to the nearest hundredth.

Course	Grade	Number of Credit-Hours (Weights)
Business Calculus	B+	3
Biology	C	4
Library Research	F	1
American Literature	A	3

46. Compute the GPA for the following grades. Round to the nearest hundredth.

Course	Grade	Number of Credit-Hours (Weights)
University Physics	C+	5
Calculus I	A	4
Computer Programming	D	3
Swimming	A	1

47. Refer to the table given in Example 7 on page 349. Replace the grade of "D" in English I with a grade of "B" and compute the GPA. How did Elmer's GPA differ with a better grade in the 4-hr English class?

Expanding Your Skills

48. There are 20 students enrolled in a 12th-grade math class. The graph displays the number of students by age. First complete the table, and then find the mean. **(See Example 7.)**

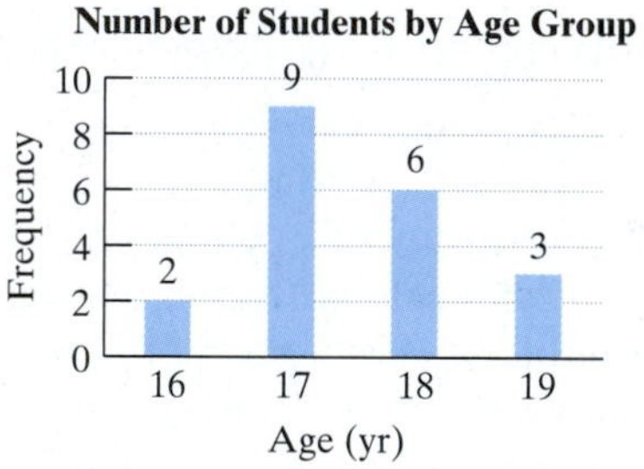

Age (yr)	Number of Students	Product
16		
17		
18		
19		
Total:		

49. A survey was made in a neighborhood of 37 houses. The graph represents the number of residents who live in each house. Complete the table and determine the mean number of residents per house.

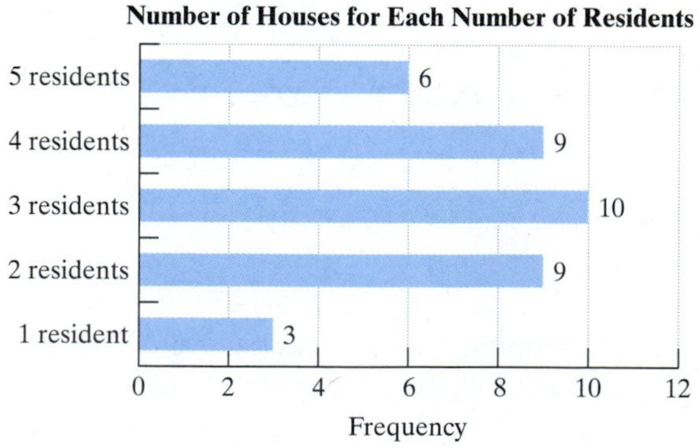

Number of Residents in Each House	Number of Houses	Product
1		
2		
3		
4		
5		
Total:		

Group Activity

Purchasing from a Catalog

Materials: Catalog (such as Amazon, Sears, JC Penney, or the like) or a computer to access an online catalog

Estimated time: 15 min

Group Size: 4

1. Each person in the group will choose an item to add to the list of purchases. All members of the group will keep a list of all items ordered and the prices.

ORDER SHEET

Item	Price Each	Quantity	Total
		SUBTOTAL	
		SHIPPING	
		TOTAL	

2. When the order list is complete, each member will find the total cost of the items ordered (subtotal) and compare the answer with the other members of the group. When the correct subtotal has been determined, find the cost of shipping from the catalog. (This is usually found on the order page of the catalog.) Now find the total cost of this order.

Chapter 5 Summary

Section 5.1 Decimal Notation and Rounding

Key Concepts

A **decimal fraction** is a fraction whose denominator is a power of 10.

Example 1

$\frac{7}{10}$, $\frac{31}{100}$, and $\frac{191}{1000}$ are decimal fractions.

Identify the place values of a decimal number.

1	2	3	4	.	5	6	7	8
thousands	hundreds	tens	ones	decimal point	tenths	hundredths	thousandths	ten-thousandths

Example 2

In the number 34.914, the 1 is in the hundredths place.

Reading a Decimal Number

1. The part of the number to the left of the decimal point is read as a whole number. *Note:* If there is not a whole-number part, skip to step 3.
2. The decimal point is read *and*.
3. The part of the number to the right of the decimal point is read as a whole number but is followed by the name of the place position of the digit farthest to the right.

Example 3

23.089 reads "twenty-three and eighty-nine thousandths."

Converting a Decimal to a Mixed Number or Proper Fraction

1. The digits to the right of the decimal point are written as the numerator of the fraction.
2. The place value of the digit farthest to the right of the decimal point determines the denominator.
3. The whole-number part of the number is left unchanged.
4. Once the number is converted to a fraction or mixed number, simplify the fraction to lowest terms, if possible.

Example 4

$$4.2 = 4\frac{\overset{1}{\cancel{2}}}{\underset{5}{\cancel{10}}} = 4\frac{1}{5}$$

Writing a Decimal Number Greater Than 1 as an Improper Fraction

1. The denominator is determined by the place position of the digit farthest to the right of the decimal point.
2. The numerator is obtained by removing the decimal point of the original number. The resulting whole number is then written over the denominator.
3. Simplify the improper fraction to lowest terms, if possible.

Example 5

$$-5.24 = -\frac{\overset{131}{\cancel{524}}}{\underset{25}{\cancel{100}}} = -\frac{131}{25}$$

Comparing Two Decimal Numbers

1. Starting at the left (and moving toward the right), compare the digits in each corresponding place position.
2. As we move from left to right, the first instance in which the digits differ determines the order of the numbers. The number having the greater digit is greater overall.

Example 6

$3.024 > 3.019$ because

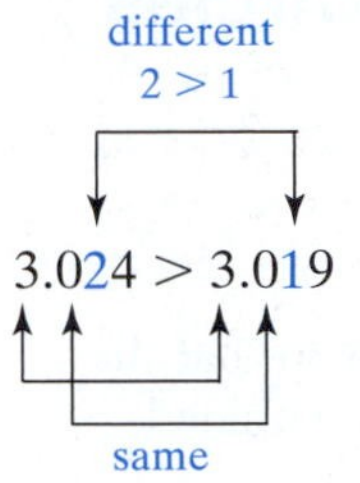

Rounding Decimals to a Place Value to the Right of the Decimal Point

1. Identify the digit one position to the right of the given place value.
2. If the digit in step 1 is 5 or greater, add 1 to the digit in the given place value. Then discard the digits to its right.
3. If the digit in step 1 is less than 5, discard it and any digits to its right.

Example 7

Round -4.8935 to the nearest hundredth.

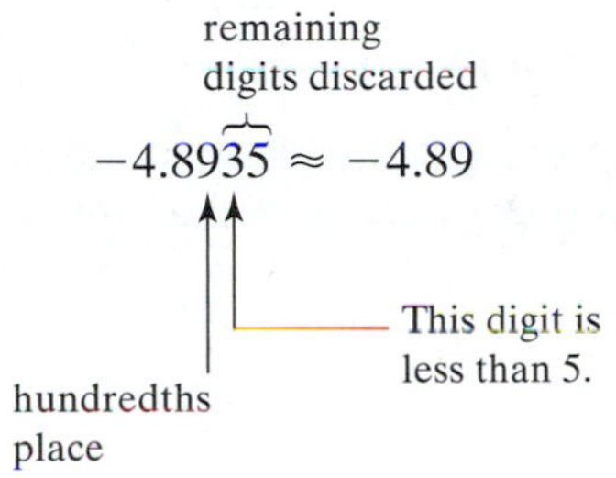

Section 5.2 Addition and Subtraction of Decimals

Key Concepts

Adding Decimals

1. Write the addends in a column with the decimal points and corresponding place values lined up.
2. Add the digits in columns from right to left as you would whole numbers. The decimal point in the answer should be lined up with the decimal points from the addends.

Examples

Example 1

Add $6.92 + 12 + 0.001$.

$$\begin{array}{r} 6.920 \\ 12.000 \\ +\ 0.001 \\ \hline 18.921 \end{array}$$

Add zeros to the right of the decimal point as placeholders.

Check by estimating:

6.92 rounds to 7 and 0.001 rounds to 0.

$7 + 12 + 0 = 19$, which is close to 18.921.

Subtracting Decimals

1. Write the numbers in a column with the decimal points and corresponding place values lined up.
2. Subtract the digits in columns from right to left as you would whole numbers. The decimal point in the answer should be lined up with the other decimal points.

Example 2

Subtract $41.03 - 32.4$.

$$\begin{array}{r} \scriptstyle 3\ \ \overset{10}{\not 0}\ \ 10 \\ \not 4\ \not 1\ .\ \not 0\ 3 \\ -\ 3\ 2\ .\ 4\ 0 \\ \hline 8\ .\ 6\ 3 \end{array}$$

Check by estimating:

41.03 rounds to 41 and 32.40 rounds to 32.

$41 - 32 = 9$, which is close to 8.63.

Section 5.3 Multiplication of Decimals and Applications with Circles

Key Concepts

Multiplying Two Decimals

1. Multiply as you would integers.
2. Place the decimal point in the product so that the number of decimal places equals the combined number of decimal places of both factors.

Multiplying a Decimal by Powers of 10

Move the decimal point to the right the same number of decimal places as the number of zeros in the power of 10.

Multiplying a Decimal by Powers of 0.1

Move the decimal point to the left the same number of places as there are decimal places in the power of 0.1.

Radius and Diameter of a Circle

$d = 2r$ and $r = \frac{1}{2}d$

Circumference of a Circle

$C = \pi d$ or $C = 2\pi r$

Area of a Circle

$A = \pi r^2$

Examples

Example 1

Multiply. 5.02×2.8

$\overset{1}{5.02}$	2 decimal places
$\times\ 2.8$	+ 1 decimal place
4016	
10040	
14.056	3 decimal places

Example 2

$-83.251 \times 100 = -8325.1$

Move 2 places to the right.

Example 3

$-149.02 \times (-0.001) = 0.14902$

Move 3 places to the left.

Example 4

Find the radius, circumference, and area. Use 3.14 for π.

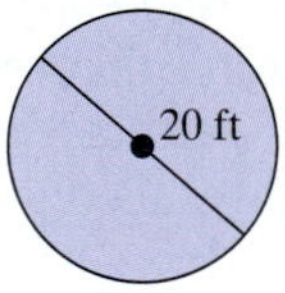

$r = \frac{20\text{ ft}}{2} = 10\text{ ft}$

$C = 2\pi r$

$= 2\pi(10\text{ ft})$

$= 20\pi\text{ ft}$ (exact circumference)

$\approx 20(3.14)\text{ ft}$

$= 62.8\text{ ft}$ (approximate value)

$A = \pi r^2$

$= \pi(10\text{ ft})^2$

$= 100\pi\text{ ft}^2$ (exact area)

$\approx 100(3.14)\text{ ft}^2$

$= 314\text{ ft}^2$ (approximate value)

Section 5.4 Division of Decimals

Key Concepts

Dividing a Decimal by a Whole Number

1. Place the decimal point in the quotient directly above the decimal point in the dividend.
2. Divide as you would whole numbers.

Examples

Example 1

62.6 ÷ 4

```
   15.65
4)62.60
 -4
  22
 -20
   26
  -24
    20
   -20
     0
```

Dividing When the Divisor Is Not a Whole Number

1. Move the decimal point in the divisor to the right to make it a whole number.
2. Move the decimal point in the dividend to the right the same number of places as in step 1.
3. Place the decimal point in the quotient directly above the decimal point in the dividend.
4. Divide as you would whole numbers. Then apply the correct sign to the quotient.

Example 2

81.1 ÷ 0.9 .9)81.1

```
   90.11...
9 ) 811.00
   -81
     01
     00
      10   <-- The pattern repeats.
      -9
      10
```

The answer is the repeating decimal $90.\overline{1}$.

To round a repeating decimal, be sure to expand the repeating digits to one digit beyond the indicated rounding place.

Example 3

Round $6.\overline{56}$ to the thousandths place.

6.5656

thousandths place

The digit 6 > 5 so increase the thousandths-place digit by 1.

$6.\overline{56} \approx 6.566$

Section 5.5 Fractions, Decimals, and the Order of Operations

Key Concepts

To write a fraction as a decimal, divide the numerator by the denominator.

These are some common fractions represented by decimals.

$\frac{1}{4} = 0.25 \quad \frac{1}{2} = 0.5 \quad \frac{3}{4} = 0.75$

$\frac{1}{9} = 0.\overline{1} \quad \frac{2}{9} = 0.\overline{2} \quad \frac{1}{3} = 0.\overline{3}$

$\frac{4}{9} = 0.\overline{4} \quad \frac{5}{9} = 0.\overline{5} \quad \frac{2}{3} = 0.\overline{6}$

$\frac{7}{9} = 0.\overline{7} \quad \frac{8}{9} = 0.\overline{8}$

To write a decimal as a fraction, first write the number as a decimal fraction and reduce.

A rational number can be expressed as a terminating or repeating decimal.

An **irrational number** cannot be expressed as a terminating or repeating decimal.

The rational numbers and the irrational numbers together make up the set of **real numbers.**

To rank decimals from least to greatest, compare corresponding digits from left to right.

$0.34501 < 0.34510$

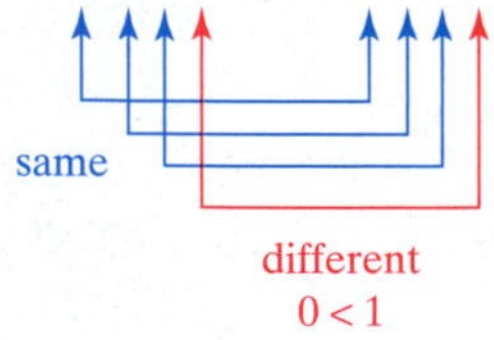

Examples

Example 1

$\frac{17}{20} = 0.85$

```
      .85
 20)17.00
   -160
    100
   -100
      0
```

Example 2

$\frac{14}{3} = 4.\overline{6}$

```
     4.66. . .
 3 ) 14.00
    -12
      20
     -18
      20
```

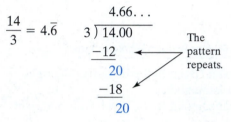

The pattern repeats.

Example 3

$-6.84 = -\frac{\overset{171}{\cancel{684}}}{\underset{25}{\cancel{100}}} = -\frac{171}{25} \quad \text{or} \quad -6\frac{21}{25}$

Example 4

Rational: $-6, \frac{3}{4}, 3.78, -2.\overline{35}$

Irrational: $\sqrt{2} \approx 1.41421356237\ldots$

$\pi \approx 3.141592654\ldots$

Example 5

Plot the decimals 2.6, $2.\overline{6}$, and 2.58 on a number line.

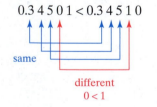

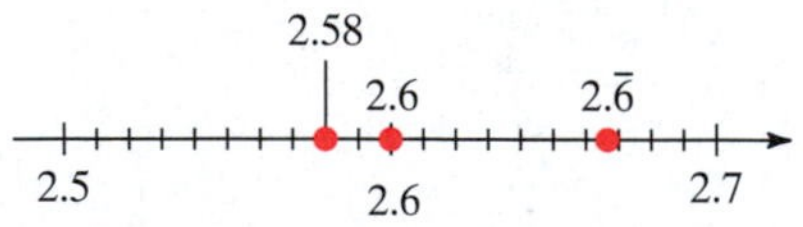

Examples 6 and 7 apply the order of operations with fractions and decimals. There are two approaches for simplifying.

Option 1: Write the expressions as decimals.

Example 6

$$\left(1.6 - \frac{13}{25}\right) \div 8 = (1.6 - 0.52) \div 8$$
$$= (1.08) \div 8$$
$$= 0.135$$

Option 2: Write the expressions as fractions.

Example 7

$$\frac{2}{3}\left(2.2 + \frac{7}{5}\right) = \frac{2}{3}\left(\frac{22}{10} + \frac{7}{5}\right)$$
$$= \frac{2}{3}\left(\frac{11}{5} + \frac{7}{5}\right)$$
$$= \frac{\overset{}{2}}{\underset{1}{\cancel{3}}}\left(\frac{\overset{6}{\cancel{18}}}{5}\right) = \frac{12}{5} = 2.4$$

Section 5.6 Solving Equations Containing Decimals

Key Concepts

We solve equations containing decimals by using the addition, subtraction, multiplication, and division properties of equality.

Examples

Example 1

Solve. $6.24 = -2(10.53 - 2.1x)$

$$6.24 = -21.06 + 4.2x$$
$$6.24 + 21.06 = -21.06 + 21.06 + 4.2x$$
$$27.3 = 4.2x$$
$$\frac{27.3}{4.2} = \frac{4.2x}{4.2}$$
$$6.5 = x$$

The solution 6.5 checks in the original equation.

We can also solve decimal equations by first clearing decimals. Do this by multiplying both sides of the equation by a power of 10 (10, 100, 1000, etc.).

Example 2

Solve by clearing decimals.

$$0.02x + 1.3 = -0.025$$
$$1000(0.02x + 1.3) = 1000(-0.025)$$
$$1000(0.02x) + 1000(1.3) = 1000(-0.025)$$
$$20x + 1300 = -25$$
$$20x + 1300 - 1300 = -25 - 1300$$
$$20x = -1325$$
$$x = -66.25$$

The solution −66.25 checks in the original equation.

Section 5.7 Mean, Median, and Mode

Key Concepts

The **mean** (or average) of a set of numbers is the sum of the values divided by the number of values.

$$\text{Mean} = \frac{\text{sum of the values}}{\text{number of values}}$$

The **median** is the "middle" number in an ordered list of numbers. For an ordered list of numbers:

- If the number of data values is *odd*, then the median is the middle number in the list.
- If the number of data values is *even*, the median is the mean of the two middle numbers in the list.

The **mode** of a set of data is the value or values that occur most often.

When data values in a list appear multiple times, we can compute a **weighted mean**.

Examples

Example 1

Find the mean test score: 92, 100, 86, 60, 90

$$\text{Mean} = \frac{92 + 100 + 86 + 60 + 90}{5}$$

$$= \frac{428}{5} = 85.6$$

Example 2

Find the median: 12 18 6 10 5

First order the list: 5 6 10 12 18

The median is the middle number, 10.

Example 3

Find the median: 15 20 20 32 40 45

The median is the average of 20 and 32:

$$\frac{20 + 32}{2} = \frac{52}{2} = 26$$ The median is 26.

Example 4

Find the mode: 7 2 5 7 7 4 6 10

The value 7 is the mode because it occurs most often.

Example 5

Compute the GPA for the following grades. Round to the tenths place.

A = 4 pts B = 3 pts C = 2 pts
D = 1 pts F = 0 pts

Course	Grade	Credit-Hours	Product
Math	C (2 pts)	4	8 pts
English	A (4 pts)	1	4 pts
Anatomy	B (3 pts)	3	9 pts
	Total:	**8**	**21 pts**

$$\text{Mean} = \frac{21}{8} \approx 2.6$$

The GPA is 2.6.

Chapter 5 Review Exercises

Section 5.1

1. Identify the place value for each digit in the number 32.16.

2. Identify the place value for each digit in the number 2.079.

For Exercises 3–6, write the word name for the decimal.

3. 5.7

4. 10.21

5. −51.008

6. −109.01

For Exercises 7–8, write the word name as a numeral.

7. Thirty-three thousand, fifteen and forty-seven thousandths.

8. Negative one hundred and one hundredth.

For Exercises 9–10, write the decimal as a proper fraction or mixed number.

9. −4.8

10. 0.025

For Exercises 11–12, write the decimal as an improper fraction.

11. 1.3

12. 6.75

For Exercises 13–14, fill in the blank with either < or >.

13. −15.032 ☐ −15.03

14. 7.209 ☐ 7.22

15. The earned run average (ERA) for five members of the American League for a recent season is given in the table. Rank the averages from least to greatest.

Player	ERA
Jon Garland	4.5142
Cliff Lee	4.3953
Gil Meche	4.4839
Jamie Moyer	4.3875
Vicente Padilla	4.5000

For Exercises 16–17, round the decimal to the indicated place value.

16. 89.9245; hundredths

17. 34.8895; thousandths

18. A quality control manager tests the amount of cereal in several brands of breakfast cereal against the amount advertised on the box. She selects one box at random. She measures the contents of one 12.5-oz box and finds that the box has 12.46 oz.

a. Is the amount in the box less than or greater than the advertised amount?

b. If the quality control manager rounds the measured value to the tenths place, what is the value?

19. Which number is equal to 571.24? Circle all that apply.

a. 571.240
b. 571.2400
c. 571.024
d. 571.0024

20. Which number is equal to 3.709? Circle all that apply.

a. 3.7
b. 3.7090
c. 3.709000
d. 3.907

Section 5.2

For Exercises 21–28, add or subtract as indicated.

21. 45.03 + 4.713

22. 239.3 + 33.92

23. 34.89 − 29.44

24. 5.002 − 3.1

25. −221 − 23.04

26. 34 + (−4.993)

27. 17.3 + 3.109 − 12.6

28. 189.22 − (−13.1) − 120.055

For Exercises 29–30, simplify by combining *like* terms.

29. $-5.1y - 4.6y + 10.2y$

30. $-2(12.5x - 3) + 11.5x$

31. The closing prices for a mutual fund are given in the graph for a 5-day period.

a. Determine the difference in price between the two consecutive days for which the price increased the most.

b. Determine the difference in price between the two consecutive days for which the price decreased the most.

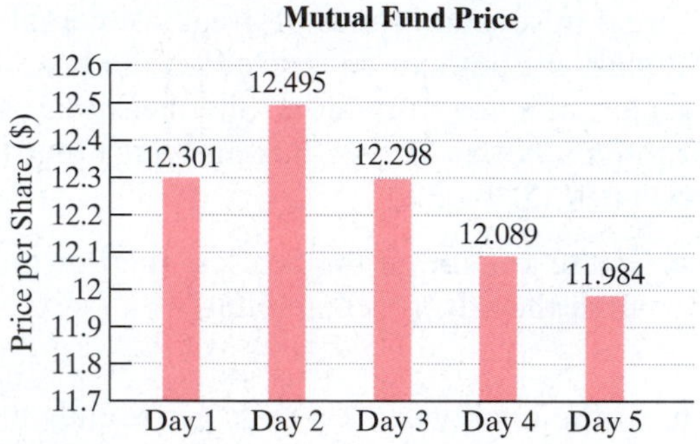

Section 5.3

For Exercises 32–39, multiply the decimals.

32. $\begin{array}{r} 3.9 \\ \times\ 2.1 \\ \hline \end{array}$

33. $\begin{array}{r} 57.01 \\ \times\ 1.3 \\ \hline \end{array}$

34. $(60.1)(-4.4)$

35. $(-7.7)(45)$

36. 85.49×1000

37. 1.0034×100

38. $-92.01 \times (-0.01)$

39. $-104.22 \times (-0.01)$

For Exercises 40–41, write the decimal number representing each word name.

40. The population of Guadeloupe is approximately 4.32 hundred-thousand.

41. A recent season premier of *American Idol* had 33.8 million viewers.

42. A store advertises a package of two 9-volt batteries for sale at $3.99.

a. What is the cost of buying 8 batteries?

b. If another store has an 8-pack for the regular price of $17.99, how much can a customer save by buying batteries at the sale price?

43. If long-distance phone calls cost $0.07 per minute, how much will a 23-min long-distance call cost?

44. Find the area and perimeter of the rectangle.

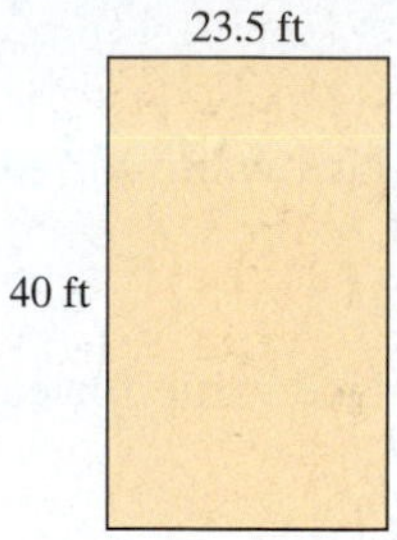

45. Population density gives the approximate number of people per square mile. The population density for Texas is given in the graph for selected years.

a. Approximately how many people would have been located in a 200-mi^2 area in 1960?

b. Approximately how many people would have been located in a 200-mi^2 area in 2006?

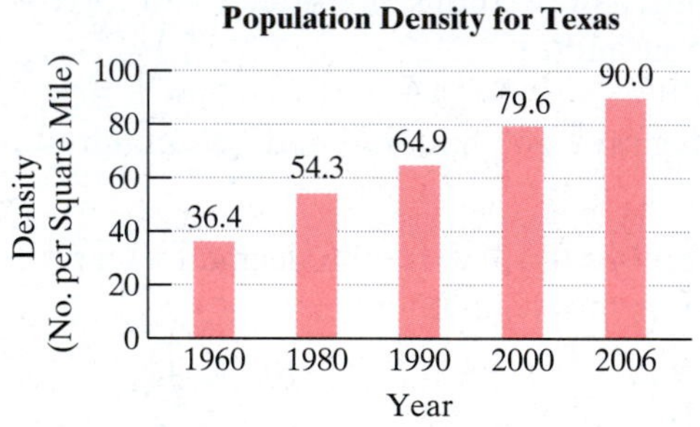

Source: U.S. Census Bureau

46. Determine the diameter of a circle with radius 3.75 m.

47. Determine the radius of a circle with diameter 27.2 ft.

48. Find the area and circumference of a circular fountain with diameter 24 ft. Use 3.14 for π.

49. Find the area and circumference of a circular garden with radius 30 yd. Use 3.14 for π.

Section 5.4

For Exercises 50–59, divide. Write the answer in decimal form.

50. $8.55 \div 0.5$

51. $64.2 \div 1.5$

52. $0.06\overline{)0.248}$

53. $0.3\overline{)2.63}$

54. $-18.9 \div 0.7$

55. $-0.036 \div 1.2$

56. $493.93 \div 100$

57. $90.234 \div 10$

58. $-553.8 \div (-0.001)$

59. $-2.6 \div (-0.01)$

60. For each number, round to the indicated place.

	$8.\overline{6}$	$52.\overline{52}$	$0.\overline{409}$
Tenths			
Hundredths			
Thousandths			
Ten-thousandths			

For Exercises 61–62, divide and round the answer to the nearest hundredth.

61. $104.6 \div (-9)$

62. $71.8 \div (-6)$

63. **a.** A generic package of toilet paper costs \$5.99 for 12 rolls. What is the cost per roll? (Round the answer to the nearest cent, that is, the nearest hundredth of a dollar.)

b. A package of four rolls costs \$2.29. What is the cost per roll?

c. Which of the two packages offers the better buy?

Section 5.5

For Exercises 64–67, write the fraction or mixed number as a decimal.

64. $2\frac{2}{5}$

65. $3\frac{13}{25}$

66. $-\frac{24}{125}$

67. $-\frac{7}{16}$

For Exercises 68–70, write the fraction as a repeating decimal.

68. $\frac{7}{12}$

69. $\frac{55}{36}$

70. $-4\frac{7}{22}$

For Exercises 71–73, write the fraction as a decimal rounded to the indicated place value.

71. $\frac{5}{17}$; hundredths

72. $\frac{20}{23}$; tenths

73. $-\frac{11}{3}$; thousandths

74. Identify the numbers as rational or irrational.

a. $6.\overline{4}$

b. π

c. $\sqrt{10}$

d. $-\frac{8}{5}$

For Exercises 75–76, write a fraction or mixed number for the repeating decimal.

75. $0.\overline{2}$

76. $3.\overline{3}$

77. Complete the table, giving the closing value of stocks as reported in the *Wall Street Journal*.

Stock	Closing Price (\$) (Decimal)	Closing Price (\$) (Fraction)
Sun	5.20	
Sony	55.53	
Verizon		$41\frac{4}{25}$

For Exercises 78–79, insert the appropriate symbol. Choose from <, >, or =.

78. $1\frac{1}{3} \;\square\; 1.33$

79. $-0.14 \;\square\; -\frac{1}{7}$

For Exercises 80–85, perform the indicated operations. Write the answers in decimal form.

80. $-7.5 \div \frac{3}{2}$

81. $\frac{1}{2}(4.6)(-2.4)$

82. $(-5.46 - 2.24)^2 - 0.29$

83. $[-3.46 - (-2.16)]^2 - 0.09$

84. $\left(\frac{1}{4}\right)^2\left(\frac{4}{5}\right) - 3.05$

85. $-1.25 - \frac{1}{3} \cdot \left(\frac{3}{2}\right)^2$

86. An audio Spanish course is available online at one website for the following prices. How much money is saved by buying the combo package versus the three levels individually?

Level	Price
Spanish I	\$189.95
Spanish II	199.95
Spanish III	219.95
Combo (Spanish I, II, III combined)	519.95

87. Marvin drives the route shown in the figure each day, making deliveries. He completes one-third of the route before lunch. How many more miles does he still have to drive after lunch?

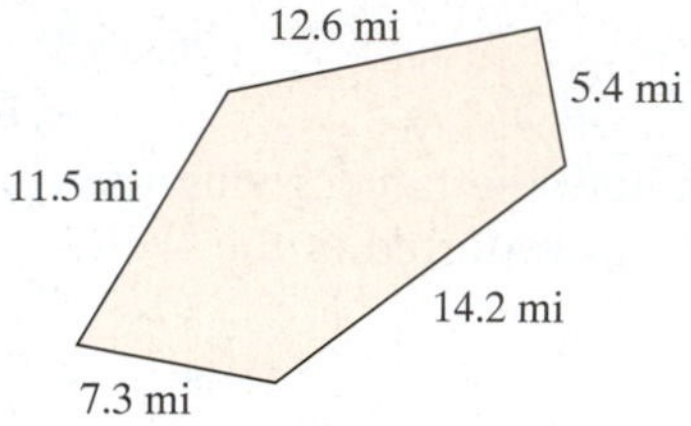

Section 5.6

For Exercises 88–96, solve the equation.

88. $x + 4.78 = 2.2$

89. $6.2 + y = 4.67$

90. $11 = -10.4 + w$

91. $-20.4 = -9.08 + z$

92. $4.6 = 16.7 + 2.2m$

93. $21.8 = 12.05 + 3.9x$

94. $-0.2(x - 6) = 0.3x + 3.8$

95. $-0.5(9 + y) = -0.6y + 6.5$

96. $0.04z - 3.5 = 0.06z + 4.8$

For Exercises 97–98, determine a number that can be used to clear decimals in the equation. Then solve the equation.

97. $2p + 3.1 = 0.14$

98. $2.5w - 6 = 0.9$

99. Four times a number is equal to 19.5 more than the number. Find the number.

100. The sum of a number and 4.8 is three times the number. Find the number.

101. The product of a number and 0.05 is equal to 80. Find the number.

102. The quotient of a number and 0.05 is equal to 80. Find the number.

103. The perimeter of a triangle is 24.8 yd. The longest side is twice the shortest side. The middle side is 2.4 yd longer than the shortest side. Find the lengths of the sides.

104. Deanna rented a video game and a DVD for $12.98. DVDs cost $1 less to rent than video games. Find the cost to rent each item.

105. The U-Store-It Company rents a 10 ft by 12 ft storage space for $59, plus $89 per month. Mykeshia wrote a check for $593. How many months of storage will this cover?

Section 5.7

106. For the list of quiz scores, find the mean, median, and mode(s).

20, 20, 18, 16, 18, 17, 16, 10, 20, 20, 15, 20

107. Juanita kept track of the number of milligrams of calcium she took each day through vitamins and dairy products. Determine the mean, median, and mode for the amount of calcium she took per day. Round to the nearest 10 mg if necessary.

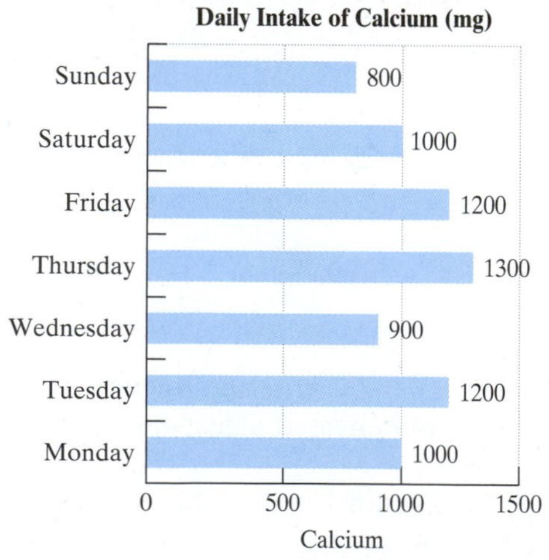

108. The seating capacity for five arenas used by the NBA is given in the table. Find the median number of seats.

Arena	Number of Seats
Phelps Arena, Atlanta	20,000
Fleet Center, Boston	18,624
Chevrolette Coliseum, Charlotte	23,799
United Center, Chicago	21,500
Gund Arena, Cleveland	20,562

109. The manager of a restaurant had his customers fill out evaluations on the service that they received. A scale of 1 to 5 was used, where 1 represents very poor service and 5 represents excellent service. Given the list of responses, determine the mode(s).

4 5 3 4 4 3 2 5 5 1 4 3 4 4 5
2 5 4 4 3 2 5 5 1 4

110. Compute the GPA for the following grades.

A = 4 pts B = 3 pts C = 2 pts
D = 1 pts F = 0 pts

Course	Grade	Credit-Hours
History	B	3
Reading	D	2
Biology	A	4

Chapter 5 Test

1. Identify the place value of the underlined digit.

a. $2\underline{3}4.17$ **b.** $234.1\underline{7}$

2. Write the word name for -509.024.

3. Write the decimal 1.26 as a mixed number and as a fraction.

4. The field goal percentages for a recent basketball season are given in the table for four NBA teams. Rank the percentages from least to greatest.

Team	Percentage
LA Lakers	0.4419
Cleveland	0.4489
San Antonio	0.4495
Utah	0.4484

5. Which statement is correct?

a. $0.043 > 0.430$ **b.** $-0.692 > -0.926$

c. $0.078 < 0.0780$

For Exercises 6–17, perform the indicated operation.

6. $-49.002 + 3.83$ **7.** $-34.09 - 12.8$

8. $28.1 \times (-4.5)$ **9.** $25.4 \div (-5)$

10. $4 - 2.78$ **11.** $12.03 + 0.1943$

12. $39.82 \div 0.33$ **13.** 42.7×10.3

14. $-45.92 \times (-0.1)$ **15.** $-579.23 \times (-100)$

16. $80.12 \div 0.01$ **17.** $2.931 \div 1000$

For Exercises 18–19, simplify by combining like terms.

18. $2.72x - 1.96x + 4.9x$

19. $2(4.1y - 3) + 6.4y + 2.7$

20. Determine the diameter of a circle with radius 12.2 ft.

21. Determine the circumference and area. Use 3.14 for π and round to the nearest whole unit.

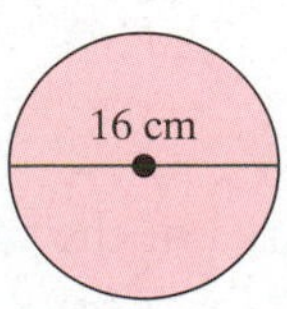

22. The temperature of a cake is recorded in 10-min intervals after it comes out of the oven. See the graph.

a. What was the difference in temperature between 10 and 20 min?

b. What was the difference in temperature between 40 and 50 min?

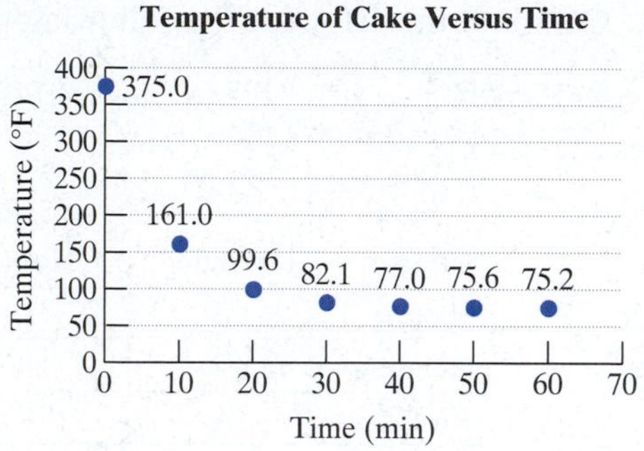

23. For a recent year, the United States consumed the most oil of any country in the world at 1.04 billion tons. China was second with 360 million tons.

a. Write a decimal number representing the amount of oil consumed by the United States.

b. Write a decimal number representing the amount of oil consumed by China.

c. What is the difference between the oil consumption in the United States and China?

24. A picture is framed and matted as shown in the figure.

a. Find the area of the picture itself.

b. Find the area of the matting *only*.

c. Find the area of the frame *only*.

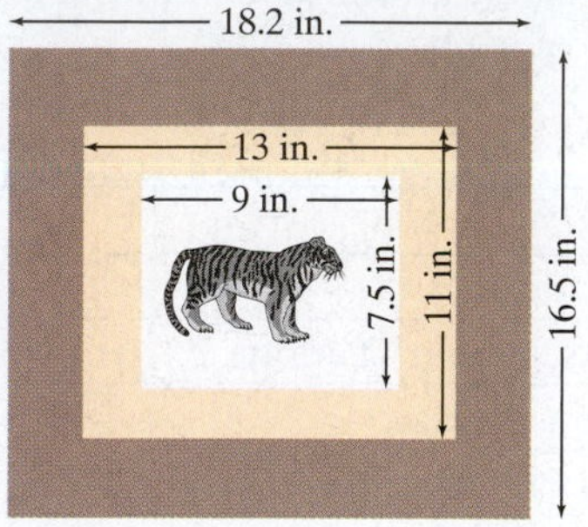

25. Jonas bought 200 shares of a stock for \$36.625 per share, and had to pay a commission of \$4.25. Two years later he sold all of his shares at a price of \$52.16 per share. If he paid \$8 for selling the stock, how much money did he make overall?

26. Dalia purchased a new refrigerator for \$1099.99. She paid \$200 as a down payment and will finance the rest over a 2-year period. Approximately how much will her monthly payment be?

27. Kent determines that his Ford Ranger pickup truck gets 23 mpg in the city and 26 mpg on the highway. If he drives 110.4 mi in the city and 135.2 mi on the highway how much gas will he use?

28. The table shows the winning times in seconds for a women's speed skating event for several years. Complete the table.

Year	Decimal	Fraction
1984		$41\frac{1}{50}$ sec
1988	39.10	
1992	40.33	
1994		$39\frac{1}{4}$

29. Rank the numbers and plot them on a number line.

$$-3\frac{1}{2}, -3.\overline{5}, -3.2$$

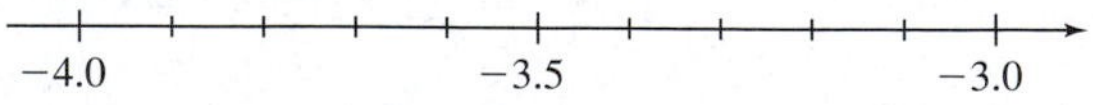

For Exercises 30–32, simplify.

30. $(8.7)\left(1.6 - \frac{1}{2}\right)$

31. $\frac{7}{3}\left(5.25 - \frac{3}{4}\right)^2$

32. $(0.2)^2 - \frac{5}{4}$

33. Identify the numbers as rational or irrational.

a. $2\frac{1}{8}$ **b.** $\sqrt{5}$ **c.** 6π **d.** $-0.\overline{3}$

For Exercises 34–39, solve the equation.

34. $0.006 = 0.014 + p$

35. $0.04y = 7.1$

36. $-97.6 = -4.3 - 5w$

37. $3.9 + 6.2x = 24.98$

38. $-0.08z + 0.5 = 0.09(4 - z) + 0.12$

39. $0.9 + 0.4t = 1.6 + 0.6(t - 3)$

40. The difference of a number and 43.4 is equal to eight times the number. Find the number.

41. The quotient of a number and 0.004 is 60. Find the number.

For Exercises 42–44, refer to the table. The table represents the heights of the Seven Summits (the highest peaks from each continent).

Mountain	Continent	Height (ft)
Mt. Kilimanjaro	Africa	19,340
Elbrus	Europe	18,510
Aconcagua	South America	22,834
Denali	North America	20,320
Vinson Massif	Antarctica	16,864
Mt. Kosciusko	Australia	7,310
Mt. Everest	Asia	29,035

42. What is the mean height of the Seven Summits? Round to the nearest whole unit.

43. What is the median height?

44. Is there a mode?

45. Mike and Darcy listed the amount of money paid for going to the movies for the past 3 months. This list represents the amount for 2 tickets. Find the mean, median, and mode.

\$11 \$14 \$11 \$16 \$15 \$16 \$12 \$16 \$15 \$20

46. David Millage runs almost every day. His distances for one week are given in the table.

a. Find the total distance that he ran.

b. Find the average distance for the 7-day period. Round to the nearest tenth of a mile.

Day	Distance (mi)
Monday	4.6
Tuesday	5.9
Wednesday	0
Thursday	8.4
Friday	2.5
Saturday	12.8
Sunday	4.6

47. Compute the GPA for the following grades. Round to the nearest hundredth. Use this scale:

A = 4.0	C = 2.0
B+ = 3.5	D+ = 1.5
B = 3.0	D = 1.0
C+ = 2.5	F = 0.0

Course	Grade	Number of Credit-Hours (Weights)
Art Appreciation	B	4
College Algebra	A	3
English II	C	3
Physical Fitness	A	1

Chapters 1–5 Cumulative Review Exercises

1. Simplify. $(17 + 12) - (8 - 3) \cdot 3$

2. Convert the number to standard form.
4 thousands + 3 tens + 9 ones

3. Add. $3902 + 34 + (-904)$

4. Subtract. $-4990 - 1118$

5. Multiply and round the answer to the thousands place. $23{,}444 \cdot 103$

6. Divide 4530 by 225. Then identify the dividend, divisor, whole-number part of the quotient, and remainder.

7. Explain how to check the division problem in Exercise 6.

8. The chart shows the retail sales for several companies. Find the difference between the greatest and least sales in the chart.

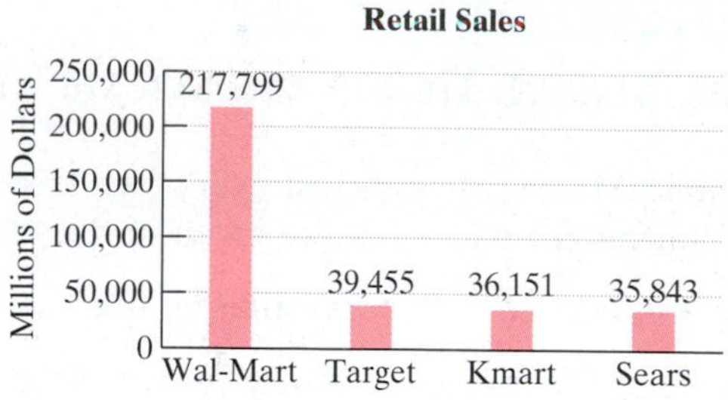

For Exercises 9–12, multiply or divide as indicated. Write the answer as a fraction.

9. $-\frac{1}{5} \cdot \left(-\frac{6}{11}\right)$

10. $\left(\frac{6}{15}\right)\left(\frac{10}{7}\right)$

11. $\left(-\frac{7}{10}\right)^2$

12. $\frac{8}{3} \div 4$

13. A settlement for a lawsuit is made for \$15,000. The attorney gets $\frac{2}{5}$ of the settlement. How much is left?

14. Simplify.

$$\frac{8}{25} + \frac{1}{5} \div \frac{5}{6} - \left(\frac{2}{5}\right)^2$$

For Exercises 15–16, add or subtract as indicated. Write the answers as fractions.

15. $\frac{5}{11} + 3$

16. $-\frac{12}{5} + \frac{9}{10}$

17. Find the area and perimeter.

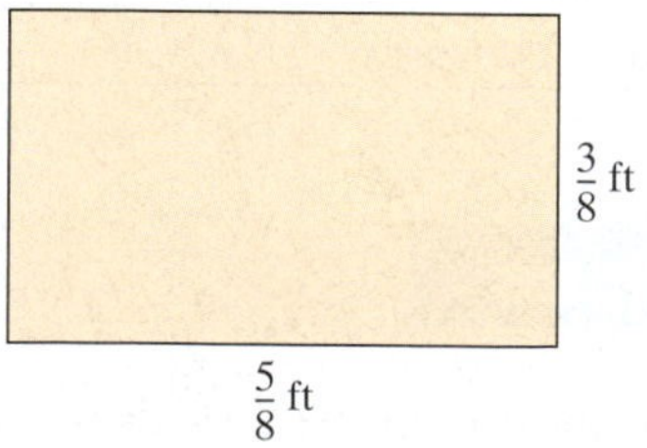

For Exercises 18–22, perform the indicated operations.

18. $50.9 + (-123.23)$

19. $700.8 - 32.01$

20. -301.1×0.25

21. $51.2 \div 3.2$

22. $\frac{4}{3}(3.14)(9)^2$

23. Divide $79.02 \div 1.7$ and round the answer to the nearest hundredth.

24. **a.** Multiply. $0.004(938.12)$

b. Multiply. $938.12(0.004)$

c. Identify the property that has been demonstrated in parts (a) and (b).

25. The table gives the average length of several bones in the human body. Complete the table by writing the mixed numbers as decimals and the decimals as mixed numbers.

Bone	Length (in.) (Decimal)	Length (in.) (Mixed Number)
Femur	19.875	
Fibula		$15\frac{15}{16}$
Humerus		$14\frac{3}{8}$
Innominate bone (hip)	7.5	

26. Simplify by combining *like* terms.

$-2.3x - 4.7 + 5.96x - 3.95$

27. Find the area and circumference of a circle with diameter 2 m. Use 3.14 for π.

For Exercises 28–30, solve the equations.

28. $-4.72 + 5x = -9.02$

29. $\frac{x}{3.9} = 4$

30. $2(x - 5) + 7 = 3(4 + x) - 9$

Ratio and Proportion

6

CHAPTER OUTLINE

Chapter 6

In this chapter, we study ratios, rates, and proportions. Unit rates are extremely important when comparing two or more items such as comparing prices at the grocery store. A proportion is an equation indicating that two rates or ratios are equal. Proportions come in handy in many applications such as converting currency (money) from one country to another.

Are You Prepared?

To prepare for this chapter, take a minute to review operations on fractions and decimals and solving equations. For each exercise on the left, find its solution on the right. Write the letter of the solution in the corresponding blank below. If you have trouble, see Sections 3.4 and 5.4.

1. What number would you divide by to isolate x in the equation $-18x = 6$?
2. Solve the equation. $4x = 28$
3. Solve the equation. $-6 = \frac{1}{2}x$
4. What is the LCD of the fractions $\frac{x}{6}$ and $\frac{5}{3}$?
5. Solve the equation. $\frac{x}{6} = \frac{5}{3}$
6. Divide. $2.36 \div 4$
7. Divide. 10 by 32
8. Divide. $24.58 \div 100$

E. 0.3125
V. 6
O. 10
P. -18
A. 0.2458
H. 7
Y. 0.59
T. -12

An improper fraction is ___ ___ ___   ___ ___ ___ ___ ___.
3 5 1   2 7 8 4 6

Section 6.1 Ratios

Objectives

1. Writing a Ratio
2. Writing Ratios of Mixed Numbers and Decimals
3. Applications of Ratios

1. Writing a Ratio

Thus far we have seen two interpretations of fractions.

- The fraction $\frac{5}{8}$ represents 5 parts of a whole that has been divided evenly into 8 pieces.
- The fraction $\frac{5}{8}$ represents $5 \div 8$.

Now we consider a third interpretation.

- The fraction $\frac{5}{8}$ represents the ratio of 5 to 8.

A **ratio** is a comparison of two quantities. There are three different ways to write a ratio.

Concept Connections

1. When forming the ratio $\frac{a}{b}$, why must b not equal zero?

DEFINITION Writing a Ratio

The ratio of a to b can be written as follows, provided $b \neq 0$.

1. a to b **2.** $a : b$ **3.** $\dfrac{a}{b}$

The colon means "to." The fraction bar means "to."

Although there are three ways to write a ratio, we primarily use the fraction form.

Skill Practice

2. For a recent flight from Atlanta to San Diego, 291 seats were occupied and 29 were unoccupied. Write the ratio of:
   a. The number of occupied seats to unoccupied seats
   b. The number of unoccupied seats to occupied seats
   c. The number of occupied seats to the total number of seats

Example 1 Writing a Ratio

In an algebra class there are 15 women and 17 men.

a. Write the ratio of women to men.

b. Write the ratio of men to women.

c. Write the ratio of women to the total number of people in the class.

Solution:

It is important to observe the *order* of the quantities mentioned in a ratio. The first quantity mentioned is the numerator. The second quantity is the denominator.

a. The ratio of women to men is $\dfrac{15}{17}$

b. The ratio of men to women is $\dfrac{17}{15}$

c. First find the total number of people in the class.

$$\text{Total} = \text{number of women} + \text{number of men}$$
$$= 15 + 17$$
$$= 32$$

Therefore the ratio of women to the total number of people in the class is $\dfrac{15}{32}$

Answers

1. The value b must not be zero because division by zero is undefined.
2. **a.** $\dfrac{291}{29}$ **b.** $\dfrac{29}{291}$ **c.** $\dfrac{291}{320}$

It is often desirable to write a ratio in lowest terms. The process is similar to simplifying fractions to lowest terms.

Example 2 **Writing Ratios in Lowest Terms**

Write each ratio in lowest terms.

a. 15 ft to 10 ft **b.** \$20 to \$10

Solution:

In part (a) we are comparing feet to feet. In part (b) we are comparing dollars to dollars. We can "cancel" the like units in the numerator and denominator as we would common factors.

a. $\frac{15\text{ ft}}{10\text{ ft}} = \frac{3 \cdot 5\text{ ft}}{2 \cdot 5\text{ ft}}$

$= \frac{3 \cdot \overset{1}{\cancel{5}}\,\cancel{\text{ft}}}{2 \cdot \underset{1}{\cancel{5}}\,\cancel{\text{ft}}}$ Simplify common factors. "Cancel" common units.

$= \frac{3}{2}$

Even though the number $\frac{3}{2}$ is equivalent to $1\frac{1}{2}$, we do not write the ratio as a mixed number. Remember that a ratio is a comparison of *two* quantities. If you did convert $\frac{3}{2}$ to the mixed number $1\frac{1}{2}$, you would write the ratio as $\frac{1\frac{1}{2}}{1}$. This would imply that the numerator is one and one-half times as large as the denominator.

b. $\frac{\$20}{\$10} = \frac{\overset{2}{\cancel{\$20}}}{\underset{1}{\cancel{\$10}}}$ Simplify common factors. "Cancel" common units.

$= \frac{2}{1}$

Although the fraction $\frac{2}{1}$ is equivalent to 2, we do not generally write ratios as whole numbers. Again, a ratio compares *two* quantities. In this case, we say that there is a 2-to-1 ratio between the original dollar amounts.

Skill Practice

Write the ratios in lowest terms.

3. 72 m to 16 m

4. 30 gal to 5 gal

Concept Connections

5. Which expression represents the ratio 4 to 1?

a. $\frac{1}{4}$ **b.** $\frac{4}{1}$

2. Writing Ratios of Mixed Numbers and Decimals

It is often desirable to express a ratio in lowest terms by using whole numbers in the numerator and denominator. This is demonstrated in Examples 3 and 4.

Example 3 **Writing a Ratio as a Ratio of Whole Numbers**

The length of a rectangular picture frame is 7.5 in., and the width is 6.25 in. Express the ratio of the length to the width. Then rewrite the ratio as a ratio of whole numbers simplified to lowest terms.

Skill Practice

Write the ratio as a ratio of whole numbers expressed in lowest terms.

6. \$4.20 to \$2.88

Answers

3. $\frac{9}{2}$ **4.** $\frac{6}{1}$ **5.** b **6.** $\frac{35}{24}$

Solution:

The ratio of length to width is $\frac{7.5}{6.25}$. We now want to rewrite the ratio, using whole numbers in the numerator and denominator. If we multiply 7.5 by 10, the decimal point will move to the right one place, resulting in a whole number. If we multiply 6.25 by 100, the decimal point will move to the right two places, resulting in a whole number. Because we want to multiply the numerator and denominator by the *same* number, we choose the greater number, 100.

$$\frac{7.5}{6.25} = \frac{7.5 \times 100}{6.25 \times 100}$$ Multiply numerator and denominator by 100.

$$= \frac{750}{625}$$ Because the numerator and denominator are large numbers, we write the prime factorization of each. The common factors are now easy to identify.

$$= \frac{2 \cdot 3 \cdot \overset{1}{\cancel{5}} \cdot \overset{1}{\cancel{5}} \cdot \overset{1}{\cancel{5}}}{5 \cdot \underset{1}{\cancel{5}} \cdot \underset{1}{\cancel{5}} \cdot \underset{1}{\cancel{5}}}$$ Simplify common factors to lowest terms.

$$= \frac{6}{5}$$ The ratio of length to width is $\frac{6}{5}$.

In Example 3, we multiplied by 100 to move the decimal point *two* places to the right. Multiplying by 10 would not have been sufficient, because $6.25 \times 10 = 62.5$, which is not a whole number.

Skill Practice

7. A recipe calls for $2\frac{1}{2}$ cups of flour and $\frac{3}{4}$ cup of sugar. Write the ratio of flour to sugar. Then rewrite the ratio as a ratio of whole numbers in lowest terms.

Example 4 Writing a Ratio as a Ratio of Whole Numbers

Ling walked $2\frac{1}{4}$ mi on Monday and $3\frac{1}{2}$ mi on Tuesday. Write the ratio of miles walked Monday to miles walked Tuesday. Then rewrite the ratio as a ratio of whole numbers reduced to lowest terms.

Solution:

The ratio of miles walked on Monday to miles walked on Tuesday is $\frac{2\frac{1}{4}}{3\frac{1}{2}}$.

To convert this to a ratio of whole numbers, first we rewrite each mixed number as an improper fraction. Then we can divide the fractions and simplify.

$$\frac{2\frac{1}{4}}{3\frac{1}{2}} = \frac{\frac{9}{4}}{\frac{7}{2}}$$ ← Write the mixed numbers as improper fractions. Recall that a fraction bar also implies division.

$$= \frac{9}{4} \div \frac{7}{2}$$

$$= \frac{9}{4} \cdot \frac{2}{7}$$ Multiply by the reciprocal of the divisor.

$$= \frac{9}{\underset{2}{\cancel{4}}} \cdot \frac{\overset{1}{\cancel{2}}}{7}$$ Simplify common factors to lowest terms.

$$= \frac{9}{14}$$ This is a ratio of whole numbers in lowest terms.

Answer

7. $\frac{2\frac{1}{2}}{\frac{3}{4}}$; $\frac{10}{3}$

3. Applications of Ratios

Ratios are used in a number of applications.

Example 5 **Using Ratios to Express Population Increase**

After the tragedy of Hurricane Katrina, New Orleans showed signs of recovery as more people moved back into the city. Three years after the storm, New Orleans had the fastest growth rate of any city in the United States. During a 1-year period, its population rose from 210,000 to 239,000. (*Source:* U.S. Census Bureau) Write a ratio expressing the increase in population to the original population for that year.

Solution:

To write this ratio, we need to know the increase in population.

$$\text{Increase} = 239{,}000 - 210{,}000 = 29{,}000$$

The ratio of the increase in population to the original number is

increase in population → $\dfrac{29{,}000}{210{,}000}$ ← original population

$$= \frac{29}{210}$$ Simplify to lowest terms.

Skill Practice

8. The U.S. Department of Transportation reported that more airports will have to expand to meet the growing demand of air travel. During a 5-year period, the Phoenix Sky Harbor Airport went from servicing 20 million passengers to 26 million passengers. Write a ratio of the increase in passengers to the original number.

Example 6 **Applying Ratios to Unit Conversion**

A fence is 12 yd long and 1 ft high.

a. Write the ratio of length to height with all units measured in yards.

b. Write the ratio of length to height with all units measured in feet.

1 ft = $\frac{1}{3}$ yd

12 yd

Solution:

a. 3 ft = 1 yd, therefore, 1 ft = $\frac{1}{3}$ yd.

Measuring in yards, we see that the ratio of length to height is $\dfrac{12 \text{ yd}}{\frac{1}{3} \text{ yd}} = \dfrac{12}{1} \cdot \dfrac{3}{1} = \dfrac{36}{1}$

b. The length is 12 yd = 36 ft. (Since 1 yd = 3 ft, then 12 yd = 12 · 3 ft = 36 ft.)

Measuring in feet, we see that the ratio of length to height is $\dfrac{36 \text{ ft}}{1 \text{ ft}} = \dfrac{36}{1}$

Notice that regardless of the units used, the ratio is the same, 36 to 1. This means that the length is 36 times the height.

Skill Practice

9. A painting is 2 yd in length by 2 ft wide.

a. Write the ratio of length to width with all units measured in feet.

b. Write the ratio of length to width with all units measured in yards.

Answers

8. $\frac{3}{10}$ **9. a.** $\frac{3}{1}$ **b.** $\frac{3}{1}$

Section 6.1 Practice Exercises

Study Skills Exercise

1. Define the key term **ratio**.

Objective 1: Writing a Ratio

2. Write a ratio of the number of females to males in your class.

For Exercises 3–8, write the ratio in two other ways.

3. 5 to 6

4. 3 to 7

5. 11 : 4

6. 8 : 13

7. $\frac{1}{2}$

8. $\frac{1}{8}$

For Exercises 9–14, write the ratios in fraction form. **(See Examples 1–2.)**

9. For a recent year, there were 10 named tropical storms and 6 hurricanes in the Atlantic. (*Source:* NOAA)

a. Write a ratio of the number of tropical storms to the number of hurricanes.

b. Write a ratio of the number of hurricanes to the number of tropical storms.

c. Write a ratio of the number of hurricanes to the total number of named storms.

10. For a recent year, 250 million persons were covered by health insurance in the United States and 45 million were not covered. (*Source:* U.S. Census Bureau)

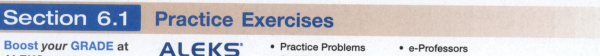

a. Write a ratio of the number of insured persons to the number of uninsured persons.

b. Write a ratio of the number of uninsured persons to the number of insured persons.

c. Write a ratio of the number of uninsured persons to the total number of persons.

11. In a certain neighborhood, 60 houses were on the market to be sold. During a 1-year period during a housing crisis, only 8 of these houses actually sold.

a. Write a ratio of the number of houses that sold to the total number that had been on the market.

b. Write a ratio of the number of houses that sold to the number that did not sell.

12. There are 52 cars in the parking lot, of which 21 are silver.

a. Write a ratio of the number of silver cars to the total number of cars.

b. Write a ratio of the number of silver cars to the number of cars that are not silver.

13. In a recent survey of a group of computer users, 21 were MAC users and 54 were PC users.

a. Write a ratio of the number of MAC users to PC users.

b. Write a ratio for the number of MAC users to the total number of people surveyed.

14. At a school sporting event, the concession stand sold 450 bottles of water, 200 cans of soda, and 125 cans of iced tea.

a. Write a ratio of the number of bottles of water sold to the number of cans of soda sold.

b. Write a ratio of the number of cans of iced tea sold to the total number of drinks sold.

For Exercises 15–26, write the ratio in lowest terms. **(See Example 2.)**

15. 4 yr to 6 yr

16. 10 lb to 14 lb

17. 5 mi to 25 mi

18. 20 ft to 12 ft

19. 8 m to 2 m

20. 14 oz to 7 oz

21. 33 cm to 15 cm

22. 21 days to 30 days

23. \$60 to \$50

24. 75¢ to 100¢

25. 18 in. to 36 in.

26. 3 cups to 9 cups

Objective 2: Writing Ratios of Mixed Numbers and Decimals

For Exercises 27–38, write the ratio in lowest terms with whole numbers in the numerator and denominator. **(See Examples 3 and 4.)**

27. 3.6 ft to 2.4 ft

28. 10.15 hr to 8.12 hr

29. 8 gal to $9\frac{1}{3}$ gal

30. 24 yd to $13\frac{1}{3}$ yd

31. $16\frac{4}{5}$ m to $18\frac{9}{10}$ m

32. $1\frac{1}{4}$ in. to $1\frac{3}{8}$ in.

33. \$16.80 to \$2.40

34. \$18.50 to \$3.70

35. $\frac{1}{2}$ day to 4 days

36. $\frac{1}{4}$ mi to $1\frac{1}{2}$ mi

37. 10.25 L to 8.2 L

38. 11.55 km to 6.6 km

Objective 3: Applications of Ratios

39. In 1981, a giant sinkhole in Winter Park, Florida, "swallowed" a home, a public pool, and a car dealership (including five Porsches). One witness said that in a single day, the hole widened from 5 ft in diameter to 320 ft.

a. Find the increase in the diameter of the sinkhole.

b. Write a ratio representing the increase in diameter to the original diameter of 5 ft. **(See Example 5.)**

40. The temperature at 8:00 A.M. in Los Angeles was 66°F. By 2:00 P.M., the temperature had risen to 90°F.

a. Find the increase in temperature from 8:00 A.M. to 2:00 P.M.

b. Write a ratio representing the increase in temperature to the temperature at 8:00 A.M.

41. A window is 2 ft wide and 3 yd in length (2 ft is $\frac{2}{3}$ yd). **(See Example 6.)**

a. Find the ratio of width to length with all units in yards.

b. Find the ratio of width to length with all units in feet.

42. A construction company needs 2 weeks to construct a family room and 3 days to add a porch.

a. Find the ratio of the time it takes for constructing the porch to the time constructing the family room, with all units in weeks.

b. Find the ratio of the time it takes for constructing the porch to the time constructing the family room, with all units in days.

For Exercises 43–46, refer to the table showing Alex Rodriguez's salary (rounded to the nearest $100,000) for selected years during his career. Write each ratio in lowest terms.

Year	Team	Salary	Position
2007	New York Yankees	$22,700,000	Third baseman
2004	New York Yankees	$22,000,000	Third baseman
2000	Seattle Mariners	$4,400,000	Shortstop
1996	Seattle Mariners	$400,000	Shortstop

Source: USA TODAY

43. Write the ratio of Alex's salary for the year 1996 to the year 2000.

44. Write a ratio of Alex's salary for the year 2004 to the year 1996.

45. Write a ratio of the increase in Alex's salary between the years 1996 and the year 2000 to his salary in 1996.

46. Write a ratio of the increase in Alex's salary between the years 2004 and 2007 to his salary in 2004.

For Exercises 47–50, refer to the table that shows the average spending per person for reading (books, newspapers, magazines, etc.) by age group. Write each ratio in lowest terms.

Age Group	Annual Average ($)
Under 25 years	60
25 to 34 years	111
35 to 44 years	136
45 to 54 years	172
55 to 64 years	183
65 to 74 years	159
75 years and over	128

Source: Mediamark Research Inc.

47. Find the ratio of spending for the under-25 group to the spending for the group 75 years and over.

48. Find the ratio of spending for the group 25 to 34 years old to the spending for the group of 65 to 74 years old.

49. Find the ratio of spending for the group under 25 years old to the spending for the group of 55 to 64 years old.

50. Find the ratio of spending for the group 35 to 44 years old to the spending for the group 45 to 54 years old.

For Exercises 51–54, find the ratio of the shortest side to the longest side. Write each ratio in lowest terms with whole numbers in the numerator and denominator.

51.

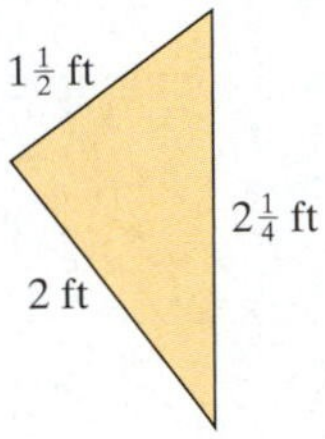

52.

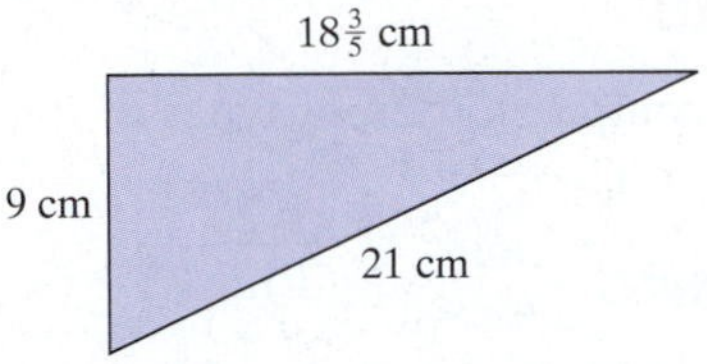

53.

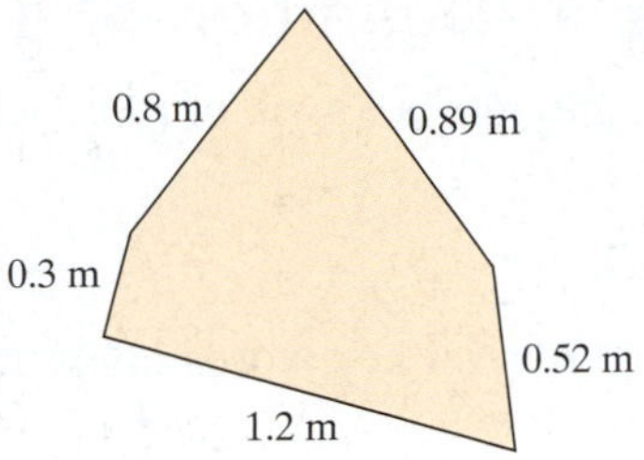

54.

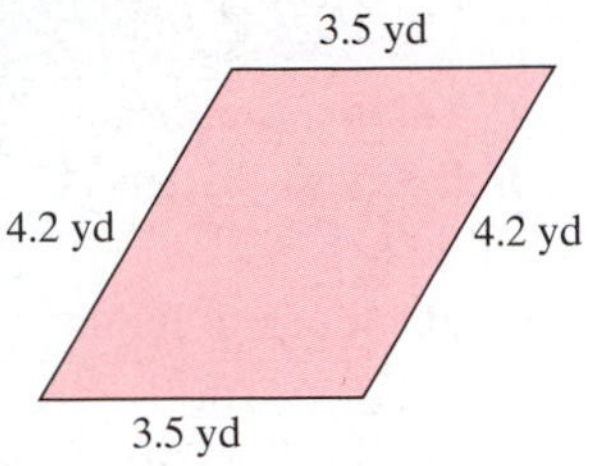

Expanding Your Skills

For Exercises 55–57, refer to the figure. The lengths of the sides for squares A, B, C, and D are given.

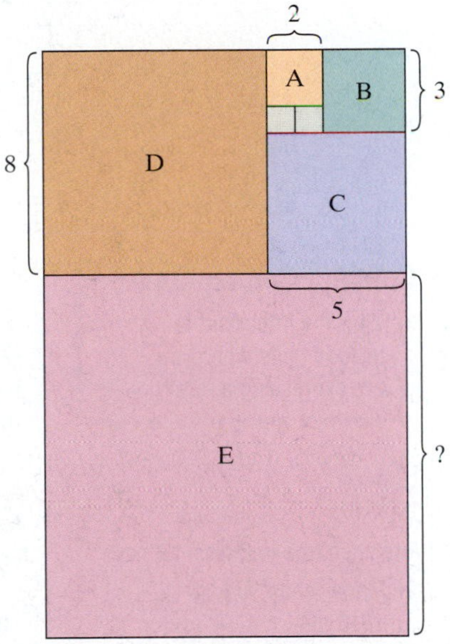

55. What are the lengths of the sides of square E?

56. Find the ratio of the lengths of the sides for the given pairs of squares.

a. Square B to square A

b. Square C to square B

c. Square D to square C

d. Square E to square D

57. Write the decimal equivalents for each ratio in Exercise 56. Do these values seem to be approaching a number close to 1.618 (this is an approximation for the *golden ratio*, which is equal to $\frac{1 + \sqrt{5}}{2}$)? Applications of the golden ratio are found throughout nature. In particular, as a result of the geometrically pleasing pattern, artists and architects have proportioned their work to approximate the golden ratio.

58. The ratio of a person's height to the length of the person's lower arm (from elbow to wrist) is approximately 6.5 to 1. Measure your own height and lower arm length. Is the ratio you get close to the average of 6.5 to 1?

59. The ratio of a person's height to the person's shoulder width (measured from outside shoulder to outside shoulder) is approximately 4 to 1. Measure your own height and shoulder width. Is the ratio you get close to the average of 4 to 1?

Section 6.2 Rates

Objectives

1. Definition of a Rate
2. Unit Rates
3. Unit Cost
4. Applications of Rates

1. Definition of a Rate

A **rate** is a type of ratio used to compare different types of quantities, for example:

$$\frac{270 \text{ mi}}{13 \text{ gal}} \quad \text{and} \quad \frac{\$8.55}{1 \text{ hr}}$$

Several key words imply rates. These are given in Table 6-1.

Table 6-1

Key Word	Example	Rate
Per	117 miles per 2 hours	$\frac{117 \text{ mi}}{2 \text{ hr}}$
For	\$12 for 3 lb	$\frac{\$12}{3 \text{ lb}}$
In	400 meters in 43.5 seconds	$\frac{400 \text{ m}}{43.5 \text{ sec}}$
On	270 miles on 12 gallons of gas	$\frac{270 \text{ mi}}{12 \text{ gal}}$

Concept Connections

1. Why is it important to include units when you are expressing a rate?

Because a rate compares two different quantities it is important to include the units in both the numerator and the denominator. It is also desirable to write rates in lowest terms.

Skill Practice

Write each rate in lowest terms.

2. Maria reads 15 pages in 10 min.
3. A Chevrolet Corvette Z06 gets 163.4 mi on 8.6 gal of gas.
4. Marty's balance in his investment account changed by −\$254 in 4 months.

Example 1 Writing Rates in Lowest Terms

Write each rate in lowest terms.

a. In one region, there are approximately 640 trees on 12 acres.

b. Latonya drove 138 mi on 6 gal of gas.

c. After a cold front, the temperature changed by −10°F in 4 hr.

Solution:

a. The rate of 640 trees on 12 acres can be expressed as $\frac{640 \text{ trees}}{12 \text{ acres}}$.

Now write this rate in lowest terms. $\frac{\overset{160}{\cancel{640}} \text{ trees}}{\underset{3}{\cancel{12}} \text{ acres}} = \frac{160 \text{ trees}}{3 \text{ acres}}$

b. The rate of 138 mi on 6 gal of gas can be expressed as $\frac{138 \text{ mi}}{6 \text{ gal}}$.

Now write this rate in lowest terms. $\frac{\overset{23}{\cancel{138}} \text{ mi}}{\underset{1}{\cancel{6}} \text{ gal}} = \frac{23 \text{ mi}}{1 \text{ gal}}$

Answers

1. The units are different in the numerator and denominator and will not "cancel."
2. $\frac{3 \text{ pages}}{2 \text{ min}}$
3. $\frac{19 \text{ mi}}{1 \text{ gal}}$ or 19 mi/gal
4. $-\frac{\$127}{2 \text{ months}}$

c. $-10°F$ in 4 hr can be represented by $\frac{-10°F}{4 \text{ hr}}$.

Writing this in lowest terms, we have: $\frac{-\overset{5}{\cancel{10}}°F}{\underset{2}{\cancel{4}} \text{ hr}} = -\frac{5°F}{2 \text{ hr}}$

2. Unit Rates

A rate having a denominator of 1 unit is called a **unit rate**. Furthermore, the number 1 is often omitted in the denominator.

$\frac{23 \text{ mi}}{1 \text{ gal}} = 23 \text{ mi/gal}$ is read as "twenty-three miles per gallon."

$\frac{52 \text{ ft}}{1 \text{ sec}} = 52 \text{ ft/sec}$ is read as "fifty-two feet per second."

$\frac{\$15}{1 \text{ hr}} = \$15/\text{hr}$ is read as "fifteen dollars per hour."

PROCEDURE Converting a Rate to a Unit Rate

To convert a rate to a unit rate, divide the numerator by the denominator and maintain the units of measurement.

Example 2 Finding Unit Rates

Write each rate as a unit rate. Round to three decimal places if necessary.

a. A health club charges $125 for 20 visits. Find the unit rate in dollars per visit.

b. In 1960, Wilma Rudolph won the women's 200-m run in 24 sec. Find her speed in meters per second.

c. During one baseball season, Barry Bonds got 149 hits in 403 at bats. Find his batting average. (*Hint:* Batting average is defined as the number of hits per the number of at bats.)

Solution:

a. The rate of $125 for 20 visits can be expressed as $\frac{\$125}{20 \text{ visits}}$.

To convert this to a unit rate, divide $125 by 20 visits.

$\frac{\$125}{20 \text{ visits}} = \frac{\$6.25}{1 \text{ visit}}$ or $6.25/visit

$$\begin{array}{r} 6.25 \\ 20\overline{)125.00} \\ \underline{-120} \\ 50 \\ \underline{-40} \\ 100 \\ \underline{-100} \\ 0 \end{array}$$

Concept Connections

5. Which rate is a unit rate?

$\frac{240 \text{ mi}}{4 \text{ hr}}$ or $\frac{60 \text{ mi}}{1 \text{ hr}}$

6. Which rate is a unit rate?

$24/1 hr or $24/hr

Skill Practice

Write a unit rate.

7. It costs $3.90 for 12 oranges.

8. A flight from Dallas to Des Moines travels a distance of 646 mi in 1.4 hr. Round the unit rate to the nearest mile per hour.

9. Under normal conditions, 98 in. of snow is equivalent to about 10 in. of rain.

Answers

5. $\frac{60 \text{ mi}}{1 \text{ hr}}$

6. They are both unit rates.

7. $0.325 per orange

8. 461 mi/hr

9. 9.8 in. snow per 1 in. of rain

b. The rate of 200 m per 24 sec can be expressed as $\dfrac{200\text{ m}}{24\text{ sec}}$.

To convert this to a unit rate, divide 200 m by 24 sec.

$$\frac{200\text{ m}}{24\text{ sec}} \approx \frac{8.333\text{ m}}{1\text{ sec}} \quad \text{or approximately } 8.333\text{ m/sec}$$

$$\begin{array}{r} 8.\overline{3} \\ 24\overline{)200.00} \\ -192 \\ \hline 80 \\ -72 \\ \hline 80 \end{array}$$

The quotient repeats.

Wilma Rudolph's speed was approximately 8.333 m/sec.

Avoiding Mistakes

Units of measurement must be included for the answer to be complete.

c. The rate of 149 hits in 403 at bats can be expressed as $\dfrac{149\text{ hits}}{403\text{ at bats}}$.

To convert this to a unit rate, divide 149 hits by 403 at bats.

$$\frac{149\text{ hits}}{403\text{ at bats}} \approx \frac{0.370\text{ hit}}{1\text{ at bat}} \quad \text{or } 0.370\text{ hit/at bat}$$

3. Unit Cost

A **unit cost** or unit price is the cost per 1 unit of something. At the grocery store, for example, you might purchase meat for \$3.79/lb (\$3.79 per 1 lb). Unit cost is useful in day-to-day life when we compare prices. Example 3 compares the prices of three different sizes of apple juice.

Skill Practice

10. Gatorade comes in several size packages. Compute the unit price per ounce for each option (round to the nearest thousandth of a dollar). Then determine which is the best buy.
    - **a.** \$2.99 for a 64-oz bottle
    - **b.** \$3.99 for four 24-oz bottles
    - **c.** \$3.79 for six 12-oz bottles

Example 3 Finding Unit Costs

Apple juice comes in a variety of sizes and packaging options. Find the unit price per ounce and determine which is the best buy.

a. \$1.69

Apple Juice
48 oz

b. \$2.39

Apple Juice
64 oz

c. \$2.99

Apple Juice
10-pack 6 oz each

Solution:

When we compute a unit cost, the cost is always placed in the numerator of the rate. Furthermore, when we divide the cost by the amount, we need to obtain enough digits in the quotient to see the variation in unit price. In this example, we have rounded to the nearest thousandth of a dollar (nearest tenth of a cent). This means that we use the ten-thousandths-place digit in the quotient on which to base our decision on rounding.

Answer

10. **a.** \$0.047/oz
    b. \$0.042/oz
    c. \$0.053/oz
    The 4-pack of 24-oz bottles is the best buy.

	Rate	Quotient	Unit Rate (Rounded)
a.	$\frac{\$1.69}{48 \text{ oz}}$	$1.69 ÷ 48 oz ≈ $0.0352/oz	$0.035/oz or 3.5¢/oz
b.	$\frac{\$2.39}{64 \text{ oz}}$	$2.39 ÷ 64 oz ≈ $0.0373/oz	$0.037/oz or 3.7¢/oz
c.	6 oz × 10 = 60 oz $\frac{\$2.99}{60 \text{ oz}}$	$2.99 ÷ 60 oz ≈ $0.0498/oz	$0.050/oz or 5.0¢/oz

From the table, we see that the most economical buy is the 48-oz size because its unit rate is the least expensive.

4. Applications of Rates

Example 4 uses a unit rate for comparison in an application.

Example 4 Computing Mortality Rates

Mortality rate is defined to be the total number of people who die due to some risk behavior divided by the total number of people who engage in the risk behavior. Based on the following statistics, compare the mortality rate for undergoing heart bypass surgery to the mortality rate of flying on the space shuttle.

a. Roughly 28 people will die for every 1000 who undergo heart bypass surgery. (*Source:* The Society of Thoracic Surgeons)

b. As of August 2007, there have been 14 astronauts killed in space shuttle missions out of 830 astronauts who have flown.

Solution:

a. Mortality rate for heart bypass surgery: $\frac{28}{1000} = 0.028$ death/surgery

b. Mortality rate for flying on the space shuttle: $\frac{14}{830} \approx 0.017$ deaths/flight

Comparing these rates shows that it is riskier to have heart bypass surgery than to fly on the space shuttle.

Skill Practice

11. In Ecuador roughly 450 out of 500 adults can read. In Brazil, approximately 1700 out of 2000 adults can read.

a. Compute the unit rate for Ecuador (this unit rate is called the *literacy rate*).

b. Compute the unit rate for Brazil.

c. Which country has the greater literacy rate?

Answer

11. a. 0.9 reader per adult
b. 0.85 reader per adult
c. Ecuador

Section 6.2 Practice Exercises

Study Skills Exercise

1. Define the key terms.

a. Rate **b.** Unit rate **c.** Unit cost

Review Exercises

2. Write the ratio 3 to 5 in two other ways.

3. Write the ratio 4:1 in two other ways.

For Exercises 4–6, write the ratio in lowest terms.

4. $6\frac{3}{4}$ ft to $8\frac{1}{4}$ ft

5. 1.08 mi to 2.04 mi

6. \$28.40 to \$20.80

Objective 1: Definition of a Rate

For Exercises 7–18, write each rate in lowest terms. **(See Example 1.)**

7. A type of laminate flooring sells for \$32 for 5 ft^2.

8. A remote control car can go up to 44 ft in 5 sec.

9. Elaine drives 234 mi in 4 hr.

10. Travis has 14 blooms on 6 of his plants.

11. Tyler earned \$58 in 8 hr.

12. Neil can type only 336 words in 15 min.

13. A printer can print 13 pages in 26 sec.

14. During a bad storm there was 2 in. of rain in 6 hr.

15. There are 130 calories in 8 snack crackers.

16. There are 50 students assigned to 4 advisers.

17. The temperature changed by −18°F in 8 hr.

18. Jill's portfolio changed by −\$2600 in 6 months.

Objective 2: Unit Rates

19. Of the following rates, identify those that are unit rates.

a. $\frac{\$0.37}{1\text{ oz}}$ **b.** $\frac{333.2\text{ mi}}{14\text{ gal}}$ **c.** 16 ft/sec **d.** $\frac{59\text{ mi}}{1\text{ hr}}$

20. Of the following rates, identify those that are unit rates.

a. $\frac{3\text{ lb}}{\$1.00}$ **b.** $\frac{21\text{ ft}}{1\text{ sec}}$ **c.** 50 mi/hr **d.** $\frac{232\text{ words}}{2\text{ min}}$

For Exercises 21–32, write each rate as a unit rate and round to the nearest hundredth when necessary. **(See Example 2.)**

21. The Osborne family drove 452 mi in 4 days.

22. The book of poetry *The Prophet* by Kahlil Gibran has estimated sales of \$6,000,000 over an 80-year period.

23. A submarine practicing evasive maneuvers descended 100 m in $\frac{1}{4}$ hr. Find a unit rate representing the change in "altitude" per hour.

24. The ground dropped 74.4 ft in 24 hr in the famous Winter Park sinkhole of 1981. Find the unit rate representing the change in "altitude" per hour.

25. If Oscar bought an easy chair for $660 and plans to make 12 payments, what is the amount per payment?

26. The jockey David Gall had 7396 wins in 43 years of riding.

27. At the market, bananas cost $2.76 for 4 lb.

28. Ceramic tile sell for $13.08 for a box of 12 tiles. Find the price per tile.

29. Lottery prize money of $1,792,000 is for seven people.

30. One WeightWatchers group lost 123 lb for its 11 members.

31. A male speed skater skated 500 m in 35 sec. Find the rate in meters per second.

32. A female speed skater skated 500 m in 38 sec. Find the rate in meters per second.

Objective 3: Unit Cost

For Exercises 33–42, find the unit costs (that is, dollars per unit). Round the answers to three decimal places when necessary. **(See Example 3.)**

33. Tide laundry detergent costs $14.99 for 100 oz.

34. Dove liquid body wash costs $3.49 for 12 oz.

35. Soda costs $1.99 for a 2-L bottle.

36. Four chairs cost $221.00.

37. A set of four tires costs $210.

38. A package of three shirts costs $64.80.

39. A package of six newborn bodysuits costs $32.50.

40. A package of eight AAA batteries costs $9.84.

41. **a.** 40 oz of shampoo for $3.00
b. 28 oz of shampoo for $2.10
c. Which is the best buy?

42. **a.** 10 lb of potting soil for $1.70
b. 30 lb of potting soil for $5.10
c. Which is the best buy?

43. Corn comes in two size cans, 29 oz and 15 oz. The larger can costs $1.59 and the smaller can costs $1.09. Find the unit cost of each can. Which is the better buy? (Round to three decimal places.)

44. Napkins come in a variety of packages. A package of 400 napkins sells for $2.99, and a package of 100 napkins sells for $1.50. Find the unit cost of each package. Which is the better buy? (Round to four decimal places.)

Objective 4: Applications of Rates

45. Carbonated beverages come in different sizes and contain different amounts of sugar. Compute the amount of sugar per fluid ounce for each soda. Then determine which has the greatest amount of sugar per fluid ounce. **(See Example 4.)**

Soda	Amount	Sugar
Coca-Cola	20 fl oz	65 g
Mello Yello	12 fl oz	47 g
Canada Dry Ginger Ale	8 fl oz	24 g

46. Compute the amount of carbohydrate per fluid ounce for each soda. Then determine which has the greatest amount of carbohydrate per fluid ounce.

Soda	Amount	Carbohydrates
Coca-Cola	20 fl oz	65 g
Mello Yello	12 fl oz	47 g
Canada Dry Ginger Ale	8 fl oz	25 g

47. Carbonated beverages come in different sizes and have a different number of calories. Compute the number of calories per fluid ounce for each soda. Then determine which has the least number of calories per fluid ounce.

Soda	Amount	Calories
Coca-Cola	20 fl oz	240
Mello Yello	12 fl oz	170
Canada Dry Ginger Ale	8 fl oz	90

48. According to the National Institutes of Health, a platelet count below 20,000 per microliter of blood is considered a life-threatening condition. Suppose a patient's test results yield a platelet count of 13,000,000 for 100 microliters. Write this as a unit rate (number of platelets per microliter). Does the patient have a life-threatening condition?

49. The number of motor vehicles produced in the United States increased steadily by a total of 5,310,000 in an 18-year period. Compute the rate representing the increase in the number of vehicles produced per year during this time period. (*Source:* American Automobile Manufacturers Association)

50. The total number of prisoners in the United States increased steadily by a total of 344,000 in an 8-year period. Compute the rate representing the increase in the number of prisoners per year.

51. **a.** The population of Mexico increased steadily by 22 million people in a 10-year period. Compute the rate representing the increase in the population per year.

    b. The population of Brazil increased steadily by 10.2 million in a 5-year period. Compute the rate representing the increase in the population per year.

    c. Which country has a greater rate of increase in population per year?

52. **a.** The price per share of Microsoft stock rose $18.24 in a 24-month period. Compute the rate representing the increase in the price per month.

    b. The price per share of IBM stock rose $22.80 in a 12-month period. Compute the rate representing the increase in the price per month.

    c. Which stock had a greater rate of increase per month?

53. A cheetah can run 120 m in 4.1 sec. An antelope can run 50 m in 2.1 sec. Compare their unit speeds to determine which animal is faster. Round to the nearest whole unit.

Calculator Connections

Topic: Applications of Unit Rates

Calculator Exercises

Don Shula coached football for 33 years. He had 328 wins and 156 losses. Tom Landry coached football for 29 years. He had 250 wins and 162 losses. Use this information to answer Exercises 54–55.

54. **a.** Compute a unit rate representing the average number of wins per year for Don Shula. Round to one decimal place.

    b. Compute a unit rate representing the average number of wins per year for Tom Landry. Round to one decimal place.

    c. Which coach had a better rate of wins per year?

55. a. Compute a unit rate representing the number of wins to the number of losses for Don Shula. Round to one decimal place.

b. Compute a unit rate representing the number of wins to the number of losses for Tom Landry. Round to one decimal place.

c. Which coach had a better win/loss rate?

56. Compare three brands of soap. Find the price per ounce and determine the best buy. (Round to two decimal places.)

a. Dove: \$7.89 for a 6-bar pack of 4.5-oz bars

b. Dial: \$2.89 for a 3-bar pack of 4.5-oz bars

c. Irish Spring: \$6.99 for an 8-bar pack of 4.5-oz bars

57. Mayonnaise comes in 32-, 16-, and 8-oz jars. They are priced at \$5.59, \$3.79, and \$2.59, respectively. Find the unit cost of each size jar to find the best buy. (Round to three decimal places.)

58. Albacore tuna comes in different-size cans. Find the unit cost of each package to find the best buy. (Round to three decimal places.)

a. \$3.99 for a 12-oz can

b. \$6.29 for a 4-pack of 6-oz cans

c. \$3.39 for a 3-pack of 3-oz cans

59. Coca-Cola is sold in a variety of different packages. Find the unit cost of each package to find the better buy. (Round to three decimal places.)

a. \$4.99 for a case of 24 twelve-oz cans

b. \$5.00 for a 12-pack of 8-oz cans

Proportions

Section 6.3

1. Definition of a Proportion

Recall that a statement indicating that two quantities are equal is called an equation. In this section, we are interested in a special type of equation called a proportion. A **proportion** states that two ratios or rates are equal. For example:

$$\frac{1}{4} = \frac{10}{40} \text{ is a proportion.}$$

We know that the fractions $\frac{1}{4}$ and $\frac{10}{40}$ are equal because $\frac{10}{40}$ reduces to $\frac{1}{4}$.

We read the proportion $\frac{1}{4} = \frac{10}{40}$ as follows: "1 is to 4 as 10 is to 40."

We also say that the numbers 1 and 4 are *proportional to* the numbers 10 and 40.

Objectives

1. Definition of a Proportion
2. Determining Whether Two Ratios Form a Proportion
3. Solving Proportions

Skill Practice

Write a proportion for each statement.

1. 7 is to 28 as 13 is to 52.
2. \$17 is to 2 hr as \$102 is to 12 hr.
3. The numbers 5 and −11 are proportional to the numbers 15 and −33.

Example 1 Writing Proportions

Write a proportion for each statement.

a. 5 is to 12 as 30 is to 72.

b. 240 mi is to 4 hr as 300 mi is to 5 hr.

c. The numbers −3 and 7 are proportional to the numbers −12 and 28.

Solution:

a. $\dfrac{5}{12} = \dfrac{30}{72}$ 5 is to 12 as 30 is to 72.

b. $\dfrac{240 \text{ mi}}{4 \text{ hr}} = \dfrac{300 \text{ mi}}{5 \text{ hr}}$ 240 mi is to 4 hr as 300 mi is to 5 hr.

c. $\dfrac{-3}{7} = \dfrac{-12}{28}$ −3 and 7 are proportional to −12 and 28.

TIP: A proportion is an equation and must have an equal sign.

2. Determining Whether Two Ratios Form a Proportion

To determine whether two ratios form a proportion, we must determine whether the ratios are equal. Recall from Section 4.2 that two fractions are equal whenever their cross products are equal. That is,

$$\frac{a}{b} = \frac{c}{d} \quad \text{implies} \quad a \cdot d = b \cdot c \quad \text{(and vice versa).}$$

Skill Practice

Determine whether the ratios form a proportion.

4. $\dfrac{4}{9} \stackrel{?}{=} \dfrac{12}{27}$
5. $\dfrac{3\frac{1}{4}}{5} \stackrel{?}{=} \dfrac{8}{12}$

Example 2 Determining Whether Two Ratios Form a Proportion

Determine whether the ratios form a proportion.

a. $\dfrac{3}{5} \stackrel{?}{=} \dfrac{9}{15}$ **b.** $\dfrac{8}{4} \stackrel{?}{=} \dfrac{10}{5\frac{1}{2}}$

Solution:

a. $\dfrac{3}{5} = \dfrac{9}{15}$

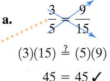

$(3)(15) \stackrel{?}{=} (5)(9)$ Cross-multiply to form the cross products.

$45 = 45$ ✓

The cross products are equal. Therefore, the ratios form a proportion.

Avoiding Mistakes

Cross multiplication can be performed only when there are two fractions separated by an equal sign.

b. $\dfrac{8}{4} = \dfrac{10}{5\frac{1}{2}}$

$(8)\left(5\dfrac{1}{2}\right) \stackrel{?}{=} (4)(10)$ Cross-multiply to form the cross products.

$\dfrac{\overset{4}{\cancel{8}}}{1} \cdot \dfrac{11}{\underset{1}{\cancel{2}}} \stackrel{?}{=} 40$ Write the mixed number as an improper fraction.

$44 \neq 40$ Multiply fractions.

The cross products are not equal. The ratios do not form a proportion.

Answers

1. $\dfrac{7}{28} = \dfrac{13}{52}$ 2. $\dfrac{\$17}{2 \text{ hr}} = \dfrac{\$102}{12 \text{ hr}}$
3. $\dfrac{5}{-11} = \dfrac{15}{-33}$ 4. Yes 5. No

Example 3 **Determining Whether Pairs of Numbers Are Proportional**

Determine whether the numbers 2.7 and −5.3 are proportional to the numbers 8.1 and −15.9.

Solution:

Two pairs of numbers are proportional if their ratios are equal.

$$\frac{2.7}{-5.3} = \frac{8.1}{-15.9}$$

$(2.7)(-15.9) \stackrel{?}{=} (-5.3)(8.1)$ Cross-multiply to form the cross products.

$-42.93 = -42.93$ ✓ Multiply decimals.

The cross products are equal. The ratios form a proportion.

Skill Practice

6. Determine whether the numbers −1.2 and 2.5 are proportional to the numbers −2 and 5.

3. Solving Proportions

A proportion is made up of four values. If three of the four values are known, we can solve for the fourth.

Consider the proportion $\frac{x}{20} = \frac{3}{4}$. We let the variable x represent the unknown value in the proportion. To solve for x, we can equate the cross products to form an equivalent equation.

$$\frac{x}{20} = \frac{3}{4}$$

$4x = 3 \cdot 20$ Cross-multiply to form the cross products.

$4x = 60$ Simplify.

$\frac{4x}{4} = \frac{60}{4}$ Divide both sides by 4 to isolate x.

$x = 15$

We can check the value of x in the original proportion.

Check: $\frac{x}{20} = \frac{3}{4}$ $\xrightarrow{\text{substitute 15 for } x}$ $\frac{15}{20} \stackrel{?}{=} \frac{3}{4}$

$(15)(4) \stackrel{?}{=} (3)(20)$

$60 = 60$ ✓ The solution 15 checks.

The steps to solve a proportion are summarized next.

PROCEDURE Solving a Proportion

Step 1 Set the cross products equal to each other.
Step 2 Solve the equation.
Step 3 Check the solution in the original proportion.

Answer

6. No

Skill Practice

Solve the proportion. Be sure to check your answer.

7. $\frac{2}{9} = \frac{n}{81}$

Example 4 Solving a Proportion

Solve the proportion. $\frac{x}{13} = \frac{6}{39}$

Solution:

$$\frac{x}{13} = \frac{6}{39}$$

$39x = (6)(13)$ Set the cross products equal.

$39x = 78$ Simplify.

$\frac{39x}{39} = \frac{78}{39}$ Divide both sides by 39 to isolate x.

$x = 2$ Check: $\frac{x}{13} = \frac{6}{39} \xrightarrow{\text{substitute } x = 2} \frac{2}{13} \stackrel{?}{=} \frac{6}{39}$

$$(2)(39) \stackrel{?}{=} (6)(13)$$

$$78 = 78 \checkmark$$

The solution 2 checks in the original proportion.

Skill Practice

Solve the proportion. Be sure to check your answer.

8. $\frac{3}{w} = \frac{21}{77}$

Example 5 Solving a Proportion

Solve the proportion. $\frac{4}{15} = \frac{9}{n}$

Solution:

$\frac{4}{15} = \frac{9}{n}$ The variable can be represented by any letter.

$4n = (9)(15)$ Set the cross products equal.

$4n = 135$ Simplify.

$\frac{4n}{4} = \frac{135}{4}$ Divide both sides by 4 to isolate n.

$n = \frac{135}{4}$ The fraction $\frac{135}{4}$ is in lowest terms.

Avoiding Mistakes

When solving a proportion, do not try to "cancel" like factors on opposite sides of the equal sign. The proportion, $\frac{4}{15} = \frac{9}{n}$ cannot be simplified.

The solution may be written as $n = \frac{135}{4}$ or $n = 33\frac{3}{4}$ or $n = 33.75$.

To check the solution in the original proportion, we may use any of the three forms of the answer. We will use the decimal form.

Check: $\frac{4}{15} = \frac{9}{n} \xrightarrow{\text{substitute } n = 33.75} \frac{4}{15} \stackrel{?}{=} \frac{9}{33.75}$

$$(4)(33.75) \stackrel{?}{=} (9)(15)$$

$$135 = 135 \checkmark$$

The solution 33.75 checks in the original proportion.

Answers

7. 18 **8.** 11

Example 6 Solving a Proportion

Solve the proportion. $\dfrac{0.8}{-3.1} = \dfrac{4}{p}$

Solution:

$\dfrac{0.8}{-3.1} = \dfrac{4}{p}$

$0.8p = 4(-3.1)$ Set the cross products equal.

$0.8p = -12.4$ Simplify.

$\dfrac{0.8p}{0.8} = \dfrac{-12.4}{0.8}$ Divide both sides by 0.8.

$p = -15.5$ The value -15.5 checks in the original equation.

The solution is -15.5.

Skill Practice

Solve the proportion. Be sure to check your answer.

9. $\dfrac{0.6}{x} = \dfrac{1.5}{-2}$

We chose to give the solution to Example 6 in decimal form because the values in the original proportion are decimal numbers. However, it would be correct to give the solution as a mixed number or fraction. The solution -15.5 is also equivalent to $-15\frac{1}{2}$ or $-\frac{31}{2}$.

Example 7 Solving a Proportion

Solve the proportion. $\dfrac{-12}{-8} = \dfrac{x}{\frac{2}{3}}$

Solution:

$\dfrac{-12}{-8} = \dfrac{x}{\frac{2}{3}}$

$(-12)\left(\dfrac{2}{3}\right) = -8x$ Set the cross products equal.

$\left(-\dfrac{\overset{4}{\cancel{12}}}{1}\right)\left(\dfrac{2}{\underset{1}{\cancel{3}}}\right) = -8x$ Write the whole number as an improper fraction.

$-8 = -8x$ Simplify.

$\dfrac{-8}{-8} = \dfrac{-8x}{-8}$ Divide both sides by -8.

$1 = x$ The value 1 checks in the original equation.

The solution is 1.

Skill Practice

Solve the proportion. Be sure to check your answer.

10. $\dfrac{\frac{1}{2}}{3.5} = \dfrac{x}{14}$

Answers

9. -0.8 10. $x = 2$

Section 6.3 Practice Exercises

Boost *your* GRADE at ALEKS.com!

ALEKS Version 3.0

- Practice Problems
- Self-Tests
- NetTutor
- e-Professors
- Videos

Study Skills Exercise

1. Define the key term **proportion**.

Review Exercises

For Exercises 2–5, write as a reduced ratio or rate.

2. 3 ft to 45 ft

3. 3 teachers for 45 students

4. 6 apples for 2 pies

5. 6 days to 2 days

For Exercises 6–8, write as a unit rate.

6. 800 revolutions in 10 sec

7. 337.2 mi on 12 gal of gas

8. 13,516 passengers on 62 flights

Objective 1: Definition of a Proportion

For Exercises 9–20, write a proportion for each statement. **(See Example 1.)**

9. 4 is to 16 as 5 is to 20.

10. 3 is to 18 as 4 is to 24.

11. −25 is to 15 as −10 is to 6.

12. 35 is to −14 as 20 is to −8.

13. The numbers 2 and 3 are proportional to the numbers 4 and 6.

14. The numbers 2 and 1 are proportional to the numbers 26 and 13.

15. The numbers −30 and −25 are proportional to the numbers 12 and 10.

16. The numbers −24 and −18 are proportional to the numbers 8 and 6.

17. \$6.25 per hour is proportional to \$187.50 per 30 hr.

18. \$115 per week is proportional to \$460 per 4 weeks.

19. 1 in. is to 7 mi as 5 in. is to 35 mi.

20. 16 flowers is to 5 plants as 32 flowers is to 10 plants.

Objective 2: Determining Whether Two Ratios Form a Proportion

For Exercises 21–28, determine whether the ratios form a proportion. **(See Example 2.)**

21. $\frac{5}{18} \stackrel{?}{=} \frac{4}{16}$

22. $\frac{9}{10} \stackrel{?}{=} \frac{8}{9}$

23. $\frac{16}{24} \stackrel{?}{=} \frac{2}{3}$

24. $\frac{4}{5} \stackrel{?}{=} \frac{24}{30}$

25. $\frac{2\frac{1}{2}}{3\frac{2}{3}} \stackrel{?}{=} \frac{15}{22}$

26. $\frac{1\frac{3}{4}}{3} \stackrel{?}{=} \frac{7}{12}$

27. $\frac{2}{-3.2} \stackrel{?}{=} \frac{10}{-16}$

28. $\frac{4.7}{-7} \stackrel{?}{=} \frac{23.5}{-35}$

For Exercises 29–34, determine whether the pairs of numbers are proportional. **(See Example 3.)**

29. Are the numbers 48 and 18 proportional to the numbers 24 and 9?

30. Are the numbers 35 and 14 proportional to the numbers 5 and 2?

31. Are the numbers $2\frac{3}{8}$ and $1\frac{1}{2}$ proportional to the numbers $9\frac{1}{2}$ and 6?

32. Are the numbers $1\frac{2}{3}$ and $\frac{5}{6}$ proportional to the numbers 5 and $2\frac{1}{2}$?

33. Are the numbers −6.3 and 9 proportional to the numbers −12.6 and 16?

34. Are the numbers −7.1 and 2.4 proportional to the numbers −35.5 and 10?

Objective 3: Solving Proportions

For Exercises 35–38, determine whether the given value is a solution to the proportion.

35. $\frac{x}{40} = \frac{1}{-8}$; $x = -5$

36. $\frac{14}{x} = \frac{12}{-18}$; $x = -21$

37. $\frac{12.4}{31} = \frac{8.2}{y}$; $y = 20$

38. $\frac{4.2}{9.8} = \frac{z}{36.4}$; $z = 15.2$

For Exercises 39–58, solve the proportion. Be sure to check your answers. **(See Examples 4–7.)**

39. $\frac{12}{16} = \frac{3}{x}$

40. $\frac{20}{28} = \frac{5}{x}$

41. $\frac{9}{21} = \frac{x}{7}$

42. $\frac{15}{10} = \frac{3}{x}$

43. $\frac{p}{12} = \frac{-25}{4}$

44. $\frac{p}{8} = \frac{-30}{24}$

45. $\frac{3}{40} = \frac{w}{10}$

46. $\frac{5}{60} = \frac{z}{8}$

47. $\frac{16}{-13} = \frac{20}{t}$

48. $\frac{12}{b} = \frac{8}{-9}$

49. $\frac{m}{12} = \frac{5}{8}$

50. $\frac{16}{12} = \frac{21}{a}$

51. $\frac{17}{12} = \frac{4\frac{1}{4}}{x}$

52. $\frac{26}{30} = \frac{5\frac{1}{5}}{x}$

53. $\frac{0.5}{h} = \frac{1.8}{9}$

54. $\frac{2.6}{h} = \frac{1.3}{0.5}$

55. $\frac{\frac{3}{8}}{6.75} = \frac{x}{72}$

56. $\frac{12.5}{\frac{1}{4}} = \frac{120}{y}$

57. $\frac{4}{\frac{1}{10}} = \frac{-\frac{1}{2}}{z}$

58. $\frac{6}{\frac{1}{3}} = \frac{-\frac{1}{2}}{t}$

For Exercises 59–64, write a proportion for each statement. Then solve for the variable.

59. 25 is to 100 as 9 is to y.

60. 65 is to 15 as 26 is to y.

61. 15 is to 20 as t is to 10

62. 9 is to 12 as w is to 30.

63. The numbers −3.125 and 5 are proportional to the numbers −18.75 and k.

64. The numbers −4.75 and 8 are proportional to the numbers −9.5 and k.

Expanding Your Skills

For Exercises 65–72, solve the proportions.

65. $\frac{x+1}{3} = \frac{5}{7}$ **66.** $\frac{2}{5} = \frac{x-2}{4}$ **67.** $\frac{x-3}{3x} = \frac{2}{3}$ **68.** $\frac{x-2}{4x} = \frac{3}{8}$

69. $\frac{x+3}{x} = \frac{5}{4}$ **70.** $\frac{x-5}{x} = \frac{3}{2}$ **71.** $\frac{x}{3} = \frac{x-2}{4}$ **72.** $\frac{x}{6} = \frac{x-1}{3}$

Problem Recognition Exercises

Operations on Fractions versus Solving Proportions

For Exercises 1–6, identify the problem as a proportion or as a product of fractions. Then solve the proportion or multiply the fractions.

1. a. $\frac{x}{4} = \frac{15}{8}$ **b.** $\frac{1}{4} \cdot \frac{15}{8}$ **2. a.** $\frac{2}{5} \cdot \frac{3}{10}$ **b.** $\frac{2}{5} = \frac{y}{10}$

3. a. $\frac{2}{7} \times \frac{3}{14}$ **b.** $\frac{2}{7} = \frac{n}{14}$ **4. a.** $\frac{m}{5} = \frac{6}{15}$ **b.** $\frac{3}{5} \times \frac{6}{15}$

5. a. $\frac{48}{p} = \frac{16}{3}$ **b.** $\frac{48}{8} \cdot \frac{16}{3}$ **6. a.** $\frac{10}{7} \cdot \frac{28}{5}$ **b.** $\frac{10}{7} = \frac{28}{t}$

For Exercises 7–10, solve the proportion or perform the indicated operation on fractions.

7. a. $\frac{3}{7} = \frac{6}{z}$ **b.** $\frac{3}{7} \div \frac{6}{35}$ **c.** $\frac{3}{7} + \frac{6}{35}$ **d.** $\frac{3}{7} \cdot \frac{6}{35}$

8. a. $\frac{4}{5} \div \frac{20}{3}$ **b.** $\frac{4}{v} = \frac{20}{3}$ **c.** $\frac{4}{5} \times \frac{20}{3}$ **d.** $\frac{4}{5} - \frac{20}{3}$

9. a. $\frac{14}{5} \cdot \frac{10}{7}$ **b.** $\frac{14}{5} = \frac{x}{7}$ **c.** $\frac{14}{5} - \frac{10}{7}$ **d.** $\frac{14}{5} \div \frac{10}{7}$

10. a. $\frac{11}{3} = \frac{66}{y}$ **b.** $\frac{11}{3} + \frac{66}{11}$ **c.** $\frac{11}{3} \div \frac{66}{11}$ **d.** $\frac{11}{3} \times \frac{66}{11}$

Applications of Proportions and Similar Figures

Section 6.4

Objectives

1. Applications of Proportions
2. Similar Figures

1. Applications of Proportions

Proportions are used in a variety of applications. In Examples 1 through 4, we take information from the wording of a problem and form a proportion.

Example 1 Using a Proportion in a Consumer Application

Linda drove her Honda Accord 145 mi on 5 gal of gas. At this rate, how far can she drive on 12 gal?

Solution:

Let x represent the distance Linda can go on 12 gal.

This problem involves two rates. We can translate this to a proportion. Equate the two rates.

distance → $\frac{145 \text{ mi}}{5 \text{ gal}} = \frac{x \text{ mi}}{12 \text{ gal}}$ ← distance

number of gallons → (denominators) ← number of gallons

Solve the proportion.

$(145)(12) = (5) \cdot x$ Cross-multiply to form the cross products.

$1740 = 5x$

$\frac{1740}{5} = \frac{\overset{1}{\cancel{5}}x}{\underset{1}{\cancel{5}}}$ Divide both sides by 5.

$348 = x$ Divide. $1740 \div 5 = 348$

Linda can drive 348 mi on 12 gal of gas.

TIP: Notice that the two rates have the same units in the numerator (miles) and the same units in the denominator (gallons).

Skill Practice

1. Jacques bought 3 lb of tomatoes for \$5.55. At this rate, how much would 7 lb cost?

In Example 1 we could have set up a proportion in many different ways.

$$\frac{145 \text{ mi}}{5 \text{ gal}} = \frac{x \text{ mi}}{12 \text{ gal}} \quad \text{or} \quad \frac{5 \text{ gal}}{145 \text{ mi}} = \frac{12 \text{ gal}}{x \text{ mi}} \quad \text{or}$$

$$\frac{5 \text{ gal}}{12 \text{ gal}} = \frac{145 \text{ mi}}{x \text{ mi}} \quad \text{or} \quad \frac{12 \text{ gal}}{5 \text{ gal}} = \frac{x \text{ mi}}{145 \text{ mi}}$$

Notice that in each case, the cross products produce the same equation. We will generally set up the proportions so that the units in the numerators are the same and the units in the denominators are the same.

Answer

1. The price for 7 lb would be \$12.95.

Skill Practice

2. It takes 2.5 gal of paint to cover 900 ft^2 of wall. How much area could be painted with 4 gal of the same paint?

Example 2 Using a Proportion in a Construction Application

If a cable 25 ft long weighs 1.2 lb, how much will a 120-ft cable weigh?

Solution:

Let w represent the weight of the 120-ft cable. Label the unknown.

$$\begin{matrix}\text{length} \rightarrow \\ \text{weight} \rightarrow\end{matrix} \frac{25 \text{ ft}}{1.2 \text{ lb}} = \frac{120 \text{ ft}}{w \text{ lb}} \begin{matrix}\leftarrow \text{length} \\ \leftarrow \text{weight}\end{matrix}$$ Translate to a proportion.

$$(25) \cdot w = (1.2)(120)$$ Equate the cross products.

$$25w = 144$$

$$\frac{\overset{1}{\cancel{25}} w}{\underset{1}{\cancel{25}}} = \frac{144}{25}$$ Divide both sides by 25.

$$w = 5.76$$ Divide. $144 \div 25 = 5.76$

The 120-ft cable weighs 5.76 lb.

Skill Practice

3. On a certain map, the distance between Shreveport, Louisiana, and Memphis, Tennessee, is represented as $4\frac{1}{2}$ in. The actual distance is 288 mi. The map distance between Seattle, Washington, and San Francisco, California, is represented as 11 in. Find the actual distance.

Example 3 Using a Proportion in a Geography Application

The distance between Phoenix and Los Angeles is 348 mi. On a certain map, this is represented by 6 in. On the same map, the distance between San Antonio and Little Rock is $8\frac{1}{2}$ in. What is the actual distance between San Antonio and Little Rock?

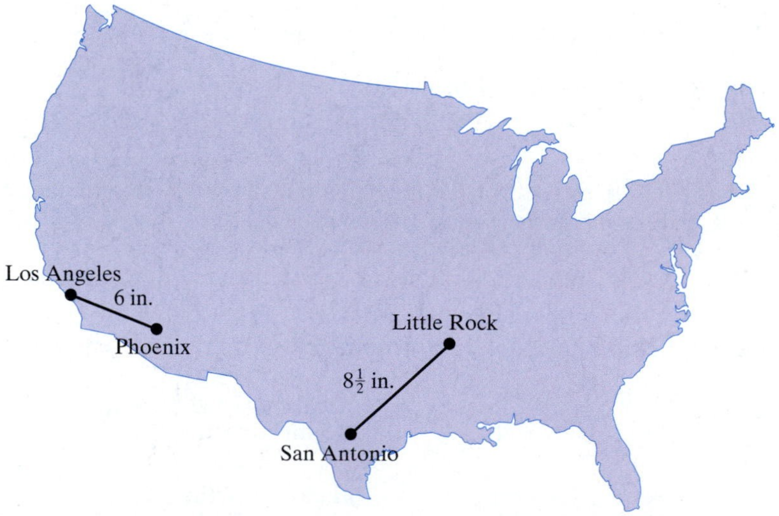

Solution:

Let d represent the distance between San Antonio and Little Rock.

$$\begin{matrix}\text{actual distance} \rightarrow \\ \text{distance on map} \rightarrow\end{matrix} \frac{348 \text{ mi}}{6 \text{ in.}} = \frac{d \text{ mi}}{8\frac{1}{2} \text{ in.}} \begin{matrix}\leftarrow \text{actual distance} \\ \leftarrow \text{distance on map}\end{matrix}$$

Translate to a proportion.

$$(348)(8\tfrac{1}{2}) = (6) \cdot d$$ Equate the cross products.

$$(348)(8.5) = 6d$$ Convert the values to decimal.

Answers

2. An area of 1440 ft^2 could be painted.
3. The distance between Seattle and San Francisco is 704 mi.

$$2958 = 6d$$

$$\frac{2958}{6} = \frac{\overset{1}{\cancel{6}}d}{\underset{1}{\cancel{6}}} \qquad \text{Divide both sides by 6.}$$

$$493 = d \qquad \text{Divide. } 2958 \div 6 = 493$$

The distance between San Antonio and Little Rock is 493 mi.

Example 4 Applying a Proportion to Environmental Science

A biologist wants to estimate the number of elk in a wildlife preserve. She sedates 25 elk and clips a small radio transmitter onto the ear of each animal. The elk return to the wild, and after 6 months, the biologist studies a sample of 120 elk in the preserve. Of the 120 elk sampled, 4 have radio transmitters. Approximately how many elk are in the whole preserve?

Solution:

Let n represent the number of elk in the whole preserve.

	Sample		Population	
number of elk in the sample with radio transmitters →	4	=	25	← number of elk in the population with radio transmitters
total elk in the sample →	120		n	← total elk in the population

$$\frac{4}{120} = \frac{25}{n}$$

$$4 \cdot n = (120)(25) \qquad \text{Equate the cross products.}$$

$$4n = 3000$$

$$\frac{\overset{1}{\cancel{4}}n}{\underset{1}{\cancel{4}}} = \frac{3000}{4} \qquad \text{Divide both sides by 4.}$$

$$n = 750 \qquad \text{Divide. } 3000 \div 4 = 750$$

There are approximately 750 elk in the wildlife preserve.

Skill Practice

4. To estimate the number of fish in a lake, the park service catches 50 fish and tags them. After several months the park service catches a sample of 100 fish and finds that 6 are tagged. Approximately how many fish are in the lake?

2. Similar Figures

Two triangles whose corresponding sides are proportional are called **similar triangles**. This means that the triangles have the same "shape" but may be different sizes. The following pairs of triangles are similar (Figure 6-1).

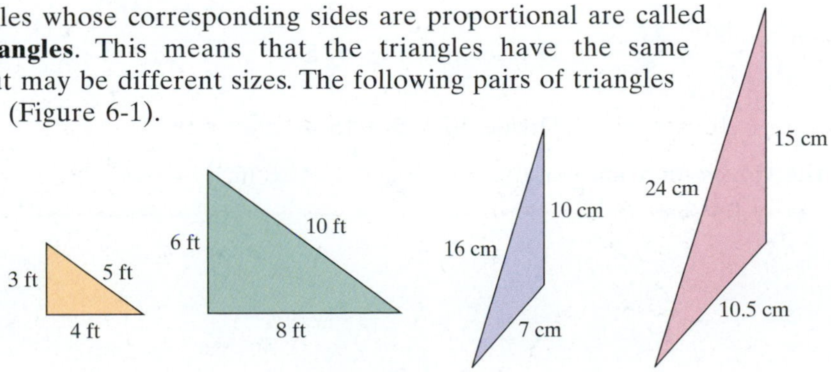

Figure 6-1

Concept Connections

5. Consider the similar triangles shown.

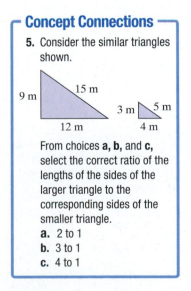

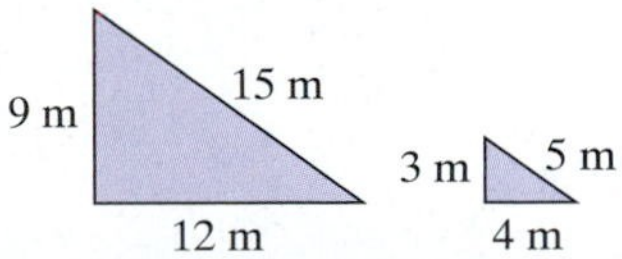

From choices **a, b,** and **c,** select the correct ratio of the lengths of the sides of the larger triangle to the corresponding sides of the smaller triangle.

a. 2 to 1
b. 3 to 1
c. 4 to 1

Answers

4. There are approximately 833 fish in the lake.
5. b

Consider the triangles:

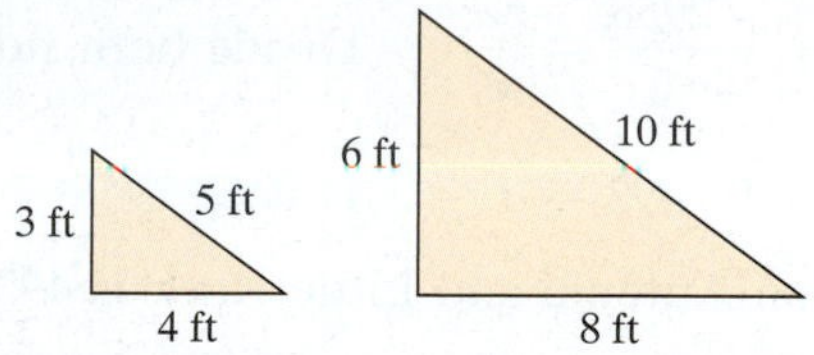

Notice that the ratio formed by the "left" sides of the triangles is $\frac{3}{6}$. This is the same as the ratio formed by the "bottom" sides, $\frac{4}{8}$, and is the same as the ratio formed by the "right" sides, $\frac{5}{10}$. Each ratio simplifies to $\frac{1}{2}$. Because all ratios are equal, the corresponding sides are proportional.

Skill Practice

6. The triangles below are similar.

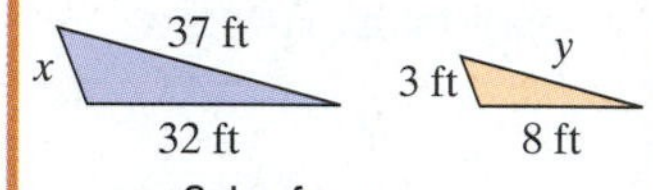

a. Solve for x.
b. Solve for y.

Example 5 Finding the Unknown Sides in Similar Triangles

The triangles in Figure 6-2 are similar.

a. Solve for x. **b.** Solve for y.

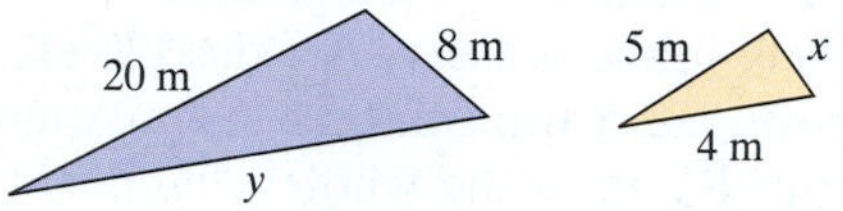

Figure 6-2

Solution:

a. Notice that the lengths of the "left" sides of both triangles are known. This forms a known ratio of $\frac{20}{5}$. Because the triangles are similar, the ratio of the other corresponding sides must be equal to $\frac{20}{5}$. To solve for x, we have

large triangle "left" side ⟶ $\frac{20}{5} = \frac{8}{x}$ ⟵ large triangle "right" side
small triangle "left" side ⟶ ⟵ small triangle "right" side

$20 \cdot x = (5)(8)$ Equate the cross products.

$20x = 40$

$\frac{\overset{1}{\cancel{20}}x}{\underset{1}{\cancel{20}}} = \frac{40}{20}$ Divide both sides of the equation by 20.

$x = 2$ Divide $40 \div 20 = 2$.

b. To solve for y, we have

large triangle "left" side ⟶ $\frac{20}{5} = \frac{y}{4}$ ⟵ large triangle "bottom" side
small triangle "left" side ⟶ ⟵ small triangle "bottom" side

$(20)(4) = 5 \cdot y$ Equate the cross products.

$80 = 5y$

$\frac{80}{5} = \frac{\overset{1}{\cancel{5}}y}{\underset{1}{\cancel{5}}}$ Divide both sides by 5.

$16 = y$ Divide. $80 \div 5 = 16$

The values for x and y are $x = 2$ m and $y = 16$ m.

TIP: In Example 5, notice that the corresponding sides of the triangles form a 4 to 1 ratio.

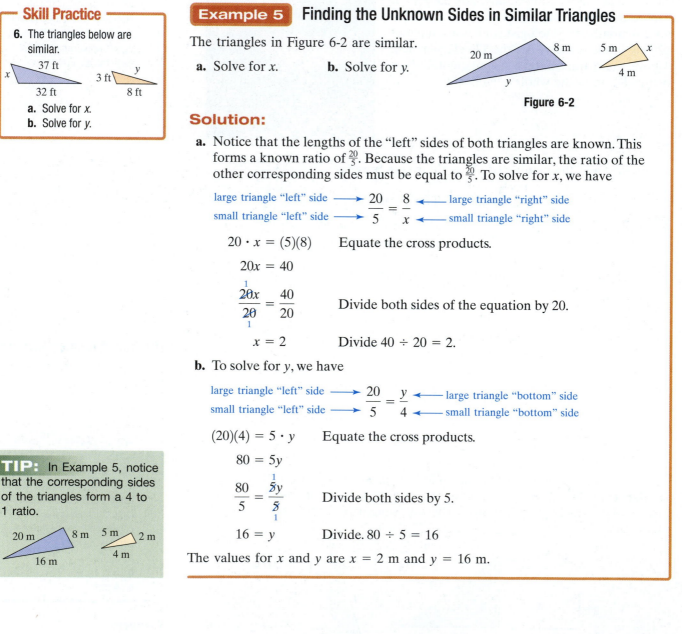

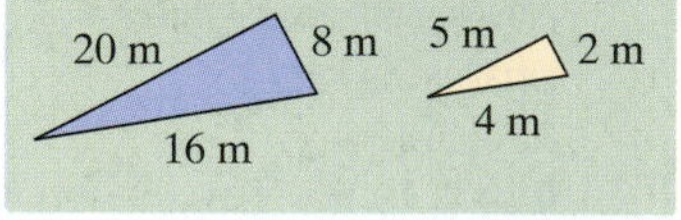

Answer

6. a. $x = 12$ ft **b.** $y = 9.25$ ft

Example 6 Using Similar Triangles in an Application

On a sunny day, a 6-ft man casts a 3.2-ft shadow on the ground. At the same time, a building casts an 80-ft shadow. How tall is the building?

Solution:

We illustrate the scenario in Figure 6-3. Note, however, that the distances are not drawn to scale.

Let h represent the height of the building in feet.

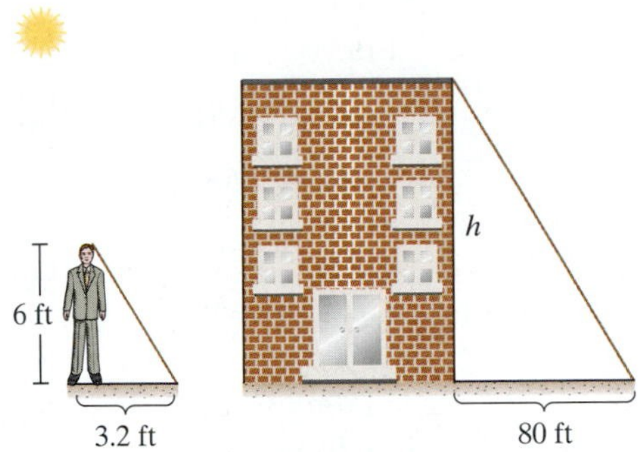

Figure 6-3

We will assume that the triangles are similar because the readings were taken at the same time of day.

$$\frac{6}{h} = \frac{3.2}{80}$$ Translate to a proportion.

$$(6)(80) = (3.2) \cdot h$$ Equate the cross products.

$$480 = 3.2h$$

$$\frac{480}{3.2} = \frac{\overset{1}{\cancel{3.2}}h}{\underset{1}{\cancel{3.2}}}$$ Divide both sides by 3.2.

$$150 = h$$ Divide. $480 \div 3.2 = 150$

The height of the building is 150 ft.

Skill Practice

7. Donna has drawn a pattern from which to make a scarf. Assuming that the triangles are similar (the pattern and scarf have the same shape), find the length of side x (in yards).

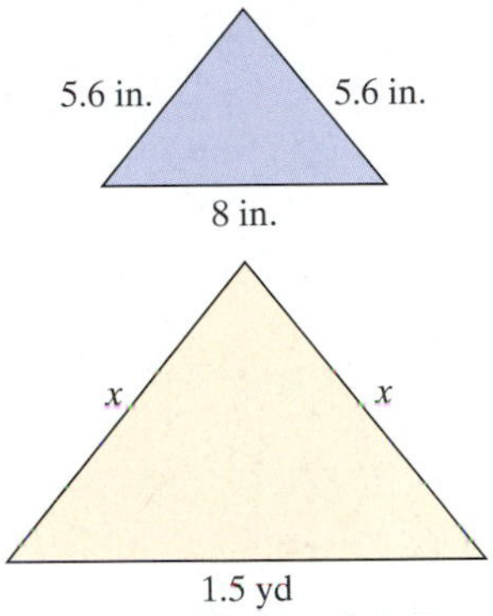

In addition to studying similar triangles, we present similar polygons. Recall from Section 1.3 that a polygon is a flat figure formed by line segments connected at their ends. If the corresponding sides of two polygons are proportional, then we have **similar polygons**.

Example 7 Using Similar Polygons in an Application

A negative for a photograph is 3.5 cm by 2.5 cm. If the width of the resulting picture is 4 in., what is the length of the picture? See Figure 6-4.

Figure 6-4

Skill Practice

8. The first- and second-place plaques for a softball tournament have the same shape but are different sizes. Find the length of side x.

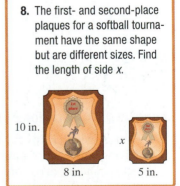

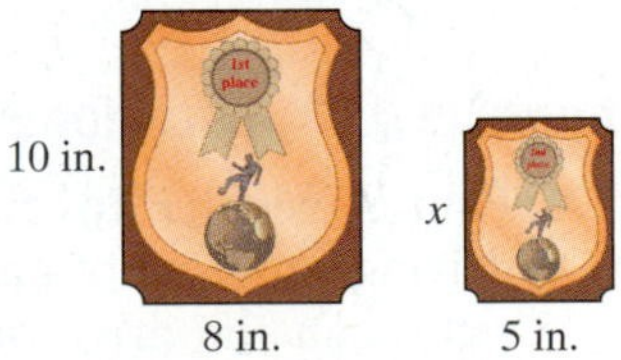

Answers

7. Side x is 1.05 yd long.
8. Side x is 6.25 in.

Solution:

Let x represent the length of the photo.

The photo and its negative are similar polygons.

$$\frac{3.5 \text{ cm}}{x \text{ in.}} = \frac{2.5 \text{ cm}}{4 \text{ in.}} \qquad \text{Translate to a proportion.}$$

$$(3.5)(4) = (2.5) \cdot x \qquad \text{Equate the cross products.}$$

$$14 = 2.5x$$

$$\frac{14}{2.5} = \frac{\overset{1}{\cancel{2.5}}x}{\underset{1}{\cancel{2.5}}} \qquad \text{Divide both sides by 2.5.}$$

$$5.6 = x \qquad \text{Divide. } 14 \div 2.5 = 5.6$$

The picture is 5.6 in. long.

Section 6.4 Practice Exercises

Study Skills Exercise

1. Define the key terms.

a. Similar triangles **b. Similar polygons**

Review Exercises

For Exercises 2–5, use = or ≠ to make a true statement.

2. $\frac{-4}{7} \square \frac{12}{-21}$ **3.** $\frac{3}{-13} \square \frac{-15}{65}$ **4.** $\frac{2}{5} \square \frac{21}{55}$ **5.** $\frac{12}{7} \square \frac{35}{19}$

For Exercises 6–12, solve the proportion.

6. $\frac{2}{7} = \frac{3}{x}$ **7.** $\frac{4}{3} = \frac{n}{5}$ **8.** $\frac{p}{9} = \frac{1}{6}$ **9.** $\frac{-3\frac{1}{2}}{k} = \frac{2\frac{1}{3}}{4}$

10. $\frac{-2.4}{3} = \frac{m}{5}$ **11.** $\frac{3}{2.1} = \frac{7}{y}$ **12.** $\frac{1.2}{4} = \frac{3}{a}$

Objective 1: Applications of Proportions

13. Pam drives her Toyota Prius 244 mi in city driving on 4 gal of gas. At this rate how many miles can she drive on 10 gal of gas? **(See Example 1.)**

14. Didi takes her pulse for 10 sec and counts 13 beats. How many beats per minute is this?

15. To cement a garden path, it takes crushed rock and cement in a ratio of 3.25 kg of rock to 1 kg of cement. If a 24 kg-bag of cement is purchased, how much crushed rock will be needed? **(See Example 2.)**

16. Suppose two adults produce 63.4 lb of garbage in one week. At this rate, how many pounds will 50 adults produce in one week?

17. On a map, the distance from Sacramento, California, to San Francisco, California, is 8 cm. The legend gives the actual distance at 91 mi. On the same map, Faythe measured 7 cm from Sacramento to Modesto, California. What is the actual distance? (Round to the nearest mile.) **(See Example 3.)**

18. On a map, the distance from Nashville, Tennessee, to Atlanta, Georgia, is 3.5 in., and the actual distance is 210 mi. If the map distance between Dallas, Texas, and Little Rock, Arkansas, is 4.75 in., what is the actual distance?

19. At Central Community College, the ratio of female students to male students is 31 to 19. If there are 6200 female students, how many male students are there?

20. Evelyn won an election by a ratio of 6 to 5. If she received 7230 votes, how many votes did her opponent receive?

21. If you flip a coin many times, the coin should come up heads about 1 time out of every 2 times it is flipped. If a coin is flipped 630 times, about how many heads do you expect to come up?

22. A die is a small cube used in games of chance. It has six sides, and each side has 1, 2, 3, 4, 5, or 6 dots painted on it. If you roll a die, the number 4 should come up about 1 time out of every 6 times the die is rolled. If you roll a die 366 times, about how many times do you expect the number 4 to come up?

23. A pitcher gave up 42 earned runs in 126 innings. Approximately how many earned runs will he give up in one game (9 innings)? This value is called the earned run average.

24. In one game Peyton Manning completed 34 passes for 357 yd. At this rate how many yards would be gained for 22 passes?

25. Pierre bought 480 Euros with \$750 American. At this rate, how many Euros can he buy with \$900 American?

26. Erik bought \$624 Canadian with \$600 American. At this rate, how many Canadian dollars can he buy with \$250 American?

27. Each gram of fat consumed has 9 calories. If a $\frac{1}{2}$-cup serving of gelato has 81 calories from fat, how many grams of fat are in this serving?

28. Approximately 24 out of 100 Americans over the age of 12 smoke. How many smokers would you expect in a group of 850 Americans over the age of 12?

29. Park officials stocked a man-made lake with bass last year. To approximate the number of bass this year, a sample of 75 bass is taken out of the lake and tagged. Then later a different sample is taken, and it is found that 21 of 100 bass are tagged. Approximately how many bass are in the lake? Round to the nearest whole unit. **(See Example 4.)**

30. Laws have been instituted in Florida to help save the manatee. To establish the number in Florida, a sample of 150 manatees was marked and let free. A new sample was taken and found that there were 3 marked manatees out of 40 captured. What is the approximate population of manatees in Florida?

31. Yellowstone National Park in Wyoming has the largest population of free-roaming bison. To approximate the number of bison, 200 are captured and tagged and then let free to roam. Later, a sample of 120 bison is observed and 6 have tags. Approximate the population of bison in the park.

32. In Cass County, Michigan, there are about 20 white-tailed deer per square mile. If the county covers 492 mi^2, about how many white-tailed deer are in the county?

Objective 2: Similar Figures

For Exercises 33–36, the pairs of triangles are similar. Solve for x and y. **(See Example 5.)**

33.

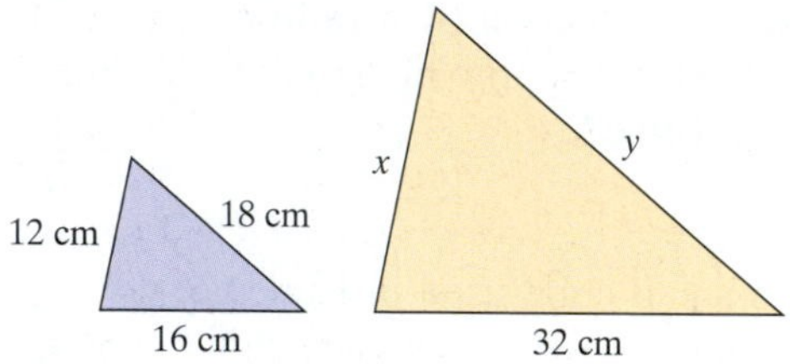

34.

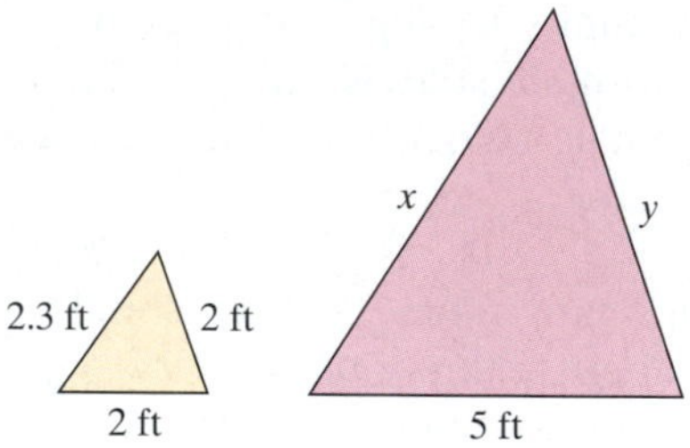

35.

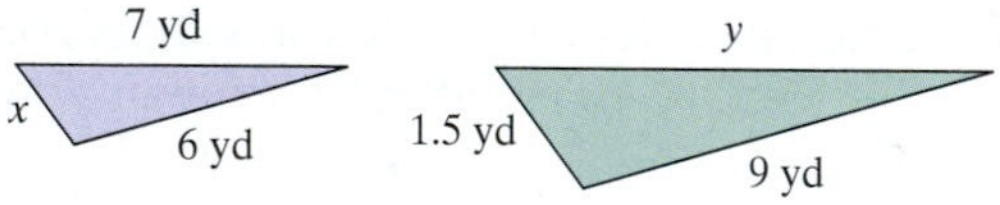

36.

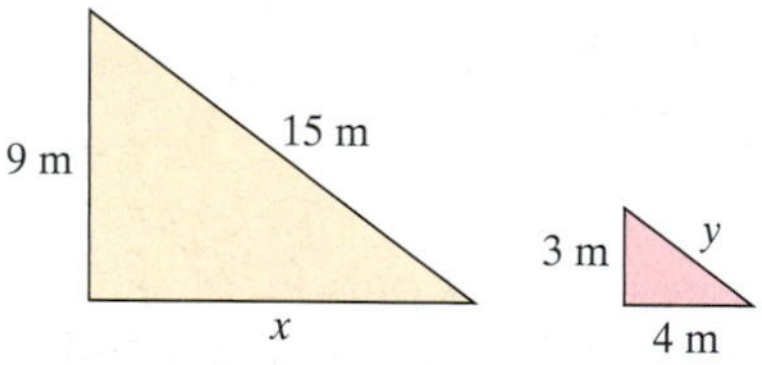

37. The height of a flagpole can be determined by comparing the shadow of the flagpole and the shadow of a yardstick. From the figure, determine the height of the flagpole. **(See Example 6.)**

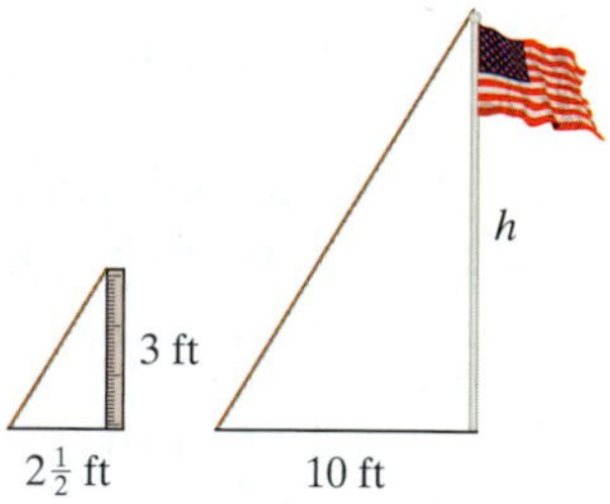

Figure for Exercise 37

38. A 15-ft flagpole casts a 4-ft shadow. How long will the shadow be for a 90-ft building?

39. A person 1.6 m tall stands next to a lifeguard observation platform. If the person casts a shadow of 1 m and the lifeguard platform casts a shadow of 1.5 m, how high is the platform?

40. A 32-ft tree casts a shadow of 18 ft. How long will the shadow be for a 22-ft tree?

For Exercises 41–44, the pairs of polygons are similar. Solve for the indicated variables. **(See Example 7.)**

41.

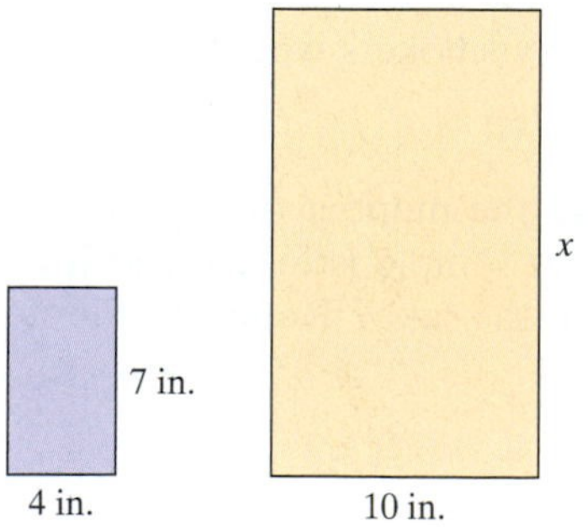

42.

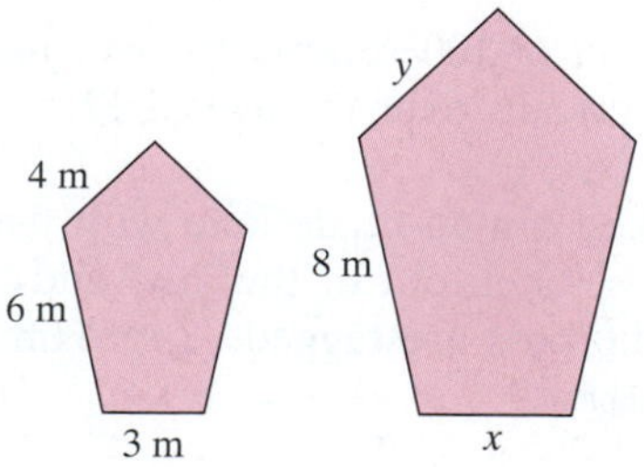

43.

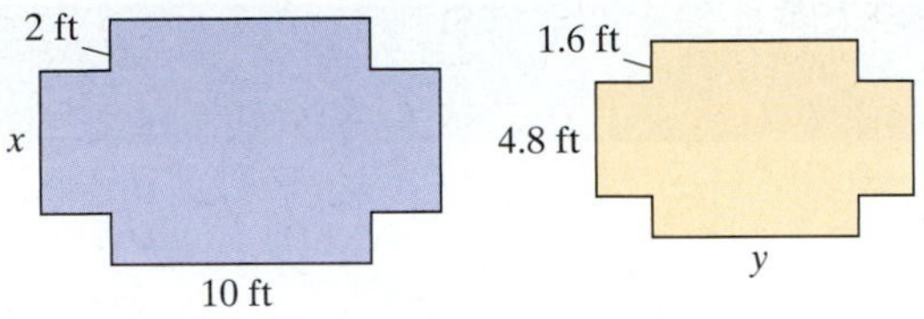

44.

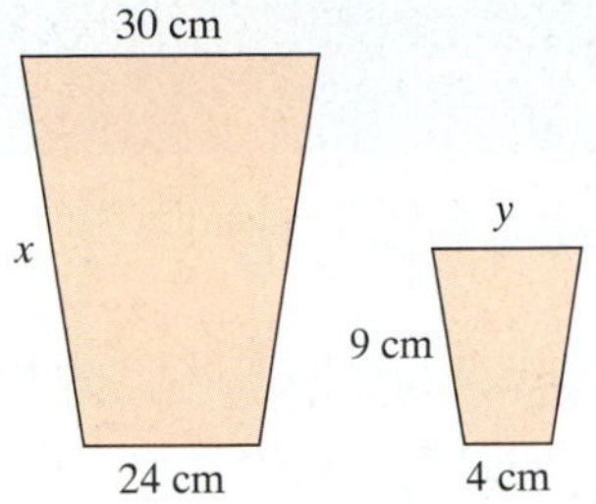

45. A carpenter makes a schematic drawing of a porch he plans to build. On the drawing, 2 in. represents 7 ft. Find the lengths denoted by x, y, and z.

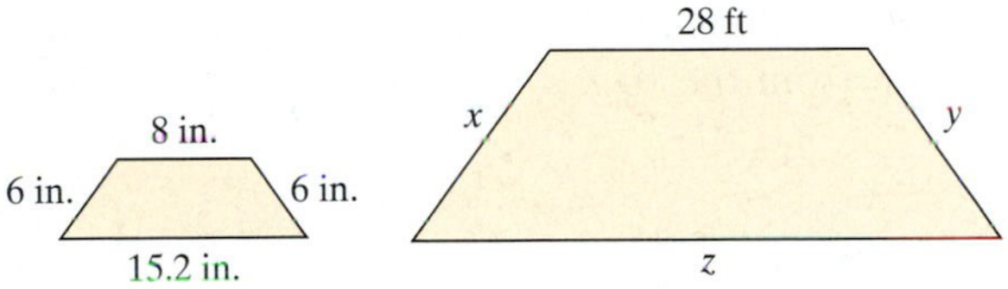

46. The Great Cookie Company has a sign on the front of its store as shown in the figure. The company would like to put a sign of the same shape in the back, but with dimensions $\frac{1}{3}$ as large. Find the lengths denoted by x and y.

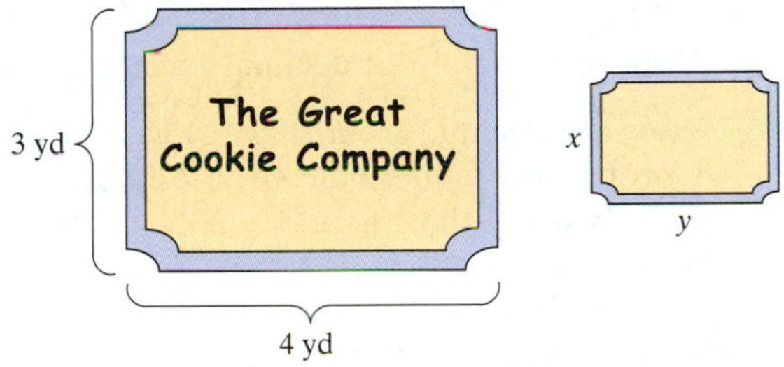

Calculator Connections

Topic: Solving Proportions

Calculator Exercises

47. In a recent year, the annual crime rate for Oklahoma was 4743 crimes per 100,000 people. If Oklahoma had approximately 3,500,000 people at that time, approximately how many crimes were committed? (*Source:* Oklahoma Department of Corrections)

48. To measure the height of the Washington Monument, a student 5.5 ft tall measures his shadow to be 3.25 ft. At the same time of day, he measured the shadow of the Washington Monument to be 328 ft long. Estimate the height of the monument to the nearest foot.

49. In a recent year, the rate of breast cancer in women was 110 cases per 100,000 women. At that time, the state of California had approximately 14,000,000 women. How many women in California would be expected to have breast cancer? (*Source:* Centers for Disease Control)

50. In a recent year, the rate of prostate disease in U.S. men was 118 cases per 1000 men. At that time, the state of Massachusetts had approximately 2,500,000 men. How many men in Massachusetts would be expected to have prostate disease? (*Source:* National Center for Health Statistics)

Group Activity

Investigating Probability

Materials: Paper bags containing 10 white poker chips, 6 red poker chips, and 4 blue poker chips.

Estimated time: 15 minutes

Group Size: 3

1. Each group will receive a bag of poker chips, with 10 white, 6 red, and 4 blue chips.

2. a. Write the ratio of red chips in the bag to the total number of chips in the bag. ____________ This value represents the *probability* of randomly selecting a red chip from the bag.

b. Write this fraction in decimal form. ____________

c. Write the decimal from step (b) as a percent. ____________
A probability value indicates the likeliness of an event to occur. For example, to interpret this probability, one might say that there is a 30% chance of selecting a red chip at random from the bag.

3. Determine the probability of selecting a white chip from the bag. Interpret your answer.

4. Determine the probability of selecting a blue chip from the bag. Interpret your answer.

5. Next, have one group member select a chip from the bag at random (without looking), and record the color of the chip. Then return the chip to the bag. Repeat this step for a total of 20 times (be sure that each student in the group has a chance to pick). Record the total number of red, white, and blue chips selected. Then write the ratio of the number selected out of 20.

	Number of Times Selected	**Ratio of Number Selected Out of 20**
Red		
White		
Blue		

6. How well do your experimental results match the theoretical probabilities found in steps 2–4?

7. The instructor will now pool the data from the whole class. Write the total number of times that red, white, and blue chips were selected, respectively, by the whole class. Then write the ratio of these values to the total number of selections made.

8. How well do the experimental results from the whole class match the theoretical probabilities of selecting a red, white, or blue chip?

Chapter 6 Summary

Section 6.1 Ratios

Key Concepts

A **ratio** is a comparison of two quantities.

The ratio of a to b can be written as follows, provided $b \neq 0$.

1. a to b
2. $a:b$
3. $\frac{a}{b}$

When we write a ratio in fraction form, we generally simplify it to lowest terms.

Ratios that contain mixed numbers, fractions, or decimals can be simplified to lowest terms with whole numbers in the numerator and denominator.

Examples

Example 1

Three forms of a ratio:

4 to 6 $\quad$ 4 : 6 $\quad$ $\frac{4}{6}$

Example 2

A hockey team won 4 games out of 6. Write a ratio of games won to total games played and simplify to lowest terms.

$$\frac{4 \text{ games won}}{6 \text{ games played}} = \frac{2}{3}$$

Example 3

$$\frac{2\frac{1}{6}}{\frac{2}{3}} = 2\frac{1}{6} \div \frac{2}{3} = \frac{13}{\underset{2}{\cancel{6}}} \cdot \frac{\overset{1}{\cancel{3}}}{2} = \frac{13}{4}$$

Example 4

$$\frac{2.1}{2.8} = \frac{2.1}{2.8} \cdot \frac{10}{10} = \frac{21}{28} = \frac{\overset{3}{\cancel{21}}}{\underset{4}{\cancel{28}}} = \frac{3}{4}$$

Section 6.2 Rates

Key Concepts

A **rate** compares two different quantities.

A rate having a denominator of 1 unit is called a **unit rate**. To find a unit rate, divide the numerator by the denominator.

A **unit cost** or unit price is the cost per 1 unit, for example, \$1.21/lb or 43¢/oz. Comparing unit prices can help determine the best buy.

Examples

Example 1

New Jersey has 8,470,000 people living in 21 counties. Write a reduced ratio of people per county.

$$\frac{8{,}470{,}000 \text{ people}}{21 \text{ counties}} = \frac{1{,}210{,}000 \text{ people}}{3 \text{ counties}}$$

Example 2

If a race car traveled 1250 mi in 8 hr during a race, what is its speed in miles per hour?

$$\frac{1250 \text{ mi}}{8 \text{ hr}} = 156.25 \text{ mi/hr}$$

Example 3

Tide laundry detergent is offered in two sizes: \$18.99 for 150 oz and \$13.59 for 100 oz. Find the unit prices to find the best buy.

$$\frac{\$18.99}{150 \text{ oz}} \approx \$0.1266/\text{oz}$$

$$\frac{\$13.59}{100 \text{ oz}} \approx \$0.1359/\text{oz}$$

The 150-oz package is the better buy because the unit cost is less.

Section 6.3 Proportions

Key Concepts

A **proportion** states that two ratios or rates are equal.

$\frac{14}{21} = \frac{2}{3}$ is a proportion.

To determine if two ratios form a proportion, check to see if the cross products are equal, that is,

$\frac{a}{b} = \frac{c}{d}$ implies $a \cdot d = b \cdot c$ (and vice versa)

To solve a proportion, solve the equation formed by the cross products.

Examples

Example 1

Write as a proportion.

56 mi is to 2 gal as 84 mi is to 3 gal.

$$\frac{56 \text{ mi}}{2 \text{ gal}} = \frac{84 \text{ mi}}{3 \text{ gal}}$$

Example 2

$$\frac{3}{8} \stackrel{?}{=} \frac{2\frac{1}{2}}{6\frac{2}{3}}$$

$$3 \cdot 6\frac{2}{3} \stackrel{?}{=} 8 \cdot 2\frac{1}{2}$$

$$\frac{3}{1} \cdot \frac{20}{3} \stackrel{?}{=} \frac{8}{1} \cdot \frac{5}{2}$$

$$20 = 20 \checkmark$$

The ratios form a proportion.

Example 3

$$\frac{5}{-4} = \frac{18}{x} \qquad 5x = -4 \cdot 18$$

$$5x = -72$$

$$\frac{5x}{5} = \frac{-72}{5}$$

$$x = -\frac{72}{5}$$

The solution is $-\frac{72}{5}$ or $-14\frac{2}{5}$ or -14.4.

Section 6.4 Applications of Proportions and Similar Figures

Key Concepts

Example 1 demonstrates an application involving a proportion. Example 2 demonstrates the use of proportions involving similar triangles.

Example 1

According to the National Highway Traffic Safety Administration, 2 out of 5 traffic fatalities involve the use of alcohol. If there were 43,200 traffic fatalities in a recent year, how many involved the use of alcohol?

Let n represent the number of traffic fatalities involving alcohol.

Set up a proportion:

$$\frac{\text{2 traffic fatalities w/alcohol}}{\text{5 traffic fatalities}} = \frac{n}{43{,}200}$$

Solve the proportion:

$$2(43{,}200) = 5n$$

$$86{,}400 = 5n$$

$$\frac{86{,}400}{5} = \frac{\overset{1}{\cancel{5}}n}{\underset{1}{\cancel{5}}}$$

$$17{,}280 = n$$

17,280 traffic fatalities involved alcohol.

Examples

Example 2

On a sunny day, a tree 6-ft tall casts a 4-ft shadow. At the same time, a telephone pole casts a 10-ft shadow. What is the height of the telephone pole?

A picture illustrates the situation. Let h represent the height of the telephone pole.

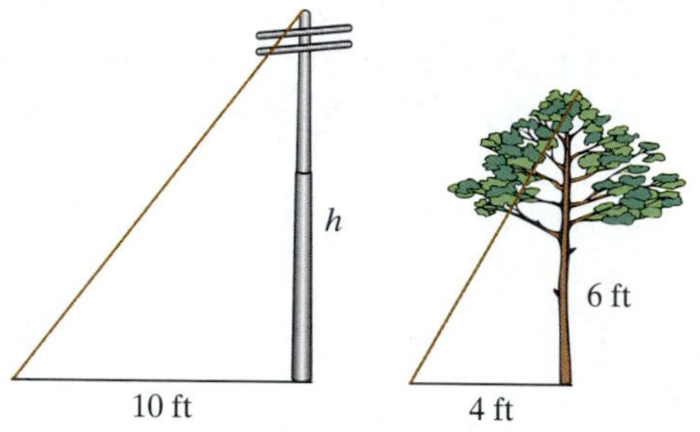

Set up a proportion:

$$\frac{10}{4} = \frac{h}{6}$$

Solve the proportion:

$$(10)(6) = 4h$$

$$60 = 4h$$

$$\frac{60}{4} = \frac{\overset{1}{\cancel{4}}h}{\underset{1}{\cancel{4}}}$$

$15 = h$ The pole is 15 ft high.

Chapter 6 Review Exercises

Section 6.1

For Exercises 1–3, write the ratios in two other ways.

1. 5 : 4 **2.** 3 to 1 **3.** $\frac{8}{7}$

For Exercises 4–6, write the ratios in fraction form.

4. Saul had three daughters and two sons.

a. Write a ratio of the number of sons to the number of daughters.

b. Write a ratio of the number of daughters to the number of sons.

c. Write a ratio of the number of daughters to the total number of children.

5. In his refrigerator, Jonathan has four bottles of soda and five bottles of juice.

a. Write a ratio of the number of bottles of soda to the number of bottles of juice.

b. Write a ratio of the number of bottles of juice to the number of bottles of soda.

c. Write a ratio of the number of bottles of juice to the total number of bottles.

6. There are 12 face cards in a regular deck of 52 cards.

a. Write a ratio of the number of face cards to the total number of cards.

b. Write a ratio of the number of face cards to the number of cards that are not face cards.

For Exercises 7–10, write the ratio in lowest terms.

7. 52 cards to 13 cards **8.** \$21 to \$15

9. 80 ft to 200 ft **10.** 7 days to 28 days

For Exercises 11–14, write the ratio in lowest terms with whole numbers in the numerator and denominator.

11. $1\frac{1}{2}$ hr to $\frac{1}{3}$ hr **12.** $\frac{2}{3}$ yd to $2\frac{1}{6}$ yd

13. \$2.56 to \$1.92 **14.** 42.5 mi to 3.25 mi

15. This year a high school had an increase of 320 students. The enrollment last year was 1200 students.

a. How many students will be attending this year?

b. Write a ratio of the increase in the number of students to the total enrollment of students this year. Simplify to lowest terms.

16. A living room has dimensions of 3.8 m by 2.4 m. Find the ratio of length to width and reduce to lowest terms.

For Exercises 17–18, refer to the table that shows the number of personnel who smoke in a particular workplace.

	Smokers	Nonsmokers	Totals
Office personnel	12	20	32
Shop personnel	60	55	115

17. Find the ratio of the number of office personnel who smoke to the number of shop personnel who smoke.

18. Find the ratio of the total number of personnel who smoke to the total number of personnel.

Section 6.2

For Exercises 19–22, write each rate in lowest terms.

19. A concession stand sold 20 hot dogs in 45 min.

20. Mike can skate 4 mi in 34 min.

21. During a period in which the economy was weak, Evelyn's balance on her investment account changed by −\$3400 in 6 months.

22. A submarine's "elevation" changed by −75 m in 45 min.

23. What is the difference between rates in lowest terms and unit rates?

For Exercises 24–27, write each rate as a unit rate.

24. A pheasant can fly 44 mi in $1\frac{1}{3}$ hr.

25. The temperature changed −14° in 3.5 hr.

26. A hummingbird can flap its wings 2700 times in 30 sec.

27. It takes David's lawn company 66 min to cut six lawns.

For Exercises 28–29, find the unit costs. Round the answers to three decimal places when necessary.

28. Body lotion costs \$5.99 for 10 oz.

29. Three towels cost \$10.00.

For Exercises 30–31, compute the unit cost (round to three decimal places). Then determine the best buy.

30. a. 48 oz of detergent for \$5.99

b. 60 oz of detergent for \$7.19

c. Which is the best buy?

31. a. 32 oz of spaghetti sauce for \$2.49

b. 48 oz of spaghetti sauce for \$3.59

c. Which is the best buy?

32. Suntan lotion costs \$5.99 for 8 oz. If Sherri has a coupon for \$2.00 off, what will be the unit cost of the lotion after the coupon has been applied?

33. A 24-roll pack of bathroom tissue costs \$8.99 without a discount card. The package is advertised at 29¢ per roll if the buyer uses the discount card. What is the difference in price per roll when the buyer uses the discount card? Round to the nearest cent.

34. In Wilmington, North Carolina, Hurricane Floyd dropped 15.06 in. of rain during a 24-hr period. What was the average rainfall per hour? (*Source:* National Weather Service)

35. For a recent year, Toyota steadily increased the number of hybrid vehicles for sale in the United States from 130,000 to 250,000.

a. What was the increase in the number of hybrid vehicles?

b. How many additional hybrid vehicles will be available each month?

36. In 1990, Americans ate on average 386 lb of vegetables per year. By 2008, this value increased to 449 lb.

a. What was the increase in the number of pounds of vegetables?

b. How many additional pounds of vegetables did Americans add to their diet per year?

Section 6.3

For Exercises 37–42, write a proportion for each statement.

37. 16 is to 14 as 12 is to $10\frac{1}{2}$.

38. 8 is to 20 as 6 is to 15.

39. The numbers -5 and 3 are proportional to the numbers -10 and 6.

40. The numbers 4 and -3 are proportional to the numbers 20 and -15.

41. \$11 is to 1 hr as \$88 is to 8 hr.

42. 2 in. is to 5 mi as 6 in. is to 15 mi.

For Exercises 43–46, determine whether the ratios form a proportion.

43. $\frac{64}{81} \stackrel{?}{=} \frac{8}{9}$

44. $\frac{3\frac{1}{2}}{7} \stackrel{?}{=} \frac{7}{14}$

45. $\frac{5.2}{3} \stackrel{?}{=} \frac{15.6}{9}$

46. $\frac{6}{10} \stackrel{?}{=} \frac{6.3}{10.3}$

For Exercises 47–50, determine whether the pairs of numbers are proportional.

47. Are the numbers $2\frac{1}{8}$ and $4\frac{3}{4}$ proportional to the numbers $3\frac{2}{5}$ and $7\frac{3}{5}$?

48. Are the numbers $5\frac{1}{2}$ and 6 proportional to the numbers $6\frac{1}{2}$ and 7?

49. Are the numbers -4.25 and -8 proportional to the numbers 5.25 and 10?

50. Are the numbers 12.4 and 9.2 proportional to the numbers -3.1 and -2.3?

For Exercises 51–56, solve the proportion.

51. $\frac{100}{16} = \frac{25}{x}$

52. $\frac{y}{6} = \frac{45}{10}$

53. $\frac{1\frac{6}{7}}{b} = \frac{13}{21}$

54. $\frac{p}{6\frac{1}{3}} = \frac{3}{9\frac{1}{2}}$

55. $\frac{2.5}{-6.8} = \frac{5}{h}$

56. $\frac{0.3}{1.2} = \frac{k}{-3.6}$

Section 6.4

57. One year of a dog's life is about the same as 7 years of a human life. If a dog is 12 years old in dog years, how does that equate to human years?

58. Lavu bought 17,120 Japanese yen with \$160 American. At this rate, how many yen can he buy with \$450 American?

59. The number of births in Alabama in a recent year was approximately 59,800. If the birthrate was about 13 per 1000, what was the approximate population of Alabama? (Round to the nearest person.)

60. If the tax on a $25.00 item is $1.20, what would be the tax on an item costing $145.00?

61. The triangles shown in the figure are similar. Solve for x and y.

62. The height of a building can be approximated by comparing the shadows of the building and of a meterstick. From the figure, find the height of the building.

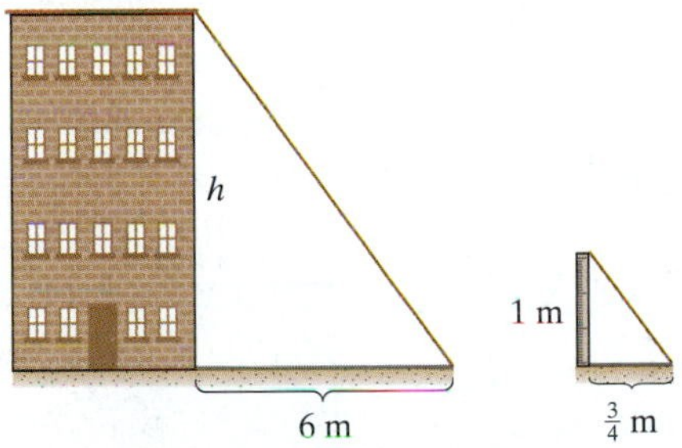

63. The polygons shown in the figure are similar. Solve for x and y.

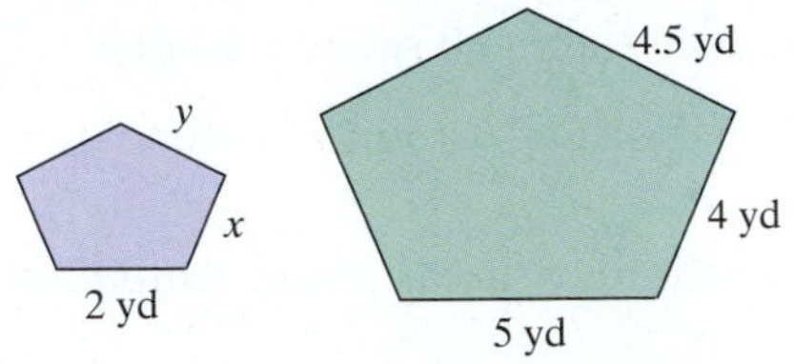

64. The figure shows two picture frames that are similar. Use this information to solve for x and y.

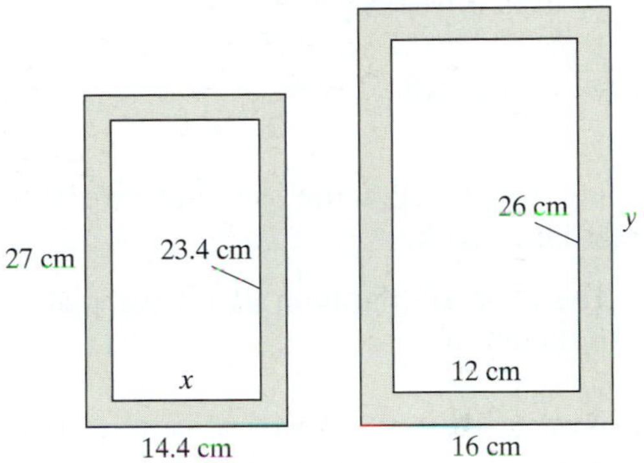

Chapter 6 Test

1. An elementary school has 25 teachers and 521 students. Write a ratio of teachers to students in three different ways.

2. The months August and September bring the greatest number of hurricanes to the eastern seaboard and gulf coast states. As of this writing, 27 hurricanes had struck the U.S. mainland in August and 44 had struck in September. (*Source:* NOAA)

a. Write a ratio of the number of hurricanes to strike in September to the number in August.

b. Write a ratio of the number of hurricanes to strike in August to the total number in these two peak months.

For Exercises 3–4, write as a reduced ratio in fraction form.

3. For a recent WNBA season, the Connecticut Sun had a win-loss ratio of 26 to 8.

4. For a recent WNBA season, the Houston Comets had a win-loss ratio of 18 to 16.

5. Find the ratio of the shortest side to the longest side. Write the ratio in lowest terms.

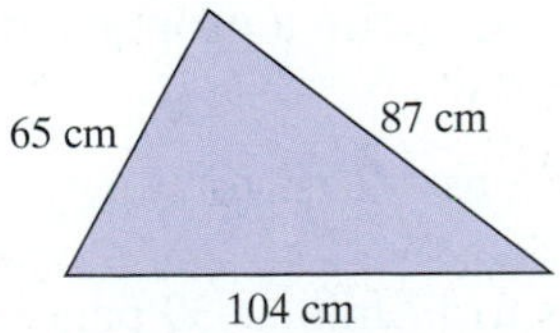

6. a. In a recent year, the number of people in New Mexico whose income was below poverty level was 168 out of every 1000. Write this as a simplified ratio.

b. The poverty level in Iowa was 72 people to 1000 people. Write this as a simplified ratio.

c. Compare the ratios and comment.

7. Write as a simplified ratio in two ways: 30 sec to $1\frac{1}{2}$ min

a. By converting 30 sec to minutes.

b. By converting $1\frac{1}{2}$ min to seconds.

For Exercises 8–10, write as a rate, simplified to lowest terms.

8. 255 mi per 6 hr

9. 20 lb in 6 weeks

10. 4 g of fat in 8 cookies

For Exercises 11–12, write as a unit rate. Round to the nearest hundredth.

11. The element platinum had density of 2145 g per 100 cm^3.

12. Approximately 104.8 oz of iron is present in 45.8 lb of rocks brought back from the moon.

13. What is the unit cost for Raid Ant and Roach spray valued at \$6.72 for 30 oz? Round to the nearest cent.

14. A package containing 3 toe rings is on sale for 2 packs for \$6.60. What is the cost of 1 toe ring?

15. A generic pain reliever is on sale for 2 bottles for \$5.49. Each bottle contains 30 tablets. Aleve pain reliever is sold in 24-capsule bottles for \$4.49. Find the unit cost of each to determine the better buy.

16. What does it mean for two pairs of numbers to be proportional?

For Exercises 17–19, write a proportion for each statement.

17. -42 is to 15 as -28 is to 10.

18. 20 pages is to 12 min as 30 pages is to 18 min.

19. \$15 an hour is proportional to \$75 for 5 hr.

20. Are the numbers 105 and 55 proportional to the numbers 21 and 10?

For Exercises 21–24, solve the proportion.

21. $\frac{25}{p} = \frac{45}{-63}$

22. $\frac{32}{20} = \frac{20}{x}$

23. $\frac{n}{9} = \frac{3\frac{1}{3}}{6}$

24. $\frac{y}{-14} = \frac{7.2}{16.8}$

25. x is to 2.5 as 12 is to 30. Solve for x.

26. A computer on dial-up can download 1.6 megabytes (MB) in 2.5 min. How long will it take to download a 4.8-MB file?

27. Cherise is an excellent student and studies 7.5 hr outside of class each week for a 3-credit-hour math class. At this rate, how many hours outside of class does she spend on homework if she is taking 12 credit-hours at school?

28. Ms. Ehrlich wants to approximate the number of goldfish in her backyard pond. She scooped out 8 and marked them. Later she scooped out 10 and found that 3 were marked. Estimate the number of goldfish in her pond. Round to the nearest whole unit.

29. Given that the two triangles are similar, solve for x and y.

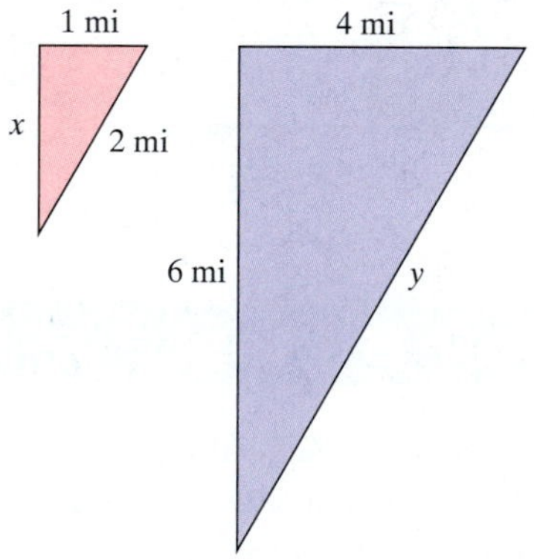

30. Maggie takes a brochure that measures 10 cm by 15 cm and enlarges it on a copy machine. If she wants the height to be 24 cm, how wide will the new image be?

Chapters 1–6 Cumulative Review Exercises

1. Write the number 503,042 in words.

2. Estimate the sum by first rounding the numbers to the nearest hundred.

$$251 + 492 + 631$$

3. Multiply. $226 \times 100{,}000$

4. Divide and write the answer with a remainder. $355 \div 16$

For Exercises 5–8, simplify.

5. $-10 + 4 + (-1) + 12$

6. $32 - (-12)$

7. $-30 - 2(3 - 7) + 2 \div (-2)$

8. $|2 - 6| - |11 - 2|$

9. Simplify the expression.

$2.1x - 3.8y + 5.04x - 4.4y$

For Exercises 10–11, solve the equations.

10. $2 = -4 + x$

11. $-5x + 3 = -12$

12. Multiply. $\frac{13}{2} \cdot \frac{3}{7}$

13. Simplify. $\left(-\frac{3}{5}\right)^2$

14. Bruce decides to share his 6-in. sub sandwich with his friend. If he gives $\frac{1}{4}$ of it to Dennis, how much of the sandwich does he have left?

15. Simplify. $\frac{7}{8} \div \frac{3}{4} + \frac{5}{6}$

16. Add. $\frac{8}{9} + 3$

17. How many ninths are in $6\frac{5}{9}$?

18. Write the number 1004.701 in words.

19. Add and subtract. $23.88 + 11.3 - 7.123$

20. Write -4.36 as an improper fraction in lowest terms.

21. Multiply. 43.923×100

22. Divide. $-237.9 \div 100$

23. Find the perimeter of the figure. Write the answer in decimal form.

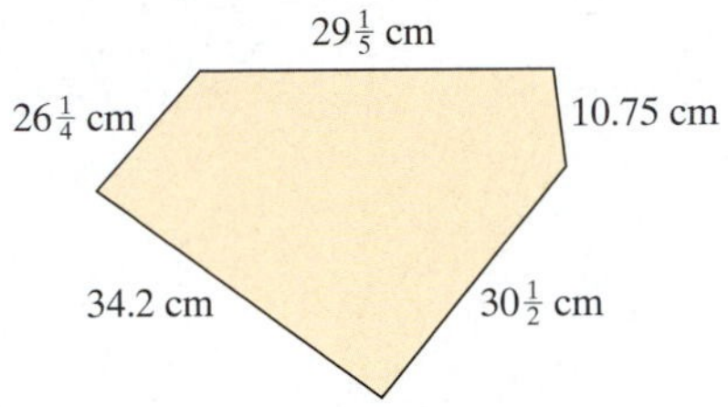

24. Americans buy 61 million newspapers each day and throw out 44 million. Write the ratio 61 to 44 in two other ways.

25. For a recent year at Southeastern Community College in North Carolina, there were 1950 students and 150 faculty. Write the student-to-faculty ratio in lowest terms.

26. In a recent study of 6000 deaths, 840 were due to cancer. Compute the ratio of deaths due to cancer to the total number of deaths studied. Simplify the ratio and interpret the answer in words.

27. Oregon has a land area of approximately 9600 mi^2. The population of Oregon is approximately 1,200,000. Compute the population density (recall that population density is a unit rate given by the number of people per square mile).

28. Determine whether the ratios form a proportion.

a. $\frac{7.5}{10} \stackrel{?}{=} \frac{9}{12}$

b. $\frac{-31}{5} \stackrel{?}{=} \frac{-33}{6}$

29. Solve the proportion. $\frac{13}{-11.7} = \frac{5}{x}$

30. Jim can drive 150 mi on 6 gal of gas. At this rate, how far can he travel on 4 gal?

Percents

7

CHAPTER OUTLINE

Chapter 7

In this chapter, we present the concept of percent. Percents are used to measure the number of parts per hundred of some whole amount. As a consumer, it is important to have a working knowledge of percents.

Are You Prepared?

To prepare for your work with percents, take a minute to review multiplying and dividing by a power of 10. Also practice solving equations and proportions. Work the problems on the left. Record the answers in the spaces on the right, according to the number of digits to the left and right of the decimal point. Then write the letter of each choice to complete the sentence below. If you need help, review Sections 5.3, 5.4, and 3.4.

1. 0.582×100

2. 0.002×100

3. $46 \times \frac{1}{100}$

4. 318×0.01

5. $0.5 \div 100$

6. Solve. $\frac{34}{160} = \frac{x}{100}$

7. Solve. $\frac{x}{4000} = \frac{36}{100}$

8. Solve. $0.3x = 16.2$

S. _ • _ _

F. _ _ • _ _

P. _ _ _ _ •

O. • _

I. _ _ •

E. • _ _ _

H. _ _ • _

U. • _ _

The mathematician's bakery was called _ _ _ _ _ _ _ _ _.
(1 2 3 4 5 2 6 7 8)

Section 7.1 Percents, Fractions, and Decimals

Objectives

1. Definition of Percent
2. Converting Percents to Fractions
3. Converting Percents to Decimals
4. Converting Fractions and Decimals to Percents
5. Percents, Fractions, and Decimals: A Summary

1. Definition of Percent

In this chapter we study the concept of percent. Literally, the word **percent** means *per one hundred.* To indicate percent, we use the percent symbol %. For example, 45% (read as "45 percent") of the land area in South America is rainforest (shaded in green). This means that if South America were divided into 100 squares of equal size, 45 of the 100 squares would cover rainforest. See Figures 7-1 and 7-2.

Figure 7-1

Figure 7-2

Consider another example. For a recent year, the population of Virginia could be described as follows.

21%	African American	21 out of 100 Virginians are African American
72%	Caucasian (non-Hispanic)	72 out of 100 Virginians are Caucasian (non-Hispanic)
3%	Asian American	3 out of 100 Virginians are Asian American
3%	Hispanic	3 out of 100 Virginians are Hispanic
1%	Other	1 out of 100 Virginians have other backgrounds

Figure 7-3 represents a sample of 100 residents of Virginia.

AA	AA	AA	C	C	C	C	C	C	C
AA	AA	C	C	C	C	C	C	C	C
AA	AA	C	C	C	C	C	C	C	C
AA	AA	C	C	C	C	C	C	C	H
AA	AA	C	C	C	C	C	C	C	H
AA	AA	C	C	C	C	C	C	C	H
AA	AA	C	C	C	C	C	C	C	A
AA	AA	C	C	C	C	C	C	C	A
AA	AA	C	C	C	C	C	C	C	A
AA	AA	C	C	C	C	C	C	C	O

AA African American
C Caucasian (non-Hispanic)
A Asian American
H Hispanic
O Other

Figure 7-3

Concept Connections

1. Shade the portion of the figure represented by 18%.

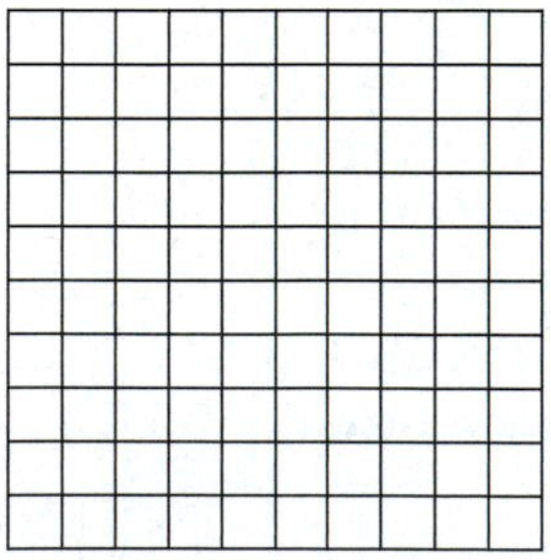

Answer

1.

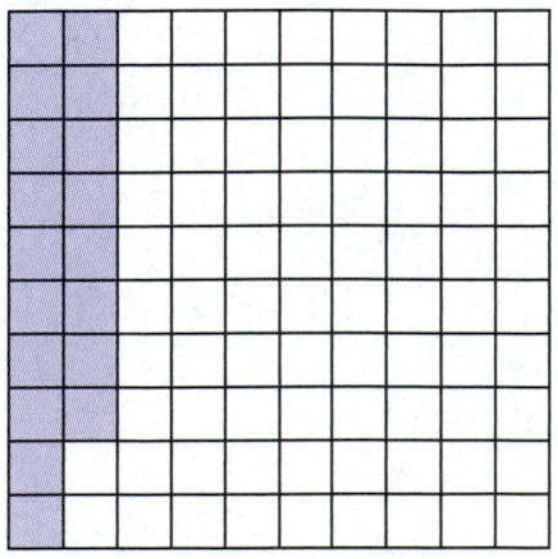

2. Converting Percents to Fractions

By definition, a percent represents a ratio of parts per 100. Therefore, we can write percents as fractions.

Percent		Fraction	Example/Interpretation
7%	=	$\frac{7}{100}$	A sales tax of 7% means that 7 cents in tax is charged for every 100 cents spent.
35%	=	$\frac{35}{100}$	To say that 35% of households own a cat means that 35 per every 100 households own a cat.

Notice that $35\% = \frac{35}{100} = 35 \times \frac{1}{100} = 35 \div 100$.

From this discussion we have the following rule for converting percents to fractions.

PROCEDURE Converting Percents to Fractions

Step 1 Replace the symbol % by $\times \frac{1}{100}$ (or by $\div 100$).

Step 2 Simplify the fraction to lowest terms, if possible.

Example 1 Converting Percents to Fractions

Convert each percent to a fraction.

a. 56% **b.** 125% **c.** 0.4%

Solution:

a. $56\% = 56 \times \frac{1}{100}$ Replace the % symbol by $\times \frac{1}{100}$.

$= \frac{56}{100}$ Multiply.

$= \frac{14}{25}$ Simplify to lowest terms.

b. $125\% = 125 \times \frac{1}{100}$ Replace the % symbol by $\times \frac{1}{100}$.

$= \frac{125}{100}$ Multiply.

$= \frac{5}{4}$ or $1\frac{1}{4}$ Simplify to lowest terms.

c. $0.4\% = 0.4 \times \frac{1}{100}$ Replace the % symbol by $\times \frac{1}{100}$.

$= \frac{4}{10} \times \frac{1}{100}$ Write 0.4 in fraction form.

$= \frac{4}{1000}$ Multiply.

$= \frac{1}{250}$ Simplify to lowest terms.

Skill Practice

Convert each percent to a fraction.

2. 55% **3.** 175%
4. 0.06%

Answers

2. $\frac{11}{20}$ **3.** $\frac{7}{4}$ **4.** $\frac{3}{5000}$

Concept Connections

Determine whether the percent represents a quantity greater than or less than 1 whole.

5. 1.92%
6. 19.2%
7. 192%

Note that 100% = $100 \times \frac{1}{100} = 1$. That is, 100% represents 1 whole unit. In Example 1(b), 125% = $\frac{5}{4}$ or $1\frac{1}{4}$. This illustrates that any percent greater than 100% represents a quantity greater than 1 whole. Therefore, its fractional form may be expressed as an improper fraction or as a mixed number.

Note that 1% = $1 \times \frac{1}{100} = \frac{1}{100}$. In Example 1(c), the value 0.4% represents a quantity less than 1%. Its fractional form is less than one-hundredth.

3. Converting Percents to Decimals

To express part of a whole unit, we can use a percent, a fraction, or a decimal. We would like to be able to convert from one form to another. The procedure for converting a percent to a decimal is the same as that for converting a percent to a fraction. We replace the % symbol by $\times \frac{1}{100}$. However, when converting to a decimal, it is usually more convenient to use the form $\times$ 0.01.

PROCEDURE Converting Percents to Decimals

Replace the % symbol by $\times$ 0.01. (This is equivalent to $\times \frac{1}{100}$ and $\div$ 100.)

Note: Multiplying a decimal by 0.01 (or dividing by 100) is the same as moving the decimal point 2 places to the left.

Skill Practice

Convert each percent to its decimal form.

8. 67%
9. 8.6%
10. 321%
11. $6\frac{1}{4}\%$
12. 0.7%

Example 2 Converting Percents to Decimals

Convert each percent to its decimal form.

a. 31% **b.** 6.5% **c.** 428% **d.** $1\frac{3}{5}\%$ **e.** 0.05%

Solution:

a. 31% = 31 $\times$ 0.01 — Replace the % symbol by $\times$ 0.01.

= 0.31 — Move the decimal point 2 places to the left.

b. 6.5% = 6.5 $\times$ 0.01 — Replace the % symbol by $\times$ 0.01.

= 0.065 — Move the decimal point 2 places to the left.

c. 428% = 428 $\times$ 0.01 — Because 428% is greater than 100% we expect the decimal form to be a number greater than 1.

= 4.28

d. $1\frac{3}{5}\%$ = 1.6 $\times$ 0.01 — Convert the mixed number to decimal form.

= 0.016 — Because the percent is just over 1%, we expect the decimal form to be just slightly greater than 0.01.

e. 0.05% = 0.05 $\times$ 0.01 — The value 0.05% is less than 1%. We expect the decimal form to be less than 0.01.

= 0.0005

TIP: In Example 2(d), convert $\frac{3}{5}$ to decimal form by dividing the numerator by the denominator.

$\frac{3}{5} = 0.6$, therefore $1\frac{3}{5} = 1.6$.

Answers

5. Less than
6. Less than
7. Greater than
8. 0.67
9. 0.086
10. 3.21
11. 0.0625
12. 0.007

4. Converting Fractions and Decimals to Percents

To convert a percent to a decimal or fraction, we replace the % symbol by $\times$ 0.01 (or $\times \frac{1}{100}$). We will now reverse this process. To convert a decimal or fraction to a percent, multiply by 100 and apply the % symbol.

PROCEDURE Converting Fractions and Decimals to Percent Form

Multiply the fraction or decimal by 100%.

Note: Multiplying a decimal by 100 moves the decimal point 2 places to the right.

Example 3 Converting Decimals to Percents

Convert each decimal to its equivalent percent form.

a. 0.62 **b.** 1.75 **c.** 1 **d.** 0.004 **e.** 8.9

Solution:

a. $0.62 = 0.62 \times 100\%$ Multiply by 100%.

$= 62\%$ Multiplying by 100 moves the decimal point 2 places to the right.

b. $1.75 = 1.75 \times 100\%$ Multiply by 100%.

$= 175\%$ The decimal number 1.75 is greater than 1. Therefore, we expect a percent greater than 100%.

c. $1 = 1 \times 100\%$ Multiply by 100%.

$= 100\%$ Recall that 1 whole is equal to 100%.

d. $0.004 = 0.004 \times 100\%$ Multiply by 100%.

$= 0.4\%$ Move the decimal point to the right 2 places.

e. $8.9 = 8.90 \times 100\%$ Multiply by 100%.

$= 890\%$

TIP: Multiplying a number by 100% is equivalent to multiplying the number by 1. Thus, the value of the number is not changed.

Skill Practice

Convert each decimal to its percent form.

13. 0.46
14. 3.25
15. 2
16. 0.0006
17. 2.5

Answers

13. 46% **14.** 325% **15.** 200%
16. 0.06% **17.** 250%

Skill Practice

Convert the fraction to percent notation.

18. $\frac{7}{10}$

Example 4 Converting a Fraction to Percent Notation

Convert the fraction to percent notation. $\frac{3}{5}$

Solution:

$$\frac{3}{5} = \frac{3}{5} \times 100\% \quad \text{Multiply by 100\%.}$$

$$= \frac{3}{5} \times \frac{100}{1}\% \quad \text{Convert the whole number to an improper fraction.}$$

$$= \frac{3}{\cancel{5}_1} \times \frac{\cancel{100}^{20}}{1}\% \quad \text{Multiply fractions and simplify to lowest terms.}$$

$$= 60\%$$

TIP: We could also have converted $\frac{3}{5}$ to decimal form first (by dividing the numerator by the denominator) and then converted the decimal to a percent.

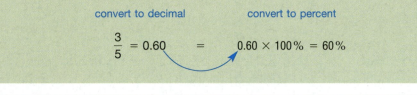

convert to decimal — convert to percent

$$\frac{3}{5} = 0.60 = 0.60 \times 100\% = 60\%$$

Skill Practice

Convert the fraction to percent notation.

19. $\frac{1}{9}$

Example 5 Converting a Fraction to Percent Notation

Convert the fraction to percent notation. $\frac{2}{3}$

Solution:

$$\frac{2}{3} = \frac{2}{3} \times 100\% \quad \text{Multiply by 100\%.}$$

$$= \frac{2}{3} \times \frac{100}{1}\% \quad \text{Convert the whole number to an improper fraction.}$$

$$= \frac{200}{3}\%$$

The number $\frac{200}{3}\%$ can be written as $66\frac{2}{3}\%$ or as $66.\overline{6}\%$.

TIP: First converting $\frac{2}{3}$ to a decimal before converting to percent notation is an alternative approach.

convert to decimal — convert to percent

$$\frac{2}{3} = 0.\overline{6} = 0.666\ldots \times 100\% = 66.\overline{6}\%$$

Answers

18. 70%

19. $\frac{100}{9}\%$ or $11\frac{1}{9}\%$ or $11.\overline{1}\%$

In Example 6, we convert an improper fraction and a mixed number to percent form.

Example 6 Converting Improper Fractions and Mixed Numbers to Percents

Convert to percent notation.

a. $2\frac{1}{4}$ **b.** $\frac{13}{10}$

Solution:

a. $2\frac{1}{4} = 2\frac{1}{4} \times 100\%$ Multiply by 100%.

$= \frac{9}{4} \times \frac{100}{1}\%$ Convert to improper fractions.

$= \frac{9}{\cancel{4}_1} \times \frac{\cancel{100}^{25}}{1}\%$ Multiply and simplify to lowest terms.

$= 225\%$

b. $\frac{13}{10} = \frac{13}{10} \times 100\%$ Multiply by 100%.

$= \frac{13}{10} \times \frac{100}{1}\%$ Convert the whole number to an improper fraction.

$= \frac{13}{\cancel{10}_1} \times \frac{\cancel{100}^{10}}{1}\%$ Multiply and simplify to lowest terms.

$= 130\%$

Skill Practice

Convert to percent notation.

20. $1\frac{7}{10}$

21. $\frac{11}{4}$

Notice that both answers in Example 6 are greater than 100%. This is reasonable because any number greater than 1 whole unit represents a percent greater than 100%.

In Example 7, we approximate a percent from its fraction form.

Example 7 Approximating a Percent by Rounding

Convert the fraction $\frac{5}{13}$ to percent notation rounded to the nearest tenth of a percent.

Solution:

$\frac{5}{13} = \frac{5}{13} \times 100\%$ Multiply by 100%.

$= \frac{5}{13} \times \frac{100}{1}\%$ Write the whole number as an improper fraction.

$= \frac{500}{13}\%$

$$\begin{array}{r} 38.46 \\ 13\overline{)500.00} \\ \underline{-39} \\ 110 \\ \underline{-104} \\ 60 \\ \underline{-52} \\ 80 \\ \underline{-78} \\ 2 \end{array}$$

To round to the nearest tenth of a percent, we must divide. We will obtain the hundredths-place digit in the quotient on which to base the decision on rounding.

$38.4\overset{1}{\cancel{6}} \approx 38.5$

Thus, $\frac{5}{13} \approx 38.5\%$.

Skill Practice

22. Write the fraction $\frac{3}{7}$ in percent notation to the nearest tenth of a percent.

Avoiding Mistakes

In Example 7, we converted a fraction to a percent where rounding was necessary. We converted the fraction to percent form *before* dividing and rounding. If you try to convert to decimal form first, you might round too soon.

Answers

20. 170% **21.** 275% **22.** 42.9%

Concept Connections

23. To convert a decimal to a percent, in which direction do you move the decimal point?

24. To convert a percent to a decimal, in which direction do you move the decimal point?

5. Percents, Fractions, and Decimals: A Summary

The diagram in Figure 7-4 summarizes the methods for converting fractions, decimals, and percents.

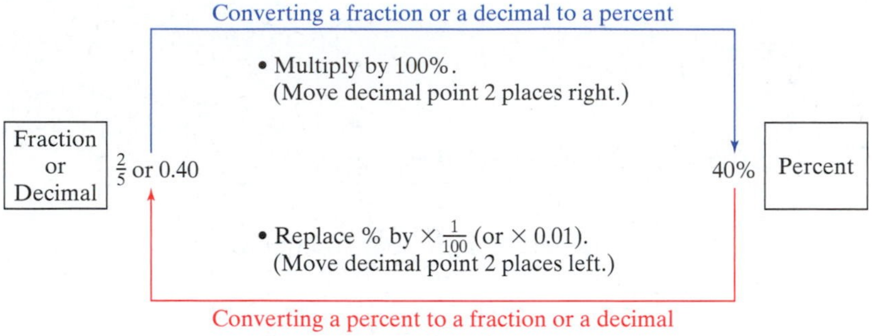

Converting a fraction or a decimal to a percent
• Multiply by 100%.
(Move decimal point 2 places right.)
Fraction or Decimal
$\frac{2}{5}$ or 0.40
40%
Percent
• Replace % by $\times \frac{1}{100}$ (or $\times$ 0.01).
(Move decimal point 2 places left.)
Converting a percent to a fraction or a decimal

Figure 7-4

Table 7-1 shows some common percents and their equivalent fraction and decimal forms.

Table 7-1

Percent	Fraction	Decimal	Example/Interpretation
100%	1	1.00	Of people who give birth, 100% are female.
50%	$\frac{1}{2}$	0.50	Of the population, 50% is male. That is, one-half of the population is male.
25%	$\frac{1}{4}$	0.25	Approximately 25% of the U.S. population smokes. That is, one-quarter of the population smokes.
75%	$\frac{3}{4}$	0.75	Approximately 75% of homes have computers. That is, three-quarters of homes have computers.
10%	$\frac{1}{10}$	0.10	Of the population, 10% is left-handed. That is, one-tenth of the population is left-handed.
20%	$\frac{1}{5}$	0.20	Approximately 20% of computers sold in stores or online are Apple computers. That is, one-fifth of the computers are Apple computers.
1%	$\frac{1}{100}$	0.01	Approximately 1% of babies are born underweight. That is, about 1 in 100 babies is born underweight.
$33\frac{1}{3}$%	$\frac{1}{3}$	$0.\overline{3}$	A basketball player made $33\frac{1}{3}$% of her shots. That is, she made about 1 basket for every 3 shots attempted.
$66\frac{2}{3}$%	$\frac{2}{3}$	$0.\overline{6}$	Of the population, $66\frac{2}{3}$% prefers chocolate ice cream to other flavors. That is, 2 out of 3 people prefer chocolate ice cream.

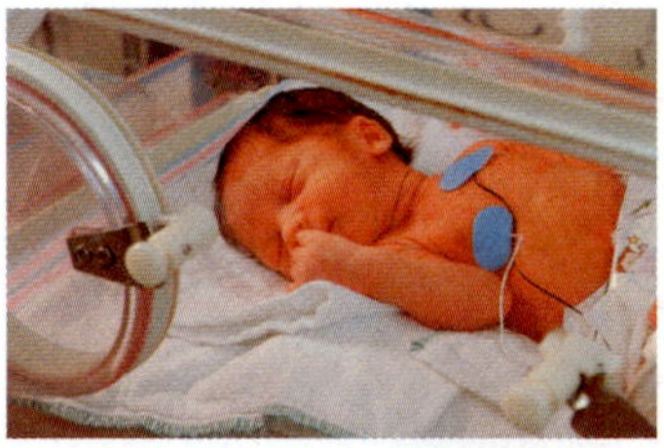

Answers

23. To the right 2 places
24. To the left 2 places

Example 8 Converting Fractions, Decimals, and Percents

Complete the table.

	Fraction	Decimal	Percent
a.		0.55	
b.	$\frac{1}{200}$		
c.			160%
d.		2.4	
e.			$66\frac{2}{3}\%$
f.	$\frac{2}{9}$		

Solution:

a. 0.55 to fraction: $0.55 = \frac{55}{100} = \frac{11}{20}$

0.55 to percent: $0.55 \times 100\% = 55\%$

b. $\frac{1}{200}$ to decimal: $1 \div 200 = 0.005$

$\frac{1}{200}$ to percent: $\frac{1}{200} \times 100\% = \frac{100}{200}\% = 0.5\%$

c. 160% to fraction: $160 \times \frac{1}{100} = \frac{160}{100} = \frac{8}{5}$ or $1\frac{3}{5}$

160% to decimal: $160 \times 0.01 = 1.6$

d. 2.4 to fraction: $\frac{24}{10} = \frac{12}{5}$ or $2\frac{2}{5}$

2.4 to percent: $2.4 \times 100\% = 240\%$

e. $66\frac{2}{3}\%$ to fraction: $66\frac{2}{3} \times \frac{1}{100} = \frac{\overset{2}{\cancel{200}}}{3} \times \frac{1}{\underset{1}{\cancel{100}}} = \frac{2}{3}$

$66\frac{2}{3}\%$ to decimal: $66\frac{2}{3} \times 0.01 = 66.\overline{6} \times 0.01 = 0.\overline{6}$

f. $\frac{2}{9}$ to decimal: $2 \div 9 = 0.\overline{2}$

$\frac{2}{9}$ to percent: $\frac{2}{9} \times 100\% = \frac{2}{9} \times \frac{100}{1}\% = \frac{200}{9}\% = 22\frac{2}{9}\%$ or $22.\overline{2}\%$

The completed table is as follows.

	Fraction	Decimal	Percent
a.	$\frac{11}{20}$	0.55	55%
b.	$\frac{1}{200}$	0.005	0.5%
c.	$\frac{8}{5}$ or $1\frac{3}{5}$	1.6	160%
d.	$\frac{12}{5}$ or $2\frac{2}{5}$	2.4	240%
e.	$\frac{2}{3}$	$0.\overline{6}$	$66\frac{2}{3}\%$
f.	$\frac{2}{9}$	$0.\overline{2}$	$22\frac{2}{9}\%$ or $22.\overline{2}\%$

Skill Practice

Complete the table.

	Fraction	Decimal	Percent
25.		1.41	
26.	$\frac{1}{50}$		
27.			18%
28.		0.58	
29.			$33\frac{1}{3}\%$
30.	$\frac{7}{9}$		

Answers

	Fraction	Decimal	Percent
25.	$\frac{141}{100}$ or $1\frac{41}{100}$	1.41	141%
26.	$\frac{1}{50}$	0.02	2%
27.	$\frac{9}{50}$	0.18	18%
28.	$\frac{29}{50}$	0.58	58%
29.	$\frac{1}{3}$	$0.\overline{3}$	$33\frac{1}{3}\%$
30.	$\frac{7}{9}$	$0.\overline{7}$	$77.\overline{7}\%$

Section 7.1 Practice Exercises

Study Skills Exercise

1. Define the key term **percent**.

Objective 1: Definition of Percent

For Exercises 2–6, use a percent to express the shaded portion of each drawing.

2.

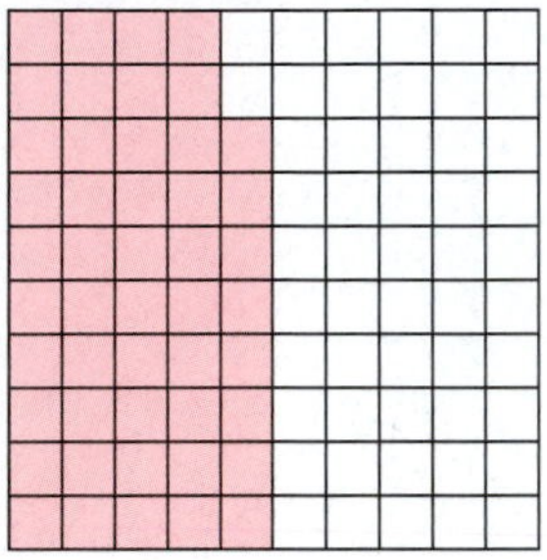

3.

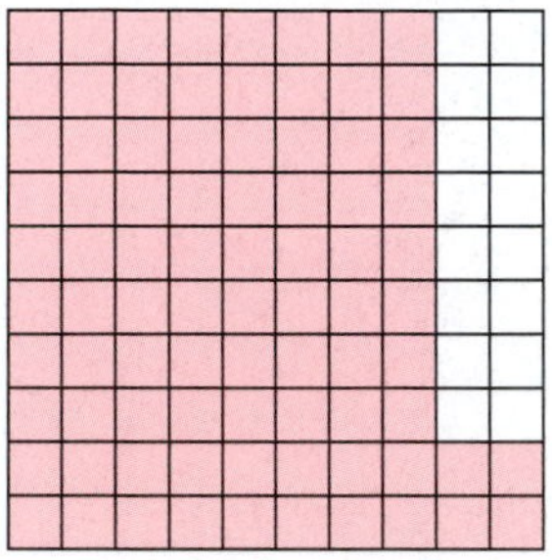

4.

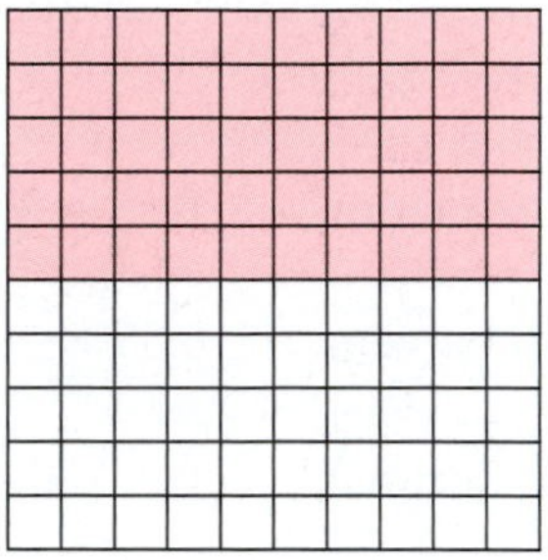

5.

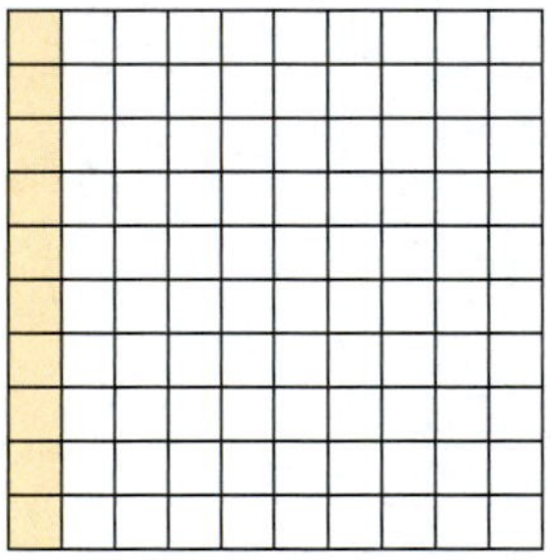

6.

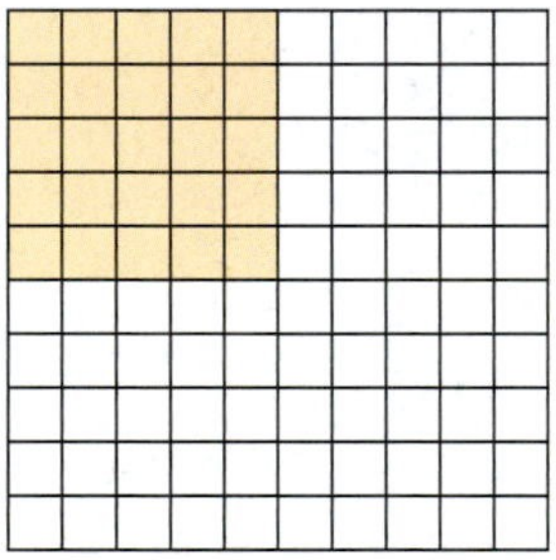

For Exercises 7–10, write a percent for each statement.

7. A bank pays \$2 in interest for every \$100 deposited.

8. In South Dakota, 5 out of every 100 people work in construction.

9. Out of 100 acres, 70 acres were planted with corn.

10. On TV, 26 out of every 100 minutes are filled with commercials.

Objective 2: Converting Percents to Fractions

11. Explain the procedure to change a percent to a fraction.

12. What fraction represents 50%.

For Exercises 13–28, change the percent to a simplified fraction or mixed number. **(See Example 1.)**

13. 3%

14. 7%

15. 84%

16. 32%

17. 3.4%

18. 5.2%

19. 115%

20. 150%

21. 0.5% **22.** 0.2% **23.** 0.25% **24.** 0.75%

25. $5\frac{1}{6}\%$ **26.** $66\frac{2}{3}\%$ **27.** $124\frac{1}{2}\%$ **28.** $110\frac{1}{2}\%$

Objective 3: Converting Percents to Decimals

29. Explain the procedure to change a percent to a decimal.

30. To change 26.8% to decimal form, which direction would you move the decimal point, and by how many places?

For Exercises 31–42, change the percent to a decimal. **(See Example 2.)**

31. 58% **32.** 72% **33.** 8.5% **34.** 12.9%

35. 142% **36.** 201% **37.** 0.55% **38.** 0.75%

39. $26\frac{2}{5}\%$ **40.** $16\frac{1}{4}\%$ **41.** $55\frac{1}{20}\%$ **42.** $62\frac{1}{5}\%$

Objective 4: Converting Fractions and Decimals to Percents

For Exercises 43–54, convert the decimal to a percent. **(See Example 3.)**

43. 0.27 **44.** 0.51 **45.** 0.19 **46.** 0.33

47. 1.75 **48.** 2.8 **49.** 0.124 **50.** 0.277

51. 0.006 **52.** 0.0008 **53.** 1.014 **54.** 2.203

For Exercises 55–66, convert the fraction to a percent. **(See Examples 4–6.)**

55. $\frac{71}{100}$ **56.** $\frac{89}{100}$ **57.** $\frac{7}{8}$ **58.** $\frac{5}{8}$

59. $\frac{5}{6}$ **60.** $\frac{5}{12}$ **61.** $1\frac{3}{4}$ **62.** $2\frac{1}{8}$

63. $\frac{11}{9}$ **64.** $\frac{14}{9}$ **65.** $1\frac{2}{3}$ **66.** $1\frac{1}{6}$

For Exercises 67–74, write the fraction in percent notation to the nearest tenth of a percent. **(See Example 7.)**

67. $\frac{3}{7}$ **68.** $\frac{6}{7}$ **69.** $\frac{1}{13}$ **70.** $\frac{3}{13}$

71. $\frac{5}{11}$ **72.** $\frac{8}{11}$ **73.** $\frac{13}{15}$ **74.** $\frac{1}{15}$

Objective 5: Percents, Fractions, and Decimals: A Summary (Mixed Exercises)

For Exercises 75–80, match the percent with its fraction form.

75. $66\frac{2}{3}\%$ **76.** 10% **a.** $\frac{3}{2}$ **b.** $\frac{3}{4}$

77. 90% **78.** 75% **c.** $\frac{2}{3}$ **d.** $\frac{1}{10}$

79. 25% **80.** 150% **e.** $\frac{9}{10}$ **f.** $\frac{1}{4}$

For Exercises 81–86, match the decimal with its percent form.

81. 0.30 **82.** $0.\overline{3}$ **a.** 5% **b.** 50%

83. 5 **84.** 0.5 **c.** 80% **d.** $33\frac{1}{3}\%$

85. 0.05 **86.** 0.8 **e.** 30% **f.** 500%

For Exercises 87–90, complete the table. **(See Example 8.)**

87.

	Fraction	Decimal	Percent
a.	$\frac{1}{4}$		
b.		0.92	
c.			15%
d.		1.6	
e.	$\frac{1}{100}$		
f.			0.8%

88.

	Fraction	Decimal	Percent
a.			0.6%
b.	$\frac{2}{5}$		
c.		2	
d.	$\frac{1}{2}$		
e.		0.12	
f.			45%

89.

	Fraction	Decimal	Percent
a.			14%
b.		0.87	
c.		1	
d.	$\frac{1}{3}$		
e.			0.2%
f.	$\frac{19}{20}$		

90.

	Fraction	Decimal	Percent
a.		1.3	
b.			22%
c.	$\frac{3}{4}$		
d.		0.73	
e.			$22.\overline{2}\%$
f.	$\frac{1}{20}$		

For Exercises 91–94, write the fraction as a percent.

91. One-quarter of Americans say they entertain at home two or more times a month. (*Source: USA TODAY*)

92. According to the Centers for Disease Control (CDC), $\frac{37}{100}$ of U.S. teenage boys say they rarely or never wear their seatbelts.

93. According to the Centers for Disease Control, $\frac{1}{10}$ of teenage girls in the United States say they rarely or never wear their seatbelts.

94. In a recent year, $\frac{2}{3}$ of the beds in U.S. hospitals were occupied.

For Exercises 95–98, find the decimal and fraction equivalents of the percent given in the sentence.

95. For a recent year, the unemployment rate in the United States was 5.8%.

96. During a slow travel season, one airline cut its seating capacity by 15.8%.

97. During a period of a weak economy and rising fuel prices, low-fare airline, Southwest, raised its average fare during the second quarter by 8.4%. (*Source: USA TODAY*)

98. One year, the average U.S. income tax rate was 18.2%.

99. Explain the difference between $\frac{1}{2}$ and $\frac{1}{2}\%$.

100. Explain the difference between $\frac{3}{4}$ and $\frac{3}{4}\%$.

101. Explain the difference between 25% and 0.25%.

102. Explain the difference between 10% and 0.10%.

103. Which of the numbers represent 125%?

a. 1.25 **b.** 0.125 **c.** $\frac{5}{4}$ **d.** $\frac{5}{4}\%$

104. Which of the numbers represent 60%?

a. 6.0 **b.** 0.60% **c.** 0.6 **d.** $\frac{3}{5}$

105. Which of the numbers represent 30%?

a. $\frac{3}{10}$ **b.** $\frac{1}{3}$ **c.** 0.3 **d.** 0.03%

106. Which of the numbers represent 180%?

a. 18 **b.** 1.8 **c.** $\frac{9}{5}$ **d.** $\frac{9}{5}\%$

Expanding Your Skills

107. Is the number 1.4 less than or greater than 100%?

108. Is the number 0.0087 less than or greater than 1%?

109. Is the number 0.052 less than or greater than 50%?

110. Is the number 25 less than or greater than 25%?

Percent Proportions and Applications

Section 7.2

Objectives

1. Introduction to Percent Proportions
2. Solving Percent Proportions
3. Applications of Percent Proportions

1. Introduction to Percent Proportions

Recall that a percent is a ratio in parts per 100. For example, $50\% = \frac{50}{100}$. However, a percent can be represented by infinitely many equivalent fractions. Thus,

$$50\% = \frac{50}{100} = \frac{1}{2} = \frac{2}{4} = \frac{3}{6} \quad \text{and infinitely many more.}$$

Equating a percent to an equivalent ratio forms a proportion that we call a **percent proportion**. A percent proportion is a proportion in which one ratio is written with a denominator of 100. For example:

$$\frac{50}{100} = \frac{3}{6} \quad \text{is a percent proportion.}$$

We will be using percent proportions to solve a variety of application problems. But first we need to identify and label the parts of a percent proportion.

A percent proportion can be written in the form:

$$\frac{\text{Amount}}{\text{Base}} = p\% \quad \text{or} \quad \frac{\text{Amount}}{\text{Base}} = \frac{p}{100}$$

For example:

4 L out of 8 L is 50% (4 = amount, 8 = base, 50 = p)

$$\frac{4}{8} = 50\% \quad \text{or} \quad \frac{4}{8} = \frac{50}{100}$$

In this example, 8 L is some total (or base) quantity and 4 L is some part (or amount) of that whole. The ratio $\frac{4}{8}$ represents a fraction of the whole equal to 50%. In general, we offer the following guidelines for identifying the parts of a percent proportion.

PROCEDURE Identifying the Parts of a Percent Proportion

A percent proportion can be written as

$$\frac{\text{Amount}}{\text{Base}} = p\% \quad \text{or} \quad \frac{\text{Amount}}{\text{Base}} = \frac{p}{100}.$$

- The **base** is the total or whole amount being considered. It often appears after the word *of* within a word problem.
- The **amount** is the part being compared to the base. It sometimes appears with the word *is* within a word problem.

Skill Practice

Identify the amount, base, and p value. Then set up a percent proportion.

1. 12 mi is 20% of 60 mi.
2. 14% of $600 is $84.
3. 32 books out of 2000 books is 1.6%.

Example 1 Identifying Amount, Base, and p for a Percent Proportion

Identify the amount, base, and p value, and then set up a percent proportion.

a. 25% of 60 students is 15 students. **b.** $32 is 50% of $64.

c. 5 of 1000 employees is 0.5%.

Solution:

For each problem, we recommend that you identify p first. It is the number in front of the symbol %. Then identify the base. In most cases it follows the word *of*. Then, by the process of elimination, find the amount.

a. 25% of 60 students is 15 students.

25: p (before % symbol); 60: base (after the word *of*); 15: amount

$$\frac{15}{60} = \frac{25}{100} \quad (\text{amount} \to 15,\ \text{base} \to 60,\ 25 \leftarrow p,\ 100 \leftarrow 100)$$

b. $32 is 50% of $64.

$32: amount; 50: p; $64: base

$$\frac{32}{64} = \frac{50}{100} \quad (\text{amount} \to 32,\ \text{base} \to 64,\ 50 \leftarrow p,\ 100 \leftarrow 100)$$

c. 5 of 1000 employees is 0.5%.

5: amount; 1000: base; 0.5: p

$$\frac{5}{1000} = \frac{0.5}{100} \quad (\text{amount} \to 5,\ \text{base} \to 1000,\ 0.5 \leftarrow p,\ 100 \leftarrow 100)$$

Answers

1. Amount = 12; base = 60; $p = 20$; $\frac{12}{60} = \frac{20}{100}$
2. Amount = 84; base = 600; $p = 14$; $\frac{84}{600} = \frac{14}{100}$
3. Amount = 32; base = 2000; $p = 1.6$; $\frac{32}{2000} = \frac{1.6}{100}$

2. Solving Percent Proportions

In Example 1, we practiced identifying the parts of a percent proportion. Now we consider percent proportions in which one of these numbers is unknown. Furthermore, we will see that the examples come in three types:

- Amount is unknown.
- Base is unknown.
- Value p is unknown.

However, the process for solving in each case is the same.

Example 2 Solving Percent Proportions—Amount Unknown

What is 30% of 180?

Solution:

What is 30% of 180? — The base and value for p are known.

(What → amount (x); 30 → p; 180 → base) — Let x represent the unknown amount.

$$\frac{x}{180} = \frac{30}{100}$$ Set up a percent proportion.

$$100 \cdot x = (30)(180)$$ Equate the cross products.

$$100x = 5400$$

$$\frac{\overset{1}{\cancel{100}}x}{\underset{1}{\cancel{100}}} = \frac{54\cancel{00}}{1\cancel{00}}$$ Divide both sides of the equation by 100.

$$x = 54$$ Simplify to lowest terms.

Therefore, 54 is 30% of 180.

TIP: We can check the answer to Example 2 as follows. Ten percent of a number is $\frac{1}{10}$ of the number. Furthermore, $\frac{1}{10}$ of 180 is 18. Thirty percent of 180 must be 3 times this amount.

$3 \times 18 = 54$ ✔

Skill Practice

4. What is 82% of 250?

Example 3 Solving Percent Proportions—Base Unknown

40% of what number is 25?

Solution:

40% of what number is 25? — The amount and value of p are known.

(40 → p; what number → base (x); 25 → amount) — Let x represent the unknown base.

$$\frac{25}{x} = \frac{40}{100}$$ Set up a percent proportion.

$$(25)(100) = 40 \cdot x$$ Equate the cross products.

$$2500 = 40x$$

$$\frac{2500}{40} = \frac{\overset{1}{\cancel{40}}x}{\underset{1}{\cancel{40}}}$$ Divide both sides by 40.

$$62.5 = x$$ Therefore, 40% of 62.5 is 25.

Skill Practice

5. 21% of what number is 42?

Answers

4. 205 5. 200

Skill Practice

6. What percent of \$48 is \$15?

Example 4 Solving Percent Proportions—*p* Unknown

12.4 mi is what percent of 80 mi?

Solution:

12.4 mi is what percent of 80 mi? (12.4 = amount; what percent = p; 80 = base)

The amount and base are known. The value of p is unknown.

$$\frac{12.4}{80} = \frac{p}{100}$$ Set up a percent equation.

$$(12.4)(100) = 80 \cdot p$$ Equate the cross products.

$$1240 = 80p$$

$$\frac{1240}{80} = \frac{\cancel{80}p}{\cancel{80}}$$ Divide both sides by 80.

$$15.5 = p$$

Therefore, 12.4 mi is 15.5% of 80.

Avoiding Mistakes

Remember that *p* represents the number of *parts* per 100. However, Example 4 asked us to find the value of *p*%. Therefore, it was necessary to attach the % symbol to our value of *p*. For Example 4, we have $p = 15.5$. Therefore, *p*% is 15.5%.

3. Applications of Percent Proportions

We now use percent proportions to solve application problems involving percents.

Skill Practice

7. In a recent year it was estimated that 24.7% of U.S. adults smoked tobacco products regularly. In a group of 2000 adults, how many would be expected to be smokers?

Example 5 Using Percents in Meteorology

Buffalo, New York, receives an average of 94 in. of snow each year. This year it had 120% of the normal annual snowfall. How much snow did Buffalo get this year?

Solution:

This situation can be translated as:

"The amount of snow Buffalo received is 120% of 94 in." (amount = amount (x); 120 = p; 94 = base)

$$\frac{x}{94} = \frac{120}{100}$$ Set up a percent proportion.

$$100 \cdot x = (120)(94)$$ Equate the cross products.

$$100x = 11{,}280$$

$$\frac{\cancel{100}x}{\cancel{100}} = \frac{11{,}280}{100}$$ Divide both sides by 100.

$$x = 112.8$$

This year, Buffalo had 112.8 in. of snow.

TIP: In a word problem, it is always helpful to check the reasonableness of your answer. In Example 5, the percent is greater than 100%. This means that the amount must be greater than the base. Therefore, we suspect that our solution is reasonable.

Answers

6. 31.25% 7. 494 people

Example 6 Using Percents in Statistics

In a recent year, Harvard University's freshman class had 18% Asian American students. If this represented 380 students, how many students were admitted to the freshman class? Round to the nearest student.

Solution:

This situation can be translated as:

"380 is 18% of what number?"

380 = amount, 18% = p, what number = base (x)

$$\frac{380}{x} = \frac{18}{100}$$ Set up a percent proportion.

$$(380)(100) = (18) \cdot x$$ Equate the cross products.

$$38{,}000 = 18x$$

$$\frac{38{,}000}{18} = \frac{\overset{1}{\cancel{18}}x}{\underset{1}{\cancel{18}}}$$ Note that $38{,}000 \div 18 \approx 2111.1$. Rounded to the nearest whole unit (whole person), this is 2111.

$$2111 \approx x$$

The freshman class at Harvard had approximately 2111 students.

Skill Practice

8. Eight students in a statistics class received a grade of A in the class. If this represents about 19% of the class, how many students are in the class? Round to the nearest student.

TIP: We can check the answer to Example 6 by substituting $x = 2111$ back into the original proportion. The cross products will not be exactly the same because we had to round the value of x. However, the cross products should be *close*.

$$\frac{380}{2111} \overset{?}{\approx} \frac{18}{100}$$ Substitute $x = 2111$ into the proportion.

$$(380)(100) \overset{?}{\approx} (18)(2111)$$

$$38{,}000 \approx 37{,}998\ ✔$$ The values are very close.

Example 7 Using Percents in Business

Suppose a tennis pro who is ranked 90th in the world on the men's professional tour earns $280,000 per year in tournament winnings and endorsements. He pays his coach $100,000 per year. What percent of his income goes toward his coach? Round to the nearest tenth of a percent.

Skill Practice

9. There were 425 donations made for various dollar amounts at an animal sanctuary. If 60 donations were made in the $100–$199 range, what percent does this represent? Round to the nearest tenth of a percent.

Answers

8. 42 students are in the class.
9. Of the donations, approximately 14.1% are in the $100–$199 range.

Solution:

This can be translated as:

"What percent of $280,000 is $100,000?"

The value of p is unknown. base (x) amount

$$\frac{100{,}000}{280{,}000} = \frac{p}{100}$$ Set up a percent proportion.

$$\frac{100{,}\cancel{000}}{280{,}\cancel{000}} = \frac{p}{100}$$ The ratio on the left side of the equation can be simplified by a factor of 10,000. "Strike through" four zeros in the numerator and denominator.

$$\frac{10}{28} = \frac{p}{100}$$

$$(10)(100) = (28) \cdot p$$ Equate the cross products.

$$1000 = 28p$$

$$\frac{1000}{28} = \frac{\overset{1}{\cancel{28}}p}{\underset{1}{\cancel{28}}}$$ Divide both sides by 28.

$$\frac{1000}{28} = p$$

$$35.7 \approx p$$ Dividing $1000 \div 28$, we get approximately 35.7.

$$\begin{array}{r} 35.71 \\ 28\overline{)1000.00} \\ -84 \\ \hline 160 \\ -140 \\ \hline 200 \\ -196 \\ \hline 40 \\ -28 \\ \hline 12 \end{array}$$

The tennis pro spends about 35.7% of his income on his coach.

Section 7.2 Practice Exercises

Study Skills Exercise

1. Define the key terms.

a. Percent proportion **b. Base** **c. Amount**

Review Exercises

For Exercises 2–4, convert the decimal to a percent.

2. 0.55 **3.** 1.30 **4.** 0.0006

For Exercises 5–7, convert the fraction to a percent.

5. $\frac{3}{8}$

6. $\frac{5}{2}$

7. $\frac{1}{100}$

For Exercises 8–10, convert the percent to a fraction.

8. $62\frac{1}{2}\%$

9. 2%

10. 77%

For Exercises 11–13, convert the percent to a decimal.

11. 82%

12. 0.3%

13. 100%

Objective 1: Introduction to Percent Proportions

14. Which of the proportions are percent proportions? Circle all that apply.

a. $\frac{7}{100} = \frac{x}{32}$ **b.** $\frac{42}{x} = \frac{15}{100}$ **c.** $\frac{2}{3} = \frac{x}{9}$ **d.** $\frac{7}{28} = \frac{10}{x}$

For Exercises 15–20, identify the amount, base, and p value. **(See Example 1.)**

15. 12 balloons is 60% of 20 balloons.

16. 25% of 400 cars is 100 cars.

17. \$99 of \$200 is 49.5%.

18. 45 of 50 children is 90%.

19. 50 hr is 125% of 40 hr.

20. 175% of 2 in. of rainfall is 3.5 in.

For Exercises 21–26, write the percent proportion.

21. 10% of 120 trees is 12 trees.

22. 15% of 20 pictures is 3 pictures.

23. 72 children is 80% of 90 children.

24. 21 dogs is 20% of 105 dogs.

25. 21,684 college students is 104% of 20,850 college students.

26. 103% of \$40,000 is \$41,200.

Objective 2: Solving Percent Proportions

For Exercises 27–36, solve the percent problems with an unknown amount. **(See Example 2.)**

27. Compute 54% of 200 employees.

28. Find 35% of 412.

29. What is $\frac{1}{2}\%$ of 40?

30. What is 1.8% of 900 g?

31. Find 112% of 500.

32. Compute 106% of 1050.

33. Pedro pays 28% of his salary in income tax. If he makes \$72,000 in taxable income, how much income tax does he pay?

34. A car dealer sets the sticker price of a car by taking 115% of the wholesale price. If a car sells wholesale at \$17,000, what is the sticker price?

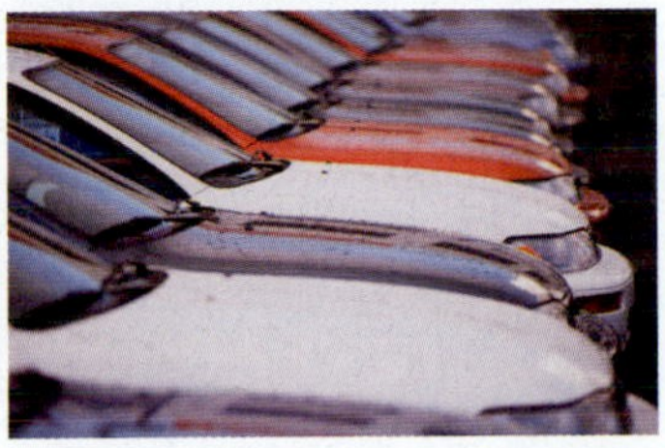

35. Jesse Ventura became the 38th governor of Minnesota by receiving 37% of the votes. If approximately 2,060,000 votes were cast, how many did Mr. Ventura get?

36. In a psychology class, 61.9% of the class consists of freshmen. If there are 42 students, how many are freshmen? Round to the nearest whole unit.

For Exercises 37–46, solve the percent problems with an unknown base. **(See Example 3.)**

37. 18 is 50% of what number?

38. 22% of what length is 44 ft?

39. 30% of what weight is 69 lb?

40. 70% of what number is 28?

41. 9 is $\frac{2}{3}$% of what number?

42. 9.5 is 200% of what number?

43. Albert saves \$120 per month. If this is 7.5% of his monthly income, how much does he make per month?

44. Janie and Don left their house in South Bend, Indiana, to visit friends in Chicago. They drove 80% of the distance before stopping for lunch. If they had driven 56 mi before lunch, what would be the total distance from their house to their friends' house in Chicago?

45. Aimee read 14 e-mails, which was only 40% of her total e-mails. What is her total number of e-mails?

46. A recent survey found that 5% of the population of the United States is unemployed. If Charlotte, North Carolina, has 32,000 unemployed, what is the population of Charlotte?

For Exercises 47–54, solve the percent problems with p unknown. **(See Example 4.)**

47. What percent of \$120 is \$42?

48. 112 is what percent of 400?

49. 84 is what percent of 70?

50. What percent of 12 letters is 4 letters?

51. What percent of 320 mi is 280 mi?

52. 54¢ is what percent of 48¢?

53. A student answered 29 problems correctly on a final exam of 40 problems. What percent of the questions did she answer correctly?

54. During his college basketball season, Jeff made 520 baskets out of 1280 attempts. What was his shooting percentage? Round to the nearest whole percent.

For Exercises 55–58, use the table given. The data represent 600 police officers broken down by gender and by the number of officers promoted.

	Promoted	Not Promoted	Total
Male	140	340	480
Female	20	100	120
Total	160	440	600

55. What percent of the officers are female?

56. What percent of the officers are male?

57. What percent of the officers were promoted? Round to the nearest tenth of a percent.

58. What percent of the officers were not promoted? Round to the nearest tenth of a percent.

Mixed Exercises

For Exercises 59–70, solve the problem using a percent proportion.

59. What is 15% of 50?

60. What is 28% of 70?

61. What percent of 240 is 96?

62. What percent of 600 is 432?

63. 85% of what number is 78.2?

64. 23% of what number is 27.6?

65. What is $3\frac{1}{2}\%$ of 2200?

66. What is $6\frac{3}{4}\%$ of 800?

67. 0.5% of what number is 44?

68. 0.8% of what number is 192?

69. 80 is what percent of 50?

70. 20 is what percent of 8?

Objective 3: Applications of Percent Proportions

71. The rainfall at Birmingham Airport in the United Kingdom averages 56 mm per month. In August the amount of rain that fell was 125% of the average monthly rainfall. How much rain fell in August? **(See Example 5.)**

72. In a recent survey, 38% of people in the United States say that gas prices have affected the type of vehicle they will buy. In a sample of 500 people who are in the market for a new vehicle, how many would you expect to be influenced by gas prices?

73. Harvard University reported that 232 African American students were admitted to the freshman class in a recent year. If this represents 11% of the total freshman class, how many freshmen were admitted? **(See Example 6.)**

74. Yellowstone National Park has 3366 mi^2 of undeveloped land. If this represents 99% of the total area, find the total area of the park.

75. At the Albany Medical Center in Albany, New York, 546 beds out of 650 are occupied. What is the percent occupancy? **(See Example 7.)**

76. For a recent year, 6 hurricanes struck the United States coastline out of 16 named storms to make landfall. What percent does this represent?

77. The risk of breast cancer relapse after surviving 10 years is shown in the graph according to the stage of the cancer at the time of diagnosis. (*Source: Journal of the National Cancer Institute*)

a. How many women out of 200 diagnosed with Stage II breast cancer would be expected to relapse after having survived 10 years?

b. How many women out of 500 diagnosed with Stage I breast cancer would *not* be expected to relapse after having survived 10 years?

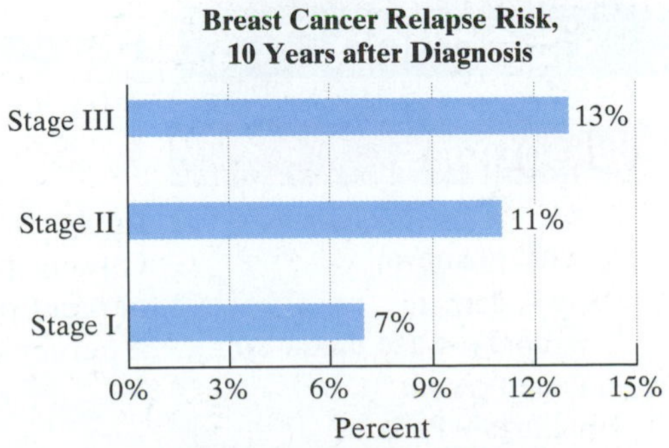

Figure for Exercise 77

78. A computer has 74.4 GB (gigabytes) of memory available. If 7.56 GB is used, what percent of the memory is used? Round to the nearest percent.

A used car dealership sells several makes of vehicles. For Exercises 79–82, refer to the graph. Round the answers to the nearest whole unit.

79. If the dealership sold 215 vehicles in 1 month, how many were Chevys?

80. If the dealership sold 182 vehicles in 1 month, how many were Fords?

81. If the dealership sold 27 Hondas in 1 month, how many total vehicles were sold?

82. If the dealership sold 10 cars in the "Other" category, how many total vehicles were sold?

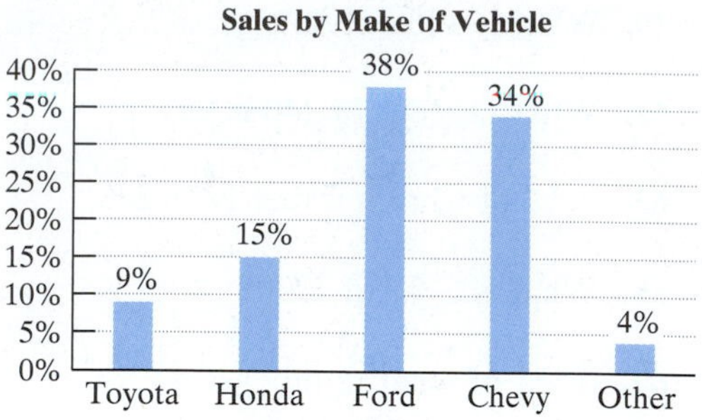

Expanding Your Skills

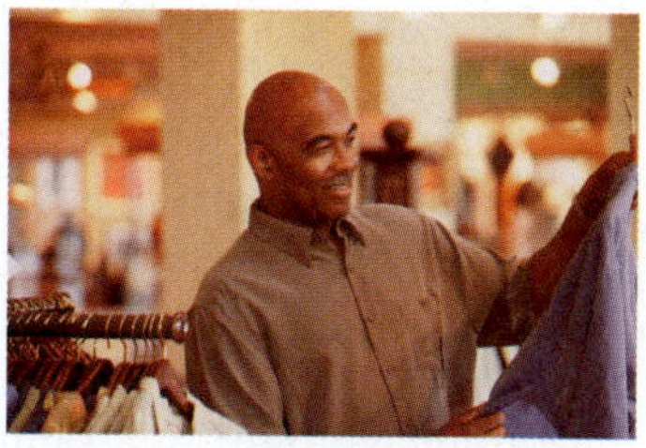

83. Carson had $600 and spent 44% of it on clothes. Then he spent 20% of the remaining money on dinner. How much did he spend altogether?

84. Melissa took $52 to the mall and spent 24% on makeup. Then she spent one-half of the remaining money on lunch. How much did she spend altogether?

It is customary to leave a 15–20% tip for the server in a restaurant. However, when you are at a restaurant in a social setting, you probably do not want to take out a pencil and piece of paper to figure out the tip. It is more socially acceptable to compute the tip mentally. Try this method.

Step 1: First, if the bill is not a whole dollar amount, simplify the calculations by rounding the bill to the next-higher whole dollar.

Step 2: Take 10% of the bill. This is the same as taking one-tenth of the bill. Move the decimal point to the left 1 place.

Step 3: If you want to leave a 20% tip, double the value found in step 2.

Step 4: If you want to leave a 15% tip, first note that 15% is $5\% + 10\%$. Therefore, add one-half of the value found in step 2 to the number in step 2.

85. Estimate a 20% tip on a bill of $57.65. (*Hint:* Round up to $58 first.)

86. Estimate a 20% tip on a bill of $18.79.

87. Estimate a 15% tip on a dinner bill of $42.00.

88. Estimate a 15% tip on a luncheon bill of $12.00.

Section 7.3 Percent Equations and Applications

Objectives

1. Solving Percent Equations—Amount Unknown
2. Solving Percent Equations—Base Unknown
3. Solving Percent Equations—Percent Unknown
4. Applications of Percent Equations

1. Solving Percent Equations—Amount Unknown

In this section, we investigate an alternative method to solve applications involving percents. We use percent equations. A **percent equation** represents a percent proportion in an alternative form. For example, recall that we can write a percent proportion as follows:

$$\frac{\text{amount}}{\text{base}} = p\% \qquad \text{percent proportion}$$

This is equivalent to writing $\quad \text{Amount} = (p\%) \cdot (\text{base}) \quad$ percent equation

To set up a percent equation, it is necessary to translate an English sentence into a mathematical equation. As you read through the examples in this section, you will notice several key words. In the phrase *percent of*, the word *of* implies multiplication. The verb *to be* (am, is, are, was, were, been) often implies = .

Example 1 Solving a Percent Equation—Amount Unknown

What is 30% of 60?

Solution:

We translate the words to mathematical symbols.

What is 30% of 60? — In this context, the word *of* means to multiply.

$x = (30\%) \cdot (60)$ — Let x represent the unknown amount.

To find x, we must multiply 30% by 60. However, 30% means $\frac{30}{100}$ or 0.30. For the purpose of calculation, we *must* convert 30% to its equivalent decimal or fraction form. The equation becomes

$$x = (0.30)(60)$$
$$= 18$$

The value 18 is 30% of 60.

TIP: The solution to Example 1 can be checked by noting that 10% of 60 is 6. Therefore, 30% is equal to (3)(6) = 18.

Skill Practice

Solve, using a percent equation.

1. What is 40% of 90?

Example 2 Solving a Percent Equation—Amount Unknown

142% of 75 amounts to what number?

Solution:

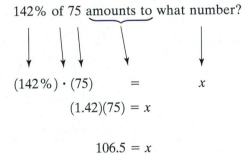

142% of 75 amounts to what number? — Let x represent the unknown amount. The word *of* implies multiplication.

$(142\%) \cdot (75) = x$ — The phrase *amounts to* implies =.

$(1.42)(75) = x$ — Convert 142% to its decimal form (1.42).

$106.5 = x$ — Multiply.

Therefore, 142% of 75 amounts to 106.5.

Skill Practice

Solve, using a percent equation.

2. 235% of 60 amounts to what number?

Examples 1 and 2 illustrate that the percent equation gives us a quick way to find an unknown amount. For example, because $(p\%) \cdot (\text{base}) = \text{amount}$, we have

$$50\% \text{ of } 80 = 0.50(80) = 40$$
$$25\% \text{ of } 20 = 0.25(20) = 5$$
$$250\% \text{ of } 90 = 2.50(90) = 225$$

2. Solving Percent Equations—Base Unknown

Examples 3 and 4 illustrate the case in which the base is unknown.

Answers

1. 36 **2.** 141

Skill Practice

Use a percent equation to solve.

3. 80% of what number is 94?

Example 3 Solving a Percent Equation—Base Unknown

40% of what number is 225?

Solution:

40% of what number is 225? — Let x represent the base number.

$$(0.40) \cdot x = 225$$

Notice that we immediately converted 40% to its decimal form 0.40 so that we would not forget.

$$0.40x = 225$$

$$\frac{0.40x}{0.40} = \frac{225}{0.40}$$ Divide both sides by 0.40.

$$x = 562.5$$ 40% of 562.5 is 225.

Skill Practice

Use a percent equation to solve.

4. 5.6 is 0.8% of what number?

Example 4 Solving a Percent Equation—Base Unknown

0.19 is 0.2% of what number?

Solution:

0.19 is 0.2% of what number? — Let x represent the base number.

$$0.19 = (0.002) \cdot x$$ Convert 0.2% to its decimal form 0.002.

$$0.19 = 0.002x$$

$$\frac{0.19}{0.002} = \frac{0.002x}{0.002}$$ Divide both sides by 0.002.

$$95 = x$$ Therefore, 0.19 is 0.2% of 95.

3. Solving Percent Equations—Percent Unknown

Examples 5 and 6 demonstrate the process to find an unknown percent.

Skill Practice

Use a percent equation to solve.

5. 16.1 is what percent of 46?

Example 5 Solving a Percent Equation—Percent Unknown

75 is what percent of 250?

Solution:

75 is what percent of 250?

$$75 = x \cdot (250)$$ Let x represent the unknown percent.

$$75 = 250x$$

$$\frac{75}{250} = \frac{250x}{250}$$ Divide both sides by 250.

$$0.3 = x$$

Answers

3. 117.5 **4.** 700 **5.** 35%

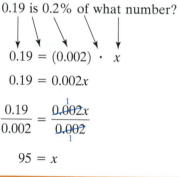

At this point, we have $x = 0.3$. To write the value of x in percent form, multiply by 100%.

$$x = 0.3$$
$$= 0.3 \times 100\%$$
$$= 30\%$$

Thus, 75 is 30% of 250.

Avoiding Mistakes

When solving for an unknown percent using a percent equation, it is necessary to convert x to its percent form.

Example 6 Solving a Percent Equation—Percent Unknown

What percent of \$60 is \$92? Round to the nearest tenth of a percent.

Solution:

What percent of \$60 is \$92?

$$x \cdot (60) = 92 \quad \text{Let } x \text{ represent the unknown percent.}$$

$$60x = 92$$

$$\frac{\overset{1}{\cancel{60}}x}{\underset{1}{\cancel{60}}} = \frac{92}{60} \quad \text{Divide both sides by 60.}$$

$$x = 1.5\overline{3} \quad \textit{Note: } 92 \div 60 = 1.5\overline{3}.$$

At this point, we have $x = 1.5\overline{3}$. To convert x to its percent form, multiply by 100%.

$$x = 1.5\overline{3}$$
$$= 1.5\overline{3} \times 100\% \quad \text{Convert from decimal form to percent form.}$$
$$= (1.53333\ldots) \times 100\%$$
$$= 153.333\ldots\%$$

The hundredths-place digit is less than 5. Discard it and the digits to its right.

Round to the nearest tenth of a percent.

$$\approx 153.3\%$$

Therefore, \$92 is approximately 153.3% of \$60. (Notice that \$92 is just over $1\frac{1}{2}$ times \$60, so our answer seems reasonable.)

Skill Practice

Use a percent equation to solve.

6. What percent of \$150 is \$213?

Avoiding Mistakes

Notice that in Example 6 we converted the final answer to percent form first *before* rounding. With the number written in percent form, we are sure to round to the nearest tenth of a percent.

4. Applications of Percent Equations

In Examples 7, 8, and 9, we use percent equations in application problems. An important part of this process is to extract the base, amount, and percent from the wording of the problem.

Answer

6. 142%

Skill Practice

7. Brianna read 143 pages in a book. If this represents 22% of the book, how many pages are in the book?

Example 7 Using a Percent Equation in Ecology

Forty-six panthers are thought to live in Florida's Big Cypress National Preserve. This represents 53% of the panthers living in Florida. How many panthers are there in Florida? Round to the nearest whole unit. (*Source:* U.S. Fish and Wildlife Services)

Solution:

This problem translates to

"46 is 53% of the number of panthers living in Florida."

$46 = (0.53) \cdot x$ Let x represent the total number of panthers.

$46 = 0.53x$

$\frac{46}{0.53} = \frac{\overset{1}{\cancel{0.53}}x}{\underset{1}{\cancel{0.53}}}$ Divide both sides by 0.53.

$87 \approx x$ *Note:* $46 \div 0.53 \approx 87$ (rounded to the nearest whole number).

There are approximately 87 panthers in Florida.

Skill Practice

8. Brandon had $60 in his wallet to take himself and a date to dinner and a movie. If he spent $28 on dinner and $19 on the movie and popcorn, what percent of his money did he spend? Round to the nearest tenth of a percent.

Example 8 Using a Percent Equation in Sports Statistics

Football player Tom Brady of the New England Patriots led his team to numerous Super Bowl victories. For one game during the season, Brady completed 23 of 30 passes. What percent of passes did he complete? Round to the nearest tenth of a percent.

Solution:

This problem translates to

"23 is what percent of 30?"

$23 = x \cdot 30$ Let x represent the unknown.

$23 = 30x$

$\frac{23}{30} = \frac{\cancel{30}x}{\cancel{30}}$ Divide both sides by 30.

$0.767 \approx x$ Divide. $23 \div 30 \approx 0.767$

The decimal value 0.767 has been rounded to 3 decimal places. We did this because the next step is to convert the decimal to a percent. Move the decimal point to the right 2 places and attach the % symbol. We have 76.7% which is rounded to the nearest tenth of a percent.

Answers

7. The book is 650 pages long.
8. Brandon spent about 78.3% of his money.

$$x \approx 0.767$$
$$= 0.767 \times 100\%$$
$$= 76.7\%$$

Tom Brady completed approximately 76.7% of his passes.

Example 9 Using a Percent Equation in Voting Statistics

Arnold Schwarzenegger received 50.5% of the votes in a California recall election for governor. If approximately 7.4 million votes were cast, how many votes did Schwarzenegger receive?

Solution:

This problem translates to

"What number is 50.5% of 7.4?"

$x \quad = 0.505 \quad \cdot \quad 7.4$ Write 50.5% in decimal form.

$x = (0.505)(7.4)$ Let x represent the number of votes for Schwarzenegger.

$x = 3.737$ Multiply.

Arnold Schwarzenegger received approximately 3.737 million votes.

Skill Practice

9. In a science class, 85% of the students passed the class. If there were 40 people in the class, how many passed?

Answer

9. 34 students passed the class.

Section 7.3 Practice Exercises

Boost *your* GRADE at ALEKS.com!

ALEKS version 3.0

- Practice Problems
- Self-Tests
- NetTutor
- e-Professors
- Videos

Study Skills Exercise

1. Define the key term **percent equation**.

Review Exercises

For Exercises 2–3, convert the decimal to a percent.

2. 0.059

3. 1.46

For Exercises 4–5, convert the percent to a decimal and a fraction.

4. 124%

5. 0.02%

For Exercises 6–10, solve the equations.

6. $3x = 27$

7. $12x = 48$

8. $\frac{62}{100} = \frac{x}{47}$

9. $\frac{924}{x} = \frac{132}{100}$

10. $\frac{43}{80} = \frac{x}{100}$

Objective 1: Solving Percent Equations—Amount Unknown

For Exercises 11–16, write the percent equation. Then solve for the unknown amount. **(See Examples 1–2.)**

11. What is 35% of 700?

12. Find 12% of 625.

13. 0.55% of 900 is what number?

14. What is 0.4% of 75?

15. Find 133% of 600.

16. 120% of 40.4 is what number?

17. What is a quick way to find 50% of a number?

18. What is a quick way to find 10% of a number?

19. Compute 200% of 14 mentally.

20. Compute 75% of 80 mentally.

21. Compute 50% of 40 mentally.

22. Compute 10% of 32 mentally.

23. Household bleach is 6% sodium hypochlorite (active ingredient). In a 64-oz bottle, how much is active ingredient?

24. One antifreeze solution is 40% alcohol. How much alcohol is in a 12.5-L mixture?

25. Hall of Fame football player Dan Marino completed 60% of his passes. If he attempted 8358 passes, how many did he complete? Round to the nearest whole unit.

26. To pass an exit exam, a student must pass a 60-question test with a score of 80% or better. What is the minimum number of questions she must answer correctly?

Objective 2: Solving Percent Equations—Base Unknown

For Exercises 27–32, write the percent equation. Then solve for the unknown base. **(See Examples 3–4.)**

27. 18 is 40% of what number?

28. 72 is 30% of what number?

29. 92% of what number is 41.4?

30. 84% of what number is 100.8?

31. 3.09 is 103% of what number?

32. 189 is 105% of what number?

33. In tests of a new anti-inflammatory drug, it was found that 47 subjects experienced nausea. If this represents 4% of the sample, how many subjects were tested?

34. Ted typed 80% of his research paper before taking a break.
   a. If he typed 8 pages, how many total pages are in the paper?
   b. How many pages does he have left to type?

35. A recent report stated that approximately 61.6 million Americans had some form of heart and blood vessel disease. If this represents 22% of the population, approximate the total population of the United States.

36. A city has a population of 245,300 which is 110% of the population from the previous year. What was the population the previous year?

Objective 3: Solving Percent Equations—Percent Unknown

For Exercises 37–42 write the percent equation. Then solve for the unknown percent. Round to the nearest tenth of a percent if necessary. **(See Examples 5–6.)**

37. What percent of 480 is 120?

38. 180 is what percent of 2000?

39. 666 is what percent of 740?

40. What percent of 60 is 2.88?

41. What percent of 300 is 400?

42. 28 is what percent of 24?

43. Of the 8079 Americans serving in the Peace Corps for a recent year, 406 were over 50 years old. What percent is this? Round to the nearest whole percent.

44. At a softball game, the concession stand had 120 hot dogs and sold 84 of them. What percent was sold?

For Exercises 45–46, refer to the table that shows the 1-year absentee record for a business.

Number of Days Missed	Number of Employees
0	4
1	2
2	14
3	10
4	16
5	18
6	10
7	6

45. **a.** Determine the total number of employees.

b. What percent missed exactly 3 days of work?

c. What percent missed between 1 and 5 days, inclusive?

46. **a.** What percent missed at least 4 days?

b. What percent did not miss any days?

Mixed Exercises

For Exercises 47–58, solve the problem using a percent equation.

47. What is 45% of 62?

48. What is 32% of 30?

49. What percent of 140 is 28?

50. What percent of 25 is 18?

51. 23% of what number is 34.5?

52. 12% of what number is 26.4?

53. What is $18\frac{1}{2}\%$ of 3000?

54. What is $\frac{1}{4}\%$ of 460?

55. 350% of what number is 2100?

56. 225% of what number is 18?

57. 1.2 is what percent of 600?

58. 10 is what percent of 2000?

Objective 4: Applications of Percent Equations

59. In a recent year, children and adolescents comprised 6.3 million hospital stays. If this represents 18% of all hospital stays, what was the total number of hospital stays? **(See Example 7.)**

60. One fruit drink advertised that it contained "10% real fruit juice." In one bottle, this was found to be 4.8 oz of real juice.

a. How many ounces of drink does the bottle contain?

b. How many ounces is something other than fruit juice?

61. Of the 87 panthers living in the wild in Florida, 11 are thought to live in Everglades National Park. To the nearest tenth of a percent, what percent is this? (*Source:* U.S. Fish and Wildlife Services) **(See Example 8.)**

62. Forty-four percent of Americans use online travel sites to book hotel or airline reservations. If 400 people need to make airline or hotel reservations, how many would be expected to use online travel sites?

63. Fifty-two percent of American parents have started to put money away for their children's college educations. In a survey of 800 parents, how many would be expected to have started saving for their children's education? (*Source: USA TODAY*) **(See Example 9.)**

64. Earth is covered by approximately 360 million km^2 of water. If the total surface area is 510 km^2, what percent is water? (Round to the nearest tenth of a percent.)

65. Brian has been saving money to buy a 61-in. Samsung Projection HDTV. He has saved \$1440 so far, but this is only 60% of the total cost of the television. What is the total cost?

66. During a weak economy, milk prices increased during a 1-year period. The average price in 2008 was \$3.90. If this value is 126% of the price from 2007, determine the price in 2007. Round to two decimal places. (*Source:* U.S. Department of Agriculture)

67. Mr. Asher made \$49,000 as a teacher in Virginia in 2007, and he spent \$8,800 on food that year. In 2008, he received a 4% increase in his salary, but his food costs increased by 6.2%.

a. How much money was left from Mr. Asher's 2007 salary after subtracting the cost of food?

b. How much money was left from his 2008 salary after subtracting the cost of food? Round to the nearest dollar.

68. The human body is 65% water. Mrs. Wright weighed 180 lb. After 1 year on a diet, her weight decreased by 15%.

a. Before the diet, how much of Mrs. Wright's weight was water?

b. After the diet, how much of Mrs. Wright's weight was water?

For Exercises 69–72, refer to the graph showing the distribution of fatal traffic accidents in the United States according to the age of the driver. (*Source:* National Safety Council)

Traffic Fatalities Distributed by Age of Driver

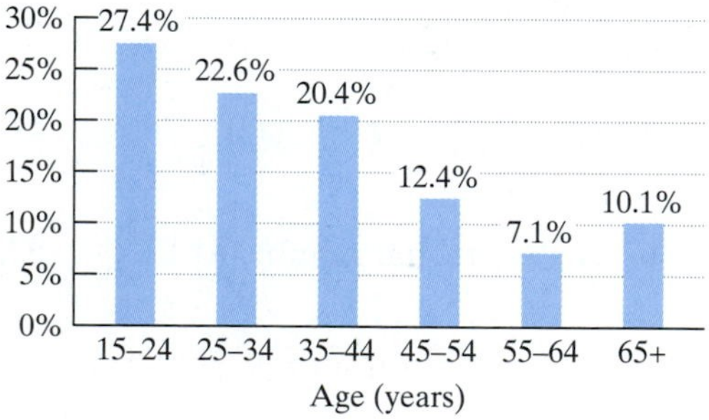

69. If there were 60,000 fatal traffic accidents during a given year, how many would be expected to involve drivers in the 35–44 age group?

70. If there were 60,000 fatal traffic accidents, how many would be expected to involve drivers in the 15–24 age group?

71. If there were 9040 fatal accidents involving drivers in the 25–34 age group, how many total traffic fatalities were there for that year?

72. If there were 3550 traffic fatalities involving drivers in the 55–64 age group, how many total traffic fatalities were there for that year?

Expanding Your Skills

The maximum recommended heart rate (in beats per minute) is given by 220 minus a person's age. For aerobic activity, it is recommended that individuals exercise at 60%–85% of their maximum recommended heart rate. This is called the aerobic range. Use this information for Exercises 73–74.

73. **a.** Find the maximum recommended heart rate for a 20-year-old.

b. Find the aerobic range for a 20-year-old.

74. **a.** Find the maximum recommended heart rate for a 42-year-old.

b. Find the aerobic range for a 42-year-old.

Problem Recognition Exercises

Percents

For Exercises 1–4, determine the percent represented by the shaded portion of the figure.

1.

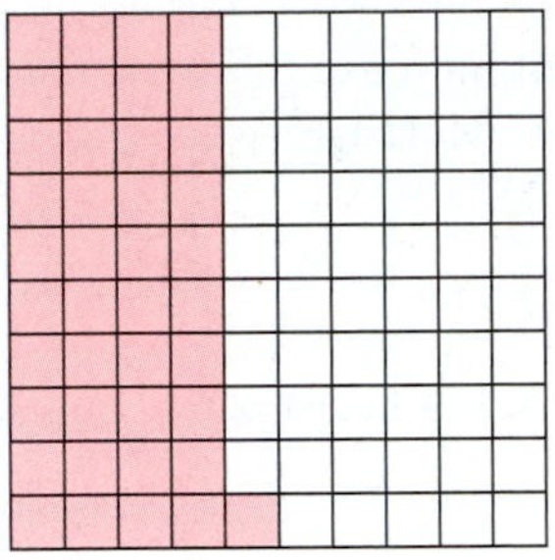

2.

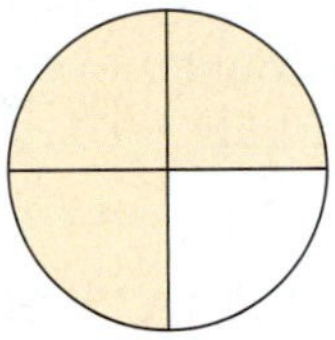

3.

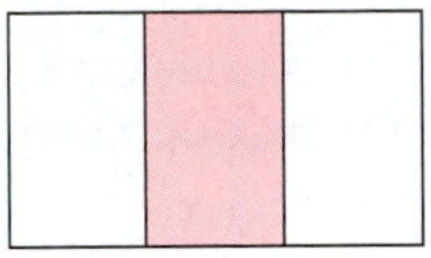

4.

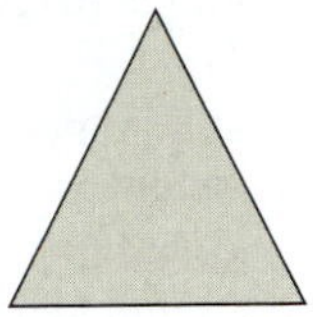

5. Is 104% of 80 less than or greater than 80?

6. Is 8% of 50 less than or greater than 5?

7. Is 11% of 90 less than or greater than 9?

8. Is 52% of 200 less than or greater than 100?

For Exercises 9–34, solve the problem by using a percent proportion or a percent equation.

9. 6 is 0.2% of what number?

10. What percent of 500 is 120?

11. 12% of 40 is what number?

12. 27 is what percent of 180?

13. 150% of what number is 105?

14. What number is 30% of 120?

15. What is 7% of 90?

16. 100 is 40% of what number?

17. 180 is what percent of 60?

18. 0.5% of 140 is what number?

19. 75 is 0.1% of what number?

20. 27 is what percent of 72?

21. What number is 50% of 50?

22. What number is 15% of 900?

23. 50 is 50% of what number?

24. 900 is 15% of what number?

25. What percent of 250 is 2?

26. 75 is what percent of 60?

27. 26 is 10% of what number?

28. 11 is 55% of what number?

29. What number is 10% of 26?

30. What number is 55% of 11?

31. 186 is what percent of 248?

32. 5 is what percent of 20?

33. 248 is what percent of 186?

34. 20 is what percent of 5?

For Exercises 35–40, perform the calculations mentally.

35. What is 10% of 82?

36. What is 5% of 82?

37. What is 20% of 82?

38. What is 50% of 82?

39. What is 200% of 82?

40. What is 15% of 82?

Section 7.4 Applications of Sales Tax, Commission, Discount, Markup, and Percent Increase and Decrease

Objectives

1. Applications Involving Sales Tax
2. Applications Involving Commission
3. Applications Involving Discount and Markup
4. Applications Involving Percent Increase and Decrease

Percents are used in an abundance of applications in day-to-day life. In this section, we investigate six common applications of percents:

- Sales tax
- Commission
- Discount
- Markup
- Percent increase
- Percent decrease

1. Applications Involving Sales Tax

The first application involves computing sales tax. **Sales tax** is a tax based on a percent of the cost of merchandise.

FORMULA Sales Tax Formula

$$\begin{pmatrix}\text{Amount of}\\ \text{sales tax}\end{pmatrix} = \begin{pmatrix}\text{Tax}\\ \text{rate}\end{pmatrix} \cdot \begin{pmatrix}\text{Cost of}\\ \text{merchandise}\end{pmatrix}$$

In this formula, the tax rate is usually given by a percent. Also note that there are three parts to the formula, just as there are in the general percent equation (see Section 7.3). The sales tax formula is a special case of a percent equation.

Skill Practice

1. A graphing calculator costs \$110. The sales tax rate is 4.5%.
   a. Compute the amount of sales tax.
   b. Compute the total cost.

Example 1 Computing Sales Tax

Suppose a Toyota Camry sells for \$20,000.

a. Compute the sales tax for a tax rate of 5.5%.

b. What is the total price of the car?

Solution:

a. Let x represent the amount of sales tax. — Label the unknown.

Tax rate $= 5.5\%$ — Identify the parts of the formula.

Cost of merchandise $= \$20{,}000$

$$\begin{pmatrix}\text{Amount of}\\ \text{sales tax}\end{pmatrix} = \begin{pmatrix}\text{Tax}\\ \text{rate}\end{pmatrix} \cdot \begin{pmatrix}\text{Cost of}\\ \text{merchandise}\end{pmatrix}$$

$$x = (5.5\%) \cdot (\$20{,}000)$$ — Substitute values into the sales tax formula.

$$x = (0.055)(\$20{,}000)$$ — Convert the percent to its decimal form.

$$x = \$1100$$

The sales tax on the vehicle is \$1100.

b. The total price is \$20,000 + \$1100 = \$21,100.

Avoiding Mistakes

Notice that we must use the decimal form of the sales tax rate in the calculation.

Answers

1. a. The tax is \$4.95.
   b. The total cost is \$114.95.

TIP: Example 1(a) can also be solved by using a percent proportion.

$$\frac{5.5}{100} = \frac{x}{20{,}000} \qquad \text{What is 5.5\% of 20,000?}$$

$$(5.5)(20{,}000) = 100x$$

$$110{,}000 = 100x$$

$$\frac{110{,}000}{100} = \frac{100x}{100} \qquad \text{Divide both sides by 100.}$$

$$1100 = x \qquad \text{The sales tax is \$1100.}$$

Example 2 Computing a Sales Tax Rate

Lindsay has just moved and must buy a new refrigerator for her home. The refrigerator costs $1200 and the sales tax is $48. Because she is new to the area, she does not know the sales tax rate. Use these figures to compute the tax rate.

Solution:

Let x represent the sales tax rate. — Label the unknown.

Cost of merchandise = $1200 — Identify the parts of the formula.

Amount of sales tax = $48

$$\begin{pmatrix}\text{Amount of}\\ \text{sales tax}\end{pmatrix} = \begin{pmatrix}\text{Tax}\\ \text{rate}\end{pmatrix}\cdot\begin{pmatrix}\text{Cost of}\\ \text{merchandise}\end{pmatrix}$$

$$48 = x \cdot (1200)$$

Substitute values into the sales tax formula.

$$48 = 1200x$$

$$\frac{48}{1200} = \frac{\overset{1}{\cancel{1200}}x}{\underset{1}{\cancel{1200}}} \qquad \text{Divide both sides by 1200.}$$

$$0.04 = x$$

The question asks for the tax rate, which is given in percent form.

$$x = 0.04$$

$$= 0.04 \times 100\%$$

$$= 4\%$$

The sales tax rate is 4%.

Skill Practice

2. A DVD sells for $15. The sales tax is $0.90. What is the tax rate?

TIP: Example 2 can also be solved by using a percent proportion.

$$\frac{48}{1200} = \frac{p}{100}$$

Example 3 Computing Cost of Merchandise

The tax on a new CD comes to $1.05. If the tax rate is 6%, find the cost of the CD before tax.

Solution:

Let x represent the cost of the CD. — Label the unknown.

Tax rate = 6% — Identify the parts of the formula.

Amount of tax = $1.05

Skill Practice

3. Sales tax on a new lawn mower is $21.50. If the tax rate is 5%, compute the price of the lawn mower before tax.

Answers

2. The tax rate is 6%.
3. The mower costs $430 before tax.

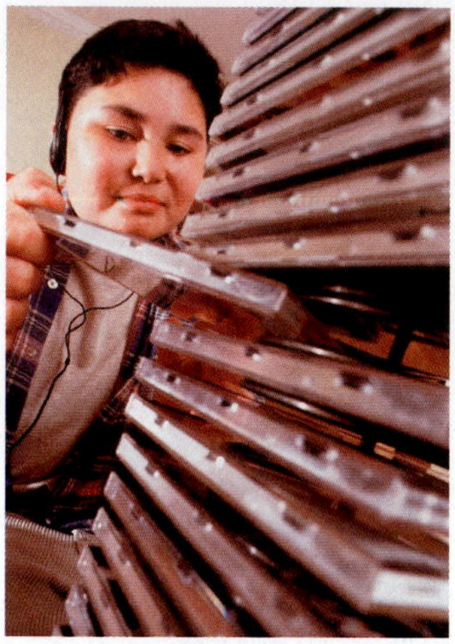

$$\begin{pmatrix}\text{Amount of}\\\text{sales tax}\end{pmatrix} = \begin{pmatrix}\text{Tax}\\\text{rate}\end{pmatrix} \cdot \begin{pmatrix}\text{Cost of}\\\text{merchandise}\end{pmatrix}$$

$$1.05 = (0.06) \cdot x$$

Notice that we immediately converted 6% to its percent form.

$$1.05 = 0.06x$$

$$\frac{1.05}{0.06} = \frac{\overset{1}{\cancel{0.06}}x}{\underset{1}{\cancel{0.06}}}$$

Divide both sides by 0.06.

$$17.5 = x$$

The CD costs $17.50.

2. Applications Involving Commission

Salespeople often receive all or part of their salary in commission. **Commission** is a form of income based on a percent of sales.

FORMULA Commission Formula

$$\begin{pmatrix}\text{Amount of}\\\text{commission}\end{pmatrix} = \begin{pmatrix}\text{Commission}\\\text{rate}\end{pmatrix} \cdot \begin{pmatrix}\text{Total}\\\text{sales}\end{pmatrix}$$

For example, if a realtor gets a 6% commission on the sale of a $200,000 home, then

$$\begin{aligned}\text{Commission} &= (0.06)(\$200{,}000)\\ &= \$12{,}000\end{aligned}$$

Skill Practice

4. Trevor sold a home for $160,000 and earned a $6400 commission. What is his commission rate?

Example 4 Computing Commission Rate

Alexis works in real estate sales.

a. If she sells a $150,000 house and earns a commission of $10,500, what is her commission rate?

b. At this rate, how much will she earn by selling a $200,000 house?

Solution:

a. Let x represent the commission rate. — Label the unknown.

Total sales = $150,000

Amount of commission = $10,500 — Identify the parts of the formula.

$$\begin{pmatrix}\text{Amount of}\\\text{commission}\end{pmatrix} = \begin{pmatrix}\text{Commission}\\\text{rate}\end{pmatrix} \cdot \begin{pmatrix}\text{Total}\\\text{sales}\end{pmatrix}$$

$$10{,}500 = x \cdot (150{,}000)$$

Substitute values into the commission formula.

$$10{,}500 = 150{,}000x$$

$$\frac{10{,}500}{150{,}000} = \frac{\overset{1}{\cancel{150{,}000}}x}{\underset{1}{\cancel{150{,}000}}}$$

Divide both sides by 150,000.

TIP: A percent proportion can be used to find the commission rate in Example 4.

$$\frac{10{,}500}{150{,}000} = \frac{p}{100}$$

Answer

4. His commission rate is 4%.

$0.07 = x$

$x = 0.07 \times 100\%$ Convert to percent form.

$= 7\%$

The commission rate is 7%.

b. The commission on a \$200,000 house is given by

Amount of commission = (0.07)(\$200,000) = \$14,000

Alexis will earn \$14,000 by selling a \$200,000 house.

Example 5 Computing Sales Base

Tonya is a real estate agent. She makes \$10,000 as her annual base salary for the work she does in the office. In addition, she makes 8% commission on her total sales. If her salary for the year amounts to \$106,000, what was her total in sales?

Solution:

First note that her commission is her total salary minus the \$10,000 for working in the office. Thus,

Amount of commission = \$106,000 − \$10,000 = \$96,000

Let x represent Tonya's total sales. Label the unknown.

Amount of commission = \$96,000 Identify the parts of the formula.

Commission rate = 8%

$$\begin{pmatrix}\text{Amount of}\\ \text{commission}\end{pmatrix} = \begin{pmatrix}\text{Commission}\\ \text{rate}\end{pmatrix} \cdot \begin{pmatrix}\text{Total}\\ \text{sales}\end{pmatrix}$$

$$96{,}000 = (0.08) \cdot x$$ Substitute values into the commission formula.

$$96{,}000 = 0.08x$$

$$\frac{96{,}000}{0.08} = \frac{\overset{1}{\cancel{0.08}}x}{\underset{1}{\cancel{0.08}}}$$ Divide both sides by 0.08.

$$1{,}200{,}000 = x$$

Tonya's sales totaled \$1,200,000 (\$1.2 million).

Skill Practice

5. A sales rep for a pharmaceutical firm makes \$50,000 as his base salary. In addition, he makes 6% commission on sales. If his salary for the year amounts to \$98,000, what were his total sales?

TIP: Example 5 can also be solved by using a percent proportion.

$$\frac{8}{100} = \frac{96{,}000}{x}$$

3. Applications Involving Discount and Markup

When we go to the store, we often find items discounted or on sale. For example, a printer might be discounted 20%, or a blouse might be on sale for 30% off. We compute the amount of the **discount** (the savings) as follows.

FORMULA Discount Formulas

$$\begin{pmatrix}\text{Amount of}\\ \text{discount}\end{pmatrix} = \begin{pmatrix}\text{Discount}\\ \text{rate}\end{pmatrix} \cdot \begin{pmatrix}\text{Original}\\ \text{price}\end{pmatrix}$$

Sale price = Original price − Amount of discount

Answer

5. He made \$800,000 in sales.

Skill Practice

6. Find the discount rate.

Adventure Kayak, ~~$600~~

On Sale Now, $480

Example 6 Computing Discount Rate

A gold chain originally priced \$500 is marked down to \$375. What is the discount rate?

Solution:

First note that the amount of the discount is given by

$$\text{Discount} = \text{Original price} - \text{Sale price}$$
$$= \$500 - \$375$$
$$= \$125$$

Let x represent the discount rate. — Label the unknown.

Original price = \$500 — Identify the parts of the formula.

Amount of discount = \$125

$$\begin{pmatrix}\text{Amount of}\\ \text{discount}\end{pmatrix} = \begin{pmatrix}\text{Discount}\\ \text{rate}\end{pmatrix} \cdot \begin{pmatrix}\text{Original}\\ \text{price}\end{pmatrix}$$

$$125 = x \cdot 500$$ Substitute values into discount formula.

$$125 = 500x$$

$$\frac{125}{500} = \frac{\overset{1}{\cancel{500}}x}{\underset{1}{\cancel{500}}}$$ Divide both sides by 500.

$$0.25 = x$$

Converting $x = 0.25$ to percent form, we have $x = 25\%$. The chain has been discounted 25%.

Retailers often buy goods from manufacturers or wholesalers. To make a profit, the retailer must increase the cost of the merchandise before reselling it. This is called **markup**.

Skill Practice

7. An importer uses a markup rate of 35%.
   a. What is the markup on a wicker chair that has a manufacturer price of \$100?
   b. What is the retail price?
   c. If there is a 5% sales tax, what is the total cost to buy the chair retail?

FORMULA Markup Formulas

$$\begin{pmatrix}\text{Amount of}\\ \text{markup}\end{pmatrix} = \begin{pmatrix}\text{Markup}\\ \text{rate}\end{pmatrix} \cdot \begin{pmatrix}\text{Original}\\ \text{price}\end{pmatrix}$$

$$\text{Retail price} = \text{Original price} + \text{Amount of markup}$$

Example 7 Computing Markup

A college bookstore marks up the price of books 40%.

a. What is the markup for a math text that has a manufacturer price of \$66?

b. What is the retail price of the book?

c. If there is a 6% sales tax, how much will the book cost to take home?

Answers

6. The kayak is discounted 20%.
7. a. The markup is \$35.
   b. The retail price is \$135.
   c. The total cost is \$141.75.

Solution:

a. Let x represent the amount of markup. — Label the unknown.

Markup rate = 40% — Identify parts of the formula.

Original price = \$66

$$\begin{pmatrix}\text{Amount of}\\ \text{markup}\end{pmatrix} = \begin{pmatrix}\text{Markup}\\ \text{rate}\end{pmatrix} \cdot \begin{pmatrix}\text{Original}\\ \text{price}\end{pmatrix}$$

$$x = (0.40) \cdot (\$66)$$ Use the decimal form of 40%.

$$x = (0.40)(\$66)$$
$$= \$26.40$$

The amount of markup is \$26.40.

b. Retail price = Original price + Markup

$$= \$66 + \$26.40$$
$$= \$92.40$$

The retail price is \$92.40.

c. Next we must find the amount of the sales tax. This value is added to the cost of the book. The sales tax rate is 6%.

$$\begin{pmatrix}\text{Amount of}\\ \text{sales tax}\end{pmatrix} = \begin{pmatrix}\text{Tax}\\ \text{rate}\end{pmatrix} \cdot \begin{pmatrix}\text{Cost of}\\ \text{merchandise}\end{pmatrix}$$

$$\text{Tax} = (0.06) \cdot (\$92.40)$$
$$\approx \$5.54$$ Round the tax to the nearest cent.

The total cost of the book is \$92.40 + \$5.54 = \$97.94.

It is important to note the similarities in the formulas presented in this section. To find the amount of sales tax, commission, discount, or markup, we multiply a rate (percent) by some original amount.

SUMMARY Formulas for Sales Tax, Commission, Discount, and Markup

Amount = Rate × Original amount

Sales tax: $$\begin{pmatrix}\text{Amount of}\\ \text{sales tax}\end{pmatrix} = \begin{pmatrix}\text{Tax}\\ \text{rate}\end{pmatrix} \cdot \begin{pmatrix}\text{Cost of}\\ \text{merchandise}\end{pmatrix}$$

Commission: $$\begin{pmatrix}\text{Amount of}\\ \text{commission}\end{pmatrix} = \begin{pmatrix}\text{Commission}\\ \text{rate}\end{pmatrix} \cdot \begin{pmatrix}\text{Total}\\ \text{sales}\end{pmatrix}$$

Discount: $$\begin{pmatrix}\text{Amount of}\\ \text{discount}\end{pmatrix} = \begin{pmatrix}\text{Discount}\\ \text{rate}\end{pmatrix} \cdot \begin{pmatrix}\text{Original}\\ \text{price}\end{pmatrix}$$

Markup: $$\begin{pmatrix}\text{Amount of}\\ \text{markup}\end{pmatrix} = \begin{pmatrix}\text{Markup}\\ \text{rate}\end{pmatrix} \cdot \begin{pmatrix}\text{Original}\\ \text{price}\end{pmatrix}$$

4. Applications Involving Percent Increase and Decrease

Two other important applications of percents are finding percent increase and percent decrease. For example:

- The price of gas increased 40% in 4 years.
- After taking a new drug for 3 months, a patient's cholesterol decreased by 35%.

When we compute **percent increase** or **percent decrease**, we are comparing the *change* between two given amounts to the *original amount.* The change (amount of increase or decrease) is found by subtraction. To compute the percent increase or decrease, we use the following formulas.

FORMULA Percent Increase or Percent Decrease

$$\begin{pmatrix}\text{Percent}\\\text{increase}\end{pmatrix} = \left(\frac{\text{Amount of increase}}{\text{Original amount}}\right) \times 100\%$$

$$\begin{pmatrix}\text{Percent}\\\text{decrease}\end{pmatrix} = \left(\frac{\text{Amount of decrease}}{\text{Original amount}}\right) \times 100\%$$

In Example 8, we apply the percent increase formula.

Skill Practice

8. After a raise, Denisha's salary increased from \$42,500 to \$44,200. What is the percent increase?

Example 8 Computing Percent Increase

The price of heating oil climbed from an average of \$1.25 per gallon to \$1.55 per gallon in a 3-year period. Compute the percent increase.

Solution:

The original price was \$1.25 per gallon. — Identify the parts of the formula.

The final price after the increase is \$1.55.

The amount of increase is given by subtraction.

$$\text{Amount of increase} = \$1.55 - \$1.25$$
$$= \$0.30$$

There was a \$0.30 increase in price.

$$\begin{pmatrix}\text{Percent}\\\text{increase}\end{pmatrix} = \left(\frac{\text{Amount of increase}}{\text{Original amount}}\right) \times 100\%$$
$$= \frac{0.30}{1.25} \times 100\%$$

Apply the percent increase formula.

$$= 0.24 \times 100\%$$
$$= 24\%$$

There was a 24% increase in the price of heating oil.

TIP: Example 8 could also have been solved by using a percent proportion.

$$\frac{p}{100} = \frac{\text{Amount of increase}}{\text{Original amount}}$$

$$\frac{p}{100} = \frac{0.30}{1.25}$$

Answer

8. Denisha's salary increased by 4%.

Example 9 **Finding Percent Decrease in an Application**

The graph in Figure 7-5 represents the closing price of Time Warner stock for a 5-day period. Compute the percent decrease between the first day and the fifth day.

Figure 7-5

Solution:

The amount of decrease is given by: $21.50 − $20.64 = $0.86

The original amount is the closing price on Day 1: $21.50

$$\left(\begin{array}{c}\text{Percent}\\ \text{decrease}\end{array}\right) = \left(\frac{\text{Amount of decrease}}{\text{Original amount}}\right) \times 100\%$$

$$= \left(\frac{\$0.86}{\$21.50}\right) \times 100\%$$

$$= (0.04) \times 100\%$$

$$= 4\%$$

The stock fell by 4%.

Skill Practice

9. Refer to the graph in Example 9. Compute the percent decrease between day 2 and day 3.

Answer

9. 2%

Section 7.4 Practice Exercises

Study Skills Exercises

1. You should not wait until the last minute to study for any test in math. It is particularly important for the final exam that you begin your review early. Check all the activities that will help you study for the final.

- ☐ Rework all old exams.
- ☐ Rework the chapter tests for the chapters you covered in class.
- ☐ Do the cumulative review exercises at the end of Chapters 2–7.

2. Define the key terms.

a. **Sales tax** **b.** **Commission** **c.** **Discount**

d. **Markup** **e.** **Percent increase** **f.** **Percent decrease**

Review Exercises

For Exercises 3–6, find the answer mentally.

3. What is 15% of 80?

4. 20 is what percent of 60?

5. 20 is 50% of what number?

6. What percent of 6 is 12?

For Exercises 7–12, solve the percent problem by using either method from Sections 7.2 or 7.3.

7. 52 is 0.2% of what number?

8. What is 225% of 36?

9. 6 is what percent of 25?

10. 18 is 75% of what number?

11. What is 1.6% of 550?

12. 32.2 is what percent of 28?

Objective 1: Applications Involving Sales Tax

For Exercises 13–14, complete the table.

13.	Cost of Item	Sales Tax Rate	Amount of Tax	Total Cost
a.	$20	5%		
b.	$12.50		$0.50	
c.		2.5%	$2.75	
d.	$55			$58.30

14.	Cost of Item	Sales Tax Rate	Amount of Tax	Total Cost
a.	$56	6%		
b.	$212		$14.84	
c.		3%	$18.00	
d.	$214			$220.42

15. A new coat costs $68.25. If the sales tax rate is 5%, what is the total bill? **(See Example 1.)**

16. Sales tax for a county in Wisconsin is 4.5%. Compute the amount of tax on a new personal MP3 player that sells for $64.

17. The sales tax on a set of luggage is $16.80. If the luggage cost before tax is $240.00, what is the sales tax rate? **(See Example 2.)**

18. A new shirt is labeled at $42.00. Jon purchased the shirt and paid $44.10.

a. How much was the sales tax?

b. What is the sales tax rate?

19. The 6% sales tax on a fruit basket came to $2.67. What is the price of the fruit basket? **(See Example 3.)**

20. The sales tax on a bag of groceries came to $1.50. If the sales tax rate is 6% what was the price of the groceries before tax?

Objective 2: Applications Involving Commission

For Exercises 21–22, complete the table.

21.	Total Sales	Commission Rate	Amount of Commission
a.	$20,000	5%	
b.	$125,000		$10,000
c.		10%	$540

22.	Total Sales	Commission Rate	Amount of Commission
a.	$540	16%	
b.	$800		$24
c.		15%	$159

23. Zach works in an insurance office. He receives a commission of 7% on new policies. How much did he make last month in commission if he sold $48,000 in new policies?

24. Marisa makes a commission of 15% on sales over $400. One day she sells $750 worth of merchandise.

a. How much over $400 did Marisa sell?

b. How much did she make in commission that day?

25. In one week, Rodney sold $2000 worth of sports equipment. He received $300 in commission. What is his commission rate? **(See Example 4.)**

26. A realtor sold a townhouse for $95,000. If he received a commission of $7600, what is his commission rate?

27. A realtor makes an annual salary of $25,000 plus a 3% commission on sales. If a realtor's salary is $67,000, what was the amount of her sales? **(See Example 5.)**

28. A salesperson receives a weekly salary of $100, plus a 5.5% commission on sales. Her salary last week was $1090. What were her sales that week?

Objective 3: Applications Involving Discount and Markup

For Exercises 29–32, complete the table.

29.	Original Price	Discount Rate	Amount of Discount	Sale Price
a.	$56	20%		
b.	$900			$600
c.			$8.50	$76.50
d.		50%	$38	

30.	Original Price	Discount Rate	Amount of Discount	Sale Price
a.	$175	15%		
b.	$900			$630
c.			$33	$77
d.		40%	$23.36	

31.	Original Price	Markup Rate	Amount of Markup	Retail Price
a.	$92	5%		
b.	$110			$118.80
c.			$97.50	$422.50
d.		20%	$9	

32.	Original Price	Markup Rate	Amount of Markup	Retail Price
a.	$25	10%		
b.	$50			$57.50
c.			$175	$875
d.		18%	$31.50	

33. Hospital employees get a 15% discount at the hospital cafeteria. If the lunch bill originally comes to $5.60, what is the price after the discount?

34. A health club membership costs $550 for 1 year. If a member pays up front in a lump sum, the member will receive a 10% discount.

a. How much money is discounted?

b. How much will the yearly membership cost with the discount?

35. A bathing suit is on sale for $45. If the regular price is $60, what is the discount rate? **(See Example 6.)**

36. A printer that sells for $229 is on sale for $183.20. What is the discount rate?

37. A business suit has a wholesale price of $150.00. A department store's markup rate is 18%. **(See Example 7.)**

a. What is the markup for this suit?

b. What is the retail price?

c. If Antonio buys this suit including a 7% sales tax, how much will he pay?

38. An import/export business marks up imported merchandise by 110%. If a wicker chair imported from Singapore originally costs $84 from the manufacturer, what is the retail price?

39. A table is purchased from the manufacturer for $300 and is sold retail at $375. What is the markup rate?

40. A $60 hairdryer is sold for $69. What is the markup rate?

41. Find the discount and the sale price of the tent in the given advertisement.

Explorer 4-person tent
On Sale 30% OFF

Was $269

42. A set of dishes had an original price of $112. Then it was discounted 50%. A week later, the new sale price was discounted another 50%. At that time, was the set of dishes free? Explain why or why not.

43. A campus bookstore adds $43.20 to the cost of a science text. If the final cost is $123.20, what is the markup rate?

44. The retail price of a golf club is $420.00. If the golf store has marked up the price by $70, what is the markup rate?

45. Find the discount and the sale price of the bike in the given advertisement.

Huffy Chopper Bike $109.99
Now 10% off.

46. Find the discount and the discount rate of the chair from the given advertisement.

Accent Chair was $235.00
and is now $188.00

Objective 4: Applications Involving Percent Increase and Decrease

47. Select the correct percent increase for a price that is double the original amount. For example, a book that originally cost $30 now costs $60.

a. 200% b. 2%

c. 100% d. 150%

48. Select the correct percent increase for a price that is greater by $\frac{1}{2}$ of the original amount. For example, an employee made $20 per hour and now makes $30 per hour.

a. 150% b. 50%

c. $\frac{1}{2}$% d. 200%

49. The number of accidents from all-terrain vehicles that required emergency room visits for children under 16 increased from 21,000 to 42,000 in an 10-year period. What was the percent increase? **(See Example 8.)**

50. The number of deaths from alcohol-induced causes rose from approximately 20,200 to approximately 20,700 in a 10-year period. (*Source:* Centers for Disease Control) What is the percent increase? Round to the nearest tenth of a percent.

51. Robin's health-care premium increased from $5000 per year to $5500 per year. What is the percent increase?

52. The yearly deductible for Diane's health-care plan rose from $800 to $1000. What is the percent increase?

53. Joel's yearly salary went from $42,000 to $45,000. What is the percent increase? Round to the nearest percent.

54. For a recent year, 67.5 million people participated in recreational boating. Sixteen years later, that number increased to 72.6 million. Determine the percent increase. Round to one decimal place.

55. During a recent housing slump, the median price of homes decreased in the United States. If James bought his house for $360,000 and the value 1 year later was $253,800, compute the percent decrease in the value of the house.

56. The U.S. government classified 8 million documents as secret in 2001. By 2003 (2 years after the attacks on 9-11), this number had increased to 14 million. What is the percent increase?

57. A stock closed at $12.60 per share on Monday. By Friday, the closing price was $11.97 per share. What was the percent decrease? **(See Example 9.)**

58. During a 5-year period, the number of participants collecting food stamps went from 27 million to 17 million. What is the percent decrease? Round to the nearest whole percent. (*Source:* U.S. Department of Agriculture)

59. Julie bought a new water-efficient toilet for her house. Her old toilet used 5 gal of water per flush. The new toilet uses only 1.6 gal of water per flush. What is the percent decrease in water per flush?

60. Nancy put new insulation in her attic and discovered that her heating bill for December decreased from $160 to $140. What is the percent decrease?

61. Gus, the cat, originally weighed 12 lb. He was diagnosed with a thyroid disorder, and Dr. Smith the veterinarian found that his weight had decreased to 10.2 lb. What percent of his body weight did Gus lose?

62. To lose weight, Kelly reduced her Calories from 3000 per day to 1800 per day. What is the percent decrease in Calories?

Expanding Your Skills

63. The retail price of four tickets to the Alicia Keys "As I Am" tour is $648. The wholesale price of each ticket is $113.

a. What is the markup amount per ticket?

b. What is the markup rate? Round to one decimal place.

Calculator Connections

Topic: Using a Calculator to Compute Percent Increase and Percent Decrease

Calculator Exercises

For Exercises 64–67, determine the percent increase in the stock price for the given companies between 2000 and 2008. Round to the nearest tenth of a percent.

	Stock	Price Jan. 2000 ($ per share)	Price Jan. 2008 ($ per share)	Change ($)	Percent Increase
64.	eBay Inc.	$16.84	$26.15		
65.	Starbucks Corp.	$6.06	$16.92		
66.	Apple Inc.	$28.66	$178.85		
67.	IBM Corp.	$118.37	$127.52		

Section 7.5 Simple and Compound Interest

Objectives

1. Simple Interest
2. Compound Interest
3. Using the Compound Interest Formula

1. Simple Interest

In this section, we use percents to compute simple and compound interest on an investment or a loan.

Banks hold large quantities of money for their customers. They keep some cash for day-to-day transactions, but invest the remaining portion of the money. As a result, banks often pay customers interest.

When making an investment, **simple interest** is the money that is earned on principal (**principal** is the original amount of money invested). When people take out a loan, the amount borrowed is the principal. The interest is a percent of the amount borrowed that you must pay back in addition to the principal.

The following formulas can be used to compute simple interest for an investment or a loan and to compute the total amount in the account.

Concept Connections

1. Consider this statement: "$8000 invested at 4% for 3 years yields $960." Identify the principal, interest, interest rate, and time.

FORMULA Simple Interest Formulas

Simple interest = Principal × Rate × Time $\quad I = Prt$

Total amount = Principal + Interest $\quad A = P + I$

where
I = amount of interest
P = amount of principal
r = annual interest rate (in decimal form)
t = time (in years)
A = total amount in an account

The time, t, is expressed in years because the rate, r, is an *annual* interest rate. If we were given a monthly interest rate, then the time, t, should be expressed in months.

Skill Practice

2. Suppose $1500 is invested in an account that earns 6% simple interest.
   a. How much interest is earned in 5 years?
   b. What is the total value of the account after 5 years?

Example 1 Computing Simple Interest

Suppose $2000 is invested in an account that earns 7% simple interest.

a. How much interest is earned after 3 years?

b. What is the total value of the account after 3 years?

Solution:

a. Principal: $P = \$2000$ — Identify the parts of the formula.

Annual interest rate: $r = 7\%$

Time (in years): $t = 3$

Let I represent the amount of interest. — Label the unknown.

$I = Prt$

$= (\$2000)(7\%)(3)$ — Substitute values into the formula.

Answers

1. Principal = $8000; interest = $960; rate = 4% (or 0.04 in decimal form); time = 3 years
2. a. $450 in interest is earned.
   b. The total account value is $1950.

$= (2000)(0.07)(3)$ Convert 7% to decimal form.

$= 420$ Multiply from left to right.

The amount of interest earned is \$420.

b. The total amount in the account is given by

$A = P + I$

$= \$2000 + \$420$

$= \$2420$

The total amount in the account is \$2420.

Avoiding Mistakes

It is important to use the decimal form of the interest rate when calculating interest.

Concept Connections

3. To find the interest earned on \$6000 at 5.5% for 6 years, what number should be substituted for r: 5.5 or 0.055?

When applying the simple interest formula, it is important that time be expressed in years. This is demonstrated in Example 2.

Example 2 Computing Simple Interest

Clyde takes out a loan for \$3500. He pays simple interest at a rate of 6% for 4 years 3 months.

a. How much money does he pay in interest?

b. How much total money must he pay to pay off the loan?

Solution:

a. $P = \$3500$ Identify parts of the formula.

$r = 6\%$

$t = 4\frac{1}{4}$ years or 4.25 years 3 months $= \frac{3}{12}$ year $= \frac{1}{4}$ year

$I = Prt$

$= (\$3500)(0.06)(4.25)$ Substitute values into the interest formula. Convert 6% to decimal form 0.06.

$= \$892.50$ Multiply.

The interest paid is \$892.50.

b. To find the total amount that must be paid, we have

$A = P + I$

$= \$3500 + \$892.50$

$= \$4392.50$ The total amount that must be paid is \$4392.50.

Skill Practice

4. Morris takes out a loan for \$10,000. He pays simple interest at 7% for 66 months.

a. Write 66 months in terms of years.

b. How much money does he pay in interest?

c. How much total money must he repay to pay off the loan?

2. Compound Interest

Simple interest is based only on a percent of the original principal. However, many day-to-day applications involve compound interest. **Compound interest** is based on both the original principal and the interest earned.

To compare the difference between simple and compound interest, consider this scenario in Example 3.

Answers

3. 0.055

4. a. 66 months = 5.5 years

b. Morris pays \$3850 in interest.

c. Morris must pay a total of \$13,850 to pay off the loan.

Skill Practice

5. Suppose $2000 is invested at 5% interest for 3 years.
   a. Compute the total amount in the account after 3 years, using simple interest.
   b. Compute the total amount in the account after 3 years of compounding interest annually.

Example 3 Comparing Simple Interest and Compound Interest

Suppose $1000 is invested at 8% interest for 3 years.

a. Compute the total amount in the account after 3 years, using simple interest.

b. Compute the total amount in the account after 3 years of compounding interest annually.

Solution:

a. $P = \$1000$

$r = 8\%$

$t = 3$ years

$I = Prt$

$= (\$1000)(0.08)(3)$

$= \$240$

The amount of simple interest earned is $240. The total amount in the account is $1000 + $240 = $1240.

b. To compute interest compounded annually over a period of 3 years, compute the interest earned in the first year. Then add the principal plus the interest earned in the first year. This value then becomes the principal on which to base the interest earned in the second year. We repeat this process, finding the interest for the second and third years based on the principal and interest earned in the preceding years. This process is outlined in a table.

Year	Interest Earned $I = Prt$	Total Amount in Account
First year	$I = (\$1000)(0.08)(1) = \$80$	$1000 + $80 = $1080
Second year	$I = (\$1080)(0.08)(1) = \$86.40$	$1080 + $86.40 = $1166.40
Third year	$I = (\$1166.40)(0.08)(1) = \$93.31$	$1166.40 + 93.31 = **$1259.71**

The total amount in the account by compounding interest annually is $1259.71.

Notice that in Example 3 the final amount in the account is greater for the situation where interest is compounded. The difference is $1259.71 − $1240 = $19.71. By compounding interest we earn more money.

Interest may be compounded more than once per year.

Annually	1 time per year
Semiannually	2 times per year
Quarterly	4 times per year
Monthly	12 times per year
Daily	365 times per year

To compute compound interest, the calculations become very tedious. Banks use computers to perform the calculations quickly. You may want to use a calculator if calculators are allowed in your class.

Answer

5. **a.** $2300 **b.** $2315.25

3. Using the Compound Interest Formula

As you can see from Example 3, computing compound interest by hand is a cumbersome process. Can you imagine computing daily compound interest (365 times a year) by hand!

We now use a formula to compute compound interest. This formula requires the use of a scientific or graphing calculator. In particular, the calculator must have an exponent key $\boxed{y^x}$, $\boxed{x^y}$, or $\boxed{\wedge}$.

Let A = total amount in an account

P = principal

r = annual interest rate

t = time in years

n = number of compounding periods per year

Then $A = P \cdot \left(1 + \frac{r}{n}\right)^{n \cdot t}$ computes the total amount in an account.

To use this formula, note the following guidelines:

- Rate r must be expressed in decimal form.
- Time t must be the total time of the investment in *years.*
- Number n is the number of compounding periods per year.

Annual	$n = 1$
Semiannual	$n = 2$
Quarterly	$n = 4$
Monthly	$n = 12$
Daily	$n = 365$

Example 4 Computing Compound Interest by Using the Compound Interest Formula

Suppose \$1000 is invested at 8% interest compounded annually for 3 years. Use the compound interest formula to find the total amount in the account after 3 years. Compare the result to the answer from Example 3(b).

Solution:

$P = \$1000$ Identify the parts of the formula.

$r = 8\%$ (0.08 in decimal form)

$t = 3$ years

$n = 1$ (annual compound interest is compounded 1 time per year)

$A = P \cdot \left(1 + \frac{r}{n}\right)^{n \cdot t}$

$= 1000 \cdot \left(1 + \frac{0.08}{1}\right)^{1 \cdot 3}$ Substitute values into the formula.

$= 1000(1 + 0.08)^3$ Apply the order of operations. Divide within parentheses. Simplify the exponent.

Skill Practice

6. Suppose \$2000 is invested at 5% interest compounded annually for 3 years. Use the formula

$A = P \cdot \left(1 + \frac{r}{n}\right)^{n \cdot t}$

to find the total amount after 3 years. Compare the answer to Skill Practice exercise 5(b).

Answer

6. \$2315.25; this is the same as the result in Skill Practice exercise 5(b).

$= 1000(1.08)^3$	Add within parentheses.
$= 1000(1.259712)$	Evaluate $(1.08)^3 = (1.08)(1.08)(1.08) = 1.259712$.
$= 1259.712$	Multiply.
≈ 1259.71	Round to the nearest cent.

The total amount in the account after 3 years is \$1259.71. This is the same value obtained in Example 3(b).

Calculator Connections

Topic: Using a Calculator to Compute Compound Interest

To enter the expression from Example 4 into a calculator, follow these keystrokes.

Expression	Keystrokes	Result
$1000 \cdot (1 + 0.08)^3$	1000 [×] [(] 1 [+] 0.08 [)] [y^x] 3 [=]	1259.712
or	1000 [×] [(] 1 [+] 0.08 [)] [^] 3 [ENTER]	1259.712

Skill Practice

7. Suppose \$5000 is invested at 9% interest compounded monthly for 30 years. Use the formula for compound interest to find the total amount in the account after 30 years.

Example 5 Computing Compound Interest by Using the Compound Interest Formula

Suppose \$8000 is invested in an account that earns 5% interest compounded quarterly for $1\frac{1}{2}$ years. Use the compound interest formula to compute the total amount in the account after $1\frac{1}{2}$ years.

Solution:

$P = \$8000$	Identify the parts of the formula.
$r = 5\%$ (0.05 in decimal form)	
$t = 1.5$ years	
$n = 4$ (quarterly interest is compounded 4 times per year)	
$A = P \cdot \left(1 + \frac{r}{n}\right)^{n \cdot t}$	
$= 8000 \cdot \left(1 + \frac{0.05}{4}\right)^{(4)(1.5)}$	Substitute values into the formula.
$= 8000(1 + 0.0125)^6$	Apply the order of operations. Divide within parentheses. Simplify the exponent.
$= 8000(1.0125)^6$	Add within parentheses.
$\approx 8000(1.077383181)$	Evaluate $(1.0125)^6$. If your teacher allows the use of a calculator, consider using the exponent key.
≈ 8619.07	Multiply and round to the nearest cent.

The total amount in the account after $1\frac{1}{2}$ years is \$8619.07.

Answer

7. The account is worth \$73,652.88 after 30 years.

Calculator Connections

Topic: Using a Calculator to Compute Compound Interest

To enter the expression from Example 5 into a calculator, follow these keystrokes.

Expression	Keystrokes	Result
$8000 \cdot \left(1 + \dfrac{0.05}{4}\right)^{(4)(1.5)}$	8000 ☒×☒ ☒(☒ 1 ☒+☒ 0.05 ☒÷☒ 4 ☒)☒ ☒y^x☒ ☒(☒ 4 ☒×☒ 1.5 ☒)☒ ☒=☒	8619.065444
	or 8000 ☒×☒ ☒(☒ 1 ☒+☒ 0.05 ☒÷☒ 4 ☒)☒ ☒∧☒ ☒(☒ 4 ☒×☒ 1.5 ☒)☒ ☒ENTER☒	8619.065444

Note: It is mandatory to insert parentheses () around the product in the exponent.

Section 7.5 Practice Exercises

Boost your GRADE at ALEKS.com!

ALEKS version 3.0

- Practice Problems
- Self-Tests
- NetTutor
- e-Professors
- Videos

Study Skills Exercise

1. Define the key terms.

 a. **Simple interest** b. **Principal** c. **Compound interest**

Review Exercises

For Exercises 2–3, write the percent in decimal form.

2. 3.5%

3. 2.25%

4. Jeff works as a pharmaceutical representative. He receives a 6% monthly commission on sales up to $60,000. He receives an 8.5% commission on all sales above $60,000. If he sold $86,000 worth of his product one month, how much did he receive in commission?

5. A paper shredder was marked down from $79 to $59. What is the percent decrease in price? Round to the nearest tenth of a percent.

6. A 19-in. computer monitor is marked down from $279 to $249. What is the percent decrease in price? Round to the nearest tenth of a percent.

Objective 1: Simple Interest

For Exercises 7–14, find the simple interest and the total amount including interest. **(See Examples 1 and 2.)**

	Principal	Annual Interest Rate	Time, Years	Interest	Total Amount
7.	$6,000	5%	3	_____	_____
8.	$4,000	3%	2	_____	_____
9.	$5,050	6%	4	_____	_____
10.	$4,800	4%	3	_____	_____
11.	$12,000	4%	$4\frac{1}{2}$	_____	_____
12.	$6,230	7%	$6\frac{1}{3}$	_____	_____

13.	$10,500	4.5%	4	______	______
14.	$9,220	8%	4	______	______

15. Dale deposited $2500 in an account that pays $3\frac{1}{2}$% simple interest for 4 years. **(See Example 2.)**

a. How much interest will he earn in 4 years?

b. What will be the total value of the account after 4 years?

16. Charlene invested $3400 at 4% simple interest for 5 years.

a. How much interest will she earn in 5 years?

b. What will be the total value of the account after 5 years?

17. Gloria borrowed $400 for 18 months at 8% simple interest.

a. How much interest will Gloria have to pay?

b. What will be the total amount that she has to pay back?

18. Floyd borrowed $1000 for 2 years 3 months at 8% simple interest.

a. How much interest will Floyd have to pay?

b. What will be the total amount that he has to pay back?

19. Jozef deposited $10,300 into an account paying 4% simple interest 5 years ago. If he withdraws the entire amount of money, how much will he have?

20. Heather invested $20,000 in an account that pays 6% simple interest. If she invests the money for 10 years, how much will she have?

21. Anne borrowed $4500 from a bank that charges 10% simple interest. If she repays the loan in $2\frac{1}{2}$ years, how much will she have to pay back?

22. Dan borrowed $750 from his brother who is charging 8% simple interest. If Dan pays his brother back in 6 months, how much does he have to pay back?

Objective 2: Compound Interest

23. If a bank compounds interest semiannually for 3 years, how many total compounding periods are there?

24. If a bank compounds interest quarterly for 2 years, how many total compounding periods are there?

25. If a bank compounds interest monthly for 2 years, how many total compounding periods are there?

26. If a bank compounds interest monthly for $1\frac{1}{2}$ years, how many total compounding periods are there?

27. Mary Ellen deposited $500 in a bank. **(See Example 3.)**

a. If the bank offers 4% simple interest, compute the amount in the account after 3 years.

b. Now suppose the bank offers 4% interest compounded annually. Complete the table to determine the amount in the account after 3 years.

Year	Interest Earned	Total Amount in Account
1		
2		
3		

28. The amount of $8000 is invested at 4% for 3 years.

a. Compute the ending balance if the bank calculates simple interest.

b. Compute the ending balance if the bank calculates interest compounded annually.

c. How much more interest is earned in the account with compound interest?

Year	Interest Earned	Total Amount in Account
1		
2		
3		

29. Fatima deposited \$24,000 in an account.

a. If the bank offers 5% simple interest, compute the amount in the account after 2 years.

b. Now suppose the bank offers 5% compounded semiannually (twice per year). Complete the table to determine the amount in the account after 2 years.

Period	Interest Earned	Total Amount in Account
1st		
2nd		
3rd		
4th		

30. The amount of \$12,000 is invested at 8% for 1 year.

a. Compute the ending balance if the bank calculates simple interest.

b. Compute the ending balance if the bank calculates interest compounded quarterly.

c. How much more interest is earned in the account with compound interest?

Period	Interest Earned	Total Amount in Account
1st		
2nd		
3rd		
4th		

Objective 3: Using the Compound Interest Formula

31. For the formula $A = P \cdot \left(1 + \frac{r}{n}\right)^{n \cdot t}$, identify what each variable means.

32. If \$1000 is deposited in an account paying 8% interest compounded monthly for 3 years, label the following variables: P, r, n, and t.

Calculator Connections

Topic: Computing Compound Interest

Calculator Exercises

For Exercises 33–40, find the total amount in an account for an investment subject to the following conditions. **(See Examples 4–5.)**

	Principal	Annual Interest Rate	Time, Years	Compounded	Total Amount
33.	\$5,000	4.5%	5	Annually	________
34.	\$12,000	5.25%	4	Annually	________
35.	\$6,000	5%	2	Semiannually	________
36.	\$4,000	3%	3	Semiannually	________
37.	\$10,000	6%	$1\frac{1}{2}$	Quarterly	________
38.	\$9,000	4%	$2\frac{1}{2}$	Quarterly	________
39.	\$14,000	4.5%	3	Monthly	________
40.	\$9,000	8%	2	Monthly	________

Group Activity

Tracking Stocks

Materials: Computer with online access or the financial page of a newspaper.

Estimated time: 15 minutes

Group Size: 4

1. Each member of the group will choose a stock that is listed on the New York Stock Exchange. You can find prices in the newspaper or online at http://finance.yahoo.com/.

   Here are some possible stocks to track.

Apple Computer	Home Depot	Best Buy	Microsoft	Walgreens
Nike	Walmart	Hershey	Exxon Mobil	Coca-Cola

2. Beginning on a Monday, record the closing price of the stock. [The closing price is the price at the end of the day at 4:00 P.M. Eastern Standard Time (EST).] Then, each day for 2 weeks, record the closing price of the stock along with the difference in price from the previous day. Record an increase in price in green with an up arrow and a decrease in price in red with a down arrow as shown below. For example:

Date	Price ($)	Increase or Decrease ($)	Percent Increase/Decrease
April 5	**34.67**		
April 6	**34.82**	0.15 ⇑	0.4%
April 7	**33.98**	0.84 ⇓	2.4%

Date	Price ($)	Increase or Decrease ($)	Percent Increase/Decrease

3. For each stock, calculate the amount of increase or decrease in price from the previous day.

4. For each stock, calculate the *percent* increase or decrease in price from the previous day.

5. Which stock was the best investment during this 2-week period? (This is generally considered to be the stock with the greatest percent increase from the original day.)

Chapter 7 Summary

Section 7.1 Percents, Fractions, and Decimals

Key Concepts

The word **percent** means *per one hundred.*

Converting Percents to Fractions

1. Replace the % symbol by $\times \frac{1}{100}$ (or by $\div$ 100).
2. Simplify the fraction to lowest terms, if possible.

Converting Percents to Decimals

Replace the % symbol by $\times$ 0.01.
(This is equivalent to $\times \frac{1}{100}$ and $\div$ 100.)

Note: Multiplying a decimal by 0.01 is the same as moving the decimal point 2 places to the left.

Converting Fractions and Decimals to Percent Form

Multiply the fraction or decimal by 100%.
(100% = 1)

Examples

Example 1

40% means 40 per 100 or $\frac{40}{100}$.

Example 2

$$84\% = 84 \times \frac{1}{100} = \frac{84}{100} = \frac{21}{25}$$

Example 3

$$24.5\% = 24.5 \times 0.01 = 0.245$$

Example 4

$$0.07\% = 0.07 \times 0.01 = 0.0007$$

(Move the decimal point 2 places to the left.)

Example 5

$$\frac{1}{5} = \frac{1}{5} \times 100\% = \frac{100}{5}\% = 20\%$$

Example 6

$$1.14 = 1.14 \times 100\% = 114\%$$

Example 7

$$\frac{2}{3} = 0.\overline{6} \times 100\% = 66.\overline{6}\% \text{ or } 66\tfrac{2}{3}\%$$

Section 7.2 Percent Proportions and Applications

Key Concepts

A **percent proportion** is a proportion that equates a percent to an equivalent ratio.

A percent proportion can be written in the form

$$\frac{\text{Amount}}{\text{Base}} = p\% \quad \text{or} \quad \frac{\text{Amount}}{\text{Base}} = \frac{p}{100}$$

The **base** is the total or whole amount being considered. The **amount** is the part being compared to the base.

To solve a percent proportion, equate the cross products and solve the resulting equation. The variable can represent the amount, base, or p. Examples 3–5 demonstrate each type of percent problem.

Example 4

Of a sample of 400 people, 85% found relief using a particular pain reliever. How many people found relief?

Solve the proportion: $\dfrac{x}{400} = \dfrac{85}{100}$

$$85 \cdot 400 = 100x$$

$$\frac{34{,}000}{100} = \frac{100x}{100}$$

$$340 = x$$

340 people found relief.

Examples

Example 1

$\dfrac{36}{100} = \dfrac{9}{25}$ is a percent proportion.

Example 2

For the percent proportion $\dfrac{12}{200} = \dfrac{6}{100}$,

12 is the amount, 200 is the base, and p is 6.

Example 3

44% of what number is 275?

Solve the proportion: $\dfrac{275}{x} = \dfrac{44}{100}$

$$44x = 275 \cdot 100$$

$$\frac{\overset{1}{\cancel{44}}x}{\underset{1}{\cancel{44}}} = \frac{27{,}500}{44}$$

$$x = 625$$

Example 5

There are approximately 750,000 career employees in the U.S. Postal Service. If 60,000 are mail handlers, what percent does this represent?

Solve the proportion: $\dfrac{60{,}000}{750{,}000} = \dfrac{p}{100}$

$$750{,}000p = 60{,}000 \cdot 100$$

$$\frac{\overset{1}{\cancel{750{,}000}}p}{\underset{1}{\cancel{750{,}000}}} = \frac{6{,}000{,}000}{750{,}000}$$

$$p = 8$$

Of career postal employees, 8% are mail handlers.

Section 7.3 Percent Equations and Applications

Key Concepts

A **percent equation** represents a percent proportion in an alternative form:

Amount = $(p\%) \cdot$ (base)

Examples 1–3 demonstrate three types of percent problems.

Example 2

Of the car repairs performed on a certain day, 21 were repairs on transmissions. If 60 cars were repaired, what percent involved transmissions?

Solve the equation: $21 = 60x$

$$\frac{21}{60} = \frac{\overset{1}{\cancel{60}}x}{\underset{1}{\cancel{60}}}$$

$$0.35 = x$$

Because the problem asks for a percent, we have

$x = 0.35$

$= 0.35 \times 100\%$

$= 35\%$

Therefore, 35% of cars repaired involved transmissions.

Examples

Example 1

Of all breast cancer cases, 99% occur in women. Out of 2700 cases of breast cancer reported, how many are expected to occur in women?

Solve the equation: $x = (99\%)(2700)$

$x = (0.99)(2700)$

$x = 2673$

About 2673 cases are expected to occur in women.

Example 3

There are 599 endangered plants in the United States. This represents 60.7% of the total number of endangered species. Find the total number of endangered species. Round to the nearest whole number.

Solve the equation: $599 = 0.607x$

$$\frac{599}{0.607} = \frac{\overset{1}{\cancel{0.607}}x}{\underset{1}{\cancel{0.607}}}$$

$$987 \approx x$$

There is a total of approximately 987 endangered species in the United States.

Section 7.4 Applications of Sales Tax, Commission, Discount, Markup, and Percent Increase and Decrease

Key Concepts

To find **sales tax**, use the formula

$$\begin{pmatrix}\text{Amount of}\\ \text{sales tax}\end{pmatrix} = \begin{pmatrix}\text{Tax}\\ \text{rate}\end{pmatrix} \cdot \begin{pmatrix}\text{Cost of}\\ \text{merchandise}\end{pmatrix}$$

Examples

Example 1

A DVD is priced at \$16.50, and the total amount paid is \$17.82. To find the sales tax rate, first find the amount of tax.

$\$17.82 - \$16.50 = \$1.32$

To compute the sales tax rate, solve: $1.32 = x \cdot 16.50$

$$\frac{1.32}{16.50} = \frac{x \cdot \overset{1}{\cancel{16.50}}}{\underset{1}{\cancel{16.50}}}$$

$0.08 = x$ The sales tax rate is 8%.

To find a **commission**, use the formula

$$\begin{pmatrix}\text{Amount of}\\\text{commission}\end{pmatrix} = \begin{pmatrix}\text{Commission}\\\text{rate}\end{pmatrix} \cdot \begin{pmatrix}\text{Total}\\\text{sales}\end{pmatrix}$$

Example 2

Fletcher makes 13% commission on the sale of all merchandise. If he sells $11,290 worth of merchandise, find how much Fletcher will earn.

$x = (0.13)(11{,}290)$

$= 1467.7$ Fletcher will earn $1467.70 in commission.

To find **discount** and sale price, use the formulas

$$\begin{pmatrix}\text{Amount of}\\\text{discount}\end{pmatrix} = \begin{pmatrix}\text{Discount}\\\text{rate}\end{pmatrix} \cdot \begin{pmatrix}\text{Original}\\\text{price}\end{pmatrix}$$

Sale price = Original price − Amount of discount

To find **markup** and retail price, use the formulas

$$\begin{pmatrix}\text{Amount of}\\\text{markup}\end{pmatrix} = \begin{pmatrix}\text{Markup}\\\text{rate}\end{pmatrix} \cdot \begin{pmatrix}\text{Original}\\\text{price}\end{pmatrix}$$

Retail price = Original price + Amount of markup

Example 3

Margaret found a ring that was originally $425 but is on sale for 30% off. To find the sale price, first find the amount of discount.

$a = (0.30) \cdot (425)$

$= 127.5$

The sale price is $425 − $127.50 = $297.50.

Percent increase or **percent decrease** compares the *change* between two given amounts to the *original amount.*

Computing Percent Increase or Decrease

$$\begin{pmatrix}\text{Percent}\\\text{increase}\end{pmatrix} = \left(\frac{\text{Amount of increase}}{\text{Original amount}}\right) \times 100\%$$

$$\begin{pmatrix}\text{Percent}\\\text{decrease}\end{pmatrix} = \left(\frac{\text{Amount of decrease}}{\text{Original amount}}\right) \times 100\%$$

Example 4

In 1 year a child grows from 35 in. to 42 in. The increase is 42 − 35 = 7. The percent increase is

$$\frac{7}{35} \times 100\% = 0.20 \times 100\%$$

$$= 20\%$$

Section 7.5 Simple and Compound Interest

Key Concepts

To find the **simple interest** made on a certain **principal**, use the formula $I = Prt$

where I = amount of interest

P = amount of principal

r = annual interest rate

t = time (in years)

The formula for the total amount in an account is $A = P + I$, where A = total amount in an account.

Examples

Example 1

Betsey deposited $2200 in her account, which pays 5.5% simple interest. To find the simple interest she will earn after 4 years, use the formula $I = Prt$ and solve for I.

$I = (2200)(0.055)(4)$

$= 484$ She will earn $484 interest.

To find the balance or total amount of her account, apply the formula $A = P + I$.

$A = 2200 + 484$

$= 2684$ Betsey's balance will be $2684.

Many day-to-day applications involve compound interest. **Compound interest** is based on both the original principal and the interest earned.

The formula $A = P \cdot \left(1 + \frac{r}{n}\right)^{n \cdot t}$ computes the total amount in an account that uses compound interest

where A = total amount in an account

P = principal

r = annual interest rate

t = time (in years)

n = number of compounding periods per year

Example 2

Gene borrows $1000 at 6% interest compounded semiannually. If he pays off the loan in 3 years, how much will he have to pay?

We are given $P = 1000$, $r = 0.06$, $n = 2$ (semiannually means twice a year), and $t = 3$.

$$A = 1000\left(1 + \frac{0.06}{2}\right)^{2 \cdot 3}$$

$$= 1000(1.03)^6$$

$$\approx 1194.05$$

Gene will have to pay $1194.05 to pay off the loan with interest.

Chapter 7 Review Exercises

Section 7.1

For Exercises 1–4, use a percent to express the shaded portion of each drawing.

1.

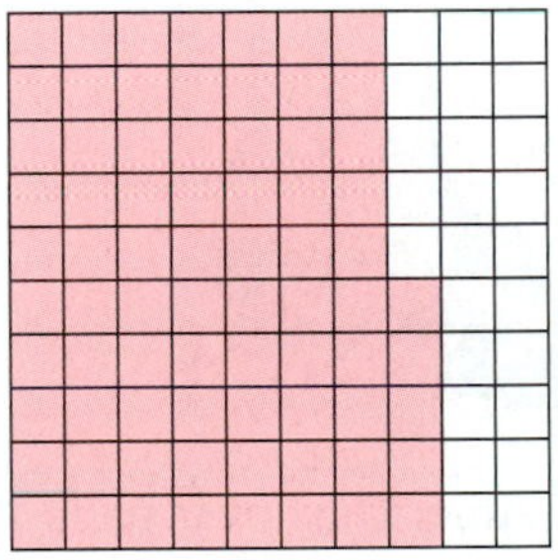

2.

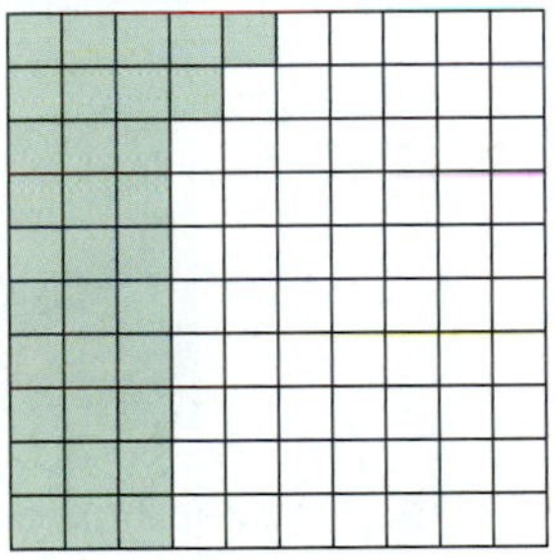

3.

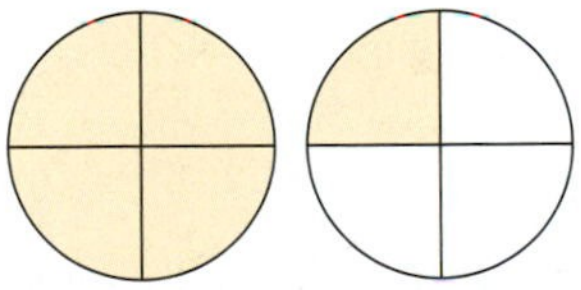

4.

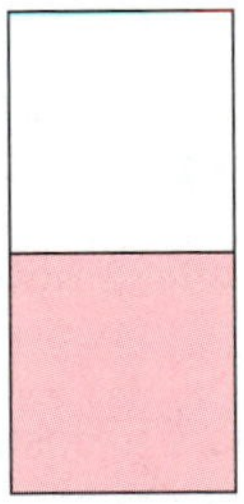

5. 68% can be expressed as which of the following forms? Identify all that apply.

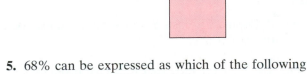

a. $\frac{68}{1000}$ **b.** $\frac{68}{100}$

c. 0.68 **d.** 0.068

6. 0.4% can be expressed as which of the following forms? Identify all that apply.

a. $\frac{4}{100}$ **b.** 0.04

c. $\frac{0.4}{100}$ **d.** 0.004

For Exercises 7–14, write the percent as a fraction and as a decimal.

7. 30% **8.** 95%

9. 135% **10.** 212%

11. 0.2% **12.** 0.6%

13. $66\frac{2}{3}\%$ **14.** $33\frac{1}{3}\%$

For Exercises 15–24, write the fraction or decimal as a percent. Round to the nearest tenth of a percent if necessary.

15. $\frac{5}{8}$ **16.** $\frac{7}{20}$

17. $\frac{7}{4}$ **18.** $\frac{11}{5}$

19. 0.006 **20.** 0.001

21. 4

22. 6

23. $\frac{3}{7}$

24. $\frac{9}{11}$

Section 7.2

For Exercises 25–28, write the percent proportion.

25. 6 books of 8 books is 75%

26. 15% of 180 lb is 27 lb.

27. 200% of \$420 is \$840.

28. 6 pine trees out of 2000 pine trees is 0.3%.

For Exercises 29–34, solve the percent problems, using proportions.

29. What is 12% of 50?

30. $5\frac{3}{4}\%$ of 64 is what number?

31. 11 is what percent of 88?

32. 8 is what percent of 2500?

33. 13 is $33\frac{1}{3}\%$ of what number?

34. 24 is 120% of what number?

35. Based on recent statistics, one airline expects that 4.2% of its customers will be "no-shows." If the airline sold 260 seats, how many people would the airline expect as no-shows? Round to the nearest whole unit.

36. In a survey of college students, 58% said that they wore their seatbelts regularly. If this represents 493 people, how many people were surveyed?

37. Victoria spends \$720 per month on rent. If her monthly take-home pay is \$1800, what percent does she pay in rent?.

38. Of the rental cars at the U-Rent-It company, 40% are compact cars. If this represents 26 cars, how many cars are on the lot?

Section 7.3

For Exercises 39–44, write as a percent equation and solve.

39. 18% of 900 is what number?

40. What number is 29% of 404?

41. 18.90 is what percent of 63?

42. What percent of 250 is 86?

43. 30 is 25% of what number?

44. 26 is 130% of what number?

45. A student buys a used book for \$54.40. This is 80% of the original price. What was the original price?

46. Veronica has read 330 pages of a 600-page novel. What percent of the novel has she read?

47. Elaine tries to keep her fat intake to no more than 30% of her total calories. If she has a 2400-calorie diet, how many fat calories can she consume to stay within her goal?

48. It is predicted that by 2010, 13% of Americans will be over the age of 65. By 2050 that number could rise to 20%. Suppose that the U.S. population is 300,000,000 in 2010 and 404,000,000 in 2050.

a. Find the number of Americans over 65 in the year 2010.

b. Find the number of Americans over 65 in the year 2050.

Section 7.4

For Exercises 49–52, solve the problem involving sales tax.

49. A Plasma TV costs \$1279. Find the sales tax if the rate is 6%.

50. The sales tax on a sofa is \$47.95. If the sofa costs \$685.00 before tax, what is the sales tax rate?

51. Kim has a digital camera. She decides to make prints for 40 photos. If the sales tax at a rate of 5% comes to \$0.44, what was the price of the photos before tax? What is the cost per photo?

52. A resort hotel charges an 11% resort tax along with the 6% sales tax. If the hotel's one-night accommodation is \$125.00, what will a tourist pay for 4 nights, including tax?

For Exercises 53–56, solve the problems involving commission.

53. At a recent auction, *Boy with a Pipe,* an early work by Pablo Picasso, sold for \$104 million. The commission for the sale of the work was \$11 million. What was the rate of commission? Round to the nearest tenth of a percent. (*Source:* The *New York Times*)

54. Andre earns a commission of 12% on sales of restaurant supplies. If he sells \$4075 in one week, how much commission will he earn?

55. Sela sells sportswear at a department store. She earns an hourly wage of \$8, and she gets a 5% commission on all merchandise that she sells over \$200. If Sela works an 8-hr day and sells \$420 of merchandise, how much will she earn that day?

56. A real estate agent earned \$9600 for the sale of a house. If her commission rate is 6%, for how much did the house sell?

For Exercises 57–60, solve the problems involving discount and markup.

57. Find the discount and the sale price of the movie if the regular price is \$28.95.

58. This notebook computer was originally priced at \$1747. How much is the discount? After the \$50 rebate, how much will a person pay for this computer?

59. A rug manufacturer sells a rug to a retail store for \$160. The store then marks up the rug to \$208. What is the markup rate?

60. Peg sold some homemade baskets to a store for \$50 each. The store marks up all merchandise by 18%. What will be the retail price of the baskets after the markup?

For Exercises 61–62, solve the problem involving percent increase or decrease.

61. The number of species of animals on the endangered species list went from 263 in 1990 to 410 in 2006. Find the percent increase. Round to the nearest tenth of a percent. (*Source:* U.S. Fish and Wildlife Services)

62. During a weak period in the economy, the stock price for Hershey Foods fell from \$50.62 per share to \$32.68 per share. Compute the percent decrease. Round to the nearest tenth of a percent.

Section 7.5

For Exercises 63–64, find the simple interest and the total amount including interest.

	Principal	Annual Interest Rate	Time, Years	Interest	Total Amount
63.	\$10,200	3%	4	_____	_____
64.	\$7000	4%	5	_____	_____

65. Jean-Luc borrowed \$2500 at 5% simple interest. What is the total amount that he will pay back at the end of 18 months (1.5 years)?

66. Kyle loaned his brother Win \$800 and charged 2.5% simple interest. If Win pays all the money back (principal plus interest) at the end of 2 yr, how much will Win pay his brother.

67. Sydney deposited \$6000 in a certificate of deposit that pays 4% interest compounded annually. Complete the table to determine her balance after 3 years.

Year	Interest	Total
1		
2		
3		

68. Nell deposited \$10,000 in a money market account that pays 3% interest compounded semiannually. Complete the table to find her balance after 2 years.

Compound Periods	Interest	Total
Period 1 (end of first 6 months)		
Period 2 (end of year 1)		
Period 3 (end of 18 months)		
Period 4 (end of year 2)		

For Exercises 69–72, find the total amount for the investment, using compound interest. Use the formula $A = P \cdot \left(1 + \frac{r}{n}\right)^{n \cdot t}$.

	Principal	Annual Interest Rate	Time in Years	Compounded	Total Amount
69.	\$850	8%	2	Quarterly	____
70.	\$2050	5%	5	Semiannually	____
71.	\$11,000	7.5%	6	Annually	____
72.	\$8200	4.5%	4	Monthly	____

Chapter 7 Test

1. Write a percent to express the shaded portion of the figure.

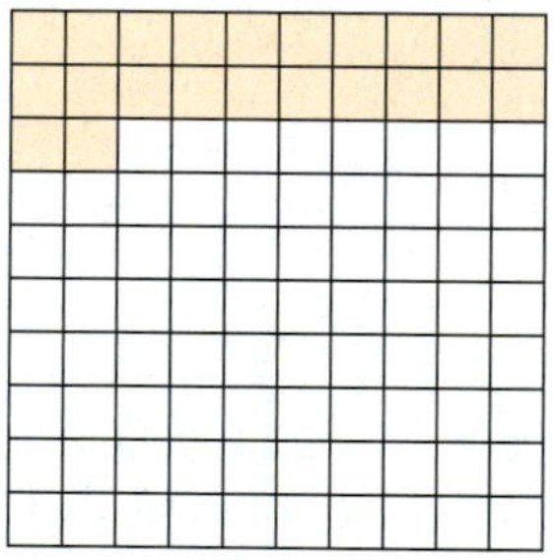

For Exercises 2–4, write the percent in decimal form and in fraction form.

2. For a recent year, the unemployment rate of Illinois was 5.4%.

3. The incidence of breast cancer increased by 0.15% between 2003 and 2005.

4. For a certain city, gas prices increased by 170% in 10 years.

5. Write the following percents in fraction form.

a. 1% **b.** 25%

c. $33\frac{1}{3}\%$ **d.** 50%

e. $66\frac{2}{3}\%$ **f.** 75%

g. 100% **h.** 150%

6. Explain the process to write a fraction as a percent.

For Exercises 7–10, write the fraction as a percent. Round to the nearest tenth of a percent if necessary.

7. $\frac{3}{5}$ **8.** $\frac{1}{250}$

9. $\frac{7}{4}$ **10.** $\frac{5}{7}$

11. Explain the process to write a decimal as a percent.

For Exercises 12–15, write the decimal as a percent.

12. 0.32 **13.** 0.052

14. 1.3 **15.** 0.006

For Exercises 16–21, solve the percent problems.

16. What is 24% of 150?

17. What is 120% of 16?

18. 21 is 6% of what number?

19. 40% of what number is 80?

20. What percent of 220 is 198?

21. 75 is what percent of 150?

22. At McDonald's, a side salad without dressing has 10 mg of sodium. With a serving of Newman's Own Low-Fat Balsamic Dressing, the sodium content of the salad is 740 mg. (*Source:* www.mcdonalds.com)

a. How much sodium is in the dressing itself?

b. What percent of the sodium content in a side salad with dressing is from the dressing? Round to the nearest tenth of a percent.

The composition of the lower level of Earth's atmosphere is given in the figure (other gases are present in minute quantities). For Exercises 23–24, use the information in the graph.

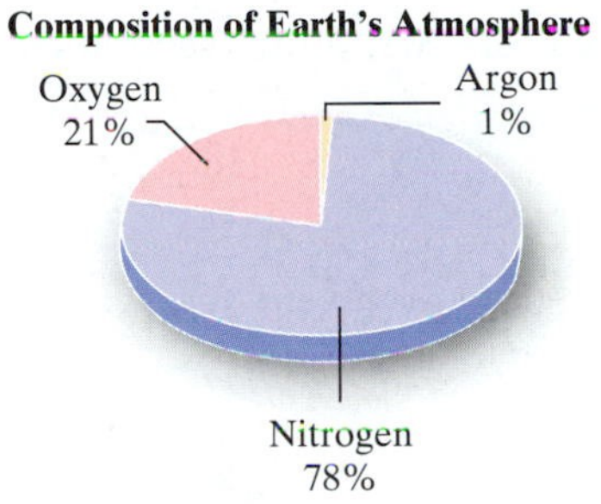

23. How much nitrogen would be expected in 500 m^3 of atmosphere?

24. How much oxygen would be expected in 2000 m^3 of atmosphere?

25. Natalia received an 8% raise in salary. If the amount of the raise is \$4160, determine her salary before the raise. Determine her new salary.

26. Brad bought a pair of blue jeans that cost \$30.00. He wrote his check for \$32.10.

a. What is the amount of sales tax that he paid?

b. What is the sales tax rate?

27. Charles earns a salary of \$400 per week and gets a bonus of 6% commission on all merchandise that he sells. If Charles sells \$3500 worth of merchandise, how much will he earn in that week?

28. Find the discount rate of the product in the advertisement.

29. A furniture store buys merchandise from the manufacturer and then marks it up by 30%.

a. If the markup on a dining room set is \$375, what was the price from the manufacturer?

b. What is the retail price?

c. If there is a 6% sales tax, what is the total cost to buy the dining room set?

30. Maury borrowed \$5000 at 8% simple interest. He plans to pay back the loan in 3 years.

a. How much interest will he have to pay?

b. What is the total amount that he has to pay back?

31. Use the formula $A = P \cdot \left(1 + \frac{r}{n}\right)^{n \cdot t}$ to calculate the total amount in an account that began with \$25,000 invested at 4.5% compounded quarterly for 5 years.

Chapters 1–7 Cumulative Review Exercises

1. What is the place value for the digit 6 in the number 26,009,235?

2. Fill in the table with either the word name for the number or the number in standard form.

	Country	Area (mi^2) Standard Form	Area (mi^2) Words
a.	United States		Three million, five hundred thirty-nine thousand, two hundred forty-five
b.	Saudi Arabia	830,000	
c.	Falkland Islands	4,700	
d.	Colombia		Four hundred one thousand, forty-four

3. Multiply. $34{,}882 \times 100$

4. Divide. $9\overline{)783}$

5. Simplify. $234 + (-44) + 6 + (-2901)$

6. Simplify. $\sqrt{16} - 6 \div 3 + 3^2$

7. Simplify. $-3(5 - 7)^2 + (-24 \div 4)$

For Exercises 8–11, multiply or divide as indicated. Simplify the fraction to lowest terms.

8. $\frac{3}{8} \cdot \frac{32}{9}$

9. $-\frac{42}{25} \div \frac{7}{100}$

10. $-\frac{21}{2} \div (-7)$

11. $16 \cdot \frac{1}{24}$

12. Find the perimeter of the figure.

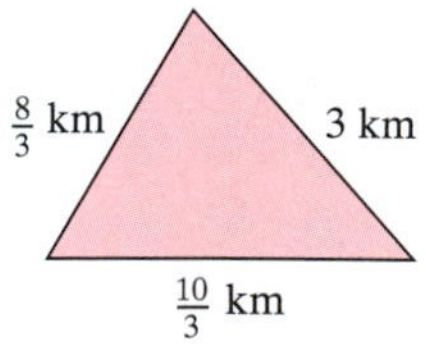

13. Add. $\frac{3}{10} + \frac{17}{100} + \frac{3}{1000}$

14. A sheet of paper has the dimensions $13\frac{1}{2}$ in. by 17 in. What is the area of the paper?

15. a. List four multiples of 18.
    b. List all factors of 18.
    c. Write the prime factorization of 18.

16. Write a fraction that represents the shaded portion of each figure.

a. 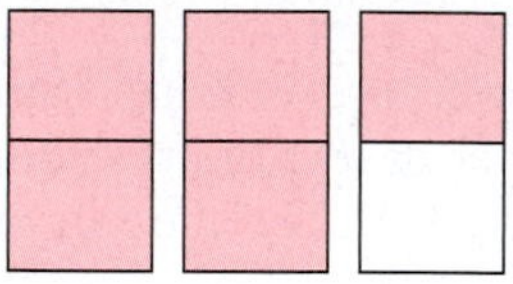

b. 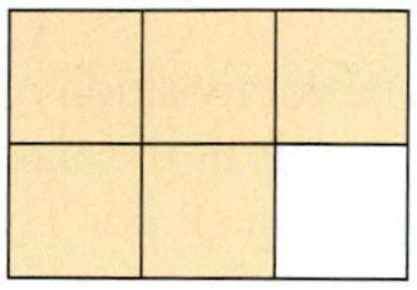

For Exercises 17–18, write the fraction as a decimal.

17. $\frac{3}{8}$

18. $\frac{4}{3}$

For Exercises 19–20, write the fraction or decimal as a percent.

19. $\frac{7}{8}$

20. 2.4

For Exercises 21–22, refer to the chart.

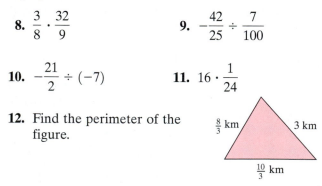

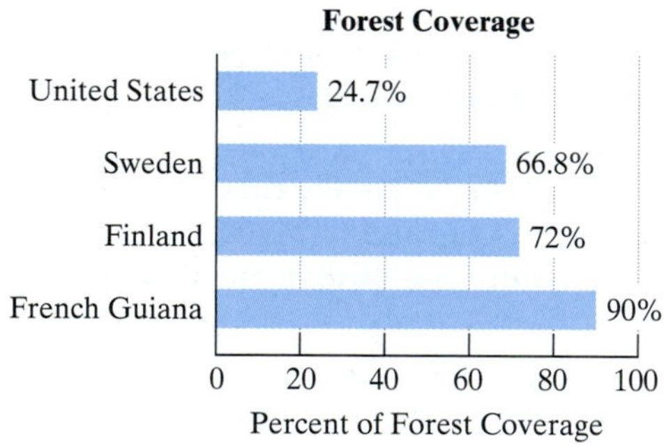

21. What is the difference in the percent of forest coverage in French Guiana and the United States?

22. If Sweden has a land area of 158,662 mi^2, how much is covered in forest? Round to the nearest square mile.

For Exercises 23–24, simplify the expression.

23. $-2x - 5(x - 6) + 9$

24. $\frac{1}{3}a - \frac{1}{5}b - \frac{5}{6}a + \frac{7}{5}b$

For Exercises 25–30, solve the equation.

25. $\frac{x}{9} = -4$

26. $0.02t = 6$

27. $-3x + 5 = 17$

28. $4y - 9 = 3(y + 2)$

29. $\frac{3}{4} = \frac{15}{p}$

30. $\frac{4\frac{1}{3}}{p} = \frac{12}{18}$

31. A computer can download 1.6 megabytes (MB) in 2.5 min. How long will it take to download a 4.6-MB file? Round to the nearest tenth of a minute.

32. A DC-10 aircraft flew 1799 mi in 3.5 hr. Find the unit rate in miles per hour.

33. Kevin deposited \$13,000 in a certificate of deposit that pays 3.2% simple interest. How much will Kevin have if he keeps the certificate for 5 years?

34. New York City has the largest population of any city in the United States. In 1900, it had 3.4 million people and in 2000, it had 8.0 million. Determine the percent increase in population. Round to the nearest whole percent.

35. A bank charges 8% interest that is compounded monthly. Use the formula $A = P \cdot \left(1 + \frac{r}{n}\right)^{n \cdot t}$ to find the total amount of interest paid for a 10-year mortgage on \$75,000.

Measurement and Geometry

8

CHAPTER OUTLINE

Chapter 8

Measurement is used when collecting data (information) for virtually all scientific experiments. In this chapter, we study both the U.S. Customary units of measure and the metric system. Then we use both systems in applications of medicine and geometry.

Are You Prepared?

To prepare for this chapter, take a minute to review some of the geometry formulas presented earlier in the text. Match the formula on the left with the figure on the right.

1. $A = \pi r^2$ ______

2. $P = 4s$ ______

3. $C = 2\pi r$ ______

4. $A = s^2$ ______

5. $A = \frac{1}{2}bh$ ______

6. $P = 2l + 2w$ ______

7. $A = lw$ ______

8. $P = a + b + c$ ______

a. Area: 

b. Circumference: 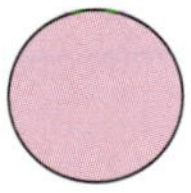

c. Area: 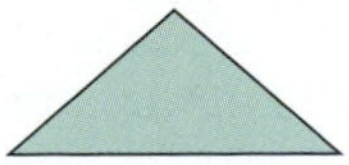

d. Perimeter: 

e. Perimeter: 

f. Area: 

g. Area: 

h. Perimeter: 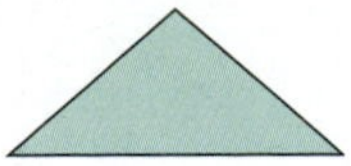

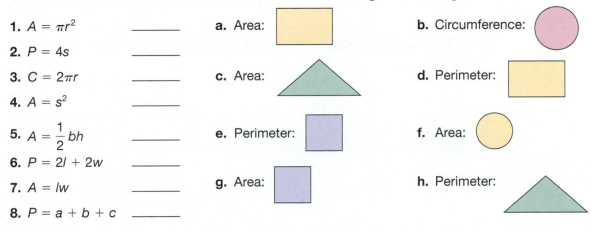

Section 8.1 U.S. Customary Units of Measurement

Objectives

1. U.S. Customary Units
2. U.S. Customary Units of Length
3. Units of Time
4. U.S. Customary Units of Weight
5. U.S. Customary Units of Capacity

1. U.S. Customary Units

In many applications in day-to-day life, we need to measure things. To measure an object means to assign it a number and a **unit of measure**. In this section, we present units of measure used in the United States for length, time, weight, and capacity. Table 8-1 gives several commonly used units and their equivalents.

TIP: Sometimes units of feet are denoted with the ′ symbol. That is, 3 ft = 3′. Similarly, sometimes units of inches are denoted with the ″ symbol. That is, 4 in. = 4″.

Table 8-1 Summary of U.S. Customary Units of Length, Time, Weight, and Capacity

Length	Time
1 foot (ft) = 12 inches (in.)	1 year (yr) = 365 days
1 yard (yd) = 3 feet (ft)	1 week (wk) = 7 days
1 mile (mi) = 5280 feet (ft)	1 day = 24 hours (hr)
1 mile (mi) = 1760 yards (yd)	1 hour (hr) = 60 minutes (min)
	1 minute (min) = 60 seconds (sec)
Capacity	**Weight**
3 teaspoons (tsp) = 1 tablespoon (T)	1 pound (lb) = 16 ounces (oz)
1 cup (c) = 8 fluid ounces (fl oz)	1 ton = 2000 pounds (lb)
1 pint (pt) = 2 cups (c)	
1 quart (qt) = 2 pints (pt)	
1 quart (qt) = 4 cups (c)	
1 gallon (gal) = 4 quarts (qt)	

2. U.S. Customary Units of Length

In Example 1, we will demonstrate how to convert between two units of measure by multiplying by a conversion factor. A **conversion factor** is a ratio of equivalent measures.

For example, note that 1 yd = 3 ft. Therefore, $\frac{1\text{ yd}}{3\text{ ft}} = 1$ and $\frac{3\text{ ft}}{1\text{ yd}} = 1$.

These ratios are called **unit ratios** (or **unit fractions**). In a unit ratio, the quotient is 1 because we are dividing measurements of equal length. To convert from one unit of measure to another, we can multiply by a unit ratio. We offer these guidelines to determine the proper unit ratio to use.

Concept Connections

Complete the unit ratio.

1. $\frac{1\text{ ft}}{\square\text{ in.}}$ 2. $\frac{\square\text{ yd}}{3\text{ ft}}$

PROCEDURE Choosing a Unit Ratio as a Conversion Factor

In a unit ratio,

- The unit of measure in the numerator should be the new unit you want to convert *to*.
- The unit of measure in the denominator should be the original unit you want to convert *from*.

Answers

1. $\frac{1\text{ ft}}{12\text{ in.}}$ 2. $\frac{1\text{ yd}}{3\text{ ft}}$

Example 1 Converting Units of Length by Using Unit Ratios

Convert the units of length.

a. 1500 ft = ____ yd **b.** 9240 yd = ____ mi **c.** 8.2 mi = ____ ft

Solution:

a. From Table 8-1, we have 1 yd = 3 ft.

$$1500\text{ ft} = 1500\text{ ft} \cdot \frac{1\text{ yd}}{3\text{ ft}}$$

← new unit to convert to (numerator); ← unit to convert from (denominator)

$$= \frac{1500\text{ ft}}{1} \cdot \frac{1\text{ yd}}{3\text{ ft}}$$

Notice that the original units of ft reduce or "cancel" in much the same way as simplifying fractions. The unit yd remains in the final answer.

$$= \frac{1500}{3}\text{ yd}$$

$$= 500\text{ yd}$$

b. From Table 8-1, we have 1 mi = 1760 yd.

$$9240\text{ yd} = 9240\text{ yd} \cdot \frac{1\text{ mi}}{1760\text{ yd}}$$

← new unit to convert to (numerator); ← unit to convert from (denominator)

$$= \frac{9240\text{ yd}}{1} \cdot \frac{1\text{ mi}}{1760\text{ yd}}$$

The units of yd "cancel," leaving the answer in miles.

$$= \frac{9240}{1760}\text{ mi}$$

Multiply fractions.

$$= 5.25\text{ mi}$$

Simplify.

c. From Table 8-1 we have 1 mi = 5280 ft.

$$8.2\text{ mi} = 8.2\text{ mi} \cdot \frac{5280\text{ ft}}{1\text{ mi}}$$

← new unit to convert to (numerator); ← unit to convert from (denominator)

$$= \frac{8.2\text{ mi}}{1} \cdot \frac{5280\text{ ft}}{1\text{ mi}}$$

The units of mi "cancel," leaving the answer in feet.

$$= 43{,}296\text{ ft}$$

TIP: It is important to write the units associated with the numbers. The units can help you select the correct unit ratio.

Skill Practice

Convert, using unit ratios.

3. 720 in. = _____ ft

4. 4224 ft = _____ mi

5. 8 mi = _____ yd

Answers

3. 60 ft **4.** 0.8 mi **5.** 14,080 yd

Skill Practice

Convert, using unit ratios.

6. 6.2 yd = ____ in.

7. 6336 in. = ____ mi

Example 2 Making Multiple Conversions of Length

Convert the units of length.

a. 0.25 mi = ____ in. **b.** 22 in. = ____ yd

Solution:

a. To convert miles to inches, we use two conversion factors. The first unit ratio converts miles to feet. The second unit ratio converts feet to inches.

(converts mi to ft; converts ft to in.)

$$0.25\text{ mi} = 0.25\text{ mi} \cdot \frac{5280\text{ ft}}{1\text{ mi}} \cdot \frac{12\text{ in.}}{1\text{ ft}}$$

$$= \frac{0.25\text{ }\cancel{\text{mi}}}{1} \cdot \frac{5280\text{ }\cancel{\text{ft}}}{1\text{ }\cancel{\text{mi}}} \cdot \frac{12\text{ in.}}{1\text{ }\cancel{\text{ft}}}$$ The units mi and ft "cancel," leaving the answer in inches.

$$= 15{,}840\text{ in.}$$

(converts in. to ft; converts ft to yd)

b. $$22\text{ in.} = 22\text{ in.} \cdot \frac{1\text{ ft}}{12\text{ in.}} \cdot \frac{1\text{ yd}}{3\text{ ft}}$$ Multiply by two conversion factors. The first converts inches to feet. The second converts feet to yards.

$$= \frac{22\text{ }\cancel{\text{in.}}}{1} \cdot \frac{1\text{ }\cancel{\text{ft}}}{12\text{ }\cancel{\text{in.}}} \cdot \frac{1\text{ yd}}{3\text{ }\cancel{\text{ft}}}$$ The units in. and ft "cancel," leaving the answer in yards.

$$= \frac{22}{36}\text{ yd}$$ Multiply fractions.

$$= \frac{11}{18}\text{ yd} \quad \text{or} \quad 0.6\overline{1}\text{ yd}$$ Simplify.

To add and subtract measurements, we must have like units. For example:

$$3\text{ ft} + 8\text{ ft} = 11\text{ ft}$$

Sometimes, however, measurements have mixed units. For example, a drainpipe might be 4 ft 6 in. long or symbolically 4′6″. Measurements and calculations with mixed units can be handled in much the same way as mixed numbers.

Skill Practice

8. Add. 8′4″ + 4′10″

9. Subtract. 6′3″ − 4′9″

Example 3 Adding and Subtracting Mixed Units of Measurement

a. Add. 4′6″ + 2′9″ **b.** Subtract. 8′2″ − 3′6″

Solution:

a. 4′6″ + 2′9″ =

$$\begin{array}{r} 4\text{ ft} + 6\text{ in.} \\ +2\text{ ft} + 9\text{ in.} \\ \hline 6\text{ ft} + 15\text{ in.} \end{array}$$ Add like units.

$$= 6\text{ ft} + 1\text{ ft} + 3\text{ in.}$$ Because 15 in. is more than 1 ft, we can write 15 in. = 12 in. + 3 in. = 1 ft + 3 in.

$$= 7\text{ ft }3\text{ in.} \quad \text{or} \quad 7'3''$$

Answers

6. 223.2 in. **7.** 0.1 mi

8. 13′2″ **9.** 1′6″

b. $8'2'' - 3'6'' = \begin{array}{r} 8\text{ ft} + 2\text{ in.} \\ -(3\text{ ft} + 6\text{ in.}) \\ \hline \end{array} = \begin{array}{r} \overset{7}{\cancel{8}}\text{ ft} + \overset{12\text{ in.}}{2}\text{ in.} \\ -(3\text{ ft} + 6\text{ in.}) \\ \hline \end{array}$ Borrow 1 ft = 12 in.

$= \begin{array}{r} 7\text{ ft} + 14\text{ in.} \\ -(3\text{ ft} + 6\text{ in.}) \\ \hline 4\text{ ft} + 8\text{ in.} \end{array}$ or $4'8''$

3. Units of Time

In Examples 4 and 5, we convert between two units of time.

Example 4 Converting Units of Time

Convert the units of time.

a. 32 hr = ___ days **b.** 36 hr = ___ sec

Solution:

a. $32\text{ hr} = \dfrac{32\text{ }\cancel{\text{hr}}}{1} \cdot \dfrac{1\text{ day}}{24\text{ }\cancel{\text{hr}}}$ ← new unit to convert to / ← unit to convert from — Recall that 1 day = 24 hr.

$= \dfrac{32}{24}\text{ days}$ Multiply fractions.

$= \dfrac{4}{3}\text{ days or } 1\dfrac{1}{3}\text{ days}$ Simplify.

b. $36\text{ hr} = \dfrac{36\text{ }\cancel{\text{hr}}}{1} \cdot \dfrac{60\text{ }\cancel{\text{min}}}{1\text{ }\cancel{\text{hr}}} \cdot \dfrac{60\text{ sec}}{1\text{ }\cancel{\text{min}}}$ (converts hr to min; converts min to sec) Multiply by two conversion factors.

$= 129{,}600\text{ sec}$ Simplify.

Skill Practice

Convert.

10. 16 hr = _____ days

11. 24 hr = _____ sec

Example 5 Converting Units of Time

After running a marathon, Dave crossed the finish line and noticed that the race clock read 2:20:30. Convert this time to minutes.

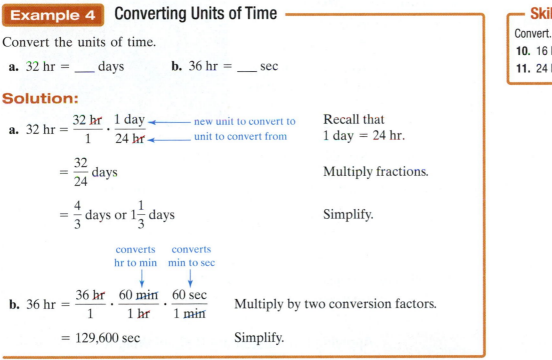

Solution:

The notation 2:20:30 means 2 hr 20 min 30 sec. We must convert 2 hr to minutes and 30 sec to minutes. Then we add the total number of minutes.

$$2\text{ hr} = \frac{2\text{ }\cancel{\text{hr}}}{1} \cdot \frac{60\text{ min}}{1\text{ }\cancel{\text{hr}}} = 120\text{ min}$$

$$30\text{ sec} = \frac{30\text{ }\cancel{\text{sec}}}{1} \cdot \frac{1\text{ min}}{60\text{ }\cancel{\text{sec}}} = \frac{30}{60}\text{ min} = \frac{1}{2}\text{ min} \quad \text{or} \quad 0.5\text{ min}$$

The total number of minutes is 120 min + 20 min + 0.5 min = 140.5 min. Dave finished the race in 140.5 min.

Skill Practice

12. When Ward Burton won the Daytona 500 his time was 2:22:45. Convert this time to minutes.

Answers

10. $\frac{2}{3}$ day **11.** 86,400 sec

12. 142.75 min

4. U.S. Customary Units of Weight

Measurements of weight record the force of an object subject to gravity. In Example 6, we convert from one unit of weight to another.

Skill Practice

13. The blue whale is the largest animal on Earth. It is so heavy that it would crush under its own weight if it were taken from the water. An average adult blue whale weighs 120 tons. Convert this to pounds.

14. A box of apples weighs 5 lb 12 oz. Convert this to ounces.

Example 6 Converting Units of Weight

a. The average weight of an adult male African elephant is 12,400 lb. Convert this value to tons.

b. Convert the weight of a 7-lb 3-oz baby to ounces.

Solution:

a. Recall that 1 ton = 2000 lb.

$$12{,}400 \text{ lb} = \frac{12{,}400 \text{ lb}}{1} \cdot \frac{1 \text{ ton}}{2000 \text{ lb}}$$

$$= \frac{12{,}400}{2000} \text{ tons} \qquad \text{Multiply fractions.}$$

$$= \frac{31}{5} \text{ tons} \quad \text{or} \quad 6.2 \text{ tons}$$

An adult male African elephant weighs 6.2 tons.

b. To convert 7 lb 3 oz to ounces, we must convert 7 lb to ounces.

$$7 \text{ lb} = \frac{7 \text{ lb}}{1} \cdot \frac{16 \text{ oz}}{1 \text{ lb}} \qquad \text{Recall that } 1 \text{ lb} = 16 \text{ oz.}$$

$$= 112 \text{ oz}$$

The baby's total weight is 112 oz + 3 oz = 115 oz.

Skill Practice

15. A set of triplets weighed 4 lb 3 oz, 3 lb 9 oz, and 4 lb 5 oz. What is the total weight of all three babies?

Example 7 Applying U.S. Customary Units of Weight

Jessica lifts four boxes of books. The boxes have the following weights: 16 lb 4 oz, 18 lb 8 oz, 12 lb 5 oz, and 22 lb 9 oz. How much weight did she lift altogether?

Solution:

$$\begin{array}{rr} 16 \text{ lb} & 4 \text{ oz} \\ 18 \text{ lb} & 8 \text{ oz} \\ 12 \text{ lb} & 5 \text{ oz} \\ + \; 22 \text{ lb} & 9 \text{ oz} \\ \hline 68 \text{ lb} & 26 \text{ oz} \end{array} \qquad \text{Add like units in columns.}$$

$$68 \text{ lb } 26 \text{ oz} = 68 \text{ lb} + 26 \text{ oz}$$

$$= 68 \text{ lb} + \overbrace{1 \text{ lb} + 10 \text{ oz}} \qquad \text{Recall that } 1 \text{ lb} = 16 \text{ oz.}$$

$$= 69 \text{ lb } 10 \text{ oz}$$

Jessica lifted 69 lb 10 oz of books.

Answers

13. 240,000 lb **14.** 92 oz
15. 12 lb 1 oz

5. U.S. Customary Units of Capacity

A typical can of soda contains 12 fl oz. This is a measure of capacity. Capacity is the volume or amount that a container can hold. The U.S. Customary units of capacity are fluid ounces (fl oz), cup (c), pint (pt), quart (qt), and gallon (gal).

One fluid ounce is approximately the amount of liquid that two large spoonfuls will hold. One cup is the amount in an average-size cup of tea. While Table 8-1 summarizes the relationships among units of capacity, we also offer an illustration (Figure 8-1).

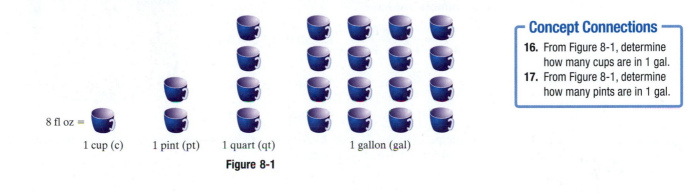

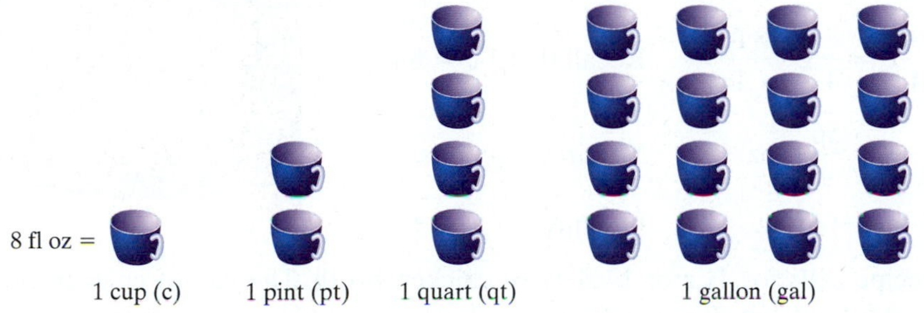

Figure 8-1

> **Concept Connections**
>
> **16.** From Figure 8-1, determine how many cups are in 1 gal.
>
> **17.** From Figure 8-1, determine how many pints are in 1 gal.

Example 8 Converting Units of Capacity

Convert the units of capacity.

a. 1.25 pt = _____ qt **b.** 2 gal = _____ c **c.** 48 fl oz = _____ gal

Solution:

a. $1.25 \text{ pt} = \dfrac{1.25 \cancel{\text{pt}}}{1} \cdot \dfrac{1 \text{ qt}}{2 \cancel{\text{pt}}}$ Recall that 1 qt = 2 pt.

$= \dfrac{1.25}{2} \text{ qt}$ Multiply fractions.

$= 0.625 \text{ qt}$ Simplify.

b. $2 \text{ gal} = 2 \text{ gal} \cdot \dfrac{4 \text{ qt}}{1 \text{ gal}} \cdot \dfrac{4 \text{ c}}{1 \text{ qt}}$ Use two conversion factors. The first unit ratio converts gallons to quarts. The second converts quarts to cups.

$= \dfrac{2 \cancel{\text{gal}}}{1} \cdot \dfrac{4 \cancel{\text{qt}}}{1 \cancel{\text{gal}}} \cdot \dfrac{4 \text{ c}}{1 \cancel{\text{qt}}}$

$= 32 \text{ c}$ Multiply.

c. $48 \text{ fl oz} = \dfrac{48 \cancel{\text{fl oz}}}{1} \cdot \dfrac{1 \cancel{\text{c}}}{8 \cancel{\text{fl oz}}} \cdot \dfrac{1 \cancel{\text{qt}}}{4 \cancel{\text{c}}} \cdot \dfrac{1 \text{ gal}}{4 \cancel{\text{qt}}}$ Convert from fluid ounces to cups, from cups to quarts, and from quarts to gallons.

$= \dfrac{48}{128} \text{ gal}$

$= \dfrac{3}{8} \text{ gal}$ or 0.375 gal

> **Skill Practice**
>
> Convert.
>
> **18.** 8.5 gal = _____ qt
>
> **19.** 2.25 qt = _____ c
>
> **20.** 40 fl oz = _____ qt

> **Avoiding Mistakes**
>
> It is important to note that ounces (oz) and fluid ounces (fl oz) are different quantities. An ounce (oz) is a measure of weight, and a fluid ounce (fl oz) is a measure of capacity. Furthermore,
>
> 16 oz = 1 lb
>
> 8 fl oz = 1 c

Answers

16. 16 c in 1 gal **17.** 8 pt in 1 gal
18. 34 qt **19.** 9 c
20. 1.25 qt

Skill Practice

21. A recipe calls for $3\frac{1}{2}$ c of tomato sauce. A jar of sauce holds 24 fl oz. Is there enough sauce in the jar for the recipe?

Example 9 Applying Units of Capacity

A recipe calls for $1\frac{3}{4}$ c of chicken broth. A can of chicken broth holds 14.5 fl oz. Is there enough chicken broth in the can for the recipe?

Solution:

We need to convert each measurement to the same unit of measure for comparison. Converting $1\frac{3}{4}$ c to fluid ounces, we have

$$1\frac{3}{4}\text{ c} = \frac{7}{4}\,\cancel{\text{c}} \cdot \frac{8\text{ fl oz}}{1\,\cancel{\text{c}}} \qquad \text{Recall that } 1\text{ c} = 8\text{ fl oz.}$$

$$= \frac{56}{4}\text{ fl oz} \qquad \text{Multiply fractions.}$$

$$= 14\text{ fl oz} \qquad \text{Simplify.}$$

The recipe calls for $1\frac{3}{4}$ c or 14 fl oz of chicken broth. The can of chicken broth holds 14.5 fl oz, which is enough.

Answer

21. No, $3\frac{1}{2}$ c is equal to 28 fl oz. The jar only holds 24 fl oz.

Section 8.1 Practice Exercises

Boost *your* GRADE at ALEKS.com!

- Practice Problems
- Self-Tests
- NetTutor
- e-Professors
- Videos

Study Skills Exercise

1. Define the key terms.

a. **Unit of measure** **b.** **Conversion factor** **c.** **Unit ratio (or unit fraction)**

Objective 2: U.S. Customary Units of Length

2. Identify an appropriate ratio to convert feet to inches by using multiplication.

a. $\dfrac{12\text{ ft}}{1\text{ in.}}$ **b.** $\dfrac{12\text{ in.}}{1\text{ ft}}$ **c.** $\dfrac{1\text{ ft}}{12\text{ in.}}$ **d.** $\dfrac{1\text{ in.}}{12\text{ ft}}$

For Exercises 3–20, convert the units of length by using unit ratios. **(See Examples 1–2.)**

3. 9 ft = ____ yd **4.** $2\frac{1}{3}$ yd = ____ ft **5.** 3.5 ft = ____ in.

6. $4\frac{1}{2}$ in. = ____ ft **7.** 11,880 ft = ____ mi **8.** 0.75 mi = ____ ft

9. 14 ft = ____ yd **10.** 75 in. = ____ ft **11.** 320 mi = ____ yd

12. $3\frac{1}{4}$ ft = ____ in. **13.** 171 in. = ____ yd **14.** 0.3 mi = ____ in.

15. 2 yd = ____ in. **16.** 12,672 in. = ____ mi **17.** 0.8 mi = ____ in.

18. 900 in. = ____ yd **19.** 31,680 in. = ____ mi **20.** 6 yd = ____ in.

21. **a.** Convert 6′4″ to inches.
b. Convert 6′4″ to feet.

22. **a.** Convert 10 ft 8 in. to inches.
b. Convert 10 ft 8 in. to feet.

23. **a.** Convert 2 yd 2 ft to feet.
b. Convert 2 yd 2 ft to yards.

24. **a.** Convert 3′6″ to feet.
b. Convert 3′6″ to inches.

For Exercises 25–30, add or subtract as indicated. **(See Example 3.)**

25. 2 ft 8 in. + 3 ft 4 in.
26. 5 ft 2 in. + 6 ft 10 in.
27. 8 ft 8 in. − 5 ft 4 in.
28. 3 ft 2 in. − 1 ft 5 in.
29. 9′2″ − 4′10″
30. 4′10″ + 6′4″

Objective 3: Units of Time

For Exercises 31–42, convert the units of time. **(See Example 4.)**

31. 2 yr = ____ days
32. $1\frac{1}{2}$ days = ____ hr
33. 90 min = ____ hr
34. 3 wk = ____ days
35. 180 sec = ____ min
36. $3\frac{1}{2}$ hr = ____ min
37. 72 hr = ____ days
38. 28 days = ____ wk
39. 3600 sec = ____ hr
40. 168 hr = ____ wk
41. 9 wk = ____ hr
42. 1680 hr = ____ wk

For Exercises 43–46, convert the time given as hr:min:sec to minutes. **(See Example 5.)**

43. 1:20:30
44. 3:10:45
45. 2:55:15
46. 1:40:30

Objective 4: U.S. Customary Units of Weight

For Exercises 47–52, convert the units of weight. **(See Example 6.)**

47. 32 oz = ____ lb
48. 2500 lb = ____ tons
49. 2 tons = ____ lb
50. 4 lb = ____ oz
51. $3\frac{1}{4}$ tons = ____ lb
52. 3000 lb = ____ tons

For Exercises 53–58, add or subtract as indicated. **(See Example 7.)**

53. 6 lb 10 oz + 3 lb 14 oz
54. 12 lb 11 oz + 13 lb 7 oz
55. 30 lb 10 oz − 22 lb 8 oz
56. 5 lb − 2 lb 5 oz
57. 10 lb − 3 lb 8 oz
58. 20 lb 3 oz + 15 oz

Objective 5: U.S. Customary Units of Capacity

For Exercises 59–70, convert the units of capacity. **(See Example 8.)**

59. 16 fl oz = ____ c
60. 5 pt = ____ c
61. 6 gal = ____ qt
62. 8 pt = ____ qt
63. 1 gal = ____ c
64. 1 T = ____ tsp
65. 2 qt = ____ gal
66. 2 qt = ____ c
67. 1 pt = ____ fl oz
68. 32 fl oz = ____ qt
69. 2 T = ____ tsp
70. 2 gal = ____ pt

Mixed Exercises

71. A recipe for minestrone soup calls for 3 c of spaghetti sauce. If a jar of sauce has 48 fl oz, is there enough for the recipe? **(See Example 9.)**

72. A recipe for punch calls for 6 c of apple juice. A bottle of juice has 2 qt. Is there enough juice in the bottle for the recipe?

73. A plumber used two pieces of pipe for a job. One piece was 4′6″ and the other was 2′8″. How much pipe was used?

74. A carpenter needs to put wood molding around three sides of a room. Two sides are 6′8″ long, and the third side is 10′ long. How much molding should the carpenter purchase?

75. If you have 4 yd of rope and you use 5 ft, how much is left over? Express the answer in feet.

76. The Blaisdell Arena football field, home of the Hawaiian Islanders football team, did not have the proper dimensions required by the arena football league. The width was measured to be 82 ft 10 in. Regulation width is 85 ft. What is the difference between the widths of a regulation field and the field at Blaisdell Arena?

77. Find the perimeter in feet.

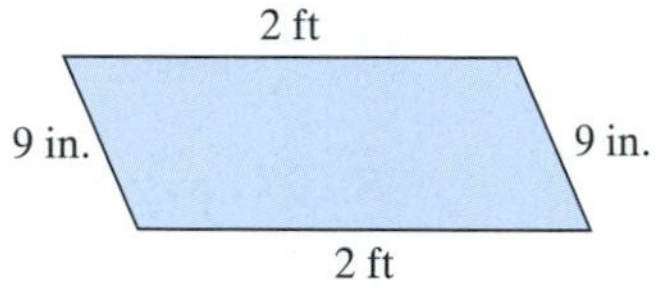

78. Find the perimeter in yards.

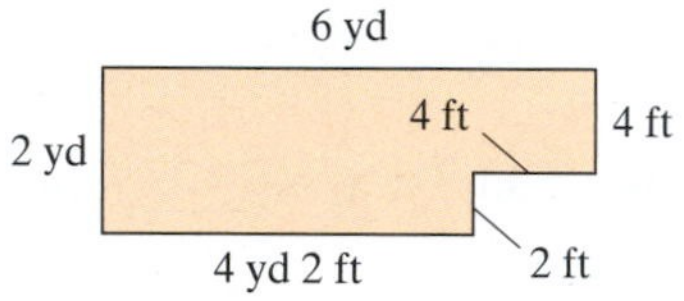

79. A 24-fl-oz jar of spaghetti sauce sells for \$2.69. Another jar that holds 1 qt of sauce sells for \$3.29. Which is a better buy? Explain.

80. Tatiana went to purchase bottled water. She found the following options: a 12-pack of 1-pt bottles; a 1-gal jug; and a 6-pack of 24-fl-oz bottles. Which option should Tatiana choose to get the most water? Explain.

81. A picnic table requires 5 boards that are 6′ long, 4 boards that are 3′3″ long, and 2 boards that are 18″ long. Find the total length of lumber required.

82. A roll of ribbon is 60 yd. If you wrap 12 packages that each use 2.5 ft of ribbon, how much ribbon is left over?

83. Byron lays sod in his backyard. Each piece of sod weighs 6 lb 4 oz. If he puts down 50 pieces, find the total weight.

84. A can of paint weighs 2 lb 4 oz. How much would 6 cans weigh?

85. The garden pictured needs a decorative border. The border comes in pieces that are 1.5 ft long. How many pieces of border are needed to surround the garden?

86. Monte fences all sides of a field with panels of fencing that are 2 yd long. How many panels of fencing does he need?

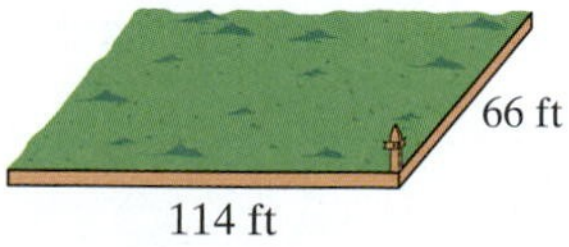

For Exercises 87–90, refer to the table.

Day	Time
Mon.	1 hr 10 min
Tues.	45 min
Wed.	1 hr 20 min
Thur.	30 min
Fri.	50 min
Sat.	Rest
Sun.	1 hr

87. Gil is a distance runner. The durations of his training runs for one week are given in the table. Find the total time that Gil ran that week, and express the answer in mixed units.

88. Find the difference between the amount of time Gil trained on Monday and the amount of time he trained on Friday.

89. In a team triathlon, Torie swims $\frac{1}{2}$ mi in 15 min 30 sec. David rides his bicycle 20 mi in 50 min 20 sec. Emilie runs 4 mi in 28 min 10 sec. Find the total time for the team.

90. Joe competes in a biathlon. He runs 5 mi in 32 min 8 sec. He rides his bike 25 mi in 1 hr 2 min 40 sec. Find the total time for his race.

Expanding Your Skills

In Section 1.5 we learned that area is measured in square units such as in.^{2}, ft^2, yd^2, and mi^2. Converting square units involves a different set of conversion factors. For example, 1 yd = 3 ft, but 1 yd^2 = 9 ft^2. To understand why, recall that the formula for the area of a rectangle is $A = l \times w$. In a square, the length and the width are the same distance s. Therefore, the area of a square is given by the formula $A = s \times s$, where s is the length of a side.

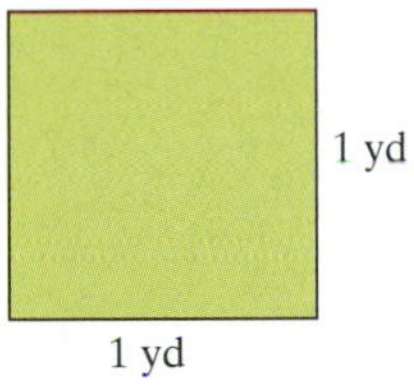

$$A = (1\text{ yd})(1\text{ yd}) = 1\text{ yd}^2$$

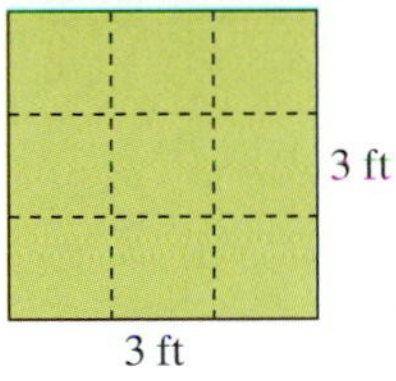

$$A = (3\text{ ft})(3\text{ ft}) = 9\text{ ft}^2$$

Instead of learning a new set of conversion factors, we can use multiples of the conversion factors that we mastered in this section.

Example: Converting Area

Convert 4 yd^2 to square feet.

Solution: We will use the unit ratio of $\frac{3\text{ ft}}{1\text{ yd}}$ twice.

$$\frac{4\text{ yd}^2}{1} \cdot \underbrace{\frac{3\text{ ft}}{1\text{ yd}} \cdot \frac{3\text{ ft}}{1\text{ yd}}}_{\text{Multiply first.}} = \frac{4\,\cancel{\text{yd}^2}}{1} \cdot \frac{9\text{ ft}^2}{1\,\cancel{\text{yd}^2}} = 36\text{ ft}^2$$

TIP: We use the unit ratio $\frac{3\text{ ft}}{1\text{ yd}}$ twice because there are two dimensions. Length and width must both be converted to feet.

For Exercises 91–98, convert the units of area by using multiple factors of the given unit ratio.

91. 54 ft^2 = ____ yd^2 $\left(\text{Use two factors of the ratio } \frac{1\text{ yd}}{3\text{ ft}}.\right)$

92. 108 ft^2 = ____ yd^2

93. 432 in.^{2} = ____ ft^2 $\left(\text{Use two factors of the ratio } \frac{1\text{ ft}}{12\text{ in.}}.\right)$

94. 720 in^2 = ____ ft^2

95. 5 ft^2 = ____ in.^{2} $\left(\text{Use two factors of } \frac{12\text{ in.}}{1\text{ ft}}.\right)$

96. 7 ft^2 = ____ in.^{2}

97. 3 yd^2 = ____ ft^2 $\left(\text{Use two factors of } \frac{3\text{ ft}}{1\text{ yd}}.\right)$

98. 10 yd^2 = ____ ft^2

Section 8.2 Metric Units of Measurement

Objectives

1. Introduction to the Metric System
2. Metric Units of Length
3. Metric Units of Mass
4. Metric Units of Capacity
5. Summary of Metric Conversions

1. Introduction to the Metric System

Throughout history the lack of standard units of measure led to much confusion in trade between countries. In 1790 the French Academy of Sciences adopted a simple, decimal-based system of units. This system is known today as the **metric system**. The metric system is the predominant system of measurement used in science.

The simplicity of the metric system is a result of having one basic unit of measure for each type of quantity (length, mass, and capacity). The base units are the *meter* for length, the *gram* for mass, and the *liter* for capacity. Other units of length, mass, and capacity in the metric system are products of the base unit and a power of 10.

2. Metric Units of Length

The **meter** (m) is the basic unit of length in the metric system. A meter is slightly longer than a yard.

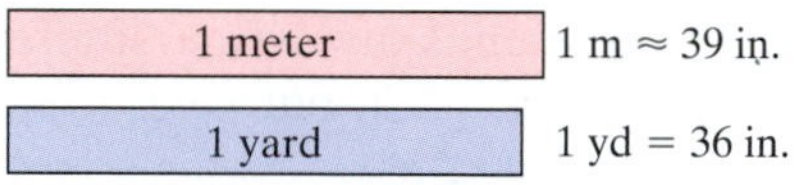

The meter was defined in the late 1700s as one ten-millionth of the distance along the Earth's surface from the North Pole to the Equator through Paris, France. Today the meter is defined as the distance traveled by light in a vacuum during $\frac{1}{299{,}792{,}458}$ sec.

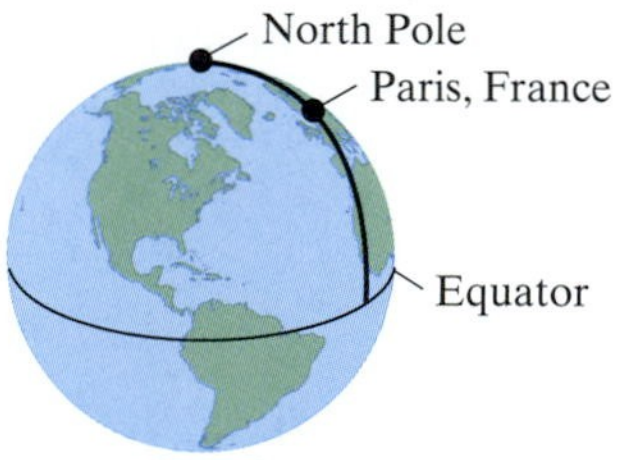

Six other common metric units of length are given in Table 8-2. Notice that each unit is related to the meter by a power of 10. This makes it particularly easy to convert from one unit to another.

TIP: The units of hectometer, dekameter, and decimeter are not frequently used.

Table 8-2 **Metric Units of Length and Their Equivalents**

1 kilometer (km) = 1000 m	
1 hectometer (hm) = 100 m	
1 dekameter (dam) = 10 m	
1 meter (m) = 1 m	
1 decimeter (dm) = 0.1 m	($\frac{1}{10}$ m)
1 centimeter (cm) = 0.01 m	($\frac{1}{100}$ m)
1 millimeter (mm) = 0.001 m	($\frac{1}{1000}$ m)

TIP: In addition to the key facts presented in Table 8-2, the following equivalences are useful.

100 cm = 1 m
1000 mm = 1 m

Notice that each unit of length has a prefix followed by the word *meter* (kilometer, for example). You should memorize these prefixes along with their multiples of the basic unit, the meter. Furthermore, it is generally easiest to memorize the prefixes in order.

kilo-	hecto-	deka-	meter	deci-	centi-	milli-
× 1000	× 100	× 10	× 1	× 0.1	× 0.01	× 0.001

As you familiarize yourself with the metric units of length, it is helpful to have a sense of the distance represented by each unit (Figure 8-2).

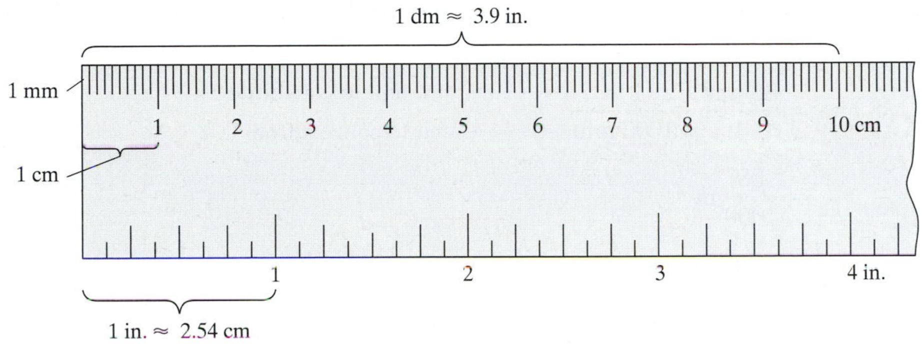

Figure 8-2

- 1 *millimeter* is approximately the thickness of five sheets of paper.
- 1 *centimeter* is approximately the width of a key on a calculator.
- 1 *decimeter* is approximately 4 in.
- 1 *meter* is just over 1 yd.
- A *kilometer* is used to express longer distances in much the same way we use miles. The distance between Los Angeles and San Diego is about 193 km.

Concept Connections

Fill in the blank with < or >.

1. 1 km ☐ 1 m
2. 1 cm ☐ 1 m
3. 1 dam ☐ 1 dm

Select the most reasonable value.

4. The length of a fork is
   a. 20 m **b.** 20 km
   c. 20 cm **d.** 20 mm
5. The length of a city block is
   a. $\frac{1}{2}$ km **b.** $\frac{1}{2}$ cm
   c. $\frac{1}{2}$ m **d.** $\frac{1}{2}$ mm

Example 1 Measuring Distances in Metric Units

Approximate the distance in centimeters and in millimeters.

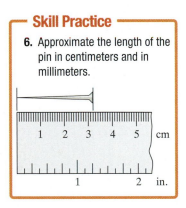

Solution:

The numbered lines on the ruler are units of centimeters. Each centimeter is divided into 10 mm. We see that the width of the penny is not quite 2 cm. We can approximate this distance as 1.8 cm or equivalently 18 mm.

Skill Practice

6. Approximate the length of the pin in centimeters and in millimeters.

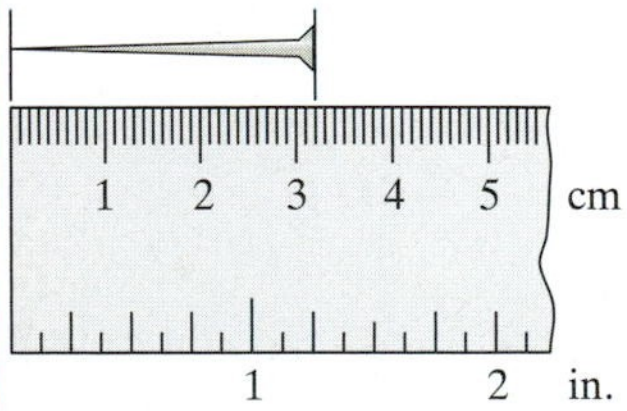

Answers

1. > **2.** < **3.** > **4.** c
5. a **6.** 3.2 cm or 32 mm

In Example 2, we convert metric units of length by using unit ratios.

Skill Practice

Convert.

7. 8.4 km = _______ m

8. 64,000 cm = _______ m

Example 2 Converting Metric Units of Length

a. 10.4 km = _______ m **b.** 88 mm = _______ m

Solution:

From Table 8-2, 1 km = 1000 m.

a. $$10.4\text{ km} = \frac{10.4\ \cancel{\text{km}}}{1} \cdot \frac{1000\text{ m}}{1\ \cancel{\text{km}}}$$ ← new unit to convert to (numerator); ← unit to convert from (denominator)

$$= 10{,}400\text{ m}$$ Multiply.

b. $$88\text{ mm} = \frac{88\ \cancel{\text{mm}}}{1} \cdot \frac{1\text{ m}}{1000\ \cancel{\text{mm}}}$$ ← new unit to convert to (numerator); ← unit to convert from (denominator)

$$= \frac{88}{1000}\text{ m}$$

$$= 0.088\text{ m}$$

Recall that the place positions in our numbering system are based on powers of 10. For this reason, when we multiply a number by 10, 100, or 1000, we move the decimal point 1, 2, or 3 places, respectively, to the right. Similarly, when we multiply by 0.1, 0.01, or 0.001, we move the decimal point to the left 1, 2, or 3 places, respectively.

Since the metric system is also based on powers of 10, we can convert between two metric units of length by moving the decimal point. The direction and number of place positions to move are based on the metric **prefix line**, shown in Figure 8-3.

Prefix Line

1000 m	100 m	10 m	1 m	0.1 m	0.01 m	0.001 m
km	hm	dam	m	dm	cm	mm
kilo-	hecto-	deka-		deci-	centi-	milli-

Figure 8-3

TIP: To use the prefix line effectively, you must know the order of the metric prefixes. Sometimes a mnemonic (memory device) can help. Consider the following sentence. The first letter of each word represents one of the metric prefixes.

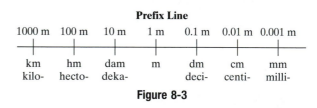

kids	**h**ave	**d**oughnuts	**u**ntil	**d**ad	**c**alls	**m**om.
kilo-	**h**ecto-	**d**eka-	**u**nit	**d**eci-	**c**enti-	**m**illi-

↑ unit: represents the main unit of measurement (meter, liter, or gram)

Answers

7. 8400 m **8.** 640 m

PROCEDURE Using the Prefix Line to Convert Metric Units

Step 1 To use the prefix line, begin at the point on the line corresponding to the original unit you are given.

Step 2 Then count the number of positions you need to move to reach the new unit of measurement.

Step 3 Move the decimal point in the original measured value the same direction and same number of places as on the prefix line.

Step 4 Replace the original unit with the new unit of measure.

Example 3 Using the Prefix Line to Convert Metric Units of Length

Use the prefix line for each conversion.

a. 0.0413 m = _______ cm **b.** 4700 cm = _______ km

Solution:

a.

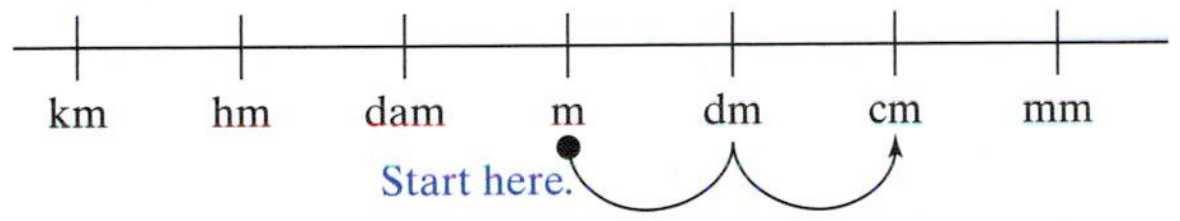

0.0413 m = 4.13 cm From the prefix line, move the decimal point two places to the right.

b.

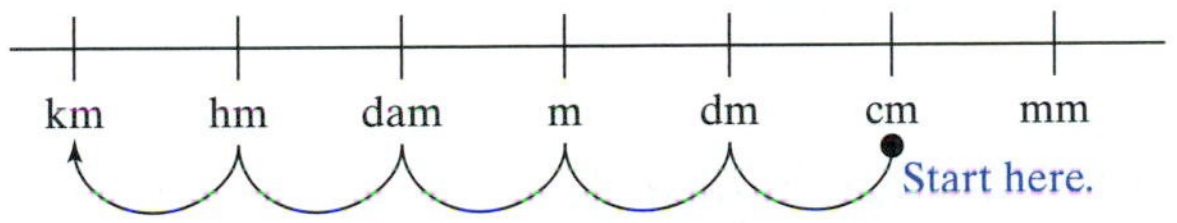

4700 cm = 0.04700 km = 0.047 km From the prefix line, move the decimal point five places to the left.

3. Metric Units of Mass

In Section 8.1 we learned that the pound and ton are two measures of weight in the U.S. Customary System. Measurements of weight give the force of an object under the influence of gravity. The mass of an object is related to its weight. However, mass is not affected by gravity. Thus, the weight of an object will be different on Earth than on the Moon because the effect of gravity is different. The mass of the object will stay the same.

The fundamental unit of mass in the metric system is the **gram** (g). A penny is approximately 2.5 g (Figure 8-4). A paper clip is approximately 1 g (Figure 8-5).

≈ 2.5 g

Figure 8-4

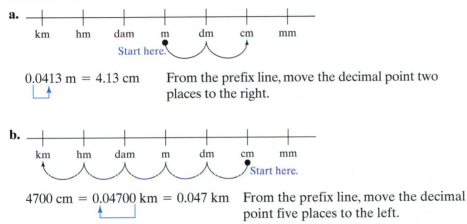

≈ 1 g

Figure 8-5

Skill Practice

Convert.

9. 864 cm = _______ m

10. 8.2 km = _______ m

Concept Connections

11. Which object could have a mass of 2 g?

a. Rubber band

b. Can of tuna fish

c. Cell phone

Answers

9. 8.64 m **10.** 8200 m

11. a

Other common metric units of mass are given in Table 8-3. Once again, notice that the metric units of mass are related to the gram by powers of 10.

Concept Connections

Fill in the blank with < or >.

12. 1 g ☐ 1 kg

13. 1 g ☐ 1 cg

Table 8-3 Metric Units of Mass and Their Equivalents

1 kilogram (kg) = 1000 g	
1 hectogram (hg) = 100 g	
1 dekagram (dag) = 10 g	
1 gram (g) = 1 g	
1 decigram (dg) = 0.1 g	$(\frac{1}{10}$ g)
1 centigram (cg) = 0.01 g	$(\frac{1}{100}$ g)
1 milligram (mg) = 0.001 g	$(\frac{1}{1000}$ g)

TIP: In addition to the key facts presented in Table 8-3, the following equivalences are useful.

100 cg = 1 g

1000 mg = 1 g

On the surface of Earth, 1 kg of mass is equivalent to approximately 2.2 lb of weight. Therefore, a 180-lb man has a mass of approximately 81.8 kg.

$$180 \text{ lb} \approx 81.8 \text{ kg}$$

The metric prefix line for mass is shown in Figure 8-6. This can be used to convert from one unit of mass to another.

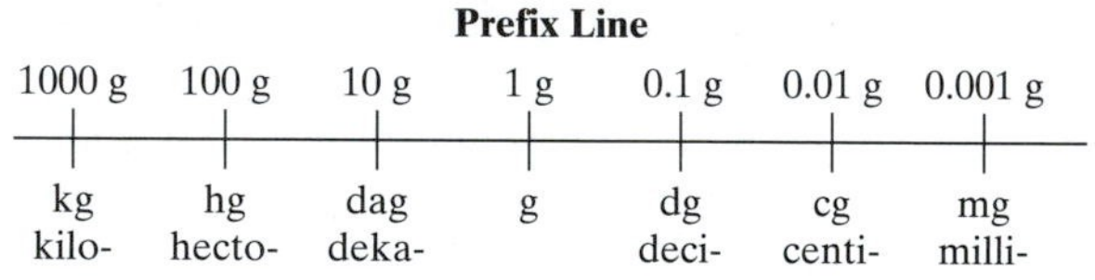

Figure 8-6

Skill Practice

Convert.

14. 80 kg = ______ g

15. 49 cg = ______ g

Example 4 Converting Metric Units of Mass

a. 1.6 kg = ______ g **b.** 1400 mg = ______ g

Solution:

a. $1.6 \text{ kg} = \dfrac{1.6 \cancel{\text{kg}}}{1} \cdot \dfrac{1000 \text{ g}}{1 \cancel{\text{kg}}}$ ← new unit to convert to / ← unit to convert from

$= 1600 \text{ g}$

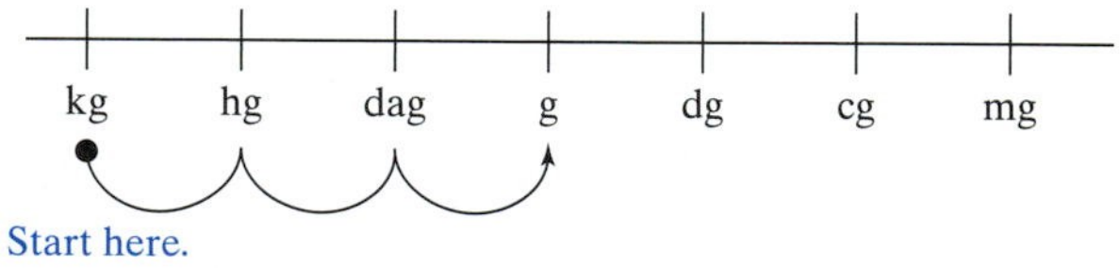

1.6 kg = 1.600 kg = 1600 g

b. $1400 \text{ mg} = \dfrac{1400 \cancel{\text{mg}}}{1} \cdot \dfrac{1 \text{ g}}{1000 \cancel{\text{mg}}}$ ← new unit to convert to / ← unit to convert from

$= \dfrac{1400}{1000} \text{ g}$

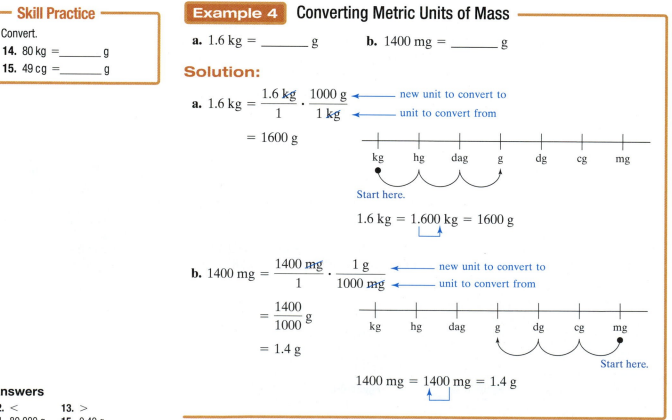

$= 1.4 \text{ g}$

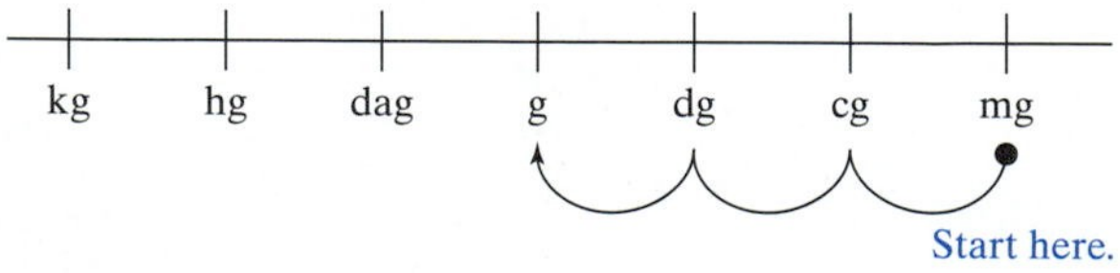

1400 mg = 1400 mg = 1.4 g

Answers

12. < **13.** >

14. 80,000 g **15.** 0.49 g

4. Metric Units of Capacity

The basic unit of capacity in the metric system is the **liter** (L). One liter is slightly more than 1 qt. Other common units of capacity are given in Table 8-4.

Table 8-4 Metric Units of Capacity and Their Equivalents

1 kiloliter (kL) = 1000 L	
1 hectoliter (hL) = 100 L	
1 dekaliter (daL) = 10 L	
1 liter (L) = 1 L	
1 deciliter (dL) = 0.1 L	($\frac{1}{10}$ L)
1 centiliter (cL) = 0.01 L	($\frac{1}{100}$ L)
1 milliliter (mL) = 0.001 L	($\frac{1}{1000}$ L)

TIP: In addition to the key facts presented in Table 8-4, the following equivalences are useful.

100 cL = 1 L

1000 mL = 1 L

1 mL is also equivalent to a **cubic centimeter** (**cc** or **cm**^{3}). The unit cc is often used to measure dosages of medicine. For example, after having an allergic reaction to a bee sting, a patient might be given 1 cc of adrenaline.

The metric prefix line for capacity is similar to that of length and mass (Figure 8-7). It can be used to convert between metric units of capacity.

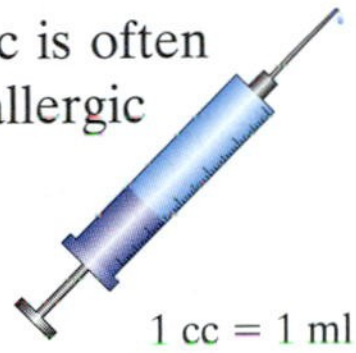

1 cc = 1 ml

Concept Connections

Fill in the blank with <, >, or =.

16. 1 kL ☐ 1 cL

17. 1 cL ☐ 1 L

18. 1 mL ☐ 1 cc

Prefix Line

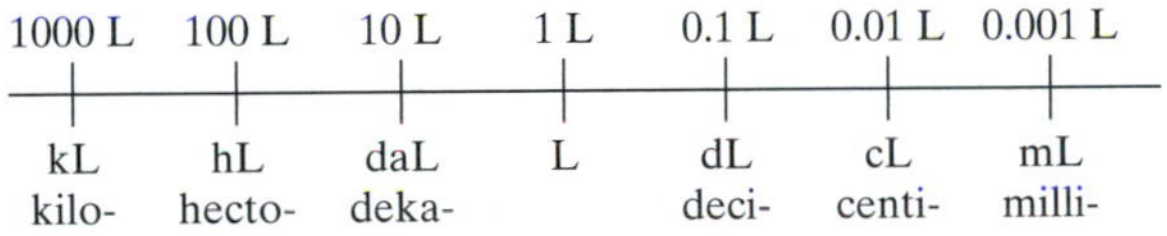

Figure 8-7

Example 5 Converting Metric Units of Capacity

a. 5.5 L = _______ mL **b.** 150 cL = _______ L

Solution:

a. $5.5\text{ L} = \frac{5.5\,\cancel{\text{L}}}{1} \cdot \frac{1000\text{ mL}}{1\,\cancel{\text{L}}}$ ← new unit to convert to / ← unit to convert from

$= 5500\text{ mL}$

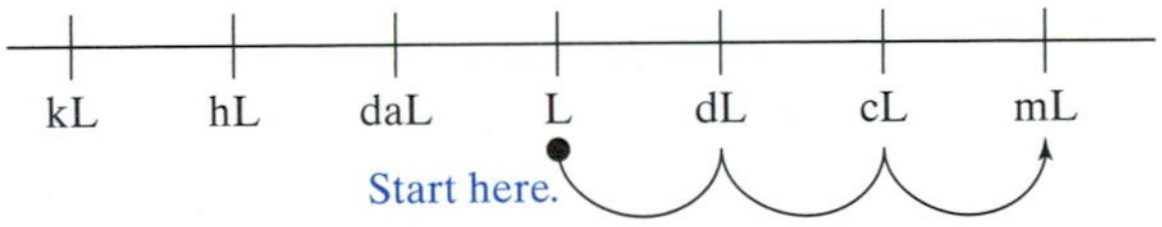

5.5 L = 5.500 L = 5500 mL

b. $150\text{ cL} = \frac{150\,\cancel{\text{cL}}}{1} \cdot \frac{1\text{ L}}{100\,\cancel{\text{cL}}}$ ← new unit to convert to / ← unit to convert from

$= \frac{150}{100}\text{ L}$

$= 1.5\text{ L}$

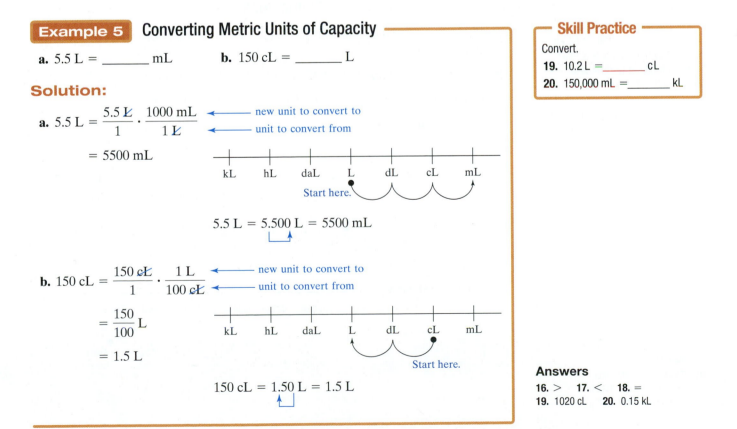

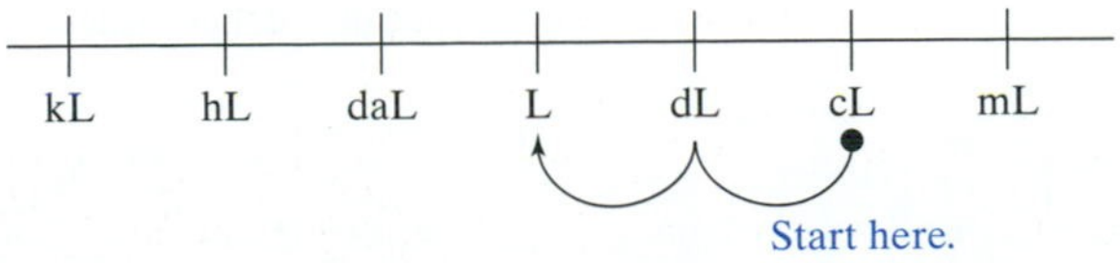

150 cL = 1.50 L = 1.5 L

Skill Practice

Convert.

19. 10.2 L = _______ cL

20. 150,000 mL = _______ kL

Answers

16. > **17.** < **18.** =

19. 1020 cL **20.** 0.15 kL

Skill Practice

Convert.

21. 0.5 cc = ______ mL

22. 0.04 L = ______ cc

Example 6 Converting Metric Units of Capacity

a. 15 cc = ______ mL **b.** 0.8 cL = ______ cc

Solution:

a. Recall that 1 cc = 1 mL. Therefore 15 cc = 15 mL.

b. We must convert from centiliters to milliliters, and then from milliliters to cubic centimeters.

$$0.8\text{ cL} = \frac{0.8\text{ cL}}{1} \cdot \frac{10\text{ mL}}{1\text{ cL}}$$

$= 8$ mL

$= 8$ cc Recall that 1 mL = 1 cc.

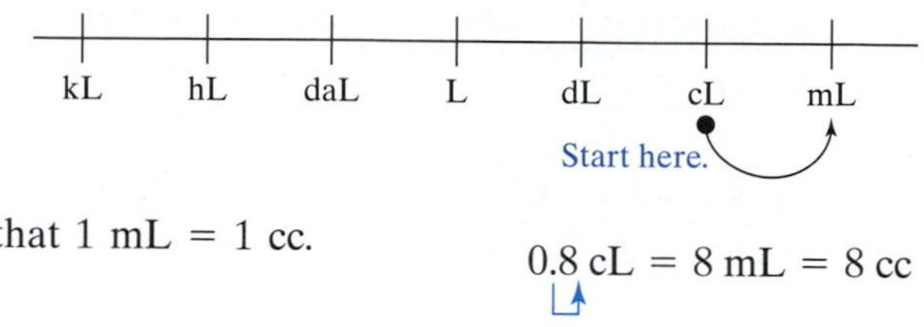

5. Summary of Metric Conversions

The prefix line in Figure 8-8 summarizes the relationships learned thus far.

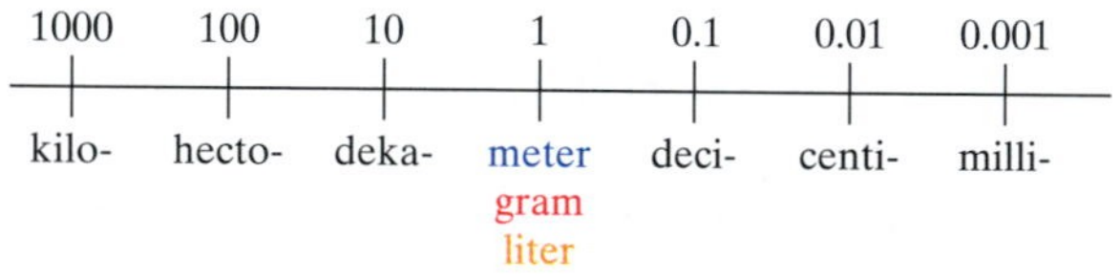

Figure 8-8

Skill Practice

23. The distance between Savannah and Hinesville is 64 km. How many meters is this?

24. A bottle of water holds 1420 mL. How many liters is this?

25. The mass of a box of cereal is 680 g. Convert this to kilograms.

26. A cat receives 1 mL of an antibiotic solution. Convert this to cc.

Example 7 Converting Metric Units

a. The distance between San Jose and Santa Clara is 26 km. Convert this to meters.

b. A bottle of canola oil holds 946 mL. Convert this to liters.

c. The mass of a bag of rice is 90,700 cg. Convert this to grams.

d. A dose of an antiviral medicine is 0.5 cc. Convert this to milliliters.

Solution:

a. 26 km = 26,000 m

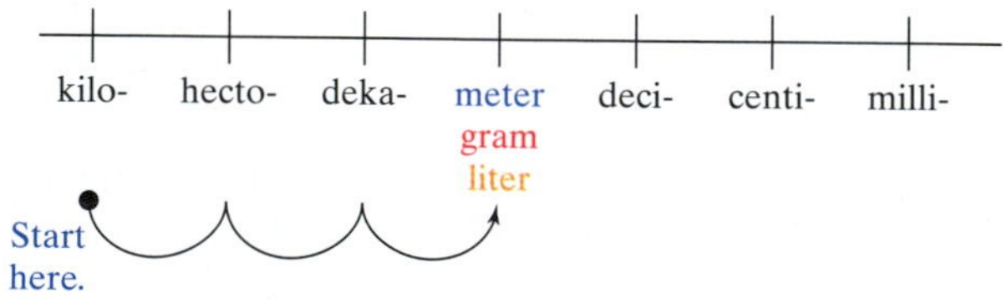

b. 946 mL = 0.946 L

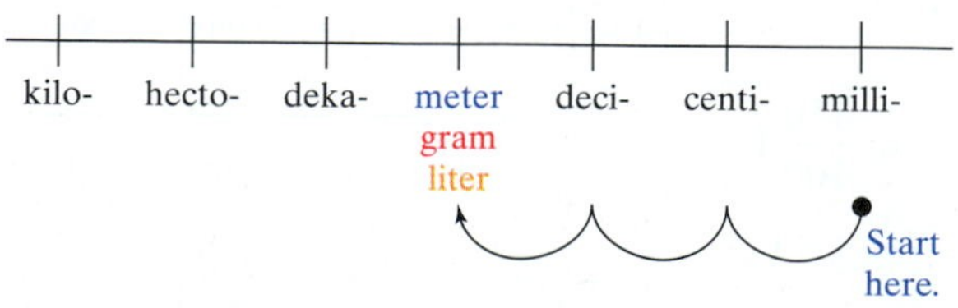

Answers

21. 0.5 mL **22.** 40 cc
23. 64,000 m **24.** 1.42 L
25. 0.68 kg **26.** 1 cc

c. 90,700 cg = 907.00 g = 907 g

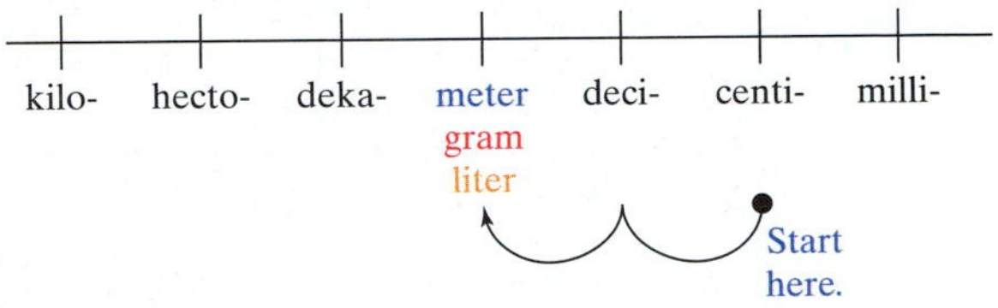

d. Recall that 1 cc = 1 mL. Therefore, 0.5 cc = 0.5 mL.

Section 8.2 Practice Exercises

Study Skills Exercise

1. Define the key terms.

a. Metric system **b. Meter** **c. Prefix line**

d. Gram **e. Liter** **f. Cubic centimeter**

Review Exercises

For Exercises 2–7, convert the units of measurement.

2. 2200 yd = _____ mi **3.** 8 c = _____ pt **4.** 48 oz = _____ lb

5. 1 day = _____ min **6.** 160 fl oz = _____ gal **7.** 3.5 lb = _____ oz

Objective 1: Introduction to the Metric System

8. Identify the units that apply to length. Circle all that apply.

a. Yard **b.** Ounce **c.** Fluid ounce **d.** Meter **e.** Quart

f. Gram **g.** Pound **h.** Liter **i.** Mile **j.** Inch

9. Identify the units that apply to weight or mass. Circle all that apply.

a. Yard **b.** Ounce **c.** Fluid ounce **d.** Meter **e.** Quart

f. Gram **g.** Pound **h.** Liter **i.** Mile **j.** Inch

10. Identify the units that apply to capacity. Circle all that apply.

a. Yard **b.** Ounce **c.** Fluid ounce **d.** Meter **e.** Quart

f. Gram **g.** Pound **h.** Liter **i.** Mile **j.** Inch

Objective 2: Metric Units of Length

For Exercises 11–12, approximate each distance in centimeters and millimeters. **(See Example 1.)**

11.

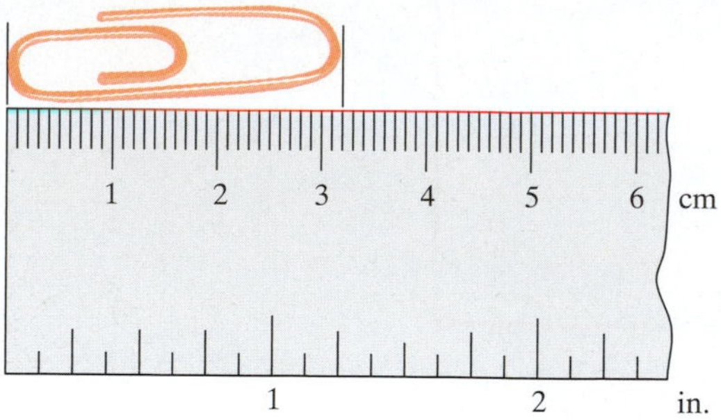

12.

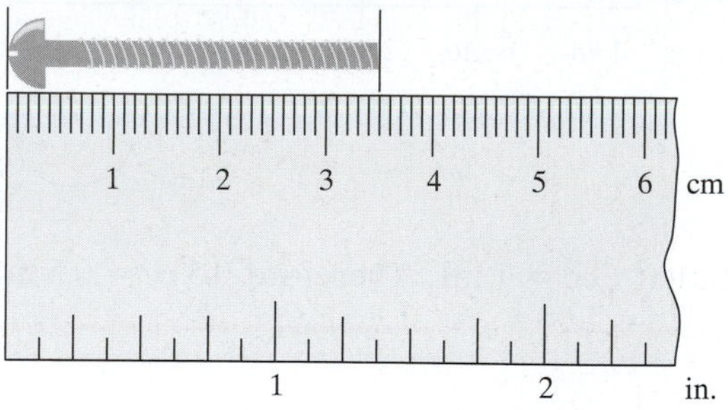

For Exercises 13–18, select the most reasonable measurement.

13. A table is ________ long.

a. 2 m **b.** 2 cm
c. 2 km **d.** 2 hm

14. A picture frame is ________ wide.

a. 22 cm **b.** 22 mm
c. 22 m **d.** 22 km

15. The distance between Albany, New York, and Buffalo, New York, is ________.

a. 210 m **b.** 2100 cm
c. 2.1 km **d.** 210 km

16. The distance between Denver and Colorado Springs is ________.

a. 110 cm **b.** 110 km
c. 11,000 km **d.** 1100 mm

17. The height of a full-grown giraffe is approximately ________.

a. 50 m **b.** 0.05 m
c. 0.5 m **d.** 5 m

18. The length of a canoe is ________.

a. 5 m **b.** 0.5 m
c. 0.05 m **d.** 500 m

For Exercises 19–30, convert metric units of length by using unit ratios or the prefix line. **(See Examples 2–3.)**

Prefix Line for Length

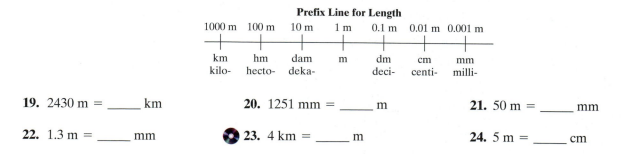

1000 m	100 m	10 m	1 m	0.1 m	0.01 m	0.001 m
km	hm	dam	m	dm	cm	mm
kilo-	hecto-	deka-		deci-	centi-	milli-

19. 2430 m = _____ km

20. 1251 mm = _____ m

21. 50 m = _____ mm

22. 1.3 m = _____ mm

23. 4 km = _____ m

24. 5 m = _____ cm

25. 4.31 cm = _____ mm

26. 18 cm = _____ mm

27. 3328 dm = _____ km

28. 128 hm = _____ km

29. 3 hm = _____ m

30. 450 mm = _____ dm

Objective 3: Metric Units of Mass

For Exercises 31–38, convert the units of mass. **(See Example 4.)**

31. 539 g = _____ kg

32. 328 mg = _____ g

33. 2.5 kg = _____ g

34. 2011 g = _____ kg

35. 0.0334 g = _____ mg

36. 0.38 dag = _____ dg

37. 409 cg = _____ g

38. 0.003 kg = _____ g

Objective 4: Metric Units of Capacity

For Exercises 39–42, fill in the blank with >, <, or =.

39. 1 cL _____ 1 L

40. 1 L _____ 1 mL

41. 1 mL _____ 1 cc

42. 1 L _____ 1 cc

43. What does the abbreviation cc represent?

44. Which of the following are measures of capacity? Circle all that apply.

a. cm **b.** cc **c.** cL **d.** cg

For Exercises 45–54, convert the units of capacity. **(See Examples 5–6.)**

45. 3200 mL = _____ L

46. 280 L = _____ kL

47. 7 L = _____ cL

48. 0.52 L = _____ mL

49. 42 mL = _____ dL

50. 0.88 L = _____ hL

51. 64 cc = _____ mL

52. 125 mL = _____ cc

53. 0.04 L = _____ cc

54. 38 cc = _____ L

Objective 5: Summary of Metric Conversions (Mixed Exercises)

55. Identify the units that apply to length.

a. mL **b.** mm **c.** hg **d.** cc **e.** kg **f.** hm **g.** cL

56. Identify the units that apply to capacity.

a. kg **b.** km **c.** cL **d.** cc **e.** hm **f.** dag **g.** mm

57. Identify the units that apply to mass.

a. dg **b.** hm **c.** kL **d.** cc **e.** dm **f.** kg **g.** cL

For Exercises 58–69, complete the table.

	Object	mm	cm	m	km
58.	Distance between Orlando and Miami				402
59.	Length of the Mississippi River				3766
60.	Thickness of a dime	1.35			
61.	Diameter of a quarter	24.3			

	Object	mg	cg	g	kg
62.	Can of tuna			170	
63.	Bag of rice			907	
64.	Box of raisins		42,500		
65.	Hockey puck		17,000		

	Object	mL	cL	L	kL
66.	1 Tablespoon	15			
67.	Bottle of vanilla extract	59			
68.	Capacity of a cooler				0.0377
69.	Capacity of a gasoline tank				0.0757

For Exercises 70–75, convert the metric units as indicated. **(See Example 7.)**

70. The height of the tallest living tree is 112.014 m. Convert this to dekameters.

71. The Congo River is 4669 km long. Convert this to meters.

72. A One-A-Day Weight Smart Multivitamin tablet contains 60 mg of vitamin C. How many tablets must be taken to consume a total of 3 g of vitamin C?

73. A can contains 305 g of soup. Convert this to milligrams.

74. A gasoline can has a capacity of 19 L. Convert this to kiloliters.

75. The capacity of a coffee cup is 0.25 L. Convert this to milliliters.

76. In one day, Stacy gets 600 mg of calcium in her daily vitamin, 500 mg in her calcium supplement, and 250 mg in the dairy products she eats. How many grams of calcium will she get in one week?

77. Cliff drives his children to their sports activities outside of school. When he drives his son to baseball practice, it is a 6-km round trip. When he drives his daughter to basketball practice, it is a 1800-m round trip. If basketball practice is 3 times a week and baseball practice is twice a week, how many kilometers does Cliff drive?

78. A gas tank holds 45 L. If it costs $74.25 to fill up the tank, what is the price per liter?

79. A can of paint holds 120 L. How many kiloliters are contained in 8 cans?

80. A bottle of water holds 710 mL. How many liters are in a 6-pack?

81. A bottle of olive oil has 33 servings of 15 mL each. How many centiliters of oil does the bottle contain?

82. A quart of milk has 130 mg of sodium per cup. How much sodium is in the whole bottle?

83. A ½-c serving of cereal has 180 mg of potassium. This is 5% of the recommended daily allowance of potassium. How many grams is the recommended daily allowance?

84. Rosanna has material 1.5 m long for a window curtain. If the window is 90 cm and she needs 10 cm for a hem at the bottom and 12 cm for finishing the top, does Rosanna have enough material?

85. Veronique has a piece of molding 1 m long. Does she have enough to cut four pieces to frame the picture shown in the figure?

86. A square tile is 110 mm in length. If they are placed side by side, how many tiles will it take to cover a length of wall 1.43 m long?

87. Two Olympic speed skating races for women are 500 m and 5 km. What is the difference (in meters) between the lengths of these races?

Expanding Your Skills

In the Expanding Your Skills of Section 8.1, we converted U.S. Customary units of area. We use the same procedure to convert metric units of area. This procedure involves multiplying by two unit ratios of length.

Example: Converting area

Convert 1000 mm² to square centimeters.

Solution:

$$\frac{1000\text{ mm}^2}{1} \cdot \underbrace{\frac{1\text{ cm}}{10\text{ mm}} \cdot \frac{1\text{ cm}}{10\text{ mm}}}_{\text{Multiply first.}} = \frac{1000\ \cancel{\text{mm}^2}}{1} \cdot \frac{1\text{ cm}^2}{100\ \cancel{\text{mm}^2}} = \frac{1000\text{ cm}^2}{100} = 10\text{ cm}^2$$

For Exercises 88–91, convert the units of area, using two factors of the given unit ratio.

88. $30{,}000\text{ mm}^2 = \_\_\_\_\_\_\text{ cm}^2$ $\left(\text{Use } \frac{1\text{ cm}}{10\text{ mm}}.\right)$

89. $65{,}000{,}000\text{ m}^2 = \_\_\_\_\_\_\text{ km}^2$ $\left(\text{Use } \frac{1\text{ km}}{1000\text{ m}}.\right)$

90. $4.1\text{ m}^2 = \_\_\_\_\_\_\text{ cm}^2$ $\left(\text{Use } \frac{100\text{ cm}}{1\text{ m}}.\right)$

91. $5600\text{ cm}^2 = \_\_\_\_\_\_\text{ m}^2$ $\left(\text{Use } \frac{1\text{ m}}{100\text{ cm}}.\right)$

In the U.S. Customary System of measurement, 1 ton = 2000 lb. In the metric system, 1 metric ton = 1000 kg. Use this information to answer Exercises 92–95.

92. Convert 3300 kg to metric tons.

93. Convert 5780 kg to metric tons.

94. Convert 10.9 metric tons to kilograms.

95. Convert 8.5 metric tons to kilograms.

Section 8.3 Converting Between U.S. Customary and Metric Units

Objectives

1. Summary of U.S. Customary and Metric Unit Equivalents
2. Converting U.S. Customary and Metric Units
3. Applications
4. Units of Temperature

1. Summary of U.S. Customary and Metric Unit Equivalents

In this section, we learn how to convert between U.S. Customary and metric units of measure. Suppose, for example, that you take a trip to Europe. A street sign indicates that the distance to Paris is 45 km (Figure 8-9). This distance may be unfamiliar to you until you convert to miles.

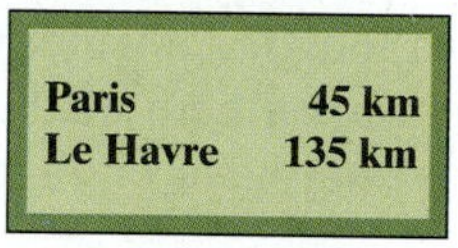

Figure 8-9

Skill Practice

1. Use the fact that 1 mi ≈ 1.61 km to convert 184 km to miles. Round to the nearest mile.

Example 1 Converting Metric Units to U.S. Customary Units

Use the fact that 1 mi ≈ 1.61 km to convert 45 km to miles. Round to the nearest mile.

Solution:

$$45\text{ km} \approx \frac{45\ \cancel{\text{km}}}{1} \cdot \frac{1\text{ mi}}{1.61\ \cancel{\text{km}}}$$ Set up a unit ratio to convert kilometers to miles.

$$= \frac{45}{1.61}\text{ mi}$$ Multiply fractions.

$$\approx 28\text{ mi}$$ Divide and round to the nearest mile.

The distance of 45 km to Paris is approximately 28 mi.

Table 8-5 summarizes some common metric and U.S. Customary equivalents.

Table 8-5

Length	Weight/Mass (on Earth)	Capacity
1 in. = 2.54 cm	1 lb ≈ 0.45 kg	1 qt ≈ 0.95 L
1 ft ≈ 0.305 m	1 oz ≈ 28 g	1 fl oz ≈ 30 mL = 30 cc
1 yd ≈ 0.914 m		
1 mi ≈ 1.61 km		

2. Converting U.S. Customary and Metric Units

Using the U.S. Customary and metric equivalents given in Table 8-5, we can create unit ratios to convert between units.

Answer

1. 114 mi

Example 2 Converting Units of Length

Fill in the blank. Round to two decimal places, if necessary.

a. 18 cm = ______ in. **b.** 15 yd ≈ ______ m **c.** 8.2 m ≈ ______ ft

Solution:

a. $18\text{ cm} = \dfrac{18\text{ cm}}{1} \cdot \dfrac{1\text{ in.}}{2.54\text{ cm}}$ From Table 8-5, we know 1 in. = 2.54 cm.

$= \dfrac{18}{2.54}\text{ in.}$ Multiply fractions.

$\approx 7.09\text{ in.}$ Divide and round to two decimal places.

b. $15\text{ yd} \approx \dfrac{15\text{ yd}}{1} \cdot \dfrac{0.914\text{ m}}{1\text{ yd}}$ From Table 8-5, we know 1 yd ≈ 0.914 m.

$= 13.71\text{ m}$ Multiply.

c. $8.2\text{ m} \approx \dfrac{8.2\text{ m}}{1} \cdot \dfrac{1\text{ ft}}{0.305\text{ m}}$ From Table 8-5, we know 1 ft ≈ 0.305 m.

$= \dfrac{8.2}{0.305}\text{ ft}$ Multiply.

$\approx 26.89\text{ ft}$ Divide and round to two decimal places.

Skill Practice

Convert. Round to one decimal place.

2. 4 m ≈ ______ ft

3. 3 in. ≈ ______ cm

4. 6500 yd ≈ ______ m

Example 3 Converting Units of Weight and Mass

Fill in the blank. Round to one decimal place, if necessary.

a. 180 g ≈ ______ oz **b.** 5.25 tons ≈ ______ kg

Solution:

a. $180\text{ g} \approx \dfrac{180\text{ g}}{1} \cdot \dfrac{1\text{ oz}}{28\text{ g}}$ From Table 8-5, we know 1 oz ≈ 28 g.

$= \dfrac{180}{28}\text{ oz}$

$\approx 6.4\text{ oz}$ Divide and round to one decimal place.

b. We can first convert 5.25 tons to pounds. Then we can use the fact that 1 lb ≈ 0.45 kg.

$5.25\text{ tons} = \dfrac{5.25\text{ tons}}{1} \cdot \dfrac{2000\text{ lb}}{1\text{ ton}}$ Convert tons to pounds.

$= 10{,}500\text{ lb}$

$\approx 10{,}500\text{ lb} \cdot \dfrac{0.45\text{ kg}}{1\text{ lb}}$ Convert pounds to kilograms.

$= 4725\text{ kg}$

Skill Practice

Convert. Round to one decimal place, if necessary.

5. 8 kg ≈ ______ lb

6. 4 tons ≈ ______ kg

Answers

2. 13.1 ft **3.** 7.6 cm **4.** 5941 m
5. 17.8 lb **6.** 3600 kg

Skill Practice

Convert. Round to one decimal place, if necessary.

7. 120 mL ≈ _______ fl oz

8. 4 qt ≈ _______ L

Example 4 Converting Units of Capacity

Fill in the blank. Round to two decimal places, if necessary.

a. 75 mL ≈ _______ fl oz **b.** 3 qt ≈ _______ L

Solution:

a. $75 \text{ mL} \approx \frac{75 \text{ mL}}{1} \cdot \frac{1 \text{ fl oz}}{30 \text{ mL}}$ From Table 8-5, we know 1 fl oz ≈ 30 mL.

$= \frac{75}{30} \text{ fl oz}$ Multiply fractions.

$= 2.5 \text{ fl oz}$ Divide.

b. $3 \text{ qt} \approx \frac{3 \text{ qt}}{1} \cdot \frac{0.95 \text{ L}}{1 \text{ qt}}$ From Table 8-5, we know 1 qt ≈ 0.95 L.

$= 2.85 \text{ L}$ Multiply.

3. Applications

Skill Practice

9. A 720-mL bottle of water sells for \$0.79. A 32-oz bottle of water sells for \$1.29. Compare the price per ounce to determine the better buy.

Example 5 Converting Units in an Application

A 2-L bottle of soda sells for \$2.19. A 32-oz bottle of soda sells for \$1.59. Compare the price per quart of each bottle to determine the better buy.

Solution:

Note that 1 qt = 2 pt = 4 c = 32 fl oz. So a 32-oz bottle of soda costs \$1.59 per quart. Next, if we can convert 2 L to quarts, we can compute the unit cost per quart and compare the results.

$2 \text{ L} \approx \frac{2 \text{ L}}{1} \cdot \frac{1 \text{ qt}}{0.95 \text{ L}}$ Recall that 1 qt = 0.95 L.

$\approx \frac{2}{0.95} \text{ qt}$ Multiply fractions.

$\approx 2.11 \text{ qt}$ Divide and round to two decimal places.

Now find the cost per quart. $\frac{\$2.19}{2.11 \text{ qt}} \approx \$1.04 \text{ per quart}$

The cost for the 2-L bottle is \$1.04 per quart, whereas the cost for 32 oz is \$1.59 per quart. Therefore, the 2-L bottle is the better buy.

Answers

7. 4 fl oz **8.** 3.8 L

9. 720 mL is 24 oz. The cost per ounce is \$0.033. The unit price for the 32-oz bottle is \$0.040 per ounce. The 720-mL bottle is the better buy.

Example 6 Converting Units in an Application

In track and field, the 1500-m race is slightly less than 1 mi. How many yards less is it? Round to the nearest yard.

Solution:

We know that 1 mi = 1760 yd. If we can convert 1500 m to yards, then we can subtract the results.

$$1500 \text{ m} \approx \frac{1500 \cancel{\text{m}}}{1} \cdot \frac{1 \text{ yd}}{0.914 \cancel{\text{m}}}$$ Recall that 1 yd = 0.914 m.

$$\approx \frac{1500}{0.914} \text{ yd}$$ Multiply fractions.

$$\approx 1641 \text{ yd}$$ Divide and round to the nearest yard.

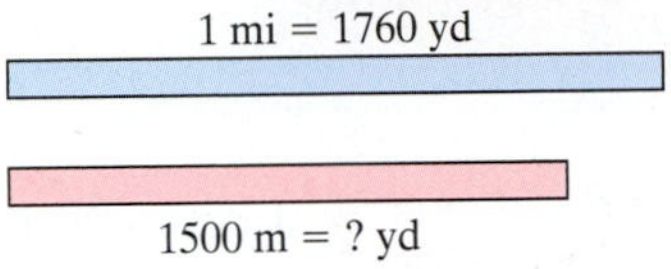

Therefore, the difference between 1 mi and 1500 m is:

(1 mi) − (1500 m)

1760 yd − 1641 yd = 119 yd

Skill Practice

10. In track and field, the 800-m race is slightly shorter than a half-mile race. How many yards less is it? Round to the nearest yard.

4. Units of Temperature

In the United States, the **Fahrenheit** scale is used most often to measure temperature. On this scale, water freezes at 32°F and boils at 212°F. The symbol ° stands for "degrees," and °F means "degrees Fahrenheit."

Another scale used to measure temperature is the **Celsius** temperature scale. On this scale, water freezes at 0°C and boils at 100°C. The symbol °C stands for "degrees Celsius."

Figure 8-10 shows the relationship between the Celsius scale and the Fahrenheit scale.

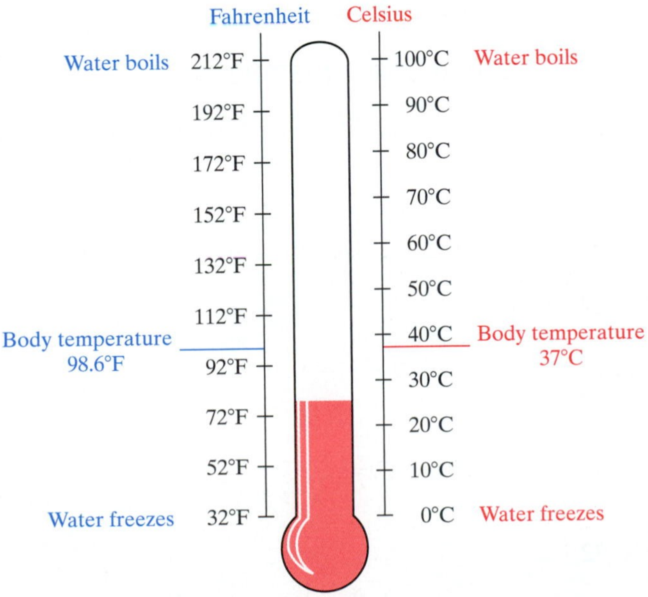

Figure 8-10

Concept Connections

Use Figure 8-10 to select the best choice.

11. The high temperature in Dallas for a day in July is
a. 34°C **b.** 34°F

12. The high temperature in Minneapolis for a day in March is
a. 34°C **b.** 34°F

Answers

10. 5 yd **11.** a **12.** b

To convert back and forth between the Fahrenheit and Celsius scales, we use the following formulas.

FORMULA Conversions for Temperature Scale

To convert from °C to °F:

$$F = \frac{9}{5}C + 32$$

To convert from °F to °C:

$$C = \frac{5}{9}(F - 32)$$

Note: Using decimal notation we can write the formulas as

$$F = 1.8C + 32 \qquad C = \frac{F - 32}{1.8}$$

Skill Practice

13. The ocean temperature in the Caribbean in August averages 84°F. Convert this to degrees Celsius and round to one decimal place.

Example 7 Converting Units of Temperature

Convert a body temperature of 98.6°F to degrees Celsius.

Solution:

Because we want to convert degrees Fahrenheit to degrees Celsius, we use the formula $C = \frac{5}{9}(F - 32)$.

$$C = \frac{5}{9}(F - 32)$$

$$= \frac{5}{9}(98.6 - 32) \qquad \text{Substitute } F = 98.6.$$

$$= \frac{5}{9}(66.6) \qquad \text{Perform the operation inside parentheses first.}$$

$$= \frac{(5)(66.6)}{9}$$

$$= 37 \qquad \text{Body temperature is 37°C.}$$

Skill Practice

14. The high temperature on a day in March for Raleigh, North Carolina, was 10°C. Convert this to degrees Fahrenheit.

Example 8 Converting Units of Temperature

Convert the temperature inside a refrigerator, 5°C, to degrees Fahrenheit.

Solution:

Because we want to convert degrees Celsius to degrees Fahrenheit, we use the formula $F = \frac{9}{5}C + 32$.

$$F = \frac{9}{5}C + 32$$

$$= \frac{9}{5} \cdot 5 + 32 \qquad \text{Substitute } C = 5.$$

$$= \frac{9}{\cancel{5}} \cdot \frac{\cancel{5}}{1} + 32$$

$$= 9 + 32$$

$$= 41 \qquad \text{The temperature inside the refrigerator is 41°F.}$$

Answers

13. 28.9°C **14.** 50°F

Section 8.3 Practice Exercises

Boost *your* GRADE at ALEKS.com!

- Practice Problems
- Self-Tests
- NetTutor
- e-Professors
- Videos

Study Skills Exercises

1. Make a list of all the section titles in the chapter that you are studying. Write each section title on a separate sheet of paper or index card. Go back and fill in the list of objectives under each section title. When you are studying for the test, try to make up an exercise that corresponds to each objective and then work the exercise. To get started, write a problem for the objective of converting metric units of capacity from Section 8.2.

2. Define the key terms.

a. **Celsius** **b.** **Fahrenheit**

Review Exercises

For Exercises 3–6, select the equivalent amounts of mass. (*Hint:* There may be more than one answer for each exercise.)

3. 500 g
4. 500 mg
5. 500 cg
6. 500 kg

a. 500,000 g
b. 5 g
c. 500,000,000 mg
d. 0.5 kg
e. 5000 mg
f. 50,000 cg
g. 0.5 g
h. 50 cg

For Exercises 7–10, select the equivalent amounts of capacity.

7. 200 L
8. 200 kL
9. 200 mL
10. 200 cL

a. 2000 mL
b. 200 cc
c. 0.2 kL
d. 20,000,000 cL
e. 200,000 L
f. 200,000 mL
g. 0.2 L
h. 2 L

Objective 1: Summary of U.S. Customary and Metric Unit Equivalents

11. Identify an appropriate ratio to convert 5 yards to meters by using multiplication.

a. $\dfrac{1\text{ yd}}{0.914\text{ m}}$ **b.** $\dfrac{0.914\text{ m}}{1\text{ yd}}$ **c.** $\dfrac{0.914\text{ yd}}{1\text{ m}}$ **d.** $\dfrac{1\text{ m}}{0.914\text{ yd}}$

12. Identify an appropriate ratio to convert 3 pounds to kilograms by using multiplication.

a. $\frac{0.45 \text{ lb}}{1 \text{ kg}}$ **b.** $\frac{1 \text{ kg}}{0.45 \text{ lb}}$ **c.** $\frac{1 \text{ lb}}{0.45 \text{ kg}}$ **d.** $\frac{0.45 \text{ kg}}{1 \text{ lb}}$

13. Identify an appropriate ratio to convert 2 quarts to liters by using multiplication.

a. $\frac{0.95 \text{ L}}{1 \text{ qt}}$ **b.** $\frac{1 \text{ qt}}{0.95 \text{ L}}$ **c.** $\frac{0.95 \text{ qt}}{1 \text{ L}}$ **d.** $\frac{1 \text{ L}}{0.95 \text{ qt}}$

14. Identify an appropriate ratio to convert 10 miles to kilometers by using multiplication.

a. $\frac{1 \text{ mi}}{1.61 \text{ km}}$ **b.** $\frac{1 \text{ km}}{1.61 \text{ mi}}$ **c.** $\frac{1.61 \text{ km}}{1 \text{ mi}}$ **d.** $\frac{1.61 \text{ mi}}{1 \text{ km}}$

Objective 2: Converting U.S. Customary and Metric Units

For Exercises 15–23, convert the units of length. Round the answer to one decimal place, if necessary. **(See Examples 1–2.)**

15. 2 in. ≈ _____ cm **16.** 120 km ≈ _____ mi **17.** 8 m ≈ _____ yd

18. 4 ft ≈ _____ m **19.** 400 ft ≈ _____ m **20.** 0.75 m ≈ _____ yd

21. 45 in ≈ _____ m **22.** 150 cm ≈ _____ ft **23.** 0.5 ft ≈ _____ cm

For Exercises 24–32, convert the units of weight and mass. Round the answer to one decimal place, if necessary. **(See Example 3.)**

24. 6 oz ≈ _____ g **25.** 6 lb ≈ _____ kg **26.** 4 kg ≈ _____ lb

27. 10 g ≈ _____ oz **28.** 14 g ≈ _____ oz **29.** 0.54 kg ≈ _____ lb

30. 0.3 lb ≈ _____ kg **31.** 2.2 tons ≈ _____ kg **32.** 4500 kg ≈ _____ tons

For Exercises 33–38, convert the units of capacity. Round the answer to one decimal place, if necessary. **(See Example 4.)**

33. 6 qt ≈ _____ L **34.** 5 fl oz ≈ _____ mL **35.** 120 mL ≈ _____ fl oz

36. 19 L ≈ _____ qt **37.** 960 cc ≈ _____ fl oz **38.** 0.5 fl oz ≈ _____ cc

Objective 3: Applications (Mixed Exercises)

For Exercises 39–54, refer to Table 8-5 on page 502.

39. A 2-lb box of sugar costs \$3.19. A box that contains single-serving packets contains 354 g and costs \$1.49. Find the unit costs in dollars per ounce to determine the better buy. **(See Example 5.)**

40. At the grocery store, Debbie compares the prices of a 2-L bottle of water and a 6-pack of bottled water. The 2-L bottle is priced at \$1.59. The 6-pack costs \$3.60 and each bottle in the package contains 24 fl oz. Compare the cost of water per quart to determine which is a better buy.

41. A cross-country skiing race is 30 km long. Is this length more or less than 18 mi? **(See Example 6.)**

42. A can of cat food is 85 g. How many ounces is this? Round to the nearest ounce.

43. Carly Patterson of the U.S. Olympic gymnastic team weighed 97 lb when she was the Olympic All-Around champion. How many kilograms is this?

44. Warren's dad ran the 100-yd dash when he was in high school. Suppose Warren runs the 100-*meter* dash on his track team.

a. Who runs the longer race?

b. Find the difference between the lengths in meters.

45. In a recent year, the price of gas in Germany was \$1.90 per liter. What is the price per gallon?

46. A jar of spaghetti sauce is 2 lb 8 oz. How many kilograms is this? Round to two decimal places.

47. The thickness of a hockey puck is 2.54 cm. How many inches is this?

48. A bottle of grape juice contains 1.9 L of juice. Is there enough juice to fill 10 glasses that hold 6 oz each? If yes, how many ounces will be left over?

49. Football player Tony Romo weighs 99,790 g. How many pounds is this? Round to the nearest pound.

50. The distance between two lightposts is 6 m. How many feet is this? Round to the nearest foot.

51. A nurse gives a patient 45 cc of saline solution. How many fluid ounces does this represent?

52. Cough syrup comes in a bottle that contains 4 fl oz. How many milliliters is this?

53. Suppose this figure is a drawing of a room where 1 cm represents 2 ft. If you were to install molding along the edge of the floor, how many feet would you need?

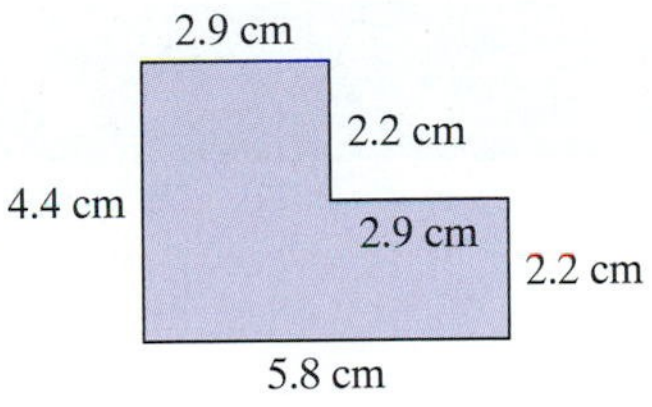

54. Suppose this figure is a drawing of a room where 1 cm represents 3 ft. If you were to install molding along the edge of the floor, how many feet would you need?

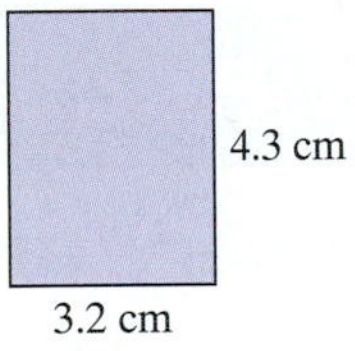

Objective 4: Units of Temperature

For Exercises 55–60, convert the temperatures by using the appropriate formula: $F = \frac{9}{5}C + 32$ or $C = \frac{5}{9}(F - 32)$. **(See Examples 7–8.)**

55. 25°C = _____ °F

56. 113°F = _____ °C

57. 68°F = _____ °C

58. 15°C = _____ °F

59. 30°C = _____ °F

60. 104°F = _____ °C

61. The boiling point of the element boron is 4000°C. Find the boiling point in degrees Fahrenheit.

62. The melting point of the element copper is 1085°C. Find the melting point in degrees Fahrenheit.

63. If the outdoor temperature is 35°C, is it a hot day or a cold day?

64. The high temperature in London, England, on a typical September day was 18°C, and the low was 13°C. Convert these temperatures to degrees Fahrenheit.

65. Use the fact that water boils at 100°C to show that the boiling point is 212°F.

66. Use the fact that water freezes at 0°C to show that the temperature at which water freezes is 32°F.

Expanding Your Skills

The USDA recommends that an adult woman get 46 g of protein per day. Peanuts are 25% protein. Use this information to answer Exercises 67–68.

67. How many grams of peanuts would an adult woman need to satisfy a day's protein requirement?

68. How many ounces of peanuts would be required for a woman's daily protein need? Round to the nearest tenth.

In the U.S. Customary System of measurement, 1 ton = 2000 lb. In the metric system, 1 metric ton = 1000 kg. Use this information to answer Exercises 69–72.

69. A Lincoln Navigator weighs 5700 lb. How many metric tons is this?

70. An elevator has a maximum capacity of 1200 lb. How many metric tons is this?

71. The average mass of a blue whale (the largest mammal in the world) is approximately 108 metric tons. How many pounds is this?

72. The mass of a Mini-Cooper is 1.25 metric tons. How many pounds is this? Round to the nearest pound.

Problem Recognition Exercises

U.S. Customary and Metric Conversions

For Exercises 1–32, convert the units as indicated.

1. 36 c = _____ qt
2. 220 cm = _____ m
3. $\frac{3}{4}$ lb = _____ oz
4. 0.3 L = _____ mL
5. 12 ft = _____ yd
6. 6.03 kg = _____ g
7. 45 dm = _____ m
8. 9 in. = _____ ft
9. $\frac{1}{2}$ mi = _____ ft
10. 6000 lb = _____ tons
11. 8 pt = _____ qt
12. 1.5 tsp = _____ T
13. 21 m = _____ km
14. 68 mg = _____ cg
15. 36 mL = _____ cc
16. 64 oz = _____ lb
17. 4322 g = _____ kg
18. 5 m = _____ mm
19. 20 fl oz = _____ c
20. 510 sec = _____ min
21. 4 pt = _____ gal
22. 26 fl oz = _____ c
23. 5.46 kg = _____ g
24. 9.02 L = _____ cL
25. 9.1 mi = _____ yd
26. 48 oz = _____ lb
27. 1.62 tons = _____ lb
28. 4.6 km = _____ m
29. 60 hr = _____ days
30. 8 cc = _____ mL
31. 8:32:24 = _____ min
32. 2 wk = _____ hr

Medical Applications Involving Measurement

Section 8.4

1. Additional Metric Units of Mass

Scientists and people in the medical community almost exclusively use the metric system. For example:

- A patient's mass may be recorded in kilograms.
- A dosage of cough syrup may be measured in milliliters.
- The active ingredient in a pain reliever is given in grams.

Sometimes doctors prescribe medicines in very small amounts. In these cases, it is sometimes more convenient to use units of **micrograms**. The abbreviation for microgram is mcg or sometimes μg. Furthermore,

1000 mcg = 1 mg — It takes 1 thousand micrograms to equal 1 milligram.

1,000,000 mcg = 1 g — It takes 1 million micrograms to equal 1 gram.

Objectives

1. Additional Metric Units of Mass
2. Medical Applications

Example 1 Converting Units of Micrograms

a. Convert. 0.85 mg = ______ mcg

b. A doctor gives a heart patient an initial dose of 200 mcg of nitroglycerin. How many milligrams is this?

Solution:

a. $$0.85\text{ mg} = \frac{0.85\text{ }\cancel{\text{mg}}}{1} \cdot \frac{1000\text{ mcg}}{1\text{ }\cancel{\text{mg}}}$$ ← new unit to convert to; ← unit to convert from

$$= 850\text{ mcg}$$

b. $$200\text{ mcg} = \frac{200\text{ }\cancel{\text{mcg}}}{1} \cdot \frac{1\text{ mg}}{1000\text{ }\cancel{\text{mcg}}}$$ ← new unit to convert to; ← unit to convert from

$$= 0.2\text{ mg}$$

Skill Practice

Convert.

1. 0.04 mg = ______ mcg
2. 95,000 mcg = ______ mg

2. Medical Applications

Example 2 Applying Metric Units of Measure to Medicine

A doctor orders the antibiotic oxacillin for a child. The dosage is 12.5 mg of the drug per kilogram of the child's body mass. This dosage is given 4 times a day.

a. How much of the drug should a 24-kg child get in one dose?

b. How much of the drug would the child get if she were on a 10-day course of the antibiotic?

Solution:

a. We need to multiply the unit rate of 12.5 mg per kilogram times the child's body mass.

$$\text{Single dose} = (12.5\text{ mg/kg})(24\text{ kg})$$

$$= 300\text{ mg}$$

Skill Practice

3. A child is to receive 0.5 mg of a drug per kilogram of the child's body mass. If the child is 27 kg, how much of the drug should the child receive?

Answers

1. 40 mcg 2. 95 mg 3. 13.5 mg

b. For a 10-day course, we need to multiply 300 g by the number of doses per day (4), and the total number of days (10).

$$\text{Total amount of drug} = (300\text{ mg})(4)(10)$$
$$= 12{,}000\text{ mg} \quad \text{or equivalently } 12\text{ g.}$$

Skill Practice

4. A doctor orders amoxicillin (an antibiotic) in a liquid suspension form. The liquid contains 75 mg of amoxicillin per 5 mL of liquid. If the doctor orders 300 mg per day for a child, how much of the suspension should be given?

Answer

4. 20 mL

Example 3 Applying Metric Units of Measure to Medicine

The drug Aldomet (used for high blood pressure) comes in a concentration of 50 mg per milliliter of solution. If a doctor prescribes 225 mg per dose, how many milliliters should be administered for a single dose?

Solution:

The statement, 50 mg per milliliter, represents the ratio, $\frac{50\text{ mg}}{1\text{ mL}}$.

$\frac{50\text{ mg}}{1\text{ mL}} = \frac{225\text{ mg}}{x}$ Use this ratio to set up a proportion.

$50x = 225(1)$ Equate the cross products.

$\frac{50x}{50} = \frac{225}{50}$ Divide by 50.

$x = 4.5$ A single dose is 4.5 mL.

Section 8.4 Practice Exercises

Boost *your* GRADE at ALEKS.com! ALEKS version 3.0

- Practice Problems
- Self-Tests
- NetTutor
- e-Professors
- Videos

Study Skills Exercise

1. Define the key term **microgram**.

Review Exercises

For Exercises 2–10, perform the conversions.

2. 9.84 m = _____ mm **3.** 4.28 km = _____ m **4.** 42 cg = _____ g

5. 80 mg = _____ cg **6.** 4 kL = _____ cL **7.** 0.009 kL = _____ mL

8. 9 kg = _____ lb **9.** 2800 g = _____ lb **10.** 18 mL = _____ cc

Objective 1: Additional Metric Units of Mass

For Exercises 11–18, perform the conversions. **(See Example 1.)**

11. 0.01 mg = _____ mcg **12.** 0.005 mg = _____ mcg **13.** 7500 mcg = _____ mg **14.** 50 mcg = _____ mg

15. 500 mcg = _____ cg **16.** 1000 mcg = _____ cg **17.** 0.05 cg = _____ mcg **18.** 0.001 cg = _____ mcg

Objective 2: Medical Applications

19. A doctor orders 0.2 mg of the drug atropine given by injection. How many micrograms is this?

20. The drug Synthroid is used to treat thyroid disease. A patient is sometimes started on a dose of 0.05 mg/day. How many micrograms is this?

21. The drug cyanocobalamin is prescribed by one doctor in the amount of 1000 mcg. How many milligrams is this?

22. An injection of naloxone is given in the amount of 800 mcg. How many milligrams is this?

23. The drug Zovirax is sometimes used to treat chicken pox in children. One doctor recommended 20 mg of the drug per kilogram of the child's body mass, 4 times daily. **(See Example 2.)**

a. How much of the drug should a 20-kg child receive for one dose?

b. How much of the drug would be given over a 5-day period?

24. The drug Amoxil is sometimes used to treat children with bacterial infections. One doctor prescribed 40 mg of the drug per kilogram of the child's body mass, 3 times daily.

a. How much of the drug should a 15-kg child receive for one dose?

b. How much of the drug would be given to the child over a 10-day period?

25. A nurse must administer 45 mg of a drug. The drug is available in a liquid form with a concentration of 15 mg per milliliter of the solution. How many milliliters of the solution should the nurse give? **(See Example 3.)**

26. A patient must receive 500 mg of medication in a solution that has a strength of 250 mg per 5 milliliter of solution. How many milliliters of solution should be given?

27. The drug amoxicillin is an antibiotic used to treat bacterial infections. A doctor orders 250 mg every 8 hr. How many grams of the drug would be given in 1 wk?

28. A blood alcohol concentration of 0.08% is the upper legal limit for operating a motor vehicle. For a typical 175-lb man, each 12-oz can of beer consumed within 1 hr will raise his blood alcohol concentration by 0.028%. If he consumes three cans of beer within 1 hr, what is his blood alcohol concentration? Is he below the legal limit for driving home?

29. Dr. Boyd gives a patient 2 cc of Zantac. How many milliliters is this?

30. If a nurse mixed 11.5 mL of sterile water with 1.5 mL of oxacillin, how many cubic centimeters will this produce?

31. A tetanus vaccine was purchased by a group of family practice doctors. They purchased 1 L of the vaccine. How many patients can be vaccinated if the normal dose is 2 cc?

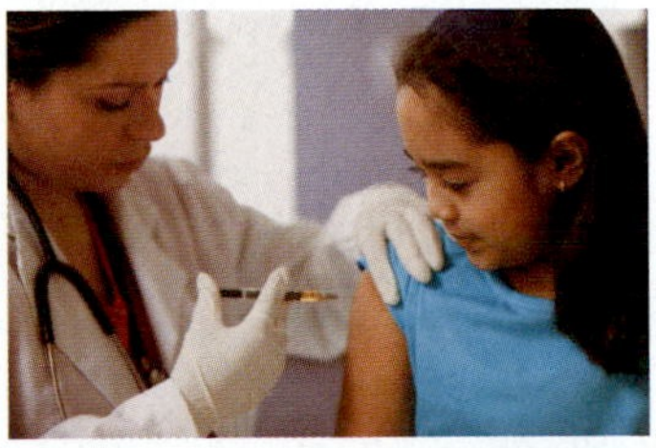

32. A pharmacist has a 1-L bottle of cough syrup. How many 20-cL bottles can she make?

33. A doctor orders 0.2 mg of a drug per kilogram of a patient's body mass. How much of the drug should be given to a patient who is 48 kg?

34. The dosage for a painkiller is 0.05 mg per kilogram of a patient's body mass. How much of the drug should be administered to a patient who is 90 kg?

Expanding Your Skills

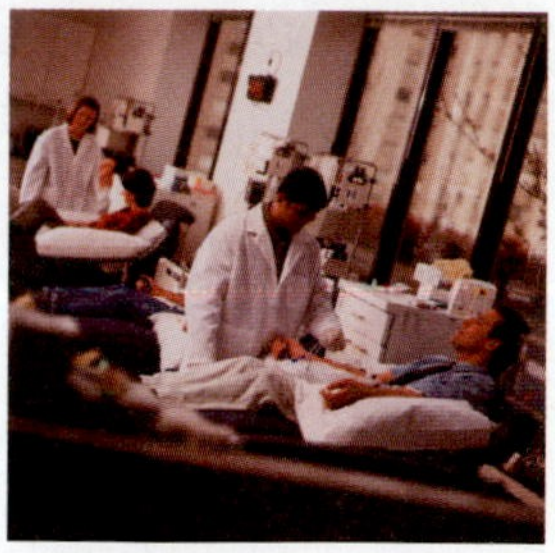

35. A normal value of hemoglobin in the blood for an adult male is 18 gm/dL (that is, 18 grams per deciliter). How much hemoglobin would be expected in 20 mL of a males's blood?

36. A normal value of hemoglobin in the blood for an adult female is 15 gm/dL (that is, 15 grams per deciliter). How much hemoglobin would be expected in 40 mL of a female's blood?

Section 8.5 Lines and Angles

Objectives

1. Basic Definitions
2. Naming and Measuring Angles
3. Complementary and Supplementary Angles
4. Parallel and Perpendicular Lines

1. Basic Definitions

In this chapter, we will introduce some basic concepts of geometry.

A **point** is a specific location in space. We often symbolize a point by a dot and label it with a capital letter such as *P*.

A **line** consists of infinitely many points that follow a straight path. A line extends forever in both directions. This is illustrated by arrowheads at both ends. Figure 8-11 shows a line through points *A* and *B*. The line can be represented as $\overleftrightarrow{AB}$ or as $\overleftrightarrow{BA}$.

Figure 8-11

A **line segment** is a part of a line between and including two distinct endpoints. A line segment with endpoints *P* and *Q* can be denoted $\overline{PQ}$ or $\overline{QP}$. See Figure 8-12.

Figure 8-12

Concept Connections

1. Are the rays $\overrightarrow{AB}$ and $\overrightarrow{BA}$ the same?
2. Are the lines $\overleftrightarrow{PQ}$ and $\overleftrightarrow{QP}$ the same?

A **ray** is the part of a line that includes an endpoint and all points on one side of the endpoint. In Figure 8-13, ray $\overrightarrow{PQ}$ is named by using the endpoint *P* and another point *Q* on the ray. Notice that the rays $\overrightarrow{PQ}$ and $\overrightarrow{QP}$ are different because they extend in different directions. See Figure 8-13.

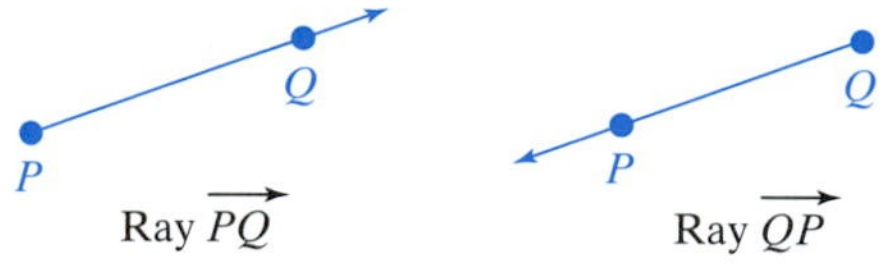

Figure 8-13

TIP: A ray has only one endpoint, which is always written first.

Skill Practice

Identify each as a point, line, line segment, or ray.

3. $\overline{RS}$ **4.** $\bullet Q$

5. $\overrightarrow{XY}$ **6.** $\overleftrightarrow{TV}$

Example 1 Identifying Points, Lines, Line Segments, and Rays

Identify each as a point, line, line segment, or ray.

a. $\overleftrightarrow{MN}$ **b.** $\overrightarrow{NM}$ **c.** $\overline{MN}$ **d.** $\bullet S$

Solution:

a. The double arrowheads indicate that $\overleftrightarrow{MN}$ is a line.

b. The single arrowhead indicates that $\overrightarrow{NM}$ is a ray with endpoint *N*.

c. The bar drawn above the letters $\overline{MN}$ indicates a line segment with endpoints *M* and *N*.

d. The dot represents a point.

Answers

1. No 2. Yes
3. Line segment
4. Point 5. Ray 6. Line

2. Naming and Measuring Angles

An **angle** is a geometric figure formed by two rays that share a common endpoint. The common endpoint is called the **vertex** of the angle. In Figure 8-14, the rays $\overrightarrow{PR}$ and $\overrightarrow{PQ}$ share the endpoint *P*. These rays form the sides of the angle and the angle is denoted $\angle QPR$ or $\angle RPQ$. Notice that when we name an angle, the vertex must be the middle letter. Sometimes a small arc) is drawn to illustrate the location of an angle. In such a case, the angle may be named by using the symbol $\angle$ along with the letter of the vertex.

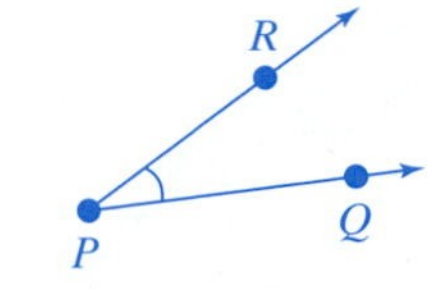

$\angle QPR$ or $\angle RPQ$ or $\angle P$

Figure 8-14

TIP: Sometimes angles are named by a number or lowercase letter between the rays. See $\angle x$ shown here.

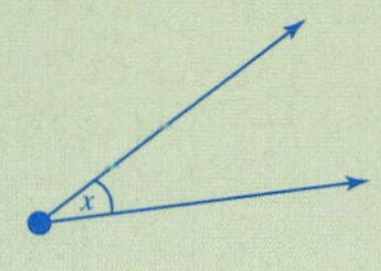

Avoiding Mistakes

The ° symbol can be used for temperature or for measuring angles. The context of the problem tells us whether temperature or angle measure is implied by the symbol.

The most common unit to measure an angle is the degree, denoted by °. To become familiar with the measure of angles, consider the following benchmarks. Two rays that form a quarter turn of a circle make a 90° angle. A 90° angle is called a **right angle** and is often depicted with a □ symbol. Two rays that form a half turn of a circle make a 180° angle. A 180° angle is called a **straight angle** because it appears as a straight line. A full circle has 360°. For example, the second hand of a clock sweeps out an angle of 360° in 1 minute. See Figure 8-15.

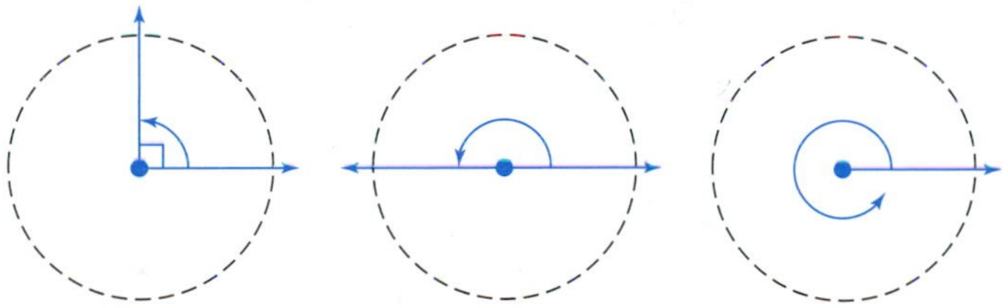

90° Right angle | 180° Straight angle | 360°

Figure 8-15

We can approximate the measure of an angle by using a tool called a *protractor*, shown in Figure 8-16. A protractor uses equally spaced tick marks around a semicircle to measure angles from 0° to 180°.

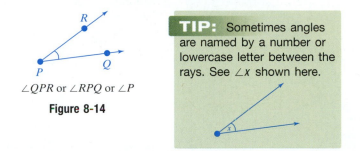

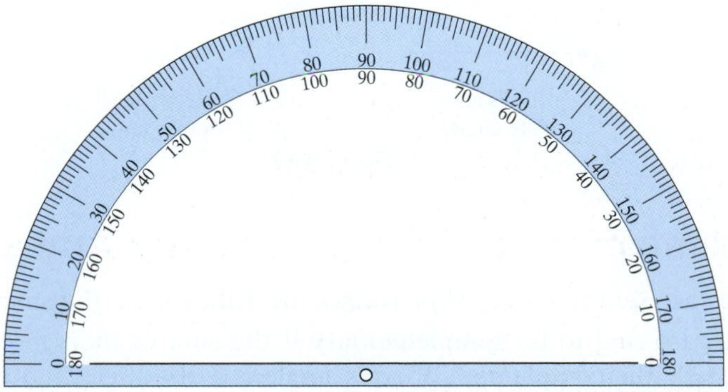

Figure 8-16

Animation

Example 2 shows how we can use a protractor to measure several angles. To denote the measure of an angle, we use the symbol *m*, written in front of the name of the angle. For example, if the measure of angle *A* is 30°, we write $m(\angle A) = 30°$.

Skill Practice

Read the protractor to determine the measure of each angle.

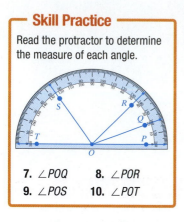

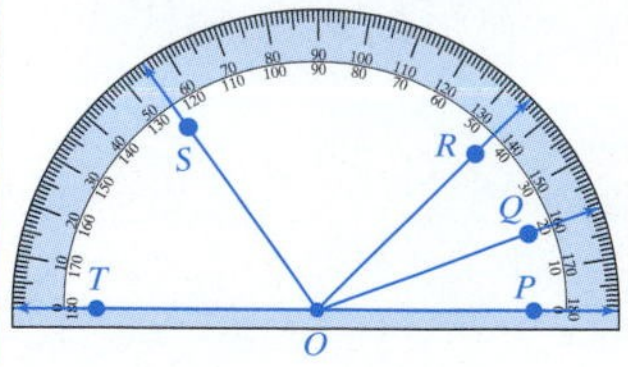

7. $\angle POQ$ **8.** $\angle POR$
9. $\angle POS$ **10.** $\angle POT$

Example 2 Measuring Angles

Read the protractor to determine the measure of each angle.

a. $\angle AOB$ **b.** $\angle AOC$ **c.** $\angle AOD$ **d.** $\angle AOE$

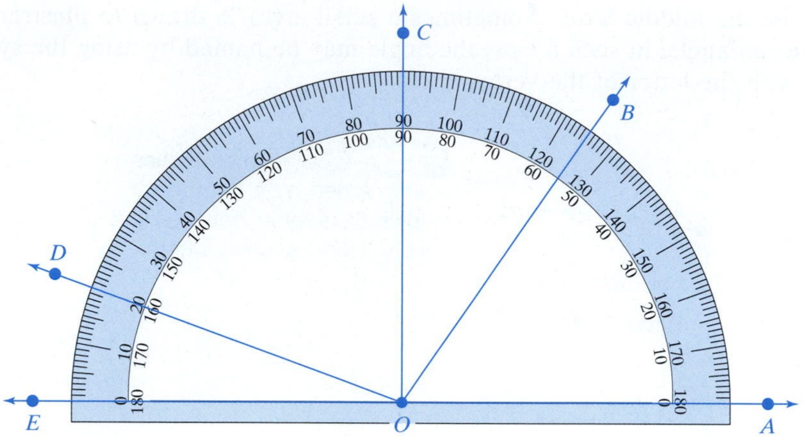

Solution:

We will use the inner scale on the protractor. This is done because we are measuring the angles in a counterclockwise direction, beginning at 0° along ray $\overrightarrow{OA}$.

a. $m(\angle AOB) = 55°$ On the inner scale, ray $\overrightarrow{OA}$ is aligned with 0° and ray $\overrightarrow{OB}$ passes through 55°. Therefore, $m(\angle AOB) = 55°$.

b. $m(\angle AOC) = 90°$ $\angle AOC$ is a right angle.

c. $m(\angle AOD) = 160°$

d. $m(\angle AOE) = 180°$ $\angle AOE$ is a straight angle.

Concept Connections

Answer true or false.

11. An angle whose measure is 102° is obtuse.

12. An angle whose measure is 98° is acute.

An angle is said to be an **acute angle** if its measure is between 0° and 90°. An angle is said to be an **obtuse angle** if its measure is between 90° and 180°. See Figure 8-17.

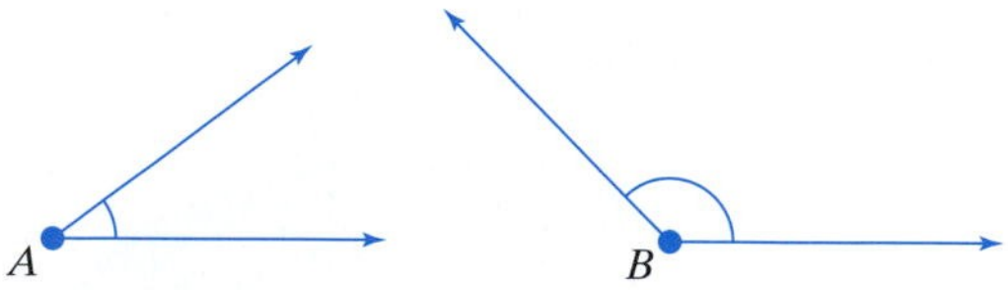

$\angle A$ is an acute angle.

$\angle B$ is an obtuse angle.

Figure 8-17

3. Complementary and Supplementary Angles

- Two angles are said to be equal or **congruent** if they have the same measure.
- Two angles are said to be **complementary** if the sum of their measures is 90°. In Figure 8-18, the complement of a 60° angle is a 30° angle, and vice versa.
- Two angles are said to be **supplementary** if the sum of their measures is 180°. In Figure 8-19, the supplement of a 60° angle is a 120° angle, and vice versa.

Answers

7. 20° **8.** 45° **9.** 125°
10. 180° **11.** True **12.** False

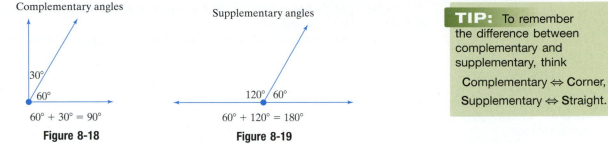

Complementary angles

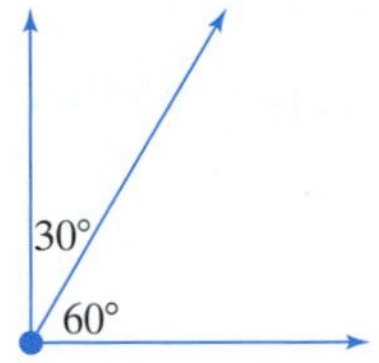

$60° + 30° = 90°$

Figure 8-18

Supplementary angles

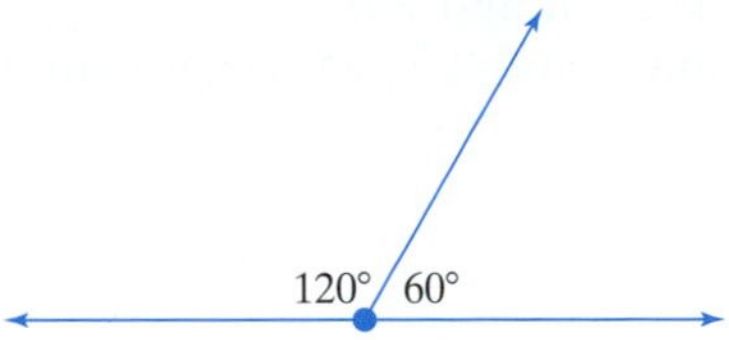

$60° + 120° = 180°$

Figure 8-19

TIP: To remember the difference between complementary and supplementary, think

Complementary ⇔ Corner,

Supplementary ⇔ Straight.

Example 3 Identifying Supplementary and Complementary Angles

a. What is the supplement of a 105° angle?

b. What is the complement of a 12° angle?

Solution:

a. Let x represent the measure of the supplement of a 105° angle.

$x + 105 = 180$	The sum of a 105° angle and its supplement must equal 180°.
$x + 105 - 105 = 180 - 105$	Subtract 105 from both sides to solve for x.
$x = 75$	The supplement is a 75° angle.

b. Let y represent the measure of the complement of a 12° angle.

$y + 12 = 90$	The sum of a 12° angle and its complement must equal 90°.
$y + 12 - 12 = 90 - 12$	Subtract 12 from both sides to solve for y.
$y = 78$	The complement is a 78° angle.

Skill Practice

13. What is the supplement of a 35° angle?

14. What is the complement of a 52° angle?

4. Parallel and Perpendicular Lines

Two lines may intersect (cross) or may be parallel. **Parallel lines** lie on the same flat surface, but never intersect. See Figure 8-20.

TIP: Sometimes we use the symbol ∥ to denote parallel lines.

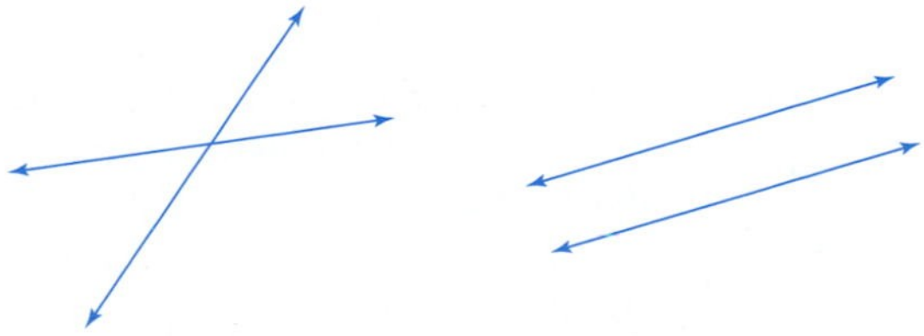

Intersecting lines Parallel lines

Figure 8-20

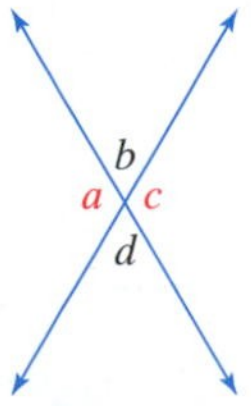

Figure 8-21

Notice that two intersecting lines form four angles. In Figure 8-21, $\angle a$ and $\angle c$ are **vertical angles**. They appear on opposite sides of the vertex. Likewise, $\angle b$ and $\angle d$ are vertical angles. Vertical angles are equal in measure. That is $m(\angle a) = m(\angle c)$ and $m(\angle b) = m(\angle d)$.

Answers

13. 145° **14.** 38°

Angles that share a side are called *adjacent* angles. One pair of *adjacent* angles in Figure 8-21 is $\angle a$ and $\angle b$.

If two lines intersect at a right angle, they are **perpendicular lines**. See Figure 8-22.

TIP: Sometimes we use the symbol $\perp$ to denote perpendicular lines.

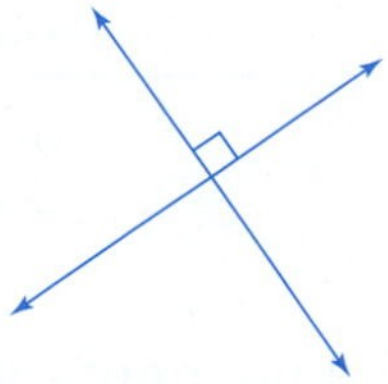

Figure 8-22

In Figure 8-23, lines L_1 and L_2 are parallel lines. If a third line m intersects the two parallel lines, eight angles are formed. Suppose we label the eight angles formed by lines L_1, L_2, and m with the numbers 1–8.

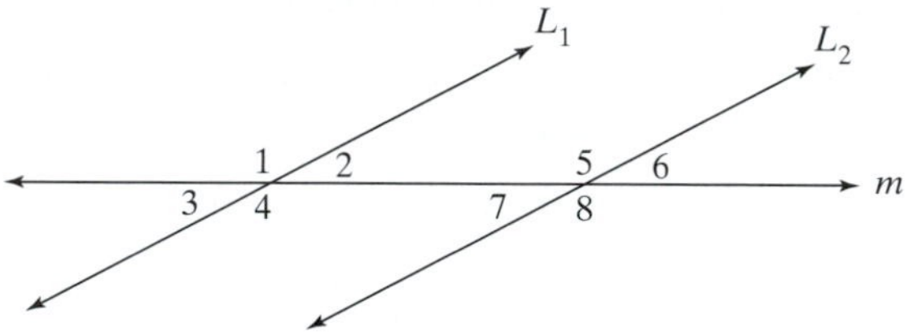

Figure 8-23

These angles have the special properties found in Table 8-6.

Table 8-6

Lines L_1 and L_2 are Parallel; Line m is an Intersecting Line	**Name of Angles**	**Property**
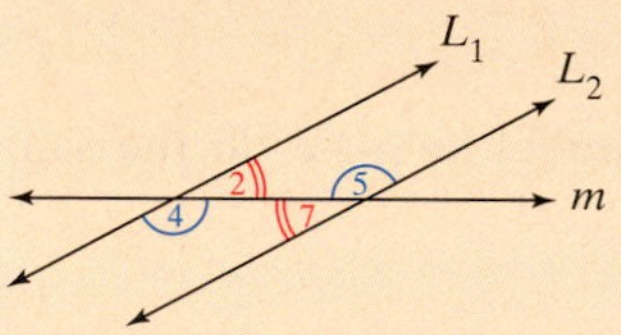	The following pairs of angles are called **alternate interior angles**: $\angle 2$ and $\angle 7$ $\angle 4$ and $\angle 5$	**Alternate interior angles are equal in measure.** $m(\angle 2) = m(\angle 7)$ $m(\angle 4) = m(\angle 5)$
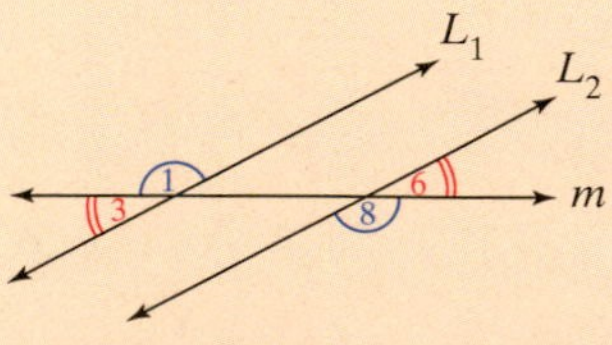	The following pairs of angles are called **alternate exterior angles**: $\angle 1$ and $\angle 8$ $\angle 3$ and $\angle 6$	**Alternate exterior angles are equal in measure.** $m(\angle 1) = m(\angle 8)$ $m(\angle 3) = m(\angle 6)$
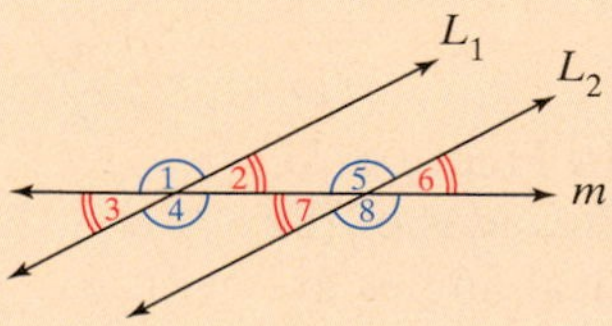	The following pairs of angles are called **corresponding angles**: $\angle 1$ and $\angle 5$ $\angle 2$ and $\angle 6$ $\angle 3$ and $\angle 7$ $\angle 4$ and $\angle 8$	**Corresponding angles are equal in measure.** $m(\angle 1) = m(\angle 5)$ $m(\angle 2) = m(\angle 6)$ $m(\angle 3) = m(\angle 7)$ $m(\angle 4) = m(\angle 8)$

Example 4 Finding the Measure of Angles in a Diagram

Assume that lines L_1 and L_2 are parallel. Find the measure of each angle, and explain how the angle is related to the given angle of 65°.

a. $\angle a$

b. $\angle b$

c. $\angle c$

d. $\angle d$

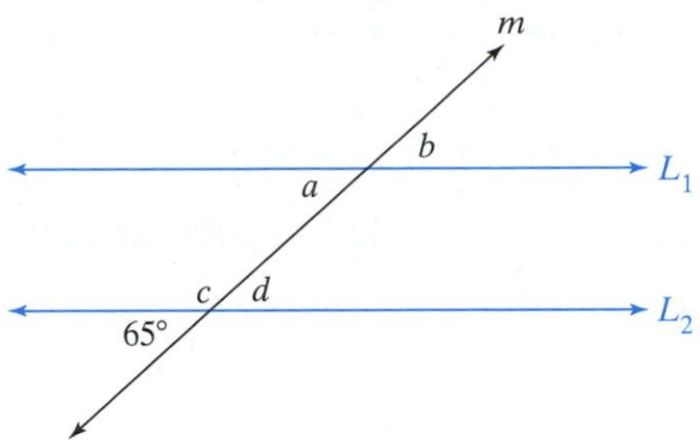

Solution:

a. $m(\angle a) = 65°$ $\angle a$ is a corresponding angle to the given angle.

b. $m(\angle b) = 65°$ $\angle b$ is an alternate exterior angle to the given angle.

c. $m(\angle c) = 115°$ $\angle c$ is the supplement to the given angle.

d. $m(\angle d) = 65°$ $\angle d$ and the given angle are vertical angles.

Skill Practice

Assume that lines L_1 and L_2 are parallel. Find the measure of each angle. Explain how the angle is related to the given angle.

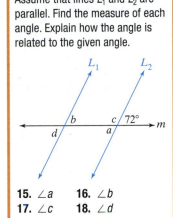

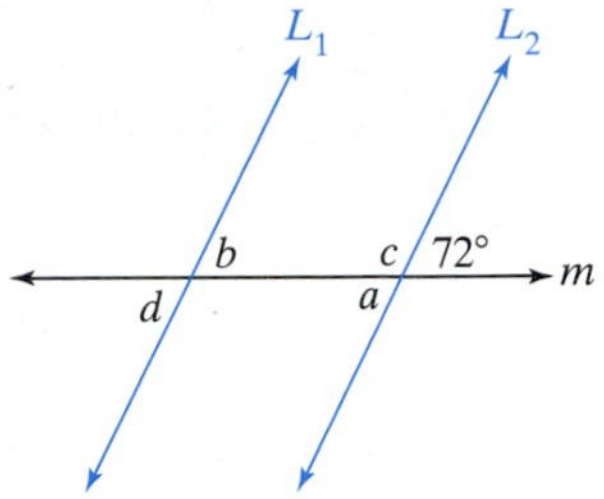

15. $\angle a$ **16.** $\angle b$
17. $\angle c$ **18.** $\angle d$

Answers

15. 72°; vertical angles
16. 72°; corresponding angles
17. 108°; supplementary angles
18. 72°; alternate exterior angles

Section 8.5 Practice Exercises

Study Skills Exercise

1. Define the key terms.

a. Point	**b. Line**	**c. Line segment**	**d. Ray**
e. Angle	**f. Vertex**	**g. Right angle**	**h. Straight angle**
i. Acute angle	**j. Obtuse angle**	**k. Congruent angles**	**l. Complementary angles**
m. Supplementary angles	**n. Parallel lines**	**o. Vertical angles**	**p. Perpendicular lines**
q. Alternate exterior angles	**r. Alternate interior angles**	**s. Corresponding angles**	

Objective 1: Basic Definitions

2. Explain the difference between a line and a line segment.

3. Explain the difference between a line and a ray.

4. Is the ray $\overrightarrow{AB}$ the same as $\overrightarrow{BA}$? Explain.

For Exercises 5–10, identify each figure as a line, line segment, ray, or point. **(See Example 1.)**

5. 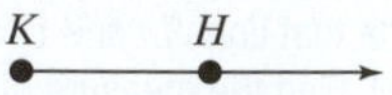

6. 

7. 

8. 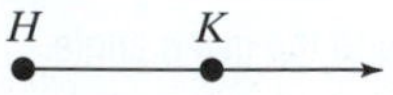

9. 

10. 

For Exercises 11–14, draw a figure that represents the expression.

11. $\overline{XY}$

12. A point named X

13. $\overleftrightarrow{YX}$

14. $\overrightarrow{XY}$

Objective 2: Naming and Measuring Angles

15. Sketch a right angle.

16. Sketch a straight angle.

17. Sketch an acute angle.

18. Sketch an obtuse angle.

For Exercises 19–24, use the protractor to determine the measure of each angle. **(See Example 2.)**

19. $m(\angle AOB)$

20. $m(\angle AOC)$

21. $m(\angle AOD)$

22. $m(\angle AOE)$

23. $m(\angle AOF)$

24. $m(\angle AOG)$

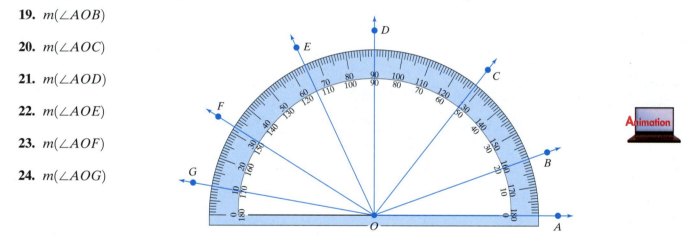

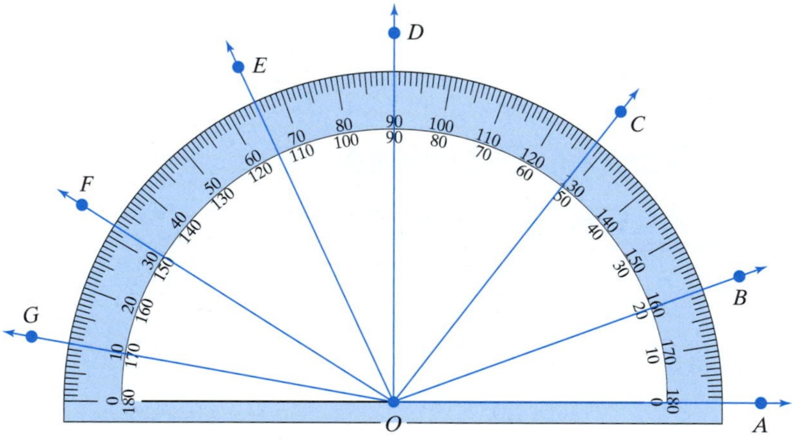

Animation

For Exercises 25–32, label each as an obtuse angle, acute angle, right angle, or straight angle.

25. $m(\angle A) = 90°$

26. $m(\angle E) = 91°$

27. $m(\angle B) = 98°$

28. $m(\angle F) = 30°$

29. $m(\angle C) = 2°$

30. $m(\angle G) = 130°$

31. $m(\angle D) = 180°$

32. $m(\angle H) = 45°$

Objective 3: Complementary and Supplementary Angles

For Exercises 33–40, the measure of an angle is given. Find the measure of the complement. **(See Example 3.)**

33. 80°

34. 5°

35. 27°

36. 64°

37. 29.5°

38. 13.2°

39. 89°

40. 1°

For Exercises 41–48, the measure of an angle is given. Find the measure of the supplement. **(See Example 3.)**

41. 80°

42. 5°

43. 127°

44. 124°

45. 37.4°

46. 173.9°

47. 179°

48. 1°

49. Can two supplementary angles both be obtuse? Why or why not?

50. Can two supplementary angles both be acute? Why or why not?

51. Can two complementary angles both be acute? Why or why not?

52. Can two complementary angles both be obtuse? Why or why not?

53. What angle is its own supplement?

54. What angle is its own complement?

Objective 4: Parallel and Perpendicular Lines

55. Sketch two lines that are parallel.

56. Sketch two lines that are *not* parallel.

57. Sketch two lines that are perpendicular.

58. Sketch two lines that are *not* perpendicular.

For Exercises 59–62, find the measure of angles *a*, *b*, *c*, and *d*.

59.

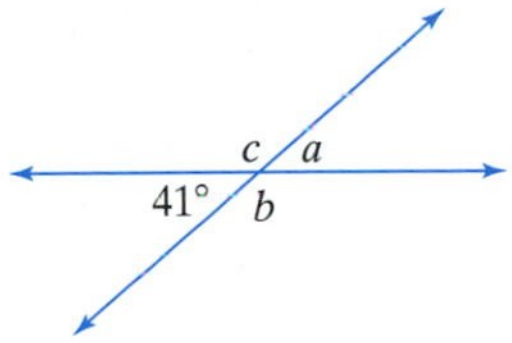

60.

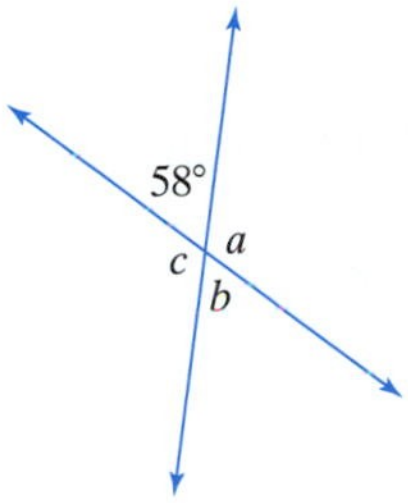

61.

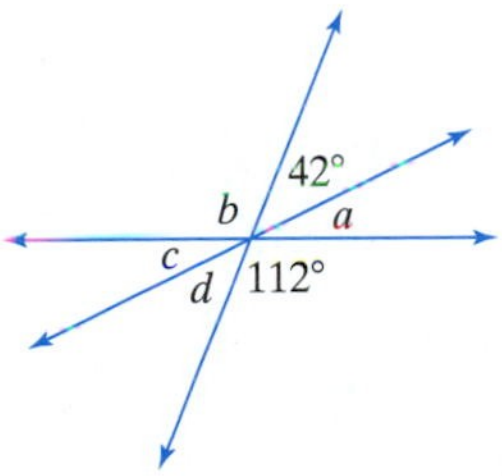

62.

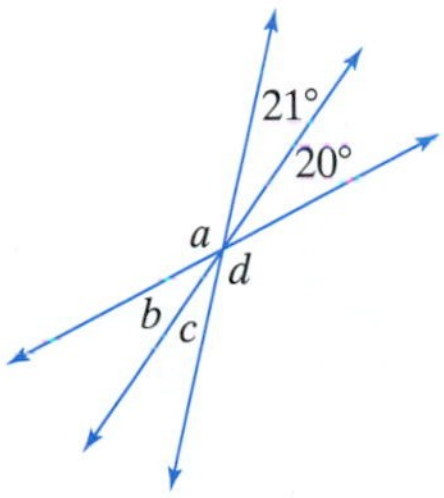

63. If two intersecting lines form vertical angles and each angle measures 90°, what can you say about the lines?

64. Can two adjacent angles formed by two intersecting lines be complementary, supplementary, or neither?

For Exercises 65–66, refer to the figure.

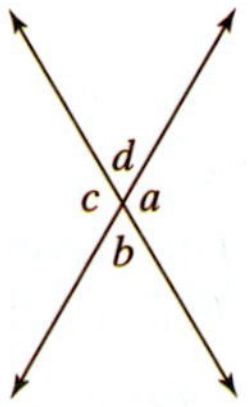

65. Describe the pair of angles, $\angle a$ and $\angle c$, as complementary, vertical, or supplementary angles.

66. Describe the pair of angles, $\angle b$ and $\angle c$, as complementary, vertical, or supplementary angles.

For Exercises 67–70, refer to the figure.

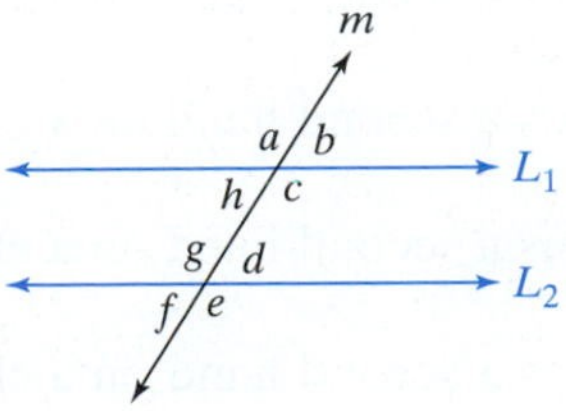

67. Identify a pair of vertical angles.

68. Identify a pair of alternate interior angles.

69. Identify a pair of alternate exterior angles.

70. Identify a pair of corresponding angles.

For Exercises 71–72, find the measure of angles a–g in the figure. Assume that L_1 and L_2 are parallel and that m is an intersecting line. **(See Example 4.)**

71.

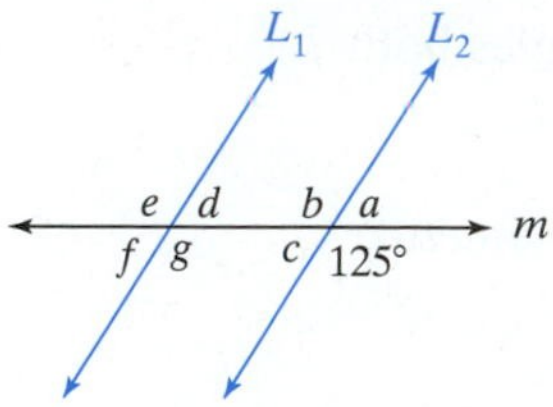

72.

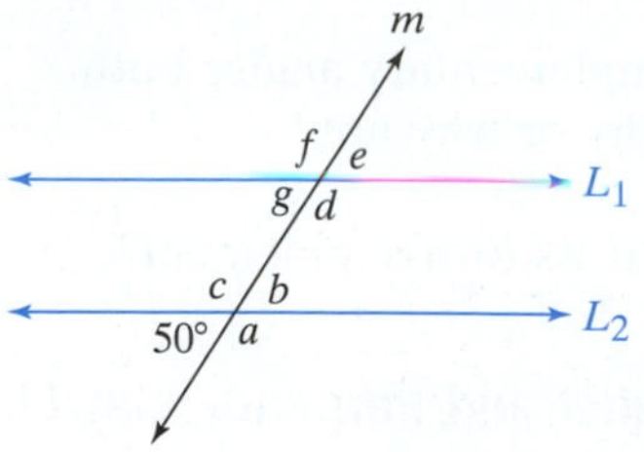

For Exercises 73–82, refer to the figure and answer true or false.

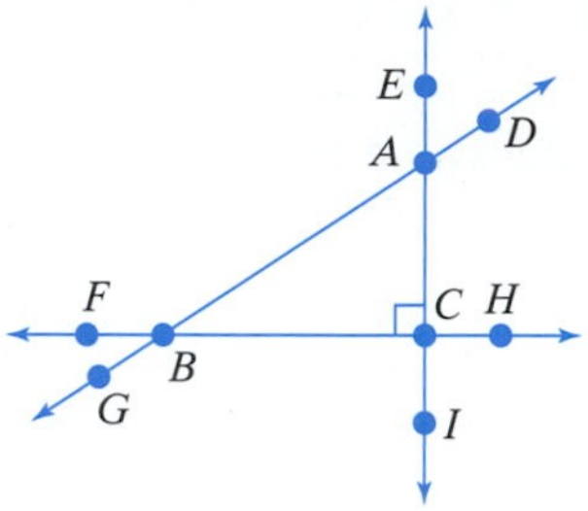

73. $\overleftrightarrow{AC}$ and $\overleftrightarrow{BC}$ are perpendicular lines.

74. $\overleftrightarrow{AB}$ and $\overleftrightarrow{AC}$ are perpendicular lines.

75. $\angle GBF$ is an acute angle.

76. $\angle EAD$ is an acute angle.

77. $\angle EAD$ and $\angle DAC$ are complementary angles.

78. $\angle GBF$ and $\angle FBA$ are complementary angles.

79. $\angle EAD$ and $\angle CAB$ are vertical angles.

80. $\angle ABC$ and $\angle FBG$ are vertical angles.

81. The point B is on $\overline{GA}$.

82. The point C is on $\overline{BH}$.

For Exercises 83–86, find the measure of $\angle XYZ$.

83.

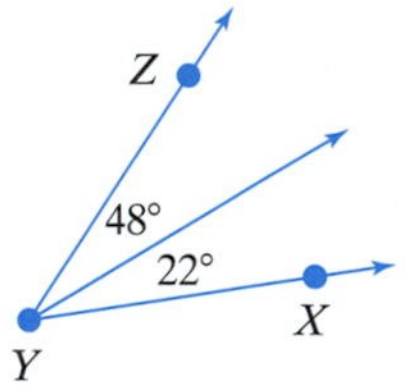

84.

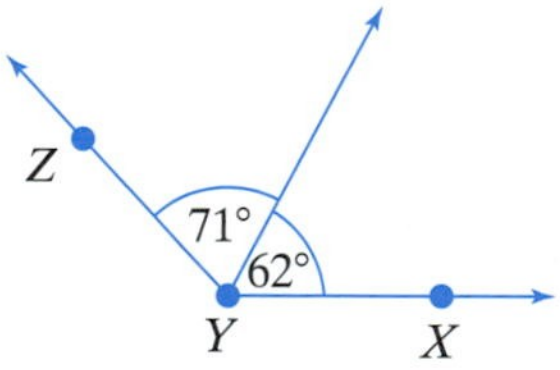

85.

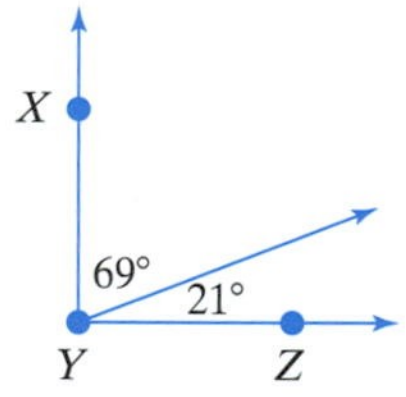

86.

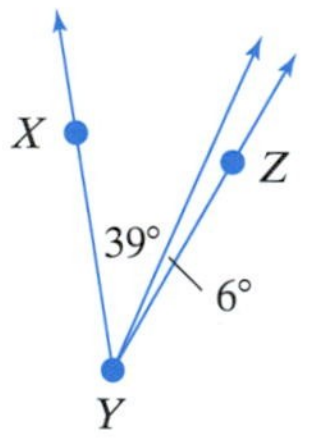

Expanding Your Skills

87. Use the figure to find the measure of each angle.

a. $\angle AOB$

b. $\angle EOD$

c. $\angle AOE$

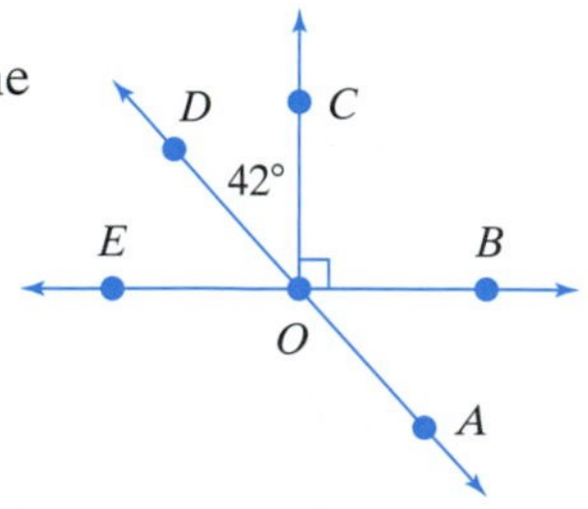

88. Use the figure to find the measure of each angle.

a. $\angle AOB$

b. $\angle EOD$

c. $\angle AOE$

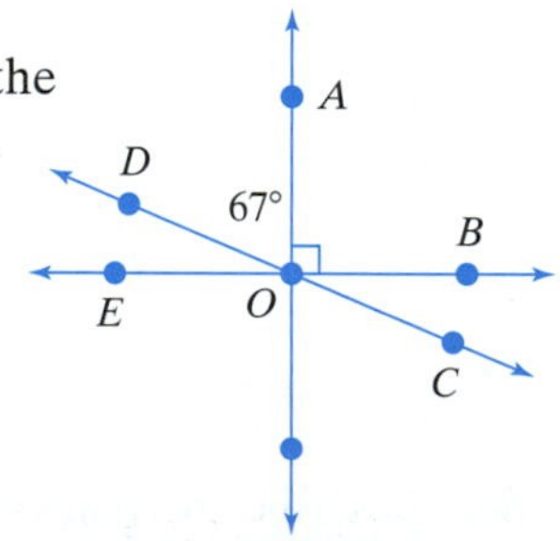

The second hand on a clock sweeps out a complete circle in 1 min. A circle forms a 360° arc. Use this information for Exercises 89–92.

89. How many degrees does a second hand on a clock move in 30 sec?

90. How many degrees does a second hand on a clock move in 15 sec?

91. How many degrees does a second hand on a clock move in 20 sec?

92. How many degrees does a second hand on a clock move in 45 sec?

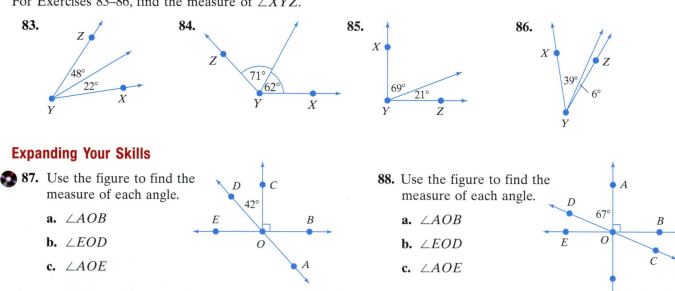

Triangles and the Pythagorean Theorem

Section 8.6

1. Triangles

A triangle is a three-sided polygon. Furthermore, the sum of the measures of the angles within a triangle is 180°. Teachers often demonstrate this fact by tearing a triangular sheet of paper as shown in Figure 8-24. Then they align the **vertices** (points) of the triangle to form a straight angle.

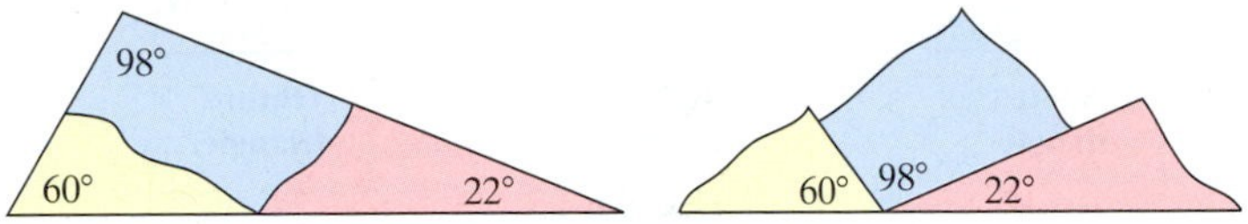

Figure 8-24

PROPERTY Angles of a Triangle

The sum of the measures of the angles of a triangle equals 180°.

Example 1 Finding the Measure of Angles Within a Triangle

Find the measure of angles a and b.

a.

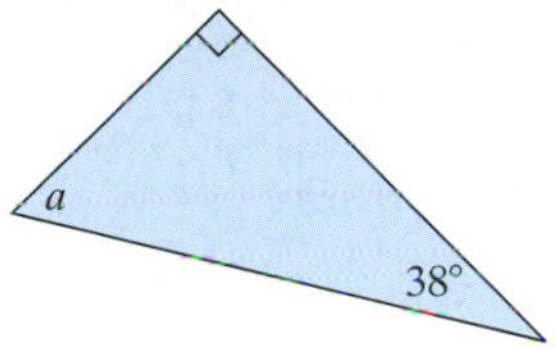

b.

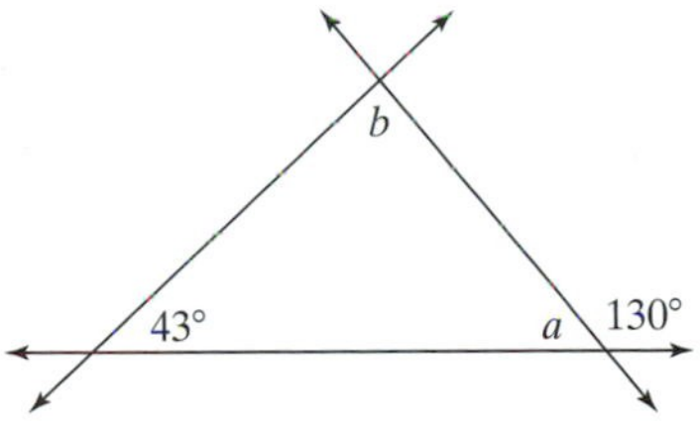

Solution:

a. Recall that the □ symbol represents a 90° angle.

$38° + 90° + m(\angle a) = 180°$	The sum of the angles within a triangle is 180°.
$128° + m(\angle a) = 180°$	Add the measures of the two known angles.
$128° - 128° + m(\angle a) = 180° - 128°$	Solve for $m(\angle a)$.
$m(\angle a) = 52°$	

b. $\angle a$ is the supplement of the 130° angle. Thus $m(\angle a) = 50°$.

$43° + 50° + m(\angle b) = 180°$	The sum of the angles within a triangle is 180°.
$93° + m(\angle b) = 180°$	Add the measures of the two known angles.
$93° - 93° + m(\angle b) = 180° - 93°$	Solve for $m(\angle b)$.
$m(\angle b) = 87°$	

Objectives

1. Triangles
2. Square Roots
3. Pythagorean Theorem

Animation

Skill Practice

Find the measures of angles a and b.

1.

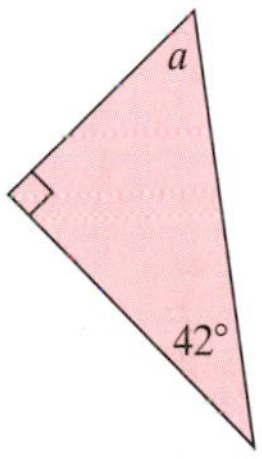

2.

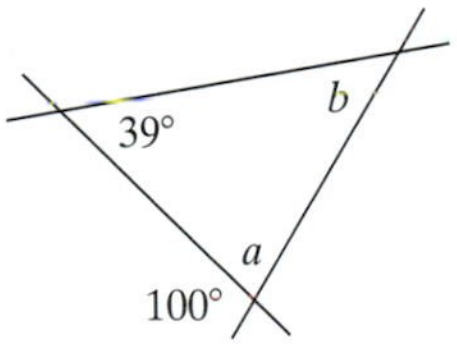

Answers

1. $m(\angle a) = 48°$
2. $m(\angle a) = 80°$
   $m(\angle b) = 61°$

Triangles may be categorized by the measures of their angles and by the number of equal sides or angles (Figures 8-25 and 8-26).

- An **acute triangle** is a triangle in which all three angles are acute.
- A **right triangle** is a triangle in which one angle is a right angle.
- An **obtuse triangle** is a triangle in which one angle is obtuse.

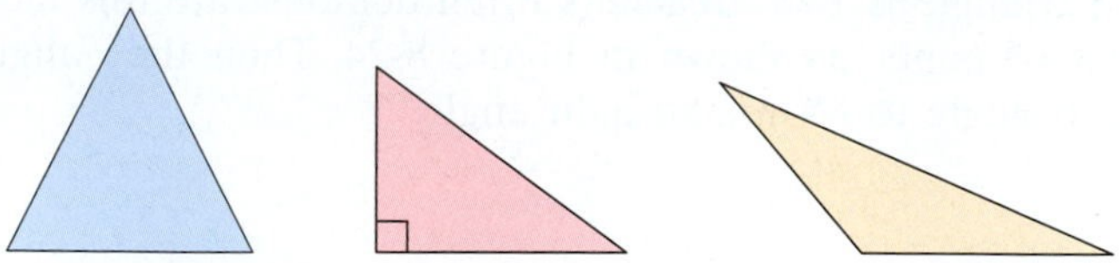

Acute triangle **Right triangle** **Obtuse triangle**

Figure 8-25

- An **equilateral triangle** is a triangle in which all three sides (and all three angles) are equal in measure.
- An **isosceles triangle** is a triangle in which two sides are equal in length (the angles opposite the equal sides are also equal in measure).
- A **scalene triangle** is a triangle in which no sides (or angles) are equal in measure.

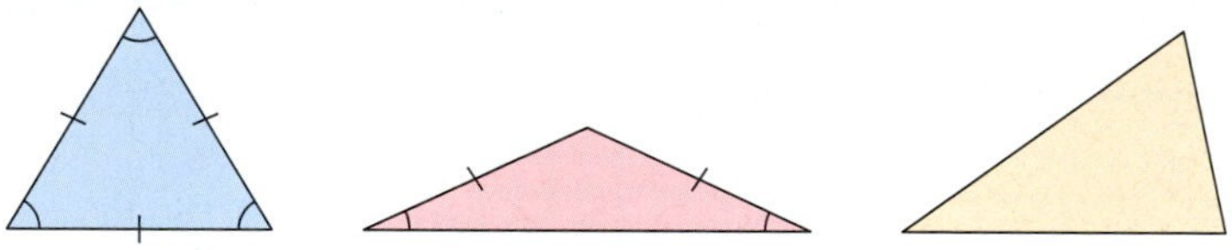

Equilateral triangle **Isosceles triangle** **Scalene triangle**

Figure 8-26

TIP: Sometimes we use tick marks / to denote segments of equal length. Similarly, we sometimes use a small arc) to denote angles of equal measure.

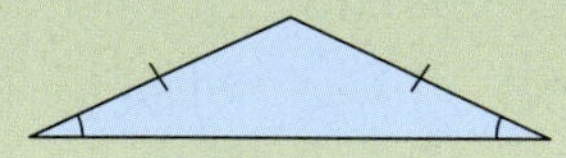

2. Square Roots

In this section we present an important theorem called the Pythagorean theorem. To understand this theorem, we first need some background definitions.

Recall from Section 1.7 that to square a number means to find the product of the number and itself. Thus, $b^2 = b \cdot b$. For example:

$$6^2 = 6 \cdot 6 = 36$$

We now want to reverse this process by finding a square root of a number. Recall that this is denoted by the radical sign $\sqrt{}$. For example, $\sqrt{36}$ reads as "the positive square root of 36." Thus,

$$\sqrt{36} = 6 \qquad \text{because } 6^2 = 6 \cdot 6 = 36.$$

Concept Connections

Match the triangle with the appropriate category, **a**, **b**, or **c**.

3. 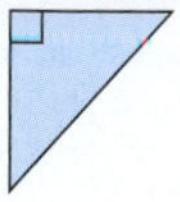

4. 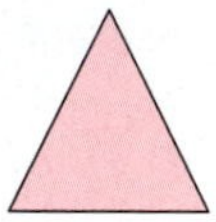

5. 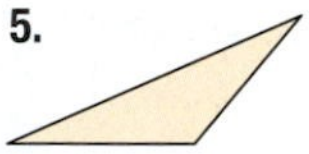

a. Acute triangle
b. Right triangle
c. Obtuse triangle

Match the triangle with the appropriate category, **a**, **b**, or **c**.

6. 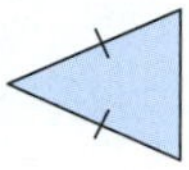

7. 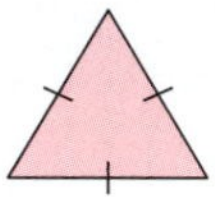

8. 

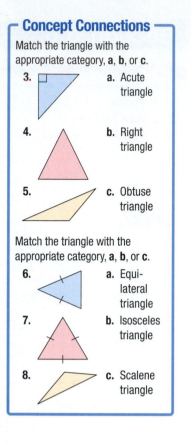

a. Equilateral triangle
b. Isosceles triangle
c. Scalene triangle

Answers

3. b **4.** a **5.** c
6. b **7.** a, b **8.** c

Example 2 Evaluating Squares and Square Roots

Simplify.

a. $\sqrt{64}$ **b.** $\sqrt{100}$ **c.** 100^2 **d.** $\sqrt{\frac{1}{4}}$

Solution:

a. $\sqrt{64} = 8$ because $8 \cdot 8 = 64$

b. $\sqrt{100} = 10$ because $10 \cdot 10 = 100$

c. $100^2 = 100 \cdot 100$

$= 10{,}000$

d. $\sqrt{\frac{1}{4}} = \frac{1}{2}$ because $\frac{1}{2} \cdot \frac{1}{2} = \frac{1}{4}$

TIP: The numbers 0, 1, 4, 9, 16, and so on are called perfect squares, because in each case, the principal square root is a whole number.

Skill Practice

Simplify.

9. $\sqrt{49}$

10. $\sqrt{9}$

11. 9^2

12. $\sqrt{\frac{1}{9}}$

3. Pythagorean Theorem

Recall that a right triangle is a triangle with a 90° angle. The two sides forming the right angle are called the **legs**. The side opposite the right angle is called the **hypotenuse**. Note that the hypotenuse is always the longest side. See Figure 8-27. We often use the letters a and b to represent the legs of a right triangle. The letter c is used to label the hypotenuse.

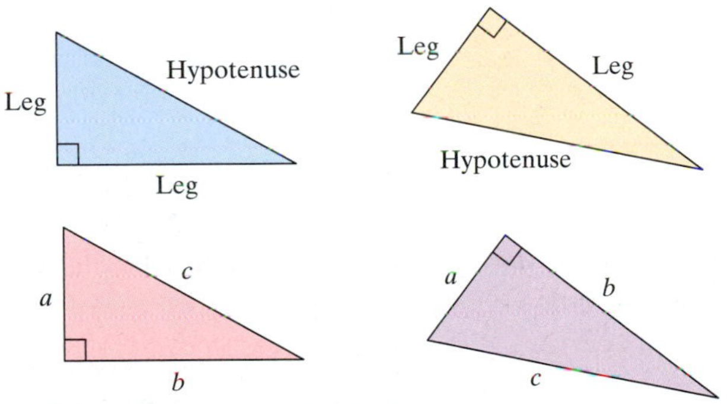

Figure 8-27

Concept Connections

13. Label the triangle with the words *leg, leg,* and *hypotenuse.*

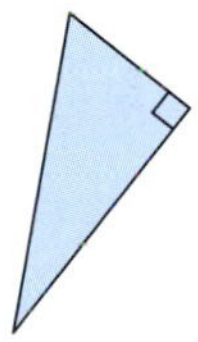

14. Label the triangle with the letters *a, b,* and *c,* where *c* represents the hypotenuse.

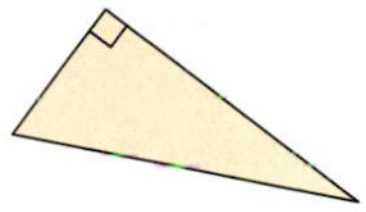

For any right triangle, the **Pythagorean theorem** gives us the following important relationship among the lengths of the sides.

PROPERTY Pythagorean Theorem

For any right triangle,

$$(\text{Leg})^2 + (\text{Leg})^2 = (\text{Hypotenuse})^2$$

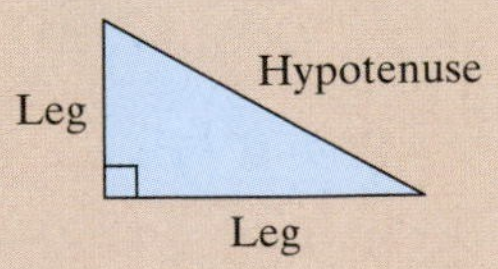

Using the letters a, b, and c to represent the legs and hypotenuse, respectively, we have

$$a^2 + b^2 = c^2$$

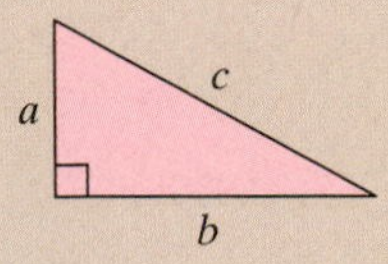

Answers

9. 7 **10.** 3 **11.** 81 **12.** $\frac{1}{3}$

13.

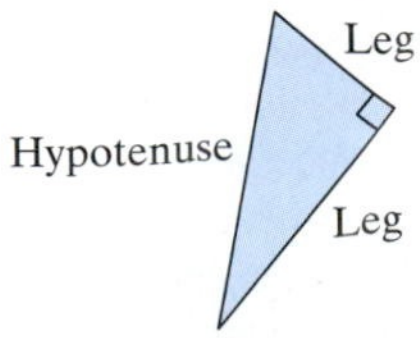

14.

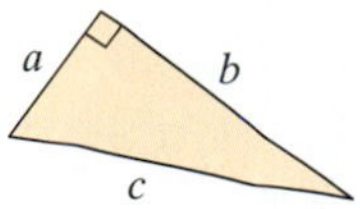

Skill Practice

15. Find the length of the hypotenuse.

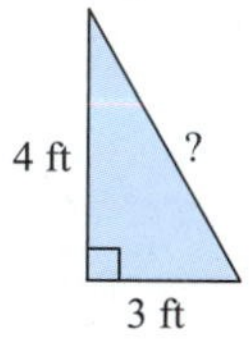

Example 3 Finding the Length of the Hypotenuse of a Right Triangle

Find the length of the hypotenuse of the right triangle.

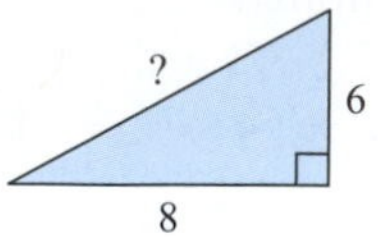

Solution:

The lengths of the legs are given.

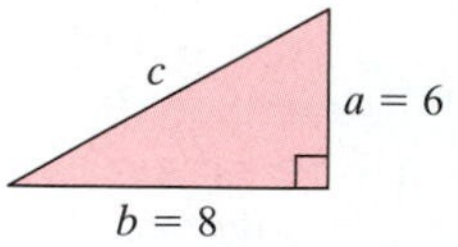

Label the triangle, using a, b, and c. It does not matter which leg is labeled a and which is labeled b.

$a^2 + b^2 = c^2$ Apply the Pythagorean theorem.

$(6)^2 + (8)^2 = c^2$ Substitute $a = 6$ and $b = 8$.

$36 + 64 = c^2$ Simplify.

$100 = c^2$ The solution to this equation is the positive number, c, that when squared equals 100.

$\sqrt{100} = c$

$10 = c$ Simplify the square root of 100.

TIP: The value $\sqrt{100} = 10$ because $10^2 = 100$.

The solution may be checked using the Pythagorean theorem.

$$a^2 + b^2 = c^2$$

$$(6)^2 + (8)^2 \stackrel{?}{=} (10)^2$$

$$36 + 64 = 100 \checkmark$$

In Example 4, we solve for one of the legs of a right triangle when the other leg and the hypotenuse are known.

Skill Practice

16. Find the length of the unknown side.

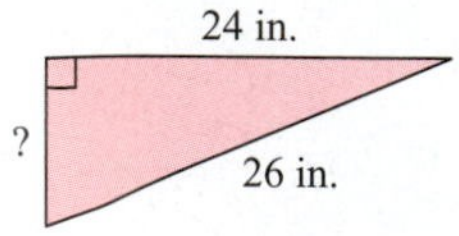

Example 4 Finding the Length of a Leg in a Right Triangle

Find the length of the unknown side of the right triangle.

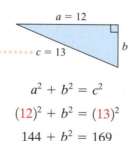

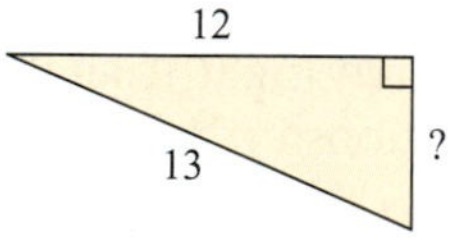

Solution:

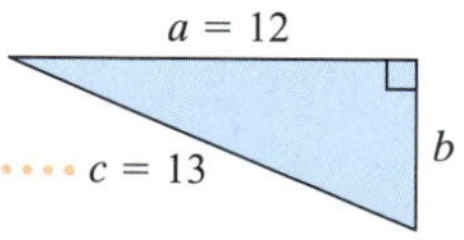

Label the triangle, using a, b, and c. One of the legs is unknown. It doesn't matter whether we call the unknown leg a or b.

$a^2 + b^2 = c^2$ Apply the Pythagorean theorem.

$(12)^2 + b^2 = (13)^2$ Substitute $a = 12$ and $c = 13$.

$144 + b^2 = 169$ Simplify.

Avoiding Mistakes

Always remember that the hypotenuse (longest side) is given the letter "c" when applying the Pythagorean theorem.

Answers

15. 5 ft 16. 10 in.

$144 - 144 + b^2 = 169 - 144$ Solve for b^2.

$b^2 = 25$ The solution to this equation is the positive number b that when squared equals 25.

$b = \sqrt{25}$

$b = 5$ Simplify the square root of 25.

The solution may be checked by using the Pythagorean theorem.

$$a^2 + b^2 = c^2$$
$$(12)^2 + (5)^2 \stackrel{?}{=} (13)^2$$
$$144 + 25 = 169 \checkmark$$

TIP: The value $\sqrt{25} = 5$ because $5^2 = 25$.

In Example 5 we use the Pythagorean theorem in an application.

Example 5 Using the Pythagorean Theorem in an Application

When Barb swam across a river, the current carried her 300 yd downstream from her starting point. If the river is 400 yd wide, how far did Barb swim?

Solution:

We first familiarize ourselves with the problem and draw a diagram (Figure 8-28). The distance Barb actually swims is the hypotenuse of the right triangle. Therefore, we label this distance c.

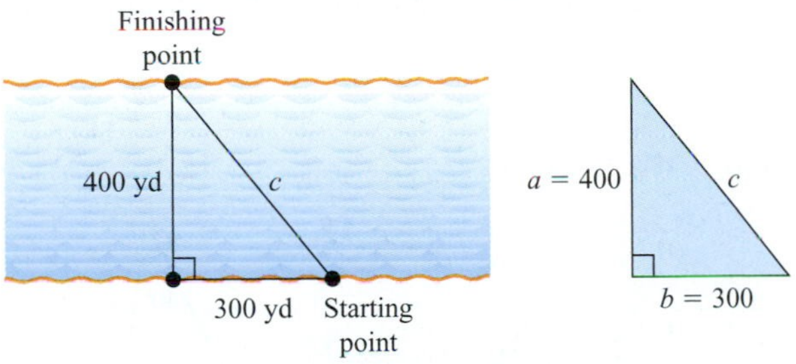

Figure 8-28

$a^2 + b^2 = c^2$ Apply the Pythagorean theorem.

$(400)^2 + (300)^2 = c^2$ Substitute $a = 400$ and $b = 300$.

$160{,}000 + 90{,}000 = c^2$ Simplify.

$250{,}000 = c^2$ Add. The solution to this equation is the positive number c that when squared equals 250,000.

$\sqrt{250{,}000} = c$

$500 = c$ Simplify.

Barb swam 500 yd.

Skill Practice

17. The bottom of a 17-ft ladder is placed 8 ft from the base of a building. How far up the building is the top of the ladder?

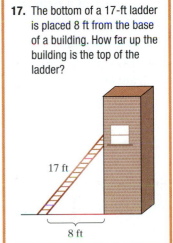

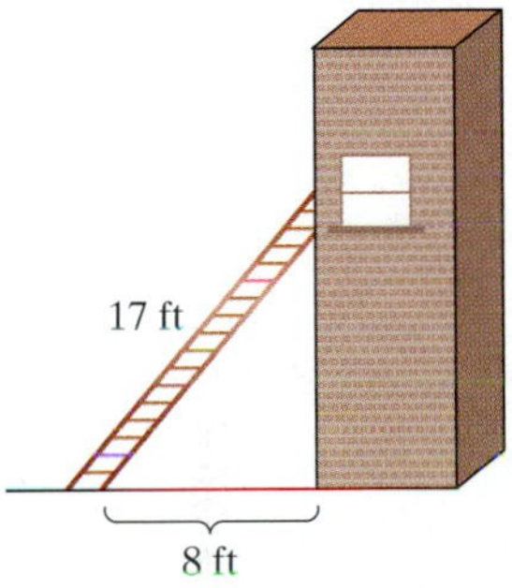

Answer

17. 15 ft

Section 8.6 Practice Exercises

Boost *your* GRADE at ALEKS.com!

ALEKS® version 3.0

- Practice Problems
- Self-Tests
- NetTutor
- e-Professors
- Videos

Study Skills Exercises

1. When solving an application involving geometry, draw an appropriate figure and label the known quantities with numbers and the unknown quantities with variables. This will help you solve the problem. After reading this section, what geometric figure do you think you will be drawing most often in this section?

2. Define the key terms.

a. **Vertices**

b. **Acute triangle**

c. **Right triangle**

d. **Obtuse triangle**

e. **Equilateral triangle**

f. **Isosceles triangle**

g. **Scalene triangle**

h. **Pythagorean theorem**

i. **Hypotenuse**

j. **Legs of a right triangle**

Review Exercises

3. Do $\angle ACB$ and $\angle BCA$ represent the same angle?

4. Is line segment $\overline{MN}$ the same as the line segment $\overline{NM}$?

5. Is ray $\overrightarrow{AB}$ the same as ray $\overrightarrow{BA}$?

6. Is the line $\overleftrightarrow{PQ}$ the same as the line $\overleftrightarrow{QP}$?

7. Is a right angle an obtuse angle?

8. Can two acute angles be supplementary?

Objective 1: Triangles

For Exercises 9–16, find the measures of angles a and b. **(See Example 1.)**

9.

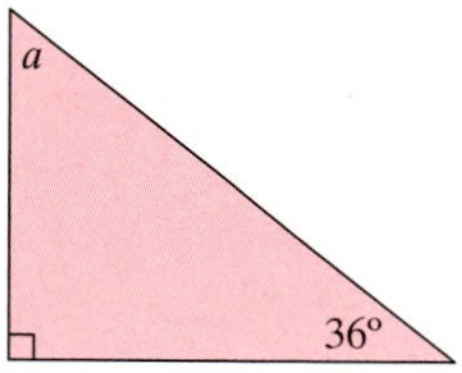

10.

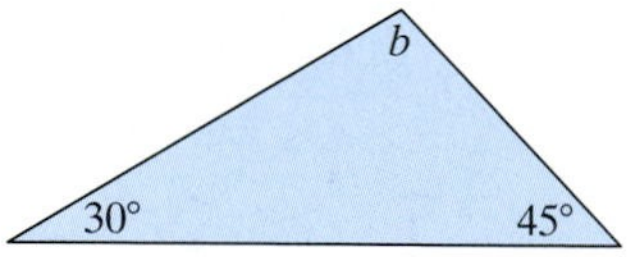

11.

12.

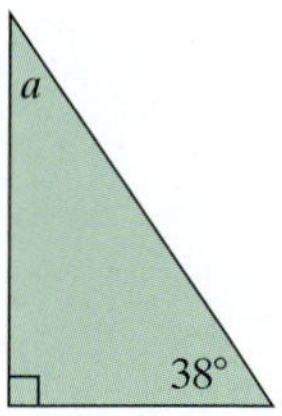

13.

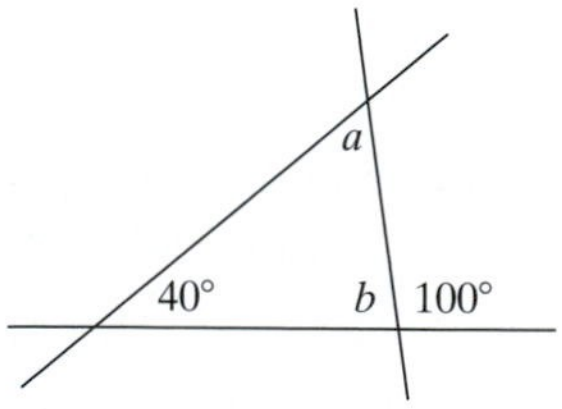

14.

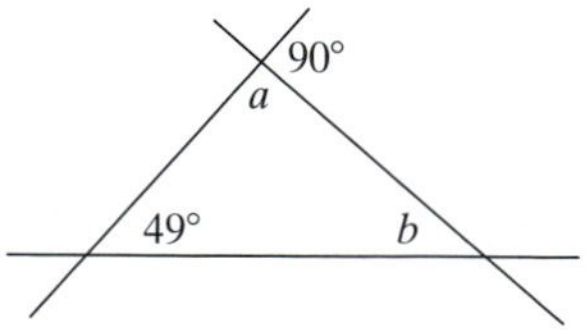

15.

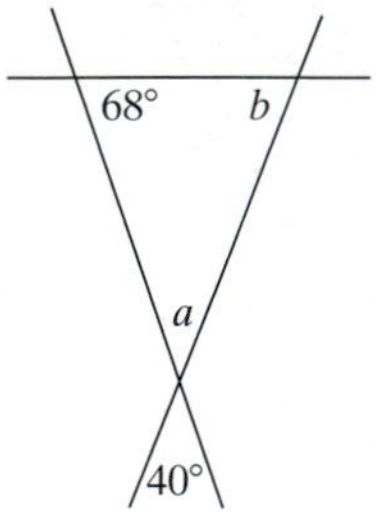

16.

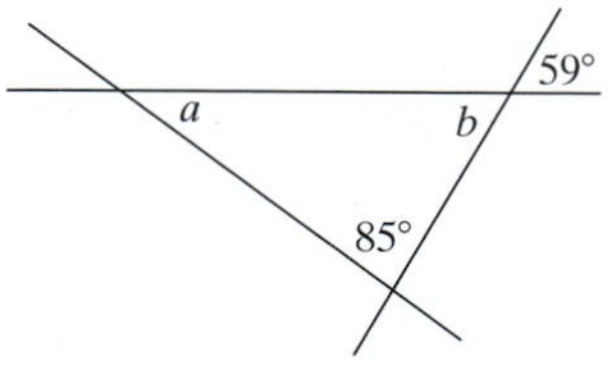

For Exercises 17–22, choose all figures that apply. The tick marks / denote segments of equal length, and small arcs) denote angles of equal measure.

17. Acute triangle

18. Obtuse triangle

19. Right triangle

20. Scalene triangle

21. Isosceles triangle

22. Equilateral triangle

a. 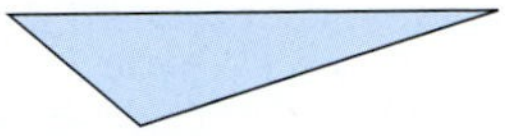

b. 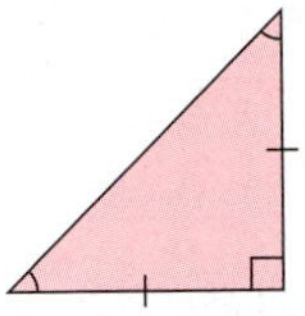

c. 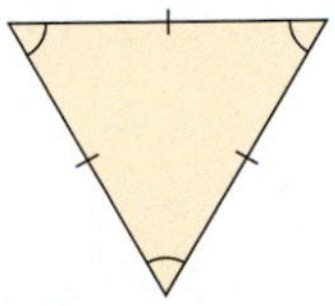

d. 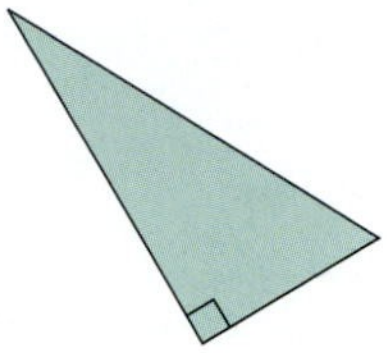

e. 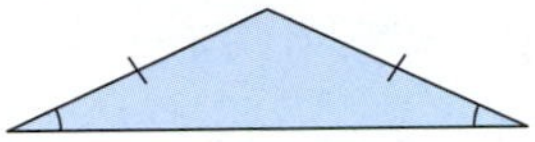

f. 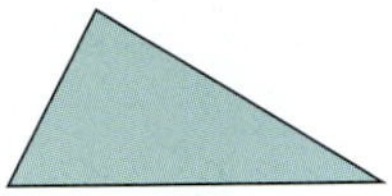

Objective 2: Square Roots

For Exercises 23–42, simplify the squares and square roots. **(See Example 2.)**

23. $\sqrt{49}$

24. $\sqrt{64}$

25. 7^2

26. 8^2

27. 4^2

28. 5^2

29. $\sqrt{16}$

30. $\sqrt{25}$

31. $\sqrt{36}$

32. $\sqrt{100}$

33. 6^2

34. 10^2

35. 9^2

36. 3^2

37. $\sqrt{81}$

38. $\sqrt{9}$

39. $\sqrt{\frac{1}{16}}$

40. $\sqrt{\frac{1}{144}}$

41. $\sqrt{0.04}$

42. $\sqrt{0.09}$

Objective 3: Pythagorean Theorem

For Exercises 43–46, find the length of the unknown side. **(See Examples 3–4.)**

43. 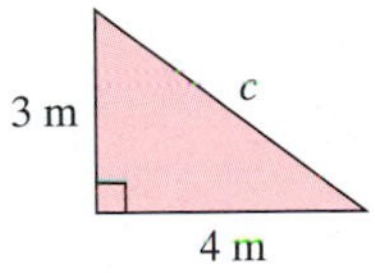

44. 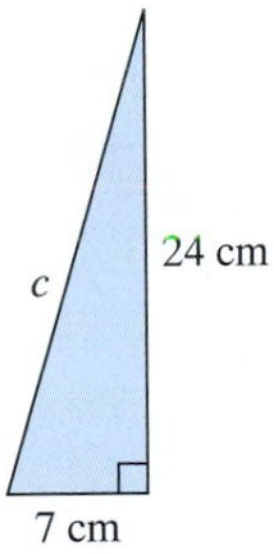

45. 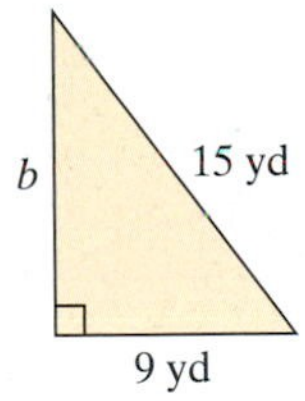

46. 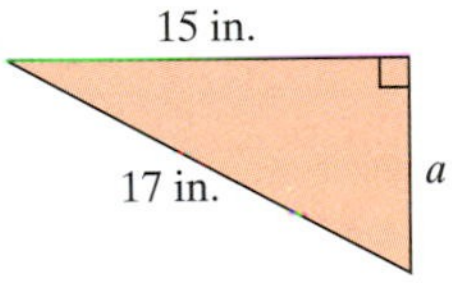

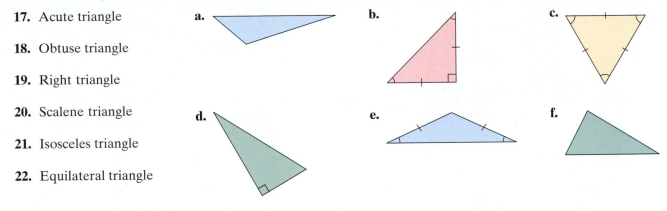

For Exercises 47–50, find the length of the unknown leg or hypotenuse.

47. Leg = 24 ft, hypotenuse = 26 ft

48. Leg = 9 km, hypotenuse = 41 km

49. Leg = 32 in., leg = 24 in.

50. Leg = 16 m, leg = 30 m

51. Find the length of the supporting brace. **(See Example 5.)**

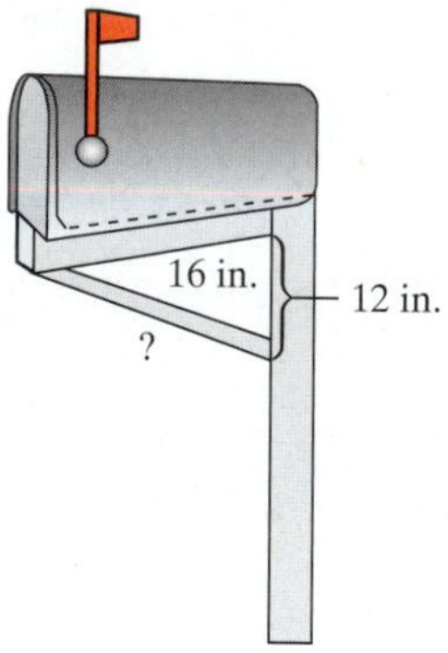

52. Find the length of the ramp.

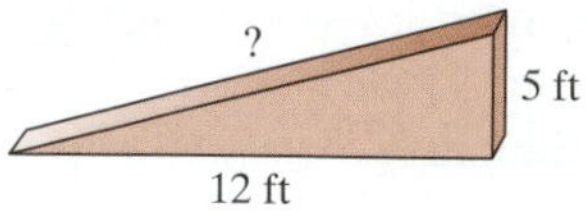

53. Find the height of the airplane above the ground.

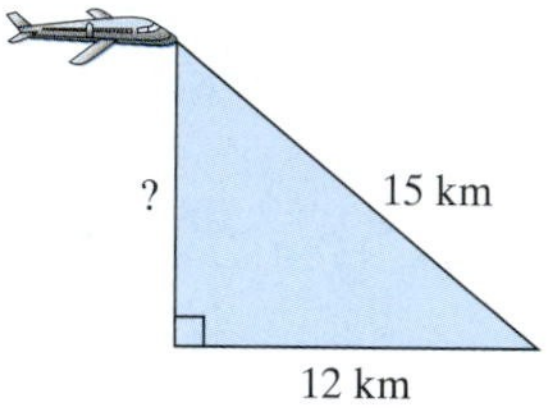

54. A 25-in. television measures 25 in. across the diagonal. If the width is 20 in., find the height.

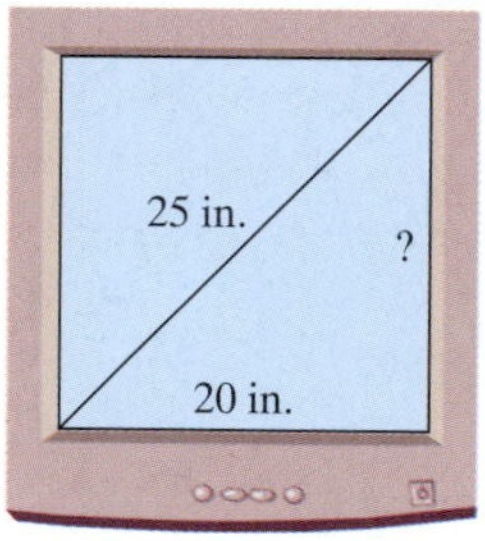

55. A car travels east 24 mi and then south 7 mi. How far is the car from its starting point?

56. A 26-ft-long wire is to be tied from a stake in the ground to the top of a 24-ft pole. How far from the bottom of the pole should the stake be placed?

Expanding Your Skills

For Exercises 57–60, find the perimeter.

57.

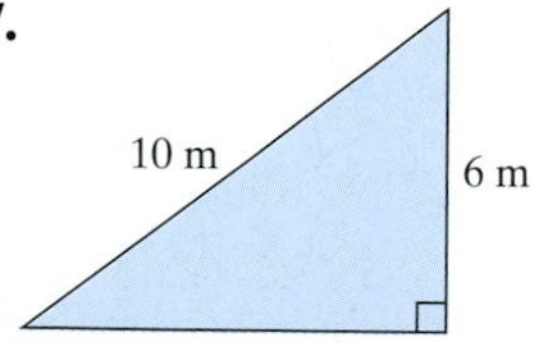

58.

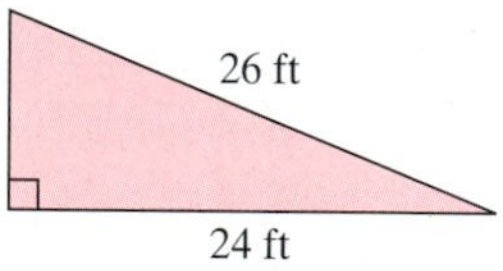

59.

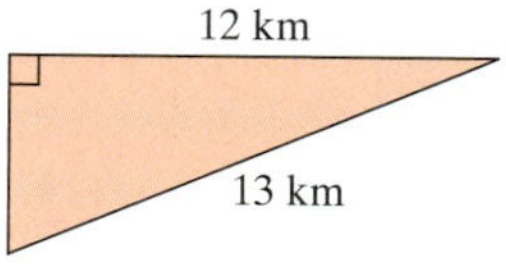

60.

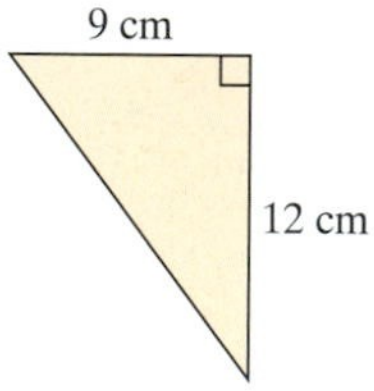

Calculator Connections

Topic: Entering Square Roots on a Calculator

In this section, we used the Pythagorean theorem to find the length of a side of a right triangle when the other two sides were given. In such problems, it is necessary to find the square root of a positive number. However, many square roots cannot be simplified to a whole number. For example, there is no whole number that when squared equals 26. However, we might speculate that $\sqrt{26}$ is a number slightly greater than 5 because $\sqrt{25} = 5$. A decimal approximation can be made by using a calculator.

$$\sqrt{26} \approx 5.099 \qquad \text{because } 5.099^2 = 25.999801 \approx 26$$

To enter a square root on a calculator, use the $\sqrt{\ }$ key. On some calculators, the $\sqrt{\ }$ function is associated with the x^2 key. In such a case, it is necessary to press 2nd or SHIFT first, followed by the x^2 key. Some calculators require the square root key to be entered first, before the number, while with others we enter the number first followed by $\sqrt{\ }$.

Expression	Keystrokes			Result
$\sqrt{26}$	26 [√]	or	[√] 26 [=]	5.099019514
$\sqrt{9325}$	9325 [√]	or	[√] 9325 [=]	96.56603958
$\sqrt{100}$	100 [√]	or	[√] 100 [=]	10

For Exercises 61–66, complete the table. For the estimate, find two consecutive whole numbers between which the square root lies. The first row is done for you.

	Square Root	Estimate	Calculator Approximation (Round to 3 Decimal Places)
	$\sqrt{50}$	is between 7 and 8	7.071
61.	$\sqrt{10}$	is between ____ and ____	
62.	$\sqrt{90}$	is between ____ and ____	
63.	$\sqrt{116}$	is between ____ and ____	
64.	$\sqrt{65}$	is between ____ and ____	
65.	$\sqrt{5}$	is between ____ and ____	
66.	$\sqrt{48}$	is between ____ and ____	

For Exercises 67–74, use a calculator to approximate the square root to three decimal places.

67. $\sqrt{427.75}$ **68.** $\sqrt{3184.75}$ **69.** $\sqrt{1{,}246{,}000}$ **70.** $\sqrt{50{,}416{,}000}$

71. $\sqrt{0.49}$ **72.** $\sqrt{0.25}$ **73.** $\sqrt{0.56}$ **74.** $\sqrt{0.82}$

Topic: Pythagorean Theorem

For Exercises 75–80, find the length of the unknown side. Round to three decimal places if necessary.

75.

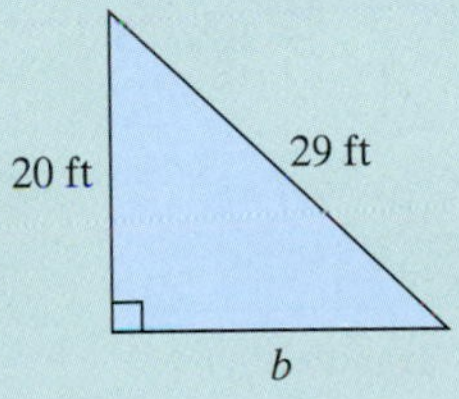

76.

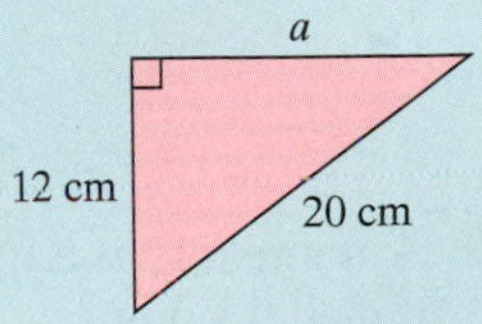

77. Leg = 5 mi, leg = 10 mi

78. Leg = 2 m, leg = 8 m

79. Leg = 12 in., hypotenuse = 22 in.

80. Leg = 15 ft, hypotenuse = 18 ft

81. A square tile is 1 ft on each side. What is the length of the diagonal? Round to the nearest hundredth of a foot.

82. A tennis court is 120 ft long and 60 ft wide. What is the length of the diagonal? Round to the nearest hundredth of a foot.

83. A contractor plans to construct a cement patio for one of the houses that he is building. The patio will be a square, 25 ft by 25 ft. After the contractor builds the frame for the cement, he checks to make sure that it is square by measuring the diagonals. Use the Pythagorean theorem to determine what the length of the diagonals should be if the contractor has constructed the frame correctly. Round to the nearest hundredth of a foot.

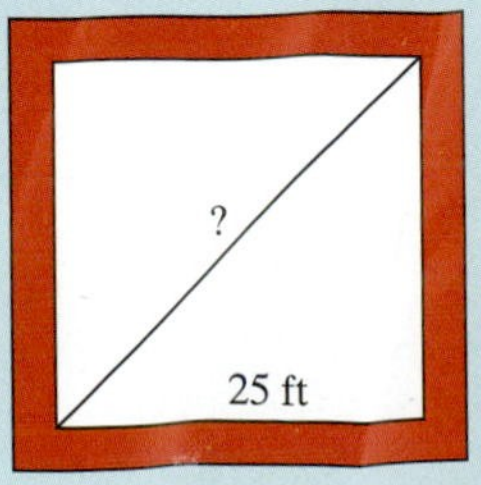

Section 8.7 Perimeter, Circumference, and Area

Objectives

1. Quadrilaterals
2. Perimeter and Circumference
3. Area

1. Quadrilaterals

At this point, you are familiar with several geometric figures. We have calculated perimeter and area of squares, rectangles, triangles and circles. In this section, we revisit these concepts and also define some additional geometric figures.

Recall that a **polygon** is a flat figure formed by line segments connected at their ends. A four-sided polygon is called a **quadrilateral**. Some quadrilaterals fall in the following categories.

A **parallelogram** is a quadrilateral with opposite sides parallel. It follows that opposite sides must be equal in length.

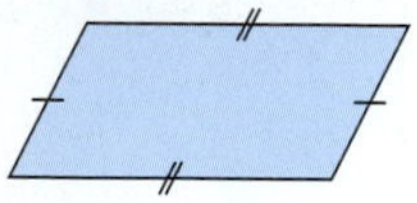

Parallelogram

A **rectangle** is a parallelogram with four right angles.

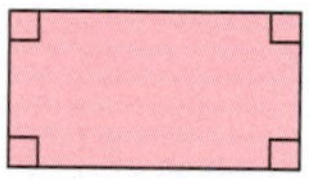

Rectangle

A **square** is a rectangle with sides of equal length.

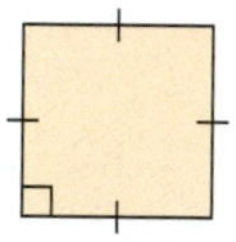

Square

A **rhombus** is a parallelogram with sides of equal length. The angles are not necessarily equal.

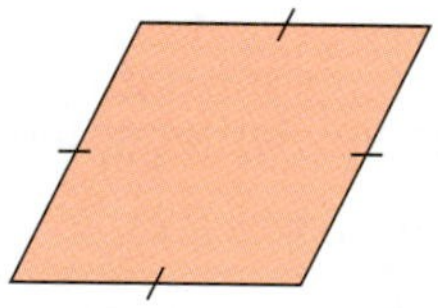

Rhombus

A **trapezoid** is a quadrilateral with one pair of parallel sides.

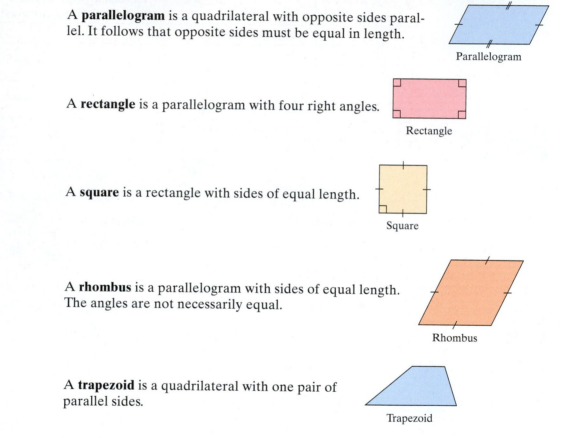

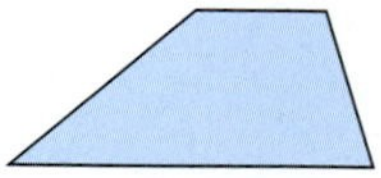

Trapezoid

Notice that some figures belong to more than one category. For example, a square is also a rectangle and a parallelogram.

2. Perimeter and Circumference

Recall that the **perimeter** of a polygon is the distance around the figure. For example, we use perimeter to find the amount of fencing needed to enclose a yard. The perimeter of a polygon is found by adding the lengths of the sides. Also recall that the "perimeter" of a circle is called the **circumference**.

We summarize some of the formulas presented earlier in the text.

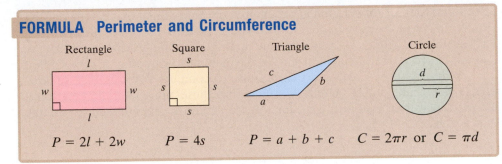

FORMULA Perimeter and Circumference

Rectangle	Square	Triangle	Circle
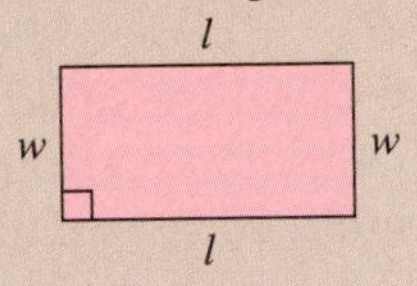	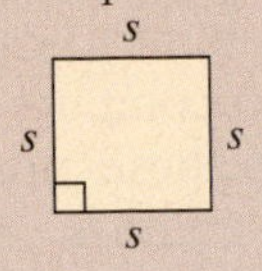	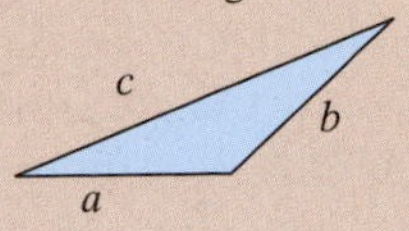	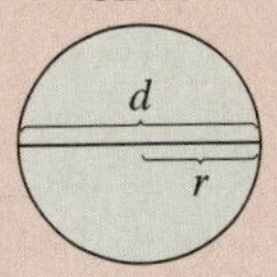
$P = 2l + 2w$	$P = 4s$	$P = a + b + c$	$C = 2\pi r$ or $C = \pi d$

TIP: The approximate circumference of a circle can be calculated by using either 3.14 or $\frac{22}{7}$ for the value of π. To express the exact circumference, the answer must be written in terms of π.

Example 1 Finding Perimeter and Circumference

Use an appropriate formula to find the perimeter or circumference. Use 3.14 for π.

a.
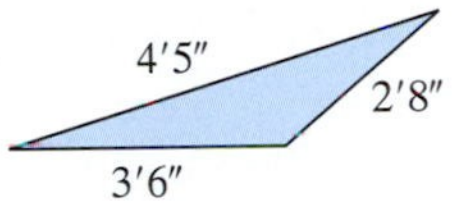

b.
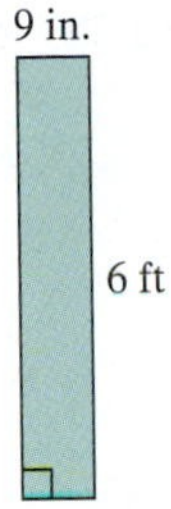

c.
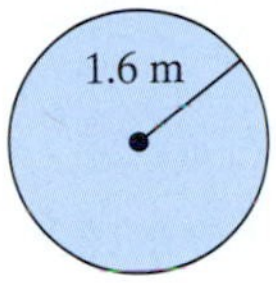

Solution:

a. To find the perimeter, add the lengths of the sides.

$P = a + b + c$

$P = 4'5'' + 2'8'' + 3'6''$ or

$$\begin{array}{rl} P = & 4\text{ ft } 5\text{ in.} \\ & 2\text{ ft } 8\text{ in.} \\ + & 3\text{ ft } 6\text{ in.} \\ \hline & 9\text{ ft } 19\text{ in.} = 9\text{ ft} + (1\text{ ft} + 7\text{ in.}) \\ & = 10\text{ ft } 7\text{ in.} \quad \text{or} \quad 10'7'' \end{array}$$

b. First note that to add the lengths of the sides, we must have like units.

$$9\text{ in.} = \frac{9\text{ in.}}{1} \cdot \frac{1\text{ ft}}{12\text{ in.}} = \frac{9}{12}\text{ ft} = \frac{3}{4}\text{ ft} \text{ or } 0.75\text{ ft}$$

$P = 2l + 2w$	The figure is a rectangle. Use $P = 2l + 2w$.
$= 2(6\text{ ft}) + 2(0.75\text{ ft})$	Substitute $l = 6$ ft and $w = 0.75$ ft.
$= 12\text{ ft} + 1.5\text{ ft}$	Simplify.
$= 13.5\text{ ft}$	

c. From the figure, $r = 1.6$ m.

$C = 2\pi(1.6\text{ m})$	Use the formula, $C = 2\pi r$.
$= 3.2\pi\text{ m}$	This is the exact value of the circumference.
$\approx 3.2(3.14)\text{ m}$	Substitute 3.14 for π.
$= 10.048\text{ m}$	The circumference is approximately 10.048 m.

Skill Practice

1. Find the perimeter.

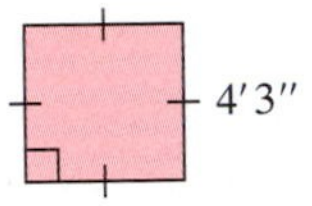

2. Find the perimeter in feet.

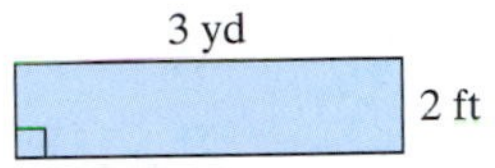

3. Find the circumference. Use $\frac{22}{7}$ for π.

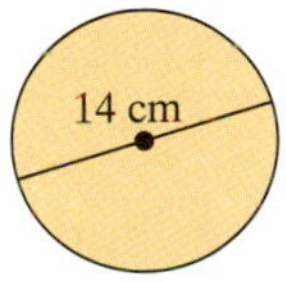

Animation

Answers

1. 17′ 2. 22 ft 3. 44 cm

3. Area

Concept Connections

4. How many square centimeters are enclosed in the figure?

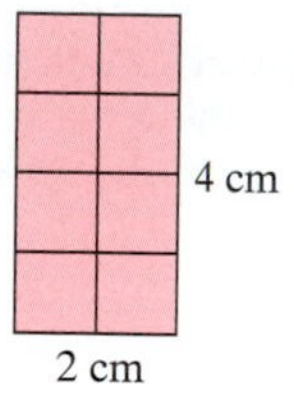

Recall that the **area** of a region is the number of square units that can be enclosed within the region. For example, the rectangle shown in Figure 8-29 encloses 6 square inches (in.^2). We would compute area in such applications as finding the amount of sod needed to cover a yard or the amount of carpeting to cover a floor.

We have already presented the formulas to compute the area of a square, a rectangle, a triangle, and a circle. We summarize these along with the area formulas for a parallelogram and trapezoid. These should be memorized as common knowledge.

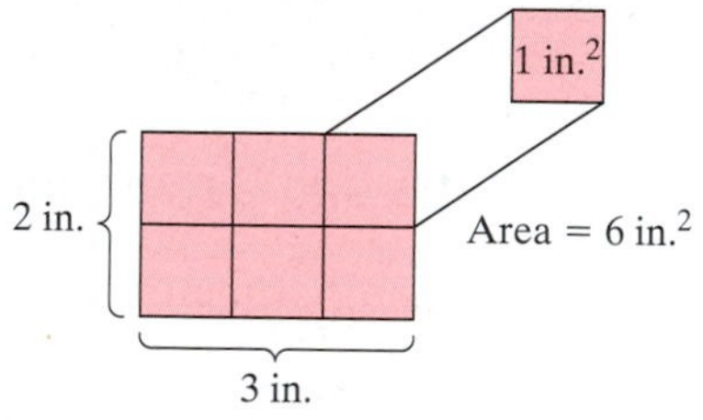

Figure 8-29

Animation

FORMULA Area Formulas

Rectangle

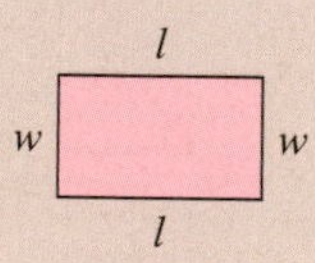

$A = lw$

Square

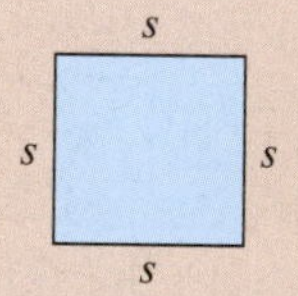

$A = s^2$

Parallelogram

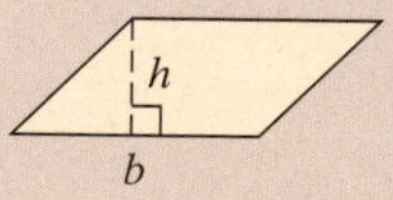

$A = bh$

Triangle

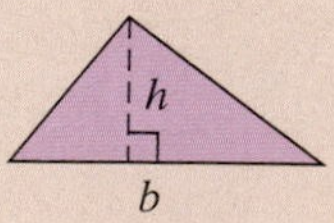

$A = \frac{1}{2}bh$

Trapezoid

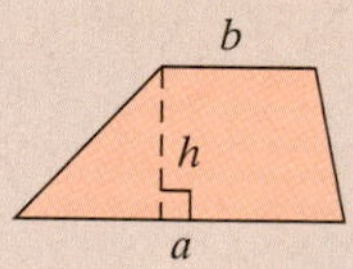

$A = \frac{1}{2}(a + b)h$

Circle

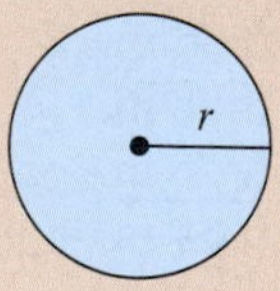

$A = \pi r^2$

TIP: Because the height of a parallelogram is perpendicular to the base, sometimes it must be drawn *outside* the parallelogram.

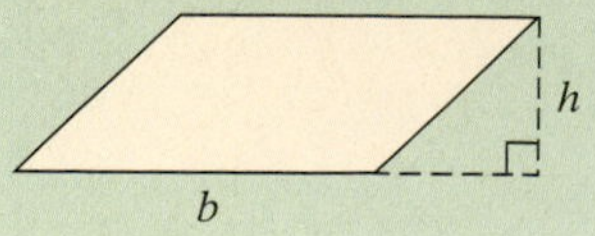

Example 2 Finding Area

a. Determine the area of the field.

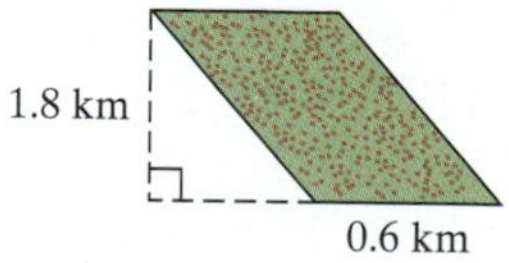

b. Determine the area of the matting.

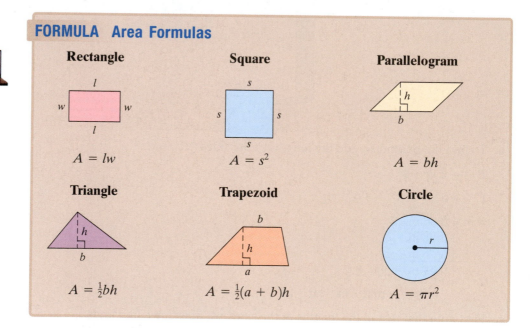

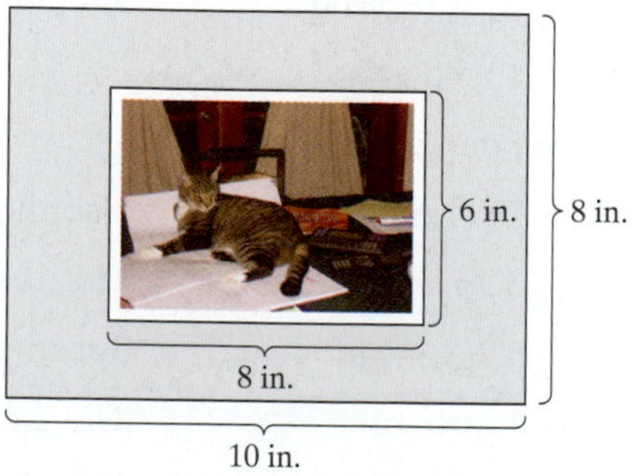

Answer

4. 8 cm²

Solution:

a. The field is in the shape of a parallelogram. The base is 0.6 km and the height is 1.8 km.

$A = bh$ — Area formula for a parallelogram.

$= (0.6 \text{ km})(1.8 \text{ km})$ — Substitute $b = 0.6$ km and $h = 1.8$ km.

$= 1.08 \text{ km}^2$

The field is 1.08 km^2.

TIP: When two common units are multiplied, such as km · km, the resulting units are square units, such as km^2.

b. To find the area of the matting only, we can subtract the inner 6-in. by 8-in. area from the outer 8-in. by 10-in. area. In each case, apply the formula, $A = lw$

$$\text{Area of matting} = \underbrace{(10 \text{ in.})(8 \text{ in.})}_{\text{outer area}} - \underbrace{(8 \text{ in.})(6 \text{ in.})}_{\text{inner area}}$$

$$= 80 \text{ in.}^2 - 48 \text{ in.}^2$$

$$= 32 \text{ in.}^2$$

The matting is 32 in.^2

Skill Practice

5. Determine the area.

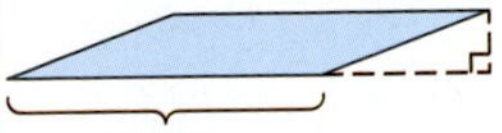

6. A side of a house must be painted. Exclude the windows to find the area that must be painted.

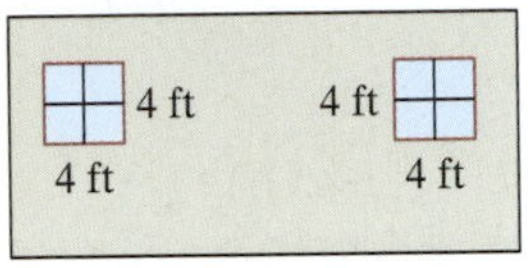

Example 3 Finding Area

Determine the area of each region.

a.

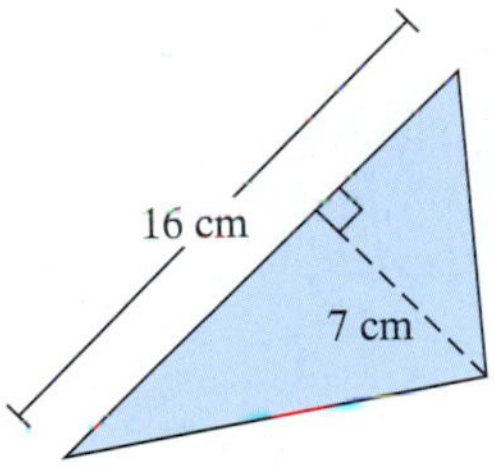

b.

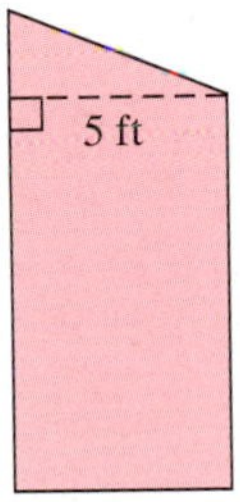

Solution:

a. $A = \frac{1}{2}bh$ — Apply the formula for the area of a triangle.

$= \frac{1}{2}(16 \text{ cm})(7 \text{ cm})$ — Substitute $b = 16$ cm and $h = 7$ cm.

$= \frac{1}{2}\left(\frac{16}{1} \text{ cm}\right)\left(\frac{7}{1} \text{ cm}\right)$ — Multiply fractions.

$= 56 \text{ cm}^2$

Skill Practice

Determine the area.

7.

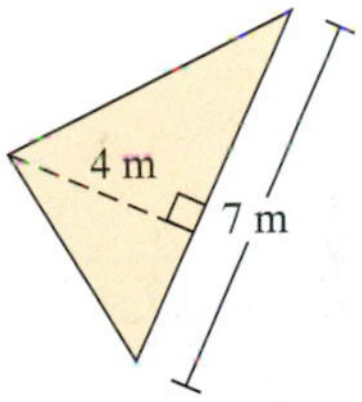

8.

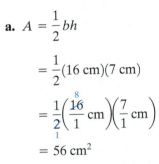

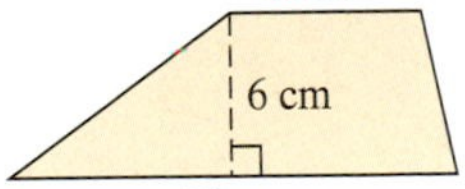

Answers

5. 32.24 in.^2 **6.** 448 ft^2
7. 14 m^2 **8.** 78 cm^2

b. $A = \frac{1}{2}(a + b)h$ Apply the formula for the area of a trapezoid.

In this case, the two parallel sides are the left-hand side and the right-hand side. Therefore, these sides are the two bases, a and b.

The "height" is the distance between the two parallel sides.

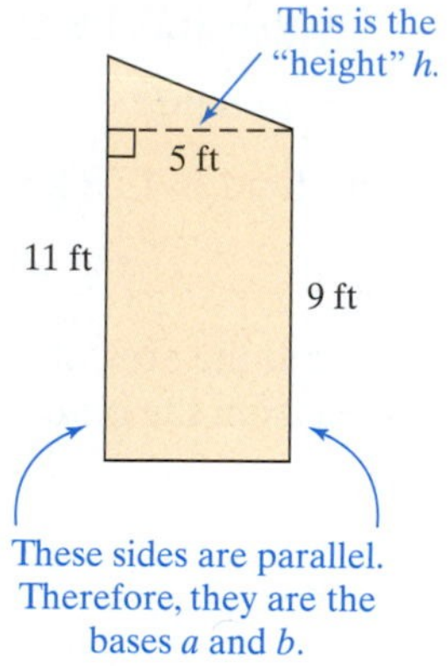

$A = \frac{1}{2}(11\text{ ft} + 9\text{ ft})(5\text{ ft})$ Substitute $a = 11$ ft, $b = 9$ ft, and $h = 5$ ft.

$= \frac{1}{2}(20\text{ ft})(5\text{ ft})$ Simplify within parentheses first.

$= \frac{1}{\cancel{2}_1}\left(\frac{\cancel{20}^{10}}{1}\text{ ft}\right)\left(\frac{5}{1}\text{ ft}\right)$ Multiply fractions.

$= 50\text{ ft}^2$

The formula for the area of a circle was given in Section 5.3. Here we show the basis for the formula.The circumference of a circle is given by $C = 2\pi r$. The length of a *semicircle* (one-half of a circle) is one-half of this amount: $\frac{1}{2}2\pi r = \pi r$. To visualize the formula for the area of a circle, consider the bottom half and top half of a circle cut into pie-shaped wedges. Unfold the figure as shown (Figure 8-30).

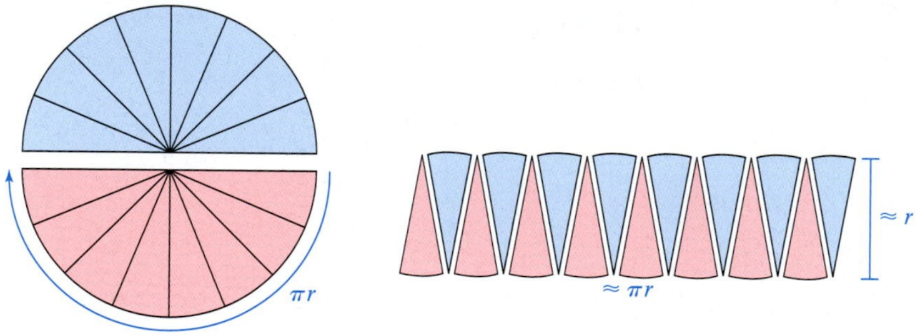

Figure 8-30

The resulting figure is nearly a parallelogram, with base approximately equal to πr and height approximately equal to the radius of the circle. The area is (base) · (height) $\approx (\pi r) \cdot r = \pi r^2$. This is the area formula for a circle.

Skill Practice

9. Determine the area of a 9-in. diameter plate. Use 3.14 for π and round to the nearest tenth.

Example 4 Computing the Area of a Circle

Determine the area of the gold medal.

Use $\frac{22}{7}$ for π.

Answer

9. 63.6 in.^2

Solution:

First note that the radius is found by $r = \frac{1}{2}d$. Therefore, $r = \frac{1}{2}(7 \text{ cm}) = \frac{7}{2}$ cm.

$A = \pi\left(\frac{7}{2}\text{ cm}\right)^2$	Substitute $r = \frac{7}{2}$ cm into the formula $A = \pi r^2$.
$= \pi\left(\frac{49}{4}\text{ cm}^2\right)$	Simplify the factor with the exponent.
$= \frac{49}{4}\pi\text{ cm}^2$	This is the exact area, which is written in terms of π.
$\approx \left(\frac{49}{4}\text{ cm}^2\right)\left(\frac{22}{7}\right)$	Substitute $\frac{22}{7}$ for π.
$= \frac{77}{2}\text{ cm}^2$	The medal has an area of approximately $\frac{77}{2}$ cm^2 or $38\frac{1}{2}$ cm^2.

TIP: The approximate area of a circle can be calculated using either 3.14 or $\frac{22}{7}$ for π. In Example 4 we could have used $\pi = 3.14$ and $r = \frac{7}{2} = 3.5$ cm.

$$A = \pi r^2$$
$$A \approx (3.14)(3.5\text{ cm})^2$$
$$A = 38.465\text{ cm}^2$$

Example 5 Finding Area for a Landscaping Application

Sod can be purchased in palettes for \$225. If a palette contains 240 ft^2 of sod, how much will it cost to cover the area in Figure 8-31?

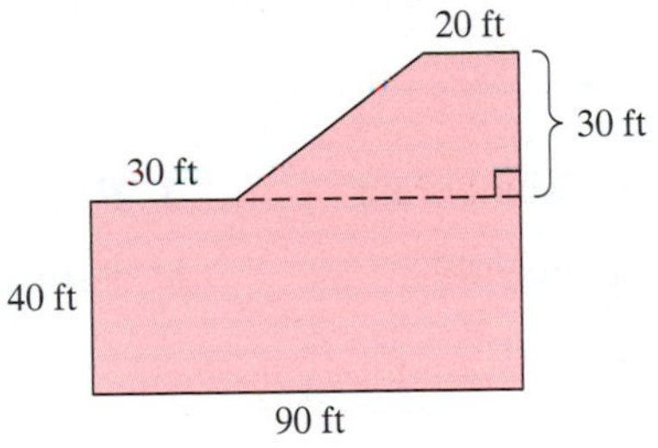

Figure 8-31

Solution:

To find the total cost, we need to know the total number of square feet. Then we can determine how many 240-ft^2 palettes are required.

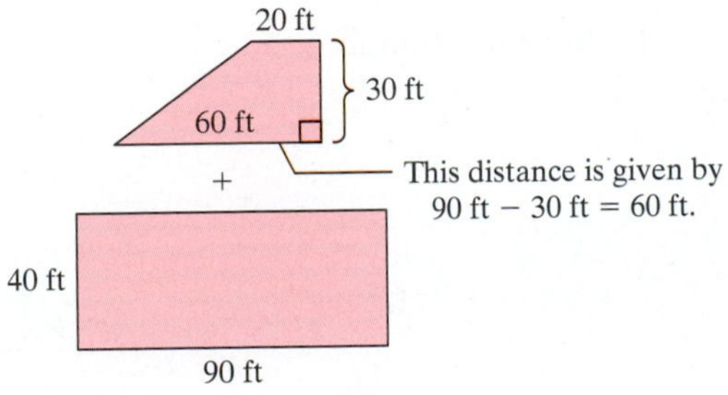

Skill Practice

10. A homeowner wants to apply water sealant to a pier that extends from the backyard to a lake. One gallon of sealant covers 160 ft^2 and sells for \$11.95. How much will it cost to cover the pier?

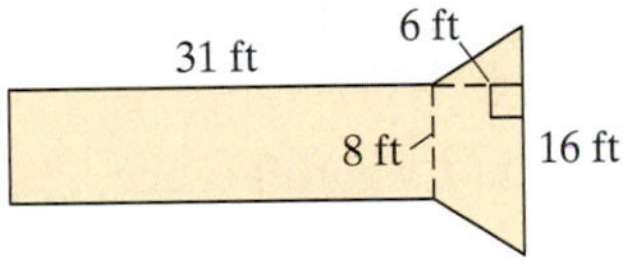

Answer

10. \$23.90

The total area is given by

$$A = \underset{\text{area of trapezoid}}{\tfrac{1}{2}(a + b)h} + \underset{\text{area of rectangle}}{lw}$$

$$= \tfrac{1}{2}(60 \text{ ft} + 20 \text{ ft})(30 \text{ ft}) + (90 \text{ ft})(40 \text{ ft})$$

$$= \tfrac{1}{2}(80 \text{ ft})(30 \text{ ft}) + 3600 \text{ ft}^2$$

$$= 1200 \text{ ft}^2 + 3600 \text{ ft}^2$$

$$= 4800 \text{ ft}^2$$

The total area is 4800 ft^2.

To determine how many 240-ft^2 palettes of sod are required, divide the total area by 240 ft^2.

Number of palettes: $4800 \text{ ft}^2 \div 240 \text{ ft}^2 = 20$

The total cost for 20 palettes is ($225 per palette) × 20 palettes = $4500

The cost for the sod is $4500.

Section 8.7 Practice Exercises

Boost *your* GRADE at ALEKS.com!

- Practice Problems
- Self-Tests
- NetTutor
- e-Professors
- Videos

Study Skills Exercises

1. It may help to remember formulas if you understand how they were derived. For example, the perimeter of a square has the formula $P = 4s$. It was derived from the fact that perimeter measures the distance around a figure.
Explain how the formula for the perimeter of a rectangle ($P = 2l + 2w$) was derived.

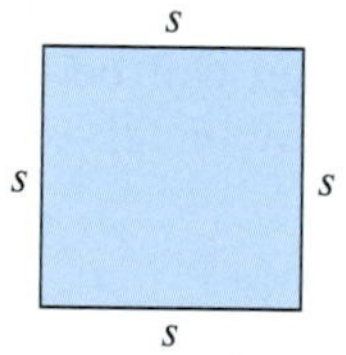

$P = s + s + s + s$
$= 4s$

2. Define the key terms.

a. Polygon **b. Quadrilateral** **c. Parallelogram** **d. Rectangle**
e. Square **f. Rhombus** **g. Trapezoid** **h. Perimeter**
i. Circumference **j. Area**

Review Exercises

For Exercises 3–8, state the characteristics of each triangle.

3. Isosceles triangle

4. Right triangle

5. Acute triangle

6. Equilateral triangle

7. Obtuse triangle

8. Scalene triangle

Objective 1: Quadrilaterals

9. Write the definition of a quadrilateral.

For Exercises 10–14, state the characteristics of each quadrilateral.

10. Square

11. Trapezoid

12. Parallelogram

13. Rectangle

14. Rhombus

Objective 2: Perimeter and Circumference

For Exercises 15–24, determine the perimeter or circumference. Use 3.14 for π. **(See Example 1.)**

15. Rectangle

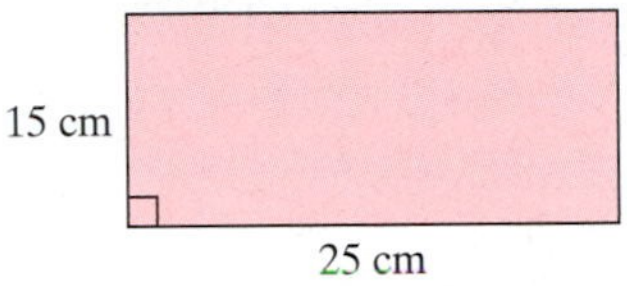

16. Square

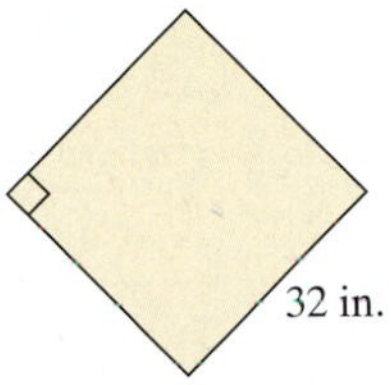

17. Square

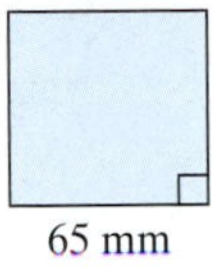

18. Rectangle

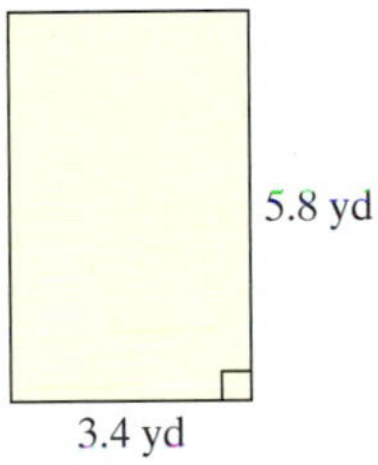

19. Trapezoid

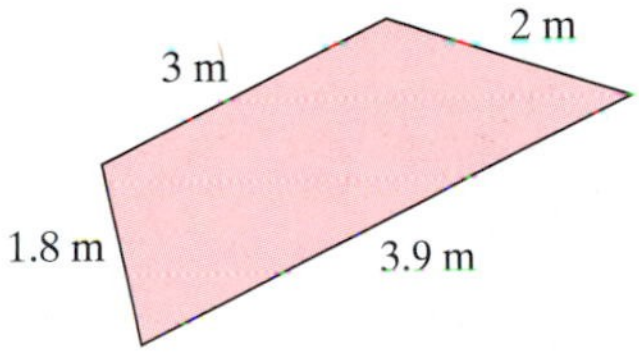

20. Trapezoid

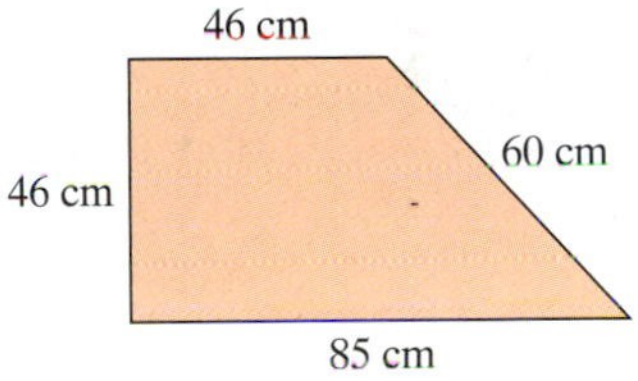

21.

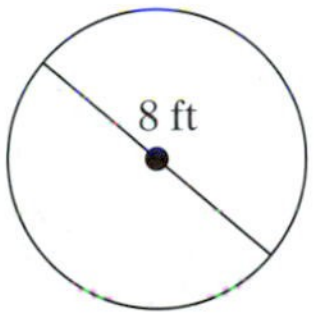

22.

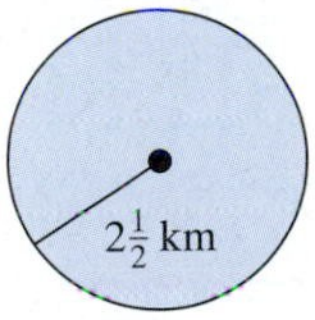

23.

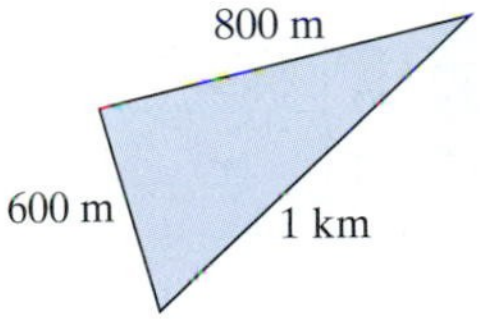

24.

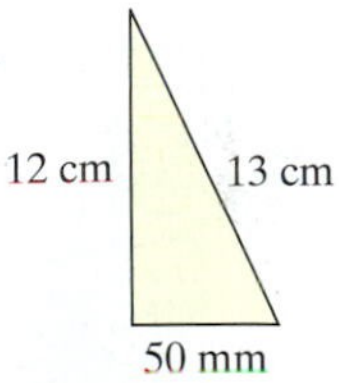

25. Find the perimeter of a triangle with sides 3 ft 8 in., 2 ft 10 in., and 4 ft.

26. Find the perimeter of a triangle with sides 4 ft 2 in., 3 ft, and 2 ft 9 in.

27. Find the perimeter of a rectangle with length 2 ft and width 6 in.

28. Find the perimeter of a rectangle with length 4 m and width 85 cm.

29. Find the lengths of the two missing sides labeled x and y. Then find the perimeter of the figure.

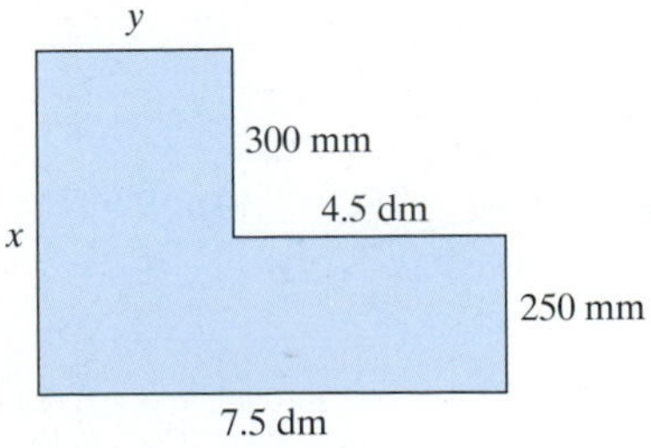

30. Find the lengths of the two missing sides labeled a and b. Then find the perimeter of the figure.

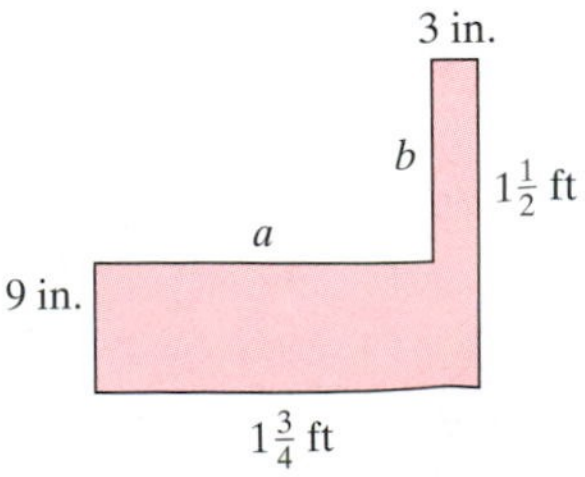

31. Rain gutters are going to be installed around the perimeter of a house. What is the total length needed?

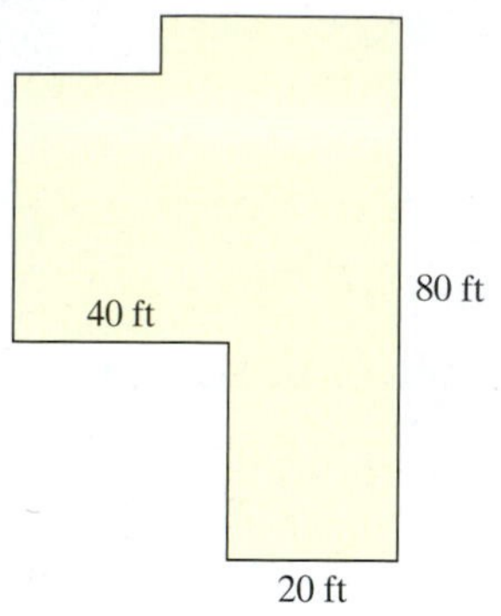

32. Wood molding needs to be installed around the perimeter of a living room floor. With no molding needed in the doorway, how much molding is needed?

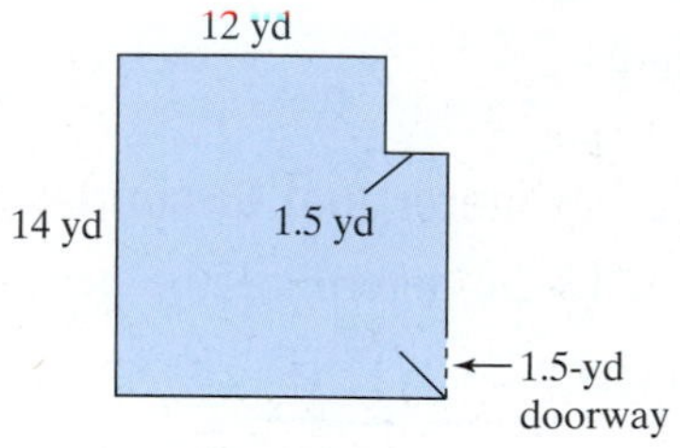

Objective 3: Area

For Exercises 33–44, determine the area of the shaded region. **(See Examples 2–3.)**

33.

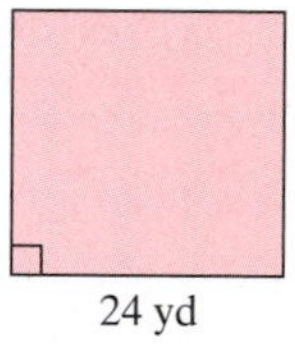

34.

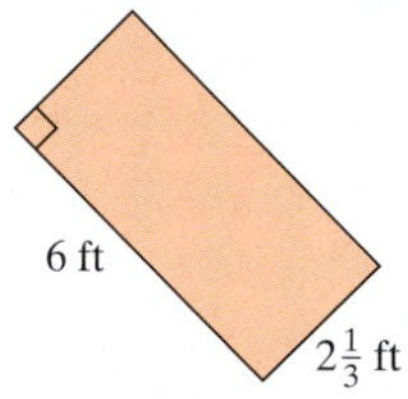

35.

36.

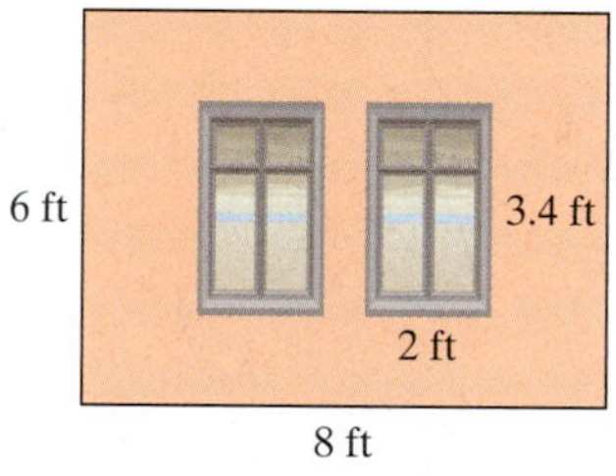

37.

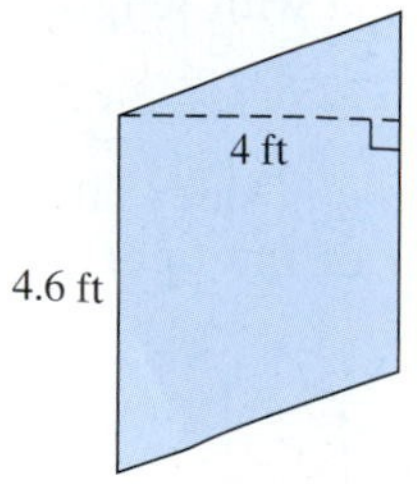

38.

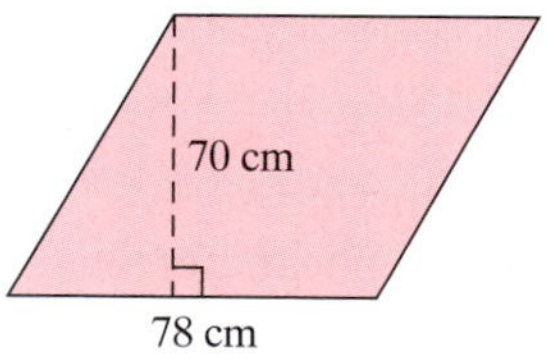

39.

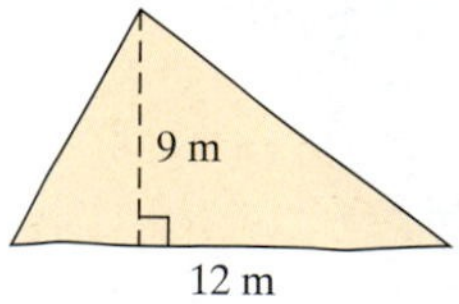

40.

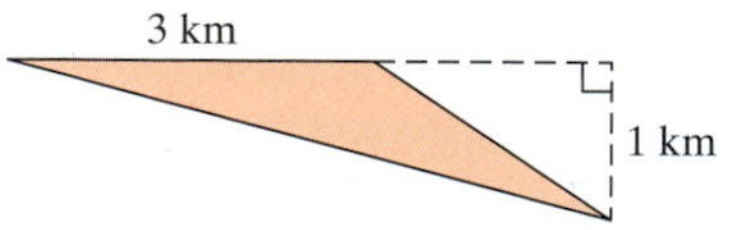

41.

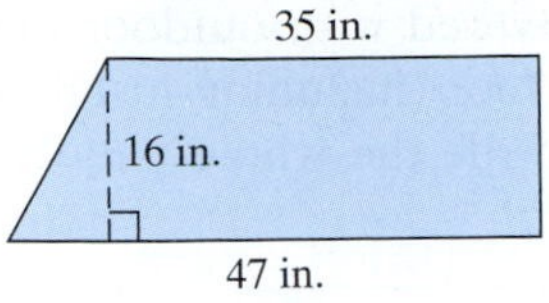

42.

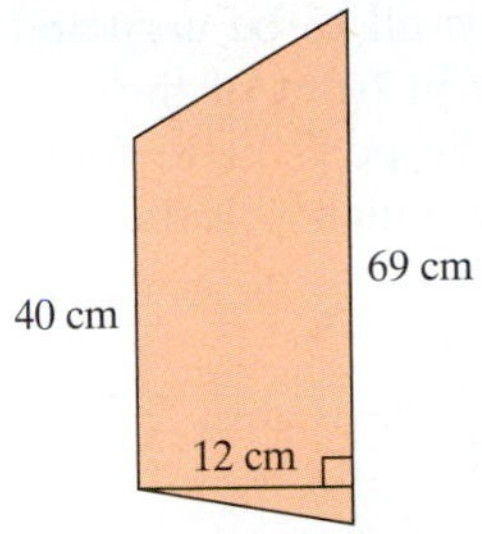

43.

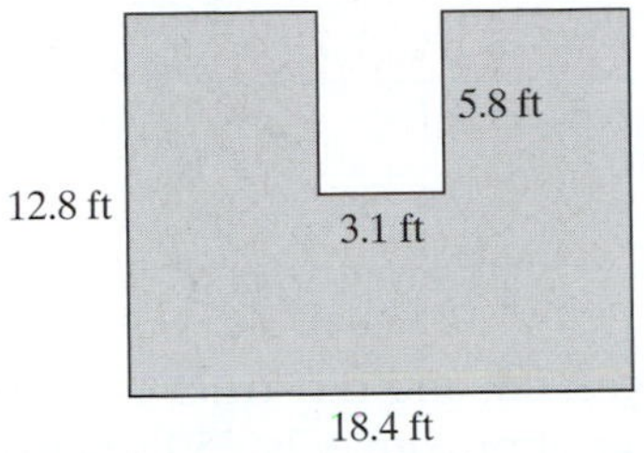

44.

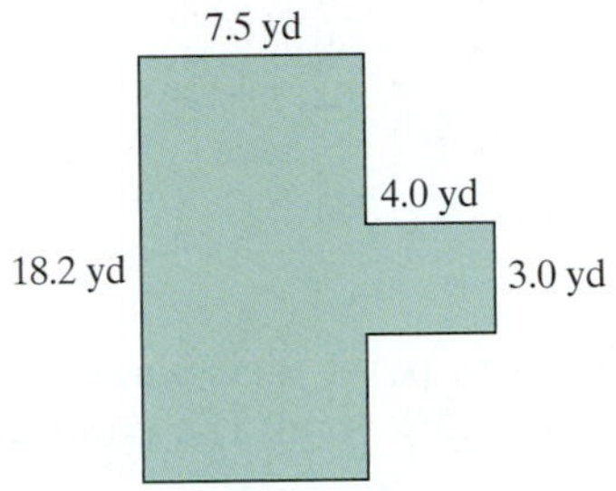

For Exercises 45–48, determine the area of the circle, using $\frac{22}{7}$ for π. **(See Example 4.)**

45.

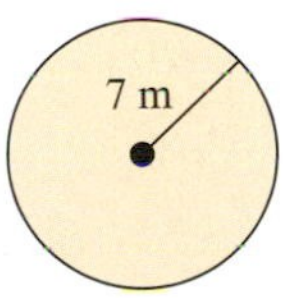

46.

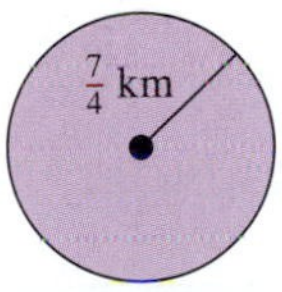

47.

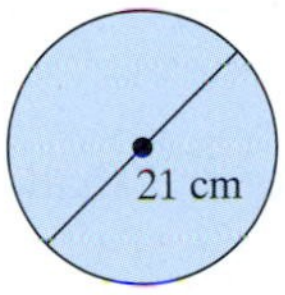

48.

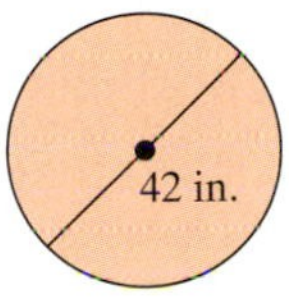

For Exercises 49–52, determine the area of the circle, using 3.14 for π. Round to the nearest whole unit.

49.

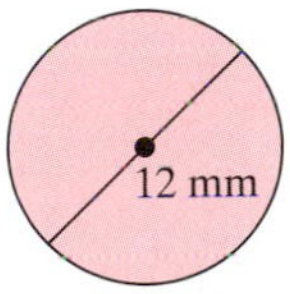

50.

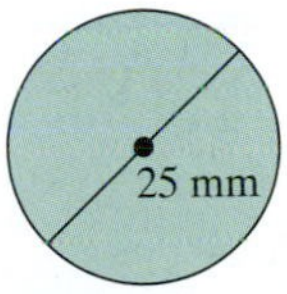

51.

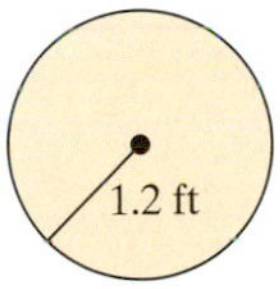

52.

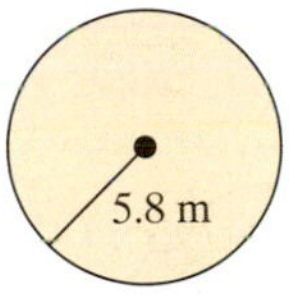

53. Determine the area of the sign.

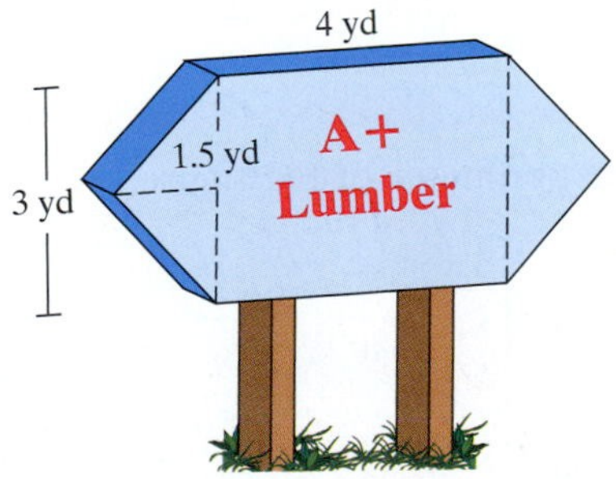

54. Determine the area of the kite.

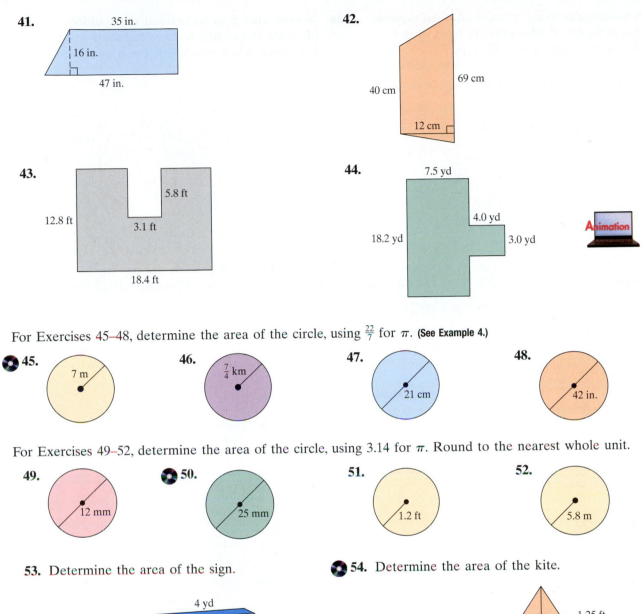

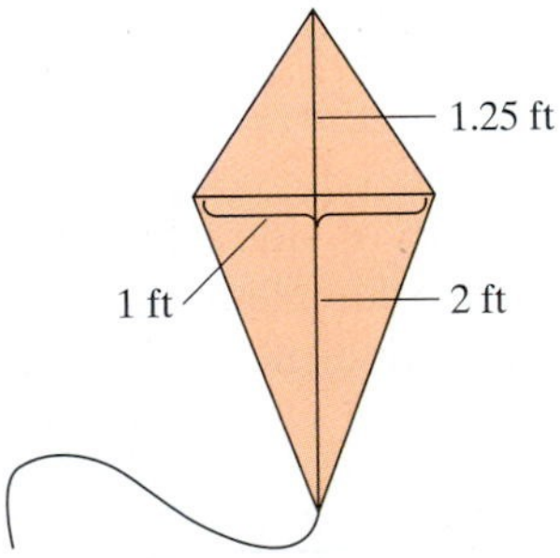

55. A rectangular living room is all to be carpeted except for the tiled portion in front of the fireplace. If carpeting is $2.50 per square foot (including installation), how much will the carpeting cost? **(See Example 5.)**

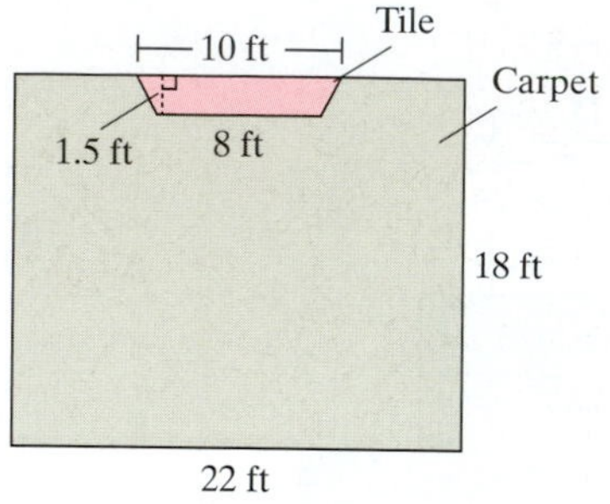

56. A patio area is to be covered with outdoor tile. If tile costs $8 per square foot (including installation), how much will it cost to tile the whole patio?

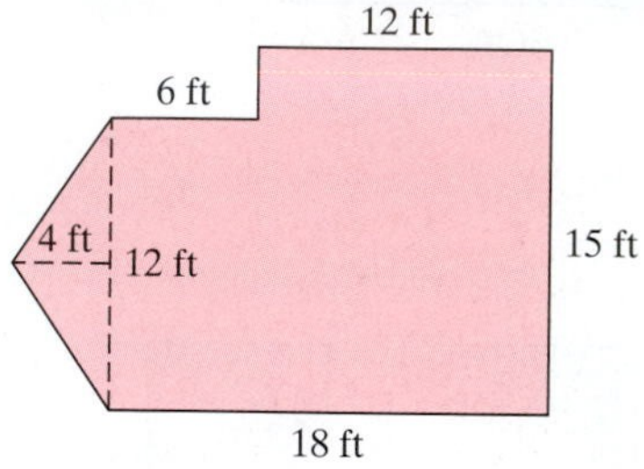

57. The Baker family plans to paint their garage floor with paint that resists gas, oil, and dirt from tires. The garage is 21 ft wide and 23 ft long. The paint kit they plan to use will cover approximately 250 square feet.

a. What is the area of the garage floor?

b. How many kits will be needed to paint the entire garage floor?

Expanding Your Skills

For Exercises 58–61, determine the area of the shaded region. Use 3.14 for π.

58.

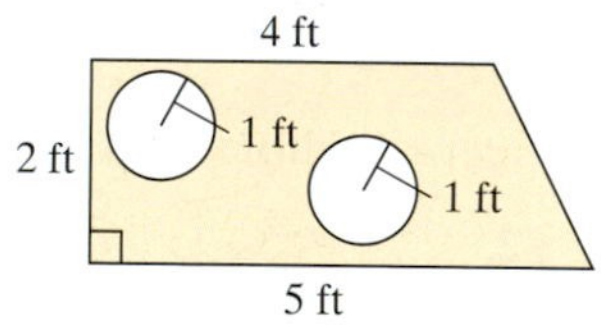

59.

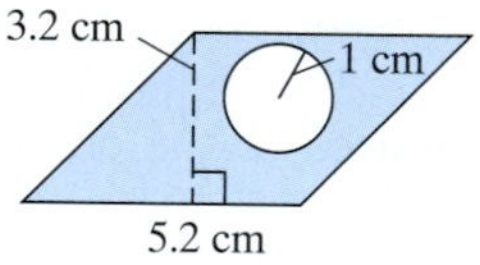

60.

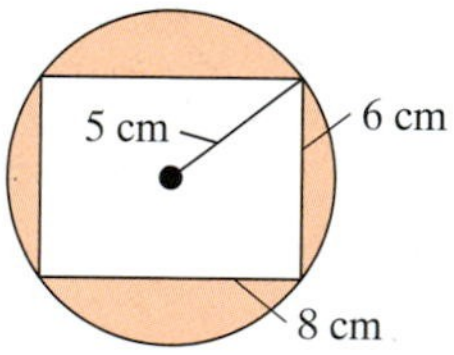

61.

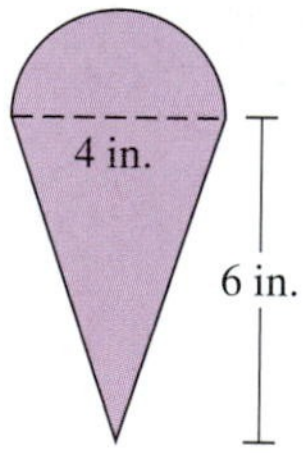

62. Find the length of side c by dividing this figure into a rectangle and a right triangle. Then find the perimeter of the figure.

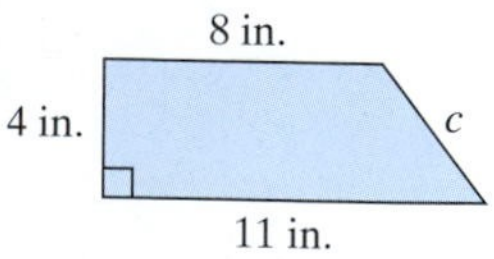

63. Find the perimeter of the figure.

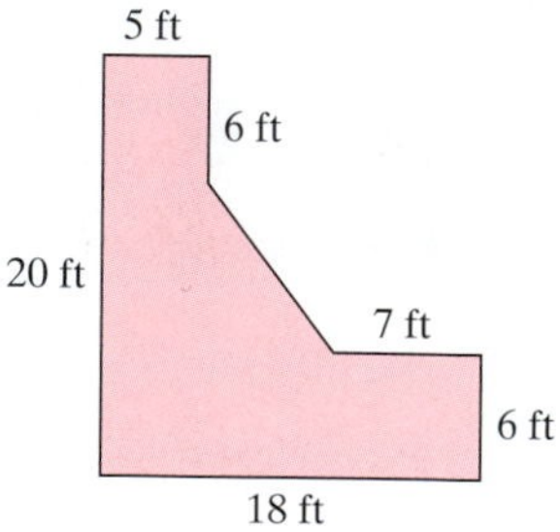

Problem Recognition Exercises

Area, Perimeter, and Circumference

For Exercises 1–14, determine the area and the perimeter or circumference for each figure.

1.

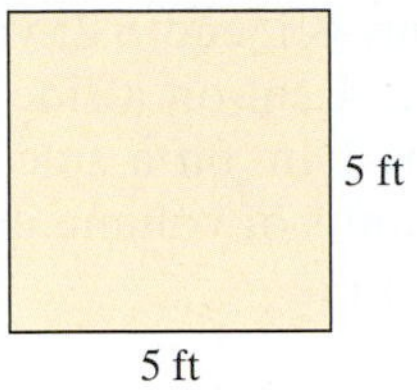

2.

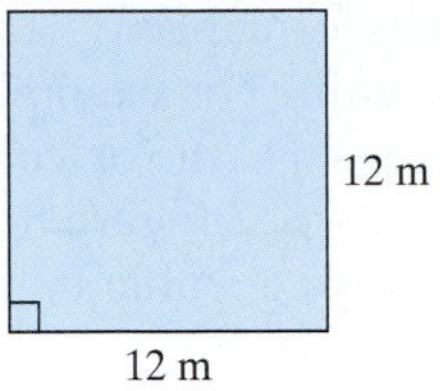

3.

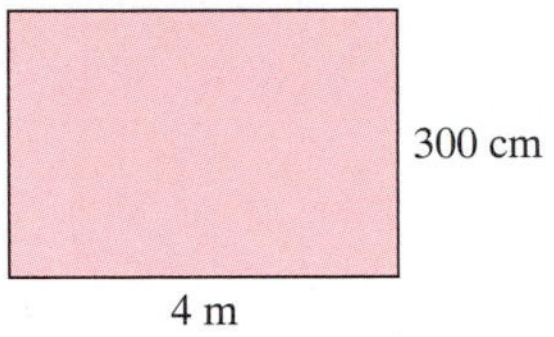

4.

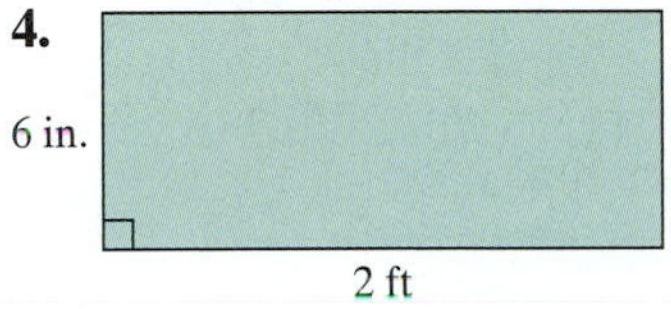

5.

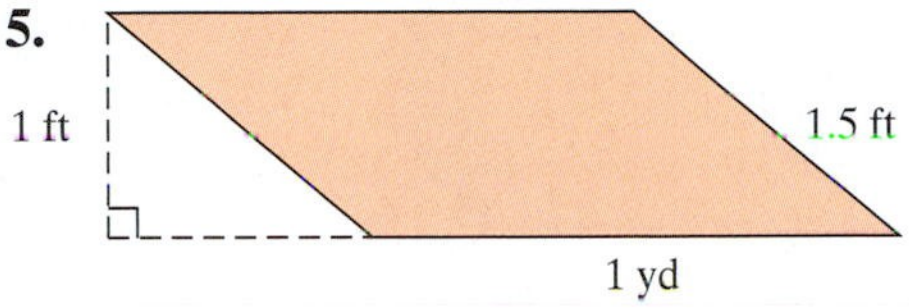

6.

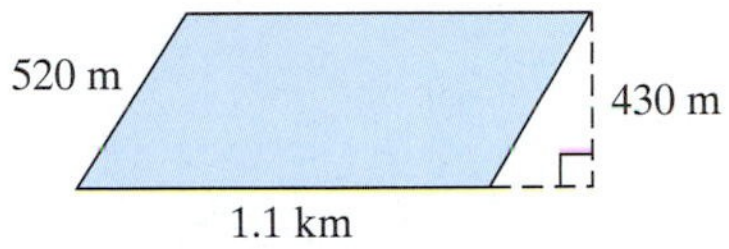

7.

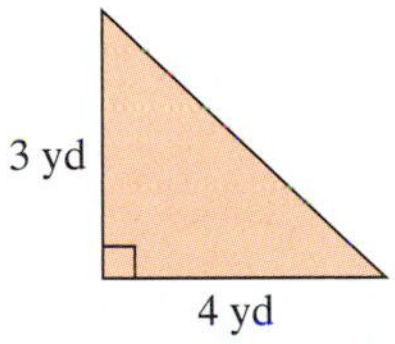

8.

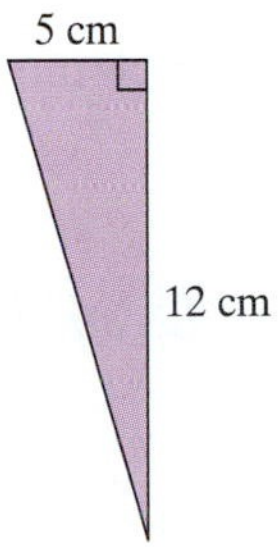

9.

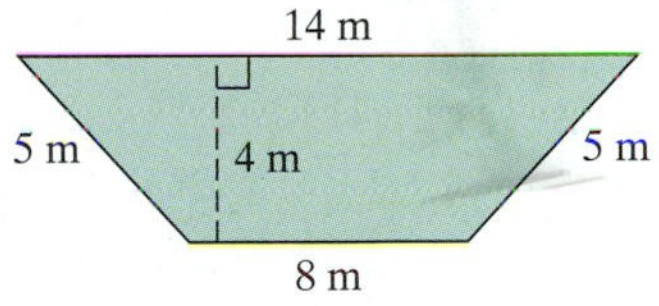

10.

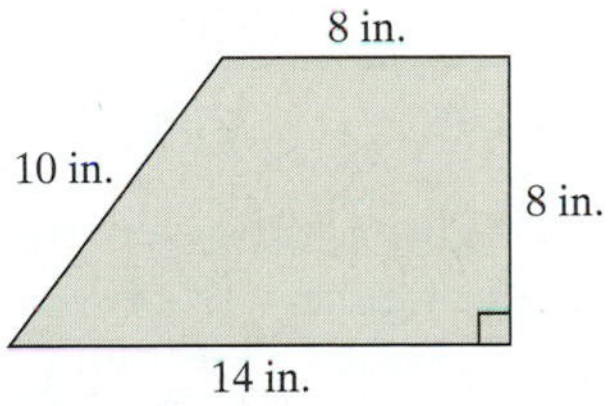

11. Use 3.14 for π.

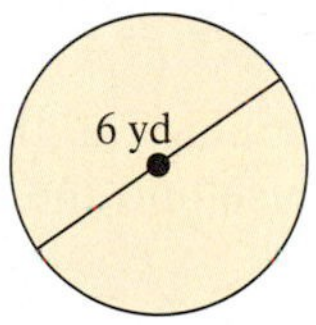

12. Use 3.14 for π.

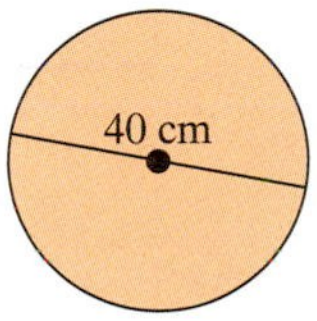

13. Use $\dfrac{22}{7}$ for π.

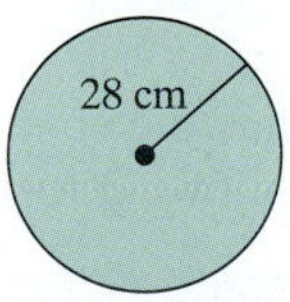

14. Use $\dfrac{22}{7}$ for π.

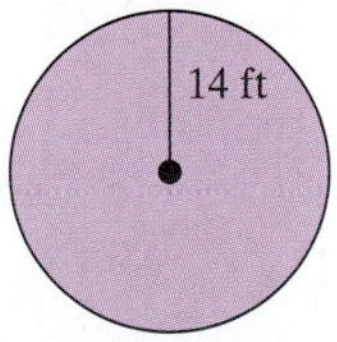

For Exercises 15–18,

a. Choose the type of formula needed to solve the problem: perimeter, circumference, or area.

b. Solve the problem.

15. Find the amount of fencing needed to enclose a square field whose side measures 16 yd.

16. Find the amount of wood trim needed to frame a circular window with diameter $1\frac{3}{4}$ ft. Use $\frac{22}{7}$ for π.

17. Find the amount of carpeting needed to cover the floor of a rectangular room that is $12\frac{1}{2}$ ft by 10 ft.

18. Find the amount of sod needed to cover a circular garden with diameter 20 m. Use 3.14 for π.

Section 8.8 Volume and Surface Area

Objectives

1. Volume
2. Surface Area

1. Volume

In this section, we learn how to compute volume. Volume is another word for capacity. We use volume, for example, to determine how much can be held in a moving van.

In addition to the units of capacity learned in Sections 8.1 and 8.2, volume can be measured in cubic units. For example, a cube that is 1 cm on a side has a volume of 1 cubic centimeter (1 cm^3 or cc). A cube that is 1 in. on a side has a volume of 1 cubic inch (1 in.^{3}). See Figure 8-32. Additional units of volume include cubic feet (ft^3), cubic yards (yd^3), cubic meters (m^3), and so on.

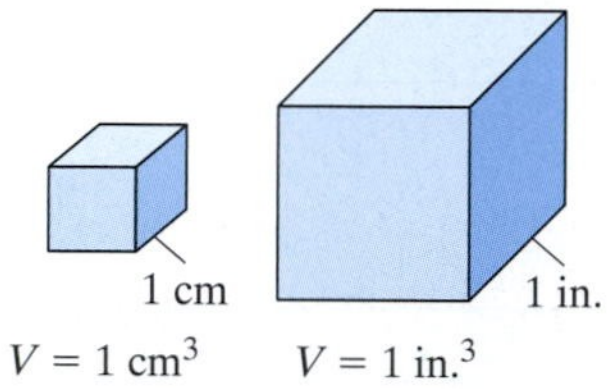

Figure 8-32

TIP: Recall that 1 cubic centimeter can also be denoted as 1 cc. Furthermore, 1 cc = 1 mL.

The formulas used to compute the volume of several common solids are given.

FORMULA

Rectangular Solid

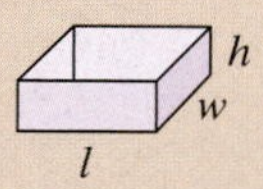

$V = lwh$

Cube

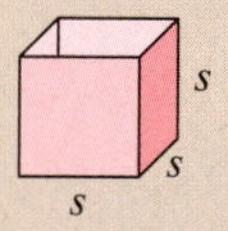

$V = s^3$

Right Circular Cylinder

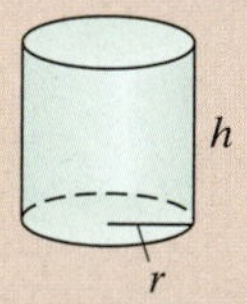

$V = \pi r^2 h$

Notice that the volume formulas for these three figures are given by the product of the area of the base and the height of the figure:

$V = lw\mathbf{h}$ — lw: area of rectangular base

$V = s \cdot s \cdot \mathbf{s}$ — $s \cdot s$: area of square base

$V = \pi r^2 \mathbf{h}$ — πr^2: area of circular base

A right circular cone has the shape of a party hat. A sphere has the shape of a ball. To compute the volume of a cone and a sphere, we use the following formulas.

TIP: Notice that the formula for the volume of a right circular cone is $\frac{1}{3}$ that of a right circular cylinder.

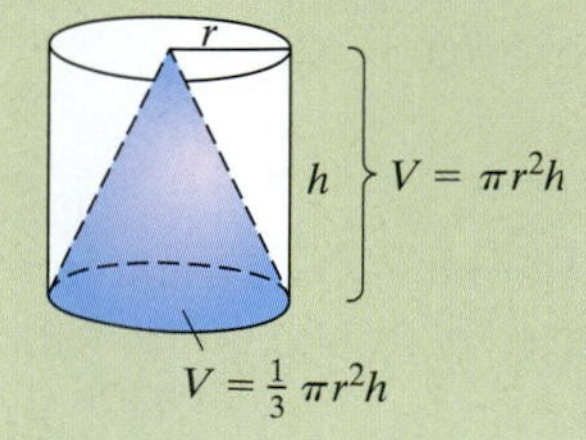

FORMULA

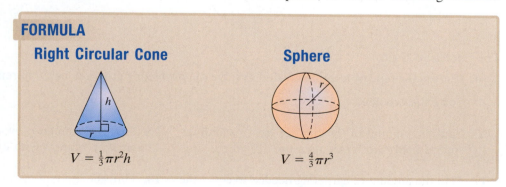

Right Circular Cone

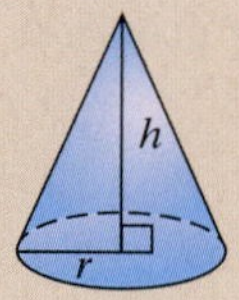

$V = \frac{1}{3}\pi r^2 h$

Sphere

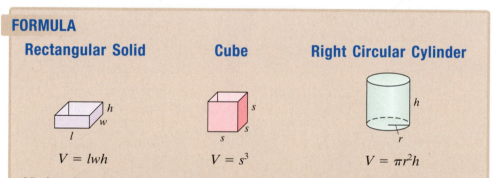

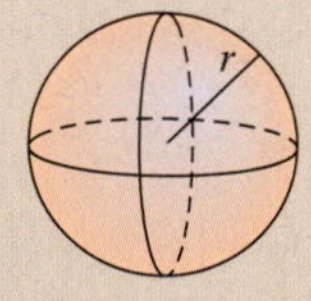

$V = \frac{4}{3}\pi r^3$

Example 1 Finding Volume

Find the volume. Round to the nearest whole unit.

Solution:

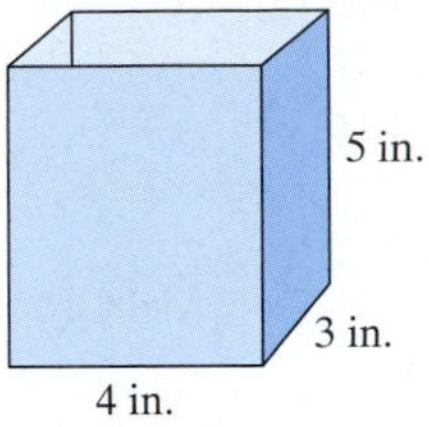

$V = lwh$ — Use the volume formula for a rectangular solid. Identify the length, width, and height.

$= (4 \text{ in.})(3 \text{ in.})(5 \text{ in.})$ — $l = 4$ in., $w = 3$ in., and $h = 5$ in.

$= 60 \text{ in.}^3$

We can visualize the volume by "layering" cubes that are each 1 in. high (Figure 8-33). The number of cubes in each layer is equal to $4 \times 3 = 12$. Each layer has 12 cubes, and there are 5 layers. Thus, the total number of cubes is $12 \times 5 = 60$ for a volume of 60 in.^{3}

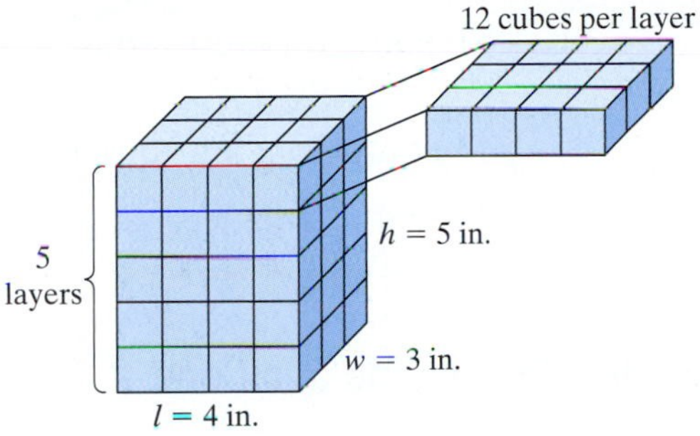

Figure 8-33

Skill Practice

Find the volume.

1.

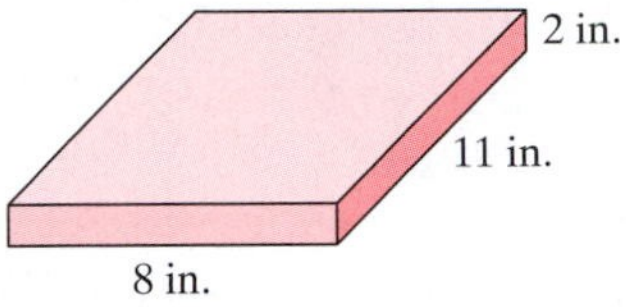

Example 2 Finding the Volume of a Cylinder

Find the volume. Use 3.14 for π. Round to the nearest whole unit.

Solution:

$V = \pi r^2 h$ — Use the formula for the volume of a right circular cylinder.

$\approx (3.14)(3.7 \text{ cm})^2(11.2 \text{ cm})$ — Substitute 3.14 for π, $r = 3.7$ cm, and $h = 11.2$ cm.

$= (3.14)(13.69 \text{ cm}^2)(11.2 \text{ cm})$ — Simplify exponents first.

$= 481.44992 \text{ cm}^3$ — Multiply from left to right.

$\approx 481 \text{ cm}^3$ — Round to the nearest whole unit.

Skill Practice

Find the volume. Use 3.14 for π.

2.

Answers

1. 176 in.^{3} **2.** $\approx$ 56.52 in.^{3}

Skill Practice

Find the volume. Use 3.14 for π.

3.

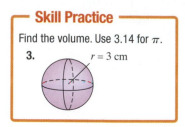

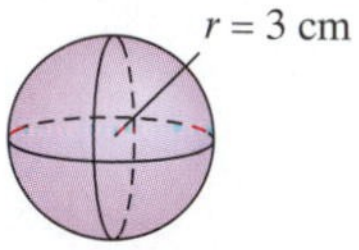

Example 3 Finding the Volume of a Sphere

Find the volume. Use 3.14 for π. Round to one decimal place.

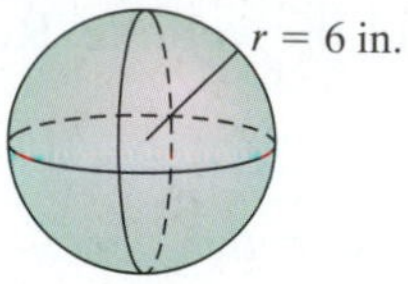

Solution:

$V = \frac{4}{3}\pi r^3$	Use the formula for the volume of a sphere.
$\approx \frac{4}{3}(3.14)(6 \text{ in.})^3$	Substitute 3.14 for π and $r = 6$ in.
$= \frac{4}{3}(3.14)(216 \text{ in.}^3)$	Simplify exponents first. $(6 \text{ in.})^3 = (6 \text{ in.})(6 \text{ in.})(6 \text{ in.}) = 216 \text{ in.}^3$
$= \frac{4}{3}\left(\frac{3.14}{1}\right)\left(\frac{216 \text{ in.}^3}{1}\right)$	Multiply fractions.
$= \frac{4}{\overset{}{\underset{1}{\cancel{3}}}}\left(\frac{3.14}{1}\right)\left(\frac{\overset{72}{\cancel{216}} \text{ in.}^3}{1}\right)$	Simplify to lowest terms.
$= 904.32 \text{ in.}^3$	Multiply from left to right.
$\approx 904.3 \text{ in.}^3$	Round to one decimal place.

Skill Practice

Find the volume. Use 3.14 for π.

4.

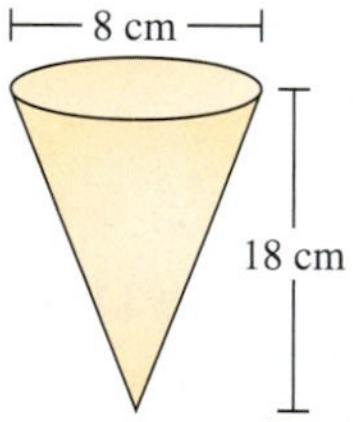

Example 4 Finding the Volume of a Cone

Find the volume. Use 3.14 for π. Round to one decimal place.

Solution:

$V = \frac{1}{3}\pi r^2 h$ Use the formula for the volume of a right circular cone.

To find the radius we have $r = \frac{1}{2}d = \frac{1}{2}(5 \text{ in.}) = 2.5 \text{ in.}$

$V \approx \frac{1}{3}(3.14)(2.5 \text{ in.})^2(8 \text{ in.})$	Substitute 3.14 for π, $r = 2.5$ in., and $h = 8$ in.
$= \frac{1}{3}\left(\frac{3.14}{1}\right)\left(\frac{6.25 \text{ in.}^2}{1}\right)\left(\frac{8 \text{ in.}}{1}\right)$	Simplify exponents first.
$= \frac{157}{3} \text{ in.}^3$	Multiply fractions.
$\approx 52.3 \text{ in.}^3$	Round to one decimal place.

Answers

3. 113.04 cm^3 **4.** 301.44 cm^3

2. Surface Area

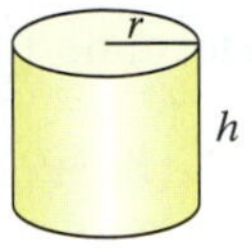

Surface area (often abbreviated SA) is the area of the surface of a three-dimensional object. To illustrate, consider the surface area of a soup can in the shape of a right circular cylinder.

If we peel off the label, we see that the label forms a rectangle whose length is the circumference of the base, $2\pi r$. The width of the rectangle is the height of the can, h. The area of the label is determined by multiplying the length and the width: $l \times w = 2\pi rh$. To determine the total surface area, add the areas of the top and bottom of the can to this rectangular area.

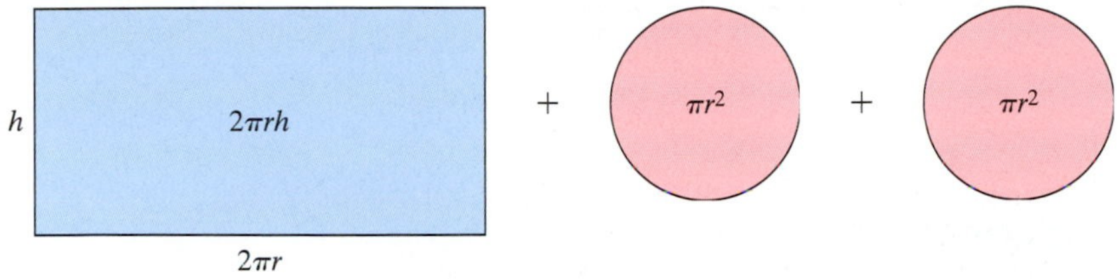

$$\text{Surface area} = 2\pi rh + \pi r^2 + \pi r^2 \quad \text{or}$$

$$\text{SA} = 2\pi rh + 2\pi r^2$$

Table 8-7 gives the formulas for the surface areas of four common solids.

Table 8-7 Surface Area

Cube	Rectangular Solid	Cylinder	Sphere
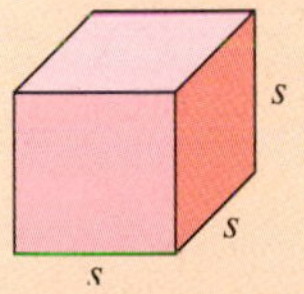	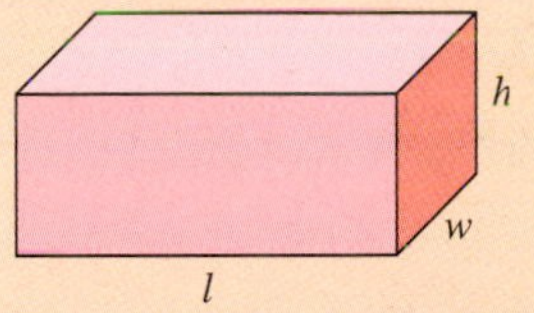	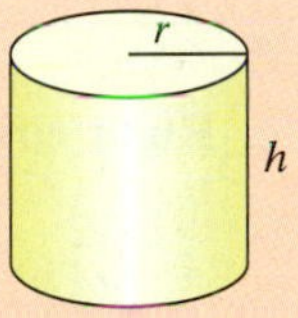	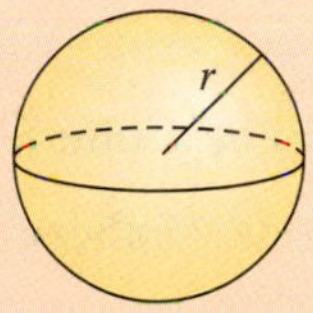
$\text{SA} = 6s^2$	$\text{SA} = 2lh + 2lw + 2hw$	$\text{SA} = 2\pi rh + 2\pi r^2$	$\text{SA} = 4\pi r^2$

Example 5 **Determining Surface Area**

Determine the surface area of the rectangular solid.

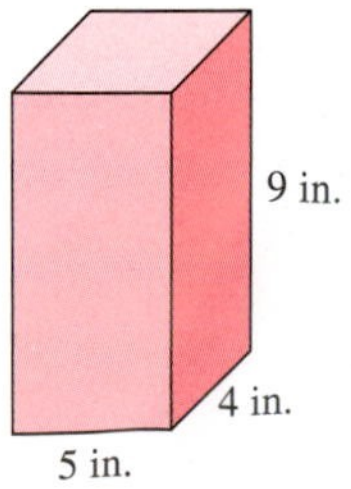

Solution:

Use the formula $\text{SA} = 2lh + 2lw + 2hw$, where $l = 5$ in., $w = 4$ in. and $h = 9$ in.

$$\text{SA} = 2(5\text{ in.})(9\text{ in.}) + 2(5\text{ in.})(4\text{ in.}) + 2(9\text{ in.})(4\text{ in.})$$

$$= 90\text{ in.}^2 + 40\text{ in.}^2 + 72\text{ in.}^2 \qquad \text{Multiply.}$$

$$= 202\text{ in.}^2 \qquad \text{Add.}$$

The surface area of the rectangular solid is 202 in.^{2}

Skill Practice

5. Determine the surface area of the cube with the side length of 3 ft.

Avoiding Mistakes

Although we are working with a three-dimensional figure, we are finding area. The answer will be in square units, not cubic units, as in calculating volume.

Answer

5. 54 ft^2

Skill Practice

6. Determine the surface area of a cylinder with radius 15 cm and height 20 cm.

Example 6 Determining Surface Area

Determine the surface area of a sphere with radius 6 m. Use 3.14 for π.

Solution:

Use the formula $SA = 4\pi r^2$, where $r = 6$ m.

$$SA = 4\pi(6\text{ m})^2$$
$$\approx 4(3.14)(36\text{ m}^2) \quad \text{Substitute 3.14 for } \pi.$$
$$\approx 452.16\text{ m}^2$$

Answer

6. 3297 cm^2

Section 8.8 Practice Exercises

Study Skills Exercises

1. For your next test, make a memory sheet. On a 3 × 5 card (or several 3 × 5 cards), write all the formulas and rules that you need to know. Memorize all this information. Then when your instructor hands you the test, write down all the information that you can remember before you begin the test. Then you can take the test without worrying that you will forget something important. This process is referred to as a "memory dump." What important definitions and concepts have you learned in this section of the text?

2. Define the key terms.

   a. **Rectangular solid** **b.** **Cube** **c.** **Right circular cylinder**

   d. **Right circular cone** **e.** **Sphere**

Review Exercises

3. Write the measure of the complement and supplement of 24°.

4. Write the measure of the complement and supplement of 79°.

5. Determine the measures of angles *a*–*g*. Assume that lines l_1 and l_2 are parallel.

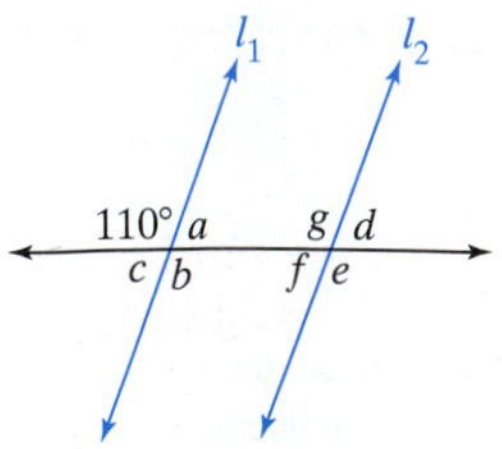

6. Determine the measures of angles *a*–*h*.

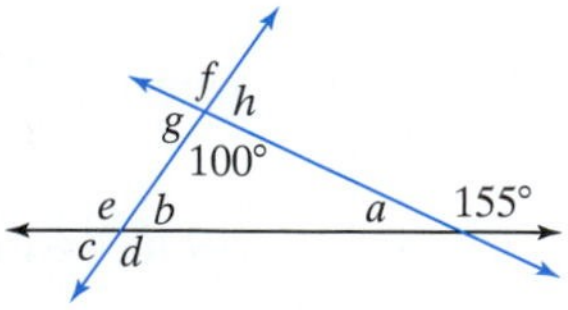

Objective 1: Volume

For Exercises 7–18, find the volume. Use 3.14 for π where necessary. **(See Examples 1–4.)**

7.

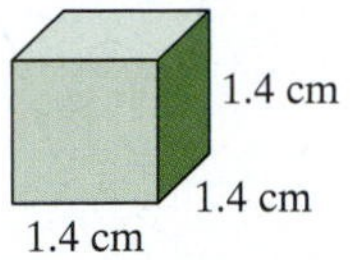

8.

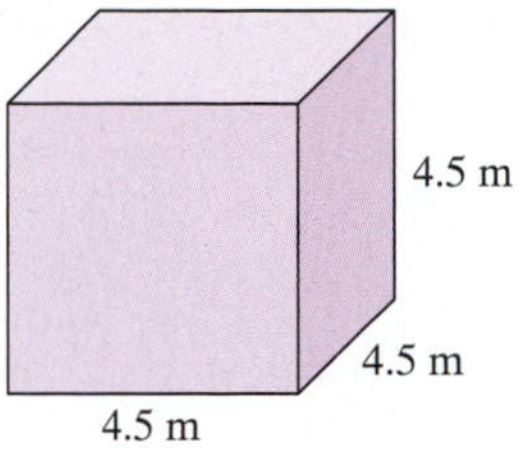

9.

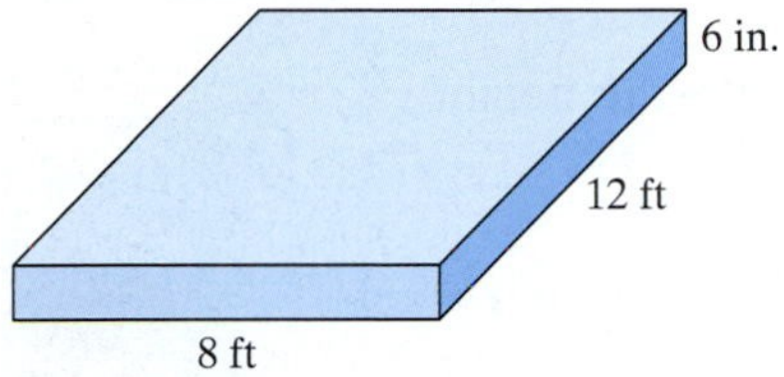

10.

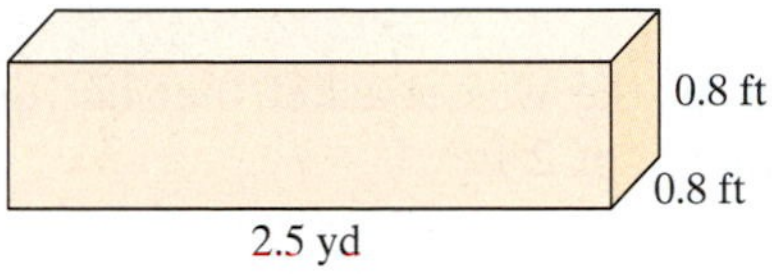

11.

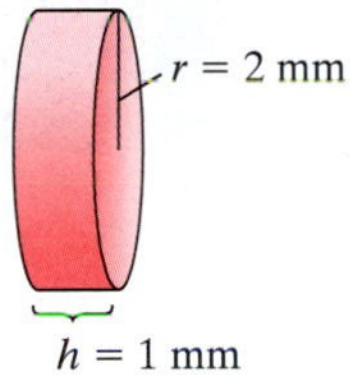

12.

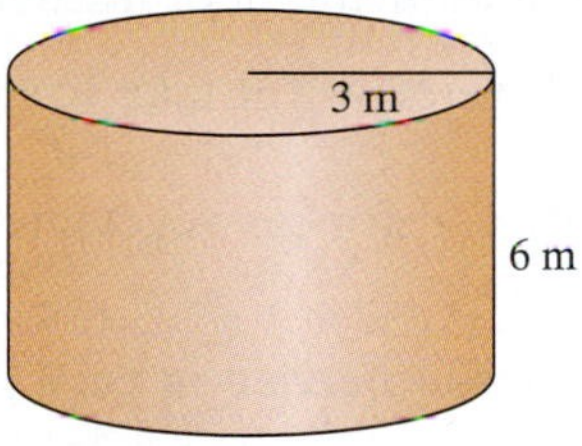

13.

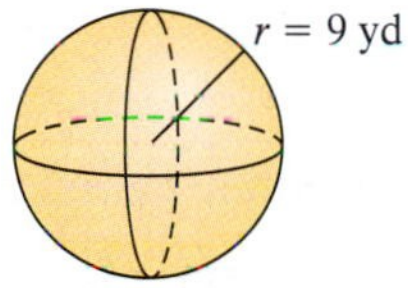

14.

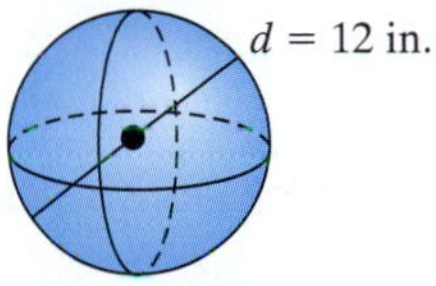

15.

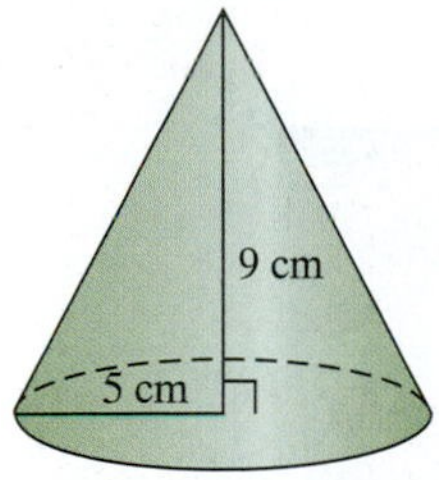

16.

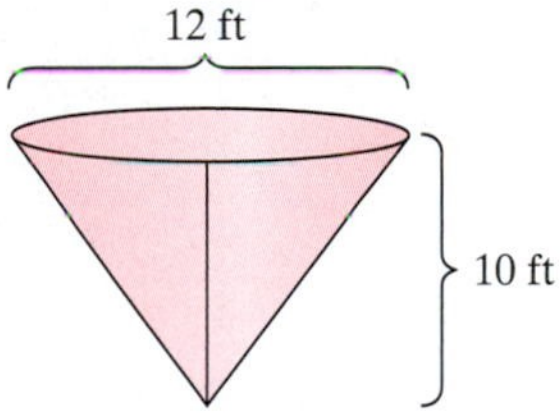

17.

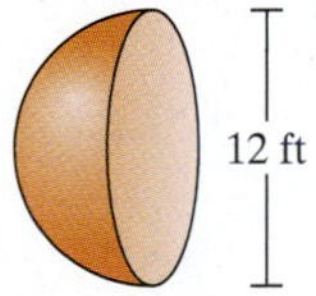

18.

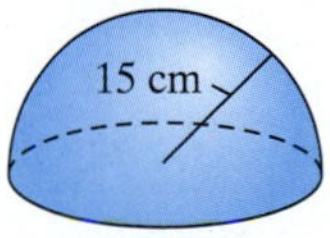

For Exercises 19–26, use 3.14 for π. Round each value to the nearest whole unit.

19. The diameter of a volleyball is 8.2 in. Find the volume.

20. The diameter of a basketball is 9 in. Find the volume.

21. Find the volume of the sand pile.

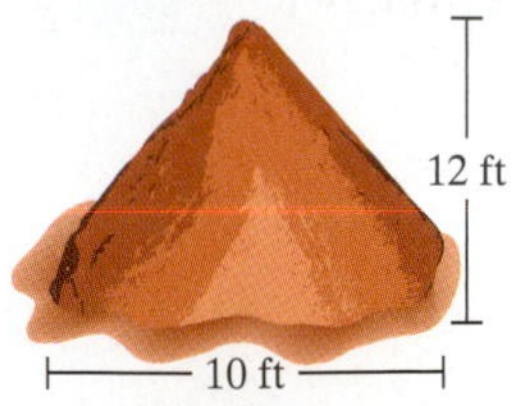

22. In decorating cakes, many people use an icing bag that has the shape of a cone. Find the volume of the icing bag.

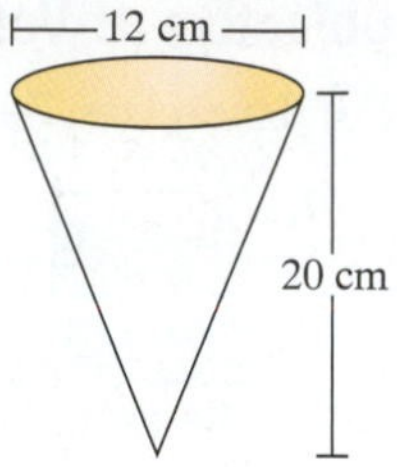

23. Find the volume of water (in cubic feet) that the pipe can hold.

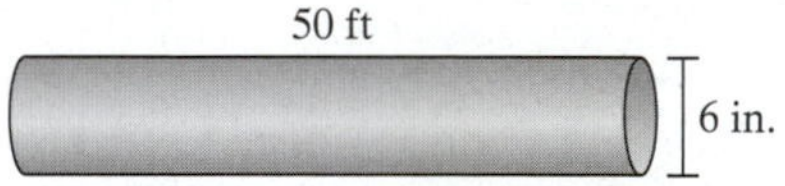

24. Find the volume of the wastebasket that has the shape of a cylinder with the height of 3 ft and diameter of 2 ft.

25. Sam bought an aboveground circular swimming pool with diameter 27 ft and height 54 in.

a. Approximate the volume of the pool in cubic feet using 3.14 for π.

b. How many gallons of water will it take to fill the pool? (*Hint:* 1 gal $\approx$ 0.1337 ft^3.)

26. Richard needs 3 in. of topsoil for his vegetable garden that is in the shape of a rectangle, 15 ft by 20 ft.

a. Find the amount of topsoil needed in cubic feet.

b. If topsoil can be purchased in bags containing 2 ft^3, how many bags must Richard purchase?

Objective 2: Surface Area

For Exercises 27–34, determine the surface area to the nearest tenth of a unit. Use 3.14 for π when necessary. **(See Examples 5–6.)**

27. Determine the amount of cardboard for the cereal box.

28. Determine the amount of cardboard for the box of macaroni.

29. Determine the surface area for the sugar cube.

30. Determine the surface area of the Sudoku cube.

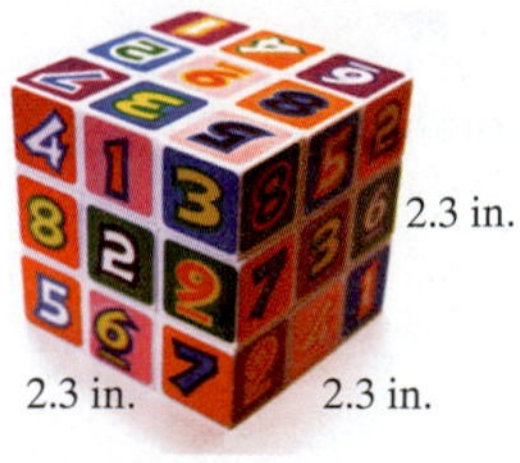

31. Determine the amount of cardboard for the container of oatmeal.

32. Determine the amount of steel needed for the can of tomato paste.

33. Determine the surface area of the volleyball.

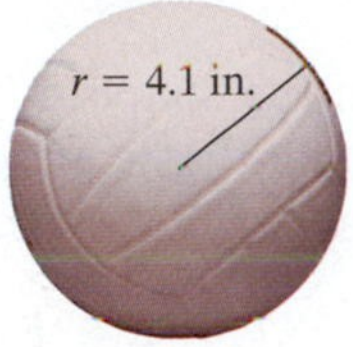

34. Determine the surface area of the golf ball.

For Exercises 35–42, determine the surface area of the object described. Use 3.14 for π when necessary. **(See Examples 5–6.)**

35. A rectangular solid with dimensions 12 ft by 14 ft by 3 ft

36. A rectangular solid with dimensions 8 m by 7 m by 5 m

37. A cube with each side 4 cm long

38. A cube with each side 10 yd long

39. A cylinder with radius 9 in. and height 15 in.

40. A cylinder with radius 90 mm and height 75 mm

41. A sphere with radius 10 mm

42. A sphere with radius 9 m

Expanding Your Skills

For Exercises 43–46, find the volume of the shaded region. Use 3.14 for π if necessary.

43. The height of the interior portion is 1 ft.

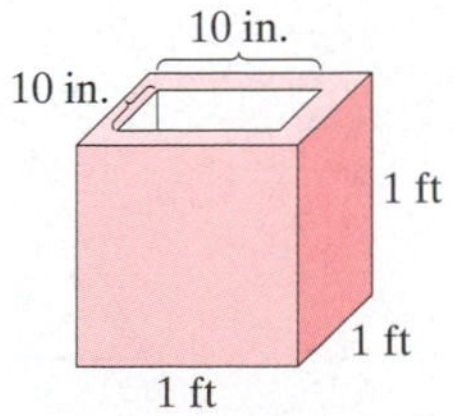

44. The height of the interior portion is 1 ft.

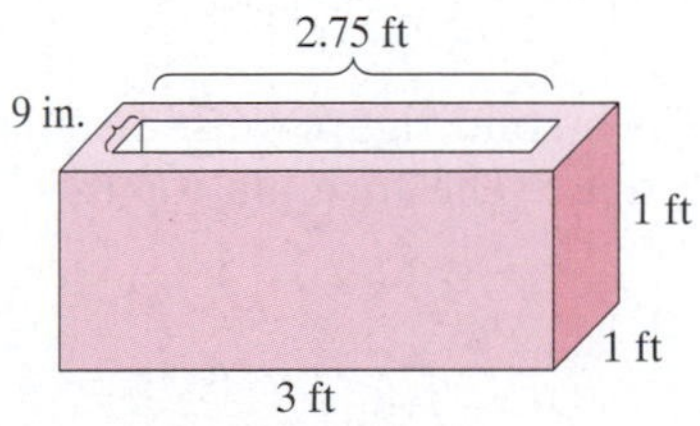

45.

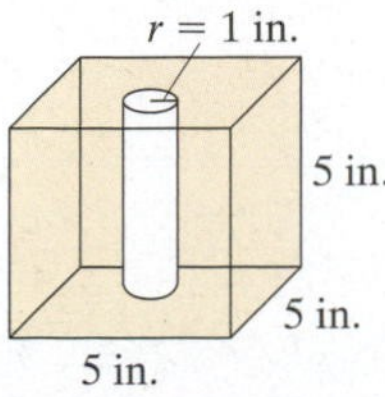

Animation

46.

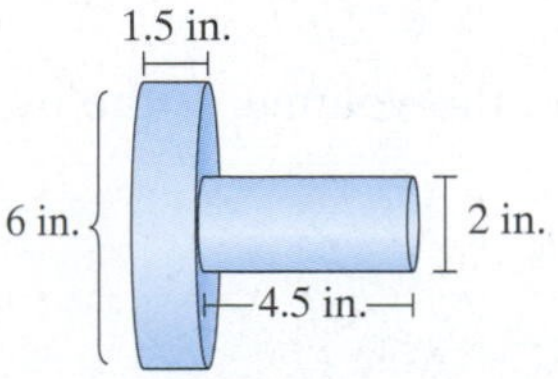

47. A machine part is in the shape of a cylinder with a hole drilled through the center. Find the volume of the machine part.

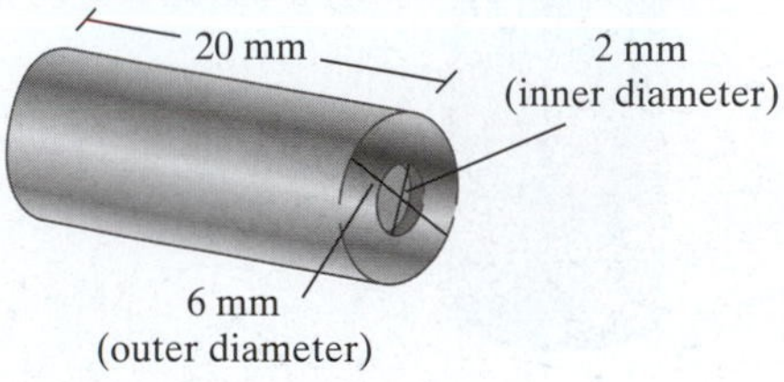

48. To insulate pipes, a cylinder of Styrofoam has a hole drilled through it to fit around a pipe. What is the volume of this piece of insulation?

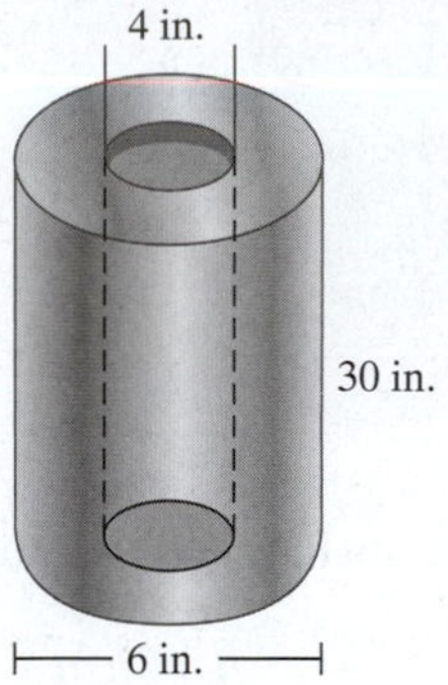

49. A silo is in the shape of a cylinder with a hemisphere on the top. Find the volume.

50. An ice cream cone is in the shape of a cone with a sphere on top. Assuming that ice cream is packed inside the cone, find the volume of the ice cream.

The volume formulas for right circular cylinders and right circular cones are the same for slanted cylinders and cones. For Exercises 51–54, find the volume. Use 3.14 for π.

51.

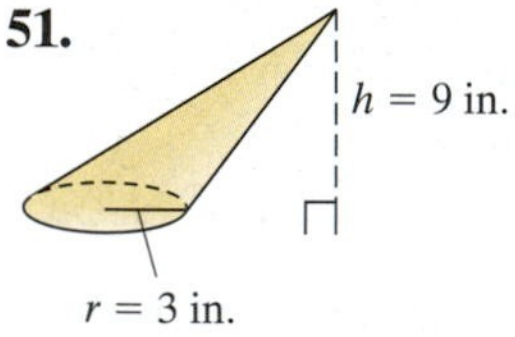

52.

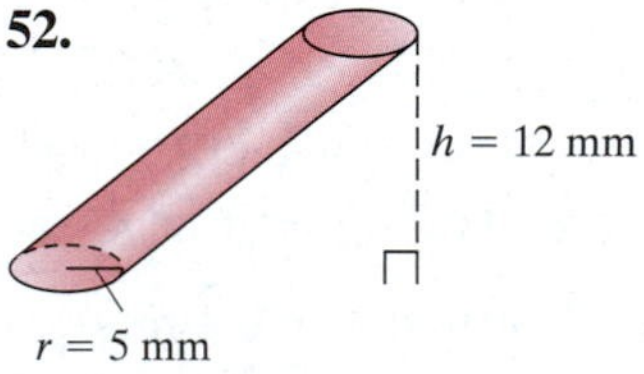

53.

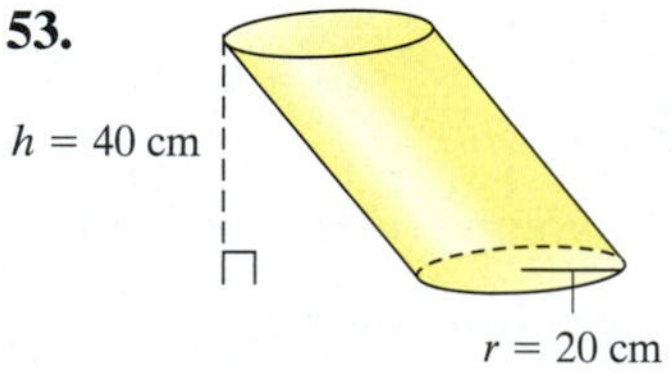

54.

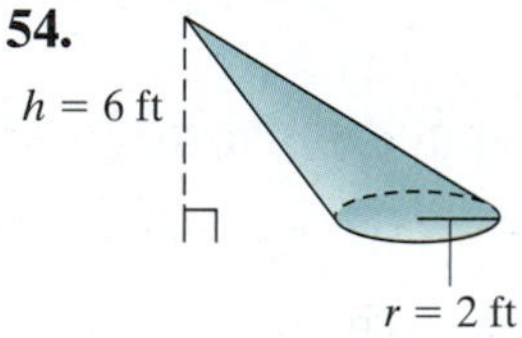

A square-based pyramid is a solid figure with a square base and triangular faces that meet at a common point called the apex. The formula for the volume is below.

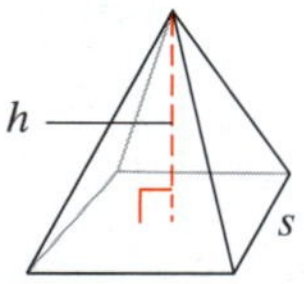

$$V = \frac{1}{3}s^2h$$

The length of the sides of the square base is s, and h is the perpendicular height from the apex to the base.

For Exercises 55–56, determine the volume of the pyramid.

55.

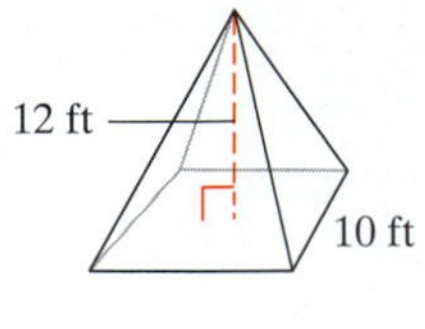

56.

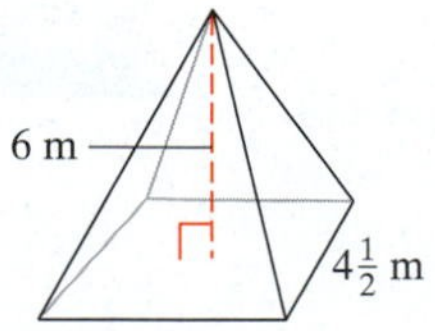

Group Activity

Remodeling the Classroom

Materials: A tape measure for measuring the size of the classroom.
Advertisements online or from the newspaper for carpet and paint.

Estimated time: 30 minutes

Group Size: 3–4

In this activity, your group will determine the cost for updating your classroom with new paint and new carpet.

1. Measure and record the dimensions of the room and also the height of the walls. You may want to sketch the floor and walls and then label their dimensions on the figure.

2. Calculate and record the area of the floor.

3. Calculate and record the total area of the walls. Subtract any area taken up by doors, windows, and chalkboards.

4. Look through advertisements for carpet that would be suitable for your classroom. You may have to look online for a better choice. Choose a carpet.

5. Calculate how much carpet is needed based on your measurements. To do this, take the area of the floor found in step 2 and add 10% of that figure to allow for waste.

6. Determine the cost to carpet the classroom. Do not forget to include carpet padding and labor to install the carpet. You may have to look online to find prices for padding and installation if they are not included in the price. Also include sales tax for your area.

7. Look through advertisements for paint that would be suitable for your classroom. You may have to look online for a better choice. Choose a paint.

8. Calculate the number of gallons of paint needed for your classroom. You may assume that 1 gal of paint will cover 400 ft^2. Calculate the cost of paint for the classroom. Include sales tax, but do not include labor costs for painting. You will do the painting yourself!

9. Calculate the total cost for carpeting and painting the classroom.

Chapter 8 Summary

Section 8.1 U.S. Customary Units of Measurement

Key Concepts

In the following lists, we give several units of measure common to the **U.S. Customary System**.

Length

$1 \text{ ft} = 12 \text{ in.}$ $\quad 1 \text{ in.} = \frac{1}{12} \text{ ft}$

$1 \text{ yd} = 3 \text{ ft}$ $\quad 1 \text{ ft} = \frac{1}{3} \text{ yd}$

$1 \text{ mi} = 5280 \text{ ft}$ $\quad 1 \text{ ft} = \frac{1}{5280} \text{ mi}$

$1 \text{ mi} = 1760 \text{ yd}$ $\quad 1 \text{ yd} = \frac{1}{1760} \text{ mi}$

Time

1 year = 365 days

1 week = 7 days

1 day = 24 hours (hr)

1 hour (hr) = 60 minutes (min)

1 minute (min) = 60 seconds (sec)

Weight

1 pound (lb) = 16 ounces (oz)

1 ton = 2000 pounds (lb)

Capacity

1 cup (c) = 8 fluid ounces (fl oz)

1 pint (pt) = 2 cups (c)

1 quart (qt) = 2 pints (pt)

1 gallon (gal) = 4 quarts (qt)

Examples

Example 1

To convert 8 yd to feet, multiply by the conversion factor.

$$8 \text{ yd} \cdot \frac{3 \text{ ft}}{1 \text{ yd}} = \frac{8 \cancel{\text{yd}}}{1} \cdot \frac{3 \text{ ft}}{1 \cancel{\text{yd}}} = 24 \text{ ft}$$

Example 2

To add 3 ft 9 in. + 2 ft 10 in., add like terms.

$$\begin{array}{r} 3 \text{ ft} + 9 \text{ in.} \\ + \, 2 \text{ ft} + 10 \text{ in.} \\ \hline 5 \text{ ft} + 19 \text{ in.} \end{array} = 5 \text{ ft} + 1 \text{ ft} + 7 \text{ in.}$$

$$= 6 \text{ ft } 7 \text{ in.}$$

Example 3

Convert. $\quad 200 \text{ min} = \underline{\qquad\qquad} \text{ hr}$

$$200 \cancel{\text{min}} \cdot \frac{1 \text{ hr}}{60 \cancel{\text{min}}} = \frac{200}{60} \text{ hr}$$

$$= \frac{10}{3} \text{ hr or } 3\frac{1}{3} \text{ hr}$$

Example 4

Convert. $\quad 6 \text{ lb} = \underline{\qquad\qquad} \text{ oz}$

$$6 \cancel{\text{lb}} \cdot \frac{16 \text{ oz}}{1 \cancel{\text{lb}}} = 96 \text{ oz}$$

Example 5

Convert. $\quad 40 \text{ c} = \underline{\qquad\qquad} \text{ gal}$

$$40 \text{ c} \cdot \frac{1 \text{ pt}}{2 \text{ c}} \cdot \frac{1 \text{ qt}}{2 \text{ pt}} \cdot \frac{1 \text{ gal}}{4 \text{ qt}}$$

$$= \frac{40 \cancel{\text{c}}}{1} \cdot \frac{1 \cancel{\text{pt}}}{2 \cancel{\text{c}}} \cdot \frac{1 \cancel{\text{qt}}}{2 \cancel{\text{pt}}} \cdot \frac{1 \text{ gal}}{4 \cancel{\text{qt}}}$$

$$= \frac{40}{16} \text{ gal} = \frac{5}{2} \text{ gal or } 2\frac{1}{2} \text{ gal}$$

Section 8.2 Metric Units of Measurement

Key Concepts

The **metric system** also offers units for measuring length, mass, and capacity. The base units are the **meter** for length, the **gram** for mass, and the **liter** for capacity. Other units of length, mass, and capacity in the metric system are powers of 10 of the base unit.

Metric units of length and their equivalents are given.

1 kilometer (km) = 1000 m

1 hectometer (hm) = 100 m

1 dekameter (dam) = 10 m

1 meter (m) = 1 m

1 decimeter (dm) = 0.1 m $\left(\frac{1}{10}\text{ m}\right)$

1 centimeter (cm) = 0.01 m $\left(\frac{1}{100}\text{ m}\right)$

1 millimeter (mm) = 0.001 m $\left(\frac{1}{1000}\text{ m}\right)$

The metric unit conversions for mass are given in **grams**.

1 kilogram (kg) = 1000 g

1 hectogram (hg) = 100 g

1 dekagram (dag) = 10 g

1 gram (g) = 1 g

1 decigram (dg) = 0.1 g $\left(\frac{1}{10}\text{ g}\right)$

1 centigram (cg) = 0.01 g $\left(\frac{1}{100}\text{ g}\right)$

1 milligram (mg) = 0.001 g $\left(\frac{1}{1000}\text{ g}\right)$

The metric unit conversions for capacity are given in **liters**.

1 kiloliter (kL) = 1000 L

1 hectoliter (hL) = 100 L

1 dekaliter (daL) = 10 L

1 liter (L) = 1 L

1 deciliter (dL) = 0.1 L $\left(\frac{1}{10}\text{ L}\right)$

1 centiliter (cL) = 0.01 L $\left(\frac{1}{100}\text{ L}\right)$

1 milliliter (mL) = 0.001 L $\left(\frac{1}{1000}\text{ L}\right)$

Note that 1 mL = 1 cc.

Example

Example 1

To convert 2 km to meters, we can use a unit ratio:

$$2\text{ km} = \frac{2\cancel{\text{ km}}}{1}\cdot\frac{1000\text{ m}}{1\cancel{\text{ km}}} \quad \begin{array}{l}\leftarrow \text{new unit}\\ \leftarrow \text{original unit}\end{array}$$

$$= 2000\text{ m}$$

Or we can use the prefix line.

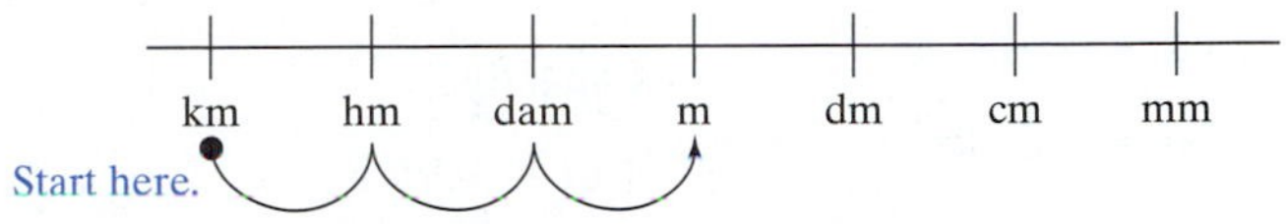

2 km = 2.000 km = 2000 m

Example 2

To convert 0.962 kg to grams, we can use a unit ratio:

$$0.962\text{ kg} = \frac{0.962\cancel{\text{ kg}}}{1}\cdot\frac{1000\text{ g}}{1\cancel{\text{ kg}}} \quad \begin{array}{l}\leftarrow \text{new unit}\\ \leftarrow \text{original unit}\end{array}$$

$$= 962\text{ g}$$

Or we can use the prefix line.

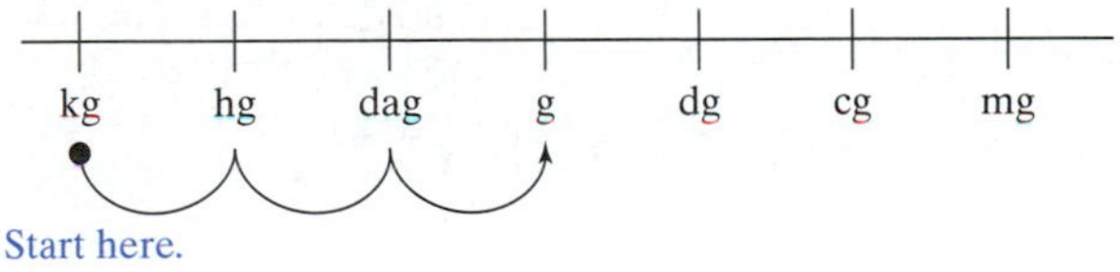

0.962 kg = 0.962 kg = 962 g

Example 3

To convert 59,000 cL to kL, we can use unit ratios:

$$59{,}000\text{ cL} = \frac{59{,}000\cancel{\text{ cL}}}{1}\cdot\frac{1\cancel{\text{ L}}}{100\cancel{\text{ cL}}}\cdot\frac{1\text{ kL}}{1000\cancel{\text{ L}}}$$

$$= 0.59\text{ kL}$$

Or we can use the prefix line.

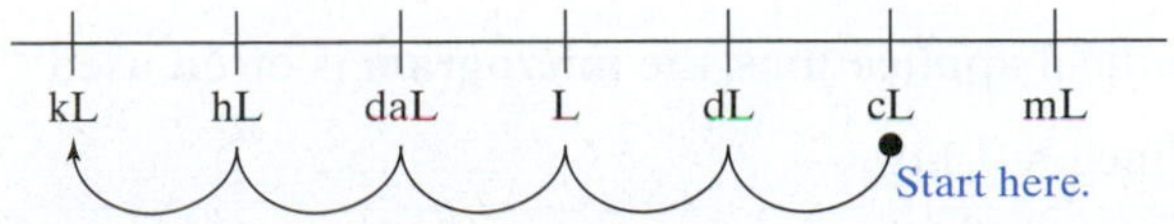

59,000 cL = 59,000 cL = 0.59 kL

Section 8.3 Converting Between U.S. Customary and Metric Units

Key Concepts

The common conversions between the U.S. Customary and metric systems are given.

Length

1 in. = 2.54 cm

1 ft ≈ 0.305 m

1 yd ≈ 0.914 m

1 mi ≈ 1.61 km

Weight/Mass (on Earth)

1 lb ≈ 0.45 kg

1 oz ≈ 28 g

Capacity

1 qt ≈ 0.95 L

1 fl oz ≈ 30 mL = 30 cc

To convert U.S. Customary units to metric or metric units to U.S. Customary units, use unit ratios.

The U.S. Customary System uses the **Fahrenheit** scale (°F) to measure temperature. The metric system uses the **Celsius** scale (°C). The conversions are given.

To convert from °C to °F: $F = \frac{9}{5}C + 32$

To convert from °F to °C: $C = \frac{5}{9}(F - 32)$

Examples

Example 1

Convert 1200 yd to meters by using a unit ratio.

$$1200 \text{ yd} \approx \frac{1200 \text{ yd}}{1} \cdot \frac{0.914 \text{ m}}{1 \text{ yd}} = 1096.8 \text{ m}$$

Example 2

To convert 900 cc to fluid ounces, recall that 1 cc = 1 mL. Therefore, 900 cc = 900 mL. Then use a unit ratio to convert to fluid ounces.

$$900 \text{ mL} \approx \frac{900 \text{ mL}}{1} \cdot \frac{1 \text{ fl oz}}{30 \text{ mL}} = 30 \text{ fl oz}$$

Example 3

The average January temperature in Havana, Cuba, is 21°C. The average January temperature in Johannesburg, South Africa, is 69°F. Which temperature is warmer?

Convert 21°C to degrees Fahrenheit:

$$F = \frac{9}{5}C + 32$$

$$= \frac{9}{5}(21) + 32$$

$$= 37.8 + 32 = 69.8$$

The value 21°C = 69.8°F, which is 0.8 degree warmer than the temperature in Johannesburg.

Section 8.4 Medical Applications Involving Measurement

Key Concepts

In medical applications, the **microgram** is often used.

1000 mcg = 1 mg

1,000,000 mcg = 1 g

Examples

Example 1

Convert.

575 mcg = ______ mg

$$575 \text{ mcg} = \frac{575 \text{ mcg}}{1} \cdot \frac{1 \text{ mg}}{1000 \text{ mcg}} = 0.575 \text{ mg}$$

The metric system is used in many applications in medicine.

Example 2

A doctor prescribes 25 mcg of a given drug per kilogram of body mass for a patient. If the patient's mass is 80 kg, determine the total amount of drug that the patient will get.

$$\text{Total amount} = \left(\frac{25\text{ mcg}}{1\text{ kg}}\right) \cdot (80\text{ kg})$$

$$= 2000\text{ mcg} \quad \text{or} \quad 2\text{ mg}$$

Section 8.5 Lines and Angles

Key Concepts

An **angle** is a geometric figure formed by two rays that share a common endpoint. The common endpoint is called the **vertex** of the angle.

An angle is **acute** if its measure is between 0° and 90°. An angle is **obtuse** if its measure is between 90° and 180°.

Two angles are said to be **complementary** if the sum of their measures is 90°. Two angles are said to be **supplementary** if the sum of their measures is 180°.

Given two intersecting lines, **vertical angles** are angles that appear on opposite sides of the vertex.

When two parallel lines are crossed by another line eight angles are formed.

Examples

Example 1

$\angle DEF$

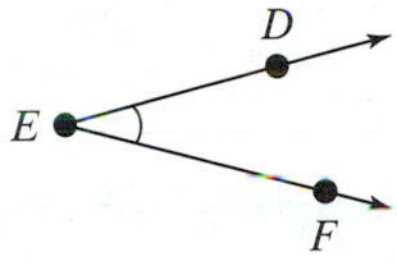

Example 2

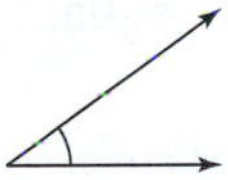

Acute angle

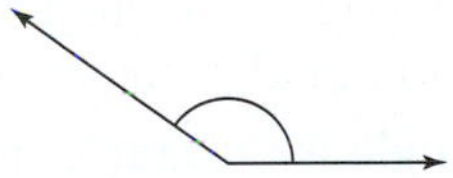

Obtuse angle

Example 3

The complement of a 32° angle is a 58° angle. The supplement of a 32° angle is a 148° angle.

Example 4

Intersecting lines:

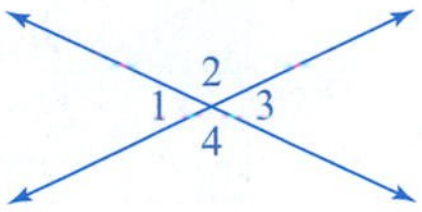

$\angle 1$ and $\angle 3$ are vertical angles and are congruent. Also, $\angle 2$ and $\angle 4$ are vertical angles and are congruent.

Example 5

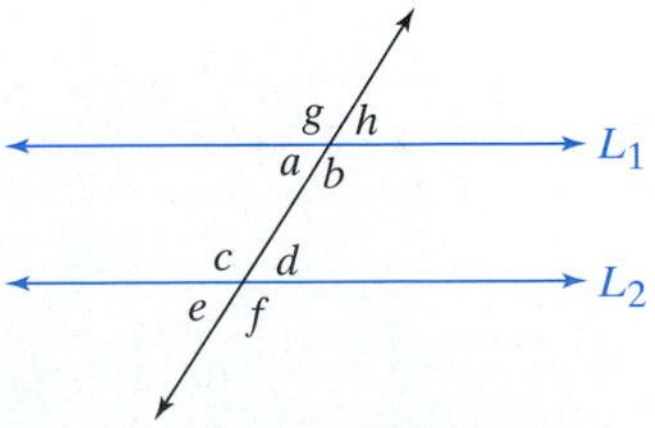

$m(\angle a) = m(\angle d)$ because they are **alternate interior angles**.
$m(\angle e) = m(\angle h)$ because they are **alternate exterior angles**.
$m(\angle c) = m(\angle g)$ because they are **corresponding angles**.

Section 8.6 Triangles and the Pythagorean Theorem

Key Concepts

The sum of the measures of the angles of any triangle is 180°.

An **acute triangle** is a triangle in which all three angles are acute.

A **right triangle** is a triangle in which one angle is a right angle.

An **obtuse triangle** is a triangle in which one angle is obtuse.

An **equilateral triangle** is a triangle in which all three sides (and all three angles) are equal in measure.

An **isosceles triangle** is a triangle in which two sides are equal in length (the angles opposite the equal sides are also equal in measure).

A **scalene triangle** is a triangle in which no sides (or angles) are equal in measure.

Pythagorean Theorem

The sum of the squares of the **legs of a right triangle** equals the square of the **hypotenuse**.

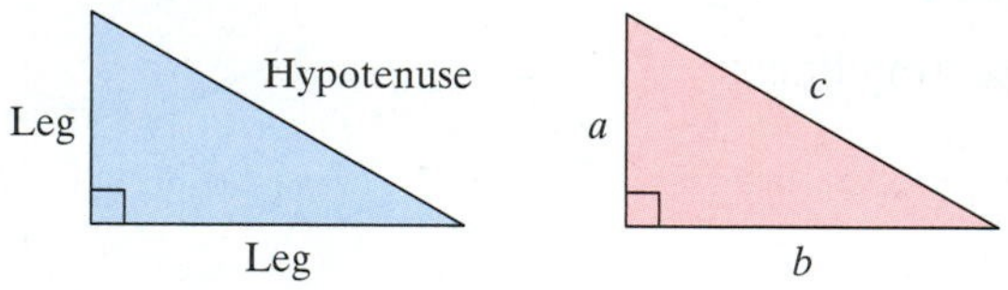

$$a^2 + b^2 = c^2$$

Examples

Example 1

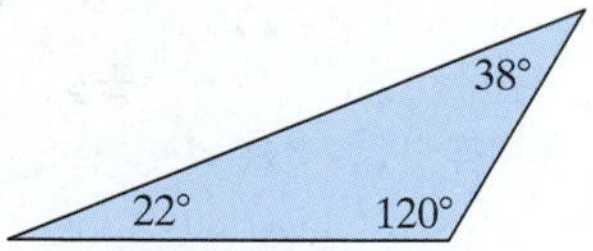

$$22° + 120° + 38° = 180°$$

Example 2

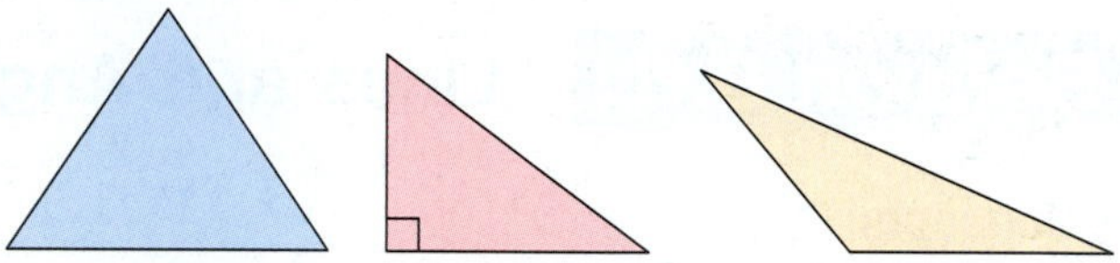

Acute triangle | Right triangle | Obtuse triangle

Example 3

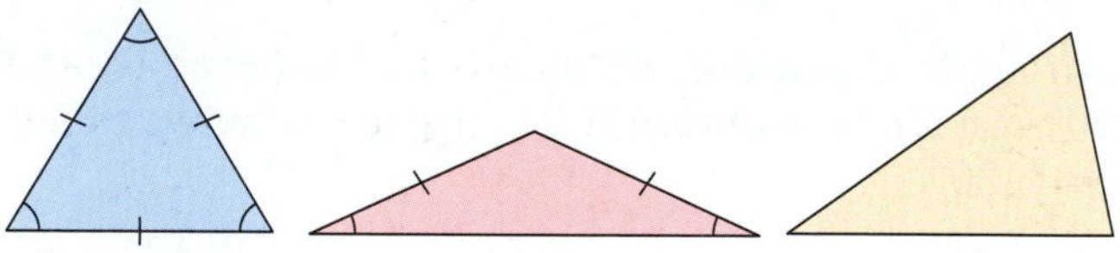

Equilateral triangle | Isosceles triangle | Scalene triangle

Example 4

To find the length of the hypotenuse, solve for c.

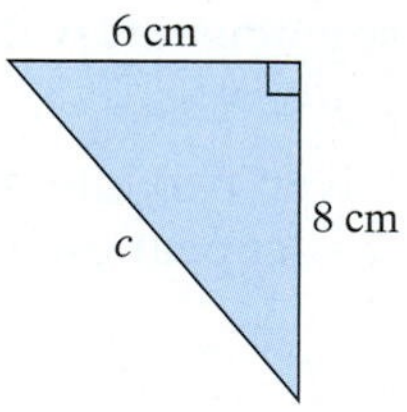

$$6^2 + 8^2 = c^2$$

$$36 + 64 = c^2$$

$$100 = c^2$$

$$\sqrt{100} = c$$

$$10 = c$$

c is the positive number that when squared equals 100.

The length of the hypotenuse is 10 cm.

Section 8.7 Perimeter, Circumference, and Area

Key Concepts

A four-sided **polygon** is called a **quadrilateral**.

A **parallelogram** is a quadrilateral with opposite sides parallel.

A **rectangle** is a parallelogram with four right angles.

A **square** is a rectangle with sides equal in length.

A **rhombus** is a parallelogram with sides equal in length.

A **trapezoid** is a quadrilateral with one pair of parallel sides.

Perimeter is the distance around a figure.

Perimeter of a triangle: $P = a + b + c$

Perimeter of a square: $P = 4s$

Perimeter of a rectangle: $P = 2l + 2w$

Circumference of a circle: $C = 2\pi r$

Area is the number of square units that can be enclosed by a figure.

Area of a rectangle: $A = lw$

Area of a square: $A = s^2$

Area of a parallelogram: $A = bh$

Area of a triangle: $A = \frac{1}{2}bh$

Area of a trapezoid: $A = \frac{1}{2}(a + b)h$

Area of a circle: $A = \pi r^2$

Examples

Example 1

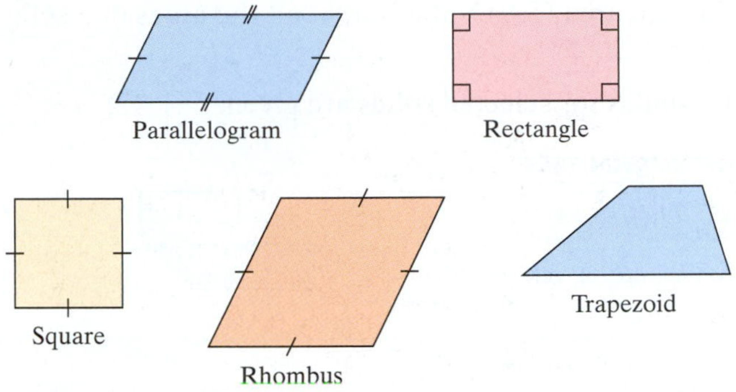

Example 2

Determine the perimeter.

First convert the length and width to the same units of measurement.

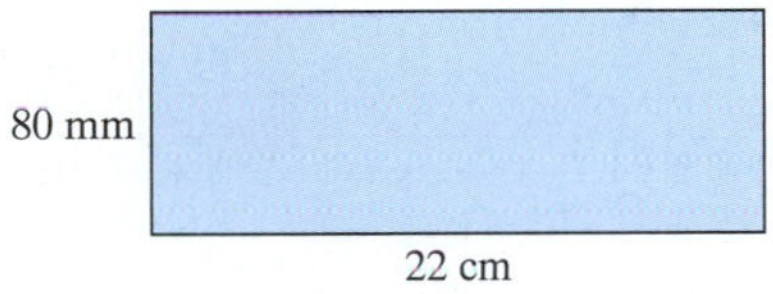

For the width: 80 mm = 8 cm.

$P = 2l + 2w$ Perimeter formula (rectangle)

$P = 2(22 \text{ cm}) + 2(8 \text{ cm})$

$= 44 \text{ cm} + 16 \text{ cm}$

$= 60 \text{ cm}$ The perimeter is 60 cm.

Example 3

Determine the area.

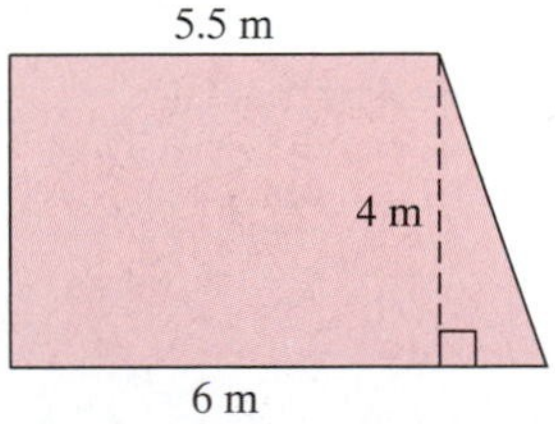

$A = \dfrac{1}{2}(a + b)h$ Area of a trapezoid

$A = \dfrac{1}{2}(6 \text{ m} + 5.5 \text{ m})(4 \text{ m})$

$= \dfrac{1}{2}(11.5 \text{ m})(4 \text{ m})$

$= 23 \text{ m}^2$

The area is 23 m^2.

Section 8.8 Volume and Surface Area

Key Concepts

Volume (V) is another word for capacity.

Surface area (SA) is the sum of all the areas of a solid figure.

Formulas for selected solids are given.

Rectangular solid

$V = lwh$

$SA = 2lh + 2lw + 2hw$

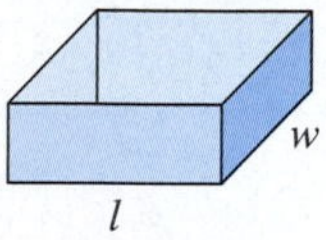

Cube

$V = s^3$

$SA = 6s^2$

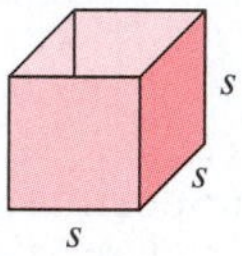

Right circular cylinder

$V = \pi r^2 h$

$SA = 2\pi rh + 2\pi r^2$

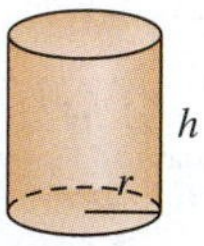

Right circular cone

$V = \frac{1}{3}\pi r^2 h$

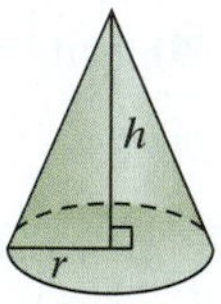

Sphere

$V = \frac{4}{3}\pi r^3$

$SA = 4\pi r^2$

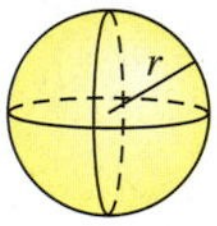

Examples

Example 1

Find the volume of the cone.

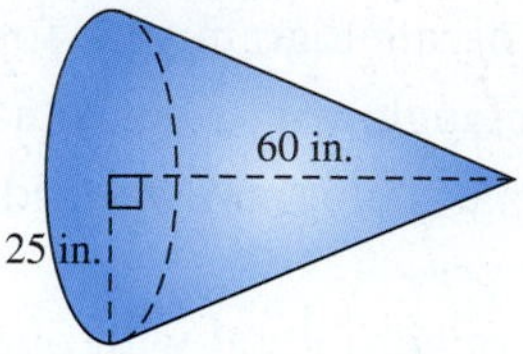

$$V = \frac{1}{3}\pi r^2 h$$

$$\approx \frac{1}{3}(3.14)(25\text{ in.})^2(60\text{ in.})$$

$$= 39{,}250\text{ in.}^3$$

The volume is approximately 39,250 in.^{3}

Example 2

Determine the surface area.

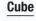

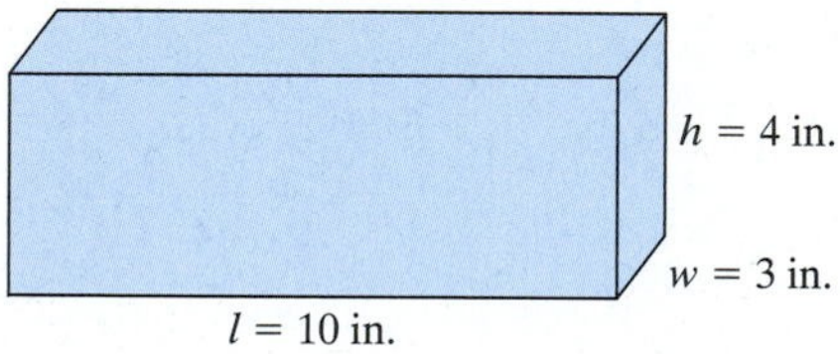

$$SA = 2lh + 2lw + 2hw$$

$$= 2(10\text{ in.})(4\text{ in.}) + 2(10\text{ in.})(3\text{ in.}) + 2(4\text{ in.})(3\text{ in.})$$

$$= 80\text{ in.}^2 + 60\text{ in.}^2 + 24\text{ in.}^2$$

$$= 164\text{ in.}^2$$

Chapter 8 Review Exercises

Section 8.1

For Exercises 1–6, convert the units of length.

1. 48 in. = _____ ft

2. $3\frac{1}{4}$ ft = _____ in.

3. 2 mi = _____ yd

4. 7040 ft = _____ mi

5. $\frac{1}{2}$ mi = _____ ft

6. 2 yd = _____ in.

For Exercises 7–10, perform the indicated operations.

7. 3 ft 9 in. + 5 ft 6 in.

8. 4′11″ + 1′5″

9. 5′3″ − 2′5″

10. 12 ft 7 in. − 8 ft 10 in.

11. Find the perimeter in feet.

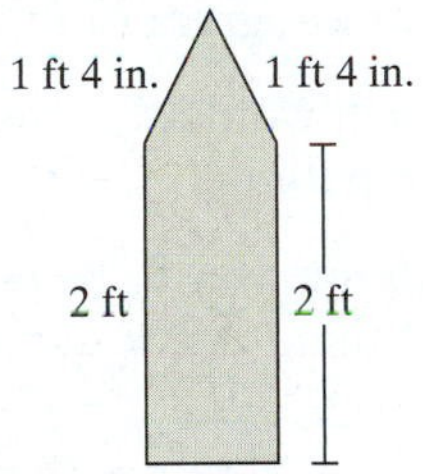

12. A roll of wire contains 50 yd of wire. If Ivan uses 48 ft, how much wire is left?

For Exercises 13–22, convert the units of time, weight, and capacity.

13. 72 hr = _____ days

14. 6 min = _____ sec

15. 5 lb = _____ oz

16. 1 wk = _____ hr

17. 12 fl oz = _____ c

18. 0.25 ton = _____ lb

19. 3500 lb = _____ tons

20. 2 gal = _____ pt

21. 12 oz = _____ lb

22. 16 qt = _____ gal

23. A runner finished a race with a time of 2:24:30. Convert the time to minutes.

24. Margaret Johansson gave birth to triplets who weighed 3 lb 10 oz, 4 lb 2 oz, and 4 lb 1 oz. What was the total weight of the triplets?

Section 8.2

For Exercises 25–26, select the most reasonable measurement.

25. A pencil is __________ long.

a. 16 mm **b.** 16 cm **c.** 16 m **d.** 16 km

26. The distance between Houston and Dallas is __________.

a. 362 mm **b.** 362 cm **c.** 362 m **d.** 362 km

For Exercises 27–32, convert the metric units of length.

27. 52 cm = _____ mm

28. 93 m = _____ km

29. 34 dm = _____ m

30. 2.1 m = _____ dam

31. 4 cm = _____ m

32. 1.2 m = _____ mm

For Exercises 33–36, convert the metric units of mass.

33. 6.1 g = _____ cg

34. 420 g = _____ kg

35. 3212 mg = _____ g

36. 0.7 hg = _____ g

For Exercises 37–40, convert the metric units of capacity.

37. 830 cL = _____ L

38. 124 mL = _____ cc

39. 225 cc = _____ cL

40. 0.49 kL = _____ L

41. The dimensions of a dining room table are 2 m by 125 cm. Convert the units to meters and find the perimeter and area of the tabletop.

42. A bottle of apple juice contains 1.2 L of juice. If a glass holds 24 cL, how many glasses can be filled from this bottle?

43. An adult has a mass of 68 kg. A baby has a mass of 3200 g. What is the difference in their masses, in kilograms?

44. From a wooden board 2 m long, Jesse needs to cut 3 pieces that are each 75 cm long. Is the 2-m length of board long enough for the 3 pieces?

Section 8.3

For Exercises 45–54, refer to page 502. Convert the units of length, capacity, mass, and weight. Round to the nearest hundredth, if necessary.

45. 6.2 in. ≈ _____ cm

46. 75 mL ≈ _____ fl oz

47. 140 g ≈ _____ oz

48. 5 L ≈ _____ qt

49. 3.4 ft ≈ _____ m

50. 100 lb ≈ _____ kg

51. 120 km ≈ _____ mi

52. 6 qt ≈ _____ L

53. 1.5 fl oz ≈ _____ cc

54. 12.5 tons ≈ _____ kg

55. The height of a computer desk is 30 in. The height of the chair is 38 cm. What is the difference in height between the desk and chair, in centimeters?

56. A bag of snack crackers contains 7.2 oz. If one serving is 30 g, approximately how many servings are in one bag?

57. The Boston Marathon is 42.195 km long. Convert this distance to miles. Round to the tenths place.

58. Write the formula to convert degrees Fahrenheit to degrees Celsius.

59. When roasting a turkey, the meat thermometer should register between 180°F and 185°F to indicate that the turkey is done. Convert these temperatures to degrees Celsius. Round to the nearest tenth, if necessary.

60. Write the formula to convert degrees Celsius to degrees Fahrenheit.

61. The average October temperature for Toronto, Ontario, Canada, is 8°C. Convert this temperature to degrees Fahrenheit.

Section 8.4

For Exercises 62–65, convert the metric units.

62. 0.45 mg = _____ mcg

63. 1.5 mg = _____ mcg

64. 400 mcg = _____ mg

65. 5000 mcg = _____ cg

66. A nasal spray delivers 2.5 mg of active ingredient per milliliter of solution. Determine the amount of active ingredient per cc.

67. A physician prescribes a drug based on a patient's mass. The dosage is given as 0.04 mg of the drug per kilogram of the patient's mass.

a. How much would the physician prescribe for an 80-kg patient?

b. If the dosage was to be given twice a day, how much of the drug would the patient take in a week?

68. A prescription for cough syrup indicates that 30 mL should be taken twice a day for 7 days. What is the total amount, in liters, of cough syrup to be taken?

69. A standard hypodermic syringe holds 3 cc of fluid. If a nurse uses 1.8 mL of the fluid, how much is left in the syringe?

70. A medication comes in 250-mg capsules. If Clayton took 3 capsules a day for 10 days, how many grams of the medication did he take?

Section 8.5

For Exercises 71–74, match the symbol with a description.

71. $\overleftrightarrow{AB}$ a. Ray AB

72. $\overrightarrow{AB}$ b. Line segment AB

73. $\overrightarrow{BA}$ c. Ray BA

74. $\overline{AB}$ d. Line AB

75. Describe the measure of an acute angle.

76. Describe the measure of an obtuse angle.

77. Describe the measure of a right angle.

78. Let $m(\angle X) = 33°$.

a. Find the complement of $\angle X$.

b. Find the supplement of $\angle X$.

79. Let $m(\angle T) = 20°$.

a. Find the complement of $\angle T$.

b. Find the supplement of $\angle T$.

For Exercises 80–84, refer to the figure. Find the measures of the angles.

80. $m(\angle a)$

81. $m(\angle b)$

82. $m(\angle c)$

83. $m(\angle d)$

84. $m(\angle e)$

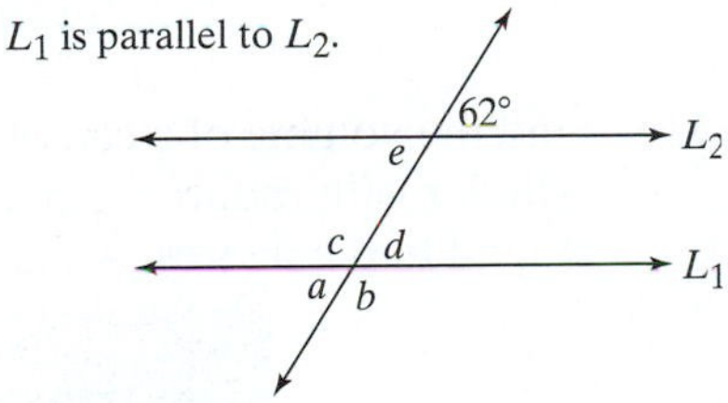

Section 8.6

For Exercises 85–86, find the measures of the angles x and y.

85.

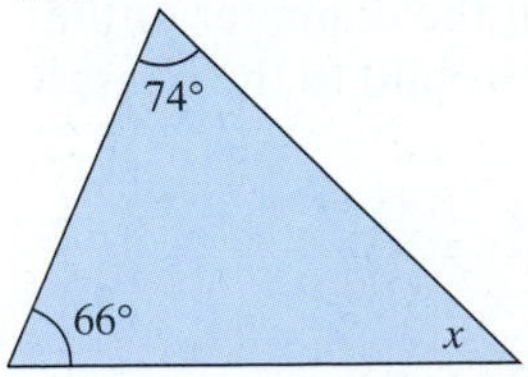

86.

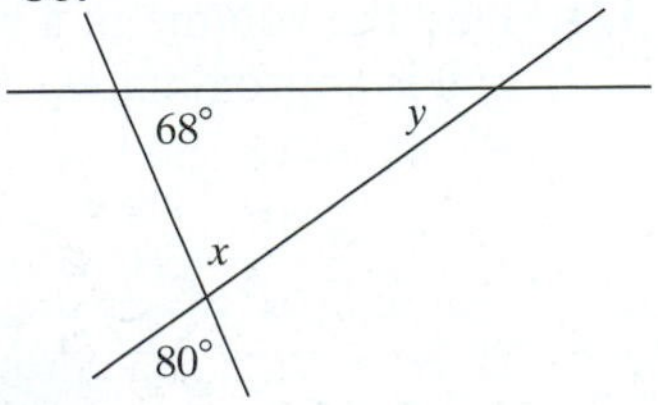

For Exercises 87–88, describe the characteristics of each type of triangle.

87. Equilateral triangle

88. Isosceles triangle

For Exercises 89–90, simplify the square roots.

89. $\sqrt{25}$

90. $\sqrt{49}$

For Exercises 91–92, find the length of the unknown side.

91.

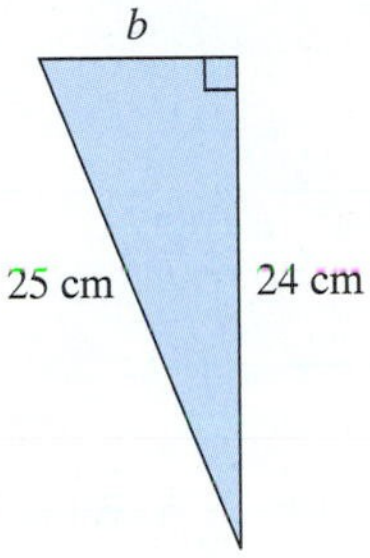

92.

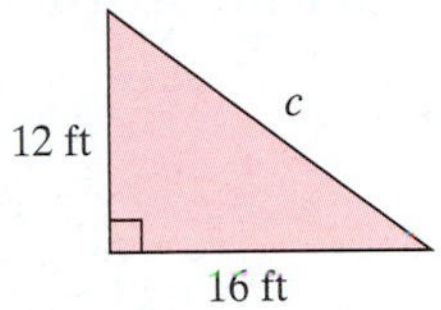

93. Kayla is flying a kite. At one point the kite is 5 m from Kayla horizontally and 12 m above her (see figure). How much string is extended? (Assume there is no slack.)

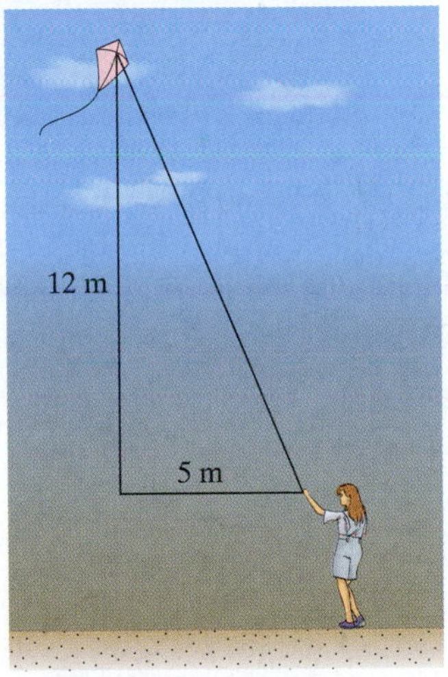

Section 8.7

94. Find the perimeter of the figure.

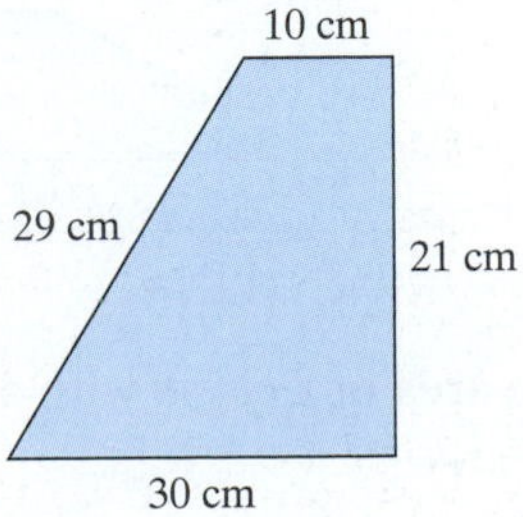

95. Find the perimeter of a triangle with sides of 4.2 m, 6.1 m, and 7.0 m.

96. How much fencing is required to put up a chain link fence around a 120-yd by 80-yd playground?

97. The perimeter of a square is 62 ft. What is the length of each side?

98. Find the circumference of a circle with a 20-cm diameter. Use 3.14 for π.

99. Find the area of the triangle.

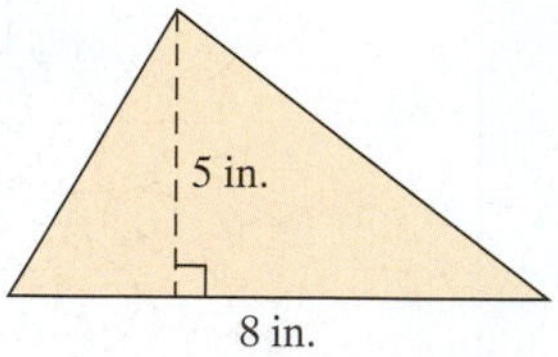

100. Fatima has a Persian rug 8.5 ft by 6 ft. What is the area?

101. A lot is 150 ft by 80 ft. Within the lot, there is a 12-ft easement along all edges. An easement is the portion of the lot on which nothing may be built. What is the area of the portion that may be used for building?

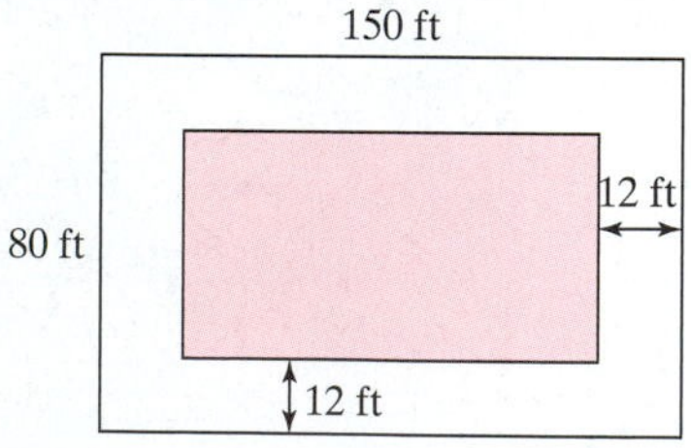

102. Find the area.

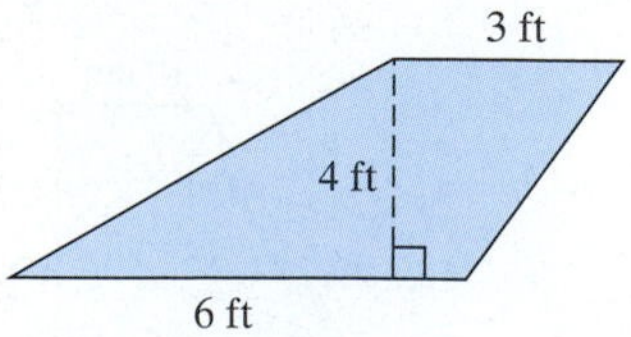

103. Find the area of a circle with a 20-cm diameter. Use 3.14 for π.

104. Find the area.

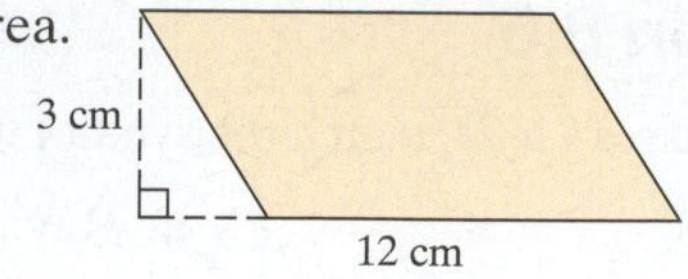

Section 8.8

For Exercises 105–108, find the volume and surface area. Use 3.14 for π.

105.

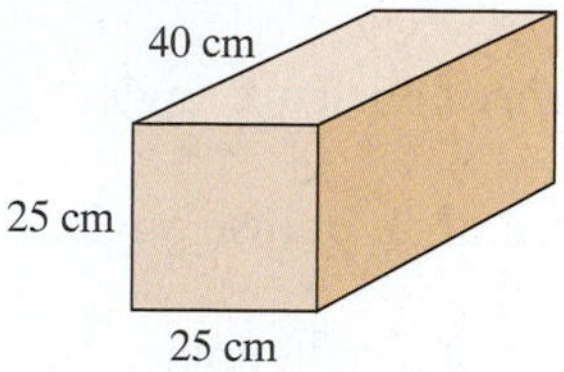

106.

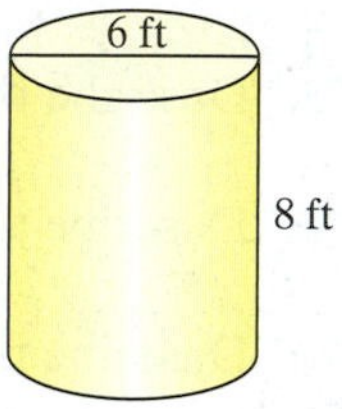

107.

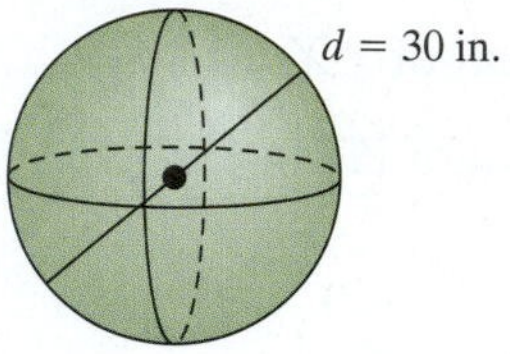

108.

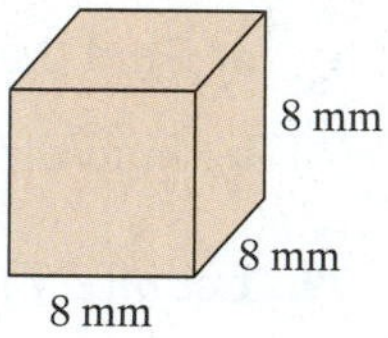

109. Find the volume. Use 3.14 for π.

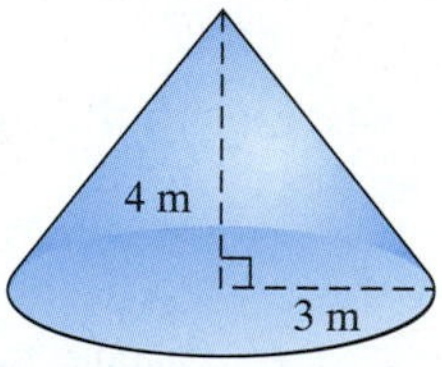

110. Find the volume of a can of paint if the can is a cylinder with radius 6.5 in. and height 7.5 in. Round to the nearest whole unit.

111. Find the volume of a ball if the diameter of the ball is approximately 6 in. Round to the nearest whole unit.

112. A microwave oven is a rectangular solid with dimensions 1 ft by 1 ft 9 in. by 1 ft 4 in. Find the volume in cubic feet.

For Exercises 113–114, find the volume of the shaded region. Use 3.14 for π if necessary. Round to the nearest whole unit.

113.

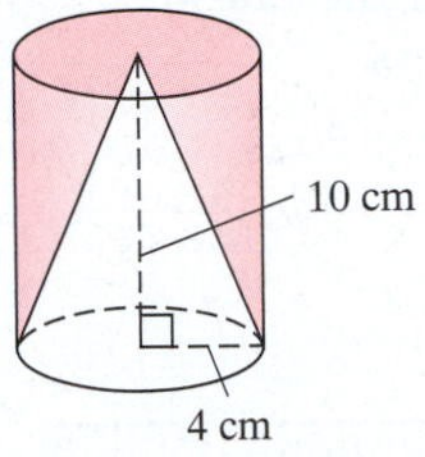

114.

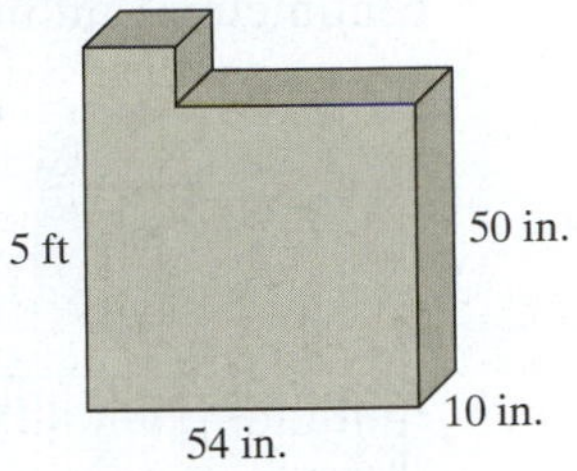

Chapter 8 Test

1. Identify the units that apply to measuring length. Circle all that apply.

a. Pound **b.** Ounce **c.** Meter
d. Mile **e.** Gram **f.** Pint
g. Feet **h.** Liter **i.** Fluid ounce
j. Kilometer

2. Identify the units that apply to measuring capacity. Circle all that apply.

a. Pound **b.** Ounce **c.** Meter
d. Mile **e.** Gram **f.** Pint
g. Foot **h.** Liter **i.** Fluid ounce
j. Kilometer

3. Identify the units that apply to measuring mass or weight. Circle all that apply.

a. Pound **b.** Ounce **c.** Meter
d. Mile **e.** Gram **f.** Pint
g. Foot **h.** Liter **i.** Fluid ounce
j. Kilometer

4. A backyard needs 25 ft of fencing. How many yards is this?

5. It's estimated that an adult *Tyrannosaurus Rex* weighed approximately 11,000 lb. How many tons is this?

6. Two exits on the highway are 52,800 ft apart. How many miles is this?

7. A recipe for brownies calls for $\frac{3}{4}$ c of milk and 4 oz of water. What is the total amount of liquid in ounces?

8. A television show has 1200 sec of commercials. How many minutes is this?

9. Find the perimeter of the rectangle in feet.

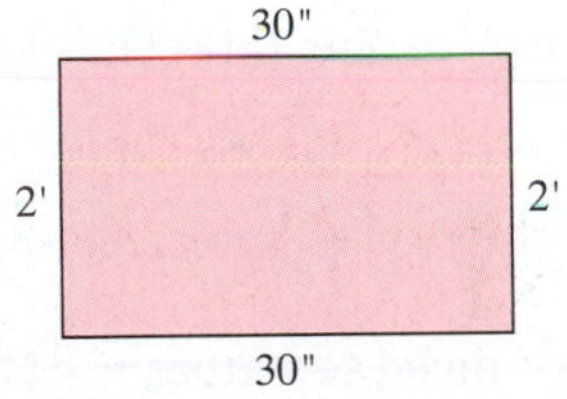

10. A decorator wraps a gift, using two pieces of ribbon. One is 1′10″ and the other is 2′4″. Find the total length of ribbon used.

11. When Stephen was born, he weighed 8 lb 1 oz. When he left the hospital, he weighed 7 lb 10 oz. How much weight did he lose after he was born?

12. A decorative pillow requires 3 ft 11 in. of fringe around the perimeter of the pillow. If five pillows are produced, how much fringe is required?

13. Josh ran a race and finished with the time of 1:15:15. Convert this time to minutes.

14. Approximate the width of the nut in centimeters and millimeters.

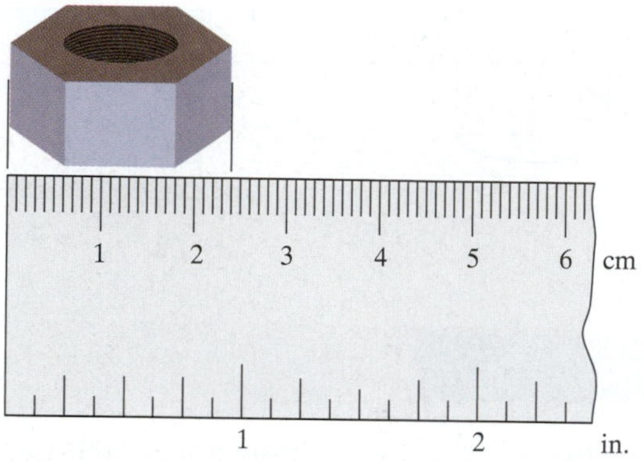

15. Select the most reasonable measurement for the length of a living room.

a. 5 mm **b.** 5 cm **c.** 5 m **d.** 5 km

16. The length of the Mackinac Bridge in Michigan is 1158 m. What is this length in kilometers?

17. A tablespoon (1 T) contains 0.015 L. How many milliliters is this?

18. **a.** What does the abbreviation cc stand for?

b. Convert 235 mL to cubic centimeters.

c. Convert 1 L to cubic centimeters.

19. A can of diced tomatoes is 411 g. Convert 411 g to centigrams.

20. A box of crackers is 210 g. If a serving of crackers is 30,000 mg, how many servings are in the box?

For Exercises 21–26, refer to Table 8-5 on page 502.

21. A bottle of Sprite contains 2 L. What is the capacity in quarts? Round to the nearest tenth.

22. Usain Bolt is one of the premier sprinters in track and field. His best race is the 100-m. How many yards is this? Round to the nearest yard.

23. The distance between two exits on a highway is 4.5 km. How far is this in miles? Round to the nearest tenth.

24. Breckenridge, Colorado, is 9603 ft above sea level. What is this height in meters? Round to the nearest meter.

25. A snowy egret stands about 20 in. tall and has a 38-in. wingspan. Convert both values to centimeters.

26. The mass of a laptop computer is 5000 g. What is the weight in pounds? Round to the nearest pound.

27. The oven temperature needed to bake cookies is 375°F. What is this temperature in degrees Celsius? Round to the nearest tenth.

28. The average January temperature in Albuquerque, New Mexico, is 2°C. Convert this temperature to degrees Fahrenheit.

29. A patient is supposed to get 0.1 mg of a drug for every kilogram of body mass, four times a day. How much of the drug would a 70-kg woman get each day?

30. Gus, the cat, had an overactive thyroid gland. The vet prescribed 0.125 mg of Methimazole every 12 hr. How many micrograms is this per week?

31. Which is a correct representation of the line shown?

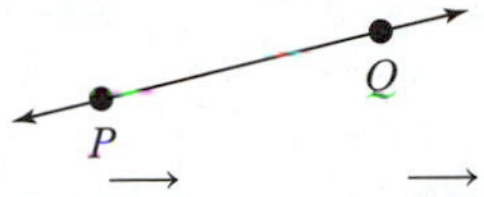

a. $\overline{PQ}$ **b.** $\overrightarrow{PQ}$ **c.** $\overrightarrow{QP}$ **d.** $\overleftrightarrow{PQ}$

32. Which is a correct representation of the ray pictured?

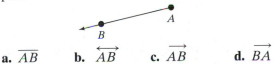

a. $\overline{AB}$ **b.** $\overleftrightarrow{AB}$ **c.** $\overrightarrow{AB}$ **d.** $\overrightarrow{BA}$

33. What is the complement of a 16° angle?

34. What is the supplement of a 147° angle?

35. Find the missing angle.

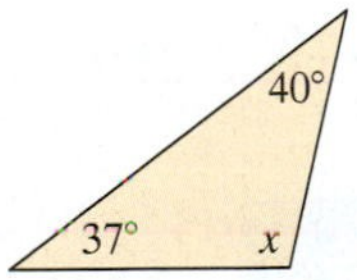

36. Determine the perimeter and area.

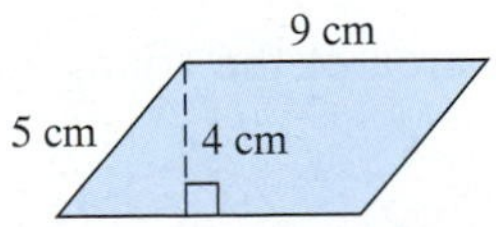

37. A farmer uses a rotating sprinkler to water his crops. If the spray of water extends 150 ft, find the area of one such region. Use 3.14 for π.

38. Find the area of the shaded region.

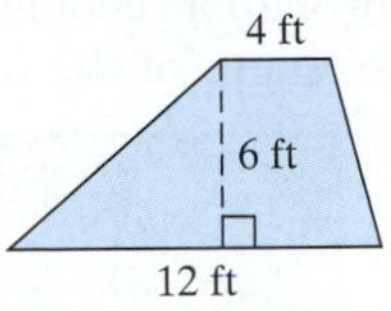

39. Determine the measure of angles x and y. Assume that line 1 is parallel to line 2.

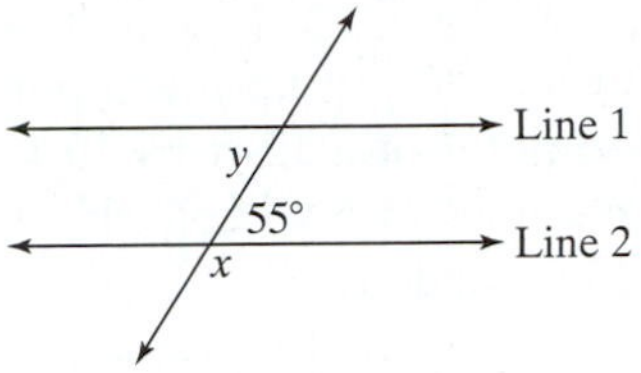

40. Given that the lengths of $\overline{AB}$ and $\overline{BC}$ are equal, what are the measures of $\angle A$ and $\angle C$?

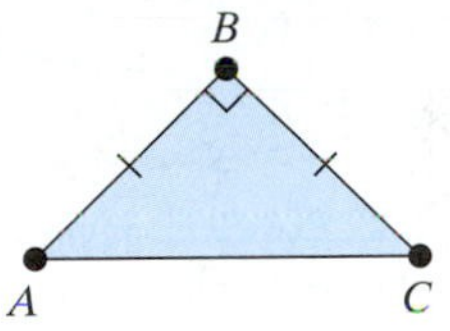

41. From the figure, determine $m(\angle S)$.

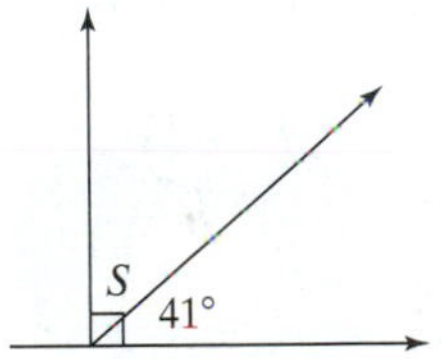

42. What is the sum of all the angles of a triangle?

43. What is the measure of $\angle A$?

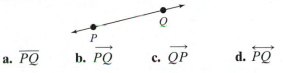

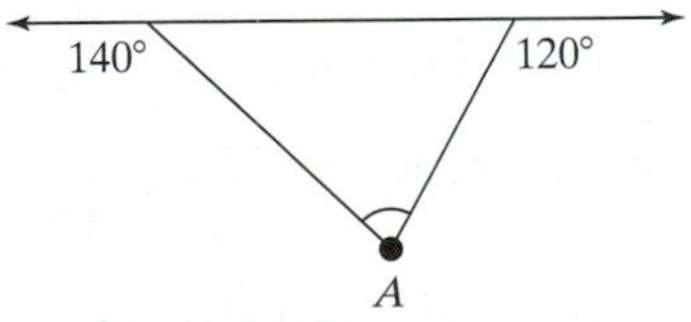

44. A firefighter places a 13-ft ladder against a wall of a burning building. If the bottom of the ladder is 5 ft from the base of the building, how far up the building will the ladder reach?

45. José is a landscaping artist and wants to make a walkway through a rectangular garden, as shown. What is the length of the walkway?

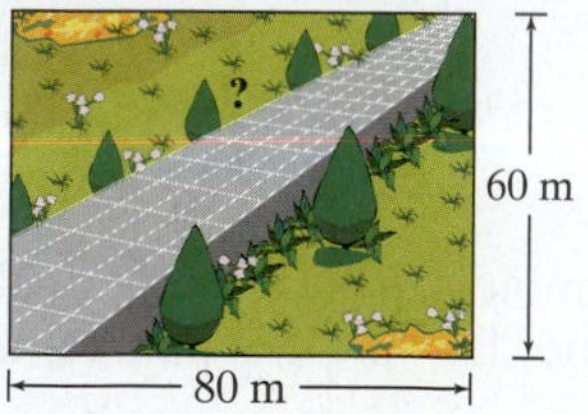

46. Jayne wants to put up a wallpaper border for the perimeter of a 12-ft by 15-ft room. The border comes in 6-yd rolls. How many rolls would be needed?

47. Find the area of the ceiling fan blade shown in the figure.

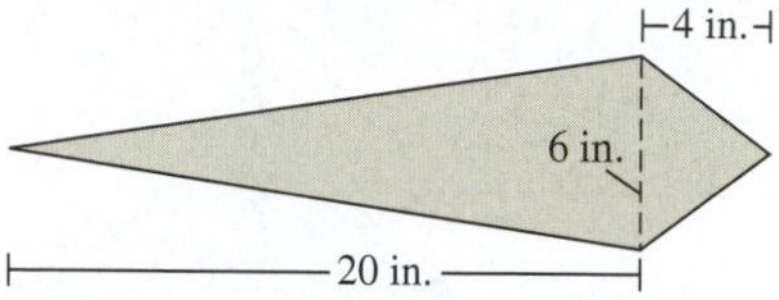

48. Find the volume of the child's wading pool shown in the figure. Use 3.14 for π and round the answer to the nearest whole unit.

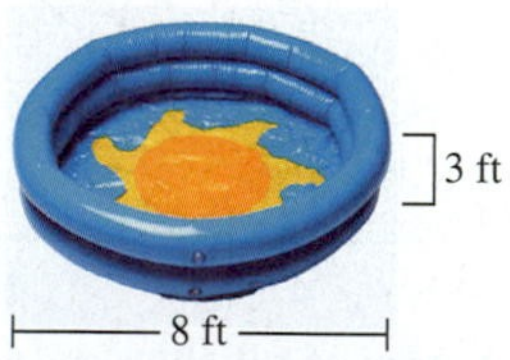

49. Find the volume of the briefcase.

For Exercises 50–51, determine the volume. Use 3.14 for π.

50.

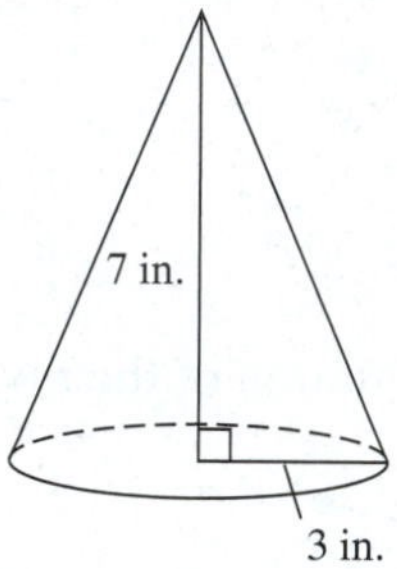

51.

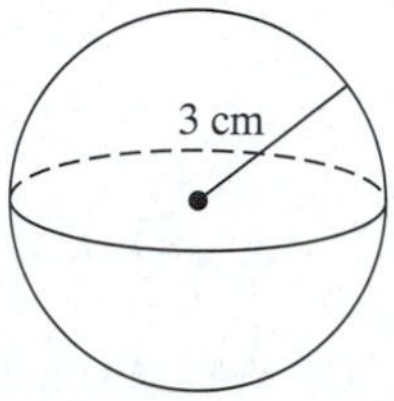

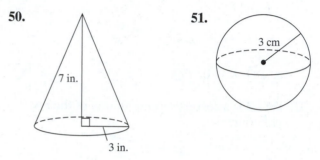

52. Determine the surface area for the sphere in Exercise 51.

53. Determine the surface area for the rectangular briefcase in Exercise 49, excluding the handle.

Chapters 1–8 Cumulative Review Exercises

1. Round each number to the indicated place.

a. 2499; thousands place **b.** 42,099; tens place

For Exercises 2–3, refer to the figure.

2. Find the perimeter.

3. Find the area.

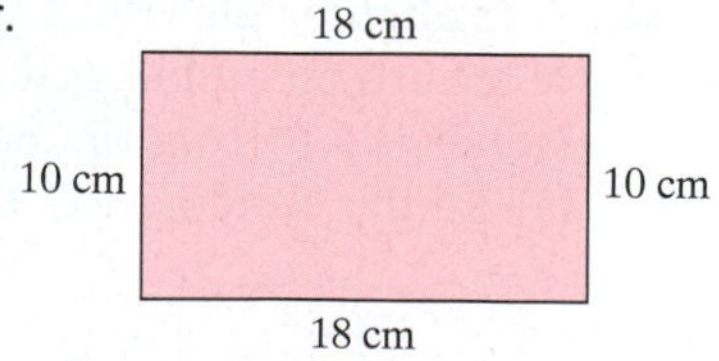

4. Simplify. $-144 \div 9 \div (17 - 3 \cdot 5)^2$

5. Is 32,542 divisible by 3? Explain why or why not.

6. Write the prime factorization of 108.

7. Find the area of the triangle.

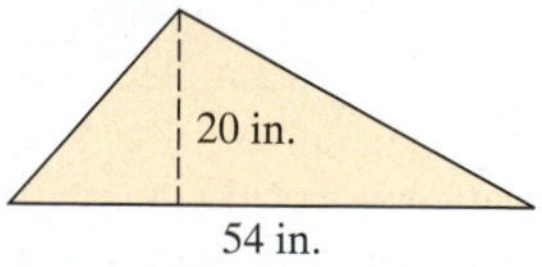

For Exercises 8–11, perform the indicated operations with mixed numbers.

8. $6\frac{2}{3} + 2\frac{5}{6}$

9. $6\frac{2}{3} \cdot 2\frac{5}{6}$

10. $6\frac{2}{3} \div 2\frac{5}{6}$

11. $6\frac{2}{3} - 2\frac{5}{6}$

For Exercises 12–15, simplify the expression.

12. $-2 - 36 \div (5 - 8)^2$

13. -6^2

14. $-5x + 3z + 8x - 4z$

15. $-7(a - 2b) + 4a + 5b$

For Exercises 16–19, solve the equation.

16. $-18 = -4 + 2x$

17. $10 = -0.4t$

18. $-\frac{2}{3}y = 12$

19. $3 - 5y = 4y - 15$

20. A pharmacist has a 1-L bottle of an antacid solution. How many smaller bottles can he fill if they contain 25 mL each?

21. If a car dealership sells 18 cars in a 5-day period, how many cars can the dealership expect to sell in 25 days?

22. A hospital ward has 40 beds and employs 6 nurses. What is the unit rate of beds per nurse? Round to the nearest tenth.

23. Four-fifths of the drinks sold at a movie theater were soda. What percent is this?

24. Assume that these figures are similar. Solve for x.

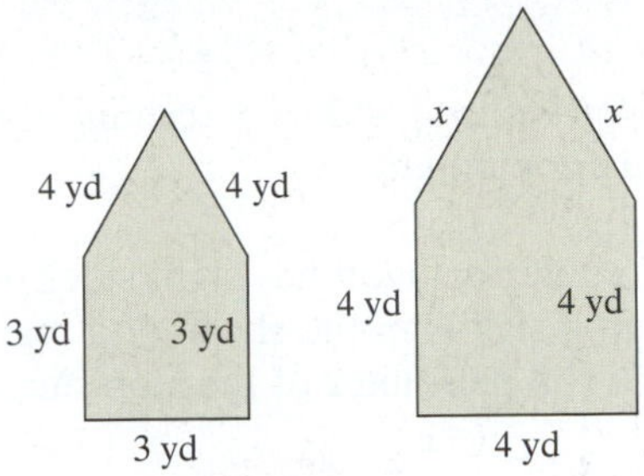

25. Of the trees in a forest, 62% were saved from a fire. If this represents 1420 trees, how many trees were originally in the forest? Round to the nearest whole tree.

26. Of 60 people, 45% had eaten at least one meal at McDonald's this week. How many people among the 60 ate at McDonald's?

27. The sales tax on a \$21 meal is \$1.26. What is the sales tax rate?

28. The commission rate that Yvonne earns is 14% of sales. If she earned \$2100 in commission, how much did she sell?

29. If \$5000 is invested at 3.4% simple interest for 6 years, how much interest is earned?

For Exercises 30–35, convert the units of capacity, length, mass, and weight. Round to one decimal place, if necessary.

30. 5800 g = _____ kg **31.** 5.8 kg ≈ _____ lb

32. 72 in. ≈ _____ cm **33.** 72 in. = _____ ft

34. $3\frac{1}{2}$ qt = _____ pt **35.** $3\frac{1}{2}$ qt ≈ _____ L

36. Given $m(\angle A) = 38°$, determine the complement and supplement.

37. A regular octagon has eight sides of equal length. A stop sign is in the shape of a regular octagon. Find the perimeter of the stop sign.

38. Determine the length of the missing side.

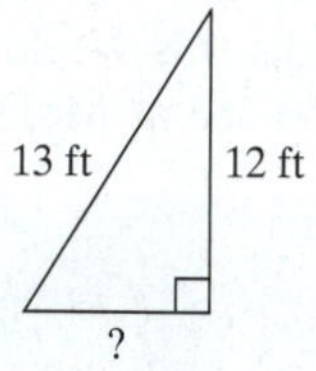

39. Determine the measure of the missing angle.

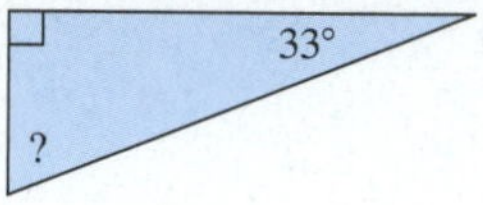

40. Find the volume and surface area.

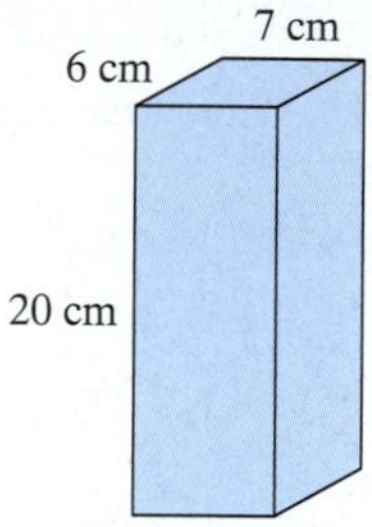

Graphs and Statistics

9

CHAPTER OUTLINE

Chapter 9

In this chapter, we study methods by which to display information graphically. We begin by plotting points to show a relationship between two variables. Then we interpret and construct different types of graphs.

Are You Prepared?

In Exercises 1–9, we review the skills of locating numbers on a number line and solving equations. Both skills will be used for graphing. For Exercises 1–5, match each number with its position on the number line. Then write the corresponding letter in the space at the bottom of the page.

1. -1 **2.** 5 **3.** -3.7

4. $-\frac{23}{4}$ **5.** $\frac{19}{3}$

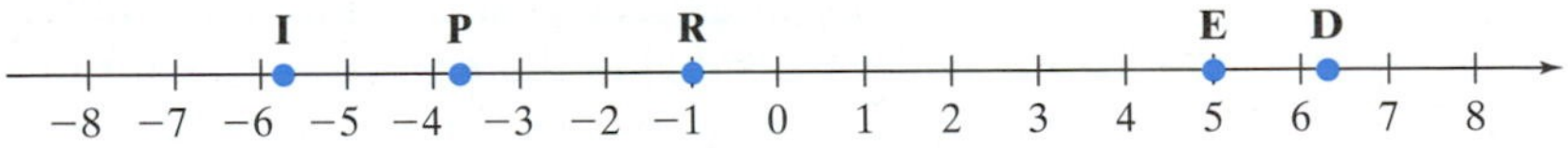

For Exercises 6–9, match the answer with the correct letter on the right. Then record the letter in the space at the bottom of the page.

6. Solve. $2x + 4 = 6$

7. Solve. $5 - 3y = 11$

8. For the equation $3x + 2y = 8$, substitute 0 for x. Then solve for y.

9. For the equation $x + 6y = -3$, substitute 0 for y. Then solve for x.

Answers

A. -2

T. -3

O. 1

N. 4

In a rectangular coordinate system, we use an ___ ___ ___ ___ ___ ___ ___   ___ ___ ___ ___
6 1 5 2 1 2 5   3 7 4 1

to represent the location of a ___ ___ ___ ___ ___.
3 6 4 8 9

Section 9.1 Rectangular Coordinate System

Objectives

1. Rectangular Coordinate System
2. Plotting Points in a Rectangular Coordinate System
3. Graphing Ordered Pairs in an Application

1. Rectangular Coordinate System

Table 9-1 represents the daily revenue for the movie *Spider-Man 3* for its opening two weeks. In table form, the information is difficult to picture and interpret. However, Figure 9-1 shows a graph of these data. From the graph, we can speculate that days 1–3 and 8–10 were weekends because revenue was up. We can also see that as time went on, revenue dropped.

Table 9-1

Day number	Revenue ($ Millions)
1	59.8
2	51.4
3	39.9
4	10.3
5	8
6	6.8
7	5.9
8	17.1
9	24.9
10	15.8
11	3.6
12	3.5
13	3.1
14	3

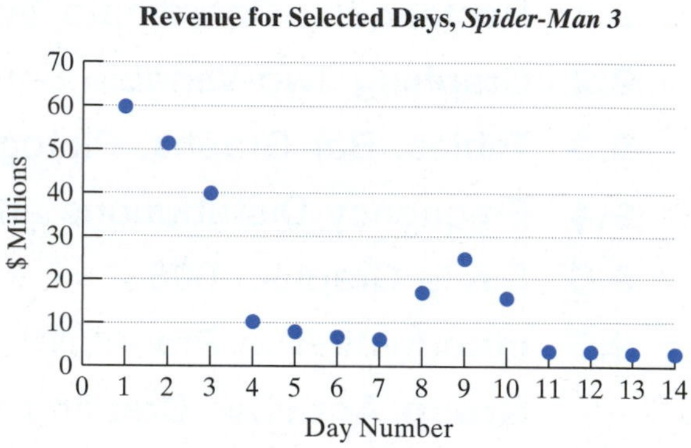

Figure 9-1

Figure 9-1 represents the variables of time and revenue. In general, to picture two variables, we use a graph with two number lines drawn at right angles to each other (Figure 9-2). This forms a **rectangular coordinate system**. The horizontal line is called the ***x*-axis**, and the vertical line is called the ***y*-axis**. The point where the lines intersect is called the **origin**. On the x-axis, the numbers to the right of the origin are positive, and the numbers to the left are negative. On the y-axis, the numbers above the origin are positive, and the numbers below the origin are negative. The x- and y-axes divide the graphing area into four regions called **quadrants**.

Animation

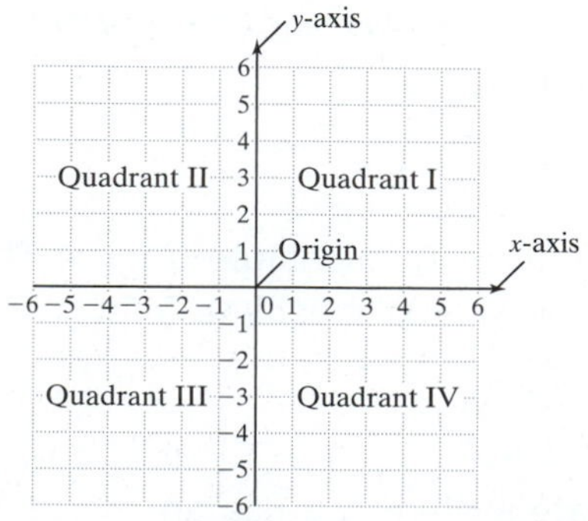

Figure 9-2

2. Plotting Points in a Rectangular Coordinate System

Points graphed in a rectangular coordinate system are defined by two numbers as an **ordered pair** (x, y). The first number (called the ***x*-coordinate** or first coordinate) is the horizontal position from the origin. The second number (called the ***y*-coordinate** or second coordinate) is the vertical position from the origin. Example 1 shows how points are plotted in a rectangular coordinate system.

Example 1 Plotting Points in a Rectangular Coordinate System

Plot the points.

a. $(3, 4)$ **b.** $(-4, 2)$ **c.** $(2, -4)$ **d.** $(-5, -3)$

Solution:

a. The ordered pair $(3, 4)$ indicates that $x = 3$ and $y = 4$.

↓ ↓
x y

Because the x-coordinate is positive, start at the origin and move 3 units in the positive x direction (to the right). Then, because the y-coordinate is positive, move 4 units in the positive y direction (upward). Draw a dot at the final location. See Figure 9-3.

The point $(3, 4)$ is located in Quadrant I.

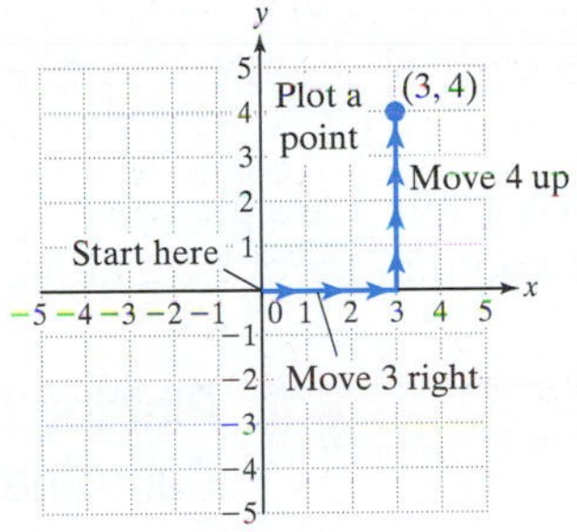

Figure 9-3

b. The ordered pair $(-4, 2)$ indicates that $x = -4$ and $y = 2$.

↓ ↓
x y

Because the x-coordinate is *negative*, we start at the origin and move 4 units in the *negative* x direction (left). Then, because the y-coordinate is positive, move 2 units in the positive y direction (upward). Draw a dot at the final location. See Figure 9-4.

The point $(-4, 2)$ is located in Quadrant II.

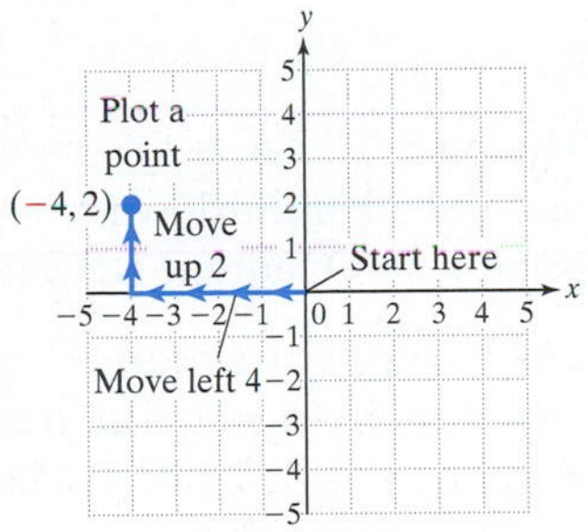

Figure 9-4

c. The ordered pair $(2, -4)$ indicates that $x = 2$ and $y = -4$.

↓ ↓
x y

Because the x-coordinate is positive, start at the origin and move 2 units in the positive x direction (to the right). Then, because the y-coordinate is *negative*, move 4 units in the *negative* y direction (downward). Draw a dot at the final location. See Figure 9-5.

The point $(2, -4)$ is located in Quadrant IV.

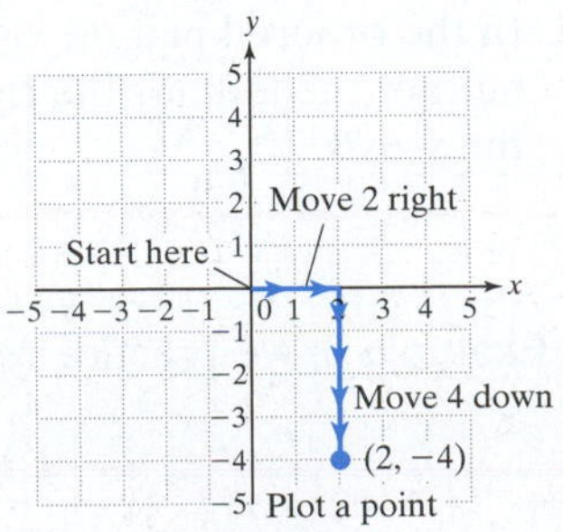

Figure 9-5

Skill Practice

Plot the points and state the quadrant in which the point lies.

1. $(-2, 3)$
2. $(4, 1)$
3. $(3, -1)$
4. $(-5, -4)$

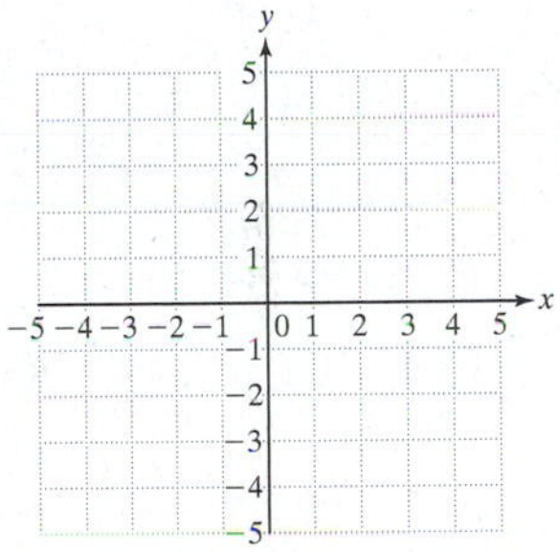

Answers

1. Quadrant II
2. Quadrant I
3. Quadrant IV
4. Quadrant III

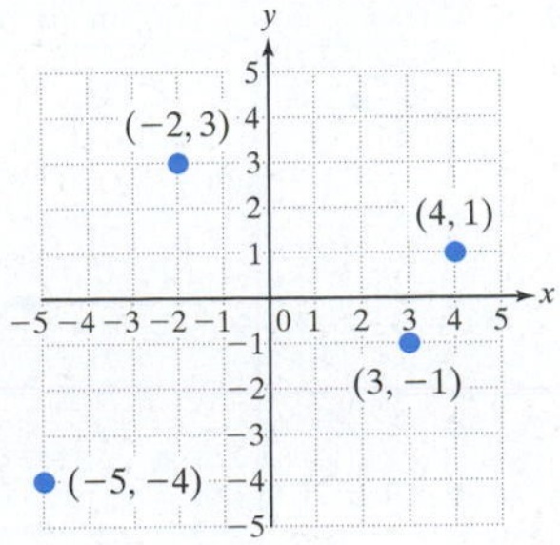

Concept Connections

5. Answer true or false: The ordered pair $(6, -1)$ is the same as $(-1, 6)$.

TIP: Notice that changing the order of the *x*- and *y*-coordinates changes the location of the point. In Example 1(b), the point $(-4, 2)$ is in Quadrant II. In Example 1(c), the point $(2, -4)$ is in Quadrant IV. This is why points are represented by *ordered* pairs. The order is important.

d. The ordered pair $(-5, -3)$ indicates that $x = -5$ and $y = -3$.

$$\underset{x}{-5}, \quad \underset{y}{-3}$$

Because the *x*-coordinate is *negative*, start at the origin and move 5 units in the *negative* *x* direction (to the left). Then, because the *y*-coordinate is *negative*, move 3 units in the *negative y* direction (downward). Draw a dot at the final location. See Figure 9-6.

The point $(-5, -3)$ is located in Quadrant III.

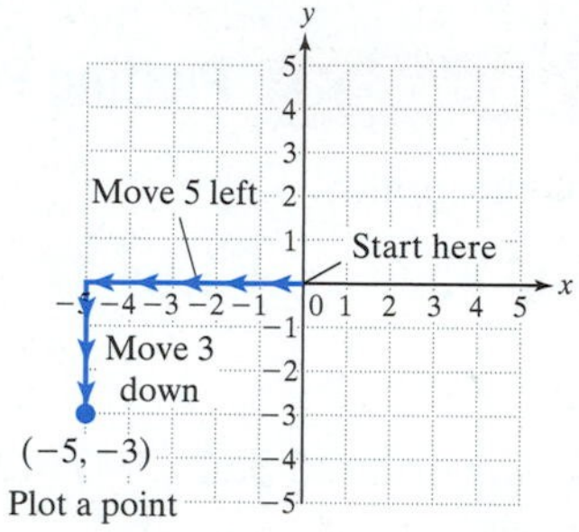

Figure 9-6

In Example 2, we plot points whose coordinates are fractions, decimals, or zero. When the coordinates of a point are not integers, their location on the graph is estimated.

Skill Practice

Plot the points.

6. $(2.1, -3.5)$
7. $\left(-\frac{9}{4}, \frac{9}{2}\right)$
8. $(5, 0)$
9. $(0, 2.5)$

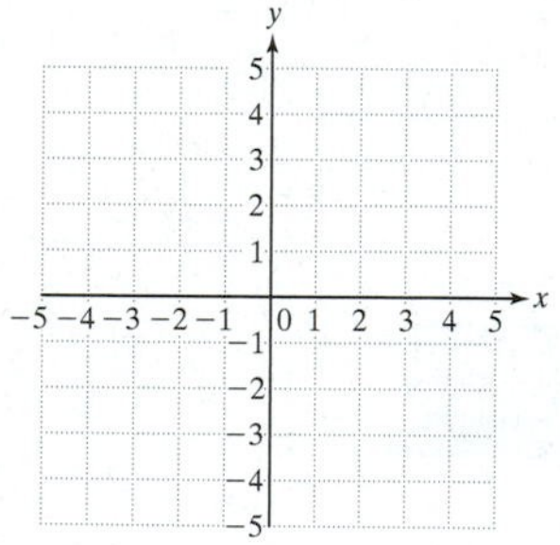

Example 2 Plotting Points in a Rectangular Coordinate System

Plot the points.

a. $(-4.6, -3.8)$ **b.** $\left(-\dfrac{5}{2}, \dfrac{13}{3}\right)$ **c.** $(3, 0)$ **d.** $(0, -4)$

Solution:

a. The point $(-4.6, -3.8)$ is located 4.6 units to the left and 3.8 units down from the origin. The point is in Quadrant III. See Figure 9-7.

b. The improper fraction $-\frac{5}{2}$ can be written as the mixed number $-2\frac{1}{2}$. The fraction $\frac{13}{3}$ can be written as $4\frac{1}{3}$. The point is in Quadrant II.

c. In the ordered pair $(3, 0)$, the *y*-coordinate is zero. Therefore, we move neither upward nor downward from the origin. This indicates that the point is on the *x*-axis.

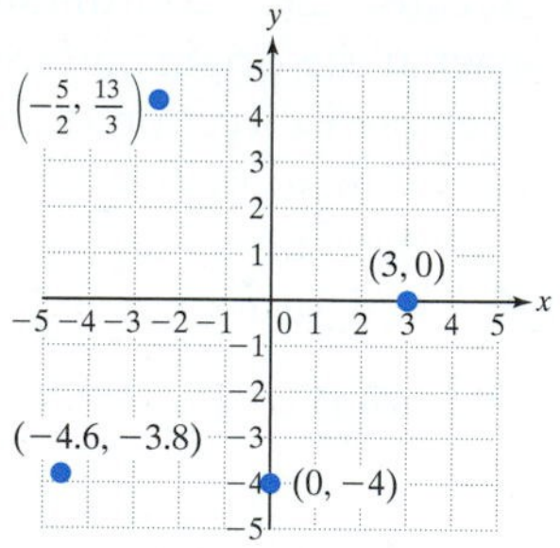

Figure 9-7

d. In the ordered pair $(0, -4)$, the *x*-coordinate is zero. Therefore, we move to neither the left nor the right of the origin. This indicates that the point is on the *y*-axis.

In Example 3, we practice reading ordered pairs from a graph.

Answers

5. False

6–9.

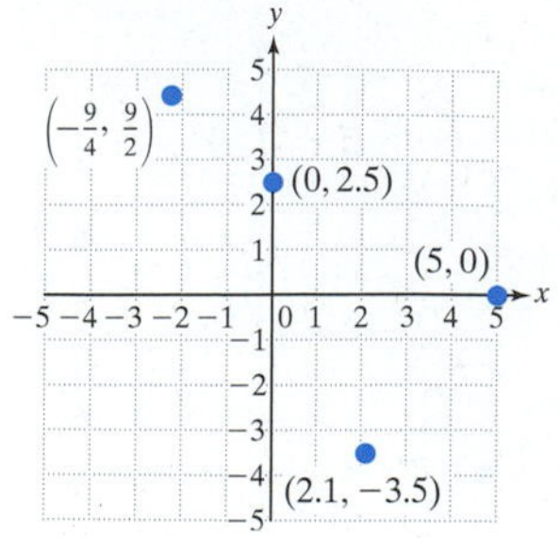

Example 3 Reading Ordered Pairs from a Graph

Estimate the coordinates of points A, B, C, D, E, and F from Figure 9-8.

Solution:

Point A is at $(-5, -1)$.

Point B is at $(-4, 4)$.

Point C is at $(2\frac{1}{2}, 0)$.

Point D is at $(5, 2)$.

Point E is at $(3\frac{1}{2}, -1\frac{1}{2})$.

Point F is at $(0, 0)$.

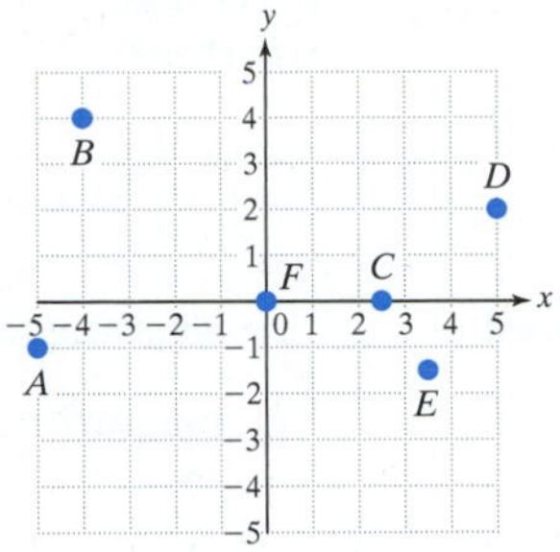

Figure 9-8

Skill Practice

Estimate the coordinates of the points in the graph.

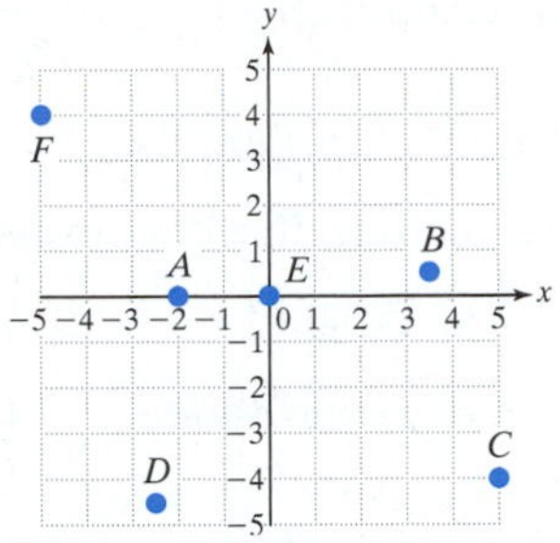

10. A **11.** B **12.** C
13. D **14.** E **15.** F

3. Graphing Ordered Pairs in an Application

In Example 4, we graph points in an application.

Example 4 Graphing Ordered Pairs in an Application

The data given in Table 9-2 represent the gas mileage for a subcompact car for various speeds. Let x represent the speed of the car, and let y represent the corresponding gas mileage.

a. Write the table values as ordered pairs.

b. Graph the ordered pairs.

Table 9-2

Speed x (mph)	Gas Mileage y (mpg)
25	27
30	29
35	33
40	34
45	35
50	33
55	30
60	27
65	24
70	21

Solution:

a. The ordered pairs are given by (25, 27), (30, 29), (35, 33), (40, 34), (45, 35), (50, 33), (55, 30), (60, 27), (65, 24), and (70, 21).

b.

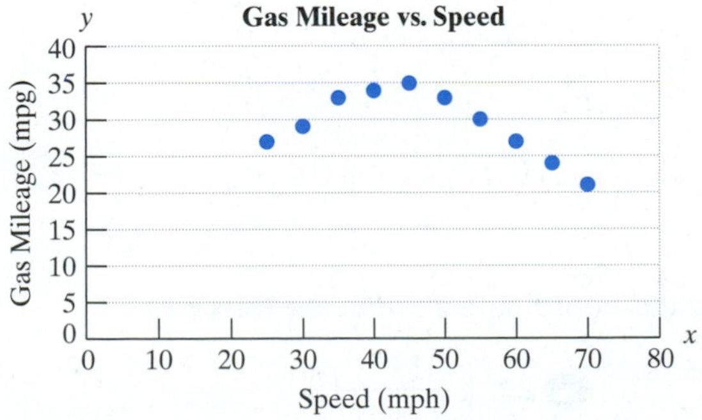

From the graph, we see that the most efficient speed is approximately 45 mph. That is where the gas mileage is at its peak.

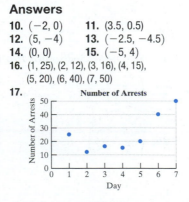

Skill Practice

For Exercises 16–17, refer to the data values given in the table representing the number of arrests made by a police department for one week.

16. Write the table values as ordered pairs.

17. Graph the ordered pairs.

Day Number x	Number of Arrests y
1	25
2	12
3	16
4	15
5	20
6	40
7	50

Answers

10. $(-2, 0)$ **11.** $(3.5, 0.5)$
12. $(5, -4)$ **13.** $(-2.5, -4.5)$
14. $(0, 0)$ **15.** $(-5, 4)$
16. (1, 25), (2, 12), (3, 16), (4, 15), (5, 20), (6, 40), (7, 50)
17.

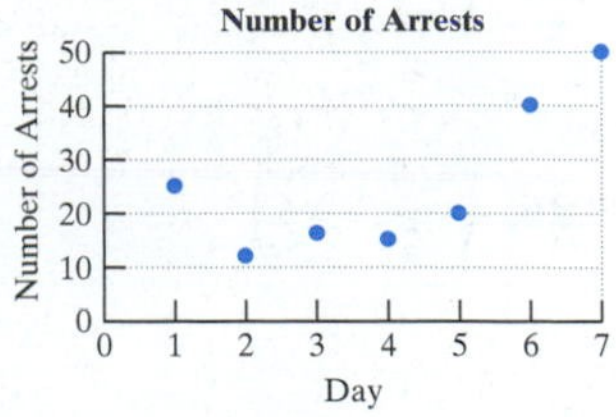

Section 9.1 Practice Exercises

Study Skills Exercise

1. Define the key terms.

a. Origin **b. Quadrant** **c. Rectangular coordinate system** **d. *x*-axis**

e. *x*-coordinate **f. *y*-axis** **g. *y*-coordinate** **h. Ordered pair**

Objective 1: Rectangular Coordinate System

For Exercises 2–8, place each key term in the appropriate location in the graph.

2. *x*-axis **3.** *y*-axis **4.** Origin

5. Quadrant I **6.** Quadrant II **7.** Quadrant III

8. Quadrant IV

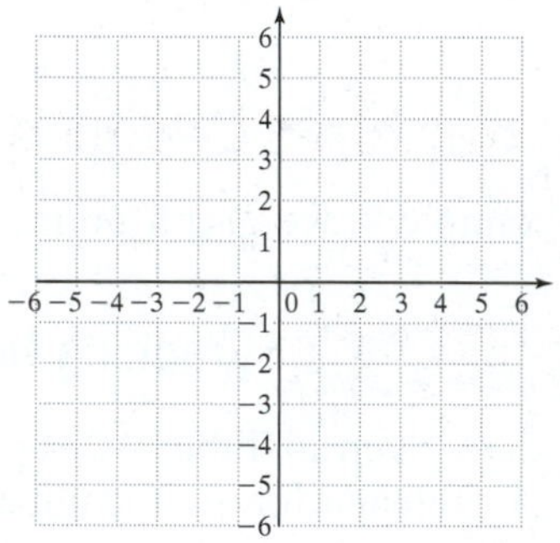

Objective 2: Plotting Points in a Rectangular Coordinate System

For Exercises 9–14, plot the points on the rectangular coordinate system. **(See Example 1.)**

9. $(-1, 4)$ **10.** $(4, -1)$ **11.** $(2, 2)$

12. $(3, -3)$ **13.** $(-5, -2)$ **14.** $(-3, 2)$

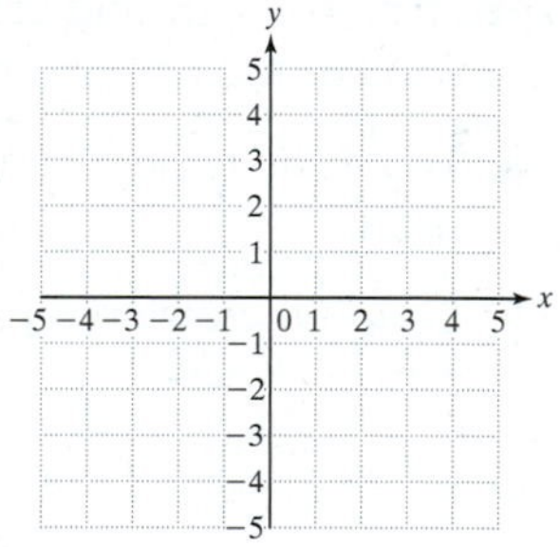

15. Explain how to plot the point $(-1.8, 3.1)$.

16. Explain how to plot the point $\left(\frac{15}{2}, \frac{15}{7}\right)$.

For Exercises 17–22, plot the points on the rectangular coordinate system. **(See Example 2.)**

17. $(-0.6, 1.1)$ **18.** $(2.3, 4.9)$ **19.** $(-1.4, 4.1)$

20. $(5.1, -3.8)$ **21.** $(0.9, -1.1)$ **22.** $(-3.3, -4.6)$

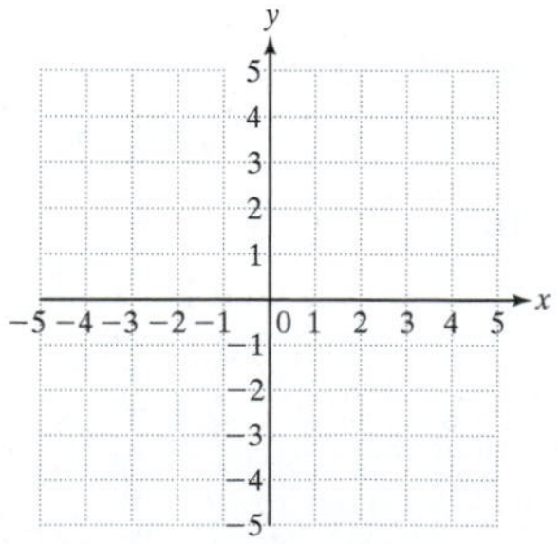

For Exercises 23–28, plot the points on the rectangular coordinate system. **(See Example 2.)**

23. $\left(\frac{1}{2}, \frac{5}{2}\right)$ **24.** $\left(-\frac{10}{3}, \frac{8}{3}\right)$ **25.** $\left(-\frac{13}{6}, -1\right)$

26. $\left(\frac{24}{5}, -\frac{8}{5}\right)$ **27.** $\left(\frac{13}{4}, \frac{29}{8}\right)$ **28.** $\left(\frac{15}{7}, -\frac{17}{6}\right)$

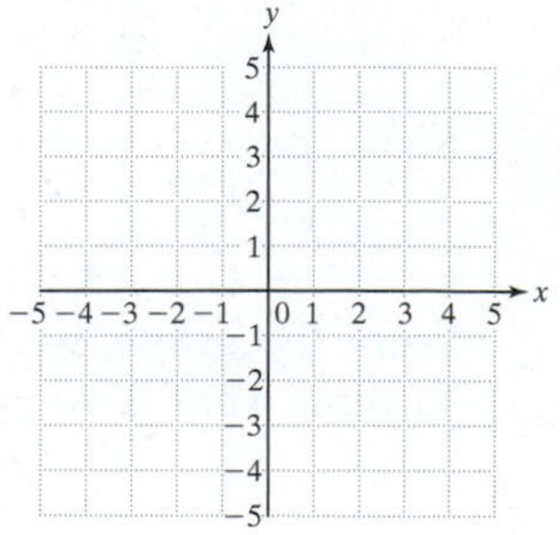

For Exercises 29–34, plot the points on the rectangular coordinate system. **(See Example 2.)**

29. $(0, -3)$ **30.** $(2, 0)$ **31.** $(4, 0)$

32. $(0, 5)$ **33.** $(0, 1)$ **34.** $(-4, 0)$

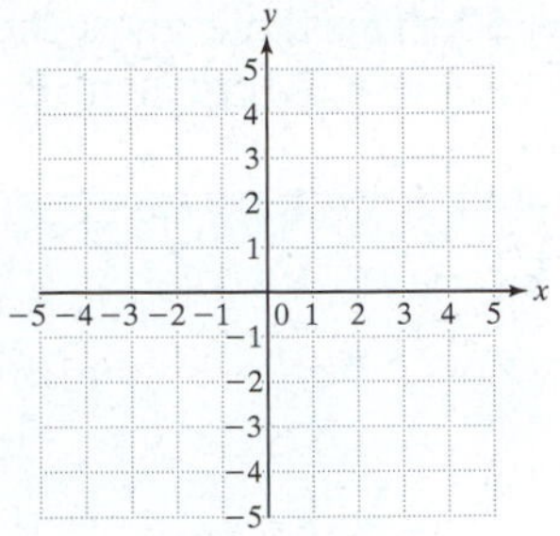

For Exercises 35–46, for each point, identify in which quadrant or on which axis it lies.

35. $(32, -44)$ **36.** $(-12, 25)$ **37.** $(-10, -5)$ **38.** $(100, 82)$

39. $(54.9, 0)$ **40.** $(0, -23.33)$ **41.** $\left(0, \frac{55}{17}\right)$ **42.** $\left(-\frac{3}{4}, 0\right)$

43. $(-27, 3)$ **44.** $(5, -75)$ **45.** $(35, 66)$ **46.** $\left(-\frac{2}{17}, -\frac{1}{50}\right)$

For Exercises 47–55, estimate the coordinate of points A, B, C, D, E, F, G, H and I. **(See Example 3.)**

47. A **48.** B **49.** C

50. D **51.** E **52.** F

53. G **54.** H **55.** I

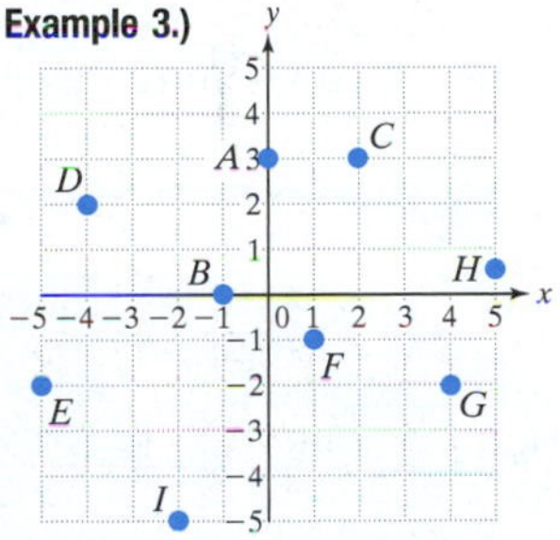

Objective 3: Graphing Ordered Pairs in an Application

For Exercises 56–61, write the table values as ordered pairs. Then graph the ordered pairs.

56. The table gives the average monthly temperature in Anchorage, Alaska, for 1 year. Let $x = 1$ represent January, $x = 2$ represent February, and so on.

Month, x	Temperature (°C), y
1	−7
2	−3
3	1
4	6
5	12
6	17
7	18
8	18
9	14
10	6
11	−1
12	−7

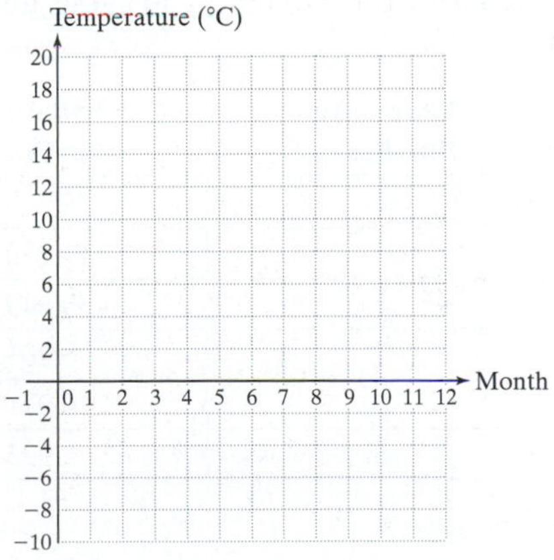

57. The table gives the average monthly temperature in Moscow for 1 year. Let $x = 1$ represent January, $x = 2$ represent February, and so on. **(See Example 4.)**

Month, x	Temperature (°C), y
1	−6
2	−5
3	1
4	11
5	18
6	22
7	24
8	22
9	15
10	8
11	1
12	−4

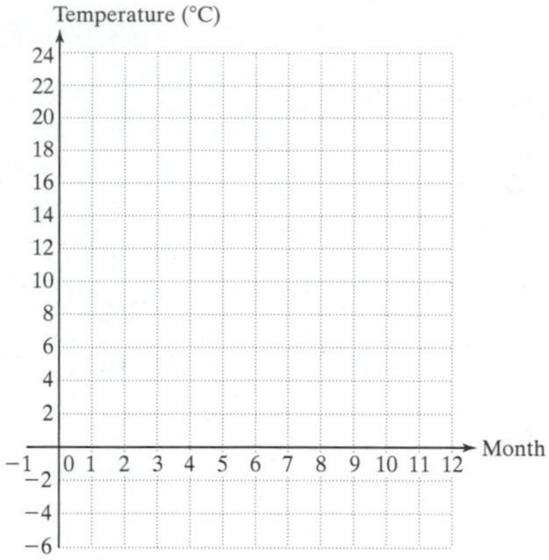

58. A car is purchased for $20,000. The table gives the amount of depreciation, x years after purchase.

Years after Purchase, x	Amount of Depreciation ($), y
1	5,000
2	6,800
3	8,384
4	10,010
5	11,509
6	12,782

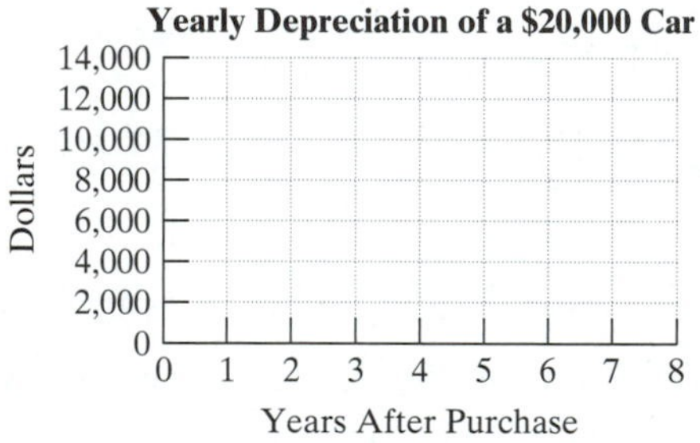

59. A car is purchased for $15,000. The table gives the value of the car, x years after purchase.

Years after Purchase, x	Value of Car ($), y
1	13,200
2	11,352
3	9,649
4	8,201
5	6,971
6	5,925

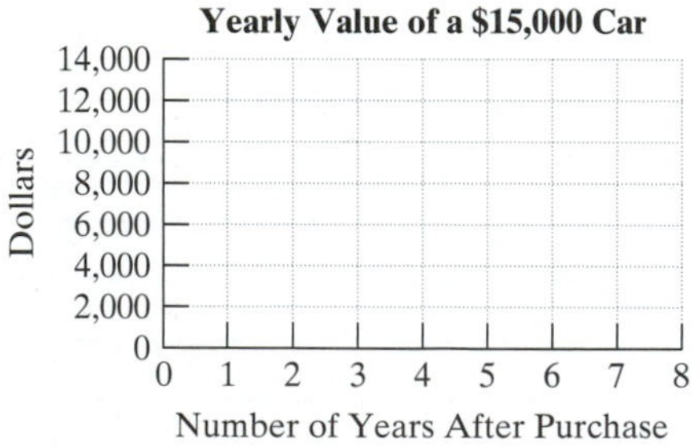

60. The table gives the gas mileage for vehicles of different weights.

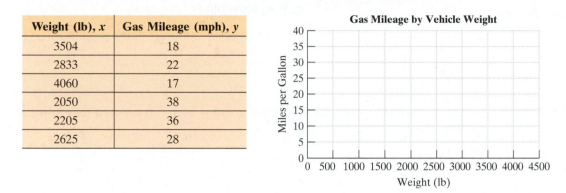

Weight (lb), x	Gas Mileage (mph), y
3504	18
2833	22
4060	17
2050	38
2205	36
2625	28

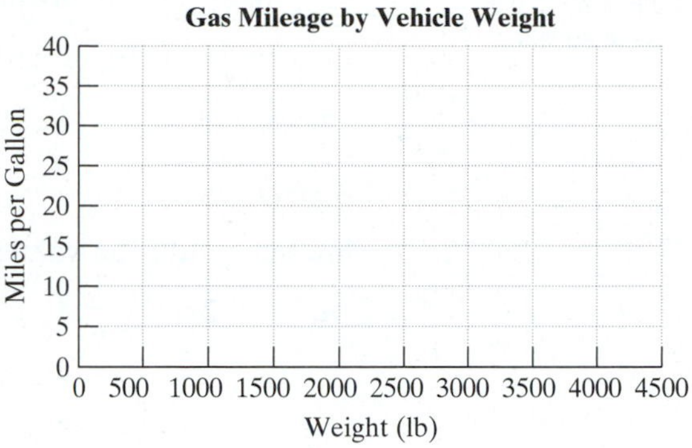

61. This table gives the gas mileage for a VW Beetle at different speeds.

Speed (mph), x	Gas Mileage (mph), y
30	33
40	35
50	40
60	37
70	30

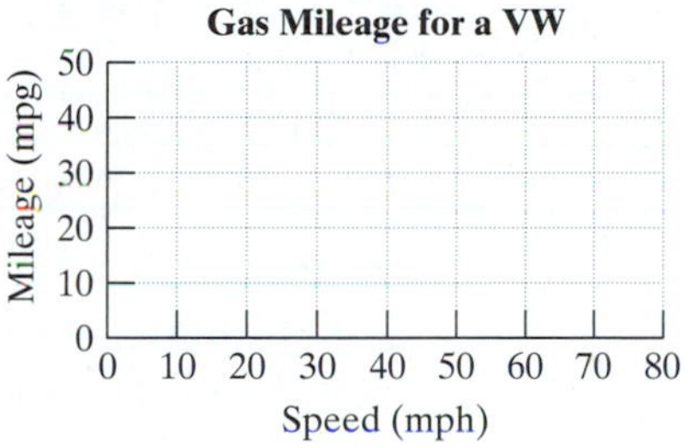

Graphing Two-Variable Equations

Section 9.2

Objectives

1. Solutions to Two-Variable Equations
2. Graphing Linear Equations in Two Variables
3. Horizontal and Vertical Lines

1. Solutions to Two-Variable Equations

We have learned how to solve linear equations in one variable such as $3x + 6 = 0$. For this equation, the solution is the single number, -2. In this section, we will find solutions to **linear equations in two variables**, such as $x + y = 4$.

A solution to $x + y = 4$ consists of two numbers, x and y, whose sum is 4. We write these numbers as an ordered pair, (x, y). Several solutions to the equation $x + y = 4$ are given:

Solution:	**Check:**
(x, y)	$x + y = 4$
$(2, 2)$	$(2) + (2) \stackrel{?}{=} 4$ ✔
$(1, 3)$	$(1) + (3) \stackrel{?}{=} 4$ ✔
$(4, 0)$	$(4) + (0) \stackrel{?}{=} 4$ ✔
$(-1, 5)$	$(-1) + (5) \stackrel{?}{=} 4$ ✔

In each case, substituting the x and y values into the equation results in a *true* statement. This means that each solution checks.

Skill Practice

Determine if the ordered pair is a solution to the equation $x + 3y = 6$.

1. $(9, -1)$
2. $(4, -2)$
3. $(0, 2)$

Example 1 Checking Solutions to a Linear Equation in Two Variables

Determine if the ordered pair is a solution to the equation $2x + y = 4$.

a. $(3, -2)$ **b.** $(1, -4)$ **c.** $(2, 0)$

Solution:

We substitute the values for x and y from the ordered pair into the equation.

a.
$$2x + y = 4$$
$$2(3) + (-2) \stackrel{?}{=} 4$$
$$6 + (-2) \stackrel{?}{=} 4$$
$$4 \stackrel{?}{=} 4 \checkmark \text{ (true)}$$

b.
$$2x + y = 4$$
$$2(1) + (-4) \stackrel{?}{=} 4$$
$$2 + (-4) \stackrel{?}{=} 4$$
$$-2 \stackrel{?}{=} 4 \text{ (false)}$$

c.
$$2x + y = 4$$
$$2(2) + (0) \stackrel{?}{=} 4$$
$$4 + 0 \stackrel{?}{=} 4$$
$$4 \stackrel{?}{=} 4 \checkmark \text{ (true)}$$

a. Substituting the values for x and y results in a *true* statement. Therefore, the ordered pair $(3, -2)$ is a solution to the equation $2x + y = 4$.

b. Substituting the values for x and y results in a *false* statement. Therefore, the ordered pair $(1, -4)$ is *not* a solution.

c. Substituting the values for x and y results in a *true* statement. Therefore, the ordered pair $(2, 0)$ is a solution.

Skill Practice

Complete the ordered pairs so that they are solutions to $2x + y = 8$.

4. $(\quad, 14)$
5. $(2, \quad)$

Example 2 Finding Solutions to a Linear Equation in Two Variables

Complete the ordered pairs so that they are solutions to the equation $x - y = 2$.

a. $(\quad, 5)$ **b.** $(3, \quad)$

Solution:

To complete an ordered pair, first substitute the given value into the equation. Then solve for the remaining variable.

a.
$$x - y = 2$$
$$x - (5) = 2 \qquad \text{Substitute 5 for } y \text{, and solve for } x.$$
$$x - 5 + 5 = 2 + 5 \qquad \text{Add 5 to both sides.}$$
$$x = 7 \qquad \text{A solution to } x - y = 2 \text{ is } (7, 5).$$

<u>Check</u>: $x - y = 2$
$$(7) - (5) \stackrel{?}{=} 2$$
$$2 \stackrel{?}{=} 2 \checkmark \text{ (true)}$$
The solution $(7, 5)$ checks.

Answers

1. Yes 2. No 3. Yes
4. $(-3, 14)$ 5. $(2, 4)$

b. $x - y = 2$

$(3) - y = 2$ Substitute 3 for x, and solve for y.

$3 - 3 - y = 2 - 3$ Subtract 3 from both sides.

$-y = -1$

$\frac{-y}{-1} = \frac{-1}{-1}$ Divide both sides by -1.

$y = 1$ A solution to $x - y = 2$ is $(3, 1)$.

Check: $x - y = 2$

$(3) - (1) \stackrel{?}{=} 2$

$2 \stackrel{?}{=} 2$ ✔ (true) The solution $(3, 1)$ checks.

Example 3 Finding Solutions to a Linear Equation in Two Variables

Determine solutions to the equation $2x + y = -3$ by completing the table.

x	y
−2	
	−5
0	

Solution:

Each entry in the table represents part of an ordered pair.

x	y	
−2		→ (−2,)
	−5	→ (, −5)
0		→ (0,)

The values in the table are selected arbitrarily. In each case, we substitute the given value into the equation. Then solve for the remaining variable.

Complete: $(-2, \ )$

$2x + y = -3$

$2(-2) + y = -3$

$-4 + y = -3$

$-4 + 4 + y = -3 + 4$

$y = 1$

$(-2, 1)$ is a solution.

Complete: $(\ , -5)$

$2x + y = -3$

$2x + (-5) = -3$

$2x - 5 + 5 = -3 + 5$

$2x = 2$

$x = 1$

$(1, -5)$ is a solution.

Complete: $(0, \ )$

$2x + y = -3$

$2(0) + y = -3$

$0 + y = -3$

$y = -3$

$(0, -3)$ is a solution.

The completed table is shown.

x	y	
−2	1	→ (−2, 1)
1	−5	→ (1, −5)
0	−3	→ (0, −3)

Skill Practice

Determine solutions to the equation $3x - y = 4$ by completing the table.

6.

x	y
1	
	5
0	

Answer

6.

x	y
1	−1
3	5
0	−4

2. Graphing Linear Equations in Two Variables

In the introduction to this section, we found several solutions to the equation $x + y = 4$. If we graph these solutions, notice that the points all line up. (See Figure 9-9.)

Equation: $x + y = 4$

Several solutions: $(2, 2)$

$(1, 3)$

$(4, 0)$

$(-1, 5)$

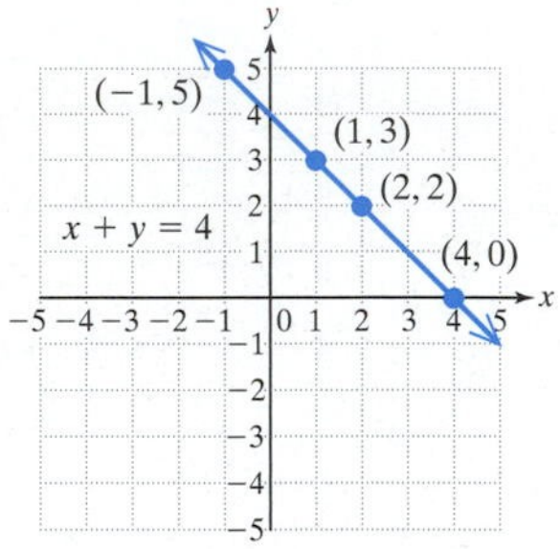

Figure 9-9

The equation actually has infinitely many solutions. This is because there are infinitely many combinations of x and y whose sum is 4. The graph of *all* solutions to this equation makes up the line shown in Figure 9-9. The arrows at each end indicate that the line extends infinitely. This is called the *graph of the equation.*

The graph of a linear equation is a line. Therefore, we need to plot at least two points and then draw the line between them. This is demonstrated in Example 4.

Skill Practice

Graph the equation.

7. $-x + y = 4$

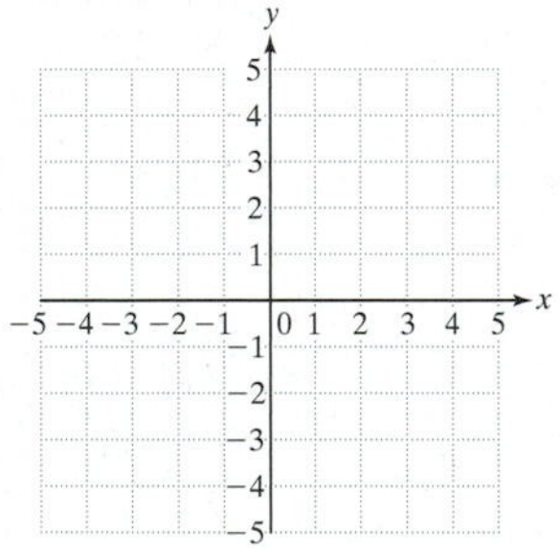

Example 4 Graphing a Linear Equation

Graph the equation. $-x + y = 2$

Solution:

We will find three ordered pairs that are solutions to $-x + y = 2$. To find the ordered pairs, choose arbitrary values for x or y, such as those shown in the table. Then complete the table.

x	y	
3		→ $(3,\ \ )$
	−2	→ $(\ \ , -2)$
−1		→ $(-1,\ \ )$

Complete: $(3,\ \ )$

$$-x + y = 2$$
$$-(3) + y = 2$$
$$-3 + 3 + y = 2 + 3$$
$$y = 5$$

Complete: $(\ \ , -2)$

$$-x + y = 2$$
$$-x + (-2) = 2$$
$$-x - 2 + 2 = 2 + 2$$
$$-x = 4$$
$$x = -4$$

Complete: $(-1,\ \ )$

$$-x + y = 2$$
$$-(-1) + y = 2$$
$$1 + y = 2$$
$$1 - 1 + y = 2 - 1$$
$$y = 1$$

Answer

7.

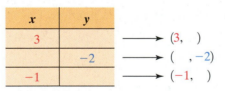

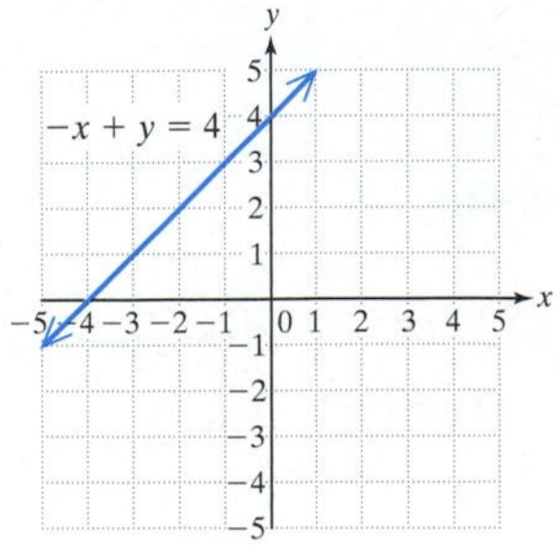

The completed table is given, and the ordered pairs are graphed in Figure 9-10. The line through the points is the graph of all solutions to the equation $-x + y = 2$.

x	y	
3	5	⟶ $(3, 5)$
−4	−2	⟶ $(-4, -2)$
−1	1	⟶ $(-1, 1)$

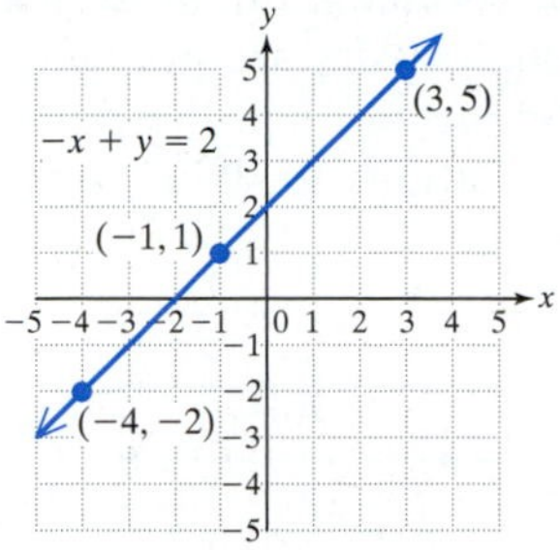

Figure 9-10

Avoiding Mistakes

Only two points are needed to graph a line. However, in Example 4, we found three points as a check. If all three points line up, we can be reasonably sure that we graphed the line correctly. If they do not line up, then we made a mistake in one or more of the ordered pairs.

PROCEDURE Graphing a Linear Equation in Two Variables

Step 1 Create a table to find at least two ordered pair solutions to the equation. For each ordered pair, arbitrarily choose a value for one of the variables.

Step 2 In each case, substitute the chosen value into the equation and solve for the remaining variable.

Step 3 Plot the points on the graph.

Step 4 Draw a line through the points. Include arrows at each end to show that the line extends infinitely.

Example 5 Graphing a Linear Equation

Graph the equation. $2x + 4y = 4$

Solution:

We will find three ordered pairs that are solutions to the equation $2x + 4y = 4$. In the table, we have selected arbitrary values for x and y. Notice that in this case, we chose zero for x and zero for y. The reason is that the resulting equation is easier to solve.

x	y	
0		⟶ $(0, \ )$
	0	⟶ $(\ , 0)$
4		⟶ $(4, \ )$

Complete: $(0, \ )$

$$2x + 4y = 4$$
$$2(0) + 4y = 4$$
$$0 + 4y = 4$$
$$4y = 4$$
$$\frac{4y}{4} = \frac{4}{4}$$
$$y = 1$$

Complete: $(\ , 0)$

$$2x + 4y = 4$$
$$2x + 4(0) = 4$$
$$2x + 0 = 4$$
$$2x = 4$$
$$\frac{2x}{2} = \frac{4}{2}$$
$$x = 2$$

Complete: $(4, \ )$

$$2x + 4y = 4$$
$$2(4) + 4y = 4$$
$$8 + 4y = 4$$
$$8 - 8 + 4y = 4 - 8$$
$$4y = -4$$
$$\frac{4y}{4} = \frac{-4}{4}$$
$$y = -1$$

Skill Practice

Graph the equation.

8. $3x + y = 3$

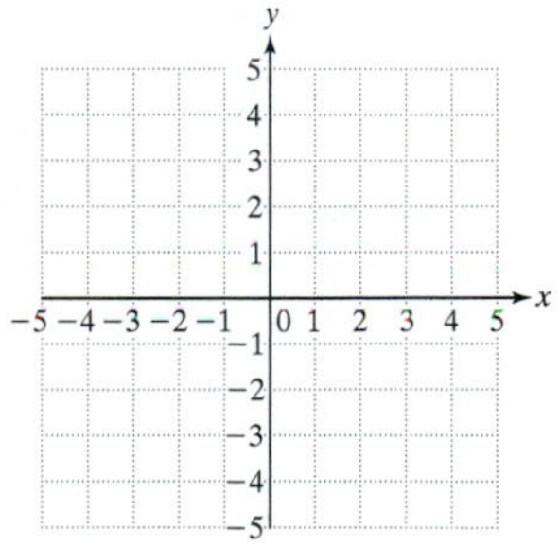

Answer

8.

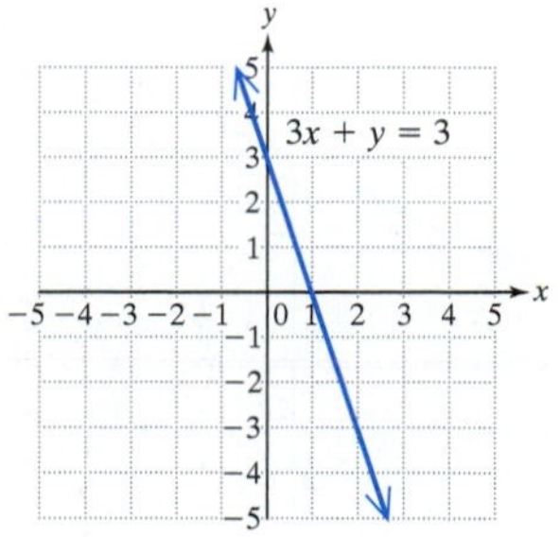

The completed table is given, and the ordered pairs are graphed in Figure 9-11. The line through the points is the graph of all solutions to the equation $2x + 4y = 4$.

x	y	
0	1	⟶ (0, 1)
2	0	⟶ (2, 0)
4	−1	⟶ (4, −1)

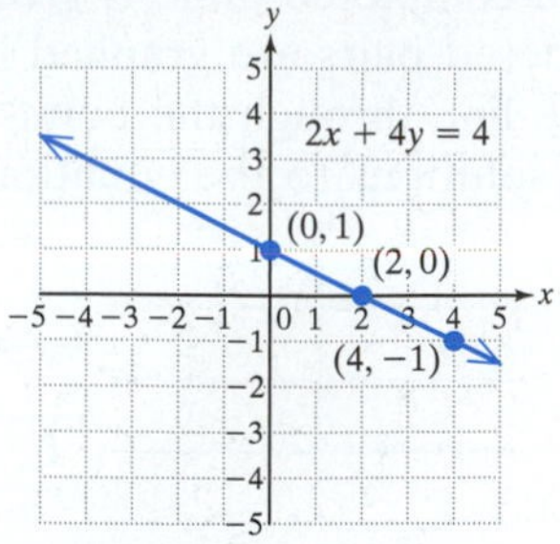

Figure 9-11

Skill Practice

Graph the equation.

9. $y = \frac{1}{2}x + 3$

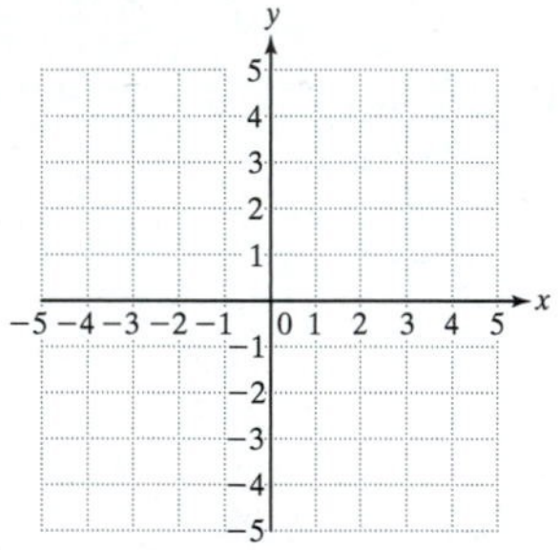

Example 6 Graphing a Linear Equation

Graph the equation. $\quad y = -\frac{1}{2}x - 2$

Solution:

In this equation, the y variable is isolated on the left-hand side. Therefore, if we substitute a value for x, then the value of y is easily found by simplifying the right-hand side of the equation. Since any number for x can be selected, we will choose numbers that are multiples of 2. These will simplify to an integer when multiplied by $-\frac{1}{2}$.

x	y	
0		⟶ (0,)
−2		⟶ (−2,)
4		⟶ (4,)

<u>**Complete:**</u> (0,)

$$y = -\frac{1}{2}x - 2$$
$$y = -\frac{1}{2}(0) - 2$$
$$y = 0 - 2$$
$$y = -2$$

<u>**Complete:**</u> (−2,)

$$y = -\frac{1}{2}x - 2$$
$$y = -\frac{1}{2}(-2) - 2$$
$$y = 1 - 2$$
$$y = -1$$

<u>**Complete:**</u> (4,)

$$y = -\frac{1}{2}x - 2$$
$$y = -\frac{1}{2}(4) - 2$$
$$y = -2 - 2$$
$$y = -4$$

The completed table is given, and the ordered pairs are graphed in Figure 9-12. The line through the points is the graph of all solutions to the equation $y = -\frac{1}{2}x - 2$.

x	y	
0	−2	⟶ (0, −2)
−2	−1	⟶ (−2, −1)
4	−4	⟶ (4, −4)

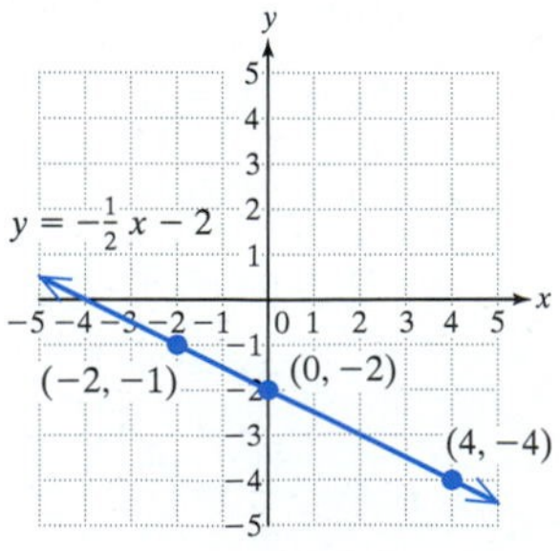

Figure 9-12

Answer

9.

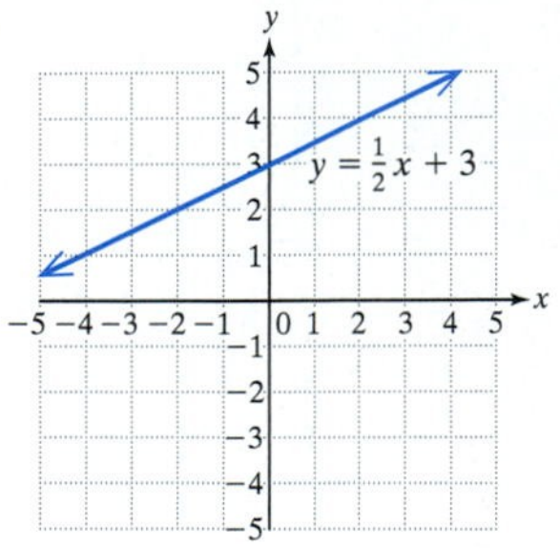

3. Horizontal and Vertical Lines

Sometimes a linear equation may have only one variable. The graph of such an equation in a rectangular coordinate system will be either a **horizontal** or **vertical line** (Figure 9-13).

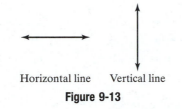

Figure 9-13

Example 7 Graphing a Horizontal Line

Graph the equation. $y = 2$

Solution:

Create a table of ordered pairs. Notice that the y-coordinates must all be 2 to satisfy the equation $y = 2$. However, the value of x may be any real number. We have arbitrarily chosen the values of x to be 1, 2, and 3. The graph is shown in Figure 9–14.

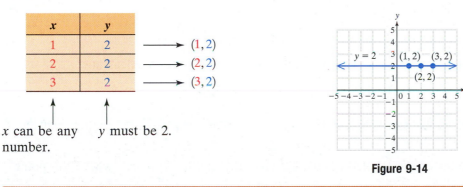

x	y	
1	2	→ (1, 2)
2	2	→ (2, 2)
3	2	→ (3, 2)

x can be any number. y must be 2.

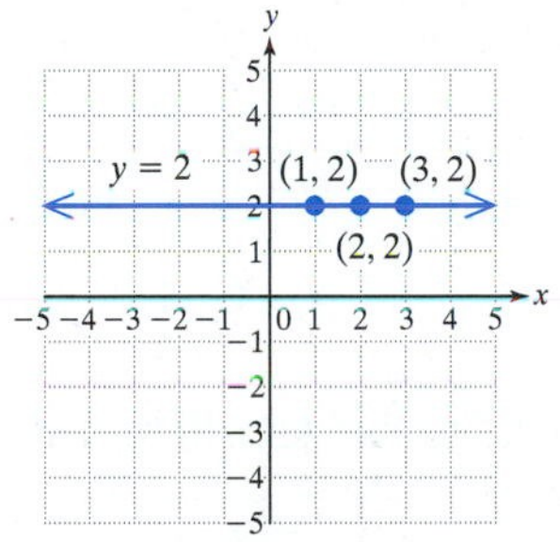

Figure 9-14

Example 8 Graphing a Vertical Line

Graph the equation. $2x = -6$

Solution:

This equation can be simplified by solving for x.

$$2x = -6$$

$$\frac{2x}{2} = \frac{-6}{2}$$

$$x = -3$$

The equation $x = -3$ will have the same graph as the original equation, $2x = -6$.

Skill Practice

Graph the equation.

10. $y = -4$

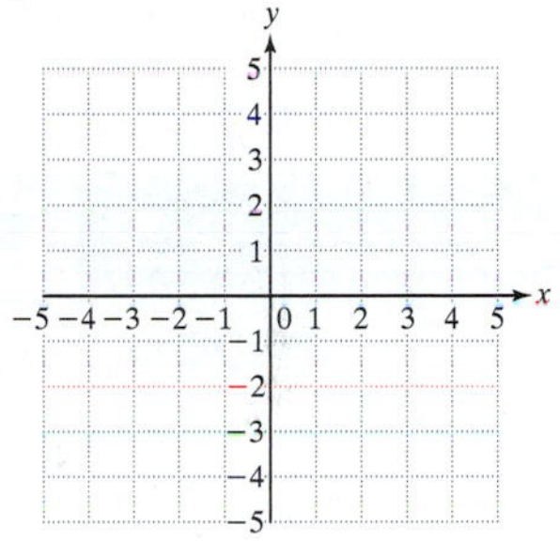

Skill Practice

Graph the equation.

11. $2x = 2$

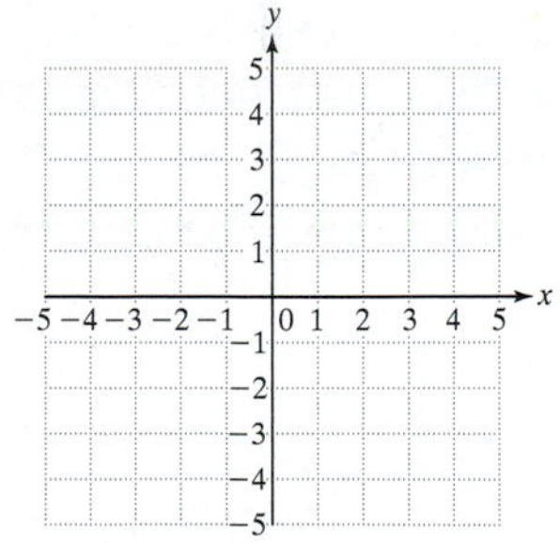

Answers

10–11.

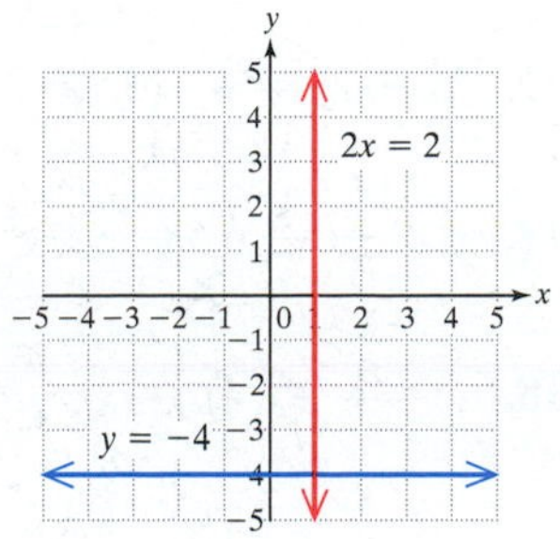

Create a table of ordered pairs. Notice that the x-coordinates must all be -3 to satisfy the equation $x = -3$. However, the value of y may be any real number. We have arbitrarily chosen the values of y to be 1, 2, and 3. The graph is shown in Figure 9–15.

x	y	
-3	1	→ $(-3, 1)$
-3	2	→ $(-3, 2)$
-3	3	→ $(-3, 3)$

x must be -3. y can be any number.

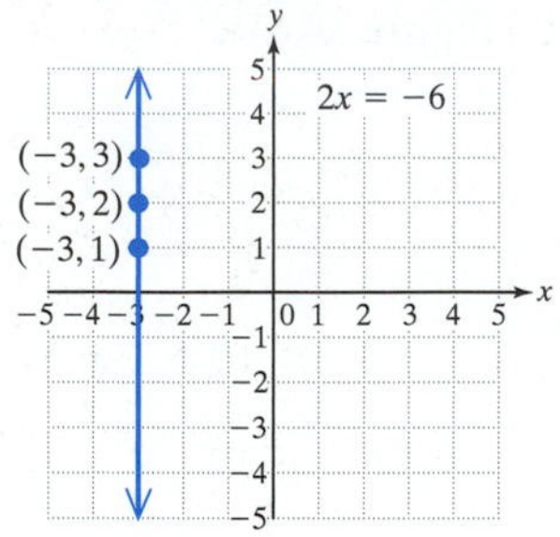

Figure 9-15

Section 9.2 Practice Exercises

Boost *your* GRADE at ALEKS.com!

ALEKS

- Practice Problems
- Self-Tests
- NetTutor
- e-Professors
- Videos

Study Skills Exercise

1. Define the key terms.

   a. **Linear equation in two variables** b. **Horizontal line** c. **Vertical line**

Review Exercises

2. The x- and y-axes divide a rectangular coordinate system into four regions. What are the regions called?

For Exercises 3–8, give the coordinates of the labeled points and state the quadrant or axis where the point is located.

3. A
4. B
5. C
6. D
7. E
8. F

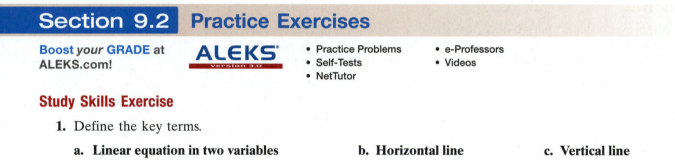

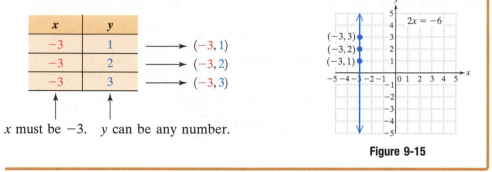

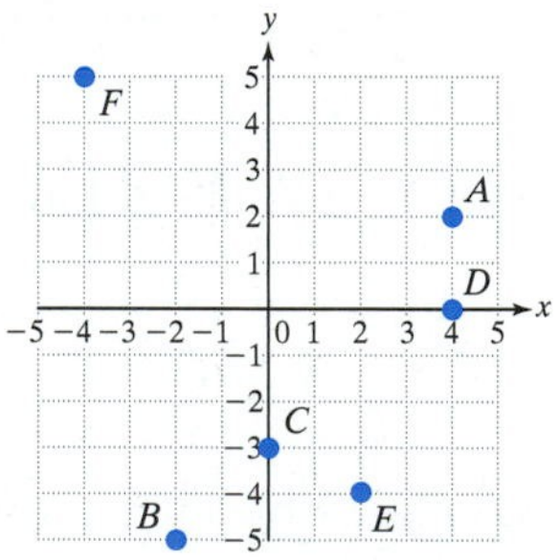

Objective 1: Solutions to Two-Variable Equations

For Exercises 9–20, determine if the ordered pair is a solution to the equation. **(See Example 1.)**

9. $x - y = 5 \quad (7, 2)$
10. $x + y = 3 \quad (-1, 4)$
11. $2x + 5y = 5 \quad (0, -1)$
12. $3x - 2y = 4 \quad (0, 2)$
13. $4x + 7y = 5 \quad (3, -1)$
14. $9x + 2y = 5 \quad (1, -2)$
15. $y = \frac{1}{3}x + 6 \quad (-3, 7)$
16. $y = -\frac{1}{4}x - 2 \quad (-8, -6)$
17. $x = -3 \quad (-3, 4)$
18. $y = 4 \quad (5, 4)$
19. $2y = 10 \quad (5, 2)$
20. $-4x = 8 \quad (3, -2)$

For Exercises 21–28, complete each ordered pair to make it a solution to the equation. **(See Example 2.)**

21. $x + y = 6$ (2,)

22. $x - y = 2$ (5,)

23. $2x + 6y = 12$ (, 1)

24. $3x + y = 5$ (, 2)

25. $3x - 8y = 0$ (, 3)

26. $5x - 2y = 0$ (, 10)

27. $y = \frac{3}{8}x + 6$ (8,)

28. $y = -\frac{2}{3}x - 4$ (6,)

For Exercises 29–34, complete the table to make solutions to the given equation. **(See Example 3.)**

29. $3x + y = 6$

x	*y*
−1	
	6
2	

30. $-x + 2y = 6$

x	*y*
	4
−4	
	−1

31. $-4x - 3y = 12$

x	*y*
	0
0	
	4

32. $-2x - 5y = 10$

x	*y*
0	
	0
10	

33. $y = \frac{3}{5}x - 3$

x	*y*
0	
5	
−10	

34. $y = -\frac{2}{3}x + 1$

x	*y*
0	
3	
−6	

Objective 2: Graphing Linear Equations in Two Variables

For Exercises 35–46, graph the equation. **(See Examples 4–6.)**

35. $x + y = 2$

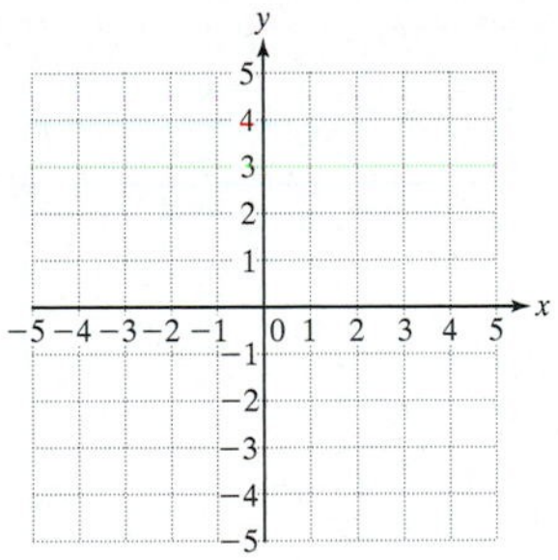

36. $x - y = 3$

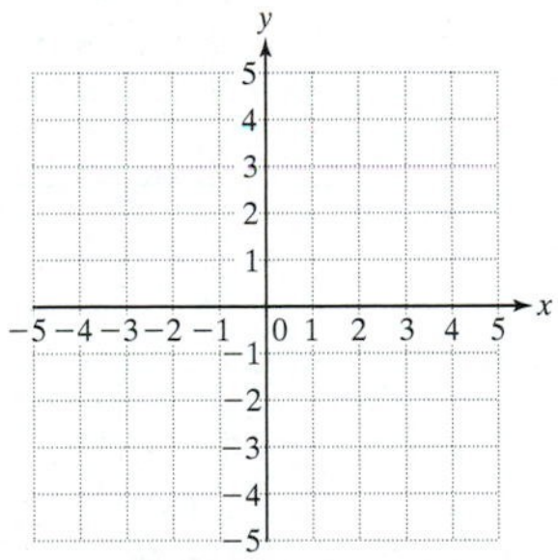

37. $2x + y = 1$

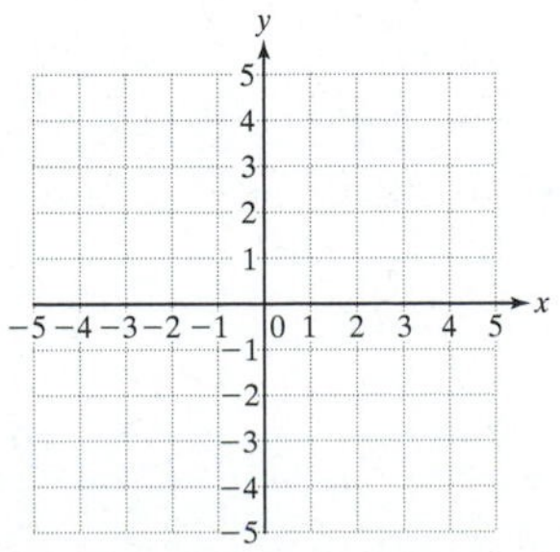

38. $x + 3y = 3$

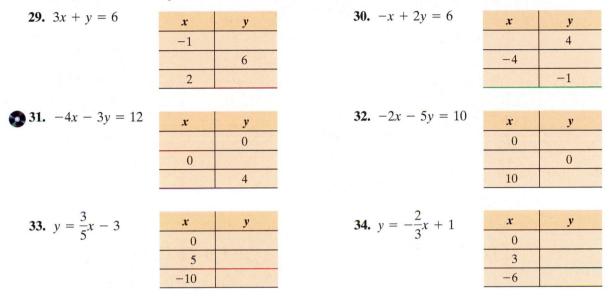

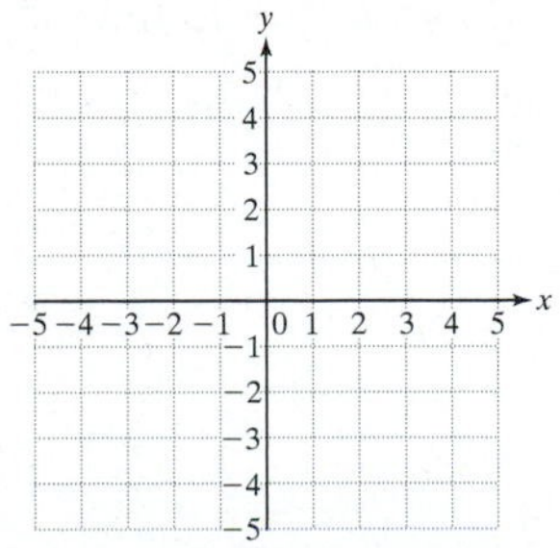

39. $y = x + 3$

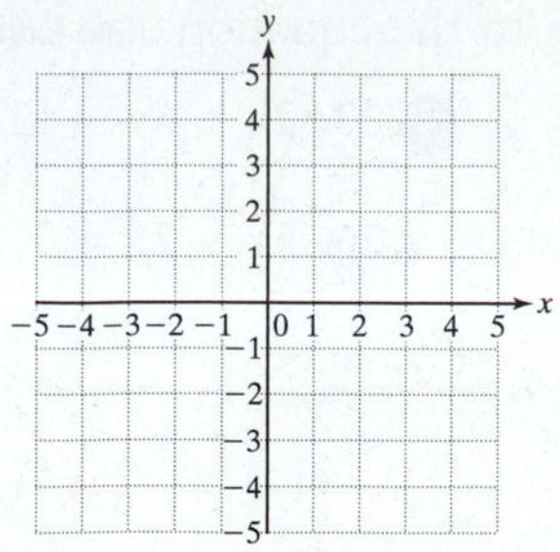

40. $y = -x - 2$

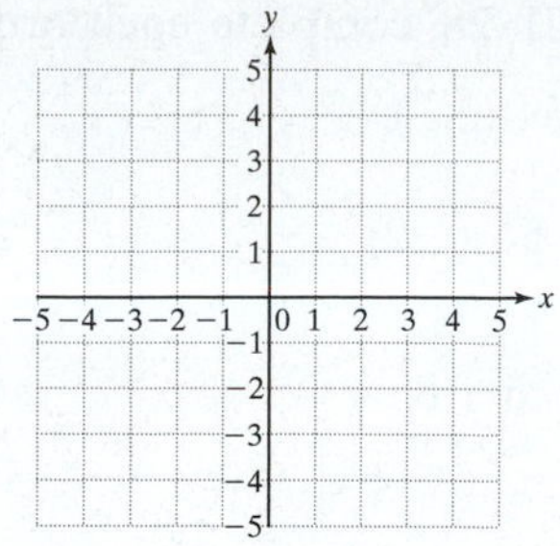

41. $y = 2x$

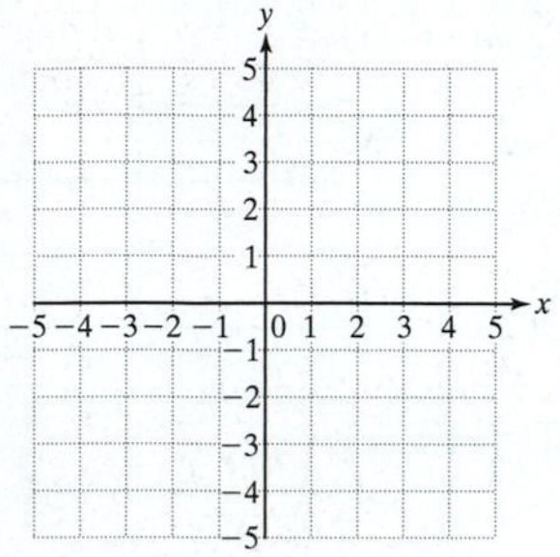

42. $y = -3x$

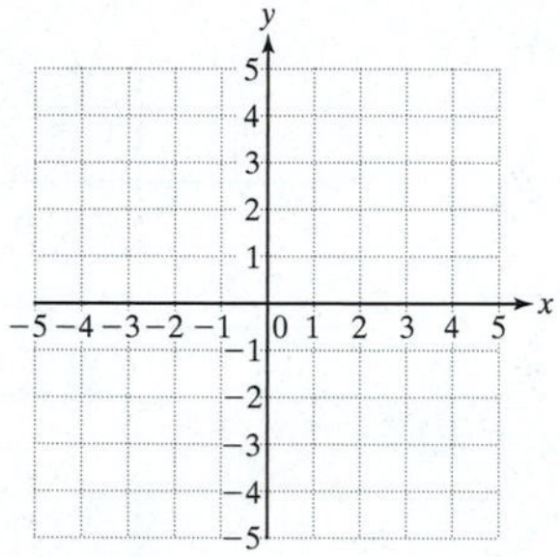

43. $y = -\frac{1}{3}x + 2$

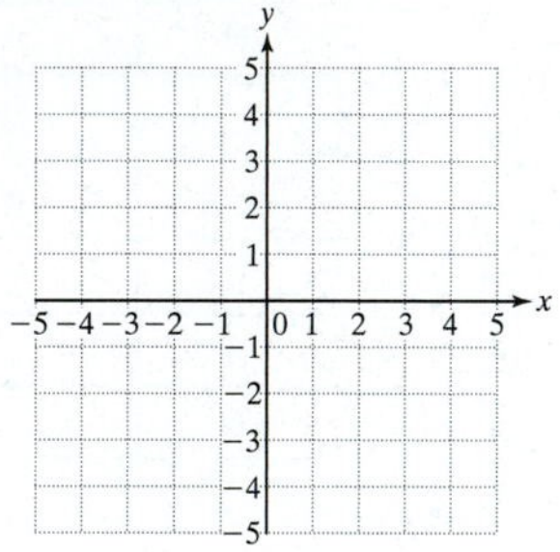

44. $y = \frac{1}{2}x - 3$

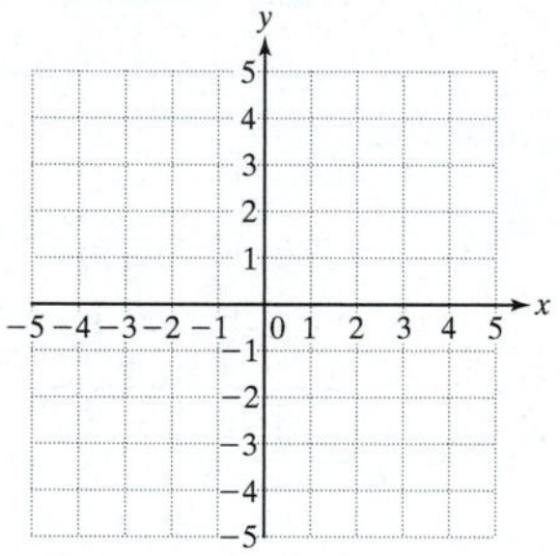

45. $3x - 6y = 12$

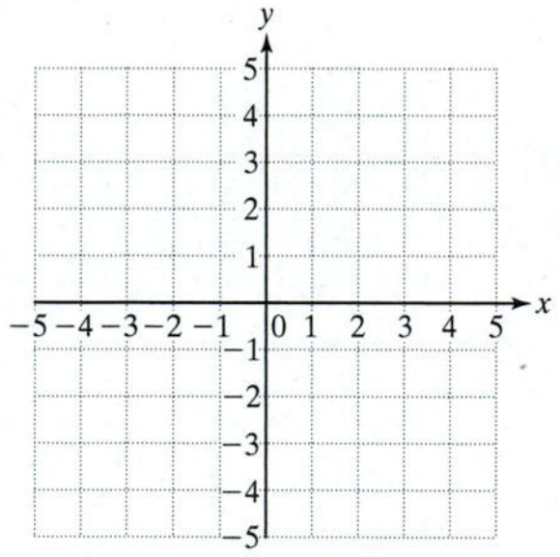

46. $4x - 3y = 12$

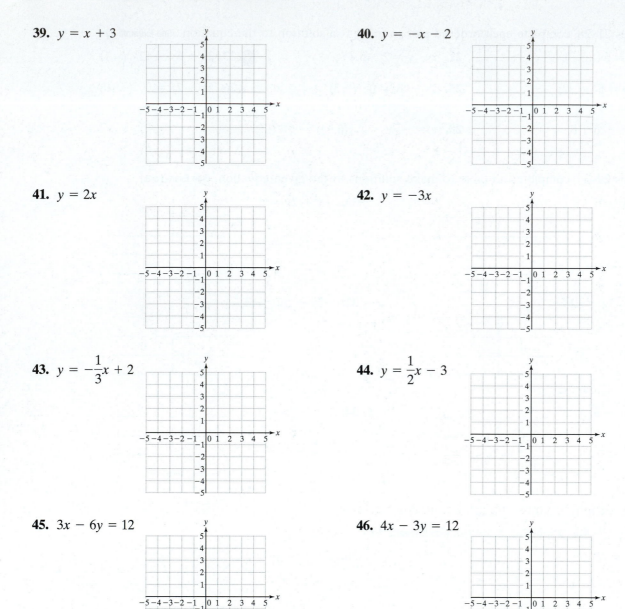

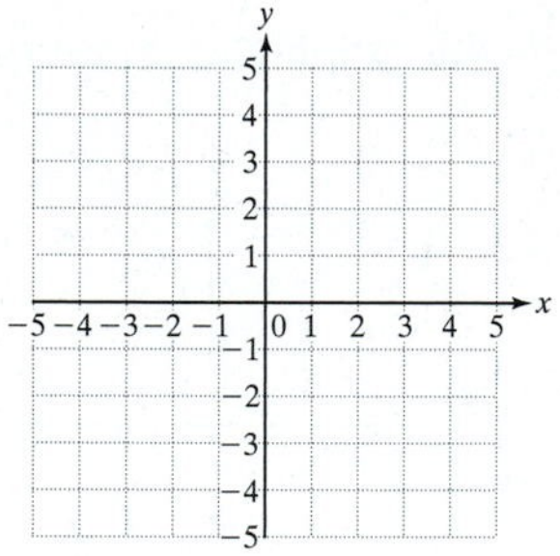

Objective 3: Horizontal and Vertical Lines

For Exercises 47–48, identify the line as horizontal or vertical.

47. 

48.

For Exercises 49–56, graph the equation. **(See Examples 7–8.)**

49. $x = -1$

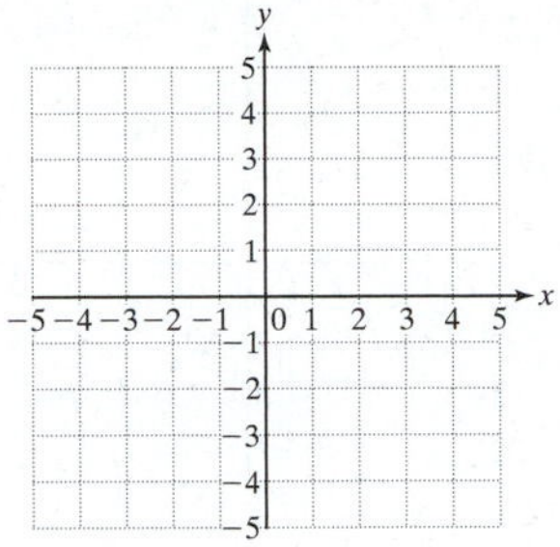

50. $y = -1$

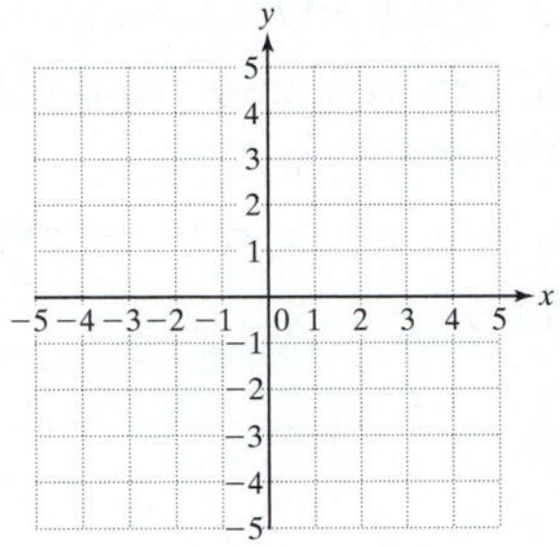

51. $y = -3$

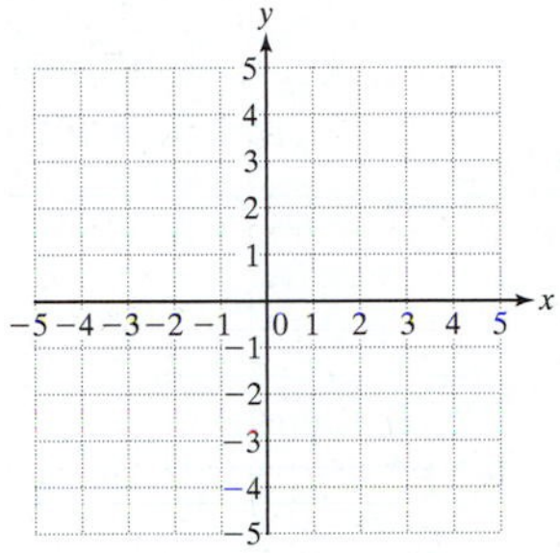

52. $x = 2$

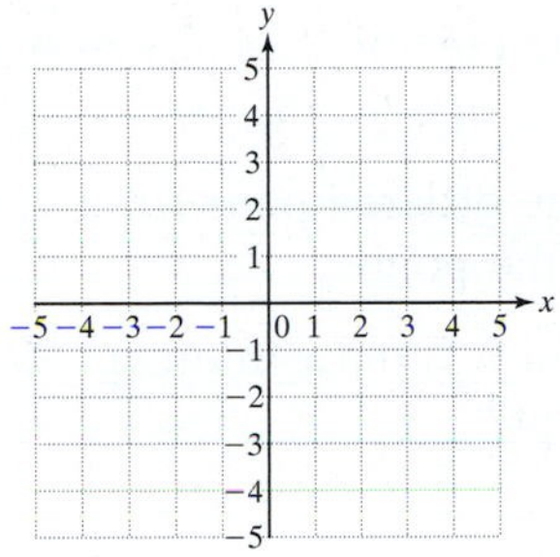

53. $5y = 20$

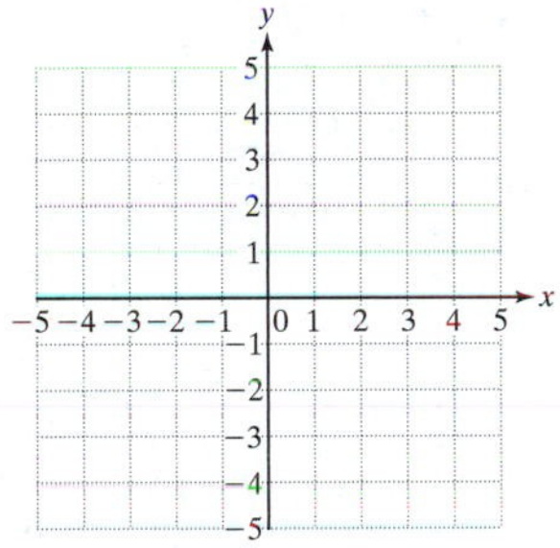

54. $4x = -8$

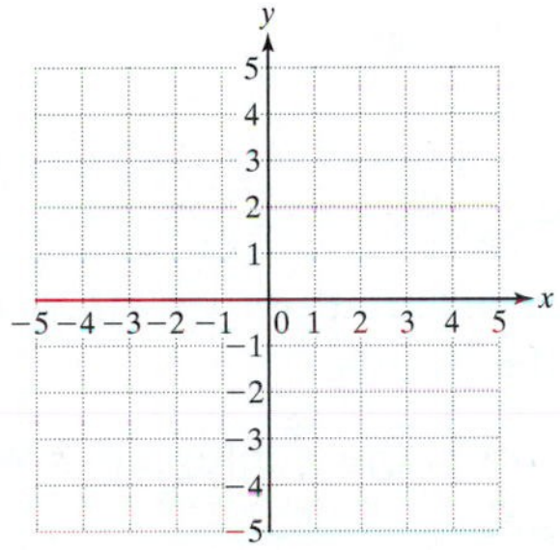

55. $-3x = 9$

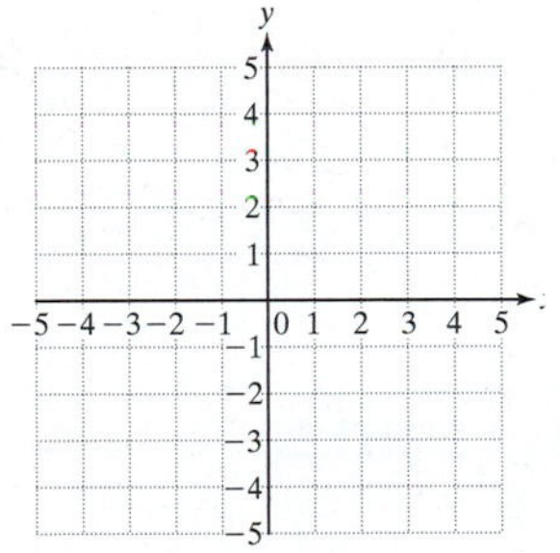

56. $-6y = 24$

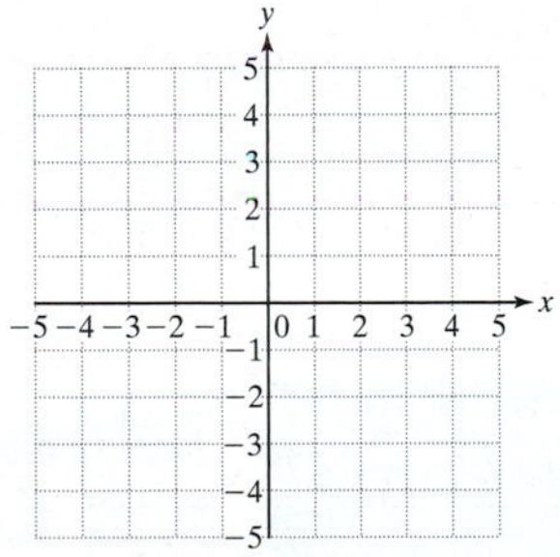

Expanding Your Skills

Suppose the points $(a, 0)$ and $(0, b)$ are solutions to an equation in two variables. Then the point $(a, 0)$ is called an ***x*-intercept**. An x-intercept is a point where a graph intersects the x-axis. The point $(0, b)$ is called a ***y*-intercept**. A y-intercept is a point where a graph intersects the y-axis. Use this information for Exercises 57–60.

57. a. Given the equation $2x + 3y = 6$, complete the ordered pairs (0,) and (, 0).

b. Graph the ordered pairs from part (a) and draw the line through the points.

c. Which point is the x-intercept? Which point is the y-intercept?

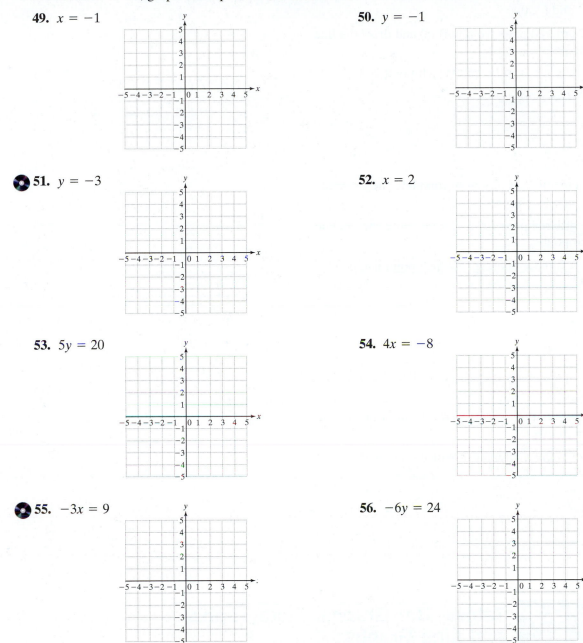

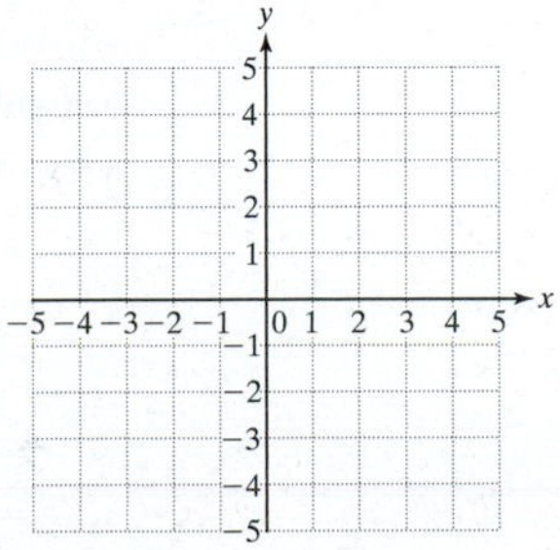

58. **a.** Given the equation $2x + y = 4$, complete the ordered pairs (0,) and (, 0).

b. Graph the ordered pairs from part (a) and draw the line through the points.

c. Which point is the x-intercept? Which point is the y-intercept?

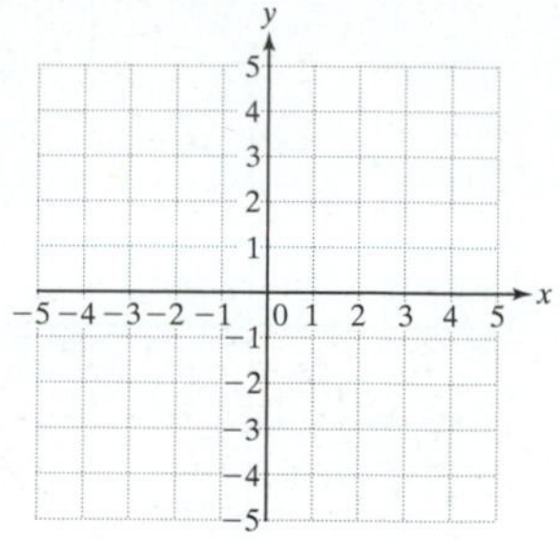

59. **a.** Given the equation $3x + y = 6$, complete the ordered pairs (0,) and (, 0).

b. Graph the ordered pairs from part (a) and draw the line through the points.

c. Which point is the x-intercept? Which point is the y-intercept?

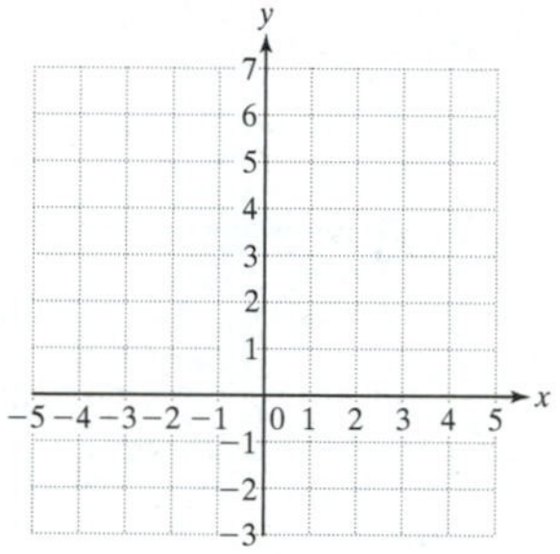

60. **a.** Given the equation $x - 4y = 8$, complete the ordered pairs (0,) and (, 0).

b. Graph the ordered pairs from part (a) and draw the line through the points.

c. Which point is the x-intercept? Which point is the y-intercept?

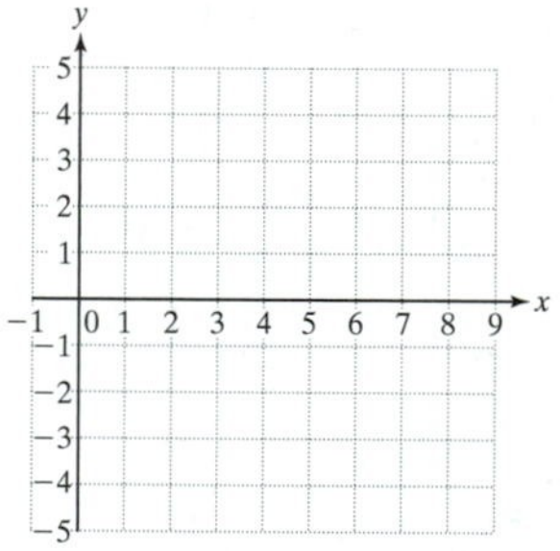

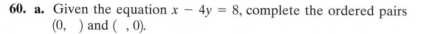

Section 9.3 Tables, Bar Graphs, Pictographs, and Line Graphs

Objectives

1. Introduction to Data and Tables
2. Bar Graphs
3. Pictographs
4. Line Graphs

1. Introduction to Data and Tables

Statistics is the branch of mathematics that involves collecting, organizing, and analyzing **data** (information). One method to organize data is by using tables. A **table** uses rows and columns to reference information. The individual entries within a table are called **cells**.

Example 1 Interpreting Data in a Table

Table 9-3 summarizes the maximum wind speed, number of reported deaths, and estimated cost for recent hurricanes that made landfall in the United States. (*Source:* National Oceanic and Atmospheric Administration)

Table 9-3

Hurricane	Date	Landfall	Maximum Sustained Winds at Landfall (mph)	Number of Reported Deaths	Estimated Cost ($ Billions)
Katrina	2005	Louisiana, Mississippi	125	1836	81.2
Fran	1996	North Carolina, Virginia	115	37	5.8
Andrew	1992	Florida, Louisiana	145	61	35.6
Hugo	1989	South Carolina	130	57	10.8
Alicia	1983	Texas	115	19	5.9

a. Which hurricane caused the greatest number of deaths?

b. Which hurricane was the most costly?

c. What was the difference in the maximum sustained winds for hurricane Andrew and hurricane Katrina?

d. How many times greater was the death toll for Hugo than for Alicia?

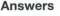

Solution:

a. The death toll is reported in the 5th column. The death toll for hurricane Katrina, 1836, is the greatest value.

b. The estimated cost is reported in the 6th column. Hurricane Katrina was also the costliest hurricane at $81.2 billion.

c. The wind speeds are given in the 4th column. The difference in the wind speed for hurricane Andrew and hurricane Katrina is 145 mph − 125 mph = 20 mph.

d. There were 57 deaths from Hugo and 19 from Alicia. The ratio of deaths from Hugo to deaths from Alicia is given by

$$\frac{57}{19} = 3$$

There were 3 times as many deaths from Hugo as from Alicia.

Skill Practice

For Exercises 1–4, use the information in Table 9-3.

1. Which hurricane had the highest sustained wind speed at landfall?
2. Which hurricane was the most recent?
3. What was the difference in the death toll for Fran and Alicia?
4. How many times greater was the cost for Katrina than for Fran?

Answers

1. Andrew
2. Katrina
3. 18 deaths
4. 14 times greater

Skill Practice

5. A political poll was taken to determine the political party and gender of several registered voters. The following codes were used.

M = male
F = female
dem = Democrat
rep = Republican
ind = Independent

Complete the table given the following results.

F–dem	F–rep	M–dem
M–rep	M–ind	M–rep
F–dem	M–rep	F–dem
F–dem	M–dem	F–ind
M–dem	M–ind	M–rep
M–rep	F–rep	F–dem

	Male	Female
Democrat		
Republican		
Independent		

Example 2 Constructing a Table from Observed Data

The following data were taken by a student conducting a study for a statistics class. The student observed the type of vehicle and gender of the driver for 18 vehicles in the school parking lot. Complete the table.

Male–car	Female–truck	Male–truck
Male–truck	Female–car	Male–truck
Female–car	Male–truck	Male–motorcycle
Female–car	Male–car	Female–car
Male–motorcycle	Female–car	Female–car
Female–motorcycle	Male–car	Male–car

Vehicle / Driver	Car	Truck	Motorcycle
Male			
Female			

Solution:

We need to count the number of data values that fall in each of the six cells. One method is to go through the list of data one by one. For each value place a tally mark | in the appropriate cell. For example, the first data value male–car would go in the cell in the first row, first column.

Vehicle / Driver	Car	Truck	Motorcycle
Male	||||	||||	||
Female	𝍸 |	|	|

To form the completed table, count the number of tally marks in each cell. See Table 9-4.

Table 9-4

Vehicle / Driver	Car	Truck	Motorcycle
Male	4	4	2
Female	6	1	1

Answer

5.

	Male	Female
Democrat	3	5
Republican	5	2
Independent	2	1

2. Bar Graphs

In Table 9-3, we see that hurricane Andrew had the greatest wind speed of those listed. This can be visualized in a graph. Figure 9-16 shows a bar graph of the wind speed for the hurricanes listed in Table 9-3. Notice that the bar showing wind speed for Andrew is the highest.

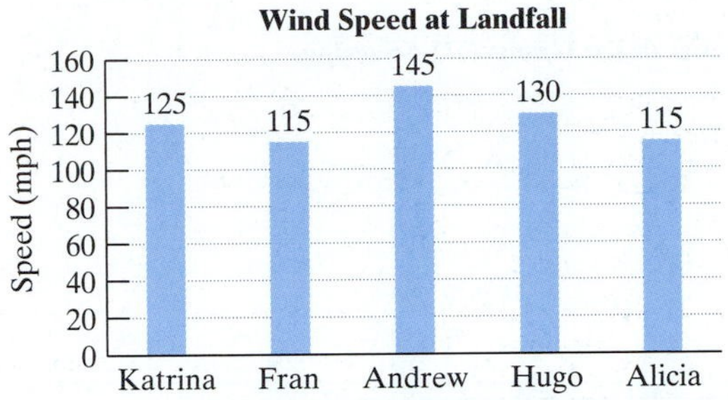

Figure 9-16

Notice that the **bar graph** compares the data values through the height of each bar. The bars in a bar graph may also be presented horizontally. For example, the double bar graph in Figure 9-17 illustrates the data from Example 2.

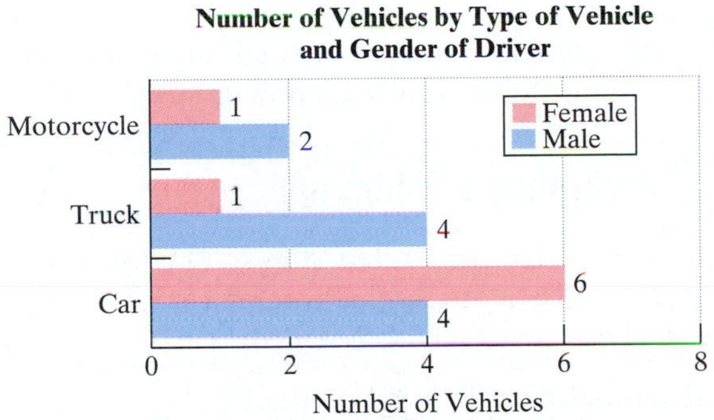

Figure 9-17

When constructing a graph or chart, be sure to include:

- A title.
- Labels on the vertical and horizontal axes.
- An appropriate range and scale.

Example 3 Constructing a Bar Graph

The number of fat grams for five different ice cream brands and flavors is given in Table 9-5. Each value is based on a $\frac{1}{2}$-c serving. Construct a bar graph with vertical bars to depict this information.

Table 9-5

Brand/Flavor	Number of Fat Grams (g) per $\frac{1}{2}$-c Serving
Breyers Strawberry	6
Edy's Grand Light Mint Chocolate Chip	4.5
Healthy Choice Chocolate Fudge Brownie	2
Ben and Jerry's Chocolate Chip Cookie Dough	15
Häagen-Dazs Vanilla Swiss Almond	20

Skill Practice

6. The amount of sodium in milligrams (mg) per $\frac{1}{2}$-c serving for three different brands of cereal is given in the table. Construct a bar graph with horizontal bars.

Brand/Flavor	Amount of Sodium (mg)
Post Grape Nuts	310
Post Cranberry Almond	190
Post Great Grains	200

Answers

6.

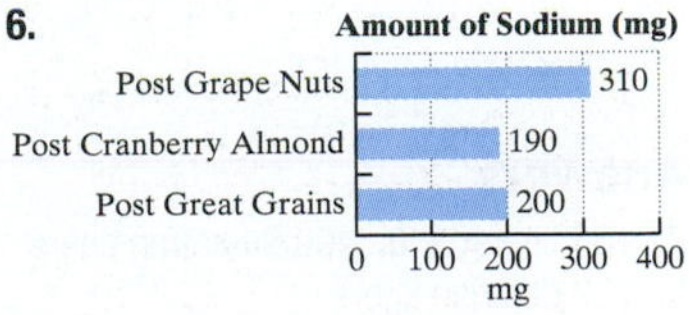

Solution:

First draw a horizontal line and label the different food categories. Then draw a vertical line on the left-hand side of the graph as in Figure 9-18. The vertical line represents the number of grams of fat. The vertical scale must extend to at least 20 to accommodate the largest value in the table. In Figure 9-18, the vertical scale ranges from 0 to 24 in multiples (or steps) of 4.

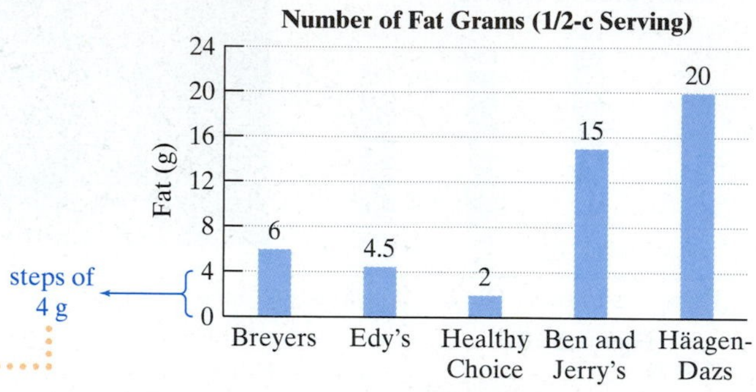

Figure 9-18

Avoiding Mistakes

The scale on the *y*-axis should begin at 0 and increase by equal intervals. We use the range of values in the data to choose a suitable scale.

3. Pictographs

Sometimes a bar graph might use an icon or small image to convey a unit of measurement. This type of bar graph is called a **pictograph**.

Skill Practice

For Exercises 7–9, refer to the pictograph showing the number of tigers living in the wild.

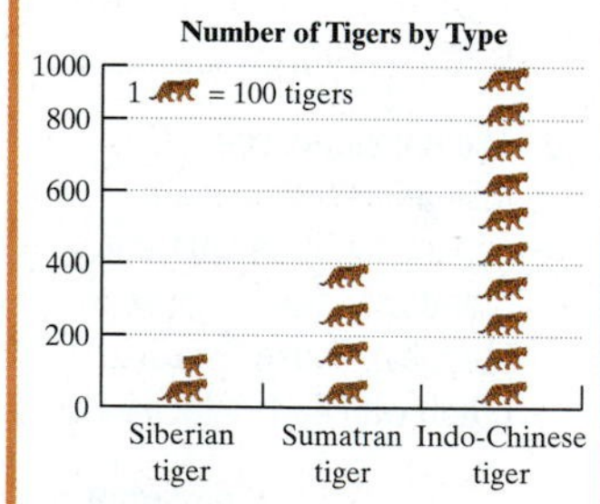

7. What is the value of each tiger icon?
8. From the graph, estimate the number of Sumatran tigers.
9. Estimate the number of Siberian tigers.

Example 4 Interpreting a Pictograph

For a recent year, California led the United States in hybrid vehicle sales (83,000 sold). The graph displays the hybrid vehicle sales for five other states for the same year (Figure 9-19). (*Source:* HybridCars.com)

a. What is the value of each car icon in the graph?

b. From the graph, estimate the number of hybrids sold in the state of Washington.

c. For which state were approximately 10,000 hybrid vehicles sold?

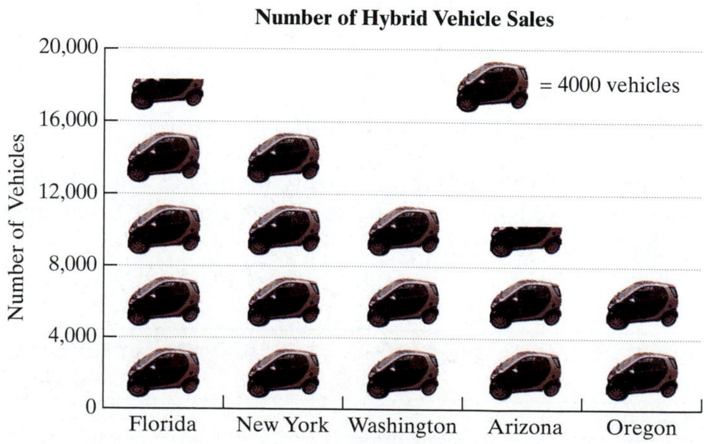

Figure 9-19

Answers

7. 100 tigers **8.** 400 Sumatran tigers
9. 150 Siberian tigers

Solution:

a. The legend indicates that 1  = 4000 vehicles sold.

b. The height of the "bar" for Washington is given by 3 car icons. This represents 3×4000 vehicles. Therefore, 12,000 hybrid vehicles were sold in Washington.

c. The bar containing $2\frac{1}{2}$ icons represents 10,000 hybrids sold. This corresponds to the state of Arizona.

4. Line Graphs

Line graphs are often used to track how one variable changes with respect to a second variable. For example, a line graph may illustrate a pattern or trend of a variable over time and allow us to make predictions.

Example 5 Interpreting a Line Graph

Figure 9-20 shows the number of bachelor's degrees earned by men and women for selected years. (*Source:* U.S. Census Bureau)

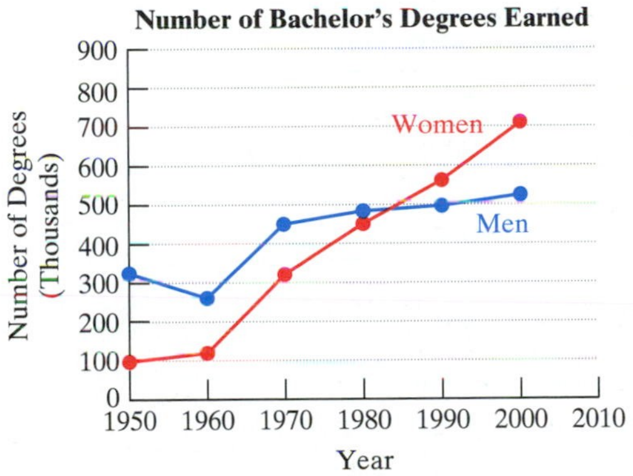

Figure 9-20

a. In 1950, who earned more bachelor's degrees, men or women?

b. In 2000, who earned more bachelor's degrees, men or women?

c. Use the trends in the graph to predict the number of bachelor's degrees earned by men and by women for the year 2010.

Solution:

a. The blue graph represents men, and the red graph represents women. In 1950, men earned approximately 320 thousand bachelor's degrees. Women earned approximately 100 thousand. Men earned more bachelor's degrees in 1950.

b. In 2000, women earned more bachelor's degrees.

Skill Practice

For Exercises 10–12, refer to the line graph showing the number of cats and dogs received by an animal shelter for the first four years since it opened.

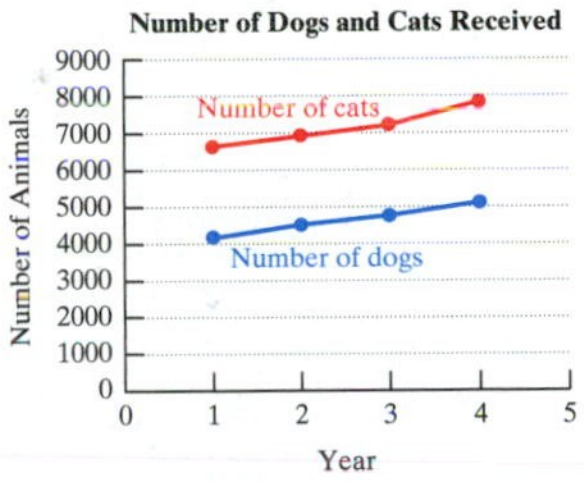

10. Which animal did the shelter receive most?

11. Approximate the number of cats received the third year after the shelter opened.

12. Use the graph to predict the number of dogs the shelter can expect for its fifth year.

Answers

10. Cats

11. Approximately 7200 cats

12. Approximately 5400 dogs

c. To predict the number of bachelor's degrees earned by men and women in the year 2010, we need to extend both line graphs. See the dashed lines in Figure 9-21. The number of bachelor's degrees earned by women is predicted to be approximately 850 thousand. The number of degrees earned by men will be approximately 550 thousand.

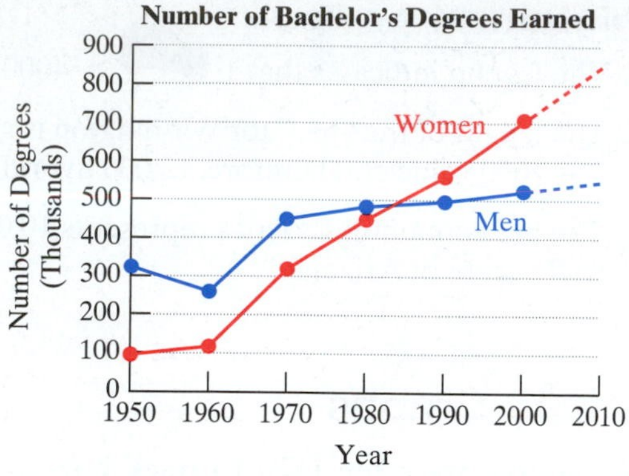

Figure 9-21

Skill Practice

13. The table gives the number of cars registered in the United States (in millions) for selected years. (*Source:* U.S. Department of Transportation) Create a line graph for this information.

Year	Number of Cars (millions)
1970	89
1980	122
1990	134
2000	134
2010	138 (projected)

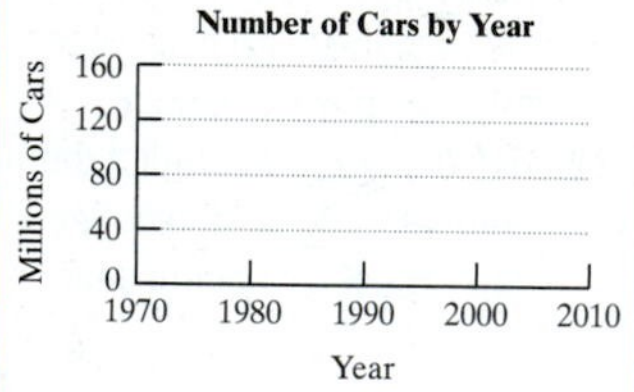

Example 6 Constructing a Line Graph

Table 9-6 gives the number of worldwide airline fatalities (excludes deaths caused by terrorism) for selected years. Use the data given in the table to create a line graph.

Table 9-6

Year	Number of Deaths
1986	641
1990	544
1994	1170
1998	904
2002	577
2006	773

Source: International Civil Aviation Administration

Solution:

First draw a horizontal line and label the year. Then draw a vertical line on the left-hand side of the graph, as in Figure 9-22. The vertical line represents the number of fatalities. In Figure 9-22, the vertical scale ranges from 0 to 1400 in multiples (or steps) of 200. For each year, plot a point corresponding to the number of deaths for that year. Then connect the points.

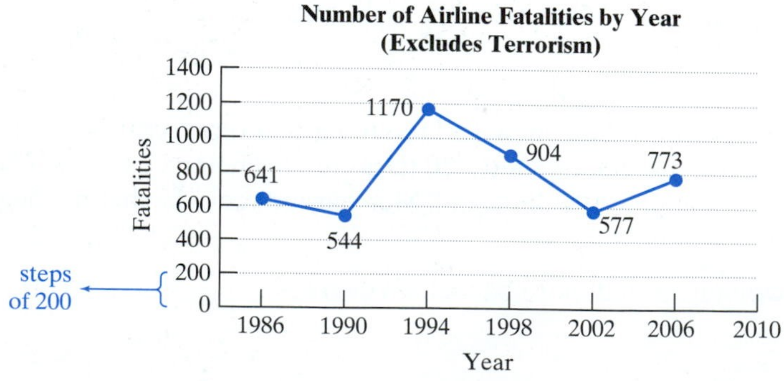

Figure 9-22

In Figure 9-22, we labeled the value at each data point because the exact values are difficult to read from the graph.

Answer

13.

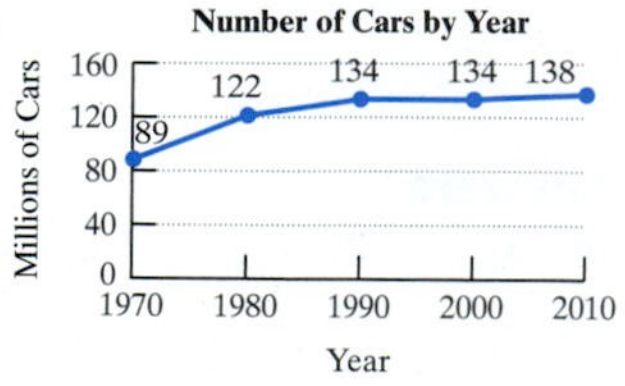

Section 9.3 Practice Exercises

Study Skills Exercise

1. Define the key terms.

a. Statistics **b. Data** **c. Table** **d. Cells**

e. Bar graph **f. Pictograph** **g. Line graph**

Review Exercises

For Exercises 2–6, graph the ordered pairs on the grid provided.

2. $(-3, -4)$ **3.** $(4, -2)$

4. $(0, 5)$ **5.** $(2, 5)$

6. $(-3, 0)$

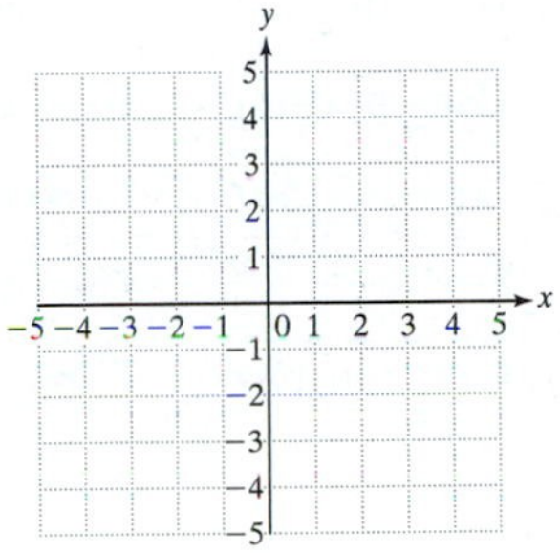

Objective 1: Introduction to Data and Tables

For Exercises 7–10, refer to the table. The table represents the Seven Summits (the highest peaks from each continent). **(See Example 1.)**

Mountain	Continent	Height (ft)
Mt. Kilimanjaro	Africa	19,340
Elbrus	Europe	18,510
Aconcagua	South America	22,834
Denali (Mt. McKinley)	North America	20,320
Vinson Massif	Antarctica	16,864
Mt. Kosciusko	Australia	7,310
Mt. Everest	Asia	29,035

7. In which continent does the highest mountain lie?

8. Which mountain among those listed is the lowest? In which continent does it lie?

9. How much higher is Aconcagua than Denali?

10. What is the difference in height between the highest mountain in Europe and the highest mountain in Australia?

For Exercises 11–16, refer to the table. The table gives the average ages (in years) for U.S. women and men married for the first time for selected years. (*Source:* U.S. Census Bureau)

	Men	Women
1940	24.3	21.5
1960	22.8	20.3
1980	24.7	22.0
2000	26.8	25.1

11. By how much has the average age for women increased between 1940 and 2000?

12. By how much has the average age for men increased between 1940 and 2000?

13. What is the difference between the men's and women's average age at first marriage in 1940?

14. What is the difference between the men's and women's average age at first marriage in 2000?

15. Which group, men or women, had the consistently higher age at first marriage?

16. Which group, men or women, had a greater increase in age between 1940 and 2000?

17. The following data were taken from a survey of a third-grade class. The survey denotes the gender of a student and whether the student owned a dog, a cat, or neither. Complete the table. Be sure to label the rows and columns. **(See Example 2.)**

Boy–dog	Boy–dog	Boy–cat	Boy–neither
Girl–dog	Girl–neither	Boy–dog	Girl–cat
Girl–neither	Girl–neither	Girl–dog	Girl–cat
Boy–dog	Girl–cat	Boy–neither	Girl–dog
Boy–neither	Girl–neither	Girl–cat	Girl–neither

	Dog	Cat	Neither
Boy			
Girl			

18. In a group of 20 women, 10 were given an experimental drug to lower cholesterol. The other 10 were given a placebo. The letter "D" indicates that the person got the drug, and the letter "P" indicates that the person received the placebo. The values "yes" or "no" indicate whether the person's cholesterol was lowered. Complete the table.

	Yes	No
Drug (D)		
Placebo (P)		

D–yes	D–yes	P–no	D–no	P–yes	D–yes	P–no	P–no	D–yes	D–no
P–yes	P–no	D–yes	P–yes	P–yes	D–no	D–yes	P–no	D–yes	P–no

Objective 2: Bar Graphs

19. The table shows the number of students per computer in U.S. public schools for selected years. (*Source:* National Center for Education Statistics) **(See Example 3.)**

a. In which year did students have the best access to a computer?

b. Draw a bar graph with vertical bars to illustrate these data.

Year	Students
1992	16
1995	10
1998	5.7
2001	4.9
2004	4.2
2007	4.0

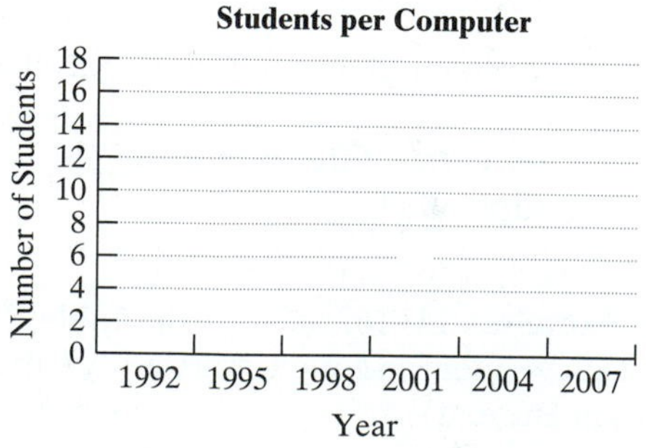

20. The number of new jobs for selected industries are given in the table. (*Source:* Bureau of Labor Statistics)

a. Which category has the greatest number of new jobs? How many new jobs is this?

b. Draw a bar graph with vertical bars to illustrate these data.

Industry	Number of New Jobs
Health care	219,400
Temporary help	212,000
Construction	173,000
Food service	167,600
Retail	78,600

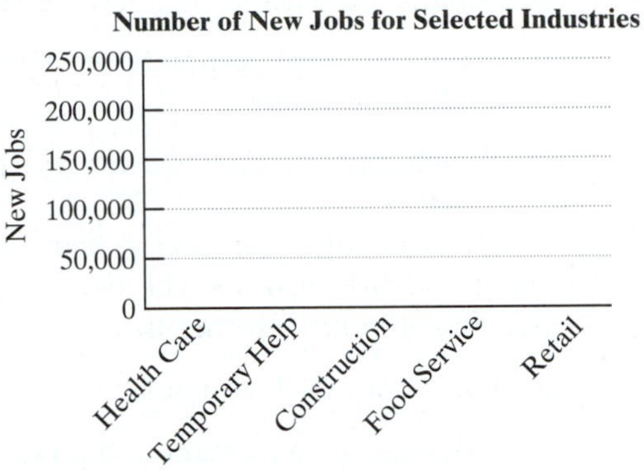

21. The table shows the number of cellular telephone subscriptions by year in the United States. (*Source:* U.S. Bureau of the Census) Construct a bar graph with horizontal bars. The length of each bar represents the number of cellular phone subscriptions for the given year.

Year	Number of Subscriptions (millions)
2002	141
2004	182
2006	233
2008	242

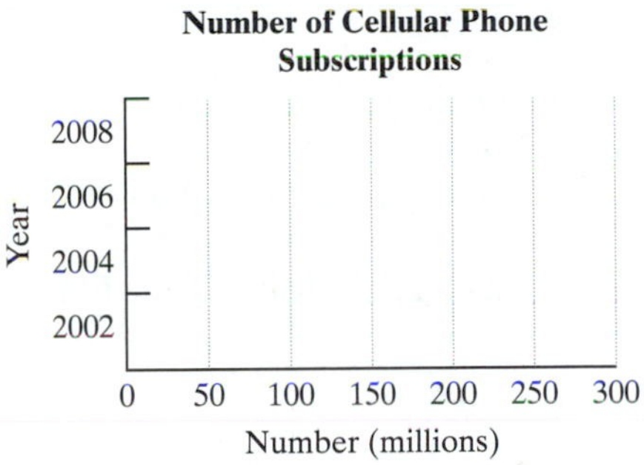

22. The table represents the world's major consumers of primary energy for a recent year. All measurements are in quadrillions of Btu. *Note:* 1 quadrillion = 1,000,000,000,000,000. (*Source:* Energy Information Administration, U.S. Department of Energy) Construct a bar graph using horizontal bars. The length of each bar gives the amount of energy consumed for that country.

Country	Amount of Energy Consumed (Quadrillions of Btu)
Germany	14
Japan	22
Russia	28
China	37
United States	99

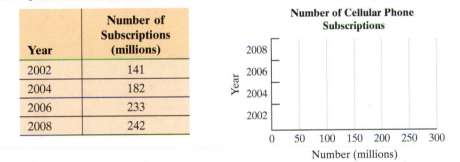

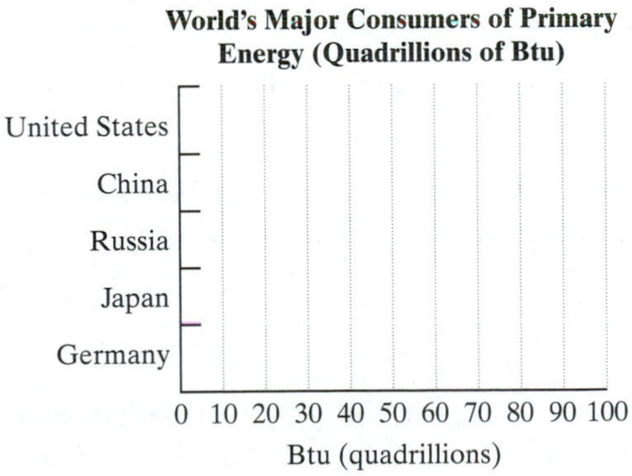

Objective 3: Pictographs

23. A local ice cream stand kept track of its ice cream sales for one weekend, as shown in the figure. **(See Example 4.)**

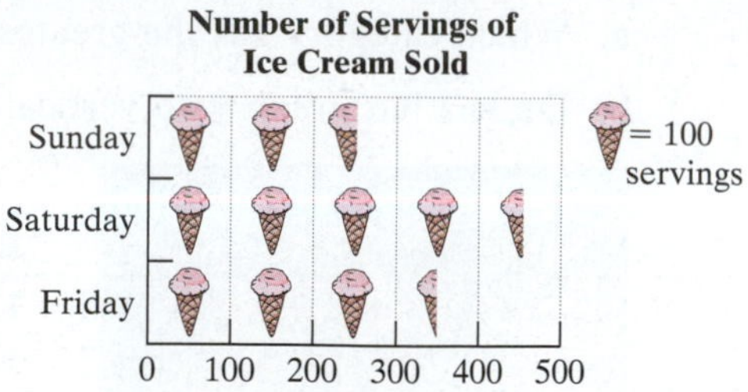

a. What does each ice cream icon represent?

b. From the graph, estimate the number of servings of ice cream sold on Saturday.

c. Which day had approximately 275 servings of ice cream sold?

24. Adults access the Internet to see weather updates and check on current news. The pictograph displays the percent of adult Internet users who access these topics.

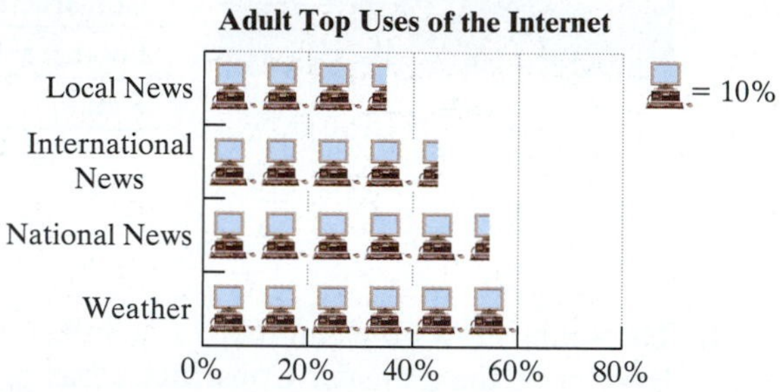

a. What does each computer icon represent?

b. From the graph, estimate the percent of adult users that access the Internet for weather.

c. Which type of news is accessed about 45% of the time?

25. The figure displays the 1-day box office gross earnings for selected films.

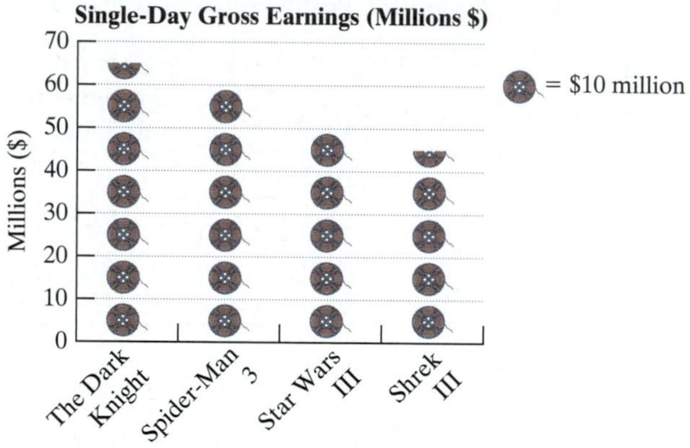

a. Estimate the earnings for *The Dark Knight*.

b. Which film grossed approximately $60 million?

c. Estimate the total earnings for all four films.

26. Recently the largest populations of senior citizens were in California, Florida, New York, Texas, and Pennsylvania, as shown in the figure. (*Source:* U.S. Bureau of the Census)

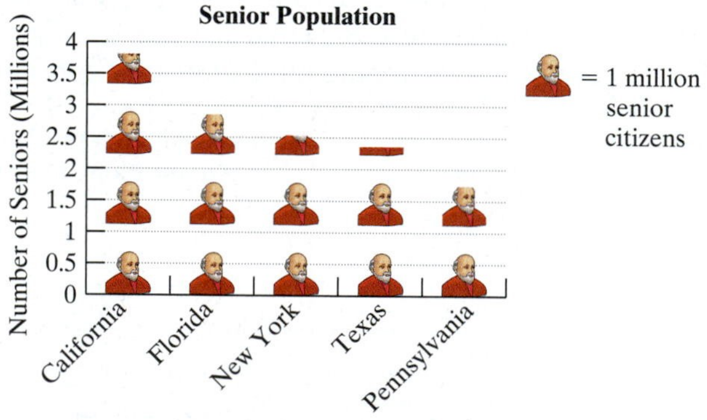

a. Estimate the number of senior citizens living in Texas.

b. How many more senior citizens are living in California than in Pennsylvania?

Objective 4: Line Graphs

For Exercises 27–32, use the graph provided. The graph shows the trend depicted by the percent of men and women over 65 years old in the labor force. (*Source:* Bureau of the Census) **(See Example 5.)**

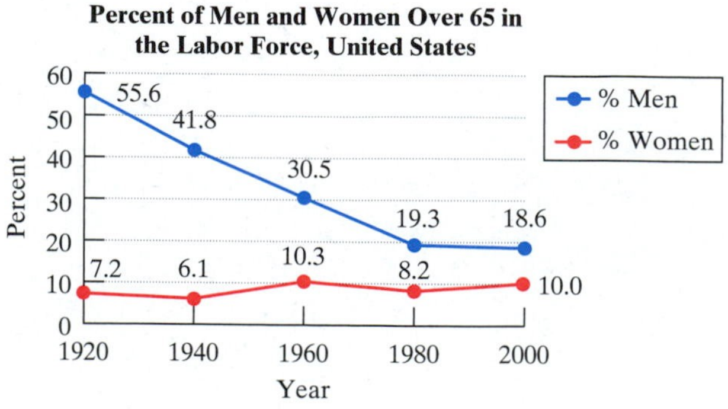

27. What was the difference in the percent of men and the percent of women over 65 in the labor force in the year 1920?

28. What was the difference in the percent of men and the percent of women over 65 in the labor force in the year 2000?

29. What was the overall trend in the percent of women over 65 in the labor force for the years shown in the graph?

30. What was the overall trend in the percent of men over 65 in the labor force for the years shown in the graph?

31. Use the graph to predict the number of men over 65 in the labor force in the year 2020. Answers will vary.

32. Use the graph to predict the number of women over 65 in the labor force in the year 2020. Answers will vary.

For Exercises 33–38, refer to the graph. The graph represents the number of kidney transplants in the United States for selected years. (*Source:* National Organ Procurement and Transplant Network)

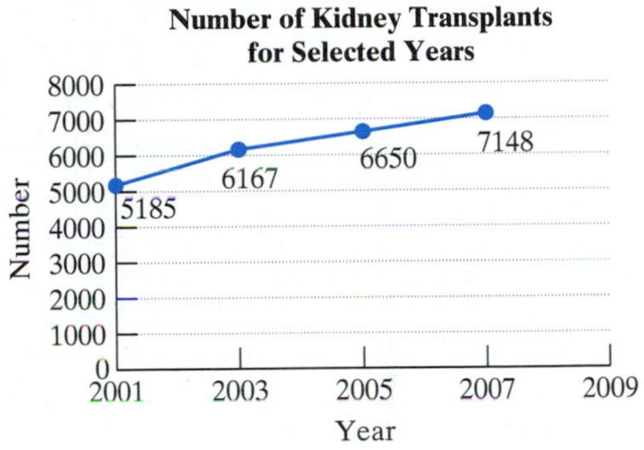

33. Which year had the greatest number of kidney transplants?

34. Which year had the least number of kidney transplants?

35. How many more kidney transplants were performed in 2003 than in 2001?

36. What is the difference in the number of transplants between 2007 and 2001?

37. Use the graph to predict the number of transplants in the year 2009. (Answers may vary.)

38. In which 2-year period was the increase in the number of transplants the greatest?

39. The data shown here give the average height for girls based on age. (*Source:* National Parenting Council) **(See Example 6.)**

a. Make a line graph to illustrate these data. That is, write the table entries as ordered pairs and graph the points.

b. Use the line graph from part (a) to predict the average height of a 10-year-old girl. (Answers may vary.)

Age, x	Height (in.), y
2	35
3	38.5
4	41.5
5	44
6	46
7	48
8	50.5
9	53

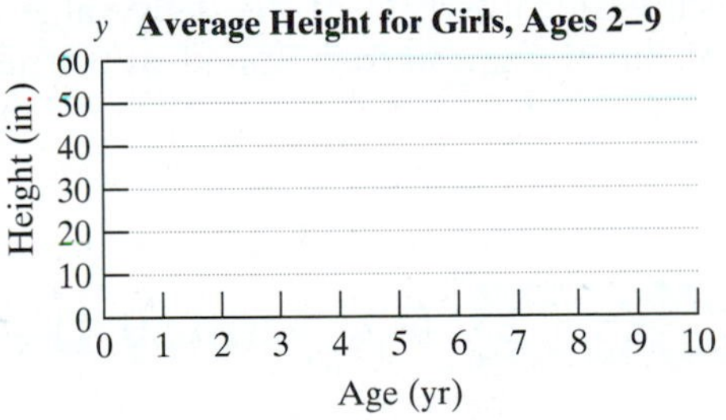

40. The data shown here give the average height for boys based on age. (*Source:* National Parenting Council)

a. Make a line graph to illustrate these data. That is, write the table entries as ordered pairs and graph the points.

b. Use the line graph from part (a) to predict the average height of a 10-year-old boy. (Answers may vary.)

Age, x	Height (in.), y
2	36
3	39
4	42
5	44
6	46.75
7	49
8	51
9	53.5

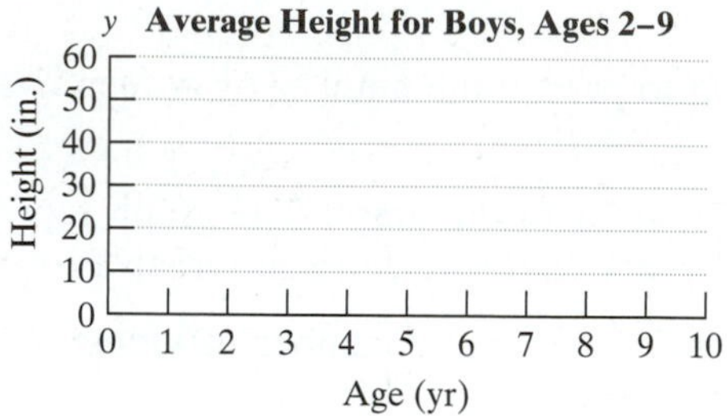

Expanding Your Skills

All packaged food items have to display nutritional facts so that the consumer can make informed choices. For Exercises 41–44, refer to the nutritional chart for Breyers French Vanilla ice cream.

41. How many servings are there per container? How much total fat is in one container of this ice cream?

42. How much total sodium is in one container of this ice cream?

43. If 8 g of fat is 13% of the daily value, what is the daily value of fat? Round to 1 decimal place.

44. If 50 mg of cholesterol is 17% of the daily value, what is the daily value of cholesterol? Round to the nearest whole unit.

Nutrition Facts
Serving Size $\frac{1}{2}$ cup (68 g)
Servings per Container 14

Amount per Serving
Calories 150 — **Calories from Fat** 80

		% Daily Value
Total Fat	8 g	13%
Saturated fat	5 g	25%
Cholesterol	50 mg	17%
Sodium	45 mg	2%
Total Carbohydrate		
Dietary fiber	0 g	
Sugars	15 g	
Protein	3 g	

Section 9.4 Frequency Distributions and Histograms

Objectives

1. Frequency Distributions
2. Histograms

1. Frequency Distributions

The ages for 36 players in the NBA (National Basketball Association) are given for a recent year. The data were taken from the rosters of the Detroit Pistons, Boston Celtics, and New York Knicks.

23	32	26	25	30	30	29	38	21	25	34	22
33	26	39	22	32	23	30	27	22	24	31	33
21	24	28	26	26	20	33	27	29	31	27	28

The youngest player is 20 years old and the oldest is 39 years old. Suppose we wanted to organize this information further by age groups. One way is to create a frequency distribution. A **frequency distribution** is a table displaying the number of values that fall within categories called **class intervals**. This is demonstrated in Example 1.

Example 1 Creating a Frequency Distribution

Complete the table to form a frequency distribution for the ages of the NBA players listed.

Class Intervals, Age (yr)	Tally	Frequency (Number of Players)
20–23		
24–27		
28–31		
32–35		
36–39		

Solution:

The classes represent different age groups. Go through the list of ages, and use tally marks to track the number of players that fall within each class. Tally marks are shown in red for the first column of data: 23, 33, and 21.

Table 9-7

Class Intervals, Age (yr)	Tally	Frequency (Number of Players)
20–23	𝍸 III	8
24–27	𝍸 𝍸 I	11
28–31	𝍸 IIII	9
32–35	𝍸 I	6
36–39	II	2

The frequency is a count of the tally marks within each class. See Table 9-7.

Skill Practice

1. The ages (in years) of individuals arrested on a certain day in Galveston, Texas, are listed.

18	20	35	46
19	26	24	32
28	25	30	34
22	29	39	19
18	19	26	40

Complete the table to form a frequency distribution.

Class (Age)	Tally	Frequency
18–23		
24–29		
30–35		
36–41		
42–47		

Example 2 Interpreting a Frequency Distribution

Consider the frequency distribution in Table 9-7.

a. Which class has the most values?

b. How many values are represented in the table?

c. What percent of the players were 32 years old or older at the time the data were recorded?

Solution:

a. The 24–27 yr class has the highest frequency (greatest number of values).

b. The number of data values is given by the sum of the frequencies.

$$\text{Total number of values} = 8 + 11 + 9 + 6 + 2$$
$$= 36$$

c. The number of players 32 years old or older is given by the sum of the frequencies in the 32–35 category and the 36–39 category. This is $6 + 2 = 8$.

There are 8 players 32 years old or older.

Therefore, $\frac{8}{36} = 0.\overline{2} \approx 22.2\%$ of the players are 32 years old or older.

Skill Practice

For Exercises 2–4, consider the frequency distribution from margin Exercise 1.

2. Which class (age group) has the most values?
3. How many values are represented in the table?
4. What percent of the people arrested are in the 42–47 age group?

Answers

1.

Class (Age)	Tally	Frequency
18–23	𝍸 II	7
24–29	𝍸 I	6
30–35	IIII	4
36–41	II	2
42–47	I	1

2. 18–23 yr 3. 20 4. 5%

When creating a frequency distribution, keep these important guidelines in mind.

- The classes should be equally spaced. For instance, in Example 1, we would not want one class to represent a 4-year interval and another to represent a 10-year interval.
- The classes should not overlap. That is, a value should belong to one and only one class.
- In general, we usually create a frequency distribution with between 5 and 15 classes.

2. Histograms

A **histogram** is a special bar graph that illustrates data given in a frequency distribution. The class intervals are given on the horizontal scale. The height of each bar in a histogram represents the frequency for each class.

Skill Practice

5. Construct a histogram for the frequency distribution in margin Exercise 1.

Example 3 Constructing a Histogram

Construct a histogram for the frequency distribution given in Example 1.

Solution:

Class Intervals, Age (yr)	Tally	Frequency (Number of Players)
20–23	𝍸 III	8
24–27	𝍸 𝍸 I	11
28–31	𝍸 IIII	9
32–35	𝍸 I	6
36–39	II	2

To create a histogram of these data, we list the classes (ages of players) on the horizontal scale. On the vertical scale we measure the frequency (Figure 9-23).

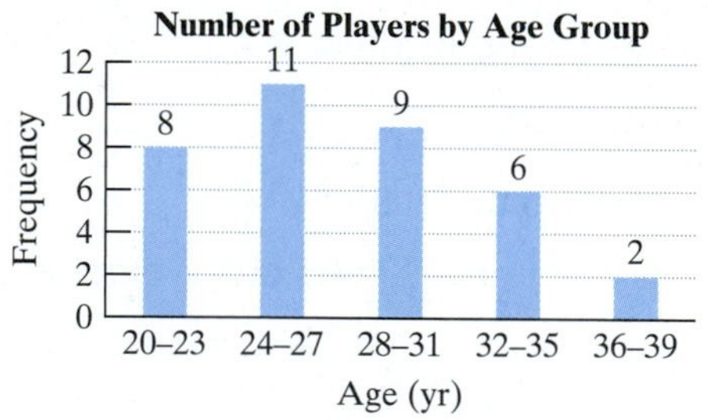

Figure 9-23

Answer

5.

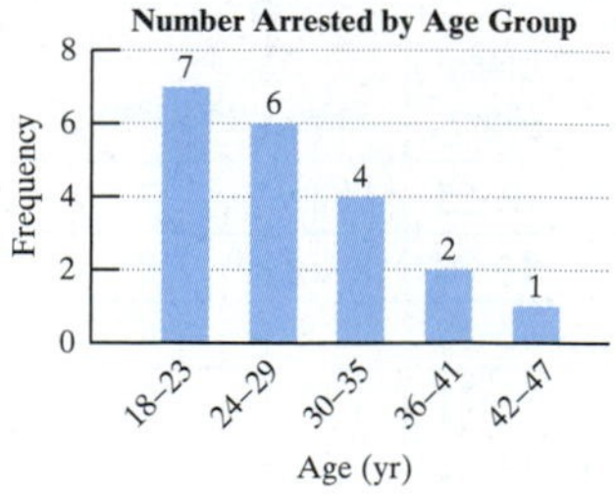

Section 9.4 Practice Exercises

Study Skills Exercise

1. Define the key terms.

a. Frequency distribution **b. Class intervals** **c. Histogram**

Review Exercises

For Exercises 2–6, answer true or false.

2. The point $(-2, 5)$ is located in Quadrant IV.

3. The point $(0, 19)$ is on the x-axis.

4. The ordered pair $(3, -6)$ is a solution to the equation $x - 2y = 15$.

5. The ordered pair $(-2, 5)$ is a solution to the equation $2x - y = -9$.

6. The ordered pair $(-3, 5)$ is the same as $(5, -3)$.

Objective 1: Frequency Distributions

7. From the frequency distribution, determine the total number of data.

Class Intervals	Frequency
1–4	14
5–8	18
9–12	24
13–16	10
17–20	6

8. From the frequency distribution, determine the total number of data.

Class Intervals	Frequency
1–50	29
51–100	12
101–150	6
151–200	22
201–250	56
251–300	60

9. For the table in Exercise 7, which category contains the most data?

10. For the table in Exercise 8, which category contains the most data?

11. The retirement age (in years) for 20 college professors is given. Complete the frequency distribution. **(See Examples 1 and 2.)**

67	56	68	70	60	65	73	72	56	65
71	66	72	69	65	65	63	65	68	70

Class Intervals (Age in Years)	Tally	Frequency (Number of Professors)
56–58		
59–61		
62–64		
65–67		
68–70		
71–73		

a. Which class has the most values?

b. How many data values are represented in the table?

c. What percent of the professors retire when they are 68 to 70 yr old?

12. The number of miles run in one day by 16 selected runners is given. Complete the frequency distribution.

2	4	7	3	8	4	5	7
4	6	4	3	4	2	4	10

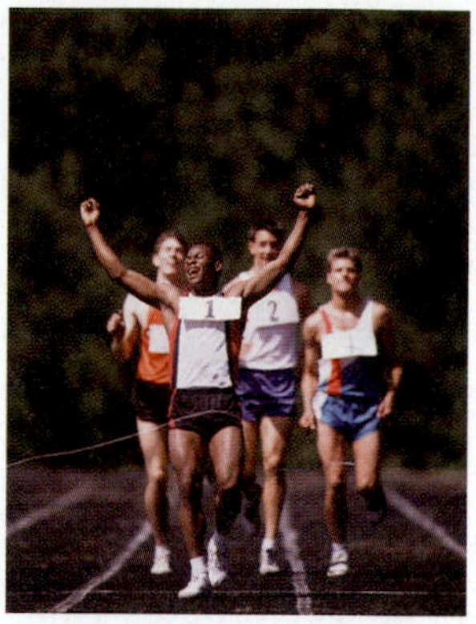

Class Intervals (Number of Miles)	Tally	Frequency (Number of Runners)
1–2		
3–4		
5–6		
7–8		
9–10		

a. Which class has the most values?

b. How many data values are represented in the table?

c. What percent of the runners ran 3 to 4 mi/day?

13. The number of gallons of gas purchased by 16 customers at a certain gas station is given. Complete the frequency distribution.

12.7	13.1	9.8	12.0	10.4	9.8	14.2	8.6
19.2	8.1	14.0	15.4	12.8	18.2	15.1	13.0

Class Intervals (Amount in Gal)	Tally	Frequency (Number of Customers)
8.0–9.9		
10.0–11.9		
12.0–13.9		
14.0–15.9		
16.0–17.9		
18.0–19.9		

a. Which class has the most values?

b. How many data values are represented in the table?

c. What percent of the customers purchaed 18 to 19.9 gal of gas?

14. The hourly wages (in dollars) for 15 employees at a department store are given. Complete the frequency distribution.

9.50	12.00	14.75	9.50	17.00
11.50	12.50	11.75	18.25	9.75
12.75	14.75	9.50	15.75	9.75

Class Intervals (Hourly Wage, $)	Tally	Frequency (Number of Employees)
9.00–10.99		
11.00–12.99		
13.00–14.99		
15.00–16.99		
17.00–18.99		

15. Explain what is wrong with the following class intervals.

Class	Tally	Frequency
0–4		
5–10		
11–17		
18–25		
26–34		

16. Explain what is wrong with the following class intervals.

Class	Tally	Frequency
1–6		
7–11		
12–17		
18–23		
24–28		

17. Explain what is wrong with the following class intervals.

Class	Tally	Frequency
1–20		
21–40		

18. Explain what is wrong with the following class intervals.

Class	Tally	Frequency
1–33		
34–66		
67–99		

19. Explain what is wrong with the following class intervals.

Class	Tally	Frequency
10–12		
12–14		
14–16		
16–18		
18–20		

20. Explain what is wrong with the following class intervals.

Class	Tally	Frequency
1–5		
5–10		
10–15		
15–20		
20–25		

21. The heights of 20 students at Daytona State College are given. Complete the frequency distribution.

70	71	73	62	65	70	69	70
64	66	73	63	68	67	69	72
64	66	67	69				

Class Interval (Height, in.)	Frequency (Number of Students)
62–63	
64–65	
66–67	
68–69	
70–71	
72–73	

22. The amount withdrawn in dollars from a certain ATM is given for 20 customers. Construct a frequency distribution.

40	50	200	200	100	120	200
50	100	60	100	100	30	40
100	100	50	200	150	200	

Class Interval (Amount, $)	Frequency (Number of Customers)
0–49	
50–99	
100–149	
150–199	
200–249	

Objective 2: Histograms

23. Construct a histogram for the frequency table in Exercise 21. **(See Example 3.)**

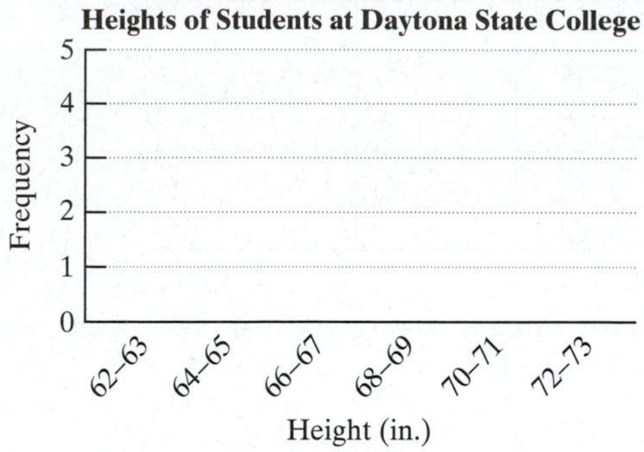

24. Construct a histogram for the frequency table in Exercise 22.

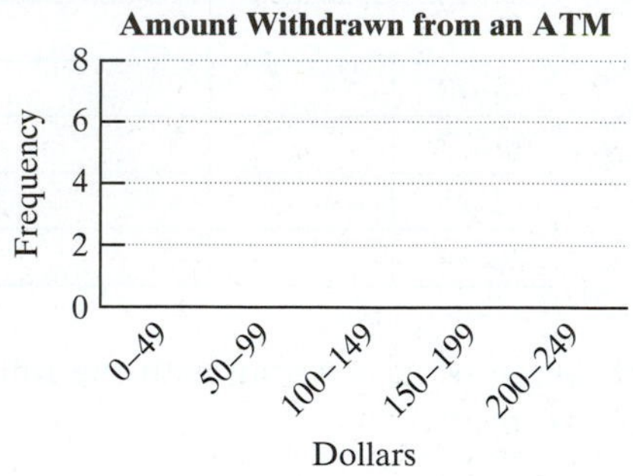

25. Construct a histogram, using the given data. Each number represents the number of Calories in a 100-g serving for selected fruits.

59	65	48	49	105	47	92	43
52	59	56	49	35	55	72	30
67	44	32	32	61	29	30	56

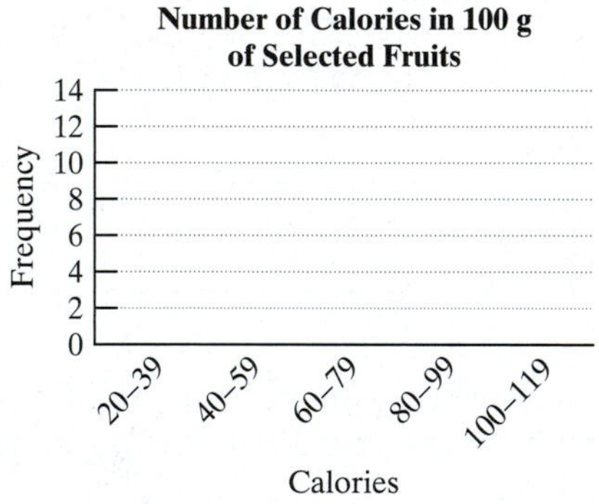

26. The list of data gives the number of children of the Presidents of the United States (in no particular order). Construct a histogram.

0	4	3	3	2	2	5	10	0	5	2
4	5	7	4	3	6	4	0	7	5	6
1	4	3	0	4	3	2	6	4	6	8
3	2	1	0	2	6	0	2	2		

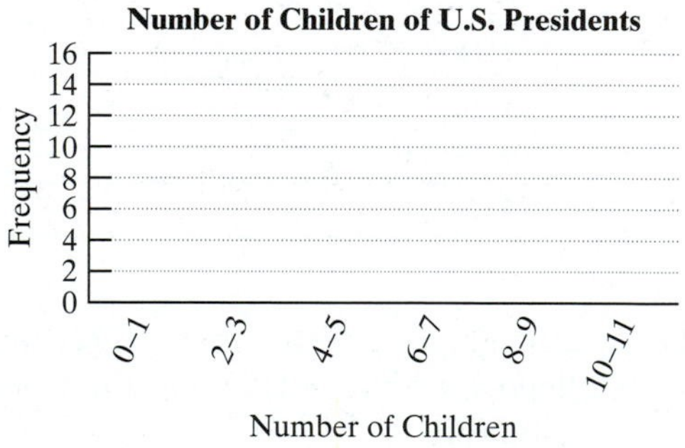

Section 9.5 Circle Graphs

Objectives

1. Interpreting Circle Graphs
2. Circle Graphs and Percents
3. Constructing Circle Graphs

1. Interpreting Circle Graphs

Thus far we have used bar graphs, line graphs, and histograms to visualize data. A **circle graph** (or pie graph) is another type of graph used to show how a whole amount is divided into parts. Each part of the circle, called a **sector**, is like a slice of pie. The size of each piece relates to the fraction of the whole it represents.

Example 1 Interpreting a Circle Graph

The grade distribution for a math test is shown in the circle graph (Figure 9-24).

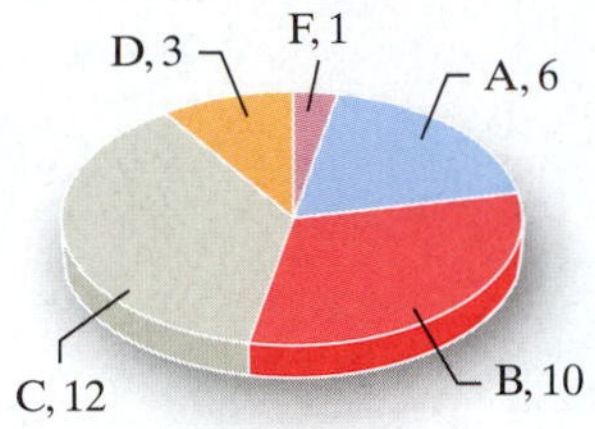

Figure 9-24

a. How many total grades are represented?

b. How many grades are B's?

c. How many times more C's are there than D's?

d. What percent of the grades were A's?

Solution:

a. The total number of grades is equal to the sum of the number of grades from each category.

$$\text{Total number of grades} = 6 + 10 + 12 + 3 + 1$$
$$= 32$$

b. The number of B's is represented by the red portion of the graph. There are 10 B's.

c. There are 12 C's and 3 D's. The ratio of C's to D's is $\frac{12}{3} = 4$. Therefore, there are 4 times as many C's as D's.

d. There are 6 A's. The percent of A's is given by

$$\frac{6}{32} = 0.1875$$

Therefore, the percent of A's is 18.75%.

Skill Practice

A used car dealership sells cars and trucks. Use the circle graph to answer Exercises 1–4.

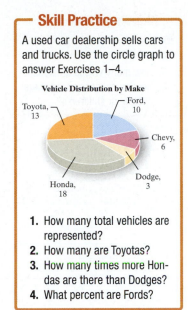

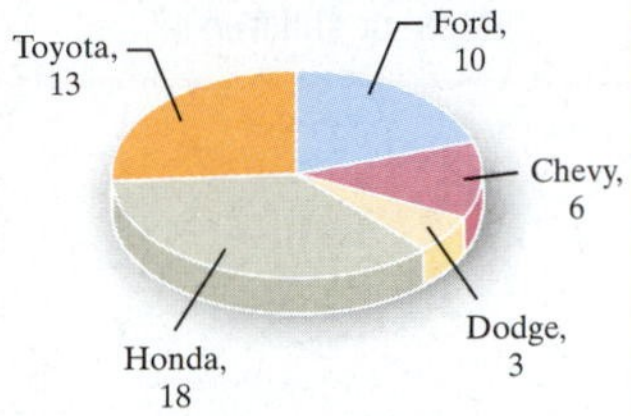

1. How many total vehicles are represented?

2. How many are Toyotas?

3. How many times more Hondas are there than Dodges?

4. What percent are Fords?

2. Circle Graphs and Percents

Sometimes circle graphs show data in percent form. This is illustrated in Example 2.

Answers

1. 50 **2.** 13

3. There are 6 times more Hondas than Dodges.

4. 20% are Fords.

Skill Practice

For Exercises 5 and 6, refer to Figure 9-25.

5. How many videos are action?
6. How many videos are general interest or children's?

Example 2 Calculating Amounts by Using a Circle Graph

A certain video rental store carries 2000 different videos. It groups its video collection by the categories shown in the graph (Figure 9-25).

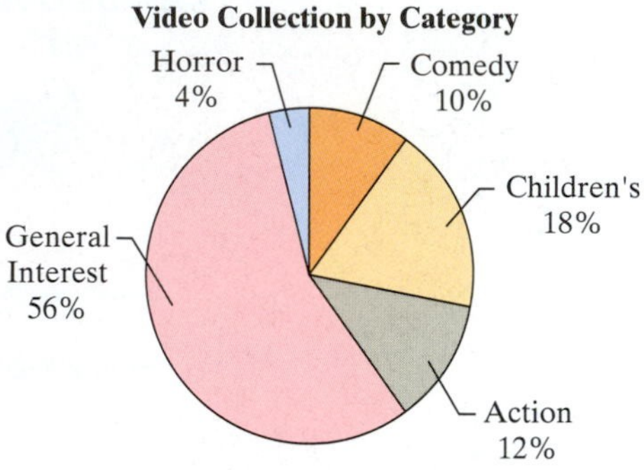

Figure 9-25

a. How many videos are comedy?

b. How many videos are action or horror?

Solution:

a. First note that the store carries 2000 different videos. From the graph we know that 10% are comedies. Therefore, this question can be interpreted as

What is 10% of 2000?

$$x = (0.10) \cdot (2000)$$

$$= 200 \qquad \text{There are 200 comedies.}$$

b. From the graph we know that 12% of the videos are action and 4% are horror. This accounts for 16% of the total video collection. Therefore, this question asks

What is 16% of 2000?

$$x = (0.16) \cdot (2000)$$

$$= 320 \qquad \text{There are 320 videos that are action or horror.}$$

3. Constructing Circle Graphs

Recall that a full circle is a 360° arc. To draw a circle graph, we must compute the number of degrees of arc for each sector. In Example 2, 10% of the videos are comedies. To draw the sector for this category, we must determine 10% of 360°.

$$10\% \text{ of } 360° = 0.10(360°) = 36°$$

The sector representing comedies should be drawn with a 36° angle. To do this, we can use a protractor (Figure 9-26).

Answers

5. There are 240 action videos.
6. There are 1480 general interest or children's videos.

To draw a sector with a 36° arc, first draw a circle. Place the hole in the protractor over the center of the circle. Using the inner scale on the protractor, place a tick mark at 0° and at 36°. Use a straightedge to draw two line segments from the center of the circle to each tick mark. See Figure 9-26.

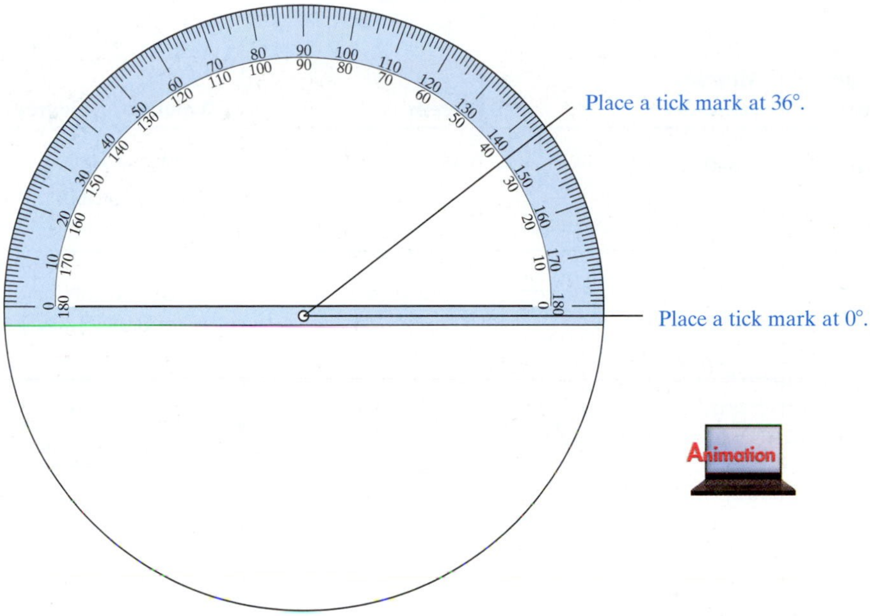

Figure 9-26

In Example 3, we use this technique to construct a circle graph.

Example 3 Constructing a Circle Graph

A teacher earns a monthly salary of $2400 after taxes. Her monthly budget is broken down in Table 9-8.

Table 9-8

Budget Item	Monthly Value ($)
Rent	840
Utilities	210
Car expenses	510
Groceries	360
Savings	300
Other	180

Construct a circle graph illustrating the information in this table. Label each sector of the graph with the percent that it represents.

Concept Connections

For Exercises 7 and 8, refer to the circle graph.

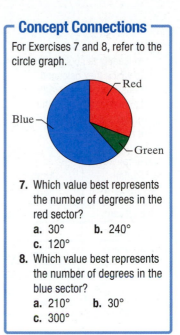

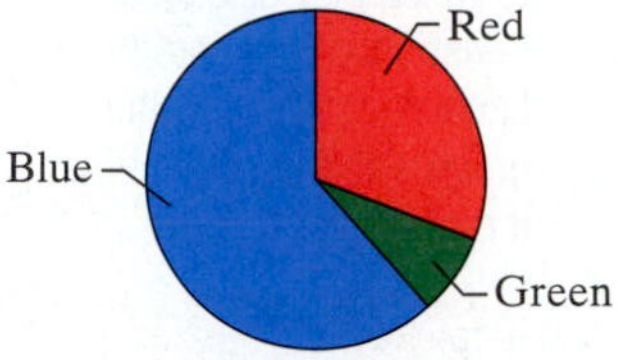

7. Which value best represents the number of degrees in the red sector?
   a. 30° **b.** 240°
   c. 120°
8. Which value best represents the number of degrees in the blue sector?
   a. 210° **b.** 30°
   c. 300°

Answers
7. c
8. a

Skill Practice

9. Voters in Oregon were asked to identify the political party to which they belonged. Construct a circle graph. Label each sector of the graph with the percent that it represents.

Political Affiliation	Number
Democrat	900
Republican	720
Libertarian	36
Green Party	144

Solution:

This problem calls for two types of calculations: (1) For each budget item, we must compute the percent of the whole that it represents. (2) We must determine the number of degrees for each category. We can use a table to help organize our calculations.

Budget Item	Monthly Value ($)	Percent	Number of Degrees
Rent	840	$= \frac{840}{2400} = 0.35$ or 35%	35% of 360° $= 0.35(360°)$ $= 126°$
Utilities	210	$= \frac{210}{2400} = 0.0875$ or 8.75%	8.75% of 360° $= 0.0875(360°)$ $= 31.5°$
Car	510	$= \frac{510}{2400} = 0.2125$ or 21.25%	21.25% of 360° $= 0.2125(360°)$ $= 76.5°$
Groceries	360	$= \frac{360}{2400} = 0.15$ or 15%	15% of 360° $= 0.15(360°)$ $= 54°$
Savings	300	$= \frac{300}{2400} = 0.125$ or 12.5%	12.5% of 360° $= 0.125(360°)$ $= 45°$
Other	180	$= \frac{180}{2400} = 0.075$ or 7.5%	7.5% of 360° $= 0.075(360°)$ $= 27°$

Now construct the circle graph. Use the degree measures found in the table for each sector. Label the graph with the percent for each sector (Figure 9-27).

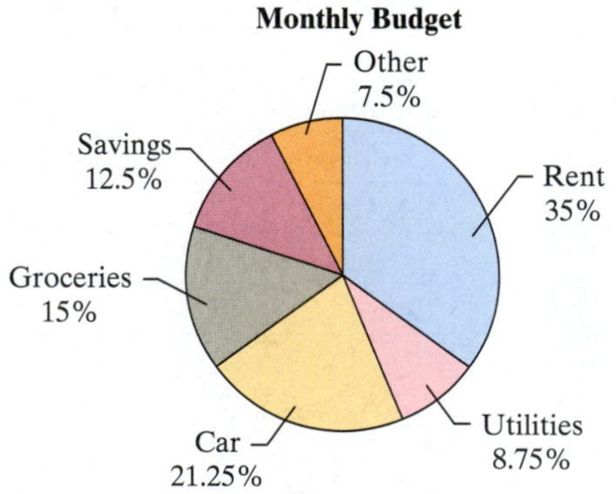

Figure 9-27

Answer

9.

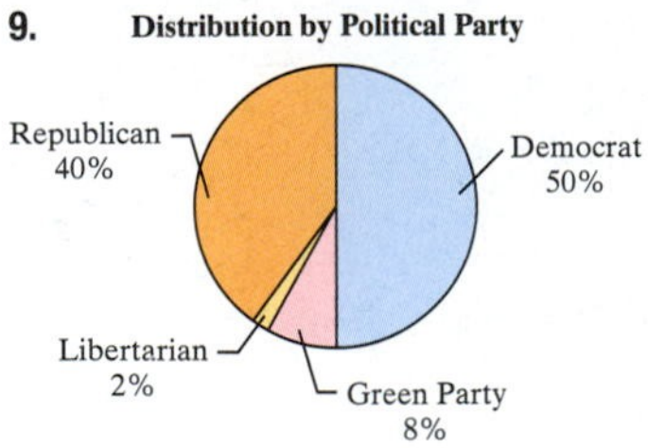

Section 9.5 Practice Exercises

Study Skills Exercise

1. Define the key terms.

a. Circle graph **b. Sector**

Review Exercises

For Exercises 2–4, graph each equation.

2. $-2x + y = -2$

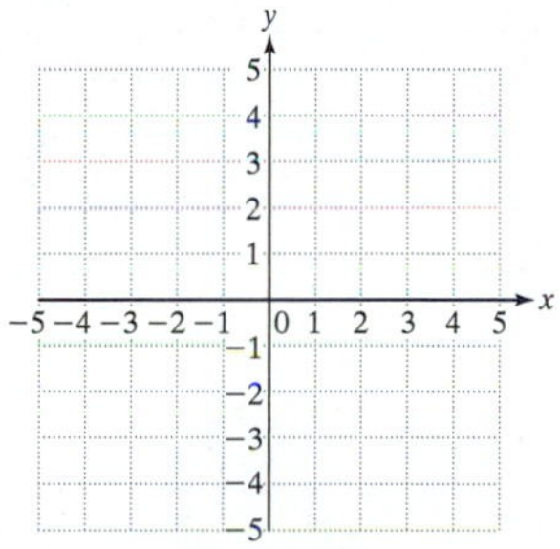

3. $y = -\frac{1}{4}x$

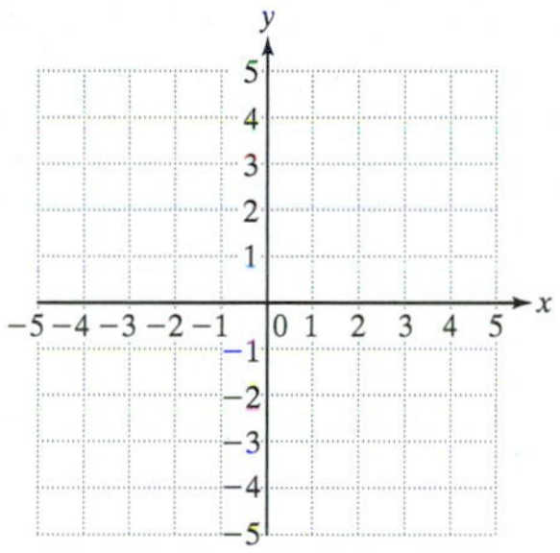

4. $y = 3$

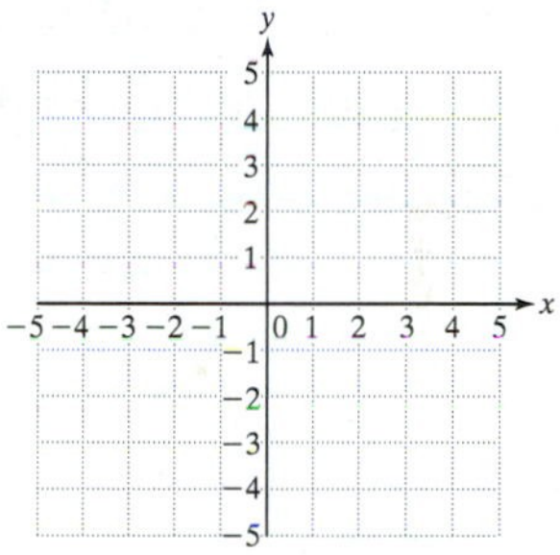

Objective 1: Interpreting Circle Graphs

For Exercises 5–12, refer to the graph. The graph represents the number of traffic fatalities by age group in the United States. (*Source:* U.S. Bureau of the Census) **(See Example 1.)**

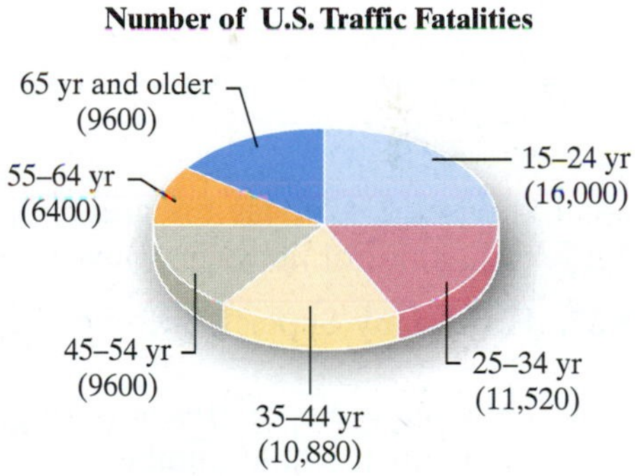

5. What is the total number of traffic fatalities?

6. Which of the age groups has the most fatalities?

7. How many more people died in the 25–34 age group than in the 35–44 age group?

8. How many more people died in the 45–54 age group than in the 55–64 group?

9. What percent of the deaths were from the 15–24 group?

10. What percent of the deaths were from the 65 and older age group?

11. How many times more deaths were from the 15–24 age group than the 55–64 age group?

12. How many times more deaths were from the 25–34 age group than from the 65 and older group?

For Exercises 13–18, refer to the figure. The figure represents the average daily number of viewers for five daytime dramas. (*Source:* Nielsen Media Research)

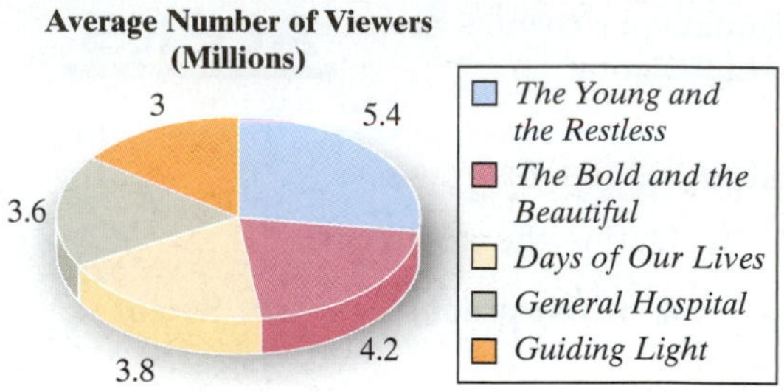

13. How many viewers are represented?

14. How many viewers does the most popular daytime drama have?

15. How many times more viewers does *The Young and the Restless* have than *Guiding Light*?

16. How many times more viewers does *The Young and the Restless* have than *General Hospital*?

17. What percent of the viewers watch *General Hospital*?

18. What percent of the viewers watch *The Bold and the Beautiful*?

Objective 2: Circle Graphs and Percents

For Exercises 19–22, use the graph representing the type of music CDs found in a store containing approximately 8000 CDs. **(See Example 2.)**

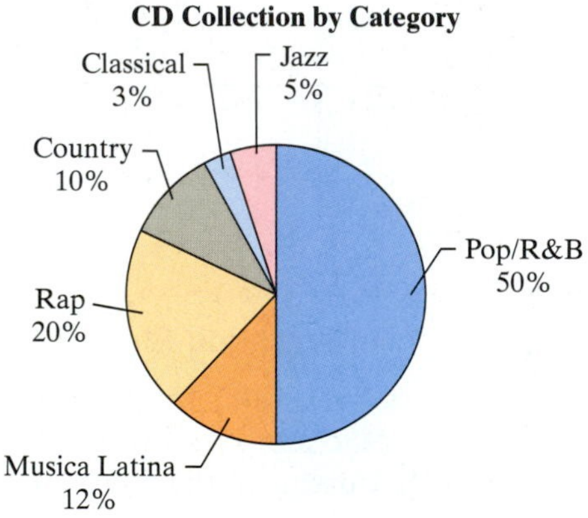

19. How many CDs are musica Latina?

20. How many CDs are rap?

21. How many CDs are jazz or classical?

22. How many CDs are *not* Pop/R&B?

For Exercises 23–26, use the graph representing the states that hosted Super Bowl I through Super Bowl XXXVI (a total of 36 Super Bowls).

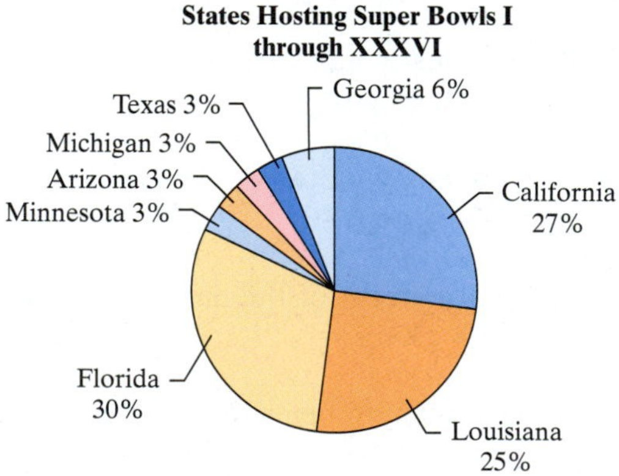

23. How many Super Bowls were played in Louisiana?

24. How many Super Bowls were played in Florida? Round to the nearest whole number.

25. How many Super Bowls were played in Georgia? Round to the nearest whole number.

26. How many Super Bowls were played in Michigan? Round to the nearest whole number.

Objective 3: Constructing Circle Graphs

For Exercises 27–34, use a protractor to construct an angle of the given measure.

27. 20° **28.** 70° **29.** 125°

30. 270° **31.** 195° **32.** 5°

33. 300° **34.** 90°

35. Draw a circle and divide it into sectors of 30°, 60°, 100°, and 170°.

36. Draw a circle and divide it into sectors of 125°, 180°, and 55°.

37. The Sunshine Nursery sells flowering plants, shrubs, ground cover, trees, and assorted flower pots. Construct a pie graph to show the distribution of the types of purchases.

Types of Purchases	Percent of Distribution
Flowering plants	45%
Shrubs	13%
Ground cover	18%
Trees	20%
Flower pots	4%

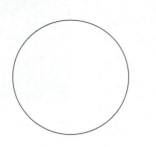

Sunshine Nursery Distribution of Sales

38. The party affiliation of registered Latino voters for a recent year is as follows:

45% Democrat   20% Republican

13% Other   22% Independent

Construct a circle graph from this information.

Party Affiliation of Latino Voters

39. The table provided gives the expenses for one semester at college. **(See Example 3.)**

a. Complete the table.

	Expenses	Percent	Number of Degrees
Tuition	$9000		
Books	600		
Housing	2400		

b. Construct a circle graph to display the college expenses. Label the graph with percents.

College Expenses for a Semester

40. The table provided gives the number of establishments of the three largest pizza chains.

a. Complete the table.

	Number of Stores	Percent	Number of Degrees
Pizza Hut	8100		
Domino's	7200		
Papa Johns	2700		

b. Construct a circle graph. Label the graph with percents.

Percent of Pizza Establishments

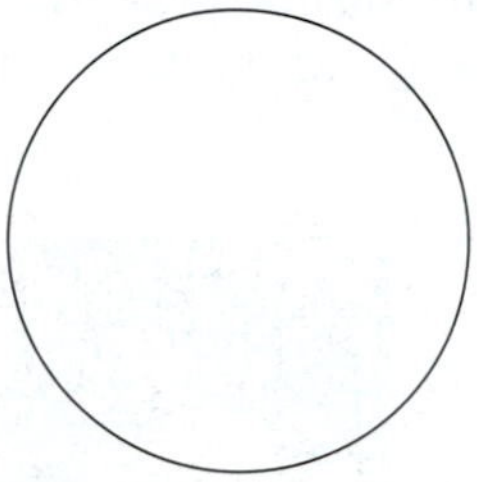

Section 9.6 Introduction to Probability

Objectives

1. Basic Definitions
2. Probability of an Event
3. Estimating Probabilities from Observed Data
4. Complementary Events

1. Basic Definitions

The probability of an event measures the likelihood of the event to occur. It is of particular interest because of its application to everyday life.

- The probability of picking the winning six-number combination for the New York lotto grand prize is $\frac{1}{45,057,474}$.
- Genetic DNA analysis can be used to determine the risk that a child will be born with cystic fibrosis. If both parents test positive, the probability is 25% that a child will be born with cystic fibrosis.

To begin our discussion, we must first understand some basic definitions.

An activity with observable outcomes such as flipping a coin or rolling a die is called an **experiment**. The collection (or set) of all possible outcomes of an experiment is called the **sample space** of the experiment.

Skill Practice

1. Suppose one ball is selected from those shown and the color is recorded. Write the sample space for this experiment.

2. For an individual birth, the gender of the baby is recorded. Determine the sample space for this experiment.

Example 1 Determining the Sample Space of an Experiment

a. Suppose a single die is rolled. Determine the sample space of the experiment.

b. Suppose a coin is flipped. Determine the sample space of the experiment.

Solution:

a. A die is a single six-sided cube on which each side has between 1 and 6 dots painted on it. When the die is rolled, any of the six sides may come up.

The sample space is {1, 2, 3, 4, 5, 6}.

Notice that the symbols { } (called *set braces*) are used to enclose the elements.

b. The coin may land as a head H or as a tail T. The sample space is {H, T}.

Answers

1. {red, green, blue, yellow}
2. {male, female}

2. Probability of an Event

Any part of a sample space is called an **event**. For example, if we roll a die, the event of rolling number 5 or a greater number consists of the outcomes 5 and 6. In mathematics, we measure the likelihood of an event to occur by its probability.

DEFINITION Probability of an Event

$$\textbf{Probability of an event} = \frac{\text{number of elements in event}}{\text{number of elements in sample space}}$$

Avoiding Mistakes

From the definition, a probability value can never be negative or greater than 1.

Example 2 Finding the Probabilities of Events

a. Find the probability of rolling a 5 or greater on a die.

b. Find the probability of flipping a coin and having it land as heads.

Solution:

a. The event can occur in 2 ways: The die lands as a 5 or 6. The sample space has 6 elements: $\{1, 2, 3, 4, 5, 6\}$.

The probability of rolling a 5 or greater: $\frac{2}{6}$ ← number of ways to roll a 5 or greater / ← number of elements in the sample space

$= \frac{1}{3}$ Simplify to lowest terms.

b. The event can occur in 1 way (the coin lands head side up). The sample space has 2 outcomes, heads or tails: $\{H, T\}$.

The probability of flipping a head on a coin: $\frac{1}{2}$ ← number of ways to get heads / ← number of elements in the sample space

Skill Practice

3. Find the probability of selecting a yellow ball from those shown.

4. Find the probability of a random birth resulting in a girl.

The value of a probability can be written as a fraction, as a decimal, or as a percent. For example, the probability of a coin landing as heads is $\frac{1}{2}$ or 0.5 or 50%. In words, this means that if we flip a coin many times, theoretically we expect one-half (50%) of the outcomes to land as heads.

Example 3 Finding Probabilities

A class has 4 freshmen, 12 sophomores, and 6 juniors. If one individual is selected at random from the class, find the probability of selecting

a. A sophomore

b. A junior

c. A senior

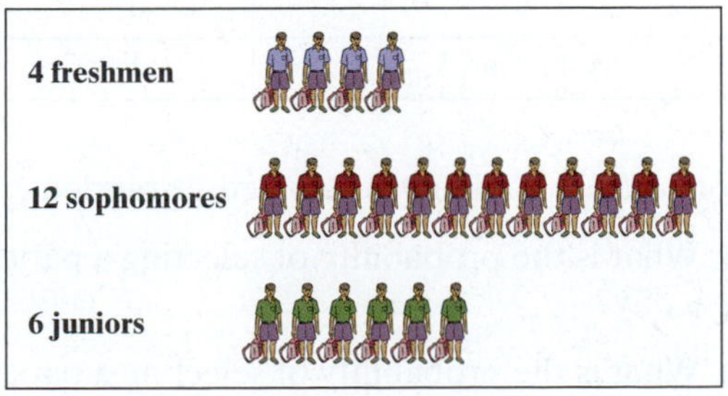

Answers

3. $\frac{1}{4}$ or 0.25 **4.** $\frac{1}{2}$ or 0.5

Skill Practice

A group of registered voters has 9 Republicans, 8 Democrats, and 3 Independents. Suppose one person from the group is selected at random.

5. What is the probability that the person is a Democrat?
6. What is the probability that the person is an Independent?
7. What is the probability that the person is registered with the Libertarian Party?

Solution:

In this case, there are 22 members of the class (4 freshmen + 12 sophomores + 6 juniors). This means that the sample space has 22 elements.

a. There are 12 sophomores in the class. The probability of selecting a sophomore is

$\frac{12}{22}$ There are 12 sophomores out of 22 people in the sample space.

$= \frac{6}{11}$ Simplify to lowest terms.

b. There are 6 juniors out of 22 people in the sample space. The probability of selecting a junior is

$$\frac{6}{22} \quad \text{or} \quad \frac{3}{11}$$

c. There are no seniors in the class. The probability of selecting a senior is

$$\frac{0}{22} \quad \text{or} \quad 0$$

A probability of 0 indicates that the event is impossible. It is impossible to select a senior from a class that has no seniors.

From the definition of the probability of an event, it follows that the value of a probability must be between 0 and 1, inclusive. An event with a probability of 0 is called an *impossible event*. An event with a probability of 1 is called a *certain event*.

3. Estimating Probabilities from Observed Data

We were able to compute the probabilities in Examples 2 and 3 because the sample space was known. Sometimes we need to collect information to help us estimate probabilities.

Skill Practice

Refer to Table 9-9 in Example 4.

8. What is the probability of selecting a patient who brushes twice a day?
9. What is the probability of selecting a patient who brushes more than twice a day?

Example 4 Estimating Probabilities from Observed Data

A dental hygienist records the number of times a day her patients say that they brush their teeth. Table 9-9 displays the results.

Table 9-9

Number of Times of Brushing Teeth per Day	Frequency
1	6
2	10
3	4
More than 3	1

If one of her patients is selected at random,

a. What is the probability of selecting a patient who brushes only one time a day?

b. What is the probability of selecting a patient who brushes more than once a day?

Answers

5. $\frac{2}{5}$ or 0.4 6. $\frac{3}{20}$ or 0.15 7. 0
8. $\frac{10}{21}$ 9. $\frac{5}{21}$

Solution:

a. The table shows that there are 6 patients who brush once a day. To get the total number of patients we add all of the frequencies $(6 + 10 + 4 + 1 = 21)$. The probability of selecting a patient who brushes only once a day is

$$\frac{6}{21} \quad \text{or} \quad \frac{2}{7}$$

b. To find the number of patients who brush more than once a day, we add the frequencies for the patients who brush 2 times, 3 times, and more than 3 times $(10 + 4 + 1 = 15)$. The probability of selecting a patient who brushes more that once a day is

$$\frac{15}{21} \quad \text{or} \quad \frac{5}{7}$$

4. Complementary Events

The events in Example 4(a) and 4(b) are called complementary events. The **complement of an event** is the set of all elements in the sample space that are not in the event. In this case, the number of patients who brush once a day and the number of patients who brush more than once a day make up the entire sample space, yet do not overlap. For this reason, the probability of an event plus the probability of its complement is 1. For Example 4, we have $\frac{2}{7} + \frac{5}{7} = \frac{7}{7} = 1$.

Example 5 **Finding the Probability of Complementary Events**

Find the indicated probability.

a. The probability of getting a winter cold is $\frac{3}{10}$. What is the probability of *not* getting a winter cold?

b. If the probability that a washing machine will break before the end of the warranty period is 0.0042, what is the probability that a washing machine will *not* break before the end of the warranty period?

Skill Practice

10. For one particular medicine, the probability that a patient will experience side effects is $\frac{1}{20}$. What is the probability that a patient will *not* experience side effects?

11. The probability that a flight arrives on time is 0.18. What is the probability that a flight will *not* arrive on time?

Solution:

a. The probability of an event plus the probability of its complement must add up to 1. Therefore, we have an addition problem with a missing addend. This may also be expressed as subtraction.

$$\frac{3}{10} + ? = 1 \qquad \text{or equivalently} \qquad 1 - \frac{3}{10} = ?$$

$$\frac{10}{10} - \frac{3}{10} = \frac{7}{10}$$ Find a common denominator and subtract.

There is a $\frac{7}{10}$ chance (70% chance) of *not* getting a winter cold.

b. The probability that a washing machine will break before the end of the warranty period is 0.0042. Then the probability that a machine will *not* break before the end of the warranty period is given by

$$1 - 0.0042 = 0.9958 \text{ or equivalently } 99.58\%$$

Answers

10. $\frac{19}{20}$ **11.** 0.82

Section 9.6 Practice Exercises

Study Skills Exercise

1. Define the key terms.

   a. Experiment **b. Sample space** **c. Event**

   d. Probability of an event **e. Complement of an event**

Review Exercises

For Exercises 2–6, determine if the ordered pair is a solution to the equation.

2. $4x + 6y = 8 \quad (-2, 0)$

3. $-3x - 5y = 10 \quad (0, 2)$

4. $-2x - 7y = 19 \quad (1, -3)$

5. $y = -7 \quad (6, -7)$

6. $x = -2 \quad (4, -2)$

Objectives 1: Basic Definitions

7. A card is chosen from a deck consisting of 10 cards numbered 1–10. Determine the sample space of this experiment. **(See Example 1.)**

8. A marble is chosen from a jar containing a yellow marble, a red marble, a blue marble, a green marble, and a white marble. Determine the sample space of this experiment.

9. Two dice are thrown, and the sum of the top sides is observed. Determine the sample space of this experiment.

10. A coin is tossed twice. Determine the sample space of this experiment.

11. If a die is rolled, in how many ways can an odd number come up?

12. If a die is rolled, in how many ways can a number less than 6 come up?

Objective 2: Probability of an Event

13. Which of the values can represent the probability of an event?

a. 1.62 **b.** $-\frac{7}{5}$ **c.** 0 **d.** 1

e. 200% **f.** 4.5 **g.** 4.5% **h.** 0.87

14. Which of the values can represent the probability of an event?

a. 1.5 **b.** 0 **c.** $\frac{2}{3}$ **d.** 1

e. 150% **f.** 3.7 **g.** 3.7% **h.** 0.92

15. If a single die is rolled, what is the probability that it will come up as a number less than 3? **(See Example 2.)**

16. If a single die is rolled, what is the probability that it will come up as a number greater than 5?

17. If a single die is rolled, what is the probability that it will come up with an even number?

18. If a single die is rolled, what is the probability that it will come up as an odd number?

For Exercises 19–22, refer to the figure. A sock drawer contains 2 white socks, 5 black socks, and 1 blue sock. **(See Example 3.)**

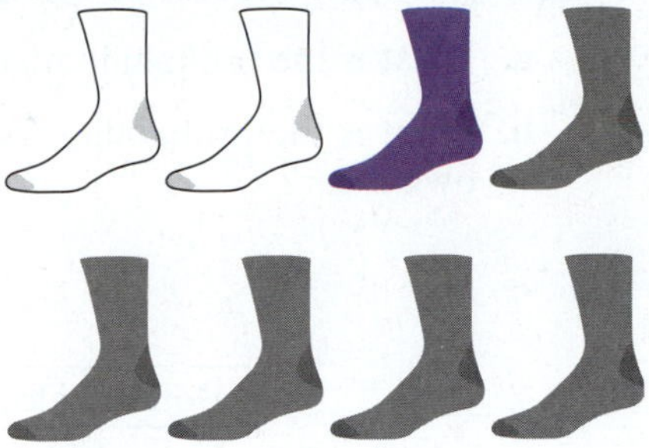

19. What is the probability of choosing a black sock from the drawer?

20. What is the probability of choosing a white sock from the drawer?

21. What is the probability of choosing a blue sock from the drawer?

22. What is the probability of choosing a purple sock from the drawer?

23. If a die is tossed, what is the probability that a number from 1 to 6 will come up?

24. If a die is tossed, what is the probability of getting a 7?

25. What is an impossible event?

26. What is the sum of the probabilities of an event and its complement?

27. In a deck of cards there are 12 face cards and 40 cards with numbers. What is the probability of selecting a face card from the deck?

28. In a deck of cards, 13 are diamonds, 13 are spades, 13 are clubs, and 13 are hearts. Find the probability of selecting a diamond from the deck.

29. A jar contains 7 yellow marbles, 5 red marbles, and 4 green marbles. What is the probability of selecting a red marble or a yellow marble?

30. A jar contains 10 black marbles, 12 white marbles, and 4 blue marbles. What is the probability of selecting a blue marble or a black marble?

Objective 3: Estimating Probabilities from Observed Data

31. The table displays the length of stay for vacationers at a small motel. **(See Example 4.)**

Length of Stay in Days	Frequency
2	14
3	13
4	18
5	28
6	11
7	30
8	6

a. What is the probability that a vacationer will stay for 4 days?

b. What is the probability that a vacationer will stay for less than 4 days?

c. Based on the information from the table, what percent of vacationers stay for more than 6 days?

32. A number of students at a large university were asked if they owned a car and if they lived in a dorm or off campus. The table shows the results.

a. What is the probability that a student selected at random lives in a dorm?

b. What is the probability that a student selected at random does not own a car?

	Number of Car Owners	Number Who Do Not Own a Car
Dorm resident	32	88
Lives off campus	59	26

33. A survey was made of 60 participants, asking if they drive an American-made car, a Japanese car, or a car manufactured in another foreign country. The table displays the results.

a. What is the probability that a randomly selected car is manufactured in America?

b. What percent of cars is manufactured in some country other than Japan?

	Frequency
American	21
Japanese	30
Other	9

34. The number of customer complaints for service representatives at a small company is given in the table. If one representative is picked at random, find the probability that the representative received

a. Exactly 3 complaints

b. Between 1 and 5 complaints, inclusive

c. At least 4 complaints

d. More than 5 complaints or fewer than 2 complaints

Number of Complaints	Number of Representatives
0	4
1	2
2	14
3	10
4	16
5	18
6	10
7	6

35. Mr. Gutierrez noted the times in which his students entered his classroom and constructed the following chart.

a. What is the probability that a student will be early to class?

b. What is the probability that a student will be late to class?

c. What percent of students arrive on time or early? Round to the nearest whole percent.

Time	Number of Students
About 10 min early	1
About 5 min early	6
On time	11
About 5 min late	7
About 10 min late	3
About 15 min late	1

36. Each person at an office party purchased a raffle ticket. The graph shows the results of the raffle.

a. How many people bought raffle tickets?

b. What is the probability of winning the TV set?

c. What percent of people won some type of prize?

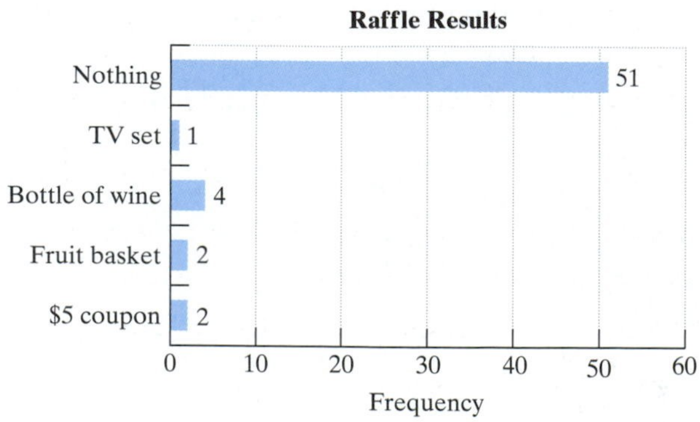

Objective 4: Complementary Events

37. If the probability of the horse Lightning Bolt to win is $\frac{2}{11}$, what is the probability that he will not win? **(See Example 5.)**

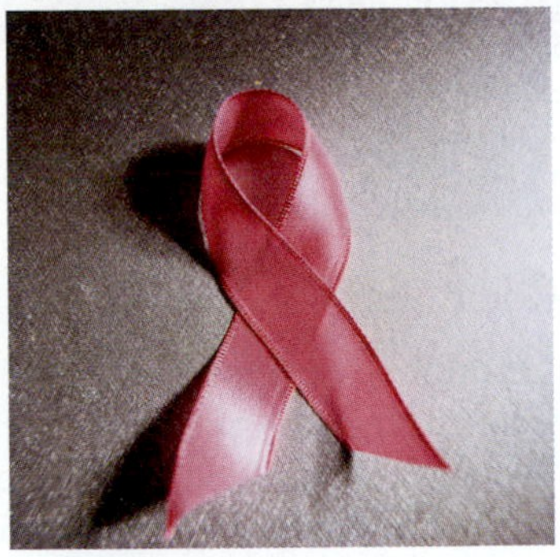

38. If the probability of being hit by lighting is $\frac{1}{1{,}000{,}000}$, what is the probability of not getting hit by lighting?

39. If the probability of having twins is 1.2%, what is the probability of not having twins?

40. The probability of a woman's surviving breast cancer is 88%. What is the probability that a woman would not survive breast cancer?

Group Activity

Creating a Statistical Report

Materials: A computer with Internet access or the local newspaper

Estimated time: 20–30 minutes

Group Size: 4

The group members will collect numerical data from the Internet or the newspaper. The data will be analyzed using the statistical techniques learned in this chapter. Here is one suggested project.

1. Record the age and gender of the individuals who were arrested in your town during the past week. This can often be found in the local section of the newspaper. For example, you can visit the website for the Daytona Beach *News-Journal* and select "local news" and then "news of record." Record 20 or 30 data values.

2. Compute the mean, median, and mode for the ages of men arrested. Compute the mean, median, and mode for the ages of women arrested. Do the statistics suggest a difference in the average age of arrest for men versus women?

3. Determine the percentage of men and the percentage of women in the sample. Does there appear to be a significant difference?

4. Organize the data by age group and construct a frequency distribution and histogram.

Note: The steps given in this project offer suggestions for organizing and analyzing the data you collect. These steps outline standard statistical techniques that apply to a variety of data sets. You might consider doing a different project that investigates a topic of interest to you. Here are some other ideas.

- Collect the weight and gender of babies born in the local hospital.
- Collect the age and gender of students who take classes at night versus those who take classes during the day.
- Collect stock prices for a 2- or 3-week period.

Can you think of other topics for a project?

Chapter 9 Summary

Section 9.1 Rectangular Coordinate System

Key Concepts

Graphical representation of numerical **data** is often helpful to study problems in real-world applications.

A **rectangular coordinate system** is made up of a horizontal line called the ***x*-axis** and a vertical line called the ***y*-axis**. The point where the lines meet is the **origin**. The four regions of the plane are called **quadrants**.

The point (x, y) is an **ordered pair**. The first element in the ordered pair is the point's horizontal position from the origin. The second element in the ordered pair is the point's vertical position from the origin.

Examples

Example 1

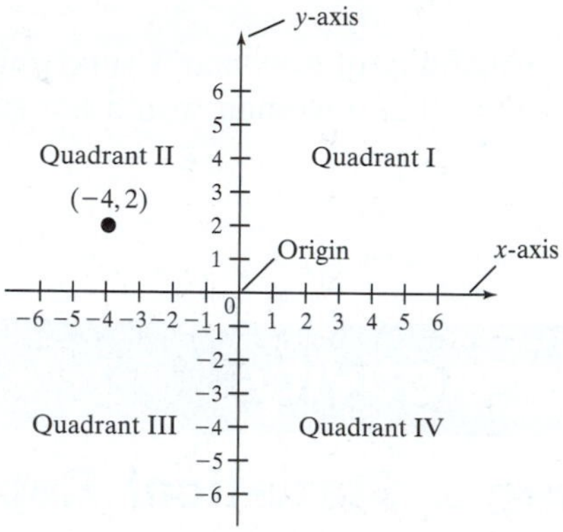

Example 2

Graph the ordered pairs.

$(2, 5)$ $(-3, -4)$ $(-2, 3)$,

$(0, -3)$, $(-4, 0)$, $(4, -5)$

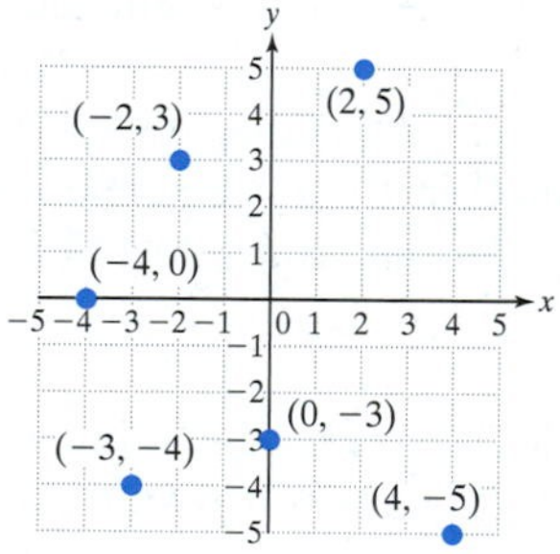

Section 9.2 Graphing Two-Variable Equations

Key Concepts

The solution to an equation in two variables is an ordered pair (x, y).

To graph a linear equation, find at least two ordered pairs that are solutions to the equation. Then graph the ordered pairs and draw the line through the points. The line represents all solutions to the equation.

If a linear equation has only one variable, the graph will be a **horizontal line** or **vertical line**.

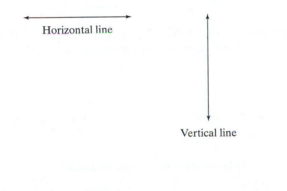

Horizontal line

Vertical line

Examples

Example 1

Determine if $(3, -2)$ is a solution to the equation $-4x + y = -14$.

$$-4(3) + (-2) \stackrel{?}{=} -14$$
$$-12 + (-2) \stackrel{?}{=} -14$$
$$-14 \stackrel{?}{=} -14 \checkmark \quad \text{(true)}$$

The ordered pair $(3, -2)$ is a solution.

Example 2

Graph the equation. $2x + y = -4$

Complete the table to find solutions to the equation. Then graph the line through the points.

x	y	
0		→ $(0, -4)$
	0	→ $(-2, 0)$
−3		→ $(-3, 2)$

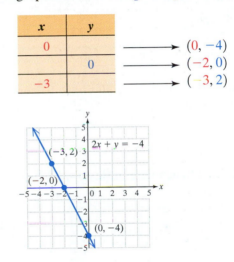

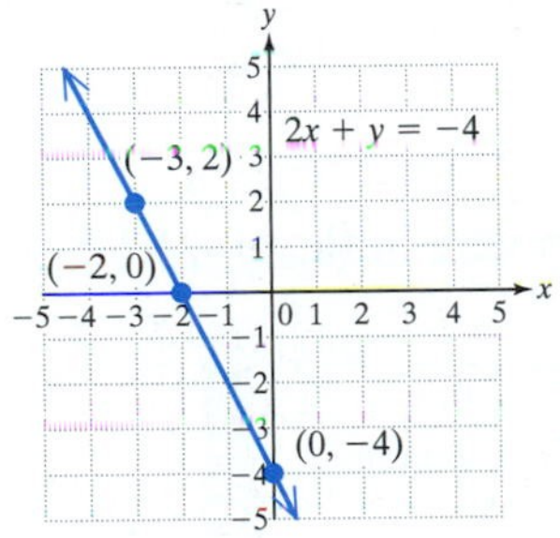

Section 9.3 Tables, Bar Graphs, Pictographs, and Line Graphs

Key Concepts

Statistics is the branch of mathematics that involves collecting, organizing, and analyzing **data** (information). Information can often be organized in tables and graphs. The individual entries within a table are called **cells**.

Example 1

The data in the table give the number of Calories for a 1-c serving of selected vegetables.

Vegetable (1 c)	Number of Calories
Corn	85
Green beans	35
Eggplant	25
Peas	125
Spinach	40

A **pictograph** uses an icon or small image to convey a unit of measurement.

Example 3

What is the value of each icon in the graph?

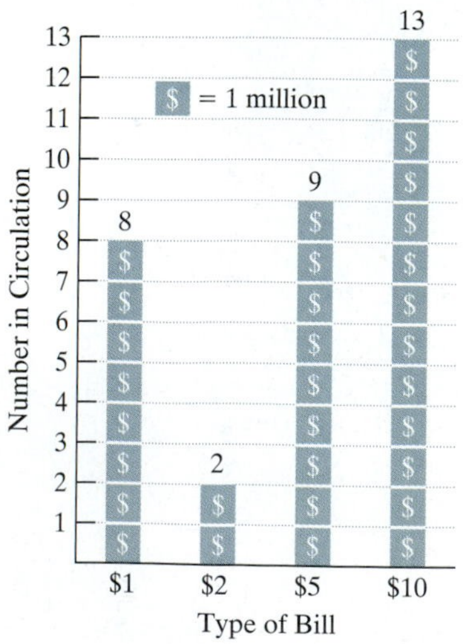

Each icon is worth 1,000,000 bills in circulation.

Examples

Example 2

Construct a bar graph for the data in Example 1.

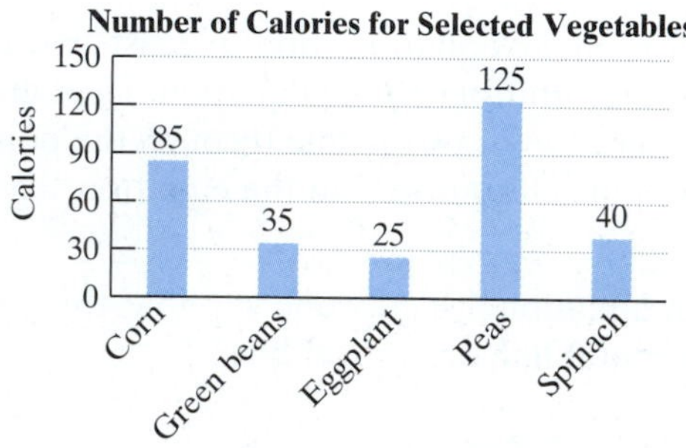

Line graphs are often used to track how one variable changes with respect to the change in a second variable.

Example 4

In what year were there 52.3 million married-couple households?

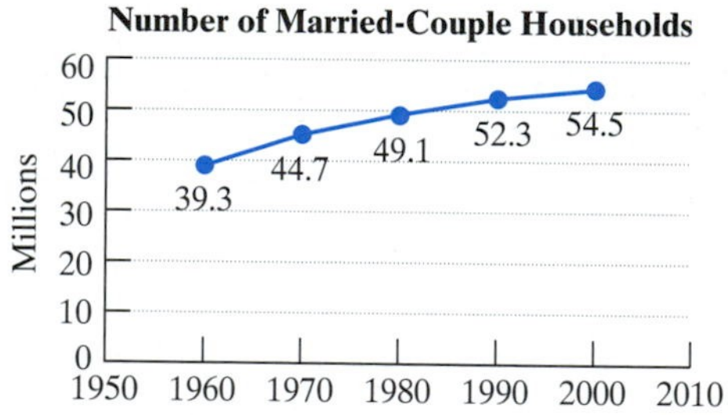

From the graph, the year 1990 corresponds to 52.3 million married-couple households.

Section 9.4 Frequency Distributions and Histograms

Key Concepts

A **frequency distribution** is a table displaying the number of data values that fall within specified intervals called **class intervals**.

When constructing a frequency distribution, keep these important guidelines in mind.

- The classes should be equally spaced.
- The classes should not overlap.
- In general, use between 5 and 15 classes.

A **histogram** is a special bar graph that illustrates data given in a frequency distribution. The class intervals are given on the horizontal scale. The height of each bar in a histogram measures the frequency for each class.

Examples

Example 1

Create a frequency distribution for the following data.

50	53
54	51
50	40
50	47
53	36
44	34
52	32
42	30

Class Intervals	Tally	Frequency
30–34	|||	3
35–39			1
40–44	|||	3
45–49			1
50–54	𝍸 |||	8

Example 2

Create a histogram for the data in Example 1.

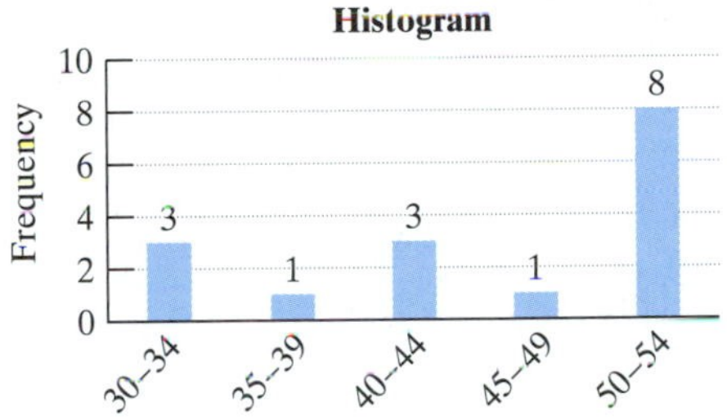

Section 9.5 Circle Graphs

Key Concepts

A **circle graph** (or pie graph) is a type of graph used to show how a whole amount is divided into parts. Each part of the circle, called a **sector**, is like a slice of pie.

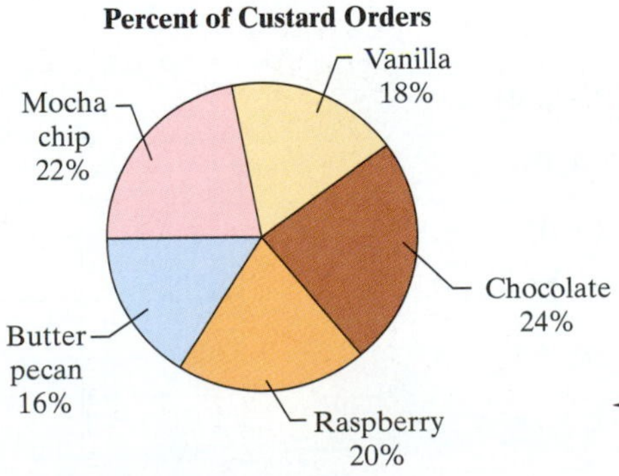

Examples

Example 1

At Ritter's Frozen Custard, the flavors for the day are given with the number of orders for each flavor.

Flavor	Number of Orders
Vanilla	180
Chocolate	240
Raspberry	200
Butter pecan	160
Mocha chip	220

Construct a circle graph for the data given. Label the sectors with percents.

Section 9.6 Introduction to Probability

Key Concepts

The collection (or set) of all possible outcomes of an experiment is called the **sample space**.

The **probability of event** is given by:

$$\frac{\text{number of elements in event}}{\text{number of elements in sample space}}$$

The probability of an event cannot be greater than 1 nor less than 0.

The **complement of an event** is the set of all elements in the sample space that are not in the event.

Examples

Example 1

Define the sample space for selecting a colored ball.

Sample space = {red, blue, yellow, green}

Example 2

What is the probability of selecting a yellow ball from Example 1?

$\frac{1}{4}$ ← number of yellow balls / ← number of balls in box

Example 3

49 CDs are in a shopping cart.

10 Rap
24 Rock
12 Latina
3 Classical

If one CD is selected at random, find the probability that

a. A rock CD is selected. $\frac{24}{49}$

b. A rock CD is *not* selected. This is the complementary event to part (a).

$$1 - \frac{24}{49} = \frac{25}{49}$$

Chapter 9 Review Exercises

Section 9.1

For Exercises 1–2, graph the ordered pairs.

1. a. $(-3, 2)$
b. $(-5, 0)$
c. $(-1, -4)$
d. $(1, 0)$
e. $(0, -4)$

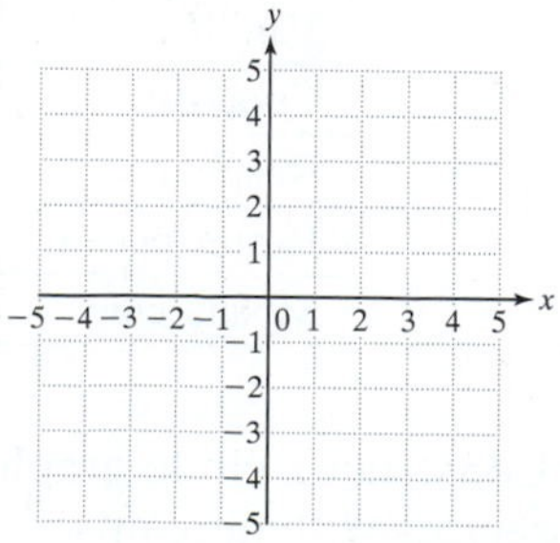

2. a. $(-2, 4)$
b. $(4, 0)$
c. $(5, 3)$
d. $(-2, -5)$
e. $(0, -1)$

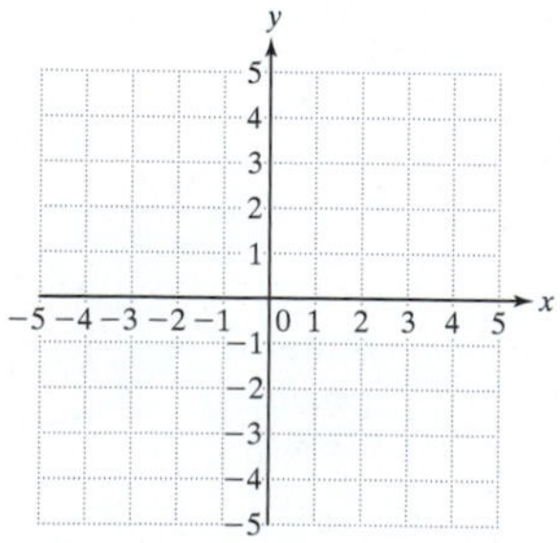

For Exercises 3–4, give the coordinates of the points A, B, C, D, and E. Also state the axis or quadrant where the point is located.

3.

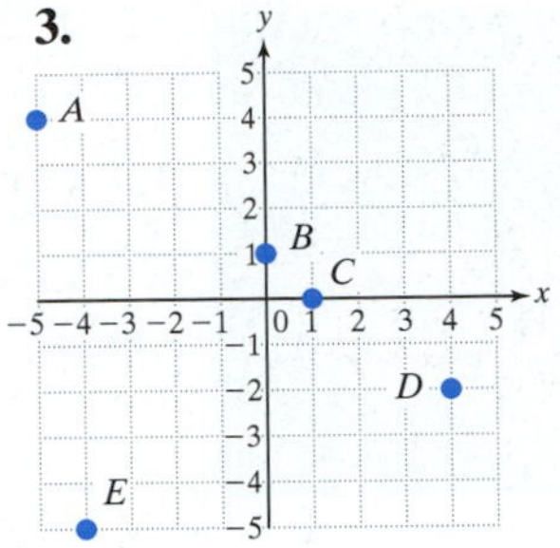

4.

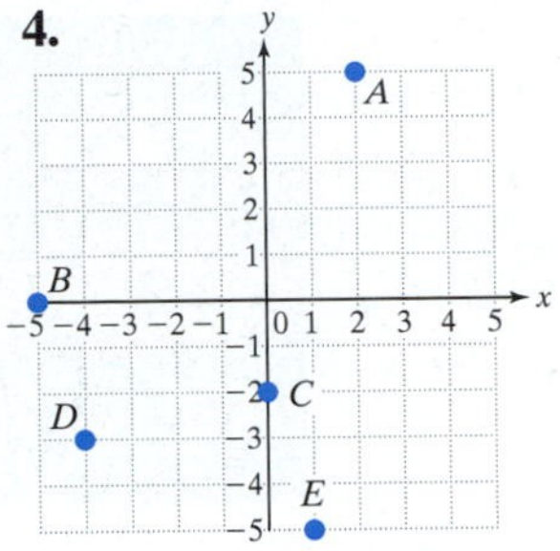

5. The average daily temperatures (in °C) are given for 5 days at the beginning of a week in January for Fargo, North Dakota. Write the table values as ordered pairs and graph the ordered pairs.

Day, x	Temperature (°C), y
1	−7
2	−6
3	4
4	2
5	0

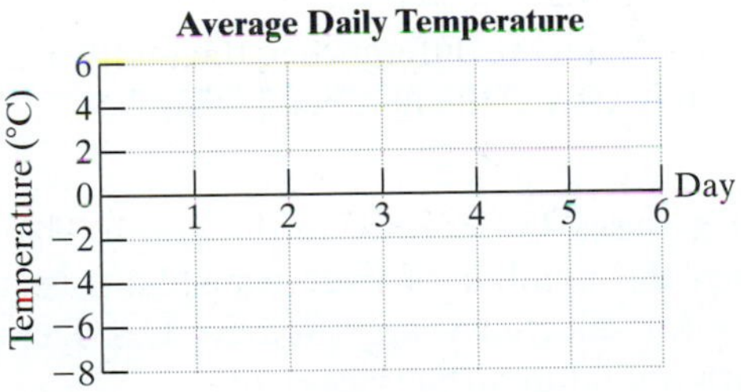

6. Are the ordered pairs $(-10, 4)$ and $(4, -10)$ the same? Explain.

Section 9.2

For Exercises 7–12, determine if the ordered pair is a solution to the equation.

7. $3x - 5y = 11 \quad (2, -1)$

8. $-6x - 2y = 4 \quad (1, -5)$

9. $-x + 4y = -7 \quad (1, 2)$

10. $-5x + y = 7 \quad (2, 3)$

11. $y = 8 \quad (4, 8)$

12. $x = -9 \quad (-9, 2)$

For Exercises 13–14, complete the ordered pairs so that they are solutions to the equation.

13. $3x - y = 2$
$(2, \quad) \; (\quad, 1)$

14. $-x + 2y = 3$
$(\quad, 4) \; (7, \quad)$

For Exercises 15–16, complete the table.

15. $-x + y = 8$

x	y
−3	
	2
5	

16. $2x - y = 16$

x	y
	−8
3	
	0

For Exercises 17–26, graph the equation.

17. $-x + y = 1$

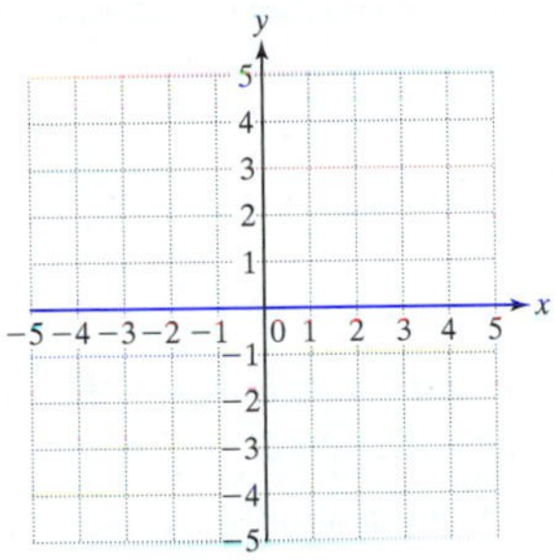

18. $x - y = -2$

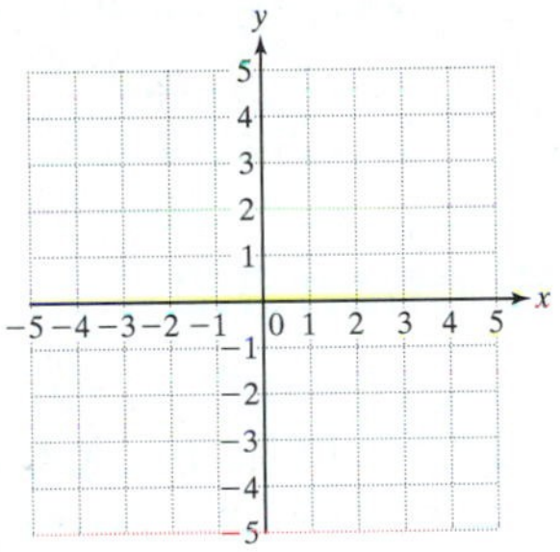

19. $y = 3x$

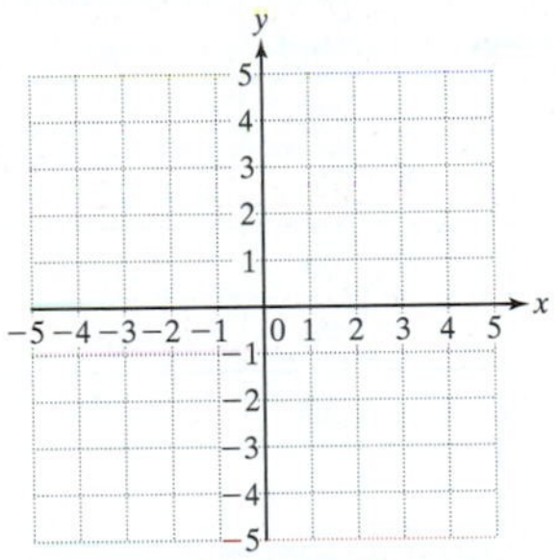

20. $y = -4x$

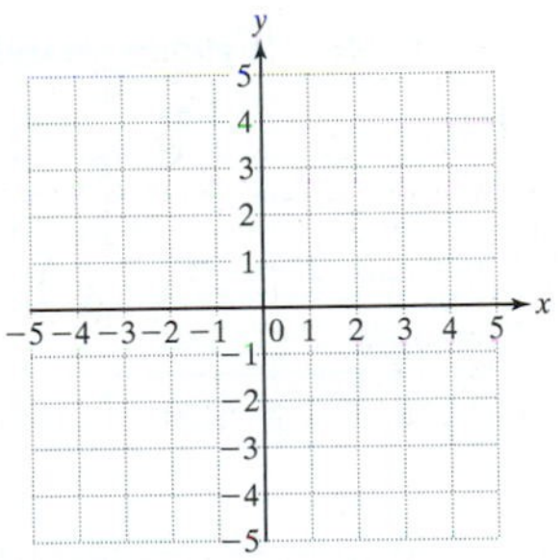

21. $y = -\frac{1}{3}x$

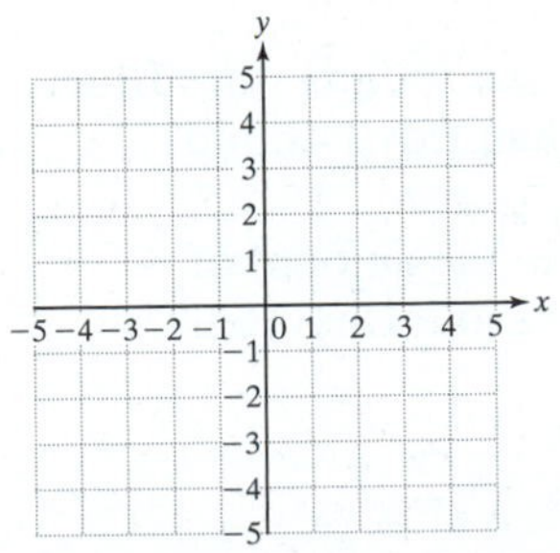

22. $y = \frac{1}{5}x$

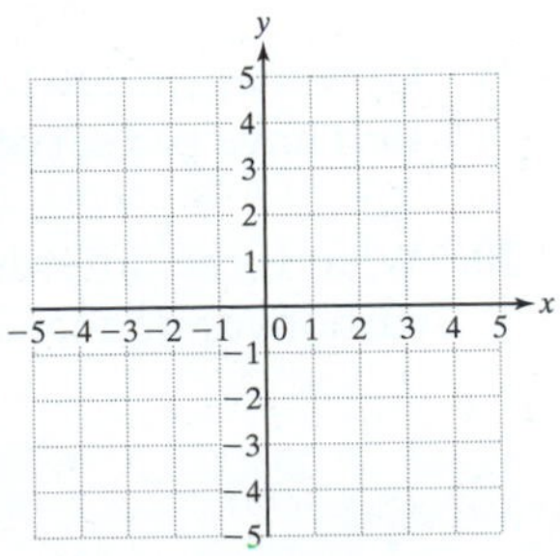

23. $x = -3$

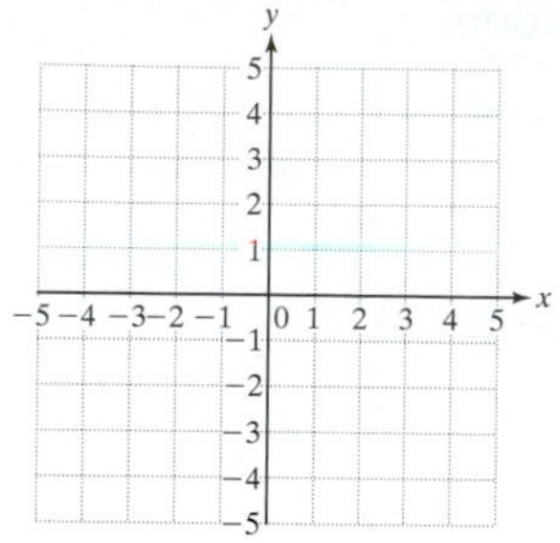

24. $y = -2$

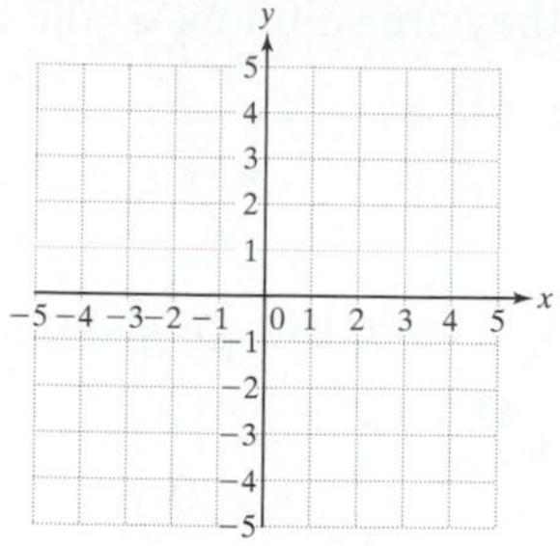

25. $y + 1 = 5$

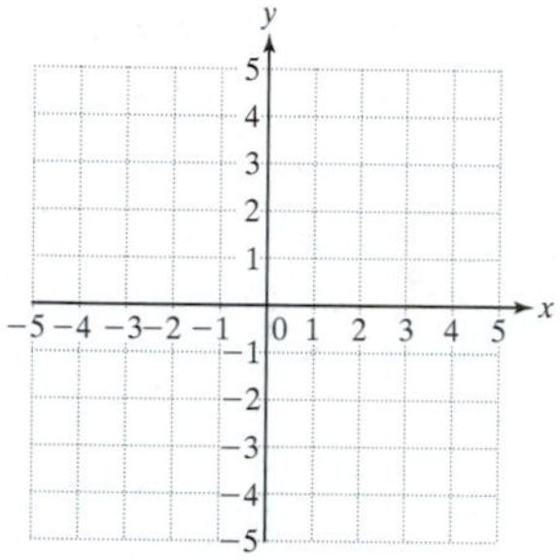

26. $x - 3 = 1$

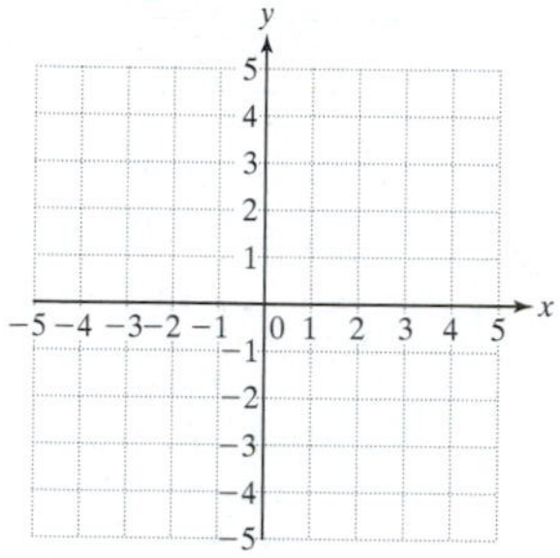

Section 9.3

For Exercises 27–30, refer to the table. The table gives the number of Calories and the amount of fat, cholesterol, sodium, and total carbohydrates for a single $\frac{1}{2}$-c serving of chocolate ice cream.

Ice Cream	Calories	Fat(g)	Cholesterol (mg)	Sodium (mg)	Carbohydrate (g)
Breyers	150	8	20	35	17
Häagen-Dazs	270	18	115	60	22
Edy's Grand	150	8	25	35	17
Blue Bell	160	8	35	70	18
Godiva	290	18	65	50	28

27. Which ice cream has the most calories?

28. Which ice cream has the least amount of cholesterol?

29. How many more times the sodium does Blue Bell have per serving than Edy's Grand?

30. What is the difference in the amount of carbohydrate for Godiva and Blue Bell?

For Exercises 31–34, refer to the pictograph. The graph represents the number of tornadoes during four months with active weather.

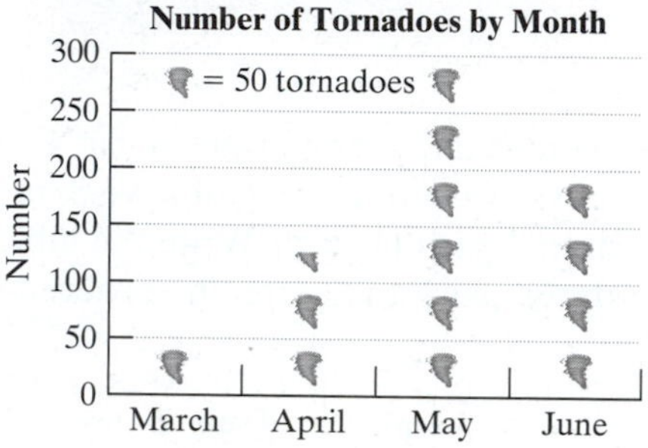

31. What does each icon represent?

32. From the graph, estimate the number of tornadoes in May.

33. Which month had approximately 200 tornadoes?

34. Estimate the difference in the number of tornadoes in April and the number in March.

For Exercises 35–38, refer to the graph. The graph represents the number of liver transplants in the United States for selected years. (*Source:* U.S. Department of Health and Human Services)

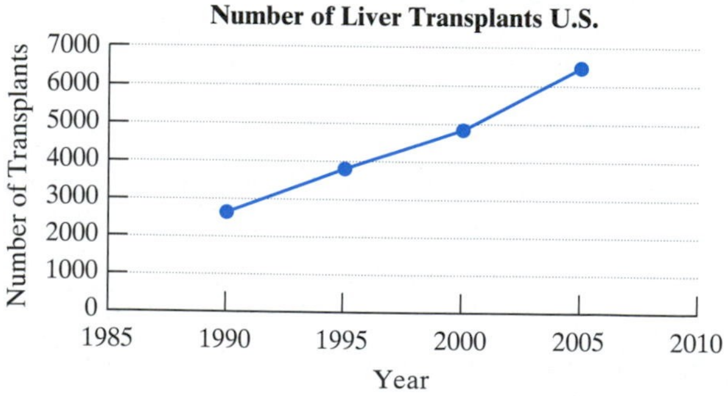

35. In which year did the greatest number of liver transplants occur?

36. Approximate the number of liver transplants for the year 2000.

37. Does the trend appear to be increasing or decreasing?

38. Extend the graph to predict the number of liver transplants for the year 2007.

39. The table shows several movies that grossed over 100 million dollars in the United States. Construct a bar graph using horizontal bars. The length of each bar should represent the amount of money in millions that each movie grossed. (*Source: Washington Post*)

Movie Title	Gross (in millions)	Year
Titanic	601	1997
Star Wars	461	1977
Shrek 2	436	2004
Lord of the Rings: The Fellowship of the Ring	314	2001
Harry Potter and the Goblet of Fire	290	2005
Wedding Crashers	209	2005

Movies that Grossed over $100 million in the U.S.

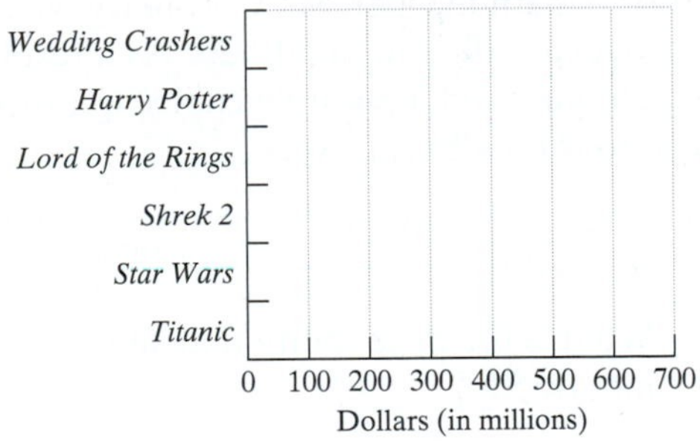

Section 9.4

The ages of students in a Spanish class are given.

18	22	19	26	31	20	40	24	43	22
29	28	35	42	29	30	24	31	23	21

Use these data for Exercises 40–41.

40. Complete the frequency table.

Class Intervals (Age)	Frequency
18–21	
22–25	
26–29	
30–33	
34–37	
38–41	
42–45	

41. Construct a histogram of the data in Exercise 40.

Number of Students in Spanish Class by Age

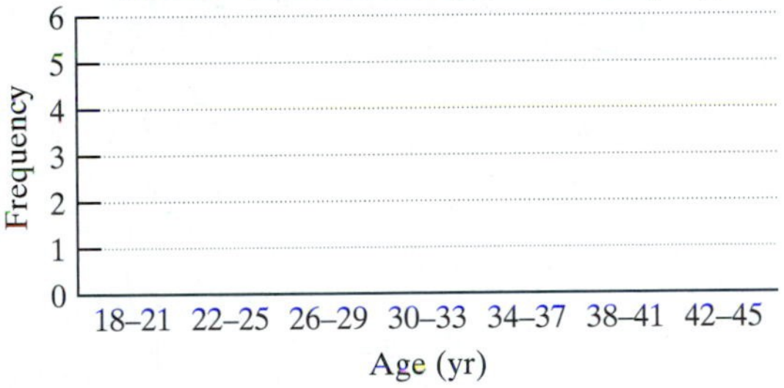

Section 9.5

The pie graph describes the types of subs offered at Larry's Sub Shop. Use the information in the graph for Exercises 42–44.

Number of Certain Types of Subs

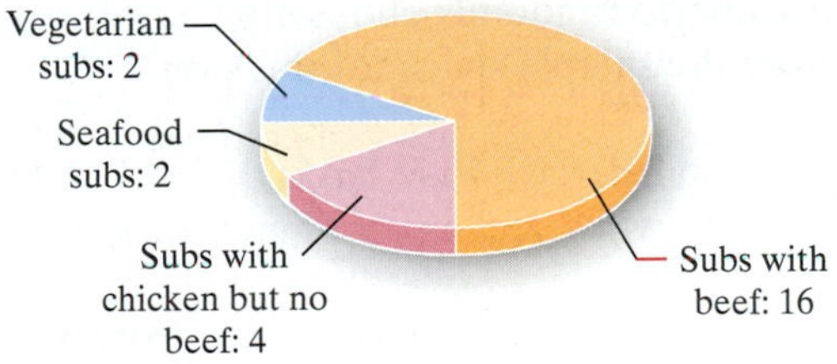

42. How many types of subs are offered at Larry's?

43. What fraction of the subs at Larry's is made with beef?

44. What fraction of the subs at Larry's is not made with beef?

45. A survey was conducted with 200 people, and they were asked their highest level of education. The results of the survey are given in the table.

a. Complete the table.

Education Level	Number of People	Percent	Number of Degrees
Grade school	10		
High school	50		
Some college	60		
Four-year degree	40		
Post graduate	40		

b. Construct a circle graph using percents from the information in the table.

Percent by Education Level

Section 9.6

46. Roberto has six pairs of socks, each a different color: blue, green, brown, black, gray, and white. If Roberto randomly chooses a pair of socks, write the sample space for this event.

47. Refer to Exercise 46. What is the probability that Roberto will select a pair of gray socks?

48. Refer to Exercise 46. What is the probability that Roberto will *not* select a blue pair of socks?

49. Which of the following numbers could represent a probability?

a. $\frac{1}{2}$ **b.** $\frac{5}{4}$ **c.** 0 **d.** 1

e. 25% **f.** 2.5 **g.** 6% **h.** 6

50. A bicycle shop sells a child's tricycle in three colors: red, blue, and pink. In the warehouse there are 8 red tricycles, 6 blue tricycles, and 2 pink tricycles.

a. If Kevin selects a tricycle at random, what is the probability that he will pick a red tricycle?

b. What is the probability that he will not pick a red tricycle?

c. What is the probability of Kevin's selecting a green tricycle?

51. Torie's math teacher, Ms. Coronel, gave a particularly difficult math test. As a result, Torie had to guess on a multiple choice question that had four possible answers.

a. What is the probability that she would guess correctly?

b. What is the probability that she would guess incorrectly?

Chapter 9 Test

1. Graph the ordered pairs.
(0, −2), (4, −4),
(−3, −5), (−2, 1), (5, 0)

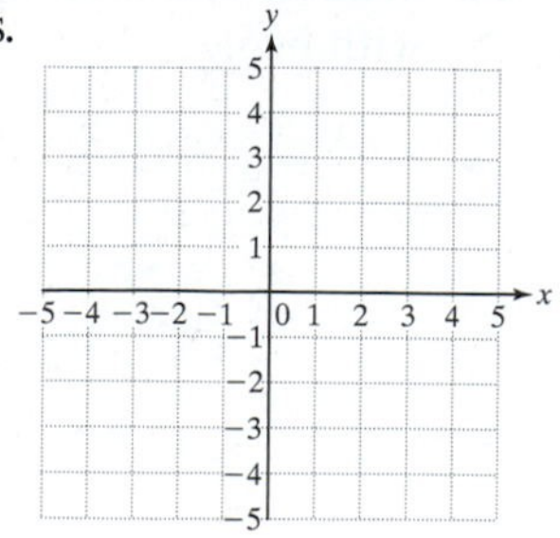

2. For each ordered pair, determine the quadrant or axis where the point is located.

a. (−15, 0) **b.** (6, −22) **c.** (−81, −12)

d. (0, −44) **e.** (−5, 17) **f.** (37, 2)

3. Fill in the blank with the word "positive" or "negative."

A point in Quadrant II has a ____________ x-coordinate and a ______________ y-coordinate.

For Exercises 4–6, determine if the ordered pair is a solution to the equation.

4. $12x - 3y = 6$ $(0, -2)$

5. $6x + 7y = 8$ $(2, -1)$

6. $y = 7$ $(7, -7)$

For Exercises 7–8, complete the ordered pairs.

7. $2x - 4y = 2$ $(5, \quad), (\quad, 0)$

8. $y = -6$ $(-3, \quad), (2, \quad)$

9. Given the equation $-3x + y = 9$, complete the table.

x	y
-3	
	9
2	

For Exercises 10–12, graph the equation.

10. $-3x + y = 6$

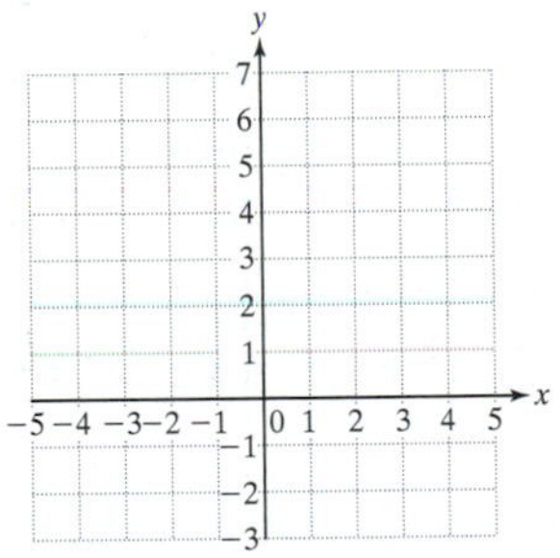

11. $y = -2x + 4$

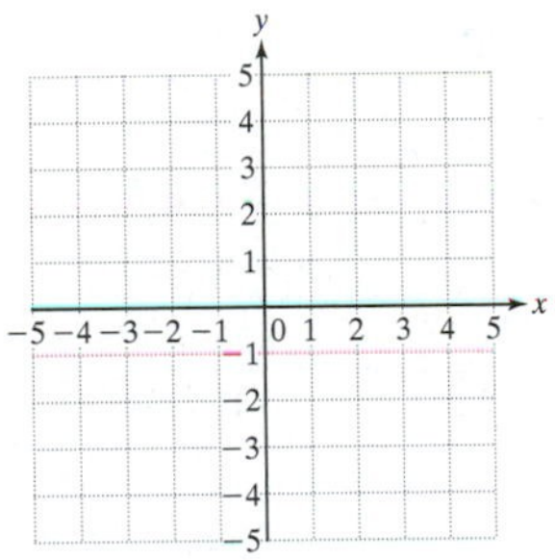

12. $y = 5$

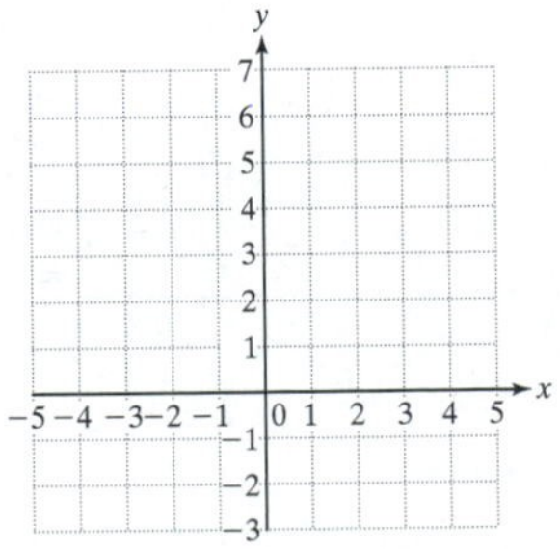

13. The table represents the world's major producers of primary energy for a recent year. All measurements are in quadrillions of Btu.

Note: 1 quadrillion = 1,000,000,000,000,000. (*Source:* Energy Information Administration, U.S. Dept. of Energy)

Country	Amount of Energy Produced (quadrillions of BTUs)
United States	72
Russia	43
China	35
Saudi Arabia	43
Canada	18

Construct a bar graph using horizontal bars. The length of each bar corresponds to the amount of energy produced for each country.

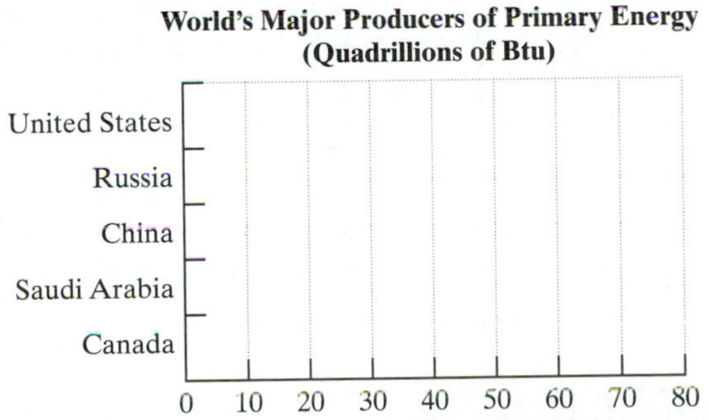

14. Of the approximately 2.9 million workers in 1820 in the United States, 71.8% were employed in farm occupations. Since then, the percent of U.S. workers in farm occupations has declined. The table shows the percent of total U.S. workers who worked in farm-related occupations for selected years. (*Source:* U.S. Department of Agriculture)

Year	Percent of U.S. Workers in Farm Occupations
1820	72%
1860	59%
1900	38%
1940	17%
1980	3%

a. Which year had the greatest percent of U.S. workers employed in farm occupations? What is the value of the greatest percent?

b. Make a line graph with the year on the horizontal scale and the percent on the vertical scale.

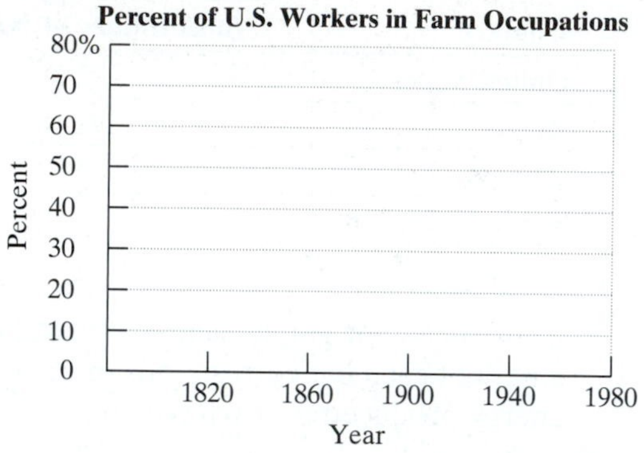

c. Based on the graph, estimate the percent of U.S. workers employed in farm occupations for the year 1960.

For Exercises 15–17, refer to the pictograph. The pictograph shows the flower sales for the first 5 months of the year for a flower shop.

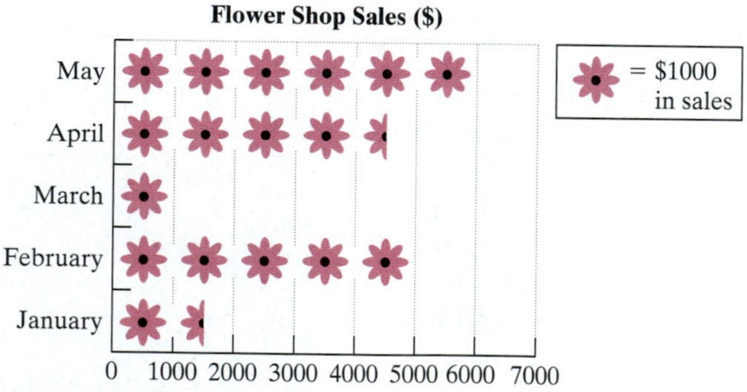

15. What is the value of each flower icon?

16. From the graph, estimate the sales for the month of April.

17. Which month brought in sales of $5000?

For Exercises 18–20, refer to the table. The rainfall amounts for Salt Lake City, Utah, and Seattle, Washington, are given in the table for selected months. All values are in inches. (*Source:* National Oceanic and Atmospheric Administration)

	April	May	June	July
Salt Lake City	2.02	2.09	0.77	0.72
Seattle	2.75	2.03	2.5	0.92

18. Which city is generally wetter?

19. What is the difference in the amount of rainfall in Seattle and Salt Lake City during June?

20. In what month does Salt Lake City have a greater rainfall amount than Seattle?

21. A cellular phone company questioned 20 people at a mall, to determine approximately how many minutes each individual spent on the cell phone each month. Using the list of results, complete the frequency distribution and construct a histogram.

100	120	250	180	300	200	250	175
110	280	330	280	300	325	60	75
100	350	60	90				

Number of Minutes Used Monthly	Tally	Frequency
51–100		
101–150		
151–200		
201–250		
251–300		
301–350		

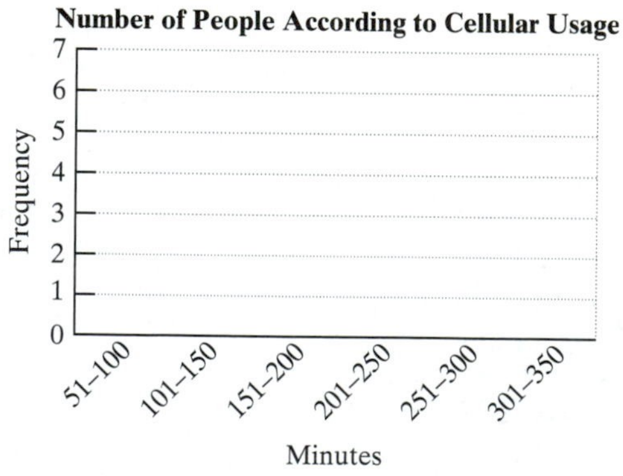

For Exercises 22–24, refer to the circle graph. The circle graph shows the percent of homes having different types of flooring in the living room area.

Percent of Types of Floor Covering

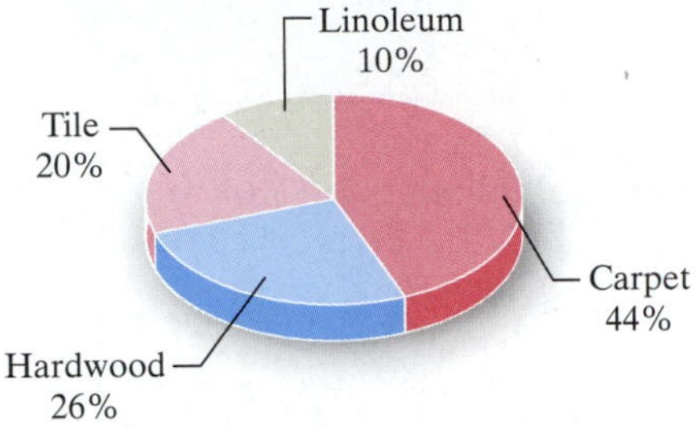

22. If 150 people were questioned, how many would be expected to have carpet on their living room floor?

23. If 200 people were questioned, how many would be expected to have tile on their living room floor?

24. If 300 people were questioned, how many would not be expected to have linoleum on their living room floor?

25. A board game has a die with eight sides with the numbers 1–8 printed on each side.

a. What is the sample space for rolling the die one time?

b. What is the probability of rolling a 6?

c. What is the probability of rolling an even number?

d. What is the probability of rolling a number less than 3?

26. At a party there is a cooler filled with ice and soft drinks: 6 cans of diet cola, 4 cans of ginger ale, 2 cans of root beer, and 2 cans of cream soda. A person takes a can of soda at random. What is the probability that the person selects a can of ginger ale?

27. Which of the following is not a reasonable value for a probability?

a. 0.36 **b.** $\frac{3}{4}$ **c.** 1.5

Chapters 1–9 Cumulative Review Exercises

1. Identify the place value of the underlined digit.

a. $23,\underline{9}04,152$ **b.** $47.2\underline{3}9$ **c.** $0.0000\underline{4}$

2. Estimate the product by first rounding each number to the nearest hundred.

$$687 \times 1243$$

For Exercises 3–13, simplify.

3. $-2087 - 53 + 10{,}499 + (-6)$

4. $-\frac{12}{7} \cdot \frac{14}{36}$

5. $\frac{5}{8} \div \left(-\frac{6}{15}\right) \cdot \left(-\frac{24}{25}\right)$

6. $\frac{3}{10} - \frac{7}{100}$

7. $\frac{1}{2} + \frac{5}{3} + \left(-\frac{1}{6}\right)$

8. $2\frac{1}{8} \div 17 + \frac{7}{12} \cdot \frac{9}{14} - \frac{1}{3}$

9. $-68.412 \div 100$

10. 68.412×0.1

11. $68.412 \div 0.001$

12. $5 - 2(6 - 10)^2 + \sqrt{9}$

13. $|-6 + 11|$

For Exercises 14–17, solve the equation.

14. $14 = -4 + 9x$

15. $-\frac{2}{3}x = 12$

16. $6 - 2(x + 7) = -4x - 12$

17. $\frac{5.5}{100} = \frac{x}{130}$

18. Quick Cut Lawn Company can service 5 customers in $2\frac{3}{4}$ hr. Speedy Lawn Company can service 6 customers in 3 hr. Find the unit rate in time per customer for both lawn companies and decide which company is faster.

19. If Rosa can type a 4-page English paper in 50 min, how long will it take her to type a 10-page term paper?

20. Find the values of x and y, assuming that the two triangles are similar.

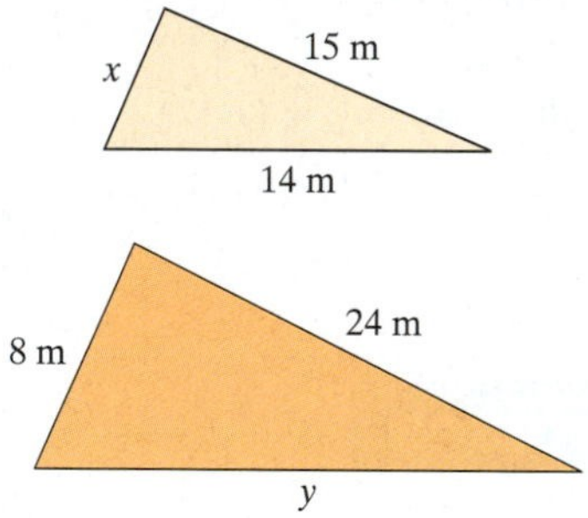

21. Out of a group of people, 95 said that they brushed their teeth twice a day. If this number represents 78% of the people surveyed, how many were surveyed? Round to the nearest whole unit.

22. The results of a survey showed that 110 people wanted to exercise more than they did in their daily routine. If this represents 44% of the sample, how many people were surveyed?

23. Out of 120 people, 78 wear glasses. What percent does this represent?

24. A savings account pays 3.4% simple interest. If \$1200 is invested for 5 years, what will be the balance?

25. Convert 2 ft 5 in. to inches.

26. Convert $4\frac{1}{2}$ gal to quarts.

27. Add. 3 yd 2 ft + 5 yd 2 ft

28. Subtract. 12 km − 2360 m

29. Find the area.

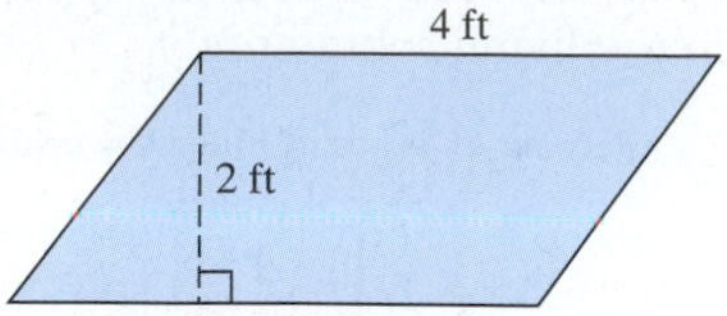

30. Find the volume. Use $\pi \approx \frac{22}{7}$.

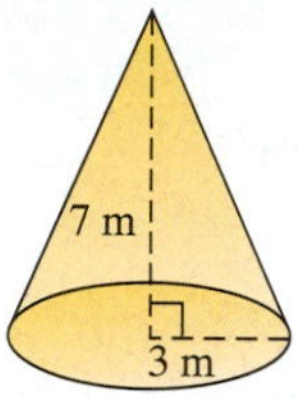

31. The data shown in the table give the average weight for boys based on age. (*Source:* National Parenting Council) Make a line graph to illustrate these data.

Age	Weight (lb)
5	44.5
6	48.5
7	54.5
8	61.25
9	69
10	74.5
11	85
12	89

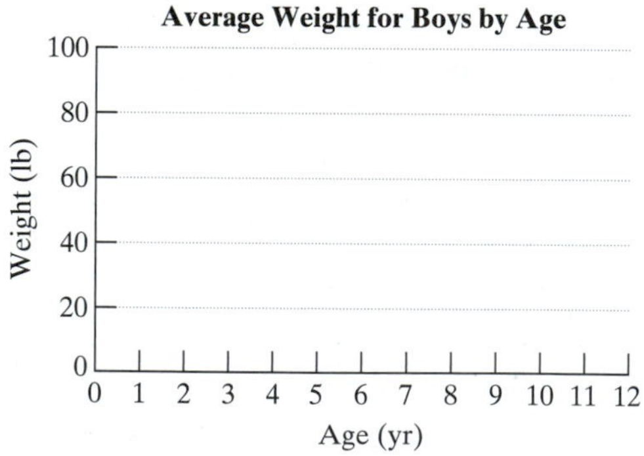

32. The monthly number of deaths resulting from tornadoes for a recent year are given. Find the mean and median. Round to the nearest whole unit.

33, 10, 62, 132, 123, 316,

123, 133, 18, 150, 26, 138

A game has a spinner with four sections of equal size. Refer to the spinner for Exercises 33–35.

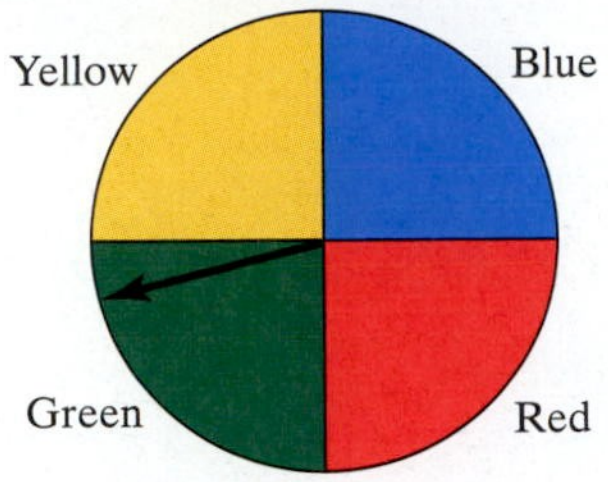

33. If a person spins the pointer once, determine the sample space.

34. What is the probability of the pointer landing on green?

35. What is the probability of the pointer *not* landing on green?

For Exercises 36–38, convert the units of measurement.

36. 49 cm = _____ m

37. 5.8 L = _____ cc

38. 510 g = _____ kg

39. For each ordered pair, identify the axis or quadrant where the point is located.

a. $(-4, -5)$

b. $(0, 6)$

c. $(4, 0)$

d. $(8, 4)$

e. $(6, -2)$

40. Graph the equation. $x - y = -5$

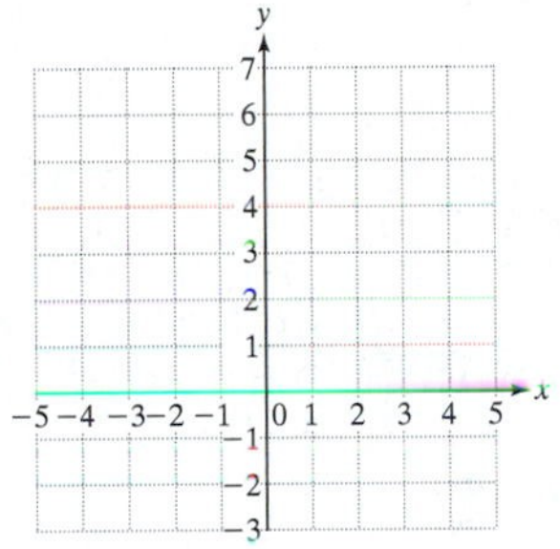

Exponents and Polynomials

10

CHAPTER OUTLINE

Chapter 10

In this chapter, we will cover additional topics in algebra, including properties of exponents, and operations on polynomials.

Are You Prepared?

To help you prepare for this chapter, try the problems below to review the related skills that you will need for this chapter. Match the exercise on the left with the answer on the right. Then use the answers from Exercises 1–11 to fill in the blank with the correct key term.

Simplify.

1. 10^2

2. 10^5

3. -5^2

4. $(-5)^2$

Clear parentheses and combine like terms.

5. $3(2x + 4)$

6. $-8(2a - 4b + c)$

7. $-6x + 8x + x$

8. $-3x + 4 - 11x + 5$

9. $3(x + 2) - (x - 5)$

10. $-6(x - 7) - 2(x + 1)$

11. $-(-16a + 32b - 8c)$

F. $6x + 12$

F. 25

N. $-8x + 40$

T. $16a - 32b + 8c$

I. $-14x + 9$

C. $3x$

C. 100

E. $2x + 11$

E. -25

I. $-16a + 32b - 8c$

O. $100{,}000$

The numerical factor within a term is called the ___ ___ ___ ___ ___ ___ ___ ___ ___ ___ ___.
1 2 3 4 5 6 7 8 9 10 11

Section 10.1 Addition and Subtraction of Polynomials

Objectives

1. Key Definitions
2. Addition of Polynomials
3. Subtraction of Polynomials
4. Evaluating Polynomials and Applications

1. Key Definitions

In Section 3.1, we learned how to add and subtract like terms. We will now extend this skill to learn how to add and subtract *polynomials*.

To begin, we need to learn some key definitions. Recall that a **term** is a number or a product or quotient of numbers and variables. The following are examples of terms:

$$5, \quad x, \quad y^2, \quad \frac{7a}{b^2}, \quad 24c^3d^5$$

A term in which the variables appear only in the numerator with whole number exponents is called a *monomial*. The terms 5, x, y^2, and $24c^3d^5$ are all monomials. The term $\frac{7a}{b^2}$ is not a monomial because the variable b appears in the denominator.

A **polynomial** is one or more monomials combined by addition or subtraction. For example the following are polynomials.

$$4x^3 - 5x^2 + 2x - 9 \qquad \text{and} \qquad 6a^2b + 4ac$$

A polynomial can be further categorized according to the number of terms that it has.

Concept Connections

1. Is the term $8x^3y$ a monomial?
2. Is the term $\frac{8y}{x^3}$ a monomial?

DEFINITION Categorizing Polynomials

- If a polynomial has exactly one term, then it is called a **monomial**.

  Example: $3xy^4$ (1 term)
- If a polynomial has exactly two terms, then it is called a **binomial**.

  Example: $5ab + 6$ (2 terms)
- If a polynomial has exactly three terms, then it is called a **trinomial**.

  Example: $6x^4 - 7x^2 - 5x$ (3 terms)

If a polynomial has only one variable, we often write the polynomial in **descending order** so that the powers on the variable decrease from left to right.

powers decrease

$5w^6 - 2w^3 + 3w^2$ This is written in descending order.

For a polynomial with one variable, the **degree** of the polynomial is the greatest power to which the variable appears. The degree of $5w^6 - 2w^3 + 3w^2$ is 6.

Concept Connections

Answer true or false.

3. The polynomial $4x^3 - 2x + 1$ is a trinomial written in descending order.
4. The degree of the term $6x^3$ is 6.

Example 1 Identifying the Characteristics of a Polynomial

Write each polynomial in descending order. Determine the degree of the polynomial, and categorize the polynomial as a monomial, binomial, or trinomial.

a. $-4t + 8t^3$ **b.** $2.3z^4 - 4.1 - 1.8z^5$ **c.** $-8p$ **d.** 5

Answers

1. Yes 2. No
3. True 4. False

Solution:

Polynomial	Descending Order	Degree	Category
a. $-4t + 8t^3$	$8t^3 - 4t$	3	binomial
b. $2.3z^4 - 4.1 - 1.8z^5$	$-1.8z^5 + 2.3z^4 - 4.1$	5	trinomial
c. $-8p$	$-8p^1$	1	monomial
d. 5	5	0	monomial

Note that the monomial in part d does not have a variable factor. Therefore, its degree is 0.

Skill Practice

Given: $-4y^3 + 7y^2 + y^6$

5. Write the polynomial in descending order.

6. Identify the degree of the polynomial.

7. Categorize the polynomial as a monomial, binomial, or trinomial.

Also recall from Section 3.1 that the **coefficient** of a term is the numerical factor of the term. For example:

$-4p^3$ The coefficient is -4.

y^5 The coefficient is 1 because $y^5 = 1y^5$.

2. Addition of Polynomials

From Section 3.1 we learned that like terms have the same variables, raised to the same powers.

$2x^5$ and $4x^5$ are like terms.

$7x^3$ and $3x^2$ are not like terms because the exponents on x are different.

In Example 2, we review the process to combine like terms by using the distributive property.

Example 2 Combining Like Terms

Combine like terms. $4x^2 + 7x^2 - 8x^2$

Solution:

$4x^2 + 7x^2 - 8x^2$

$= (4 + 7 - 8)x^2$ Apply the distributive property.

$= 3x^2$

Skill Practice

Combine like terms.

8. $-6z^3 + 9z^3 + 4z^3$

The process of combining like terms can be shortened by adding or subtracting the coefficients and leaving the variable factor unchanged.

$4x^2 + 7x^2 - 8x^2 = 3x^2$ This method will be used throughout the chapter.

To add two polynomials, first use the associative and commutative properties of addition to group like terms. Then combine like terms.

Answers

5. $y^6 - 4y^3 + 7y^2$ **6.** 6
7. Trinomial **8.** $7z^3$

Skill Practice

Add.

9. $(5a^4 - 6a^2) + (10a^4 + 3a^2)$

Example 3 Adding Polynomials

Add the polynomials. $(2x^3 + 5x) + (7x^3 - x)$

Solution:

$(2x^3 + 5x) + (7x^3 - x)$

$= 2x^3 + 7x^3 + 5x - x$ Regroup and collect like terms.

$= \underbrace{2x^3 + 7x^3} + \underbrace{5x - 1x}$ Note that $-x$ is the same as $-1x$.

$= 9x^3 + 4x$ Combine like terms.

TIP: Polynomials can also be added vertically. Begin by lining up like terms in the same column.

$$\begin{array}{r} 2x^3 + 5x \\ + \; 7x^3 - x \\ \hline 9x^3 + 4x \end{array}$$

Skill Practice

10. Add $(8t^3 - 9t + 12)$ to $(3t^3 + 7t^2 + 4t - 2)$.

Example 4 Adding Polynomials

Add $(2y^3 + 4y^2 - 9)$ to $(4y^3 - 6y^2 - 11y + 4)$.

Solution:

$(4y^3 - 6y^2 - 11y + 4) + (2y^3 + 4y^2 - 9)$

$= \underbrace{4y^3 + 2y^3} \underbrace{- 6y^2 + 4y^2} \underbrace{- 11y} + \underbrace{4 - 9}$ Regroup and collect like terms.

$= 6y^3 - 2y^2 - 11y - 5$

The sum is $6y^3 - 2y^2 - 11y - 5$.

TIP: To add the polynomials vertically, insert place holders for "missing" terms.

$$\begin{array}{r} 4y^3 - 6y^2 - 11y + 4 \\ + \; 2y^3 + 4y^2 + 0y - 9 \\ \hline 6y^3 - 2y^2 - 11y - 5 \end{array}$$

3. Subtraction of Polynomials

Recall that subtraction of real numbers is defined as $a - b = a + (-b)$. That is, we add the opposite of the second number to the first number. We will use the same strategy to subtract polynomials. To find the opposite of a polynomial, take the opposite of each term.

Skill Practice

Write the opposite of the polynomial.

11. $-6y^3 + 4y - 9$

Example 5 Finding the Opposite of a Polynomial

Write the opposite of the polynomial. $2x^2 - 7x + 5$

Solution:

The opposite of a real number a is written as $-(a)$. We can apply the same notation to find the opposite of a polynomial.

The opposite of $2x^2 - 7x + 5$ is $-(2x^2 - 7x + 5)$. Apply the distributive property.

$= -2x^2 + 7x - 5$

Answers

9. $15a^4 - 3a^2$
10. $11t^3 + 7t^2 - 5t + 10$
11. $6y^3 - 4y + 9$

To subtract two polynomials, add the opposite of the second polynomial to the first polynomial. This is demonstrated in Example 6.

Example 6 Subtracting Polynomials

Subtract. $(15x^4 - 8x^2 + 5) - (-11x^4 + 4x^2 + 3)$

Solution:

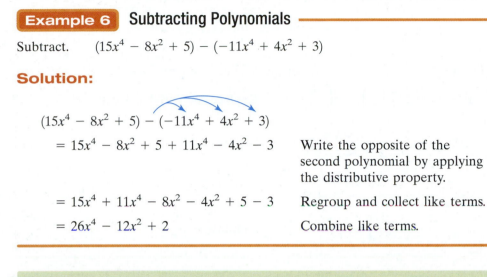

$(15x^4 - 8x^2 + 5) - (-11x^4 + 4x^2 + 3)$

$= 15x^4 - 8x^2 + 5 + 11x^4 - 4x^2 - 3$ Write the opposite of the second polynomial by applying the distributive property.

$= 15x^4 + 11x^4 - 8x^2 - 4x^2 + 5 - 3$ Regroup and collect like terms.

$= 26x^4 - 12x^2 + 2$ Combine like terms.

Skill Practice

Subtract.

12. $(8p^5 + 10p^3 - p) - (7p^5 - 3p^3 - 4p)$

TIP: Subtraction of polynomials can also be performed vertically. To do so, add the opposite of the second polynomial to the first polynomial.

$$\begin{array}{r} 15x^4 - 8x^2 + 5 \\ -\ (-11x^4 + 4x^2 + 3) \\ \hline \end{array} \xrightarrow{\text{Add the opposite.}} \begin{array}{r} 15x^4 - \ 8x^2 + 5 \\ +\ 11x^4 - \ 4x^2 - 3 \\ \hline 26x^4 - 12x^2 + 2 \end{array}$$

Example 7 Subtracting Polynomials

Subtract. $(8w^2z + 4wz - 7z^2)$ from $(10w^2z - 9wz + 6)$

Solution:

Recall that to subtract b from a, we write $a - b$. The order is important.

$(10w^2z - 9wz + 6) - (8w^2z + 4wz - 7z^2)$ Apply the distributive property to write the opposite of the second polynomial.

$= 10w^2z - 9wz + 6 - 8w^2z - 4wz + 7z^2$ Regroup and collect like terms.

$= \underbrace{10w^2z - 8w^2z}_{} - \underbrace{9wz - 4wz}_{} + \underbrace{7z^2}_{} + \underbrace{6}_{}$

$= \quad 2w^2z \quad\quad -13wz \quad + 7z^2 + 6$ Combine like terms.

The difference is $2w^2z - 13wz + 7z^2 + 6$.

Skill Practice

13. Subtract. $(4xy + 9x^2y - 3y)$ from $(9xy - 4x^2y + 8x)$

4. Evaluating Polynomials and Applications

A polynomial is an algebraic expression. Evaluating a polynomial for a value of the variable is the same as evaluating an expression.

Answers

12. $p^5 + 13p^3 + 3p$
13. $5xy - 13x^2y + 8x + 3y$

Skill Practice

14. Evaluate the polynomial for the given value of the variable.

$2y^2 + y - 6$ for $y = \frac{2}{3}$

Example 8 Evaluating a Polynomial

Evaluate the polynomial for the given value of the variable:

$$x^2 - 5x + 2 \quad \text{for} \quad x = \frac{1}{2}$$

Solution:

$x^2 - 5x + 2 = \left(\frac{1}{2}\right)^2 - 5\left(\frac{1}{2}\right) + 2$ Replace the variable x with $\frac{1}{2}$.

$= \frac{1}{4} - \left(\frac{5}{2}\right) + 2$ Simplify.

$= \frac{1}{4} - \frac{10}{4} + \frac{8}{4}$ The LCD is 4.

$= -\frac{1}{4}$ Simplify.

Avoiding Mistakes

Use parentheses to enclose the value for the variable.

In Example 9, we present an application of a polynomial. We evaluate the polynomial for different values of the variable and interpret the meaning in the context of the problem.

Skill Practice

The cost (in dollars) to rent a kayak for x hours is given by $6x + 15$.

15. Evaluate the polynomial for $x = 2$ and interpret the result.

16. Evaluate the polynomial for $x = 8$ and interpret the result.

Example 9 Evaluating a Polynomial in an Application

The cost (in dollars) to rent storage space for x months is given by the polynomial.

$$49.99x + 129$$

a. Evaluate the polynomial for $x = 3$ and interpret the result in the context of the problem.

b. Evaluate the polynomial for $x = 12$ and interpret the result in the context of the problem.

Solution:

a. $49.99(3) + 129$ Evaluate $49.99x + 129$ for $x = 3$.

$= 149.97 + 129$ Multiply before adding.

$= 278.97$ This means that the cost to rent storage space for 3 months is \$278.97.

b. $49.99(12) + 129$ Evaluate $49.99x + 129$ for $x = 12$.

$= 599.88 + 129$ Multiply before adding.

$= 728.88$ This means that the cost to rent storage space for 12 months (1 year) is \$728.88.

Answers

14. $-\frac{40}{9}$

15. 27; The cost to rent a kayak for 2 hours is \$27.

16. 63; The cost to rent a kayak for 8 hours is \$63.

Section 10.1 Practice Exercises

Study Skills Exercises

1. List three benefits of successfully completing this course.

2. Define the key terms.

a. Binomial **b. Coefficient** **c. Degree** **d. Descending order**

e. Monomial **f. Polynomial** **g. Term** **h. Trinomial**

Objective 1: Key Definitions

For Exercises 3–10, complete the table. Write the polynomial in descending order. Determine the degree of the polynomial, and categorize the polynomial as a monomial, a binomial, or trinomial. **(See Example 1.)**

		Descending Order	Degree	Category
3.	$-7 + 2x^4$			
4.	$3 + 4p^3$			
5.	$2y^3 - 4y^5 + 9y$			
6.	$7m - 6m^4 + 5m^3$			
7.	$12z^7$			
8.	$11n^2$			
9.	$9k - 4$			
10.	$13w + 8$			

For Exercises 11–14, determine if the term is a monomial.

11. $18p^2q$ **12.** $-4a^9c$ **13.** $\frac{7a^2}{d^3}$ **14.** $\frac{15n^4}{m}$

15. Identify the coefficient of each term.

a. $-5x^2y$ **b.** $12a^5$ **c.** 6 **d.** wz

16. Identify the coefficient of each term.

a. $9ab$ **b.** $-4y^7$ **c.** -8 **d.** cd

Objective 2: Addition of Polynomials

For Exercises 17–20, determine if the terms are like terms.

17. $-6x$ and $6x^2$ **18.** $9y^4$ and $9y^2$ **19.** $-2x^3y$ and $5x^3y$ **20.** a^2b^4 and $-4a^2b^4$

For Exercises 21–26, combine like terms. **(See Example 2.)**

21. $-8p^2 - 2p^2$ **22.** $-16w^4 - 2w^4$ **23.** $12y^3 - 4y^3 + 6y^3$

24. $9z^2 + 11z^2 - 8z^2$ **25.** $2.4x^4 - 1.6x^4 + 5.1x^4$ **26.** $-6.8w^3 + 2.1w^3 - 1.5w^3$

For Exercises 27–38, add the polynomials. **(See Examples 3–4.)**

27. $(4t - 8) + (3t + 11)$

28. $(14m - 9) + (12m + 2)$

29. $(2y^2 + 6y) + (-8y^2 + 2y)$

30. $(15x^4 + 10x^3) + (-6x^4 + 9x^3)$

31. $\left(-6z^3 - 2z^2 + \frac{3}{4}\right) + \left(4z^3 + 10z^2 - \frac{1}{3}\right)$

32. $\left(-w^2 + 6w - \frac{5}{4}\right) + \left(9w^2 - 5w + \frac{1}{2}\right)$

33. $(8x^2y + 2xy - 3xy^2) + (-9x^2y - xy + 6xy^2)$

34. $(4a^3b^2 + 2a^2b^2 - 9ab^2) + (6a^3b^2 - 11a^2b^2 + 4ab^2)$

35.
$$\begin{array}{r} 18x^3 - x^2 + 9x - 3 \\ \underline{+\ 5x^3 + 6x^2 - 14x + 11} \end{array}$$

36.
$$\begin{array}{r} 14z^3 + 7z^2 - 13z + 2 \\ \underline{+\ 6z^3 - 9z^2 + 10z - 20} \end{array}$$

37.
$$\begin{array}{r} 6y^3 \qquad\qquad + 8y + 5 \\ \underline{+\ (-9y^3 + 16y^2 \qquad\quad - 3)} \end{array}$$

38.
$$\begin{array}{r} 12p^3 + 7p^2 \qquad\quad - 9 \\ \underline{+\ (-p^3 \qquad\quad - 8p - 4)} \end{array}$$

39. Add $(-12x^4 - 2x^2 + 6x)$ to $(9x^4 + 2x^3 + 4x)$. **(See Example 4.)**

40. Add $(5a^3 + 2a - 3)$ to $(6a^2 + 9a + 7)$.

Objective 3: Subtraction of Polynomials

For Exercises 41–44, write the opposite of the polynomial. **(See Example 5.)**

41. $9x^2 - 2x - 3$

42. $12x^3 + 4x^2 - 8$

43. $-1.3y + z$

44. $2.4m - 5.1n$

For Exercises 45–54, subtract the polynomials. **(See Examples 6–7.)**

45. $(9p^3 - 4p^2 + 2p) - (2p^3 + 6p^2 - 3p)$

46. $(12c^2 + 6c - 3) - (4c^2 - 3c + 8)$

47. $(-19a^5 - 6a^3 + 2a^2 + 7) - (4a^5 + 2a^3 + 6a^2 - 3)$

48. $(-8y^4 + 2y^2 - 3y + 10) - (6y^4 + y^3 + 11y - 9)$

49. $(2a^2b^2 - 7ab^3 + 6ab - 8b^3) - (3a^2b^2 + 9ab - 4b^3)$

50. $(2mn^3 + 6m^2n^2 + 9mn^2 - 3mn) - (5mn^3 - 2m^2n^2 - 7mn)$

51.
$$\begin{array}{r} 5x^2 + 9x - 6 \\ \underline{-\ (3x^2 - 2x + 1)} \end{array}$$

52.
$$\begin{array}{r} 15a^3 - 2a^2 + 4a \\ \underline{-\ (2a^3 + 6a^2 - 12a)} \end{array}$$

53.
$$\begin{array}{r} 13z^4 + 2z^3 \qquad\quad + 5z - 3 \\ \underline{-\ (-10z^4 \qquad + 5z^2 - 6z \qquad)} \end{array}$$

54.
$$\begin{array}{r} 9m^4 \qquad\qquad - 2m^2 + 6m + 8 \\ \underline{-\ (-4m^4 - 5m^3 \qquad\qquad - 10m \qquad)} \end{array}$$

55. Subtract $(9x^2 + 16x - 4)$ from $(2x^2 - 4x - 9)$. **(See Example 7.)**

56. Subtract $(-5w^2 + 6w - 3)$ from $(11w^2 - 20w + 4)$.

Objective 4: Evaluating Polynomials and Applications

For Exercises 57–60, evaluate the polynomial for the given values of the variable. **(See Example 8.)**

57. $t^2 - 6t - 1$

a. for $t = -2$

b. for $t = \frac{1}{4}$

58. $x^2 + x + 2$

a. for $x = -10$

b. for $x = \frac{2}{5}$

59. $y^3 - y^2 - 1$

a. for $x = -2$

b. for $x = 0.2$

60. $3c^3 - c + 8$

a. for $c = -1$

b. for $c = 0.1$

61. The cost (in dollars) to rent a beach house for x days is given by the polynomial $250 + 219x$. **(See Example 9.)**

a. Evaluate the polynomial for $x = 3$ and interpret the answer in the context of this problem.

b. Evaluate the polynomial for $x = 14$ and interpret the answer in the context of this problem.

62. The cost (in dollars) for a speeding ticket is given by the polynomial $110 + 15x$. In this context, x is the number of miles per hour a motorist travels over the speed limit.

a. Evaluate the polynomial for $x = 15$ and interpret the answer in the context of this problem.

b. Evaluate the polynomial for $x = 25$ and interpret the answer in the context of this problem.

63. The cost (in dollars) to rent a car is given by the polynomial $29.99x + 0.40y$. In this context, x is the number of days that the car is rented and y is the number of miles driven.

a. Evaluate the polynomial for $x = 3$ and $y = 350$. Interpret the answer in the context of this problem.

b. Evaluate the polynomial for $x = 7$ and $y = 720$. Interpret the answer in the context of this problem.

64. The cost (in dollars) for a cab ride is given by the polynomial $2 + 1.25x + 0.25y$. In this context, x is the number of miles driven and y is the time in minutes for the ride.

a. Evaluate the polynomial for $x = 10$ and $y = 20$. Interpret the answer in the context of this problem.

b. Evaluate the polynomial for $x = 22$ and $y = 40$. Interpret the answer in the context of this problem.

Mixed Exercises

For Exercises 65–68, write a polynomial that represents the perimeter of the region.

65.

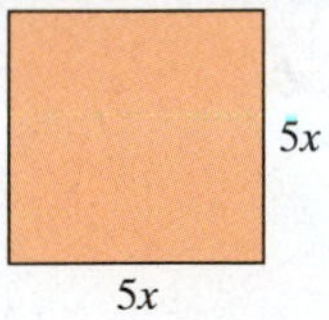

66.

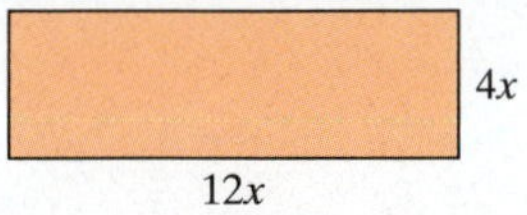

67.

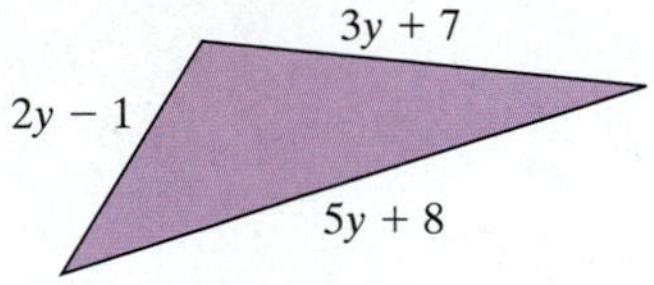

68.

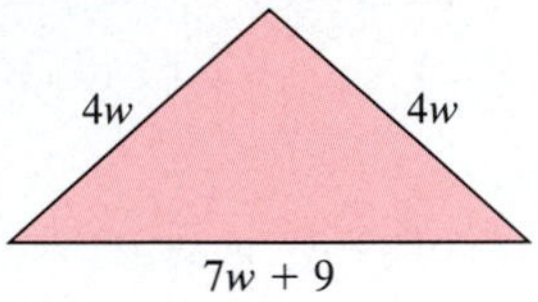

69. Find the missing polynomial. $(4x + 5) + (\quad\quad) = 2x - 1$

70. Find the missing polynomial. $(9y - 3) + (\quad\quad) = -4y + 7$

For Exercises 71–78, perform the indicated operations.

71. $(2x^3 - 5x + 8) - (4x^2 + 2x - 3) + (-7x^3 + 8x^2)$

72. $(-9m^4 - 7m^2 + 8) - (3m^4 + 9m^3 - 11) - (8m^3 + m^2 - 3)$

73. $(-5a^2 - 9a^3 + 10a^4) + (2a^3 - a^4) - (8a + 4a^3 + 10a^4)$

74. $(-t + 8t^3 - 3t^4) - (5t + 4t^4 + 6t^5) + (15t^3 + 2t^4 + 6t^5)$

75. $(1.2x^2 + 3.5x - 9.6) + (2.9x^2 - 6.7x - 3.5)$

76. $(-5.1y^2 + 4.8y + 2.3) + (-1.1y^2 - 8.9y + 3.0)$

77. $\left(-\frac{5}{4}t^3 + \frac{1}{8}t + 1\right) - \left(\frac{3}{2}t^3 - \frac{1}{2}t + \frac{2}{3}\right)$

78. $\left(\frac{7}{8}a - 4b - \frac{1}{2}c\right) - \left(-\frac{1}{4}a + \frac{1}{2}b + \frac{3}{8}c\right)$

Expanding Your Skills

79. Write a binomial with degree 4.

80. Write a binomial with degree 5.

81. Write a trinomial with degree 2.

82. Write a trinomial with degree 6.

Multiplication Properties of Exponents

Section 10.2

Objectives

1. Multiplication of Like Bases: $a^m \cdot a^n = a^{m+n}$
2. Multiplying Monomials
3. Power Rule of Exponents: $(a^m)^n = a^{m \cdot n}$
4. The Power of a Product and the Power of a Quotient

1. Multiplication of Like Bases: $a^m \cdot a^n = a^{m+n}$

In this section, we investigate the effect of multiplying two exponential quantities with the same base. For example, consider the expression: $x^2 \cdot x^4$.

$$x^2 \cdot x^4 = \underbrace{(x \cdot x)}_{x^2}\underbrace{(x \cdot x \cdot x \cdot x)}_{x^4} = \underbrace{x \cdot x \cdot x \cdot x \cdot x \cdot x}_{x^6} = x^6$$

This example suggests the following property of exponents.

PROPERTY Multiplication of Factors with Like Bases

Assume that $a \neq 0$ and m and n represent integers. Then,

$$a^m \cdot a^n = a^{m+n}$$

This rule implies that when multiplying like bases, leave the base unchanged and add the exponents.

Example 1 Multiplying Factors with the Same Base

Multiply. **a.** $z^8 \cdot z^4$ **b.** $2^5 \cdot 2^3 \cdot 2^7$

Solution:

a. $z^8 \cdot z^4 = z^{8+4}$ The base remains unchanged. Add the exponents.

$= z^{12}$

b. $2^5 \cdot 2^3 \cdot 2^7 = 2^{5+3+7}$ The base remains unchanged. Add the exponents.

$= 2^{15}$

Skill Practice

Multiply.

1. $x^9 \cdot x^{11}$
2. $5^7 \cdot 5^6 \cdot 5^2$

Avoiding Mistakes

In Example 1(b), do not be tempted to multiply the 2's. The base should remain unchanged.

2. Multiplying Monomials

In Example 2, we demonstrate how to multiply monomials. We use the commutative and associative properties of multiplication to regroup factors and multiply like bases.

Example 2 Multiplying Monomials

Multiply. $(2x)(5x)$

Solution:

$(2x)(5x) = (2 \cdot 5)(x \cdot x)$ Regroup factors.

$= 10x^2$ Multiply.

Skill Practice

Multiply.

3. $(6p)(7p)$

TIP: If a variable does not have an exponent explicitly written, then the exponent is 1. That is, $x = x^1$.

Answers

1. x^{20} 2. 5^{15} 3. $42p^2$

The product in Example 2 can be visualized as an area of a rectangle. Suppose that $2x$ represents the width of a rectangle and $5x$ represents the length. The area is given by $A = lw$. (See Figure 10-1.)

$$A = lw$$
$$A = (2x)(5x)$$
$$= 10x^2$$

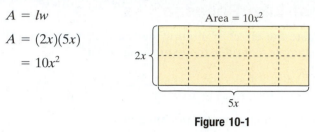

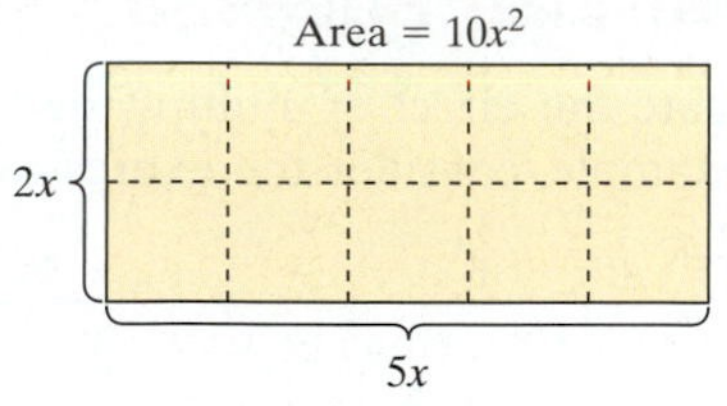

Figure 10-1

Skill Practice

Multiply.

4. $(4z^5)(8z^3)$

5. $(-5cd^7)(2c^2d^5)$

Example 3 **Multiplying Monomials**

Multiply. **a.** $(2y^2)(5y^6)$ **b.** $(-3a^2b)(4a^3b^4)$

Solution:

a. $(2y^2)(5y^6) = (2 \cdot 5)(y^2 \cdot y^6)$ Regroup factors.

$= 10y^8$ Multiply.

b. $(-3a^2b)(4a^3b^4) = (-3 \cdot 4)(a^2 \cdot a^3)(b^1 \cdot b^4)$ Regroup factors. Note that $b = b^1$.

$= -12a^5b^5$ Multiply.

3. Power Rule of Exponents: $(a^m)^n = a^{m \cdot n}$

The expression $(x^4)^2$ indicates that the expression x^4 is squared.

$$(x^4)^2 = \underbrace{(x \cdot x \cdot x \cdot x)}_{x^4}\underbrace{(x \cdot x \cdot x \cdot x)}_{x^4} = x^8$$

This example illustrates the power rule for exponents.

PROPERTY Power Rule for Exponents

Assume that $a \neq 0$ and that m and n represent integers. Then,

$$(a^m)^n = a^{m \cdot n}$$

This rule tells us to leave the base unchanged and multiply the exponents.

Skill Practice

Simplify.

6. $(m^4)^7$ **7.** $(8^4)^3$

Example 4 **Applying the Power Rule for Exponents**

Simplify the expressions. **a.** $(p^4)^5$ **b.** $(5^2)^3$

Solution:

a. $(p^4)^5 = p^{4 \cdot 5} = p^{20}$ The base is unchanged. Multiply the exponents.

b. $(5^2)^3 = 5^{2 \cdot 3} = 5^6$ The base is unchanged. Multiply the exponents.

Note that 5^6 is equal to $5 \cdot 5 \cdot 5 \cdot 5 \cdot 5 \cdot 5 = 15{,}625$. However, when such a large number is involved, we usually leave the answer in exponential form.

Answers

4. $32z^8$ **5.** $-10c^3d^{12}$

6. m^{28} **7.** 8^{12}

Avoiding Mistakes

Take notice of the difference between the power rule and multiplication of like bases:

$(p^4)^5 = p^{4 \cdot 5} = p^{20}$ power rule—multiply the exponents.

$p^4 \cdot p^5 = p^{4+5} = p^9$ multiplication of like bases—add the exponents.

4. The Power of a Product and the Power of a Quotient

Consider the expressions $(xy)^3$ and $\left(\frac{a}{b}\right)^3$. These can be expanded and simplified.

$$(xy)^3 = (xy)(xy)(xy) = (x \cdot x \cdot x)(y \cdot y \cdot y) = x^3y^3$$

$$\left(\frac{a}{b}\right)^3 = \left(\frac{a}{b}\right)\left(\frac{a}{b}\right)\left(\frac{a}{b}\right) = \frac{a \cdot a \cdot a}{b \cdot b \cdot b} = \frac{a^3}{b^3}$$

These examples suggest that to raise a product or quotient to a power, we can raise each individual factor to the power.

PROPERTY Power of a Product and the Power of a Quotient

Assume that a and b are nonzero numbers. Let m represent an integer. Then,

$$(ab)^m = a^mb^m \quad \text{and} \quad \left(\frac{a}{b}\right)^m = \frac{a^m}{b^m}$$

Example 5 Simplifying a Power of a Product or Quotient

Simplify. **a.** $(5x)^2$ **b.** $\left(\frac{2a}{b}\right)^3$ **c.** $(-3c^2d^3)^4$

Solution:

a. $(5x)^2 = 5^2x^2$ Square each factor within the parentheses (power of a product).

$= 25x^2$

b. $\left(\frac{2a}{b}\right)^3 = \frac{2^3a^3}{b^3}$ Cube each factor within the parentheses (power of a quotient).

$= \frac{8a^3}{b^3}$ Note that $2^3 = 2 \cdot 2 \cdot 2 = 8$.

c. $(-3c^2d^3)^4 = (-3)^4(c^2)^4(d^3)^4$ Raise each factor to the fourth power (power of a product rule).

$= 81c^{2 \cdot 4}d^{3 \cdot 4}$ Note that $(-3)^4 = (-3)(-3)(-3)(-3) = 81$. Apply the power rule to the factors c and d.

$= 81c^8d^{12}$

Skill Practice

Simplify.

8. $(7n)^2$ **9.** $\left(\frac{3u}{v}\right)^4$

10. $(-5x^3y)^2$

Avoiding Mistakes

Do not forget to apply the exponent to the numerical coefficient, 5.

Answers

8. $49n^2$ **9.** $\frac{81u^4}{v^4}$ **10.** $25x^6y^2$

Skill Practice

Simplify.

11. $(t^8)^4(t^2)^{10}$

12. $(-3m^4)^2(m^5n^2)^3$

Example 6 Simplifying Expressions Involving Exponents

Simplify. **a.** $(x^3)^5(x^2)^6$ **b.** $(-2y^2)^3(y^7z^2)^5$

Solution:

a. $(x^3)^5(x^2)^6 = x^{3\cdot 5} \cdot x^{2\cdot 6}$ Apply the power rule to each factor in parentheses. Operations with exponents are performed before multiplication.

$= x^{15} \cdot x^{12}$

$= x^{15+12}$ When multiplying factors with the same base, leave the base unchanged and add the exponents.

$= x^{27}$

b. $(-2y^2)^3(y^7z^2)^5 = (-2)^3(y^2)^3 \cdot (y^7)^5(z^2)^5$ Apply the power rule to each factor within parentheses.

$= -8 \cdot y^6 \cdot y^{35} \cdot z^{10}$ Note that $(-2)^3 = (-2)(-2)(-2) = -8$.

$= -8y^{6+35}z^{10}$ Add the exponents on the common base, y.

$= -8y^{41}z^{10}$

Answers

11. t^{52} **12.** $9m^{23}n^6$

Section 10.2 Practice Exercises

Boost *your* GRADE at ALEKS.com!

ALEKS version 3.0

- Practice Problems
- Self-Tests
- NetTutor
- e-Professors
- Videos

Review Exercises

1. a. Answer true or false. $3x + 4x = 7x^2$ **b.** Answer true or false. $3x + 4x = 7x$

For Exercises 2–3, add the polynomials.

2. $(-3y^5 - 4y^2 + 6y) + (2y^5 + 6y^2 - 9y)$

3. $(-2a^2 + 5a - 3) + (a^2 - 11a - 7)$

4. Subtract $(-7z^3 - 3z^2 + 8z)$ from $(4z^3 + 9z - 5)$.

5. Subtract $(-12w^4 - 3w + 14)$ from $(3w^4 + 9w^2 - 7)$.

Objective 1: Multiplication of Like Bases: $a^m \cdot a^n = a^{m+n}$

6. Fill in the blanks. $a^7 \cdot a = a^{\square+\square} = a^8$

7. Fill in the blanks. $m \cdot m^3 = m^{\square+\square} = m^4$

8. Fill in the blanks. $n^5 \cdot n = n^{\square+\square} = n^6$

For Exercises 9 – 18, multiply. **(See Example 1.)**

9. y^7y^5

10. z^3z^8

11. $w^{11} \cdot w$

12. $p \cdot p^4$

13. $c^3c^8c^{10}$

14. $d^2d^3d^4$

15. $3^4 \cdot 3^2$

16. $2^5 \cdot 2^6$

17. $5 \cdot 5^2 \cdot 5^9$

18. $4 \cdot 4^3 \cdot 4^{10}$

Objective 2: Multiplying Monomials

For Exercises 19–30, multiply. **(See Examples 2–3.)**

19. $(3y)(4y)$

20. $(9z)(2z)$

21. $(-w)(-10w^3)$

22. $(-p)(4p^2)$

23. $(6x^2)(2x^4)$

24. $(8t^3)(3t^5)$

25. $2p(p^5)(4p^7)$

26. $6m(m^2)(3m^3)$

27. $a^5b^7 \cdot a^4b^2$

28. $c^4d^{10} \cdot c^3d^6$

29. $(-4x^3y^4)(2xy^2)$

30. $(-8mn^5)(5m^2n^3)$

Objective 3: Power Rule of Exponents: $(a^m)^n = a^{m \cdot n}$

31. Fill in the blanks. $(x^3)^5 = x^{\square \cdot \square} = x^{15}$

32. Fill in the blanks. $(y^2)^7 = y^{\square \cdot \square} = y^{14}$

For Exercises 33–36, simplify the expression. **(See Example 4.)**

33. $(x^2)^{10}$

34. $(y^3)^6$

35. $(2^4)^7$

36. $(9^3)^8$

37. Which is greater? $(2^3)^2$ or $(2^2)^3$

38. Which is greater? $(4^4)^5$ or $(4^5)^4$

Objective 4: The Power of a Product and the Power of a Quotient

39. Fill in the blanks. $(xy)^3 = x^{\square}y^{\square}$

40. Fill in the blanks. $\left(\frac{x}{y}\right)^4 = \frac{x^{\square}}{y^{\square}}$

For Exercises 41–60, simplify the expression. **(See Examples 5–6.)**

41. $(2w)^5$

42. $(3t)^2$

43. $\left(\frac{10}{y}\right)^2$

44. $\left(\frac{z}{2}\right)^3$

45. $(4xy)^3$

46. $(7cd)^4$

47. $\left(\frac{2w}{p}\right)^4$

48. $\left(\frac{9u}{t}\right)^2$

49. $(6a^2b)^2$

50. $(4xy^3)^3$

51. $(-2tv^3)^4$

52. $(-10w^2z^3)^3$

53. $\left(\frac{9a^4}{b}\right)^2$

54. $\left(\frac{7c}{d^3}\right)^2$

55. $(y^2)^3(y^4)^2$

56. $(z^3)^5(z^2)^3$

57. $(2p^2)^3(pq)^2$

58. $(-4a^5)^2(ab)^3$

59. $(2xy^4)^3(4x^2y^3)$

60. $(3tv^2)^3(2t^2v^4)$

Mixed Exercises

For Exercises 61–66, write a polynomial that represents the area of the region.

61.

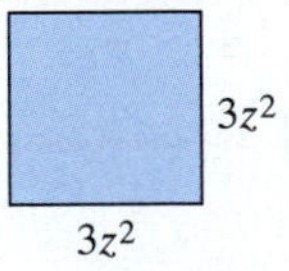

62.

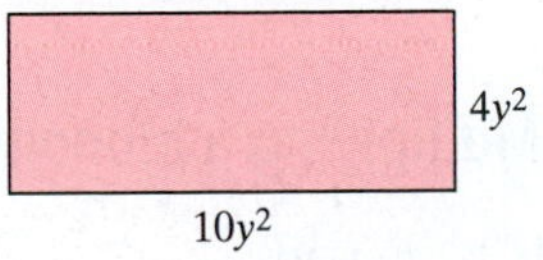

63.

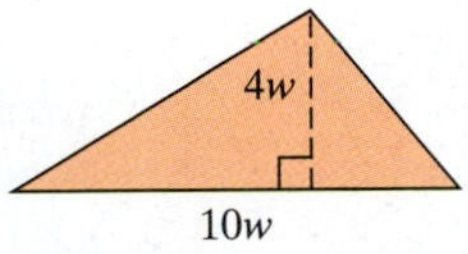

64.

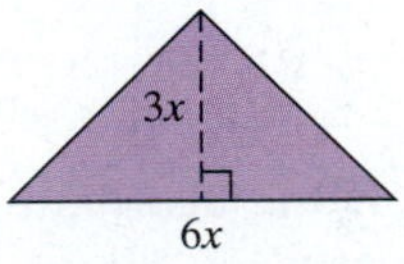

65.

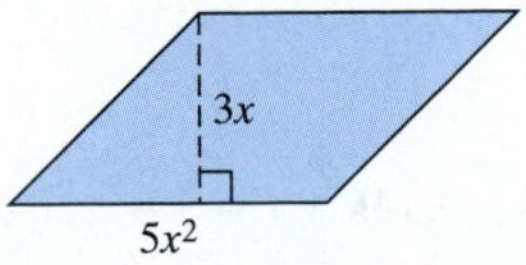

66.

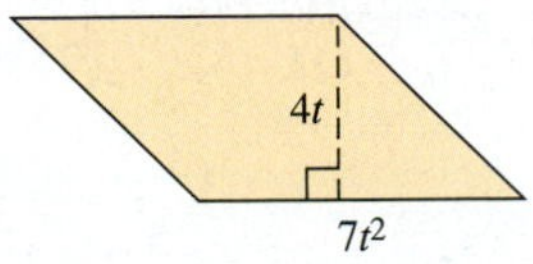

Expanding Your Skills

For Exercises 67–72, simplify.

67. $\left(\frac{2}{3}\right)^2\left(\frac{2}{3}\right)^8$

68. $\left(\frac{3}{4}\right)^3\left(\frac{3}{4}\right)^4$

69. $(x^{60}y^{50})^2$

70. $(p^{44}z^{32})^3$

71. $k^{24}k^{36}k^{50}$

72. $v^9v^{30}v^{55}$

Section 10.3 Multiplication of Polynomials

Objectives

1. Multiplying a Monomial by a Polynomial
2. Multiplying a Polynomial by a Polynomial

1. Multiplying a Monomial by a Polynomial

In Section 10.2, we practiced multiplying two monomials. In this section, we will extend this skill to multiplying polynomials with many terms. To multiply a monomial by a polynomial with more than one term, use the distributive property.

Skill Practice

Multiply.

1. $8m(4m + 7)$

Example 1 Multiplying a Monomial by a Polynomial

Multiply. $6a(3a - 2)$

Solution:

$6a(3a - 2) = (6a)(3a) + (6a)(-2)$ Apply the distributive property.

$= 18a^2 - 12a$ Simplify each term using the properties of exponents.

Skill Practice

Multiply.

2. $-5z(4z^2 + 8z - 3)$

Example 2 Multiplying a Monomial by a Polynomial

Multiply. $-2w(5w^2 + 4w - 9)$

Solution:

$-2w(5w^2 + 4w - 9)$ Apply the distributive property.

$= (-2w)(5w^2) + (-2w)(4w) + (-2w)(-9)$

$= -10w^3 - 8w^2 + 18w$ Simplify each term.

Skill Practice

Multiply.

3. $5a^3b(-2ab + 4a^2b^4)$

Example 3 Multiplying a Monomial by a Polynomial

Multiply. $4x^2y(8x^4y^3 + 3xy)$

Solution:

$4x^2y(8x^4y^3 + 3xy) = (4x^2y)(8x^4y^3) + (4x^2y)(3xy)$ Apply the distributive property.

$= 32x^6y^4 + 12x^3y^2$ Simplify each term.

Answers

1. $32m^2 + 56m$
2. $-20z^3 - 40z^2 + 15z$
3. $-10a^4b^2 + 20a^5b^5$

2. Multiplying a Polynomial by a Polynomial

We offer the following general rule for multiplying polynomials.

> **PROCEDURE Multiplying Polynomials**
>
> **Step 1** To multiply two polynomials, use the distributive property to multiply each term in the first polynomial by each term in the second polynomial.
>
> **Step 2** Combine like terms if possible.

In Examples 4–7, we demonstrate this process by multiplying two binomials.

Example 4 Multiplying a Binomial by a Binomial

Multiply. $(x + 1)(x + 4)$

Solution:

$(x + 1)(x + 4) = (x)(x) + (x)(4) + (1)(x) + (1)(4)$ Multiply each term in the first binomial by each term in the second binomial.

$= x^2 + 4x + x + 4$ Simplify each term.

$= x^2 + \underbrace{5x} + 4$ Combine like terms: $4x + x$ is $5x$.

The product is $x^2 + 5x + 4$.

> **Skill Practice**
>
> Multiply.
>
> **4.** $(y + 6)(y + 3)$

TIP: Notice that the product of two *binomials* equals the sum of the products of the **F**irst, **O**uter, **I**nner, and **L**ast terms.

The word "FOIL" can be used as a memory device to multiply two binomials.

Outer terms

First terms

First Outer Inner Last

$(x + 1)(x + 4) = (x)(x) + (x)(4) + (1)(x) + (1)(4)$

Inner terms

Last terms

Note that FOIL only works when multiplying *binomials*.

Answer

4. $y^2 + 9y + 18$

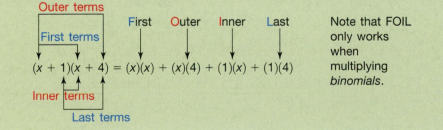

Skill Practice

Multiply.

5. $(4z - 5)(2y + 3)$

Example 5 Multiplying a Binomial by a Binomial

Multiply. $(2y + 5)(3x - 7)$

Solution:

$$(2y + 5)(3x - 7)$$

$$= (2y)(3x) + (2y)(-7) + (5)(3x) + (5)(-7)$$

Multiply each term in the first binomial by each term in the second binomial.

$$= 6xy - 14y + 15x - 35$$

Simplify each term. Notice that there are no like terms to combine.

Skill Practice

Multiply.

6. $(8t - 3)(8t + 3)$

Example 6 Multiplying a Binomial by a Binomial

Multiply. $(7m - 5)(7m + 5)$

Solution:

$$(7m - 5)(7m + 5) = (7m)(7m) + (7m)(5) + (-5)(7m) + (-5)(5)$$

$$= 49m^2 + \underbrace{35m - 35m}_{0m} - 25$$

Simplify each term. Notice that the difference, $35m - 35m = 0m$.

$$= 49m^2 - 25$$

Skill Practice

Square the binomial.

7. $(2w + 5)^2$

Example 7 Squaring a Binomial

Square the binomial. $(4x + 3)^2$

Solution:

$$(4x + 3)^2 = (4x + 3)(4x + 3)$$

To square a quantity, multiply the quantity times itself.

$$(4x + 3)(4x + 3) = (4x)(4x) + (4x)(3) + (3)(4x) + (3)(3)$$

$$= 16x^2 + \underbrace{12x + 12x} + 9$$

Simplify each term.

$$= 16x^2 + 24x + 9$$

Combine like terms.

The product is $16x^2 + 24x + 9$.

Avoiding Mistakes

The power rule can be applied to a product but cannot be applied to a sum. $(4x + 3)^2 \neq (4x)^2 + (3)^2$

Answers

5. $8zy + 12z - 10y - 15$
6. $64t^2 - 9$
7. $4w^2 + 20w + 25$

Example 8 **Multiplying a Polynomial by a Polynomial**

Multiply. $(x + 2)(x^2 - 4x + 5)$

Solution:

Apply the distributive property. Multiply each term in the first polynomial by each term in the second.

$$(x + 2)(x^2 - 4x + 5) = (x)(x^2) + (x)(-4x) + (x)(5) + (2)(x^2) + (2)(-4x) + (2)(5)$$

$= x^3 - 4x^2 + 5x + 2x^2 - 8x + 10$ Simplify each term.

$= x^3 \underbrace{- 4x^2 + 2x^2} + \underbrace{5x - 8x} + 10$ Group like terms.

$= x^3 \quad - 2x^2 \quad - 3x \quad + 10$ Combine like terms.

$= x^3 - 2x^2 - 3x + 10$

Skill Practice

Multiply.

8. $(y + 3)(y^2 - 6y + 1)$

Avoiding Mistakes

It is important to note that the acronym FOIL does not apply to Example 8 because the product does not involve two binomials.

TIP: Multiplication of polynomials can also be performed vertically. Here we have the product from Example 8 compared to the product of two real numbers.

$$\begin{array}{r} 235 \\ \times\ 21 \\ \hline 235 \\ 4700 \\ \hline 4935 \end{array} \qquad \begin{array}{r} x^2 - 4x + 5 \\ \times \qquad x + 2 \\ \hline 2x^2 - 8x + 10 \\ x^3 - 4x^2 + 5x + 0 \\ \hline x^3 - 2x^2 - 3x + 10 \end{array}$$

Note: When multiplying by the column method, it is important to align like terms vertically before adding terms.

Answer

8. $y^3 - 3y^2 - 17y + 3$

Section 10.3 Practice Exercises

Boost *your* GRADE at ALEKS.com! ALEKS version 3.0

- Practice Problems
- Self-Tests
- NetTutor
- e-Professors
- Videos

Review Exercises

1. Complete the rule for multiplying like bases. $a^m a^n =$ ______

For Exercises 2–7, simplify.

2. $x^3x^{12}x$

3. $6^2 6^5 6^3$

4. $(5^2)^{11}$

5. $(x^3)^{10}$

6. $(-3x^2z^3)^2$

7. $(-4cd^5)^2$

For Exercises 8–10, add or subtract as indicated.

8. $(-5x^2 + 7x - 3) + (2x^2 + 9)$

9. $(-3y^2 + 5y - 7) - (8y^3 + 2y^2 + 9y)$

10. $(-3cd - 4d^2 + 5c) - (12cd + 2d^2 - 4c)$

Objective 1: Multiplying a Monomial by a Polynomial

For Exercises 11–22, multiply. **(See Examples 1–3.)**

11. $2x(3x + 4)$
12. $3w(5w + 7)$
13. $-2y(y - 8)$
14. $-5z(z - 1)$
15. $6a(a^2 - 2a + 3)$
16. $-4t(t^2 + 5t - 7)$
17. $8w(-3w^2 - 2w + 4)$
18. $5m(-2m^2 - 4m + 1)$
19. $2a^2b(5a^2b + 4ab)$
20. $3c^3d(2cd^2 + 6cd)$
21. $-9xy(x^2y - 2xy + xy^2)$
22. $-6wz(w^2z - 4wz - wz^2)$

Objective 2: Multiplying a Polynomial by a Polynomial

For Exercises 23–46, multiply the polynomials. **(See Examples 4–7.)**

23. $(y + 5)(y + 3)$
24. $(z + 6)(z + 2)$
25. $(m - 4)(m + 7)$
26. $(n - 5)(n + 8)$
27. $(x - 4)(y - 10)$
28. $(w - 3)(z - 9)$
29. $(4y + 6)(2y - 3)$
30. $(6x + 5)(3x - 1)$
31. $(5t - 4)(2t + 6)$
32. $(10w - 3)(4w + 3)$
33. $(2a + b)(5a - 3)$
34. $(10m + 2n)(m + 4)$
35. $(x - 4)(x + 4)$
36. $(y + 5)(y - 5)$
37. $(8p + 3)(8p - 3)$
38. $(9v + 4)(9v - 4)$
39. $(2 - m)(2 + m)$
40. $(3 - n)(3 + n)$
41. $(x + y)^2$
42. $(m + n)^2$
43. $(3x + 5)^2$
44. $(2y + 4)^2$
45. $(7z - 1)^2$
46. $(8p - 1)^2$
47. Explain why FOIL will not work when you multiply $(x + 4)(x^2 - 2x + 3)$.
48. Explain why FOIL will not work when you multiply $(y + 7)(y^2 - 3y + 1)$.

For Exercises 49–60, multiply the polynomials. **(See Example 8.)**

49. $(a + 5)(a^2 - 2a + 4)$
50. $(w + 4)(w^2 - 5w + 6)$
51. $(2z - 3)(3z^2 + 5z + 7)$
52. $(5y - 1)(6y^2 + 2y + 5)$
53. $(c + 3)(c^2 - 3c + 9)$
54. $(d + 2)(d^2 - 2d + 4)$
55. $$\begin{array}{r} x^2 - 5x + 1 \\ \times \qquad 2x - 3 \\ \hline \end{array}$$
56. $$\begin{array}{r} y^2 - 7y + 3 \\ \times \qquad 3y - 5 \\ \hline \end{array}$$

57. $\begin{array}{r} 9a^2 + 2a - 4 \\ \times\ 4a^2 + a + 3 \\ \hline \end{array}$

58. $\begin{array}{r} 10c^2 + 3c - 6 \\ \times\ 2c^2 + c + 2 \\ \hline \end{array}$

59. $(4x^2 - 3x + 2)(2x^2 - 5x - 1)$

60. $(5y^2 - 3y - 4)(y^2 + 4y + 7)$

Expanding Your Skills

For Exercises 61–66, fill in the missing polynomial.

61. $(x + 3)(\quad\quad) = x^2 + 8x + 15$

62. $(y + 2)(\quad\quad) = y^2 + 9y + 14$

63. $(\quad\quad)(z - 5) = z^2 - 3z - 10$

64. $(\quad\quad)(w - 7) = w^2 - 4w - 21$

65. $(2x + 5)(\quad\quad) = 6x^2 + 23x + 20$

66. $(3y + 2)(\quad\quad) = 15y^2 + 13y + 2$

Problem Recognition Exercises

Operations on Polynomials and Exponential Expressions

For Exercises 1–12, perform the indicated operation and simplify if possible.

1. a. $12x - 4x$
b. $12x \cdot 4x$
c. $12x + 4x$

2. a. $6y \cdot 2y$
b. $6y - 2y$
c. $6y + 2y$

3. a. $(ab)^2$
b. $(a + b)^2$
c. $(a - b)^2$

4. a. $(c + d)^2$
b. $(cd)^2$
c. $(c - d)^2$

5. a. $4x^3 + 3x^2$
b. $4x^3 \cdot 3x^2$
c. $4x^3 - 3x^2$

6. a. $6y^4 \cdot 3y^3$
b. $6y^4 + 3y^3$
c. $6y^4 - 3y^3$

7. a. $2x^3y^5 - 5x^3y^5$
b. $2x^3y^5 + 5x^3y^5$
c. $2x^3y^5 \cdot 5x^3y^5$

8. a. $9a^2b + 7a^2b$
b. $9a^2b - 7a^2b$
c. $9a^2b \cdot 7a^2b$

9. a. $(2x - 3)(4x + 7)$
b. $(2x - 3) - (4x + 7)$
c. $(2x - 3) + (4x + 7)$

10. a. $(3w + 4) - (6w - 9)$
b. $(3w + 4)(6w - 9)$
c. $(3w + 4) + (6w - 9)$

11. a. $(t + 2) + (t^2 - 3t + 4)$
b. $(t + 2) - (t^2 - 3t + 4)$
c. $(t + 2)(t^2 - 3t + 4)$

12. a. $(z - 5)(z^2 + 4z - 3)$
b. $(z - 5) - (z^2 + 4z - 3)$
c. $(z - 5) + (z^2 + 4z - 3)$

Section 10.4 Introduction to Factoring

Objectives

1. Greatest Common Factor
2. Factoring Out the Greatest Common Factor

1. Greatest Common Factor

In Section 4.2, we learned how to factor a number into prime factors. For example:

$$10 = 2 \cdot 5 \qquad \text{2 and 5 are } \textit{factors} \text{ of 10.}$$

$$15 = 3 \cdot 5 \qquad \text{3 and 5 are } \textit{factors} \text{ of 15.}$$

From this we see that 5 is a common factor of 10 and 15. In fact, it is the greatest factor that divides evenly into both 10 and 15.

The **greatest common factor** (denoted **GCF**) of two or more integers is the greatest factor that divides evenly into each integer.

PROCEDURE Finding the GCF of Two or More Integers

Step 1 Factor each integer into prime factors.

Step 2 Determine the prime factors common to each integer, including repeated factors.

Step 3 The product of the factors from step 2 is the GCF.

Skill Practice

Determine the GCF.

1. 28 and 44
2. 42, 70, and 56

Example 1 Determining the Greatest Common Factor

Determine the GCF. **a.** 18 and 45 **b.** 45, 30, and 105

Solution:

a. Factor each number into prime factors.

$$18 = 2 \cdot 3 \cdot 3$$

$$45 = 3 \cdot 3 \cdot 5 \qquad \text{The GCF is } 3 \cdot 3 = 9.$$

The factor 3 appears twice in each factorization. Therefore, the GCF is $3 \cdot 3 = 9$.

b. Factor each number into prime factors.

$$45 = 3 \cdot 3 \cdot 5$$

$$30 = 2 \cdot 3 \cdot 5$$

$$105 = 3 \cdot 5 \cdot 7 \qquad \text{The GCF is } 3 \cdot 5 = 15.$$

The factors 3 and 5 are common to all three numbers. Therefore, the GCF is $3 \cdot 5 = 15$.

Some expressions have variable factors. To factor an expression such as $15x^3y^2$, factor the integer coefficient and expand the powers of the variables. For example:

$$15x^3y^2 = 3 \cdot 5 \cdot x \cdot x \cdot x \cdot y \cdot y$$

Answers

1. 4 2. 14

Example 2 **Determining the Greatest Common Factor**

Determine the GCF. **a.** y^4, y^6, and y^3 **b.** $6a^2b$ and $12a^3c^2$

Solution:

a. Factor the expressions.

$$y^4 = y \cdot y \cdot y \cdot y$$
$$y^6 = y \cdot y \cdot y \cdot y \cdot y \cdot y$$
$$y^3 = y \cdot y \cdot y \qquad \text{GCF} = y \cdot y \cdot y = y^3.$$

Three factors of y are common to each expression. Therefore, the GCF is y^3.

b. Factor the expressions.

$$6a^2b = 2 \cdot 3 \cdot a \cdot a \cdot b$$
$$12a^3c^2 = 2 \cdot 2 \cdot 3 \cdot a \cdot a \cdot a \cdot c \cdot c \qquad \text{GCF} = 2 \cdot 3 \cdot a \cdot a = 6a^2.$$

The expressions each have a factor of 2, a factor of 3, and two factors of a. The GCF is $6a^2$.

Skill Practice

Determine the GCF.

3. t^6, t^8, and t^4

4. $5x^3y^2$ and $15xy^4$

TIP: The GCF is the product of common factors, where each factor is raised to the lowest exponent to which it appears. This is demonstrated in Example 2.

2. Factoring Out the Greatest Common Factor

In addition to factoring whole numbers, we can also factor algebraic expressions. We first revisit the process to multiply a monomial and a polynomial. Consider the product:

$3(x + 2y)$ Multiply each term in parentheses by 3.

$= 3(x) + 3(2y)$

$= 3x + 6y$ Simplify.

Reversing this process is called *factoring out the greatest common factor*. To factor out a 3 from the polynomial $3x + 6y$, we have

$3x + 6y$

$= 3(x) + 3(2y)$ Write each term as a product of 3 and another factor. This can be determined by dividing the original terms by the GCF.

$$\frac{3x}{3} = x \quad \text{and} \quad \frac{6y}{3} = 2y$$

$= 3(x + 2y)$ Apply the distributive property "in reverse."

The expression $3(x + 2y)$ is the factored form of $3x + 6y$. Notice that the terms within parentheses are determined by dividing the original terms by the GCF.

Answers

3. t^4 **4.** $5xy^2$

Skill Practice

Factor out the GCF.

5. $8a + 12b + 24c$

Example 3 Factoring Out the GCF

Factor out the GCF from $15x + 10y + 30z$.

Solution:

The GCF of $15x$, $10y$, and $30z$ is 5.

$15x + 10y + 30z = 5(3x) + 5(2y) + 5(6z)$ Write each term as a product of the GCF, 5, and another factor.

$\frac{15x}{5} = 3x$, $\frac{10y}{5} = 2y$, and $\frac{30z}{5} = 6z$

$= 5(3x + 2y + 6z)$ Apply the distributive property "in reverse."

The factored form of $15x + 10y + 30z$ is $5(3x + 2y + 6z)$.

The answer to a factoring problem can be checked by multiplication. To check the solution to Example 3, we have

$$5(3x + 2y + 6z) = 15x + 10y + 30z \checkmark$$

Skill Practice

Factor out the GCF.

6. $28w^5 + 7w^2$

Example 4 Factoring Out the GCF

Factor out the GCF from $10y^4 + 5y^3$.

Solution:

The GCF of $10y^4$ and $5y^3$ is $5y^3$.

$10y^4 + 5y^3 = 5y^3(2y) + 5y^3(1)$ Write each term as a product of the GCF, $5y^3$, and another factor. Notice $5y^3 = 5y^3(1)$.

$= 5y^3(2y + 1)$ Apply the distributive property "in reverse."

The factored form of $10y^4 + 5y^3$ is $5y^3(2y + 1)$.

Avoiding Mistakes

Do not forget to write the 1 in the second term of the parentheses.

Skill Practice

Factor out the GCF.

7. $15ab + 5a^2b^2 - 25a^3b^3$

Example 5 Factoring Out the GCF

Factor out the GCF from $12xy^3 + 4x^2y^2 - 8x^3y$.

Solution:

The GCF of $12xy^3$, $4x^2y^2$, and $-8x^3y$ is $4xy$.

$12xy^3 + 4x^2y^2 - 8x^3y$

$= 4xy(3y^2) + 4xy(xy) - 4xy(2x^2)$ Write each term as a product of $4xy$ and another factor.

$= 4xy(3y^2 + xy - 2x^2)$ Apply the distributive property "in reverse."

The factored form of $12xy^3 + 4x^2y^2 - 8x^3y$ is $4xy(3y^2 + xy - 2x^2)$.

Answers

5. $4(2a + 3b + 6c)$
6. $7w^2(4w^3 + 1)$
7. $5ab(3 + ab - 5a^2b^2)$

Section 10.4 Practice Exercises

Study Skills Exercises

1. The material learned in this chapter will help you build a foundation for future courses in algebra. Take a minute to reflect on your educational and career goals. Do you plan to take any other algebra courses beyond this course?

2. Define the key term, **greatest common factor (GCF)**.

Review Exercises

For Exercises 3–10, perform the indicated operations.

3. $(2x^2y)^4$ **4.** $(x^4y^3)^3$ **5.** w^4w^5w **6.** p^3p^9p

7. $(4x^3 - 6x^2 - 5x) - (6x^3 + 10x^2 - 5x)$ **8.** $(-9y^2 + 4y - 3) - (-8y^2 + 4y + 11)$

9. $(2x - 3)(x - 10)$ **10.** $(4x + 1)(x - 7)$

Objective 1: Greatest Common Factor

For Exercises 11–28, determine the greatest common factor. **(See Examples 1–2.)**

11. 12 and 18 **12.** 15 and 20 **13.** 21 and 28 **14.** 24 and 40

15. 75 and 25 **16.** 36 and 72 **17.** 12, 18, and 24 **18.** 30, 42, and 54

19. 30, 45, and 75 **20.** 16, 40, and 48 **21.** x^5, x^3, and x^7 **22.** z^5, z^9, and z^2

23. w^3, w^2, and w **24.** p^4, p, and p^3 **25.** $3x^2y^3$ and $6xy^4$ **26.** $15c^3d^3$ and $10c^4d^2$

27. $5ab^3$ and $10a^2b$ **28.** $20xy^2$ and $28x^2y$

29. a. Identify the GCF for the terms t^3, t^5, and t^4.

b. In the GCF, was the power on t selected from the exponent of greatest value or least value?

30. a. Identify the GCF for the terms x^4, x^5, x^2.

b. In the GCF, was the power on x selected from the exponent of greatest value or least value?

Objective 2: Factoring Out the Greatest Common Factor

For Exercises 31–52, factor out the greatest common factor. **(See Examples 3–5.)**

31. $4x - 20$ **32.** $5y - 30$ **33.** $2p + 2$ **34.** $7w + 7$

35. $5m + 40n$ **36.** $6t + 18v$ **37.** $4a + 8b + 2c$ **38.** $5m - 15n + 25p$

39. $12x - 20y - 8z$ **40.** $20b - 40c + 50d$ **41.** $x^4 + x^3$ **42.** $y^8 + y^5$

43. $5y^3 - 15y^2$ **44.** $6z^4 - 12z^3$ **45.** $3x^3 + 5x^2 + 7x$ **46.** $7y^4 + 9y^3 - 11y^2$

47. $12z^5 - 6z^3 + 3z^2$ **48.** $15p^4 - 10p^3 + 5p^2$ **49.** $4x^4y^3 - 3x^3y^4 + 5x^2y^5$ **50.** $2a^2b^3 - 3a^3b^2 + 4a^4b$

51. $15c^3d^5 + 30c^2d^4 - 45cd^3$ **52.** $24m^4n^2 + 36m^3n - 48m^2n^3$

Section 10.5 Negative Exponents and the Quotient Rule for Exponents

Objectives

1. Division of Like Bases
2. Definition of b^0
3. Definition of b^{-n}
4. Properties of Exponents: A Summary

1. Division of Like Bases

In Section 10.2, we learned that $a^m a^n = a^{m+n}$. That is, to multiply expressions with the same base, leave the base unchanged and add the exponents.

Now consider the expression $\dfrac{x^5}{x^2} = \dfrac{\not{x} \cdot \not{x} \cdot x \cdot x \cdot x}{\not{x} \cdot \not{x}} = \dfrac{x \cdot x \cdot x}{1} = x^3$

This example suggests the following rule for *dividing* expressions with the same base.

PROPERTY Division of Like Bases

Assume that $a \neq 0$ and that m and n represent integers. Then,

$$\frac{a^m}{a^n} = a^{m-n}$$

This property indicates that to divide expressions with the same base, leave the base unchanged and subtract the denominator exponent from the numerator exponent.

Skill Practice

Simplify.

1. $\dfrac{p^{10}}{p^4}$
2. $\dfrac{5^7}{5^2}$
3. $\dfrac{m^{12}m}{m^3}$

Example 1 Dividing Like Bases

Simplify.

a. $\dfrac{w^6}{w^4}$ **b.** $\dfrac{7^3}{7^2}$ **c.** $\dfrac{z^4 z}{z^3}$

Solution:

a. $\dfrac{w^6}{w^4} = w^{6-4} = w^2$ The base is unchanged. Subtract the exponents.

b. $\dfrac{7^3}{7^2} = 7^{3-2} = 7^1 = 7$ The base is unchanged. Subtract the exponents.

Avoiding Mistakes

Recall that the order of operations requires us to simplify the numerator before dividing.

c. $\dfrac{z^4 z}{z^3} = \dfrac{z^{4+1}}{z^3}$ First simplify the numerator. When multiplying expressions with the same base, add the exponents.

$= \dfrac{z^5}{z^3}$

$= z^{5-3}$ The base is unchanged. Subtract the exponents.

$= z^2$

Answers

1. p^6 2. 5^5 3. m^{10}

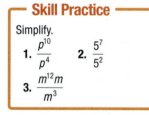

2. Definition of b^0

The definition of b^0 must be consistent with the other properties of exponents learned thus far. For example, we know that $1 = \frac{5^3}{5^3}$. If we subtract exponents, the result is 5^0.

Subtract exponents.

$$1 = \frac{5^3}{5^3} = 5^{3-3} = 5^0.$$ Therefore, 5^0 must be defined as 1.

DEFINITION b^0

Let b be a nonzero number. Then $b^0 = 1$.

Example 2 Simplifying Expressions with a Zero Exponent

Simplify.

a. z^0 **b.** $4x^0$ **c.** $(6wz)^0$ **d.** $t^0 + 5$

Solution:

a. $z^0 = 1$ By definition

b. $4x^0$ Notice that the exponent of 0 applies only to the factor x. The factor of 4 has an implied exponent of 1. Thus, $4x^0 = 4^1x^0$.

$= 4(1)$ Note that $x^0 = 1$. The operation involving exponents is performed before multiplying.

$= 4$

c. $(6wz)^0 = 1$ By definition

d. $t^0 + 5 = 1 + 5$ Note that $t^0 = 1$.

$= 6$

Skill Practice

Simplify.

4. x^0 **5.** $9y^0$

6. $(7ab)^0$ **7.** $-3 + t^0$

3. Definition of b^{-n}

The definition of a base raised to a negative exponent must be consistent with the other properties of exponents. For example:

$$\frac{x^3}{x^5} = \frac{\overset{1}{\cancel{x}} \cdot \overset{1}{\cancel{x}} \cdot \overset{1}{\cancel{x}}}{\cancel{x} \cdot \cancel{x} \cdot \cancel{x} \cdot x \cdot x} = \frac{1}{x^2} \quad \text{and} \quad \frac{x^3}{x^5} = x^{3-5} = x^{-2}$$

These must be the same.

This suggests the following definition involving negative exponents.

DEFINITION b^{-n}

Let n be an integer and b be a nonzero number. Then,

$$b^{-n} = \left(\frac{1}{b}\right)^n \quad \text{or} \quad \frac{1}{b^n}$$

Answers

4. 1 **5.** 9 **6.** 1 **7.** -2

To evaluate b^{-n}, take the reciprocal of the base and change the sign of the exponent.

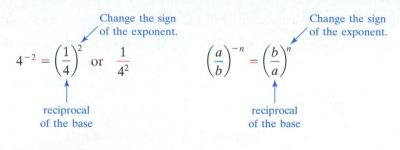

$$4^{-2} = \left(\frac{1}{4}\right)^2 \text{ or } \frac{1}{4^2} \qquad \left(\frac{a}{b}\right)^{-n} = \left(\frac{b}{a}\right)^n$$

Change the sign of the exponent. — reciprocal of the base

Skill Practice

Simplify.

8. d^{-6} 9. 4^{-1}

10. $(-7)^{-2}$ 11. $\left(\frac{1}{3}\right)^{-2}$

Example 3 Simplifying Expressions Containing Negative Exponents

Simplify. Write the answer with positive exponents only.

a. c^{-3} **b.** 5^{-1} **c.** $(-3)^{-4}$ **d.** $\left(\frac{1}{5}\right)^{-2}$

Solution:

a. $c^{-3} = \left(\frac{1}{c}\right)^3$ Take the reciprocal of the base. Change the sign of the exponent.

$= \frac{1}{c^3}$

b. $5^{-1} = \left(\frac{1}{5}\right)^1$ Take the reciprocal of the base. Change the sign of the exponent.

$= \frac{1}{5}$

Avoiding Mistakes

In Example 3(b), $5^{-1} \neq -5$. A negative exponent does not affect the sign of the base.

c. $(-3)^{-4} = \left(\frac{1}{-3}\right)^4$ Take the reciprocal of the base. Change the sign of the exponent.

$= \frac{(1)^4}{(-3)^4}$ Raise the numerator and denominator to the fourth power.

$= \frac{1}{81}$ Note that $(-3)^4 = (-3)(-3)(-3)(-3) = 81$.

d. $\left(\frac{1}{5}\right)^{-2} = \left(\frac{5}{1}\right)^2$ Take the reciprocal of the base. Change the sign of the exponent.

$= 25$

Skill Practice

Simplify.

12. $(3x)^{-4}$ 13. $3x^{-4}$

Example 4 Simplifying Expressions Containing Negative Exponents

Simplify. Write the answer with positive exponents only.

a. $(5x)^{-2}$ **b.** $5x^{-2}$

Answers

8. $\frac{1}{d^6}$ 9. $\frac{1}{4}$ 10. $\frac{1}{49}$

11. 9 12. $\frac{1}{81x^4}$ 13. $\frac{3}{x^4}$

Solution:

a. $(5x)^{-2} = \left(\frac{1}{5x}\right)^2$ Take the reciprocal of the base. Change the sign of the exponent.

$= \frac{(1)^2}{(5x)^2}$ Square the numerator and square the denominator.

$= \frac{1}{25x^2}$

b. $5x^{-2} = 5^1x^{-2}$ There are no parentheses to group the 5 and x as a single base. Therefore, the exponent of -2 applies only to x. The factor of 5 has an implied exponent of 1.

$= 5 \cdot \left(\frac{1}{x}\right)^2$

$= 5 \cdot \frac{1}{x^2}$

$= \frac{5}{x^2}$

Notice that a negative exponent changes the position of the base within a fraction. For example:

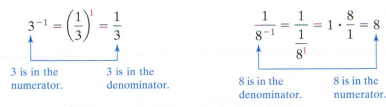

$$3^{-1} = \left(\frac{1}{3}\right)^1 = \frac{1}{3}$$

3 is in the numerator. 3 is in the denominator.

$$\frac{1}{8^{-1}} = \frac{1}{\frac{1}{8^1}} = 1 \cdot \frac{8}{1} = 8$$

8 is in the denominator. 8 is in the numerator.

Example 5 Simplifying Expressions Containing Negative Exponents

Simplify. Write the answer with positive exponents only.

a. $\frac{1}{6^{-1}}$ **b.** $\frac{x^4}{y^{-5}}$ **c.** $\frac{a^{-3}b}{c^{-4}}$

Solution:

a. $\frac{1}{6^{-1}} = 6^1 = 6$

b. $\frac{x^4}{y^{-5}} = \frac{x^4y^5}{1}$ The negative exponent on y changes the position of y within the fraction. Notice that x does not change position because its exponent is positive.

$= x^4y^5$

c. $\frac{a^{-3}b}{c^{-4}} = \frac{bc^4}{a^3}$ The factor of a and the factor of c have negative exponents. Change their positions within the fraction. Notice that b does not change position because its exponent is positive.

Skill Practice

Simplify.

14. $\frac{1}{7^{-1}}$ **15.** $\frac{m^6}{n^{-2}}$

16. $\frac{x^2y^{-6}}{z^{-5}}$

Answers

14. 7 **15.** m^6n^2 **16.** $\frac{x^2z^5}{y^6}$

4. Properties of Exponents: A Summary

The properties of exponents that we have learned thus far are summarized in Table 10-1.

Table 10-1 Properties of Integer Exponents

Assume that $a \neq 0$, $b \neq 0$, and m and n represent integers.	
1. Multiplication of Like Bases: $a^m a^n = a^{m+n}$	Example: $x^2x^4 = x^{2+4} = x^6$
2. Division of Like Bases: $\dfrac{a^m}{a^n} = a^{m-n}$	Example: $\dfrac{x^8}{x^3} = x^{8-3} = x^5$
3. The Power Rule: $(a^m)^n = a^{m \cdot n}$	Example: $(x^5)^3 = x^{5 \cdot 3} = x^{15}$
4. Power of a Product: $(ab)^m = a^m b^m$	Example: $(2y)^3 = 2^3y^3$ or $8y^3$
5. Power of a Quotient: $\left(\dfrac{a}{b}\right)^m = \dfrac{a^m}{b^m}$	Example: $\left(\dfrac{6}{w}\right)^2 = \dfrac{6^2}{w^2}$ or $\dfrac{36}{w^2}$
6. Definition of b^0: $b^0 = 1$	Example: $11^0 = 1$
7. Definition of b^{-n}: $b^{-n} = \left(\dfrac{1}{b}\right)^n = \dfrac{1}{b^n}$	Example: $p^{-3} = \left(\dfrac{1}{p}\right)^3 = \dfrac{1}{p^3}$

In Example 6, we practice applying all the properties of exponents.

Skill Practice

Simplify.

17. $y^{-5}y^7$ **18.** $\dfrac{t^8}{t^{-5}}$

19. $(w^{-4})^3$

20. $(4a^4b^{-8})(a^{-4}b^9)$

Example 6 Simplifying Expressions Containing Exponents

Simplify. Write the answer with positive exponents only.

a. $x^{-3}x^8$ **b.** $\dfrac{y^4}{y^{-6}}$ **c.** $(x^{-2})^3$ **d.** $(3x^{-2}y^{-5})(x^6y^5)$

Solution:

a. $x^{-3}x^8 = x^{-3+8}$ The base is unchanged. Add the exponents (property 1).

$= x^5$

b. $\dfrac{y^4}{y^{-6}} = y^{4-(-6)}$ The base is unchanged. Subtract the exponents (property 2).

$= y^{10}$

c. $(x^{-2})^3 = x^{-2 \cdot 3}$ Apply the power rule (property 3).

$= x^{-6}$

$= \dfrac{1}{x^6}$ Simplify the negative exponent.

d. $(3x^{-2}y^{-5})(x^6y^5)$

$= 3(x^{-2}x^6)(y^{-5}y^5)$ Regroup like factors.

$= 3x^{-2+6}y^{-5+5}$ Add the exponents (property 1).

$= 3x^4y^0$

$= 3x^4(1)$ The expression $y^0 = 1$.

$= 3x^4$

Answers

17. y^2 **18.** t^{13} **19.** $\dfrac{1}{w^{12}}$

20. $4b$

Section 10.5 Practice Exercises

Study Skills Exercise

1. Define the key terms.

a. b^0 **b.** b^{-n}

Review Exercises

2. Simplify. $-3x^2 + 4y - 7x^2 - 10y$

For Exercises 3–10, simplify.

3. $x^2 \cdot x^5$ **4.** $y^3 \cdot y^9$ **5.** $(x^2)^5$ **6.** $(y^3)^9$

7. $(4x^2y)^3$ **8.** $(2a^5b)^4$ **9.** $\left(\dfrac{3x}{y^2}\right)^2$ **10.** $\left(\dfrac{5c^2}{d}\right)^3$

Objective 1: Division of Like Bases

For Exercises 11–18, simplify. **(See Example 1.)**

11. $\dfrac{m^5}{m^2}$ **12.** $\dfrac{k^7}{k^3}$ **13.** $\dfrac{10^{10}}{10^2}$ **14.** $\dfrac{9^4}{9^3}$

15. $\dfrac{x^3}{x}$ **16.** $\dfrac{y^7}{y}$ **17.** $\dfrac{w^5w^2}{w^6}$ **18.** $\dfrac{d^7d^6}{d^{12}}$

Objective 2: Definition of b^0

For Exercises 19–30, simplify. **(See Example 2.)**

19. a^0 **20.** t^0 **21.** 55^0 **22.** 48^0

23. $6x^0$ **24.** $7y^0$ **25.** $(6x)^0$ **26.** $(7y)^0$

27. $\left(\dfrac{1}{3}\right)^0$ **28.** $\left(\dfrac{4}{5}\right)^0$ **29.** $3 - n^0$ **30.** $6 - w^0$

Objective 3: Definition of b^{-n}

For Exercises 31–54, simplify. Write the answer with positive exponents only. **(See Examples 3–5.)**

31. a^{-4} **32.** t^{-7} **33.** 6^{-1} **34.** 8^{-1}

35. $(-5)^{-2}$ **36.** $(-4)^{-2}$ **37.** $\left(\dfrac{1}{4}\right)^{-2}$ **38.** $\left(\dfrac{1}{8}\right)^{-2}$

39. $(2x)^{-3}$ **40.** $(6x)^{-2}$ **41.** $2x^{-3}$ **42.** $6x^{-2}$

43. $16x^{-1}y$ **44.** $12a^{-1}b$ **45.** $\dfrac{1}{8^{-1}}$ **46.** $\dfrac{1}{9^{-1}}$

47. $\dfrac{c^5}{d^{-4}}$ **48.** $\dfrac{b^2}{c^{-5}}$ **49.** $\dfrac{x^{-2}y^3}{z^{-4}}$ **50.** $\dfrac{m^{-1}n^7}{p^{-3}}$

51. $2x^{-2}y^{-3}$ **52.** $5a^{-2}b^{-4}$ **53.** $\dfrac{2}{x^{-2}y^{-3}}$ **54.** $\dfrac{5}{a^{-2}b^{-4}}$

Objective 4: Properties of Exponents: A Summary (Mixed Exercises)

For Exercises 55–78, simplify. Write the answers with positive exponents only. **(See Example 6.)**

55. $y^{-12}y^{14}$ **56.** $z^{-4}z^7$ **57.** $p^{-8}p^3$ **58.** w^4w^{-7}

59. $\dfrac{x^5}{x^{-7}}$ **60.** $\dfrac{a^3}{a^{-5}}$ **61.** $\dfrac{t^4}{t^9}$ **62.** $\dfrac{m^5}{m^7}$

63. $(k^{-7})^2$ **64.** $(p^4)^{-5}$ **65.** $(k^{-7})^{-2}$ **66.** $(p^{-4})^{-5}$

67. $(a^{-2}b)^3$ **68.** $(xy^{-4})^3$ **69.** $(w^{-2}z^{-3})^{-2}$ **70.** $(p^{-7}q^{-4})^{-2}$

71. $5^4 5^{-7}$ **72.** $8^2 8^{-4}$ **73.** $\dfrac{4^2}{4^{-1}}$ **74.** $\dfrac{2^3}{2^{-2}}$

75. $(-3c^2d^{-3})(6c^3d^{-4})$ **76.** $(4m^3n^{-5})(m^{-4}n^6)$ **77.** $(x^2)^4(x^5)^{-2}$ **78.** $(w^{-7})^2(w^3)^4$

Section 10.6 Scientific Notation

Objectives

1. Scientific Notation
2. Converting to Scientific Notation
3. Converting Scientific Notation to Standard Form

1. Scientific Notation

In many applications in mathematics, business, and science, it is necessary to work with very large or very small numbers. For example:

- The space shuttle travels approximately 17,000 mph.
- A piece of paper is 0.002 in. thick.

Scientific notation is a means by which we can write very large numbers and very small numbers without having to write numerous zeros in the number. To understand scientific notation, we first review the patterns associated with powers of 10.

$$\begin{aligned}10^4 &= 10 \cdot 10 \cdot 10 \cdot 10 = 10{,}000\\ 10^3 &= 10 \cdot 10 \cdot 10 = 1000\\ 10^2 &= 10 \cdot 10 = 100\\ 10^1 &= 10\\ 10^0 &= 1\\ 10^{-1} &= 0.1\\ 10^{-2} &= 0.01\\ 10^{-3} &= 0.001\\ 10^{-4} &= 0.0001\end{aligned}$$

TIP: For negative powers of 10, we have

$10^{-1} = \dfrac{1}{10^1} = 0.1$

$10^{-2} = \dfrac{1}{10^2} = \dfrac{1}{100} = 0.01$ and so on.

Therefore,

- The value 17,000 can be expressed as $1.7 \times 10{,}000 = 1.7 \times 10^4$.
- The value 0.002 can be expressed as $2.0 \times 0.001 = 2.0 \times 10^{-3}$.

The numbers 1.7×10^4 and 2.0×10^{-3} are written in scientific notation. Using **scientific notation** we express a number as the product of two factors. One factor is a number greater than or equal to 1, but less than 10. The other factor is a power of 10.

> **DEFINITION Scientific Notation**
>
> A positive number written in scientific notation is written as $a \times 10^n$, where a is a number greater than or equal to 1, but less than 10, and n is an integer.

2. Converting to Scientific Notation

To write a number in scientific notation, follow these guidelines.

Move the decimal point so that its new location is to the right of the first nonzero digit. Count the number of places that the decimal point is moved. Then

1. If the original number is *greater than or equal to 10:*

   The exponent for the power of 10 is *positive* and is equal to the number of places that the decimal point was moved.

   $7{,}500{,}000 = 7.5 \times 10^6$ — Move the decimal point 6 places to the *left*.

   $30{,}000 = 3 \times 10^4$ — Move the decimal point 4 places to the *left*.

2. If the original number is *between 0 and 1:*

   The exponent for the power of 10 is *negative*. Its absolute value is equal to the number of places the decimal point was moved.

   $0.000089 = 8.9 \times 10^{-5}$ — Move the decimal point 5 places to the *right*.

   $0.00000004 = 4 \times 10^{-8}$ — Move the decimal point 8 places to the *right*.

3. If the original number is *between 1 and 10*

   The exponent on 10 is 0.

   $2.5 = 2.5 \times 10^0$ *Note:* $10^0 = 1$.

   For this case, scientific notation is not needed.

Example 1 Writing Numbers in Scientific Notation

Write the number in scientific notation.

a. 93,000,000 mi (the distance between Earth and the Sun)

b. 0.000 000 000 753 kg (the mass of a dust particle)

c. 300,000,000 m/sec (the speed of light)

d. 0.00017 m (length of the smallest insect in the world)

Skill Practice

Write the numbers in scientific notation.

1. The amount of income brought in by the movie *Shrek the Third* in its first weekend was $121,630,000.
2. The time required for light to travel 1 mi is 0.00000534 sec.
3. During the peak of her career, Bette Midler earned $31,000,000 in a single year.
4. The diameter of a grain of sand is approximately 0.0025 in.

Solution:

a. 93,000,000 mi — The number is greater than 10. Move the decimal point left 7 places.

$= 9.3 \times 10^7$ mi — For a number greater than 10, the exponent is *positive*.

b. 0.000 000 000 753 kg — The number is between 0 and 1. Move the decimal point to the right 10 places.

$= 7.53 \times 10^{-10}$ kg — For a number between 0 and 1, the exponent is *negative*.

c. 300,000,000 m/sec — The number is greater than 10. Move the decimal point to the left 8 places.

$= 3 \times 10^8$ m/sec — For a number greater than 10, the exponent is *positive*.

d. 0.00017 m — The number is between 0 and 1. Move the decimal point to the right 4 places.

$= 1.7 \times 10^{-4}$ m — For a number between 0 and 1, the exponent is *negative*.

3. Converting Scientific Notation to Standard Form

To convert from scientific notation to standard form, follow these guidelines.

1. If the exponent on 10 is *positive*, move the decimal point to the right the same number of places as the exponent. Add zeros as necessary.
2. If the exponent on 10 is *negative*, move the decimal point to the left the same number of places as the exponent. Add zeros as necessary.

Skill Practice

Convert to standard form.

5. 2.79×10^{-8}
6. 8.603×10^5
7. 1×10^{-6}
8. 6×10^1

Example 2 Converting Scientific Notation to Standard Form

Convert to decimal notation.

a. 3.52×10^{-5} **b.** 4.6×10^4 **c.** 9×10^{-12} **d.** 1×10^{15}

Solution:

a. $3.52 \times 10^{-5} = 0.0000352$ — The exponent is negative. Move the decimal point to the left 5 places.

b. $4.6 \times 10^4 = 46{,}000$ — The exponent is positive. Move the decimal point to the right 4 places.

c. $9 \times 10^{-12} = 0.000\,000\,000\,009$ — The exponent is negative. Move the decimal point to the left 12 places.

d. $1 \times 10^{15} = 1{,}000{,}000{,}000{,}000{,}000$ — The exponent is positive. Move the decimal point to the right 15 places.

Answers

1. 1.2163×10^8 2. 5.34×10^{-6} sec
3. 3.1×10^7 4. 2.5×10^{-3} in.
5. 0.0000000279 6. 860,300
7. 0.000001 8. 60

Section 10.6 Practice Exercises

Study Skills Exercise

1. Define the key term, **scientific notation**.

Review Exercises

2. Apply the distributive property and combine like terms. $-4(3t - w) + 5(t + 2w)$

For Exercises 3–12, simplify the expression. Write the answers with positive exponents only.

3. $\dfrac{p^2p^9}{p^4}$

4. $\dfrac{z^{10}z^{20}}{z^5}$

5. $(x^6y^{-7})^{-2}$

6. $(c^{-5}d^9)^{-3}$

7. $5m^0$

8. $(5m)^0$

9. $\left(\dfrac{8x}{y^{12}}\right)^2$

10. $\left(\dfrac{6m^7}{p}\right)^2$

11. $(5xy + 9y^2 - 3x^2) - (-7x^2 - 8y^2 + 5xy)$

12. $(-8x^2 - 9x + 4) - (-3x^2 + 12x - 1)$

Objective 1: Scientific Notation

For Exercises 13–20, write each power of 10 in exponential notation.

13. 10,000

14. 100,000

15. 1000

16. 1,000,000

17. 0.001

18. 0.01

19. 0.0001

20. 0.1

For Exercises 21–28, identify which of the expressions are in correct scientific notation. (Answer yes or no.)

21. 43×10^3

22. 82×10^{-4}

23. 6.1×10^{-1}

24. 2.34×10^4

25. 2×10^{10}

26. 8×10^{-5}

27. 0.02×10^4

28. 0.052×10^{-3}

Objective 2: Converting to Scientific Notation

For Exercises 29–32, convert the number to scientific notation. **(See Example 1.)**

29. For a recent year the national debt was approximately \$7,455,000,000,000.

30. The gross national product is approximately \$13,247,000,000,000.

31. The diameter of an atom is 0.000 000 2 mm.

32. The length of a flea is 0.0625 in.

For Exercises 33–48, convert to scientific notation. **(See Example 1.)**

Animation

33. 20,000,000

34. 5000

35. 8,100,000

36. 62,000

37. 0.003

38. 0.0009

39. 0.025

40. 0.58

41. 142,000

42. 25,500,000

43. 0.0000491

44. 0.000116

45. 0.082

46. 0.15

47. 4920

48. 13,400

Objective 3: Converting Scientific Notation to Standard Form

For Exercises 49–64, convert to standard form. **(See Example 2.)**

49. 6×10^3 **50.** 3×10^4 **51.** 8×10^{-2} **52.** 2×10^{-5}

53. 4.4×10^{-1} **54.** 2.1×10^{-3} **55.** 3.7×10^4 **56.** 5.5×10^3

57. 3.26×10^2 **58.** 6.13×10^7 **59.** 1.29×10^{-2} **60.** 4.04×10^{-4}

61. 2.003×10^{-6} **62.** 5.02×10^{-5} **63.** 9.001×10^8 **64.** 7.07×10^6

Expanding Your Skills

For Exercises 65–72, write the mass of the planet in scientific notation if it is not already in scientific notation.

	Planet	Mass (kg)	Scientific Notation
65.	Mercury	0.33×10^{24}	
66.	Venus	4.87×10^{24}	
67.	Earth	5.98×10^{24}	
68.	Mars	0.64×10^{24}	
69.	Jupiter	1899×10^{24}	
70.	Saturn	586.5×10^{24}	
71.	Uranus	86.8×10^{24}	
72.	Neptune	102.4×10^{24}	

Group Activity

Evaluating and Interpreting a Polynomial Model

Estimated Time: 15 minutes

Group Size: 3

Polynomials can be used to model phenomena in the world around us. For example, suppose a softball player throws a ball from right field to third base. The height of the ball (in ft) can be computed by the polynomial $-16t^2 + 32t + 5$. That is:

$$\text{Height} = -16t^2 + 32t + 5$$

In this problem, the variable t is the time (in seconds) after the ball is released from the player's hand.

1. Evaluate the polynomial for the given values of t in the table. You can divide the workload by having each member of the group evaluate three or four of the values in the polynomial.

Time (sec)	Height (ft)
0	
0.2	
0.4	
0.6	
0.8	
1.0	
1.2	
1.4	
1.6	
1.8	
2.0	

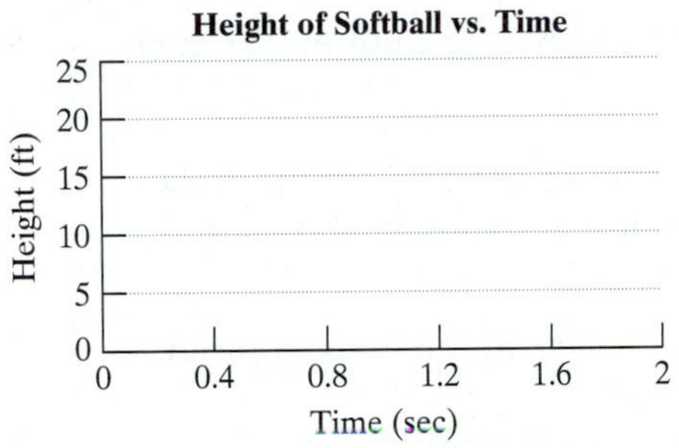

2. Plot the ordered pairs from the completed table. Plot time on the horizontal axis and height on the vertical axis.

3. From your results, how high will the ball be at the time of release?

4. How high will the ball be 0.6 sec after release?

5. Approximate the maximum height of the ball.

6. If the third baseman catches the ball 2 sec after the right-fielder threw the ball, how high was the ball when it was caught?

Chapter 10 Summary

Section 10.1 Addition and Subtraction of Polynomials

Key Concepts

A **term** is a number or a product or quotient of numbers and variables.

A term in which the variables appear only in the numerator with whole number exponents is called a **monomial**.

A **polynomial** is one or more monomials combined by addition or subtraction.

- If a polynomial has exactly one term, then it is called a **monomial**.
- If a polynomial has exactly two terms, then it is called a **binomial**.
- If a polynomial has exactly three terms, then it is called a **trinomial**.

A polynomial is written in **descending order** if the powers on the variable decrease from left to right. The **degree** of the polynomial is the greatest power to which the variable appears.

Adding Polynomials

To add polynomials, combine like terms.

Subtracting Polynomials

To subtract two polynomials, add the opposite of the second polynomial to the first polynomial.

Examples

Example 1

The term $5x^2y$ is a monomial.

The term $\frac{5x^2}{y}$ is not a monomial because the variable y appears in the denominator.

Example 2

	Descending Order	Degree
$-7x^2 + 5x^3$	$5x^3 - 7x^2$ binomial	3
$4y - 9y^2 + 5$	$-9y^2 + 4y + 5$ trinomial	2
$-16p$	$-16p$ monomial	1

Example 3

$$(-4x^2y + 5xy^2) + (6x^2y - 12xy^2)$$
$$= -4x^2y + 6x^2y + 5xy^2 - 12xy^2$$
$$= 2x^2y - 7xy^2$$

Example 4

$$(8m^2 - 4m + 3) - (2m^2 - 9m + 3)$$
$$= 8m^2 - 4m + 3 - 2m^2 + 9m - 3$$
$$= 8m^2 - 2m^2 - 4m + 9m + 3 - 3$$
$$= 6m^2 + 5m$$

Section 10.2 Multiplication Properties of Exponents

Key Concepts

Multiplication of Like Bases

Assume that $a \neq 0$ and m and n represent integers. Then, $a^m \cdot a^n = a^{m+n}$.

Multiplying Monomials

Use the commutative and associative properties of multiplication to regroup factors and multiply like bases.

Power Rule for Exponents

Assume that $a \neq 0$ and that m and n represent integers. Then, $(a^m)^n = a^{m \cdot n}$.

Power of a Product and Power of a Quotient

Assume that a and b are nonzero numbers. Let m represent an integer. Then,

$$(ab)^m = a^m b^m \quad \text{and} \quad \left(\frac{a}{b}\right)^m = \frac{a^m}{b^m}$$

Examples

Example 1

$x^8 x^7 x = x^{8+7+1} = x^{16}$

Example 2

$(-3y^5)(8y^7) = (-3 \cdot 8)(y^5 y^7) = -24y^{12}$

Example 3

$(m^4)^9 = m^{4 \cdot 9} = m^{36}$

Example 4

$(3x^4y)^2 = (3)^2(x^4)^2(y)^2 = 9x^8y^2$

$\left(\frac{5c}{d}\right)^3 = \frac{5^3c^3}{d^3}$ or $\frac{125c^3}{d^3}$

Section 10.3 Multiplication of Polynomials

Key Concepts

Multiplying a Monomial by a Polynomial

Apply the distributive property.

Multiplying a Polynomial by a Polynomial

Multiply each term in the first polynomial by each term in the second polynomial. Combine like terms if possible.

Examples

Example 1

$3y(7y^2 + 6y - 5)$

$= 3y(7y^2) + 3y(6y) + 3y(-5)$

$= 21y^3 + 18y^2 - 15y$

Example 2

$(2x + 3)(4x - 7)$

$= (2x)(4x) + (2x)(-7) + (3)(4x) + 3(-7)$

$= 8x^2 - 14x + 12x - 21$

$= 8x^2 - 2x - 21$

Section 10.4 Introduction to Factoring

Key Concepts

Greatest Common Factor (GCF)

The **GCF** is the product of prime factors (and repeated prime factors) common to each tern.

Factoring Out the GCF

Apply the distributive property "in reverse."

Examples

Example 1

Determine the GCF of $18y^3$ and $30y^5$.

$18y^3 = 2 \cdot 3 \cdot 3 \cdot y \cdot y \cdot y$

$30y^5 = 2 \cdot 3 \cdot 5 \cdot y \cdot y \cdot y \cdot y \cdot y$

The GCF is $2 \cdot 3 \cdot y \cdot y \cdot y = 6y^3$

Example 2

Factor out the GCF.

$25a^3b^2 + 10ab^3 + 5ab^2$ The GCF is $5ab^2$.

$= 5ab^2(5a^2) + 5ab^2(2b) + 5ab^2(1)$

$= 5ab^2(5a^2 + 2b + 1)$

Section 10.5 Negative Exponents and the Quotient Rule for Exponents

Key Concepts

Division of Like Bases

Assume that $a \neq 0$ and that m and n represent integers. Then,

$$\frac{a^m}{a^n} = a^{m-n}$$

Key Definitions

Let b be a nonzero number. Then, $\boldsymbol{b^0} = 1$

Let b be a nonzero number. Then,

$$\boldsymbol{b^{-n}} = \left(\frac{1}{b}\right)^n \quad \text{or} \quad \frac{1}{b^n}$$

Examples

Example 1

$$\frac{x^{13}}{x^4} = x^{13-4} = x^9$$

Example 2

a. $m^0 = 1$
b. $(4m)^0 = 1$
c. $4m^0 = 4(1) = 4$

Example 3

a. $d^{-6} = \left(\frac{1}{d}\right)^6 = \frac{1}{d^6}$

b. $\frac{x^{-4}y}{z^{-6}} = \frac{yz^6}{x^4}$

Section 10.6 Scientific Notation

Key Concepts

A positive number written in **scientific notation** is written as $a \times 10^n$, where a is a number greater than or equal to 1 and less then 10, and n is an integer.

Examples

Example 1

Write the numbers in scientific notation.

a. $130{,}000 = 1.3 \times 10^5$

b. $0.00943 = 9.43 \times 10^{-3}$

Example 2

Write the numbers in standard form.

a. $5.2 \times 10^7 = 52{,}000{,}000$

b. $4.1 \times 10^{-5} = 0.000041$

Chapter 10 Review Exercises

Section 10.1

For Exercises 1–4,

a. Write the polynomial in descending order.

b. Determine the degree.

c. Categorize the polynomial as a monomial, binomial, or trinomial.

1. $-13y^2 + 7y^4 - 5$

2. $-7x + 4x^2 - 2x^3$

3. $7.8 - 4.2a$

4. $-\frac{2}{3}$

For Exercises 5–6, combine like terms.

5. $-8.4z^3 - 4.1z^3 + 7.2z^3$

6. $5.8p^4 - 6.2p^4 - p^4$

For Exercises 7–10, add or subtract as indicated.

7. $(5x^2 - 3x) + (7x^2 - 4x)$

8. $(9w^4 + 2w^2) + (7w^4 - 3w^2)$

9. $(4x^2 + 8xy - y^2) - (2xy - 6x^2 + 2y^2)$

10. $(6ab - 9c^2 - 8a^2) - (-2ab + 6a^2 + 5c^2)$

11. Subtract $(-11p^2 - 4p + 8)$ from $(14p^3 - 2p^2 - 5)$.

12. Subtract $(-6m^3 - 14m - 3)$ from $(3m^3 + 2m^2 - 15)$.

13. The membership cost (in dollars) at a health club is given by $350 + 60x$, where x is the number of months for the membership.

a. Evaluate the polynomial for $x = 3$ and interpret the answer in the context of this problem.

b. Evaluate the polynomial for $x = 12$ and interpret the answer in the context of this problem.

14. Kris rents space at an art show to set up a lemonade stand. The cost (in dollars) for her to produce x cups of lemonade is given by the polynomial $100 + 0.45x$.

a. Evaluate the polynomial for $x = 300$ and interpret the answer in the context of this problem.

b. If Kris sells 300 lemonades for \$2 each, will she make a profit?

Section 10.2

For Exercises 15–30, simplify.

15. $a^3a^7a^2$
16. $8^2 8^4 8^{10}$
17. $(-3t)(7t)$
18. $(-5x)(-12x)$
19. $(-4y^2)(7y^3)$
20. $(-24a)(2a^6)$
21. $(8pq)(7p^2q^3)$
22. $(10c^2d)(-2cd)$
23. $(10^2)^3$
24. $(q^2)^{10}$
25. $\left(\frac{xy}{z}\right)^5$
26. $\left(\frac{pq}{n}\right)^4$
27. $(x^3)^4(x)^5$
28. $(z^3)^2(z)^{10}$
29. $(-4a^2b)^3$
30. $(3mn^6)^2$

Section 10.3

For Exercises 31–46, multiply.

31. $-5a(a^2 + 2a - 3)$
32. $-6t(t^3 + 4t + 4)$
33. $5cd(9c^2 + 2d)$
34. $3m^3n(m - 2mn^2)$
35. $(z + 9)(z + 3)$
36. $(p + 7)(p + 4)$
37. $(2y + 8)(3y - 4)$
38. $(7x + 3)(2x - 11)$
39. $(x - 4)(x^2 - 3x + 9)$
40. $(y - 5)(y^2 + 8y - 3)$
41. $(x - 1)(x^2 + x + 1)$
42. $(x + 6)(x^2 - 6x + 36)$
43. $(9z - 5)(9z + 5)$
44. $(10w + 4)(10w - 4)$
45. $(a + 2)^2$
46. $(z + 4)^2$

Section 10.4

For Exercises 47–50, determine the GCF.

47. 21, 28, and 35
48. 20, 28, and 40
49. $6p^3$ and $4p^5$
50. $18x^2$ and $9x^3$

For Exercises 51–58, factor out the GCF.

51. $18p + 9$
52. $14w + 2$
53. $3x^4 - 2x^3 + 5x^2$
54. $7y^3 + 6y^4 + 8y^5$
55. $81m^2 + 36m + 45$
56. $12n^2 - 48n + 36$
57. $6c^4d^2 - 3c^3d^3 + 9c^2d^4$
58. $8w^2z^3 - 4w^3z^2 + 12w^4z$

Section 10.5

For Exercises 59–82, simplify. Write the answers with positive exponents only.

59. $\frac{h^9}{h^2}$
60. $\frac{7^{10}}{7^4}$
61. $\frac{p^{15}p^2}{p}$
62. $\frac{q^7q^4}{q}$
63. $x + 14^0$
64. $w^0 + 4$
65. $(14x)^0$
66. $(12y)^0$
67. $14x^0$
68. $12y^0$
69. r^{-5}
70. z^{-8}
71. $\left(\frac{4}{3}\right)^{-2}$
72. $\left(\frac{3}{5}\right)^{-2}$
73. $\frac{1}{a^{-10}}$
74. $\frac{1}{4^{-2}}$
75. $\frac{x^2}{y^{-4}}$
76. $\frac{a^5}{b^{-7}}$
77. x^4x^{-10}
78. $y^{12}y^{-14}$

79. $\dfrac{w^3}{w^4}$

80. $\dfrac{z^5}{z^6}$

81. $(p^3)^{-6}$

82. $(q^{-4})^2$

Section 10.6

For Exercises 83–90, write the numbers in scientific notation.

83. The world wheat production is approximately 720,000,000 tons.

84. The number of cats owned in the United States is approximately 85,100,000.

85. The smallest spider species (*Patu marplesi*) resides in Western Samoa. Its average length is 0.017 in.

86. The probability of selecting 4 numbers from 20 numbers in no particular order is approximately 0.0002.

87. 9,456,000

88. 43,100

89. 0.0000456

90. 0.00367

For Exercises 91–98, write the numbers in standard notation.

91. The lightest gas is hydrogen at 5.612×10^{-3} lb/ft^3.

92. The most unstable isotope is lithium 5, which decays in 4.4×10^{-22} sec.

93. Metallic tungsten melts at a temperature of 1.055×10^4 °F.

94. The distance between the Earth and Moon is approximately 3.825×10^5 km.

95. 4.59×10^8

96. 3.589×10^2

97. 7.8×10^{-6}

98. 4.79×10^{-2}

Chapter 10 Test

1. Write the polynomial in descending order and determine the degree.

$-9y^2 + 3y - 7y^4 + 2$

2. Define the terms monomial, binomial, and trinomial.

3. Combine like terms. $-11m^3 - 6m^3 + 4m^3$

For Exercises 4–13, perform the indicated operations.

4. $(2x + 3) + (5x - 7)$

5. $(2x + 3)(5x - 7)$

6. $(2x + 3) - (5x - 7)$

7. $-4x^2(5x^2 - 3x + 7)$

8. $5a^2b^3(2ab - 4b^2 + 7a^2)$

9. $(-6x^3 - 4x^2 - 7x + 2) + (9x^3 + 10x)$

10. $(5x^2y + 15xy + 3y^2) + (-4x^2y - 3xy + x^2)$

11. $(3x + 6)(x^2 + 2x - 5)$

12. $(3x - 5)(3x + 5)$

13. $(8y + 3)^2$

For Exercises 14–29, simplify. Write the answers with positive exponents only.

14. $2^2 2^3 2^7$

15. $(x^3)^{10}$

16. $(6y^4)^2$

17. $\left(\dfrac{5a}{b^2}\right)^2$

18. $9z^0$

19. $(7d)^0 + 2$

20. k^{-4}

21. $\left(\frac{1}{5}\right)^{-3}$

22. $\dfrac{a^{-3}b^4}{c^{-10}}$

23. $\dfrac{x^{15}}{x^7}$

24. $\dfrac{x^7}{x^{15}}$

25. $y^{-9}y^3$

26. $(-2x^{-4}y)(3x^3y^6)$

27. $(x^3)^4(x^7)^2$

28. $(2a^2b^5)^2$

29. $(5x^{-4}y)^3$

For Exercises 30–31, determine the greatest common factor.

30. 12, 18, and 36

31. $4p^2$ and $12p^3$

For Exercises 32–35, factor out the greatest common factor.

32. $16m + 8$

33. $15p - 5$

34. $12x^7 - 14x^3 + 6x^4$

35. $2x^5y + 3x^3y^2 + 5x^2y^3$

For Exercises 36–37, write the number in scientific notation.

36. 42,500,000

37. 0.00004134

For Exercises 38–39, write the number in standard form.

38. 2.7×10^{-2}

39. 9.52×10^6

Chapters 1–10 Cumulative Review Exercises

1. Identify the place value of the underlined digit.

a. $\underline{4}$,976.215

b. 4,976.21$\underline{5}$

2. Write the word name for the number. 300,300.03

For Exercises 3–11, simplify the expression.

3. $60 - 2 \cdot 12 \div 6$

4. $-3^2 - 4^2$

5. $-5 - 3 - 4 - 1$

6. $(-5)(-3)(-4)(-1)$

7. $\dfrac{1}{3} \div \dfrac{5}{9}$

8. $\dfrac{4}{15} + \dfrac{1}{3} \div 10$

9. $4\dfrac{3}{10} - 2\dfrac{3}{5}$

10. $8\dfrac{7}{8} + 4\dfrac{1}{4}$

11. $6\dfrac{2}{3} - 10\dfrac{1}{3}$

For Exercises 12–17, solve the equation.

12. $-9 = -5 + 2x$

13. $7(x - 2) - 6x = 18$

14. $3(x - 5) + 2x = 7x - 9 + 2x$

15. $x + \dfrac{1}{3} = \dfrac{1}{2}$

16. $-\dfrac{3}{5}y = 12$

17. $4.7x - 2.1 = 40.2$

18. The following data give the ages of 12 students at a college graduation. Find the mean, median, and mode. Round to one decimal place if necessary.

22	24	24	28	29	29
52	32	29	40	29	21

19. Find the diameter, circumference, and area of the circle. Use 3.14 for π.

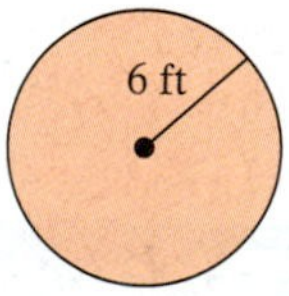

For Exercises 20–21, find the volume. Use 3.14 for π.

20.

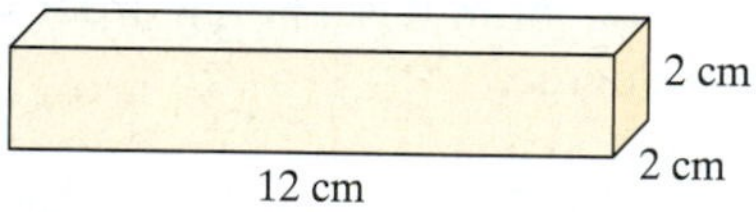

21.

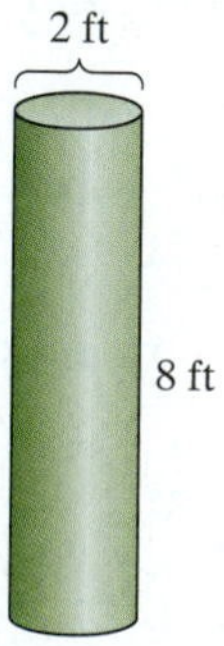

22. An angle measures 71°.

a. Find its complement.

b. Find its supplement.

23. Find the length of the missing side.

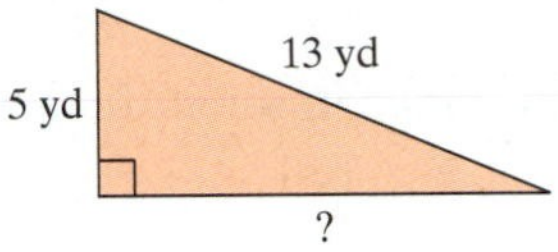

24. Find the measure of $\angle a$ and $\angle b$.

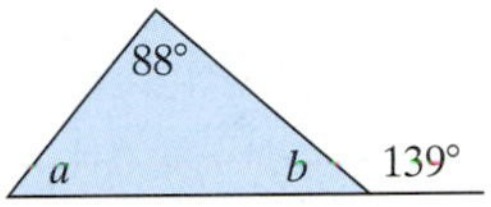

25. Determine if the numbers 6 and 15 are proportional to the numbers 10.8 and 27.

26. If 1,610,000 people travel through Atlanta's Hartsfield Airport in 1 week (7 days), how many people would travel through the airport in a 30-day month?

27. The triangles are similar. Find the missing value of x.

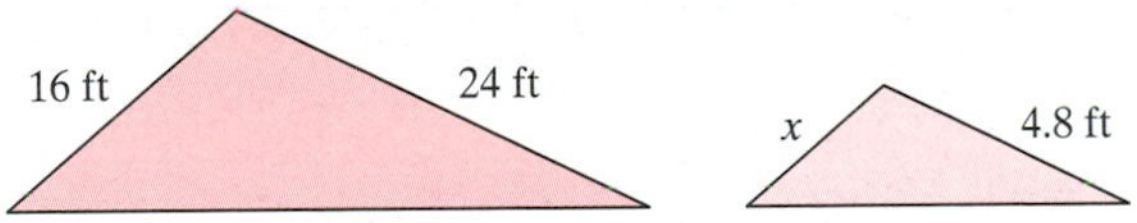

28. Julie's favorite dark chocolate comes in 12-oz packages for $3.89. Find the unit cost to the nearest tenth of a cent.

29. A bottle of fertilizer requires 2 fl oz for 2 qts of water. How much is needed for 5 gal of water?

For Exercises 30–33, convert the units of measurement.

30. 180 m = _____ km **31.** 5.2 g = _____ mg

32. 12 pt = _____ gal **33.** 27 ft = _____ yd

For Exercises 34–35, find the area and perimeter.

34.

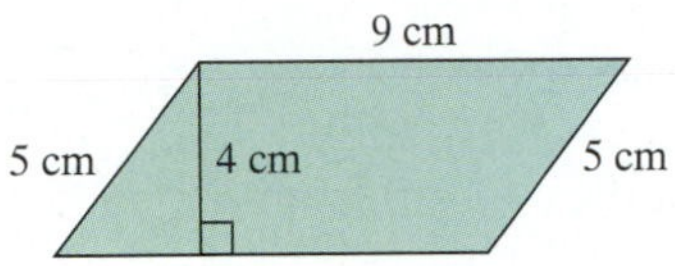

35.

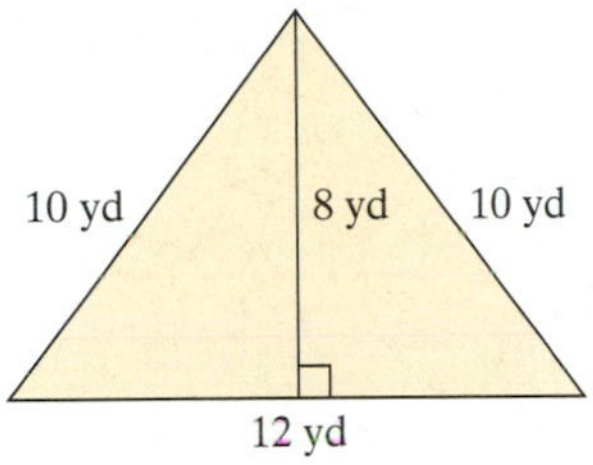

36. A realtor receives a 6% commission on a house that sold for $235,000. How much was the commission?

37. In a sample of college students, 384 reported that they watch less than 4 hr of television a week. If this represents 32% of the sample, how many students were surveyed?

38. The tax on a $44 pair of slacks was $2.20. What is the tax rate?

39. The table is a frequency distribution showing the gas mileage for a random sample of vehicles (cars, SUVs, and light trucks).

a. How many vehicles were in the survey?

b. What percent of vehicles got less than 15 mpg?

c. What percent got 25 mph or more?

Gas Mileage (mpg)	Frequency
10–14	6
15–19	12
20–24	11
25–34	8
35–39	2
40–44	1

For Exercises 40–43, perform the indicated operations.

40. $-4a^2b(2a^2 - 3ab + 7b^2)$

41. $(4a + 2)(4a - 2)$

42. $(x + 7)^2$

43. $(w - 4)(w^2 - 3w + 8)$

44. Subtract $(9x^2 - 3y^2)$ from $(2x^2 + 5xy + 6y^2)$

For Exercises 45–48, simplify the expressions. Write the answers with positive exponents only.

45. $\dfrac{w^9w^4}{w^6}$

46. $(a^{-5}b^2)^3$

47. $(x^2)^{10}(x^3)^{-6}$

48. $15k^0$

49. Write the number in scientific notation.
852,400

50. Write the number in standard form.
3.15×10^{-3}

Student Answer Appendix

Chapter 1

Chapter Opener Puzzle

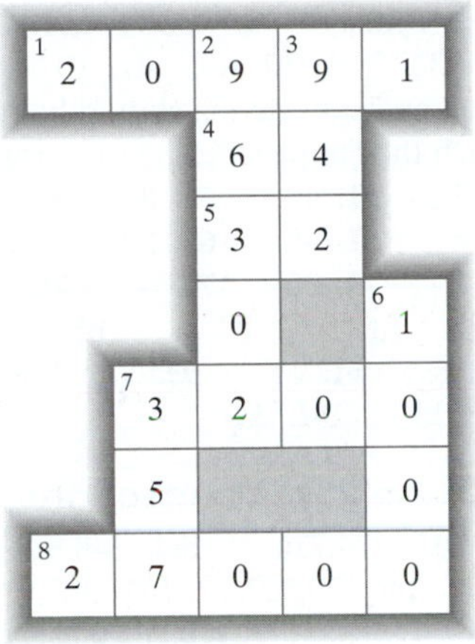

Section 1.2 Practice Exercises, pp. 9–12

3. 7: ones; 5: tens; 4: hundreds; 3: thousands; 1: ten-thousands; 2: hundred-thousands; 8: millions
5. Tens **7.** Ones **9.** Hundreds **11.** Thousands
13. Hundred-thousands **15.** Billions
17. Ten-thousands **19.** Millions **21.** Ten-millions
23. Billions **25.** 5 tens + 8 ones
27. 5 hundreds + 3 tens + 9 ones
29. 5 thousands + 2 hundreds + 3 ones
31. 1 ten-thousand + 2 hundreds + 4 tens + 1 one
33. 524 **35.** 150 **37.** 1906 **39.** 85,007
41. Ones, thousands, millions, billions
43. Two hundred forty-one **45.** Six hundred three
47. Thirty-one thousand, five hundred thirty
49. One hundred thousand, two hundred thirty-four
51. Nine thousand, five hundred thirty-five
53. Twenty thousand, three hundred twenty
55. Five hundred ninety thousand, seven hundred twelve
57. 6005 **59.** 672,000 **61.** 1,484,250
63.

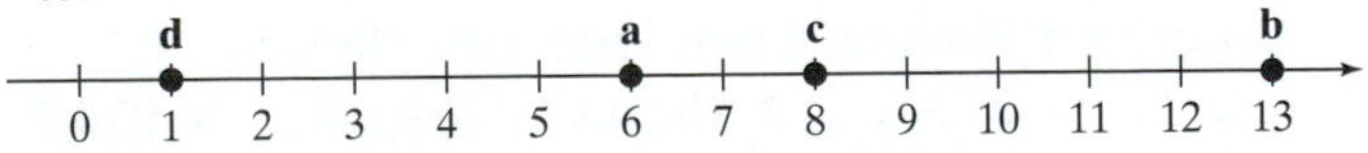

65. 10 **67.** 4
69. 8 is greater than 2, or 2 is less than 8
71. 3 is less than 7, or 7 is greater than 3
73. $<$ **75.** $>$ **77.** $<$ **79.** $>$ **81.** $<$ **83.** $<$
85. False **87.** 99 **89.** There is no greatest whole number. **91.** 7 **93.** 964

Section 1.3 Practice Exercises, pp. 22–26

3. 3 hundreds + 5 tens + 1 one
5. 4012
7.

+	0	1	2	3	4	5	6	7	8	9
0	0	1	2	3	4	5	6	7	8	9
1	1	2	3	4	5	6	7	8	9	10
2	2	3	4	5	6	7	8	9	10	11
3	3	4	5	6	7	8	9	10	11	12
4	4	5	6	7	8	9	10	11	12	13
5	5	6	7	8	9	10	11	12	13	14
6	6	7	8	9	10	11	12	13	14	15
7	7	8	9	10	11	12	13	14	15	16
8	8	9	10	11	12	13	14	15	16	17
9	9	10	11	12	13	14	15	16	17	18

9. Addends: 1, 13, 4; sum: 18 **11.** 75 **13.** 59
15. 997 **17.** 119 **19.** 121 **21.** 111
23. 889 **25.** 701 **27.** 203 **29.** 15,203
31. 40,985 **33.** 44 + 101 **35.** $y + x$
37. 23 + (9 + 10) **39.** $(r + s) + t$
41. The commutative property changes the order of the addends, and the associative property changes the grouping.
43. Minuend: 12; subtrahend: 8; difference: 4
45. 18 + 9 = 27 **47.** 27 + 75 = 102 **49.** 5
51. 3 **53.** 1126 **55.** 1103 **57.** 17 **59.** 521
61. 4764 **63.** 1,403 **65.** 2217 **67.** 713
69. 30,941 **71.** 5,662,119 **73.** The expression 7 − 4 means 7 minus 4, yielding a difference of 3. The expression 4 − 7 means 4 minus 7 which results in a difference of −3. (This is a mathematical skill we have not yet learned.)
75. 13 + 7; 20 **77.** 7 + 45; 52 **79.** 18 + 5; 23
81. 1523 + 90; 1613 **83.** 5 + 39 + 81; 125
85. 422 − 100; 322 **87.** 1090 − 72; 1018
89. 50 − 13; 37 **91.** 103 − 35; 68
93. 74,283,000 viewers **95.** 423 desks
97. Denali is 6074 ft higher than White Mountain Peak.
99. 7748 **101.** 195,489 **103.** 821,024 nonteachers
105. 4256 ft **107.** Jeannette will pay $29,560 for 1 year.
109. 104 cm **111.** 42 yd **113.** 288 ft **115.** 13 m

Section 1.3 Calculator Connections, p. 27

117. 192,780 **118.** 21,491,394 **119.** 5,257,179
120. 4,447,302 **121.** 897,058,513 **122.** 2,906,455
123. 49,408 mi^2 **124.** 17,139 mi^2 **125.** 96,572 mi^2
126. 224,368 mi^2

Section 1.4 Practice Exercises, pp. 32–34

3. 26 **5.** 5007 **7.** Ten-thousands
9. If the digit in the tens place is 0, 1, 2, 3, or 4, then change the tens and ones digits to 0. If the digit in the tens place is 5, 6, 7, 8, or 9, increase the digit in the hundreds place by 1 and change the tens and ones digits to 0.
11. 340 **13.** 730 **15.** 9400 **17.** 8500
19. 35,000 **21.** 3000 **23.** 10,000 **25.** 109,000

27. 490,000 **29.** $148,000,000 **31.** 239,000 mi **33.** 160 **35.** 220 **37.** 1000 **39.** 2100 **41.** $151,000,000 **43.** $11,000,000 more **45.** $10,000,000 **47. a.** 2003; $3,500,000 **b.** 2005; $2,000,000 **49.** Massachusetts; 79,000 students **51.** 71,000 students **53.** 10,000 mm **55.** 440 in.

Section 1.5 Practice Exercises, pp. 43–46

3. 1,010,000 **5.** 5400 **7.** 6×5; 30 **9.** 3×9; 27 **11.** Factors: 13, 42; product: 546 **13.** Factors: 3, 5, 2; product: 30 **15.** For example: 5×12; $5 \cdot 12$; $5(12)$ **17.** d **19.** e **21.** c **23.** $8 \cdot 14$ **25.** $(6 \cdot 2) \cdot 10$ **27.** $(5 \cdot 7) + (5 \cdot 4)$ **29.** 144 **31.** 52 **33.** 655 **35.** 1376 **37.** 11,280 **39.** 23,184 **41.** 378,126 **43.** 448 **45.** 1632 **47.** 864 **49.** 2431 **51.** 6631 **53.** 19,177 **55.** 186,702 **57.** 21,241,448 **59.** 4,047,804 **61.** 24,000 **63.** 2,100,000 **65.** 72,000,000 **67.** 36,000,000 **69.** 60,000,000 **71.** 2,400,000,000 **73.** $1000 **75.** $1,370,000 **77.** 4000 min **79.** $1665 **81.** 144 fl oz **83.** 287,500 sheets **85.** 372 mi **87.** 276 ft^2 **89.** 5329 cm^2 **91.** 105,300 mi^2 **93. a.** 2400 in.^2 **b.** 42 windows **c.** 100,800 in.^2 **95.** 128 ft^2

Section 1.6 Practice Exercises, pp. 54–56

3. 4944 **5.** 1253 **7.** 664,210 **9.** 902 **11.** 9; the dividend is 72; the divisor is 8; the quotient is 9 **13.** 8; the dividend is 64; the divisor is 8; the quotient is 8 **15.** 5; the dividend is 45; the divisor is 9; the quotient is 5 **17.** You cannot divide a number by zero (the quotient is undefined). If you divide zero by a number (other than zero), the quotient is always zero. **19.** 15 **21.** 0 **23.** Undefined **25.** 1 **27.** Undefined **29.** 0 **31.** $2 \cdot 3 = 6$, $2 \cdot 6 \neq 3$ **33.** Multiply the quotient and the divisor to get the dividend. **35.** 13 **37.** 41 **39.** 486 **41.** 409 **43.** 203 **45.** 822 **47.** Correct **49.** Incorrect; 253 R2 **51.** Correct **53.** Incorrect; 25 R3 **55.** 7 R5 **57.** 10 R2 **59.** 27 R1 **61.** 197 R2 **63.** 42 R4 **65.** 1557 R1 **67.** 751 R6 **69.** 835 R2 **71.** 479 R9 **73.** 43 R19 **75.** 308 **77.** 1259 R26 **79.** 22 **81.** 35 R1 **83.** 229 R96 **85.** 302 **87.** $497 \div 71$; 7 **89.** $877 \div 14$; 62 R9 **91.** $42 \div 6$; 7 **93.** 14 classrooms **95.** 5 cases; 8 cans left over **97.** There will be 120 classes of Beginning Algebra. **99.** It will use 9 gal. **101.** $1200 \div 20 = 60$; approximately 60 words per minute **103.** Yes, they can all attend if they sit in the second balcony.

Section 1.6 Calculator Connections, p. 57

105. 7,665,000,000 bbl **106.** 13,000 min **107.** $211 billion **108.** Each crate weighs 255 lb.

Chapter 1 Problem Recognition Exercises, p. 57

1. a. 120 **b.** 72 **c.** 2304 **d.** 4 **2. a.** 575 **b.** 525 **c.** 13,750 **d.** 22 **3. a.** 946 **b.** 612 **4. a.** 278 **b.** 612 **5. a.** 1201 **b.** 5500 **6. a.** 34,855 **b.** 22,718 **7. a.** 20,000 **b.** 400 **8. a.** 34,524 **b.** 548 **9. a.** 230 **b.** 5060 **10. a.** 15 **b.** 1875 **11. a.** 328 **b.** 4 **12. a.** 8 **b.** 547 **13. a.** 4180 **b.** 41,800 **c.** 418,000 **d.** 4,180,000 **14. a.** 35,000 **b.** 3500 **c.** 350 **d.** 35 **15.** 506 **16.** 230 R4 **17.** 22,761 **18.** 6561

Section 1.7 Practice Exercises, pp. 63–65

3. True **5.** False **7.** True **9.** 9^4 **11.** 3^6 **13.** $4^4 \cdot 2^3$ **15.** $8 \cdot 8 \cdot 8 \cdot 8$ **17.** $4 \cdot 4 \cdot 4 \cdot 4 \cdot 4 \cdot 4 \cdot 4 \cdot 4$ **19.** 8 **21.** 9 **23.** 27 **25.** 125 **27.** 32 **29.** 81 **31.** The number 1 raised to any power is 1. **33.** 1000 **35.** 100,000 **37.** 2 **39.** 6 **41.** 10 **43.** 0 **45.** No, addition and subtraction should be performed in the order in which they appear from left to right. **47.** 26 **49.** 1 **51.** 49 **53.** 3 **55.** 2 **57.** 53 **59.** 8 **61.** 45 **63.** 24 **65.** 4 **67.** 40 **69.** 5 **71.** 26 **73.** 4 **75.** 50 **77.** 2 **79.** 0 **81.** 5 **83.** 6 **85.** 3 **87.** 201 **89.** 6 **91.** 15 **93.** 32 **95.** 24 **97.** 75 **99.** 400 **101.** 3

Section 1.7 Calculator Connections, p. 65

102. 24,336 **103.** 174,724 **104.** 248,832 **105.** 1,500,625 **106.** 79,507 **107.** 357,911 **108.** 8028 **109.** 293,834 **110.** 66,049 **111.** 1728 **112.** 35 **113.** 43

Section 1.8 Practice Exercises, pp. 69–73

3. $71 + 14$; 85 **5.** $2 \cdot 14$; 28 **7.** $102 - 32$; 70 **9.** $10 \cdot 13$; 130 **11.** $24 \div 6$; 4 **13.** $5 + 13 + 25$; 43 **15.** Lucio's monthly payment was $85. **17.** The interstate will take 4 hr, and the back roads will take 5 hr. The interstate will take less time. **19.** Area **21.** It will cost $1650. **23.** The cost is $720. **25.** There will be $2676 left in Jose's account. **27.** The amount of money required is $924. **29. a.** Shevona's paycheck is worth $320. **b.** She will have $192 left. **31.** There will be 2 in. of matte between the pictures. **33. a.** The difference between the number of male and female doctors is 424,400. **b.** The total number of doctors is 836,200. **35. a.** The distance is 320 mi. **b.** 15 in. represents 600 mi. **37.** 354 containers will be filled completely with 9 eggs left over. **39. a.** Byron needs six $10 bills. **b.** He will receive $6 in change. **41.** The total amount paid was $2748. **43.** 109 **45.** 18 **47.** 78¢ per pound **49.** 121 mm per month

Chapter 1 Review Exercises, pp. 78–82

1. Ten-thousands **2.** Hundred-thousands **3.** 92,046 **4.** 503,160 **5.** 3 millions + 4 hundred-thousands + 8 hundreds + 2 tens **6.** 3 ten-thousands + 5 hundreds + 5 tens + 4 ones **7.** Two hundred forty-five **8.** Thirty thousand, eight hundred sixty-one **9.** 3602 **10.** 800,039

11. 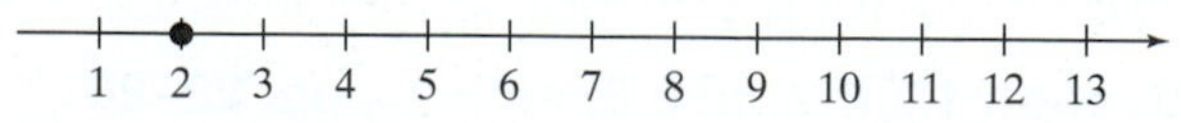

12. 

13. True **14.** False **15.** Addends: 105, 119; sum: 224

16. Addends: 53, 21; sum: 74 **17.** 71 **18.** 54 **19.** 17,410 **20.** 70,642 **21. a.** Commutative property **b.** Associative property **c.** Commutative property **22.** Minuend: 14; subtrahend: 8; difference: 6 **23.** Minuend: 102; subtrahend: 78; difference: 24 **24.** 26 **25.** 20 **26.** 121 **27.** 1090 **28.** 31,019 **29.** 34,188 **30.** 403 + 79; 482 **31.** 44 + 92; 136 **32.** 38 − 31; 7 **33.** 111 − 15; 96 **34.** 36 + 7; 43 **35.** 23 + 6; 29 **36.** 251 − 42; 209 **37.** 90 − 52; 38 **38. a.** 96 cars **b.** 66 Fords **39.** 45,797 thousand seniors **40.** 7709 thousand **41.** 45,096 thousand **42.** 71,893,000 tons **43.** $21,772,784 **44.** 177 m **45.** 5,000,000 **46.** 9,330,000 **47.** 800,000 **48.** 1500 **49.** 17,000,000 people **50.** 163,000 m^3 **51.** Factors: 33, 40; product: 1320 **52. a.** Yes **b.** Yes **c.** No **53.** c **54.** e **55.** d **56.** a **57.** b **58.** 6106 **59.** 52,224 **60.** 3,000,000 **61.** $429 **62.** 7714 lb **63.** 7; divisor: 6, dividend: 42, quotient: 7 **64.** 13; divisor: 4, dividend: 52, quotient: 13 **65.** 3 **66.** 1 **67.** Undefined **68.** 0 **69.** Multiply the quotient and the divisor to get the dividend. **70.** Multiply the whole number part of the quotient and the divisor, and then add the remainder to get the dividend. **71.** 58 **72.** 41 R7 **73.** 52 R3 **74.** $\frac{72}{4}$; 18 **75.** $9\overline{)108}$; 12 **76.** 26 photos with 1 left over **77. a.** 4 T-shirts **b.** 5 hats **78.** 8^5 **79.** $2^4 \cdot 5^3$ **80.** 125 **81.** 256 **82.** 1 **83.** 1,000,000 **84.** 8 **85.** 12 **86.** 7 **87.** 75 **88.** 90 **89.** 15 **90.** 12 **91.** 55 **92.** 3 **93.** 42 **94.** 0 **95.** 4 **96.** 100 **97. a.** 21 mi **b.** 840 mi **98.** He received $19,600,000 per year. **99. a.** She should purchase 48 plants. **b.** The plants will cost $144. **c.** The fence will cost $80. **d.** Aletha's total cost will be $224. **100.** 8 **101.** $89 **102.** 8 houses per month

Chapter 1 Test, pp. 82–83

1. a. Hundreds **b.** Thousands **c.** Millions **d.** Ten-thousands **2. a.** 4,065,000 **b.** Twenty-one million, three hundred twenty-five thousand **c.** Twelve million, two hundred eighty-seven thousand **d.** 729,000 **e.** Eleven million, four hundred ten thousand **3. a.** 14 > 6 **b.** 72 < 81 **4.** 129 **5.** 328 **6.** 113 **7.** 227 **8.** 2842 **9.** 447 **10.** 21 R9 **11.** 546 **12.** 8103 **13.** 20 **14.** 1,500,000,000 **15.** 336 **16.** 0 **17.** Undefined **18. a.** The associative property of multiplication; the expression shows a change in grouping. **b.** The commutative property of multiplication; the expression shows a change in the order of the factors. **19. a.** 4900 **b.** 12,000 **c.** 8,000,000 **20.** There were approximately 1,430,000 people. **21.** 4 **22.** 24 **23.** 48 **24.** 33 **25.** 57 **26.** 9 **27.** Jennifer has a higher average of 29. Brittany has an average of 28. **28. a.** 23,418 thousand subscribers **b.** The largest increase was between the years 2004 and 2005. **29.** The North Side Fire Department is the busiest with five calls per week. **30.** 156 mm **31.** Perimeter: 350 ft; area: 6016 ft^2 **32.** 4,560,000 m^2

Chapter 2

Chapter Opener Puzzle

1	$^{A}5$	2	$^{B}3$	6	4
$^{C}6$	3	$^{D}4$	2	5	1
$^{E}4$	2	3	5	$^{F}1$	6
5	$^{G}1$	6	4	3	$^{H}2$
3	4	1	$^{I}6$	2	$^{J}5$
2	$^{K}6$	$^{L}5$	1	$^{M}4$	3

Section 2.1 Practice Exercises, pp. 89–92

3. −86 m **5.** $3800 **7.** −$500 **9.** −14 lb **11.** 1,400,000 **13.**

15. −2 **17.** > **19.** > **21.** < **23.** < **25.** 2 **27.** 2 **29.** 427 **31.** 100,000 **33. a.** −8 **b.** $|-12|$ **35. a.** 7 **b.** $|7|$ **37.** $|-5|$ **39.** Neither, they are equal. **41.** −5 **43.** 12 **45.** 0 **47.** 1 **49.** 15 **51.** −15 **53.** −15 **55.** 15 **57.** 36 **59.** −107 **61. a.** 6 **b.** 6 **c.** −6 **d.** 6 **e.** −6 **63.** −8 **b.** 8 **c.** −8 **d.** 8 **e.** 8 **65.** −6 **67.** −(−2) **69.** $|7|$ **71.** $|-3|$ **73.** $-|14|$ **75.** = **77.** > **79.** > **81.** < **83.** < **85.** 25°F **87.** −22°F **89.** −2°F **91.** 44°F **93.** September **95.** −60, $-|-46|$, $|-12|$, $-(-24)$, 5^2 **97.** Positive **99.** Negative

Section 2.2 Practice Exercises, pp. 97–99

3. > **5.** = **7.** < **9.** 2 **11.** −2 **13.** −8 **15.** 6 **17.** −7 **19.** −4 **21.** To add two numbers with the same sign, add their absolute values and apply the common sign. **23.** 15 **25.** −16 **27.** −124 **29.** 89 **31.** −3 **33.** 5 **35.** −24 **37.** 45 **39.** 0 **41.** 0 **43.** 9 **45.** −26 **47.** −41 **49.** −150 **51.** −17 **53.** −41 **55.** 2 **57.** −30 **59.** 10 **61.** −8 **63.** 0 **65.** −12 **67.** −23 + 49; 26 **69.** 3 + (−10) + 5; −2 **71.** (−8 + 6) + (−5); −7 **73.** 7 in.; Marquette had above average snowfall. **75.** −8 **77.** 8°F **79.** $333 **81.** $600 **83.** 0 **85.** For example: −12 + 2 **87.** For example: −1 + (−1)

Section 2.2 Calculator Connections, p. 99

89. −120 **90.** −566 **91.** −28,414 **92.** −24,325 **93.** 711 **94.** 339

Section 2.3 Practice Exercises, pp. 102–105

3. −47 **5.** −26 **7.** −8 **9.** 2 + (−9); −7 **11.** 4 + 3; 7 **13.** −3 + (−15); −18 **15.** −11 + 13; 2

17. 52 **19.** −33 **21.** −12 **23.** 8 **25.** 0 **27.** 161 **29.** −34 **31.** −22 **33.** −26 **35.** −1 **37.** 32 **39.** −15 **41.** 0 **43.** −1 **45.** 52 **47.** minus, difference, decreased, less than, subtract from **49.** 14 − 23; −9 **51.** 105 − 110; −5 **53.** 320 − (−20); 340 **55.** 5 − 12; −7 **57.** −34 − 21; −55 **59.** −35 − 24; −59 **61.** 6423°F **63.** His balance is −$375. **65.** The balance is $18,084 **67.** 164 **69.** −112 **71.** The range is 3° − (−8°) = 11°. **73.** For example: 4 − 10 **75.** −11, −15, −19 **77.** Positive **79.** Positive **81.** Negative **83.** Negative

Section 2.3 Calculator Connections, p. 105

85. −413 **86.** −433 **87.** −14,623 **88.** 19,906 **89.** 916,450 **90.** 129,777 **91.** 64,827 ft **92.** 4478 m

Section 2.4 Practice Exercises, pp. 110–113

3. 19 **5.** −44 **7.** 17 **9.** −15 **11.** 40 **13.** −21 **15.** 48 **17.** −45 **19.** −72 **21.** 0 **23.** 95 **25.** −3(−1); 3 **27.** −5 · 3; −15 **29.** 3(−5); −15 **31.** 400 **33.** −88 **35.** 0 **37.** 1 **39.** 32 **41.** −100 **43.** 100 **45.** −1000 **47.** −1000 **49.** −625 **51.** 625 **53.** 1 **55.** −1 **57.** −20 **59.** 7 **61.** −3 **63.** 21 **65.** Undefined **67.** 0 **69.** 4 **71.** −34 **73.** −100 ÷ 20; −5 **75.** −64 ÷ (−32); 2 **77.** −52 ÷ 13; −4 **79.** −6 ft/min **81.** −13°F **83.** −$235 **85.** −18 ft **87.** −108 **89.** −3 **91.** 108 **93.** 15 **95.** 0 **97.** Undefined **99.** 40 **101.** 49 **103. a.** 7 **b.** −7 **105.** +1 **107.** −2 **109.** Negative **111.** Negative **113.** Positive

Section 2.4 Calculator Connections, p. 113

115. −359,723 **116.** 594,125 **117.** 54 **118.** −629

Chapter 2 Problem Recognition Exercises, p. 114

1. a. 48 **b.** −22 **c.** −26 **d.** 12 **2. a.** −36 **b.** 15 **c.** 9 **d.** −4 **3.** −5 + (−3); −8 **4.** 9(−5); −45 **5.** −3 − (−7); 4 **6.** $\frac{28}{-4}$; −7 **7.** −23(−2); 46 **8.** −4 − 18; −22 **9.** $\frac{42}{-2}$; −21 **10.** −18 + (−13); −31 **11.** 10 − (−12); 22 **12.** $\frac{-21}{-7}$; 3 **13.** −6(−9); 54 **14.** −7 + 4 + 8 + (−16) + (−5); −16 **15. a.** 20 **b.** −75 **c.** 10 **d.** −3 **16. a.** 72 **b.** −34 **c.** 18 **d.** −38 **17. a.** −80 **b.** 80 **c.** −80 **d.** 80 **18. a.** −16 **b.** −90 **19. a.** −5 **b.** 25 **20. a.** 24 **b.** 24 **c.** −24 **d.** −24 **21. a.** 50 **b.** 50 **c.** 50 **d.** −50 **22.** 0 **23.** Undefined **24.** 90 **25.** −593 **26.** −30 **27.** 40 **28.** 0 **29.** 0 **30.** 81 **31.** −32 **32.** −81 **33.** −32 **34.** Undefined **35.** 0 **36.** −10,149 **37.** −85,048

Section 2.5 Practice Exercises, pp. 119–121

3. 25 **5.** 400 **7.** 144 **9.** −17 **11.** 120 **13.** 1 **15.** 16 **17.** −44 **19.** −150 **21.** 56 **23.** 110 **25.** −3 **27.** −4 **29.** 16 **31.** 13 **33.** 11 **35.** 2 **37.** −5 **39.** 8 **41.** −16 **43.** 6 **45.** −1 **47.** 1 **49.** 14 **51.** −5 **53.** 38 **55.** $15x$ **57.** $t + 4$ **59.** $v - 6$ **61.** $2g$ **63.** $-12n$ **65.** $-9 - x$ **67.** $\frac{t}{-2}$ **69.** $y + (-14)$ **71.** $2(c + d)$ **73.** $x - (-8)$ **75.** −37 **77.** 17 **79.** 7 **81.** −48 **83.** 9 **85.** −4 **87.** 9 **89.** −9 **91.** −52 **93.** −5 **95.** −2° **97.** −3

Chapter 2 Review Exercises, pp. 125–127

1. −4250 ft **2.** −$3,000,000 **3–6.**

7. Opposite: 4; absolute value: 4 **8.** Opposite: −6; absolute value: 6 **9.** 3 **10.** 1000 **11.** 74 **12.** 0 **13.** 9 **14.** 28 **15.** −20 **16.** −45 **17.** < **18.** < **19.** > **20.** = **21.** 4 **22.** 3 **23.** −5 **24.** −3 **25.** See page 93. **26.** See page 94. **27.** 13 **28.** −15 **29.** −70 **30.** −140 **31.** 3 **32.** −15 **33.** 23 + (−35); −12 **34.** 57 + (−10); 47 **35.** −5 + (−13) + 20; 2 **36.** −42 + 12; −30 **37.** −12 + 3; −9 **38.** −89 + (−22); −111 **39.** −2 in.; Caribou had below average snowfall. **40.** −5 **41.** See page 100. **42.** 27 **43.** −25 **44.** 22 **45.** −419 **46.** −8 **47.** 40 **48.** −17 **49.** 100 **50. a.** 8 − 10; −2 **b.** 10 − 8; 2 **51.** For example: 14 subtracted from −2. **52.** For example: Subtract −7 from −25. **53.** The temperature rose 5°F. **54.** Sam's new balance is $92. **55.** The average is 1 above par. **56.** 3450 ft **57.** −18 **58.** −3 **59.** 15 **60.** 56 **61.** −4 **62.** −12 **63.** 96 **64.** −32 **65.** Undefined **66.** 0 **67.** −125 **68.** −125 **69.** 36 **70.** −36 **71.** 1 **72.** −1 **73.** Negative **74.** Positive **75.** −45 ÷ (−15); 3 **76.** −4 · 19; −76 **77.** −3°F **78.** −$90 **79.** 38 **80.** 57 **81.** −11 **82.** 8 **83.** −7 **84.** −2 **85.** −2 **86.** 3 **87.** 17 **88.** −13 **89.** $a + 8$ **90.** $3n$ **91.** $-5x$ **92.** $p - 12$ **93.** $(a + b) + 2$ **94.** $\frac{w}{4}$ **95.** $y - (-8)$ **96.** $2(5 + z)$ **97.** −23 **98.** −55 **99.** −18 **100.** −30 **101.** −2 **102.** −5 **103.** −10 **104.** 5

Chapter 2 Test, pp. 128–129

1. −$220 **2.** 26 **3.** < **4.** > **5.** < **6.** = **7.** < **8.** < **9.** 10 **10.** 10 **11.** −5 **12.** −28 **13.** 9 **14.** −41 **15.** 6 **16.** −23 **17.** −72 **18.** 88 **19.** 2 **20.** −18 **21.** Undefined **22.** 0 **23.** −3(−7); 21 **24.** −13 + 8; −5 **25.** 18 −(−4); 22 **26.** 6 ÷ (−2); −3 **27.** −8 + 5; −3 **28.** −3 + 15 + (−6) + (−1); 5 **29.** −7 in.; Atlanta had below average rainfall. **30.** −7°F **31. a.** 64 **b.** −64 **c.** −64 **d.** −64 **32.** 3 **33.** −60 **34.** 0 **35.** −19 **36.** −55 **37.** 3 **38.** $18m$ **39.** −15 **40.** 12

Chapters 1–2 Cumulative Review Exercises, p. 129

1. Ten-thousands place **2.** One hundred thirty is less than two hundred forty-four. **3.** 1200 ft **4.** 114 **5.** 42 **6.** −196 **7.** 766 **8.** −248 **9.** 193

10. 105 R2 **11.** 93,252 **12.** 0 **13.** Undefined **14.** 140 m^2 **15. a.** −4 **b.** 4 **c.** −16 **d.** 16 **16.** 18 **17.** 21 **18.** 5 **19. a.** Torie can take the herb for 40 days if she takes 3 a day. **b.** Torie can take the herb for 60 days if she takes 2 a day. **20.** −4°F

Chapter 3

Chapter Opener Puzzle

1. Yes **2.** No **3.** Yes **4.** No **5.** Yes **6.** Yes **7.** No **8.** Yes

Section 3.1 Practice Exercises, pp. 138–140

3. −16 **5.** 6 **7.** −144 **9.** $2a$: variable term; $5b^2$: variable term; 6: constant term **11.** 8: constant term; $9a$: variable term **13.** 6, −4 **15.** −1, −12 **17.** *Like* terms **19.** Unlike terms; different variables **21.** Unlike terms; different powers of y **23.** Unlike terms; different variables **25.** $w + 5$ **27.** $2r$ **29.** $-st$ **31.** $7 - p$ **33.** $(3 + 8) + t$; $11 + t$ **35.** $(-2 \cdot 6)b$; $-12b$ **37.** $(3 \cdot 6)x$; $18x$ **39.** $[-9 + (-12)] + h$; $-21 + h$ **41.** $4x + 32$ **43.** $4a + 16b - 4c$ **45.** $-2p - 8$ **47.** $-3x - 9 + 5y$ **49.** $-12 + 4n^2$ **51.** $15q + 6s + 9t$ **53.** $12x$ **55.** $12 + 6x$ **57.** $-4 - p$ **59.** $-32 + 8p$ **61.** $14r$ **63.** $7h$ **65.** $-2a^2b$ **67.** $6x - 15y + 9$ **69.** $-3k - 4$ **71.** $4uv + 6u$ **73.** $5t - 28$ **75.** $-6x - 16$ **77.** $6y - 14$ **79.** $7q$ **81.** $-2n - 4$ **83.** $4x + 23$ **85.** $2z - 11$ **87.** $-w - 4y + 9$ **89.** $8a - 9b$ **91.** $-5m + 6n - 10$ **93.** $12z^2 + 7$ **95. a.** 6 **b.** 6 **97. a.** 5 **b.** 5 **99. a.** −45 **b.** −45 **101. a.** 36 **b.** 36

Section 3.2 Practice Exercises, pp. 145–146

3. $-13a + 16b$ **5.** $8h - 2k + 13$ **7.** $-3z + 4$ **9.** Yes **11.** No **13.** Yes **15.** No **17.** Equation **19.** Expression **21.** Equation **23.** 0 **25.** 7 **27.** 3 **29.** 37 **31.** 16 **33.** −15 **35.** 16 **37.** −78 **39.** 34 **41.** 52 **43.** 0 **45.** 100 **47.** −28 **49.** 3 **51.** −46 **53.** 61 **55.** −55 **57.** 42 **59.** −1 **61.** −6 **63.** 12 **65.** −5 **67.** −1 **69.** 11 **71.** −1 **73.** 10 **75.** 3 **77.** 28

Section 3.3 Practice Exercises, pp. 151–152

3. $6x + 4y$ **5.** $2m - 10n$ **7.** 45 **9.** −36 **11.** 3 **13.** −7 **15.** 2 **17.** 13 **19.** −5 **21.** −21 **23.** 30 **25.** −28 **27.** 4 **29.** 0 **31.** 0 **33.** 6 **35.** If the operation between the variable and a number is subtraction, use the addition property to isolate the variable. **37.** If the operation between the variable and a number is multiplication, use the division property to isolate the variable. **39.** −16 **41.** −3 **43.** −8 **45.** −48 **47.** 2 **49.** 30 **51.** −15 **53.** −57 **55.** 0 **57.** 20 **59.** −31 **61.** 5 **63.** 13 **65.** 3 **67.** −2 **69.** 1 **71.** −5 **73.** −1

Section 3.4 Practice Exercises, pp. 156–157

3. −12 **5.** 1 **7.** −19 **9.** 0 **11.** 6 **13.** 6 **15.** 5 **17.** 2 **19.** 9 **21.** −12 **23.** −60 **25.** −21 **27.** −3 **29.** 2 **31.** 1 **33.** 4 **35.** −3 **37.** −4 **39.** −13 **41.** 3 **43.** −12 **45.** −2 **47.** 0 **49.** 15 **51.** 6 **53.** 1 **55.** 5

Chapter 3 Problem Recognition Exercises, pp. 157–158

1. Equation **2.** Expression **3.** Expression **4.** Equation **5.** Equation **6.** Expression **7.** 4 **8.** 6 **9.** $-15t$ **10.** $-30x$ **11.** $5w - 15$ **12.** $6x - 12$ **13.** 7 **14.** 8 **15.** 15 **16.** 30 **17.** $t + 25$ **18.** $x + 42$ **19.** −1 **20.** 9 **21.** 2 **22.** 3 **23.** $11u + 74$ **24.** $61w + 129$ **25.** $-3x + 26$ **26.** $-7x + 21$ **27.** 14 **28.** −7 **29.** 8 **30.** −34

Section 3.5 Practice Exercises, pp. 164–167

3. −3 **5.** −45 **7.** 6 **9. a.** $x + 6 = 19$ **b.** The number is 13. **11. a.** $x - 14 = 20$ **b.** The number is 34. **13. a.** $\frac{x}{3} = -8$ **b.** The number is −24. **15. a.** $-6x = -60$ **b.** The number is 10. **17. a.** $-2 - x = -14$ **b.** The number is 12. **19. a.** $13 + x = -100$ **b.** The number is −113. **21. a.** $60 = -5x$ **b.** The number is −12. **23. a.** $3x + 9 = 15$ **b.** The number is 2. **25. a.** $5x - 12 = -27$ **b.** The number is −3. **27. a.** $\frac{x}{4} - 5 = -12$ **b.** The number is −28. **29. a.** $8 - 3x = 5$ **b.** The number is 1. **31. a.** $3(x + 4) = -24$ **b.** The number is −12. **33. a.** $-4(3 - x) = -20$ **b.** The number is −2. **35. a.** $-12x = x + 26$ **b.** The number is −2. **37. a.** $10(x + 5) = 80$ **b.** The number is 3. **39. a.** $3x = 2x - 10$ **b.** The number is −10. **41.** $5x$ **43.** $6n$ **45.** $A - 30$ **47.** $p + 1481$ **49.** $5r$ **51.** $2c$ **53.** The pieces should be 2 ft and 6 ft. **55.** The distance between Minneapolis and Madison is 240 mi. **57.** The Beatles had 19 and Elvis had 10. **59.** The Giants scored 17 points and the Patriots scored 14 points. **61.** Charlene's rent is $650 a month and the security deposit is $300. **63.** Stefan worked 6 hr of overtime.

Chapter 3 Review Exercises, pp. 172–174

1. $3a^2$ is a variable term with coefficient 3; $-5a$ is a variable term with coefficient −5; 12 is a constant term with coefficient 12. **2.** $-6xy$ is a variable term with coefficient −6; $-y$ is a variable term with coefficient −1; $2x$ is a variable term with coefficient 2. **3.** Unlike terms **4.** *Like* terms **5.** *Like* terms **6.** Unlike terms **7.** $-5 + t$ **8.** $3h$ **9.** $(-4 \cdot 2)p$; $-8p$ **10.** $m + (10 - 12)$; $m - 2$ **11.** $6b + 15$ **12.** $20x + 30y - 15z$ **13.** $-4c + 6d$ **14.** $4k - 8w + 12$ **15.** $2x$ **16.** $-9y$ **17.** $6x + 4y + 10$ **18.** $7a + 9b + 9$ **19.** $-2x + 26$ **20.** $-2z - 2$ **21.** $-u - 17v$ **22.** $p - 16$ **23.** −3 is a solution. **24.** −3 is not a solution. **25.** −35 **26.** −12 **27.** 14 **28.** 18 **29.** 13 **30.** 15 **31.** 2 **32.** −14 **33.** −7 **34.** 4 **35.** 26 **36.** 35 **37.** 6 **38.** −48 **39.** −1 **40.** 3 **41.** 8 **42.** −12 **43.** 32 **44.** 5 **45.** −28 **46.** −7 **47.** −4 **48.** 2 **49.** −1 **50.** 2 **51.** −3 **52.** −6 **53.** $x - 4 = 13$; The number is 17. **54.** $\frac{x}{-7} = -6$; The number is 42.

55. $-4x + 3 = -17$; The number is 5. **56.** $3x - 7 = -22$; The number is -5. **57.** $2(x + 10) = 16$; The number is -2. **58.** $3(4 - x) = -9$; The number is 7. **59.** $9n$ **60.** $4x$ **61.** $x + 2$ **62.** $h + 6$ **63.** Monique drove 120 mi, and Michael drove 360 mi. **64.** Angela ate 4 pieces and Joel ate 8 pieces. **65.** Tom Hanks starred in 35 films and Tom Cruise starred in 30 films. **66.** Raul took 16 hours in the fall and 12 hours in the spring.

Chapter 3 Test, pp. 174–175

1. d **2.** a **3.** c **4.** e **5.** b **6.** $-7x$ **7.** $5a + b$ **8.** $4a + 24$ **9.** $-13b + 8$ **10.** $12y + 39$ **11.** $-5w - 3$ **12.** An expression is a collection of terms. An equation has an equal sign that indicates that two expressions are equal. **13.** Expression **14.** Equation **15.** Equation **16.** Expression **17.** 21 **18.** 18 **19.** -3 **20.** 7 **21.** -7 **22.** -10 **23.** -21 **24.** 32 **25.** -2 **26.** 18 **27.** -72 **28.** -2 **29.** 7 **30.** -3 **31.** -84 **32.** -14 **33.** 4 **34.** 3 **35.** 0 **36.** 2 **37.** The number is -5. **38.** The number is -3. **39.** $15m$ **40.** $a + 5$ **41.** Phil makes \$252 and Monica makes \$504. **42.** The computer costs \$570 and the monitor costs \$329.

Chapters 1–3 Cumulative Review Exercises, pp. 175–176

1. a. hundreds **b.** ten-thousands **c.** hundred-thousands **2.** 46,000 **3.** 1,290,000 **4.** 25,400 **5.** Dividend is 39,190; divisor is 46; quotient 851; remainder is 44. **6.** Sarah makes \$20 per room. **7.** 10 **8.** 365 ft **9.** 20 **10.** -5 **11.** -25 **12.** -9 **13.** Undefined **14.** -12 **15.** $-5x + 29$ **16.** $-2y - 18$ **17.** -4 **18.** -8 **19.** 4 **20.** Kanye received 10 nominations and Alicia Keys received 8 nominations.

Chapter 4

Chapter Opener Puzzle

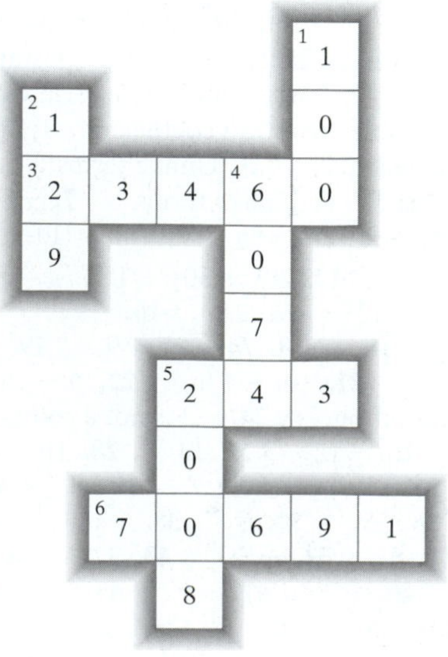

Section 4.1 Practice Exercises, pp. 184–188

3. Numerator: 8; denominator: 9 **5.** Numerator: $7p$; denominator: $9q$ **7.** $\frac{5}{9}$ **9.** $\frac{3}{8}$ **11.** $\frac{4}{7}$ **13.** $\frac{1}{8}$ **15.** $\frac{41}{103}$ **17.** $\frac{10}{21}$ **19.** -13 **21.** 1 **23.** 0 **25.** Undefined **27.** $\frac{9}{10}$ **29.** b, c **31.** a, b, c **33.** Proper **35.** Improper **37.** Improper **39.** $\frac{5}{2}$ **41.** $\frac{12}{4}$ **43.** $\frac{7}{4}$; $1\frac{3}{4}$ **45.** $\frac{13}{8}$ in.; $1\frac{5}{8}$ in. **47.** $\frac{7}{4}$ **49.** $-\frac{38}{9}$ **51.** $-\frac{24}{7}$ **53.** $\frac{27}{4}$ **55.** $\frac{137}{12}$ **57.** $-\frac{171}{8}$ **59.** 30 **61.** 19 **63.** 7 **65.** $4\frac{5}{8}$ **67.** $-7\frac{4}{5}$ **69.** $-2\frac{7}{10}$ **71.** $5\frac{7}{9}$ **73.** $12\frac{1}{11}$ **75.** $-3\frac{5}{6}$ **77.** $44\frac{1}{7}$ **79.** $1056\frac{1}{5}$ **81.** $810\frac{3}{11}$ **83.** $12\frac{7}{15}$

85. 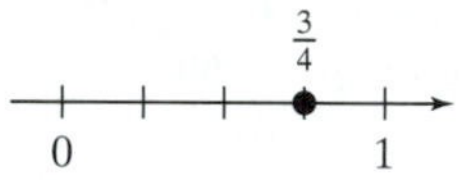

87. 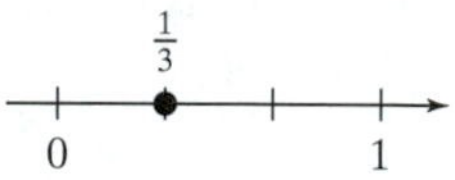

89. 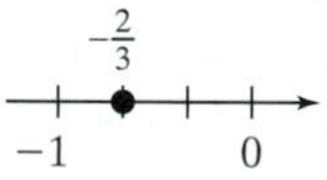

91. 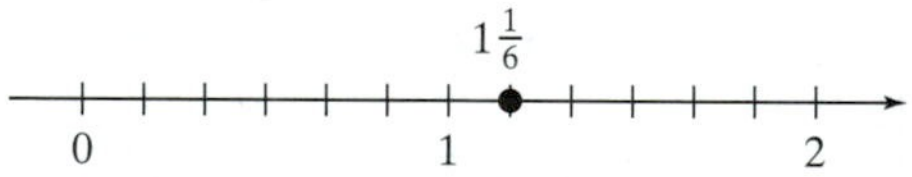

93. 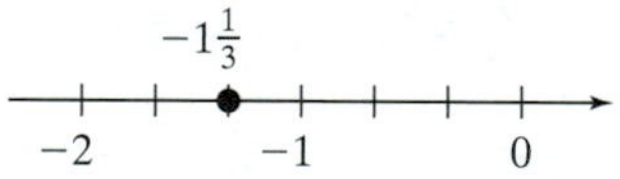

95. $\frac{3}{4}$ **97.** $\frac{1}{10}$ **99.** $-\frac{7}{3}$ **101.** $\frac{7}{3}$ **103.** False **105.** True

Section 4.2 Practice Exercises, pp. 196–200

3. $\frac{5}{2}$; $\frac{1}{2}$ **5.** $4\frac{3}{5}$ **7.** For example: $2 \cdot 4$ and $1 \cdot 8$ **9.** For example: $4 \cdot 6$ and $2 \cdot 2 \cdot 2 \cdot 3$ **11.** A whole number is divisible by 2 if it is an even number. **13.** A whole number is divisible by 3 if the sum of its digits is divisible by 3. **15. a.** No **b.** Yes **c.** Yes **d.** No **17. a.** No **b.** No **c.** No **d.** No **19. a.** Yes **b.** Yes **c.** No **d.** No **21. a.** Yes **b.** No **c.** Yes **d.** Yes **23.** Yes **25.** Prime **27.** Composite **29.** Neither **31.** Prime **33.** 2, 3, 5, 7, 11, 13, 17, 19, 23, 29, 31, 37, 41, 43, 47 **35.** False **37.** No, 9 is not a prime number. **39.** Yes **41.** $2 \cdot 5 \cdot 7$ **43.** $2 \cdot 2 \cdot 5 \cdot 13$ or $2^2 \cdot 5 \cdot 13$ **45.** $3 \cdot 7 \cdot 7$ or $3 \cdot 7^2$ **47.** $2 \cdot 2 \cdot 2 \cdot 7 \cdot 11$ or $2^3 \cdot 7 \cdot 11$

49. 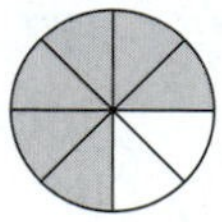

51. False **53.** $\neq$ **55.** $=$ **57.** $=$ **59.** $\neq$ **61.** $\frac{1}{2}$ **63.** $\frac{1}{3}$ **65.** $-\frac{9}{5}$ **67.** $-\frac{5}{4}$ **69.** 1 **71.** $\frac{3}{4}$ **73.** -3 **75.** $\frac{7}{10}$ **77.** $\frac{77}{39}$ **79.** $\frac{2}{5}$ **81.** $\frac{3}{4}$ **83.** $-\frac{5}{3}$ **85.** $\frac{21}{11}$ **87.** $\frac{17}{100}$ **89.** $\frac{8b}{5}$ **91.** $2xy$ **93.** $\frac{x}{3}$ **95.** $-\frac{1}{2c^2}$ **97.** Heads: $\frac{5}{12}$; tails: $\frac{7}{12}$ **99. a.** $\frac{3}{13}$ **b.** $\frac{10}{13}$ **c.** $\frac{3}{25}$ **101. a.** Jonathan: $\frac{5}{7}$; Jared: $\frac{6}{7}$ **b.** Jared sold the greater fractional part. **103. a.** Raymond: $\frac{10}{11}$; Travis: $\frac{9}{11}$ **b.** Raymond read the greater fractional part. **105. a.** 300,000,000 **b.** 36,000,000 **c.** $\frac{3}{25}$ **107.** For example: $\frac{6}{8}, \frac{9}{12}, \frac{12}{16}$ **109.** For example: $-\frac{6}{9}, -\frac{4}{6}, -\frac{2}{3}$

Section 4.2 Calculator Connections, p. 200

111. $\frac{8}{9}$ **112.** $\frac{13}{14}$ **113.** $\frac{41}{51}$ **114.** $\frac{21}{10}$ **115.** $\frac{29}{30}$ **116.** $\frac{13}{7}$ **117.** $\frac{3}{2}$ **118.** $\frac{31}{19}$

Section 4.3 Practice Exercises, pp. 210–213

3. Numerator: 2100; denominator: 7000; $\frac{3}{10}$ **5.** $\frac{23}{8}$

7.

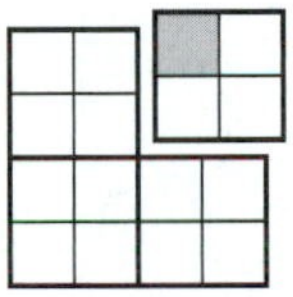

9. $\frac{3}{16}$ **11.** $\frac{24}{35}$ **13.** $\frac{8}{11}$ **15.** $-\frac{24}{5}$ **17.** $\frac{2}{15}$ **19.** $\frac{5}{8}$ **21.** $\frac{35}{4}$ **23.** $-\frac{8}{3}$ **25.** $\frac{4}{5}$ **27.** $\frac{30}{7}$ **29.** $\frac{3}{2}$ **31.** $-\frac{a}{10}$ **33.** $\frac{1}{20y}$ **35.** $4m^2$

37.

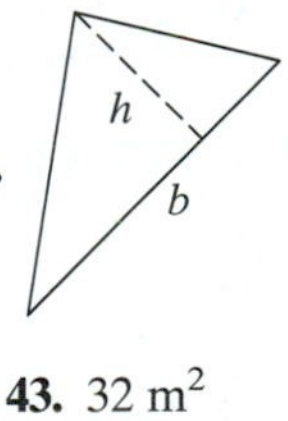

39.

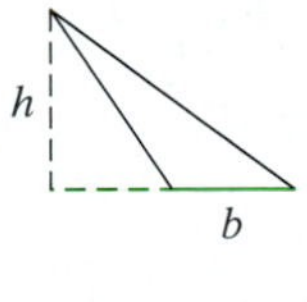

41. 44 cm^2 **43.** 32 m^2 **45.** 4 yd^2 **47.** $\frac{8}{7}$ **49.** $-\frac{9}{10}$ **51.** $-\frac{1}{4}$ **53.** No reciprocal exists. **55.** $\frac{1}{3}$ **57.** multiplying **59.** $\frac{8}{25}$ **61.** $\frac{35}{26}$ **63.** $\frac{35}{9}$ **65.** -5 **67.** 1 **69.** $-\frac{21}{2}$ **71.** $\frac{3}{5}$ **73.** $-\frac{90}{13}$ **75.** $\frac{6y}{7}$ **77.** $-4c^2d$ **79.** $\frac{7}{2}$ **81.** $\frac{5}{36}$ **83.** -8 **85.** $\frac{2}{5}$ **87.** 2 **89.** $\frac{3}{2}$ **91.** $-\frac{xy}{2}$ **93.** $\frac{2}{d}$ **95.** Li wrapped 54 packages. **97.** 24 cups of juice **99.** The stack will be 12 in. high. **101. a.** 27 commercials in 1 hr **b.** 648 commercials in 1 day **103. a.** Ricardo's mother will pay \$16,000. **b.** Ricardo will have to pay \$8000. **c.** He will have to finance \$216,000. **105.** Frankie mowed 960 yd^2. He has 480 yd^2 left to mow. **107.** $\frac{1}{10}$ of the sample has O negative blood. **109.** She can prepare 14 samples. **111. a.** 455 gal was required to transport the 130 gal. **b.** A total of 585 gal was used. **113.** 18 **115.** $\frac{2}{25}$ **117.** 12 ft, because $30 \div \frac{5}{2} = 12$. **119.** $\frac{1}{32}$ **121.** They are the same.

Section 4.4 Practice Exercises, pp. 219–222

3. $\frac{3y}{2x^2}$ **5.** $-\frac{10}{7}$ **7.** $-\frac{23}{4}$ **9. a.** 48, 72, 240 **b.** 4, 8, 12 **11. a.** 72, 360, 108 **b.** 6, 12, 9 **13.** 50 **15.** 48 **17.** 72 **19.** 60 **21.** 210 **23.** 120 **25.** 60 **27.** 240 **29.** 180 **31.** 180 **33.** The shortest length of floor space is 60 in. (5 ft). **35.** It will take 120 hr (5 days) for the satellites to be lined up again. **37.** $\frac{14}{21}$ **39.** $\frac{10}{16}$ **41.** $-\frac{12}{16}$ **43.** $\frac{-12}{15}$ **45.** $\frac{49}{42}$ **47.** $\frac{121}{99}$ **49.** $\frac{20}{4}$ **51.** $\frac{11{,}000}{4000}$ **53.** $-\frac{55}{15}$ **55.** $\frac{-15}{24}$ **57.** $\frac{16y}{28}$ **59.** $\frac{3y}{8y}$ **61.** $\frac{15p}{25p}$ **63.** $\frac{2x}{x^2}$ **65.** $\frac{8b^2}{ab^3}$ **67.** $>$ **69.** $<$ **71.** $=$ **73.** $>$ **75.** $\frac{7}{8}$ **77.** $\frac{2}{3}, \frac{3}{4}, \frac{7}{8}$ **79.** $-\frac{3}{8}, -\frac{5}{16}, -\frac{1}{4}$ **81.** $-\frac{4}{3}, -\frac{13}{12}, \frac{17}{15}$ **83.** The longest cut is above the left eye. The shortest cut is on the right hand. **85.** The least amount is $\frac{3}{4}$ lb of cheddar, and the greatest amount is $\frac{7}{8}$ lb of Swiss. **87.** a and b **89.** 336 **91.** 540

Section 4.5 Practice Exercises, pp. 229–231

3. $-\frac{12}{14}$ **5.** $\frac{25}{5}$ **7.** $\frac{4t^2}{t^3}$

9.

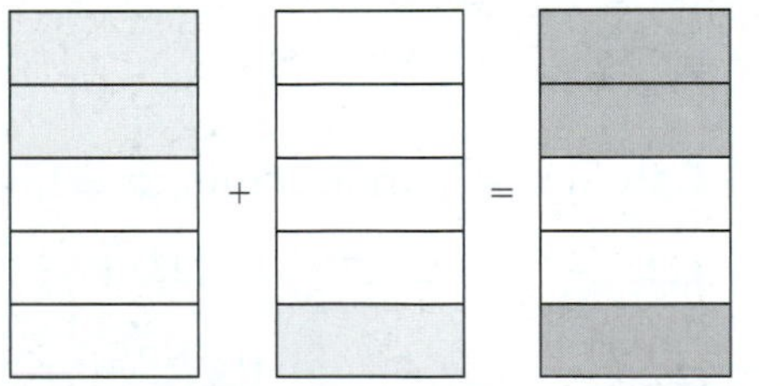

11.

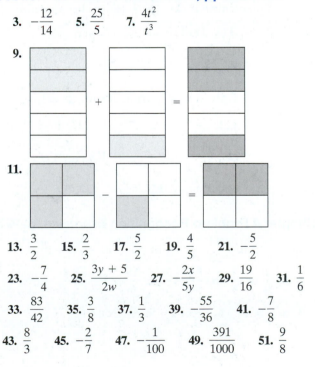

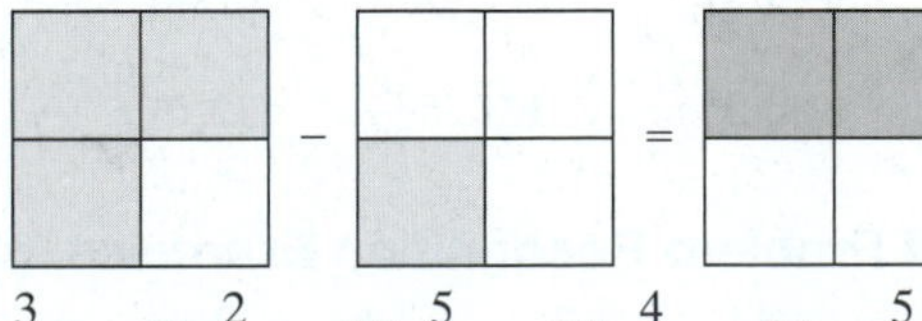

13. $\frac{3}{2}$ **15.** $\frac{2}{3}$ **17.** $\frac{5}{2}$ **19.** $\frac{4}{5}$ **21.** $-\frac{5}{2}$ **23.** $-\frac{7}{4}$ **25.** $\frac{3y+5}{2w}$ **27.** $-\frac{2x}{5y}$ **29.** $\frac{19}{16}$ **31.** $\frac{1}{6}$ **33.** $\frac{83}{42}$ **35.** $\frac{3}{8}$ **37.** $\frac{1}{3}$ **39.** $-\frac{55}{36}$ **41.** $-\frac{7}{8}$ **43.** $\frac{8}{3}$ **45.** $-\frac{2}{7}$ **47.** $-\frac{1}{100}$ **49.** $\frac{391}{1000}$ **51.** $\frac{9}{8}$

53. $-\frac{1}{8}$ **55.** $\frac{9}{16}$ **57.** $\frac{3x+8}{4x}$ **59.** $\frac{10y+7x}{xy}$
61. $\frac{10x-2}{x^2}$ **63.** $\frac{5-2x}{3x}$ **65.** Inez added $\frac{9}{8}$ cups or $1\frac{1}{8}$ cups.
67. The storm delivered $\frac{5}{32}$ in. of rain. **69. a.** $\frac{13}{36}$ **b.** $\frac{23}{36}$
71. $\frac{13}{5}$ m or $2\frac{3}{5}$ m **73.** Perimeter: 3 ft **75.** b

Section 4.6 Practice Exercises, pp. 241–245

1. $\frac{24}{5}$ **3.** $\frac{13}{6}$ **5.** $\frac{12}{11}$ **7.** $\frac{1}{2}$ **9.** $\frac{17}{5}$
11. $-12\frac{5}{6}$ **13.** $7\frac{2}{5}$ **15.** $15\frac{2}{5}$ **17.** -38 **19.** $27\frac{2}{3}$
21. $72\frac{1}{2}$ **23.** 2 **25.** $-4\frac{5}{12}$ **27.** $2\frac{6}{17}$ **29.** $\frac{3}{5}$
31. $-2\frac{3}{4}$ **33.** $7\frac{4}{11}$ **35.** $15\frac{3}{7}$ **37.** $15\frac{9}{16}$ **39.** $10\frac{13}{15}$
41. 5 **43.** 2 **45.** $3\frac{1}{5}$ **47.** $8\frac{2}{3}$ **49.** $14\frac{1}{2}$
51. $23\frac{1}{8}$ **53.** $19\frac{17}{48}$ **55.** $9\frac{5}{12}$ **57.** $12\frac{19}{24}$ **59.** $9\frac{7}{8}$
61. $171\frac{1}{2}$ **63.** $11\frac{3}{5}$ **65.** $12\frac{1}{6}$ **67.** $11\frac{1}{2}$ **69.** $1\frac{3}{4}$
71. $7\frac{13}{14}$ **73.** $3\frac{1}{6}$ **75.** $2\frac{7}{9}$ **77.** $\frac{11}{16}$ **79.** $\frac{32}{35}$
81. $-7\frac{11}{14}$ **83.** $2\frac{7}{8}$ **85.** $5\frac{3}{4}$ **87.** $-2\frac{5}{6}$ **89.** $-\frac{36}{5}$
91. $\frac{7}{24}$ **93.** $\frac{50}{13}$ **95.** $\frac{61}{30}$ **97.** $7\frac{3}{4}$ in.
99. The index finger is longer. **101.** Tabitha earned \$38.
103. $642\frac{1}{2}$ lb **105.** The total is $16\frac{11}{12}$ hr.
107. a. 7 weeks old **b.** $8\frac{1}{2}$ weeks old **109. a.** Lucy earned \$72 more than Ricky. **b.** Together they earned \$922.
111. $3\frac{5}{12}$ ft **113.** The printing area width is 6 in.
115. a. $3\frac{3}{8}$ L **b.** $\frac{5}{8}$ L **117.** $2\frac{2}{3}$ **119.** $2\frac{1}{6}$

Section 4.6 Calculator Connections, p. 246

121. $318\frac{1}{4}$ **122.** $3\frac{1}{15}$ **123.** $17\frac{18}{19}$ **124.** $466\frac{1}{5}$
125. $1\frac{43}{168}$ **126.** $\frac{11}{30}$ **127.** $\frac{37}{132}$ **128.** $\frac{137}{391}$
129. $46\frac{25}{54}$ **130.** $25\frac{71}{84}$ **131.** $5\frac{17}{77}$ **132.** $3\frac{9}{68}$

Chapter 4 Problem Recognition Exercises, p. 247

1. a. -1 **b.** $-\frac{14}{25}$ **c.** $-\frac{7}{2}$ or $-3\frac{1}{2}$ **d.** $-\frac{9}{5}$ or $-1\frac{4}{5}$
2. a. $\frac{10}{9}$ or $1\frac{1}{9}$ **b.** $\frac{8}{5}$ or $1\frac{3}{5}$ **c.** $\frac{13}{6}$ or $2\frac{1}{6}$ **d.** $\frac{1}{2}$
3. a. $\frac{5}{4}$ or $1\frac{1}{4}$ **b.** $\frac{17}{4}$ or $4\frac{1}{4}$ **c.** $-\frac{11}{6}$ or $-1\frac{5}{6}$
d. $-\frac{33}{8}$ or $-4\frac{1}{8}$ **4. a.** $\frac{221}{18}$ or $12\frac{5}{18}$ **b.** $\frac{26}{17}$ or $1\frac{9}{17}$
c. $\frac{3}{2}$ or $1\frac{1}{2}$ **d.** $\frac{43}{6}$ or $7\frac{1}{6}$ **5. a.** $-\frac{35}{8}$ or $-4\frac{3}{8}$
b. $-\frac{3}{2}$ or $-1\frac{1}{2}$ **c.** $-\frac{32}{3}$ or $-10\frac{2}{3}$ **d.** $-\frac{29}{8}$ or $-3\frac{5}{8}$
6. a. $\frac{11}{6}$ or $1\frac{5}{6}$ **b.** $\frac{5}{3}$ or $1\frac{2}{3}$ **c.** $\frac{17}{3}$ or $5\frac{2}{3}$ **d.** $\frac{22}{3}$ or $7\frac{1}{3}$
7. a. $-\frac{53}{15}$ or $-3\frac{8}{15}$ **b.** $-\frac{73}{15}$ or $-4\frac{13}{15}$ **c.** $\frac{14}{5}$ or $2\frac{4}{5}$
d. $\frac{63}{10}$ or $6\frac{3}{10}$ **8. a.** $\frac{25}{18}$ or $1\frac{7}{18}$ **b.** $\frac{50}{9}$ or $5\frac{5}{9}$ **c.** $\frac{7}{9}$
d. $\frac{43}{9}$ or $4\frac{7}{9}$ **9. a.** -1 **b.** $-\frac{56}{45}$ or $-1\frac{11}{45}$
c. $-\frac{81}{25}$ or $-3\frac{6}{25}$ **d.** $-\frac{106}{45}$ or $-2\frac{16}{45}$
10. a. 1 **b.** 1 **c.** 1 **d.** 1

Section 4.7 Practice Exercises, pp. 252–254

3. $-\frac{47}{30}$ **5.** $-\frac{12}{35}$ **7.** $-9\frac{5}{24}$ **9.** $\frac{1}{81}$ **11.** $\frac{1}{81}$
13. $-\frac{27}{8}$ **15.** $-\frac{27}{8}$ **17.** $\frac{1}{1000}$ **19.** $\frac{1}{1{,}000{,}000}$
21. $-\frac{1}{1000}$ **23.** -27 **25.** $4\frac{1}{4}$ **27.** 42
29. -2 **31.** $\frac{2}{9}$ **33.** 3 **35.** $2\frac{3}{8}$ **37.** $\frac{5}{3}$
39. $\frac{1}{36}$ **41.** $\frac{23}{24}$ **43.** $1\frac{3}{7}$ **45.** $\frac{25}{3}$ **47.** $5\frac{1}{4}$
49. -7 **51.** $7\frac{7}{9}$ **53.** $\frac{5}{6}$ **55.** $-\frac{7}{4}$ **57.** $\frac{x}{28}$
59. $-\frac{3}{5}$ **61.** $\frac{7}{4}$ **63.** $\frac{28}{11}$ **65.** -11 **67.** $\frac{2}{9}$
69. $2y$ **71.** $\frac{5}{8}a$ **73.** $\frac{17}{30}x$ **75.** $\frac{4}{3}y - \frac{7}{4}z$
77. a. $\frac{1}{36}$ **b.** $\frac{1}{6}$ **79.** $\frac{1}{5}$ **81.** $\frac{8}{9}$

Section 4.8 Practice Exercises, pp. 259–261

3. $\frac{25}{36}$ **5.** $-\frac{65}{36}$ **7.** $-10\frac{1}{24}$ **9.** $\frac{7}{6}$ **11.** $-\frac{13}{10}$
13. $\frac{13}{12}$ **15.** $\frac{13}{8}$ **17.** 2 **19.** $\frac{7}{6}$ **21.** $-\frac{2}{5}$
23. -24 **25.** 21 **27.** -21 **29.** 30 **31.** 0
33. $\frac{2}{5}$ **35.** $\frac{1}{18}$ **37.** $\frac{35}{8}$ **39.** -5 **41.** $\frac{1}{4}$
43. 1 **45.** $-\frac{7}{2}$ **47.** $\frac{2}{3}$ **49.** $\frac{10}{9}$ **51.** 7
53. 5 **55.** -48 **57.** $\frac{1}{3}$ **59.** $-\frac{21}{10}$ **61.** $-\frac{1}{12}$
63. $\frac{7}{10}$ **65.** -12 **67.** $\frac{1}{3}$ **69.** $\frac{8}{5}$ **71.** $-\frac{44}{15}$
73. 6 **75.** $\frac{9}{5}$

Chapter 4 Problem Recognition Exercises, pp. 261–262

1. Equation; $\frac{1}{10}$ **2.** Equation; $\frac{6}{7}$ **3.** Expression; $\frac{3}{2}$
4. Expression; $\frac{1}{8}$ **5.** Expression; $\frac{7}{5}$ **6.** Expression; $\frac{9}{5}$
7. Equation; $-\frac{2}{5}$ **8.** Equation; $-\frac{4}{5}$ **9.** Equation; 4
10. Equation; $\frac{5}{3}$ **11.** Expression; $\frac{4}{9}$
12. Expression; 0 **13.** Equation; 6 **14.** Equation; $-\frac{5}{2}$
15. Expression; $6x - 24$ **16.** Expression; $2x + 30$
17. Equation; $\frac{8}{3}$ **18.** Equation; 7 **19.** Expression; $\frac{3}{2}$
20. Expression; $\frac{25}{7}$ **21.** Equation; $\frac{2}{7}$ **22.** Equation; $\frac{3}{4}$
23. Expression; $\frac{18}{7}c$ **24.** Expression; $\frac{21}{4}d$

Chapter 4 Review Exercises, pp. 272–275

1. $\frac{1}{2}$ **2.** $\frac{4}{7}$ **3. a.** $\frac{5}{3}$ **b.** Improper **4. a.** $\frac{1}{6}$
b. Proper **5. a.** $\frac{3}{8}$ **b.** $\frac{2}{3}$ **c.** $\frac{4}{9}$ **6.** $\frac{23}{8}$ or $2\frac{7}{8}$
7. $\frac{7}{6}$ or $1\frac{1}{6}$ **8.** $\frac{43}{7}$ **9.** $\frac{57}{5}$ **10.** $5\frac{2}{9}$ **11.** $1\frac{2}{21}$
12., 13., 14., 15.

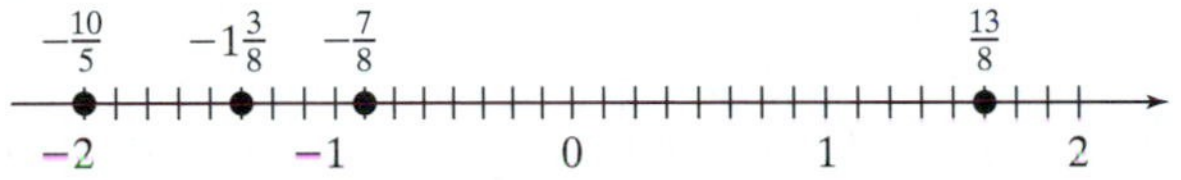

16. $134\frac{3}{7}$ **17.** $60\frac{11}{13}$ **18.** 21, 51, 1200
19. 55, 140, 260, 1200 **20.** 2, 53, 113
21. 12, 27, 51, 63, 130 **22.** $2 \cdot 3 \cdot 5 \cdot 11$
23. $2 \cdot 2 \cdot 3 \cdot 3 \cdot 5 \cdot 5$ or $2^2 \cdot 3^2 \cdot 5^2$ **24.** $\neq$ **25.** $=$
26. $\frac{1}{4}$ **27.** $\frac{1}{5}$ **28.** $-\frac{3}{2}$ **29.** $-\frac{7}{3}$ **30.** $\frac{2}{25}$
31. $\frac{7}{1000}$ **32.** $\frac{2a}{5c}$ **33.** $\frac{4t^2}{5}$ **34.** $\frac{14}{15}; \frac{1}{15}$
35. a. $\frac{3}{5}$ **b.** $\frac{2}{5}$ **36.** $-\frac{3}{7}$ **37.** $-\frac{3}{2}$ **38.** 63
39. 15 **40.** $\frac{4}{x}$ **41.** $\frac{3y^2}{2}$ **42.** $A = \frac{1}{2}bh$ **43.** 51 ft^2
44. 1 **45.** 1 **46.** $\frac{2}{7}$ **47.** $-\frac{1}{7}$ **48.** $\frac{16}{9}$ **49.** $\frac{7}{5}$
50. $-\frac{1}{21}$ **51.** -14 **52.** $8a$ **53.** $\frac{3y^2}{2}$
54. 36 bags of candy **55.** Yes. $9 \div \frac{3}{8} = 24$ so he will have 24 pieces, which is more than enough for his class.
56. There are 900 African American students. **57.** There are 300 Asian American students. **58.** Amelia earned \$576.
59. a. $2^2 \cdot 5^2$ **b.** $5 \cdot 13$ **c.** $2 \cdot 5 \cdot 7$
60. 420 **61.** 96 **62.** They will meet on the 12th day.
63. $\frac{15}{48}$ **64.** $\frac{63}{35}$ **65.** $\frac{35y}{60y}$ **66.** $\frac{-28}{4x}$ **67.** $<$
68. $>$ **69.** $=$ **70.** $-\frac{27}{35}, -\frac{7}{10}, -\frac{72}{105}, -\frac{8}{15}$
71. $\frac{3}{2}$ **72.** $\frac{2}{3}$ **73.** $\frac{29}{100}$ **74.** $\frac{1}{25}$ **75.** $-\frac{47}{11}$
76. $-\frac{117}{20}$ **77.** $\frac{17}{40}$ **78.** $\frac{12}{7}$ **79.** $\frac{17}{5w}$
80. $\frac{11b + 4a}{ab}$ **81. a.** $\frac{35}{4}$ m or $8\frac{3}{4}$ m **b.** $\frac{315}{128}\text{ m}^2$ or $2\frac{59}{128}\text{ m}^2$
82. a. $\frac{23}{3}$ yd or $7\frac{2}{3}$ yd **b.** $\frac{7}{2}\text{ yd}^2$ or $3\frac{1}{2}\text{ yd}^2$
83. $23\frac{7}{15}$ **84.** $23\frac{2}{3}$ **85.** $\frac{10}{11}$ **86.** $4\frac{1}{2}$ **87.** $-2\frac{3}{11}$
88. $-\frac{3}{5}$ **89.** 50; $50\frac{9}{40}$ **90.** 23; $22\frac{71}{75}$ **91.** $11\frac{11}{63}$
92. $14\frac{7}{16}$ **93.** $2\frac{5}{8}$ **94.** $1\frac{11}{12}$ **95.** $3\frac{2}{5}$ **96.** $3\frac{3}{14}$
97. $63\frac{15}{16}$ **98.** $50\frac{1}{2}$ **99.** $-2\frac{5}{6}$ **100.** $-3\frac{7}{8}$
101. Corry drove a total of $8\frac{1}{6}$ hr.
102. Denise will have $\frac{7}{8}$ acre left. **103.** It will take $3\frac{1}{8}$ gal.
104. There will be 10 pieces. **105.** $\frac{9}{64}$ **106.** $\frac{9}{64}$
107. $-\frac{1}{100{,}000}$ **108.** $\frac{1}{10{,}000}$ **109.** $-\frac{29}{10}$ **110.** $\frac{31}{15}$
111. $\frac{1}{6}$ **112.** $\frac{7}{16}$ **113.** $\frac{14}{5}$ **114.** $\frac{2x}{9}$ **115.** $-\frac{3}{7}$
116. $\frac{1}{2}$ **117.** $-\frac{4}{3}$ **118.** $\frac{3}{5}$ **119.** $\frac{5}{16}$ **120.** $11\frac{2}{3}$
121. $-\frac{17}{12}x$ **122.** $-\frac{13}{10}y$ **123.** $\frac{2}{3}a + \frac{1}{6}c$ **124.** $\frac{9}{10}w + \frac{5}{3}y$
125. $\frac{19}{15}$ **126.** $-\frac{5}{14}$ **127.** $\frac{10}{9}$ **128.** $\frac{3}{8}$ **129.** $\frac{3}{5}$
130. $\frac{3}{4}$ **131.** -1 **132.** 2 **133.** -15 **134.** 12

Chapter 4 Test, pp. 276–277

1. a. $\frac{5}{8}$ **b.** Proper **2. a.** $\frac{7}{3}$ **b.** Improper
3. a. $3\frac{2}{3}$ **b.** $\frac{34}{9}$

4.

5.

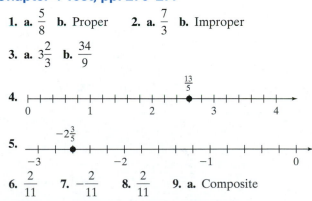

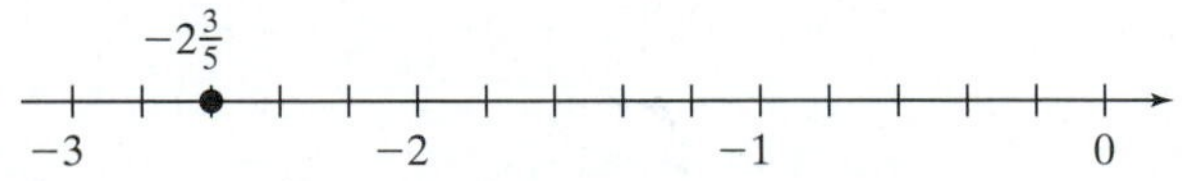

6. $\frac{2}{11}$ **7.** $-\frac{2}{11}$ **8.** $\frac{2}{11}$ **9. a.** Composite
b. Neither **c.** Prime **d.** Neither **e.** Prime **f.** Composite
10. $3 \cdot 3 \cdot 5$ or $3^2 \cdot 5$ **11.** Add the digits of the number. If the sum is divisible by 3, then the original number is divisible by 3. **b.** Yes **12. a.** No **b.** Yes **c.** Yes **d.** No
13. $=$ **14.** $\neq$ **15.** $\frac{10}{7}$ or $1\frac{3}{7}$ **16.** $\frac{2}{7b}$

17. a. Christine: $\frac{3}{5}$; Brad: $\frac{4}{5}$
b. Brad has the greater fractional part completed.

18. $\frac{19}{69}$ **19.** $\frac{25}{2}$ or $12\frac{1}{2}$ **20.** $\frac{4}{9}$ **21.** $-\frac{1}{2}$ **22.** $\frac{2y}{3}$

23. $-5bc$ **24.** $\frac{44}{3}\text{ cm}^2$ or $14\frac{2}{3}\text{ cm}^2$ **25.** $20 \div \frac{1}{4}$

26. 48 quarter-pounders

27. They can build on a maximum of $\frac{2}{5}$ acre.

28. a. 24, 48, 72, 96 **b.** 1, 2, 3, 4, 6, 8, 12, 24
c. $2 \cdot 2 \cdot 2 \cdot 3$ or $2^3 \cdot 3$

29. 240 **30.** $\frac{35}{63}$ **31.** $\frac{22w}{42w}$ **32.** $-\frac{5}{3}, -\frac{4}{7}, -\frac{11}{21}$

33. When subtracting like fractions, keep the same denominator and subtract the numerators. When multiplying fractions, multiply the denominators as well as the numerators.

34. $\frac{9}{16}$ **35.** $\frac{1}{3}$ **36.** $-\frac{1}{3}$ **37.** $\frac{12y-6}{y^2}$ **38.** $-1\frac{21}{25}$

39. $9\frac{3}{5}$ **40.** $17\frac{3}{8}$ **41.** $2\frac{1}{11}$ **42.** $-7\frac{4}{9}$ **43.** $3\frac{9}{10}$

44. 1 lb is needed. **45.** Area: $25\frac{2}{25}\text{ m}^2$; perimeter: $20\frac{1}{5}$ m

46. $\frac{36}{49}$ **47.** $-\frac{1}{1000}$ **48.** $-\frac{4}{15}$ **49.** $\frac{9}{7}$ **50.** $-\frac{1}{17}$

51. $\frac{3}{2}$ **52.** $\frac{8}{15}m$ **53.** $\frac{11}{9}$ **54.** $-\frac{6}{5}$ **55.** $-\frac{2}{11}$

56. 16 **57.** $-\frac{7}{2}$ **58.** $\frac{20}{3}$

Chapters 1–4 Cumulative Review Exercises, p. 278

1. Ten-thousands place **2.** One hundred thirty is less than two hundred forty-four

3. $2 \cdot 2 \cdot 2 \cdot 3 \cdot 3 \cdot 5$ or $2^3 \cdot 3^2 \cdot 5$ **4.** 42 **5.** $-\frac{3}{8}$

6. -92 **7.** 0 **8.** $\frac{1}{8}$ **9.** $\frac{2}{5}$ **10.** $\frac{26}{17}$ or $1\frac{9}{17}$

11. $\frac{10}{3}$ or $3\frac{1}{3}$ **12.** $\frac{1}{5}$

13. 140 m^2

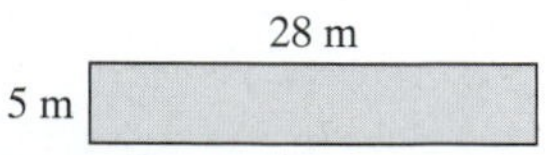

14. a. -4 **b.** 4 **c.** -16 **d.** 16 **15.** 18
16. $-11x - 3y - 5$ **17.** $-7x + 18$ **18.** 24

19. $\frac{1}{4}$ **20.** 14

Chapter 5

Chapter Opener Puzzle

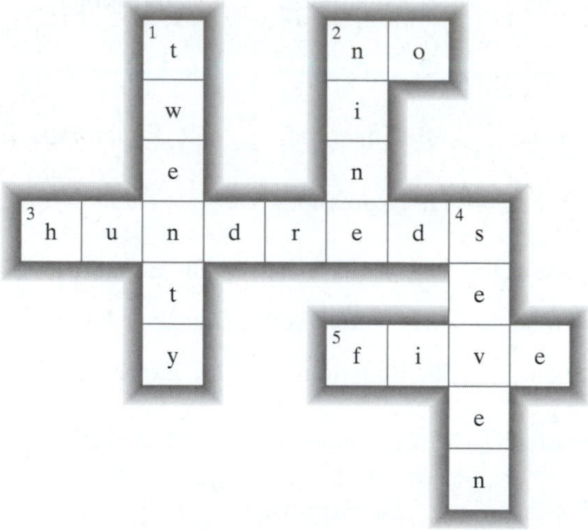

Section 5.1 Practice Exercises, pp. 287–289

3. 100 **5.** 10,000 **7.** $\frac{1}{100}$ **9.** $\frac{1}{10{,}000}$
11. Tenths **13.** Hundredths **15.** Tens
17. Ten-thousandths **19.** Thousandths **21.** Ones
23. Nine-tenths **25.** Twenty-three hundredths
27. Negative thirty-three thousandths
29. Four hundred seven ten-thousandths
31. Three and twenty-four hundredths
33. Negative five and nine-tenths
35. Fifty-two and three-tenths
37. Six and two hundred nineteen thousandths
39. -8472.014 **41.** 700.07 **43.** $-2{,}469{,}000.506$

45. $3\frac{7}{10}$ **47.** $2\frac{4}{5}$ **49.** $\frac{1}{4}$ **51.** $-\frac{11}{20}$ **53.** $20\frac{203}{250}$

55. $-15\frac{1}{2000}$ **57.** $\frac{42}{5}$ **59.** $\frac{157}{50}$ **61.** $-\frac{47}{2}$

63. $\frac{1191}{100}$ **65.** $<$ **67.** $>$ **69.** $<$

71. $>$ **73.** a, b **75.** 0.3444, 0.3493, 0.3558, 0.3585, 0.3664 **77.** These numbers are equivalent, but they represent different levels of accuracy. **79.** 7.1 **81.** 49.9
83. 33.42 **85.** -9.096 **87.** 21.0 **89.** 7.000
91. 0.0079 **93.** 0.0036 mph

	Number	Hundreds	Tens	Tenths	Hundredths	Thousandths
95.	971.0948	1000	970	971.1	971.09	971.095
97.	21.9754	0	20	22.0	21.98	21.975

99. 0.972

Section 5.2 Practice Exercises, pp. 295–298

1. a, c **3.** 23.5 **5.** 8.603 **7.** 2.8300 **9.** 63.2 **11.** 8.951 **13.** 15.991 **15.** 79.8005 **17.** 31.0148 **19.** 62.6032 **21.** 100.414 **23.** 128.44 **25.** 82.063 **27.** 14.24 **29.** 3.68 **31.** 12.32 **33.** 5.677 **35.** 1.877 **37.** 21.6646 **39.** 14.765 **41.** 159.558 **43.** 0.9012 **45.** −422.94 **47.** −1.359 **49.** 50.979 **51.** −3.27 **53.** −4.432 **55.** 1.4 **57.** −111.2 **59.** 0.5346 **61.**

Check No.	Description	Debit	Credit	Balance
				$ 245.62
2409	Electric bill	$ 52.48		193.14
2410	Groceries	72.44		120.70
2411	Department store	108.34		12.36
	Paycheck		$1084.90	1097.26
2412	Restaurant	23.87		1073.39
	Transfer from savings		200	1273.39

63. 1.35 million cells per microliter **65. a.** The water is rising 1.7 in./hr. **b.** At 1:00 P.M. the level will be 11 in. **c.** At 3:00 P.M. the level will be 14.4 in. **67.** The pile containing the two nickels and two pennies is higher. **69.** $x = 8.9$ in.; $y = 15.4$ in.; the perimeter is 98.8 in. **71.** $x = 2.075$ ft; $y = 2.59$ ft; the perimeter is 22.17 ft. **73.** 27.2 mi **75.** $3.87t$ **77.** $-13.2p$ **79.** $0.4y$ **81.** $0.845x + 0.52y$ **83.** $c - 5d$

Section 5.2 Calculator Connections, p. 299

85. IBM increased by $5.90 per share. **86.** FedEx decreased by $3.66 per share. **87.** Between February and March, FedEx increased the most, by $2.27 per share. **88.** Between April and May, IBM increased the most, by $7.96 per share. **89.** Between March and April, FedEx decreased the most, by $8.09 per share. **90.** Between February and March, IBM decreased the most, by $6.73 per share.

Section 5.3 Practice Exercises, pp. 306–311

3. 50.0 **5.** −0.003 **7.** 7.958 **9.** 0.4 **11.** 3.6 **13.** 0.18 **15.** 17.904 **17.** 37.35 **19.** 4.176 **21.** −4.736 **23.** 2.891 **25.** 114.88 **27.** 2.248 **29.** −0.00144 **31. a.** 51 **b.** 510 **c.** 5100 **d.** 51,000 **33. a.** 0.51 **b.** 0.051 **c.** 0.0051 **d.** 0.00051 **35.** 3490 **37.** 96,590 **39.** −0.933 **41.** 0.05403 **43.** 2,600,000 **45.** 400,000 **47.** $20,549,000,000 **49. a.** 201.6 lb of gasoline **b.** 640 lb of CO_2 **51.** The bill was $312.17. **53.** $2.81 can be saved. **55.** 0.00115 km^2 **57.** The area is 333 yd^2. **59.** 0.16 **61.** 1.69 **63.** 0.001 **65.** −0.04 **67.** The length of a radius is one-half the length of a diameter. **69.** 12.2 in. **71.** 83 m **73.** 62.8 cm **75.** 15.7 km **77.** 18.84 cm **79.** 14.13 in. **81.** 314 mm^2 **83.** 121 ft^2 **85.** 16.642 mi **87.** 2826 ft^2 **89. a.** 62.8 in. **b.** 23 times **91.** 69,080 in. or 5757 ft

Section 5.3 Calculator Connections, p. 311

93. Area ≈ 517.1341 cm^2; circumference ≈ 80.6133 cm
94. Area ≈ 81.7128 ft^2; circumference ≈ 32.0442 ft
95. Area ≈ 70.8822 $in.^2$; circumference ≈ 29.8451 in.
96. Area ≈ 8371.1644 mm^2; circumference ≈ 324.3380 mm

Section 5.4 Practice Exercises, pp. 319–322

3. −10.203 **5.** −101.161 **7.** −0.00528 **9.** 314 ft^2 **11.** 0.9 **13.** 0.18 **15.** 0.53 **17.** 21.1 **19.** 1.96 **21.** 0.035 **23.** 16.84 **25.** 0.12 **27.** −0.16 **29.** $5.\overline{3}$ **31.** $3.1\overline{6}$ **33.** $2.\overline{15}$ **35.** 503 **37.** 9.92 **39.** −56 **41.** 2.975 **43.** $208.\overline{3}$ **45.** 48.5 **47.** 1100 **49.** 42,060 **51.** The decimal point will move to the left two places. **53.** 0.03923 **55.** −9.802 **57.** 0.00027 **59.** 0.00102 **61. a.** 2.4 **b.** 2.44 **c.** 2.444 **63. a.** 1.8 **b.** 1.79 **c.** 1.789 **65. a.** 3.6 **b.** 3.63 **c.** 3.626 **67.** 0.26 **69.** −14.8 **71.** 20.667 **73.** 35.67 **75.** 111.3 **77.** Unreasonable; $960 **79.** Unreasonable; $140,000 **81.** The monthly payment is $42.50. **83. a.** 13 bulbs would be needed (rounded up to the nearest whole unit). **b.** $9.75 **c.** The energy efficient fluorescent bulb would be more cost effective. **85.** Babe Ruth's batting average was 0.342. **87.** 2.2 mph **89.** −47.265 **91.** b, d

Section 5.4 Calculator Connections, pp. 322-323

93. 1149686.166 **94.** 3411.4045 **95.** 1914.0625 **96.** 69,568.83693 **97.** 95.6627907 **98.** 293.5070423 **99.** Answers will vary. **100.** Answers will vary. **101. a.** 0.27 **b.** Yes the claim is accurate. The decimal, 0.27 is close to 0.25, which is equal to $\frac{1}{4}$. **102.** 272 people per square mile **103. a.** 1,600,000 mi per day **b.** $66{,}666.\overline{6}$ mph **104.** When we say that 1 year is 365 days, we are ignoring the 0.256 day each year. In 4 years, that amount is $4 \times 0.256 = 1.024$, which is another whole day. This is why we add one more day to the calendar every 4 years.

Chapter 5 Problem Recognition Exercises, pp. 323–324

1. a. 223.04 **b.** 12,304 **c.** 23.04 **d.** 1.2304 **e.** 123.05 **f.** 1.2304 **g.** 12,304 **h.** 123.03 **2. a.** 6078.3 **b.** 5,078,300 **c.** 4078.3 **d.** 5.0783 **e.** 5078.301 **f.** 5.0783 **g.** 5,078,300 **h.** 5078.299 **3. a.** −7.191 **b.** 7.191 **4. a.** −730.4634 **b.** 730.4634 **5. a.** 52.64 **b.** 52.64 **6. a.** 59.384 **b.** 59.384 **7. a.** 86.4 **b.** −5.4 **8. a.** 185 **b.** −46.25 **9. a.** −80 **b.** −448 **10. a.** −54 **b.** −496.8 **11.** 1 **12.** 1 **13.** 4000 **14.** 6,400,000 **15.** 200,000 **16.** 2700 **17.** 1,350,000,000

18. 1,700,000 **19.** 4.4001 **20.** 76.7001
21. 5.73 **22.** 0.055 **23.** −12.67 **24.** −0.139

Section 5.5 Practice Exercises, pp. 333–336

3. 0.0225 **5.** 6.4 **7.** −0.756 **9.** $\frac{4}{10}$; 0.4
11. $\frac{98}{100}$; 0.98 **13.** 0.28 **15.** −0.632 **17.** −3.2
19. −5.25 **21.** 0.75 **23.** 7.45 **25.** $3.\overline{8}$
27. $0.52\overline{7}$ **29.** $-0.\overline{126}$ **31.** $1.1\overline{36}$ **33.** 0.9
35. 0.143 **37.** 0.08 **39.** −0.71
41. **a.** $0.\overline{1}$ **b.** $0.\overline{2}$ **c.** $0.\overline{4}$ **d.** $0.\overline{5}$

If we memorize that $\frac{1}{9} = 0.\overline{1}$, then $\frac{2}{9} = 2 \cdot \frac{1}{9} = 2 \cdot 0.\overline{1} = 0.\overline{2}$, and so on.

43.

	Decimal Form	Fraction Form
a.	0.45	$\frac{9}{20}$
b.	1.625	$1\frac{5}{8}$ or $\frac{13}{8}$
c.	$-0.\overline{7}$	$-\frac{7}{9}$
d.	$-0.\overline{45}$	$-\frac{5}{11}$

45.

	Decimal Form	Fraction Form
a.	$0.\overline{3}$	$\frac{1}{3}$
b.	−2.125	$-2\frac{1}{8}$ or $-\frac{17}{8}$
c.	$-0.8\overline{63}$	$-\frac{19}{22}$
d.	1.68	$\frac{42}{25}$

47. Rational **49.** Rational **51.** Rational
53. Irrational **55.** Irrational **57.** Rational
59. = **61.** < **63.** > **65.** >
67. 1.75, $1.\overline{7}$, 1.8

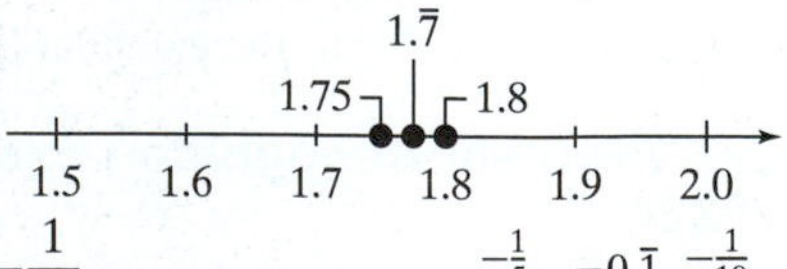

69. $-\frac{1}{5}$, $-0.\overline{1}$, $-\frac{1}{10}$

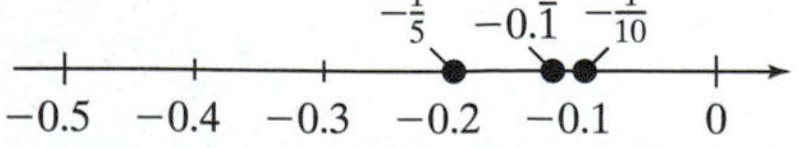

71. 6.25 **73.** 10 **75.** 8.77 **77.** 25.75 **79.** −2
81. −8.58 **83.** 67.35 **85.** 25.05 **87.** 23.4
89. 1.28 **91.** −10.83 **93.** 2.84 **95.** 43.46 cm^2
97. 84.78 **99.** −78.4 **101.** **a.** 471 mi **b.** 62.8 mph
103. Jorge will be charged $98.75. **105.** She has 24.3 g left for dinner. **107.** Hannah should get $4.77 in change.
109. 3.475 **111.** 0.52

Section 5.5 Calculator Connections, p. 337

113. **a.** 237 shares **b.** $13.90 will be left.
114. **a.** Approximately 921,800 homes could be powered. **b.** Approximately 342,678 additional homes could be powered. **115.** **a.** The BMI is approximately 30.1. The person is at high risk. **b.** The BMI is approximately 19.8. The person has a lower risk. **116.** **a.** Marty will have to finance $120,000. **b.** There are 360 months in 30 yr. **c.** He will pay $287,409.60 **d.** He will pay $167,409.60 in interest.
117. Each person will get approximately $13,410.10.

Section 5.6 Practice Exercises, pp. 342–344

1. area: 0.0314 m^2; circumference: 0.628 m **3.** 2.25
5. −2.23 **7.** $0.51x$ **9.** $-22.1z + 6.2$ **11.** 0.86
13. −4.78 **15.** −49.9 **17.** 0.095 **19.** 42.78
21. −0.12 **23.** −2.9 **25.** 19 **27.** 5 **29.** 9
31. −24 **33.** 6.72 **35.** 50 **37.** −4
39. −600 **41.** 10 **43.** The number is 4.5.
45. The number is 2.7. **47.** The number is −9.4.
49. The number is 7.6. **51.** The sides are 4.6 yd, 7.7 yd, and 9.2 yd. **53.** Rafa made $65.25, Toni made $43.20, and Henri made $59.35 in tips. **55.** The painter rented the pressure cleaner for 3.5 hr. **57.** The previous balance is $104.75 and the new charges amount to $277.15.
59. Madeline rode for 54.25 min and Kim rode for 62.5 min.

Section 5.7 Practice Exercises, pp. 350–354

3. 5 **5.** 6 **7.** −15.8 **9.** 5.8 hr
11. **a.** 397 Cal **b.** 386 Cal **c.** There is an 11-Cal difference in the means.
13. **a.** 86.5% **b.** 81% **c.** The low score of 59% decreased Zach's average by 5.5%. **15.** 17 **17.** 110.5
19. −52.5 **21.** 3.93 deaths per 1000 **23.** 0
25. 51.7 million passengers **27.** 4 **29.** −21 and −24
31. No mode **33.** $600 **35.** 5.2%, 5.8%
37. Mean: 85.5%; median: 94.5%; The median gave Jonathan a better overall score. **39.** Mean: $250; median: $256; mode: There is no mode. **41.** Mean: $942,500; median: $848,500; mode: $850,000 **43.** 2.38 **45.** 2.77
47. 3.3; Elmer's GPA improved from 2.5 to 3.3.
49.

Number of Residents in Each House	Number of Houses	Product
1	3	3
2	9	18
3	10	30
4	9	36
5	6	30
Total:	37	117

The mean number of residents is approximately 3.2.

Chapter 5 Review Exercises, pp. 363–367

1. The 3 is in the tens place, 2 is in the ones place, 1 is in the tenths place, and 6 is in the hundredths place.
2. The 2 is in the ones place, 0 is in the tenths place, 7 is in the hundredths place, and 9 is in the thousandths place.
3. Five and seven-tenths **4.** Ten and twenty-one hundredths **5.** Negative fifty-one and eight thousandths
6. Negative one hundred nine and one-hundredth
7. 33,015.047 **8.** −100.01 **9.** $-4\frac{4}{5}$ **10.** $\frac{1}{40}$
11. $\frac{13}{10}$ **12.** $\frac{27}{4}$ **13.** < **14.** <
15. 4.3875, 4.3953, 4.4839, 4.5000, 4.5142 **16.** 89.92
17. 34.890 **18. a.** The amount in the box is less than the advertised amount. **b.** The amount rounds to 12.5 oz.
19. a, b **20.** b, c **21.** 49.743 **22.** 273.22
23. 5.45 **24.** 1.902 **25.** −244.04 **26.** 29.007
27. 7.809 **28.** 82.265 **29.** $0.5y$ **30.** $-13.5x + 6$
31. a. Between days 1 and 2, the increase was \$0.194. **b.** Between days 3 and 4, the decrease was \$0.209.
32. 8.19 **33.** 74.113 **34.** −264.44 **35.** −346.5
36. 85,490 **37.** 100.34 **38.** 0.9201 **39.** 1.0422
40. 432,000 **41.** 33,800,000 **42. a.** Eight batteries cost \$15.96 on sale. **b.** A customer can save \$2.03.
43. The call will cost \$1.61. **44.** Area = 940 ft^2, perimeter = 127 ft **45. a.** 7280 people **b.** 18,000 people
46. 7.5 m **47.** 13.6 ft **48.** Area: 452.16 ft^2; circumference: 75.36 ft **49.** Area: 2826 yd^2; circumference: 188.4 yd **50.** 17.1 **51.** 42.8
52. $4.1\overline{3}$ **53.** $8.7\overline{6}$ **54.** −27 **55.** −0.03
56. 4.9393 **57.** 9.0234 **58.** 553,800 **59.** 260
60.

	$8.\overline{6}$	$52.\overline{52}$	$0.\overline{409}$
Tenths	8.7	52.5	0.4
Hundredths	8.67	52.53	0.41
Thousandths	8.667	52.525	0.409
Ten-thousandths	8.6667	52.5253	0.4094

61. −11.62 **62.** −11.97 **63. a.** \$0.50 per roll **b.** \$0.57 per roll **c.** The 12-pack is better.
64. 2.4 **65.** 3.52 **66.** −0.192 **67.** −0.4375
68. $0.58\overline{3}$ **69.** $1.52\overline{7}$ **70.** $-4.3\overline{18}$ **71.** 0.29
72. 0.9 **73.** −3.667 **74. a.** Rational **b.** Irrational **c.** Irrational **d.** Rational
75. $\frac{2}{9}$ **76.** $3\frac{1}{3}$
77.

Stock	Closing Price (\$) (Decimal)	Closing Price (\$) (Fraction)
Sun	5.20	$5\frac{1}{5}$
Sony	55.53	$55\frac{53}{100}$
Verizon	41.16	$41\frac{4}{25}$

78. > **79.** > **80.** −5 **81.** −5.52 **82.** 59
83. 1.6 **84.** −3 **85.** −2 **86.** \$89.90 will be saved by buying the combo package. **87.** Marvin must drive 34 mi more. **88.** −2.58 **89.** −1.53 **90.** 21.4
91. −11.32 **92.** −5.5 **93.** 2.5 **94.** −5.2
95. 110 **96.** −415
97. Multiply by 100; solution: −1.48
98. Multiply by 10; solution: 2.76 **99.** The number is 6.5.
100. The number is 2.4.
101. The number is 1600. **102.** The number is 4.
103. The sides are 5.6 yd, 8 yd, and 11.2 yd.
104. DVDs cost \$5.99 and video games cost \$6.99.
105. Mykeshia can rent space for 6 months.
106. Mean: 17.5; median: 18; mode: 20 **107.** Mean: 1060 mg; median: 1000 mg; modes: 1000 mg and 1200 mg
108. The median is 20,562 seats. **109.** 4 **110.** 3.0

Chapter 5 Test, pp. 367–369

1. a. Tens place **b.** Hundredths place
2. Negative five hundred nine and twenty-four thousandths
3. $1\frac{13}{50}; \frac{63}{50}$ **4.** 0.4419, 0.4484, 0.4489, 0.4495
5. b is correct. **6.** −45.172 **7.** −46.89
8. −126.45 **9.** −5.08 **10.** 1.22 **11.** 12.2243
12. $120.\overline{6}$ **13.** 439.81 **14.** 4.592 **15.** 57,923
16. 8012 **17.** 0.002931 **18.** $5.66x$ **19.** $14.6y - 3.3$
20. 24.4 ft **21.** Circumference: 50 cm; area: 201 cm^2
22. a. 61.4°F **b.** 1.4°F **23. a.** 1,040,000,000 tons **b.** 360,000,000 tons **c.** 680,000,000 tons **24. a.** 67.5 in.^2 **b.** 75.5 in.^2 **c.** 157.3 in.^2
25. He made \$3094.75. **26.** She will pay approximately \$37.50 per month. **27.** He will use 10 gal of gas.
28.

Year	Decimal	Fraction
1984	41.02 sec	$41\frac{1}{50}$ sec
1988	39.10	$39\frac{1}{10}$
1992	40.33	$40\frac{33}{100}$
1994	39.25	$39\frac{1}{4}$

29. $-3.\overline{5}, -3\frac{1}{2}, -3.2$

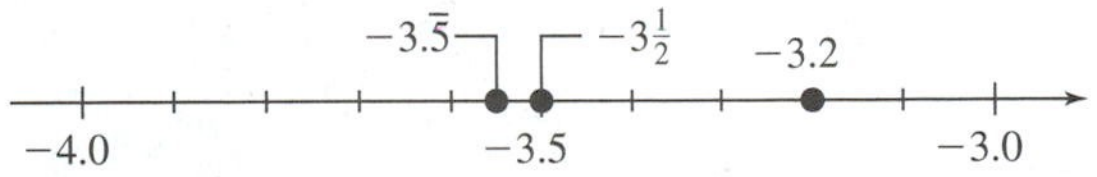

30. 9.57 **31.** 47.25 **32.** −1.21 **33. a.** Rational **b.** Irrational **c.** Irrational **d.** Rational
34. −0.008 **35.** 177.5 **36.** 18.66 **37.** 3.4
38. −2 **39.** 5.5 **40.** The number is −6.2.
41. The number is 0.24. **42.** 19,173 ft **43.** 19,340 ft
44. There is no mode. **45.** Mean: \$14.60; median: \$15; mode \$16 **46. a.** 38.8 mi **b.** 5.5 mi/day **47.** 3.09

Chapters 1–5 Cumulative Review Exercises, pp. 369–370

1. 14 **2.** 4039 **3.** 3032 **4.** −6108
5. 2,415,000 **6.** Dividend: 4530; divisor: 225; whole-number part of the quotient: 20; remainder: 30

7. To check a division problem, multiply the whole-number part of the quotient and the divisor. Then add the remainder to get the dividend. That is, $20 \times 225 + 30 = 4530$.

8. The difference between sales for Wal-Mart and Sears is $181,956 million.

9. $\frac{6}{55}$ **10.** $\frac{4}{7}$ **11.** $\frac{49}{100}$ **12.** $\frac{2}{3}$

13. There is $9000 left. **14.** $\frac{2}{5}$ **15.** $\frac{38}{11}$ **16.** $-\frac{3}{2}$

17. Area: $\frac{15}{64}$ ft^2; perimeter: 2 ft **18.** -72.33

19. 668.79 **20.** -75.275 **21.** 16 **22.** 339.12 **23.** 46.48 **24. a.** 3.75248 **b.** 3.75248 **c.** Commutative property of multiplication

25.

Bone	Length (in.) (Decimal)	Length (in.) (Mixed Number)
Femur	19.875	$19\frac{7}{8}$
Fibula	15.9375	$15\frac{15}{16}$
Humerus	14.375	$14\frac{3}{8}$
Innominate bone (hip)	7.5	$7\frac{1}{2}$

26. $3.66x - 8.65$ **27.** Circumference: 6.28 m; area: 3.14 m^2 **28.** -0.86 **29.** 15.6 **30.** -6

Chapter 6

Chapter Opener Puzzle

T	O	P	H	E	A	V	Y
3	5	1	2	7	8	4	6

Section 6.1 Practice Exercises, pp. 376–379

3. 5 : 6 and $\frac{5}{6}$ **5.** 11 to 4 and $\frac{11}{4}$ **7.** 1 : 2 and 1 to 2

9. a. $\frac{5}{3}$ **b.** $\frac{3}{5}$ **c.** $\frac{3}{8}$ **11. a.** $\frac{2}{15}$ **b.** $\frac{2}{13}$

13. a. $\frac{7}{18}$ **b.** $\frac{7}{25}$ **15.** $\frac{2}{3}$ **17.** $\frac{1}{5}$ **19.** $\frac{4}{1}$ **21.** $\frac{11}{5}$

23. $\frac{6}{5}$ **25.** $\frac{1}{2}$ **27.** $\frac{3}{2}$ **29.** $\frac{6}{7}$ **31.** $\frac{8}{9}$ **33.** $\frac{7}{1}$

35. $\frac{1}{8}$ **37.** $\frac{5}{4}$ **39. a.** 315 ft **b.** $\frac{63}{1}$ **41. a.** $\frac{2}{9}$ **b.** $\frac{2}{9}$

43. $\frac{1}{11}$ **45.** $\frac{10}{1}$ **47.** $\frac{15}{32}$ **49.** $\frac{20}{61}$ **51.** $\frac{2}{3}$

53. $\frac{1}{4}$ **55.** 13 units **57. a.** 1.5 **b.** $1.\overline{6}$ **c.** 1.6 **d.** 1.625; yes **59.** Answers will vary.

Section 6.2 Practice Exercises, pp. 383–386

3. 4 to 1 and $\frac{4}{1}$ **5.** $\frac{9}{17}$ **7.** $\frac{\$32}{5 \text{ ft}^2}$ **9.** $\frac{117 \text{ mi}}{2 \text{ hr}}$

11. $\frac{\$29}{4 \text{ hr}}$ **13.** $\frac{1 \text{ page}}{2 \text{ sec}}$ **15.** $\frac{65 \text{ calories}}{4 \text{ crackers}}$ **17.** $-\frac{9°\text{F}}{4 \text{ hr}}$

19. a, c, d **21.** 113 mi/day **23.** -400 m/hr **25.** $55 per payment **27.** $0.69/lb **29.** $256,000 per person **31.** 14.29 m/sec **33.** $0.150 per oz **35.** $0.995 per liter **37.** $52.50 per tire **39.** $5.417 per bodysuit **41. a.** $0.075/oz **b.** $0.075/oz **c.** Both sizes cost the same amount per ounce. **43.** The larger can is $0.055 per ounce. The smaller can is $0.073 per ounce. The larger can is the better buy. **45.** Coca-Cola: 3.25 g/fl oz; MelloYello: 3.92 g/fl oz; Ginger Ale: 3 g/fl oz; MelloYello has the greatest amount per fluid oz. **47.** Coca-Cola: 12 cal/fl oz; MelloYello: 14.2 cal/fl oz; Ginger Ale: 11.25 cal/fl oz; Ginger Ale has the least number of calories per fluid oz.

49. 295,000 vehicles/year **51. a.** 2.2 million per year **b.** 2.04 million per year **c.** Mexico **53.** Cheetah: 29 m/sec; antelope: 24 m/sec. The cheetah is faster.

Section 6.2 Calculator Connections, pp. 386–387

54. a. 9.9 wins/yr **b.** 8.6 wins/yr **c.** Shula

55. a. 2.1 wins/loss **b.** 1.5 wins/loss **c.** Shula

56. a. $0.29 per ounce **b.** $0.21 per ounce **c.** $0.19 per ounce; The best buy is Irish Spring. **57.** The unit prices are $0.175 per ounce, $0.237 per ounce, and $0.324 per ounce. The best buy is the 32-oz jar. **58. a.** $0.333 per ounce **b.** $0.262 per ounce **c.** $0.377 per ounce The best buy is the 4-pack of 6-oz cans for $6.29. **59. a.** $0.017 per ounce **b.** $0.052 per ounce The case of 24 twelve-oz cans for $4.99 is the better buy.

Section 6.3 Practice Exercises, pp. 392–394

3. $\frac{1 \text{ teacher}}{15 \text{ students}}$ **5.** $\frac{3}{1}$ **7.** 28.1 mpg **9.** $\frac{4}{16} = \frac{5}{20}$

11. $\frac{-25}{15} = \frac{-10}{6}$ **13.** $\frac{2}{3} = \frac{4}{6}$ **15.** $\frac{-30}{-25} = \frac{12}{10}$

17. $\frac{\$6.25}{1 \text{ hr}} = \frac{\$187.50}{30 \text{ hr}}$ **19.** $\frac{1 \text{ in.}}{7 \text{ mi}} = \frac{5 \text{ in.}}{35 \text{ mi}}$ **21.** No

23. Yes **25.** Yes **27.** Yes **29.** Yes
31. Yes **33.** No **35.** Yes **37.** No

39. 4 **41.** 3 **43.** -75 **45.** $\frac{3}{4}$

47. $-\frac{65}{4}$ or $-16\frac{1}{4}$ or -16.25 **49.** $\frac{15}{2}$ or $7\frac{1}{2}$ or 7.5

51. 3 **53.** 2.5 **55.** 4 **57.** $-\frac{1}{80}$ **59.** 36

61. 7.5 **63.** 30 **65.** $\frac{8}{7}$ **67.** -3 **69.** 12

71. -6

Chapter 6 Problem Recognition Exercises, p. 394

1. a. Proportion; $\frac{15}{2}$ **b.** Product of fractions; $\frac{15}{32}$

2. a. Product of fractions; $\frac{3}{25}$ **b.** Proportion; 4

3. a. Product of fractions; $\frac{3}{49}$ **b.** Proportion; 4

4. a. Proportion; 2 **b.** Product of fractions; $\frac{6}{25}$

5. a. Proportion; 9 **b.** Product of fractions; 32

6. a. Product of fractions; 8 **b.** Proportion; $\frac{98}{5}$

7a. 14 **b.** $\frac{5}{2}$ **c.** $\frac{3}{5}$ **d.** $\frac{18}{245}$ **8a.** $\frac{3}{25}$ **b.** $\frac{3}{5}$
c. $\frac{16}{3}$ **d.** $-\frac{88}{15}$ **9a.** 4 **b.** $\frac{98}{5}$ **c.** $\frac{48}{35}$
d. $\frac{49}{25}$ **10a.** 18 **b.** $\frac{29}{3}$ **c.** $\frac{11}{18}$ **d.** 22

Section 6.4 Practice Exercises, pp. 400–403

3. = **5.** ≠ **7.** $\frac{20}{3}$ or $6\frac{2}{3}$ or $6.\overline{6}$ **9.** −6
11. 4.9 **13.** Pam can drive 610 mi on 10 gal of gas. **15.** 78 kg of crushed rock will be required. **17.** The actual distance is about 80 mi. **19.** There are 3800 male students. **21.** Heads would come up about 315 times. **23.** There would be approximately 3 earned runs for a 9-inning game. **25.** Pierre can buy 576 Euros. **27.** 9 g **29.** There are approximately 357 bass in the lake. **31.** There are approximately 4000 bison in the park. **33.** $x = 24$ cm, $y = 36$ cm **35.** $x = 1$ yd, $y = 10.5$ yd **37.** The flagpole is 12 ft high. **39.** The platform is 2.4 m high. **41.** $x = 17.5$ in. **43.** $x = 6$ ft, $y = 8$ ft **45.** $x = 21$ ft; $y = 21$ ft; $z = 53.2$ ft

Section 6.4 Calculator Connections, p. 403

47. There were approximately 166,005 crimes committed. **48.** The Washington Monument is approximately 555 ft tall. **49.** Approximately 15,400 women would be expected to have breast cancer. **50.** Approximately 295,000 men would be expected to have prostate disease.

Chapter 6 Review Exercises, pp. 408–411

1. 5 to 4 and $\frac{5}{4}$ **2.** 3 : 1 and $\frac{3}{1}$ **3.** 8 : 7 and 8 to 7
4. a. $\frac{2}{3}$ **b.** $\frac{3}{2}$ **c.** $\frac{3}{5}$ **5. a.** $\frac{4}{5}$ **b.** $\frac{5}{4}$ **c.** $\frac{5}{9}$
6. a. $\frac{12}{52}$ or $\frac{3}{13}$ **b.** $\frac{12}{40}$ or $\frac{3}{10}$ **7.** $\frac{4}{1}$ **8.** $\frac{7}{5}$ **9.** $\frac{2}{5}$
10. $\frac{1}{4}$ **11.** $\frac{9}{2}$ **12.** $\frac{4}{13}$ **13.** $\frac{4}{3}$ **14.** $\frac{170}{13}$
15. a. This year's enrollment is 1520 students. **b.** $\frac{4}{19}$
16. $\frac{19}{12}$ **17.** $\frac{1}{5}$ **18.** $\frac{24}{49}$ **19.** $\frac{4 \text{ hot dogs}}{9 \text{ min}}$
20. $\frac{2 \text{ mi}}{17 \text{ min}}$ **21.** $-\frac{\$1700}{3 \text{ months}}$ **22.** $-\frac{5 \text{ m}}{3 \text{ min}}$
23. All unit rates have a denominator of 1, and reduced rates may not. **24.** 33 mph **25.** −4° per hour **26.** 90 times/sec **27.** 11 min/lawn **28.** $0.599 per ounce **29.** $3.333 per towel **30. a.** $0.125/oz **b.** $0.120/oz **c.** The 60-oz bottle is the best buy. **31. a.** $0.078/oz **b.** $0.075/oz **c.** The 48-oz jar is the best buy. **32.** $0.499 per ounce **33.** The difference is about 8¢ per roll or $0.08 per roll. **34.** 0.6275 in./hr **35. a.** There was an increase of 120,000 hybrid vehicles. **b.** There will be 10,000 additional hybrid vehicles per month. **36. a.** There was an increase of 63 lb. **b.** Americans increased the amount of vegetables in their diet by 3.5 lb per year.

37. $\frac{16}{14} = \frac{12}{10\frac{1}{2}}$ **38.** $\frac{8}{20} = \frac{6}{15}$ **39.** $\frac{-5}{3} = \frac{-10}{6}$
40. $\frac{4}{-3} = \frac{20}{-15}$ **41.** $\frac{\$11}{1 \text{ hr}} = \frac{\$88}{8 \text{ hr}}$ **42.** $\frac{2 \text{ in.}}{5 \text{ mi}} = \frac{6 \text{ in.}}{15 \text{ mi}}$
43. No **44.** Yes **45.** Yes **46.** No **47.** Yes **48.** No **49.** No **50.** Yes **51.** 4 **52.** 27 **53.** 3 **54.** 2 **55.** −13.6 **56.** −0.9 **57.** The human equivalent is 84 years. **58.** He can buy 48,150 yen. **59.** Alabama had approximately 4,600,000 people. **60.** The tax would be $6.96. **61.** $x = 10$ in., $y = 62.1$ in. **62.** The building is 8 m high. **63.** $x = 1.6$ yd, $y = 1.8$ yd **64.** $x = 10.8$ cm, $y = 30$ cm

Chapter 6 Test, pp. 411–412

1. 25 to 521, 25 : 521, $\frac{25}{521}$ **2. a.** $\frac{44}{27}$ **b.** $\frac{27}{71}$ **3.** $\frac{13}{4}$
4. $\frac{9}{8}$ **5.** $\frac{5}{8}$ **6. a.** $\frac{21}{125}$ **b.** $\frac{9}{125}$
c. The poverty ratio was greater in New Mexico.
7. a. $\frac{\frac{1}{2}}{1\frac{1}{2}} = \frac{1}{3}$ **b.** $\frac{30}{90} = \frac{1}{3}$ **8.** $\frac{85 \text{ mi}}{2 \text{ hr}}$ **9.** $\frac{10 \text{ lb}}{3 \text{ weeks}}$
10. $\frac{1 \text{ g}}{2 \text{ cookies}}$ **11.** 21.45 g/cm^3 **12.** 2.29 oz/lb
13. $0.22 per ounce **14.** $1.10 per ring **15.** Generic: $0.092/tablet; Aleve: $0.187/capsule. The generic pain reliever is the better buy. **16.** They form equal ratios or rates.
17. $\frac{-42}{15} = \frac{-28}{10}$ **18.** $\frac{20 \text{ pages}}{12 \text{ min}} = \frac{30 \text{ pages}}{18 \text{ min}}$
19. $\frac{\$15}{1 \text{ hr}} = \frac{\$75}{5 \text{ hr}}$ **20.** No **21.** −35 **22.** 12.5
23. 5 **24.** −6 **25.** 1 **26.** It will take 7.5 min. **27.** Cherise spends 30 hr each week on homework outside of class. **28.** There are approximately 27 fish in her pond. **29.** $x = 1\frac{1}{2}$ mi, $y = 8$ mi **30.** 16 cm

Chapters 1–6 Cumulative Review Exercises, p. 413

1. Five hundred three thousand, forty-two **2.** Approximately 1400 **3.** 22,600,000 **4.** 22 R 3 **5.** 5 **6.** 44 **7.** −23 **8.** −5 **9.** $7.14x - 8.2y$
10. 6 **11.** 3 **12.** $\frac{39}{14}$ **13.** $\frac{9}{25}$
14. Bruce has $4\frac{1}{2}$ in. of sandwich left. **15.** 2 **16.** $\frac{35}{9}$
17. There are 59 ninths. **18.** One thousand four and seven hundred one thousandths.
19. 28.057 **20.** $-\frac{109}{25}$ **21.** 4392.3 **22.** −2.379
23. 130.9 cm **24.** $\frac{61}{44}$ or 61 : 44 **25.** $\frac{13}{1}$
26. $\frac{7}{50}$; Approximately 7 out of 50 deaths are due to cancer.
27. 125 people/mi^2
28. a. Yes **b.** No **29.** −4.5 **30.** Jim can drive 100 mi on 4 gal.

Chapter 7

Chapter Opener Puzzle

H	O	U	S	E		O	F		P	I
1	2	3	4	5		2	6		7	8

Section 7.1 Practice Exercises, pp. 424–427

3. 84% **5.** 10% **7.** 2% **9.** 70% **11.** Replace the symbol % by $\times \frac{1}{100}$ (or ÷ 100). Then reduce the fraction to lowest terms. **13.** $\frac{3}{100}$ **15.** $\frac{21}{25}$ **17.** $\frac{17}{500}$ **19.** $\frac{23}{20}$ or $1\frac{3}{20}$ **21.** $\frac{1}{200}$ **23.** $\frac{1}{400}$ **25.** $\frac{31}{600}$ **27.** $\frac{249}{200}$ **29.** Replace the % symbol by × 0.01 (or ÷ 100). **31.** 0.58 **33.** 0.085 **35.** 1.42 **37.** 0.0055 **39.** 0.264 **41.** 0.5505 **43.** 27% **45.** 19% **47.** 175% **49.** 12.4% **51.** 0.6% **53.** 101.4% **55.** 71% **57.** 87.5% or $87\frac{1}{2}$% **59.** $83.\overline{3}$% or $83\frac{1}{3}$% **61.** 175% **63.** $122.\overline{2}$% or $122\frac{2}{9}$% **65.** $166.\overline{6}$% or $166\frac{2}{3}$% **67.** 42.9% **69.** 7.7% **71.** 45.5% **73.** 86.7% **75.** c **77.** e **79.** f **81.** e **83.** f **85.** a

87.

	Fraction	Decimal	Percent
a.	$\frac{1}{4}$	0.25	25%
b.	$\frac{23}{25}$	0.92	92%
c.	$\frac{3}{20}$	0.15	15%
d.	$\frac{8}{5}$ or $1\frac{3}{5}$	1.6	160%
e.	$\frac{1}{100}$	0.01	1%
f.	$\frac{1}{125}$	0.008	0.8%

89.

	Fraction	Decimal	Percent
a.	$\frac{7}{50}$	0.14	14%
b.	$\frac{87}{100}$	0.87	87%
c.	1	1	100%
d.	$\frac{1}{3}$	$0.\overline{3}$	$33.\overline{3}$% or $33\frac{1}{3}$%
e.	$\frac{1}{500}$	0.002	0.2%
f.	$\frac{19}{20}$	0.95	95%

91. 25% **93.** 10% **95.** 0.058; $\frac{29}{500}$ **97.** 0.084; $\frac{21}{250}$ **99.** The fraction $\frac{1}{2} = 0.5$ and $\frac{1}{2}\% = 0.5\% = 0.005$. **101.** 25% = 0.25 and 0.25% = 0.0025 **103.** a, c **105.** a, c **107.** 1.4 > 100% **109.** 0.052 < 50%

Section 7.2 Practice Exercises, pp. 432–436

3. 130% **5.** 37.5% or $37\frac{1}{2}$% **7.** 1% **9.** $\frac{1}{50}$ **11.** 0.82 **13.** 1 **15.** Amount: 12; base: 20; $p = 60$ **17.** Amount: 99; base: 200; $p = 49.5$ **19.** Amount: 50; base: 40; $p = 125$ **21.** $\frac{10}{100} = \frac{12}{120}$ **23.** $\frac{80}{100} = \frac{72}{90}$ **25.** $\frac{104}{100} = \frac{21{,}684}{20{,}850}$ **27.** 108 employees **29.** 0.2 **31.** 560 **33.** Pedro pays $20,160 in taxes. **35.** Jesse Ventura received approximately 762,200 votes. **37.** 36 **39.** 230 lb **41.** 1350 **43.** Albert makes $1600 per month. **45.** Aimee has a total of 35 e-mails. **47.** 35% **49.** 120% **51.** 87.5% **53.** She answered 72.5% correctly. **55.** 20% **57.** 26.7% **59.** 7.5 **61.** 40% **63.** 92 **65.** 77 **67.** 8800 **69.** 160% **71.** 70 mm of rain fell in August. **73.** Approximately 2110 freshmen were admitted. **75.** The hospital is filled to 84% occupancy. **77. a.** 22 women would be expected to relapse. **b.** 465 would not be expected to relapse. **79.** 73 were Chevys. **81.** There were 180 total vehicles. **83.** $331.20 **85.** $11.60 **87.** $6.30

Section 7.3 Practice Exercises, pp. 441–444

3. 146% **5.** 0.0002; $\frac{1}{5000}$ **7.** 4 **9.** 700 **11.** $x = (0.35)(700)$; $x = 245$ **13.** $(0.0055)(900) = x$; $x = 4.95$ **15.** $x = (1.33)(600)$; $x = 798$ **17.** 50% equals one-half of the number. So multiply the number by $\frac{1}{2}$. **19.** $2 \cdot 14 = 28$ **21.** $\frac{1}{2} \cdot 40 = 20$ **23.** There is 3.84 oz of sodium hypochlorite. **25.** Marino completed approximately 5015 passes. **27.** $18 = 0.4x$; $x = 45$ **29.** $0.92x = 41.4$; $x = 45$ **31.** $3.09 = 1.03x$; $x = 3$ **33.** There were 1175 subjects tested. **35.** At that time, the population was about 280 million. **37.** $x \cdot 480 = 120$; $x = 25\%$ **39.** $666 = x \cdot 740$; $x = 90\%$ **41.** $x \cdot 300 = 400$; $x = 133.3\%$ **43.** 5% of American Peace Corps volunteers were over 50 years old. **45. a.** There are 80 total employees. **b.** 12.5% missed 3 days of work. **c.** 75% missed 1 to 5 days of work. **47.** 27.9 **49.** 20% **51.** 150 **53.** 555 **55.** 600 **57.** 0.2% **59.** There were 35 million total hospital stays that year. **61.** Approximately 12.6% of Florida's panthers live in Everglades National Park. **63.** 416 parents would be expected to have started saving for their children's education. **65.** The total cost is $2400. **67. a.** $40,200 **b.** $41,614 **69.** 12,240 accidents involved drivers 35–44 years old. **71.** There were 40,000 traffic fatalities. **73. a.** 200 beats per minute **b.** Between 120 and 170 beats per minute

Chapter 7 Problem Recognition Exercises, p. 445

1. 41% **2.** 75% **3.** $33\frac{1}{3}$% **4.** 100% **5.** Greater than **6.** Less than **7.** Greater than **8.** Greater than **9.** 3000 **10.** 24% **11.** 4.8 **12.** 15% **13.** 70 **14.** 36 **15.** 6.3 **16.** 250 **17.** 300% **18.** 0.7 **19.** 75,000 **20.** 37.5% **21.** 25 **22.** 135 **23.** 100 **24.** 6000 **25.** 0.8% **26.** 125% **27.** 260 **28.** 20 **29.** 2.6 **30.** 6.05 **31.** 75% **32.** 25% **33.** $133\frac{1}{3}$% **34.** 400% **35.** 8.2 **36.** 4.1 **37.** 16.4 **38.** 41 **39.** 164 **40.** 12.3

Section 7.4 Practice Exercises, pp. 453–457

3. 12 **5.** 40 **7.** 26,000 **9.** 24% **11.** 8.8

13.	Cost of Item	Sales Tax Rate	Amount of Tax	Total Cost
a.	\$20	5%	\$1.00	\$21.00
b.	\$12.50	4%	\$0.50	\$13.00
c.	\$110	2.5%	\$2.75	\$112.75
d.	\$55	6%	\$3.30	\$58.30

15. The total bill is \$71.66. **17.** The tax rate is 7%. **19.** The price is \$44.50.

21.	Total Sales	Commission Rate	Amount of Commission
a.	\$20,000	5%	\$1000
b.	\$125,000	8%	\$10,000
c.	\$5400	10%	\$540

23. Zach made \$3360 in commission. **25.** Rodney's commission rate is 15%. **27.** Her sales were \$1,400,000.

29.	Original Price	Discount Rate	Amount of Discount	Sale Price
a.	\$56	20%	\$11.20	\$44.80
b.	\$900	$33\frac{1}{3}$%	\$300	\$600
c.	\$85	10%	\$8.50	\$76.50
d.	\$76	50%	\$38	\$38

31.	Original Price	Markup Rate	Amount of Markup	Retail Price
a.	\$92	5%	\$4.60	\$96.60
b.	\$110	8%	\$8.80	\$118.80
c.	\$325	30%	\$97.50	\$422.50
d.	\$45	20%	\$9	\$54

33. The discounted lunch bill is \$4.76. **35.** The discount rate is 25%. **37. a.** The markup is \$27.00. **b.** The retail price is \$177.00. **c.** The total price is \$189.39. **39.** The markup rate is 25%. **41.** The discount is \$80.70 and the sale price is \$188.30. **43.** The markup rate is 54%. **45.** The discount is \$11.00, and the sale price is \$98.99. **47.** c **49.** 100% **51.** 10% **53.** 7% **55.** 29.5% **57.** 5% **59.** 68% **61.** 15% **63. a.** \$49 per ticket **b.** 43.4%

Section 7.4 Calculator Connections, p. 457

	Stock	Price Jan. 2000 (\$ per share)	Price Jan. 2008 (\$ per share)	Change (\$)	Percent Increase
64.	eBay Inc.	\$16.84	\$26.15	\$9.31	55.3%
65.	Starbucks Corp.	\$6.06	\$16.92	\$10.86	179.2%
66.	Apple Inc.	\$28.66	\$178.85	\$150.19	524.0%
67.	IBM Corp.	\$118.37	\$127.52	\$9.15	7.7%

Section 7.5 Practice Exercises, pp. 463–465

3. 0.0225 **5.** 25.3% **7.** \$900, \$6900 **9.** \$1212, \$6262 **11.** \$2160, \$14,160 **13.** \$1890.00, \$12,390.00 **15. a.** \$350 **b.** \$2850 **17. a.** \$48 **b.** \$448 **19.** \$12,360 **21.** \$5625 **23.** There are 6 total compounding periods. **25.** There are 24 total compounding periods. **27. a.** \$560 **b.**

Year	Interest Earned	Total Amount in Account
1	\$20.00	\$520.00
2	20.80	540.80
3	21.63	**562.43**

29. a. \$26,400 **b.**

Period	Interest Earned	Total Amount in Account
1st	\$600	\$24,600
2nd	615	25,215
3rd	630.38	25,845.38
4th	646.13	**26,491.51**

31. A = total amount in the account; P = principal; r = annual interest rate; n = number of compounding periods per year; t = time in years

Section 7.5 Calculator Connections, p. 465

33. \$6230.91 **34.** \$14,725.49 **35.** \$6622.88 **36.** \$4373.77 **37.** \$10,934.43 **38.** \$9941.60 **39.** \$16,019.47 **40.** \$10,555.99

Chapter 7 Review Exercises, pp. 471–474

1. 75% **2.** 33% **3.** 125% **4.** 50% **5.** b, c **6.** c, d **7.** $\frac{3}{10}$; 0.3 **8.** $\frac{19}{20}$; 0.95 **9.** $\frac{27}{20}$; 1.35 **10.** $\frac{53}{25}$; 2.12 **11.** $\frac{1}{500}$; 0.002 **12.** $\frac{3}{500}$; 0.006 **13.** $\frac{2}{3}$; $0.\overline{6}$ **14.** $\frac{1}{3}$; $0.\overline{3}$ **15.** 62.5% **16.** 35% **17.** 175% **18.** 220% **19.** 0.6% **20.** 0.1% **21.** 400% **22.** 600% **23.** 42.9% **24.** 81.8% **25.** $\frac{6}{8} = \frac{75}{100}$ **26.** $\frac{27}{180} = \frac{15}{100}$ **27.** $\frac{840}{420} = \frac{200}{100}$ **28.** $\frac{6}{2000} = \frac{0.3}{100}$ **29.** 6 **30.** 3.68 **31.** 12.5% **32.** 0.32% **33.** 39 **34.** 20 **35.** Approximately 11 people would be no-shows. **36.** 850 people were surveyed. **37.** Victoria spends 40% on rent. **38.** There are 65 cars. **39.** $0.18 \cdot 900 = x$; $x = 162$ **40.** $x = 0.29 \cdot 404$; $x = 117.16$ **41.** $18.90 = x \cdot 63$; $x = 30\%$ **42.** $x \cdot 250 = 86$; $x = 34.4\%$ **43.** $30 = 0.25 \cdot x$; $x = 120$ **44.** $26 = 1.30 \cdot x$; $x = 20$ **45.** The original price was \$68.00. **46.** Veronica read 55% of the novel. **47.** Elaine can consume 720 fat calories. **48. a.** 39,000,000 **b.** 80,800,000 **49.** The sales tax is \$76.74. **50.** The sales tax rate is 7%. **51.** The cost before tax was \$8.80. The cost per photo is \$0.22. **52.** The total amount for 4 nights will be \$585.00. **53.** The commission rate was approximately 10.6%.

54. Andre earned \$489 in commission. **55.** Sela will earn \$75 that day. **56.** The house sold for \$160,000. **57.** The discount is \$8.69. The sale price is \$20.26. **58.** The discount is \$174.70. The final price is \$1522.30. **59.** The markup rate is 30%. **60.** The baskets will sell for \$59 each. **61.** 55.9% **62.** 35.4% **63.** \$1224, \$11,424 **64.** \$1400, \$8400 **65.** Jean-Luc will have to pay \$2687.50. **66.** Win will pay \$840.

67.

Year	Interest	Total
1	\$240.00	\$6240.00
2	249.60	6489.60
3	259.58	**6749.18**

68.

Compound Periods	Interest	Total
Period 1 (end of first 6 months)	\$150.00	\$10,150.00
Period 2 (end of year 1)	152.25	10,302.25
Period 3 (end of 18 months)	154.53	10,456.78
Period 4 (end of year 2)	156.85	10,613.63

69. \$995.91 **70.** \$2624.17 **71.** \$16,976.32 **72.** \$9813.88

Chapter 7 Test, pp. 474–475

1. 22% **2.** 0.054; $\frac{27}{500}$ **3.** 0.0015; $\frac{3}{2000}$ **4.** 1.70; $\frac{17}{10}$

5. a. $\frac{1}{100}$ **b.** $\frac{1}{4}$ **c.** $\frac{1}{3}$ **d.** $\frac{1}{2}$ **e.** $\frac{2}{3}$ **f.** $\frac{3}{4}$ **g.** 1 **h.** $\frac{3}{2}$

6. Multiply the fraction by 100%. **7.** 60% **8.** 0.4% **9.** 175% **10.** 71.4% **11.** Multiply the decimal by 100%. **12.** 32% **13.** 5.2% **14.** 130% **15.** 0.6% **16.** 36 **17.** 19.2 **18.** 350 **19.** 200 **20.** 90% **21.** 50% **22. a.** 730 mg **b.** 98.6% **23.** 390 m^3 **24.** 420 m^3 **25.** Her salary before the raise was \$52,000. Her new salary is \$56,160. **26. a.** The amount of sales tax is \$2.10. **b.** The sales tax rate is 7%. **27.** Charles will earn \$610. **28.** The discount rate of this product is 60%. **29. a.** The dining room set is \$1250 from the manufacturer. **b.** The retail price is \$1625. **c.** The cost after sales tax is \$1722.50. **30. a.** \$1200 **b.** \$6200 **31.** \$31,268.76

Chapters 1–7 Cumulative Review Exercises, pp. 476–477

1. Millions place **2.**

	Country	Area (mi^2) Standard Form	Words
a.	United States	3,539,245	Three million, five hundred thirty-nine thousand, two hundred forty-five
b.	Saudi Arabia	830,000	Eight hundred thirty thousand
c.	Falkland Islands	4,700	Four thousand, seven hundred
d.	Colombia	401,044	Four hundred one thousand, forty-four

3. 3,488,200 **4.** 87 **5.** −2705 **6.** 11 **7.** −18 **8.** $\frac{4}{3}$ or $1\frac{1}{3}$ **9.** −24 **10.** $\frac{3}{2}$ or $1\frac{1}{2}$ **11.** $\frac{2}{3}$

12. 9 km **13.** $\frac{473}{1000}$ **14.** $229\frac{1}{2}$ in.^{2} or 229.5 in.^{2}

15. a. 18, 36, 54, 72 **b.** 1, 2, 3, 6, 9, 18 **c.** $2 \cdot 3^2$

16. a. $\frac{5}{2}$ **b.** $\frac{5}{6}$ **17.** 0.375 **18.** $1.\overline{3}$ **19.** 87.5%

20. 240% **21.** 65.3% **22.** 105,986 mi^2

23. $-7x + 39$ **24.** $-\frac{1}{2}a + \frac{6}{5}b$ **25.** −36 **26.** 300

27. −4 **28.** 15 **29.** 20 **30.** $6\frac{1}{2}$ **31.** It will take about 7.2 min. **32.** The DC-10 flew 514 mph. **33.** Kevin will have \$15,080. **34.** 135% **35.** There is \$91,473.02 paid in interest.

Chapter 8

Chapter Opener Puzzle

1. f **2.** e **3.** b **4.** g **5.** c **6.** d **7.** a **8.** h

Section 8.1 Practice Exercises, pp. 486–489

3. 3 yd **5.** 42 in. **7.** $2\frac{1}{4}$ mi **9.** $4\frac{2}{3}$ yd

11. 563,200 yd **13.** $4\frac{3}{4}$ yd **15.** 72 in.

17. 50,688 in. **19.** $\frac{1}{2}$ mi **21. a.** 76 in. **b.** $6\frac{1}{3}$ ft

23. a. 8 ft **b.** $2\frac{2}{3}$ yd **25.** 6 ft **27.** 3 ft 4 in.

29. 4′4″ **31.** 730 days **33.** $1\frac{1}{2}$ hr **35.** 3 min

37. 3 days **39.** 1 hr **41.** 1512 hr **43.** 80.5 min **45.** 175.25 min **47.** 2 lb **49.** 4000 lb **51.** 6500 lb **53.** 10 lb 8 oz **55.** 8 lb 2 oz **57.** 6 lb 8 oz **59.** 2 c

61. 24 qt **63.** 16 c **65.** $\frac{1}{2}$ gal **67.** 16 fl oz

69. 6 tsp **71.** Yes, 3 c is 24 oz, so the 48-oz jar will suffice. **73.** The plumber used 7′2″ of pipe. **75.** 7 ft is left over.

77. $5\frac{1}{2}$ ft **79.** The unit price for the 24-fl-oz jar is about \$0.112 per ounce, and the unit price for the 1-qt jar is about \$0.103 per ounce; therefore the 1-qt jar is the better buy. **81.** The total length is 46′. **83.** The total weight is 312 lb 8 oz. **85.** 18 pieces of border are needed. **87.** Gil ran for 5 hr 35 min. **89.** The total time is 1 hr 34 min. **91.** 6 yd^2 **93.** 3 ft^2 **95.** 720 in.^{2} **97.** 27 ft^2

Section 8.2 Practice Exercises, pp. 497–501

3. 4 pt **5.** 1440 min **7.** 56 oz **9.** b, f, g **11.** 3.2 cm or 32 mm **13.** a **15.** d **17.** d **19.** 2.43 km **21.** 50,000 mm **23.** 4000 m **25.** 43.1 mm **27.** 0.3328 km **29.** 300 m **31.** 0.539 kg **33.** 2500 g **35.** 33.4 mg **37.** 4.09 g **39.** $<$ **41.** $=$ **43.** Cubic centimeter **45.** 3.2 L **47.** 700 cL **49.** 0.42 dL **51.** 64 mL **53.** 40 cc **55.** b, f **57.** a, f

	Object	mm	cm	m	km
59.	Length of the Mississippi River	3,766,000,000	376,600,000	3,766,000	3766
61.	Diameter of a quarter	24.3	2.43	0.0243	0.0000243

	Object	mg	cg	g	kg
63.	Bag of rice	907,000	90,700	907	0.907
65.	Hockey puck	170,000	17,000	170	0.17

	Object	mL	cL	L	kL
67.	Bottle of vanilla extract	59	5.9	0.059	0.000059
69.	Capacity of a gasoline tank	75,700	7570	75.7	0.0757

71. 4,669,000 m **73.** 305,000 mg **75.** 250 mL
77. Cliff drives 17.4 km per week.
79. 8 cans hold 0.96 kL. **81.** The bottle contains 49.5 cL.
83. 3.6 g **85.** No, she needs 1.04 m of molding.
87. The difference is 4500 m. **89.** 65 km^2
91. 0.56 m^2 **93.** 5.78 metric tons **95.** 8500 kg

Section 8.3 Practice Exercises, pp. 507–510

3. d, f **5.** b, e **7.** c, f **9.** b, g **11.** b
13. a **15.** 5.1 cm **17.** 8.8 yd **19.** 122 m
21. 1.1 m **23.** 15.2 cm **25.** 2.7 kg **27.** 0.4 oz
29. 1.2 lb **31.** 1980 kg **33.** 5.7 L **35.** 4 fl oz
37. 32 fl oz **39.** The box of sugar costs $0.100 per ounce, and the packets cost $0.118 per ounce. The 2-lb box is the better buy. **41.** 18 mi is about 28.98 km. Therefore the 30-km race is longer than 18 mi. **43.** 97 lb is approximately 43.65 kg. **45.** The price is approximately $7.22 per gallon. **47.** A hockey puck is 1 in. thick. **49.** Tony weighs about 222 lb. **51.** 45 cc is 1.5 fl oz. **53.** 40.8 ft
55. 77°F **57.** 20°C **59.** 86°F
61. 7232°F **63.** It is a hot day. The temperature is 95°F.
65. $F = \frac{9}{5}C + 32 = \frac{9}{5} \cdot 100 + 32 = 9(20) + 32$
$= 180 + 32 = 212$ **67.** 184 g **69.** The Navigator weighs approximately 2.565 metric tons. **71.** The average weight of the blue whale is approximately 240,000 lb.

Chapter 8 Problem Recognition Exercises, p. 510

1. 9 qt **2.** 2.2 m **3.** 12 oz **4.** 300 mL
5. 4 yd **6.** 6030 g **7.** 4.5 m **8.** $\frac{3}{4}$ ft
9. 2640 ft **10.** 3 tons **11.** 4 qt **12.** $\frac{1}{2}$ T
13. 0.021 km **14.** 6.8 cg **15.** 36 cc **16.** 4 lb
17. 4.322 kg **18.** 5000 mm **19.** 2.5 c **20.** 8.5 min
21. 0.5 gal **22.** 3.25 c **23.** 5460 g **24.** 902 cL
25. 16,016 yd **26.** 3 lb **27.** 3240 lb **28.** 4600 m
29. 2.5 days **30.** 8 mL **31.** 512.4 min **32.** 336 hr

Section 8.4 Practice Exercises, pp. 512–514

3. 4280 m **5.** 8 cg **7.** 9000 mL **9.** $6.\overline{2}$ lb
11. 10 mcg **13.** 7.5 mg **15.** 0.05 cg **17.** 500 mcg
19. 200 mcg **21.** 1 mg **23. a.** 400 mg **b.** 8000 mg or 8 g **25.** 3 mL **27.** 5.25 g of the drug would be given in 1 wk. **29.** 2 mL **31.** 500 people **33.** 9.6 mg
35. 3.6 g

Section 8.5 Practice Exercises, pp. 519–522

3. A line extends forever in both directions. A ray begins at an endpoint and extends forever in one direction.
5. Ray **7.** Point **9.** Line
11. For example:

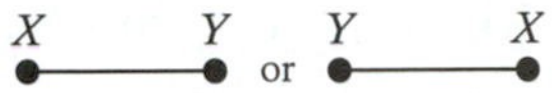

13. For example:

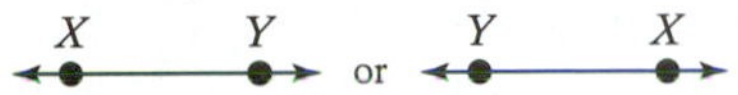

15.

17.

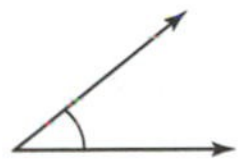

19. 20° **21.** 90° **23.** 148° **25.** Right
27. Obtuse **29.** Acute **31.** Straight
33. 10° **35.** 63° **37.** 60.5° **39.** 1°
41. 100° **43.** 53° **45.** 142.6° **47.** 1°
49. No, because the sum of two angles that are both greater than 90° will be more than 180°. **51.** Yes. For two angles to add to 90°, the angles themselves must both be less than 90°. **53.** A 90° angle
55.

57.

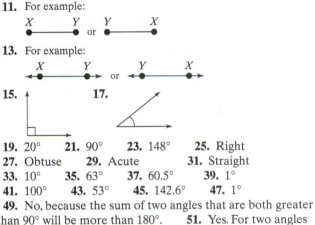

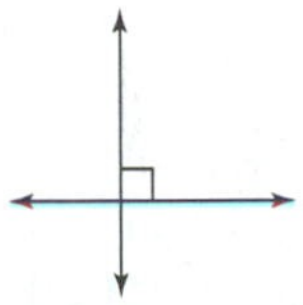

59. $m(\angle a) = 41°; m(\angle b) = 139°; m(\angle c) = 139°$
61. $m(\angle a) = 26°; m(\angle b) = 112°;$
$m(\angle c) = 26°; m(\angle d) = 42°$
63. The two lines are perpendicular. **65.** Vertical angles
67. *a*, *c* or *b*, *h* or *e*, *g* or *f*, *d* **69.** *a*, *e* or *f*, *b*
71. $m(\angle a) = 55°; m(\angle b) = 125°;$
$m(\angle c) = 55°; m(\angle d) = 55°; m(\angle e) = 125°; m(\angle f) = 55°;$
$m(\angle g) = 125°$
73. True **75.** True **77.** False **79.** True
81. True **83.** 70° **85.** 90° **87. a.** 48° **b.** 48°
c. 132° **89.** 180° **91.** 120°

Section 8.6 Practice Exercises, pp. 528–530

3. Yes **5.** No **7.** No **9.** $m(\angle a) = 54°$
11. $m(\angle b) = 78°$ **13.** $m(\angle a) = 60°, m(\angle b) = 80°$
15. $m(\angle a) = 40°, m(\angle b) = 72°$ **17.** c, f **19.** b, d
21. b, c, e **23.** 7 **25.** 49 **27.** 16 **29.** 4
31. 6 **33.** 36 **35.** 81 **37.** 9 **39.** $\frac{1}{4}$ **41.** 0.2
43. $c = 5$ m **45.** $b = 12$ yd **47.** Leg = 10 ft
49. Hypotenuse = 40 in. **51.** The brace is 20 in. long.
53. The height is 9 km. **55.** The car is 25 mi from the starting point. **57.** 24 m **59.** 30 km

Section 8.6 Calculator Connections, pp. 530–531

	Square Root	Estimate	Calculator Approximation (Round to 3 Decimal Places)
	$\sqrt{50}$	is between 7 and 8	7.071
61.	$\sqrt{10}$	is between 3 and 4	3.162
62.	$\sqrt{90}$	is between 9 and 10	9.487
63.	$\sqrt{116}$	is between 10 and 11	10.770
64.	$\sqrt{65}$	is between 8 and 9	8.062
65.	$\sqrt{5}$	is between 2 and 3	2.236
66.	$\sqrt{48}$	is between 6 and 7	6.928

67. 20.682 **68.** 56.434 **69.** 1116.244 **70.** 7100.423 **71.** 0.7 **72.** 0.5 **73.** 0.748 **74.** 0.906 **75.** $b = 21$ ft **76.** $a = 16$ cm **77.** Hypotenuse = 11.180 mi **78.** Hypotenuse = 8.246 m **79.** Leg = 18.439 in. **80.** Leg = 9.950 ft **81.** The diagonal length is 1.41 ft. **82.** The length of the diagonal is 134.16 ft. **83.** The length of the diagonal is 35.36 ft.

Section 8.7 Practice Exercises, pp. 538–542

3. An isosceles triangle has two sides of equal length. **5.** An acute triangle has all acute angles. **7.** An obtuse triangle has an obtuse angle. **9.** A quadrilateral is a polygon with four sides. **11.** A trapezoid has one pair of opposite sides that are parallel. **13.** A rectangle has four right angles. **15.** 80 cm **17.** 260 mm **19.** 10.7 m **21.** 25.12 ft **23.** 2.4 km or 2400 m **25.** 10 ft 6 in. **27.** 5 ft or 60 in. **29.** $x = 550$ mm; $y = 3$ dm; perimeter = 26 dm or 2600 mm **31.** 280 ft of rain gutters is needed. **33.** 576 yd^2 **35.** 23 in.^2 **37.** 18.4 ft^2 **39.** 54 m^2 **41.** 656 in.^2 **43.** 217.54 ft^2 **45.** 154 m^2 **47.** $346\frac{1}{2}$ cm^2 **49.** 113 mm^2 **51.** 5 ft^2 **53.** The area of the sign is 16.5 yd^2. **55.** $956.25 **57. a.** The area is 483 ft^2. **b.** They will need 2 paint kits. **59.** 13.5 cm^2 **61.** 18.28 in.^2 **63.** Perimeter = 72 ft

Chapter 8 Problem Recognition Exercises, p. 543

1. Area = 25 ft^2; perimeter = 20 ft **2.** Area = 144 m^2; perimeter = 48 m **3.** Area = 12 m^2 or 120,000 cm^2; perimeter = 14 m or 1400 cm **4.** Area = 1 ft^2 or 144 in.^2; perimeter = 5 ft or 60 in. **5.** Area = $\frac{1}{3}$ yd^2 or 3 ft^2; perimeter = 3 yd or 9 ft **6.** Area = 0.473 km^2 or 473,000 m^2; perimeter = 3.24 km or 3240 m **7.** Area = 6 yd^2; perimeter = 12 yd **8.** Area = 30 cm^2; perimeter = 30 cm **9.** Area = 44 m^2; perimeter = 32 m **10.** Area = 88 in.^2; perimeter = 40 in. **11.** Area ≈ 28.26 yd^2; circumference ≈ 18.84 yd **12.** Area ≈ 1256 cm^2; circumference ≈ 125.6 cm **13.** Area ≈ 2464 cm^2; circumference ≈ 176 cm **14.** Area ≈ 616 ft^2; circumference ≈ 88 ft **15. a.** perimeter **b.** 64 yd **16. a.** circumference **b.** $\frac{11}{2}$ ft or $5\frac{1}{2}$ ft **17. a.** area **b.** 125 ft^2 **18. a.** area **b.** 314 m^2

Section 8.8 Practice Exercises, pp. 548–552

3. Complement: 66°; supplement: 156° **5.** $m(\angle a) = 70°$, $m(\angle b) = 110°$, $m(\angle c) = 70°$, $m(\angle d) = 70°$, $m(\angle e) = 110°$, $m(\angle f) = 70°$, $m(\angle g) = 110°$ **7.** 2.744 cm^3 **9.** 48 ft^3 **11.** 12.56 mm^3 **13.** 3052.08 yd^3 **15.** 235.5 cm^3 **17.** 452.16 ft^3 **19.** 289 in.^3 **21.** 314 ft^3 **23.** 10 ft^3 **25. a.** 2575 ft^3 **b.** 19,260 gal **27.** 342 in.^2 **29.** 13.5 cm^2 **31.** 113.0 in.^2 **33.** 211.1 in.^2 **35.** 492 ft^2 **37.** 96 cm^2 **39.** 1356.48 in.^2 **41.** 1256 mm^2 **43.** $\frac{11}{36}$ ft^3 or 0.306 ft^3 or 528 in.^3 **45.** 109.3 in.^3 **47.** 502.4 mm^3 **49.** 621.72 ft^3 **51.** 84.78 in.^3 **53.** 50,240 cm^3 **55.** 400 ft^3

Chapter 8 Review Exercises, pp. 561–565

1. 4 ft **2.** 39 in. **3.** 3520 yd **4.** $1\frac{1}{3}$ mi **5.** 2640 ft **6.** 72 in. **7.** 9 ft 3 in. **8.** 6′4″ **9.** 2′10″ **10.** 3 ft 9 in. **11.** $7\frac{1}{2}$ ft **12.** There is 102 ft or 34 yd of wire left. **13.** 3 days **14.** 360 sec **15.** 80 oz **16.** 168 hr **17.** $1\frac{1}{2}$ c **18.** 500 lb **19.** $1\frac{3}{4}$ tons **20.** 16 pt **21.** $\frac{3}{4}$ lb **22.** 4 gal **23.** 144.5 min **24.** The total weight was 11 lb 13 oz. **25.** b **26.** d **27.** 520 mm **28.** 0.093 km **29.** 3.4 m **30.** 0.21 dam **31.** 0.04 m **32.** 1200 mm **33.** 610 cg **34.** 0.42 kg **35.** 3.212 g **36.** 70 g **37.** 8.3 L **38.** 124 cc **39.** 22.5 cL **40.** 490 L **41.** Perimeter: 6.5 m; area: 2.5 m^2 **42.** 5 glasses can be filled. **43.** The difference is 64.8 kg. **44.** No, the board is 25 cm too short. **45.** 15.75 cm **46.** 2.5 fl oz **47.** 5 oz **48.** 5.26 qt **49.** 1.04 m **50.** 45 kg **51.** 74.53 mi **52.** 5.7 L **53.** 45 cc **54.** 11,250 kg **55.** The difference in height is 38.2 cm. **56.** There are approximately 6.72 servings. **57.** The marathon is approximately 26.2 mi. **58.** $C = \frac{5}{9}(F - 32)$ **59.** 82.2°C to 85°C **60.** $F = \frac{9}{5}C + 32$ **61.** 46.4°F **62.** 450 mcg **63.** 1500 mcg **64.** 0.4 mg **65.** 0.5 cg **66.** 2.5 mg/cc **67. a.** 3.2 mg **b.** 44.8 mg **68.** The total amount of cough syrup is approximately 0.42 L. **69.** There is 1.2 cc or 1.2 mL of fluid left. **70.** Clayton took 7.5 g. **71.** d **72.** a **73.** c **74.** b **75.** The measure of an acute angle is between 0° and 90°. **76.** The measure of an obtuse angle is between 90° and 180°. **77.** The measure of a right angle is 90°. **78. a.** 57° **b.** 147° **79. a.** 70° **b.** 160° **80.** 62°

81. 118° **82.** 118° **83.** 62° **84.** 62°
85. $m(\angle x) = 40°$ **86.** $m(\angle x) = 80°$; $m(\angle y) = 32°$
87. An equilateral triangle has three sides of equal length and three angles of equal measure. **88.** An isosceles triangle has two sides of equal length and two angles of equal measure. **89.** 5 **90.** 7 **91.** $b = 7$ cm
92. $c = 20$ ft **93.** 13 m of string is extended.
94. 90 cm **95.** 17.3 m **96.** 400 yd **97.** 15.5 ft
98. 62.8 cm **99.** 20 in.^{2} **100.** 51 ft^2 **101.** 7056 ft^2
102. 18 ft^2 **103.** 314 cm^2 **104.** 36 cm^2
105. Volume: 25,000 cm^3; SA: 5250 cm^2
106. Volume: 226.08 ft^3; SA: 207.24 ft^2
107. Volume: 14,130 in.^{3}; SA: 2826 in.^{2}
108. Volume: 512 mm^3; SA: = 384 mm^2
109. 37.68 m^3 **110.** 995 in.^{3} **111.** 113 in.^{3}
112. $2\frac{1}{3}$ ft^3 **113.** 335 cm^3 **114.** 28,500 in.^{3}

Chapter 8 Test, pp. 565–568

1. c, d, g, j **2.** f, h, i **3.** a, b, e **4.** $8\frac{1}{3}$ yd
5. 5.5 tons **6.** 10 mi **7.** 10 oz of liquid
8. 20 min **9.** 9′ **10.** 4′2″ **11.** He lost 7 oz.
12. 19 ft 7 in. **13.** 75.25 min **14.** 2.4 cm or 24 mm
15. c **16.** 1.158 km **17.** 15 mL
18. a. Cubic centimeters **b.** 235 cc **c.** 1000 cc
19. 41,100 cg **20.** 7 servings **21.** 2.1 qt **22.** 109 yd
23. 2.8 mi **24.** 2929 m **25.** 50.8 cm tall and 96.52-cm wingspan **26.** 11 lb **27.** 190.6°C **28.** 35.6°F
29. 28 mg **30.** 1750 mcg per week **31.** d
32. c **33.** 74° **34.** 33° **35.** 103° **36.** Perimeter: 28 cm; area 36 cm^2 **37.** 70,650 ft^2 **38.** 48 ft^2
39. $m(\angle x) = 125°$, $m(\angle y) = 55°$
40. They are each 45°. **41.** $m(\angle S) = 49°$ **42.** 180°
43. $m(\angle A) = 80°$ **44.** 12 ft **45.** 100 m
46. 3 rolls are needed. **47.** The area is 72 in.^{2}
48. The volume is about 151 ft^3.
49. The volume is 1260 in.^{3} **50.** 65.94 in.^{3}
51. 113.04 cm^3 **52.** 113.04 cm^2 **53.** 824 in.^{2}

Chapters 1–8 Cumulative Review Exercises, pp. 568–570

1. a. 2000 **b.** 42,100 **2.** 56 cm **3.** 180 cm^2
4. −4 **5.** The number 32,542 is not divisible by 3 because the sum of the digits (16) is not divisible by 3.
6. $2 \cdot 2 \cdot 3 \cdot 3 \cdot 3$ **7.** 540 in.^{2} **8.** $9\frac{1}{2}$ **9.** $18\frac{8}{9}$
10. $2\frac{6}{17}$ **11.** $3\frac{5}{6}$ **12.** −6 **13.** −36
14. $3x - z$ **15.** $-3a + 19b$ **16.** −7
17. −25 **18.** −18 **19.** 2 **20.** 40 bottles
21. 90 cars **22.** 6.7 beds per nurse **23.** 80%
24. $x = \frac{16}{3}$ yd or $5\frac{1}{3}$ yd **25.** 2290 trees **26.** 27 people
27. 6% **28.** $15,000 in sales **29.** $1020 in interest
30. 5.8 kg **31.** 12.9 lb **32.** 182.9 cm **33.** 6 ft
34. 7 pt **35.** 3.3 L **36.** Complement: 52°; supplement: 142° **37.** 96 in. **38.** 5 ft **39.** 57°
40. Volume: 840 cm^3; SA: 604 cm^2

Chapter 9

Chapter Opener Puzzle

In a rectangular coordinate system, we use an

O	R	D	E	R	E	D		P	A	I	R
6	1	5	2	1	2	5		3	7	4	1

to represent the location of a

P	O	I	N	T
3	6	4	8	9

Section 9.1 Practice Exercises, pp. 576–579

2.–8.

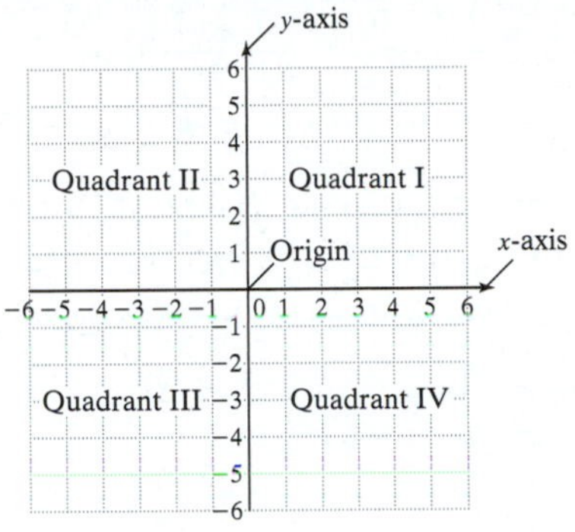

9.–14.

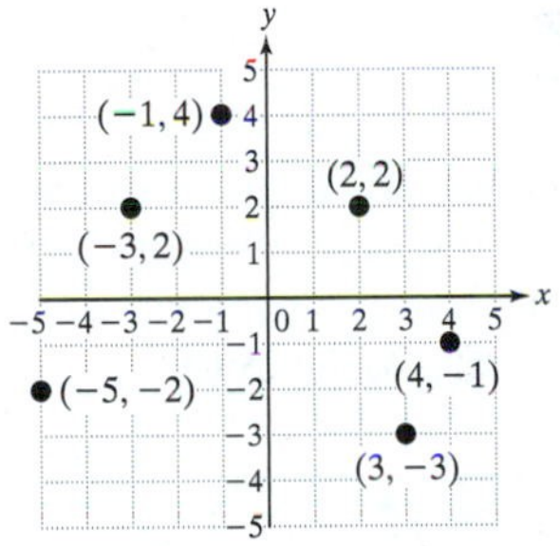

15. First move to the left 1.8 units from the origin. Then go up 3.1 units. Place a dot at the final location. The point is in Quadrant II.

17.–22.

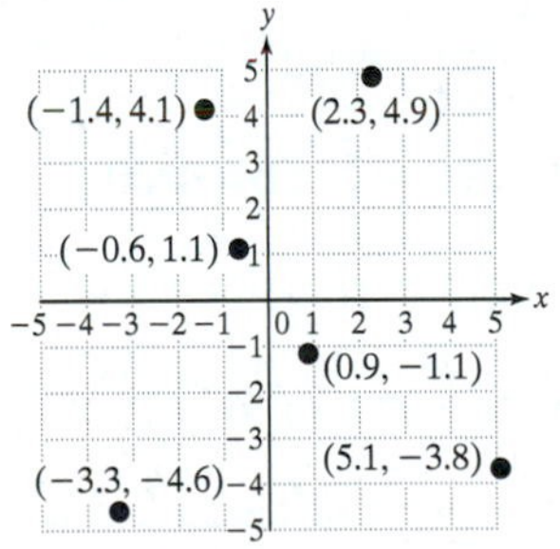

23.–28.

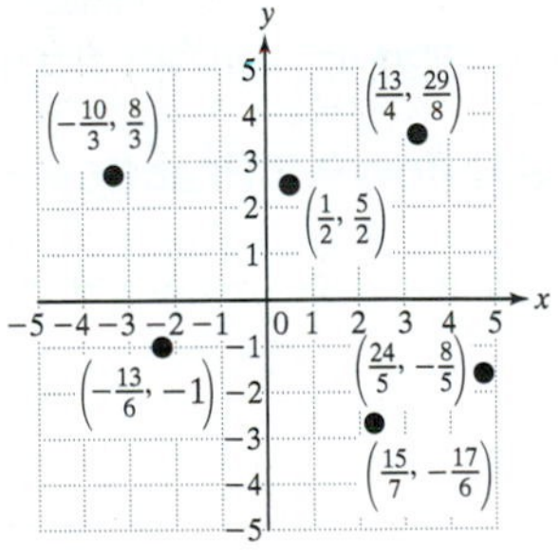

29.–34.

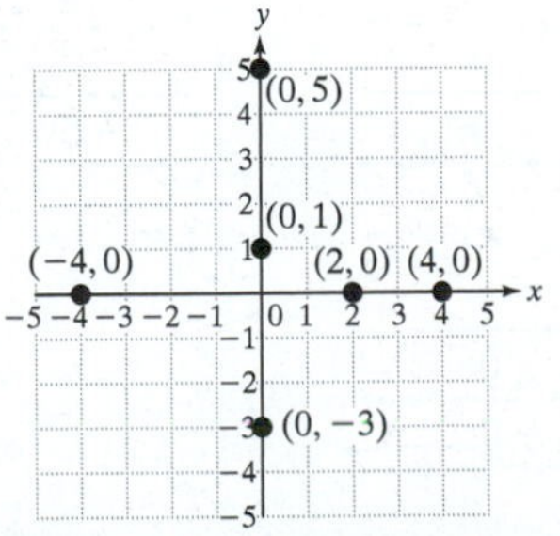

35. Quadrant IV **37.** Quadrant III **39.** x-axis
41. y-axis **43.** Quadrant II **45.** Quadrant I
47. (0, 3) **49.** (2, 3) **51.** (−5, −2) **53.** (4, −2)
55. (−2, −5)

57.

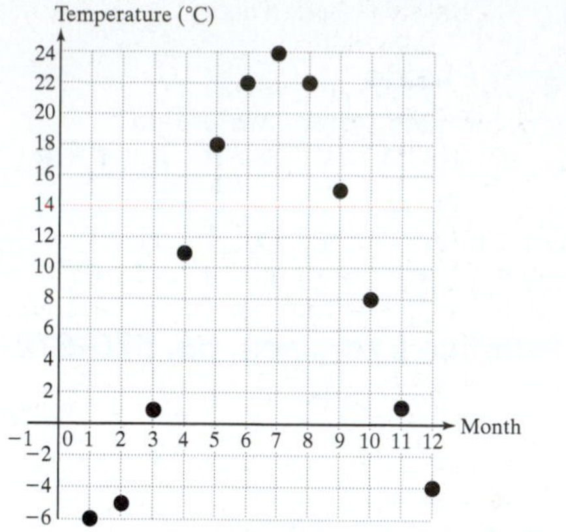

(1, −6), (2, −5), (3, 1), (4, 11), (5, 18), (6, 22), (7, 24), (8, 22), (9, 15), (10, 8), (11, 1), (12, −4)

59.

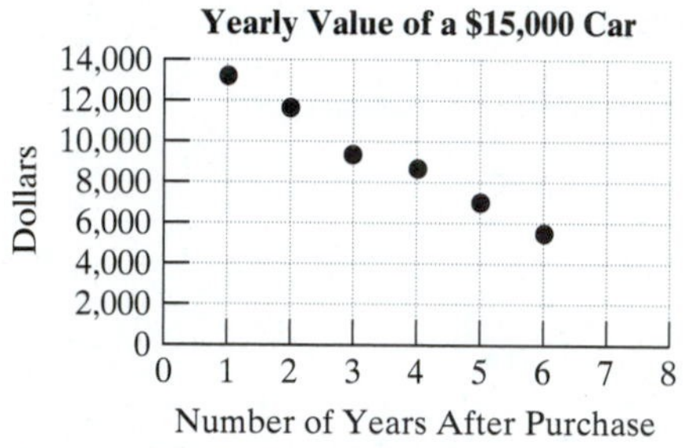

(1, 13,200), (2, 11,352), (3, 9649), (4, 8201), (5, 6971), (6, 5925)

61.

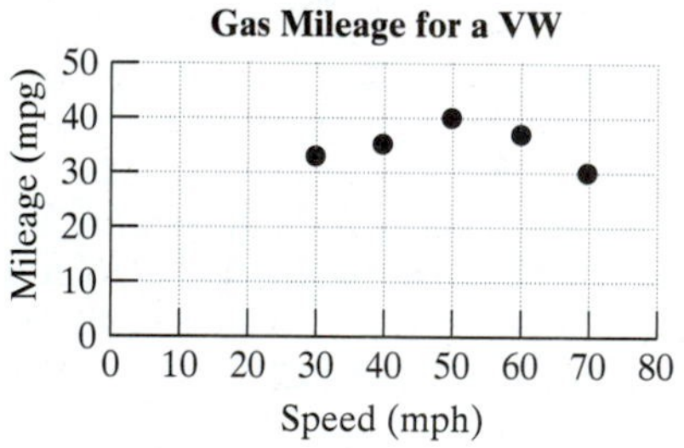

(30, 33), (40, 35), (50, 40), (60, 37), (70, 30)

Section 9.2 Practice Exercises, pp. 586–590

3. (4, 2); Quadrant I **5.** (0, −3); *y*-axis **7.** (2, −4); Quadrant IV **9.** Yes **11.** No **13.** Yes **15.** No **17.** Yes **19.** No **21.** (2, 4) **23.** (3, 1) **25.** (8, 3) **27.** (8, 9)

29.

x	y
−1	9
0	6
2	0

31.

x	y
−3	0
0	−4
−6	4

33.

x	y
0	−3
5	0
−10	−9

35.

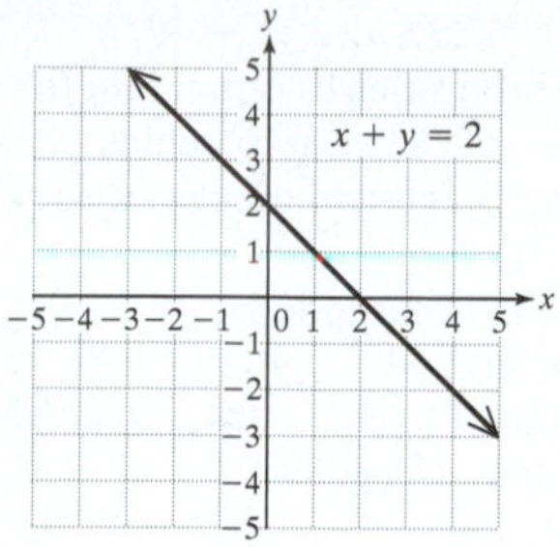

37.

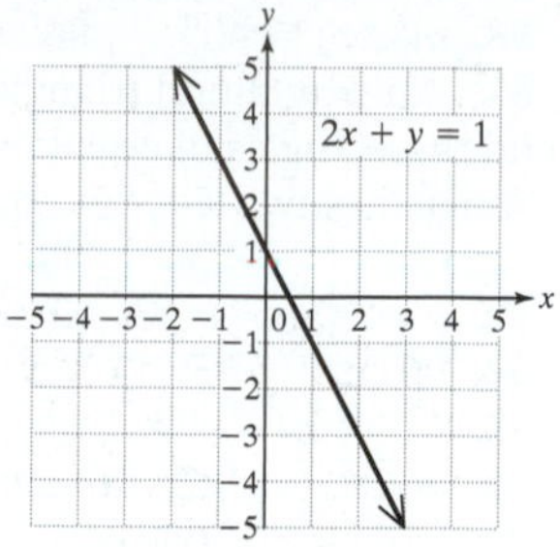

39.

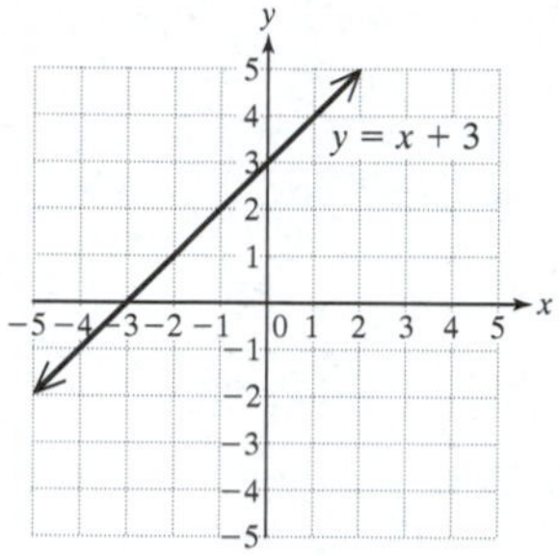

41.

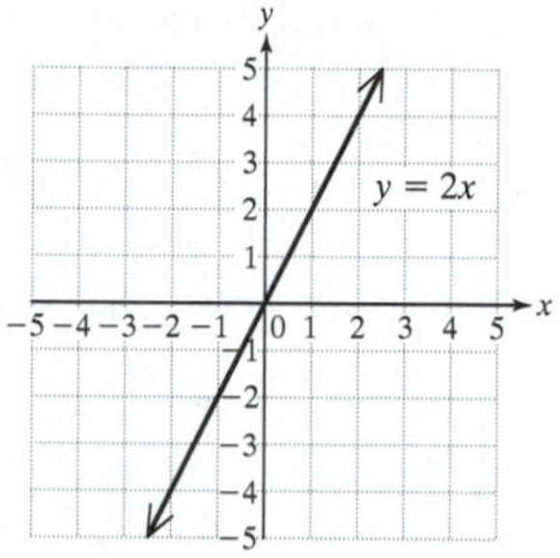

43.

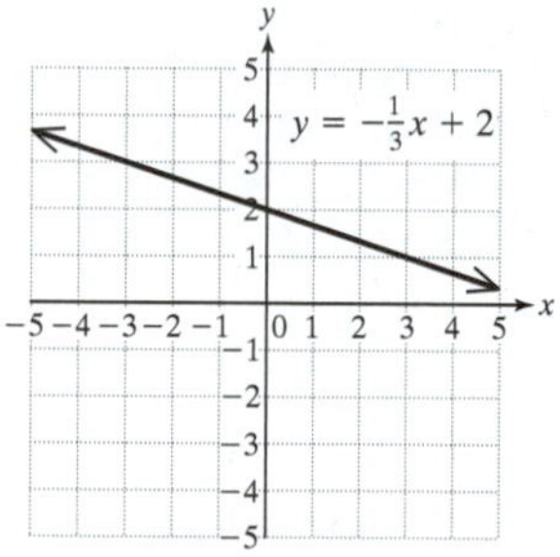

45.

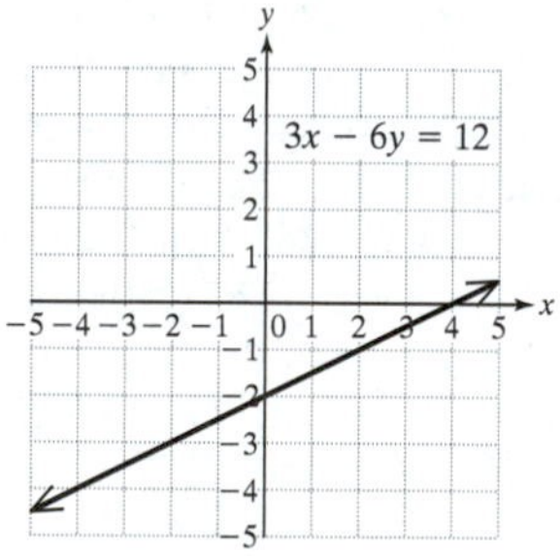

47. Vertical

49.

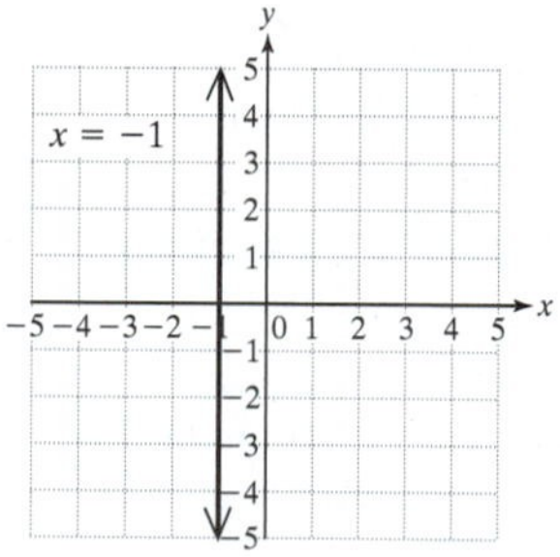

51.

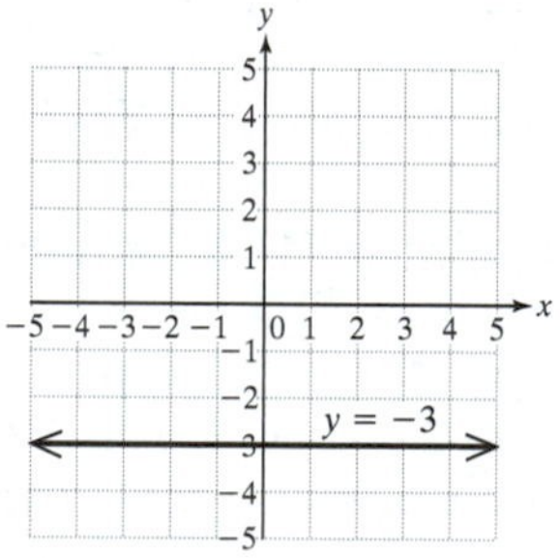

53.

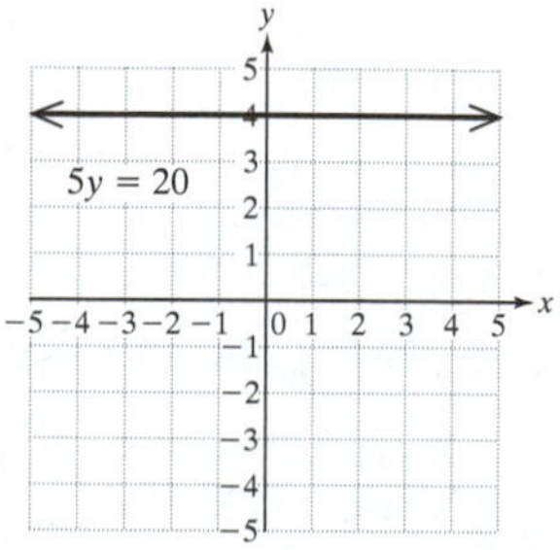

55.

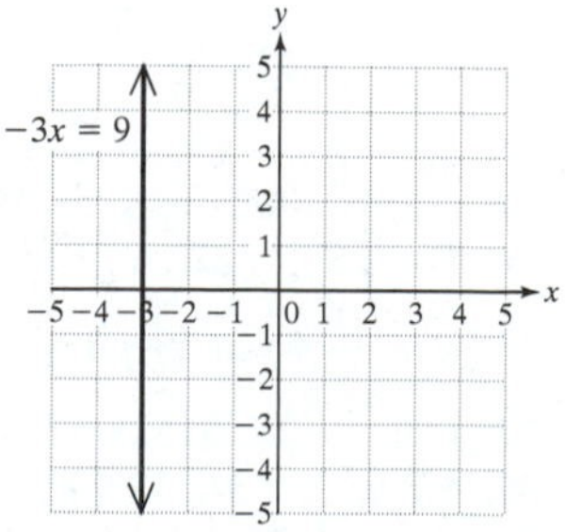

57. a. (0, 2) and (3, 0)

b.

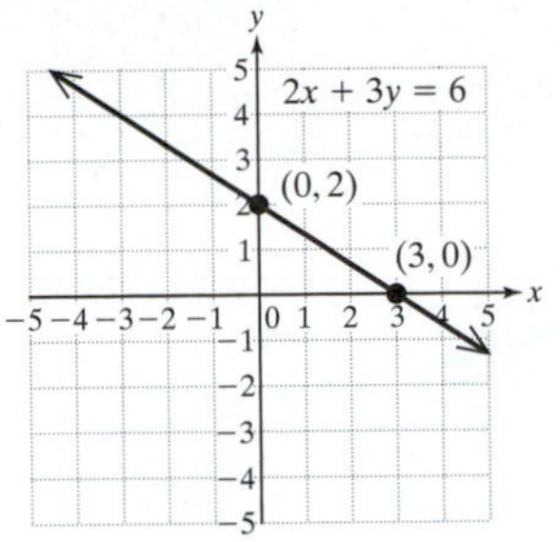

c. (3, 0) is the x-intercept; (0, 2) is the y-intercept.

59. a. (0, 6) and (2, 0)

b.

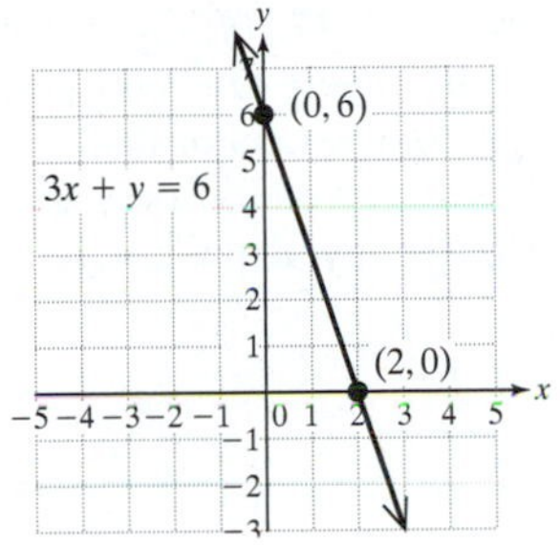

c. (2, 0) is the x-intercept; (0, 6) is the y-intercept.

Section 9.3 Practice Exercises, pp. 597–602

2.–6.

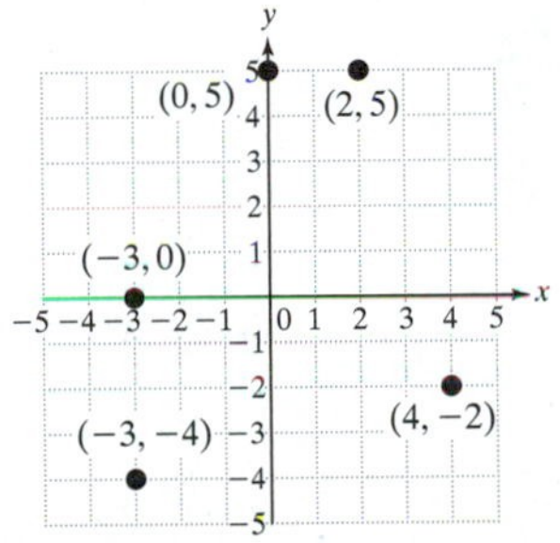

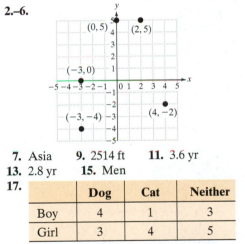

7. Asia **9.** 2514 ft **11.** 3.6 yr **13.** 2.8 yr **15.** Men

17.

	Dog	Cat	Neither
Boy	4	1	3
Girl	3	4	5

19. a. Students had the best access to a computer in 2007. Only four students were sharing a computer.

b. **Students per Computer**

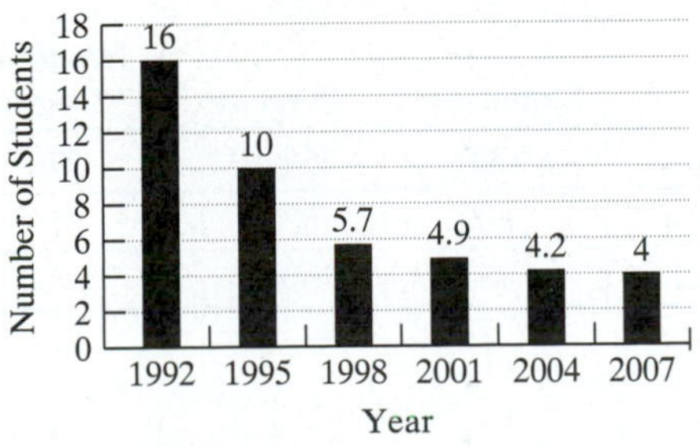

21. **Number of Cellular Phone Subscriptions**

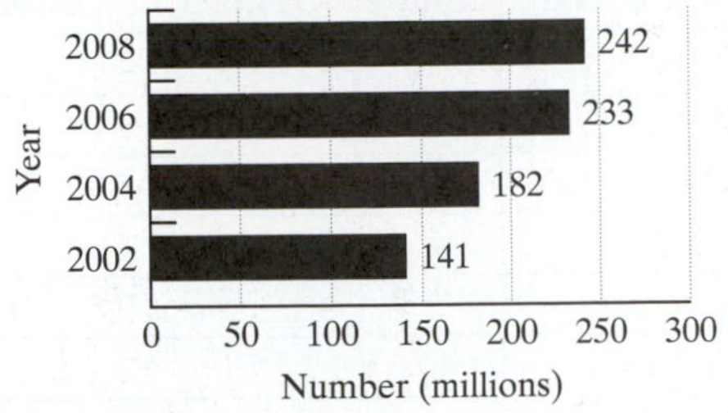

23. a. One icon represents 100 servings sold. **b.** About 450 servings **c.** Sunday **25. a.** \$65 million **b.** *Spider-Man 3* **c.** Approximately \$220 million **27.** 48.4% **29.** The trend for women over 65 in the labor force shows a slight increase. **31.** For example: 18% **33.** The year 2007 had the greatest number of kidney transplants. **35.** There were 982 more kidney transplants in 2003 than in 2001. **37.** Approximately 7600 transplants would be performed in the year 2009.

39. a. **Average Height for Girls, Ages 2–9**

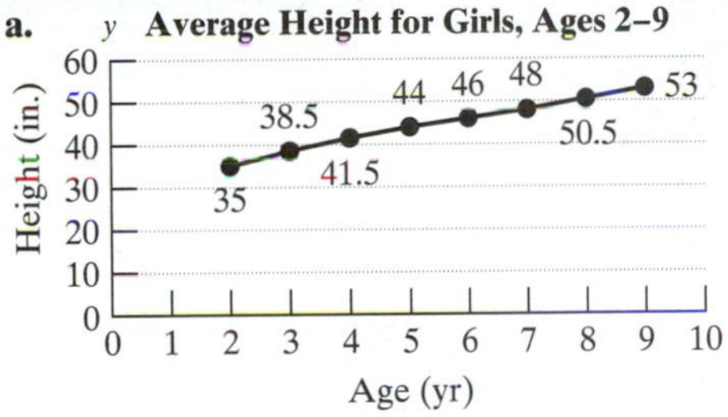

b. ≈ 55 in.

41. There are 14 servings per container, which means that there is 8 g × 14 = 112 g of fat in one container. **43.** The daily value of fat is approximately 61.5 g.

Section 9.4 Practice Exercises, pp. 605–608

3. False **5.** True **7.** There are 72 data values. **9.** 9–12

11.

Class Intervals (Age in Years)	Tally	Frequency (Number of Professors)
56–58	II	2
59–61	I	1
62–64	I	1
65–67	~~IIII~~ II	7
68–70	~~IIII~~	5
71–73	IIII	4

a. The class of 65–67 yr has the most values. **b.** There are 20 values represented in the table. **c.** Of the professors, 25% retire when they are 68 to 70 yr old.

13.

Class Intervals (Amount in Gal)	Tally	Frequency (Number of Customers)
8.0–9.9	||||	4
10.0–11.9	|	1
12.0–13.9	~~||||~~	5
14.0–15.9	||||	4
16.0–17.9		0
18.0–19.9	||	2

a. The 12.0–13.9 gal class has the highest frequency.
b. There are 16 data values represented in the table.
c. Of the customers, 12.5% purchased 18 to 19.9 gal of gas.

15. The class widths are not the same. **17.** There are too few classes. **19.** The class intervals overlap. For example, it is unclear whether the data value 12 should be placed in the first class or the second class.

21.

Class Interval (Height, in.)	Frequency (Number of Students)
62–63	2
64–65	3
66–67	4
68–69	4
70–71	4
72–73	3

23. Heights of Students at Daytona State College

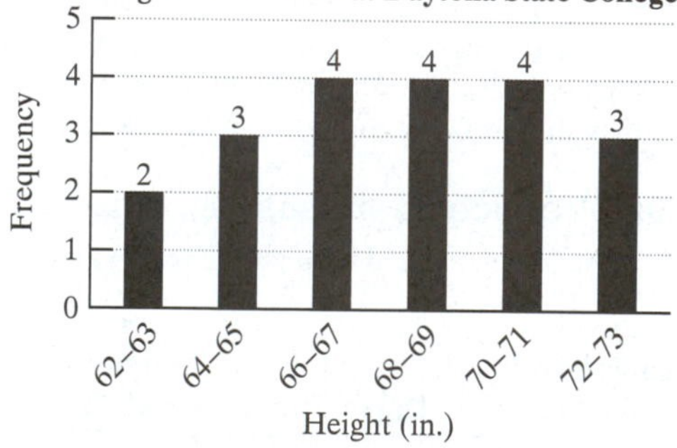

25. Number of Calories in 100 g of Selected Fruits

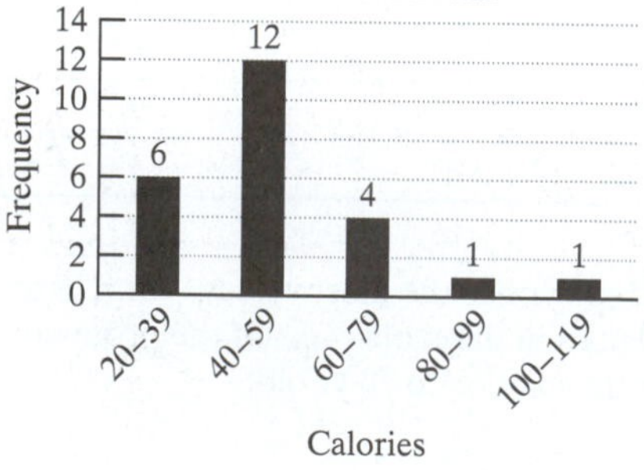

Section 9.5 Practice Exercises, pp. 613–616

3.

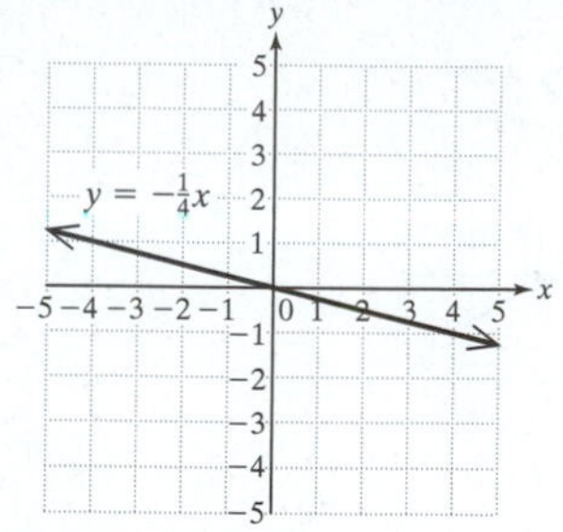

5. 64,000 **7.** 640 **9.** 25% **11.** 2.5 times
13. There are 20 million viewers represented.
15. 1.8 times as many viewers **17.** Of the viewers, 18% watch *General Hospital.* **19.** There are 960 Latina CDs.
21. There are 640 CDs that are classical or jazz.
23. There were 9 Super Bowls played in Louisiana.
25. There were 2 Super Bowls played in Georgia.

27.

29.

31.

33.

35.

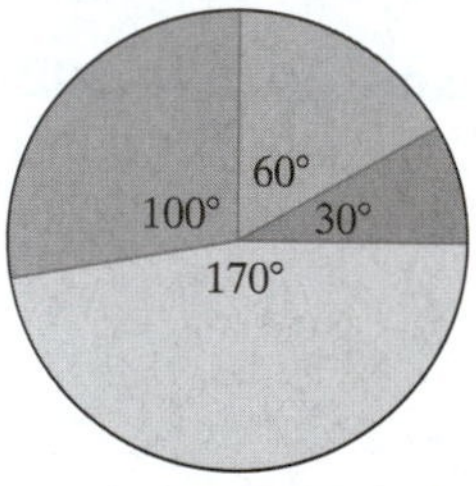

37. Sunshine Nursery Distribution of Sales

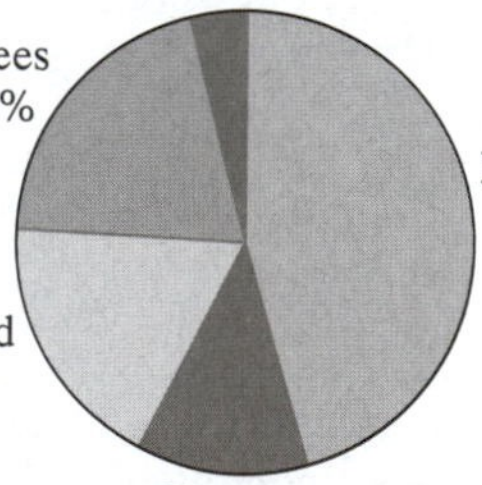

39. a.

	Expenses	Percent	Number of Degrees
Tuition	$9000	75%	270°
Books	600	5%	18°
Housing	2400	20%	72°

b.

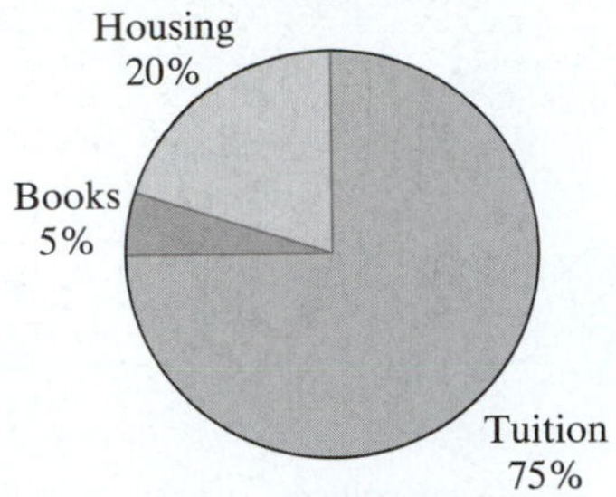

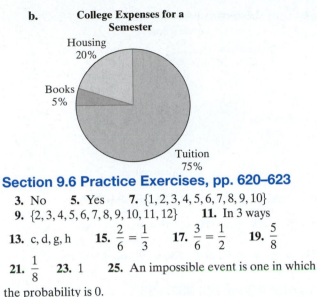

Section 9.6 Practice Exercises, pp. 620–623

3. No **5.** Yes **7.** {1, 2, 3, 4, 5, 6, 7, 8, 9, 10} **9.** {2, 3, 4, 5, 6, 7, 8, 9, 10, 11, 12} **11.** In 3 ways **13.** c, d, g, h **15.** $\frac{2}{6} = \frac{1}{3}$ **17.** $\frac{3}{6} = \frac{1}{2}$ **19.** $\frac{5}{8}$ **21.** $\frac{1}{8}$ **23.** 1 **25.** An impossible event is one in which the probability is 0. **27.** $\frac{12}{52} = \frac{3}{13}$ **29.** $\frac{12}{16} = \frac{3}{4}$ **31. a.** $\frac{18}{120} = \frac{3}{20}$ **b.** $\frac{27}{120} = \frac{9}{40}$ **c.** 30% **33. a.** $\frac{21}{60} = \frac{7}{20}$ **b.** 50% **35. a.** $\frac{7}{29}$ **b.** $\frac{11}{29}$ **c.** 62% **37.** $1 - \frac{2}{11} = \frac{9}{11}$ **39.** 100% − 1.2% = 98.8%

Chapter 9 Review Exercises, pp. 628–632

1.

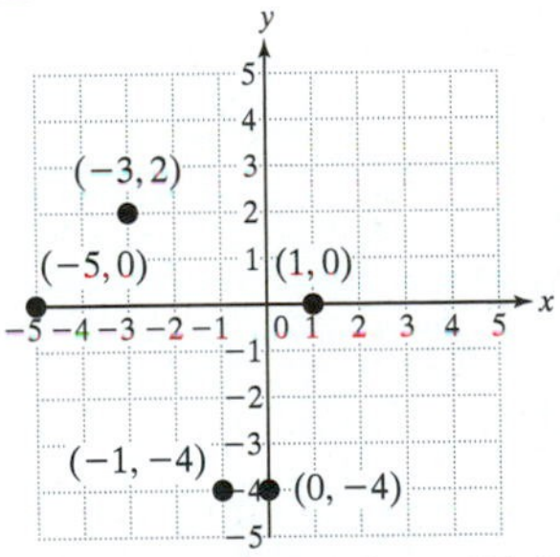

2.

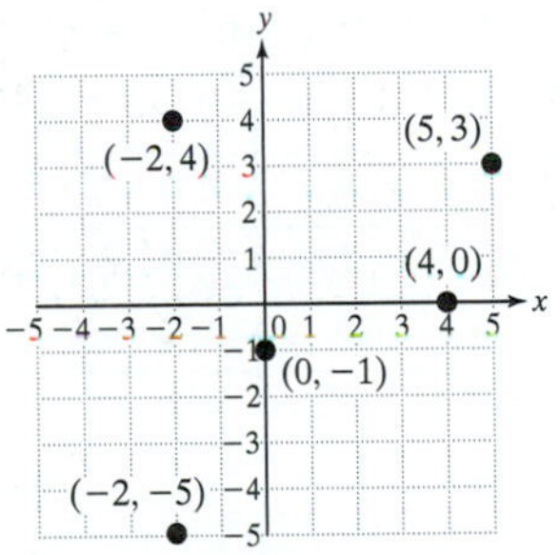

3. *A* (−5, 4) Quadrant II; *B* (0, 1) *y*-axis; *C* (1, 0) *x*-axis; *D* (4, −2) Quadrant IV; *E* (−4, −5) Quadrant III
4. *A* (2, 5) Quadrant I; *B* (−5, 0) *x*-axis; *C* (0, −2) *y*-axis; *D* (−4, −3); Quadrant III; *E* (1, −5) Quadrant IV
5. (1, − 7), (2, −6), (3, 4), (4, 2), (5, 0)

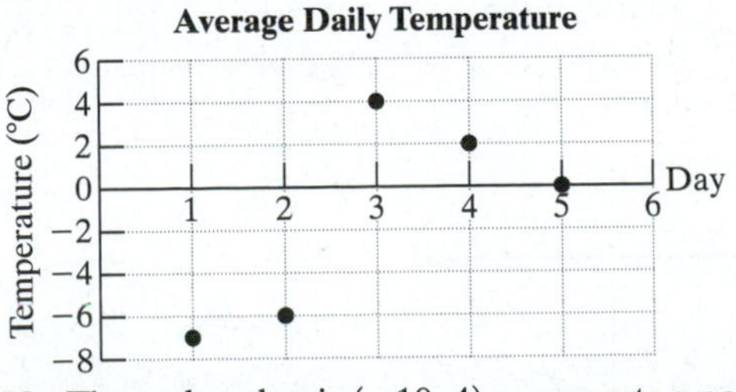

6. No. The ordered pair (−10, 4) represents a point in Quadrant II. The ordered pair (4, −10) represents a point in Quadrant IV. **7.** Yes **8.** Yes **9.** No **10.** No **11.** Yes **12.** Yes **13.** (2, 4), (1, 1) **14.** (5, 4), (7, 5)

15.

x	y
−3	5
−6	2
5	13

16.

x	y
4	−8
3	−10
8	0

17.

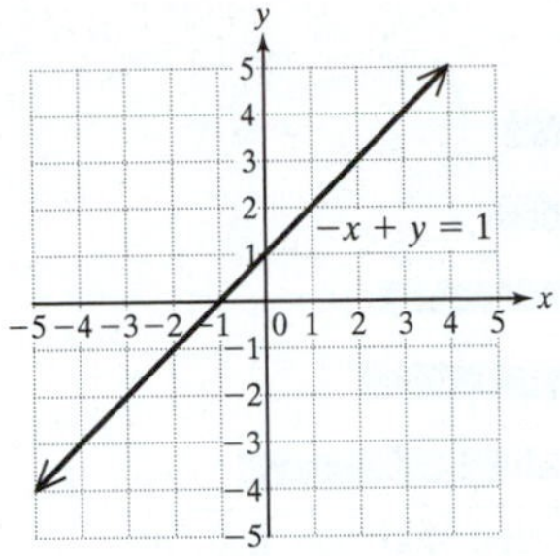

18.

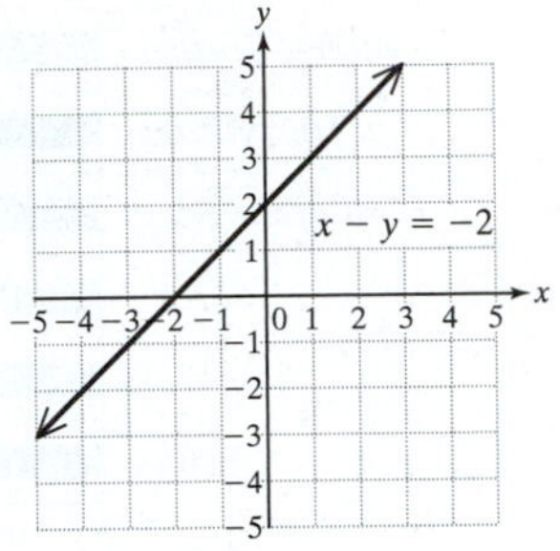

19.

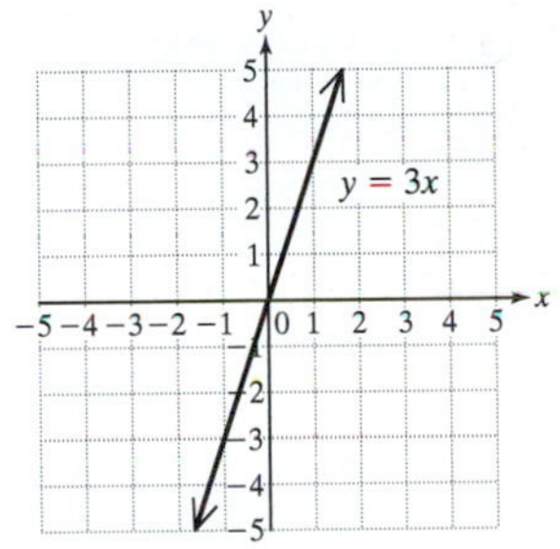

20.

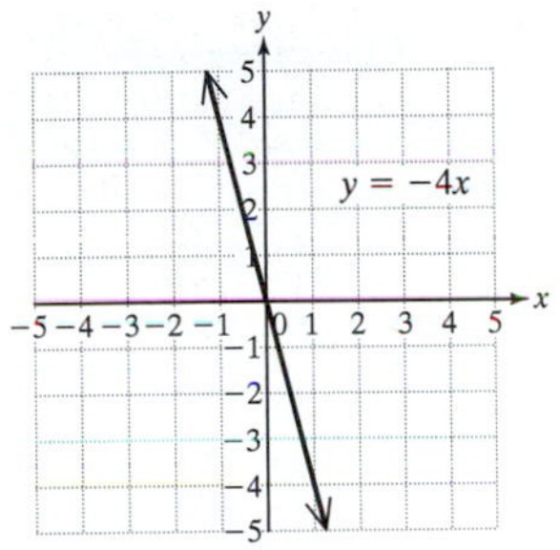

21.

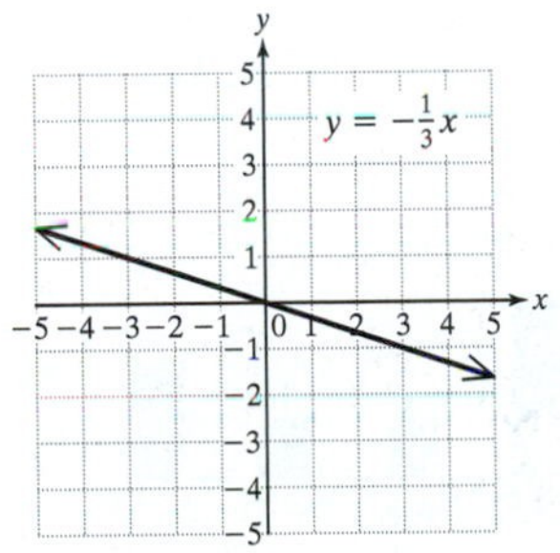

22.

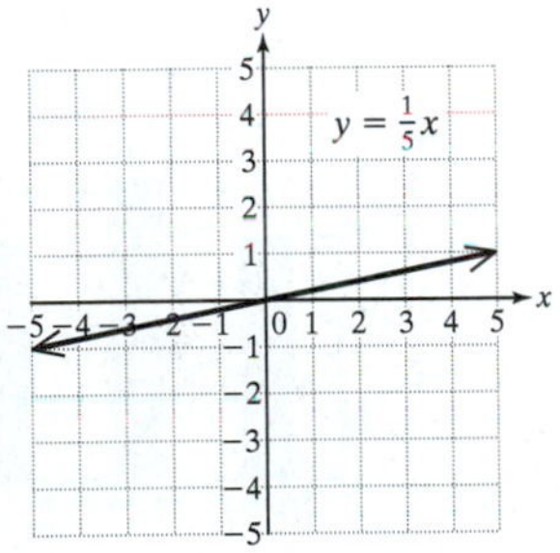

23.

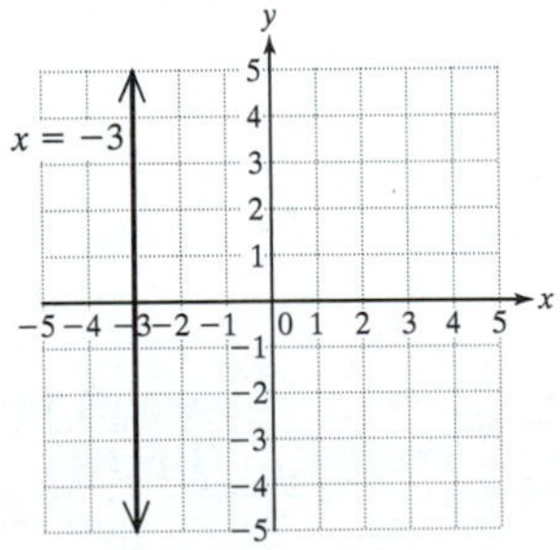

24.

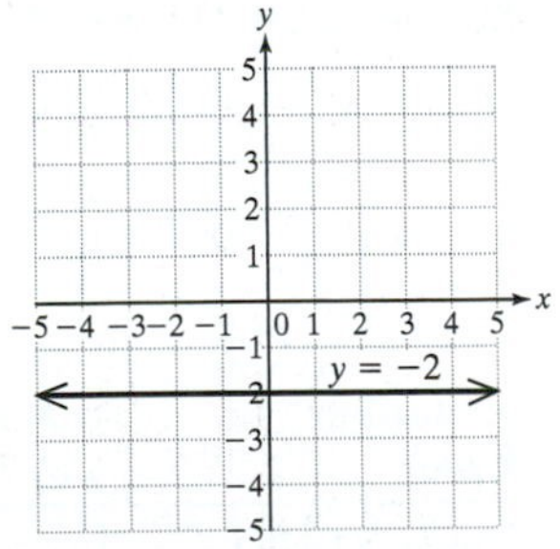

25.

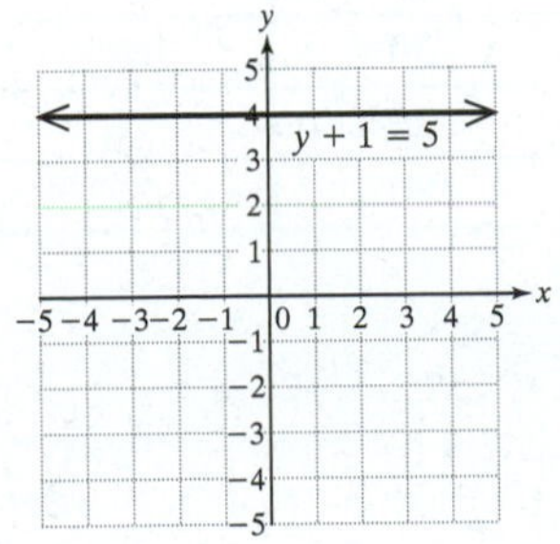

26.

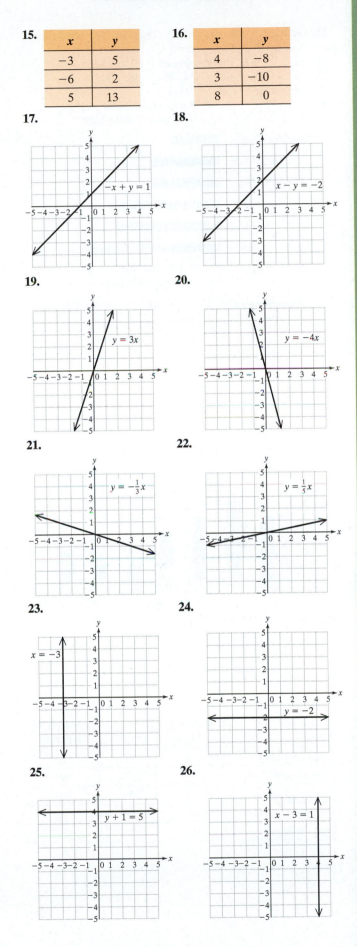

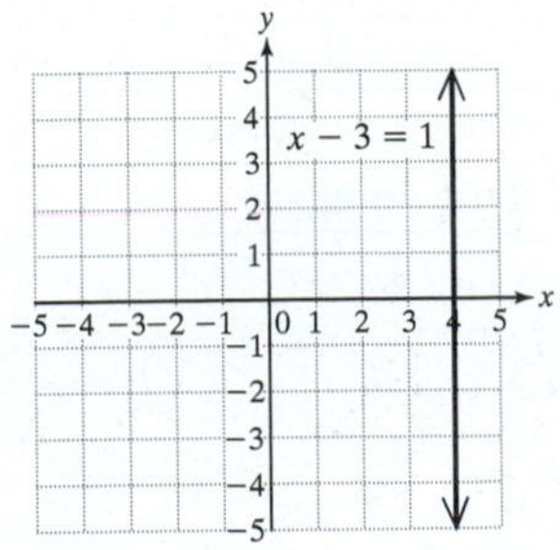

27. Godiva **28.** Breyers **29.** Blue Bell has 2 times more sodium than Edy's Grand. **30.** There is a 10-g difference. **31.** 1 icon represents 50 tornadoes.
32. 300 **33.** June **34.** 75 **35.** 2005 **36.** 4900
37. Increasing **38.** ≈ 7000
39.

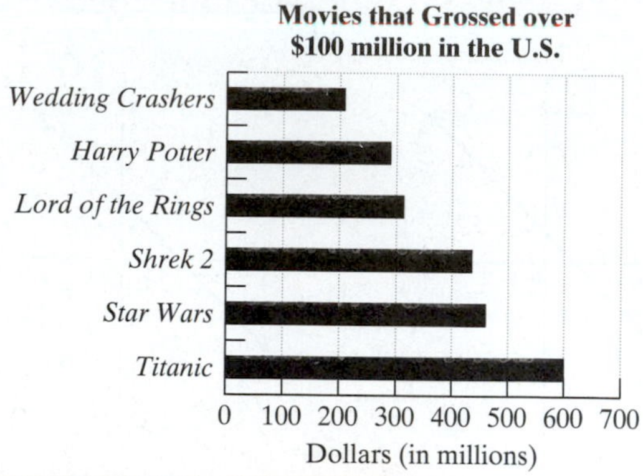

40.

Class Intervals (Age)	Frequency
18–21	4
22–25	5
26–29	4
30–33	3
34–37	1
38–41	1
42–45	2

41.

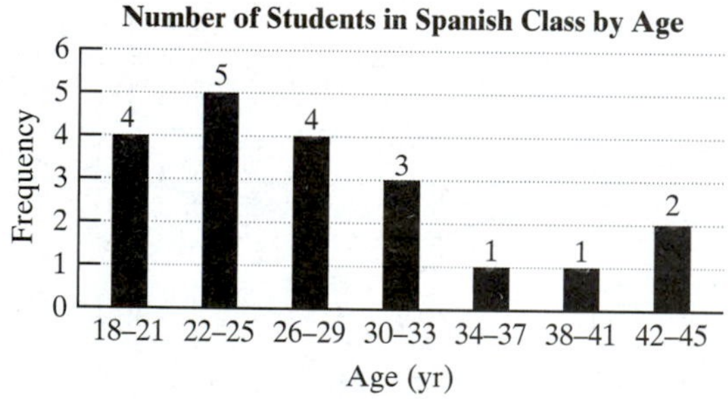

42. There are 24 types of subs.

43. $\frac{2}{3}$ of the subs contain beef.

44. $\frac{1}{3}$ of the subs do not contain beef.

45. a.

Education Level	Number of People	Percent	Number of Degrees
Grade school	10	5%	18°
High school	50	25%	90°
Some college	60	30%	108°
Four-year degree	40	20%	72°
Post graduate	40	20%	72°

b.

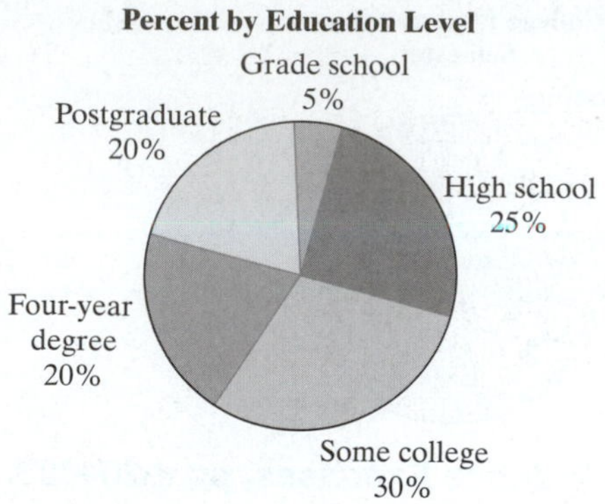

46. {blue, green, brown, black, gray, white}

47. $\frac{1}{6}$ **48.** $\frac{5}{6}$ **49.** a, c, d, e, g

50. a. $\frac{1}{2}$ **b.** $\frac{1}{2}$ **c.** 0 **51. a.** $\frac{1}{4}$ **b.** $\frac{3}{4}$

Chapter 9 Test, pp. 632–635

1.

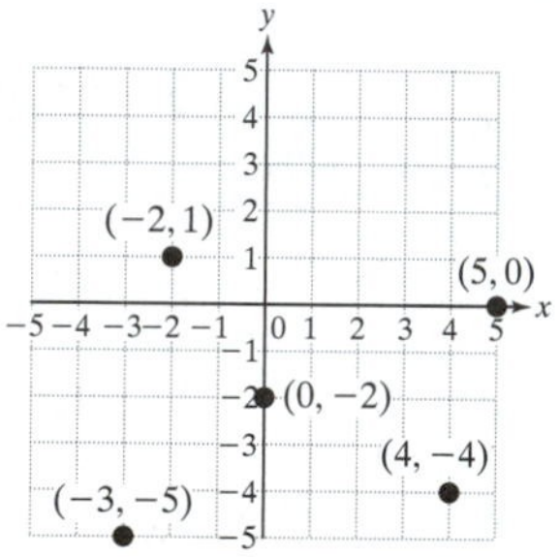

2. a. x-axis **b.** Quadrant IV **c.** Quadrant III **d.** y-axis **e.** Quadrant II **f.** Quadrant I
3. Negative; positive **4.** Yes **5.** No **6.** No
7. (5, 2), (1, 0) **8.** (−3, −6), (2, −6)
9.

x	y
−3	0
0	9
2	15

10.

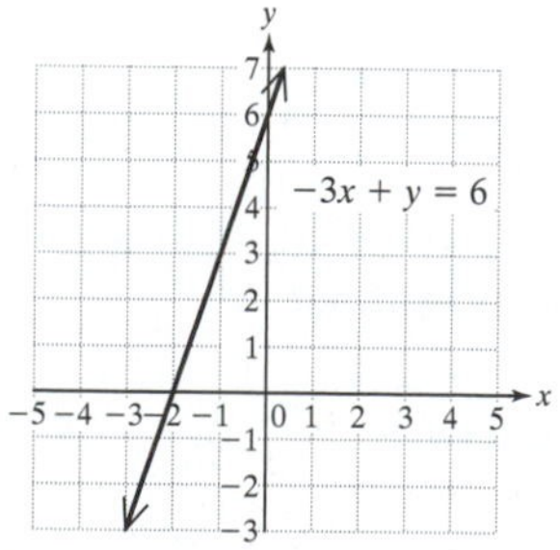

11.

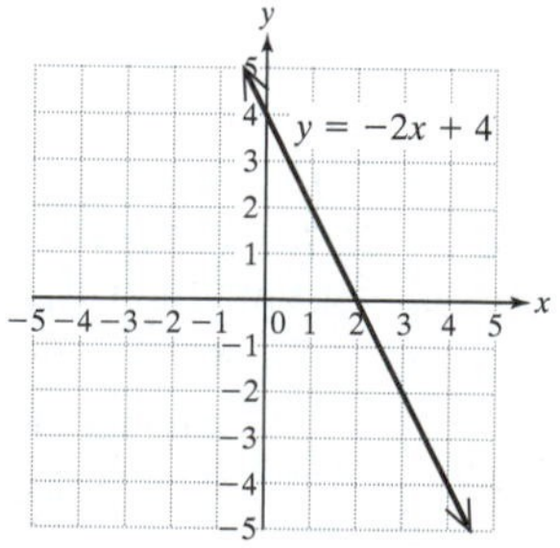

12.

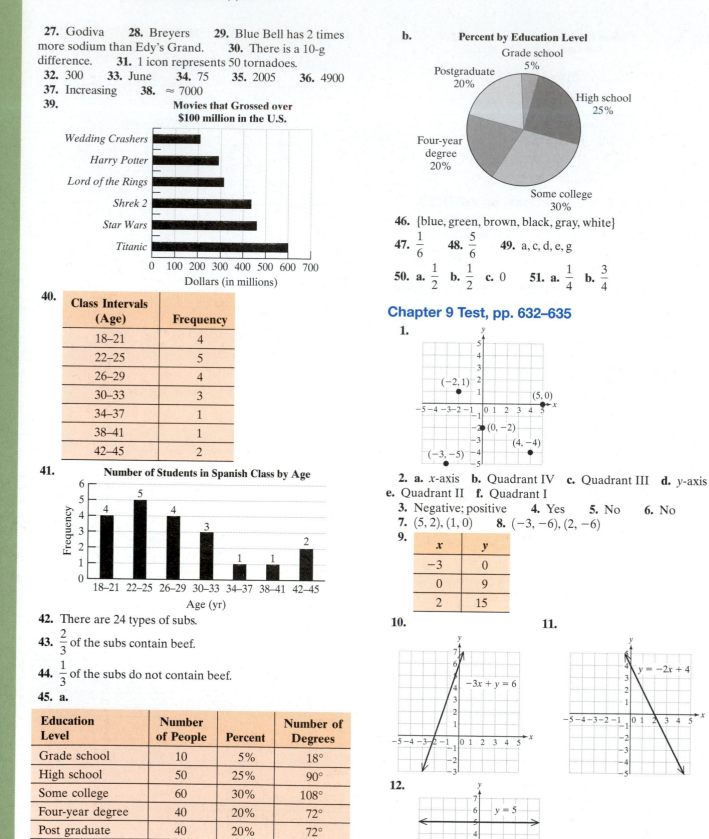

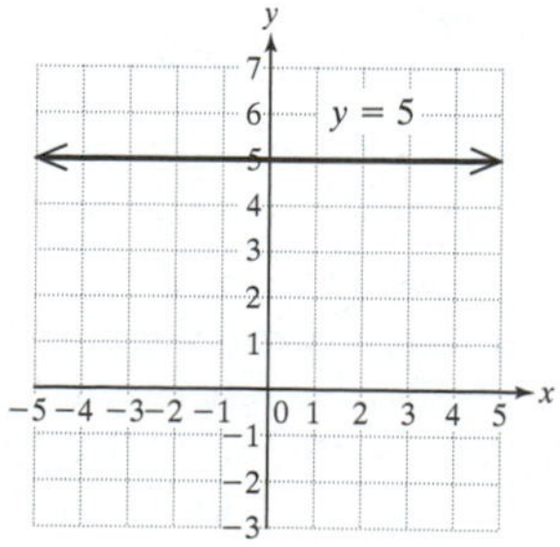

13.

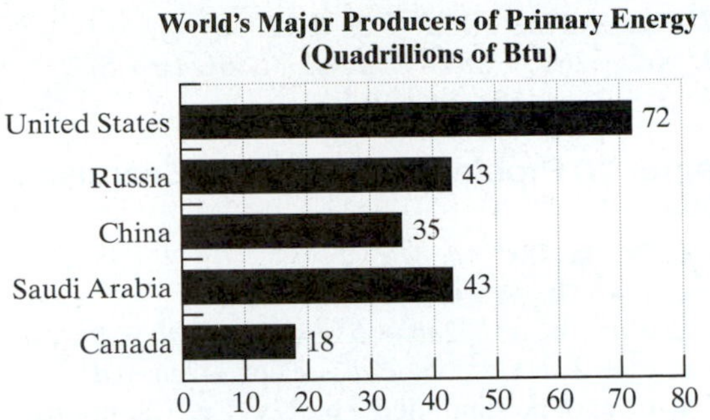

14. a. The year 1820 had the greatest percent of workers employed in farm occupations. This was 72%.
b.

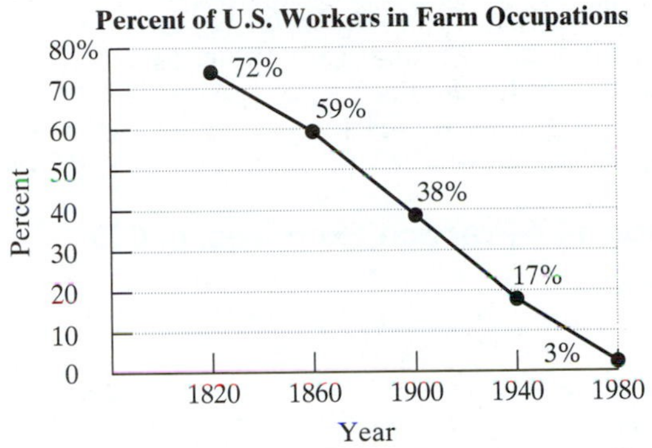

c. Approximately 10% of U.S. workers were employed in farm occupations in the year 1960.
15. \$1000 **16.** \$4500 **17.** February **18.** Seattle
19. 1.73 in. **20.** May
21.

Number of Minutes Used Monthly	Tally	Frequency
51–100	~~IIII~~ I	6
101–150	II	2
151–200	III	3
201–250	II	2
251–300	IIII	4
301–350	III	3

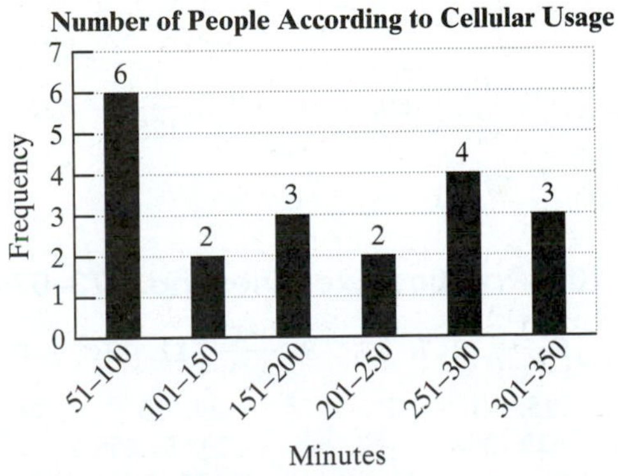

22. 66 people would have carpet.
23. 40 people would have tile. **24.** 270 people would have something other than linoleum.
25. a. $\{1, 2, 3, 4, 5, 6, 7, 8\}$ **b.** $\frac{1}{8}$ **c.** $\frac{1}{2}$ **d.** $\frac{1}{4}$
26. $\frac{2}{7}$ **27.** c

Chapters 1–9 Cumulative Review Exercises, pp. 635–637

1. a. Hundred-thousands **b.** Hundredths **c.** Hundred-thousandths **2.** $700 \times 1200 = 840{,}000$ **3.** 8353
4. $-\frac{2}{3}$ **5.** $\frac{3}{2}$ **6.** $\frac{23}{100}$ **7.** 2 **8.** $\frac{1}{6}$
9. -0.68412 **10.** 6.8412 **11.** 68,412 **12.** -24
13. 5 **14.** 2 **15.** -18 **16.** -2 **17.** 7.15
18. Quick Cut Lawn Company's rate is 0.55 hr per customer. Speedy Lawn Company's rate is 0.5 hr per customer. Speedy Lawn Company is faster.
19. 125 min or 2 hr 5 min **20.** $x = 5$ m, $y = 22.4$ m
21. 122 people **22.** 250 people **23.** 65%
24. \$1404 **25.** 29 in. **26.** 18 qt **27.** 9 yd 1 ft
28. 9.64 km or 9640 m **29.** 8 ft^2 **30.** 66 m^3
31.

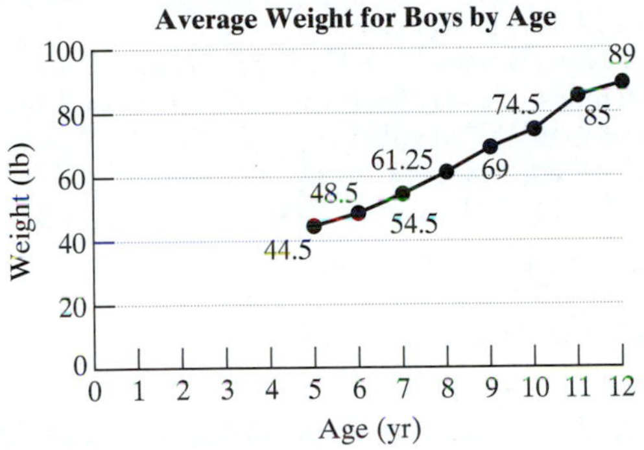

32. Mean: 105; median: 123
33. {yellow, blue, red, green}
34. $\frac{1}{4}$ **35.** $\frac{3}{4}$ **36.** 0.49 m **37.** 5800 cc
38. 0.51 kg **39. a.** Quadrant III **b.** *y*-axis **c.** *x*-axis **d.** Quadrant I **e.** Quadrant IV
40.

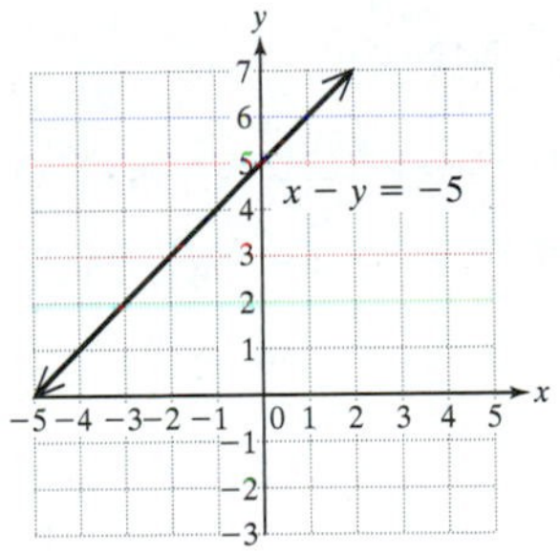

Chapter 10

Chapter Opener Puzzle

The numerical factor within a term is called the C O E F F I C I E N T.

C	O	E	F	F	I	C	I	E	N	T
1	2	3	4	5	6	7	8	9	10	11

Section 10.1 Practice Exercises, pp. 645–648

		Descending Order	Degree	Category
3.	$-7 + 2x^4$	$2x^4 - 7$	4	binomial
5.	$2y^3 - 4y^5 + 9y$	$-4y^5 + 2y^3 + 9y$	5	trinomial
7.	$12z^7$	$12z^7$	7	monomial
9.	$9k - 4$	$9k - 4$	1	binomial

11. Yes **13.** No **15. a.** -5 **b.** 12 **c.** 6 **d.** 1 **17.** No **19.** Yes **21.** $-10p^2$ **23.** $14y^3$ **25.** $5.9x^4$ **27.** $7t + 3$ **29.** $-6y^2 + 8y$ **31.** $-2z^3 + 8z^2 + \frac{5}{12}$ **33.** $-x^2y + xy + 3xy^2$ **35.** $23x^3 + 5x^2 - 5x + 8$ **37.** $-3y^3 + 16y^2 + 8y + 2$ **39.** $-3x^4 + 2x^3 - 2x^2 + 10x$ **41.** $-9x^2 + 2x + 3$ **43.** $1.3y - z$ **45.** $7p^3 - 10p^2 + 5p$ **47.** $-23a^5 - 8a^3 - 4a^2 + 10$ **49.** $-a^2b^2 - 7ab^3 - 3ab - 4b^3$ **51.** $2x^2 + 11x - 7$ **53.** $23z^4 + 2z^3 - 5z^2 + 11z - 3$ **55.** $-7x^2 - 20x - 5$ **57. a.** 15 **b.** $-\frac{39}{16}$ **59. a.** -13 **b.** -1.032 **61. a.** 907; It costs \$907 to rent the house for 3 days. **b.** 3316; It costs \$3316 to rent the house for 14 days (2 weeks).

63. a. 229.97; It costs \$229.97 to rent a car for 3 days and to drive 350 miles. **b.** 497.93; It costs \$497.93 to rent a car for 7 days and to drive 720 miles.

65. $20x$ **67.** $10y + 14$ **69.** $-2x - 6$ **71.** $-5x^3 + 4x^2 - 7x + 11$ **73.** $-a^4 - 11a^3 - 5a^2 - 8a$ **75.** $4.1x^2 - 3.2x - 13.1$ **77.** $-\frac{11}{4}t^3 + \frac{5}{8}t + \frac{1}{3}$ **79.** For example: $x^4 + x^3$ **81.** For example: $w^2 + w + 1$

Section 10.2 Practice Exercises, pp. 652–654

1. a. False **b.** True **3.** $-a^2 - 6a - 10$ **5.** $15w^4 + 9w^2 + 3w - 21$ **7.** m^{1+3} **9.** y^{12} **11.** w^{12} **13.** c^{21} **15.** 3^6 **17.** 5^{12} **19.** $12y^2$ **21.** $10w^4$ **23.** $12x^6$ **25.** $8p^{13}$ **27.** a^9b^9 **29.** $-8x^4y^6$ **31.** $x^{3 \cdot 5}$ **33.** x^{20} **35.** 2^{28} **37.** Neither. They are both equal to 2^6. **39.** x^3y^3 **41.** $32w^5$ **43.** $\frac{100}{y^2}$ **45.** $64x^3y^3$ **47.** $\frac{16w^4}{p^4}$ **49.** $36a^4b^2$ **51.** $16t^4v^{12}$ **53.** $\frac{81a^8}{b^2}$ **55.** y^{14} **57.** $8p^8q^2$ **59.** $32x^5y^{15}$ **61.** $9z^4$ **63.** $20w^2$ **65.** $15x^3$ **67.** $\left(\frac{2}{3}\right)^{10}$ or $\frac{2^{10}}{3^{10}}$ **69.** $x^{120}y^{100}$ **71.** k^{110}

Section 10.3 Practice Exercises, pp. 657–659

1. a^{m+n} **3.** 6^{10} **5.** x^{30} **7.** $16c^2d^{10}$ **9.** $-8y^3 - 5y^2 - 4y - 7$ **11.** $6x^2 + 8x$ **13.** $-2y^2 + 16y$ **15.** $6a^3 - 12a^2 + 18a$ **17.** $-24w^3 - 16w^2 + 32w$ **19.** $10a^4b^2 + 8a^3b^2$ **21.** $-9x^3y^2 + 18x^2y^2 - 9x^2y^3$ **23.** $y^2 + 8y + 15$ **25.** $m^2 + 3m - 28$ **27.** $xy - 10x - 4y + 40$ **29.** $8y^2 - 18$ **31.** $10t^2 + 22t - 24$ **33.** $10a^2 - 6a + 5ab - 3b$ **35.** $x^2 - 16$ **37.** $64p^2 - 9$ **39.** $4 - m^2$ **41.** $x^2 + 2xy + y^2$ **43.** $9x^2 + 30x + 25$ **45.** $49z^2 - 14z + 1$

47. FOIL only applies to a product of two binomials. The product given here has a trinomial and its middle term must also be accounted for in the product.

49. $a^3 + 3a^2 - 6a + 20$ **51.** $6z^3 + z^2 - z - 21$ **53.** $c^3 + 27$ **55.** $2x^3 - 13x^2 + 17x - 3$ **57.** $36a^4 + 17a^3 + 13a^2 + 2a - 12$ **59.** $8x^4 - 26x^3 + 15x^2 - 7x - 2$ **61.** $(x + 5)$ **63.** $(z + 2)$ **65.** $(3x + 4)$

Chapter 10 Problem Recognition Exercises, p. 659

1. a. $8x$ **b.** $48x^2$ **c.** $16x$
2. a. $12y^2$ **b.** $4y$ **c.** $8y$
3. a. a^2b^2 **b.** $a^2 + 2ab + b^2$ **c.** $a^2 - 2ab + b^2$
4. a. $c^2 + 2cd + d^2$ **b.** c^2d^2 **c.** $c^2 - 2cd + d^2$
5. a. Cannot be simplified **b.** $12x^5$ **c.** Cannot be simplified **6. a.** $18y^7$ **b.** Cannot be simplified **c.** Cannot be simplified **7. a.** $-3x^3y^5$ **b.** $7x^3y^5$ **c.** $10x^6y^{10}$ **8. a.** $16a^2b$ **b.** $2a^2b$ **c.** $63a^4b^2$
9. a. $8x^2 + 2x - 21$ **b.** $-2x - 10$ **c.** $6x + 4$
10. a. $-3w + 13$ **b.** $18w^2 - 3w - 36$ **c.** $9w - 5$
11. a. $t^2 - 2t + 6$ **b.** $-t^2 + 4t - 2$ **c.** $t^3 - t^2 - 2t + 8$
12. a. $z^3 - z^2 - 23z + 15$ **b.** $-z^2 - 3z - 2$ **c.** $z^2 + 5z - 8$

Section 10.4 Practice Exercises, p. 663

3. $16x^8y^4$ **5.** w^{10} **7.** $-2x^3 - 16x^2$ **9.** $2x^2 - 23x + 30$ **11.** 6 **13.** 7 **15.** 25 **17.** 6 **19.** 15 **21.** x^3 **23.** w **25.** $3xy^3$ **27.** $5ab$ **29. a.** t^3 **b.** Least value **31.** $4(x - 5)$ **33.** $2(p + 1)$ **35.** $5(m + 8n)$ **37.** $2(2a + 4b + c)$ **39.** $4(3x - 5y - 2z)$ **41.** $x^3(x + 1)$ **43.** $5y^2(y - 3)$ **45.** $x(3x^2 + 5x + 7)$ **47.** $3z^2(4z^3 - 2z + 1)$ **49.** $x^2y^3(4x^2 - 3xy + 5y^2)$ **51.** $15cd^3(c^2d^2 + 2cd - 3)$

Section 10.5 Practice Exercises, pp. 669–670

3. x^7 **5.** x^{10} **7.** $64x^6y^3$ **9.** $\frac{9x^2}{y^4}$ **11.** m^3 **13.** 10^8 **15.** x^2 **17.** w **19.** 1 **21.** 1 **23.** 6 **25.** 1 **27.** 1 **29.** 2 **31.** $\frac{1}{a^4}$ **33.** $\frac{1}{6}$ **35.** $\frac{1}{25}$ **37.** 16 **39.** $\frac{1}{8x^3}$ **41.** $\frac{2}{x^3}$ **43.** $\frac{16y}{x}$ **45.** 8 **47.** c^5d^4 **49.** $\frac{y^3z^4}{x^2}$ **51.** $\frac{2}{x^2y^3}$ **53.** $2x^2y^3$ **55.** y^2 **57.** $\frac{1}{p^5}$ **59.** x^{12} **61.** $\frac{1}{t^5}$ **63.** $\frac{1}{k^{14}}$ **65.** k^{14} **67.** $\frac{b^3}{a^6}$ **69.** w^4z^6 **71.** $\frac{1}{125}$ **73.** 64 **75.** $-\frac{18c^5}{d^7}$ **77.** $\frac{1}{x^2}$

Section 10.6 Practice Exercises, pp. 673–674

3. p^7 **5.** $\frac{y^{14}}{x^{12}}$ **7.** 5 **9.** $\frac{64x^2}{y^{24}}$ **11.** $17y^2 + 4x^2$ **13.** 10^4 **15.** 10^3 **17.** 10^{-3} **19.** 10^{-4} **21.** No **23.** Yes **25.** Yes **27.** No **29.** $\$7.455 \times 10^{12}$ **31.** 2×10^{-7} mm **33.** 2×10^7 **35.** 8.1×10^6 **37.** 3×10^{-3} **39.** 2.5×10^{-2} **41.** 1.42×10^5 **43.** 4.91×10^{-5} **45.** 8.2×10^{-2} **47.** 4.92×10^3 **49.** 6000 **51.** 0.08 **53.** 0.44 **55.** 37,000 **57.** 326 **59.** 0.0129 **61.** 0.000002003 **63.** 900,100,000

	Planet	Mass (kg)	Scientific Notation
65.	Mercury	0.33×10^{24}	3.3×10^{23}
67.	Earth	5.98×10^{24}	Already in scientific notation
69.	Jupiter	1899×10^{24}	1.899×10^{27}
71.	Uranus	86.8×10^{24}	8.68×10^{25}

Chapter 10 Review Exercises, pp. 679–681

1. a. $7y^4 - 13y^2 - 5$ **b.** 4 **c.** Trinomial
2. a. $-2x^3 + 4x^2 - 7x$ **b.** 3 **c.** Trinomial
3. a. $-4.2a + 7.8$ **b.** 1 **c.** Binomial
4. a. $-\frac{2}{3}$ **b.** 0 **c.** Monomial
5. $-5.3z^3$ **6.** $-1.4p^4$ **7.** $12x^2 - 7x$
8. $16w^4 - w^2$ **9.** $10x^2 + 6xy - 3y^2$
10. $-14a^2 + 8ab - 14c^2$ **11.** $14p^3 + 9p^2 + 4p - 13$
12. $9m^3 + 2m^2 + 14m - 12$ **13. a.** \$530; The cost for a 3-month membership is \$530. **b.** \$1070; The cost for a 12-month membership is \$1070. **14. a.** \$235; The cost to produce 300 lemonades is \$235. **b.** Yes. Her gross sales would be \$600. If she subtracts her cost of \$235, her profit is \$365. **15.** a^{12} **16.** 8^{16} **17.** $-21t^2$ **18.** $60x^2$
19. $-28y^5$ **20.** $-48a^7$ **21.** $56p^3q^4$ **22.** $-20c^3d^2$
23. 10^6 **24.** q^{20} **25.** $\frac{x^5y^5}{z^5}$ **26.** $\frac{p^4q^4}{n^4}$ **27.** x^{17}
28. z^{16} **29.** $-64a^6b^3$ **30.** $9m^2n^{12}$
31. $-5a^3 - 10a^2 + 15a$ **32.** $-6t^4 - 24t^2 - 24t$
33. $45c^3d + 10cd^2$ **34.** $3m^4n - 6m^4n^3$
35. $z^2 + 12z + 27$ **36.** $p^2 + 11p + 28$
37. $6y^2 + 16y - 32$ **38.** $14x^2 - 71x - 33$
39. $x^3 - 7x^2 + 21x - 36$ **40.** $y^3 + 3y^2 - 43y + 15$
41. $x^3 - 1$ **42.** $x^3 + 216$ **43.** $81z^2 - 25$
44. $100w^2 - 16$ **45.** $a^2 + 4a + 4$ **46.** $z^2 + 8z + 16$
47. 7 **48.** 4 **49.** $2p^3$ **50.** $9x^2$ **51.** $9(2p + 1)$
52. $2(7w + 1)$ **53.** $x^2(3x^2 - 2x + 5)$
54. $y^3(7 + 6y + 8y^2)$ **55.** $9(9m^2 + 4m + 5)$
56. $12(n^2 - 4n + 3)$ **57.** $3c^2d^2(2c^2 - cd + 3d^2)$
58. $4w^2z(2z^2 - wz + 3w^2)$ **59.** h^7 **60.** 7^6
61. p^{16} **62.** q^{10} **63.** $x + 1$ **64.** 5 **65.** 1
66. 1 **67.** 14 **68.** 12 **69.** $\frac{1}{r^5}$ **70.** $\frac{1}{z^8}$
71. $\frac{9}{16}$ **72.** $\frac{25}{9}$ **73.** a^{10} **74.** 16 **75.** x^2y^4
76. a^5b^7 **77.** $\frac{1}{x^6}$ **78.** $\frac{1}{y^2}$ **79.** $\frac{1}{w}$ **80.** $\frac{1}{z}$
81. $\frac{1}{p^{18}}$ **82.** $\frac{1}{q^8}$ **83.** 7.2×10^8 tons **84.** 8.51×10^7
85. 1.7×10^{-2} in. **86.** 2.0×10^{-4} **87.** 9.456×10^6
88. 4.31×10^4 **89.** 4.56×10^{-5}
90. 3.67×10^{-3} **91.** 0.005612 lb/ft³
92. 0.000 000 000 000 000 000 000 44 sec **93.** 10,550°F
94. 382,500 km **95.** 459,000,000 **96.** 358.9
97. 0.0000078 **98.** 0.0479

Chapter 10 Test, pp. 681–682

1. $-7y^4 - 9y^2 + 3y + 2$; degree 4 **2.** The terms monomial, binomial, and trinomial refer to polynomials with one, two, and three terms, respectively. **3.** $-13m^3$
4. $7x - 4$ **5.** $10x^2 + x - 21$ **6.** $-3x + 10$
7. $-20x^4 + 12x^3 - 28x^2$ **8.** $10a^3b^4 - 20a^2b^5 + 35a^4b^3$
9. $3x^3 - 4x^2 + 3x + 2$ **10.** $x^2y + 12xy + 3y^2 + x^2$
11. $3x^3 + 12x^2 - 3x - 30$ **12.** $9x^2 - 25$
13. $64y^2 + 48y + 9$ **14.** 2^{12} **15.** x^{30} **16.** $36y^8$
17. $\frac{25a^2}{b^4}$ **18.** 9 **19.** 3 **20.** $\frac{1}{k^4}$ **21.** 125
22. $\frac{b^4c^{10}}{a^3}$ **23.** x^8 **24.** $\frac{1}{x^8}$ **25.** $\frac{1}{y^6}$
26. $-\frac{6y^7}{x}$ **27.** x^{26} **28.** $4a^4b^{10}$ **29.** $\frac{125y^3}{x^{12}}$
30. 6 **31.** $4p^2$ **32.** $8(2m + 1)$ **33.** $5(3p - 1)$
34. $2x^3(6x^4 - 7 + 3x)$ **35.** $x^2y(2x^3 + 3xy + 5y^2)$
36. 4.25×10^7 **37.** 4.134×10^{-5} **38.** 0.027
39. 9,520,000

Chapter 10 Cumulative Review Exercises, pp. 682–684

1. a. Thousands **b.** Thousandths **2.** Three hundred-thousand, three hundred and three hundredths
3. 56 **4.** −25 **5.** −13 **6.** 60 **7.** $\frac{3}{5}$
8. $\frac{3}{10}$ **9.** $1\frac{7}{10}$ **10.** $13\frac{1}{8}$ **11.** $-3\frac{2}{3}$
12. −2 **13.** 32 **14.** $-\frac{3}{2}$ **15.** $\frac{1}{6}$ **16.** −20
17. 9 **18.** Mean: 29.9 yr; median: 29 yr; mode: 29 yr
19. Diameter: 12 ft; circumference: 37.68 ft; area: 113.04 ft²
20. 48 cm³ **21.** 25.12 ft³ **22. a.** 19° **b.** 109°
23. 12 yd **24.** $m(\angle a) = 51°$ and $m(\angle b) = 41°$
25. Yes **26.** 6,900,000 people **27.** 3.2 ft
28. \$0.324 per oz **29.** 20 fl oz **30.** 0.18 km
31. 5200 mg **32.** 1.5 gal **33.** 9 yd
34. Area: 36 cm²; perimeter: 28 cm
35. Area: 48 yd²; perimeter: 32 yd
36. The realtor received \$14,100 in commission.
37. 1200 students were surveyed. **38.** The tax rate is 5%.
39. a. 40 **b.** 15% **c.** 27.5%
40. $-8a^4b + 12a^3b^2 - 28a^2b^3$ **41.** $16a^2 - 4$
42. $x^2 + 14x + 49$ **43.** $w^3 - 7w^2 + 20w - 32$
44. $-7x^2 + 5xy + 9y^2$ **45.** w^7 **46.** $\frac{b^6}{a^{15}}$ **47.** x^2
48. 15 **49.** 8.524×10^5 **50.** 0.00315

Photo Credits

Chapter 1

Page 2: © Getty Images/Blend Images RF; p. 3: © Photodisc/Getty RF; p. 6: © RF/Corbis; p. 7: © SMSU-WP Photo: Vickie Driskell; p. 10: © BrandX Pictures/Punchstock RF; p. 11(left): © Royalty-Free/Corbis; p. 11(right): © BrandX Pictures/Punchstock RF; p. 15: © The McGraw-Hill Companies, Inc./Jill Braaten, photographer; p. 25(left): © Corbis RF; p. 25(right): © The McGraw-Hill Companies, Inc./Jill Braaten, photographer; p. 28: © Julie Miller; p. 33(top): © Corbis RF; p. 33(bottom): © Banana Stock/agefotostock RF; p. 40: © Corbis RF; p. 45(top): © Veer RF; p. 45(bottom left): © Vol. 43 PhotoDisc/Getty RF; p. 45(bottom right): © The McGraw-Hill Companies, Inc./Jill Braaten, photographer; p. 50: © BrandX/Jupiter; p. 52:© PhotoLink/Getty Images; p. 53: © Corbis RF; p. 56: © Corbis RF; p. 66: © Corbis/Jupiterimages; p. 71(top): © Getty RF; p. 71(bottom): © Julie Miller; p. 72(top): © Vol. 43 PhotoDisc/Getty; p. 72(bottom): © Getty RF; p. 73: © Vol. 156/Corbis; p. 80(top): © IMS/Capstone Design RF; p. 80(bottom): © Lynette King; p. 81: © Comstock/PunchStock RF; p. 82: © Photodisc/Getty Images RF; p. 83: © Buzzshotz/Alamy RF.

Chapter 2

Page 86: © 2006 Glowimages, Inc. - All Rights Reserved RF; p. 89: © Doug Sherman/Geofile RF; p. 90(top): © Imagestate Media (John Foxx)/Imagestate RF; p. 90(bottom): © Fancy Photography/Veer RF; p. 98: National Park Service, photo by DL Coe; p. 99: © Ilene MacDonald/Alamy RF; p. 104(left): © Getty Images/Purestock RF; p. 104(right): © Ingram Publishing/SuperStock RF; p. 105: © J. Luke/PhotoLink/Getty Images RF; p. 112(top): © ImageState/Punchstock RF; p. 112(bottom): © StockTrek/SuperStock RF; p. 117: © BananaStock/Punchstock RF; p. 120(top): © Imagestate Media (John Foxx)/Imagestate RF; p. 120(bottom): © Somos Images/Corbis RF; p. 121: © BananaStock/Punchstock RF; p. 122: © Siede Preis/Getty Images RF; p. 126: © Stockdisc RF; p. 127: © Randy Faris/Corbis RF; p. 128: © Getty Images/Digital Vision RF; p. 129: © Fancy Photography/Veer RF.

Chapter 3

Page 161: © Purestock/PunchStock RF; p. 162: © Alamy RF; p. 166(top): © Comstock Images/SuperStock RF; p. 166(bottom): © Brand X Pictures RF; p. 167(top): © Corbis RF; p. 167(bottom): © Ingram Publishing/SuperStock RF; p. 174(left): © Stock 4B RF; p. 174(right): © Dynamic Graphics/JupiterImages RF; p. 175(top): © Tetra Images/Alamy RF; p. 175(bottom): © Burke/Triolo/Brand X Pictures RF.

Chapter 4

Page 178: © The McGraw-Hill Companies, Inc./Mark Dierker, photographer; p. 182: © The McGraw-Hill Companies, Inc./Mark Dierker, photographer; p. 185(top): © Royalty-Free/ Corbis; p. 185(bottom): © Liquidlibrary/PictureQuest RF; p. 187: © The McGraw-Hill Companies, Inc./Jill Braaten, photographer; p. 196: © Image100 Ltd. RF; p. 199(top left): © BrandX/Punchstock RF; p. 199(top right): © Banana Stock/ Punchstock RF; p. 199(bottom top): © Banana Stock/Jupiterimages RF; p. 199 (bottom): © PhotoAlto/ PunchStock RF; p. 212: © Corbis RF; p. 213(top): © Digital Vision/Getty Images RF; p. 213(middle): © C Squared Studies/Getty Images RF; p. 213(bottom): © BrandX/Punchstock RF; p. 221: © Vol. 85/PhotoDisc/Getty RF; p. 230: © RF;/Corbis RF; p. 244(top): © Corbis RF; p. 244(bottom): © Comstock/Alamy RF; p. 245(top): © BrandX Pictures/ Jupiter RF; p. 245(bottom): Photo by Benjamin Schott; p. 274: © Tanya Constantine/Blend Images/Getty Images RF; p. 275(top): © Stockbyte/SuperStock RF; p. 275(bottom): © Brand X Pictures RF; p. 277(left): © Dynamic Graphics/PunchStock RF; p. 275(right): © Brand X Pictures/PunchStock RF.

Chapter 5

Page 288(left): © Mastefile RF; p. 288(right): U.S. Air Force Photo by Master Sgt. Michael A. Kaplan; p. 289: © EP4 PhotoDisc/Getty RF; p. 297(top): © NASA; p. 297(bottom): © Bob Jacobson/Corbis RF; p. 303(top): © Creatas/Punchstock RF; p. 303(bottom): © The McGraw-Hill Companies, Inc./Ken Karp, photographer; p. 308: © The McGraw-Hill Companies, Inc./Lars A. Niki, photographer; p. 309(left): © The McGraw-Hill Companies, Inc./Mark Dierker, photographer; p. 309(right): © The McGraw-Hill Companies, Inc./Mark Dierker, photographer; p. 310(left): © Art Vandalay/Getty Images RF; p. 310(right): © Stockbyte/PunchStock RF; p. 318: © Vol. 190/Getty RF; p. 321(top): © Getty RF; p. 321(bottom): © Brownstock Inc./Alamy RF; p. 332: © The McGraw-Hill Companies, Inc./Gary He, photographer; p. 336: © Vol. 49/Getty RF; p. 342: © BananaStock/Jupiterimages RF; p. 344(top): © Brand X Pictures/PunchStock RF; p. 344(bottom): © Corbis - All Rights Reserved RF; p. 347(bottom): © Brand X Pictures/Punchstock RF; p. 348: © David Buffington/Blend Images/Getty Images RF; p. 349: © Stockdisc/PunchStock RF; p. 351(left): © Stockdisc RF; p. 351(right): © Royalty-Free/CORBIS RF; p. 366: © Isabelle Rozenbaum & Frederic Cirou/PhotoAlto/PunchStock RF; p. 367: © BananaStock/PunchStock RF; p. 368: © Getty Images/Digital Vision RF.

Chapter 6

Page 372: © moodboard/Corbis RF; p. 373: © Molly O'Neill; p. 374: © The McGraw-Hill Companies, Inc./L.Niki; p. 375(top): © Brand X Pictures/PunchStock RF; p. 375(bottom): © Corbis RF; p. 376(left): © Corbis RF; p. 376(right): © Kent Knudson/PhotoLink/Getty Images RF; p. 377: © Photograph courtesy of USGS, Water Resources Division; p. 378(top): © Vol. 77/PhotoDisc/Getty RF; p. 378(bottom): © MIXA/PunchStock RF; p. 382a: © The McGraw-Hill Companies, Inc./Jill Braaten, photographer; p. 382b: © The McGraw-Hill Companies, Inc./Jill Braaten, photographer; p. 382c: © The McGraw-Hill Companies, Inc./Jill Braaten, photographer; p. 384(top): © Getty RF; p. 384(bottom): © bildagentur-online.com/th-foto/Alamy RF; p. 385: © Vol. 101 PhotoDisc/Getty RF; p. 386(top): © Digital Vision/Punchstock RF; p. 386(bottom): © EP073/Getty RF; p. 397: © Corbis RF; p. 399: © Linda Waymire; p. 401(top): © Corbis RF; p. 401(bottom): © Don Farrall/Getty Images RF; p. 401(middle): © The McGraw-Hill Companies, Inc./Mark Dierker, photographer; p. 401: © Corbis RF; p. 409: © David Tietz/Editorial

Image, LLC p. 410(left): © Imagestate Media (John Foxx)/Imagestate RF; p. 410(right): © Jarvell Jardey/Alamy RF; p. 413: © The McGraw-Hill Companies, Inc./Lars A. Niki, photographer.

Chapter 7

Page 422(top): © Royalty-Free/CORBIS RF; p. 422(bottom): © C Squared Studios/Getty Images RF; p. 426: © Vol. 76/PhotoDisc/Getty RF; p. 431(top): © Stockbyte/Punchstock RF; p. 431(bottom): © Stockbyte/Getty Images RF; p. 433: © Vol. 21 PhotoDisc/Getty RF; p. 434: © BrandX RF; p. 435: © Medioimages/Punchstock RF; p. 436: © Blend Images/Getty Images RF; p. 440: © Digital Vison/Getty RF; p. 441: © The McGraw-Hill Companies, Inc./Jill Braaten, photographer; p. 442: © The McGraw-Hill Companies, Inc./Jill Braaten, photographer; p. 443: © Creatas/Punchstock RF; p. 444: © RF/Getty; p. 448: © Image100/Corbis RF; p. 454(top): © OS23/ Getty RF; p. 454(bottom): © Creatas/PictureQuest RF; p. 455(top): © RF/Corbis RF; p. 455(bottom): © Digital Vision/Getty Images RF; p. 457: © Julie Miller RF; p. 464: © RF/Getty RF; p. 472: © Big Cheese Photo/ PunchStock RF; p. 477: © JUPITERIMAGES/Thinkstock/Alamy RF.

Chapter 8

Page 483: © LCPL Casey N. Thurston, USMC/DoD Media RF; p. 484: © Vol. 1/PhotoDisc/Getty RF; p. 486: © The McGraw-Hill Companies, Inc./Jill Braaten, photographer; p. 488(left): The McGraw-Hill Companies, Inc./Mark Dierker, photographer; p. 488(right): The McGraw-Hill Companies, Inc./Mark Dierker, photographer; p. 489(top): © RF/Corbis RF; p. 489(bottom): © Corbis-All Rights Reserved RF; p. 491: © Image Club RF; p. 496(left): © Ingram Publishing/Alam RF; p. 496(right): © Medioimages/Superstock RF; p. 501(top): © RF/Corbis p. 501(bottom): © BrandX/Punchstock RF; p. 504: The McGraw-Hill Companies, Inc./Mark Dierker, photographer; p. 505: © JUPITERIMAGES/Brand X/Alamy RF; p. 508: © Vol. 27 PhotoDisc/Getty RF; p. 509: © Vol. 10 PhotoDisc/Getty RF; p. 510: © Vol. 89 PhotoDisc/Getty RF; p. 511: © Vol. 83 PhotoDisc/Getty RF; p. 513: © Blend Images/Jupiterimages RF; p. 514: © Vol. 59 PhotoDisc/Getty RF; p. 534: © Linda Waymire RF; p. 569(right): © The McGraw-Hill Companies, Inc./John Flournoy, photographer RF; p. 522: © Image Club RF; p. 540: © Julie Miller; p. 545: © Studiohio RF; p. 546: © The McGraw-Hill Companies; p. 550(top left): © The McGraw-Hill Companies, Inc./Jacques Cornell photographer; p. 550(top right): © The McGraw-Hill Companies, Inc./Mark Dierker, photographer; p. 550(bottom left): © The McGraw-Hill Companies, Inc./Jacques Cornell photographer; p. 550(bottom right): © The McGraw-Hill Companies, Inc./Mark Dierker, photographer; p. 551(top left): © The McGraw-Hill Companies, Inc./Mark Dierker, photographer; p. 550(top right): © The McGraw-Hill Companies, Inc./Mark Dierker, photographer; p. 551(bottom left): © Comstock Images/Alamy RF; p. 551(bottom right): © Photodisc/Getty Images RF; p. 552(left): © Pixtal/age fotostock RF; p. 552(right): © OS23 PhotoDisc/Getty RF; p. 561: © Image Source/Alamy RF; p. 562(left): © Ernie Friedlander/Cole Group/Getty Images RF; p. 562(right): © PhotoDisc/Getty Images RF; p. 564(top): © Getty Images/DAJ, Inc. RF; p. 562(middle): © Lawrence Manning/Corbis RF; p. 562(bottom): © Stockbyte/Alamy RF; p. 563(left): © Royalty-Free/CORBIS; p. 563(right): © Brand X Pictures/PunchStock RF; p. 566(top): © Image Ideas/PictureQuest RF; p. 566(middle): U.S. Fish & Wildlife Service/David Hall; p. 566(bottom): © Digital Vision RF; p. 567: © Julie Miller; p. 568(left): © Stockdisc/Getty Images RF; p. 568(right): © Image Club RF; p. 569(left): © Blend Images/Getty Images RF.

Chapter 9

Page 578(top): © Neil Beer/Getty Images RF; p. 578(bottom): © Getty Images/SW Productions RF; p. 591: © U.S. Air Force photo by Master Sgt. Michael A. Kaplan; p. 593: © Royalty-Free/CORBIS p. 594: © Erica Simone Leeds RF; p. 595: © Corbis/PictureQuest RF; p. 597: © RF/Corbis; p. 598(top): © Creatas/Punchstock RF; p. 598(bottom): © Ian Shaw/Alamy RF; p. 599: © Getty Images RF; p. 605: © PhotoDisc/Getty RF; p. 606(top): © PhotoDisc/Getty RF; p. 606(bottom): © RF/Corbis; p. 607: © Image Source/Getty Images RF; p. 610: © The McGraw-Hill Companies, Inc./Andrew Resek, photographer; p. 611: © Terry Vine/Blend Images/Punchstock RF; p. 615: © Getty Images/Photodisc RF; p. 616: © The McGraw-Hill Companies, Inc./Barry Barker, photographer; p. 617: © BrandX/PunchStock RF; p. 618(bottom): © The McGraw-Hill Companies, Inc./Mark Dierker, photographer; p. 618(top): © Erica Simone Leeds RF; p. 621: © Getty RF; p. 622: © Stockbyte/Getty RF; p. 623: © PhotoDisc/Getty RF; p. 630: © Comstock Images/Jupiterimages RF; p. 632: © Ingram Publishing/Alamy RF; p. 633: Photo by Jack Dykinga/USDA; p. 636: © Liquidlibrary/Dynamic Graphics/Jupiterimages RF.

Chapter 10

Page 647(top): © Big Cheese Photo/PunchStock RF; p. 647(middle): © Moodboard/Corbis RF; p. 647(right): © fstop/Punchstock RF; p. 679: © Corbis - All Rights Reserved RF; p. 680: © John A. Rizzo/Getty RF; p. 681(top): USDA; p. 681(bottom): © Nancy Hyde; p. 683(top left): © Getty Images/Digital Vision RF; p. 683(bottom left): © The McGraw-Hill Companies, Inc./Mark Dierker, photographer; p. 683(right): © BananaStock/Jupiterimages RF.

Application Index

Distance/Speed/Time

Education

Entertainment/Recreation

Geography/Meteorology

Health/Medicine

Physics/Engineering

Politics/Government

Sports

Statistics/Demographics

Travel/Transportation

Subject Index

G

H

I

K

L

M